International Perspectives on Aging

The study of aging is continuing to increase rapidly across multiple disciplines. This wide-ranging series on International Perspectives on Aging provides readers with much-needed comprehensive texts and critical perspectives on the latest research, policy, and practical developments. Both aging and globalization have become a reality of our times, yet a systematic effort of a global magnitude to address aging is yet to be seen. The series bridges the gaps in the literature and provides cutting-edge debate on new and traditional areas of comparative aging, all from an international perspective. More specifically, this book series on International Perspectives on Aging puts the spotlight on international and comparative studies of aging.

More information about this series at http://www.springer.com/series/8818

Marisa Cordella • Aldo Poiani

Fulfilling Ageing

Psychosocial and Communicative Perspectives on Ageing

Springer

Marisa Cordella
School of Languages and Cultures
University of Queensland
St.Lucia, QLD, Australia

Aldo Poiani
School of Biological Sciences
Monash University
Clayton, VIC, Australia

ISSN 2197-5841 ISSN 2197-585X (electronic)
International Perspectives on Aging
ISBN 978-3-030-60069-3 ISBN 978-3-030-60071-6 (eBook)
https://doi.org/10.1007/978-3-030-60071-6

This Springer imprint is published by the registered company Springer Nature Switzerland AG
The registered company address is: Gewerbestrasse 11, 6330 Cham, Switzerland

To all of us,
travelling along different roads,
on a common human journey.

Preface

Now,
time, I wrap you up,
I lay you inside my
wild box
and I go fishing,
with your long line,
the fishes of my dawn!
Pablo Neruda, *Ode to Age*[1]

Pablo Neruda was not only a great poet (Nobel Prize winner in 1971) but also a great *bon vivant*, lover, and political activist right until the end of his life in 1973, at the age of 69. His poems and life exemplify an optimistic view of old age, a period of new challenges, and never-ending personal growth and fulfillment. This contrasts with a different view of old age as a time for retirement from work, physical and mental decay, personal identity crisis, and financial challenges; rupture of social bonds, potential economic load on the rest of society, and an overall problematic period for both the individual and family. Neruda experienced what we will call in this book a *fulfilling ageing*, or *the ability to effectively express capability and agency, leading to a personally satisfactory level of well-being, contentment, and adaptation within the self and with our surrounding environment as we become older*.

Understanding our elders requires exploring all aspects of old age: the positive and the negative, the psychological, the social, the biological; but in our opinion such exploration should be enlightened by a broader view that digs deep into what

[1] *Ahora,*
tiempo, te enrollo,
te deposito en mi
caja silvestre
y me voy a pescar
con tu hilo largo
los peces de la aurora!
Pablo Neruda, *Oda a la Edad*

makes life worth living. We believe that being fulfilled makes not only our life worth living, but it also makes the inevitable hurdles of our ageing more bearable.

Greater life fulfillment in old age brings greater meaning and satisfaction to our existence despite the inevitable shortcomings of getting old. Life fulfillment is experienced through the development of all our human capabilities, from the cognitive to the emotional, to the existential and the spiritual. It is a dynamic process and one that engages not only the self but also others in social interactions and our broader environment. Fulfilling ageing involves finding personal meaning in the experience of living in the world.

Developing a sense of fulfillment makes life worth living even when not everything runs smoothly or as we wish. Fulfillment also means feeling comfortable with our legacy work, no matter what that might be and how far it might reach.

Life fulfillment should not be equated with "success." Life is not a race, it is not a contest. As we age, we may change, develop, expand, or contract; but for as long as we feel fulfilled, we will be fine. Over time, we may develop a greater integration of the self, a better coherence, a clearer self-identity which may allow us to also develop our wisdom. Through greater life fulfillment, we may even regard ourselves as happy people.

In order to live a fulfilling old age, however, it helps to retain a degree of good physical and mental health, possess some level of material security (whether it is of personal origin or provided by government programs), enjoy positive interactions with others, and be able to engage in personally satisfying activities that give meaning to our life.

Everybody can experience a fulfilling old age. Life fulfillment in old age is a personal experience and so anything goes as far as the specific path towards it is concerned. We may remain cognitively active through using the natural plastic capabilities of our brain in different ways. Building cognitive reserve throughout life (e.g. through education and intellectual curiosity) will certainly help in old age, by supporting our capacity for personal growth. The same is true for broader physical capabilities, but let us not forget that we can also find life fulfillment within the reality of specific physical disabilities and even some mental disorders. A level of optimism will help in all circumstances, as through optimism we may be able to overcome the difficulties encountered throughout our life and incorporate those difficult experiences into our personal growth. Nurturing social relationships with family, friends, and the local community of residence can also promote fulfilling ageing, especially with regard to securing emotional and material support.

Throughout our life, we may often confront important choices. For as long as the choice, whatever it is, responds to strong urgencies rather than superficial whims, it may open a door to fulfillment. Regrets are always possible, nobody is perfect, but it is the overall result that matters.

Our life story can be expressed into a meaningful narrative in old age and such life story will not only help shape our personal identity, from which an integrated self may emerge; but it will also provide the raw material for our legacy and generative work. Meaningful narratives help put the picture of our life into a sharper focus, thus contributing to our ability to develop and transmit the wisdom of our life.

A greater sense of fulfillment and meaning in life should better prepare us to withstand inevitable levels of suffering and, eventually, to face the end of our life with dignity.

St.Lucia, QLD, Australia Marisa Cordella
Clayton, VIC, Australia Aldo Poiani

Acknowledgments

This book is dedicated to Franco Poiani, who passed away in 2020 at the age of 89. He continues his journey in our memories.

We are grateful to Janice Stern from Springer for supporting this project. Many thanks also to Lilith Dorko who eventually replaced Janice at Springer upon Janice's retirement, and who took over the responsibility for the publication of this book, and to Brinda Megasyamalan and Kala Palanisamy for overseeing the final stages of production.

The University of Queensland granted a study leave to M. Cordella in 2017 that allowed her a much-needed reprieve from usual academic duties to focus on the writing of two chapters. Thanks to Dr. Andrés Losada from the Department of Psychology of the Universidad Rey Juan Carlos, Madrid, Spain, for hosting M. Cordella during her study leave and for providing feedback on one chapter and to Ms. Vilma Masini who hosted us in Madrid.

We are grateful to various publishers and sources for their permission to reproduce some of the figures and tables.

Many thanks to Elsevier for granting permission to reproduce:

Figure 1.1, which is Figure 1 from Christensen, K., Doblhammer, G., Rau, R. and Vaupel, J.W. 2009. Ageing populations: the challenges ahead. Lancet 374: 1196–1208. Figure 3.4, which is Figure 1 from Sung, K.-t. 2004. Elder respect among young adults: A cross-cultural study of Americans and Koreans. Journal of Aging Studies 18: 215–230. Figure 3.7, which is Figure 1 from De Groot, C.P.G.M., Schlettwein-Gsell, D., Schroll-Bjørnsbo, K. and van Staveren, W.A. 1998. Meal patterns and food selection of elderly people from six European towns. Food Quality and Preference 9: 479–486. Figure 4.1, which is Figure 2 from Cabeza, R., Anderson, N.D., Locantore, J.K. and McIntosh, A.R. 2002. Aging gracefully: Compensatory brain activity in high-performing older adults. NeuroImage 17: 1394–1402. Figure 4.5, which is Figure 3 from Binder, L.I., Guillozet-Bongaarts, A.L., Garcia-Sierra, F. and Berry, R.W. 2005. Tau, tangles, and Alzheimer's disease. Biochimica et Biophysica Acta (BBA) - Molecular Basis of Disease 1739: 216–223. Figure 4.7, which is Figure 1 from Fox, N.C. and Schott, J.M. 2004. Imaging cerebral atrophy: normal ageing to Alzheimer's disease. Lancet 363: 392–394. Figure 4.9, which is

Figure 1 from Gott, M. and Hinchliff, S. 2003a. How important is sex in later life? The views of older people. Social Science and Medicine 56: 1617–1628. Figure 4.12, which is Figure 3 from Wilmoth, J.R. 2000. Demography of longevity: Past, present, and future trends. Experimental Gerontology 35: 1111–1129. Figure 5.6, which is Figure 3 from Mather, M. and Carstensen, L.L. 2005. Aging and motivated cognition: the positivity effect in attention and memory. TRENDS in Cognitive Sciences 9: 496–502. Figure 5.7, which is Figure 1 from Lachman, M.E., Agrigoroaei, S., Murphy, C. and Tun, P.A. 2010. Frequent cognitive activity compensates for education differences in episodic memory. American Journal of Geriatric Psychiatry 18: 4–10. Figure 6.11, which is Figure 4 from Robine, J.-M. and Cubaynes, S. 2017. Worldwide demography of centenarians. Mechanisms of Ageing and Development 165 (Part B): 59–67. Figure 6.15, which is Figure 1 from Schulz, R. and Martire, L.M. 2004. Family caregiving of persons with dementia: Prevalence, health effects, and support strategies. American Journal of Geriatric Psychiatry 12: 240–249. Figure 8.7, which is Figure 1 from Hunter, E.G. and Rowles, G.D. 2005. Leaving a legacy: Toward a typology. Journal of Aging Studies 19: 327–347. Figures 8.17 and 8.18, which are, respectively, Figures 1 and 2 from Turecki, G. and Brent, D.A. 2016. Suicide and suicidal behavior. Lancet 387: 1227–1239. Figures 8.19 and 8.20, which are, respectively, Figures 2 and 3 from Conwell, Y., Van Orden, K. and Caine, E.D. 2011. Suicide in older adults. Psychiatric Clinics of North America 34: 451–468. Figure 9.6, which is Figure 4 from Sauer, M.V. 2015. Reproduction at an advanced maternal age and maternal health. Fertility and Sterility 103: 1136–1143. Figure 9.10, which is Figure 1 from Griffin, A.S. 2008. Naked mole-rat. Current Biology 18: R844–R845.

We are also grateful to both the authors of the article and Elsevier for allowing the publication of Figure 9.5 in this book, which is Figure 1 from Croft, D.P., Brent, L.J.N., Franks, D.W. and Cant, M.A. 2015. The evolution of prolonged life after reproduction. Trends in Ecology and Evolution 30: 407–416. The article is under a Creative Commons Attribution License (CC BY) and therefore a specific permission is not required for reuse of its contents.

We are thankful to Springer Nature for granting permission to reproduce: Figure 1.2, which is Figure 1 from Finkel, T. and Holbrook, N.J. 2000. Oxidants, oxidative stress and the biology of ageing. Nature 408: 239–247. Figure 1.5, which is Figure 3 from Cabeza, R., Albert, M., Belleville, S., Craik, F.I.M., Duarte, A., Grady, C.L., Lindenberger, U., Nyberg, L., Park, D.C., Reuter-Lorenz, P.A., Rugg, M.D., Steffener, J. and Rajah, M.N. 2018. Maintenance, reserve and compensation: the cognitive neuroscience of healthy ageing. Nature Reviews Neuroscience 19: 701–710. Figure 4.8, which is Figure 3 from Li, Y., Rinne, J.O., Mosconi, L., Pirraglia, E., Rusinek, H., DeSanti, S., Kemppainen, N., Någren, K., Kim, B.-C., Tsui, W. and de Leon, M.J. 2008. Regional analysis of FDG and PIB-PET images in normal aging, mild cognitive impairment, and Alzheimer's disease. European Journal of Nuclear Medicine and Molecular Imaging 35: 2169–2181. Figure 4.13, which is Figure 2 from Vaupel, J.W. 2010. Biodemography of human ageing. Nature 464: 536–542. Figure 5.2, which is Figure 1 from Hedden, T. and Gabrieli, J.D.E. 2004. Insights into the ageing mind: A view from cognitive neuroscience.

Nature Reviews Neuroscience 5: 87–96. Figure 8.14, which is Figure 1 from Wink, P. and Dillon, M. 2002. Spiritual development across the adult life course: Findings from a longitudinal study. Journal of Adult Development 9: 79–94. Figure 8.22, which is Figure 1 from Larsson, K., Kåreholt, I. and Thorslund, M. 2008. Care utilisation in the last years of life in relation to age and time to death: results from a Swedish urban population of the oldest old. European Journal of Ageing 5: 349–357.

The Springer Publishing Company also kindly allowed the publication of Figure 1.3, which is Figure 1 from van den Beld, A.W., Kaufman, J.-M., Zillikens, M.C., Lamberts, S.W.J., Egan, J.M. and van der Lely, A.J. 2018. The physiology of endocrine systems with ageing. The Lancet Diabetes and Endocrinology 6: 647–658. Figure 4.10, which is Figure 1 from Laumann, E.O., Paik, A., Glasser, D.B., Kang, J.-H., Wang, T., Levinson, B., Moreira, Jr., E.D., Nicolosi, A. and Gingell, C. 2006. A cross-national study of subjective sexual well-being among older women and men: Findings from the Global Study of Sexual Attitudes and Behaviors. Archives of Sexual Behavior 35: 145–161. Figure 6.4, which is Figure 3 from Dykstra, P.A. 2009. Older adult loneliness: myths and realities. European Journal of Ageing 6: 91–100. Figure 6.7, which is Figure 1 from Daatland, S.O. and Lowenstein, A. 2005. Intergenerational solidarity and the family–welfare state balance. European Journal of Ageing 2: 174–182.

We are appreciative to John Wiley and Sons for allowing the publication of Figure 1.4, which is Figure 1 from Weiskopf, D., Weinberger, B. and Grubeck-Loebenstein, B. 2009. The aging of the immune system. Transplant International 22: 1041–1050. Figure 1.6, which is Figure 1 from Middleton, L.E., Barnes, D.E., Lui, L.-Y. and Yaffe, K. 2010. Physical activity over the life course and its association with cognitive performance and impairment in old age. Journal of the American Geriatrics Society 58: 1322–1326. Figure 8.4, which is Figure 1 from Park, N., Park, M. and Peterson, C. 2010. When is the search for meaning related to life satisfaction? Applied Psychology: Health and Well-Being 2: 1–13. Figure 8.21, which is Figure 4 from Snowdon, J. and Hunt, G.E. 2002. Age, period and cohort effects on suicide rates in Australia, 1919–1999. Acta Psychiatrica Scandinavica 105: 265–270. Figure 9.2, which is Figures 1 and 10 from Kaplan, H., Hill, J., Lancaster, J. and Hurtado, A.M. 2000. A theory of human life history evolution: diet, intelligence, and longevity. Evolutionary Anthropology 9: 156–185.

Taylor & Francis kindly allowed us to reproduce the following figures: Figure 7.3, which is Figure 2 from Mayer-Smith, J., Bartosh, O. and Peterat, L. 2007. Teaming children and elders to grow food and environmental consciousness. Applied Environmental Education and Communication 6: 77–85. Figure 8.1, which is Figure 1 from Sjöberg, M., Beck, I., Rasmussen, B.H. and Edberg, A.-K. 2018. Being disconnected from life: meanings of existential loneliness as narrated by frail older people. Aging and Mental Health 22: 1357–1364. Figure 8.2, which is Figure 1 from Steger, M.F., Oishi, S. and Kashdan, T.B. 2009. Meaning in life across the life span: Levels and correlates of meaning in life from emerging adulthood to older adulthood. The Journal of Positive Psychology 4: 43–52. Figure 8.10, which is Figure 1 from Chopik, W.J. 2017. Death across the lifespan: Age differences in death-related thoughts and anxiety. Death Studies 41: 69–77. Figure 8.13, which is Figure 1 from Hui, V.K.-Y. and Fung, H.H. 2008. Mortality anxiety as a function of intrinsic religiosity and perceived purpose in life. Death Studies 33: 30–50.

We are grateful to Oxford University Press for permitting the publication of Figure 4.2, which is Figure 2 from Frotscher, M., Drakew, A. and Heimrich, B. 2000. Role of afferent innervation and neuronal activity in dendritic development and spine maturation of fascia dentata granule cells. Cerebral Cortex 10: 946–951. Figures 4.15, 4.16, and 4.17, which are, respectively, Figure 1, 2, and 3 from Gavrilova, N.S. and Gavrilov, L.A. 2019. Are we approaching a biological limit to human longevity? Journals of Gerontology Series A, Biological Sciences and Medical Sciences, glz164, https://doi.org/10.1093/gerona/glz164. Figure 8.12, which is Figure 1 from Wink, P. and Scott, J. 2005. Does religiousness buffer against the fear of death and dying in late adulthood? Findings from a longitudinal study. Journal of Gerontology 60B: P207-P214. Figure 8.16, which is Figure 1 from Kirby, S.E., Coleman, P.G. and Daley, D. 2004. Spirituality and well-being in frail and nonfrail older adults. Journal of Gerontology 59B: P123–P129.

The Nature Publishing Group allowed the publication of Figure 4.14, which is Figure 6 from Dong, X., Milholland, B. and Vijg, J. 2016. Evidence for a limit to human lifespan. Nature 538: 257–259.

SAGE-Hindawi and SAGE Publications, respectively, allowed the publication of Figure 5.1, which is Figure 3 from Tse, M.M.Y., Lo, A.P.K., Cheng, T.L.Y., Chan, E.K.K., Chan, A.H.Y. and Chung, H. S.W. 2010. Humor therapy: Relieving chronic pain and enhancing happiness for older adults. Journal of Aging Research Volume 2010, Article ID 343574, 9 pages doi:https://doi.org/10.4061/2010/343574 and Figure 5.3, which is Figure 1 from Li, S.-C., Lindenberger, U., Hommel, B., Aschersleben, G., Prinz, W., and Baltes, P.B. 2004. Transformations in the couplings among intellectual abilities and constituent cognitive processes across the life span. Psychological Science 15: 155–163.

We thank the American Psychological Association for granting permission to publish Figure 5.9, which is Figure 1 from Roberts, B.W., Walton, K.E. and Viechtbauer, W. 2006. Patterns of mean-level change in personality traits across the life course: A meta-analysis of longitudinal studies. Psychological Bulletin 132: 1–25. Figure 5.11, which is Figure 3 from Pinquart, M. and Schindler, I. 2007. Changes of life satisfaction in the transition to retirement: A latent-class approach. Psychology and Aging 22: 442–455. Figure 5.12, which is Figure 3 from Moen, P. 1996. A life course perspective on retirement, gender, and well-being. Journal of Occupational Health Psychology 1: 131–144.

Cambridge University Press permitted the publication of Figure 6.8, which is Figure 3 from Gray, A. 2009. The social capital of older people. Ageing and Society 29: 5–31. We are also thankful to Diane Sullenberger, PNAS Executive Editor, for granting permission to publish Figure 9.1, which is Figure 3 from Alberts, S.C., Altmann, J., Brockman, D.K., Cords, M., Fedigan, L.M., Pusey, A., Stoinski, T.S., Strier, K.B., Morris, W.F. and Bronikowski, A.M. 2013. Reproductive aging patterns in primates reveal that humans are distinct. Proceedings of the National Academy of Sciences, USA 110: 13440–13445. The British Ecological Society permitted the publication of Figure 9.4, which is Figure 1 from Bonduriansky, R., Maklakov, A., Zajitschek, F. and Brooks, R. 2008. The evolutionary ecology of senescence: Sexual selection, sexual conflict and the evolution of ageing and life span. Functional Ecology 22: 443–453. We also acknowledge Wolters Kluwer Health, Inc. for allowing us to publish Figure 9.7, which is Figure 1 From Hollier, L.M., Leveno, K.J., Kelly, M.A., McIntire, D.D. and Cunningham, F.G. 2000. Maternal age and malformations in singleton births. Obstetrics and Gynecology 96:701–706.

Figure 2.1 is in the public domain, having been downloaded from Wikipedia. There are no known restrictions for publication in the USA. Figure 2.6 is also in the public domain, having been downloaded from Wikipedia https://commons.wikimedia.org/wiki/File:Lucas_Cranach__Der_Jungbrunnen_(Gem%C3%A4ldegalerie_Berlin).jpg, as it is Figure 2.8, having been also downloaded from Wikipedia Hubert_von_Herkomer_1878_-_Eventide.jpg.

Figure 2.2 was downloaded from the USA Library of Congress website http://www.loc.gov/pictures/item/2006686268/ on 7 December 2018. Rights Advisory: No known restrictions on publication. Figure 2.4 was downloaded from the American School of Classical Studies Digital Collections website: http://ascsa.net/research?v=default on 26 July 2016. Material that has been published is made completely available to the public. We are also grateful to the Trustees of the British Museum to allow us to publish Figures 2.5 and 2.7 given the academic purpose of this publication.

Figure 2.10 is in the public domain, having been downloaded from https://www.kurrimine.com.au/facilities/info-for-grey-nomads on 7 December 2018.

Figure 3.2 is available for reproduction in academic publications. We are grateful to Kasturi & Sons Ltd., publishers of *The Hindu*. Figure 3.6: the image is in the public domain and it was downloaded from: https://commons.wikimedia.org/wiki/File:Tai_Chi1.jpg on 16 August 2016.

The following figures were redrawn from the original, sometimes with minor modifications. The modifications did not affect the relevant information appearing in the original figure that concern this book: 1.7, 2.3, 2.9, 3.1, 3.5, 3.8, 3.9, 4.3, 4.4, 4.6, 4.18, 5.4, 5.5, 5.8, 5.10, 6.1, 6.2, 6.5, 6.6, 6.9, 6.10, 6.12, 6.13, 6.14, 6.16, 7.1, 7.2, 7.4, 7.5, 7.7, 8.2, 8.4, 8.5, 8.8, 8.11, 8.15, and 9.3. The acknowledgment of the original source is provided in the figure caption.

Contents

1 Introduction 1
1.1 The Multiple Dimensions of Ageing: An Overview 8
References 46

2 Old Ages in History 57
2.1 The Ages of Life 59
2.2 Retirement Age and Old Age Pension 67
2.3 Before Greece and Rome 70
2.4 Ancient Greece and Rome 71
2.5 Middle Ages and Renaissance 79
2.6 The Seventeenth–Eighteenth Centuries 86
2.7 The Nineteenth Century 91
2.8 The Twentieth and Twenty-First Centuries 96
2.9 The Safety Net 103
2.10 Older Ages and New Meanings of Life 105
2.11 Conclusions 112
References 113

3 Old People Across Cultures 119
3.1 The Cultural Context of Old Age 122
3.2 Fulfilling Ageing Across Cultures 127
3.3 Inter-Generational Relationships Across Cultures 137
3.4 Perceptions of Old People's Health Across Cultures 143
3.5 Old-Age Stereotypes: A Cross-Cultural View 151
3.6 The Ethnic Dimension of Elder Abuse and Neglect 155
3.7 Caring for Older People 159
3.8 Food Choices of the Elders 164
3.9 Life After Retirement 170
3.10 Old Age and End of Life Across Cultures 183
3.11 Conclusions 188
References 189

4 Ageing in Better Mental Health 201
4.1 Body Changes with Age 202
4.2 Plasticity of the Ageing Brain 206
4.3 Mental Health 215
4.4 Mental Disorders in the Elderly 226
4.5 Neurocognitive Disorders: Causes, Prevention, and Psychological Interventions 238
4.6 Ageing, Gender, and Sexual Life 270
4.7 Ageing with a Disability 291
4.8 Physical Disabilities 293
4.9 Intellectual Disabilities 296
4.10 Reaching Very Old Ages 301
4.11 Conclusions 322
References 322

5 The Psychology of Older Ages 355
5.1 Psychological Ageing 355
5.2 Self-Identity 359
5.3 The Emotional Dimension 368
5.4 The Cognitive Dimension 381
5.5 Memory 392
5.6 Personality and Ageing 404
5.7 Ageing and Psychological Stress 414
5.8 Retirement and Fulfilling Life 420
5.8.1 Preparing for Retirement 428
5.8.2 Transition to Retirement 430
5.8.3 Post-Transition Adjustment 431
5.8.4 Gender and Retirement 434
5.9 Conclusions 437
References 438

6 The Social Dimension of Older Ages 461
6.1 Social Integration of the Elderly 462
6.2 Loneliness 476
6.3 Abuse 487
6.4 Family 493
6.5 Community 504
6.6 The Ethics of Ageing: Stereotypes, Stigma, and Ageism 514
6.6.1 Old Age and Ethics 514
6.6.2 Stereotypes of Old Age 517
6.6.3 Stigma 526
6.6.4 Ageism and Discrimination 532
6.7 Oldest-Old in Society 540
6.8 Older Immigrants 546
6.9 Older Gay, Lesbian, Bisexual, Transsexual/Transgender(trans), and Intersex People 558

6.10 Ageing Prison Inmates . . . 572
6.11 Caring for Older People and Caregiver Burden . . . 580
6.11.1 Caring for Older People . . . 580
6.11.2 Caregiver Burden . . . 587
6.12 Conclusions . . . 599
References . . . 600

7 Language and Communication . . . 633
7.1 Ageing and Communication: An Overview . . . 634
7.2 A Cross-Cultural Perspective on Communication in Old People . . . 647
7.3 Intra-Generational Communication . . . 654
7.4 Inter-Generational Communication . . . 657
7.4.1 Improving Inter-Generational Communication: Intervention Programmes . . . 668
7.5 Narratives of the Elders . . . 677
7.6 Elderly and the Media . . . 698
7.7 Translating and Interpreting for Older Immigrants . . . 699
7.8 Doctor–Patient Communication in the Elderly . . . 704
7.9 Conclusions . . . 708
References . . . 708

8 Journey Towards the End of Life . . . 721
8.1 The Existential Dimension . . . 728
8.2 Life Meaning and Legacy . . . 734
8.2.1 Life Meaning . . . 735
8.2.2 The Study of Meaning . . . 745
8.2.3 Meaning and Life Satisfaction . . . 749
8.2.4 Life Meaning and Life Traumas . . . 753
8.2.5 Meaning Making . . . 758
8.2.6 Legacy . . . 768
8.3 Fear of Death and Death Anxiety . . . 774
8.3.1 Factors Affecting Fear of Death and Death Anxiety . . . 786
8.3.2 Decreasing Fear of Death and Death Anxiety . . . 790
8.4 Religion and Spirituality . . . 797
8.5 Hastening Own Death: Euthanasia and Suicide . . . 812
8.5.1 Euthanasia . . . 819
8.5.2 Euthanasia, Life Fatigue, and Existential Suffering . . . 825
8.5.3 Suicide . . . 828
8.5.4 Will to Live and Suicide . . . 842
8.5.5 Attempted Suicide, Protective Factors, and Interventions . . . 845
8.6 End of Life in Multicultural Societies . . . 848
8.7 End-of-Life Care . . . 856
8.8 Conclusions . . . 864
References . . . 865

9 Ageing in Evolutionary Perspective 891
9.1 Evolution and Lifespan 896
9.2 Evolution and Senescence 902
9.2.1 Evolution, Sex, and Ageing 907
9.2.2 Kin Selection and Ageing 912
9.3 Neoteny 921
9.4 Conclusions 927
References 927

10 Fulfilling Ageing 937
10.1 Experiencing Fulfilling Ageing 938
10.2 Spontaneous Personal and Community Initiatives 940
10.3 Institutional and Professionally Guided Initiatives 942
References 944

Index 947

About the Authors

Marisa Cordella is Associate Professor in linguistics in the School of Languages and Cultures at the University of Queensland. She is an expert in discourse analysis in the areas of intercultural and doctor–patient communication, language studies, and inter-generational and aging studies. She is currently the Director of Research in the School of Languages and Cultures at the University of Queensland. She is the author of *The Dynamic Consultation: A Discourse Analytical Study of Doctor-Patient Communication* (Benjamins, The Netherlands, 2004), *Behavioural Oncology: Psychological, Communicative, and Social Dimensions* (Springer, New York, 2014; with A. Poiani), and co-editor of the book *Rethinking Second Language Learning: Using Intergenerational Community Resources* (Multilingual Matters, Bristol, United Kingdom, 2016; with Hui Huang).

Website: https://researchers.uq.edu.au/researcher/2812

Aldo Poiani is adjunct research associate at the School of Biological Sciences of Monash University. He is a professional biologist expert in behaviour and evolution and has carried out research in social behaviour, immunology, endocrinology, and disease, adopting an evolutionary perspective. He is author of the books *Behavioural Oncology: Psychological, Communicative, and Social Dimensions* (Springer, New York, 2014; with M. Cordella), *Animal Homosexuality: A Biosocial Perspective* (Cambridge University Press, Cambridge, 2010), and editor of the books *Pragmatic Evolution: Applications of Evolutionary Theory* (Cambridge University Press, Cambridge, 2012) and *Floods in an Arid Continent* (Academic Press, Amsterdam, 2006).

Website: http://www.researchgate.net/profile/Aldo_Poiani

Chapter 1
Introduction

Abstract *Introduction* provides an overview of the book, introducing the major biological, psychological, social, cultural, historical, and existential themes that will be further developed in the rest of the chapters. In particular, the concept of fulfilling ageing is defined and explained here as a central idea that links together the various aspects of ageing addressed in this book.

When should a person be considered "old"? What does define old age, and is it possible to determine a specific point in time for the transition from middle to old age? Even if defining a specific threshold point is possible, would such accepted threshold vary over the years and decades, and would it change across cultures as well?

Regardless of some intuitive views of who should be considered old and who should not, determining a clear set of criteria to distinguish between the two is not easy. Although there is an established twentieth century tradition to identify old age with qualification for the old-age pension or with retirement from work—that is currently variable across countries ranging between 60 and 67 years old, with retirement age for men and women being either the same or women retiring at a somewhat younger age than men—the question remains: What are the criteria to choose those retirement ages rather than others?

In Elizabethan England, the old were identified on the ground of not only age but also the level of deterioration of their body and mind:

> The aged were those who were infirm, frail, and suffering incapacities of body or mind to the extent that they could no longer fully support or take care of themselves, and who also gave the appearance of being old. The assumption was that people could be advancing in years, or they could be incapable of supporting themselves, but it was only when the two conditions came together in one person that that person was considered 'old' by the authorities (Roebuck 1979, p. 417).

The combination of two main factors: "look" and "incapacity" produced a range of ages that even in the eighteenth century could qualify a person as being "old" if she or he was from 40-year-old to 80 plus. The advent of pensions required a stricter

M. Cordella, A. Poiani, *Fulfilling Ageing*, International Perspectives on Aging 30, https://doi.org/10.1007/978-3-030-60071-6_1

limit to the beginning of old age that necessarily shifted the threshold to older ages. In the *Friendly Societies Act* of 1875 in England (and its amendment of 1887) old age started after 50, although pensions tended to begin at the age of 65 (Roebuck 1979). In the late nineteenth and early twentieth century Germany, age of pension eligibility started at 70 (Denton and Spencer 2002). Interestingly, in Australian traditional Aboriginal societies a person is also considered old at the age of 55–60, a broad confluence of views with the Western tradition, that is presumably guided by the capacity to perform basic activities of daily living, and also some biological considerations, such as the onset of menopause in women, for instance. In such societies, adult children are expected to support their old and increasingly incapacitated parents in retribution for what they received from them when they were younger, whereas other members of the extended family also contribute to the care of older people (Berndt and Berndt 1992).

According to the *World Health Organization* (WHO), most developed countries currently consider 65 years old as the beginning of old age, although the age of 60 is used by the *United Nations* (Millane 2013). Further distinctions have been introduced within the population of old people based on age. Individuals aged 60–69 are known as the "young-old", those aged 70–79 are the "old-old", whereas those 80 years old and older are referred to as the "oldest-old". Centenarians and ultracentenarians are currently the subject of specific studies, being regarded as categories on their own.

The original introduction of age pensions was a consequence of the difficulties that people of older ages had in finding work. According to Janet Roebuck, in the late nineteenth century people older than 50 who were looking for work had difficulties in finding it and this could vary with the kind of work the person was applying for. However, pensions for those people did cost a significant amount of money to the government, which led to the inevitable push to try to shift the pensionable age to older ages. The argument about the age pension was well alive in the early twentieth century, and it is resurfacing around the world in the twenty-first century, with an ageing population (Denton and Spencer 2002) and in current discussions about what is the best and most sustainable system of pensions (e.g. James 1998). The issue is made even more urgent now given that the ratio of people at work over those in retirement is decreasing, thus raising concerns about the sustainability of pension systems (e.g. Millane 2013). However, there is a counter-argument against this negative view of old age and pensions, as when the rates of unemployment are increasing then it could be better to lower the pensionable age, so that more opportunities are open for the young, an issue that is also relevant at the beginning of this third decade of the twenty-first century. We will see in the next chapter that the conundrum between early retirement to create job opportunities for the young and sustainability of the pensions system can be resolved by the proper taxation of profits from the increasingly mechanised/computerised systems of production.

Older people are currently confronted with additional financial challenges such as their increasing level of debt, that in developed countries may have been accrued in part to finance a better lifestyle, whilst their income after retirement may not be enough to sustain such level of debt in the long term. This has put some pressure on the individual to continue working even when she or he may be officially "retired", although such person may face barriers of discrimination when applying for a job (see Millane 2013).

Emily Millane reports the results of a survey carried out by the *Australian Bureau of Statistics* with data for the years 2002–2011 where interviewees were asked about the reasons older people are discouraged from seeking work; the vast majority of respondents (62–78%, depending on the year) was concerned about being considered too old by a potential employer, 10–18% complained about lack of jobs in the area they live in, and 4–8% had issues with lack of qualifications. Such concerns notwithstanding, older people willing and able to work should be given a chance.

The difficulties that older people may confront in their quest to find a job have put to centre stage the issue of superannuation. Given that the recent trend has been to move away from the *defined benefits system* of superannuation where the retirees are expected to receive a predictable income no matter what the fluctuations of the market are, and for as long as they live, and towards a more risky system relying on an amount of accumulated funds that are contributed by the employer—perhaps with some additional voluntary contribution by the employee through salary sacrifice—and that are invested in the stock market or in a secure but usually low-yield cash deposit, the question of whether the retiree will have enough financial means to support him/herself in retirement is also becoming more central than ever. To increase the degree of financial certainty in retirement more and better planning is required as governments reconsider their level of expenditure on age pensions, and superannuation systems are also modified (Feldman and Beehr 2011).

But ageing is first and above all characterised by body and mental changes. Although most of the older people experience a relative decline in their cognitive capacities (e.g. Verhaegen and Salthouse 1997; Skirbekk 2003), the rate of decline can vary considerably across both individuals and specific cognitive skills, and it can be somewhat improved by appropriate interventions. The same variability in decline can be seen in some body changes associated with old age such as wrinkles, grey hair, age spots, health conditions such as arthritis, osteoporosis, and so forth, which can also affect body image in old people (Hurd Clarke and Korotchenko 2011). Deterioration in such variables may be predictive of mortality (e.g. Lee 2000); therefore, recent attempts have been undertaken to include in the definition of old age also the probability of mortality, whereby the limits of what is considered a "young-old", "old-old", and "oldest-old" can change following changes in life expectancy (Denton and Spencer 2002).

In traditional societies the concept of old person is also intermingled with that of "elder". An elder could be defined as "someone who has been sought by their peers for spiritual and cultural leadership and who has knowledge of some aspect of tradition" (Stiegelbauer 1996, p. 39). The elders are the repository of history, beliefs, traditional thoughts, knowledge, and practices that are passed on from one generation to another (e.g. Berndt and Berndt 1992). In a statement produced by an Innu delegation from the *Sheshatshiu Native Canadian Centre* of Toronto, reported by Stiegelbauer (1996, p. 39), elders:

> should be role models for everyone else. Elders should be teachers to the grandchildren and all young people because of their wisdom. Elders should be advisors, law-givers, dispensers of justice. Elders should be open to everyone. Elders should be knowledgeable in all aspects of Innu culture. Elders should be teachers for everyone of the past history of Innu people. Elders should be recorders of history, not only orally but to be preserved in print. Elders

> should be teachers of values important to Innu to be passed on from generation to generation. Elders should be teachers of language and oral history. Elders should be teachers of Innu medicine. We place great importance in our Elders. Their directions for us will guide our lives.

In other words, the elders are "experts on traditional life". In some traditional societies such expertise is kept somewhat secret and it also functions as a way to exert a degree of control over the youth, and to protect useful knowledge from being appropriated by outsiders to the family or the group (e.g. see the *Kpelle* of West Africa described by Murphy (1980), see also Berndt and Berndt (1992)). In these "gerontocratic societies" the level of inter-generational tensions may be particularly high. However, transmission of knowledge from old to young is more often a positive feature of inter-generational relationships across cultures (Maccoby 2007).

In modern societies the traditional cultural role of old people as elders is being compromised by easy access to recorded information from the past and by a fast-changing world that may challenge the capacity of many old people to stay abreast with new developments. Still, they remain living witnesses of a time past and their accumulated life experience may provide useful guidance to the solution of those problems that remain relevant from one generation to the next. We will see throughout this book how older people could continue to contribute to society in various ways, and for that reason, we will use the words *old* and *elder* interchangeably.

Our world is undergoing a *demographic transition*; that is, it is changing and ageing as the global population moves from a regime of high fertility and mortality to one of low fertility and mortality. People 65 years old and older represented 8.2% of the global population in 2013; this is expected to increase to at least 19% by 2050 (Lutz and Kc 2010; Draper 2014). From the information available in Table 1.1, it can be seen how all continents, including the two most populous countries in the world: China and India, are trending towards an increased proportion of older people over time, although some continents are ageing faster (Europe) than others (Africa). This trend towards an ageing world population is the result of increased life expectancy, especially since the twentieth century, leading to an increase in the

Table 1.1 Proportion of the population above the age of 65 and projections into the future for the world and various specific areas (UN Scenario of IIASA education projections)

Area	2000 (%)	2010 (%)	2020 (%)	2030 (%)	2040 (%)	2050 (%)
World	7	8	10	13	16	19
Africa	3	3	4	5	5	7
Asia	6	7	9	13	17	21
Europe	15	16	19	23	25	28
Latin	6	7	9	12	15	19
America and Caribbean North America	12	13	16	20	21	21
Oceania	10	11	14	16	18	19
China	7	8	12	17	24	27
India	5	5	7	9	11	14

Simplified from Lutz and Kc (2010)

number of people expected to live to 75 and beyond (e.g. past 100 years), with the oldest-old group expanding rapidly, especially in economically developed countries (Christensen et al. 2009; HelpAge International 2013). In addition, the global ageing population can be also explained by slower overall population growth caused by reduced fertility. Indeed, around the world the section of the population aged 60 years old or older is the one growing the fastest, ranging from 1 to 3.7% annually depending on the region (United Nations, Department of Economic and Social Affairs, Population Division 2013a, b).

Currently, the five countries with the highest percentage of people older than 65 are Japan (21.6% of the population), Italy (20%), Germany (20%), Greece (19.1%), and Sweden (18.3%) (Sokolovsky 2009).

Life expectancy has been dramatically increasing since the mid-nineteenth century in economically developed countries (see Fig. 1.1). In recent decades this increase in life expectancy has been mainly accounted for by increased probability of surviving to old age due to improved general life conditions and medical interventions. Across more than 30 developed countries, probability of survival from age 80 to 90 years increased from 15–16% for women and 12% for men in 1950, to 37% and 25%, respectively, in 2002 (Christensen et al. 2009). This seems to be a result of more people reaching their older ages in better health than it was the case in the past (Vaupel 2010). Women in general tend to live longer than men, with 60-year-old or older women outnumbering men by a ratio of 1.19 in Africa, 1.07 in Asia, 1.56 in Europe, and 1.16 in Latin America and the Caribbean, this is also true for the section of the elderly population aged 70 and older (Lloyd-Sherlock 2000), in spite of the well-known world trend for more males than females to be born (105–107 male for every 100 female births, Hesketh and Xing 2006). In 2008 the sex ratio of the 67–79 years old around the world was 1.31 women/men, whereas for the ≥80 years old it was 2.08 women/men (Cauley 2012). The bias in life expectancy in favour of older women has been consistently increasing throughout the twentieth and the beginning of the twenty-first centuries in most countries, with only few exceptions such as the United Kingdom, Australia, and the USA among developed countries, which in 2007 saw the female bias in life expectancy decrease somewhat in comparison with the 1950s (Kinsella 2009).

Increased ageing, however, also comes at a cost in terms of increased susceptibility to diseases, both infectious and non-infectious, as the immune system becomes more dysregulated. In particular, the oldest-olds may become frailer and less capable of effectively caring for themselves in their daily necessities. Although women are more likely to survive to older ages than men, physical deterioration in old age seems to be more accentuated among women than men (Christensen et al. 2009), although this issue is also ground for much debate (Kinsella 2009, p. 22).

Despite the negative effects of increased age on health, in recent times older people have also become better able to take care of themselves than it was the case in previous generations. This is because they have enjoyed the aid of technology, better housing conditions, and public services, including healthcare services; changing gender roles, better education, better conditions in their workplace, and overall improved economic standards (Christensen et al. 2009).

Although women tend to live longer than men, a pattern that, as we have seen, is broadly common around the world, they also tend to have less material protection

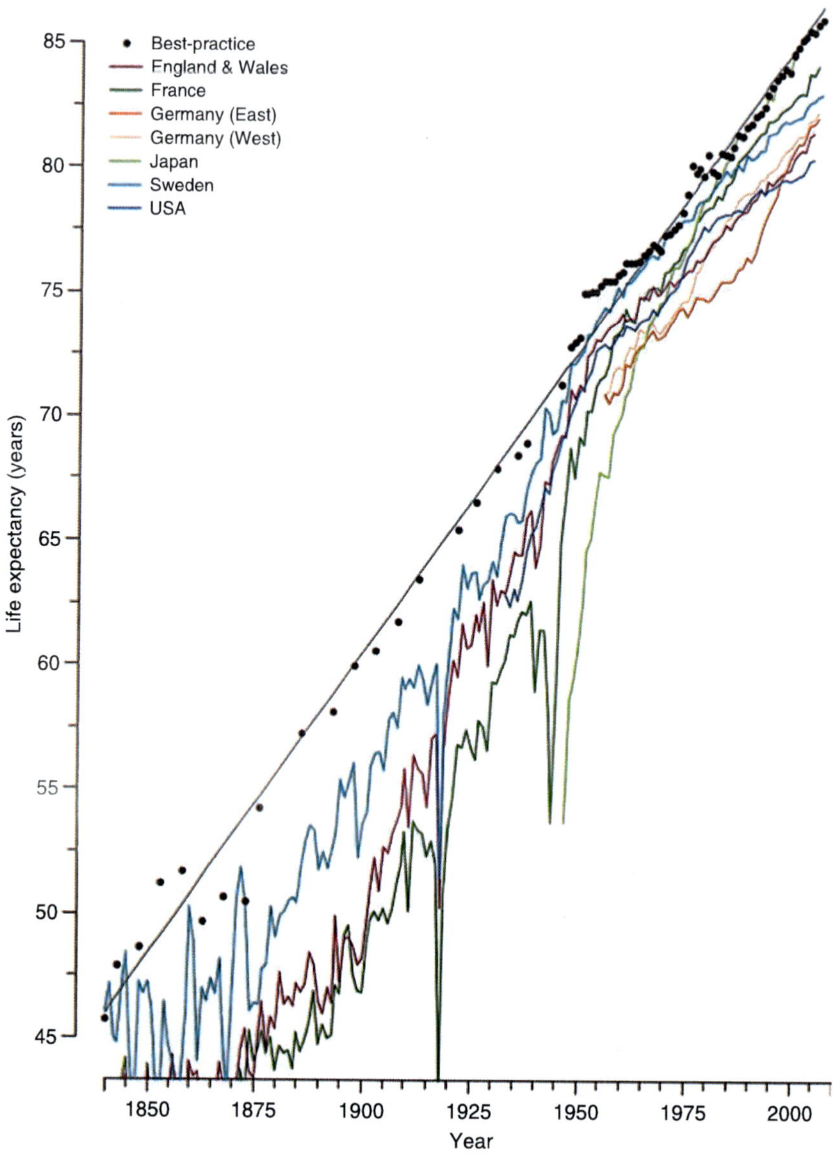

Fig. 1.1 Increase in life expectancy in various economically developed countries since the mid-nineteenth century (from Christensen et al. 2009)

than men in old age, but they tend to enjoy greater social support from a broader network of family and acquaintances than men do.

In terms of the capacity of various countries to secure the well-being and to provide good quality of life to their older citizens, the *Global AgeWatch Index* (HelpAge International 2013) opens a window into the ability of 91 countries to afford their older citizens enough income, ability to maintain good health, and access to quality healthcare services, along with a capacity to participate in various activities within their communities. Northern European and North American countries top the list, along with New Zealand in number 7, Japan (number 10) and Australia (number 14). The best South American country is Chile (number 19) and, not surprisingly, at the bottom of the list are poor countries ravaged by war and civil unrest. It should be noted that although the ranking reflects to a great extent the material wealth of the country, it is also dependent on the specific policies and social factors affecting the well-being of older people, that can be modulated by historical processes and specific political priorities. For instance, although Chile ranks number 19, the much wealthier Brazil (as measured through the country's gross domestic product (GDP): https://data.worldbank.org/indicator/NY.GDP.MKTP.CD), also a South American country, ranks 31. Incidentally, as we revise the text of this chapter in November 2019, we have to note that Chile is in the grip of significant social unrest that was initially motivated by a broad dissatisfaction with the costs of living and wealth inequality in that country: not all that seems to shine is actually gold.

The *Human Development Index* (HDI), an indicator of social progress, is linearly associated with the values of the *Global AgeWatch Index*, but countries such as Canada, Switzerland, The Netherlands, Germany, and Sweden perform much better than the expected from the linear trend due to their national social policies in favour of the elderly.

An important issue to consider when we study older people is that not all cohorts of, say, 60–70 years olds are the same. Even within a single country, each cohort will have experienced different historical processes throughout their lifetime that may mark the way in which they reach and live their old age (Millane 2013). For instance, old people in the 1980s–1990s had been young adults at the time of World War II and the fall of European colonialism that followed across the world; on the other hand, in the 2010s old people had been young adults in the 1970s, when the world was dominated by the Cold War, social unrest, unseen-before levels of consumerism, the rise of feminism, and the emerging conscience of the human impact on the environment. This phenomenon is known as *generational imprinting* (Dorfman et al. 2004). Therefore, both similarities and differences between cohorts of equally aged old people can be explained by a combination of biological, psychological, cultural/social, and historical factors. For instance, the decrease in the number of children per family observed in recent decades may produce a cohort effect in future generations of elders, by decreasing the probability that younger generations may directly help their ageing parents, thus shifting the responsibility of aged care for future cohorts of elderly to the individual (through savings and lifetime investments) and society at large (pension systems and public aged care initiatives) and away from the direct family.

Older generations are better able of taking care of themselves after retirement and enjoy a greater level of well-being the more they have built their capabilities over their lifetime. Following the views of Amartya Sen, capabilities in old age should not only be seen in terms of financial backup, although this is certainly an important issue, but also in terms of the mental, existential, social, and physical scaffolding that allows individuals to build a personally satisfactory level of functioning in their later life:

> The capability of a person reflects the alternative combinations of functionings the person can achieve, and from which he or she can choose one collection. The approach is based on a view of living as a combination of various 'doings and beings', with quality of life to be assessed in terms of the capability to achieve valuable functionings (Sen 1993, p. 271).

> The functionings relevant for well-being vary from such elementary ones as escaping morbidity and mortality, being adequately nourished, having mobility, etc., to complex ones such as being happy, achieving self-respect, taking part in the life of the community, appearing in public without shame ... The claim is that the functionings make up a person's being, and the evaluation of a person's well-being has to take the form of an assessment of these constituent elements (Sen 1993, p. 276).

We agree with Sen that the ability to effectively express personal capabilities and agency is at the core of human freedom and the achievement of a satisfactory level of well-being, this remains true throughout the life of an individual, but it is especially so at older ages. We will introduce and develop later in this chapter our concept of *fulfilling ageing* that encapsulates such notions as those mentioned by Sen and others into a comprehensive view of ageing. The concept will be further developed throughout the book and synthesised in the last chapter.

Personal capabilities may be also translated into material productivity. Material productivity is dependent on many specific factors such as physical and mental abilities, formal education, experience. Although experience is likely to increase with age some abilities that are currently of use in many jobs may in fact decline as technology changes (Skirbekk 2003).

Capabilities that do not require constant retraining and updates, but just maintenance and reinforcement of well-established routines (*crystallised abilities*) may not decline with age; whereas *fluid abilities*—or the capacity to solve new problems and identify new patterns—are relatively more susceptible to age-dependent decline (Horn and Cattell 1967). Whenever constant adaptation to a fast-changing world is required, some older workers may find themselves at a relative disadvantage, leading to higher levels of unemployment or sub-employment, and decreased income (Skirbekk 2003). At that point in their life they may hope to have real choices: going ahead alone if they have sufficient personal resources, fully relying on the support from a good welfare system if they lack personal resources, or enjoying the benefits of a combination of both kinds of resources. Not having any adequate choice is the worst situation an elderly person can expect.

In what follows we provide a general overview of some major areas of interest in ageing studies that will be analysed in greater detail in the remaining chapters.

1.1 The Multiple Dimensions of Ageing: An Overview

Many theories have been proposed over the years to explain the process of ageing at several levels of analysis. Table 1.2 summarises some of the main biological theories that consider mechanisms spanning from the evolutionary to the more proximate molecular, cellular, and systemic ones. Such theories encompass two main broad perspectives of the biological processes of ageing: the "programmed ageing"

Table 1.2 Classification and brief description of the main biological theories of ageing

Biological level/theory	Description
Evolutionary	
Mutation accumulation	Mutations that affect health at older ages are not selected against
Disposable soma	Somatic cells are maintained only to ensure continued reproductive success; after reproduction, soma becomes disposable
Antagonistic pleiotropy	Genes beneficial at younger age become deleterious at older ages
Molecular	
Gene regulation	Ageing is caused by changes in the expression of genes regulating both development and ageing
Codon restriction	Fidelity/accuracy of mRNA translation is impaired due to inability to decode codons in mRNA
Error catastrophe	Decline in fidelity of gene expression with ageing results in increased fraction of abnormal proteins
Somatic mutation	Molecular damage accumulates, primarily to DNA/genetic material
Dysdifferentiation	Gradual accumulation of random molecular damage impairs regulation of gene expression
Cellular	
Cellular senescence–Telomere theory	Phenotypes of ageing are caused by an increase in frequency of senescent cells. Senescence may result from telomere loss (replicative senescence) or cell stress (cellular senescence)
Free radical	Oxidative metabolism produces highly reactive free radicals that subsequently damage lipids, proteins, and DNA
Wear and tear	Accumulation of normal injury
Apoptosis	Programmed cell death from genetic events or genome crisis
System	
Neuroendocrine	Alterations in neuroendocrine control of homeostasis results in ageing-related physiological changes
Immunologic	Immune function dysregulation with ageing results in increased incidence of autoimmunity

From Weinert and Timiras (2003), with slight modifications

view, whereby intrinsic molecular clocks regulate the timing of biological events that describe the ageing process, and the "error accumulation ageing" view that includes inputs from the external environment (stresses, for instance) that can progressively damage aspects of our biology, eventually leading to ageing (see Weinert and Timiras 2003 for a review).

From a genetic perspective, genes that affect phenotypic traits that are damaging to the organism at old ages, and that underlie the process of senescence, are usually expressed after the individual has already reproduced, and therefore they escape the filtering process of *natural selection*; this allows those genes to remain in the population from one generation to the next (Medawar 1946, 1952). In addition, the longer an individual survives, the more likely it is that it may accumulate mutations in the DNA of cells in various tissues (*somatic mutations*), that could eventually lead to genomic instability and eventually ageing, as first suggested by Leo Szilard (1959, see also Trifunovic et al. 2005). Therefore, ageing is likely to be a combination of "programme" and "error accumulation".

Studies carried out on human centenarians support a degree of heritable variation in the ability to survive to oldest ages (Weinert and Timiras 2003). Heritability, however, tends to be lower than 50% as seen in studies of identical twins (Finch and Ruvkun 2001). One gene that has been shown to have clear effects on ageing is the *apolipoprotein E* (*APOE*) gene. Apolipoprotein E, the gene product, is a major cholesterol carrier that is involved in the transport of lipids and injury repair in the brain, along with being involved in the repair of arterial lesions and protection against atherosclerosis. In particular, the *APOE2* allele is especially prevalent among centenarians (Finch and Ruvkun 2001), whereas *APOE4* is associated with dementia. Caloric restriction also increases longevity in a variety of mammals, this could be mediated by insulin and it could therefore be affected by mutations in genes of the growth factor system. The growth factor system includes the hormones insulin, insulin-like growth factor-I (IGF-I) and IGF-II. Recent studies have identified an insulin-like signalling pathway that can regulate lifespan in both invertebrates and vertebrates (Cheng et al. 2005).

Apart from heritable genetic aspects of ageing, there are also modes of inheritance that can affect ageing but that do not involve any change in the primary structure of the genetic material, the DNA. These are known as *epigenetic mechanisms*. Such epigenetic mechanisms include random events in gene expression that affect the phenotype, and DNA methylation. In this case older people may display a variable degree of methylation in diverse regions of the DNA of various cells, which affects the expression of the genes involved (Fraga and Esteller 2007).

Other specific molecular mechanisms have been also proposed that could affect the process of ageing. Ageing at the cellular level can be affected by the accumulated effects of repeated cell replications and stresses suffered by the cell, alongside the effects of the shortening of telomeres following cell replications when telomerase becomes less active. Telomeres are repetitive sequences of DNA found at the tips of the chromosomes that protect the integrity of chromosomes. They are shortened by each cell replication but are reconstituted thanks to the action of the enzyme telomerase. The longer the telomeres are, the more likely it is that the person can survive beyond the age of 60 years old (Muller et al. 2007). Telomeres can be shortened by life events such as long-term chronic psychological stresses and infections (e.g. Effros 2011).

It is also known that oxidative damage to DNA—and to proteins and lipids—by reactive oxygen species (ROS, such as hydroxyl radicals, hydrogen peroxide, and superoxide anions that are mainly produced by mitochondria) accumulates with age. This can have a direct effect on the ageing process (Martin et al. 1996; Finkel and Holbrook 2000; Weinert and Timiras 2003; Beal 2005; Muller et al. 2007), a suggestion that was first made by Denham Harman in the 1950s (Harman 1957). Genes which products can defend tissues against the effect of free radicals, as it is the case of superoxide dismutase (SOD), can delay the process of ageing. In addition, reduced availability of calories can reduce the capacity of mitochondria to produce ROS, thus also delaying cellular ageing (Finkel and Holbrook 2000; Wallace 2005).

ROS can be produced endogenously within the cell by organelles such as mitochondria or peroxisomes. They can have physiological functions in homeostasis, helping in normal growth and metabolism, but they can also have damaging effects on the organism through their roles in ageing, cell death and disease if they are in

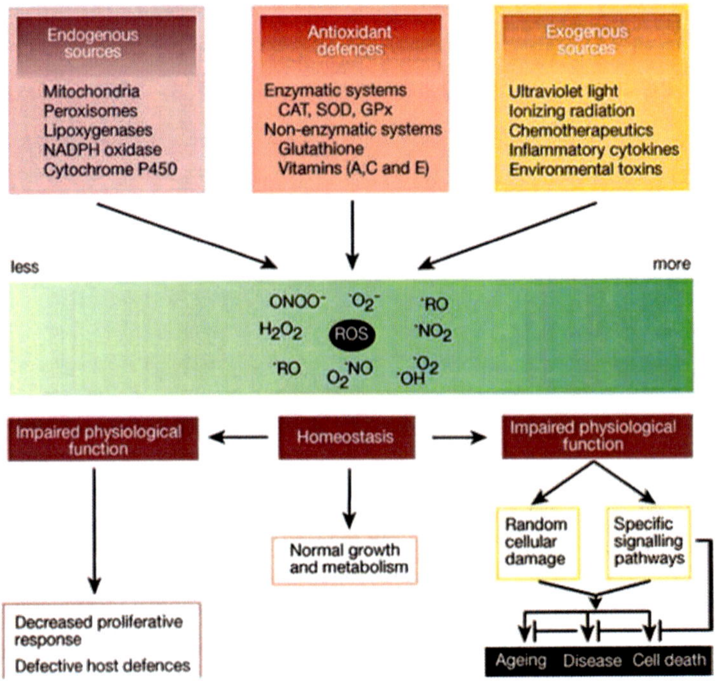

Fig. 1.2 Various sources of reactive oxygen species (ROS) and antioxidant defences, along with effects of ROS on the organism's physiology, including ageing (from Finkel and Holbrook 2000)

high concentration. At lower concentrations than normal, they can impair the immune response. Production of ROS is regulated by antioxidant molecules such as SOD and various vitamins as summarised in Fig. 1.2.

Intracellular degradation of macromolecules occurs in the lysosomes as part of the normal functioning of cells. This happens through a process known as *autophagy*, which literally means "eating itself", but autophagy decreases in effectiveness with age. Age-dependent failure in autophagy leads to waste accumulation within cells and a decrease in the normal functioning of such cells. In turn such malfunction will affect the tissues those cells are part of and the physiological processes the tissues are involved in (e.g. immunity, brain activity), leading to altered health, physiological performance, and behaviours (Cuervo et al. 2005). *Proteasomes* are intracellular multi-subunit assemblies of proteases that can specifically contribute to the ageing process through this autophagy mechanism in the cells of the nervous system (Keller et al. 2002).

Apoptosis, an internal mechanism of cellular self-destruction, that can be very adaptive in the defence against cancer, can also have the negative side effect (*antagonistic pleiotropic* effect) of promoting tissue senescence (Campisi 2003).

At the systemic level, ageing can be expressed in changes to endocrinological, immunological, and neurological functions that in fact form a closely integrated and interacting biological network, thus what affects some parts of this neuro–immune–endocrine system may also affect others.

Table 1.3 Some endocrine changes with ageing

Hormone	Change from younger to older ages	Reference
Norepinephrine	Increases	Noth and Mazzaferri (1985)
Renin	Decreases	Noth and Mazzaferri (1985)
Aldosterone	Decreases	Noth and Mazzaferri (1985)
Thyroid hormone	Unchanged	Noth and Mazzaferri (1985)
Thyroxine	Decreased metabolism	Noth and Mazzaferri (1985)
Vasopressin	Increases	Noth and Mazzaferri (1985)
Glucocorticoids	Unchanged	Noth and Mazzaferri (1985)
Delta-5 adrenal steroids	Decreases	Noth and Mazzaferri (1985)
Heat shock proteins	Decrease	Weinert and Timiras (2003)
Dehydroepiandrosterone	Decreases	Weinert and Timiras (2003)
Growth hormone	Decreases	Clegg et al. (2013)
Insulin-like growth factor-1 (IGF1)	Decreases	Clegg et al. (2013)
Oestradiol	Decreases	Clegg et al. (2013)
Testosterone	Decreases	Clegg et al. (2013)
Luteinising hormone	Increases	Clegg et al. (2013)
Follicle-stimulating hormone	Increases	Clegg et al. (2013)

Table 1.3 summarises some major hormonal alterations described in older people, such alterations contribute to physiological changes in the elderly such as hypertension, carbohydrate intolerance (due to changed responsiveness to insulin), hyponatraemia (lower sodium in plasma) (Noth and Mazzaferri 1985); modification of "biological clocks" and of the hypothalamo–pituitary–adrenal/gonadal axes (Weinert and Timiras 2003; Shaw et al. 2010; Clegg et al. 2013). Endocrinological alterations that occur with ageing are also co-contributors to the general condition of frailty observed in older people, that is expressed in lower neuronal plasticity and skeletal muscle strength, reduced energy expenditure and decreased body mass, increased weakness and fatigue, among other changes (Clegg et al. 2013). The *frailty syndrome* is a progressive decline in the performance of multiple body systems with ageing, leading to greater vulnerability to disease and eventually death (Espinoza and Walston 2005). More specifically, it is "a state of muscular weakness and other secondary widely distributed losses in function and structure" (Bortz 2002, p. M284). The insulin-like growth factor-1 (IGF1) in particular, has been linked to the development of frailty (Chen et al. 2014), and it is also involved in mechanisms that regulate the lifespan (e.g. Russell and Kahn 2007).

A more detailed description of the hormonal changes occurring with ageing in various endocrine organs is provided in Fig. 1.3 (see also van den Beld et al. 2018). The thyroid function is highly variable in old age and such variability can be affected by increased autoimmunity within the thyroid in older individuals. Levels of the growth hormone (somatotropin) produced in the liver tend to decrease as we age, as it does ghrelin, a hormone that controls appetite and food intake. Lower levels of ghrelin in old age can explain reduction of appetite and therefore body mass reduction among the elderly. Other hormones can also contribute to the development of anorexia in old

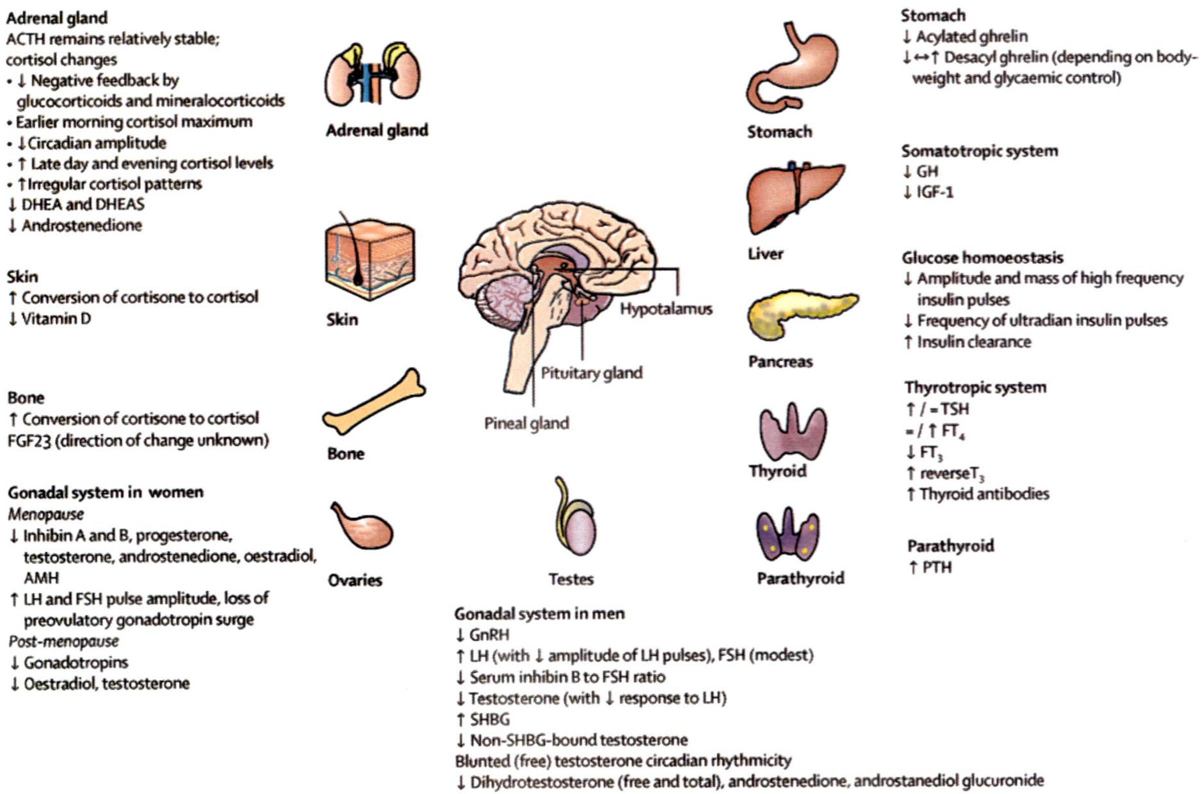

Fig. 1.3 Most common changes in circulating hormone levels and action occurring with ageing in different organs. Reproduced, with permission, from van den Beld et al. (2018)

people, such as cholecystokinin, leptin, and some cytokines that increase their activity with ageing. Cortisol secreted by the adrenal glands shows some complex patterns with ageing becoming irregular in its changes in concentration. Circulating levels of gonadal hormones decrease in old age, especially in women after menopause (oestradiol and testosterone), but also in men (testosterone), due to lower levels of gonadotropins. Lower concentrations of gonadal hormones in circulation negatively affect bone mass and strength, and muscle mass and function (van den Beld et al. 2018) along with increasing skin ageing (Jung et al. 2008). Glucocorticoids levels are maintained with ageing although in old age the daily fluctuation in circulating glucocorticoids is lower and the levels later in the day tend to be higher than in younger individuals (Ferrari et al. 2001). Levels of circulating parathyroid hormone increase with ageing, leading to decreased bone mineral density and therefore increased likelihood of experiencing bone fractures. Glucose metabolism is also altered in old age being affected by a progressive decline in insulin action which, in turn, is affected by obesity, high caloric intake, and decreased physical activity among other factors. This means that the risk of developing diabetes increases with age, although the risk does not increase further beyond the age of 85 (van den Beld et al. 2018).

An especially important alteration of bodily functions with increasing age is that concerning immunity, a process that is also referred to as *immunosenescence*. Given that the immune system is crucially involved in the defence of our body against infectious and other diseases, a decrease in the competence of such system can only increase the chances of serious bodily harm, eventually leading to death. In addition, dysregulation of the immune system with age can also lead to increased probability of autoimmune diseases.

Our immune system is comprised of two major types of organs: (1) the primary lymphoid organs (e.g. bone marrow and thymus), where the monocytes, natural killer (NK) cells, polymorphonuclear granulocytes, and T-lymphocytes complete their maturation, and (2) the secondary lymphoid organs (e.g. lymph glands, mucus-associated lymphoid tissue, and spleen), where antigens or foreign molecules interact with cells of the immune system and where the B-lymphocytes complete their maturation.

The response of such immune organs can be classified into two broad categories: an innate immune response that can be non-specific and is carried out by leukocytes (or white blood cells) such as granulocytes, lymphocytes, monocytes, natural killer cells, mast cells and phagocytic cells (macrophages, dendritic cells, neutrophils) and a more specific or adaptive (or acquired) immune response that mainly involves T-lymphocytes (CD4 "helper" and CD8 cytotoxic T-cells) and B-lymphocytes. Cellular immunity involves the action of T-lymphocytes, whereas humoral immunity implicates B-lymphocytes.

The immune reaction following an infection can be local, in which case inflammation occurs caused by lymphocytes and other immune cells through the release of cytokines (e.g. interleukin-1 (IL-1), IL-6 and tumour necrosis factor (TNF)) or it can be systemic, a reaction involving cytokines released by immune cells in distant tissues such as those of the nervous and endocrine systems. When the immune system becomes dysregulated, as it happens in the process of ageing (Li et al. 2011; Shaw et al. 2013), it may turn against the tissues of the body leading to autoimmune diseases.

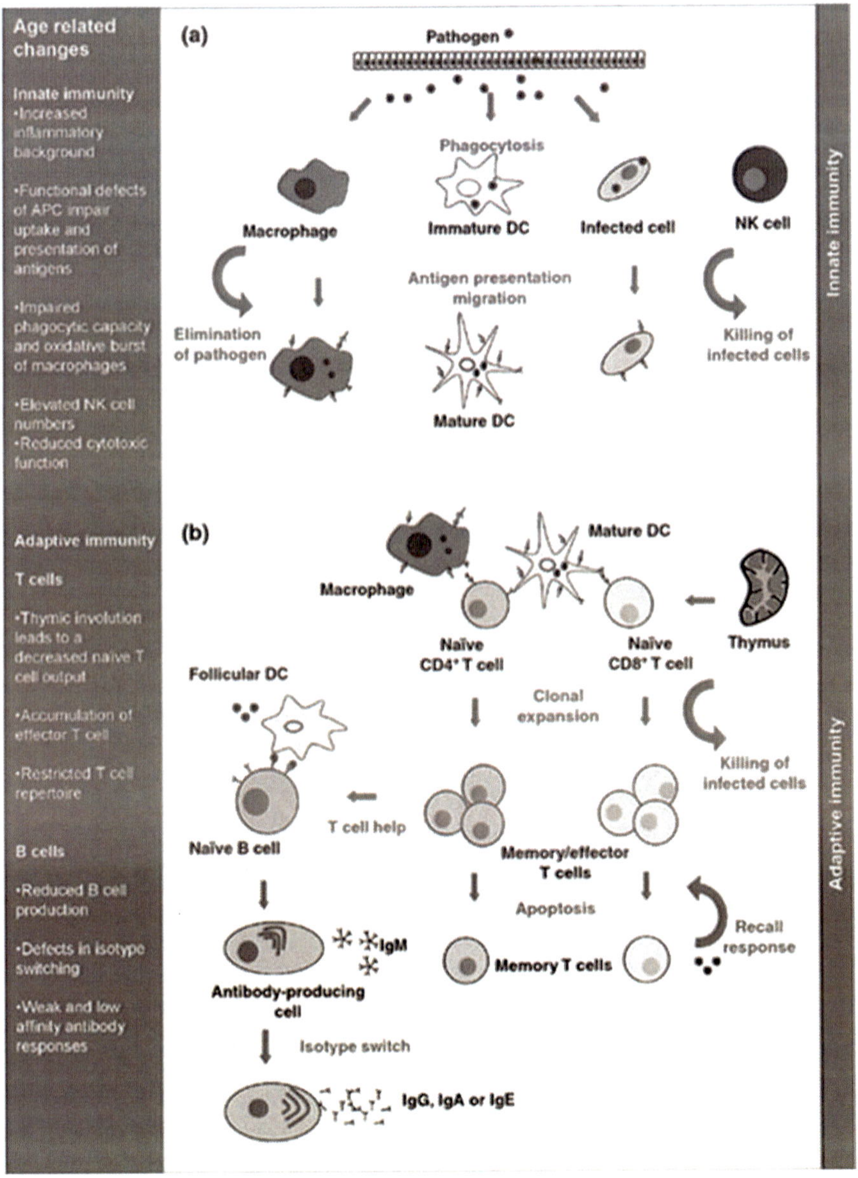

Fig. 1.4 The ageing immune system. (a) Innate immunity; (b) Adaptive immunity (from Weiskopf et al. 2009)

As we age, both innate and adaptive immune capacities decrease, as do cellular and humoral immune responses (Solana et al. 2006). Figure 1.4 summarises the main immune changes that occur with ageing. The innate immune response triggered by an infectious pathogen becomes less efficient at older ages through a functional impairment of antigen presenting cells (APC), impaired phagocytic and oxidative capacity of macrophages and reduced cytotoxicity of NK cells (on a per-cell basis), only partially compensated by increased numbers of NK cells in older people. Increased dysregulation of immunity with age causes increased levels of inflammatory background known as *inflamm-ageing* (e.g. Gomez et al. 2005; Liberale et al. 2020). This can eventually lead to autoimmune disorders in the elderly such as neurodegeneration. Cellular adaptive immunity is reduced by thymic involution and a more restricted repertoire in the T-lymphocytes populations (see also Agarwal and Busse 2010), whereas decreased humoral adaptive immunity is seen, for instance, in the reduced production of B-lymphocytes and their impaired affinity to antibodies.

In particular, it is the naïve lymphocyte populations that are affected by ageing. As we age, and the thymus decreases in size and productivity (e.g. Hakim and Gress 2007), fewer functional and smaller numbers of naïve T-lymphocytes are produced. Naïve B-lymphocytes also decrease in numbers with age, and antibodies become of lower affinity, as we already mentioned (see Weiskopf et al. 2009, for a review). In fact, changes in humoral immunity in the elderly involve not only a reduced response of antibodies, with the period of protective immunity shortened and lower antibody affinity, but also an increase in the number of autoantibodies that can potentially cause autoimmune disorders (Prelog 2006; Hakim and Gress 2007; Sansoni et al. 2008).

It has been suggested that immune senescence could be also mediated, at least partially, by epigenetic processes involving DNA methylation (Issa 2003). In addition, both T- and B-lymphocytes suffer shortening of their telomeres with ageing which compromises their functioning (Weng 2008).

Many other aspects of the biology of the organism are also affected by ageing. For instance, sex differences have been described in ageing. Although survival is greater in females than males, with females accounting for 59% of the 65-year-old or older population (Strawbridge et al. 1993; Peace et al. 2007), males tend to reach older ages with a better capacity to perform daily activities, but this is especially the case for cohorts where women have lower education, practice less exercise, have higher rate of smoking, and are more likely to be separated or divorced (Strawbridge et al. 1993).

With regard to general health conditions, although they have been improving for the elderly in recent decades thus extending life expectancy, given that individuals live longer morbidity has also increased, such as vision and hearing impairments, hypertension, back and neck and other neuromuscular problems, loss of muscle mass (sarcopenia), osteoporosis, heart pathologies, diabetes mellitus and other autoimmune conditions, incontinence and increased physical functional limitations among others, the latter are especially prevalent among women (Okada et al. 2009; Doherty 2008; Sjölund et al. 2010). In addition, thermal regulation is also impaired in the elderly along with other autonomous body functions (Doherty et al. 1993; Roos et al. 1997; Verdú et al. 2000).

Declining health can jeopardise the ability of old people to carry out their daily activities. Researchers often measure the ability to perform activities that are necessary for daily living (ADL)—such as dressing, bathing, ability to eat, and using the toilet—in their studies to understand the effects of ageing on general well-being; but instrumental activities of daily living (IADL) that are not strictly fundamental for daily functioning (such as cleaning the house), are also measured. In the last two decades results of such studies have shown various trends regarding values of ADL and IADL in the elderly, some indicating deterioration over time, others suggesting no change, and some aspects of daily living may even improve with age, perhaps due to improved standards of living in many countries and also improved education. In general ADL disabilities are more pronounced in older women than men (Parker and Thorslund 2007).

Body changes with age can be also estimated through anthropometric measurements (total body weight, fat-free mass, fat mass, height, for instance). Body weight decreases with age (at a higher rate in males), especially since the age of 75-year-old. In spite of this, obesity is common in some elderly, among women in particular, with subcutaneous fat tending to increase in older people (Hughes et al. 2004; Perissinotto et al. 2002), and there is redistribution of fat tissue and accumulation in the trunk and visceral areas (Perissinotto et al. 2002). But there is also an overall reduction of fat-free mass with age (Perissinotto et al. 2002), which explains the broad trend towards an overall decrease in body weight observed in most elderly people. Lean mass loss, also known as *sarcopenia*, is more likely to occur in elderly men than women (Newman et al. 2005).

Although body height decreases with age, older men are on average taller than women and because of this they are also heavier, as it is the case for younger ages as well (Perissinotto et al. 2002). Body shape measurements such as the mean value of waist circumference decreases with age, as it does hip circumference in both sexes, although in the latter case the hip circumference is higher in women than men. Waist-to-hip ratio is significantly higher in males, where it decreases with age, whereas in females it increases with age (Perissinotto et al. 2002; Hughes et al. 2004). Body mass index (BMI) measured in kg/m^2 is higher in females than males and it decreases with age in both sexes (Perissinotto et al. 2002).

Older ages are also associated with increased probability of tumour and cancer development, in part due to immune dysregulation occurring with ageing (e.g. Vasto et al. 2009). Cancer developing in old individuals may or may not be detected. In a study of 3535 necropsies of 66-year-old and older patients (47.8% females) carried out over an 11-year period at the Kings County Hospital Center in Brooklyn, New York, Suen et al. (1974) found an overall prevalence of cancer of 32.5% (743 males and 406 females), and the prevalence of incidental, that is undetected cancer increased with age in the population as a proportion of total cancers (Table 1.4).

Sensorimotor capacities such as visual and auditory acuity, grip and lower limbs strength, expiratory volume, and tactile abilities, also decrease with age, which may affect cognitive capacities in older people. In fact, exercises that improve motor functions also translate into better cognitive capacities in old age (Schäfer et al. 2006; Wickremaratchi and Llewelyn 2006).

Table 1.4 Incidental (undetected) cancers by age group in female and male elderly patients

Age group	Total number of cancers	Number of incidental cancers	Percent
66–75	670	139	20.7
76–85	469	138	29.4
86 and over	130	47	36.2

The difference is statistically significant: $X^2 = 20.1$, $P < 0.001$. Incidental cancers are those that were not suspected during the life of the individual and did not contribute to the patients' death. From Suen et al. (1974)

So far, we have seen the effects of ageing on various aspects of body functioning and hinted at cognitive issues in the previous paragraph. But if we want to understand the biological substrate of behavioural change with ageing, we must turn our attention to the nervous system. However, we should also note that the nervous system does not function in isolation, instead it is closely interlinked with other systems (e.g. endocrine, immune). Inter-systemic communication is facilitated by molecules such as cytokines, neurotransmitters, and hormones and such complex network of molecular and cellular relationships is characterised by a high degree of plasticity, which is more prevalent at younger ages but which, to some extent, could be also extended to older ages. Such plasticity may be put to good use in interventions that can promote fulfilling ageing, characterised by better health and mental capacities (Mazzoccoli et al. 1997; Weinert and Timiras 2003). The immune system has been implicated in the modulation of learning and memory processes via neurogenesis and neural plasticity. Such functions are mediated by neurotransmitters, hormones such as glucocorticoids and prostaglandins, cytokines such as IL-1, IL-6, and TNF-α, and neurotrophins, the latter modulate survival, growth, and differentiation of neurons (e.g. Yirmiya and Goshen 2011).

Regarding the nervous system more specifically, both the peripheral and the central nervous systems can be affected by ageing. Two major changes occurring with ageing that affect peripheral neurons are the loss of nerve fibres and alterations in the myelination patterns of nerve cell axons, such as loss of myelin sheath that can then affect conduction of the nerve impulse. Axonal atrophy may also occur and capacity for nerve regeneration decreases. All this can affect the muscular function leading to motor weakness and lower coordination, there is also decreased sensory capacity. Overall, those changes contribute to the *frailty syndrome* that we mentioned above in this section.

Complex mental and behavioural faculties, however, are critically dependent on the activity of the central nervous system, especially the brain and its functions. Therefore, we now turn our focus on this issue.

The central nervous system is composed of the brain and the spinal cord. Studies carried out in rodents indicate that ageing of the spinal cord is characterised by a decrease in the number of terminal connections of motoneurons—which vary according to the area of the cord (Kullberg et al. 1998)—and a decrease in the intensity of capillaries, or the fine branchings of blood vessels (Qiu and Zhu 2004). Early studies in humans based on preserved tissues noted that the spinal cord of older people shows calcareous deposits, thickening of arterial walls, increased proliferation of astrocytes—a type of glial cells—and atrophy of the myelin sheath in various regions (Bailey 1953), the latter is a structural change that decreases the efficiency of nerve impulse transmission.

Ageing is also associated with degenerative alterations of sensory and motor fibres (Lindberg et al. 2010, and references therein) and broader progressive degeneration of various areas along the spinal cord (e.g. Mufson and Stein 1980; Lexell 1997; Okada et al. 2009). Such changes, however, are variable and different individuals may express different degrees of plastic capacity to compensate for the effects of degenerative processes (Wolpaw 2006).

The brain, and the functions controlled by the various brain regions also change with age. Memory clearly declines as we become older and this is certainly the case for delayed recall of verbal information, working and spatial memory, and short-term recall, but also aspects of long-term memory and the speed of processing information. Our spatial navigation ability tends to decline with age as it does trace eyeblink conditioning, whereas there is better preservation of memory of some specific long-term life events, world knowledge, and unconscious responses as we age. There is also an improvement in attention span—but with inter-individual variability, Zanto and Gazzaley (2014), and some aspects of verbal knowledge (Park and Reuter-Lorenz 2009). Interestingly, some aspects of emotionality may also improve with age such as the emotional aspects of memory and emotional stability. All these changes mirror functional changes in both the prefrontal cortex and the hippocampus, two areas of the brain that in combination control memory and the production and interactions of emotions and cognition. As the human brain ages, more areas are involved in the control of various behaviours and mental processes, indicating a trend to compensate for loss of brain tissue and functionality across some areas of the organ over time (Burke and Barnes 2006; Yankner et al. 2008).

Past studies of the ageing brain included old individuals with dementia that obviously biased the conclusions when they were compared with healthy young brains. Subsequent studies of both healthy young and older human brains have shown that reduction in the number of neurons with age is not as widespread a phenomenon across all areas of the brain as previously thought. On the other hand, the amount of white matter—that is mainly formed by neuronal axons and glial cells—decreases with ageing in areas of the brain such as the prefrontal cortex and the corpus callosum. Although neuronal loss is variable across some areas of the brain it may be compensated by increased dendritic branching of the surviving neurons, but older individuals may also experience a decline in the synaptic function of those neurons. That is, neurons of older brains may decrease their functional connections with other neurons and therefore their ability to transmit impulses. Synaptic plasticity may be reduced in the elderly by alterations in their calcium metabolism (Burke and Barnes 2006; Yankner et al. 2008). In addition, the number of dopaminergic receptors on neurons also declines in older people thus affecting cognitive capacities, whereas the number of neurofibrillary plaques and tangles—which have been associated, although not in a straightforward manner, with onset of dementia—increases (Park and Reuter-Lorenz 2009).

In a study carried out in the USA, Raz et al. (2005) compared 72 healthy individuals ranging from 20 to 77 years old—both men and women—focusing on some aspects of their brain characteristics, using magnetic resonance imaging (MRI) scans. They found that although there was a general trend for brain shrinkage over time, the extent is variable across different areas of the brain, being higher in the cerebellum, hippocampus (both of which displayed acceleration of shrinkage with age), caudatum, entorhinal and tertiary association cortices, but smaller in the fusiform and entorhinal cortices, whereas the visual cortex remained unchanged (see also Park and Reuter-Lorenz 2009, although Salat et al. 2004 did find thinning of the visual cortex with age). The calcarine cortex, an area close to the visual cortex, also shrinks with age (Salat et al. 2004). These patterns, however, were variable inter-individually. The shrinkage seems to start in the mid-1950s, it is sensitive to hypertension and it can be explained by a combination of factors including loss of neurons, myelin, and dendritic arborisation.

Changes in brain structure and function with ageing are affected by the *brain reserve capacity*. Brain reserve capacity is a concept first proposed by early students of dementia in the 1960s and 1970s and it is broadly associated with brain size, architecture, and function (Satz 1993). Such reserve capacity just expresses the ability of the brain to respond to environmental and internal challenges, including traumas and injuries, but also ageing, in a functionally adaptive manner. Although some studies (e.g. Raz et al. 2005; Christensen et al. 2007) have not supported the *Brain Reserve Capacity hypothesis*, others have (e.g. Valenzuela and Sachdev 2006; Sumowski et al. 2014). More recently, the maintenance of brain capacity with age has inspired the *Scaffolding Theory of Ageing and Cognition* (STAC) of Park and Reuter-Lorenz (2009):

> After reviewing the range of data examining brain and behavioral function with age, we believe that the corpus of these data suggests that the brain is a dynamic organism seeking to maintain homeostatic cognitive function...We argue that the brain responds to these neural insults by engaging in continuous functional reorganization and functional repairs that result in self-generated support of cognitive function (p. 174).

By *scaffolding* they mean "the recruitment of additional circuitry that shores up declining structures whose functioning has become noisy, inefficient, or both". Such dynamic process of brain plasticity has been reported by various authors, being illustrated by the differential activation of alternative areas of the brain in the performance of a similar task as the individual ages. Figure 1.5d shows an example of high-performing older adults improving their positive results in the performance of an episodic memory retrieval task through increased bilateral brain activation in the frontal lobe (Cabeza et al. 2018). Plasticity may help the brain of older people perform better, especially when the task is not exceedingly demanding, as suggested in Fig. 1.5a, b. Recruitment of other brain regions by older people, in order to compensate for decreased efficiency of the main brain areas devoted to specific tasks, may just be a best possible job performed under constraining circumstances (Fig. 1.5c).

Therefore, brain compensation in the face of increasing challenges due to ageing can be achieved in different ways, as illustrated in Fig. 1.5. As task-demands increase the brain may compensate through *upregulation* by increasing neuronal activity; or compensation may occur through *selection* by using brain areas that are also available to young adults but that they do not use for those specific tasks. Finally, compensation may operate through *reorganisation* in which older people use neural mechanisms that are not available to younger adults (Cabeza et al. 2018).

Although the plastic ability of the brain can somewhat buffer our mental capacities from the changes coming with ageing changes do occur, and they can simultaneously affect various interrelated functions such as sensory perception and cognition (Baltes and Lindenberger1997). Interestingly, Park and Reuter-Lorenz (2009) stress how the increased de-differentiation—that is, less specialisation—occurring in areas of the brain as we age can be at the core of greater integration between sensory and cognitive faculties in older individuals: a change that can in fact be quite adaptive.

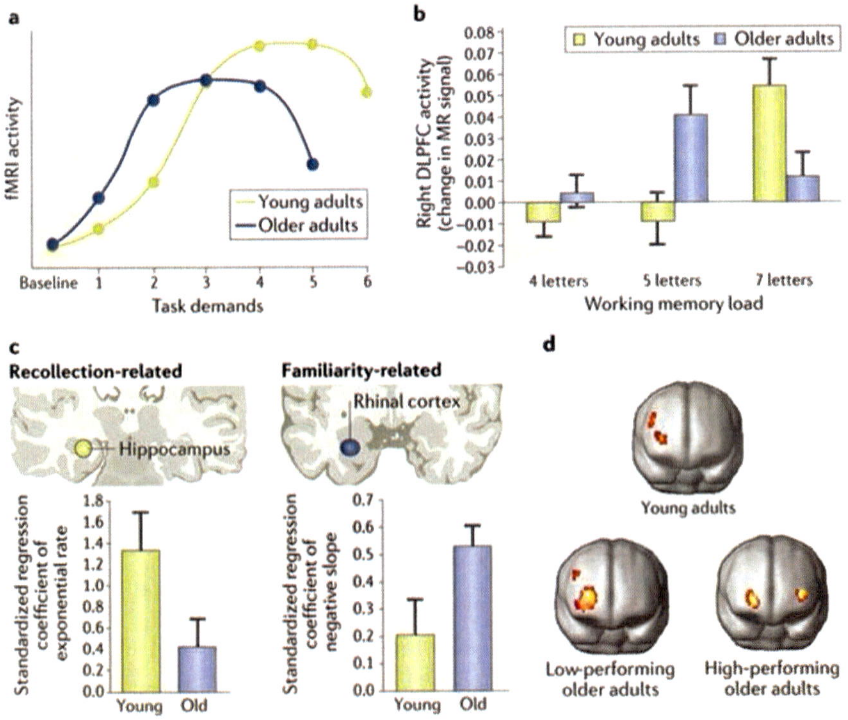

Fig. 1.5 (**a**) Theoretical patterns of compensation by upregulation. (**b**) An example of compensation by upregulation involving a working memory-related activity in the right dorsolateral prefrontal cortex (DLPFC). (**c**) Compensation by selection shown in a fMRI study comparing ageing effects on recollection and familiarity (a less precise form of memory). Older adults compensated for reduced recollection-related activity in the hippocampus with increased familiarity-related activity in the rhinal cortex. (**d**) Compensation by reorganisation is shown during an episodic memory retrieval task where young adults and low-performing older adults displayed unilateral frontal activity, whereas high-performing older adults responded with bilateral frontal activity. Reproduced from Cabeza et al. (2018) with permission

Working memory is defined as the "brain system that provides temporary storage and manipulation of the information necessary for such complex cognitive tasks as language comprehension, learning, and reasoning" (Baddeley 1992, p. 556). Shorter-term memory is referred to as *primary memory*. Memory is a specific cognitive ability that not only tends to decline with age as we have already mentioned, but when it does it is often noticed by the individual and it can be cause for concerns. The degree of decline in memory, however, is not uniform across various tasks. For instance, some aspects of short-term memory seem to be resilient to ageing, and the mnemonic task can be strengthened by repetition (Grady and Craik 2000). The process of memorising may be interfered with in older people by multi-tasking, but it can be helped by verbalising. In addition, older people are more prone to experience false memories and, in general, they also display lower capacities for *prospective memory* (i.e. remembering to carry out a task in the future) than younger individuals. Such memory difficulties can be improved, to some extent and in some individuals, through a compensatory role of a variety of brain regions such as the prefrontal cortex. The prefrontal cortex is an area of the brain that can be somewhat resilient to ageing in healthy individuals, although the pattern is not homogeneous across its ventral, dorsal, and anterior regions (Rajah and D'Esposito 2005).

Apart from changes in memory, both emotions and various cognitive processes may be also affected by ageing. It is broadly recognised that affective well-being and emotional stability are greater at older ages, especially in the oldest-old. This trend can be explained by the combined effect of both ability and motivation to regulate emotional states as we become older.

As we age, existential issues regarding end of life also come to the fore more frequently, thus affecting our emotional states, a trend that has been encapsulated in the *Socioemotional Selectivity theory* (Carstensen 2006). In older ages, both positive and optimistic views tend to be emphasised in response to existential challenges. Greater optimism in older ages can also result from a compensatory response to decreased cognitive capacities with age, as purported by the *Dynamic Integration theory* of Labouvie-Vief (2003). Decreased cognitive capacities may also affect the ability of the individual to control the events of life. One way to recover a degree of control (at least in a psychological sense) is by using *emotion regulation* (see the *Life-span theory of control* of Heckhausen and Schulz 1995). These views taken together point to decreased cognitive abilities at older ages which can be compensated by emotional regulation. Both effects have been synthesised into a *Selective Optimisation with Compensation* (SOC) model (Baltes and Baltes 1990) which states that:

> selective changes in emotional preferences, compensatory efforts to adapt to declining cognitive and control capacities, and the optimization of emotional behavior through lifelong learning and practice can all be expected to drive changes in emotional experience and regulation across adulthood (Scheibe and Carstensen 2010, p. 136).

As applied to old age in particular, the SOC model simply emphasises the well-known interaction between emotion and cognition, where the *selection* considered in the model refers to an "increasing restriction of life domains as a consequence or in anticipation of changes in personal and environmental resources. In old age, these changes are often losses. Selection can entail the avoidance of one domain altogether or it can mean a restriction in tasks and goals within one or more domains" (Baltes and Carstensen 1996, p. 406). In other words, old people, over time, may prefer to use those faculties that remain most functional. *Compensation* refers to the use of reserve capacities to compensate for the loss of abilities in other domains with age, with the objective to maintain an adequate level of functioning. *Optimisation* refers to the "enrichment and augmentation of reserves or resources and, thus, the enhancement of functioning and adaptive fitness in selected life domains" (p. 412).

Cognition varies with age and such changes run in parallel with functional changes in our brain. Both the brain and its cognitive capacities, however, also express a variable degree of plasticity which can be functionally adaptive. Some cognitive theories of ageing tend to mainly focus on the cognitive deficits experienced by older people, seeing them as inevitable, irreversible, and universal, an approach that has been labelled *nihilistic* (e.g. Eisdorfer and Cohen 1980). Such theories tend to emphasise deficits in cognitive processing, frontal cortex function, and decision speed. This in turn may lead to the adoption of *therapeutic nihilism*, which simply accepts the inevitability of the negative effects of old age and allows nature to follow its course. On the other hand, other authors suggest that such deficits could be at least partially overcome with the help of environmental and therapeutical support (see Greenwood 2007). Greenwood's (2007) *Functional Plasticity in Cognitive Ageing hypothesis* postulates—similarly to Park and Reuter-Lorenz's (2009) *Scaffolding Theory of Ageing and Cognition* (STAC) and Satz's (1993) *Brain Reserve Capacity hypothesis* mentioned above—that as regions of the brain may lose volume with advancing age a plastic shift may occur whereby other areas of the brain increase their control of some of those threatened functions. In the words of Pamela Greenwood (2007): "I argue for the existence of an interaction between volume loss and plasticity, with regional decline in brain integrity accompanied by both atrophy-driven and strategy-driven plasticity in the same and adjacent regions".

The process of compensation for decreased cognitive abilities in older ages may be, therefore, helped by training (e.g. Lustig et al. 2009). Moreover, resilience to brain changes in old age starts at younger ages, not only in terms of education and opportunities to develop cognitive capacities, but also in terms of protecting the individual from traumas such as those caused by infections that can negatively affect cognitive development (e.g. Case and Paxson 2009).

In a broader approach to the study of ageing it is essential that we adopt a *life course paradigm*, which could be defined as *the developmental pathways that individuals go through in their life as they are influenced by individual, social, and other environmental factors*. This is relevant whether we are interested in cognition or any other specific physiological or psychological aspect of our life experience.

Individuals go through their life following an individual life course trajectory which may comprise a variable number of identifiable stages punctuated by periods of transitions from one stage to another. In this historical process of development, previous stages of the life course are likely to affect later stages. Glen Elder Jr. (1994, p. 5) identifies four major themes of the life course paradigm:

(a) the interplay of human lives and historical times,
(b) the timing of lives,
(c) linked or independent lives, and
(d) human agency in choice making.

Historical contingencies affect our life course starting from our very birth year, which will expose us to a specific historical world characterised by a set of social issues, technological advancements, broader environmental conditions, prevailing worldviews, morals, and so forth. Throughout development, the individual life is punctuated by social and cultural events that affect the timing and duration of life stages, the sequence of roles, expectations, and beliefs that are socially marked based on the specific life stage. Such life course trajectory is also affected by individual characteristics that can change over time. Individuals do try to control and direct the course of their life despite cultural and social pressures, but such *agency* and its results are variable across individuals. The external world, and the social world in particular, may constrain or help the capacity of individuals to exert agency and achieve their goals.

Various models have been proposed to make sense of the broader process of the life course. The *Generation-based model*, for instance, focuses on reproductive events and inter-generational processes of socialisation, whereas the *Age-based model* puts more emphasis on the effect of external social structures throughout the life span. Such "generational" and "age" perspectives are just two aspects of a broader process that combines the two (Elder 1994). A more promising approach to understand the dynamics of the life course must necessarily be integrative of various dimensions: the individual, the social, and the cultural/historical.

The life course approach can also inspire early interventions through education or social programmes that can use physical activity, cognitive development, and optimisation of emotional behaviours to favour a better ageing experience (Middleton et al. 2010). In fact, as shown in Fig. 1.6, women who had been involved in greater physical activities during their teenage years have relatively better cognitive capacities at all adult ages, tendentially including the older ages, than those who were not physically active as teenagers. Although Fig. 1.6 emphasises the life-long cognitive benefits of early involvement in physical activities, a life course approach must also consider all other aspects of the personal experience, including the negative effects of past traumas.

In a study of Holocaust survivors carried out in Israel by Stessman et al. (2008), as part of the *Jerusalem Longitudinal Cohort Study*, they interviewed 605

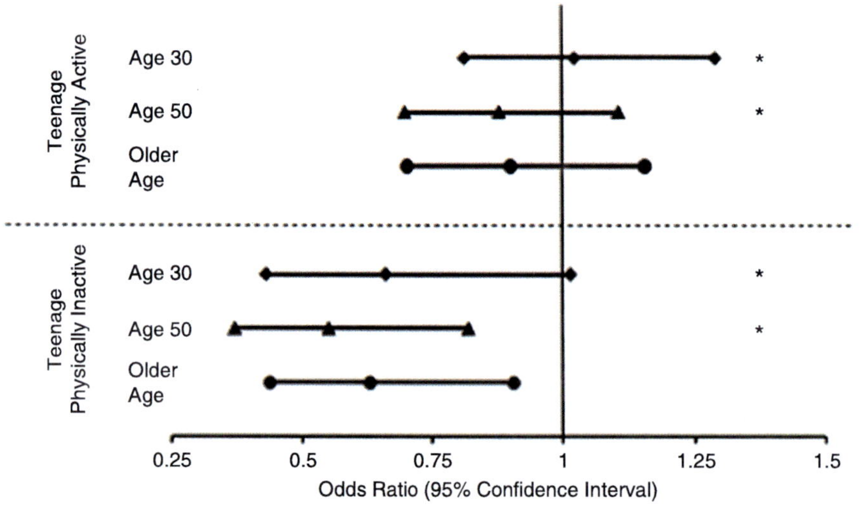

Fig. 1.6 Cognitive impairment over the life course in older women who were physically active versus those who were inactive as teenagers. Physically inactive women are significantly more cognitively impaired at 30 years old ($P = 0.004$) and 50 years old ($P = 0.011$). The trend remains the same at older ages, but it is not statistically significant ($P = 0.26$). From Middleton et al. (2010)

participants who were living in Jerusalem and were born in 1920/21. The group was subdivided into people who survived extreme experiences in extermination, concentration, or slave labour camps (Camps group) and those who had lived under Nazi occupation but were not interned in camps (Exposure group). Although from a psychological perspective the two groups appeared similar, sedatives and psychiatric medication were taken significantly more frequently by the Camps survivor group, suggesting that they tried to compensate for the effects of past traumas through medication. In general, similar studies have detected greater long-term psychological problems in ageing Holocaust survivors than in those who were not exposed to such traumatic experience in their youth. Psychological problems include post-traumatic stress disorder (PTSD), ambivalence over emotional expression, altered appreciation of own health and well-being, greater levels of somatization and depression, often leading to higher frequency of suicidal ideation and attempted suicide. Camps survivors in Stessman et al.'s work also tended to have more difficulties with, and to be more dependent on others in one or more activities of daily living (ADL), but their experience in camps was not a predictor of mortality at older ages. Of course, the experience of a concentration camp is a rather extreme case of psychological trauma, but other intense traumatic or stressful experiences can also have negative psychological effects that can drag on into old ages (Lupien et al. 2009). The point is that the life course approach needed to understand the elders requires attention to both the positive and the negative experiences in life.

Physical and psychological changes unfolding with age may finally affect the *quality of life* of older people and their family, especially for those who depend on assistance from others. Quality of life is a multidimensional construct that is dependent on individual, social, and broader environmental factors along with their interactions. Although quality of life is ultimately a subjective experience, it is also affected by variables that can be measured objectively: e.g. muscle tone, blood pressure, autoimmune conditions, access to quality medical care, and the performance of various ADLs (Borglin et al. 2005).

In a study carried out in Spain, Prieto-Flores et al. (2010) identified four major themes in relation to quality of life after conducting semi-structured interviews with 24, 64 to 90-year-old participants (45.8% women): (1) adaptation to health limitations; (2) seeking balance; (3) the experience of place in centres for older people; and (4) health and family interrelated dimensions of quality of life. Living a good quality life, a fulfilling life, when older people experience health limitations can be a challenge, but a degree of adaptation to such limitations can be achieved whenever the individual has an "active mind", even when it is expressed by "a limited body". A positive attitude towards ageing is essential to better adapt to the changes. Among the participants in Prieto-Flores et al.'s work a positive attitude was helped by good family relationships, the freedom of doing things that before retirement they could not do, the realisation that the accumulated life experience was a useful asset in life, and a satisfactory and fulfilling experience as parents. On the negative side, the participants also stated that as partners and relatives declined in health or passed away their own quality of life suffered; as it did when they lost a degree of their own independence and with the increasing realisation of the proximity of death. The shift in older ages towards focusing on the deterioration of own health is a widespread phenomenon (Borglin et al. 2005).

In a study carried out in Sweden, Borglin et al. (2005) analysed conversations with six women and five men (80- to 85-year-old) to conclude that their quality of life was rooted in the preservation of both self and life meaning which, in turn, was also affected by interactions with others. As in the Spanish study of Prieto-Flores et al., these Swedish respondents were also concerned about deteriorating health, the issue of getting closer to their inevitable death and that of dear people, increasing dependency, and they also had concerns about their material wealth. The better all those factors were perceived the better their quality of life was.

Therefore, quality of life at older ages is crucially dependent on individual factors such as health, personality, and life history, which in turn can be affected by socio-demographic factors and social support (Pinquart and Sorensen 2000).

The *belief in a just world* (Lerner 1980)—that is, that we all get what we deserve in life—can also modify the dynamics of quality of life in old age by producing some form of adaptation in response to perceived past failures or injustice. Some strategies that old people use to assimilate past injustices by means of a belief in a just world include apportioning a degree of self-culpability, decreasing the importance of the injustice, move on and avoid excessive rumination on the issue.

The belief in a just world is a defence mechanism to protect the individual's subjective well-being from the long-term psychological impact of distressing experiences (see Dzuka and Dalbert 2006, and references therein). In a study of the relationship between beliefs in a just world and subjective well-being carried out in Slovakia, Dzuka and Dalbert (2006) interviewed 122 elderly people (74-year-old on average; 60.6% females) who were living in public senior residences. They measured belief in a just world through the *Personal Belief in a Just World Scale* and used the *Positive* and *Negative Affect Scales* to measure subjective emotional well-being. The *Brunstein's five-item scale* was used to measure social contacts outside the family. Belief in a just world correlated with dimensions of well-being. Older people's subjective health status was positively associated with social contacts and with belief in a just world. Personal life satisfaction was also positively correlated with positive affect, whereas negative affect was inversely related to better subjective health status and stronger beliefs in a just world. That is, it seems that the belief in a just world affords the elderly person a better chance to adapt to adversities, and it should be regarded as a coping strategy that some individuals may wish to adopt.

Coping mechanisms such as the belief in a just world, taken together, can be organised into a set of strategies to experience fulfilling ageing. Such strategies emphasise an active stance of the individual in navigating and directing his/her ageing experience, as far as it is possible of course. Such proactive coping can be *preventive*—in preparation of expected stresses—or *corrective* when it is adopted in reaction to experienced stresses (Kahana and Kahana 2001).

We have now accumulated enough preliminary understanding of old age in this chapter to shift our focus to one of the central themes of this book, *fulfilling ageing*. Our analysis starts with a critique of the concept of *successful ageing*, leading to the need to focus on what is, in our view, the vastly more useful concept of *fulfilling ageing*.

Successful ageing (often also referred to as *healthy ageing*, *productive ageing*, *optimal ageing*, and *ageing well*; Chapman 2005; van Wagenen et al. 2013) implies the avoidance of disease, the maintenance of cognitive and physical functioning, and sustained engagement with life (Rowe and Kahn 1998). In the words of Rowe and Kahn (1998, p. 433):

> We define successful aging as including three main components: low probability of disease and disease-related disability, high cognitive and physical functional capacity, and active engagement with life.

The above definition has been criticised on the ground that it lacks a specific reference to the importance of social support and, we may also add, a degree of material/financial security. The lack of a specific social dimension to successful ageing was criticised by Holstein and Minkler (2003) among others, which subsequently led Rowe and Khan (2015) to acknowledge the importance of social relationships in a *successful ageing 2.0* revision of their model. Despite the improvement, Toni Calasanti and Neal King remain concerned that, in the social context, there are still possible links between successful ageing and ageism that should be obviously avoided (Calasanti and King 2020).

In principle, *successful ageing* offers some worthwhile objectives in life for anyone experiencing old age. However, even beyond the issues of social interactions and ageism mentioned above, successful ageing is laden with meanings of "competitiveness", "achievement", and therefore also "failure", characteristics that have even led Rubinstein and de Medeiros (2015) to link successful ageing to the tenets of neoliberalism.

For those various reasons Hubble and Tew (2013, p. 199) have stressed the need to replace the concept of successful ageing with one that prioritises satisfaction in life and psychological well-being:

> it is possible to see the basis for the emergence of a new social narrative of ageing, which both allows for long active post-retirement years and a gradual acceptance of old age, seen not as a decrepitude or social problem, but as the attainment of a self acceptance that transcends any purely medical concept of well-being. In other words, our public and social concept of 'successful ageing' has to be revised.

We agree that successful ageing is a concept that is too limited, and it can in fact lead to negative discrimination of the elders, frustration, and unhappiness. We, therefore, offer the alternative concept of *fulfilling ageing*, as ageing only needs to be fulfilling and meaningful to the person experiencing it. There is no reason to divide the elders between the "successful ones" and the "failures", those who reached the top and those who lay at the bottom, the winners and the losers in the competitive race of life. Fulfilling ageing is not a *competitive race*, rather it is a *life journey*.

Although we will see below that Viktor Frankl combined both concepts of successful and fulfilling ageing into a two-dimensional space describing a diverse view of the human experience, we regard fulfilment in life as a central aspect of ageing in all people, that can embrace a very diverse set of life experiences whether they can be regarded as being "successful" or not under any set of criteria.

In the *Preface* we defined fulfilling ageing as:

- *The ability to effectively express capability and agency, leading to a personally satisfactory level of well-being, contentment, and adaptation within the self and with our surrounding environment as we become older.*

In the definition, the "surrounding environment" includes all the environmental components relevant to each individual both the human (social) ones and not.

We can trace the historical antecedents of *fulfilling ageing* back to the classic Greek concept of *eudaimonia*. In the *Nicomachean ethics* Aristotle (fourth century b.c.e.) distinguished between *hedonism* (or the search for pleasure in life) and *eudaimonia* ("happiness that arises from good works" Kashdan et al. 2008, p. 219). Therefore, the concept of eudaimonia to a great extent contrasts with the hedonistic tradition of the interpretation of human life. The hedonistic tradition stresses personal happiness, success, and their usual external expressions, whereas in the eudaimonic tradition life is expected to be lived "in a full and deeply satisfying way" (Deci and Ryan 2008, p. 1).

Eudaimonia is "concerned with living well or actualizing one's human potentials. This conceptualization maintains that well-being is not so much an outcome or end state as it is a process of fulfilling or realizing ones daimon or true nature—that is, of fulfilling one's virtuous potentials and living as one was inherently intended to live" (Deci and Ryan 2008, p. 1).

In Nagel's (1972, p. 252) words:

> *eudaimonia* essentially involves not just the activity of the theoretical intellect, but the full range of human life and action, in accordance with the broader excellences of moral virtue and practical wisdom. This view connects *eudaimonia* with the conception of human nature as composite, i.e. as involving the interaction of reason, emotion, perception, and action ... (italics original)

In a comparison between the eudaimonic and hedonistic traditions in the understanding of human well-being, Ryan and Deci (2001) noted that hedonism has a long historical rooting in Western culture as a source of happiness derived from pleasure and self-interest. Ancient hedonistic philosophers included Aristippus (c.435–c.356 b.c.e.) and Epicurus (341–270 b.c.e.) (Shmotkin and Shrira 2012a, b). Eudaimonic living is also expected to lead to personal happiness and enjoyment of life, but not all instances of enjoyment are eudaimonic, some are hedonistic and grounded on selfishness (e.g. Waterman et al. 2008). The difference between the two concepts is clear in Ryan et al.'s (2008, p. 139) *model of eudaimonic living*, that is characterised by the pursue of personal growth, relationships, consensus and mindfulness, rather than hedonistic wealth, fame, and power:

1. Pursuing intrinsic goals and values for their own sake, including personal growth, relationships, community, and health, rather than extrinsic goals and values, such as wealth, fame, image, and power.
2. Behaving in autonomous, volitional, or consensual ways, rather than heteronomous or controlled ways.
3. Being mindful and acting with a sense of awareness.

4. Behaving in ways that satisfy basic psychological needs for competence, relatedness, and autonomy.

This model encompasses many of the aspects of eudaimonia listed by Nagel, although the emotional aspects of human life are mainly contained in the reference to relationships and relatedness, which are not fully descriptive of all our human emotional needs. *Fulfilling ageing*, on the other hand, incorporates all those positive emotional aspects of life that contribute to making it worth living, such as joy, love, amusement, pride, hope, gratitude, serenity, interest, awe, inspiration. Positive emotions build personal resilience against the stresses of life and facilitate social cooperation, and in so doing they can increase our well-being (Fredrickson 2001; Cohn et al. 2009). But we can derive fulfilment also from transforming experiences that initially elicit negative emotions (e.g. sadness) into instances of personal growth (Frankl 1988).

Ryff (2014) to an extent echoes the list provided by Ryan et al., but again her conception of eudaimonia seems to underestimate the emotional aspects of well-being. In Ryff's view, the eudaimonic perspective provides a focus on self-acceptance, purpose in life, environmental mastery, positive relationships, personal growth, autonomy.

Experiencing well-being and fulfilment in life is undoubtedly a dynamic process (Kashdan et al. 2008) that expresses the adaptation of the self within him/herself and with the rest of his/her environment. Such dynamic process encompasses all aspects of our human reality and therefore it should include not only cognitive and existential aspects, but also our emotional needs, as Nagel (1972) did.

Therefore, *fulfilling ageing* takes into consideration the various psychological and social aspects of eudaimonia, including the emotional aspects of being, that inspire and motivate us throughout our life. Neugarten et al.'s (1961, pp. 137–138) *Life Satisfaction Index* offers a comprehensive list of variables to be considered in the measurement of life fulfilment. The index broadly includes what Ryff (2014) and Ryan et al. (2008) regard as the main aspects of eudaimonic living, and it also emphasises Nagel's (1972) four major areas of eudaimonic experience: reason, emotion, perception, and action. In Neugarten et al.'s view life satisfaction can be described in terms of:

- Zest: enthusiasm and degree of personal involvement in various activities, relationships with persons, or ideas that the individual finds of interest.
- Resolution and fortitude: these involve acceptance of personal responsibility in life and regarding life as being meaningful.
- Congruence between desired and achieved goals: extent to which the individual has achieved his/her goals in life, whatever they might be; regarding what was achieved as personally important.
- Self-concept: an individual's concept of self which encapsulates physical as well as psychological and social attributes. In fulfilling ageing an elder only needs to be satisfied with him/herself and what has been achieved in life.
- Mood tone: High mood tone individuals express happy, optimistic attitudes and mood; use spontaneous, positively-toned affective terms for people and things, and take pleasure from life, expressing it.

Ultimately, it is the combined cognitive, emotional, and existential aspects of our self that co-contribute to experiencing fulfilment in life (e.g. Fredrickson and Joiner 2002).

In addition, and complementing the eudaimonic view, fulfilling ageing is not just limited to expressing our own self at any point in time; rather, it is also open to the development of a new self over time. Fulfilling ageing is the expression of a dynamic personal experience that has roots in the past, expressions in the present, and projections into the future.

Thus, fulfilling ageing remains wide open and accommodative to the variable and specific needs of individuals, their creativity and desires, who are trying to live the best life possible despite the challenges of ageing. The zest expressed in various activities is determined more by the initiative of the individual than by cultural expectations imposing standards of "success", which may be too constraining and often frustrating. The individual seeks meaning and meaning will be valid once it is found, whether such meaning makes or does not make sense to some others. But meaning is also open to dynamic change according to personal experiences and circumstances. In addition, the social aspects of the self are determined by the specific relationship of the individual with specific others, which may also change over time.

Fulfilling ageing also implies taking pleasure in life, even to the extent of feeling content in the expectation that life will inevitably come to an end, with our personal fulfilment at the end of life being also linked to our legacy, whatever that may be and whoever may benefit from it. Fulfilling ageing should be a good path to building up a legacy.

Fulfilling ageing retains some aspects of "successful ageing", such as the desirability of avoiding disease and disability, but also maintaining good physical and cognitive capacities—to the extent that such capacities can be controlled—self-acceptance, personal growth, emotional and existential satisfaction, along with productive engagement in social activities and life, and living an active life in general (e.g. Ryff 1986; Rowe and Kahn 1997, 2015; Byles 2017); but without the pressure imposed by the pursue of a competitive edge. Above all, fulfilling ageing is a personal path that, however, we do not always walk alone. Such personal path is built through constructively living all the experiences in life, those that are positive and those that are negative.

Viktor Frankl (1988) provided a basic, two-dimensional model of the human experience. One dimension of the model spans from despair to fulfilment in life and another dimension ranges from failure to success in life (see Fig. 1.7).

A life can be obviously both successful—however we may define "success"—and fulfilling; or it can be one where despair prevails in the midst of perceived failure. But other outcomes are also possible. We agree with Frankl that fulfilment could be found in a life that, according to some criteria, may be considered a "failure". There is no reason to believe that a person with modest financial means who is living a simple life out of personal choice, for instance, cannot find fulfilment in such a life, no matter how much others may regard him/her as a "failure". In fact, we all can experience life fulfilment, no matter what. Frankl describes this situation

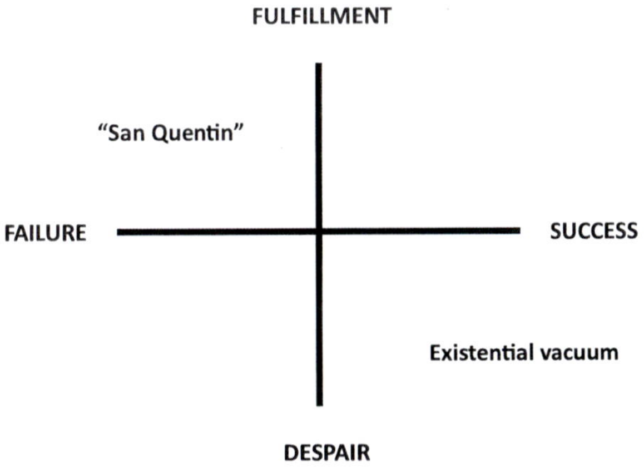

Fig. 1.7 Frankl model of life meaning with success and fulfilment in life being independent. Redrawn with slight modifications from Frankl (1988)

as "San Quentin" (Fig. 1.7), from an experience he had with some inmates in the notorious jail of San Quentin, who were able to find life fulfilment even in such a situation. On the other hand, experiencing despair despite having achieved social and economic success is seen in some cases of *existential vacuum*, or loss of life meaning, as suggested in Fig. 1.7. Existential vacuum expresses a profound lack of life meaning even in situations where an external observer may find it difficult to understand why such life meaning is missing. Here, however, organic aspects of depression may also play a role in some circumstances.

Fulfilling ageing may be helped by better education, financial resources, physical condition, contentment, social connectedness and, in various contemporary societies, it may require some government intervention in order to be equally experienced by all men and women. Fulfilling ageing is also helped by perceptions of wellness, satisfaction, and by the ability to cope with and recover from adverse circumstances. All those are important issues to keep in mind because although life fulfilment is a personal and idiosyncratic experience, the cynic in us warns us against the danger of manipulative intents such as telling the poor to be "happy and fulfilled with what you have", no need for any social support from government to provide better services and standards of living, and hence no need to properly tax the wealthy. We regard that to be a gross misuse of the concept of fulfilling ageing.

The *Disengagement theory* of Cumming and collaborators (Cumming et al. 1960; Cumming and Henry 1961) considers fulfilling ageing as a gradual process of disengagement from active life and personal preparation for the inevitable death. This may be regarded as a somewhat passive and perhaps even pessimistic approach that is currently losing traction in modern societies, although the original intent was optimistic, seeing such disengagement as a "graceful withdrawal from society" (Cumming et al. 1960, p. 7). On the other hand, *Activity theory* emphasises the continuation of the active lifestyle that is usually typical of middle ages into the older ages, if the ageing process is to be experienced as a fulfilling one (Nussbaum 1983). This is perhaps an over-optimistic approach that may go to the opposite extreme. In the end both views can be subsumed into a compromise concept of fulfilling ageing as a process of development of a sense of satisfaction with our life (Havighurst 1963)—no matter whether we find satisfaction in contemplation, action or a combination of both—that may better prepare us for whatever additional developments we may experience into the future (Ryff 1989).

In a broader perspective, fulfilling ageing relies on reaching old age in good enough general health and mental conditions that promote a personally satisfying engagement with life (Rowe and Kahn 1987). But alternative routes to fulfilling ageing are also possible, that improve satisfaction with life at older ages under a set of various circumstances and constraints (e.g. in the case of older people with physical and/or mental handicaps, for instance). Such multi-way approach to experiencing fulfilling ageing is encapsulated in the *Selective Optimisation with Compensation* (SOC) model (Baltes and Baltes 1990). As we have explained previously in this chapter, *selection* in the SOC model refers to the directional paths of development, outcomes and goals that are selected by the individual among what she/he recognises as the available alternatives. *Optimisation* refers to the strategies used to achieve the desired goals and levels of functioning in life; whereas *compensation* points to the acquisition and then activation of alternative means to counteract the loss or the decline of those abilities used to maintain a certain level of functioning (Baltes and Smith 2003). Baltes and Smith (2003, p. 132) beautifully illustrate their model with a nice example inspired by the then 80-year-old Polish-American pianist Arthur Rubinstein:

> When Rubinstein was asked how he continued to be such an excellent concert pianist, he named three reasons. He played fewer pieces, but practiced them more often, and he used contrasts in tempo to simulate faster playing than he in the meantime could master. Rubinstein reduced his repertoire (i.e., *selection*). This gave him the opportunity to practice each piece more (i.e., *optimisation*). And finally, he used contrasts in speed to hide his loss in mechanical finger speed, a case of *compensation* (italics ours).

Rubinstein's case exemplifies an adaptable approach to life in old age, implying an engagement with *competent living*, or the ability to somehow continue the activities associated with the most important aspects of daily life despite the growing challenges of old age (Baltes 1998). Moreover, reaching older ages in a better condition can be helped by adopting better lifestyles at younger ages (Fries 1980). That is, we can prepare for a more fulfilling old age by nurturing ourselves since younger ages and by using all the resources at our disposal to help adjust to the challenges of ageing.

Therefore, as we age we may optimise the resources available to us (both internal to our organism and external) to experience our fulfilment in life, and when some of those resources decline or are missing, we may adopt strategies to compensate, in order to keep ourselves on track to live a fulfilling old age. Goals may be also adapted to changing circumstances, of course (Freund and Baltes 2002). Through a process of disengagement from unachievable goals and the construction of new ones, the individual can better accommodate to a changing reality (Brandtstädter and Greve 1994; Schulz and Heckhausen 1996; Brandtstädter and Rothermund 2002). Thus, by adopting a dynamic stance on ageing we either take actions to modify our external environment, or we adapt to it and to our changing body and mind, by modifying ourselves. Ouwehand et al. (2007) appropriately refer to this dynamic process as *accommodative flexibility*.

Accommodative flexibility involves the use of the plasticity of our body and mind to decrease the impact and the limitations of old age (Bowling 2008). *Active ageing* is one result of accommodative flexibility, being a process that optimises opportunities for better health, greater participation, and security over time, with the result of enhancing quality of life in old age, against the obstacles imposed by various constraints. But accommodative flexibility can be also used to disengage ourselves from some undesirable aspects of life. That is, accommodative flexibility is a process that is common to both the *disengagement* and *activity* approaches mentioned above; and it is therefore also a process characterising *fulfilling ageing*.

Following Bowling (2008, p. 294), the term "active" in *active ageing* points to a "continuing participation in society, retaining the social, mental and physical health to enable this, maintenance of dignity, self-efficacy, human rights and age-friendly physical environments facilitating autonomy and independence". Therefore, active ageing aims at maintaining a good and satisfactory quality of life and, according to Bowling (2008), this is achieved by focusing on four major themes:

(a) Health
(b) Participation
(c) Security
(d) Fulfilment

These themes are regarded by Bowling as the pillars of active ageing in that they stress the physical, material, social, and existential aspects of the ageing person that should lead to better quality of life in old age. Active ageing may be helped by *ageing in place*, that is by "remaining living in the community, with some level of independence, rather than in residential care" (Davey et al. 2004, p. 133). Ageing in place affords a sense of *existential continuity* and above all maintenance of well-established social networks (Wiles et al. 2011). Although life fulfilment in old age can be achieved through a specific active ageing path, we remind that the paths to fulfilment are many and each one of us will choose his/her own. In this context, we do not regard *life fulfilment* as just one aspect of "active ageing", as Bowling does; rather, we see it as a central and encompassing theme of ageing.

Continuity may also help in experiencing fulfilling ageing through the maintenance of consistency in personality (Nussbaum et al. 2000). *Continuity theory* (Neugarten et al. 1968) states that personality establishes a continuity thread into old age. Different personalities, therefore, will experience different styles of ageing. In this case, some personalities may be more likely to disengage in old age, or to maintain a consistent long-term stance; whereas others may be more likely to remain active and embrace change, and so on.

The *Socioemotional Selectivity theory* (Carstensen 1991) is another attempt to synthesise disengagement and activity theories, as the "aging individuals reduce their number of overall interaction partners, but maintain those relationships that provide the most satisfaction" (Carstensen 1991, p. 10). That is, the focus here is on fewer but closer relationships as people become older. Again, such selection process is fully compatible with adaptive adjustments aimed at experiencing fulfilling ageing under changing circumstances.

Finally, *Social-environmental theory* (Hendricks and Hendricks 1986) focuses on the combined individual and social contexts in which the life of elderly people unfolds. Social-environmental theory incorporates aspects of various theories (e.g. disengagement, activity, continuity, socioemotional selectivity, and the selective optimisation with compensation) in the search of life fulfilment (Nussbaum et al. 2000).

Over time the body inevitably deteriorates, despite our attempts to remain engaged with life into older ages, and some of those changes may also affect the mental health of older people. At this point the meaning of life fulfilment will also require appropriate re-adjustment. Chief among the major mental disorders typical of older ages is dementia. Prevalence of dementia is low for people younger than 65 years, but it increases exponentially thereafter ranging from 0.6 to 3.7% in the 65–69 years old to 25.2–75% in people aged 90 or more years (Riedel-Heller et al. 2006). Prevalence of dementia in the elder population is also variable geographically around the world (Draper 2014). The most common form of dementia is Alzheimer's disease (38.5–78% of dementia cases), followed by vascular dementia (15.8% of cases). Other mental disorders can also appear with ageing such as delusional disorders, mood, and stress-related disorders among others (Riedel-Heller et al. 2006).

Maintaining a physically and mentally active life during old ages is a protective factor against the onset of dementia and other mental disorders. Such protection of mental health can be specifically achieved through increasing our *cognitive reserve* whilst decreasing both vascular risk factors and stress (Wang et al. 2002; Fratiglioni et al. 2004).

In order to experience fulfilment in life in old age, however, a degree of planning is necessary, especially as old ages also coincide with retirement from work. What to do with the time available in retirement is a major issue for older people who were used to strict routines at work that are suddenly discontinued, often also with the loss of part of their social network and income (Moffic 2012). These days retirement could encompass up to a quarter of the life of an individual and planning for it is essential, especially when public support for retired people does not come in the form of appropriate pensions and social benefits such as healthcare (Adams and Rau 2011; Shultz and Wang 2011). Retirement can last 20, perhaps even more years depending on circumstances, and therefore a sound preparation for it is becoming critical. Although such preparation may in fact be a source of some anxiety for some individuals, due to the uncertainties about the future and whether assets and income will be enough to sustain an acceptable lifestyle, it is known that it can also be a source of relief and better adjustment (Adams and Rau 2011, and references therein).

According to Albert Bandura (2001) the preparation for retirement is an *agentic process* that involves a degree of intentionality and forethought, activity and self-regulation. In this process the individual addresses financial concerns that will affect health, the chances of survival and lifestyle, social issues including relationships with family, existential and *generative needs* that involve the passing on of knowledge and experience to younger generations, and the establishment of own *legacy* (e.g. Mor-Barak 1995).

We have mentioned the importance of the social sphere for older people. In fact, old age itself could be regarded as a social construct superimposed on biological changes (Paris 2008; Mintz 2009). Although, as a social species, our network of inter-individual relationships is important in our life since we are born, it becomes ever more relevant as we age whenever we are in strife and require a degree of help, be it financial, emotional, physical, or cognitive. Old age brings with it a degree of increasing frailty that can be compensated, to some extent, through social support, which is especially critical after retirement.

The objective of "successful" ageing specifically espoused by the *New Gerontology* approach tends to over-emphasise the individual sphere, but through *Critical Gerontology* we may better approach fulfilling ageing by also giving proper prominence to the social dimension, including those aspects that impinge on social inequality, justice, access to resources and services, and also the effects of social class, race, and gender that can interfere with life fulfilment (see Holstein and Minkler 2003). What the precise individual position within the social sphere is—his/her *contextual embeddedness*—it becomes rather relevant at an age when social support can have major effects on well-being (Szinovacz et al. 2001).

Sociality is strongly dependent on attachment between individuals and attachment figures obviously change over time as we grow up and then become old. *Socioemotional Selectivity theory* tends to emphasise the reduction of social interactions with old age to those that are more emotionally fulfilling to each individual elder (Carstensen et al. 1999), usually involving close family members (Cicirelli 2010); and yet older people do realise that social relationships, whether they are numerous or not, may become more loose at older ages and less reliable, especially among peers. In this case a tendency towards a *dismissing attachment style* may develop in some elderly people (Magai 2008), perhaps as a strategy to safeguard their emotional stability against demoralisation or even depression in the face of losses within their social network.

The social network of a person is defined as:

> a structure of individuals with a designed relationship to the focal person, as well as an average frequency of contact and a specified geographical proximity to that person. Social networks can be thought of as a key resource over the life course, a form of social capital that potentially influences the exchange of supports (Ajrouch et al. 2001, p. S112).

As the *social network* of an elder starts to loosen up, there is an increasing risk of greater levels of loneliness. Loneliness could be seen, following Peplau and Perlman (1982), as the subjective experience of a negative, aversive emotional state resulting from the perception of unfulfilled intimate and social needs. The lack of attachment figures and the decreased number of contacts with others all conspire to increase the sense of loneliness in older people (Heylen 2010).

Two major types of loneliness have been described: *social loneliness* that results from lack of social integration within a community; and *emotional loneliness* that derives from the lack of a reliable attachment figure (see Ó Luanaigh and Lawlorz 2008). We agree with Ó Luanaigh and Lawlorz, however, that loneliness has a complex and multicausal basis that includes a potential biological predisposition that may be associated with personality traits, and it can be also precipitated by grief or depression. In addition, Ó Luanaigh and Lawlorz also distinguish between the concepts of loneliness and *aloneness*, the latter being defined as a personal desire to be alone. The distinction between those two concepts is important because being alone is not always seen as negative by older individuals depending on personality, culture, and current priorities. Being alone can indeed be desirable to foster creativity, facilitate self-reflection, self-regulation, concentration, and learning (Luanaigh and Lawlorz 2008). Such individual desire for aloneness may even reach a stage whereby both old person and those initially closer to him/her may undertake a process of partial (sometimes even total) mutual *disengagement* (Mutran 1987). Social disengagement derived from a desire for aloneness is usually accompanied by experiences of what Mihaly Csikszentmihalyi (1990) calls *flow*, or full immersion of the individual in the performance of a specific activity. We will return to the concept of flow in subsequent chapters.

The caveat of aloneness notwithstanding, it remains true that loneliness can have negative effects on both the psychology and the physiology of the older person, being often associated with vascular and blood pressure problems, increased activation of the hypothalamo–pituitary–adrenocortical axis, immune dysregulation, depression, altered circadian rhythms, decreased memory, and other cognitive declines, all ultimately leading to a greater probability of mortality (Luo et al. 2012).

The partner/spouse is a major figure of attachment for many older individuals that can decrease loneliness, even today and in spite of the long-term trend to increasing rates of divorce (Stevenson and Wolfers 2007); and when the relationship either fails or one of the partners passes away, it is not uncommon for couplehood to be restored within a new partnership in order to avoid loneliness: "Couplehood fulfills needs such as friendship, intimacy, sexuality, love, compassion, emotional support, and mutual help" (Koren 2011, p. 687). The social network of older individuals also extends beyond the couple to usually embrace direct family members, such as the offspring, in an inter-generational network of relationships (White Riley 1983). Adult offspring can be important providers of assistance to elderly parents, thus helping in decreasing their loneliness; but they may also receive help from parents in return, usually in the form of advice, emotional support, money, childcare, and household assistance (Eggebeen 1992).

The relationship of the elderly person with direct family members can be also affected by broader cultural expectations that may change over time. If prevailing cultural views about the elderly are negative, this may affect the intra-family relationship to the point where the elders may not only receive less support but also assimilate such negative views into a negative self-image (Arnold-Cathalifaud et al. 2008), thus perhaps contributing to their own demoralisation and even depression. From the perspective of the family members, on the other hand, they may experience a degree of *caregiver burden* as they assist an ageing relative who may be experiencing serious health deterioration (Spillman and Pezzin 2000). It is not unusual, however, that the older persons may also feel a degree of pressure or even coercion into providing help to their family. This pressure may be either accepted or resisted.

The potential for caregiver burden leading to burnout makes the maintenance of a broader social network of central importance to the elderly, against the backdrop of the various difficulties to retain, let alone enlarge, such broader network of social relationships as they age. The larger and committed the network is, the more likely it is that a replacement will be available if a caregiver may feel burnt out thus withdrawing from helping.

Social network types can vary among the elderly according to their circumstances, Litwin (2001) described social networks among 60-year-old and older people living in Israel as diverse, mainly composed of friends, or neighbours, family, or they could be overall restricted. By nurturing a reliable social network of family and acquaintances the elders can improve their quality of life and well-being (Cantor 1980; Nilsson et al. 2006), which can be of great help especially when they are facing financial and other material constraints (Lloyd-Sherlock 2000). When the older person is physically impeded to nurture links across a potential social network, she or he could be helped by learning how to use technology such as the Internet. Informal online networks could complement the existing network of family and close friends (Stoller and Pugliesi 1991), whereas social programmes of home assistance may ensure that the elder is properly cared for.

Through nurturing positive social relationships old people may also improve their subjective age. *Subjective age* is the personal perception of ones' feelings and categorisation within any specific age group. Subjective age may be close to actual age or, alternatively, one may feel older or younger than one is. In this respect, the psychology of the individual may not necessarily match her/his external physical look or even actual biological capabilities (Kleinspehn-Ammerlahn et al. 2008). Positive social relationships help old people feel younger (e.g. Stephan et al. 2013).

Nevertheless, not all social relationships are positive. Despite recent trends towards a better perception of ourselves and our capacities in old age, an older-looking person often faces social barriers in the form of stigma and discrimination that usually derive from stereotypes about the elderly. Such stereotypes can even persist against attempts by the elderly to resist them by openly demonstrating their real capabilities.

Negative discrimination against older people, also referred to as *ageism*, derives from—and in turn produces—inter-generational conflicts. It is often caused by clashes of interests such as competition between the young and the old for access to resources, like jobs or government support. *Ageistic* discrimination is reinforced by *stereotypes* that are "exaggerated views or conceptions about a person or a group based on simplistic generalizations, stereotypes do not consider individual characteristics" (Ory et al. 2003, p. 165).

Stereotypes are expressed in the form of myths about specific groups of people and they can be variable across societies and through time. Some common myths about old people include that old age is inevitably synonymous with poor health, learning difficulties, low libido, poor levels of initiative and engagement with the rest of the community; it is also a common belief that good health in old age is just a matter of genetic good luck (Ory et al. 2003). These myths are often perpetuated through the media. Although the myth may draw a degree of justification from scientific biological and psychological knowledge, and also population studies; not all individuals are the same and their differences may respond to their personal circumstances, agency, and the tendency in society to stereotype the elders. How good the mental and physical capacities of each elder are, can only be ascertained once the older person has willingly engaged in various activities. Limitations should be proven in facts, not assumed. The latter is what stereotypes do.

Although stereotypes may lead to *stigma* and discrimination (see Goffman 1963), the targets of such stigma may also express a degree of *resistance* (Hurd 1999) that could be either productive in terms of decreasing the level of discrimination, or counterproductive if it elicits a retaliatory response from the rest of the community (see Cordella and Poiani 2014). The situation may become even worse for older people who are also ill, especially if their medical condition affects their mental capacities. In such cases they could be "doubly-stigmatised" (Thomas and Shute 2006): they are both old and impaired.

When social bonds are weakened by stigma or other factors, the elderly are more likely to suffer higher mortality rates (Avlund et al. 1998).

The structure and intensity of social networks can also change according to other variables that may interact with age, such as personality, race, ethnicity, and socio-cultural factors. Social networks may vary with time throughout our lifespan, including old and oldest-old ages, thus calling for a *life course perspective* in the study of the relationship between sociality and ageing (Granovetter 1973; Ajrouch et al. 2001).

One social aspect of ageing that is gaining importance at present, following an increasing level of movement of people around the world, is the experience of older migrants. Immigration produces greater cultural diversity within countries. In addition, immigrants are confronted with the challenges of adapting to the host country in terms of language, culture, and the law; whether they migrate in old age or whether they age in their adopted country after they have migrated as a young person (White 2006). Old immigrants may be at a disadvantage, compared to native locals, in terms of using and benefiting from the social and institutional support available in the country. As a result, old immigrants tend to rely more heavily on family members, which often exacts a degree of caregiver burden upon the latter (Treas and Mazumdar 2002; White 2006). In a study of 28 elderly immigrants (72 year old on average, 75% females, mainly from Asia) carried out in California, USA, Treas and Mazumdar (2002) identified the following major themes that were recurrent in the discourse of the interviewees: loneliness and therefore the need to cultivate a social network, geographic constraints that complicated contacts with others and that finally led to prefer interactions with family over people outside the family circle; hanging onto their cultural values in a way that impeded a better integration within the host community, this was in part a result of fear of acculturation into values that they did not necessarily share. Fulfilling ageing in older immigrants requires a combined effort of society (with emphasis on tolerance and provision of services), family but also of the immigrant him/herself who must be ready not just to "teach" but also to "learn".

Whether they are immigrants or not, some old people may continue to age until they become oldest-old. The oldest-old age is also referred to as the *fourth age* and the number of people in such age category has been increasing over the years as a proportion of the overall elderly population around the world (Vaupel 2010). This category of old people is currently the focus of great attention from the part of researchers. The transition between the third and fourth age is currently 75–80 years old (Baltes 1998; Baltes and Smith 2003) based on a definition of the fourth age as "the chronological age at which 50% of the birth cohort are no longer alive" (Baltes and Smith 2003, p. 125). Another definition in use is: "the age at which 50% of the people who attained age 50 or 60 have died subsequently" (Baltes and Smith 2003, p. 125), the latter definition would further push the start of the fourth age to about 80–85 years old.

Although the oldest-old experience an increasing degree of body frailty over time, that can become an impediment to some activities, other activities may still be possible provided that the mind remains alert and active; and the oldest-old could even experience further developments and improvements as they continue to age, leading to what is known as the *paradox of well-being* (Isaacowitz and Smith 2003), where purpose in life can remain high or even increase in spite of advancing age and deteriorating physical abilities (Hedberg et al. 2013). Baltes and Smith (2003, p. 124) recount a beautiful story about the famous Catalan cellist Pablo Casals who "as an 80-year-old, was asked by a young student why he continued to practice so hard. 'Why?', Casals answered. 'This is simple. Because I want to get better!'".

The overall experiences of the elderly, whether they are perceived as positive (support received, life satisfaction) or negative (burden, discrimination), are expressed in their discourse. Carrying out a *discourse analysis* of older persons' narratives is a way to open a window into their mind and to hopefully access their ideas, beliefs, expectations, emotions. This is obviously true for all kinds of old people, but narratives can be especially revealing in the case of old people who may be going through intense traumas (e.g. immigrants, refugees; Becker and Beyene 1999) that can be better uncovered through long and in-depth conversations with them, rather than through the analysis of brief answers to a questionnaire.

The words we choose to express ourselves encapsulate our view of the subject matter that can be coloured by our world view, philosophy of life, beliefs, prejudices, and previous experiences. The discourse on ageing has been dominated by the concepts of what is *natural* (usually understood as "inevitable"), what is *normal* (prevalent) but also what is *normative* (socially expected) (Rees Jones and Higgs 2010). Therefore, through discourse we express what we see as prevalent but also what we expect on the ground of our beliefs. Such beliefs, expectations, and observations are dynamic and do change over time, and their changing is in turn affected by the level of our agency. Today the discourse on ageing contrasts, in a dynamic and creative way, the issues of the individual responsibility to prepare for old age with the responsibility of society not to leave anyone behind, and instead provide both equality of opportunities and access to support when needed.

It is such a diversity of perspectives that is at the core of modern *Narrative gerontology*. Narrative gerontology challenges the idea of a single and static model of fulfilling ageing—that is expected to apply to everybody—embracing instead a plurality of pathways, an approach that can better satisfy the personal reality of each elder (Holstein and Minkler 2003). The life paths to fulfilment are diverse because each elder is different and the social influences he or she is exposed to may also vary.

The diversity of personal pathways that can be followed throughout the process of ageing is not only determined by the diversity of individualities, but also by the space of possible changes that is available to any individual at any given moment. In a way, at each point in time, there is only a partially known space of possible selves that is opened to any person to both explore and create. The anchoring of such *possible selves* into the past and the external constraints of the present potentially limit the possibilities for the future, but there is still a region of unexplored realities that can only be discovered once one decides to undertake a life journey into it. Indeed, such possible selves may be cognitively and emotionally constructed as a set of aims that can then drive the exploratory action. In the words of Smith and Freund (2002, p. 492):

> Possible selves are highly personalized hoped-for and feared images of the self that function as incentives for action… these images contain information about self-related concerns that guide decisions about what goals to work on, where to expend time and effort, what to avoid or resist, and what to abandon. In this sense, possible selves can be regarded as guides for selection processes… Possible selves also reflect an individual's motivation to try to control the direction of her or his future life: motivations such as self-improvement, self-maintenance, or efforts to minimize loss. It has been suggested that individuals change and recalibrate their possible selves in response to changes in life circumstances in order to maintain well-being.

In old age, however, narratives may also become constraining to further developments of the self. Biggs (2004, p. 46), for instance, mention the production of *masquerades* and how they can become crystallised into a limited and limiting self-identity:

> Masquerade draws on the idea that identity is performative: put on, so to speak, in a particular context and for a specific audience, even if that audience exists in the inner world of the self. It also reflects a psychodynamic understanding of identity as something that is inevitably layered, having depth as well as appearance. Both narrative and masquerade, I argue, hold the danger of becoming inward-looking, and thus solipsistic. They may cease to be part of an interactive, dialogical process and end up as fixed, yet ungrounded psychic positions for scholars and individual older people alike.

Story-making and social masquerading are strategies that older persons may use to buffer themselves from the effects of stereotypes and stigma, but in so doing they may create their own stereotypes. For instance, it is not uncommon for the elderly to refer to "in the good old days" myths (Lundgren 2010) that are in fact narratives rooted in memory, personal biases, and socially constructed stereotypes.

Thus, through identity narratives older persons can foster either "fixity" (reinforcement of a given identity) or "flux" (promotion of a changing identity) of the self in relationship to time and the environment (Biggs 2004). The social environment is especially important to either stress fixity through stereotypes and reinforcement of specific views, or to foster flux when the environment itself is changing (e.g. in multicultural societies). Flux can be seen in the changing patterns of identity that may even occur in a hierarchical manner, whereby some identities may be stronger or more pervasive than others depending on the personal and social circumstances of the elder (Mutran 1987; Osborne 2012).

Maintaining a sense of identity into old age (whether it is fixed or more fluid, provided it is not a masquerade), is expected to nurture our existential well-being.

Following Erickson (1997) we may say that the elderly face a set of challenges that encompass the maintenance of an identity, their ability to operate in life, their individual freedom but also their need for social interactions and *generativity*. Generativity can be defined as a process used to achieve meaningfulness and transcendence through the ability to guide the next generation. In addition, Rubinstein (2002) also expands the meaning of generativity to include social, political, or cultural activities that may contribute to alter the broader environmental and social "settings" in which one's children and grandchildren will live. Such challenges and motivation to focus on our contribution to the next generation may fall into a sharper focus upon retirement. Generativity relies on the elders offering to others the gifts of their own identity.

Identity develops over the course of a lifetime, but it is seriously challenged by changes in roles and in the activities that give life much of its meaning. Retirement may involve such a challenge to our identity. In the words of John William Osborne (2012, p. 9), after retirement: "A worker who was a somebody can suddenly feel like a nobody". But retirement is not always such a major shake up of previously held certainties. Sometimes retirement is a relief from the stress of a job that may have become intellectually or emotionally draining and unrewarding, but still required for financial reasons. If the job is still cherished and fulfilling, on the other hand, obligatory retirement may indeed come with a significant shake up of the world of the individual. In the end, the life of an older person after retirement can be existentially fulfilling for as long as it is lived *authentically*, meaningfully, and in a way that strengthens our identity (Osborne 2012).

Authenticity should not be confused with selfishness. A life can be lived authentically without causing pain and suffering to others. Living an authentic life in old age, when we also tend to be more in need of help from others, will obviously require a degree of adjustment and the reaching of some compromises. On the other hand, the more self-sufficient a person is, the more easily that person will be able to live her/his life authentically without social constraints; the less self-sufficiency, the more need there is for the individual to reach compromises and to adjust to others, in order to live a fulfilling life. The need for a compromise between the self and others is likely the answer to Golomb's (1995, p. 4) question about authenticity:

> Must we be content with postulating authenticity as an ideal to aspire to rather than a valuable social norm—an ideal that is nevertheless necessary if we are to become what we are, given a cultural and social context that undermines this authentic selfhood?

The answer is probably "yes", barring total solipsism, in most cases a balance will be needed between the independence of the self and the need for social support that implies social compromises.

Although in their quest for *authenticity* some older persons may desire to maintain a degree of continuity with their pre-retirement life, or even with their life at much younger ages, the reality is that old age and retirement open up the possibilities for a diversity of identities to manifest themselves, as the shackles of work obligations disappear, along with some associated expectations. In this context, some personality characteristics such as Extraversion may facilitate the finding of an appropriate new niche in a changed external environment (Lockenhoff et al. 2009a, b, c).

Authenticity can be achieved through a process of *assimilation* or the interpretation of one's experience in the world as being compatible with own identity; or even *accommodation*, which requires a more proactive stance whereby the individual adjusts both behaviours and expectations to make self-identity more consistent with the experience in the world (Whitbourne 1986).

Older persons often explore their spiritual dimension in their path to living their life authentically. In our book "Behavioural Oncology" we defined *spirituality* as "the search a human being undertakes to find a sense of peace, satisfaction and meaning in life" (Cordella and Poiani 2014, p. 240). This definition encapsulates most of the major aspects of spirituality usually considered by various authors (see Narayanasamy et al. 2004, and Velasco-Gonzalez and Rioux 2013, for brief reviews). Although spirituality is not a specific characteristic of old ages, it may take an increasingly important existential role in the process of seeking meaning in life past, present, and future, as we inevitably move closer to the end of our life.

Spirituality is the mean to achieve *spiritual well-being*. In the words of Velasco-Gonzalez and Rioux (2013, p. 1125):

> spiritual well-being refers to a positive state and feelings, to behavior and cognition, to relationships with oneself and with others, and to a transcendent dimension, giving the individual a sense of identity, integrity, satisfaction, beauty, love, respect, positive attitudes, inner peace and harmony, a goal, and a direction in life.

This complements the more basic concept of *existential well-being* (or *subjective well-being*; see Baltes 1991; Freund and Baltes 1998) that could be defined as "a subjective sense of happiness and vigour in life" (Cordella and Poiani 2014, p. 226).

It is often also the case that spirituality may take a more specific shape in the form of *religiosity*. Following Brady et al. (1999, p. 418) religiosity could be seen "as 'participation in the particular beliefs, rituals and activities of traditional religion'… It can serve as a 'nurturer and channel of expression' for spirituality …" Religiosity can be either *intrinsic* (which refers to both religious practices and the more individualised spiritual aspects of religion) or *extrinsic* (associated with the rituals and practices of institutionalised religions) (Cordella and Poiani 2014).

Spirituality, meaning-seeking and religiosity have a diversity of effects on life fulfilment in older people. On the one hand, spiritual changes that occur in an individual over time may culminate in the attainment of greater levels of existential well-being, wisdom, and spiritual development at older ages that Tornstam (1989, see also Erikson and Erikson 1997) call *gerotranscendence*. On the other hand, spirituality and religiosity may develop as a specific psychological defence against the challenges and constraints of life associated with advancing age (Wilber 1999).

Spirituality may also interact with cognitive functions in a process known as *spiritual intelligence*, or the mental processes involved in the search for meaning, purpose in life, and the production of unifying concepts. This may lead to higher levels of self-awareness, seeing connections among seemingly separated entities that may improve our capacity to overcome suffering (e.g. Sijuwade 2013).

The existential well-being afforded by greater life meaning, spirituality, or religiosity is an important pillar to fulfilling ageing, and it may even feedback on our body and physiological processes through the biological links between the activities of our nervous system that generate existential well-being and the rest of our biology (e.g. immunity, endocrinology), even to the point of decreasing mortality rate, at least to a degree (e.g. Zhang 2008).

Alas, over time, even old people in the best body and mental condition eventually deteriorate. Whether we like it or not, gradually the oldest ages become an inevitable transition to the end of life. The optimistic discourse on ageing, that strives for the maintenance of high quality of life, cannot ignore the reality of the inevitability of death (see Whitaker 2010, and references therein). Regardless of our best efforts, the body will deteriorate, and the mind may also fail us. Eventually we may reach a stage where the level of suffering may overcome our hope and optimism. As death inevitably nears, gerotranscendence and a sense of life fulfilment may help in smoothing the transition over; but sometimes physical pain, demoralisation, and depression may overcome any attempt to control ourselves, leading to thoughts about euthanasia or even the consideration of suicide (e.g. Harwood et al. 2000). These are real issues that we must confront as a society and not brush aside, dismiss, or hide under a stiff ideological carpet.

In the remaining chapters we explore all the above issues, plus others, in greater detail, ending with the last chapter where we provide a synthesis grounded in the concept of *fulfilling ageing*.

Throughout the book the readers will notice many references to political issues concerning old age, including critical arguments against *neoliberalism*. We regard that a subject such as old age, that relies so much on policies for the elders to ensure their well-being and their ability to live a fulfilling life, cannot be analysed properly whilst ignoring the politics of ageing. The term neoliberalism was introduced by Alexander Rüstow at *The Walter Lippmann Colloquium* in Paris in 1938. His objective was to offer a new liberalism as an alternative to collectivism, socialism, and laissez-faire liberalism. By neoliberalism Rüstow meant an economic system that prioritises price mechanisms, free enterprise, competition, and a strong and impartial state, to which it was subsequently added deregulation, privatisation, and inequality. In 1951, Milton Friedman wrote that:

> Neo-liberalism would accept the nineteenth century liberal emphasis on the fundamental importance of the individual, but it would substitute for the nineteenth century goal of laissez faire as a means to this end, the goal of the competitive order. (Friedman 1951)

Historically, neoliberalism spread throughout the capitalist world since the end of the 1970s–beginning of the 1980s, to be also adopted by former communist countries in Eastern Europe after the fall of the Soviet Union. Here, we use the term to emphasise the neoliberal policies adopted by many countries that give priority to individualism, self-reliance, efficiency, government cost-cutting on services and pensions, and competitiveness that can seriously affect the social and financial situation of old retirees.

The term modernisation also appears frequently in this book. By modernisation we simply mean "the processes of democratization, industrialization, urbanization, immigration, secularization, and bureaucratization" (Achenbaum 1982, p. 348; see also Achenbaum and Stearns 1978) that developed especially since the nineteenth century.

We also wish to note that although gender issues will be the specific focus of some sections in this book (e.g. *Ageing, gender, and sexual life* in Chap. 4) they will be also mentioned, as appropriate, throughout the various chapters and sections. After all, gender permeates most topics of interest to older people. Although there is a long-standing tradition to distinguish *gender* from *sex*, in that gender refers to cultural and psychological factors, whereas sex is more associated to biological differences between individuals, many authors use the terms interchangeably (Muehlenhard and Peterson 2011). Given that feminine behaviours are also present in biological males and masculine behaviours can be observed in biological females (e.g. Poiani 2010), the two terms are indeed somewhat blurred. Still, when in this book we refer to *sex* (male/man and female/woman) we will be pointing to a biological distinction. When we use the term *gender*, we will be implicating both biological and cultural/psychological aspects.

We conclude this introductory chapter with a brief summary of the remaining chapters.

In Chap. 2 *Old ages in history* we review how older people have been regarded in the past, whether their roles in society have changed, how and why. We will also highlight what kind of lessons we can learn from the historical past that can be useful today, not only to understand the reality of older people, but also to perhaps improve our relationship with them.

Chapter 3 *Old people across cultures* focuses on the cross-cultural similarities and differences in the meaning of being old, the behaviour of older people, and the cultural traditions that derive from the process of ageing but that also shape it in the view of the members of an ethnic group.

Ageing in better health is the topic of Chap. 4. Here we return to the biology of ageing, that was first introduced in this chapter, but with a special emphasis on brain plasticity, a very important topic that is the focus of a fast-developing research programme, and we will also review psychological health in old age. Initially mentally healthy persons may be at risk of experiencing serious deterioration of their mental capacities as a result of ageing; therefore, we will devote a section in this chapter to review mental pathologies in the elderly. Special emphasis will be given to the analysis of the various forms of dementia, including Alzheimer's disease. From there we proceed to address those aspects of ageing that affect sexual behaviour. The challenges experienced by people with a disability (mental or physical) who are getting older will be also the topic of a section in this chapter. We conclude the chapter with a section devoted to older people who reach very advanced ages: the centenarians, semi-supercentenarians, and supercentenarians.

We next move to *The psychology of older ages* in Chap. 5 where we start with an analysis of self-identity and then review emotions and cognition in the elderly. Aspects of memory and personality and how they can affect the psychology of older people are addressed next, to then conclude the chapter with a section devoted to an analysis of psychological stress in the elderly, followed by one on retirement and how the experience of retired people can be improved through an active life.

Chapter 6 addresses *The social dimension of older ages* starting with the many factors affecting the social integration of the elderly and the psychological complexities associated with loneliness. Partners and the family are first in line as potential social factors that can contribute to elders' well-being. But the community at large, including friends and acquaintances can also become important, especially if direct family members are not available. Social interactions can be also negative and therefore we analyse the important issue of stigma affecting the elderly, and the various ethical issues, including those associated with the abuse of old people, that are relevant to old age. The oldest-old represent a special case as they face unique social challenges, not only due to their physical frailty, but also because at very advanced ages their peer network of social relationships weakens considerably. In multi-ethnic societies older immigrants experience specific problems of adaptation that cannot be ignored, and thus a section will be also devoted to this topic. Other minorities are also important, and they may have unique needs associated with their ageing. Here we will address the social issues relevant to older gay, lesbian, bisexual, transsexual/transgender (trans), and intersex people, and in a separate section we focus on the experience of ageing prison inmates. We conclude the chapter with an analysis of the challenges associated with caring for older people and the consequences of such care in terms of caregiver burden.

We then shift our focus to *Language and communication* in Chap. 7, starting with intra-generational communication or the ways in which old people communicate with each other. Intra-generational communication tends to differ from communicating with younger generations (inter-generational communication) which, in turn, poses specific challenges to both the old and the young. Communication is expressed through narratives that can unravel our priorities, concerns, emotions, beliefs, ways of processing life events, and understanding of the relationships we establish with other people. The narratives of the elderly are to some extent a window into their mind. Such narratives can however remain more hidden in multicultural societies due to language barriers, in which case older people will require the services of bilingual family members or specifically trained interpreters and translators to communicate more effectively, especially in institutional and medical settings.

Chapter 8, *Journey towards the end of life*, deals with the existential aspects of old age, including changes in life meaning and concerns about legacy. Existential well-being often benefits from the development of spirituality which, in turn, can be expressed in several ways, including through specific religious beliefs and practices. Existential issues are brought into sharper focus in the life of an old person by the obvious fact that inevitably, as time passes, death nears. This brings to the fore questions about what constitutes a "good death", and thus the debate over euthanasia and suicide will be the focus of a specific section. We conclude this chapter with an exploration of the challenges of the end of life in multicultural societies, and the issues that affect end-of-life care.

In the following Chap. 9 *Ageing in evolutionary perspective* we delve into an area of inquiry that is still in development but that it promises to be of great utility in the study of ageing. In this chapter we aim to lay down the foundations for an *evolutionary gerontology*.

The book ends with Chap. 10 *Fulfilling ageing* where we provide our synthesis of old age and review some approaches to helping older people live a more fulfilling life. We emphasise the intermingled individual and social dimensions of fulfilling ageing, through the idea of finding personal meaning in the experience of living with others.

References

Achenbaum, W. A. (1982). Further perspectives on modernization and aging: A (p)review of the historical literature. *Social Science History, 6*, 347–368.

Achenbaum, W. A., & Stearns, P. N. (1978). Old age and modernization. *The Gerontologist, 18*, 307–312.

Adams, G. A., & Rau, B. L. (2011). Putting off tomorrow to do what you want today: Planning for retirement. *American Psychologist, 66*, 180–192.

Agarwal, S., & Busse, P. J. (2010). Innate and adaptive immunosenescence. *Annals of Allergy, Asthma and Immunology, 104*, 183–190.

Ajrouch, K. J., Antonucci, T. C., & Janevic, M. R. (2001). Social networks among blacks and whites: The interaction between race and age. *Journal of Gerontology, 56B*, S112–S118.

Arnold-Cathalifaud, M., Thumala, D., Urquiza, A., & Ojeda, A. (2008). Young people's images of old age in Chile: Exploratory research. *Educational Gerontology, 34*, 105–123.

Avlund, K., Trab Damsgaard, M., & Holstein, B. E. (1998). Social relations and mortality. An eleven year follow-up study of 70-year-old men and women in Denmark. *Social Sciences and Medicine, 47*, 635–643.

Baddeley, A. (1992). Working memory. *Science, 255*, 556–559.

Bailey, A. A. (1953). Changes with age in the spinal cord. *AMA Archives of Neurology and Psychiatry, 70*, 299–309.

Baltes, M. M. (1998). The psychology of the oldest-old: The fourth age. *Current Opinion in Psychiatry, 11*, 411–415.

Baltes, M. M., & Carstensen, L. L. (1996). The process of successful aging. *Ageing and Society, 16*, 397–422.

Baltes, P. (1991). The many faces of human ageing: Toward a psychological culture of old age. *Psychological Medicine, 21*, 837–854.

Baltes, P. B., & Baltes, M. M. (1990). Psychological perspectives on successful aging: The model of selective optimization with compensation. In P. B. Baltes & M. M. Baltes (Eds.), *Successful aging: Perspectives from the behavioral sciences* (pp. 1–34). Cambridge: Cambridge University Press.

Baltes, P. B., & Smith, J. (2003). New frontiers in the future of aging: From successful aging of the young old to the dilemmas of the fourth age. *Gerontology, 49*, 123–135.

Bandura, A. (2001). Social cognitive theory: An agentic perspective. *Annual Review of Psychology, 52*, 1–26.

Beal, M. F. (2005). Mitochondria take center stage in aging and neurodegeneration. *Annals of Neurology, 58*, 495–505.
Becker, G., & Beyene, Y. (1999). Narratives of age and uprootedness among older Cambodian refugees. *Journal of Aging Studies, 13*, 295–314.
Berndt, R. M., & Berndt, C. H. (1992). The world of the first Australians. In *Aboriginal traditional life: Past and present*. Canberra, ACT: Aboriginal Studies Press.
Biggs, S. (2004). Age, gender, narratives, and masquerades. *Journal of Aging Studies, 18*, 45–58.
Borglin, G., Edberg, A.-K., & Rahm Hallberg, I. (2005). The experience of quality of life among older people. *Journal of Aging Studies, 19*, 201–220.
Bortz, W. M. I. I. (2002). A conceptual framework of frailty: A review. *Journal of Gerontology, 57A*, M283–M288.
Bowling, A. (2008). Enhancing later life: How older people perceive active ageing? *Aging and Mental Health, 12*, 293–301.
Brady, M. J., Peterman, A. H., Fitchett, G., Mo, M., & Cella, D. (1999). A case for including spirituality in quality of life measurement in oncology. *Psycho-Oncology, 8*, 417–428.
Brandtstädter, J., & Greve, W. (1994). The aging self: Stabilizing and protective processes. *Developmental Review, 14*, 52–80.
Brandtstädter, J., & Rothermund, K. (2002). The life-course dynamic of goal pursuit and goal adjustment: A two-process framework. *Developmental Review, 22*, 117–150.
Burke, S. N., & Barnes, C. A. (2006). Neural plasticity in the ageing brain. *Nature Reviews Neuroscience, 7*, 30–40.
Byles, J. (2017). Health and ageing. In K. O'Loughlin, C. Browning, & H. Kendig (Eds.), *Ageing in Australia: Challenges and opportunities* (pp. 63–81). New York: Springer.
Cabeza, R., Albert, M., Belleville, S., Craik, F. I. M., Duarte, A., Grady, C. L., et al. (2018). Maintenance, reserve and compensation: The cognitive neuroscience of healthy ageing. *Nature Reviews Neuroscience, 19*, 701–710.
Calasanti, T., & King, N. (2020). Beyond successful aging 2.0: Inequalities, ageism, and the case for normalizing old ages. *The Journals Gerontology B*. https://doi.org/10.1093/geronb/gbaa037
Campisi, J. (2003). Cancer and ageing: Rival demons? *Nature Reviews Cancer, 3*, 339–349.
Cantor, M. H. (1980). The informal support system: Its relevance in the lives of the elderly. In E. F. Borgatta & N. G. McCluskey (Eds.), *Aging and society: Current research and policy perspectives* (pp. 131–144). Beverly Hills, CA: Sage.
Carstensen, L. L. (1991). Socioemotional activity theory: Social activity in life-span context. *Annual Review of Gerontology and Geriatrics, 11*, 195–217.
Carstensen, L. L. (2006). The influence of a sense of time on human development. *Science, 312*, 1913–1915.
Carstensen, L. L., Isaacowitz, D. M., & Charles, S. T. (1999). Taking time seriously:A theory of socioemotional selectivity. *American Psychologist, 54*, 165–181.
Case, A., & Paxson, C. (2009). Early life health and cognitive function in old age. *American Economic Review: Papers and Proceedings, 99*, 104–109.
Cauley, J. A. (2012). The demography of aging. In A. B. Newman & J. A. Cauley (Eds.), *The epidemiology of aging* (pp. 3–14). Dordrecht: Springer.
Chapman, S. A. (2005). Theorizing about aging well: Constructing a narrative. *Canadian Journal on Aging, 24*, 9–18.
Chen, X., Mao, G., & Leng, S. X. (2014). Frailty syndrome: An overview. *Clinical Interventions in Aging, 9*, 433–441.
Cheng, C.-L., Gao, T.-Q., Wang, Z., & Li, D.-D. (2005). Role of insulin/insulin-like growth factor 1 signaling pathway in longevity. *World Journal of Gastroenterology, 11*, 1891–1895.
Christensen, H., Anstey, K. J., Parslow, R. A., Maller, J., Mackinnon, A., & Sachdev, P. (2007). The brain reserve hypothesis, brain atrophy and aging. *Gerontology, 53*, 82–95.
Christensen, K., Doblhammer, G., Rau, R., & Vaupel, J. W. (2009). Ageing populations: The challenges ahead. *Lancet, 374*, 1196–1208.

Cicirelli, V. G. (2010). Attachment relationships in old age. *Journal of Social and Personal Relationships, 27*, 191–199.

Clegg, A., Young, J., Iliffe, S., Rikkert, M. O., & Rockwood, K. (2013). Frailty in elderly people. *The Lancet, 381*, 752–762.

Cohn, M. A., Fredrickson, B. L., Brown, S. L., Mikels, J. A., & Conway, A. M. (2009). Happiness unpacked: Positive emotions increase life satisfaction by building resilience. *Emotion, 9*, 361–368.

Cordella, M., & Poiani, A. (2014). *Behavioural oncology: Psychological , communicative, and social dimensions*. New York: Springer.

Csikszentmihályi, M. (1990). *Flow: The psychology of optimal experience*. New York: Harper & Row.

Cuervo, A. M., Bergamini, E., Brunk, U. T., Dröge, W., Ffrench, M., & Terman, A. (2005). Autophagy and aging: The importance of maintaining "clean" cells. *Autophagy, 1*, 131–140.

Cumming, E., Dean, L. R., Newell, D. S., & McCaffrey, I. (1960). Disengagement - A tentative theory of aging. *Sociometry, 1*, 23–35.

Cumming, E., & Henry, W. E. (1961). *Growing old: The process of disengagement*. New York: Basic Books.

Davey, J., Nana, G., de Joux, V., & Arcus, M. (2004). Accommodation options for older people in Aotearoa/New Zealand. In *NZ Institute for Research on Ageing/Business and Economic Research Ltd, for Centre for Housing Research Aotearoa/New Zealand*. Wellington: New Zealand.

Deci, E. L., & Ryan, R. M. (2008). Hedonia, eudaimonia, and well-being: An introduction. *Journal of Happiness Studies, 9*, 1–11.

Denton, F. T., & Spencer, B. G. (2002). Some demographic consequences of revising the definition of "old age" to reflect future changes in life table probabilities. *Canadian Journal on Aging, 21*, 349–356.

Doherty, T. J. (2008). Aging and sarcopenia. *Journal of Applied Physiology, 95*, 1717–1727.

Doherty, T. J., Vandervoort, A. A., & Brown, W. F. (1993). Effects of ageing on the motor unit: A brief review. *Canadian Journal of Applied Physiology, 18*, 331–358.

Dorfman, L. T., Murty, S. A., Evans, R. J., Ingram, J. G., & Power, J. R. (2004). History and identity in the narratives of rural elders. *Journal of Aging Studies, 18*, 187–203.

Draper, B. (2014, February 14). Preparing for the graying of the world: Meeting the needs of older adults. *Psychiatric Times*. Downloaded on March 20, 2014, from http://www.psychiatrictimes.com/geriatric-psychiatry/preparing-graying-world- meeting-needs-older-adults/page/0/1

Dzuka, J., & Dalbert, C. (2006). The belief in a just world and subjective well-being in old age. *Aging and Mental Health, 10*, 439–444.

Effros, R. B. (2011). Telomere/telomerase dynamics within the human immune system: Effect of chronic infection and stress. *Experimental Gerontology, 46*, 135–140.

Eggebeen, D. J. (1992). Family structure and intergenerational exchanges. *Research on Aging, 14*, 427–447.

Eisdorfer, C., & Cohen, D. (1980). The issue of biological and psychological deficits. In E. F. Borgatta & N. G. McCluskey (Eds.), *Aging and society: Current research and policy perspectives* (pp. 49–70). Beverly Hills, CA: Sage.

Elder, G. H., Jr. (1994). Time, human agency, and social change: Perspectives on the life course. *Social Psychology Quarterly, 57*, 4–15.

Erikson, E. H. (1997). *The life cycle completed*. New York: Norton.

Erikson, E. H., & Erikson, J. (1997). *The life cycle completed: Extended version*. New York: Norton.

Espinoza, S., & Walston, J. D. (2005). Frailty in older adults: Insights and interventions. *Cleveland Clinic Journal of Medicine, 72*, 1105–1112.

Feldman, D. C., & Beehr, T. A. (2011). A three-phase model of retirement decision making. *American Psychologist, 66*, 193–203.

Ferrari, E., Cravello, L., Muzzoni, B., Casarotti, D., Paltro, M., Solerte, S. B., et al. (2001). Age-related changes of the hypothalamic-pituitary-adrenal axis: Pathophysiological correlates. *European Journal of Endocrinology, 144*, 319–329.

Finch, C. E., & Ruvkun, G. (2001). The genetics of aging. *Annual Review of Genomics and Human Genetics, 2*, 435–462.

Finkel, T., & Holbrook, N. J. (2000). Oxidants, oxidative stress and the biology of ageing. *Nature, 408*, 239–247.

Fraga, M. F., & Esteller, M. (2007). Epigenetics and aging: The targets and the marks. *Trends in Genetics, 23*, 423–418.

Frankl, V. E. (1988). *The will to meaning: Foundations and applications of logotherapy*. New York: A Plume Book.

Fratiglioni, L., Paillard-Borg, S., & Winblad, B. (2004). An active and socially integrated lifestyle in late life might protect against dementia. *Lancet Neurology, 3*, 343–353.

Fredrickson, B. L. (2001). The role of positive emotions in positive psychology: The broaden-and-build theory of positive emotions. *American Psychologist, 56*, 218–226.

Fredrickson, B. L., & Joiner, T. (2002). Positive emotions trigger upward spirals toward emotional well-being. *Psychological Science, 13*, 172–175.

Freund, A., & Baltes, P. B. (1998). Selection, optimization and compensation as strategies of life management: Correlations with subjective indicators of successful aging. *Psychology of Aging, 13*, 531–543.

Freund, A. M., & Baltes, P. B. (2002). Life-management strategies of selection, optimization, and compensation: Measurement by self-report and construct validity. *Journal of Personality and Social Psychology, 82*, 642–662.

Friedman, M. (1951). Neo-liberalism and its prospects. *Farmand, 1951*, 89–93.

Fries, J. F. (1980). Aging, natural death, and the compression of morbidity. *New England Journal of Medicine, 303*, 130–135.

Goffman, E. (1963). *Stigma*. Englewood Cliffs, NJ: Prentice-Hall.

Golomb, J. (1995). *In search of authenticity*. New York: Routledge.

Gomez, C. R., Boehmer, E. D., & Kovacs, E. J. (2005). The aging innate immune system. *Current Opinions in Immunology, 17*, 457–462.

Grady, C. L., & Craik, F. I. M. (2000). Changes in memory processing with age. *Current Opinion in Neurobiology, 10*, 224–231.

Granovetter, M. S. (1973). The strength of weak ties. *American Journal of Sociology, 78*, 1360–1380.

Greenwood, P. M. (2007). Functional plasticity in cognitive aging: Review and hypothesis. *Neuropsychology, 21*, 657–673.

Hakim, F. T., & Gress, R. E. (2007). Immunosenescence: Deficits in adaptive immunity in the elderly. *Tissue Antigens, 70*, 179–189.

Harman, D. (1957). Aging: A theory based on free radical and radiation chemistry. *Journal of Gerontology, 2*, 298–300.

Harwood, D. M. J., Hawton, K., Hope, T., & Jacoby, R. (2000). Suicide in older people: Mode of death, demographic factors, and medical contact before death. *International Journal of Geriatric Psychiatry, 15*, 736–743.

Havighurst, R. J. (1963). Successful aging. In R. H. Williams, C. Tibbitts, & W. Donahue (Eds.), *Processes of aging* (pp. 299–320). New York: Atherton Press.

Heckhausen, J., & Schulz, R. (1995). A life-span theory of control. *Psychological Review, 102*, 284–304.

Hedberg, P., Gustafson, Y., Brulin, C., & Aléx, L. (2013). Purpose in life among very old men. *Advances in Aging Research, 2*, 100–105.

HelpAge International. (2013). *Global AgeWatch Index 2013: Insight report*. London: HelpAge International.

Hendricks, J., & Hendricks, C. D. (1986). *Aging in mass society: Myths and realities* (3d ed.). Brown, Boston: Little.

Hesketh, T., & Xing, Z. W. (2006). Abnormal sex ratios in human populations: Causes and consequences. *Proceedings of the National Academy of Sciences of the United States of America, 103*, 13271–13275.

Heylen, L. (2010). The older, the lonelier? Risk factors for social loneliness in old age. *Ageing and Society, 30*, 1177–1196.

Holstein, M. B., & Minkler, M. (2003). Self, society, and the "new gerontology". *The Gerontologist, 43*, 787–796.

Horn, J. L., & Cattell, R. B. (1967). Age differences in fluid and crystallized intelligence. *Acta Psychologica, 26*, 107–129.

Hubble, N., & Tew, P. (2013). *Ageing, narrative and identity: New qualitative social research.* Hampshire: Palgrave Macmillan.

Hughes, V. A., Roubenoff, R., Wood, M., Frontera, W. R., Evans, W. J., & Fiatarone Singh, M. A. (2004). Anthropometric assessment of 10-y changes in body composition in the elderly. *American Journal of Clinical Nutrition, 80*, 475–482.

Hurd Clarke, L., & Korotchenko, A. (2011). Aging and the body: A review. *Canadian Journal of Aging, 30*, 495–510.

Hurd, L. C. (1999). "WE'RE NOT OLD!": Older women's negotiation of aging and oldness. *Journal of Aging Studies, 13*, 419–439.

Isaacowitz, D. M., & Smith, J. (2003). Positive and negative affect in very old age. *Journal of Gerontology, 58B*, P143–P152.

Issa, J.-P. (2003). Age-related epigenetic changes and the immune system. *Clinical Immunology, 109*, 103–108.

James, E. (1998). New systems for old age security: Experiments, evidence and unanswered questions. *World Bank Research Observer, 13*, 271–301.

Jung, B. H., Jeon, M. J., & Bai, S. W. (2008). Hormone-dependent aging problems in women. *Yonsei Medical Journal, 49*, 345–351.

Kahana, E., & Kahana, B. (2001). Successful aging among people with HIV/AIDS. *Journal of Clinical Epidemiology, 54*, S53–S56.

Kashdan, T. B., Biswas-Diener, R., & King, L. A. (2008). Reconsidering happiness: The costs of distinguishing between hedonics and eudaimonia. *The Journal of Positive Psychology, 3*, 219–233.

Keller, J. N., Gee, J., & Ding, Q. (2002). The proteasome in brain aging. *Ageing Research Reviews, 1*, 279–293.

Kinsella, K. (2009). Global perspectives on the demography of aging. In J. Sokolovsky (Ed.), *The cultural context of aging: Worldwide perspectives* (pp. 13–29). Westport, CT: Praeger.

Kleinspehn-Ammerlahn, A., Kotter-Grühn, D., & Smith, J. (2008). Self-perceptions of aging: Do subjective age and satisfaction with aging change during old age? *Journal of Gerontology, 63B*, P377–P385.

Koren, C. (2011). Continuity and discontinuity: The case of second couplehood in old age. *The Gerontologist, 51*, 687–698.

Kullberg, S., Ramírez-León, V., Johnson, H., & Ulfhake, B. (1998). Decreased axosomatic input to motoneurons and astrogliosis in the spinal cord of aged rats. *Journal of Gerontology, 53A*, B369–B379.

Labouvie-Vief, G. (2003). Dynamic integration: Affect, cognition, and the self in adulthood. *Current Directions in Psychological Science, 12*, 201–206.

Lee, Y. (2000). The predictive value of self assessed general, physical, and mental health on functional decline and mortality in older adults. *Journal of Epidemiology and Community Health, 54*, 123–129.

Lerner, M. J. (1980). *The belief in a just world: A fundamental delusion.* New York: Plenum Press.

Lexell, J. (1997). Evidence for nervous system degeneration with advancing age. *The Journal of Nutrition, 127*, 1011S–1013S.

Li, H., Manwani, B., & Leng, S. X. (2011). Frailty, inflammation, and immunity. *Aging and Disease, 2*, 466–473.

Liberale, L., Montecucco, F., Tardif, J.-C., Libby, P., & Camici, G. G. (2020). Inflamm- ageing: The role of inflammation in age-dependent cardiovascular disease. *European Heart Journal, 41*(31), 2974–2982. https://doi.org/10.1093/eurheartj/ehz961

Lindberg, P. G., Feydy, A., & Maier, M. A. (2010). White matter organization in cervical spinal cord relates differently to age and control of grip force in healthy subjects. *Journal of Neuroscience, 30*, 4102–4109.

Litwin, H. (2001). Social network type and morale in old age. *The Gerontologist, 41*, 516–524.

Lloyd-Sherlock, P. (2000). Old age and poverty in developing countries: New policy challenges. *World Development, 8*, 2157–2168.

Löckenhoff, C. E., De Fruyt, F., Terracciano, A., McCrae, R. R., De Bolle, M., Costa, P. T., Jr., et al. (2009a). Perceptions of aging across 26 cultures and their culture-level associates. *Psychology and Aging, 24*, 941–954.

Löckenhoff, C. E., Terracciano, A., & Costa, P. T. (2009b). Five-factor model personality traits and the retirement transition: Longitudinal and cross-sectional associations. *Psychology and Aging, 24*, 722–728.

Löckenhoff, C. E., Terracciano, A., Patriciu, N. S., Eaton, W. W., & Costa, P. T., Jr. (2009c). Self-reported extremely adverse life events and longitudinal changes in Five- Factor Model personality traits in an urban sample. *Journal of Traumatic Stress, 22*, 53–59.

Luanaigh, O. C., & Lawlorz, B. A. (2008). Loneliness and the health of older people. *International Journal of Geriatric Psychiatry, 23*, 1213–1221.

Lundgren, A. S. (2010). "In the good old days": Insidious nostalgia and the constitution of old age identity. *Journal of Aging Studies, 24*, 248–256.

Luo, Y., Hawkley, L. C., Waite, L. J., & Cacioppo, J. T. (2012). Loneliness, health, and mortality in old age: A national longitudinal study. *Social Sciences and Medicine, 74*, 907–914.

Lupien, S. J., McEwen, B. S., Gunnar, M. R., & Heim, C. (2009). Effects of stress throughout the lifespan on the brain, behavior and cognition. *Nature Reviews Neuroscience, 10*, 434–445.

Lustig, C., Shah, P., Seidler, R., & Reuter-Lorenz, P. A. (2009). Aging, training, and the brain: A review and future directions. *Neuropsychology Review, 19*, 504–522.

Lutz, W., & Kc, S. (2010). Dimensions of global population projections: What do we know about future population trends and structures? *Philosophical Transactions of the Royal Society B, 365*, 2779–2791.

Maccoby, E. E. (2007). Historical overview of socialization research and theory. In J. E. Grusec & P. D. Hastings (Eds.), *Handbook of socialization: Theory and research* (pp. 13–41). New York: The Guilford Press.

Magai, C. (2008). Attachment in middle and later life. In J. Cassidy & P. R. Shaver (Eds.), *Handbook of attachment: Theory, research, and clinical applications* (2nd ed., pp. 532–551). New York: Guilford Press.

Martin, G. M., Austad, S. N., & Johnson, T. E. (1996). Genetic analysis of ageing: Role of oxidative damage and environmental stresses. *Nature Genetics, 13*, 25–34.

Mazzoccoli, G., Correra, M., Bianco, G., De Cata, A., Balzanelli, M., Giuliani, A., et al. (1997). Age-related changes of neuro-endocrine-immune interactions in healthy humans. *Journal of Biological Regulatons and Homeostatic Agents, 11*, 143–147.

Medawar, P. B. (1946). Old age and natural death. *Modern Quarterly, 1*, 30–56.

Medawar, P. B. (1952). *An unsolved problem of biology*. London: H.K. Lewis.

Middleton, L. E., Barnes, D. E., Lui, L.-Y., & Yaffe, K. (2010). Physical activity over the life course and its association with cognitive performance and impairment in old age. *Journal of the American Geriatrics Society, 58*, 1322–1326.

Millane, E. (2013). Longevity and ageing: The demographic climate change of our time. *Still Kicking, 2013*, 1–37.

Mintz, S. (2009). Reflections on age as a category of historical analysis. *The Journal of the History of Childhood and Youth, 1*, 91–94.

Moffic, H. S. (2012, June 25). Mental bootcamp: Today is the first day of your retirement! *Psychiatric Times*, Downloaded on 15 February 15, 2014, from http://www.psychiatrictimes.com/blogs/couch-crisis/mental- bootcamp-today-first-day-your-retirement

Mor-Barak, M. E. (1995). The meaning of work for older adults seeking employment. *International Journal of Aging and Human Development, 41*, 325–344.

Muehlenhard, C. L., & Peterson, Z. D. (2011). Distinguishing between sex and gender: History, current conceptualizations, and implications. *Sex Roles, 64*, 791–803.

Mufson, E. J., & Stein, D. G. (1980). Degeneration in the spinal cord of old rats. *Experimental Neurology, 70*, 179–186.

Muller, F. L., Lustgarten, M. S., Jang, Y., Richardson, A., & Van Remmen, H. (2007). Trends in oxidative aging theories. *Free Radical Biology and Medicine, 43*, 477–503.

Murphy, W. P. (1980). Secret knowledge as property and power in Kpelle society: Elders versus youth. *Africa, 50*, 193–207.

Mutran, E. (1987). Family, social ties and self-meaning in old age: The development of an affective identity. *Journal of Social and Personal Relationships, 4*, 463–480.

Nagel, T. (1972). Aristotle on eudaimonia. *Phronesis, 17*, 252–259.

Narayanasamy, A., Clissett, P., Parumal, L., Thompson, D., Annasamy, S., & Edge, R. (2004). Responses to the spiritual needs of older people. *Journal of Advanced Nursing, 48*, 6–16.

Neugarten, B. L., Havighurst, R., & Tobin, S. (1961). The measurement of life satisfaction. *Journal of Gerontology, 16*, 134–143.

Neugarten, B. L., Havinghurst, R. J., & Tobin, S. S. (1968). Personality and patterns of aging. In B. L. Neugarten (Ed.), *Middle age and aging* (pp. 173–177). Chicago: University of Chicago Press.

Newman, A. B., Lee, J. S., Visser, M., Goodpaster, B. H., Kritchevsky, S. B., Tylavsky, F. A., et al. (2005). Weight change and the conservation of lean mass in old age: The Health, Aging and Body Composition Study. *American Journal of Clinical Nutrition, 82*, 872–878.

Nilsson, J., Masud Rana, A. K. M., & Nahar Kabir, Z. (2006). Social capital and quality of life in old age: Results from a cross-sectional study in rural Bangladesh. *Journal of Aging and Health, 18*, 419–434.

Noth, R. H., & Mazzaferri, E. L. (1985). Age and the endocrine system. *Clinics in Geriatric Medicine, 1*, 223–250.

Nussbaum, J. F. (1983). Perception of communication content and life satisfaction among the elderly. *Western Journal of Speech Communication, 47*, 229–243.

Nussbaum, J. F., Pecchioni, L. L., Robinson, J. D., & Thompson, T. L. (2000). *Communication and aging* (2nd ed.). Mahwah, NJ: Lawrence Erlbaum Associates Publishers.

Okada, E., Matsumoto, M., Ichihara, D., Chiba, K., Toyama, Y., Fujiwara, H., et al. (2009). Aging of the cervical spine in healthy volunteers : A 10-year longitudinal magnetic resonance imaging study. *Spine, 34*, 706–712.

Ory, M., Hoffman, M. K., Hawkins, M., Sanner, B., & Mockenhaupt, R. (2003). Challenging aging stereotypes strategies for creating a more active society. *American Journal of Preventative Medicine, 25*, 164–171.

Osborne, J. W. (2012). Existential and psychological aspects of the transition to retirement. *European Journal of Psychotherapy and Counselling, 2012*, 1–15.

Ouwehand, C., de Ridder, D. T. D., & Bensing, J. M. (2007). A review of successful aging models: Proposing proactive coping as an important additional strategy. *Clinical Psychology Review, 27*, 873–884.

Paris, L. (2008). Through the looking glass: Age, stages, and historical analysis. *The Journal of the History of Childhood and Youth, 1*, 106–113.

Park, D. C., & Reuter-Lorenz, P. (2009). The adaptive brain: Aging and neurocognitive scaffolding. *Annual Review of Psychology, 60*, 173–196.

Parker, M. G., & Thorslund, M. (2007). Health trends in the elderly population: Getting better and getting worse. *The Gerontologist, 47*, 150–158.

Peace, S., Dittmann-Kohli, F., Westerhof, G. J., & Bond, J. (2007). The ageing world. In J. Bond, S. Peace, F. Dittmann-Kohli, & G. J. Westerhof (Eds.), *Ageing in society* (pp. 1–14). London: Sage.

Peplau, L., & Perlman, D. (1982). Perspectives on loneliness. In L. A. Peplau & D. Perlman (Eds.), *Loneliness: A sourcebook of current theory, research and therapy*. New York: Wiley.

Perissinotto, E., Pisent, C., Sergi, G., Grigoletto, F., Enzi, G., & The ILSA Working Group. (2002). Anthropometric measurements in the elderly: Age and gender differences. *British Journal of Nutrition, 87*, 177–186.

Pinquart, M., & Sörensen, S. (2000). Influences of socioeconomic status, social network, and competence on subjective well-being in later life: A meta-analysis. *Psychology and Aging, 15*, 187–224.

Pisent, C., Sergi, G., Grigoletto, F., Enzi, G., & the ILSA Working Group. (2002). Anthropometric measurements in the elderly: Age and gender differences. *British Journal of Nutrition, 87*, 177–186.

Poiani, A. (2010). *Animal homosexuality: A biosocial perspective*. Cambridge: Cambridge University Press.

Prelog, M. (2006). Aging of the immune system: A risk factor for autoimmunity? *Autoimmunity Reviews, 5*, 136–139.

Prieto-Flores, M.-E., Fernandez-Mayoralas, G., Rosenberg, M. W., & Rojo-Perez, F. (2010). Identifying connections between the subjective experience of health and quality of life in old age. *Qualitative Health Research, 20*, 1491–1499.

Qiu, M.-G., & Zhu, X.-H. (2004). Aging changes of the angioarchitecture and arterial morphology of the spinal cord in rats. *Gerontology, 50*, 360–365.

Rajah, M. N., & D'Esposito, M. (2005). Region-specific changes in prefrontal function with age: A review of PET and fMRI studies on working and episodic memory. *Brain, 128*, 1964–1983.

Raz, N., Lindenberger, U., Rodrigue, K. M., Kennedy, K. M., Head, D., Williamson, A., et al. (2005). Regional brain changes in aging healthy adults: General trends, individual differences and modifiers. *Cerebral Cortex, 15*, 1676–1689.

Rees Jones, I., & Higgs, P. F. (2010). The natural, the normal and the normative: Contested terrains in ageing and old age. *Social Science and Medicine, 71*, 1513–1519.

Riedel-Heller, S. G., Busse, A., & Angermeyer, M. C. (2006). The state of mental health in old-age across the old European Union - A systematic review. *Acta Psychiatrica Scandinavica, 113*, 388–401.

Roebuck, J. (1979). When does "old age" begin?: The evolution of the English definition. *Journal of Social History, 12*, 416–428.

Roos, M. R., Rice, C. L., & Vandervoort, A. A. (1997). Age-related changes in motor unit function. *Muscle and Nerve, 20*, 679–690.

Rowe, J. W., & Kahn, R. L. (1987). Human aging: Usual and successful. *Science, 237*, 143–149.

Rowe, J. W., & Kahn, R. L. (1997). Successful aging. *The Gerontologist, 37*, 433–440.

Rowe, J. W., & Kahn, R. L. (1998). *Successful aging*. New York: Pantheon Books.

Rowe, J. W., & Kahn, R. L. (2015). Successful aging 2.0: Conceptual expansions for the 21st century. The Journals of Gerontology. *Series B: Psychological Sciences and Social Sciences, 70*, 593–596.

Rubinstein, R. L. (2002). The third age. In R. S. Weiss & S. A. Bass (Eds.), *Challenges of the third age: Meaning and purpose in later life* (pp. 29–40). Oxford: Oxford University Press.

Rubinstein, R. L., & de Medeiros, K. (2015). "Successful aging," gerontological theory and neoliberalism: A qualitative critique. *The Gerontologist, 55*, 34–42.

Russell, S. J., & Kahn, C. R. (2007). Endocrine regulation of ageing. *Nature Reviews Molecular Cell Biology, 8*, 681–691.

Ryan, R. M., & Deci, E. L. (2001). On happiness and human potentials: A review of research on hedonic and eudaimonic well-being. *Annual Review of Psychology, 52*, 141–166.

Ryan, R. M., Huta, V., & Deci, E. L. (2008). Living well: A self-determination theory perspective on eudaimonia. *Journal of Happiness Studies, 9*, 139–170.

Ryff, C. D. (1986). *The failure of successful aging research.* Paper presented at the annual meeting of the Gerontological Society of America, Chicago.
Ryff, C. D. (1989). Beyond Ponce de Leon and life satisfaction: New directions in quest of successful ageing. *International Journal of Behavioural Development, 12*, 35–55.
Ryff, C. D. (2014). Psychological well-being revisited: Advances in the science and practice of eudaimonia. *Psychotherapy and Psychosomatics, 83*, 10–28.
Salat, D. H., Buckner, R. L., Snyder, A. Z., Greve, D. N., Desikan, R. S., Busa, E., et al. (2004). Thinning of the cerebral cortex in aging. *Cerebral Cortex, 14*, 721–730.
Sansoni, P., Vescovini, R., Fagnoni, F., Biasini, C., Zanni, F., Zanlari, L., et al. (2008). The immune system in extreme longevity. *Experimental Gerontology, 43*, 61–65.
Satz, P. (1993). Brain reserve capacity on symptom onset after brain injury: A formulation and review of evidence for threshold theory. *Neuropsychology, 3*, 273–295.
Schäfer, S., Huxhold, O., & Lindenberger, U. (2006). Healthy mind in healthy body? A review of sensorimotor - Cognitive interdependencies in old age. *European Review of Aging and Physical Activity, 3*, 45–54.
Scheibe, S., & Carstensen, L. L. (2010). Emotional aging: Recent findings and future trends. *Journal of Gerontology, 65B*, 135–144.
Schulz, R., & Heckhausen, J. (1996). A life span model of successful aging. *American Psychologist, 51*, 702–714.
Sen, A. (1993). Capability and well-being. Reprinted in: D. M. Hausman (Ed.) (2007), *The philosophy of economics: An antology* (3rd ed., pp. 270–293). Cambridge: Cambridge University Press.
Shaw, A. C., Goldstein, D. R., & Montgomery, R. R. (2013). Age-dependent dysregulation of innate immunity. *Nature Reviews Immunology, 13*, 875–887.
Shaw, A. C., Joshi, S., Greenwood, H., Panda, A., & Lord, J. M. (2010). Aging of the innate immune system. *Current Opinion in Immunology, 22*, 507–513.
Shmotkin, D., & Shrira, A. (2012a). Happiness and suffering in the life story: An inquiry into conflicting expectations concerning the association of perceived past with present subjective well-being in old age. *Journal of Happiness Studies, 13*, 389–409.
Shmotkin, D., & Shrira, A. (2012b). On the distinction between subjective well-being and meaning in life: Regulatory versus reconstructive functions in the face of a hostile world. In P. T. P. Wong (Ed.), *The human quest for meaning: Theories, research, and applications* (pp. 143–163). London: Routledge.
Shultz, K. S., & Wang, M. (2011). Psychological perspectives on the changing nature of retirement. *American Psychologist, 66*, 170–179.
Sijuwade, P. O. (2013). Spiritual intelligence, living status and general health of the elderly. *Pakistan Journal of Social Sciences, 10*, 135–138.
Sjölund, B.-M., Nordberg, G., Wimo, A., & von Strauss, E. (2010). Morbidity and physical functioning in old age: Differences according to living area. *Journal of the American Geriatrics Society, 58*, 1855–1862.
Skirbekk, V. (2003). *Age and individual productivity: A literature survey* (MPIDR Working Paper WP 2003-028). Max Planck Institute for Demographic Research, Rostock, Germany.
Smith, J., & Freund, A. M. (2002). The dynamics of possible selves in old age. *The Journals of Gerontology, 57B*, P492–P500.
Sokolovsky, J. (2009). A global vision of aging, culture and context. In J. Sokolovsky (Ed.), *The cultural context of aging: Worldwide perspectives* (pp. 1–12). Westport, CT: Praeger.
Solana, R., Pawelec, G., & Tarazona, R. (2006). Aging and innate immunity. *Immunity, 24*, 491–494.
Spillman, B. C., & Pezzin, L. E. (2000). Potential and active family caregivers changing networks and the "sandwich generation". *The Milbank Quarterly, 78*, 347–374.
Stephan, Y., Chalabaev, A., Kotter-Grühn, D., & Jaconelli, A. (2013). 'Feeling younger, being stronger': An experimental study of subjective age and physical functioning among older adults. *Journal of Gerontology B, 68*, 1–7.

Stessman, J., Cohen, A., Hammerman-Rozenberg, R., Bursztyn, M., Azoulay, D., Maaravi, Y., et al. (2008). Holocaust survivors in old age: The Jerusalem Longitudinal Study. *Journal of the American Geriatrics Soiety, 56*, 470–477.

Stevenson, B., & Wolfers, J. (2007). *Marriage and divorce: Changes and their driving forces* (IZA Discussion Papers, No. 2602). Downloaded on April 26, 2014, from http://nbn-resolving.de/urn:nbn:de:101:1-20080422181

Stiegelbauer, S. M. (1996). What is an elder? What do elders do?: First nation elders as teachers in culture-based urban organizations. *The Canadian Journal of Native Studies, 16*, 37–66.

Stoller, E. P., & Pugliesi, K. L. (1991). Size and effectiveness of informal helping networks: A panel study of older people in the community. *Journal of Health and Social Behavior, 32*, 180–191.

Strawbridge, W. J., Camacho, T. C., Cohen, R. D., & Kaplan, G. A. (1993). Gender differences in factors associated with change in physical functioning in old age: A 6-year longitudinal study. *The Gerontologist, 33*, 603–609.

Suen, K. C., Lau, L. I., & Yermako, V. (1974). Cancer and old age: An autopsy study of 3,535 patients over 65 years old. *Cancer, 33*, 1164–1168.

Sumowski, J. F., Rocca, M. A., Leavitt, V. M., Dackovic, J., Mesaros, S., Drulovic, J., et al. (2014). Brain reserve and cognitive reserve protect against cognitive decline over 4.5 years in MS. *Neurology, 82*, 1776–1783.

Szilard, L. (1959). On the nature of the aging process. *Proceedings of the National Academy of Sciences of the United States of America, 45*, 35–45.

Szinovacz, M. E., DeViney, S., & Davey, A. (2001). Influences of family obligations and relationships on retirement: Variations by gender, race, and marital status. *Journal of Gerontology, 56B*, S20–S27.

Thomas, K., & Shute, R. (2006). The old and mentally ill in Australia: Doubly stigmatized. *Australian Psychologist, 41*, 186–192.

Tornstam, L. (1989). Gero-transcendence; a reformulation of the disengagement theory. *Aging: Clinical and Experimental Research, 1*, 55–63.

Treas, J., & Mazumdar, S. (2002). Older people in America's immigrant families: Dilemmas of dependence, integration, and isolation. *Journal of Aging Studies, 16*, 243–258.

Trifunovic, A., Hansson, A., Wredenberg, A., Rovio, A. T., Dufour, E., Khvorostov, I., et al. (2005). Somatic mtDNA mutations cause aging phenotypes without affecting reactive oxygen species production. *Proceedings of the National Academy of Sciences of the United States of America, 102*, 17993–17998.

United Nations, Department of Economic and Social Affairs, Population Division. (2013a). *World population ageing 2013* (ST/ESA/SER.A/348).

United Nations, Department of Economic and Social Affairs, Population Division. (2013b). *World population prospects: The 2012 revision, key findings and advance tables* (Working Paper No. ESA/P/WP.227).

Valenzuela, M. J., & Sachdev, P. (2006). Brain reserve and dementia: A systematic review. *Psychological Medicine, 36*, 441–454.

van den Beld, A. W., Kaufman, J.-M., Zillikens, M. C., Lamberts, S. W. J., Egan, J. M., & van der Lely, A. J. (2018). The physiology of endocrine systems with ageing. *The Lancet Diabetes and Endocrinology, 6*, 647–658.

van Wagenen, A., Driskell, J., & Bradford, J. (2013). "I'm still raring to go": Successful aging among lesbian, gay, bisexual, and transgender older adults. *Journal of Aging Studies, 27*, 1–14.

Vasto, S., Carruba, G., Lio, D., Colonna-Romano, G., Di Bona, D., Candore, G., et al. (2009). Inflammation, ageing and cancer. *Mechanisms of Ageing and Development, 130*, 40–45.

Vaupel, J. W. (2010). Biodemography of human ageing. *Nature, 464*, 536–542.

Velasco-Gonzalez, L., & Rioux, L. (2013). The spiritual well-being of elderly people: A study of a French sample. *Journal of Religion and Health, 53*, 1123–1137.

Verdú, E., Ceballos, D., Vilches, J. J., & Navarro, X. (2000). Influence of aging on peripheral nerve function and regeneration. *Journal of the Peripheral Nervous System, 5*, 191–208.

Verhaegen, P., & Salthouse, T. A. (1997). Meta-analyses of age-cognition relations in adulthood: Estimates of linear and nonlinear age effects and structural models. *Psychological Bulletin, 122*, 231–249.

Wallace, D. C. (2005). A mitochondrial paradigm of metabolic and degenerative diseases, aging, and cancer: A dawn for evolutionary medicine. *Annual Review of Genetics, 39*, 359–407.

Wang, H.-X., Karp, A., Winblad, B., & Fratiglioni, L. (2002). Late-life engagement in social and leisure activities is associated with a decreased risk of dementia: A longitudinal study from the Kungsholmen project. *American Journal of Epidemiology, 155*, 1081–1087.

Waterman, A. S., Schwartz, S. J., & Conti, R. (2008). The implications of two conceptions of happiness (hedonic enjoyment and eudaimonia) for the understanding of intrinsic motivation. *Journal of Happiness Studies, 9*, 41–79.

Weinert, B. T., & Timiras, P. S. (2003). Theories of aging. *Journal of Applied Physiology, 95*, 1706–1716.

Weiskopf, D., Weinberger, B., & Grubeck-Loebenstein, B. (2009). The aging of the immune system. *Transplant International, 22*, 1041–1050.

Weng, N. (2008). Telomere and adaptive immunity. *Mechanisms of Ageing and Development, 129*, 60–66.

Whitaker, A. (2010). The body as existential midpoint—The aging and dying body of nursing home residents. *Journal of Aging Studies, 24*, 96–104.

Whitbourne, S. K. (1986). *The me I know: A study of adult identity*. New York: Springer.

White, P. (2006). Migrant populations approaching old age: Prospects in Europe. *Journal of Ethnic and Migration Studies, 32*, 1283–1300.

White Riley, M. (1983). The family in an aging society: A matrix of latent relationships. *Journal of Family Issues, 4*, 439–454.

Wickremaratchi, M. M., & Llewelyn, J. G. (2006). Effects of ageing on touch. *Postgraduate Medical Journal, 82*, 301–304.

Wilber, K. (1999). Spirituality and developmental lines: Are there stages? *Journal of Transpersonal Psychology, 31*, 1–10.

Wiles, J. L., Leibing, A., Guberman, N., Reeve, J., & Allen, R. E. S. (2011). The meaning of "aging in place" to older people. *The Gerontologist, 52*, 357–366.

Wolpaw, J. R. (2006). The education and re-education of the spinal cord. *Progress in Brain Research, 157*, 261–280.

Yankner, B. A., Lu, T., & Loerch, P. (2008). The aging brain. *Annual Review of Pathology: Mechanisms of Disease, 3*, 41–66.

Yirmiya, R., & Goshen, I. (2011). Immune modulation of learning, memory, neural plasticity and neurogenesis. *Brain, Behavior, and Immunity, 25*, 181–213.

Zanto, T. P., & Gazzaley, A. (2014). Attention and ageing. In A. C. Nobre & S. Kastner (Eds.), *The Oxford handbook of attention* (pp. 927–971). Oxford: Oxford University Press.

Zhang, W. (2008). Religious participation and mortality risk among the oldest old in China. *Journal of Gerontology, 63B*, S293–S297.

Chapter 2
Old Ages in History

Abstract *Old ages in history* is a review of how older people have been regarded throughout history, whether their roles in society have changed, how and why. We will also highlight what kind of lessons we can learn from the historical past that can be useful today, not only to understand the reality of older people, but also to perhaps improve our relationship with them.

Humans have demonstrated the capacity to reach old ages for a very long time, but the probability of doing so has been quite variable in the human population throughout history. Although the elders do not appear frequently in the ancient literature or on records that have survived to our days (Charles 1977), and when references are made to old people they usually mention men rather than women (Parkin 2003), we know from the available information that perception of old age and the relationship of the elders with the rest of society have changed over time. In oral, pre-literary cultures the old tended to be highly regarded as repositories of wisdom and local knowledge that was then passed on to the younger generations. This was a source of social relevance and prestige for the elders that was intermittently lost and regained over time in the Western world, as knowledge that could be transferred through written documents did not necessitate the intermediacy of older generations, but the old could still find some niches of relevance as the circumstances changed.

The relative importance of older people in society has also varied in response to their numerical representation in the population. How has the percentage of older people changed throughout the centuries? Tim Parkin (2003) estimates that around 6–8% of the European population was older than 60 years during the Roman Empire (6–7% according to Cokayne 2003), a similar percentage of 5–8% was maintained in the Middle Ages (Shahar 2005), with the exception of periods of Black Death. During epidemics of bubonic plague that ravaged Europe in the second half of the fourteenth century (1361–1400), and that especially killed those younger than 60 (Razi 1980), the percentage of the population older than 60 reached peaks of 15%

M. Cordella, A. Poiani, *Fulfilling Ageing*, International Perspectives on Aging 30,
https://doi.org/10.1007/978-3-030-60071-6_2

(Fleming et al. 2003). Such figures of about 6–8% of old people during both the Roman Empire and most of the Middle Ages in Europe are quite remarkable if we consider Pat Thane's (2003) point that in England the percentage of the population older than 60 was just 3.4% in the nineteenth century and 6% in 1911, and it was 8–9% in the eighteenth century according to Troyansky (2005), declining to 6–7% by the end of the eighteenth century (Cole and Edwards 2005). For Victorian England Cokayne (2003) provides a figure of about 7.2%. But by 1951 the percentage had reached 14% of the population in England and then 18% in 1991. That is, it took 600 years for the percentage of old people to reach levels in Europe that were first observed during the epidemics of bubonic plague in the fourteenth century.

In France, the percentage of the population that was older than 60 was 7% in the eighteenth century (6–10% according to Troyansky 2005), 10% in the nineteenth century (9% is given by Thane (2005a) for 1836), 12% in the early twentieth century, 14% in 1946 (Thane 2003), and 18% in 1976 (Thane 2005a). Cole and Edwards (2005) provide figures of 8.7% in 1801 and 12.6% in 1906 also for France. In Sweden the population over 60 was 9% in 1876, doubling to 18% in 1962 (Thane 2005a). In the seventeenth century more Europeans started to live to older ages because of improved dietary and economic conditions, compounded by the introduction of some social welfare programs. In Austria in 1632 a total of 5.5% of the population was older than 60, which increased to 6.9% in 1671 and then again to 8.6% in 1779 (Botelho 2005).

In the USA people aged over 60 years comprised 2% of the population from 1650 until the beginning of the nineteenth century (Fleming et al. 2003), 4% of the population in 1830, and 6.5% in 1906 (Cole and Edwards 2005). The same increase in the transition from the nineteenth to the twentieth century was seen in the USA among the 65 year old and older who represented 2.9% of the population in 1870 and 4.6% in 1920 (Achenbaum 1974). By 1935 those 60 and older reached 8% of the population, with immigration, that usually involves the transfer of relatively young individuals from one area or country to another, being responsible for slowing down the ageing trend across the USA population as compared to Europe (Thane 2005a).

In Australia 0.9% of the population was 65 years old and older in 1861, the percentage had increased to 2.4% by 1881, it was 4.3% in 1911, 8% in 1947, 9.8% in 1981, and 14% in 2011 (Jalland 2015). Just as it was the case in the USA, immigration also slowed down the level of ageing in the Australian population compared to Europe.

The above data suggest that for many centuries the percentage of old people in some Western countries remained remarkably stable (roughly between 6% and 9%), increasing significantly only in relatively recent times, especially in the post-World War II period when the percentage of people older than 60 jumped to 14% immediately after the war and then again to 18% in the 1960s. In 2013 the world population of 60 year old and older was 11.7%, but in Europe—our main source of historical data—it had increased to 22.9%; in North America it was similarly up to 19.9%, whereas in Africa it was 5.4%, which is similar to the pre-twentieth century levels recorded in Western countries (United Nations 2013).

In antiquity, there was high mortality at young ages, but for people who could reach old age their probability of survival from then on was similar to the one observed today. The recent increase in the proportion of the elderly population is mainly due to medical advances and improved living conditions that allow many more people to reach older ages than it was the case in the past.

The sex ratio among the older segments of the population has also changed. Men were more frequent than women among the old in Roman times, especially due to high women mortality during child delivery. In the Middle Ages also more women died young than men, thus maintaining the male-biased sex ratio in the adult population, even though surviving older women lived longer than older men. The male-biased sex-ratio in the broader population reversed in the Renaissance when women became overall more frequent than men, as health conditions improved.

Hence the relative impact of older people in society has been changing over time due to various demographic factors.

In this chapter we focus on how the views on old age have varied throughout the history of Western culture. We start by reviewing the historical changes in the division of the lifetime into specific stages: the ages of life. We then review old age in antiquity, with emphasis on the classical Greek and Roman worlds, to be followed by a brief analysis of old age in the Middle Ages. We subsequently move to the Renaissance and later centuries up to the present. Throughout the chapter we explore the continuous tensions between competition and cooperation that have animated the relationship between the older and the younger generations throughout history.

2.1 The Ages of Life

Observable change is what marks our perception of time and therefore the ageing of our self and others. Change may be perceived as continuous or punctuated, but even within a continuum of variation there are transitions in our development that tend to mark specific points of departure from the past. Such transitions have been the base for the establishment of ages of life systems, and in some cultures some of those transitions are marked by rites of passage (Settersten 1997).

In general, in antiquity there was not a strong need to know the exact age of a person (Parkin 2003), and so the beginning of old age throughout history has been somewhat variable. Since the times of ancient Greece and Rome the Western tradition has been to place the beginning of old age between 60 and 70 years old, although various sources of variability could extend the range, with old age (*senectus* in Latin) sometimes commencing in the 40s, whereas other individuals were regarded as quite fit even in their late 70s or 80s (Parkin 2003). In the Middle Ages old age started at 50 (Minois 1989), which in reality only represented a mid-point between a larger range of ages for the old (between 40 and 65) that was also recognised in the Renaissance (Gilbert 1967).

The judgement whether a person had reached old age was based mainly on external physical characteristics, rather than on the probability of being close to death or on date of birth, which usually was not recorded with exactitude. Criteria based on the interactions of the individual with society, and the capacity to effectively perform some specific tasks, were also used to determine whether a person was old or not. The physical limitations usually associated with old age have also led old individuals to be exempted from some public services since antiquity (Thane 2005a, b). Such judgement regarding ageing based on broad body condition and capacities tended to reflect more negatively on women than men. It is only more recently that chronological age as such has been a defining characteristic of old age, due to the introduction of the retirement age and pensions (Covey 1992).

Traditionally the ages of life from childhood to old age have been described from the perspective of men, although there are also some authors who have mentioned women. In general, women were regarded as ageing faster than men and they were also described as losing their social status faster once they reached menopause and were beyond child-bearing age.

"Youth", "maturity", and "old age" have been the common basic subdivision of the ages of life (the ages of men and women), although in more recent times categories such as "young-old age" and "old-old age" have been added to provide a more precise description of "old age". Overall, this latter classification forms a schema of first, second, third, and fourth age (Thane 2005a, b). The concept of the "fourth age" was introduced in 1991 by the British demographer Peter Laslett to describe those over 80 (Jalland 2015) and it tends to be defined as a period of dependence, decrepitude, and eventually death (Hubble and Tew 2013). Throughout history, however, the ages of life have been remarkably variable, with subdivisions being based on a combination of biological changes, social roles, but also myths, prejudice, and personal perceptions. At the individual level, the limit of what one regards as own "old age" may be shifted at ever later ages by self-perception, especially when the individual still feels active and well, both mentally and physically, but in part also in order to avoid the stigma of old age.

Given that in most ancient classifications the ages of life focused on the male experience, they have been referred to as the "ages of man" (e.g. Chojnacka and Wiesner-Hanks 2002). We will retain this nomenclature following the original use in the texts cited, when appropriate; otherwise, we will use the broader concept of the "ages of life".

In ancient Greece, Solon (the Athenian statesman and legislator, seventh to sixth century b.c.e.) divided man's lifespan into ten periods of 7 years each (Payne 2015), with old age being the stage when individuals have acquired the knowledge and experience to be able to counsel others. Whereas Pythagoras (sixth–fifth century b.c.e) saw the ages of life as an equivalent to the four seasons, a concept that subsequently became popular in Western culture (Minois 1989). Each "season" in the four ages of live was of 20 years, and old age begun at 60 in this Pythagorean system (Payne 2015). Old age was associated with the cold winter, and any person was supposed to feel better in the season that is contrary to his/her age; thus, for the elderly, summer and autumn would be the best seasons (Parkin 2005).

Aristotle (fourth century b.c.e.) recognised three ages of man, whereas the Roman Horace (first century b.c.e.) identified four: boy, young man, man, old man. The child is focused on playing and is changeable and capricious, "the beardless youth, enjoying new-found freedom, lavish, impressionable, and irresponsible; the mature man, wealth-seeking, ambitious, conservative in action" (Coffman 1934, pp. 249–250). Old age is a period of desire for gain, miserliness, lack of energy, a desire to have a longer life, quarrelsomeness, a praise of the "good old days" when he was younger, and condemnation of the younger generations (Coffman 1934, p. 250).

Cicero (first century b.c.e.) also used the Pythagorean seasonal metaphor to describe the ages of life, each one of which is associated with specific aspects of character: the dependent childhood, ardour in youth, seriousness in middle age, and maturity in old age.

In classical antiquity, the characterisation of different stages of ageing was mainly based on external characteristics of the body, as even a simple accounting task such as describing the number of years a person lived was not a straightforward matter, and it was also affected by cultural beliefs. For instance, most ages given in ancient Roman tombs were multiples of five (Payne 2015), which presumably did not reflect the exact age of the deceased. Variability was also recognised in the population, Seneca (first century c.e.) observed that different individuals experience old age in different ways: "There is not one type of old age for all people" (Parkin 2005, p. 31).

In the Middle Ages, life expectancy at birth was 40 years, mainly due to high infant mortality and the effect of epidemics, but much older ages could be achieved by some individuals. Old age was usually regarded to start between 60 and 70 (for both men and women), but specific individuals could be regarded as old at 35 (Shahar 2005).

The ages of life in the Middle Ages were not only descriptive but also prescriptive of specific sets of age-appropriate behaviours. Such behaviours were supposed to conform to a natural order (Burrow 1986; Covey 1989), although in reality not all individuals followed such prescriptions (Dove 1986). For instance, old age was expected to conduce to wisdom, hence any childish behaviour performed by older people was frowned upon. Similarly, some wise younger people could be also seen as elders.

Although in Medieval times the ages of life were deterministic and their sequence had a somewhat mechanical character (Covey 1989), they could be divided in different ways according to various traditions, ranging from three, to four, six, seven, or even 12 ages. A less popular tradition considered five ages of life. The intervals of time determining the ages of life were constant within each system and varied from 7 years to 10 or 20, which fixed the specific transitions from one age to the next.

A widespread subdivision of the ages of life in the Middle Ages involved three periods: youth, middle age, and old age. The four ages of life also became more common in the late Middle Ages, which also saw the organisation of other phenomena into groups of four: the four seasons, four elements, four humours, continuing a tradition that harked back to ancient Greece and Rome. During this period, old age was regarded as the winter of life (Covey 1989). Dante (fourteenth century) also divided life into four ages: Adolescence, youth (starting at 25), old age (starting at 45), and senility (starting at 70) (Gilbert 1967).

Beginning in the twelfth century, the seven ages of life became popular, with the number seven being associated with other systems of belief such as the seven virtues, the seven planets known at the time and the seven vices. This also dovetailed with the Biblical description of the lifespan as "three score and ten" (70 years) (Covey 1989). The number seven also recurs frequently in the Old Testament in various contexts. Figure 2.1 shows an example of the seven ages of life from an engraving published in a 1486 edition of Bartholomeus Anglicus' *De proprietatibus rerum* (On the Properties of Things).

Fig. 2.1 The seven ages of life, published in a 1486 edition of *De proprietatibus rerum* by Bartholomeus Anglicus

This is how Covey (1989, p. 697) synthesised the various concepts of the ages of life in the Middle Ages, with special reference to old age:

> Viewed simply as heuristic devices used to make sense out of the aging process, ages-of-life models have elements of both continuity and abrupt change, of ambiguity and inconsistency. What is clear is the consistency of what was defined as age-appropriate behavior in all of the models. On the positive side, the ages of life characterize old age as a time for contemplation, spiritual restoration, repentance, and wisdom. On the negative side, decay, miserliness, lust, foolishness, childlike behavior, dementia, and poor health were also attributed to older people.

Thus, old age although representing a period with some obvious negative aspects to it, it could be rescued in the prevailing view of society through the wisdom and experience the elders had acquired and could share with younger generations.

During the Renaissance old age was considered to start at 50 (Minois 1989), but in fact some variability was also recognised. Just like in the Middle Ages, the most common subdivisions of the ages of life in this period were: 3, 4, 6, and 7. Three was usually directly associated with the Holy Trinity but also with other triads such as the Augustinian view of the soul that was composed of three parts: understanding, will (*voluntas*), and memory. Understanding, will, and memory were in turn directly related to the Trinity, and the Trinity represented unity in diversity, just as the three parts of the soul were united into one soul. Four was also a common subdivision for the ages of life, with classic associations to the four seasons, complexions, and elements. Six were the ages of the world and seven were the planets known until then (Patrides 1958).

In the seventeenth century the systems of classification of the ages of life ranged from 3, to 4, 7, 10, or 12 age periods (Botelho 2005). An example of a ten-stage classification used in Germany during the sixteenth and seventeenth centuries is provided by Thane (2005a, p. 119):

- 10 years—a child
- 20 years—youth
- 30 years—a man
- 40 years—standing still
- 50 years—settled and prosperous
- 60 years—departing
- 70 years—protect your soul
- 80 years—the world's fool
- 90 years—the scorn of children
- 100 years—God have mercy

Note how in this classification it seems that there was no real role for anybody older than 70 years, presumably because the number of people who could live to such ages was limited at the time. The advice to "protect your soul" at the age of 70 seems to be a preparation for imminent death, with those living longer becoming a puzzle of sorts. The description of the oldest-old stages apparently reflects a difficulty to understand who the oldest-old are and what to do with them.

In the eighteenth century the ages of man and woman were illustrated as running in strict parallel. Troyansky (2005) provides an example from France that contemplates 12 age periods from infancy (newborn to younger than 4) to childhood (4 years old), adolescence (10 years), youth (20 years), manhood (30 years; "womanhood" in the case of females), discretion (40 years), maturity (50 years), decline (60 years), decadence (70 years), age of frailty ("l'age caduc", 80 years), decrepitude (90 years), age of imbecility and childishness (100 years). After the age of 60 the description of the ages of life is overwhelmingly negative.

In the nineteenth century the relative determinism of the ages of life was challenged, at least to an extent, by the notion that individual choices and lifestyle can affect the length, the quality of life, and the definitions of young and old. In 1853, in his book *On the Decline of Life*, Barnard van Oven expressed the view that:

> the duration of the period of maturity is greatly within our control; and that, although the termination of the journey of human life is absolute and certain, yet that not only the length of that journey, but the manner of its division into various stages, and the degree of ease and pleasure with which we may travel, depend essentially on ourselves (Cole and Edwards 2005, p. 246).

Romanticism added a sense of security by idealising the journey of life with a spiritual undertone of religious inspiration. Interestingly, some illustrations of the ages of life of this period maintain a large number of stages (i.e. 12) in the case of men, whereas for women the stages seem to be reduced in number and running at irregular intervals (Cole and Edwards 2005) (see Fig. 2.2): from a minimum of 6 years (12 years old to 18 years old) to 25 years (50 years old to 75 years old). Perhaps the shortest intervals between stages at younger ages in women reflected the importance of achieving reproductive condition, whereas after the age of

Fig. 2.2 The ages of women's life: from infancy to the brink of the grave. Alden (1835), Library of Congress (http://www.loc.gov/pictures/item/2006686268/)

reproduction the role of women relaxed, and longer intervals could be described for subsequent stages. The social roles of men remained somewhat diverse into older ages.

In the twentieth century old age was initially further subdivided into a young-old period (55–75 years old) and the old-old period (older than 75), which eventually led to the classification that we mentioned in Chap. 1: young-old (60–69 years old), old-old (70–79 years old), oldest-old (80 years old and older). The young-old are characterised by the initiation of their official retirement from work. Although young-old women were more likely to be widowed than men (30% vs 7% in the USA), the rate of home ownership was high during this period, thus giving greater security to old women compared to previous generations. In addition, this generation of young-old overlapped with the generation of their parents (the old-old), due to a general increase in the average lifespan (Neugarten 1974).

In this century, a trend developed towards giving greater importance no so much to the chronological age but to reaching older ages in a "youthful" mental and physical condition (Neugarten 1979), although we will see below that other alternative views of old age also emerged. In addition, the increasing economic uncertainties facing individuals in the late twentieth and early twenty-first centuries compounded with more people reaching older ages, the relaxation of some social traditions, better education, and greater consciousness of our individual capabilities have made the patterns of old age more complex and variegated (part of a "fluid life cycle", Hirschhorn 1977; see also Settersten 1997), in contrast to the more rigid schemes of the ages of life of the past:

> There are also changes in the rhythm of the life cycle for the increasing number of men and women who are marrying, divorcing, and then remarrying, and, most striking in young women, for the number who are rearing children first in two-parent, then one-parent, and then in two-parent households. There are increasing numbers of women and men who enter and reenter the labor force, change jobs, undertake new careers, or return to school (Neugarten 1979, p. 889).

During the first half of the twentieth century and under the influence of psychoanalysis, especially in the West, new views on the ages of life emerged. The psychoanalyst Erik H. Erikson saw the development of the human being as a dynamic process unfolding over time through eight stages. Each stage was the product of the combined effect of "ego-syntonic" (i.e. defining of the self) and "ego-dystonic" (i.e. disrupting of the self) factors determining a life stage-specific tension or crisis. Table 2.1 shows Erikson's eight stages of life model. Infancy, for instance, is characterised by hope and the infant must resolve the tension between trust and mistrust. Adulthood is the stage of provision of care—to offspring, partner, old parents—that confronts the existential tension between *generativity* (self-fulfilment through guiding the next generation) and stagnation. Old age, on the other hand, is supposed to bring wisdom and be affected by the crisis of conserving *ego integrity* without falling into the grip of despair (Agronin 2014). Here Erikson points to a fundamental existential "contradiction" of old age: just when we may be reaching a stage of greater knowledge and wisdom, we start our irreversible physical decay that will eventually lead to death (Elkhonon Goldberg calls this the *Wisdom Paradox*, Goldberg 2007). The conflict is real, and *acceptance* could be regarded as a possible resolution of it: at one point we have done all what we could, and it is up to others to do the rest from then on. The sense of acceptance of the physical limitations of the self may also drive older people away from self and into caregiving and role-modelling activities towards younger generations, something that Erikson did include in his definition of wisdom (Agronin 2014). Through generativity, the achievements of the old could be a springboard for further achievements by the young.

In the 1990s, Erikson's wife, Joan, added a ninth stage to the original eight stages of life. Such stage would include people 80 year old and older. This stage is characterised by serious decay of both body and mental faculties leading to hopelessness. By then Erik had developed dementia and this might have affected the view Joan had of this final stage.

Towards the end of the twentieth century and continuing into the twenty-first century the ages of life have lost their traditional structured and staged progression in time, as mentioned above, to be blurred into a more dynamic process (a set of "intertwined life trajectories", Cattell and Albert 2009). For instance, George Vaillant refers not to stages but to "developmental tasks" (Agronin 2014): to acquire skills to obtain a job, further develop skills and expertise to advance in one's career,

Table 2.1 The eight stages of life according to Erik Erikson. Each successive stage is listed on the top, the diagonal shows the central tension or crisis characterising each stage of life, and the bottom row indicates the stage-specific strengths. Modified from Agronin (2014)

Infancy	Early Childhood	Play Age	School Age	Adolescence	Young Adulthood	Adulthood	Old Age
Trust vs Mistrust							
	Autonomy vs Shame						
		Initiative vs Guilt					
			Industry vs Inferiority				
				Identity vs Confusion			
					Intimacy vs Isolation		
						Generativity vs Stagnation	
							Integrity vs Dispair
Hope	**Will**	**Purpose**	**Competence**	**Fidelity**	**Love**	**Care**	**Wisdom**

establish positive social relationships and break negative ones, develop an identity, accumulate knowledge. Such tasks are not strictly sequential in time, and they can be achieved at different speeds and at different points in life by different individuals. Such dynamism of the human experience that somewhat blurs the classification of human life into strict stages has been nicely backed up by current research on the plasticity of the human brain and its ability for continued change until advanced age in healthy individuals (e.g. Cohen 2005). We will address this issue in some detail in Chap. 4: *Ageing in Better Mental Health.*

In addition, stronger emphasis has been put recently into defining the experience of personal development through time also from the perspective of women. Women, for instance, have a stronger sense of social interconnectivity than men (Gilligan 1982), a characteristic that is profoundly entrenched in our biology and that is shared with females of many other social mammals (Poiani 2010). Therefore, the characterisation of the stages of life should also consider the factor of social interconnectivity. Still, the tendency of classifying according to age categories is alive and well, see, for instance, the current classification of various ages into *Millennials* (18–33 years old), *Generation X* (34–45), *Younger Boomers* (46–55), *Older Boomers* (56–64), the *Silent Generation* (65–73), and the *G.I. Generation* (74+; G.I. stands for Government Issue, a military expression) (Zickuhr 2010).

2.2 Retirement Age and Old Age Pension

The structure and meaning of retirement have also changed over the centuries. Starting from the age of 70, Roman citizens were not required to fulfil any duties in the form of compulsory public service, either via financial contribution or actual physical work (Payne 2015), although the age was progressively decreased in Roman Egypt (Parkin 2003). In the Roman Senate retirement was voluntary and 65 years old Senators (the age was perhaps subsequently lowered to 60 years) were allowed to attend the sessions if they wanted to, but there was no compulsion (Parkin 2003), although if they decided to attend the Senate sessions they were expected to make an active contribution.

All Roman citizens were compelled to contribute *munera* (duties) and to hold *honores* (public offices)—although many did all they could to avoid them—but people older than 70 were exempt (Parkin 2003). The *munera personalia* (personal duties) in particular involved physical labour. In the third–fourth centuries c.e., during the rule of Diocletian and Maximian, citizens older than 55 were not expected to perform personal munera, at least duties such as being a member of the city Senate or a local administrative officer (munera of decurion) (Parkin 2003).

In Medieval England compulsory public service ceased at the age of 60 (Thane 2003). Although in the Middle Ages some old clerics did enjoy a modest "age pension" after retirement, this was not a common occurrence across society (Shahar 2005).

In the USA, around 1790, public officers had to retire at 60 or 70 years old in many states (Fleming et al. 2003). But it was only in the nineteenth century when the concept and practice of large-scale retirement started to take hold (Sargent et al. 2012). The age of 70 was established as the age of retirement in Germany in 1889 (Payne 2015) entitling the retiree to receive a government pension. Two years later, Denmark also introduced an old-age pension scheme for retirees (the *Danish Act on Old Age Relief*, Petersen 1990), but starting at 60 years old, and in England *The Friendly Societies Act* of 1875 established old age as any age after 50 (Kirk 1992), but pensions tended to start at 65. Pensions were directly available for "deserving poor" from friendly societies and subsequently from a government initiative as established in the *Old Age Pensions Act* of 1908 (Gilbert 1965). The "deserving poor" were defined in various ways according to the act. For instance, the Danish act, similarly to the English act, listed:

> (a) the pensioner must not have undergone sentence for any act accounted dishonourable, and in respect of which he had not received rehabilitation, (b) his 'need' must not be caused by any actions by which, for the benefit of his children or others, he had deprived himself of the means of subsistence, or be caused by a disorderly and extravagant mode of life, or in any other way be brought about by his own default, and (c) he must not have received poor law relief for the 10 years prior to his application (Petersen 1990, p. 72).

The German system relied on some contribution of the individual during his working life and it was therefore more like some form of insurance. Character was not a factor that was taken into consideration for a pension.

A system of pensions for the aged was introduced in Australia following the disastrous effects of the 1890s depression, after which the country suffered a long drought; this led the state of New South Wales to take the initiative with the design of a support system for the elderly (Jalland 2015). A debate took place in 1896 to introduce an old age pension, which was eventually made reality in 1908.

However, it was only in the twentieth century when retirement became a distinctive phase of the life cycle for the majority of the elderly population in many countries around the world, through government and enterprise-based contributions to retirement income. Two major models of retirement pensions dominated the initial two-thirds of the twentieth century: the *Bismarckian model*, where the government provided benefits that changed according to the pre-retirement income and grew thanks to contributions, whereas the *Beveridgean model* was based on arrangements with the employer with added state pensions for workers in disadvantaged situations (Sargent et al. 2012). More broadly, during most of the twentieth century the most widespread system of pensions in retirement was based on *defined benefits*: an annuity that was received for life after retirement and that was calculated from a formula that considered both the last salary of the worker and the years worked (Sargent et al. 2012). This changed with the spread of neoliberalism since the 1970s, as the *accumulation* of retirement income shifted more and more towards the ability not only of the employer but also of the worker to both contribute and manage his/her retirement capital as an investment (*defined contribution*). Greater longevity has added the complication of both how to make the retirement money last and what kind of lifestyle could the retiree afford with the money available. Given that most accumulation capital under the new schemes is currently invested in the financial market, one big Damocles sword hanging over all retirees is the potential effect of financial markets downturns on the capital invested. Hence, the freedom afforded by retirement can be tainted by the financial uncertainties coming with it.

The complex historical development of retirement, with its positive and negative aspects, has been neatly summarised by Sargent et al. (2012) as seen in Fig. 2.3. Indeed, modern retirement as a concept and a practice has not only been invented in the twentieth century but also significantly reinvented, especially under the influence of neoliberalism and other factors, such as the increase in the average lifespan and the emergence of women as a strong social force in many areas of employment, including the professions. It is currently a mixture of situations where financially well-supported retirees can enjoy great freedom and security, whereas a growing proportion of the population of retirees is under increasing stress in spite of enjoying a longer lifespan, as they face the combined challenges of looking for work in an environment of ageism and discrimination against the old, fast-developing technologies, decreasing need for manpower with the advance of such technologies and mechanisation of production, and continued shrinking of government active intervention on social issues under the influence of neoliberalism. Starting a new career after 50 is more likely to work for those who plan to start their own enterprise or business. Seeking to be employed by somebody else in order to supplement an insufficient pension, on the other hand, is an endeavour that will likely bump into a strong brick wall of prejudice, as old people directly compete with younger ones for any

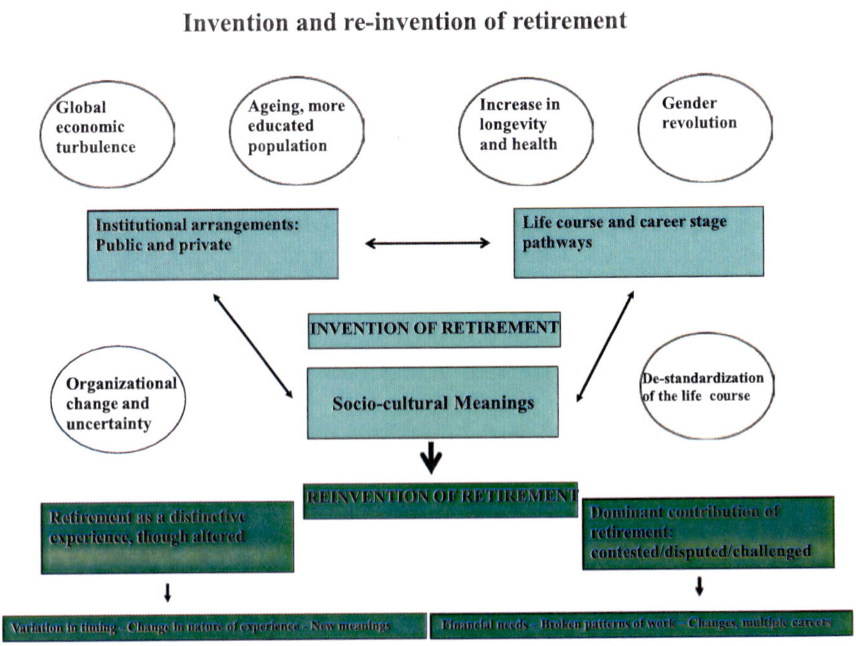

Fig. 2.3 Changes in the experience of retirement in the second half of the twentieth and the first half of the twenty-first century. The major transition in the structure of retirement during this period has involved the shift from a secure path to a financially well-supported retirement after a life of usually continuous work, to a situation where retirement becomes more active but at the same time more financially uncertain. Redrawn from Sargent et al. (2012) with minor modifications

job vacancy. The increased level of discontinuity in the experience of paid work can also seriously affect the ability of individuals to accumulate enough personal savings and funds in their retirement pension scheme. In contrast, for those with enough financial support the sky is the limit and they can orient their life in whatever direction they want, and for as long as their physical and mental abilities can allow it. Sargent et al. (2012) mention a good intermediate solution that is common practice in Germany and that has been an initiative pursued by the German trade unions: protect routes to early retirement across all sectors (public and private) for those who wish to retire whilst also negotiating age equality policies in the country in order to safeguard the opportunities to work for those who are willing and able to do so.

Among the professionals, paths to *bridge employment* could be also available as a transition to retirement (Sargent et al. 2012).

Currently, retirement age around the world is about 65 years for men and 60 for women. Most countries also have a retirement pensions system, although some systems are better than others, with the exception of about 54 countries, mainly in Africa but including also some East Asian and Latin American countries, where retirement is uncommon (World Bank 2000). We will return to the historical aspects of old age pensions in other sections of this chapter.

2.3 Before Greece and Rome

Records of the life and social roles of old people in pre-classic times are scanty and therefore the history of human old age during its dawn is very difficult to reconstruct. However, we may perhaps obtain a glimpse of old age in that period by studying current and some past pre-literary societies, which traditions have been recorded. In societies where knowledge is orally transmitted from one generation to the next, older people tend to acquire a role of repositories of the group's wisdom, experience, and memory, fulfilling roles as educators and judges that are often reflected in their significant position of political power. This important role for the elders is what characterises the system known as *gerontocracy*. Gerontocracy was already formally organised in some ancient cultures through a "council of the elders". Even in ancient Egypt village elders participated in the *kenbet*, a local council, in charge of settling disputes between villagers (Aldred 1972). In the Middle East, the *Zekenim* or the Hebrew council of the elders: "… began in the most remote period of the history of the nomadic Hebrews as a tribal council of the heads of the patriarchal families" (Amram 1900, p. 34).

Such roles tended to be appreciated by the younger generations in those societies, despite the obvious physical decline and mental changes experienced by older people. In fact, the usefulness of old people to others could even help the elders in terms of decreased prejudice, as some of their decaying faculties, such as increased somnolence, may have been interpreted positively as "meditative contemplation" (Minois 1989). This important social function of older people as repositories of knowledge and wisdom tended to decline with the onset of writing, as knowledge could be also transmitted through books rather than just orally. Importantly, George Minois (1989) also points to a decline in the relevance of the elders in those societies that are more democratic and where religion does not directly condition the political organisation.

From the above, however, we should not expect the life of the elders to have been one of privilege in all pre-literary societies. If older people became a drag on the capacity of the group to survive and prosper, then they might have been killed, abandoned, or encouraged to commit suicide. But more generally, old people could survive within a society for as long as they were able to contribute to the group. In totalitarian civilisations such as the Incas "[j]udges, called *ilactamayu*, would walk into the houses and check that everyone was engaged in useful work; the blind, the lame, the deaf and dumb, all had their tasks to accomplish, according to their capacity" (Minois 1989, p. 13) in return, the rest of the community cared for the elders (and others) providing food and shelter.

Ageing was described as a form of physical and mental decay, if not an illness, in both ancient Egypt and China. The ancient Indian text *Sushruta Samhita* (sixth century b.c.e.) points out that "ageing is a natural process, which weakens resistance to illness" (Minois 1989, p. 16). This negative view of old age gave raise to "myths of rejuvenation" in antiquity, as a strategy to ward off the physical decay coming with the passing years. In addition, a link was made between longevity and living a just life, thus establishing an association between the realms of the moral and the physical.

Interestingly, it seems that in the pre-Hellenic world, old women may have also been repository of wisdom, just like men (e.g. Pratt 1978), with their physical decay being less important than it has become in more recent historical periods.

2.4 Ancient Greece and Rome

We have seen that in pre-literary societies the elders played an important positive role as the guardians of knowledge and repositories of wisdom, and they also represented the living link with the family ancestors. In ancient Greek literature the association of old age with wisdom is seen in the Homer's character of Nestor, the old warrior respected for his experience and whose advice was always worth listening to, see also Fig. 2.4 for a representation of the Athenian people as an old, wise man.

But inter-generational tensions started to permeate both culture and politics in ancient Greece since at least the fifth century b.c.e., especially in Athens, whereas in Sparta the elders (*gerontes*) continued to wield significant political power (Finley 1981; Minois 1989; Parkin 2005), although a *gerontocratic* view was also supported

Fig. 2.4 Old man Demos representing the Athenian people is crowned by Democracy (338 b.c.e.). The figure shows the top section of the Eukrates' motion stele. The stele also contains the text of a decree forbidding plots against democracy. The figure is reproduced with permission from the American School of Classical Studies

by the Athenian Plato in *The Republic*. Sparta's *gerontocracy* relied on the rule of a council of the elders known as *Gerousia* that had both legislative and judicial powers.

The same broad pattern showing a loss of social power by the elders over time is also seen in Hebrew society around the same time: "the young deride them (Job 30.1) and instruct them, because wisdom no longer depends on age" (Minois 1989, p. 35). Concomitantly, a new stereotype of old people started to take over: "the old man was now de-sacralized and trivialized… His longevity, previously the source of his prestige, now only served to increase his culpability when he committed faults. The type of odious, garrulous, senile, disgusting and lascivious old man…" (Minois 1989, p. 38). This negative view of old people also pervaded theatre through comedies that expressed the competitive tensions between generations. On the other hand, note how both the Jewish God and also the Greek Zeus are depicted as older men in traditional iconography: "Thus, just as the old man lost his prestige in human society, he made a symbolic entry into eternity by personifying divine wisdom and permanence" (Minois 1989, p. 39). The knowledgeable old man also became a "sage". A characteristic of aristocratic education in Greece was a teacher–pupil liaison between an old teacher and a young pupil, whereas the unlearned old man was regarded as "decrepit".

During the Hellenistic period (fourth–first centuries b.c.e.) Greek society became more open to external influences, and prejudices against the elders tended to be more subdued. Everybody was allowed to compete in the political arena according to personal capacities.

The Western biological tradition of the understanding of old age was arguably initiated by Hippocrates (fifth–fourth century b.c.e.) who believed that each person receives a specific amount of energy at birth, called "internal heat" which would be eventually consumed throughout life. The gradual loss of such heat or energy with time constitutes the process of ageing. This idea of internal heat that is gradually lost through ageing was subsequently taken up by Aristotle (fourth century b.c.e.) (Minois 1989) who believed that ageing was an illness, a concept that was also maintained by Seneca (first century c.e.) (Payne 2015).

According to Plutarch (first–second century c.e.), the beginning of old age in classic Greece was 50 years old. During this period a common recommendation to help the individual better cope with ageing involved practising physical exercises, which produce heat, and mental exercises, alongside following a healthy diet. In his *Nine Books of Table Talk* Plutarch emphasises negative physical and psychological aspects of old age by stating that old men are "characterized by trembling limbs and stammering tongue, excessive talkativeness, irascible temper, forgetfulness and wandering minds" (Minois 1989, p. 72).

With the praise of youth during the Roman Empire there was a strong rejection of the ugliness of the body of old people, the old-female body in particular. The same attitude would reappear during the Renaissance.

In Cicero's (first century b.c.e.) opinion, there are four reasons to consider old age as an unhappy period: the old person is forced to withdraw from active public life, the body becomes more feeble, physical pleasures are considerably diminished, death is approaching (Minois 1989; Parkin 2005). The same pessimistic stance on old age was maintained by Seneca with his view of ageing as disease (Payne 2015).

Although Cicero also points to some positive aspects of old age such as the freedom to pursue specific intellectual interests and the acquisition of prestige and respect if the old man had led a productive public life (Parkin 2003). In Cicero's times wrinkles were celebrated in men, perhaps as an expression of character and masculinity, but they were not in women.

Celsus (first century c.e.) described how "[i]n older people there occur difficulty of breathing, catarrhal coughs, strangury, painful urination, arthritis, kidney disease, dizziness, apoplexy, cachexy, violent itching over the whole body, insomnia, watery discharges from the bowels, eyes, and nostrils, failing sight, blindness from glaucoma, and deafness" (Parkin 2003, p. 254), which are all physical symptoms of a body that is losing its ability to work effectively.

To add to Celsus' list, according to Juvenal (first–second centuries c.e.) the main physical characteristics of old age comprised: "wrinkles, tremulousness, baldness, a runny nose, and loss of teeth" (Parkin 2003, p. 81). Juncus (a second century c.e. philosopher) complemented the negative physical aspects of old age with psychological and social ones: "physical disabilities and illnesses, loss of pleasures and senses, fear of death, impotence, the ridicule of younger generations, exclusion from political and military affairs—and … the specter of poverty in old age" (Parkin 2003, p. 68). Among the positives of old age, Juncus mentioned wisdom and self-control, whereas the natural fear of approaching death could be relieved by older people through believing in an afterlife or by thinking that the suffering experienced in this life will soon be over. For Juvenal the objective of each person should not be so much that of living a long life but living a healthy life in both mind and body: *mens sana in corpore sano* (Parkin 2005).

Following Aristotle and in contrast to Hippocrates (fifth–fourth century b.c.e.)—who believed that old persons are characterised by coldness, humidity, and moisture—Galen (second–third century c.e.) considered ageing to be a process of drying up, due to loss of fluids, and of cooling, but he regarded it as a natural change and therefore not an illness. Such a process was regarded as both natural and inevitable, although each elder experienced it in a specific way. Galen thought that the elders' coughing and runny nose are just evidence of residual humidity, not an innate condition of older ages (Parkin 2005). Galen called his art the *gerocomic* art (from the Greek *geron* = old and *komeo* = to care), a precursor of what we now refer to as geriatrics.

The speed of ageing was determined by our humours, or more specifically it was due to the balance between the wet, dry, hot, and cold elements within our bodies. Hence the performance of regular physical exercises would help decrease cooling, to a degree, whereas taking regular baths would help against drying. Cooling could be also attenuated with the right kind of food. Other recommendations suggested by Galen to help the ageing body included drinking undiluted wine, being carried in a litter, practising the game of *harpastum* (involving snatching a ball), and engaging in conversation (Payne 2015).

Pliny described the fulfilling ageing process of a first–second century c.e. Roman patrician (Spurinna, a name that suggests an Etruscan origin) that he allegedly achieved through the following recipe: eat good but moderate food, perform regular physical exercises and relaxation that keep the stress away and the mind and body active (Kebric 1983).

The mind of old people attracted special concerns. Old age was associated with the potential onset of some form of dementia since at least Pythagoras (sixth century b.c.e.) who saw old age as a regression to an infantile state of mind (Payne 2015). Galen attributed dementia in old age to "coldness", which is supposed to damage the activities of the "soul" (Parkin 2005). Referring to dementia, Juvenal wrote: "but worse than all bodily ills is madness" (Parkin 2003, p. 228).

More generally, before the fifth century b.c.e. it was recognised that there is a natural, mental decay associated with old age, but the elders' capacities could be rescued through wisdom and a relevant role for old people in society, although such mental capacities could be worsened by low social status (Minois 1989). Later, Aristotle thought that whereas the young are typically hot and tempestuous, older people have the unflattering characteristics of being cold and lethargic (Parkin 2003). Tim Parkin (2003, p. 77) lists the following negative psychological characteristics of old people according to Aristotle (see the *Rhetoric*): "wary", "overly cautious", "overly pessimistic, distrustful, malicious, suspicious, and small minded", "they lack generosity, are cowardly and always anticipating danger", "and yet they also love life to excess, especially on their last day of life", they are self-centred, utilitarian, shameless, "caring more for profit than for honor", disregard others' opinions, they live in the past, "depend on memory rather than hope", are "prone to fits of anger", display "pettiness", and they care for others only out of self-convenience. Such stereotypes were associated with older men in particular.

Early on it was recognised by Plato—in *The Republic*, for instance—that the psychological characteristics of old people are determined not just by the ageing process but also by more individual aspects of character (Parkin 2003). In Plato's dialogue, Cephalus replies to Socrates: "In all this, and in the lack of respect their families show them, there is only one thing to blame; and that is not their old age, Socrates, but their character. For if men are sensible and good-tempered, old age is easy enough to bear: if not, youth as well as age is a burden" (cit. in Thane 2005a, p. 14). The same point was made later by the Roman Cicero who in his *Cato Maior de Senectute* (Cato the Elder on Old Age) wrote that many problems of older individuals "are a consequence not of old age but of his own character" (Parkin 2003, p. 61). Aristotle is also less understanding of the elders, seeing them as not necessarily wiser because of their age, especially when their life has been just an accumulation of mistakes that are increased by their old mind's lack of flexibility.

Plutarch described old people as having a "propensity to chatter interminably, their unstable temper and loss of memory, their excessive attachment to this life and material possessions…The old are also vainglorious, exaggerating their former deeds and prowess… less acquisitive [have less desire] and more reasonable" (Minois 1989, p. 75).

In the first century b.c.e. the Roman philosopher and politician Cicero described old people as "morose, troubled, prone to anger, and hard to please…. they are also misers" (Parkin 2003, p. 76). But he also pointed to the existential benefits of a life review for the elders: "The fruit of old age is the remembering and amassing of fine accomplishments" (Payne 2015, p. 18). Such accumulated experience over a lifetime coming with old age also made the Roman Seneca consider old age as a period of wisdom, despite all its physical and mental limitations. Seneca also provided some consolation to the ageing individual who is getting closer to the end of life, by seeing the whole lifespan as a continuous process of birth, death, and rebirth of an ever changing self: "I'm amazed at our madness, because we love so dearly something so fleeting as the body, and we fear that some time we will die, when every moment there is a death of our previous self" (Payne 2015, p. 89). For Seneca the pains associated with an ageing body should not lead a person to desire to die, as this would be a sign of weakness, although he regarded as "stupid to live for pain's sake" (Payne 2015, p. 92).

The Roman poet Maximianus (sixth century c.e.) left us a poem where he laments the ills of old age, especially physical and sexual impotence. He regards old age as the age of "living death", but he ends the poem with a touch of stoicism by recommending to be "reconciled to old age; be content that you are coming to the inevitable: death and the grave" (Coffman 1934, p. 251). By then the Roman Empire was also coming to an end.

Sexual desires in women were seen in a particularly negative way in classic antiquity, but not so much in men. The satirical and comic tradition of Roman theatre and literature was harsh against old age but especially against old women: "The stereotyped old woman is, in sum, a disgusting, haggard, stinking, toothless, and sex-crazed fellatrix" (Parkin 2003, p. 86). Old women were portrayed as brothel madams, alcoholics, and even as "evil witches". But we will see below that the *mater familias* did enjoy some position of power.

Old men were also regarded as sexual individuals, to the point that homosexual behaviour between an old man and a much younger one was accepted, with the old man being seen as an experienced teacher to the young on matters of life.

Welfare programs for the elderly were not readily available in antiquity, although old people in ancient Greece (e.g. during Pericles' rule in the fifth century b.c.e.) did receive some assistance through programs that also involved a degree of payback from the elders through various forms of public service (e.g. jury duties) (Kebric 1988). In general, a broader sense of duty and the need to have everybody contributing to the various tasks required in society—given the relatively low technological advances of the time—can explain why old people engaged in various communal activities (e.g. both Greek and Roman military units did have some soldiers older than 60 years). In ancient Greece the approach to activities performed by the elderly and those with some handicaps seems to have been a bit more humane (as judged from our modern standards) than that prevalent in Rome.

In the fourth century b.c.e. old men were positively regarded for public office because they were "less subject to envy and blind criticism, because their reputations were already firmly established" (Minois 1989, pp. 73–74), and their network of relations made them harder to uproot by younger competitors. According to Plutarch, old age should be a period of active participation in life because idleness and slackness in the elders are sources of baseness (Minois 1989). In addition, he suggested that older people could contribute a sense of "justice, moderation and wisdom" (Payne 2015). According to Cicero, as young people overthrow states, the old can put them back together again.

The old could have some access to power, as it was the case in the Roman Senate, in order to provide an input of prudence. Incidentally, the word Senate derives from the Latin word *senex*, meaning "old man", although not all members of the Roman Senate were old. Given the importance of rhetorical declamation in ancient Rome, especially in public life, the old were also regarded as models for the younger orator-in-training (Holmes 2012). But Plutarch also suggested that the old should share some of the duties with the younger and upcoming in order to diminish inter-generational conflicts. As a result of such conflicts old age could be the target of hatred and resentfulness from the part of the younger generations.

The Roman world is characterised by Minois as "history's first melting pot" which, however, was organised within a basic social structure. The *pater familias* had absolute power over all the members of his family, along with the duty of fulfilling specific rituals such as ceremonies in honour of the *lares*, *penates*, *genius* (or household gods) and the ancestors (Parkin 2005; Scheid 2007). Such power was cause of great conflict between father and sons as the father grew older. This led the old to be despised and detested by the younger generations. The *mater familias* exerted some power over her family but more subtly. Throughout the imperial period the *pater familias* gradually lost his power, whereas the *mater familias* gained it, even to the point of being able to become her children's guardian. Children were supposed to care for their elders, but this was not a duty that was necessarily taken up with gusto by the young, some needing state coercion (Payne 2015). The duty of care for the old parent from the part of children was expressed in the Roman concept of *pietas*, which was engrained deep in Roman mythology, as described by Virgil in the *Aeneid*, through the image of Aeneas carrying his old father Anchises during their escape from Troy. *Pietas* was the ethical principle that underpinned inter-generational reciprocity and the duty of care: children were helped by their parents during the process of growing up, and then, once they become adult, the children reciprocate by caring for the ageing parent. Such reciprocity, however, was engrained in tradition especially for the benefit of the father who, as *pater familias*, exerted a control on his dependents. The rest of the family could also receive support in old age, but the expectation was not straightforward (Parkin 2003, 2005).

With regard to the duty of care that the family should feel towards their elders, the stoic philosopher Hierokles (second century c.e.) wrote: "For our parents, therefore, we should provide food freely, and such is fitting for the weakness of old age; besides this, a bed, sleep, oil, a bath, and clothing—in short, general physical necessities, so that they should never lack any of these things; thus we imitate the care they took in rearing ourselves when we were infants" (Parkin 2005, p. 45). Such dependency of old people on family probably also helped give rise to the idea of old age as a second childhood. In Athens the responsibilities of children towards their old parents were legislated, not so in Rome where support for the elders was left to the judgement of the younger generations, although more strict legal regulations were introduced since the second century c.e. (Parkin 2005).

The elders who lived long enough were likely to remain within a multi-generational household, especially among the more socially disadvantaged classes and in rural areas. There, grandparents interacted with their grandchildren, child-minding, and often even legally adopting them (Parkin 2003, 2005).

A sad chapter of the history of old age in Roman times concerns the old slaves regarded by their masters as being no longer capable of rendering useful services, many of them were unceremoniously dumped on the Tiberina Island in Rome and abandoned to their fate (Parkin 2003).

During the Republic the elders had great social but also political power, especially through the Senate, whereas during the Empire such power decreased considerably. This loss of influence in society by the older generations led to a pessimistic view of old age and an increase in suicides among the elders in the second half of the first century and beginning of the second century c.e. Old men could retain a degree of dignity in retirement, provided that they were "rich, cultivated and in perfect health" (Minois 1989, p. 91), whereas older senators could still play a role as advisors to the emperor in their role of *amici principis* (Parkin 2003).

Although ancient Romans were worried that people receiving welfare could become idle (*otiosus*), the Athenian Pericles apparently replied to the news that an older slave had broken a leg with the words: "There's another teacher" (Payne 2015). However, the Romans also had the concept of *otium honestum* or the "pursuing of worthwhile and rewarding activities" (Parkin 2003, p. 72) which was an acceptable behaviour in old age, although it was usually the prerogative of the patrician class, as poorer people had to continue working until very old ages in order to sustain themselves. Pliny the Younger (first–second century c.e.) laments exactly such incapacity of many people, potentially including himself, to be able to live old age in comfortable *otium honestum* (Kebric 1983).

Pliny, in his *Letters*, also mentions evidence of some sort of "retirement communities" during the Roman Empire: places located in temperate climate regions where the elderly who could afford it could retire and enjoy a relaxed life in the company of others. In such communities, emphasis was put on an orderly and predictable life, which Pliny regarded as important during old age. This, however, did not impede older patricians to stay in touch with family matters and to also participate in some professional activities (e.g. teaching).

Otium honestum was fine, inertia was not. The ideal form of retirement in the Roman world involved enjoying tranquil and comfortable surroundings where one could continue developing the individual potential to reach wisdom, thanks to the enjoyment of good physical and mental health in spite of the advancing years, and after having lived a distinguished life to the service of the Roman state. As already mentioned, realising this ideal was not something that could be done by all citizens.

The concept of *otium honestum* points to the desirability of an active old age in retirement. The elders could still contribute to society if they wished, but at their own pace if they could afford it, and in proportion to their capacities. Such active life in retirement was also expected to help the elderly maintain their self-respect and the respect from the rest of society. In the words of Cicero (quoted in Parkin 2003, p. 235): "Old age will only be respected if it fights for itself, maintains its own rights, avoids dependence on anyone, and asserts control over its own to the last breath".

Just as it is today, death was also a matter of concern for the elderly in classic antiquity, but the Stoic Seneca was more interested in the quality of life than its length. In fact, he advocated a process of dignified dead under specific conditions:

> So the question we need to consider is whether one should shrink from the final stages of old age and not await the end, but bring it artificially… But if the body is useless for service, why should one not free the struggling soul? Perhaps you ought to do this a little before the debt is due, lest, when it falls due, you may be unable to perform the act… I shall not abandon old age, if old age preserves me intact for myself, and intact as regards the better part of myself; but if old age begins to shatter my mind, and to pull its various faculties to pieces, if it leaves me, not life, but only the breath of life, I shall leap from a building that is crumbling and tottering. I shall not avoid illness through death so long as the illness is curable and does not impede my mind… *He who dies just because he is in pain is a weakling and a coward; but he who lives merely to brave out this pain is a fool* (italics ours) (cit. in Parkin 2003, pp. 70–71, see also Agronin 2014).

Suicide was not unusual in the Roman world among elderly individuals suffering the chronic pains of old age, and it was regarded as a proper way of ending own life by the Stoic philosophy (Kebric 1983). Stoicism was especially prevalent among the political class during the imperial period in Rome.

For Cicero it is the achievements of life that give meaning to death: "When the end comes what has passed has flowed away, and all that is left is what you have achieved by virtue and good deeds" (Agronin 2014, p. 31).

In the ancient world the end of life of old people could be also hastened by external intervention. For instance, there were cases of senicide—with sometimes added anthropophagy—and also cases of imposed (or voluntary) euthanasia/suicide (Parkin 2003). In the Greek myth of Herakles and Geras (a god of old age), as depicted in a sequence of Greek vases from the fifth century b.c.e., mostly coming from excavations carried out in Etruria (central Italy), some ancient negative stereotypes of old age leading to threat of death are on display. The full story has not been transmitted but it could be reconstructed by looking at the figures in different vases as conforming to a sequence. In one vase, a scene shows Herakles engaging in a conversation with old Geras, who is presumably tempting Herakles with wine, or more likely sex as in two vases—one at the Etruscan Museum of Villa Giulia in Rome and the other at the Louvre in Paris—Geras is depicted with an enlarged penis. This is consistent with the stereotype of old men as lascivious individuals. In another vase, Herakles lifts his club in a threatening attitude towards Geras, presumably showing his disgust towards Geras' insinuations, and finally in the vase shown in Fig. 2.5, Geras runs away whilst still extending his inviting arms towards Herakles, with the latter following in pursue holding his club, presumably indicating his violent rejection of Geras' offer.

In sum, the picture that emerges of the elderly in antiquity is that their position in society changed according to their actual capacity to exert power. For as long as they had been able to show their usefulness to others they could be in control, but as soon as younger individuals felt that they did not need the counsel of the elders, then

Fig. 2.5 Attic neck-amphora (fifth century b.c.e.) displaying Herakles in pursue of Geras. The image is reproduced with permission from the British Museum

the young did not hesitate to take over. All along, however, older individuals who had achieved success in life were able to maintain the respect of the rest of society and could retire in tranquillity, if they had the material means to do so, but such cases were the exception, not the rule. Prejudice against old age, but especially against old women, was widespread and filled with stereotypes. The physical and mental limitations of old age were explicitly recognised and acknowledged as, however, was also the elders' experience.

2.5 Middle Ages and Renaissance

The Middle Ages can be roughly subdivided into three major periods: Early Middle Ages (fifth–tenth centuries), the High Middle Ages (eleventh–thirteenth centuries), and the Late Middle Ages (fourteenth–fifteenth centuries). Three major themes that prevailed about old age throughout the Middle Ages were: an ambivalent attitude about old age, the issue of physical decay—with women seen as ageing faster than men—and age-appropriate behaviour. The ambivalent attitude was due to the contrast between a positive view of the elderly in terms of experience and wisdom and a negative view regarding both their loss of physical capacities (de Beauvoir 1972) and the inevitable progression towards death. Links between old age and various beliefs in the supernatural and the natural world were also of concern in the Middle Ages.

During this period older people were not supposed to display avarice and greed or engage in sex and war, rather they were expected to devote more of their time to contemplation. The old were supposed to abandon the competitive attitude of their younger years in order to enter a period of reflection and spiritual restoration, in preparation for death.

In the Christian tradition of the Early Middle Ages "old age" was equated with wisdom, but not necessarily with actual age, hence a wise young man could be considered old. The influence of Christianity was also seen in the interpretation of the decaying body functions in people advanced in age as a painful legacy of the original sin. Hence, the belief that being a good Christian in life brings the promise of eternal youth in Heaven. Religion, being a pervasive personal and social referent throughout the Middle Ages, affected various aspects of the perception of old people. For instance, it was considered that sinning was worse if the individual was old than if he was young, as the young can be excused on virtue of their youth and inexperience, but the old have no excuse. Therefore, it was regarded as inappropriate for old people to act as if they were still young (Minois 1989).

Old women suffered from a double handicap: their perceived physical unattractiveness and their lost reproductive capacity. This presumably contributed to their association with evil in the prevailing culture, and the common stereotype of a "witch" combined both old age and evil throughout this period, as it did also in later literary works such as Chaucer's, for instance. According to Minois (1989), "in some villages the effigy of an old woman would be burnt in order to drive away old age". Such negative stereotypes about old women have persisted to these days. A. Poiani, growing up in the 1960s in a small rural town in Central Italy still remembers a wooden mask representing an old woman, a witch, hanging on the door of a country house. The mask was supposed to scare evil spirits away. Paradoxically, although in the Late Middle Ages the menstrual blood was considered as impure and possessing a destructive power, post-menopausal women were considered even more dangerous than pre-menopausal women.

Poor old people relied on family solidarity for survival and the general goodwill of society, whereas the wealthier ones introduced a form of retirement since the sixth century c.e., by spending the last part of their life in a monastery. This was regarded as a preparation for "eternal life" after death: "Old men, being cut off from the life of this world, were in transit; they were preparing for eternal life. They were no longer completely of this world but were not yet of the next. As the antechamber to life everlasting, retirement to a monastery provided old age with its chief preoccupation: that of ensuring salvation" (Minois 1989, p. 137).

During the High Middle Ages the concept of old age varied according to social classes. Retirement was a privilege of the upper classes only; the peasantry essentially never retired and worked the land for as long as they physically could. Old age was considered to have both negative and positive aspects. Negative aspects of old age included ugliness, bad character, and physical and mental frailty, whereas the positive aspects included a tendency to help others, wisdom, and experience. Vincent of Beauvais (thirteenth century) mentioned some positive aspects of old age such as increased wisdom and decreased passion; but absence of energy, forgetfulness, and gullibility were regarded as downsides of ageing (Shahar 2005).

The pervasiveness of Christian spirituality during the Middle Ages went as far as to minimise the relevance of the material body in favour of the "soul", thus providing a cultural niche for an optimistic view of old age. Given that the body is bound to decay for biological reasons, hope was offered to the elderly through the separation of body and soul. Being different in their fundamental nature, even if the body was decaying the soul could still be beautiful. In fact, religion and God were offered as sources of "rejuvenation" in the form of spiritual renewal: "In a life of grace, man will always be renewed, and renewal is life" (Shahar 2005, p. 93). Although the body always ages, the soul can remain young provided that it stays close to God. Everlasting happiness was promised to old people in the continuation of their soul into the afterlife.

The ageing body was interpreted as an opportunity for the soul to take centre stage. In the words of George Minois (1989, p. 169): "Old age's opportunity lay in its bodily decrepitude which allowed a man's soul to raise more easily towards the celestial realities. The weaker the body grows, the stronger the soul becomes". St Bernard further stressed this view: "A healthy body, which leads man to infirmity of soul, is a bad thing: but an infirm body, which leads man to the health of his soul, is a good thing" (Minois 1989, p. 170). Contrast this with the more materialistic view of the Roman Juvenal's *mens sana in corpore sano*, where the body is the fundamental repository of our mind and thus it must be protected and nurtured to remain healthy, in doing so the mind will be healthy too.

But clearly the body did remain important in the real lives of people in the Middle Ages—how it could not be—and the physical limitations coming with old age were indeed a concern leading to seeking remedies to escape the decay of old age. Remedies to counteract the physical ravages of ageing included those that were inherited from the Greco-Roman tradition: living in a good climate, following an appropriate diet, performing regular physical exercise, taking baths, and drinking wine. Psychologically, old people also tried to keep themselves optimistic, especially against the inevitable social pressure coming from the young. Pope Innocent III (twelfth–thirteenth centuries) famously wrote in his *De Comtemptu Mundi sive de Miseria Humanae Conditionis* (The Contempt of Worldly Concerns or the Misery of the Human Condition):

> Young men, be not proud in the presence of a decaying old man; he was once that which you are; he now is that which you in turn will be. (cit. in Coffman 1934, p. 255).

The inevitability of old age was also expressed beautifully by the poet Petrarch (fourteenth century) through tactfully playing with the concept of the youthful body being similar to a flower. Shulamith Shahar (2005, p. 85) paraphrases Petrarch so: "The young body is like a flower, but a flower whose fading is inherent in its blossoming". Old age is just the inevitable consequence of once having been young and of living long enough.

During this period the number of adult men still exceeded the number of women, as it was the case in ancient Greece and Rome, this was likely the result of elevated women mortality during childbirth, which led to high levels of male celibacy (Minois 1989), although women who survived the perils of childbirth were capable of living longer than men (Thane 2003).

Just like in classic antiquity, old men in the Middle Ages enjoyed greater respect in society if they had been successful warriors or a prominent member of privileged classes. But more broadly, it is in this period when the *conjugal family*, where power resides with the married children and old parents are seen as somewhat of a burden, replaced the old *patriarchal family*, in which the old man was in charge to the very end. Patriarchal power, however, was retained among the upper classes. The potential negative effect of the conjugal family on the elderly was compensated by the emergence of family names that reinforced inter-generational solidarity and support for older relatives, in spite of their loss of social and decision-making power (Minois 1989). Moreover, during the thirteenth century some early initiatives of public support for indigent elderly people were implemented, and the network of social relationships within a Medieval village was such that older people did retain some role in society in terms of telling stories to the younger generations and as links with the past and with cultural traditions, the village square being the main place for the elders to exchange memories, opinions, news, and gossips.

In the Late Middle Ages, Europe was ravaged by a series of epidemics of the bubonic plague that resulted in a relative increase in the number of old people in the population, with surviving old people displaying higher life expectancy (Minois 1989). Larger relative numbers of elders resulted in big age gaps between the father and his children. As a consequence, old men regained control of the wealth of the family and they also married much younger women, thus resulting in both more widows (Shahar 2005) and in the mother feeling closer to her children than her husband, leading to new inter-generational conflicts. The tensions derived from such inter-generational conflicts during this period are clearly described by Shahar (2005, p. 103):

> It is in the literary description of old men in epics and romances that the raw intergenerational tension is exposed. The old man is an object of contempt. He tries to do things he no longer can: he wants war when he himself can no longer fight; he is feeble and senile, but tyrannizes over his sons, and his commands are disastrous. Above all, he still holds his fief and can marry for the second or third time, taking to wife a young girl, while the young men in his kinship group have no fief and cannot marry.

The resolution of inter-generational conflicts, however, varied geographically across Europe. In Southern Europe the elder continued to be the head of family and remained in control of both property and the lives of the members of the household, whereas in Northern Europe, the elder male eventually handed over the control of the family property to an offspring at some stage through a written agreement, which also made the duties of the offspring clear regarding the care of the old father (Shahar 2005).

Old men were described by Aegidius Romanus (thirteenth–fourteenth centuries), following Aristotle, as "cowardly, suspicious, shameless (because advantage is more important than honour), miserly and without hope", but they also show "pity and avoid passing judgments hurriedly" (Shahar 2005, p. 88). Bernardino da Siena (fourteenth–fifteenth centuries) characterised the elderly as: impatient, melancholic, gloomy, ignorant, perverse, mentally blind, and stupid. But mercifully he added that these characteristics were mainly typical of "bad elderly" (Shahar 2005).

Old women continued to be treated particularly harshly during this period. For instance, Guillaume de Deguileville wrote in the fourteenth century in his *Pèlerinage de la vie humaine* (The Pilgrimage of Human Life) that young women were characterised by the virtues of reason, penitence, charity, mercy, diligence and they also benefited from God's grace. On the other hand, old women apart from being depicted as ugly were also characterised by pride, flattery, sloth, envy, treachery, hypocrisy, avarice, gluttony, lust, anger and heresy, disease, and tribulation (Shahar 2005, p. 87). Gabriele Zerbi in the fifteenth century went as far as to regard women as "deformed" and "inferior" versions of men. But ultimately, apart from the general aesthetic decay, which is shared with old men, it is the incapacity of old women to reproduce that negatively affected their standing in society, whereas old men, by retaining their biological capacity to procreate for longer, could at least cling their sense of "superiority" onto something that had the resemblance of a fact.

At the time, the myth of the *Fountain of Youth* also became popular. The *Fountain of Youth* had old roots in stories already told by Herodotus (fifth century b.c.e.) and the myth survived to later centuries. We may remember that Galen considered the process of ageing as one of drying, hence bathing was recommended to stave off old age. The *Fountain of Youth* is certainly consistent with such Galenic principle. Although there are artistic representations of the *Fountain of Youth* from the Late Middle Ages (see, for instance, the fifteenth century fresco from the Castle of Manta in Italy, Fig. 1a in Demontis 2015), Fig. 2.6 shows a later, sixteenth century representation by Lucas Cranach. In this painting, as in the one from the Castle of Manta, old people both men and women enter the fountain from the left of the observer and as soon as the process of rejuvenation starts, they engage in heterosexual sexual acts whilst they bathe in the "miraculous" water—a sign of recovered vital vigour and energy—to finally exit from the right side, dress up, and leave no longer as old people but as strong, healthy young ones.

The *Fountain of Youth* reflects a concern of the elderly for their failing bodies as they advance in age. Another Medieval myth that expressed the same concern is shown in the Walsperger map, dated 1448, depicting an island in the Atlantic Ocean where it was believed both ageing and death were unknown (Demontis 2015). But the Zeitz map of 1470 gives a grimmer view of this island (the island of Jupiter) as ageing and death were unknown there apparently only because old people were expelled from the island (Winter 1953).

A well-known fifteenth century authority on the healthcare for the elderly was Gabriele Zerbi who in 1489 published the book on old age *Gerontocomia Scilicet*

Fig. 2.6 Lucas Cranach, *The Fountain of Youth* (1546)

de Senum Cura atque Victu (Gerontocomia, namely the care and diet for the elderly). Zerbi introduced the word *gerontocomi* to refer to medical specialists on old age (Freeman 1965). For him a medical intervention was still valuable in the elderly simply because although old age is inevitable, the actual time of death remains uncertain and therefore it could be delayed if appropriate actions are taken. His approach was based on Greco-Roman beliefs and his method of extending longevity relied on the retention of heat and moisture (Warren 2012), given that the old were considered to be cold and dry in the Galenic tradition. Demontis (2015) lists other Medieval works that provide remedies to combat ageing, such as the lord of the castle of Goet's *De retardatione accidentium senectutis* (The Slowing of the Accidents of Old Age), Arnoldus de Villanova's *De conservatione juventutis et retardatione senectutis* (Preservation of Youth and Old Age Retardation), Marsilio Ficino's *De vita longa* (On Long Life), and several others.

The elderly among the commoners did not enjoy many privileges during this period, apart from the exemption from some duties. In fact, older individuals were rather marginalised in the Middle Ages, although older women were more so than men. During this period, only the Venetian Republic gave more power to older people in public life (Venetian *doges*—the most senior elected officials in Venice—were usually quite old: about 72 years when they were elected into office) (Shahar 2005), whereas Roman Catholic popes' median age at death was 66 years between 1200 and 1599 (Carrieri and Serraino 2005).

One area where the elderly (mainly men, but sometimes also women) did have a public role was as witnesses in legal disputes and, given the still significant level of illiteracy among the general population, they also retained their ancient role of storytellers and repositories of local traditions.

Cossar (2012) describes some aspects of ageing among the clergy and the notariate in the fourteenth century in the Italian city of Bergamo. Such persons held a position of authority in their late years, which was expressed through honorific titles such as *venerabile* (worthy of reverence) and *discretus* (distinguished) and they continued working and performing their duties until the very end of their life. An interesting analysis of the handwriting of notary Gerardo suggests a consistent style that only changed a short time before his death. Presumably it was the social status brought by his profession that kept Gerardo and his colleagues working to the very end, with no sign of retirement. The poor, on the other hand, most likely continued working up to the very end of their lives simply out of basic material need.

At the time, the only available system of pensions was created by the Church for retired clergy. The elderly poor also received some material support from church institutions, but it was never quite sufficient.

The European Late Middle Ages transitioned into the Renaissance in the fifteenth century, with the Renaissance proper blossoming in the sixteenth century. In the words of George Minois (1989, p. 249): "The Renaissance, like every time of renewal and rebirth, celebrated youth, the fullness of life, beauty and novelty. It abhorred everything that presaged decline, decrepitude and death… the Renaissance railed against old age, blackening, demeaning and damning it". This was a direct result of a return to the principles and aesthetic preferences of classic Greece and Rome. Older men were rejected even in more practical roles such as leading an army, where younger commanders were preferred by the soldiers.

Just like in the Middle Ages, older women attracted a particular level of rejection from Renaissance man. For instance, they were associated with witchcraft far more frequently than young women.

Although knowledge and wisdom, and therefore accumulated life experience—all qualities that were praised by the Renaissance humanists—were still associated with older ages during this period, Erasmus of Rotterdam put old men in his "gallery of fools", as there was also a widespread belief that old people reverted to infantile behaviours, which was considered foolish. For Niccolò Machiavelli the "ideal" old man should be: "resolved, reserved, honourable, grave, honest, serious, regular and exemplary" (Minois 1989, p. 280). The same old man was expected to be wise and faultless in the view of William Shakespeare. Although the flaws of old age were a matter of concern, the oldest of old men were still regarded as worthy of respect.

In the utopian literature of the time, the old (considered so since the age of 50) were ultimately recommended to hasten the end of their life by committing suicide. The view of such authors was that those who are of no use to the rest of society have no right to live. This idea harks back to antiquity and it is still found in some contemporary traditional societies (see Chap. 3). This form of hastened death of the elders is a social imposition (a "duty to die"), and it is not a precursor of our current concept of "right-to-die", which is a personal choice. Old age was indeed an annoying and stubborn truth that ultimately threatened to ruin the utopian Renaissance vision of eternal youth. The suggestion of having fewer old people around may have been a simplistic ploy to live in the delusion of "eternal youth", by restricting exposure to the old: a reminder of the story of the Island of Jupiter.

It was during the Renaissance that the consistent male-biased sex ratio in the population, that had been in place since antiquity and that was maintained throughout the Middle Ages, started to change towards a female bias. The transition unfolded over two centuries between the fifteenth and the seventeenth century (Minois 1989). The pattern of greater number of females than males in the population has been subsequently maintained to these days.

The first truly modern treatise on medical approaches to ageing appeared in the Renaissance, published in Padua (Italy) in 1558 and authored by Alvise (Luigi) Cornaro (known as the "Apostle of Senescence") (Freeman 1965). Cornaro introduced a method of tackling the process of ageing that was based on diet and that we know today as calorie restriction. In his *Discorsi della vita sobria* (Discourses on the Sober Life) he described how he restricted himself to eating 350 g of food each day (including sources of carbohydrates and proteins) and 414 mL of wine.

To recap, in the Middle Ages the power of control of the Church over society was pervasive, reinforcing a patriarchal model with emphasis on the soul and the afterlife. The ravages of the epidemics of the bubonic plague also helped change the demography to older ages, thus shifting the power position towards old men, especially whenever they were in possession of property. However, the situation for older women did not change. In the Renaissance, for the first time the sex ratio in the population favoured women, a situation that has been maintained to these days. Yet, the position of low power for older women, and women in general, was not modified from what it had been in previous centuries. Old individuals in the less wealthy classes (both men and women) remained socially marginalised throughout both the Middle Ages and the Renaissance.

2.6 The Seventeenth–Eighteenth Centuries

Being old in the seventeenth century simply meant "looking old", and due to a widespread perception dating back centuries, women were considered to be already "old" at a much younger age than men. As a result of generally poor dietary standards and living conditions women were already regarded as old at the age of 50 (Botelho 2005), but the difference between men and women decreased in the higher social classes that enjoyed better standards of living. People in such better-off sections of society were also regarded to enter old age at least a decade later than the poor. Continuing the pattern started in the Renaissance, during this period a greater percentage of women lived to reach old age than men, also dying at older ages than men. This is the reverse of the pattern found in antiquity and in the Middle Ages. Widows tended not to remarry and preferred to live alone and, in any event, after they turned

50 the chances of remarrying were slim. Some could have moved with their adult children, but children not always lived nearby. During this period there was also an increase in the number of men and women who had never married (Botelho 2005).

Just like in previous centuries, old women were still more likely to be subject to accusations of witchcraft than men and younger women, such accusations were particularly directed to poor women whose external ageing looked more advanced than that of more wealthy women of the same age. A seventeenth century German doctor, however, explained the association of old women with witchcraft with the suggestion that it could have been a consequence of social alienation: "They are so unfairly despised and rejected by everyone, enjoy no-one's protection, much less their affection and loyalty… so that it is no wonder that, on account of poverty and need, misery and faint-heartedness, they often… give themselves to the devil and practice of witchcraft" (Thane 2005a, p. 14). Thane adds an interesting comment by pointing out that the knowledge (e.g. on matters medical, remedies to cure illnesses) acquired by older people throughout their lives could be seen either as a positive contribution to the community or, more ominously, as "proof of witchcraft" according to circumstances. In the seventeenth century North American colony of Massachusetts women most likely to be accused of witchcraft were old, poor, single, or widowed and they were regarded as odd or a nuisance by their neighbours (Sokolovsky 2009b).

Older widows in particular were also subject to the stereotype of being lustful which in this case was considered repugnant, given that sexual intercourse was supposed to only have the purpose of reproduction and once a woman was beyond child-bearing age sexual acts were understood to have no meaning anymore (Botelho 2005). Talking too much or wearing makeup was also regarded as inappropriate for old women.

Other widespread stereotypes about old people included that they are greedy, melancholic, self-important, meddling, and talkative individuals; such stereotypes were the main themes of various jokes about the elderly (Botelho 2005). But there were also some prescriptive characteristics for old people: "The old were to be wise and somber; they were to make peace with their lives and with their souls; and they were to be mocked for clinging too tightly to the ways of youth" (Botelho 2005, p. 144). Such prescriptions contributed to entrenching societal stereotypes into the minds of old and young people alike, thus reinforcing both the behaviour and the stereotype itself.

Characters in the Italian *Commedia dell'Arte* such as Pantaloon, the Physician, and the Doctor reflected such stereotypes. Pantaloon is the old man pretending to be younger than he is and behaving accordingly, the Doctor is the allegedly erudite person. He is excessively talkative but either lacking in logic or otherwise simply stating the bleeding obvious. The Physician is what we could today call a medical doctor. Botelho (2005, p. 140) quotes an exchange between the Doctor and the Physician that could have been taken from a sketch by the Marx brothers:

> Physician:…. In short, I wage such cruel and relentless warfare against all forms of illness that when I see a disorder becoming ineradicable in a patient I even go as far as to kill the patient in order to relieve him of his disorder.
> Doctor: That is an excellent cure.

Table 2.2 Alignment of the *Ages of Man* with the four humours, elements, qualities, and seasons according to a seventeenth century European tradition with roots into classical antiquity

Four humours	Yellow bile[a,b]	Blood	Phlegm	Black bile[c]
Four elements	Fire[d]	Air	Water	Earth
Four qualities or building blocks	Hot[b]/dry	Hot/moist	Cold/moist	Cold/dry[e]
Four seasons	Summer	Spring	Autumn	Winter
Four ages of man	Childhood	Youth	Adulthood	Old age

Modified from Botelho (2005)
[a]Also known as choler, which was associated with irascible temperament
[b]From Hippocrates, subsequently maintained by Aristotle
[c]Melancholy
[d]From Aristotle
[e]From Galen

The issues of stereotypes and stigma of old age will be further analysed in the section *The Ethics of Ageing: Stereotypes, Stigma, and Ageism* of Chap. 6.

By the seventeenth century various alignments between life stages, psychological traits, and environmental factors such as elements and seasons, that harked back to traditions from classical antiquity, were established in the prevailing culture across Europe, and they were expressed into a parallel system of four components as seen in Table 2.2.

The seventeenth century system shown in Table 2.2 is in fact a summation of ancient Greek and Roman beliefs that synthesises the views of Aristotle, who suggested the four elements as fundamental building blocks of all what is around us: earth, water, air, and fire; Hippocrates, who introduced the idea of the four humours: blood, yellow bile, black bile, and phlegm; and Galen, who suggested that temperament is determined by combinations of hot/cold and moist/dry. Heat was associated with action, cold with inaction, and whereas choler or yellow bile meant courage, being a coward was linked to phlegm. Old age could therefore be regarded as *the cold and dry winter of life, full of earthly melancholy.*

The medical advice of the day was more aimed at slowing down the process of ageing than bathing into the *Fountain of Youth* to reverse the effects of advancing age. Ageing could not be reversed, but it could be slowed down. To achieve that, people were advised to take care of their eating and drinking habits and to follow Cornaro's suggestion to drink red wine. Eating in small amounts but often during the day was also believed to be beneficial (Botelho 2005).

Family responsibility for the care of the elderly varied culturally across Europe, with Eastern Europe and some sections of Southern Europe preferring the communal life of three-generation households, whereas life for the elderly was more independent in Northern Europe (Botelho 2005). Old women could live alone, or if they were widow they could organise themselves into all-women cooperative groups, whereas old men tended to live in—and usually headed—a household. For the poor there were hospitals and workhouses, which required the able poor to contribute some work in exchange for accommodation. The wealthy old, especially in Roman Catholic areas, tended to find care and accommodation in monasteries and convents.

According to Lynn Botelho, at the end of the seventeenth century the modern "spoil-the-child" form of grandparenting started to emerge: "… the idea of the sentimental grandparent arose as part of a search by active and leisured elderly people to find 'a positive, active and integrated' role for themselves in a stage of life that had previously been viewed as socially marginalised. It was the very wealthy that could afford this type of old age; it was also women who were most likely to experience it; and, consequently, it was grandmothers who first redefined its meaning" (Botelho 2005, p. 172).

In the seventeenth century colony of New England in North America, 60 years was regarded as the beginning of old age, and the elderly were supposed to be deserving of "esteem, and reverence, and veneration", a view that was directly derived from the role of old age in the Old Testament (Demos 1978). But outside the strict religious tradition, the opinions about old age were less than flattering.

In 1679 William Bridge wrote about his views of older people in the colony, describing old age as a "dry and barren ground", a "state of weakness and of much infirmity". Older people are "too drowsy and remiss in the things of God, … too covetous and tenacious for the things of this world, … too timorous and fearful, … too touchy, peevish, angry, and forward, … very unteachable … [since] they think they know more than others, … hard to be pleased, and as hard to please others, … full of complaints of the present times, … [and] full of suspicions, and very apt to surmise, suspect, and fear the worst…"; and Cotton Mather added:

> The sun, the light, the moon, and the stars begin to be darkened with you; that is, your parts are under a decay; your fancy, your judgment are failing you … Your hands now shake and shrink, and must lean upon a staff … Your thighs and legs now buckle under you … Your teeth grow weak and few, and are almost all rotted out … Your eyes become dim, and clouds disturb the visive powers in them … You become deaf and thick of hearing … You can't without some difficulty go up a pair of stairs, and are in danger of stumbling at every stone in the street … Your backs are so feeble that instead of carrying anything else they can scarce bear themselves… (quotes from Bridge and Mather are cit. in Demos 1978, pp. S251–S252).

This is certainly not the description of an old age that is the target of much veneration. Quite likely, the prescriptions coming from Biblical recommendations regarding the relationship with old people were not followed in society at large. Instead, old people were highly recommended to practise the virtues of sobriety, gravity, temperance, orthodoxy, charity, and patience, as if it was something in urgent need of improvement in society.

An inter-generational network of mutual responsibilities was in place in the early New England colony where grandparents actively helped their grandchildren and vice versa. Older people also tended to gradually decrease their levels of commitment with work, but they continued working for as long as they could, including members of the public service and the clergy. In part this could be explained by the local lack of workforce in the colony. Demos (1978) describes early New England as a place that was "land rich and labor poor". In spite of the social "usefulness" of older people, they were also the target of both negative stereotypes, being regarded as irritable, suspicious, and complaining and derision ("staggering, spewing, old drunkard") (Demos 1978).

The eighteenth century opened a new era for old age, as life expectancy increased and modernity started to emerge in the culture of Europe, with revolutionary zeal. In France in particular, "old age underwent changes in cultural representations which had their own impacts on how people lived their lives. They included a shift from the religiosity of previous centuries to greater worldliness, and a literary and artistic turn from comic derision aimed at aged characters to greater respect and sentimentality" (Troyansky 2005, p. 175).

In England the *Old Poor Law* that was initially introduced in the sixteenth century was further improved in the eighteenth century whereby local parishes were expected to provide some material assistance to the elderly, in spite of some societal expectations that the elderly themselves and their family should be also expected to take their own responsibilities. Public service pensions were also introduced during this century in England, Germany, and France (Troyansky 2005).

In revolutionary France in particular, a new reappraisal of the values of old age emerged. David Troyansky (2005, p. 185) describes what was known as the *fête de la vieillesse* (festival of old age, Fig. 2.7): "one of a series of secular holidays that sought to anchor society in such natural phenomena as the seasons and the ages of life". This involved celebrations for the local elders who were regarded as the living testimony of "natural authority and honour". In this way

Fig. 2.7 France *fête de la vieillesse*, 1795 etching, reproduced with permission from the British Museum

the ideas of the *Enlightenment* such as that of the "social good" and also social cohesion and mutual responsibilities were intended to replace the religious ethics of "charity".

2.7 The Nineteenth Century

The modernisation that spread during the nineteenth century, as the industrial revolution unfolded at an increasing pace, was also accompanied with a broader trend to lower levels of mortality (infant mortality in particular) and increased birth rate, leading to higher growth in the population. Caring for older people was still largely a family duty with some contribution from the local community (including the churches). Whereas grandparents took child-caring roles in the family, when both parents were out at work; the family reciprocated by providing various forms of help to the elderly, such as accommodation, financial assistance, food or just company, nursing and help with household chores (Cole and Edwards 2005). In the late nineteenth century, most extended families in the USA included at least one grandparent (Dahlin 1980). This changed, especially for men, when government-funded social support programs were introduced, initially in Germany by Chancellor Otto von Bismarck in 1883 (Achenbaum and Stearns 1978), thus providing some financial relief to both elders and their family. Speaking in March 1884, von Bismarck stated:

> The social significance of a general insurance of the propertyless appears immeasurable to me; it is essential to generate a basic conservative conviction among the vast majority of unpropertied people through the feelings that come with the right to a pension. Why should the soldier of work not have a pension like the soldier of the civil service? That is state socialism, that is the legitimate operation of practical Christianity (Cole and Edwards 2005, p. 261).

In the nineteenth century older workers were over-represented in the less modern economic activities such as agriculture, fishing, or handicrafts. But even with income from their work or some form of government pension, many old people still required institutionalised support which was only a last option if nothing better was available, as conditions in institutions such as the workhouses and hospices tended to be rather dreadful, even though, paradoxically, they were intended only for the "deserving poor" (see Fig. 2.8 reproducing a scene from the Westminster Union workhouse in London):

> … for the most they were considered their final choice as it involved a complete loss of autonomy, of personal belongings, contact with friends and family (they might even be separated from their spouses in the same institution), and life in often crowded and unsanitary conditions, with little means of distraction, poor or nonexistent nursing arrangements and monotonous diets (Cole and Edwards 2005, p. 233)

The "deserving poor" were those who were genuinely incapacitated to work (because of sickness or disability, for instance), but being poor was not sufficient

Fig. 2.8 "Eventide: A Scene in the Westminster Union" painted by Sir Hubert von Herkomer in 1878

reason to receive support, as it was commonly believed that most persons were poor as a result of their immorality, alcohol addiction, or laziness (the "undeserving poor"). Those who could still work were forced into work (Fleming et al. 2003). In the nineteenth century the concept of "undeserving poor" was widespread in countries such as the UK, USA, and Australia that were experiencing the industrial revolution and the spreading of ideologies such as laissez-faire capitalism. The competitive spirit underlying the spreading of modern capitalism clearly affected the kind of stereotype the elderly were subject to, and in fact the stereotype could be even positive for some old people provided that: "they occupied their positions by merit, achievement and aggression" (Jalland 2015, p. 60). In addition, an entrenched individualism tended to view the person as the source of his/her own problems and the individual was expected to be the main provider of his/her own solutions to those problems (Randall 1965).

In spite of such deplorable conditions prevalent in the workhouses, with residents being often physically abused and, to make things worse, they also suffered severe social stigma, institutionalisation increased in the second half of the nineteenth century for both women and men, and by 1901 almost 10% of men and 6% of women aged 75 and older were hosted in workhouses in England. In France the percentage of people older than 75 living in hospices was lower: 2.1% (Cole and Edwards 2005).

Retirement in a workhouse for the poor contrasted with the far better conditions of retirement for the better off:

> The economic and social realities towards the end of the lifecycle were shaped by class, gender and personal wealth, ability to work, access to kin and availability of communal

> resources for support, and were reflected most obviously in one's place of living, appearance and scope for social interaction. At one end of the spectrum, emaciated, infirm and isolated elders lived out their last days in a public institution, separated from family and friends, deprived of personal belongings, autonomy and dignity. At the other end, well-to-do heads of family commanded respect and maintained control over their personal wealth, surrounded (if not necessarily cherished) by a circle of family, staff and acquaintances, living out their last days in their own homes, retired from work and with sufficient time, energy and intellect to prepare for the inevitable. (Cole and Edwards 2005, p. 256)

The condition of old women was not much better in the nineteenth century than in previous centuries. Women's old age was regarded to begin at menopause (Stearns 1980), whereas men's decline occurred 15–20 years later. Menopause and loss of both sexual interest and aesthetic appeal were the defining characteristics of old age in women, giving raise to negative stereotypes mixed with mockery and sarcasm. Some French authors, for instance, were of the opinion that: "In general one can divide the life of women into three periods: in the first, they dream of love; in the second, they make it; in the third, they regret its passing" (Reveillé-Parise 1853, p. 63). Thus sex ("love") becomes the stereotypical definer of a young woman, who therefore reaches old age when her sexual desire and sexual appeal to men wane. In other words, this is the definition of ageing in women from a man's perspective.

Although women lived longer than men, their role in society outside the family was very limited compared to older men, and they also retired younger from work. In the nineteenth century it was clear that women outlived men on average, but they did so in an environment of social discrimination against them in general and against old women in particular, thus creating the problem of an underclass. This paved the way to the development of the modern cosmetics industry, products that women used to try to slow down the appearance of ageing. However, old women enjoyed a better level of social support from family and others than men, which may have played an important role in decreasing mortality in old women compared to men.

Although widows increased in numbers during this century, in most cases they remained financially unprotected due to lack of pension allowances. This situation changed only in the twentieth century. Interestingly, becoming a widower was associated with an increase in mortality in men, but the effect was much less dramatic in widows. Apparently, greater emotional ties with children and neighbours helped old widows survive better than widowers. There may also be an effect of religion and spirituality that could have decreased the levels of emotional distress in older women. But women were also better cushioned existentially in retirement, given that in this century most of them did not have a continuous job, thus retirement did not mean the drastic life change that it meant to men (Stearns 1980). The scene depicted in Fig. 2.8 although showing old women in retirement may be equally representative of the same group of women thirty years earlier: busy producing handcrafts, reading, and talking to each other.

In the late 1800s in Australia, the sex ratio in asylums for the elderly and the poor was biased towards male elders, among other things this was due to older women being better able to build networks of social support when they needed it, especially with family, a daughter in particular, thus finding themselves in a better position to obtain accommodation outside the public institutions system (Jalland 2015).

Support received by older women from their family was further strengthened by the social prejudice that saw women as naturally dependent throughout their entire life. As expressed in an almshouse report: "... the world recognizes the inevitable dependence of women by considering it a most disgraceful thing for relatives or children to allow an old woman to go to the almshouse. On the other hand, men are supposed to have had their chance to lay up money, and if they have not done so they must take the consequences". (Dahlin 1980, p. 103).

In Victorian England negative stereotypes affecting the elderly of any sex were similar to those affecting women of any age (young and old): they were regarded has having poor physical strength, inferior intelligence, economic dependency, and being childlike and sensitive (Roebuck and Slaughter 1979).

During the period of their revolution of independence and until the early nineteenth century, Americans tended to praise youth, with older people facing serious difficulties with little help, especially if they did not have a family or did not live in a community with established traditions to care for those in need. For the less lucky ones who lived in communities under economic stress the situation could be dreadful in the extreme:

> In response to rising relief taxes, a number of towns rid themselves of poor, elderly residents by auctioning them off to farms for labor. This practice continued until the 1820s, frequently accompanied by neglect, cruelty, and even torture. Some communities denied refuge to nonresidents, forcing the old and sick to wander from town to town in search of assistance, even in the middle of winter (Fleming et al. 2003, p. 916).

In the second half of the nineteenth century, and presumably under the influence of the emerging new theory of evolution by natural selection introduced by Charles Darwin and Alfred Russell Wallace, old people were seen as "survivors" and therefore the "fittest" within their generation, they were also seen as contributing positively to society through their various social roles (e.g. helping the family), wisdom, and experience. In the USA the emergence of evolutionary thinking also coincided with the post-Civil War period which made it easier to see the elderly as "survivors" and "winners" in the "battles of life", deserving of appropriate recognition, respect, and support (Fleming et al. 2003). The idealistic views of Romanticism may have also played a role in the construction of a positive image of the elderly. Andrew Achenbaum (1974, p. 49) quotes Ralph Waldo Emerson, noting that: "We do not count a man's years until he has nothing else to count".

The optimistic interpretation of evolutionary thinking, as applied to the social sphere, would soon be overturned by the emergence of *Social Darwinism* that interpreted the demise of the weak, the sick, the poor, that is the "losers" in the competitive scramble for survival, as good for society at large (Haraven 1976). An illustrative example of *Social Darwinism* in action in the early twentieth century is given by Tamara Hareven (1976, p. 20):

> Contemplating the 'industrial scrap heap' in one's middle years became one of the nightmares of industrial society. The Amoskeag Corporation, the world's largest textile manufacturer, began to lay off, rather than reassign, its slow workers beginning about 1920. When a middle-aged woman weaver protested her dismissal, after twenty years of work, before the union's grievance committee, the management upheld the overseer's rule: 'In deciding who

we should keep and who we should lay off, we follow the simple rule of the 'survival of the fittest''. This same overseer, who had risen from bobbin boy to that much coveted position, himself ended up as a sweeper.

Social Darwinism spread across many countries towards the end of the nineteenth century and further developed into the twentieth century in association with socio-economic theories such as laissez-faire and then neoliberalism, but also eugenics and racism (Claeys 2000). In Australia: "[o]ld people suffering from dementia, paralysis, cancer and incurable diseases were too often viewed as useless and could be subjected to humiliation and torment in asylums supposed to care for them" (Jalland 2015, p. 19). In Hareven's (1982) words: "Advanced old age, which had earlier been regarded as a manifestation of survival of the fittest was, by the late nineteenth century, denigrated as a condition of dependency and inexorable deterioration".

Views about the elderly continued to change in the transition into the twentieth century when a cult of youth re-emerged and compulsory retirement at 65 was more generally introduced, alongside a system of pensions. At the time, the chance of those older than 65 to reenter the workforce was slim.

Under the influence of technological progress, towards the end of the nineteenth century thinking about old age became more realistic and pragmatic. Then all the negative aspects of old age, including physical and mental decay became even more visible. The old years were regarded as a time of dependency and decay, including dementia, with old age as such believed to be a disease. This, of course, did not help older individuals find a job if they required an income (Fleming et al. 2003). Old people were then described as being tactless, not necessarily polite; careless with their appearance and too prone to complain (Achenbaum 1974). Even the songs of the time reflected a negative view of old age (Cohen and Kruschwitz 1990). The classic poem by George Catlin "Over the Hill to the Poor House" (made into a song in 1874) reflects this pessimism:

"What? No! It can't be that they've driven
Their father so helpless and old,
Oh God may their crime be forgiven,
To perish out there in the cold.
Oh heavens, I am saddened and weary,
See the tears how they course down my cheeks,
Oh this world it is lonely and dreary,
And my heart for relief vainly seeks.
For I'm old and I'm helpless and feeble.
The days of my youth have gone by,
Then over the hill to the poor house,
I wander alone there to die".

This produced a cultural schism whenever the ageing body was nevertheless associated with a capacity of the individual to still operate mentally in a perfectly productive way, a "marked ambiguity, ambivalence, and conflict about feelings about growing old and being old" (Achenbaum 1978).

At the time, a more scientific approach to the understanding of old age was introduced in the USA by Ignatz Leo Nascher who coined the term *geriatrics* in 1909, a new medical discipline. Geriatrics initially considered old age as "disease" a view that, according to Kirk (1992), has ancient roots in Terence (161 b.c.e.), but also Aristotle and Seneca (as we have already mentioned in this chapter) and more recent humanist writers since the sixteenth century. During this period the Belgian mathematician Adolphe Quetelet introduced the body mass index that was used to determine the beginning of old age in a more scientific manner, based on measurable biological characteristics. Other contemporary studies also suggested a decline in brain size with age, thus leading to a biological underpinning of some observable aspects of "senility" (Kirk 1992). Some of the early geriatrists, however, rejected the view inspired by the Latin maxim *Senectus ipsa morbus est* that old age is a disease, among them was the German geriatrician Lorenz Geist who suggested that old age rather than being a disease in itself is a natural process of change (an idea that arks back to Galen and before him to the sixth century b.c.e. Indian text *Sushruta Samhita*) that simply increases the probability of contracting diseases.

At the dawn of the twentieth century youth was back in charge. To quote Andrew Achenbaum (1974, p. 59): "The young refused the old admittance into their world. Youth no longer revered age; they excluded it. The stage of life romantics once characterized as the season of rest had become one of discontent".

2.8 The Twentieth and Twenty-First Centuries

At the beginning of the twentieth century old age was culturally relegated to a position of less prominence than youth, but this changed as the century progressed and then again into the twenty-first century, as the view about the elders flip-flopped between negative and positive over the decades following socio-economic and cultural changes that affected the relationship between younger and older generations.

In Australia, those born in the 1920s are known as the *Lucky Generation* because they were adults just when World War II ended and a period of social democratic and unprecedented economic growth in greater social equality started in most capitalistic countries (Mackay 1997). This generation was followed by the *Baby Boomers*, that is, people born between 1946 and 1964; those born in 1946–1955 are known as "leading edge boomers", the others are "shadow boomers"; pre-boomers are also referred to as "stoic" (Butler 2015).

As the *Baby Boomers* are entering their old age, they are reclaiming the social relevance that they themselves denied to their elders. Given the serious financial stresses the younger adult generations are in at the beginning of the twenty-first century, the old can indeed exert a degree of power and social control, a reminder of the situation of power the elders found themselves in during the periods of bubonic plague in the Late Middle Ages. Schneider and Ingram (1993) summarise the characteristics of various generations according to a combination of positive or negative social constructions or perceptions of their condition and their true strong or weak social power (see Fig. 2.9). Many old *Baby Boomers* are socially *advantaged* compared to younger generations, because they enjoy a *positive social image* and a

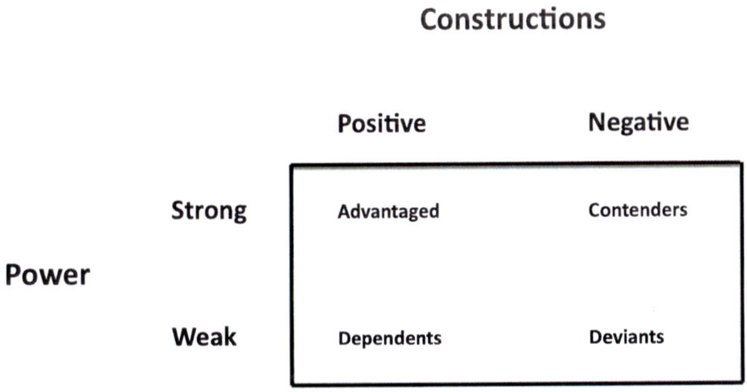

Fig. 2.9 Typology of target generations according to social perception and actual power (adapted from Schneider and Ingram 1993). *Baby Boomers* can be seen as *advantaged* from a combination of *positive* image and actual *strong* material *power*

strong power due to their accumulated wealth (see also Hudson and Gonye 2012), but not all is rosy for all *Baby Boomers*. Guy Standing, in his book *The Precariat*, describes the split of the population of old people under neoliberalism into *grinners* (i.e. "those with adequate pension and health care coverage, who can do odd jobs for the pleasure of activity or to earn money for extras") and *groaners* (i.e. those "without a reasonable pension, who face competition from more energetic youth and less needy old agers") (Standing 2016, p. 69).

The *Baby Boomers*' optimism was overshadowed by the *Cold War* and the threat of nuclear war, leading to nihilism and hedonism and also a strong focus on individualism and personal growth starting in the 1970s, that challenged the sense of community and social responsibility that prevailed among the members of the *Lucky Generation*. In addition, they were confronted with a life of uncertainties at work and in the family since the 1980s, which is at the root of widespread levels of stress and stress-related medical conditions in this generation (Quine and Carter 2006). This has made divorce far more frequent than in previous generations and it has also required a blurring of traditional gender roles as any partner (male or female in heterosexual couples) can become unemployed. The *Baby Boomers* also experienced the raise of neoliberal political thinking since at least the 1980s, which has led to a societal demand for greater "flexibility" from young and old and an expectation to embrace uncertainty as an "exciting" challenge, rather than a burden. But the grim reality of increasing unemployment, sub-employment, and uncertainty in the 2000s, especially uncertainty in the material future, is not as "exciting" for the common people as some ideologues would like it to be. In the 2000s we are seeing the emergence of the *Options Generation*. The *Options Generation* is less materially independent than previous generations precisely due to the uncertainty and instability of life, especially at work (Mackay 1997): the "option" of being more insecure? What kind of "option" is that?

In the twentieth century the study of old age also became more scientific and less intuitive. Both psychological and physiological aspects of old age begun to be better understood and integrated (Birren 1961a). Research centres were built, journals published, and international conferences organised, for instance, by the *Gerontological Society*, founded in 1945 (Birren 1961b).

The view of old age as pathology that started to take hold in the nineteenth century further developed during the first half of the twentieth through the bio-medicalisation of old age (Hirshbein 2001; Vincent 2006). Although old age is ultimately an inescapable period of physiological decay, this should not discourage old people from taking actions and performing activities that could help them improve their health and mental capacities. Medications and surgical interventions may be necessary in specific cases, but much can be also done about old age outside the traditional medical paradigm, once entrenched stereotypes are abandoned (Vertinsky 1991).

In both the twentieth and twenty-first centuries the trend has continued for older people to behave as providers of care for both family and others. Pat Thane (2005a, p. 12) lists: "bringing up orphaned grandchildren, giving a home to a widowed or deserted daughter and her children, caring for sick or disabled children, spouses, neighbours". But such level of support is also mutual, inter-generational.

In Australia in the late 1960s grandparents played a significant role within their extended family by acting as babysitters, for instance, whereas neighbours also fulfilled a vital role providing support for old people who were living alone (Jalland 2015). But the burden of providing care for the elders usually fell on the family, especially a daughter, or sometimes even a daughter-in-law, often leading to stress and burn out. Caregivers "often felt guilty, helpless and exhausted and suffered silently 'in the belief that care giving is their responsibility and theirs alone'" (Jalland 2015, p. 187).

In urban settings, families with elderly members tended to be biased towards the wife's old parents, whereas the opposite relationship was found in rural settings where the old parents in residence tended to be the husband's parents. This is known as *Sweetser's Rule* (Sweetser 1964).

In the 1970s a spur towards greater individualism and a quest for privacy in their living arrangements characterised the older generations. In the USA they explicitly preferred to live alone or with their spouse rather than with any of their children or relatives (Dahlin 1980). In spite of this, women retained their ability to build and maintain broader social networks than men, to the extent that the death of his wife was especially devastating for an older man, whereas older women tended to be better protected in widowhood by their ability to establish social bonds with others, both within and outside their family.

In the social sphere old people were also the target of stereotypes, prejudice, and discrimination during this period. In Australia, the social stigma of being dependent on charity, that was widespread in society and was felt by old people living on an old age pension, started to abate only after 1945, in concomitance with an increased social interest in the needs of older people (Jalland 2015). However, stereotypes endured in some sectors of society and indeed they are still maintained today in contemporary Australia.

In a letter to *The Sydney Morning Herald* newspaper that was published in April 1952, the author, a 22-year-old man with wife and two children, wondered why his taxpayer money should be spent on old pensioners who neglected to save during their working life, rather than being used for the benefit of his own family (Jalland 2015). The same argument can be seen repeated in the 2010s in posts published in various blogs and in newspaper articles, virtually unchanged down to almost each word. As a counterpoint, most other letters to the 1952 newspaper pointed out that people on a pension had paid taxes throughout their lives and contributed to the progress of the country, and they deserved at least such minimum support. Not surprisingly, the same reply is repeated today by progressive citizens. It seems that on this issue we have been culturally running in circles for the past 60 years or so.

In the 1960s negative stereotyping of the old was encapsulated in the new term *ageism*, coined by the American geriatrician Robert Butler in 1968, which he defined as "'a process of systematic stereotyping or discrimination against people because they are old'—categorizing them as senile, unproductive and frail" (Jalland 2015, p. 254). At the same time the term *gerontophobia* was also introduced as: "the unreasonable fear and or irrational hatred of older people by society and by themselves" (Bunzel 1972, p. 116).

The combination of individualism and withdrawal into private life by the emerging middle-class living in the suburbs of big cities not only diminished contacts with older people but further entrenched stereotypes that were already common towards the end of the previous century, giving rise to a "spoiled identity" (Goffman 1963) for the old: a view of the old as useless, unattractive, senile, temperamental, inefficient.

Social discrimination, prejudice, stereotypes, and misconceptions about old age and the capabilities of older people may feedback on the individual, who may incorporate such views into his/her own behaviours or concept of self, thus causing a potential runaway process where external prejudice ends up becoming internal reality. This may lead to further prejudice, followed by accelerated mental and physical decay. Goffman pointed that this runaway process underpins the development of any stigma in society (Goffman 1963). The runaway process of stigmatisation may lead to myth becoming reality, a self-fulfilling prophecy (Friedan 1993; see also Cordella and Poiani 2014).

Ageism makes the ability of older workers to find a job more difficult, and the feedback mechanism that escalates the likelihood of discrimination only makes things worse. In the words of Betty Friedan (1993, pp. 208–209): "Hence the vicious circle: stereotypes regarding older employees' lower abilities, motivation, and learning capacity lead to restricted opportunities for training and development, and thus to 'obsolescence' and career stagnation, to blocked promotion ladders and career plateaus…. And the critical juncture, leading to continued growth with age for some and 'obsolescence, stagnation, and career decline' for others, is the period around age forty".

A common way of expressing prejudice against older people is by using *overaccommodation* when communicating with the old; this consists of assuming that old people are thick to understand, not flexible, and limited in their sensory abilities, thus others start talking to them in an excessively simple language, slowly and loudly (Butler 2015). But the most serious prejudice is arguably that of age-negative discrimination at work and to find work, that we mentioned above, which can seriously and negatively affect not only self-confidence and quality of life but the capacity of survival itself, especially when the old person cannot access a helping social network or she/he lives in a country with an inadequate social welfare system.

The cult of youth that developed in post-World War II in Australia was reflected in 1984 in the official replacement of the British *God Save the Queen* with *Advance Australia Fair* as the national anthem, a song written in 1878 but that also reflected the spirit of youth pervading society in the late twentieth century: "Australians all let us rejoice, for we are young and free…". This spirit of youth has fostered a widespread prejudice against older Australians that they are "warm but incompetent", and older people resisting the stereotype do not necessarily gain recognition on the competence side but certainly lose on the warmth side (Butler 2015).

But inter-generational conflicts are certainly not new in the twentieth century. In the 1970s, a period that was at the top-end of over two decades of post-World War II international economic growth and social policies that, at least in the industrialised West, saw high levels of employment and low levels of income inequality, generations engaged in tense relationships. The inter-generational tension derived not only from the young having a worldview that clashed with that of their parents, but also and perhaps more importantly opportunities for personal growth and fulfilment started to shrink from then on for everybody, and therefore the level of inter-generational competition increased. Unemployment rates across the industrialised world were lower in the 1960s and started to increase in the early 1970s to continue, with ups and downs, until present. This occurred in spite of a new definition of "employed" being adopted in the 1980s whereby a person is regarded as "employed" if she/he is engaged in paid work for at least 1 h per week. Such definition clearly masks the levels of real unemployment and under-employment. Following that definition, the *Australian Bureau of Statistics* gives a March 2020 official level of unemployment of 5.2% in Australia, whereas *Roy Morgan Research* gives 12.2% for the same period (http://www.roymorgan.com/morganpoll/unemployment/estimates-detailed), after regarding as unemployed any person who during the survey period was looking for work, which is certainly a far more realistic approach.

The antagonism between generations that gave rise to the concept of *ageism* is in fact bi-directional: discrimination of the young against the old and vice versa. This inter-generational tension—that, as we have seen, has very deep historical roots—is also compounded by physical segregation into age-specific ghettos and it also involves middle-age generations who are in the midst of their work productive life and tend to see taxation as a burden they have to bear in order to support both the old and the young (Neugarten 1974). With regard to retirees, this conflict has given rise to the issue of the *burden of dependency* and some myths associated with it.

The demographic trend towards an increased percentage of people older than 60 has produced what is ultimately a myth, the neoliberal myth of the *greedy geezers* (or the increasing ratio of *leaners* over *lifters*, Butler 2015; Standing 2016) that aims at constructing a view of retirees as less "deserving" (e.g. Hudson and Gonye 2012), or some sort of social parasites. In the sarcastic words of Caryl Rivers, Professor of

Journalism at Boston University (USA) and also a senior: greedy geezers are all current seniors who are supposed to be "enjoying the good life spending all our money on ourselves while siphoning off public money because we vote for our 'entitlements.' We are stealing from our children, increasing the debt—maybe we even created the financial crisis with our overly affluent lifestyles" (Rivers 2011). This is supposed to establish the conditions for a *burden of dependency*, the burden of retirees on society, but we argue here that, among other suggestions, such alleged burden of dependency disappears as soon as we realise that technological advances will tend to produce higher revenues at lowers costs to enterprises. The escalating profit could be used to help sustain the increasing population of retirees (and working-age unemployed people as well) at no real cost to economic growth, via appropriate taxation. Those retirees and unemployed people will then return the money to the economy through consumption.

The myth of the high cost to society of an increasing number of retirees, who are supposed to impose an unbearable financial burden on the younger generations, thus creating a significant inter-generational conflict, is based on the false premise that social services and government pensions are too expensive and are a major reason for increasing government debt. The myth has been debunked by many but here we focus on the argument put forward by Holstein and Pavle (2013).

Holstein and Pavle start by establishing the "moral foundation" of their critique of the *burden of dependency*. This is important, because all sides of the argument are appealing to some sort of morality: it is immoral for the "greedy geezers" to jeopardise the future of the younger generations, and so the counter-argument goes that it is also immoral to negatively affect the future of those who have worked hard throughout their life and contributed to society with their taxes. In the words of the authors: "An acceptable moral foundation will support broad risk-sharing that involves both the public and the private sector and will accept no changes that harm the least well off" (p. 720). It should be immoral to be insensitive to the *cumulative disadvantage* of the less well off when it is clearly detected in society. This moral stance is of course not new but in this case it is formally inspired by the work of John Rawls (1971): "any changes in the status quo must work to benefit the least advantaged so that they can sustain dignity, continue to be moral agents, and flourish" (p. 729). This is especially relevant to old people (but not only) whose dignity and ability to flourish can be seriously jeopardised by material stress, and they often find themselves in the physical incapacity to significantly change their condition without help.

The alternative neoliberal argument rests on the fundamental assumption that the more government support is provided for the elderly the more incentive there is for the family to save on support, and therefore relax. Hence, withdrawing government support for the elderly is expected to provide an incentive for the families to take care of their older members (Sokolovsky 2009d). More specifically, the neoliberal tenet could be stated in this manner: The reality of an ageing population, the concern about the future sustainability of Social Security and other public services, the need for investment in the younger generations (even though, we should note, they are increasingly required to pay for services such as education out of their own or their family's pockets), the unwarranted gifts to the "greedy geezers", and the public deficit are justifications for reforming Social Security by modifying its benefit structure (Holstein and Pavle 2013). Such changes in the benefit structure of Social Security are suggested to include an increase in the retirement age (but see how this will immediately create increased inter-generational competition for jobs) and modifying the way in which costs of living are calculated. Note, however, that lifespan has not increased homogeneously across society, wealthier people live longer than poorer ones, and so if by following the neoliberal paradigm the age of retirement is lifted for everyone, society will find itself trying to answer this question, originally asked by Paul Krugman: "Why should janitors work longer because lawyers are living longer?" (cit. in Holstein and Pavle 2013, p. 737). Indeed! Moreover, revenues raised in the above manners are not capable of addressing the issue of deficit and sustainability. The reality is that the long-term sustainability of the Social Security system can be guaranteed into the future provided that some additional sources of revenues are added (Holstein and Pavle 2013). Where are the additional revenues going to come from? In the USA context, Holstein and Pavle correctly argue that the additional revenues can come from raising the FICA (*Federal Insurance Contributions Act*) tax and by raising the cap on taxable earnings. Similar solutions are being considered by progressive political parties around the world.

Therefore what is required is a change in the political model, from neoliberalism to social democracy, for instance, in order to shift the burden of support for an increasingly ageing population from the individual to the broader society through better levels of taxation on companies—such as large and multinational companies—and on individuals, especially those in the top 1% of the income distribution. This requires a re-evaluation of the role of government as an agent for the redistribution of wealth and promoter of greater social equality:

> Thus it is ideology, and changed assumptions about the role of the public sector with regard to social need, including material deprivation and health and social care, which underlie the current economic pessimism about ageing emanating from some governments and international agencies (Walker 1990, p. 386).

The perception of a crisis is therefore a result of the application of the wrong policies that are ideologically driven: (1) the increasing reliance on private-sector solutions to the risks of ageing, especially the economic risks, (2) the increasing gulf between the rich and everybody else that only deepens in the case of retirees, (3) lower or stagnant wages that have seriously incapacitated large groups in society from building a safety net of savings, (4) the shift from defined benefit plans for pensions to plans characterised by defined contributions which are more subject to the vagaries of the financial markets (Holstein and Pavle 2013).

Therefore, ethically and economically sustainable solutions for the viability of social security and pensions into the future can only come from increasing revenues from, and cutting benefits to, the highest earning sectors of society (individuals and companies). This will also minimise inter-generational conflicts. Automation and technological development could be transformed from the currently perceived threat to older people (Randall 1965) to an asset, via appropriately targeted taxation.

This brings us back to the issue of the long-term sustainability of our modern, industrialised societies and their system of support for the elderly. Such societies can indeed be sustainable, with no or little "burden of dependency" and in a way that takes into account humanitarian values and also issues of environmental sustainability, even against the fears and doubts of some progressive politicians around the world (e.g. Butler 2015). The increased technological advances that make our life easier but that create unemployment and early de facto retirement, along with the potential for negative environmental effects such as pollution, can be addressed by not only introducing environmentally sustainable technologies (e.g. sources of energy alternative to fossil fuels) but also appropriate forms of taxation that can better spread the wealth created across society as suggested, for instance, by Holstein and Pavle (2013) above. Population growth is another important issue affecting sustainability that must be addressed in a democratic way. Although the ageing world population could be interpreted as a symptom of a slowing population growth, our species keeps exerting significant pressure on the environment across the globe, not only because of our lifestyle, but also by the sheer effect of our population size (Ehrlich and Ehrlich 2009).

2.9 The Safety Net

At the beginning of the twentieth century, more often than not old people found themselves still in the necessity to work, in order to provide for their needs and those of their families; but in addition, there was also cultural pressure to remain visibly active as idleness was regarded harmful to both the individual and society at large (Dahlin 1980). In the USA in 1900, about 70% of men who were 65 year old and older still remained in the workforce out of material need.

A better social safety net for the elderly and destitute was put in place after 1918 in the USA to improve the conditions of the poor, but the economic turmoil of the decades following 1910, that finally led to the *Great Depression* of 1929, produced alarming levels of poverty: from 30% in the late 1920s to 50% in 1935 and about 66% in 1940, thus putting enormous pressure on social welfare initiatives due to costs (Fleming et al. 2003). In Australia, about 20% of the workforce was unemployed between 1930 and 1934, reaching a peak of 30% in 1932, with the elderly not always seeking help, out of pride (Jalland 2015). This social tragedy, however, revived the initiatives of the movement for old-age pension, thus leading the American President Franklin D. Roosevelt to sign the *Social Security Act* of 1935, on which occasion he said:

> We can never insure one-hundred percent of the population against one-hundred percent of the hazards and vicissitudes of life. But we have tried to frame a law which will give some measure of protection to the average citizen and to his family against the loss of a job and against poverty-ridden old age. (Fleming et al. 2003, p. 917).

This eventually led to the signing into law of *Medicare* and *Medicaid* by the American President Lyndon B. Johnson in 1965. Medicare and an improved level of social security were major boosts to the ability of older people to afford a safer retirement.

In Australia, the distinction between "deserving" and "undeserving" poor was culturally entrenched as it was in the USA and the UK, but the concept of an aged pension being received because it was a right, rather than a charity, finally took hold, thanks to the activism of the *Australian Pensioners League* that was established in 1932 (Jalland 2015).

In the 1950s retirement was associated with loss of purpose and meaning in life, especially in men: "A sense of uselessness quickly developed into apathy and disinterest in the world around, and they lived 'a vegetable existence'" (Jalland 2015, p. 180). Three basic pillars of retirement, health, income, and home ownership, were not always within the reach of everybody and the situation would only deteriorate in Australia and around the world in the twenty-first century, as wealth inequality has increased alongside the costs of living, thus making housing affordability a challenge for most. Job insecurity and low pay have just compounded the overall sense of crisis for both young and old.

During the so-called *social democratic phase* of economic and social development of western democracies, that started soon after the end of World War II (Dumenil and Levy 2013), employment started to raise, opportunities were better spread across the population, and wealth inequality decreased. Fewer older people were living below the poverty line (15% in 1979 in the USA compared to 35% in 1959) and pensions coverage improved, thus making older people more independent, with an overall better social security system also playing an important role. But these improved conditions for old people were also the result of greater activism from the part of organisations of pensioners since the late 1960s and early 1970s (Hudson and Gonye 2012). More old people were supported in retirement by a pension system that allowed acceptable living standards, and this was also seen as a solution to youth unemployment: the removal of older people from the workforce created opportunities for the younger ones (Randall 1965).

The situation started to change in the 1980s (Dumenil and Levy 2013), reaching a current crisis point in the second decade of the twenty-first century, as the level of socio-economic inequality continues to increase. In the words of Mark Butler (2015, p. 9) who refers to the current situation in Australia: "Older Australians are generally living on the edge financially, and are very nervous about the constant chatter about the unsustainability of existing pension entitlements. They are resentful—and often shocked—by the way in which the rest of society treats them as they age. And they are frightened about what confronts them in their 'fourth age'; the period of decline, dependence and, eventually, death". But even middle age workers are living on the edge. Employers increasingly regard an employee to be past his/her prime at about 45 years old, which is seriously affecting jobs opportunities for this category of people, at an age that is still two decades away from retirement.

The dramatic situation for retirees that started to unfold with the onset of neoliberalism is clearly described by Thane (2005a, p. 277):

> So, at the very time that most people were living longer and remaining fitter to later ages, instead of working later in life they left the labour market at earlier ages than ever before. This became particularly evident in the 1980s, when the recurrence of high levels of unemployment internationally again led to earlier retirements. By late 1990s almost one-third of West European workers had retired permanently from paid work by the age of 60. Some left willingly, on comfortable pensions, to enjoy relaxation, travel, consumption. A new generation of WOOPIES (Well-Off Older Persons) was identified by advertisers and welcomed by politicians. *But they were a distinct minority*. Others gave up work more reluctantly. (italics ours).

2.10 Older Ages and New Meanings of Life

Once retirement has become a reality, then the issue is what to do with the rest of our life. Two basic questions acquire a degree of urgency at that point in time: Where to live? How to live? The answers to those questions will affect the meaning of our life into retirement.

From the twentieth century on, there has been an increasing trend towards old people to live alone (e.g. Cattell 2009), a trend that has been especially strong in northern European cultures (Thane 2005a). This may be an issue of choice for many old people, but for many others it may just be a consequence of constraints, having no option (Kaida et al. 2009). In Australia in the 1950s, in spite of older women being more numerous in the population than older men, fewer women than men lived in their adult children's households (Jalland 2015); the same pattern is observed in today's USA, where women over 60 tend to live by themselves two to three times more frequently than old men (Sokolovsky 2009e). Table 2.3 shows data from the USA indicating the progressive detachment of older people from living with their children (or even with a spouse, see "Unmarried" living "Alone" in Table 2.3) throughout the initial three-quarters of the twentieth century.

The increase in the number of elders living alone is a relatively new phenomenon. Until the beginning of the twentieth century, elders may have lived alone, in the home of their adult children, or in their home with their adult children. Dahlin (1980, p.100) notes that living in their home with their own adult children was the most common arrangement for old people at the beginning of the twentieth century, at least for those younger than 75: "In 1900, most old people were either heads or wives of heads in their own homes, not dependents in the homes of their children or grandchildren. In other words, the old had their grown children living with them more often than they went to live with their children". In the specific case of parents older than 75, however, living with children was a necessity at the beginning of the twentieth century, due to their increasing frailty and lack of acceptable elder care facilities outside the family.

A new phenomenon also started to emerge in the twentieth century: that of old widowed women who remarried. In France the frequency of remarried widows in the population started to increase since 1906, and "by the 1960s 25% of all single,

Table 2.3 Living arrangements of the elderly by marital status in 1900 and 1975

	Married		Unmarried	
	1900	1975	1900	1975
Living arrangement				
Alone or with spouse only	29	84	11	66
With married child	16	2	38	7
With unmarried child	42	10	27	10
With kin	7	4	13	13
With non-kin	6	0	11	3

From Dahlin (1980), data are from the USA and are expressed in percentages

divorced, and widowed women over 50 would marry, a rate well below the male but increasing far more rapidly" (Stearns 1980, p. 55).

The trend towards living alone observed among the elderly in the second half of the twentieth century (Table 2.3) developed hand in hand with an increasing social change towards individualism. The independence of old people was also helped by changes in society that made them less essential to their adult children, such as the greater availability of information through better education and various media that to a great extent replaced the knowledge that the elders formerly provided, and the institutionalised childcare that took off especially after the *Great Depression*. Although living alone may have brought increased independence for some elders, it also contributed to increased social isolation and loneliness in some others (Pantell et al. 2013; Klinenberg 2016).

Alternative forms of retirement also appeared in the second half of the twentieth century that combined the need for privacy with some access to company: the retirement villages, that superseded the old retirement homes of the past. Retirement homes were not much better during the early decades of the twentieth century than their predecessors of the nineteenth century. One inspector described one of such institutions in the state of Pennsylvania (USA) in 1920 so:

> There was no genuine homelike spirit. Most of the inmates looked sullen and wore depressed and downcast mien. Practically all were eager to get out of the place.... This feeling of depression is augmented by the fact that in no homes is an attempt made to segregate the old people—who have been compelled to go to the almshouse through no fault of their own—from the feeble-minded, and in some cases even the partially insane. In many places they are compelled to eat at the same table and sleep in the same dormitories with the latter groups. (Hareven 1976, p. 24).

The stigma associated with the nineteenth century almshouse to some extent continued into the twentieth century with retirement villages, at least initially. But more modern villages—that can be usually afforded mainly by wealthier retirees—tend to make some effort to project an image of homely normality in a safe and attractive environment. In Australia a spur of activity in the building of retirement villages started in the 1970s (Jalland 2015).

In most cases, however, elders prefer to remain in their home for as long as possible. Over the years, older people may develop a level of attachment to their home and neighbourhood that could be important in the maintenance of their self-identity, see, for instance, the phenomenon of the *Downtowners* or communities of interconnected elders living in their own house or apartment in large city centres where they have easy access to amenities and services (Cattell 2009). Such attachment could be dramatically disrupted by retirement into a village where old people may share the communal living spaces with complete strangers, in an age-segregated social environment where all memory of past life-history and achievements, that may be an important part of personal identity, may be lost (Keith 2009). Still, there is diversity among older people. Some may appreciate staying within a familiar, multi-generational environment in their home and neighbourhood, others may appreciate the safety provided by a good quality retirement village, and still others may prefer to unleash their sense of adventure and move on to other places, especially if they are still healthy and can afford the costs of their lifestyle. Indeed, communities of old people may dwell on specific yet changeable *elderscapes* (Katz 2009a, b), that is specific environs used and created by such communities of elders in the performance of their activities, and as they do so they may also create new cultural spaces and scripts (Sokolowski 2009a). In other words, what we are witnessing is the development of multiple *landscapes of ageing* (Cohen 1988).

The lifestyle of moving to places has produced a particularly dynamic elderscape such as that of the *Grey Nomads* in Australia and the *RVers* in the USA (the "R" stands for recreational and the "V" for vehicle). Also in the USA, the so-called *Snowbirds* follow a "seasonally migratory" lifestyle rather than a "nomadic" one. Together, these groups of elders give life to what Stephen Katz (2009a, 470) calls a *gerontology of mobility*, which includes "the transculturality of both people and places as they age and change while adding a dynamic sense of retirement 'flow' to the more static tradition of retirement 'time'". The concept of *transculturality* describes a process of combining local and global cultures, as a result of simultaneous trends to retain what is traditional and what is becoming internationally widespread. Physical mobility also affords older people an opportunity to cultivate their plastic mental capabilities as they continue learning and incorporating more experiences into their lives.

The *Grey Nomads* are a community of elders who like life on the move, driving their caravans and pursuing both change and stability at the same time. They seek the thrill of adventure combined with sociability (communication, reassurance from other "nomads" and plain good fun) in a friendly and cooperative environment (Fig. 2.10), sometimes they even fulfil a useful social role of protectors of the natural environment (e.g. Holloway et al. 2011). The wandering of the *Grey Nomads* or the more predictable "migratory" activity of the American *Snowbirds*, who may spend the cold months of the year away from their residence to then return during the warmer months, provide them with a variety of landscapes, environments, and climates whilst decreasing stress and increasing attention and engagement. Katz (2009a, b) describes the case of Canadian *Snowbirds* spending the winter months in Florida, whereas the octogenarian parents of A. Poiani were, until recently, *Snowbirds* in their native Italy, spending the cold months of the year in the relatively mild winter climate of Rome to then escape the hot and humid Roman summer in

Fig. 2.10 A group of *Grey Nomads* enjoying "happy hour" at the Kurrimine Beach Holiday Park in Queensland, Australia

their refuge on the Adriatic side of the peninsula at 535 m. above sea level, where they could enjoy a fresher summer weather, but it snows there in winter.

Snowbirds create a complex culture and identity that can embrace two different regions of a country or indeed two different countries. As Katz (2009a, p. 480) put it regarding Canadian *Snowbirds* in the USA: "snowbirds are quite comfortable with their bi-national identities and that travelling in both directions feels like 'going home'".

In the case of the *Grey Nomads*, Onyx and Leonard (2005) identify a variety of motivations that lead this group of old people to adopt their specific lifestyle: "reunion" (returning to the places of their past), "recuperation" (rest, seeking peace, and quiet), "circulation" (indulgence, looking for comfort, and pampering), "reduction" (go through a list of see-it-before-I-die), and "exploration" (active involvement). Although the vast majority of *Grey Nomads* travel with their spouse, they tend to regard meeting other people as one of the most positive aspects of their experience; some of those acquaintances may become life-long friends.

The *Grey Nomads* lifestyle is an example of continuous change and growth into old age. Seeing "Age as Adventure", as Friedan puts it, can be literal, as in the case of the *Grey Nomads*, or metaphorical: an inner adventure exploring artistic capabilities, for instance, such as painting, writing, and so forth. New meanings in life emerge from new meaningful activities, whichever they may be. In the words of Cattell and Albert (2009, p. 131), *RVing* elders "have created a rootless, independent lifestyle of deep personal meaning and have become experts at building community over time and space and creating instant communities wherever they park their rigs".

The nomads' lifestyle seems to better accommodate the adventurous types who would probably not accept living in the artificial environment even of well-organised retirement villages—such as "Leisure World" a retirement community in Laguna Hills in California, USA—that these days tend to be built by multinationals of the retirement villages business. In Friedan's (1993, p. 371) words: "In a certain sense, the official Leisure World attitude was a Peter Pan denial of the very ability to keep changing". People who left the village also mentioned a sense of disconnect with real life that they recovered within their community of origin once they returned. A real town or neighbourhood affords the elders the capacity to interact with individuals of different ages within a natural social setting. Interacting with people of all ages can be beneficial to the well-being of older people, especially when interactions occur with meaningful others who are fully engaged emotionally with the elder.

Retirement also affords both men and women an opportunity to behave in ways that are less constrained by social expectations. For instance, older men may feel freer to develop their feminine side, older women their masculine side (Gutmann 1987). Once the pressures of work with all its duties and deadlines come to an end, the self is finally freer to focus on life and its deeper meaning. It is to this issue that we turn next.

For many retirees, life after work presents some challenges not only in terms of what to do with so much time available, but also how to solve the material issues of sustenance, and how to maintain life meaning. Becoming more obvious in the late twentieth and early twenty-first centuries, the older person who is retired from work is confronted with a deceptively simple existential question: Now what?

Throughout our working life each one of us produces a "life script" that provides some form of meaning to our cumulative life experience. Relevant memories are organised into a story and we also give them emotional valence (the good, the bad, the ugly). Grounded in our past experiences, we establish the conditions that can improve our well-being and that help achieve our self-realisation, allowing us to cope with the negatives of life (Dittmann-Kohli 1990). Retirement from work that may socially mark the "beginning" of old age, at least to those men and women who have been working full-time for a considerable part of their life, can in some cases disrupt the life script that had been constructed until then. This poses a challenge to the individual in terms of what Dittmann-Kohli refers to as the "re-organisation of the personal meaning system". A similar challenge is also confronted by survivors of life-threatening diseases such as cancer (Cordella and Poiani 2014).

Dittmann-Kohli (1990) has shown that old people can adopt specific strategies to face the uncertainties of the future: they could relax thinking that there are some events that are just beyond their control, this may help decrease self-criticism and lead them to become more open and accepting of themselves as they are. This is a contemplative stance that has been embraced within a religious framework for centuries, both in the West and in the East. We will return to this issue below with Elaine Cumming's concept of "disengagement".

More proactive approaches to ageing (e.g. Moody 2009a, b) have been inspired by the work of Carl Gustav Jung and they became more common during the second half of the twentieth century. They have been embraced especially by the *Baby Boomers* who have adopted a more optimistic and independent stance towards retirement and old age (Quine and Carter 2006). The Jungian view of old age suggests that old people actively seek new meaning and purpose in life through specific aims, spirituality, and a communion with nature that could be helped by introspective self-awareness and a life review (Barry et al. 1986). Therefore the old can still experience what Jung called the *developmental imperative* which spurs us to continue to grow and move on to other phases in our life. Throughout this process the elders can not only use but also spread the knowledge acquired in their life, by means of mentoring the new generations. Our own concept of *fulfilling ageing* is also consistent with this view, but we will see in Chap. 8 that fulfilling ageing often requires an extension of our sources of meaning in life and learning also towards the broader social sphere.

Unleashed creativity and learning from others can lead to a greater sense of wonder—as a core element of growing wisdom—and a continuous process of development of the self, what Jung called *individuation*. When this process of learning and growing through activities and life experiences becomes natural and pleasurable, we reach what Csikszentmihalyi (1990) calls a state of *flow*.

The proactive approach to old age that spread especially across industrialised countries in the 1990s is clearly exemplified in Betty Friedan's book *The Fountain of Age* (the title is a mocking reference to the Medieval *Fountain of Youth*). In her book, Friedan (1993) develops an optimistic view of the potential for further personal development in old age: "Why are we not looking at age as a new, evolving stage of human life—not merely as a decline from youth, but as an open-ended development in its own terms, which, in fact, may be *uniquely ours to define*?" (Friedan 1993, p. 75, italics original). The ability to continue to grow as an individual, in pursue of self-fulfilment, even throughout old age, whilst sharing what has been learned with younger generations, is known as *generativity* (Erikson 1997). Therefore, through a continuous state of *flow* we can keep *generating* our own *individual* self into old age, as we keep interacting with (and perhaps contributing to) our social and natural environment.

Friedan argues against the negativity and passivity associated with the traditional understanding of old age, what she calls "compassionate ageism", that regards older people as sick and dependent individuals. Old age can be an active, creative, and fulfilling stage of life, with older people being perfectly capable to continue enjoying new experiences and to maintain their ability to grow personally. *Connectedness*, involving social support and caring ties or networks, can be an additional help in delivering the beneficial effects of an active life on the psychology and general health of older people. On the other hand, she also rejects the "denial of age": pretending that time can be stopped and that ageing can be averted, arguing instead to embrace (and indeed create) the opportunities that old age can offer. From this it comes to no surprise that she rejects *Disengagement Theory* (Cumming 1975), which suggests that it is beneficial to both the individual and society for older people to retire, retreat, and disengage from active participation in social life. Sarcastically, Friedan asks: "[disengage] in their sixties—with a quarter of their possible life span still to live?" (Friedan 1993, p. 52). Friedan's criticism aside, disengagement is a legitimate path to old age, as we will see below, when it leads to gerotranscendence.

Elders who keep developing in a coherent manner are better able to experience a fulfilling old age: "individuals over sixty-five who do not decline seem to become more *integrated* in their various characteristics as they get older. They also become *increasingly individual* and different from others who decline; they continue to develop on the basis of their own accumulated and increasingly divergent experience. In this increasing integration, traumas, declines, or deficiencies in one sphere are compensated for by another" (Friedan 1993, p. 85, italics original).

Activity can be beneficial not only for the mind, but also for longevity. However, Friedan derides the literal understanding of old age as a "second childhood": you do not create a constructive experience of old age by reverting to a clueless and infantile state. Indeed, experience matters, and life past can be used to help in the continued growth of the individual throughout what remains of the future, rather than to block new developments.

Friedan concludes that: "The exercise of our unique human capacity for mindful control is key to vital age versus decline". But there are some types of social environments that foster the development of mindfulness, whereas others favour mindlessness (an automatic performance of behaviour out of habit and repetition). Mindlessness is what could eventually deliver the traits of senility and mental deterioration: "Since 'mindfulness' can only take place in situations which are new or require effort to master, older people who accept their 'deteriorated' status may indeed be reduced to mindlessness; given no opportunities for new and challenging experience, they sink into brain-numbing passivity" (Friedan 1993, p. 89). Through a process of mindlessness older people may gradually lose control over their own life which may lead them to become helpless. The depression and loneliness that are commonly experienced by older people, especially in large urban centres, can only compound the above problems.

Contrasting the proactive approach to old age espoused by Friedan, Jung, and others (Katz 2005) is Lars Tornstam's concept of *gerotranscendence* (Tornstam 1989), which is based on a view of old age that is characterised by increased detachment, as described by Cumming et al. (1960). Elaine Cumming refers to this process of increased detachment in the old as *disengagement*. Tornstam's *gerotranscendence* describes what is supposed to be a natural alteration of consciousness with old age that leads to increased disengagement with (or lack of interest in) some aspects of the external environment (see also Jönson and Magnusson 2001; Tornstam 2011).

Although a level of disengagement and re-prioritising does occur with old age, care should be taken to understand the exact situation of each individual, as disengagement is also expected as a reaction to processes of discrimination and marginalisation (a form of giving up), which may require some kind of intervention. If such concerns can be ruled out, then *gerotranscendence* can be useful in bringing greater peace of mind to specific individuals, who may be otherwise under undue stress if they were subject to the same expectations of youth or middle age. Some old people may achieve a greater level of happiness by simply relaxing (in the ancient Roman sense of *otium honestum*), play, create, or develop wisdom in whichever way they can (Tornstam 1992; Jönson and Magnusson 2001).

In reality, it seems clear to us that in the twenty-first century the Jungian tradition of *developmental imperative* and Tornstam's *gerotranscendence* are simply two different, yet equally valid paths to fulfilling old age that can be followed by different old people with different personalities: the active, the contemplative; and if any of them is not enough, we can always try a combination of both and add to that the enrichment of our self through social interactions and contact with nature, for greater completeness of our life experience.

2.11 Conclusions

Given the opportunity to live long enough, human beings have always aged biologically, but being old has not meant the same throughout the history of humankind. Societies have changed, systems of belief appeared and disappeared or underwent cultural evolution, political circumstances have varied over time, economic factors came and went, and all that affected how the old felt, how they were seen by others, and how they lived their life at any point in time. Throughout history there have been intermittent periods when old age was seen favourably, and others when youth was seen in a better light, but for as long as old people had the security that comes with material possessions, they could at least compensate for any changed view over their physical appearance, knowledge, or mental capacity by retaining a degree of power and control. The struggle for control has generated inter-generational tensions between old and young throughout history since antiquity, but inter-generational cooperation has also been present: through transmission of knowledge and inheritance, for instance.

Major historical changes tend to unfold over a period of several generations, but on a smaller time scale, older individuals have always lived their past childhood, adolescence, and maturity in specific historical periods, and their past experiences would influence how they behave and think as old people. Each generation acquires specific sets of "life scripts" that are affected by life circumstances (Dittmann-Kohli 1990). A 60 year old in the 1970s was in his/her 30s during World War II, a 60 year old in 2016 was 33 when the Berlin Wall fell. Those born during the Great Depression were young adults during World War II and had children during or after the war, and those children became the post-war *Baby Boomers* who enjoyed a period of great economic prosperity even if it was in the shadows of the Cold War. Such period transitioned from a social democratic environment, with policies that favoured the spreading of opportunities across socio-economic strata, to a neoliberal environment characterised by an increasing level of inequality, individualism, competition, and entrenched unemployment. In the words of Pat Jalland (2015, p. 104): "Old age is powerfully influenced by the historical period in which people lived their adult lives before sixty". Due to this historical effect, the understanding of old age can lead to different conclusions whether we study old age synchronically: when a cross sectional group of elders, a cohort, is observed at a specific point in time and they are compared with a younger cohort in the same population; or diachronically, if a group of people is followed up over time until they reach old age and they are compared with themselves when they were younger (White Riley 1994).

Neugarten (1974, p. 197) asked in the 1970s what kind of elders would the young American *Baby Boomers*, who went through "the civil rights movement, the Women's Liberation movement and the antipollution movement" make. We have seen above how the behaviour of *Baby Boomers* as old people is different from that of old people of the previous generation. Thus each older generation is also the product of their times (Mackay 1997).

Hareven (1976, p. 16, quoting Hill 1970) has noted how "in periods of rapid social change, each cohort 'encounters at marriage a unique set of historical constraints and incentives which influence the timing of its crucial life decisions, making for marked generational dissimilarities in life-cycle career patterns'". Generations that have been used to a safer and more predictable job environment would likely feel under far more psychological pressure in a situation of early retirement than a generation that has experienced greater job uncertainty over a long period. But the former are also likely to be financially better off than the latter. This is the current reality that will influence the lives of old people in the near future.

As a result of the historicity of life, understanding old age at any given time can be better achieved if a *life course* approach is adopted (Cattell and Albert 2009). In Chap. 1 we defined the life course as *the developmental pathways that individuals go through in their life as they are influenced by individual, social, and other environmental factors*. If we reach advanced old ages, our life course encompasses a long period that has exposed and it continues to expose us to unexpected historical events and challenges for as long as we live. To quote Jay Sokolovsky (2009c, p. 277) who was referring to a group of old East European Jewish immigrants living in California in the 1970s: "the scripts they anticipated for late life and the spaces they expected to inhabit once they got there, had largely vanished. It was up to them to create new possibilities for their old age". With personal resources and commitment and with social support, elders may have a chance to adapt to unexpected historical circumstances and find their best way to live a fulfilling old age.

References

Achenbaum, W. A. (1974). The obsolescence of old age in America, 1865–1914. *Journal of Social History, 8*, 48–62.

Achenbaum, W. A. (1978). *Old age in the new land: The American experience since 1790*. Baltimore: Johns Hopkins University Press.

Achenbaum, W. A., & Stearns, P. N. (1978). Old age and modernization. *The Gerontologist, 18*, 307–312.

Agronin, M. E. (2014). From Cicero to Cohen: Developmental theories of aging, from antiquity to the present. *The Gerontologist, 54*, 30–39.

Alden, A. (1835). The life and age of woman: Stages of woman's life from the infancy to the brink of the grave. *Library of Congress*. Downloaded on December 7, 2018, from http://www.loc.gov/pictures/item/2006686268/

Aldred, C. (1972). *Akhenaten: Pharaoh of Egypt*. London: Abacus, Sphere Books Ltd..

Amram, D. W. (1900). The Zekenim, or Council of Elders. *Journal of Biblical Literature, 19*, 34–52.
Barry, J. B., Bozarth, J., Myers, J. E., & Heyn, J. (1986). Jungian analytical psychology and old age. *Journal of Applied Gerontology, 4*, 105–110.
Birren, J. E. (1961a). A brief history of the psychology of aging—Part I. *The Gerontologist, 1*, 69–77.
Birren, J. E. (1961b). A brief history of the psychology of aging—Part II. *The Gerontologist, 1*, 127–134.
Botelho, L. A. (2005). The 17th century. In P. Thane (Ed.), *The long history of old age* (pp. 113–173). London: Thames & Hudson Ltd.
Bunzel, J. H. (1972). Note on the history of a concept: Gerontophobia. *The Gerontologist, 12*, 116. (cont. on p. 203).
Burrow, J. A. (1986). *The ages of man: A study in medieval writing and thought.* Oxford: Clarendon Press.
Butler, M. (2015). *Advanced Australia: The politics of ageing*. Carlton: Melbourne University Press.
Carrieri, M. P., & Serraino, D. (2005). Longevity of popes and artists between the 13th and the 19th century. *International Journal of Epidemiology, 34*, 1435–1436.
Cattell, M. G. (2009). Global perspectives on widowhood and aging. In J. Sokolovsky (Ed.), *The cultural context of aging: Worldwide perspectives* (pp. 155–172). Westport, CT: Praeger.
Cattell, M. G., & Albert, S. M. (2009). Elders, ancients, ancestors and the modern life course. In J. Sokolovsky (Ed.), *The cultural context of aging: Worldwide perspectives* (pp. 115–133). Westport, CT: Praeger.
Charles, D. C. (1977). Literary old age: A browse through history. *Educational Gerontology, 2*, 237–253.
Chojnacka, M., & Wiesner-Hanks, M. E. (2002). *Ages of woman, ages of man. Sources in European social history, 1400–1750.* London: Longman.
Claeys, G. (2000). The "survival of the fittest" and the origins of Social Darwinism. *Journal of the History of Ideas, 61*, 223–240.
Coffman, G. E. (1934). Old age from Horace to Chaucer: Some literary affinities and adventures on an idea. *Speculum, 9*, 249–277.
Cohen, E. S., & Kruschwitz, A. L. (1990). Old age in America represented in nineteenth and twentieth century popular sheet music. *The Gerontologist, 30*, 345–354.
Cohen, G. D. (1988). Do health science concepts influence care?: The case for a new landscape of aging. *The American Journal of Geriatric Psychiatry, 6*, 273–276.
Cohen, G. D. (2005). *The mature mind: The positive power of the aging brain.* New York: Basic Books.
Cokayne, K. (2003). *Experiencing old age in ancient Rome*. London: Routledge.
Cole, T. R., & Edwards, C. (2005). The 19th century. In P. Thane (Ed.), *The long history of old age* (pp. 211–261). London: Thames & Hudson Ltd.
Cordella, M., & Poiani, A. (2014). *Behavioural oncology: Psychological , communicative, and social dimensions*. New York: Springer.
Cossar, R. (2012). Portraits of aging men in late Medieval Italy. *The Gerontologist, 52*, 553–560.
Covey, H. C. (1989). Old age portrayed by the ages-of-life models from the Middle Ages to the 16th century. *The Gerontologist, 29*, 692–698.
Covey, H. C. (1992). The definitions of the beginning of old age in history. *The International Journal of Aging and Human Development, 34*, 325–337.
Csikszentmihályi, M. (1990). *Flow: The psychology of optimal experience.* New York: Harper & Row.
Cumming, E. (1975). Engagement with an Old Theory. *International Journal of Aging and Human Development, 6*, 187–191.
Cumming, E., Dean, L. R., Newell, D. S., & McCaffrey, I. (1960). Disengagement - A tentative theory of aging. *Sociometry, 1*, 23–35.

Dahlin, M. (1980). Perspectives on the family life of the elderly in 1900. *The Gerontologist, 20*, 99–107.

de Beauvoir, S. (1972). *The coming of age*. New York: Putnam.

Demontis, L. (2015). Comparative analysis of medieval and modern scientific research on ageing reveals many conceptual similarities. *Journal of Gerontology and Geriatric Research, 4*, 216. https://doi.org/10.4172/2167-7182.1000216

Demos, J. (1978). Old age in early New England. *American Journal of Sociology, 84*, S248–S287.

Dittmann-Kohli, F. (1990). The construction of meaning in old age: Possibilities and constraints. *Ageing and Society, 10*, 279–294.

Dove, M. (1986). *The perfect age of man's life*. Cambridge: Cambridge University Press.

Duménil, G., & Lévy, D. (2013). *The crisis of neoliberalism*. Cambridge, MA: Harvard University Press.

Ehrlich, P. R., & Ehrlich, A. H. (2009). The population bomb revisited. *The Electronic Journal of Sustainable Development, 1*, 5–13.

Erikson, E. H. (1997). *The life cycle completed*. New York: Norton.

Finley, M. I. (1981). The elderly in classical antiquity. *Greece and Rome, 28*, 156–171.

Fleming, K. C., Evans, J. M., & Chutka, D. S. (2003). A cultural and economic history of old age in America. *Mayo Clinic Proceedings, 78*, 914–921.

Freeman, J. T. (1965). Medical perspectives in aging (12th-19th century). *The Gerontologist, 5*, 1–24.

Friedan, B. (1993). *The fountain of age*. New York: Simon & Schuster.

Gilbert, B. B. (1965). The decay of nineteenth-century provident institutions and the coming of old age pensions in Great Britain. *The Economic History Review, 17*, 551–563.

Gilbert, C. (1967). When did a man in the Renaissance grow old? *Studies in the Renaissance, 14*, 7–32.

Gilligan, C. (1982). *In a different voice: Psychological theory and women's development*. Cambridge, MA: Harvard University Press.

Goffman, E. (1963). *Stigma*. Englewood Cliffs, NJ: Prentice-Hall.

Goldberg, E. (2007). *The wisdom paradox: How your mind can grow stronger as your brain grows older*. London: Pocket Books.

Gutmann, D. (1987). *Reclaimed powers: Men and women in later life*. Illinois: Northern University Press.

Hareven, T. K. (1976). The last stage: Historical adulthood and old age. *Daedalus, 105*, 13–27.

Hareven, T. K. (1982). The life course and aging in historical perspective. In T. K. Hareven & K. J. Adams (Eds.), *Aging and life course transitions: An interdisciplinary perspective* (pp. 1–26). London: Tavistock Publications.

Hill, R. (1970). *Family development in three generations*. Cambridge, MA: Schenkman Pub..

Hirschhorn, L. (1977). Social policy and the life cycle: A developmental perspective. *Social Service Review, 51*, 434–450.

Hirshbein, L. D. (2001). Popular views of old age in America, 1900-1950. *Journal of the American Geriatrics Society, 49*, 1555–1560.

Holloway, D., Green, L., & Holloway, D. (2011). The intratourist gaze: Grey nomads and 'other tourists'. *Tourist Studies, 11*, 235–252.

Holmes, B. (2012). *Gender: Antiquity and its legacy*. New York: Oxford University Press.

Holstein, M., & Pavle, K. (2013). The "crisis" in retirement security: Social security is the answer, not the problem. *John Marshall Law Reveview, 46*, 719–747.

Hubble, N., & Tew, P. (2013). *Ageing, narrative and identity: New qualitative social research*. Hampshire: Palgrave Macmillan.

Hudson, R. B., & Gonye, J. G. (2012). Baby boomers and the shifting political construction of old age. *The Gerontologist, 52*, 272–282.

Jalland, P. (2015). *Old age in Australia: A history*. Carlton: Melbourne University Press.

Jönson, H., & Magnusson, J. A. (2001). A new age of old age? Gerotranscendence and the re-enchantment of aging. *Journal of Aging Studies, 15*, 317–331.

Kaida, L., Moyser, M., & Park, S. Y. (2009). Cultural preferences and economic constraints: The living arrangements of elderly Canadians. *Canadian Journal on Aging/La Revue canadienne du vieillissement, 28*, 303–313.
Katz, R. (2009a). Intergenerational family relations and subjective well-being in old age: A cross-national study. *European Journal of Ageing, 6*, 79–90.
Katz, S. (2005). *Cultural aging: Life course, life style and senior worlds*. Peterborough, ON: Broadview Press.
Katz, S. (2009b). Spaces of age, snowbirds and the gerontology of mobility: The elderscapes of Charlotte County, Florida. In J. Sokolovsky (Ed.), *The cultural context of aging: Worldwide perspectives* (pp. 463–486). Westport, CT: Praeger.
Kebric, R. B. (1983). Aging in Pliny's letters: A view from the second century A.D. *The Gerontologist, 23*, 538–545.
Kebric, R. B. (1988). Old age, the ancient military, and Alexander's army: Positive examples for a graying America. *The Gerontologist, 28*, 298–302.
Keith, J. (2009). When old is new: Cultural spaces and symbolic meaning in late life. In J. Sokolovsky (Ed.), *The cultural context of aging: Worldwide perspectives* (pp. 145–154). Westport, CT: Praeger.
Kirk, H. (1992). Geriatric medicine and the categorisation of old age - The historical linkage. *Ageing and Society, 12*, 483–497.
Klinenberg, E. (2016). Social isolation, loneliness, and living alone: Identifying the risks for public health. *American Journal of Public Health, 106*, 786–787.
Mackay, H. (1997). *Generations: Baby boomers, their parents and their children*. Sydney, NSW: Macmillan.
Minois, G. (1989). *History of old age*. Cambridge: Polity Press.
Moody, H. (2009a). From successful aging to conscious aging. In J. Sokolovsky (Ed.), *The cultural context of aging: Worldwide perspectives* (pp. 67–76). Westport, CT: Praeger.
Moody, H. R. (2009b). Eco-elders: Legacy and environmental advocacy. *Generations, 33*, 70–74.
Neugarten, B. L. (1974). Age groups in American society and the rise of the young-old. *Annals of the American Academy of Political and Social Science, 415*, 187–198.
Neugarten, B. L. (1979). Time, age, and the life cycle. *The American Journal of Psychiatry, 136*, 887–894.
Onyx, J., & Leonard, R. (2005). Australian Grey Nomads and American Snowbirds: Similarities and differences. *The Journal of Tourism Studies, 16*, 61–68.
Pantell, M., Rehkopf, D., Jutte, D., Syme, S. L., Balmes, J., & Adler, N. (2013). Social isolation: A predictor of mortality comparable to traditional clinical risk factors. *American Journal of Public Health, 103*, 2056–2062.
Parkin, T. (2005). The ancient Greek and Roman worlds. In P. Thane (Ed.), *The long history of old age* (pp. 31–69). London: Thames & Hudson Ltd.
Parkin, T. G. (2003). *Old age in the Roman world: A cultural and social history*. Baltimore and London: The Johns Hopkins University Press.
Patrides, C. A. (1958). The numerological approach to cosmic order during the English Renaissance. *Isis, 49*, 391–397.
Payne, T. (2015). *The ancient art of growing old*. London: Vintage Books.
Petersen, J. H. (1990). The Danish 1891 Act on Old Age Relief: A response to agrarian demand and pressure. *Journal of Social Policy, 19*, 69–91.
Poiani, A. (2010). *Animal homosexuality: A biosocial perspective*. Cambridge: Cambridge University Press.
Pratt, A. (1978). Aunt Jennifer's tigers: Notes toward a preliminary history of women's archetypes. *Feminist Studies, 4*, 163–194.
Quine, S., & Carter, S. (2006). Australian baby boomers' expectations and plans for their old age. *Australasian Journal on Ageing, 25*, 3–8.
Randall, O. A. (1965). Some historical developments of social welfare aspects of aging. *The Gerontologist, 5*, 40–49.

Rawls, J. (1971). *A theory of justice*. Cambridge: Harvard University Press.

Razi, Z. (1980). *Life, marriage, and death in a medieval parish : Economy, society, and demography in Halesowen, 1270–1400*. Cambridge: Cambridge University Press.

Reveillé-Parise, J. H. (1853). *Traité de la vieillesse hygienique, médical et philosophique*. Paris: J. B. Baillière.

Rivers, C. (2011). The greedy geezers libel. *The Huffington Post*. Downloaded on July 13, 2016, from http://www.huffingtonpost.com/caryl-rivers/the-greedy-geezers-libel_b_867034.html

Roebuck, J., & Slaughter, J. (1979). Ladies and pensioners: Stereotypes and public policy affecting old women in England, 1880-1940. *Journal of Social History, 13*, 105–114.

Sargent, L. D., Lee, M. D., Martin, B., & Zikic, J. (2012). Reinventing retirement: New pathways, new arrangements, new meanings. *Human Relations, 66*, 3–21.

Scheid, J. (2007). Sacrifices for gods and ancestors. In J. Rüpke (Ed.), *A companion to Roman religion* (pp. 263–271). Malden, MA: Blackwell Publishing Ltd..

Schneider, A., & Ingram, H. (1993). The social construction of target populations. *American Political Science Review, 87*, 334–347.

Settersten, R. A., Jr. (1997). The measurement of age, age structuring, and the life course. *Annual Reviews of Sociology, 23*, 233–261.

Shahar, S. (2005). The middle ages and renaissance. In P. Thane (Ed.), *The long history of old age* (pp. 71–111). London: Thames & Hudson Ltd.

Sokolovsky, J. (2009a). Introduction: Human maturity and global aging in cultural context. In J. Sokolovsky (Ed.), *The cultural context of aging: Worldwide perspectives* (pp. xv–xxxv). Westport, CT: Praeger.

Sokolovsky, J. (2009b). A global vision of aging, culture and context. In J. Sokolovsky (Ed.), *The cultural context of aging: Worldwide perspectives* (pp. 1–12). Westport, CT: Praeger.

Sokolovsky, J. (2009c). The ethnic dimension in aging: Culture, context and creativity. In J. Sokolovsky (Ed.), *The cultural context of aging: Worldwide perspectives* (pp. 277–287). Westport, CT: Praeger.

Sokolovsky, J. (2009d). Ethnic elders and the limits of family support in a globalizing world. In J. Sokolovsky (Ed.), *The cultural context of aging: Worldwide perspectives* (pp. 289–301). Westport, CT: Praeger.

Sokolovsky, J. (2009e). Families, communities and elderscapes: Transforming cultural spaces for aging. In J. Sokolovsky (Ed.), *The cultural context of aging: Worldwide perspectives* (pp. 371–382). Westport, CT: Praeger.

Standing, G. (2016). *The precariat: The new dangerous class*. London: Bloomsbury.

Stearns, P. N. (1980). Old women: Some historical observations. *Journal of Family History, 5*, 44–57.

Sweetser, D. A. (1964). Urbanization and the patrilineal transmission of farms in Finland. *Acta Sociologica, 7*, 215–228.

Thane, P. (2003). Social histories of old age and aging. *Journal of Social History, 37*, 93–111.

Thane, P. (2005a). The age of old age. In P. Thane (Ed.), *The long history of old age* (pp. 9–29). London: Thames & Hudson Ltd.

Thane, P. (2005b). The 20th century. In P. Thane (Ed.), *The long history of old age* (pp. 263–300). London: Thames & Hudson Ltd.

Tornstam, L. (1989). Gero-transcendence; a reformulation of the disengagement theory. *Aging: Clinical and Experimental Research, 1*, 55–63.

Tornstam, L. (1992). The quo vadis of gerontology; on the gerontological research paradigm. *The Gerontologist, 32*, 318–326.

Tornstam, L. (2011). Maturing into gerotranscendence. *The Journal of Transpersonal Psychology, 43*, 166–180.

Troyansky, D. G. (2005). The 18th century. In P. Thane (Ed.), *The long history of old age* (pp. 175–209). London: Thames & Hudson Ltd.

United Nations, Department of Economic and Social Affairs, Population Division. (2013). *World population prospects: The 2012 revision, key findings and advance tables* (Working Paper No. ESA/P/WP.227).

Vertinsky, P. (1991). Old age, gender and physical activity: The biomedicalization of aging. *Journal of Sport History, 18*, 64–80.

Vincent, J. A. (2006). Ageing contested: Anti-ageing science and the cultural construction of old age. *Sociology, 40*, 681–698.

Walker, A. (1990). The economic 'burden' of ageing and the prospect of intergenerational conflict. *Ageing and Society, 10*, 377–396.

Warren, C. A. B. (2012). Assisted living in 1489. *The Gerontologist, 52*, 698–702.

White Riley, M. W. (1994). The 1993 Kent Lecture. Aging and society: Past, present, and future. *The Gerontologist, 34*, 436–446.

Winter, H. (1953). A circular map in a Ptolemaic MS. *Imago Mundi: The International Journal of History and Cartography, 10*, 15–22.

World Bank. (2000). *Retirement: Can pension reform reverse the trend to earlier retirement?* Downloaded on 17 July, 2016, from https://openknowledge.worldbank.org/handle/10986/11444

Zickuhr, K. (2010). *Generations 2010*. Washington DC: Pew Internet and American Life Project, Pew Research Center. Retrieved March 23, 2018, from http://pewinternet.org/Reports/2010/Generations-2010.aspx

Chapter 3
Old People Across Cultures

Abstract *Old people across cultures* focuses on the cross-cultural similarities and differences in the meaning of being old, the behaviour of older people, and the cultural traditions that derive from the process of ageing and that also shape it in the view of the members of an ethnic group.

Specific groups of people can be identified by their culture, and cultures are distinguished not only for their commonalities, that are typical of our species, such as the capacity of individuals to develop a language, but also for their diversity. Languages, systems of beliefs, traditions can be very different from one culture to another and they can change over time.

Culture can be defined as a dynamic "set of shared symbols, beliefs, and customs that shapes individual and/or group behaviour" (Dilworth-Anderson and Gibson 2002, p. S56). Although individuals process their experiences according to their cultural background, they also do so in an idiosyncratic manner, which is dynamic and may be subject to additional external influences. Thus, each one of us creates a personal "cultural frame", which interacts with those of other individuals through our social relationships (Goodenough 1981).

For any specific human group, the cultural system of beliefs and traditions that characterises it and gives it its cultural distinctiveness constitutes its *ethnicity*. In the words of the anthropologist Jay Sokolovsky (2009f, pp. 277–278): "Ethnicity is the manifestation of a cultural tradition in a heterogeneous societal framework … Ethnicity is therefore typically a creative act, meshing ancestral 'native' patterns with restraints imposed by the broader society and the demands of the local environment".

The comparative study of old age across ethnicities is known as *ethnogerontology* or *ethnogeriatrics*, whereas the anthropology of the ageing process is known as *geroanthropology* (Cohen 1994a, b). One broad conclusion of the studies carried out in those specialised fields is that across cultures there are three traditional ways

M. Cordella, A. Poiani, *Fulfilling Ageing*, International Perspectives on Aging 30,
https://doi.org/10.1007/978-3-030-60071-6_3

of identifying a person as being old: through experiencing a change in socio-economic role (e.g. retirement from work), a change in chronological age, and/or changes in physical characteristics (Sokolovsky 2009b). As we have seen in Chap. 2, the identification of old age with qualification for the age pension is a relatively recent phenomenon.

Cultural perceptions of ageing combine both positive and negative views of old people which may be objective (e.g. increasing physical frailty with old age) or mythical (e.g. association of old people with witchcraft). Combinations of such positive, negative, objective, and mythical views establish a multidimensional space that defines the perception of old age as it is recorded across cultures (Löckenhoff et al. 2009a).

Factors such as *modernisation*—with its negative correlation between economic development and the status of older citizens (Cowgill 1986)—religion and cultural traditions, all combine to explain how the elders are currently seen and treated by society and how, in turn, they see society and respond to the challenges they are confronted with.

Modernisation and globalisation are modifying the cultural landscape across the world, by increasing the flow of migrants, goods, services, and information, and by also accelerating cultural change within countries. This has led to an increased level of cultural diversity within countries, and a relative decrease in cultural diversity between countries, that have affected old people both in the positive (e.g. making it easier for the elders to move from one country to another one) and in the negative (e.g. increasing some inter-generational conflicts).

Although increased ethnic diversity within local populations of elders comes with some constraints to adaptation, acceptance, and ability to navigate the various complexities of a heterogeneous society, it can be also a source of reassurance, social support, and identity that can help older people carve a specific niche within a new social environment, whether such environment is a country of immigration or the native country that is undergoing major social and cultural changes. That is, ethnicity can be a source of either *deprivation* of opportunities for the elders (e.g. wherever ethnic minorities are discriminated against) or *compensation* for what the current environment does not provide in terms of elders' care, according to circumstances (Sokolovsky 2009g).

A traditional ethnic divide in the study of old age has been that between Western and Eastern/Asian cultures, where Asian cultures are expected to show a higher degree of *filial piety* and worship of the ancestors that lead to holding the elders in higher esteem and to (traditionally) treating them with greater respect than it is the case in modern Western societies; the latter, in turn, are expected to be more individualistic and youth-focussed (Löckenhoff et al. 2009a). For instance, Tannahill (1988, pp. 130–131) states that:

> (a)mong the spirits to whom the Chinese gentleman might find himself making sacrifice were 'the ancestors'—and when a man reached the age at which his conversion into an ancestor was becoming imminent, he was treated with the greatest consideration by both his family and the state.

Such considered treatment of old people in China was even reflected in the high-quality food served to the elderly in the family (Tannahill 1988). Filial piety *Xiao*

(孝) (also written *Hsiao*) was formally enshrined in Chinese society through the rules of behaviour dictated by Confucius (fifth century b.c.e.) in his *Analects*:

> Mêng I Tzu asked about the treatment of parents. The Master said, Never disobey! ... Behave in such a way that your father and mother have no anxiety about you, except concerning your health ... 'Filial sons' nowadays are people who see to it that their parents get enough to eat. But even dogs and horses are cared for to that extent. If there is no feeling of respect, wherein lies the difference? ... Filial piety does not consist merely in young people undertaking the hard work, when anything has to be done, or serving their elders first with wine and food. It is something much more than that (Confucius 1989, pp. 88–89).

The show of respect towards the elders is central to Confucian philosophy in all aspects of behaviour. When a disciple was waiting for the master in a sprawling position that was considered disrespectful "The Master said, Those who when young show no respect to their elders achieve nothing worth mentioning when they grow up. And merely to live on, getting older and older, is to be a useless pest. And he struck him across the shins with his stick" (Confucius 1989, p. 192).

In Confucian philosophy filial piety derives at least in part from the guiding principles of *Ren* (仁, "humanism"), *Yi* (仪, "righteousness"), *Li* (礼, "propriety") and *Chong* (聪, "wisdom") (Yum 1988).

Given the formal importance of the concept of *filial piety* in Asian cultures (and in other cultures as well, Laidlaw et al. 2010), we should also note that, following Yeh (2003), filial piety can be subdivided into *reciprocal filial piety* and *authoritarian filial piety*, the former being "positive" and reflecting inter-generational reciprocation: parents provide love and material support to their young children who return the favour when they are adult and their parents are old; whereas authoritarian filial piety reflects a degree of social control against which the young generations may rebel. Sung (1995) described six fundamental dimensions of filial piety following an analysis of Korean culture: harmonising the family, showing respect, making a repayment, fulfilling a responsibility, showing affection, and making a sacrifice.

In the West, filial piety has been traditionally more of the authoritarian kind, as represented by the concept of *patria potestas*, for instance. *Patria potestas* emphasises specific relationships between parent and child, whereas the Chinese concept of *Xiao* points to the importance of roles. In the words of Gary Hamilton (1990, p. 92): "*Patria potestas* identifies the father's authoritative power, whereas *hsiao* identifies the son's filial duty". *Xiao* is closer to the Western concept of *pietas*, which is rooted in a "complex of reverent regard, moral norms, ritual observance and material duty in the relationship between parent and child, more particularly of son to father, both during the lifetime and after the death of the parent" (Fortes 1961, p. 178).

Filial piety has tended to wane faster in Western than Eastern cultures in recent decades, but we will see in this chapter how contemporary research is showing that the current family relationships in various Asian countries are becoming more complex and less attached to ancient traditions.

In this chapter we review some important cross-cultural aspects of old age, with emphasis on the factors that we regard relevant to fulfilling ageing. Culture can affect fulfilling ageing both at the level of the individual conceptions of life and its meaning, and in terms of the relationships between the individual and the rest of society. Social support can be a significant aid in experiencing life fulfilment, and social support can be modulated, at the most basic level, by inter-generational relationships within the family. The more positive social relationships are, the more and better support the elder will receive, and the more fulfilling the ageing experience can be. We then review some of the perceptions about old people's health across cultures which may also contribute to strengthening some old-age stereotypes. When stereotypes are negative, they may lead to discrimination against the elders, which is sometimes expressed as abuse. Although abuse against the elders may come from within the family, the family and others are mainly a source of much needed care especially to the frail and dependent elders. Such care may also come at a cost to the carer in the form of burnout. On the other hand, if the elder is in good physical and mental conditions, old age is not necessarily a time of increased dependence but rather it could be a creative and active period of life, with opportunities opening up to embrace new experiences. But all comes to an end eventually, and so we conclude this chapter with a cross-cultural analysis of the end-of-life process in old age.

3.1 The Cultural Context of Old Age

In one of the first reviews of old age across cultures Leo Simmons concluded that in so-called primitive societies elders tend to remain integrated into the group, actively contributing to group life. They are also regarded as repositories and transmitters of traditional knowledge, as well as mediators between the living and the spiritual world (Simmons 1945, 1946).

Most cultures see old age not in isolation but as part of a more complex *Life Course* (Cattell and Albert 2009). We have seen in Chap. 2 how the life course has been organised throughout history into various *Ages of Life* systems, such tendency to subdivide the life course has been a common feature of most cultures to these days. Among the *Samburu* of northern Kenya, for instance, stages of life are organised into intervals of 12–14 years: from *moran* (unmarried warrior), to various categories of elder, for a total of six stages. The last stage: *father-of-moran* starts at 60 years old. Each passage from one stage to the next is demarcated by specific ceremonies, a characteristic common to many cultures. Interestingly, women's life stages are far more simplified: girl (unmarried) or woman (married, widow), compared with those of men, a characteristic that has also pervaded Western culture during some historical periods, as we mentioned in Chap. 2 (see Fig. 2.2). It is unclear, however, whether in this case gender differences across cultures are factual or simply derive from a difficulty of the usually male ethnographer/anthropologist to understand the ages of life of women in traditional societies.

In most cultures the life course encompasses not only the entire life of the individual, but it usually extends beyond death, into the *afterlife* (i.e. life after death, e.g. Legare et al. 2012), as in a spiritual life as ancestor. In some cultures, ancestors are believed to even "help" the living through "communicating" with them, in dreams for instance. On the other hand, in Buddhist cultures the extended life course is expressed as a continuous and cyclical return, rather than a linear progression projecting into eternity, as in the Judeo-Christian tradition.

Some cultures see the life course, and the process of ageing, as a constant loss of some property (e.g. heat in ancient western cultures, *raka* in the *Bororo* of Brazil), whereas yet other cultures see the life course as a relentless acquisition of knowledge, "spiritual power" or both (e.g. some Aboriginal cultures of Australia).

The tradition that associates ageing with increasing knowledge and experience is ethnically widespread. Although in such cultures knowledge can be transferred from the old to others with no loss to the donor, among the *Hua* of Papua New Guinea the transference of a vital substance from women to men known as *nu* occurs to the detriment of the donor who loses *nu* whilst the recipient acquires it until an equilibrium of *nu* content is achieved between men and women in old age, thus making them almost indistinguishable. Eating the body of a deceased person also helps in transferring *nu* (Cattell and Albert 2009).

In Australian Aboriginal societies old age is assumed to be a stage of lower stress and loneliness, bringing greater accumulated experience to both men and women that allows them to act as advisers to other, less experienced persons (Berndt and Berndt 1992). Old men become "elders" at the age of 55–60, assuming the role of advisers to the young, repositories of oral traditions, and of knowledge about food and medicinal herbs, experts in spiritual matters and knowledge of the land. They also provide both guidance and support to the overall community, and directly participate in decision-making, various cultural activities and ceremonies; they share their stories, dispense their advice and pass key knowledge on to others (Clapham and Duncan 2017). Given the central importance of connection with the land in Aboriginal cultures, the elders also become linchpins for the stability of such connection with the land for the rest of the community, thus providing both cultural continuity and a sense of identity. But the elders also rely on their kin and in-laws for the provision of food, thus having children is of vital importance for survival through old age. Although Aboriginal societies are nominally gerontocratic, elders have a voice on various matters not only on the grounds of their age but especially because they have something of interest to offer to the community from their accumulated knowledge (see the specific case of the *Tiwi* people from the Tiwi Islands of Northern Australia described by Sokolovsky 2009a; see also Clapham and Duncan 2017 and references therein). A gender role differentiation is also evident among older Aborigines, with older men assuming political and ceremonial roles, whereas older women actively carve their personal sphere of authority and influence (Clapham and Duncan 2017).

Similar roles of service to the group are also found among the elders in the *Ju/'hoansi* of Botswana (Sokolovsky 2009b), where old people transmit their knowledge to the younger generations, care for the children, participate in decision-making, and officiate rituals as masters of spiritual forces such as the so-called *num*. They also perform roles of "knowledge repository about kin ties and natural resource management; transcendental curing and performance; entertainment through clowning and dancing; and being an emotional focus for community integration" (Sokolovsky 2009b, p. 6). Interestingly, the value of caring for the elders is reinforced by the elders themselves through constant complaining, with caregiving duties for the elders being equally spread across both sexes within the kin group.

Among the elderly Druse of Lebanon, Syria and Galilean Israel, there is a cultural expectation of greater "passivity" in old age mixed with great social prestige and the establishment of specific social roles for old people within a socially constructive attitude of *passive mastery*:

> The old man's emerging humility and submissiveness fits him to live on the dangerous interface between the gods and the mundane community, and, through his prayers, to bring life-sustaining forces into the world so as to maintain and increase children, flocks, and crops. Through him, the world becomes vivid with the aura of the gods. The passivity that could lead to vulnerability, depression and psychosomatic illness in other settings becomes in these settings—of which the traditional Druse society is one—the very core and pivot of the older man's social prestige and personal identity (Gutmann 2002, p. 130).

Sokolovsky (2009e) describes the situation of the elders in a small village near Mexico City, *Amatango*, that is inhabited by people of Aztec ethnicity, and that has undergone recent development and modernisation. In *Amatango* people are regarded to be old between the ages of 60 and 79 and, from then on, they are not expected to contribute heavy communal work, although they continue helping the family with less physically demanding tasks such as childcaring. Women carry on with their domestic work for the family until the very end, and able elders contribute to the economic activity of the town by producing household items (such as wooden crates) to be sold in the market in Mexico City. In general, older people are shown respect, expressed through a hand-kissing ritual, and receive assistance from the rest of the family.

In *Amatango* women enjoy greater supportive social networks than men, whereas in other societies it is the elders of both sexes who take the initiative to support each other. Among the Australian Aborigines, self-supportive social networks are actively fostered by the elders, and old men and women who do not have a partner (or other sources of social support) live together in monosexual groups: "The 'old women's camp' may be a centre for visiting women who have left their husbands behind, or even for unmarried girls. It may be no more than a temporary association between people who have come together as a matter of convenience; but while it lasts it is a 'household' unit, one of the many which make up the total camp" (Berndt and Berndt 1992, p. 209).

The views on old age and how older people see themselves and try to live their life have been significantly affected by recent globalisation trends. Globalisation increases the inter-connectivity of different cultures via both communication and immigration, simultaneously creating a *centripetal homogenising force* that leads to modification of some local cultural traditions, and a *centrifugal reactive force* towards the maintenance of such traditions, with the often-seen consequence of the emergence of new cultures that have roots both into the old and the new. Immigration and the melting pot of ethnicities that results from it creates what Sokolovsky calls *glocalisation*: "transforming a cultural pattern that travelled over global boundaries into a new variant of its original form" (Sokolovsky 2009f). In a situation of immigration, ethnicity can be either a handicap to full integration into the mainstream of society or an asset to sustain oneself despite the difficulties of adaptation and survival in a foreign environment. In general, successful immigrants are able to combine both aspects of immigration (they *glocalise*): keep aspects of their roots to sustain themselves, whilst changing and adapting to the extent that is needed, in response to the requirements for survival in the new environment. The capacity for cultural adaptation, however, tends to be relatively less developed in older people who may have not been exposed to such level of cultural change and variability in the past. This can be felt especially strongly when new developments interfere with established cultural relationships between old and young people (respect, filial piety, duties). In such circumstances it is often the case that old people find themselves at the losing end of cultural change (Sokolovsky 2009d) and they either look for new ways to adapt (which often requires some external help) or suffer the consequences of frustration, depression, and isolation.

Presumably reflecting the combined influence of personal experience and cultural factors that are affected by globalisation, a study of the opinion about old age in 3435 college students from 26 cultures found a general consensus in that old age is perceived negatively in society regarding aspects of physical attractiveness, capacity to perform daily tasks and to acquire new knowledge; but from a more positive perspective, there is a perceived increase in wisdom in old age which should command a degree of respect and some authority, at least in the context of the family (Löckenhoff et al. 2009a). Variations around those themes also reflected the national diversity in cultures and stereotypes, prevailing levels of education, local values but also demographic aspects, with people from countries where old people are more frequent in the population expressing more negative views of old age. Interestingly, although most respondents in Löckenhoff et al.'s study considered that old age was mainly viewed negatively in their culture, with the most negative views found in Serbia, Argentina, the Czech Republic, and the UK; there were some exceptions: Mainland China, India, Malaysia, Russia, and New Zealand, where participants reported neutral or slightly positive views of old age. In a comparison between Western and Eastern countries, participants from the latter displayed more positive views of ageing, but their perceptions of age-related changes in wisdom were less favourable than those of Western participants. This is likely a result of a recent shift in traditional views of the old in Eastern countries.

Yun and Lachman (2006) also investigated cultural variations in the beliefs about ageing, with special focus on anxiety about getting older. They administered an *Anxiety about Ageing* scale to 153 American and 167 South Korean men and women: young (18–39 years old), middle age (40–59), and older (60–91). In general, Koreans were more anxious about ageing, and they also displayed greater fear of old people and concerns about physical appearance. However, whereas in the case of Americans it was the older respondents who were more fearful of old people, in the South Korean sample it was the younger respondents who were more fearful (see Fig. 3.1). That is, Koreans may be more affected by their thoughts about the future, hence the young fearing the old, whereas Americans seem to be more concerned about the present, hence the old being more fearful of old people.

Greater anxiety about old age was also shown by women as compared to men in the American sample (Yun and Lachman 2006). Fear of being devalued may be at the root of such anxiety in women. This is a broader phenomenon that is similar

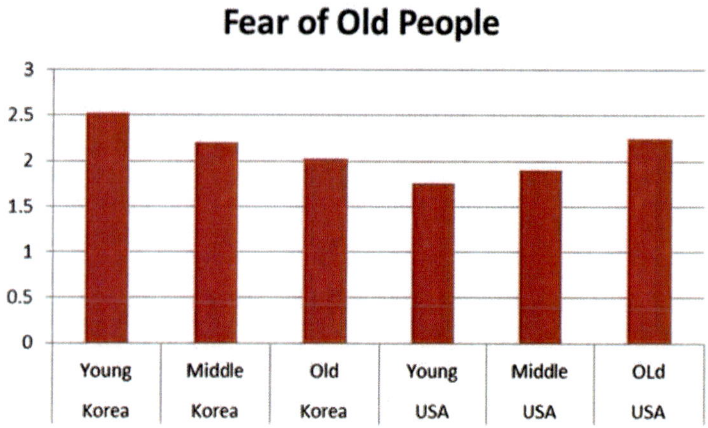

Fig. 3.1 Fear of old people as expressed by young, middle age, and old Korean and USA respondents. Fear was measured on a five-point Likert scale. Redrawn from Yun and Lachman (2006)

across many cultures where females of any age are less valued than males, they have less access to resources and property, earn less and tend to be poorer than men, and they are also more dependent on the aid provided by kin, their children in particular (Cattell 2009).

The above studies have focused on the perception of old age, and they indicate that there are some ethnic differences but also some similarities, especially regarding the physical aspects of old people. An important study also emphasising cross-cultural similarities, but in the psychological characteristics of the elders, is that of McCrae et al. (1999) who compared the Costa and McCrae's (1992) five personality traits: Neuroticism, Extraversion, Openness, Agreeableness, and Conscientiousness from the *Revised NEO Personality Inventory* (NEO-PI-R) among four age classes: 18–21 years old; 22–29; 30–49; 50 and over, in a large sample of 7361 respondents, both males and females, from five ethnic groups: German, Italian, Portuguese, Croatian, and Korean.

Each personality trait reflects some broad psychological characteristics:

- *Neuroticism*: anxiety, angry hostility, depression, self-consciousness, impulsiveness, vulnerability.
- *Extraversion*: warmth, gregariousness, assertiveness, activity, excitement seeking, positive emotions.
- *Openness*: Fantasy, aesthetics, feelings, actions, ideas, values.
- *Agreeableness*: trust, straightforwardness, altruism, compliance, modesty, tender mindedness.
- *Conscientiousness*: Competence, order, dutifulness, achievement striving, self-discipline, deliberation.

Some cross-cultural trends found by McCrae et al. (1999) regarding the 50 year old and older category included:

1. They tend to be lower in neuroticism (less anxious) across ethnic groups compared with the younger age classes; with the exception of the Portuguese sample where the older category is rather high in neuroticism.
2. They tend to be lower in extraversion (less active) across all groups.
3. They tend to be lower in openness (less intellectually dynamic) across all groups.
4. They tend to be higher in agreeableness (more cooperative) except for the Croatian sample where the middle-age group was also agreeable, like the older group.
5. They tend to be higher in conscientiousness (more self-disciplined, dutiful) except for the Croatian sample where the middle-age group was relatively more conscientious than the older group.

That is, although healthy old people tend to take their life easier, they can still offer much to others thanks to both their experience and their broader willingness to help. The limitations of old age are also clearly perceived across cultures, thus making older people more dependent on family and society at large. Those trends, however, can be also variable across ethnicities.

3.2 Fulfilling Ageing Across Cultures

What fulfilling ageing means can be broadly variable across ethnic groups. For instance, Lewis (2011) mentions how the concept of self-sufficiency is important for elders in many Western cultures, whereas elders in traditional Asian cultures tend not to regard it as a central issue, expecting some helping support from family members. However, given that fulfilling ageing is first and above all a personal experience, it is also variable among individuals within any ethnic group. Fulfilling ageing for an extraverted elder, who values physical engagement in outdoors activities and the thrill of travelling and knowing new places, is likely to take a different shape to fulfilling ageing for a more contemplative, inward-looking elder who enjoys the familiarity of local places and people; even if both individuals could be otherwise regarded as members of the same ethnicity. Gender and other differences may further diversify the paths to fulfilling ageing.

In this section we will focus on both the features of fulfilling ageing that are common to various ethnicities, and on those aspects that are idiosyncratic to some ethnic groups. In doing so, we will refer not only to *fulfilling ageing*, but also to other variables that are associated with it such as *ageing well*, *subjective well-being,* and *psychological well-being*, following their use in specific works.

In an international cross-cultural study of *ageing well*, Fernández-Ballesteros et al. (2010) analysed responses to questionnaires obtained from 1189 participants (58.4% women, mean age was 68 year old with an age range of 50–100). Participants were residents in 10 countries: Brazil, Chile, Colombia, Cuba, Ecuador, Mexico, and Uruguay in Latin America and Greece, Portugal, and Spain in Europe. Results indicate that there is a broad consistency across nationalities in the answers referred to various aspects of ageing well. Four items in the questionnaire received consistently high scores across countries: "remaining in good health until death", "being able to take care of myself", "having family and friends who are there for me", and "feeling good about myself". That is, better health, capacity for independent living, social support, and overall satisfaction with life are features of ageing well that are valid cross-culturally in this sample. Two items received the lowest scores: "living a very long time" and "being able to work after normal retirement age" indicating that greater longevity per se was not regarded as one of the most important aspects of ageing well; whereas retirement was considered to be a period in life when the individual is not expected to work.

The item "continuing to learn new things" was only regarded important by the younger participants, matching similar results from previous studies carried out in North America and Japan (Phelan et al. 2004; Matsubayashi et al. 2006). Thus, it seems that at least some aspects of fulfilling ageing are currently shared across various cultures, especially those aspects concerning good health, psychological, and social well-being. Still, the cross-cultural agreement is not perfect and there are some differences that can be attributed to specific cultural views on ageing (Fernández-Ballesteros et al. 2010). In addition, in this sample of elders from cultures that are directly or indirectly associated with Southern Europe, fulfilling ageing is not seen in an active and participative way, but in a more passive and perhaps contemplative way, with no urgency to keep working or learning new skills into old age.

Elders living in Northern European countries, on the other hand, tend to have a more active and participative stance in their retirement, as it is seen, for instance, in their stronger tendency to become involved in voluntary and charity work (e.g. Haski-Leventhal 2009).

Another psychological and social element of fulfilling ageing that is observed across cultures is the importance of maintaining existential relevance as recognised by the community. Hence the concept of *eldership* that contains an element of community respect for old people (Lewis 2011). Eldership involves becoming a repository of traditional knowledge that may be recognised as an essential component for the maintenance of cultural identity. Such role can become critically important in the case of ethnic native groups embedded within a matrix of mainstream culture in countries that were former colonies, as it is the case in the Americas, Australia, and New Zealand, for instance; but it can also acquire special relevance in the case of immigrant ethnic minority groups living in multicultural societies.

In his study of Alaskan native elders from the Bristol Bay region (*Aleut*, *Athabascan*, and *Yup'ik Eskimo* native cultural groups), Lewis (2011) found that ageing well means not only achieving personal good health and well-being into old age but also contributing to the well-being of both family and the community at large. Achieving such combined personal and social goals attracts the respect of the rest of the community, thus elevating the old person to the status of *elder*. Elders who are ageing well display good emotional well-being, engagement with the community, solid spirituality, and good physical health. They are also better integrated within their kin group and, in the process, they strengthen specific *kinscripts* or family-centred cultural scripts that connect older people with both their kin and their broader cultural networks (Stack and Burton 1993).

The social context of fulfilling ageing is also important in traditional Asian cultures. Although the traditional view of ageing well in Asia implies living old age within an extended family (*multigenerational co-residential family*, Lamb 2009), things have been changing in some countries in recent decades under the pressure of modernisation. Lamb (2009), for instance, describes such change in the case of India, where there is a growing degree of social isolation among the elders, against the cultural tradition of multi-generational co-residential family-living of the past. In the specific case of India, however, both the social and the individual aspects of fulfilling ageing have coexisted for a long time (Lamb 2020), as in the traditional Hindu four stages of life the last two are: disengaged forest dweller (*vanaprastha*) and wandering ascetic renouncer (*sannyasi*) (e.g. Savishinsky 2004), which involve a greater degree of focus on the self. This cultural aspect notwithstanding, most old parents in India have always been cared for in a family home, this being a natural consequence of seeing life as a "reciprocal inter-generational cycle". Caring for the elders is traditionally regarded as part of an inter-generational debt repayment, a form of *inter-generational reciprocity*. More recently, this crucial aspect of traditional ageing in Indian society has been challenged by modernisation, as we have already mentioned. There are many specific factors explaining the change. For start, the son's wife tends to be better educated, hence she is likely to be a professional and working, and thus in command of her own source of income, which makes her more reluctant to act as a servant to her husband's parents, as expected from tradition. In addition, there is a greater level of emigration away from both family and native town, including emigration overseas, due to professional reasons. Lamb (2009, p. 425) also mentions the important issue of neoliberalism (as it is expressed in current trends of globalisation, Westernisation, and modernisation) with its focus on "individualism, materialism, consumerism, … nuclear families, small flats, a pervasive lack of time…" and the outsourcing of elders' care, especially into old age homes and retirement villages. Such changes do not always help achieve fulfilling ageing. In fact, they often interfere with it as retirement villages attract social stigma in India, being regarded as places of abandonment of elderly parents by their children or evidence that children do not love their old parents, given that they are not caring for them. Yet, the alternative of living alone also attracts social stigma as loneliness is perceived as a problem. Indian elders are thus trying to find new ways of remaining associated with a supportive social network, in the face of these cultural transformations and constraints. For instance, new social-cultural spaces are being created, such as laughing clubs (see Fig. 3.2) which have become an expression of a *naturally occurring retirement community* (NORC, Sokolovsky 2009h).

Although co-habitation with a child's family is a traditional expectation for old people also in China, recent modernisation trends have led some older parents to prefer living alone rather than sharing the same house with a child's family. Co-habiting with a child's family may lead to conflicts in the current environment where individualism is becoming more prominent (Sokolovsky 2009g and references therein). Ip et al. (2007) refer to this situation as "veiled entrapment", where

Fig. 3.2 Group laughter at Central Park in Jaipur, India (*The Hindu* online, 6 May 2013)

ethnic traditions of the past clash with current cultural changes often to the detriment of the elders. Such situation of "entrapment" suggests that the *compensation* role of ethnicity as a source of support for the elders that we mentioned in the introductory section of this chapter has its own limits (Sokolovsky 1985, 2009g). A similar condition of *entrapment* was described by Hegland and Associates (2009) for the case of Iranian immigrants in the USA.

In New York (USA) modernisation has affected different ethnic groups differently, especially in concomitance with factors related to immigration. For instance, rates of living alone are much lower among Hispanics and Asians (most of whom are relatively recent migrants) and slightly lower among African Americans than Caucasians. The same pattern is observed in London (United Kingdom) where a higher proportion of Caucasians live alone than immigrant populations of Black Caribbeans, Indians, and Bangladeshis (Gusmano 2009).

In the case where the elder is not sufficiently independent to incorporate him/herself into a NORC, it is either the family that usually takes direct care of the elder or if the family is prevented from (or unwilling of) providing such care, children (or close relatives) may outsource or sub-contract "filial piety", as it is the case in some Chinese communities in the USA, where caregivers from a similar ethnic background are contracted to care for old parents at home (Lan 2002). Alternatively, the elder would transfer or be transferred to an aged care facility.

Among old Cuban immigrants in Florida (USA), the original multi-generational household that was prevalent in Cuba and that was maintained in the early phases of immigration (1960s) is now being replaced with the nuclear family and older persons living alone. The traditional Hispanic concept of *familism*, or the centrality of family for social support and as an aid to life fulfilment for the elders, is upheld in theory but in practice it is declining as a source of help for older Cuban immigrants, thus putting pressure on such elders to find new ways to experience fulfilling ageing. The demographic transition that is reflected in smaller families among the descendants of the original immigrants also decreases the chances of social contacts and family support, although in the Cuban community the elders are still regarded as a valuable source of child caring, especially during the periods of school holidays (Martinez 2009).

Perhaps a better mechanism of ethnic compensation in the experience of fulfilling ageing, in the face of current social changes can be found not so much in the direct family but in peers, such as a local community of same-age elders from a similar ethnic background. For instance, Sokolovsky (2009g) describes the case of Italian immigrants in the USA, many of whom were able to engage with extra-familial elders of the same ethnicity, as in the case of men and women becoming members of Italian social clubs or nurturing relationships with a stable group of friends. A similar response is described by Hegland and Associates (2009) for the case of older Iranian immigrants in the USA, who maintained the tradition of the *doreh*, that is regular meetings with the same group of friends carried out once a week or once a month, each time changing the venue (a private home or a public space). Likewise, old Cuban immigrants in Florida, who mostly fled Cuba escaping Castro's revolution, have organised seniors' centres where they can nurture their social networks whilst retaining their sense of *cubanidad*, or Cuban ethnic identity (Martinez 2009). Such localised NORCs are widespread across ethnic groups and although they could be perceived as "ghettos", it is ultimately the level of satisfaction of the participating old people that will determine whether such communities are significant promoters of fulfilling ageing or not.

The previous examples from various ethnicities suggest that although the concept of fulfilling old age is rooted into the individuality of each elder, the relevance of a supportive social network remains intact (Eriksson et al. 1999), not only in terms of providing company, stimulation, exchange of knowledge, but also emotional and sometimes material support that can significantly improve the ability of any individual to live a fulfilling life into old age. Such supportive social network can be intra-generational, inter-generational or a combination of both.

The more stable social relationships are and the more beneficial they also are for the elder, the more likely it is that they may help older people better face the challenges of ageing, and experience fulfilling ageing. Social relationships can be better stabilised the more the interacting people share a common worldview (Goffman 1974; Verhoeven 1985) or at least agree on a basic framework for coexistence despite their differences.

The experiencing of positive relationships within their social network can increase the elders' well-being and lead to an overall more fulfilling ageing (Eriksson et al. 1999). It is the positivity of the relationship that matters, not the fact that the elder is engaged in social interactions per se, as some social interactions can have negative effects on the individual (e.g. Wellman 1981). Positive social interactions that improve well-being may involve family, the direct community of residence and supportive government institutions.

Katz (2009) carried out a study of subjective well-being among 400 urban people 75 year old and older from Norway, England, Germany, Spain, and Israel. These countries represent a diversity of socio-economic and political traditions that affect their national welfare system, with Norway being categorised as social democratic, Germany and Spain are characterised as mainly conservative, England as liberal (and arguably neoliberal since the late 1970s), whilst Israel represents a "mixed" model. The level of solidarity was relatively high and similar across all countries, but help received from children—although being relatively low across countries—was higher in Germany, England, and Spain than in Norway and Israel, presumably because the latter two countries have a better government support for the elderly (Katz 2009). Interestingly, although older people may improve their well-being by receiving social support, the positive effect of sociality for the older person can also apply if they are the donors of support: by giving support to others older people can improve their own well-being through a sense of personal satisfaction (see also Krause and Shaw 2000).

A positive effect of social networks on fulfilling ageing was also shown cross-culturally by Eriksson et al. (1999) who compared two populations of elders: one in Gothenburg, Sweden; where data were collected in the context of the *H70 Longitudinal Gerontological and Geriatric Population Study*, comprising respondents who were initially 70 year old, and were then interviewed again regularly until some were 95 years old; and the other population was in Missouri, USA, with data gathered within the *Rural Missouri Elders* longitudinal study that included people 65 years old and older who were followed up over a period of 21 years. Three major types of social networks were analysed: (1) *secondary formal*, which are subject to a system of explicit formal rules, with goals and administrative decisions that follow a hierarchical organisation; (2) *primary friendships*, which are more informal and which rules tend to be negotiated directly by interactants; and (3) *primary children and relatives* groups.

American elders were more inclined to participate in organisations (*secondary formal* social networks) and kept more regular contact with their friends (*primary friendships*) than their Swedish counterparts, Americans also reported better health than the Swedes. Whereas old American women tended to live alone more frequently than men, the opposite trend was detected among Swedish elders, with men living more frequently alone than women. Being married was the single most important factor explaining survival in both samples, with the Swedes also showing a positive effect of higher frequency of visits to and from children (*primary children and relatives* groups) on survival. Therefore, it seems that although older Americans and Swedes seek support within social networks, the kind of social network is different, with the Americans seeking support outside the immediate family more frequently than the Swedes, who tend to seek support within the family. For as long as each source of support has a positive effect on the elder well-being, they can be equally valid.

Different social sources of support: within and outside the family, can occur together in specific cases, as suggested by Sokolovsky (2009h), thus co-contributing to sustain better health and psychological well-being among older people. In turn,

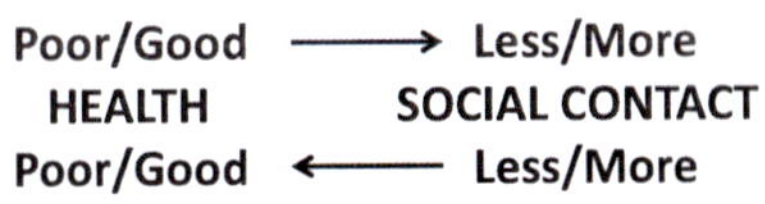

Fig. 3.3 Mutually reinforcing effects of social contact and health in older people. Poor health depresses the probability of social contacts, and lower social contacts, in turn, tend to negatively affect health. The same mechanism works in the positive: good health promotes more and better social contacts which, in turn, help in maintaining good health

with better health, social life can be also more easily enriched. Indeed, sociality and health are inter-dependent in complex ways, with both positive and negative feedback mechanisms operating (Fig. 3.3).

Toni Antonucci modelled the social network that helps sustain fulfilling ageing as a "Social Support Bank" (Antonucci 1985). In Antonucci's view social support becomes like a "bank account" subject to the flow of transactions of "debits" and "credits" that define every individual "account balance". The accounting of who owes to whom and how much, determines a system of expectations that, if unfulfilled may destabilise the social network, otherwise the stability of the network is increased. Although a certain degree of accounting of the favours given and received over time does play a role in the stability of social networks, the reality is a bit more complex and other factors do also play a role, such as the willingness to accept asymmetrical relationships in the flow of cooperative acts between individuals (e.g. a poor old parent and a wealthy adult child). Akiyama et al. (2009) show how in Japan fulfilling ageing is helped by such a system of inter-generational reciprocity; that in this case is expressed as high levels of co-residence between elder parents and their adult children.

Fulfilling ageing is much more difficult to experience when the overall network of local social support fails without any compensation (e.g. in circumstances when there is negligible or no government assistance). Such a situation is described by Indrizal et al. (2009) in the *Minangkabau* of West Sumatra, Indonesia; an ethnic group displaying a matrilineal social organisation. In the absence of government support, *Minangkabau* elders become especially vulnerable if they are childless (or de facto childless if children emigrate, although in this case children could send some remittance and keep in touch through modern means of communication) or in the case of men if they do not have a wife, given the matrilineal system of wealth inheritance prevailing in this population (Stark 2013).

Although fulfilling ageing can be helped in small communities by a local network of relationships with kin and non-kin; in larger urban communities, stable social networks are not always easy to establish and maintain. Nevertheless, in such larger urban communities the establishment and maintenance of social networks could be aided by modern communication technologies, and in any event older people must also weigh additional benefits of urban life such as access to better quality health care and other services, cultural opportunities, and so on. In this regard some large cities (*world cities*, *global cities*) are better than others (Gusmano 2009). World cities that display better *collective efficacy* or "social cohesion among neighbors combined with their willingness to intervene on behalf of the common good" (Sampson et al. 1997, p. 918) can offer a good environment to live a fulfilling old age.

Another important cross-cultural aspect of fulfilling ageing is for old people to feel that they are valued and are shown a degree of respect by the rest of their community. Respect is central to the maintenance of human dignity in the elderly and it is an important consideration in providing effective care, but levels of respect for old people are variable across cultures. Different authors have identified a diversity of forms of respect shown to elders in studies carried out in East Asia. These forms of elders' respect include (Sung 2004, see also Silverman and Maxwell 1978):

- Care (e.g. showing concern for the elders)
- Acquiescent (e.g. following elders' advice)
- Consulting (e.g. asking for their opinion on personal matters)
- Precedential (e.g. precedence to access seats, food)
- Salutatory (e.g. greeting them in specific ways)
- Linguistic (e.g. using honorifics to address them)
- Victual (e.g. preparing them their preferred dish)
- Gift (e.g. providing them with material or non-material things that they would appreciate or need)
- Presentational (e.g. showing politeness)
- Public (e.g. display of respect to elders at large in public spaces)
- Celebrative (e.g. celebrating the elder's birthday)
- Spatial (e.g. giving them a central spatial position, such as a head seat)
- Ancestor (e.g. visiting the grave of deceased elders)
- Funeral (e.g. solemn funeral)

Many of those expressions of respect towards the elders are seen not just in Eastern cultures but also in traditional forms of Western cultures. In a study of surveys of university students carried out in the USA and Korea, Sung (2004) found that in both cultures respect for the elders is shown as engaging behaviours (e.g. caring, asking for advice) and also symbolic displays (e.g. expression of respect, greetings, courtesy, demonstrations of obedience, and others). Figure 3.4 compares the frequency in which Korean and American respondents cited the forms of respect that are used in their country. Overall it is clear that both ethnic groups use most forms of respect in the list. Koreans, however, use them more frequently than Americans and two forms of respect were cited only by Koreans: Ancestor respect (commemorating the anniversary of ancestors' death, and visiting ancestors' resting place) and Funeral respect (mourning and burying deceased elders with respect, and holding funeral rites), which involve deceased people but that could affect the relationships within the family if they are ignored. This suggests a large degree of similarity but also some quantitative and qualitative differences between Western and Eastern traditions in treating old people with respect.

Respect for the elderly in four Asian countries (the Philippines, Singapore, Taiwan, and Thailand) was also studied by Ingersoll-Dayton and Saengtienchai (1999). In those countries respect for the elders is traditionally expressed as a sense

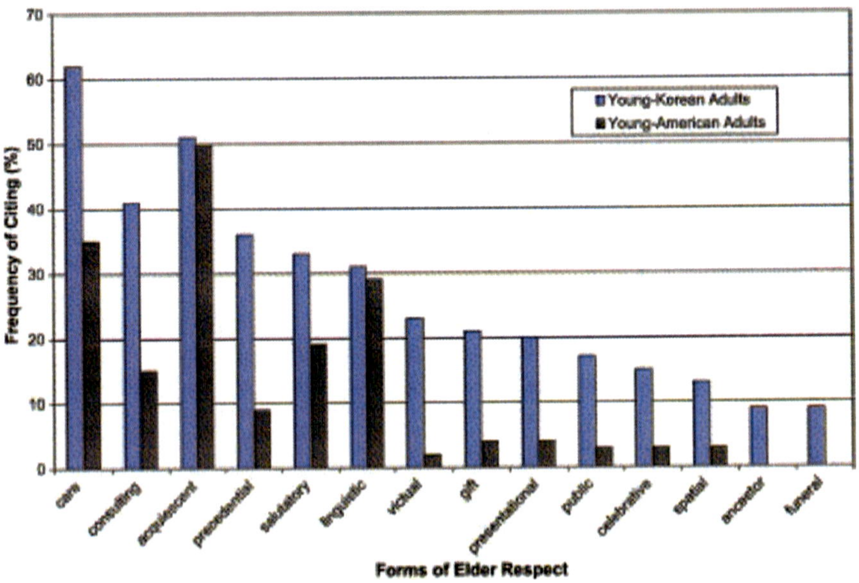

Fig. 3.4 Forms of respect towards elders mentioned by American and Korean university students; reproduced from Sung (2004)

of gratitude and repayment of debt for their efforts and sacrifices in bringing up the younger generations. Such gratitude is manifested in the language used to refer to the elders and it is expected to at least lead to genuine attempts of repayment. Ingersoll-Dayton and Saengtienchai (1999) used focus groups in each of those four Asian countries to understand the patterns of respect. Participants included old parents who were aged 60 and over and adult children aged 30–55 years old. The forms of respect detected in this study are listed in Table 3.1 and they are a sub-sample of the larger list provided by Sung (2004).

Thus, the elders, through those traditional forms of respect are not only better integrated into the family group but they are so in a manner that maintains their dignity, despite the unavoidable negative effects of the ageing process. Ingersoll-Dayton and Saengtienchai (1999) mentions how in Thailand elders are addressed using respectful terms such as "grandma" or "aunty" even if they are not biological relatives. By doing so the individual is bestowing a status of virtual family member upon the elder, indicating not only respect but also willingness to help if help is required. Growing up in Chile, M. Cordella and her family referred to a good old friend of their grandmother also as "grandmother": "abuelita" Estefanía. The Spanish word "abuelita" denotes not only the status of "grandmother" ("abuela") but it also adds the diminutive "ita" which gives a loving character to the relationship. Showing respect to older people is an important factor promoting their psychological well-being across cultures (e.g. Cheng and Chan 2006; Cheng 2009).

We have mentioned that the traditional collectivistic aspects of culture, including respect for the elders, have been changing in recent years in Asia under the influence

Table 3.1 Forms and examples of respect shown to the elderly in a group of participants from the Philippines, Singapore, Taiwan, and Thailand

Gestures and manners
Ritualised gestures: Hands joined in front of bent head, kissing elder's hand
Respectful terms: "Third aunty"
Helpful behaviours: Carrying heavy objects, offering a seat on the bus to an elder
Tokens
Food: Preparing special food, saving the best portion of the meal for the elder
Money: Providing for living expenses, facilitating the elder's merit making
Customs and rituals
Celebrations focused on older people: Thai new year, surprise gift-giving
Celebrations involving older people: Making the marriage bed party
Advice
Major purchases: Asking the elder before purchasing a television, car, or house
Auspicious dates: Asking elders about when to plan a wedding or plant rice
Consultants in family quarrels: Relying on elders as peacemakers
Obedience
Leaving home: Waiting for permission to leave the house
Marriage decisions: Following advice of elders about whom to marry

From: Ingersoll-Dayton and Saengtienchai (1999)

of Western culture and individualism. Ingersoll-Dayton and Saengtienchai (1999) report some of those changes as they were expressed by their respondents: changes in family structure and function, education, income, and modernisation, the latter pointing to the direct influence of Western culture on Asian ways of life. In spite of such changes and as we have also seen in our historical analysis of old age in Chap. 2, wealthy elders in modernised cultures still command a great deal of respect and freedom to pursue their own path to self-fulfilment, whereas elders who are less well-off experience a more challenging path to fulfilling ageing, being a likely target of abuse, disrespect, and neglect (e.g. Moon and Williams 1993).

In multicultural and multi-ethnic societies elders can be also shown respect by regarding their knowledge of their native language, culture, and traditions with interest and through willingness to learn from them. Indeed, many elders are willing to transmit their language to younger generations, but they often have to confront the negative stance of the rest of society that may tend to discourage the young from learning a second language (Meek 2007), especially if the language is not highly regarded in the cultural context of the country. In some circumstances social and cultural disruption of traditional societies may even discourage elders from passing on knowledge to unworthy young people, as it has been observed among the Nunggubuyu speakers of Numbulwar, an Aboriginal community resident in Arnhem Land, Australia (Burbank 2011).

Thus, although somewhat different strategies to experiencing fulfilling ageing may be adopted across cultures (e.g. individualistic vs collectivistic cultures), there are some cross-cultural commonalities too, which include: good health, constructive engagement with the community, emotional and existential/spiritual well-being, expressions of respect and demonstrations of worthiness by others. These traditional aspects of fulfilling ageing have been challenged by recent social changes introduced by modernisation in a variety of countries. The changes have shifted the cultural emphasis towards the individual and to an expectation of self-reliance. In response, elders are inventing new ways of experiencing a fulfilling life by organising themselves and creating new *elderscapes* and social networks, where individuals who share common interests and cultural background come together to support each other.

3.3 Inter-Generational Relationships Across Cultures

Although in contemporary Western societies—and increasingly so also in Asian societies—old people are creating their own *elderscapes*, where it is the relationships among elders that are preferentially nurtured, it remains true that interactions between different generations are not only a persistent reality, but when they are supportive they could be an important aid to experiencing fulfilling ageing.

Inter-generational *solidarity*, or mutual agreement between generations in the need for action, and *cooperation*, or the coordinated action between generations in the pursue of a common objective, can be culturally based on established codes of conduct of respect and ethics (Silverman and Maxwell 1978; Ingersoll-Dayton and Saengtienchai 1999; Sung 2004). But they can also emerge from more immediate transactions of convenience, implying the giving and taking of material goods and favours, like those suggested in Antonucci's *Social Support Bank* model; or they may simply result from emotional attachment between young and old. It is often the case, however, that inter-generational solidarity and cooperation are grounded on a variety of motivations and constraints.

Silverstein et al. (1998) investigated inter-generational solidarity between older parents and their adult children comparing rural North Wales (UK) and urban areas in the USA. Table 3.2 shows their results on variables reflecting inter-generational solidarity.

In spite of an overall degree of similarity between the two countries, rural Welsh parent-adult child dyads tended to live closer to each other, and thus older parents were somewhat more in contact with and also received more support from their offspring than it was the case with the American urban samples. A more detailed analysis of spatial proximity also suggests that when parent and offspring live closer, they

Table 3.2 Parent-adult child solidarity variables in three samples of old parents from North Wales, UK (NW), Los Angeles, USA (LA) and USA National (USAN)

	NW	LA	USAN
Receives instrumental help	38	35	32
Lives less than 50 km away	60	49	–
Lives less than 1 h drive away	58	–	55
In contact at least weekly	72	65	55
More than somewhat emotionally close	85	88	84

Values are percentages; adapted from Silverstein et al. (1998)

also tend to be more emotionally attached. This is true across samples, but the effect is stronger in North Wales. Overall, however, the American samples show values of "More than somewhat emotionally close" in Table 3.2 that are similar to North Wales, which may simply indicate that urban American old parents and their children can still maintain some contact, even at a distance, thus nurturing their emotional closeness through the use of modern means of communication, which indeed have become even more sophisticated now than it was the case when Silverstein et al.'s published their article in the late 1990s. Thus, although inter-generational solidarity is favoured by physical proximity and, conversely, physical proximity may be also a manifestation of inter-generational solidarity (e.g. the old parent and/or the adult child taking the initiative to live close in order to better help), modern means of communication can still allow the expression of solidarity at a distance.

The agreement for action implicit in the concept of *solidarity*, and the coordinated supportive action expressed in the term *cooperation*, are conflated by Schans and Komter (2010, pp. 194–195) into a more nuanced categorisation of inter-generational solidarity: *normative solidarity* ("the obligations adult children feel towards their parents"), *associational solidarity* ("the frequency of contact between adult children and their parents"), and *functional solidarity* ("the degree of assistance provided by adult children to their parents"). They studied those three categories of inter-generational solidarity in The Netherlands comparing native Dutch participants (adult children and their old parents) with equivalent samples of migrants from Turkey, Morocco, Suriname, and the Dutch Antilles. Normative solidarity (filial responsibility) is more common in adult child-old parent relationships among migrants than in the native Dutch sample. This may be a combined result of cultural differences and the effect of migration, in fact the difference between native Dutch and immigrants decreases in the immigrant sample that is better educated and presumably also better integrated into the mainstream of society. There is a difference in associational solidarity between the Dutch and migrant samples but it is mainly due to the effect of daughters, who reported higher levels of contact with parents than sons, and most categories of migrant daughters also have more contact with parents than Dutch (and Antillean) daughters. More solidarity expressed by daughters also explains greater practical support (functional solidarity) for migrant than Dutch parents.

Therefore, cooperation between generations is not only possible, but it is indeed more necessary than ever, following the demographic transition that the world is experiencing (Chap. 1). Inter-generational cooperation can still regulate a degree of familial flow of goods and support, especially in terms of care from grandparents to grandchildren and other forms of support offered to adult children. Conversely, children may keep in touch with old parents and provide advice (to obtain government assistance, for instance), or some material aid if it is required (Sokolovsky 2009c). Such inter-generational cooperation has deep roots into our social biology, and it is found in human societies across the world. Hill and Hurtado (1996), for instance, describe how post-reproductive women among the *Aché* of Eastern Paraguay help their adult children by collecting food, attending to household chores according to their physical capacities, and babysitting grandchildren. Old men continue to reproduce if they can and provide care to family into their sixties, but they transition to an old-age pattern when they cannot hunt anymore. They then shift to gathering food that it is not difficult to collect and making tools for younger men. They also babysit their grandchildren just like grandmothers do.

Becker et al. (2003) investigated mutual assistance between old and young members of the family in the USA. The sample comprised 270 participants, 49–97-year-old women and men (with a women bias), from four ethnic groups: 59 African Americans, 85 Latinos, 78 Filipino Americans, and 48 Cambodian Americans. Mutual assistance between generations was recognised as a crucial aspect of social relationships in all ethnic groups. Such cooperative stance, in fact, extended to all members of the family both within and across generations. Mutual assistance, however, adopted various expressions according to the specific cultural characteristics of the group: African Americans tended to balance their individual autonomy with the diverse duties within the family; Latinos showed a priority for emotional interdependence, with *familism* being the underlying value supporting such web of relationships among kin. *Compadrazgo* (co-parenthood, Mintz and Wolf 1950) extended the web of support also to non-kin. Among Filipino Americans mutual assistance was especially centred on economic support, and elders who can contribute to the family financial well-being are held in higher esteem. For Cambodian Americans the central aspect of mutual assistance was co-residence where they could care for and receive help from old relatives (e.g. parents). In the provision of such care family members are expected to demonstrate selflessness, cooperation, and respect. Thus, each ethnic group showed specific expressions of mutual assistance. A common feature of the cooperative interactions between elders and other family members across ethnic groups was the expectation from elders to be treated with deference. They could help the family in very practical ways (providing childcare, for instance) and in so doing they also enhanced their self-esteem, but their help was conditional to being treated with respect.

Inter-generational cooperation can specifically contribute to *family stability*. In a study carried out in Evergreen, an area in Chicago of low socio-economic background but with a surprisingly high degree of stability within families, Iris (2009) reported how African American older adults are seen as an asset to the family, playing an important role in maintaining stability within the family:

> We heard numerous stories of how these African American elders provided both material and nonmaterial support: they contributed to their grandchildren's financial and family stability, paid for the education of grandchildren and assisted with childcare (p. 366).

Grandparents in fact may go even further than just providing the occasional child-minding or financial help. Some may go as far as to become *custodial grandparents*, thus legally committing themselves to raising their grandchildren. In the USA, for instance, custodial grandparenting has been on the raise, affecting communities of various ethnic backgrounds: Caucasians, 2% custodial grandparents; 6% among Asians; 8% American Indians and Alaska Natives; 8% African Americans; 8% Hispanics; 10% Pacific Islanders (Hayslip 2009). Given the usually advanced age of grandparents, custodial grandparenting already involves a series of challenges, but the situation can become even worse whenever custodial grandparents in some ethnic groups are culturally reluctant (or may be communicatively incapable) of seeking help, thus putting themselves and their grandchildren through unnecessary strain. Being unaware about the local social support system can also affect the likelihood that they miss on the help available, especially the help provided by government agencies. This may lead custodial grandparents to seek help elsewhere, which may expose them to unscrupulous service providers (Hayslip 2009). In addition, there may be also instances in which immigrant grandparents may suffer additional stress and frustration due to their incapacity to help their grandchildren as a result of the compounding effects of both cultural and generational gap, leading to conflicts and misunderstanding. Language difficulties can only make things worse. Hence, among ethnic minorities custodial grandparenting may be constrained by the degree of acculturation of the grandparent, the ability to transmit values inter-generationally, and how flexible are the roles within the family. Despite such constraints, custodial grandparents in some ethnic groups still commit themselves to the task of caring and raising their grandchild or grandchildren, out of love and duty. This can produce a situation of *role captivity* (Hayslip 2009) that could potentially conduce to the development of depression unless the grandparent receives external help.

An important contribution that older members of the family can make to the younger generations—that is especially relevant in multicultural/multi-ethnic societies—is *cultural continuity*. This is an issue that can be of great importance especially to ethnic minority groups as they try to retain a link with their cultural roots amidst a whirlwind of change and external influences. Moreover, across ethnicities elders can also act as "family historians" (Becker et al. 2003). Given the current growing interest in family trees, with genealogy being referred to as "one of the world's most popular hobbies" (Veale 2004), elders can help younger generations by providing background information on the history of the family, which may help the young in the construction of their own identity.

Following "post-welfare" neoliberal policies, families have found themselves under the stress of filling the gaps left by a retreating government. This has led grandparents to seek more support from family sources and *vice versa*. The ethnic background of the family, however, can affect the extent in which support is distributed within the family. In a study carried out in New Zealand, Ng et al. (2000) administered questionnaires to adult grandchildren from 100 Chinese and 103 European families in the Wellington area. Figure 3.5 shows that across various aspects of support for the elders: obey, financial help, providing care, please, respect,

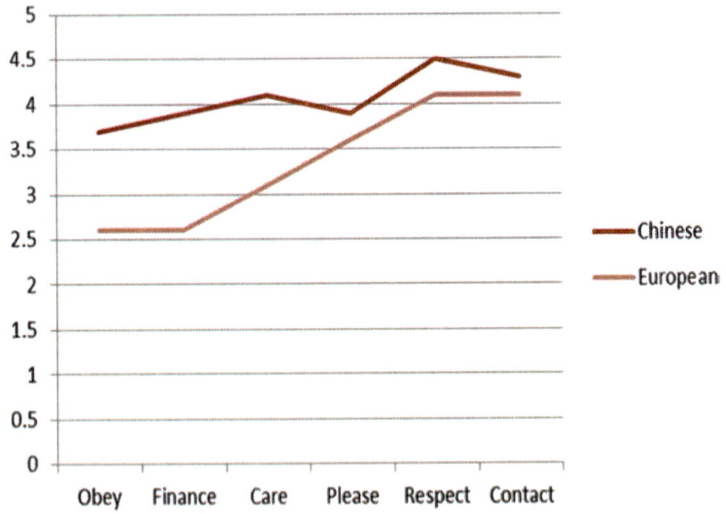

Fig. 3.5 Family obligations towards grandparents according to Chinese and Europeans living in New Zealand, values on the *y*-axis are scores from a questionnaire on a scale from 1 to 5 (redrawn from Ng et al. 2000)

and keeping contact, the Chinese tend to be more supportive of grandparents than people of European descent, but the broad distribution of priorities is relatively similar, with "obeying" being less frequent, presumably as a result of current modern trends in Western societies, and "respect" and "contact" being the most frequent. That is, grandparents are not necessarily neglected, but their authority over the family is not perceived to be significant, the specific pattern being dependent on ethnicity.

Respecting older people and keeping in contact with them may directly result from traditional filial piety, but it is likely that both behaviours could be enhanced if the older person is also making a contribution (e.g. giving valuable advice), even if such contribution is not subject to strict "accounting" requirements as seen in Antonucci's "Social Support Bank" and as expected from *Social exchange theory* more broadly (Dowd 1983).

Inter-generational interactions can be also negative as it is the case in some cultures where elders form socially powerful organised groups, that are aimed at controlling the younger members of society both politically and economically (see for instance Murphy 1980). Alternatively, the younger generations may abuse their elders, as we will see in section *The ethnic dimension of elder abuse and neglect* below in this chapter.

Notwithstanding the current ability of old and young to stay in touch through modern technologies, relationships between generations can be nurtured only to the extent that a shared system of beliefs and communicative strategies favour such relationship. If they hinder it, then the inter-generational relationship may become strained and each generation may recoil inside the social confines of their own cohort. In a study of young adults' beliefs about intra- and inter-generational communication carried out with participants from South Vietnam, North Vietnam, and the USA (30 years old or younger, 66.9% female respondents), McCann et al. (2004) concluded that overall, the young adults that they surveyed perceived their relationships with older people as being both "more problematic and more pleasing" than those with other young adults. Problems arose from older people being perceived as more negative, more self-centred, and less positive in their communicative strategies than peers. However, older people were also perceived as being more *accommodative* (i.e. they tend to "inhibit impulses to respond in kind, instead reacting in a constructive manner" (Yovetich and Rusbult 1994, p. 138)) than peers, this pattern was especially strong among women respondents, across cultures. In response, young adults may even *overaccommodate*, as when they are exceedingly polite or warm and use a simple language to address older people (e.g. Edwards and Noller 1993).

However, not always older people are regarded as accommodative when they communicate with the young. Sometimes they are perceived as rigid/inflexible, or they may be judged as "patronising" or even disapproving, "not listening" and "incoherently verbose" (McCann et al. 2004 and references therein), in other words they can be *underaccommodating* (Giles et al. 2003). The overall pattern was similar across cultures in McCann et al.'s (2004) work, perhaps in part because the sample of university students they used may be more sensitive to the equalising effects of globalisation and modernisation.

A similar study was carried out by Giles et al. (2003) who administered questionnaires to young adults from three Western countries: Canada, USA, and New Zealand; and three East Asian countries: The Philippines, South Korea, and Japan. In this case, only Westerners regarded old relatives as more accommodating than their same-age peers, whereas East Asian respondents did not see any difference between the two in terms of accommodation. Respondents from both groups also coincided in being most deferential in their interactions with non-related elders than with related elders, and finally with same-age peers. Although intra-generational communication is seen as less problematic than inter-generational communication in both groups, the pattern is stronger in the East Asian than the Western sample, thus again suggesting that younger Asians are somewhat distancing themselves from traditional forms of interactions with older people, such as the Confucian concept of filial piety, although a traditional sense of filial responsibility is still somewhat persistent especially among young Asian women (Gallois et al. 1999). The departure from traditional expressions of filial piety can be also affected by the reaction of older people to the cultural changes being experienced by the young. This is seen, for instance, when old people stereotype the young as "irresponsible" and "naïve" (Giles et al. 2003).

To recap, inter-generational interactions can vary across ethnic groups and they can be either positive or negative, and accommodative, overaccommodative or underaccommodative. However, a common theme that prevails cross-culturally is that there are positive aspects of such interactions that can be beneficial to all those involved: the elder, the young members of the family, and beyond. Such beneficial interactions are defined by solidarity and cooperation, with elders receiving love and support and in return they can contribute their love and care, their time, experience, and sometimes even material resources to their relatives. Modernisation is affecting the specific expressions of inter-generational solidarity and cooperation around the world, with the neoliberal emphasis on individualism and self-reliance pushing older people to seek more support beyond the family and within the social network of their own peers. This pattern is also affecting societies that have been traditionally characterised by a strong sense of filial piety and respect towards the elders.

3.4 Perceptions of Old People's Health Across Cultures

In this section we focus on how age-related health changes in old age are perceived across cultures. We start with some general health topics: What are the conceptions of illness and good health in old age, and how do they vary from one culture to another? We then analyse some aspects of menopause across cultures, to conclude with a specific focus on mental health in old age. Cross-cultural studies have emphasised both the similarities and the differences in the perception of the elders' health across ethnic groups, with some of those differences and similarities being affected, among other factors, by recent patterns of migration and modernisation.

Following Boorse (1977, p. 555), given a reference class of individuals (e.g. the elders) it should be possible to define a *normal function* of a body part or process within the members of that reference class as a "statistically typical contribution by it to their individual survival and reproduction". Therefore, "(*h*)*ealth* in a member of the reference class is *normal functional ability*: the readiness of each internal part to perform all its normal functions on typical occasions with at least typical efficiency". From this it follows that a *disease* "is a type of internal state which impairs health, i.e. reduces one or more functional abilities below typical efficiency". In the definition of *normal function*, survival broadly remains a valid component of it for all ages, including old age, but reproduction decreases in the elderly—especially among women, following menopause—as a "normal function" not as a "disease", although the process of menopause can be also associated with health problems, as we will see in this section.

Therefore, a healthy elder can function normally within an expected statistical range of variability of body functions, and a disease condition is suspected when such abilities depart from the statistical range. Further analyses of the condition of the elder may produce a specific diagnosis.

Illness, on the other hand, "refers to a person's first-hand experience of the disease" (Carel and Cooper 2012, p. 8). Illness is therefore a concept that stresses the personal perceptions of own body and its functioning, and such perceptions can be influenced not only by the organic aspects of a disease but also by psychological and cultural factors. Therefore, a person may feel ill without being diseased, or she/he may suffer from a disease without necessarily feeling ill (e.g. attributing tiredness to hard work rather than a developing cancer) or, she/he may both feel ill and be diagnosed with a disease.

Although current biomedical paradigms exert great influence on the perception of illness and the treatments required to return to a healthy condition; traditional views of health and disease (folk medicine; e.g. Hatfield 2004) are still common and they tend to coexist with biomedical approaches, often complementing them, but sometimes interfering with them.

For instance, in a study comparing the Chamorros of Guam with a sample of Chinese Americans from Houston and Philadelphia, Torsch and Ma (2000) noted that in both cultures traditional holistic forms of healing/medicine are still practiced, despite much Western influence. Elderly Chinese explained illness as an imbalance between *yin* and *yang*, whereas balance in life in general is regarded as the best route to achieving a long and healthy life. This springs naturally from the traditional belief that mind and body are interconnected and influence each other, with good health being dependent on behaviour, the effects of emotions and lifestyle. In the view of Chinese elders, the loss of harmony between mind and body leads to a weakening of the positive energy of life or *qi*. Healing can therefore also be aided by actions performed on the body such as massage (*taijiquan*) and exercises (*qi gong*). For the Chamorros, healthy ageing simply meant the avoidance (or decrease) of the usual bodily changes experienced by old people (e.g. loss of strength and decrease in the levels of activity), but they also included mental changes. If achieved, this avoidance/decrease allows a more harmonious participation into family life. We can see that a difference between the two cultures is the more "abstract" approach to health of the Chinese (e.g. *yin* and *yang*) as compared to the more "concrete" approach of the Chamorros who focused on body functionality. Interestingly, the influences of modernisation on traditional cultural values have become a cause of anxiety among elders in both groups, leading to the retention of some traditional beliefs through a process that Torsch and Ma describe as *adaptation with continuity*, where older people may mix the new with the old. Through *adaptation with continuity* elders may also become an inter-generational linchpin to maintain a degree of cultural continuity, despite the changes that may occur over time.

Although traditional ethnological studies can provide an in-depth analysis of the health perceptions across cultures, a complementary approach that has been adopted for rapid accumulation of comparative data is the use of *self-rated health* (SRH) instruments (e.g. Bombak 2013). Self-rating of health is a surprisingly good descriptor of the actual health condition of an individual, despite being potentially influenced by emotions, levels of psychological distress, personality, past experiences and of course age, ethnicity, socio-economic status, education, and gender (Bardage et al. 2005). SRH has been shown to be even a good predictor of mortality across various ethnic groups (Benyamini et al. 2000).

Bardage et al. (2005) used the information from the *Comparison of Longitudinal European Studies on Aging* (CLESA) to analyse data on health perception in a sample of 11,557 respondents, aged 65–89 years old, from Finland, Israel, The Netherlands, Spain, and Sweden. Their evaluation of SRH considered only a choice between a "good" and "fair and poor" options. In a cross-cultural convergence, older respondents (75 years old and older) rated their health as poorer than that of the younger categories of old people, with women reporting worse conditions than men in most countries. Marriage was associated with better self-reported health, as it was better education. But some cross-cultural differences in SRH have been also found. Jylhä et al. (1998), for instance, compared SRH for people 60 years old and older between Finland (Tampere) and Italy (Florence). The trend was for both men and women to report better self-rated health in Florence than Tampere. SRH was more capable of predicting mortality in men than women in both areas; and in both areas, self-rated "very good" health was associated with a relatively lower risk of mortality.

Menec et al. (2007) investigated SRH in Manitoba, Canada, as part of the *Aging in Manitoba* (AIM) study, comparing the results of interviews carried out between 1983 ($n = 5278$) and 1996 ($n = 1868$) (55% women; 60–104 years old). Ethnicities included: British/Canadian, Central/Northern European (French, German, and Scandinavian), Eastern European (Polish, Russian/Ukrainian), and Other (other Europeans, Middle Eastern, Asia Oceanic, Native American or Inuit, and U.S./ Western hemisphere). The British/Canadian ethnicity was the reference group. SRH significantly varied among ethnic groups in both periods of the study. The most striking difference was observed between the British/Canadians and the Eastern Europeans, with the former rating their health significantly higher than the latter: 63% of British/Canadians regarded their health as "good or excellent" in 1983 (61.8% in 1996), whereas only 44.9% of Eastern Europeans did so in 1983 (39.4% in 1996). The trend between those two ethnicities was maintained after controlling for some confounding variables. The difference in SRH could be at least in part explained by some socio-economic differences, such as education and income, which were lower among Eastern Europeans; but also the level of acculturation, as Eastern Europeans who spoke better English also scored higher in SRH.

In other words, although old people perceive their health to be poorer the older they are, such perception can vary between ethnic groups of elders; with an added effect of immigration, especially when recent immigrants are included in the sample.

Given that elders do perceive a degree of health deterioration with advancing age: Are they prepared to do something about it in order to slow down the rate of decay over time? Wilcox et al. (2009) carried out focus groups in the USA with 50–90-year-old African Americans ($n = 10$), Caucasians (19), Hispanics (2), American Indians (3), Chinese (4), and Vietnamese (3)—the sample is overall female biased—to determine their beliefs regarding the role of physical activity and nutrition in the maintenance of brain health in old people. There was an agreement on the role of physical activity (e.g. walking, dancing, gardening, and so forth, including practicing Tai Chi and yoga for the Chinese) as an important factor in keeping a healthy brain. Some believed that the role of physical activity in maintaining a healthy brain may be mediated by learning and mental challenge, breathing faster inhaling more oxygen, and keeping a steady blood circulation. Performing such activities in the company of others, rather than alone, was also mentioned as positive. A Chinese respondent emphasised the importance of being mentally, physically, and socially active, but without overdoing it, balance was the key concept for this respondent, to which a Vietnamese participant also agreed. Eating well and the appropriate kind of food were also regarded as a contributing factor to good brain health.

In another study carried out in the USA as part of the *Women's Health and Aging Study* (WHAS) that focused on old women's mobility and SRH, Jylhä et al. (2001) also found that being able to mobilise independently was central to good SRH. Motility had a strong positive effect, but negative effects on SRH were noted when the elders experienced depression (see also Ormel et al. 1998). It is known that keeping a reasonable physical fitness through exercise helps in the maintenance of independent motility and it improves SRH (e.g. Shirom et al. 2008).

Therefore, when it comes to general health it is remarkable to see how much old people across cultures converge in their opinions. This can be in part due to our shared biology but there may also be an effect of increased flow of medical information. Differences based on traditional beliefs, however, are also noted as exemplified by the Chinese concepts of *qi*, *ying,* and *yang*.

In recent years there has been an increased interest in cross-cultural studies of health perceptions in old women, in part because they undergo distinctive age-related biological changes such as menopause.

Around the world, the age at onset of menopause can be somewhat variable between late forties and early fifties, ranging from 50.9 years in Italy, 50.4 in Iran and Slovenia, 49.3 in Korea and Lebanon, 49.1 in Singapore, 48.9 in the case of Korean migrants living in China, 48.7 in Greece, 48.4 in Morocco, and 48 in Mexico, Turkey and also for the Han Chinese living in Taiwan (Melby et al. 2005).

Across cultures, menopause tends to be seen more as a process than an abrupt change, as it is expressed in the English word *climacteric* (from the Greek *klimakterikos* "of a critical period") or the Japanese *kônenki* and the Chinese *gengnianqi* (Lock and Kaufert 2001). Menopause is broadly manifested through a variety of symptoms that conform to a *menopausal syndrome*: hot flashes, sweats, vaginal dryness, depression, irritability, anxiety, palpitations, headaches, insomnia, lack of energy, difficulty in concentrating, spells of dizziness (Bromberger et al. 2001; Avis 2005). The syndrome, however, is not manifested in the same manner across ethnic groups, and variability is especially accentuated with respect to psychological symptoms (Avis 2005). Lock (2002, p. 132) correctly points out that there is an "inevitable ambivalence associated with the end of reproductive life, the implications of which vary enormously depending upon local attitudes towards ageing in general, older women in particular, and their place in society".

The cross-cultural study of menopause is not necessarily easy, as in some traditional societies it is difficult to know whether a reported absence of symptoms (e.g. hot flashes) is a genuine description of a biological fact or whether information is being withdrawn, being considered taboo. We refer the reader to Lock (2002), for a discussion of the potential interference of cultural beliefs on the reporting of menopause symptoms. With this caveat in mind, Beyene (2009) states that menopausal symptoms are reported more often in Western women (see also Obermeyer 2000; Melby et al. 2005). For instance, hot flashes and night sweats are reported less frequently among women in India, Indonesia, Africans living in Israel, Taiwan, Hong Kong, Japan, Singapore, China, Korea, Thailand, and Malaysia compared to North

American and European women (Melby et al. 2005). In cultures where many of the menopausal symptoms are not reported, such as in Japan, there are also better attitudes towards menopause (Bromberger et al. 2001).

Beyene (2009) compared Mayan and rural Greek women to find that both groups see menstruation negatively: bad omen, unclean, can cause contamination. After menopause, women from both ethnicities felt that they gained more freedom to participate in social activities. In addition, both also felt more relaxed about sexual activities after menopause and reported improvements in their relationship with husband. Relief was also felt by post-menopausal Thai women who saw the end of menstruation as equivalent to entering a more spiritual phase of their life (Lock and Kaufert 2001).

In both Mayan and Greek cultures, the social status of a woman increases with age, and older women are believed to have special healing capacities. However, Mayan women reported cessation of menstruation at menopause but not hot flashes or cold sweats, or any anxiety, stress or additional health concerns; whereas among the Greek participants in Beyene's study pre-menopausal women saw menopause negatively, due to the anxiety and potential health problems associated with it, the indication that menopause marks a clear step towards growing old, and the decreasing levels of energy experienced. Unlike Mayan women, hot flashes and cold sweats are reported by Greek women, in addition to insomnia, irritability, and "melancholia" (depression). Ayers et al. (2010) suggest that women who hold a more negative attitude towards menopause also report more symptoms during the transition to menopause.

In a multi-ethnic study carried out in the USA, Sommer et al. (1999) interviewed 16,065 women: Caucasians, African Americans, Japanese Americans, Chinese Americans, and Hispanics to understand their attitudes towards ageing and menopause. Across all ethnicities, menopause was greeted with positive attitudes, but especially among African American women. The positive attitude increased with the degree of Western acculturation, especially in Asian women. African American women also had the most positive attitude towards ageing in general. However, Bromberger et al. (2001) also reported a 24.1% of women manifesting psychological distress during menopause, with Caucasian, Hispanic, and African American women expressing the highest rates of psychological distress, followed by the Chinese and Japanese. This result is consistent with other studies that found that across cultures, depression in menopausal women is lower in Asia than in Europe and the USA (Obermeyer 2000).

Thus, the general response of women to the onset of menopause is more positive than negative across cultures, especially once the symptoms of menopause are over or at least under control. Differences do occur, however, across ethnic groups and they are more likely to be explained by a combination of biological and cultural factors (see also Melby et al. 2005).

We have already briefly mentioned the increased risk of developing dementia with advanced age in Chap. 1. We will review various forms of dementia—Alzheimer's disease in particular—in greater detail in Chap. 4, here we focus on some cross-cultural aspects of it.

Alzheimer's disease (AD) and vascular dementia have been on the increase for the past 30 years, especially in East Asia (Traphagan 2009), and within countries some ethnic groups are at relatively higher risk of developing dementia than others. For instance, in the multi-ethnic USA older African Americans are at a greater risk of developing AD than older Caucasians (Mahoney et al. 2005).

In the past, AD, vascular dementia, and the more general frailty syndrome that usually characterises old age were subsumed into the common term of *senility* (Henderson and Traphagan 2005). This term has lost a great deal of its former academic usefulness, but it can still be relevant as a current concept in some cultures. The challenge we face when trying to understand cognitive changes associated with ageing around the world is to be able to integrate both the biological and the cultural aspects of those changes. This integrative approach can not only conduce to a better understanding of dementia, but it can be also used to better intervene on specific individuals.

After the symptoms of dementia become evident in an older person, the causation of the mental deterioration can be understood very differently by family and acquaintances according to their culture. Dementia can be culturally seen as a necessary outcome of the normal process of ageing or as a disease that could potentially be prevented first and then hopefully treated once it develops (Downs 2000). The hallucinatory aspects of dementia can be interpreted in traditional cultures in a religious or transcendental manner, with people experiencing dementia seen as going through a "supernormal" state that makes sense in relation to a prevailing spiritual worldview and belief in the afterlife (Henderson and Henderson 2002).

Indeed, the entire construct of the meaning and causation of dementia can be affected by cultural beliefs (Zeilig 2014). Among African Americans, old people's behaviours associated with dementia may be regarded as evidence of "worration" (a mixture of worry and stress causing brain damage), or "falling out" and "high blood" (Dilworth-Anderson and Gibson 2002). Hispanic and Asian American cultures in the USA tend to refer to dementia as "being crazy" or having "bad blood".

La Fontaine et al. (2007) studied perceptions about ageing and dementia in the United Kingdom with focus on Indian migrants of Punjabi origin using a focus group methodology. Given the prevailing stigma affecting old people with dementia in this immigrant population, the immediate expectation was to hope that the old person would put an effort into self-helping, aided by the family and doctors. Regarding the causations of dementia, there was a belief that it could be caused by psychological factors such as anxiety and depression, especially resulting from unstable family relationships, or simply by the natural wear and tear of ageing, with some added personality factors such as some people actually liking others to do things for them, an attitude that suggests a view that old people may take advantage of the condition or even go as far as to fake helplessness. The latter belief can be especially damaging during the initial phases of development of a cognitive disorder, as it is likely to cause frustration in the older person affected by it and it may delay medical intervention.

In traditional Japanese culture the ability of the individual to maintain working relationships with the social network is an essential part of the defining characteristics of personal identity. Age-related loss of cognitive capacities and the ability to interact socially can significantly affect the position of the individual in society, hence the tendency to hide or conveniently reinterpret any cognitive decline. In traditional Japanese culture it is expected that old individuals do put an effort into avoiding the development of *boke* (*senility*, Henderson and Traphagan 2005), by remaining physically and mentally active and contributing to family life. Indeed, there are four moral pillars that old Japanese people are culturally expected to consider as they age: a duty to remain active, keep their effort up, not to be a burden to their family, and being willing to reciprocate (Traphagan 2009). Old people who are mentally incapacitated from fulfilling those functions may experience an added level of distress.

Although some cultural expectations may be detrimental to the old person developing dementia, others may be helpful. A social network is perceived to be useful to the extent that it becomes an effective source of social stimulation and support. Functioning social networks can provide material help and safety to the elder but also a challenging cognitive environment, by stimulating memory, for instance, thus potentially protecting the individual from mental decay later in life (Pillai and Verghese 2009 and references therein, but see also counter-examples such as Karp et al. 2006). Therefore, social networks can offer the appropriate environment to help protect the cognitive capacities of ageing people, for as long as relationships are positive, and they decrease stress.

Receiving good social support from family and friends is of course important for old people suffering from dementia, but such social support may be jeopardised by stigma. A stigma may already kick in, and affect the elder directly, during the initial stages of development of dementia when the elder is still fully conscious of the reactions of others. Liu et al. (2008) studied the interactions between stigma and dementia after interviewing Chinese and Vietnamese caregivers. Participants frequently mentioned the issue of stigma in the interviews (91% of all interviews included the issue of stigma). The sources of stigma are double in this case: stigma against the elderly, and stigma against mental illness. The issue with multiple stigmas is that their effects may be compounded, thus seriously compromising the ability of the old person to operate in life, in circumstances that can be already quite challenging even in the absence of stigma.

Stigma can affect not only the older persons developing dementia, but also their caregivers, such as spouse and other family members, who may become the recipient of what Goffman (1963) called a *courtesy stigma*. As older people from a diversity of ethnic backgrounds who are affected by dementia lose their capacity to discern, it is the caregivers (families and professionals) who are left with the entire burden of care, often in the ungrateful circumstance where the elder has lost the capacity to recognise the carers, including partner and children. Caring for dementia patients is not only taxing on caregivers' physical and mental resources but the commitment could extend over a long period of time, sometimes years, following the usual course of the disease, so much so that caregivers often refer to dementia as the "long goodbye" (Henderson 2009).

In an early review of caregiving for old dementia patients, Connell and Gibson (1997) concluded that among Caucasian caregivers the spouse is the most frequent person taking care of the patient, whereas among non-Caucasian caregivers the immediate and extended family, and also the circle of friends, are more likely to share the burden of caregiving. In turn, caregivers expressed high levels of stress and depression derived from the burden of caregiving, especially among Caucasians. The burden can be increased by cultural factors such as a relatively weaker sense of filial piety in some ethnic groups compared to others that may restrict the caring effort on just a small number of individuals. Those patterns were confirmed by Janevic and Connell (2001) in a review published 4 years later.

In a study carried out in the USA, Mahoney et al. (2005) interviewed 22 family caregivers of older people who were in their early stages of Alzheimer's disease. Most informants were Latinos, but there were also African Americans and Chinese. The mean age of caregivers was 53 years old, most of them were women (see also Henderson 2009), married and well educated, and they were caring for parents or spouses. Caregivers stated that the early signs of the disease, such as forgetfulness, were easily dismissed and attributed to just a normal ageing process, there was also a tendency to minimise the condition: "I would say to her, you know I don't remember either"; "(s)he never had a good memory for the start"; a Chinese respondent said: "Friends suggested us to relocate our house ... they thought mom's craziness was caused by bad feng sui". Derived from an ancient Chinese tradition, *fengshui* is a property of the structure and surroundings of dwellings, that affects the vital energy or *qi* of those living there. The Chinese were also concerned about the social stigma expected if their family member was diagnosed with a mental illness; their fear was that other members of the community would blame them for what happened to their old relative. Among African Americans their respect for older people tended to minimise concerns in response to elders' behaviours that deviated from the norm, a similar attitude was adopted by Latino families who tried to avoid upsetting the old person until the situation became untenable and some action had to be taken.

Once the disease was diagnosed, family members from the three ethnic groups turned to a diversity of sources for advice: African Americans approached their church (ministers), Latinos sought advice from friends, and the Chinese obtained information directly from public health centres.

Social factors contributed to caregivers' decisions about where to seek practical help from: African American caregivers were preoccupied about racism, even from some medical doctors; as already noted above, the Chinese were concerned about stigmatisation and therefore sought support from the family, but this could also be a problem as the fear of stigma permeated the family too: "As soon as my brother heard [that it was Alzheimer's disease] he and his wife walked away scared ...[that it was] contagious. I had to move in to stay with my mom" (Mahoney et al. 2005, p. 789); whereas Latinos saw the family as the main provider of support, this was further reinforced by mistrust of institutionalised care. The Chinese also sought to develop a personal and trusting relationship with medical personnel, in accordance to the Chinese value of *guanxi*, which came with an expectation from doctors to take their time to clearly explain the situation, provide information on the steps to follow into the future, so that family members could share their knowledge with the rest of their social network thus hopefully increasing the level of overall support: "We need education about AD [Alzheimer's disease] in the Chinese community so that people can help one another, not walk away".

The cultural tradition among Latin Americans is that the females of the family (wife, daughter, even daughter-in-law) will take care of the old relatives who are developing dementia (Henderson 2009); such tradition is shared with other ethnic groups (e.g. Henderson and Traphagan 2005; see also the case of Italian Australians in Benedetti et al. 2013). However, this attitude is embedded in the far more profound and complex concept of *familism* (Losada et al. 2008), as we have already mentioned before, that encompasses a sense of duty of care from all members of the family to all other members of the family, irrespective of sex and age. But when family caregivers are alone and unaided, especially in circumstances of immigration into a foreign country, adherence to familism may add an additional level of burden and stress on the caregivers. Losada et al. (2006) examined the effects of caregiving burden and its relationship with familism in 48 Hispanic caregivers of dementia patients from Los Angeles (USA; the group included people with ethnic roots in Mexico, Puerto Rico, Cuba, El Salvador, Guatemala, Argentina, and Nicaragua) and 60 caregivers from Madrid (Spain). All groups embraced the concept of familism and displayed a positive relationship between burden of care and depressive symptoms, thus pointing to the costs of caring for a dementia patient. But whereas the burden was somewhat decreased in the USA sample under the effect of familism, it was increased in Spain, although the effect disappeared after controlling for level of education and years of caregiving. Thus, although familism is a shared cultural value among these Hispanic participants, it is expressed in somewhat different ways according to their specific situation. We agree with Losada et al. (2006) that the fact that the American sample is mainly composed of immigrants to the USA may indeed make a difference, as immigrants would be more likely to regard the family as a major referent for support when experiencing stressful circumstances.

In sum, given the above effects of ethnicity on perceptions of health, it is clear that for health programmes for the elders to be effective, especially in multi-ethnic societies, they must be specifically adapted to the cultural needs of those elders, in a process known as *cultural tailoring*. Kreuter et al. (2002, p. 137) define cultural tailoring as "any combination of information or change strategies intended to reach one specific person, based on characteristics that are unique to that person, related to the outcome of interest, and have been derived from an individual assessment". Thus, the elder can be better assisted to the extent that care is delivered in a personally and culturally appropriate manner. Doing so requires the understanding of how relevant cultural and social variables (e.g. ethnicity, immigration), as well as specific individual factors such as personality and life history, affect perceptions of health.

3.5 Old-Age Stereotypes: A Cross-Cultural View

Prejudice, stereotypes, discrimination, and stigma can have serious consequences when they produce a devaluation of the worthiness, if not humanity, of a target such as an old person. Social status is lost, or never gained, and it can be only recovered (or attained) with difficulty, if at all. In the words of Arnhoff et al. (1964, pp. 41–42):

> prestige and high status are accorded to those who perform functions the society values highly or finds necessary for its survival; if, for example, the society possesses no written language, the older person becomes the repository of, and transmission agent for, societal information, and is therefore valued. In our current society, such is not the case, and since the culture has been in a rapidly changing cycle with major technological advances and a chronically saturated labor market, the information and skills of the older person tend to be obsolete, unessential, and devalued. Low status of the aged results.

Such statement assumes that older people who can indeed prove themselves to be still "functional", "technologically savvy", "relevant" would automatically be able to access all the opportunities and respect that they deserve; the problem with stereotypes leading to *ageism* is that a negative judgment is often made about an old person simply from the fact of being old, thus denying the individual even the chance of proving his/her worth in deeds.

Stereotypes, despite their fame as crystallised opinions, can be also somewhat dynamic, being subject to reinforcement but also transformation through communication and mutual influence among the members of society (Cordella and Poiani 2014). When consensus is reached on a specific stereotype, such stereotypical view becomes *hegemonic* thus transforming itself into an "established reality", rather than the arbitrary social construct that it really is (Moscovici 1981). We have also mentioned before that stereotypes leading to stigma can be subject to social self-reinforcement in a positive feedback loop, whereby the stereotype is incorporated by the target into his/her own behaviour thus further reinforcing the reality of the stereotype (e.g. Goffman 1963). This process is also referred to by Cuddy et al. (2005) as the *oppositional stereotype enhancement effect*. This social dynamic can have serious negative consequences for the targets of stigma, being potentially detrimental to the cognitive abilities of older people. By decreasing elders' *self-efficacy* (Lachman et al. 1992)—that is, their conviction of being capable of achieving a specific goal—negative stereotypes can potentially impair the cognitive capacities of older people, such as memory (e.g. Ryan et al. 1992; Albert et al. 1995); although this is variable (Yoon et al. 2000). Negative stereotypes of old people could also contribute to decreasing their *mastery* (Surbone et al. 2007), which is their ability to exert control over life circumstances.

The behaviour of those using stereotypes in their interactions with the elders is reinforced by a *just-world* view, or the belief that "those who suffer have brought about their own misery" (Cuddy et al. 2005, p. 267). A just-world view simply legitimises the prevailing social structure with its inequalities and social hierarchies by deflecting the responsibility for the discriminatory behaviour from the actor to the victim. Jost and Banaji's (1994) *System-justification theory* suggests that stereotypes are further reinforced when individuals who are regarded socially subordinate are, however, granted some "warmth stereotypes" (e.g. the elder who is judged "ignorant" may be also seen as "cute") in order to decrease the sense of responsibility in the dominant group for the discrimination and inequalities imposed on the subordinate group. For instance, Cuddy et al. (2005) mention how the stereotypical "incompetent" elder can be still regarded as a person full of "warmth". Whether the elders themselves are happy to be considered "warm" but also rather "incompetent" is an entirely different matter.

Stereotypes, discrimination and their negative effects on old people are often criticised, but counter-arguments against such criticisms are also routinely put forward in the form of accusations of elders' "victimism" or "playing the victim", leading to an anti-victim stance in the rest of the community (see Rohrer 2008 for a critical analysis).

We will further expand the general analysis of stereotypes affecting old people in the section *The ethics of ageing: stereotypes, stigma and ageism* of Chap. 6. In what follows we comment on some cross-cultural aspects of old-age stereotypes.

In contemporary Western societies there are widespread positive and negative stereotypes about the elders. For instance, in a study carried out in the USA that investigated the prevailing stereotypes about old people among young, middle age, and old participants, Hummert et al. (1994a, b) described a set of seven prevailing negative and positive stereotypes: Perfect Grandparent, Golden Ager, John Wayne Conservative, Severely Impaired, Shrew/Curmudgeon, Despondent, and Recluse. In general, however, negative stereotypes tend to prevail over the positive ones (e.g. Boduroglu et al. 2006).

Arnhoff et al. (1964) carried out a cross-cultural study of age-stereotyping among six groups of university students from various countries: the USA ($n = 423$), Japan ($n = 184$), Puerto Rico ($n = 246$), Sweden ($n = 305$), Greece ($n = 336$), and England ($n = 245$). They administered the *Attitudes Towards Old People Questionnaire* which provides a list of statements about old people. The tendency to stereotype old people varied across countries from 46% to 66.5%: USA 46.0%, Sweden 49.4%, England 52.2%, Puerto Rico 56.6%, Japan 57.3%, Greece 66.5%. High levels of stereotyping can start a self-reinforcing process that could difficult changing the views about the elders in society. This work was carried out in the 1960s and it is interesting to see how the countries that here show the lowest levels of stereotyping of the elders (e.g. USA, Sweden) are also those that subsequently have been at the forefront in the creation of new and more proactive ways of living in old age (see Chap. 2).

We have seen in Chap. 2 how the negative stereotyping of old people has ancient historical roots, emerging whenever the young pressed ahead in order to achieve greater power, against the interests of the old. The ancient competitive tension between the young and the old continues to these days. Current cultures that are still formally *gerontocratic* may become fertile ground for the emergence of negative stereotypes against old people under the influence of modernisation. This was shown, for instance, by Sharps et al. (1998) in a comparison between rural northern Thailand (with a strong traditional culture of filial piety and respect for the elders) and the USA. In this case, young adult Thai respondents displayed a relatively more negative attitude towards old people than their counterparts in the USA. Sharps et al. correctly interpreted these results in terms of increased inter-generational conflicts in the formally gerontocratic society of rural Thailand, in the face of recent cultural changes; a phenomenon that has repeated itself periodically in history since at least ancient Greece and Rome.

Negative stereotyping of the old could be counteracted by further developing the core of positive stereotypes that may be already present in society. In a study carried out in New Zealand focusing on Chinese immigrants and respondents of European descent, Liu et al. (2003) described some positive stereotypes about old people in the Chinese sample that reflected a sense of filial piety. For instance, the Chinese to a degree retained the image of the "revered elder, wise in the ways of the world", which was not present among New Zealanders of European descent. European New Zealanders, on the other hand, shared the stereotype of the old person as "political animal" (a person full of contradictions and paradoxes). The "nurturing" view of the old person was common to both ethnic groups, as it was the "curmudgeon" elder stereotype.

Liu et al.'s (2003) work reinforces the idea that across ethnic groups, elders may live in a precarious balance between positive and negative stereotypes (see also Cuddy et al. 2005; Boduroglu et al. 2006): wise vs. curmudgeon. But in this game, the dice are loaded against the elders, as negative stereotypes are very easily reinforced (e.g. old people seen as incompetent, not competitive, incapable of keeping up with the pace of technological change, out-dated in their views), whereas it takes a considerable amount of effort for the elder to gain and retain a positive stereotype (e.g. wisdom, experience, balanced calm, focus on the issues that truly matter in life, and so forth).

Negative stereotypes against older people can be especially costly to their well-being when they are still below the legal age of retirement and yet find themselves in a situation of looking for work. The difficulties for older workers to find a job have been increasing around the world in recent times under the influence of neoliberalism, with its emphasis on efficiency, cost-cutting, and competitiveness; whether the country does possess anti-discrimination laws (which may be circumvented by some employers) or not (Numhauser-Henning 2018). In terms of stereotypes, although there are some views of older workers that can be seen as positive: they have more experience, are more loyal and reliable, more interpersonally skilled, more conscientious, more confident and effective, think more before acting and work better within a team; they are also burdened with negative stereotypes such as being allegedly less flexible and motivated, possessing outdated skills, poor health and fitness, they are supposed to be less open to new technology, less adaptable to change and less interested in training (Chiu et al. 2001).

In sum, although negative stereotypes leading to discrimination against older people are a cross-cultural reality and may have become hegemonic within society, they are not necessarily immutable. Changes can occur within cultures in situ and through the process of immigration. We believe that the following extract from Cuddy et al. (2005, p. 27) nicely summarises the challenges and potential solutions to the stereotyping of old people across cultures: "While noticing someone's age is not inherently offensive, acting on age-based stereotypes clearly is; sadly, prejudice against older people often goes unchallenged by mainstream society. Nevertheless, certain avenues may decrease ageist beliefs and behaviors. High-quality intergenerational contact … and exposure to stereotype-inconsistent … and individuating … information could help to undo the elderly stereotype by restoring a sense of esteem for older people". The "incompetence" stereotype might be challenged by simply allowing the elders to prove themselves, and then judge their performance on the ground of the results achieved. There are also some positive stereotypes about the elders, such as being wise, which may help the elders find fulfilment in old age. In a recent study of ageing stereotypes comparing 26 cultures, Löckenhoff et al. (2009a) concluded that some broad cross-cultural trends are revealed where elders are perceived as declining in physical attractiveness, their ability to perform everyday tasks, and capacity to acquire new learning; whereas on the positive side they are perceived as individuals with increased wisdom, knowledge, and deserving of respect.

3.6 The Ethnic Dimension of Elder Abuse and Neglect

Discrimination associated with negative stereotypes of the elders could go as far as leading to elder abuse and neglect and even death. We have already seen in Chap. 2 how throughout history neglect and abuse of the elderly may have led to the hastening of their death. Anthony Glascock (2009) defines death hastening behaviour as: "killing, abandoning and forsaking of the elderly" (p. 84); where forsaking is defined as a "nonsupportive treatment that leads directly to the death of aged individuals… it includes the denial of sufficient food, medical care, clothing and shelter" (p. 85).

Glascock provides various anthropological examples of violence and death hastening behaviours against the elderly. For instance, the *Chukchee* of North-eastern Siberia allowed ill old persons who were becoming a burden to the group to be killed by a nearest relative. The *Lau* Pacific Islanders took the old and weak to an islet in a lagoon and abandoned them there to their fate, whereas the *Bororo* of Central Brazil at one point just stopped bringing food to the old and frail. The *Yakut* of north-central Siberia also stopped feeding the old and may have beaten them and often forced them to leave their house. Among the *Ju/'hoansi* there were cases in the past of frail elderly suffering a hastened end to their life (Sokolovsky 2009b). Sokolovsky (2009b) also reports that in Ghana both men and women practice witchcraft, but only women—the old ones in particular—are banished from their village as a consequence of it, whereas men tend to be feared. Once an old woman has been banished, she becomes stigmatised and risks to be lynched if she comes back. In the male-dominated gerontocratic society of the *Tiwi* people from the Tiwi Islands of Northern Australia, some old and frail women may have been killed by their sons and brothers (Sokolovsky 2009a); whereas among the *Aché* of Eastern Paraguay, Hill and Hurtado (1996) described how extremely old people who could not be carried from one place to another were left behind (old men) or killed (old women). A very gruesome description of the mode of killing old women is given in pp. 236–237 of Hill and Hurtado's book, as recounted by a 75-year-old man recollecting memories of his youth in a tape recorded in 1985. In the Italian island of Sardinia there was an ancient tradition of death hastening of terminally ill old people through the services of a woman, specialised in such practice, known as *accabadora* (Tulli 2016). The accabadora euthanised the old infirm by asphyxiation (by using a rod known as *su jualeddu*) or, after caressing the head of the elder and whilst reciting the Holy Rosary she struck a blow on the head with a hammer known as *su mazzolu*.

In his review, Glascock (2009) found death hastening behaviours in 21 of 41 societies studied. Death of old people is most frequently hastened by killing (14 of the 21 societies where death hastening behaviours occur). Nine societies use forsaking, whereas in eight societies old people are just abandoned to their fate. A combination of killing and abandoning is found in five cultures. Interestingly, both old men and old women receive the same treatment in these cases. In such circumstances the decision to hasten the death of the old person was made by the family; the reason for this decision was mainly because the old person had become a burden to the group. In fact, in the same societies, older people who are not a serious burden (they are not decrepit) do receive support from their family. Such abuse of the elderly—sometimes leading to death—also occurs in modern Western societies, where it is often associated with caregiver psychological morbidity (Cooper et al. 2008).

Death hastening behaviours, however, are at the extreme end of a spectrum involving a variety of forms of elder abuse which can be physical or psychological, sexual, and financial (Malley-Morrison et al. 2006, Cooper et al. 2008, see also Table 3.3). With the caveat that there is a real potential for under-reporting abuse, prevalence values of elder abuse tend to range between 3.2% and 27.5% in population studies according to Cooper et al. (2008); but there are cases where that maximum has been exceeded. Patterson and Malley-Morrison (2006) quote estimated levels of abuse ranging from a low 0.5% to as high as 18.4% in the general population of old people in Israel, with high levels of abuse reported especially among old Arab women. The range in Germany goes from 2% among elders living independently to as high as 68% in the case of those with dementia. In Japan a rate of 17.9%

Table 3.3 Worst things that a family member can do to an old person among four ethnic groups in the USA

	Ethnic group			
Worst things	European-Americans ($N = 14$)	African-Americans ($N = 15$)	Puerto Ricans ($N = 8$)	Japanese-Americans ($N = 6$)
Psychological abuse	7 (50.0%)	8 (53.3%)	3 (37.5%)	5 (83.3%)
Psychological neglect	11 (78.6%)	7 (46.7%)	5 (62.5%)	2 (33.3%)
Physical neglect	3 (21.4%)	7 (46.7%)	5 (62.5%)	1 (16.7%)
Unmet expectations	1 (7.1%)	8 (53.3%)	1 (12.5%)	2 (33.3%)
Physical abuse	4 (28.6%)	2 (13.3%)	1 (12.5%)	2 (33.3%)
Exploitation	0	5 (33.3%)	0	0
Violation of rights	0	0	0	1 (6.7%)
Multiple abuse forms	0	0	0	1 (6.7%)

From Anetzberger et al. (1996)

has been reported. According to Lindert et al. (2013) psychological abuse of the elderly in Europe spanned from 10.4% in Italy to 29.7% in Sweden in their study. Financial abuse, however, was low in Sweden (1.8%) and high in Portugal (7.8%). In general, physical abuse was not very common and ranged from 1% in Italy to 4% in Sweden. The least common form of elder abuse was sexual which ranged from 0.3% in Italy and Spain to 1.5% in Greece. Anetzberger et al. (1996) provide additional examples of elder abuse among various ethnic groups.

In Patterson and Malley-Morrison's (2006) review, old women tend to report more abuse in general than men, and more psychological and economic abuse in particular. The perpetrators of abuse against older people are often family members and other carers of such elders, this includes a small number of cases of sadistic acts against elders occurring in situations of institutionalised care, such as nursing homes (Patterson and Malley-Morrison 2006; Cooper et al. 2008).

Recent studies have been also reporting abuse of the elderly perpetrated by daughters-in-law and also daughters in particular, perhaps in part because young females in the family are subject to higher burden of care of the old and therefore are more likely to suffer burnout, that could find relief through violent or abusive behaviours. Elder abuse can also result directly from financial pressures on the family; a situation that seems equivalent to the reports of elder abuse, neglect and even hastened death occurring under environmentally stressful conditions coming from the anthropological literature, as mentioned above in this section.

Barker (2009) studied abuse inflicted on elders in Niue, a western Polynesian island. In Polynesia, elders who have established their respected social position by middle age do not lose status as they become older. Such elders are respected not only due to their personal accomplishments in life or family background, but also as a result of their advanced age. The local cultural tradition establishes that older people should be "obeyed, respected, served and emulated" and in return, those elders will "nurture, teach, love and protect". This establishes a form of "social contract" of rights and duties, meaning that when the elders cannot fulfil their side of the bargain they could be neglected by the rest of the community. For instance, this was the case of old people who "yelled frequently, especially at night; swore at neighbors and kin; fought all time, hitting out at all and sundry; forgot people's names or what they were doing; wandered away at all times of day and night; talked only of events in the remote past, or conversed with absent friends and long-dead relatives; and stared vacantly about them, constantly drooled or were incontinent" (Barker 2009, pp. 609–610). Such symptomatology could well fit a possible diagnosis of dementia. In response, the family neglected their hygiene duties towards such elders (bathing, cleaning their homes), and did not provide them with appropriate food and other forms of material comfort. In particular, elders who were decrepit (they were mainly males) "received minimal attention or inadequate care". Neglect resulting from the onset of decrepitude can be minimised by the old persons by shifting their communal work to simpler and easier house cores, adopt new and valued social roles such as those of "storyteller", "clown", "traditional healer" and help raise grandchildren as their changing capacities allow, but some individuals may not be in the condition to provide any help at all.

In a study of perception of elder abuse carried out among two generations of European-Americans, African-Americans, Puerto Ricans, and Japanese-Americans (60 years old and older persons and their *Baby Boomer* caregiver), Anetzberger et al. (1996) found that European-Americans and Puerto Ricans considered "psychological neglect" as the worst thing that a family member could do to an old person, and for Puerto Ricans "physical neglect" was equally important. "Psychological abuse" was highest in the list for Japanese Americans, whereas for African Americans it was both "psychological abuse" and "unmet expectations" ("not being there when needed, no gifts, and disregarding wishes or opinions") (Table 3.3).

Thus, what in a technologically advanced Western society, such as the USA, various ethnic groups consider to be the worst things that one can do against older people, coincide with what is actually done against the elders across various kinds of societies, as we have seen in this section. The perception of open physical abuse against the elders seems to be more contained in the USA, compared with other forms of abuse, perhaps because physical abuse usually does not occur in public spaces, becoming less visible. A legal framework protecting the elderly from the worst forms of abuse may also help in decreasing violence and abuse across society.

Chang and Moon (1997) also studied perceptions of elder abuse but they did so among Korean Americans ($n = 100$, 60–94 years old, 65% women). They recorded 46 cases of abuse that included: financial abuse (36% of cases), psychological abuse (24%), culturally specific abuse (17%), neglect (15%), physical abuse (4%), and other (4%). Psychological abuse included "lack of respect", "inappropriate treatment of elders by family members", and "failure to employ Korean language usage that denotes respect" which includes disagreeing with the elder and failing to show politeness. Rude treatment was also mentioned (e.g. accusing grandmother of "smelling bad" and even of "eating Korean food"). Interestingly, Chang and Moon (1997) remark that more than half of the perpetrators of psychological abuse were daughters-in-law, but across all forms of abuse sons are the greatest perpetrators (33% of all cases of abuse) followed by the daughter-in-law (24% of cases). Such cases of perceived elder abuse are all seen as transgressions of the culturally expected filial piety (*Hyo* in Korean), an expectation that these elder immigrants have obviously maintained through the process of emigration to the USA.

Given that the family itself can be the source of elder abuse, programmes have been put in place to address this issue, especially among ethnic minorities. For instance, the *Family Care Conference* (FCC) that originated in New Zealand as an intervention for Maori families to prevent abuse of elders, has been also used in other countries such as the USA to intervene in Native American communities. The idea is to use the FCC to take advantage of traditional family, religious, and cultural values among ethnic minorities in order to overcome elders' abuse. Issues are discussed and problems about the welfare of the elder identified in meetings attended by family members, family-nominated supportive members of the community, a spiritual leader if requested, and health and social service providers. All participants bring their concerns to the meetings and eventually solutions are offered, with the engagement of both the immediate and extended family. The results of the intervention are subsequently evaluated in a follow up.

Finally, abuse grounded on age discrimination can be compounded with abuse based on other factors such as socio-economic status and ethnicity. Ethnic minorities being targeted with discrimination and abuse can in turn react with coping strategies. For instance, Mossakowski (2003) showed how Filipino Americans used ethnic identity and pride to buffer themselves against the negative effects of discrimination, with positive consequences on mental health. In a work studying negative discrimination against the *Mapuche* Native Americans in Chile, respondents expressed how such discrimination was "psychologically wounding" which led to a response of "anger, undifferentiated bad feelings, shame and sense of powerlessness". To safeguard their well-being against such discrimination the *Mapuche* adopted various coping strategies such as "self-protective, self-controlled or confronting actions", with resilience being enhanced by increased cooperation, ethnic-reaffirmation, and social bonds within the *Mapuche* community (Mellor et al. 2009). Group identification is also a strategy that increases well-being in the face of discrimination among African Americans (Branscombe et al. 1999). In general, finding social support is a major coping strategy for individuals experiencing discrimination and other sources of psychological stress (e.g. Knight and Sayegh 2010).

To recap, abuse and neglect of elderly people are not uncommon around the world, especially when old people become a drag on the rest of the family. Although cultural, educational, and legal actions can help prevent some cases of abuse, the core issue is to provide relief to the families, thus making the care of the elderly less of a burden.

3.7 Caring for Older People

Older people who are independent, in control of their life and well supported financially may still need a caring social environment that allows them to live a fulfilling old age. Others may prefer a more solitary lifestyle. But it is those who are not in full control even of the basic requirements of their daily living that should be the main concern for their immediate family, entourage of friends and acquaintances and, ultimately, government. This is obviously true if the elders are living in their native country, but some additional complications may emerge for the care of old migrants.

In multi-ethnic societies proper care of old people from various ethnic backgrounds (also known as people from *culturally and linguistically diverse* (CALD) backgrounds) should take into account the specific needs of those elders, whether care is provided at home or in institutional facilities (Ebrahim 1996; see also Radermacher et al. 2009; Federation of Ethnic Communities' Councils of Australia 2015, for a review of the situation in Australia). Moreover, care of CALD elders implies the availability of a caregiver who should be culturally competent, in order to provide effective support. This, in turn, also calls for the provision of meaningful assistance to the caregiver, in order to prevent caregiver burnout.

Traditionally, the first source of care for the elders has been their immediate family. Such care of the elderly provided by the family is still common practice in various parts of the world, whether the elders live on their own or are co-resident with their family, and it has led to increased international concern for the improvement of home care for old people (WHO 2015).

Among the *Ju/'hoansi* of Botswana the elders (*na*, considered to be so as early as in their mid and late forties) are persons with both physical limitations and generative and healing powers, and they expect to receive some help from their relatives. Rosenberg (2009) explains how older *Ju/'hoansi* tend to complain vociferously whenever they perceive that they are not receiving an appropriate level of care, the objective of their complaining is to "constantly lubricate" the system of mutual responsibility and caregiving. This is especially understandable in view of the traditional nomadic lifestyle of the *Ju/'hoansi* that in the past may have implied leaving the old and frail behind, whenever the group could not afford to carry the individual in their journeys. By reminding the family of their duties, the elders may try to keep the interest in their well-being alive. Older people see the care from the family as an "unquestionable right" and they do not perceive themselves as a burden. But of course, they can be a burden, hence the system of inter-generational care is actively maintained not only by the reminders from the elders but also through education, with the children being taught to care for their old parents in the future. Rosenberg (2009, p. 42) reported one of them, Gai Koma, saying:

> We feel under an obligation [to care for our parents] because they brought us up. We've drained all their energies. After they die, we would be left with bad luck if we had not cared for them. We could fall ill.

The burden of caring for the elders among the *Ju/'hoansi* is expected to be spread between genders, with women having a more natural propensity to establish "caregiving networks". Caregiving is also extended to the rest of the members of the community, not just the immediate family. Spontaneously, four to eight people take caregiving responsibilities towards a single elder. In Rosenberg's (2009, p. 45) words, old people and their caregivers "are rooted in a social matrix that undoubtedly eased the burden of care and perhaps enhanced the quality of this very frail person's last days".

The family is also a traditional core source of support for the elderly in India. Describing the deteriorating conditions of his grandfather, Atul Gawande (Gawande 2015, p. 15) wrote:

> In America, he would almost certainly have been placed in a nursing home. Health professionals have a formal classification system for the level of function a person has. If you cannot, without assistance, use the toilet, eat, dress, bathe, groom, get out of bed, get out of a chair, and walk—the eight 'Activities of Daily Living'—then you lack the capacity for basic physical independence. If you cannot shop for yourself, prepare your own food, maintain your housekeeping, do your laundry, manage your medications, make phone calls, travel on your own, and handle your finances—the eight 'Independent Activities of Daily Living'—then you lack the capacity to live safely on your own. My grandfather could per-

> form only some of the basic measures of independence, and few of the more complex ones. But in India, this was not of any dire consequence… It was clear that the family would ensure my grandfather could continue to live as he desired… The arrangement allowed him to maintain a way of life that few elderly people in modern societies can count on.

In Italy there is a strong tradition for the care of the elderly to be first and above all a family responsibility. Although in small villages care for the elders may also involve other members of the community, especially if the elder is a popular person. Weibel-Orlando (2009, p. 544) recounts the story of 95-year-old Zita:

> Her health and continued safety have become the concern and responsibility of the entire village. Her neighbors regularly check her whereabouts by knocking on the door of her tiny house on the piazza whenever Zita swerves from her well-known daily round of activities.

In Japan, the country with the highest percentage of old people in the world (see Chap. 1) and with still some traditional sense of filial piety, family-centred caregiving is expressed in words such as *funerai* (warm contact) and the concept of *skinship*, or physical touch by kin. Kin are expected to provide care for the elderly in the form of physical assistance with movement or bathing but also through emotional support and family relationships that should be empathetic (Jenike and Traphagan 2009). In return, the elders try not to be an excessive burden on their family, to the point that they may even pray for the end of their life. Jenike and Traphagan (2009) write about the *pokkuri* (or "swift" death) temples where older Japanese go to pray for a timely and peaceful death, preceded by a healthy old age. This may also explain why the rate of suicide among older Japanese is high for international standards.

Following the contemporary trend shifting the family away from multi-generational households, a trend that is seen in various industrialised countries (e.g. Gawande 2015), in Japan there has been an increase in the number of elders living alone, or with their spouse only, since the mid-1980s. This trend has somewhat affected the distribution of responsibilities for the care of the elders away from the family. China has also been going through very important social and economic transformations following the pro-market reforms initiated in the late 1970s. The traditional Chinese concept of filial piety was under attack during the Communist era of Mao Zedong, only to be replaced by a government-based system of care for the elders; but with the introduction of market reforms by Deng Xiaoping since 1978, economic inequality has been increasing and the care of the elders is back to be a responsibility of the family, but with a "modern" twist: filial piety applies only if parents had treated their children well, if not the child feels in no obligation to help the old parent. Zhang (2009) refers to this situation as one of *inter-generational exchange*. This has led to a disruption of traditional family structure, resulting in an increase in the number of elders living alone. In addition, market economy reforms have aggravated the competition for limited family resources between son and elderly parents, as the son tends to prioritise the needs of his wife and children and spends less time and resources providing care for his parents (Zhang 2009).

On the other hand, economic reforms have also introduced greater individual autonomy among some elders in China, especially those who are financially better

Fig. 3.6 Outdoor practice of *Tai Chi* in Beijing's Temple of Heaven (photo by Craig Nagy)

off. Such elders use their own resources to purchase eldercare as a user-pay service and, in cities, old people organise to help themselves, such as congregating in public parks to exercise (e.g. through *Tai Chi*; see Fig. 3.6) or dance. Spontaneous activities may even start as an initiative by a single person, who may then be joined by others. Hung and Crompton (2006) mention elders using the *Tsuen Wan Jockey Club Tak Wah Park* in Hong Kong to practice *Tai Chi* and Chinese *Kung Fu*. A 70–79-year-old male park user recounted that:

> I visit this park every morning. It is open at 6:30 am. I always come here around that time. I usually practice Tai Chi and Chinese Ku Fu, walk and talk to others. I learned Tai Chi and Chinese Ku Fu from an old person who is from mainland China. He was here to visit his relatives. He practiced Tai Chi in this park during his stay. Since I was retired and was spending my time in the park, I started to learn Tai Chi and Chinese Ku Fu with him. I have practiced them every day since then. While I am practicing, some people also follow my actions. I teach them Tai Chi and Chinese Ku Fu if they are interested in it. It is good for health and can cultivate the mind … This park is close to my residence. I have nothing to do, and have nowhere to go. I come here to practice Ku Fu and Tai Chi … walking or doing a little exercise here every day can improve my health, pass my time, and cultivate my mind (Hung and Crompton 2006, p. 305).

Such activities also help create social bonds among elders that could compensate for the decreased level of social contact with and support from the immediate family. Although this form of "active ageing" can be seen favourably from a current Western perspective, it is the actual level of fulfilment experienced by the elders that is the final test of its success.

Older people in need of care require the services of a caregiver of some sort, but caregivers—especially family members—do also need some support for themselves, in order to avoid burning-out. In recent times, much research has focused on the caregivers' perspective, the challenges they face and their needs as they care for older members of their family.

Carers of elders affected by Alzheimer's disease and other forms of dementia face the double challenge of both providing the required help to the old person and doing so with a decreasing level of reward, as the elder gradually loses mental faculties to the point of not recognising close family members anymore. The situation can become even more critical in the case of minority groups of immigrants who may have limited access to support from the extended family.

Traditional expectations among ethnic groups may bias the burden of care on a specific sex (usually women; e.g. Hispanics, Asians) which may reach breaking point when the family group is under cultural pressure not to expose the mental condition of their older members to the outside world. Unless the family opens up to external help, and they are also assisted in navigating the bureaucratic intricacies to access the services of government agencies, then the burden of the overall situation may become unbearable to the main carers. Dilworth-Anderson and Gibson (2002) also add additional cultural aspects that may interfere with seeking help in some ethnic groups of immigrants: low levels of education, poor access to information and/or resources, some religious beliefs.

Dilworth-Anderson and Gibson (2002) carried out a study in the USA with the collaboration of 121 caregivers of old people with dementia (African, Chinese, European, and Hispanic descent). Caregivers were interviewed and they assigned meaning to the illness of their family members by using three major cues: the *ageing process*, the *disease process*, and *pressure*. By *ageing process* caregivers meant that dementia is a "normal or natural" outcome of ageing, not a disease. Chinese Americans used this category of meaning to avoid seeking help from outside the family; and yet, following the effects of modernisation and acculturation, adult children were not as keen to provide help as it would have been expected in the same situation in China. This situation only made the helping effort of other adult members of the family harder. The *disease process* implies the construction of dementia as a biomedical phenomenon that caregivers may use to gain a better understanding of the situation and to lay down a plan for action. Western caregivers gave priority to this approach. The *pressure* category emphasised "illness definitions and meanings of dementia that are related to the sufferers' experiences of pressure, stress, worry, anxiety, trauma, and/or family discord" (Dilworth-Anderson and Gibson 2002, p. S59). This category was widespread across ethnic groups. Social pressure was especially evident among the Chinese caregivers who felt under the cultural obligation from family to adhere to the Asian concepts of filial piety and maintaining harmony in the family. Psychological pressure on caregivers also resulted from the Chinese conception of "worry" (*danxin*), that when it is accumulated in excess it can lead to physical and psychological imbalance. Psychological pressure/stress was also believed to be a co-contributing factor in the development of dementia across cultures.

We have already mentioned that caring for older members of the family tends to be gender-biased around the world, with women providing, or being expected to provide most of the care (e.g. Brewer 2001; Surbone et al. 2007, and references therein). Such caregiving carries a cost for women in terms of increased stress, fatigue, financial hardship, social isolation, and loneliness. The burden of care is further increased by emotions such as guilt, anger, and grief (Brewer 2001). This often leads to deterioration of both the health and financial situation of the female caregivers, unless they receive external support.

To recap, caring for older people is traditionally considered a responsibility of the immediate family across ethnic groups, with an extension to the broader community of friends and acquaintances depending on ethnic background and the specific situation (e.g. migration, living in a small town). The burden of care can be significant, especially if the elders suffer from serious health deterioration and if they are migrants; but alleviating such burden can be helped by assistance from family, acquaintances, community and the provision of appropriate government services. In the case of older people who still enjoy good health and live on their own, they often self-organise as a group of neighbours, thus decreasing the burden of care on their family.

3.8 Food Choices of the Elders

Food is one of the many important aspects of elder care that is subject to cultural variability. Carers of old people, especially institutional carers in nursing homes or in hospitals, should take questions such as: "What kind of food are older people willing to eat?" and "how does their choice contrast with the food that they should eat?" very seriously. Depending on the answer to those questions, the type of elder care provided could be effective or not.

The ageing process involves the dysregulation of some physiological functions that affect both the absorption of nutrients in the digestive system and some sensory capacities that can affect taste and smell and therefore decrease food intake (Morley 2001). In addition, the level of activity decreases in older individuals who therefore necessitate less food (Buchowski and Sun 1996). This eventually leads to loss of weight, although some behavioural/social factors can further accelerate weight loss (e.g. depression) or alternatively improve weight gain (e.g. social contact). For instance, eating alone tends to be associated with lower levels of food intake compared to eating in company (see Roberts and Rosenberg 2006 for a review).

But food intake is also regulated by more complex cognitive functions, as food can acquire diverse connotations according to the culture and contexts within culture. Food can have religious meanings, it can be a social mediator of hospitality or status, and there may be an expectation to consume some traditional foods during specific festivities or occasions. Some food items may not be regarded as suitable or desirable, whereas others may be regarded as indispensable irrespective of their nutritional value. Food may bring back good memories, and for immigrants it may be an important link with life in their native country and part of their ethnic identity.

Hence, any initiative to improve the quality of nutrition in older people must be informed by their cultural traditions (Bermudez and Tucker 2004).

Krondl and Lau (1982) proposed a general model of food selection based on perception. Our perceptions of food are both sensory: feeling of satiety, tolerance, and taste preferences; and cognitive: familiarity with specific ingredients, prestige conferred to some food items, price, convenience, health beliefs, and knowledge. We organise this overall system of perceptions about food into memory schemata that can then automatically direct our behaviour (e.g. Olson 1981; Barker 1982), in this case towards preferring specific kinds of food.

The choice of food by older people is therefore modulated by a set of factors that can vary in their specific relevance both culturally and individually. Herne (1995, p. 13) provided a list of such factors affecting food choice that we reproduce here verbatim:

- Personal factors: expectation of food, familiarity, personality, influence of others, appetites, moods, and emotions, meanings attached to food.
- Socio-economic factors: household income, cost of food, meaning and status of food, security, society.
- Educational factors: general education, nutrition education.
- Biological, physiological, and psychological factors: age, sex, physiology, psychology, biology.
- Cultural, religious, and regional factors: cultural origins, religious background, beliefs, and traditions of culture and race, geographical region.
- Extrinsic factors: environment, advertising, merchandising, time and seasonal variation.
- Intrinsic factors: appearance, odour, texture, colour, flavour, quantity, quality, preparation, and presentation.

In this section we will focus on the effects of ethnicity and its interaction with other factors to understand the choice of food among elderly people.

Preferences made on the ground of expected biological effects of food were shown by Wilcox et al. (2009) in their work with old African Americans and Caucasians in the USA. Old African Americans preferred eating fruit, fish, and chicken, whereas vegetables were a preferred food item for both African Americans and Caucasians. Caucasians explicitly indicated low-fat food as part of their diet. The reasons expressed by the elders for preferring such food items included expected effects on the functioning of the body, such as fish and blueberries being good for the brain, and pineapple being good for memory in particular. Consumption of dietary supplements and vitamins was also mentioned as a strategy to maintain brain function.

Whether older people exert their choice of food on the ground of cultural expectations to improve their health or just out of routine or acquired tastes, their choice of food does necessarily affect the ingestion of specific nutrients needed for their daily physiological functions. Bermúdez et al. (2000) studied this issue in a sample of 937, 60+ year old in the USA (majority women). Participants were distributed among four major ethnic groups: Puerto Ricans, Dominicans, Other Hispanics (Cubans, Central and South Americans), and Non-Hispanic Whites. The information was gathered as part of the *Massachusetts Hispanic Elders Study* (MAHES). Puerto Ricans tended to have high intake of carbohydrates, whereas Mexicans tended to display high intake of polyunsaturated fatty acids, which have an important role in nervous system functioning, immunity and the regulation of other essential aspects of physiology (e.g. Benatti et al. 2004). Cultural orientation affects these choices among older immigrants more than any other consideration, including socio-economic status.

In all three Hispanic groups there was higher intake of complex carbohydrates (which have several physiological functions in human health, e.g. Topping (2007)) and lower intake of monounsaturated fatty acids as compared with the Non-Hispanic Whites group. Puerto Rican and Dominican men also reported lower intake of simple sugars and alcohol and higher consumption of polyunsaturated fatty acids, the latter pattern coincided with the preference of Mexicans in general. In addition, Puerto Rican men reported lower intake of saturated fatty acids and Dominican men less cholesterol intake than Non-Hispanic Whites. Other Hispanics reported higher fibre and lower total fat intake than Non-Hispanic Whites.

With regard to the proportion of macronutrients to total energy: there was a higher contribution to energy intake from total carbohydrates and polyunsaturated fatty acids, and lower energy contribution from simple sugars and alcohol in Puerto Rican and Dominican men than in Non-Hispanic Whites. All Hispanic groups had lower intake of monounsaturated fatty acids than Non-Hispanic White men.

Ethnic differences detected among women were similar to those observed in men. Energy intake was lower for Puerto Ricans and Dominicans than Non-Hispanic White women. Women from all Hispanic groups reported higher intake of complex carbohydrates and lower intake of monounsaturated and saturated fatty acids than Non-Hispanic White women.

With increased level of acculturation, the diet of Hispanics becomes more similar to that of Non-Hispanic Whites. This was positively correlated with the time of residence in the USA. For instance, high-fibre cereals were consumed by the most acculturated Hispanics.

For Hispanics rice was the major single source of energy, whereas for the Non-Hispanic Whites it was bread. Total fat came mainly from cooking oil for all groups. However, it was used twice as much by Hispanics than Non-Hispanic Whites. Beans and plantains were major sources of complex carbohydrates for Hispanics, whereas for Non-Hispanic Whites it was bread, pasta, cakes, and others.

Therefore, old people of different ethnic groups living in the same country (USA) not only differ in terms of the distribution of the nutrients they consume as broad categories, but also in terms of the specific sources of such nutrients. Some of those food preferences can in fact be detrimental to the health of old people, especially when those food items are consumed in excess, such as the Non-Hispanic preferences for fatty foods and sources of carbohydrates such as cakes. It is possible that, at least in part, such differences in dietary preferences among elderly Hispanics and non-Hispanics can explain what in the USA is known as the *Hispanic paradox*, which points to better survival rates among elder Hispanics, compared with non-Hispanic Caucasians, in spite of the former being generally poorer and less educated (e.g. Markides and Eschbach 2005). Of course, other factors can also contribute to explain the *Hispanic paradox*, such as an initial bias among people who can migrate successfully towards those who are in a relatively good health condition (e.g. Crimmins et al. 2007); such initial bias could then be maintained into old age. Alternatively, old Hispanics who are in worse health condition may prefer to return to their native country (Markides and Eschbach 2005).

Total daily intake of energy and nutrients can be also distributed differently throughout the day among old people of different ethnic backgrounds. De Groot et al. (1998) carried out a study on food eaten and health in 681 Europeans, 74–79 years old (51.6% women) who were part of the prospective *Survey in Europe on Nutrition and the Elderly, a Concerted Action* (SENECA) programme started in 1988. The countries surveyed included: Denmark, France, Italy, The Netherlands, and Switzerland.

The distribution of meals throughout the day differed between countries but it was broadly similar between men and women, in accordance to the local cultural pattern. In most areas—Italy and The Netherlands being exceptions—more than 50% of participants started the day with a continental breakfast that included a drink, bread, butter, and jam/honey. In the Netherlands most men added cheese, eggs, fruits, cereals and/or cold cuts, thus their breakfast was rather more substantial; whereas in Italy it was much simpler, comprising of just a drink and some bread. At midday elders in French and Italian locations had their main cooked meal of the day to which they added a second cooked meal in the evening. In Switzerland, the men had a bread-meal in the evening. In both Denmark and The Netherlands men and women mainly had a bread-meal at midday and then a cooked meal in the evening. Overall the range of number of meals consumed in the day went from a minimum of 3 in France to about 7 in The Netherlands.

Energy intake varied at different times of the day and between ethnic groups. Figure 3.7 summarises the results obtained by De Groot et al. (1998). For both men and women there is a general trend for total energy intake to decrease at lower latitudes as one would expect from the effect of geographic changes in temperature (i.e. warmer climates in Southern Europe), although the trend is somewhat stronger for women.

Grain products and vegetables were consumed by all participants in the study, and the vast majority also consumed milk, meat, fruit, and sources of fat. Eggs and sugar were consumed only in small amounts. The ethnic groups, however, varied in the relative amount of those items in their diet and how they consumed them. Vegetable oils, for instance, were mainly consumed by the French and Italians.

Ethnic differences in the patterns of consumption of food can therefore reflect not only physiological and environmental factors but also the way older people perceive food, and their choice can eventually affect some aspects of their general health. Once we understand the motivation for food choice among the elders, we can then better design programmes that can improve the quality and quantity of the nutrients they consume, whilst maintaining their cultural preferences for specific food items as much as possible.

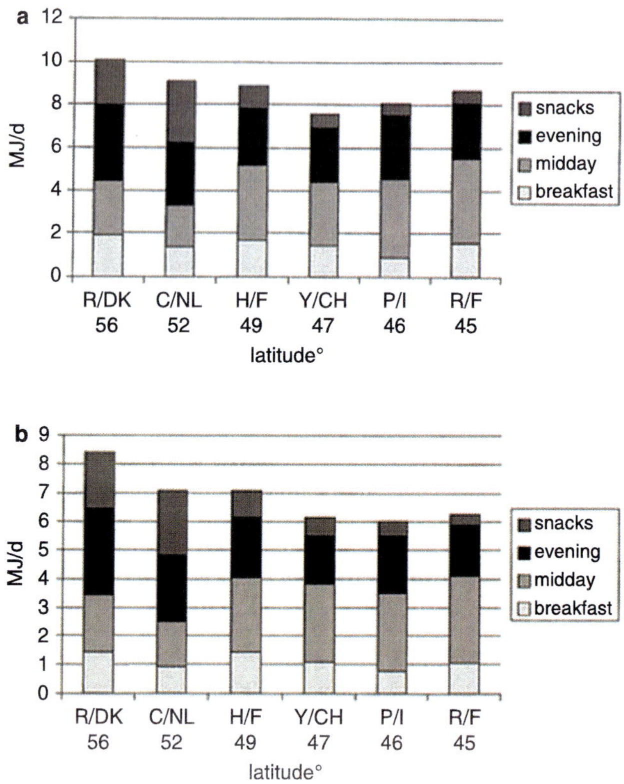

Fig. 3.7 Energy intake at different times of the day (expressed in MJ/d) by 74–79 years old: (**a**) Men, (**b**) Women in six different locations in Europe. *R/DK* Roskilde/Denmark, *C/NL* Culemborg/The Netherlands, *H/F* Haguenau/France, *Y/CH* Yverdon/Switzerland, *P/I* Padua/Italy, *R/F* Romans/France. The latitude of each location is also shown in the *x*-axis (from De Groot et al. 1998)

Although old people who migrated to other countries may continue to nurture their established food tastes and overall eating habits, some degree of change and adaptation may also occur through the process of *acculturation*, as we have already seen above for the case of Hispanics living in the USA (Bermúdez et al. 2000; see also Lin et al. 2003).

Both ethnic preferences for specific food items and capacity to acculturate, and therefore to accommodate to new diets, can become especially important in institutionalised situations, such as nursing homes, where meals may be mainly provided rather than personally cooked by the elder or her/his family. If the nursing home fails to meet the taste requirements of the ethnically diverse population of elders under their care, especially if the elders retain many aspects of their cultural food preference, the institution may risk some old people to become malnourished, with escalating consequences for their general health (Kayser-Jones 2001). Focusing on Chinese residents in a nursing home in the USA, Wu and Barker (2008) noted how mealtime in Chinese culture is a moment of sharing and socialising with the family, rather than a solitary affair. In the southern Chinese tradition, the family (or a group of friends) seat around a table, each with a bowl where they put rice and morsels of food (e.g. vegetables, fish, meat) communally available from larger dishes in the table. In this case food is served like a small smorgasbord, providing diversity of choice from which one selects the appropriate variety and amounts according to personal taste and current needs. The main component for breakfast is *juk*, a rice porridge that is flavoured in a diversity of ways according to local traditions and it is served with hot tea available from a common pot and poured into individual small tea cups.

Wu and Barker (2008) studied the food served to Chinese residents in a nursing home in the USA. Participants were 81 year old on average and mainly women. They had been living in the USA for 25 years (on average) but none of them spoke English fluently, thus presumably, their level of acculturation was limited. The nursing home took special care to focus on the nutritional value of food and did make some concessions to ethnic preferences such as providing *juk* and rice. They put together a so-called Asian diet which consisted of *juk* served with all three daily meals, rice at lunch and dinner, and to drink they served hot tea or water. Four Chinese entrees were also alternated in a cycle, such as tofu stir-fry and a dim sum platter. Despite such efforts, the Chinese residents did not find the menu appealing, sometimes refusing to eat their food, according to family members, which forced the family to bring their own appropriately cooked food from home. In fact, most residents did not even recognise the "Asian diet" as real Chinese food at all.

Regarding her mother's food preferences, one family member stated:

> No yogurt or ice cream. She doesn't like cold foods. She is afraid of cold food. The box milk [carton] she doesn't like. I think it's a habit. The elderly Chinese people; they don't like cold. Mom will spit it out (p. 51).

In traditional Chinese medicine the property of *qi* (a concept that we have already introduced in this chapter) determines the individual's health, whereas regarding food the *yin/yang* duality is a major organising principle, where *yin* is associated with cold and *yang* with hot. The link between *yin* (a negative principle) and cold also implies an association with darkness, passivity, and disintegration. Hence the negative reaction to cold food noted in the former quote is not surprising.

Food in nursing homes is therefore not always the highlight of the day for residents belonging to ethnic minorities. Non-institutionalised elders, on the other hand, can better arrange their meals according to their preferences of taste, location and their choice of companionship. In a study carried out in Hawaii, Cheang (2002) interviewed 26, 57–83 years old, second generation Japanese Americans who were regular clients in a restaurant where they had meals on average four times a week with their friends. Conversations with the elders revealed three major reasons for their regular gatherings around food: sociability, play, and laughter. Their voluntary association in a non-institutionalised setting allowed them to freely interact and, in their words, just "hang out with my friends" as they consumed the food of their choice. Whether their diet was nutritionally optimal given their age may have not been a high priority in their mind.

In sum, food is an important factor affecting both the general health of old people and their psychological well-being. Ingestion of food is modulated not only by needs but also habits—including social habits—tastes and cultural traditions. Optimal nutritional care of the elders, especially in institutionalised settings, should be able to provide a diet that combines both nutrition and sensory appeal, according to cultural preferences.

3.9 Life After Retirement

Widespread retirement from work in old age, and the leisure time it affords, is a recent phenomenon that in the past was confined only to the wealthiest social classes (see Chap. 2), whereas now it involves a large percentage of the workforce, especially in industrialised countries. Some aspects of the new lifestyles afforded to retirees who are in good health and enjoy a degree of financial security were mentioned in Chap. 2 (e.g. *Grey Nomads*, *RVers*, *Snowbirds*). Here we expand our exploration of life after retirement within a cross-cultural perspective.

One of the most basic forms of life after retirement can be found among the *Samia* of Kenya where frail elders will stop working and "just sit and eat" for as long as their family and community will help them, which can be seen as a form of "retirement" (Cattell and Albert 2009). How widespread are this and other forms of retirement for the elderly across cultures?

Luborsky and Leblanc (2003) have compiled a varied list of views on this issue from the literature that range from blunt statements such as "retirement is rare" to "all societies retire the elderly", to more nuanced descriptions of the reality of retirement that recognise both historical patterns (e.g. our current understanding of retirement that comes with the expectation of some form of pension is a relatively recent phenomenon), and cross-cultural variability (e.g. full retirement from work is uncommon in tribal societies). Although work can be ongoing and gradual shifts in the type of work and activities can occur with age in many traditional societies, in some others such as the *Fulani* of West Africa a married couple will abruptly face forced retirement and loss of their property (cattle) upon the marriage of their last child, their eldest son being in charge of caring for them. The couple is regarded to be "socially dead" from then on and are forced into sleeping in sexually segregated quarters.

The kind of activities the elderly can shift to after a certain age is also variable from society to society. In Asia it is not uncommon for older people to focus on religious matters, both public such as assuming active roles in religious ceremonies, and private, such as meditation. A degree of altruistic commitment to others is also expected in the traditional view of retirement in old age in China, as expected from the stage of old age, known as *lao-nien*, which starts at about 55 years old. But such commitment also comes with disengagement from the most arduous aspects of work, including housework for old women, which are taken over by younger members of the family (Luborsky and Leblanc 2003).

Wherever there is a cultural expectation for older people to experience some form of retirement at one point in their life, then the issue of preparing for retirement comes to the fore. Many practical and psychological considerations combine in the planning for retirement, such as: How to use the time available in retirement? Are finances under control? Are health issues being taken care of?

Winston and Barnes (2007) explored the preparation for retirement in the experience of a sample of academics. They analysed 21 interviews conducted with academics in the USA and 11 in New Zealand. All interviewees were women of the *Baby Boomers* generation. The *Baby Boomers*, especially the older members of the generation, are probably the best equipped individuals in history to be able to properly plan for retirement, as they enjoyed an unprecedented period of not only world economic expansion but also high employment, better education (making them also financially more savvy), and good remuneration for their work during the social democratic phase of the evolution of capitalism (from the end of World War II to the 1970s; e.g. Duménil and Lévy 2013), whilst still accessing considerable support from government services, including good pension schemes. This situation has changed quite dramatically for subsequent generations.

Many *Baby Boomers* have not only been anticipating their retirement and preparing for a "retiree role", but in fact they have been redefining the "role" of retiree altogether. The proactive stance of many *Baby Boomers* makes their retirement look not like the "roleless role" envisaged by Burgess (1960) but an almost extravaganza of potential roles, including aspects of continuity with their professional work. A recurrent theme of concern for the academics interviewed by Winston and Barnes was to see retirement as an opportunity to do things that their current work obligations made them difficult to achieve. Paradoxically, many such activities are supposed to be part of the job description of an academic, such as doing research and writing. Retirement for them was the chance to do the true academic work they could not do in their professional job as a result of the current constraining reality of academia, this is especially true in Anglo-Saxon countries that have experienced the full brunt of neoliberal reforms in the past 30 years. A. Poiani vividly recalls the case of Richard Zann, a professional ornithologist and academic, who on the verge of retirement was simply planning to spend some time in tropical north Queensland, Australia, to survey bird populations in the same areas he had surveyed decades before, to see what changes had occurred. Unfortunately, a tragic accident deprived him of the experience. Wishing for *continuity* with the past work life into retirement is expected to be especially prevalent among people who see their profession as a vocation. Stebbins (2000) regards such retirees as being in a state of "flow" (see Chap. 2).

In Winston and Barnes' work 55% of the academics that they interviewed would have liked to continue doing research and writing after retirement. One of their respondents said: "I don't see it (retirement) as a major change in how I'm going to live my life because I will continue to be active intellectually and active in research and writing—I don't see that changing!" (p. 147). The same percentage anticipated travelling and 41% was expecting to devote some time to gardening. Teaching (or other types of work) was an activity envisaged by 38% of respondents. Spending time with family and friends was mentioned by 31%, the same percentage as those who expected doing volunteering work in their community. Other activities that were mentioned included physical activities such as excursions, engaging in the arts, learning to better use modern technologies and other forms of retraining, doing various works at home.

Among the main concerns regarding retirement, the respondents mentioned healthcare/health and financial matters/pension. Indeed, the issue of financial security into retirement is a widespread worry for those approaching retirement age. Being on top of the financial security game demands a considerable degree of planning from the part of the individual, which should start early-on in working life. Hershey et al. (2007) compared the patterns of financial planning for retirement between two samples of respondents from the USA and The Netherlands (age range between 25 and 64 years old). In the USA, but increasingly also in The Netherlands, the responsibility for financial security in retirement has been shifting from government to the individual under the influence of neoliberalism. This shift has significantly increased the impact of personal circumstances on the ability to accumulate enough resources for a safe and satisfactory life in retirement. Even for workers who have experienced parallel careers, the combined effects of differences in personality, general education, specific knowledge of financial matters, clarity in their planning for the future, not to speak of the usual component of "luck" can produce a diversity of outcomes that will see some individuals better off and others worse off upon commencing their retirement, producing social inequality.

In a study carried out in multi-ethnic Singapore, Mehta (1999) surveyed 95 adults, 50–59 years old, and recorded their opinions and interactions in 13 focus groups. The ethnicities of the participants were Chinese, Malay, or Indian. Most of the participants were still employed. Although 65% of those who had not retired yet were against fixing a specific age for retirement, they thought that if an age for retirement must be set it should be older than it currently is: 60 years. They wished for 65 and over. Financial needs and maintaining employer-provided medical benefits were important reasons for delaying retirement, but they also added usefulness to society, avoiding boredom and remaining active. Despite these concerns about retirement, only around 50% of the sample had planned financially for retirement, as they assumed that a sense of filial piety would lead their children to help them. The better educated participants, however, had made some plans and were in a better financial position for retirement. There was a trend towards Chinese and Indians to make better preparations for their future, being also motivated by the need to pass some inheritance on to their children. A less materialistic approach was taken by the Muslim Malays who were more focused on transmitting ethical and religious values to their children. The Chinese did not mention any spiritual aspects in the preparation for retirement, whereas the Indians, like the Malays, did.

A good preparation for retirement is expected to smooth the transition into retirement proper. A common form of transition is to switch from the commitments of full-time employment to the less demanding *bridge employment*, which can be defined as "employment that takes place after a person's retirement from a full-time position but before the person's permanent withdrawal from the workforce" (Kim and Feldman 2000, p. 1195). This form of transition ensures a degree of continuity and smooth, rather than abrupt change, giving the individual the time to adjust and find new roles that will provide meaning to his/her life into the future (Atchley 1989). A high degree of continuity with the usual working life, can be also maintained throughout the transition into retirement (as already described for some research academics, see also Stebbins 2000) through volunteer work and leisure activities for instance.

Kim and Feldman (2000) studied bridge employment in a sample of academics from the University of California system who had decided to retire early. As an incentive to early retirement the university allowed them to still engage in part-time or temporary employment there. Accepting bridge employment was favoured by academics who were in good health, tenured, married to someone who was still working and had dependent children. Satisfaction was also better for those academics accepting bridge employment. But adjustment to retirement was further facilitated by also engaging in other activities outside work, such as leisure and volunteering.

Hershey and Henkens (2014) studied transition to retirement in The Netherlands with an emphasis on the distinction between voluntary and involuntary retirement. Indeed, it does make a difference whether retirement is a decision of the retiree, who is in full control of his/her life, or whether retirement is imposed on the retiree by external circumstances, such as retrenchment. In fact, as one would expect, there are consequences for life satisfaction depending on the type of retirement. Figure 3.8 summarises some of the results obtained by Hershey and Henkens (2014). Retirees who left work voluntarily had the highest mean score of life satisfaction, and the score did not change in the time interval considered in this study. A slightly lower degree of life satisfaction was found in those who were still working, and such satisfaction decreased with time. But the group that was in the worst situation were those who had been retired involuntarily. They were not only the least satisfied ones with their life, but they dropped their score more sharply with time than any of the other groups.

Once the individual is officially retired from work, then the focus shifts to whether she or he is adjusting well to the new situation. Wong and Earl (2009), after studying retirees in Australia, pointed out that better adjustment to retirement is predicted by higher income, being married, possessing better health, and better psychological well-being. However, adjustment may be compromised by the loss of social role and identity. This is not surprising as sociality is a central characteristic of our species and losing one's former place in society, without replacement with a new social role, can be disconcerting and frustrating. To better understand this issue Linda George proposed what is known as *Role theory* (George 1993).

A *social role* is defined by both the status of an individual and the behaviours (expected and expressed) associated with it within a specific social environment. Life transitions can be expressed as changes in social roles (e.g. within the professional and family contexts, for instance). Such role changes can be challenging to the retiree, in terms of some loss of life meaning; or they can be liberating, when the individual has already lost enthusiasm for his/her work and is looking forward to retirement.

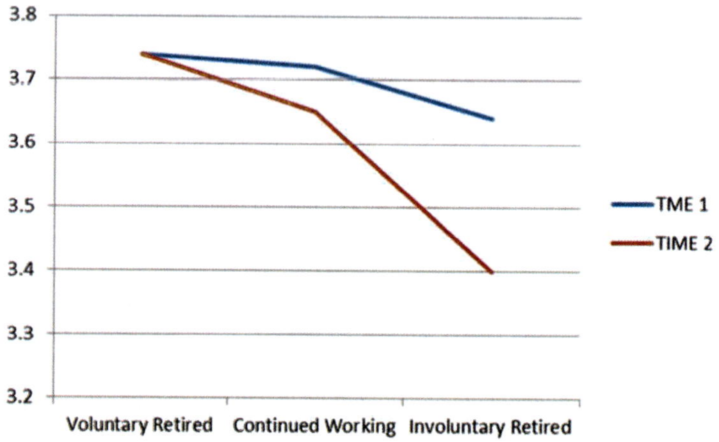

Fig. 3.8 Mean values of life satisfaction scores for participants who retired voluntarily, continued working or retired involuntarily. Values were taken at two times (T1: 2001 when respondents were 50 years old and older, and at T2: 2006–2007); redrawn in a different format from the data in Figure 1a of Hershey and Henkens (2014)

Role theory thus provides a perspective to understand the transition and adjustment to retirement. Wang et al. (2011) also reviewed other theoretical frameworks such as *Continuity theory* that emphasises the maintenance of personal identity throughout the transition to retirement—in spite of the obvious external changes—which should help in making such transition less stressful (see Atchley 1989). We have seen how continuity is displayed by some retiring academics, for instance.

The *bridge employment* mentioned above affords the individual an opportunity to extend the process of retirement in a more staged manner (*Stage theory*, Atchley 1976), without a drastic change in social role and providing greater continuity, thus allowing the individual to better adapt. But staged changes may also occur post-retirement: such as initial enthusiasm for a specific activity followed by disenchantment, perhaps because of limited financial resources or unmet expectations. Disenchantment may lead to a re-evaluation stage and a new adjustment to retirement into a different role, potentially leading to greater stability and fulfilment in life. Another approach to adjustment into retirement that we have already mentioned in this chapter is the *Life Course* perspective. In the words of Wang et al. (2011, p. 206): "The life course perspective thus emphasizes that life transitions are contextually embedded, which implies that the experiences of retirement adjustment are contingent on the specific circumstances under which the adjustment occurs". That is, adjustment will be a function of the combined effects of external circumstances and individual agency. Adjustment will be easier for people who have already developed an ability to be flexible and adapt to changing circumstances throughout their lives.

Finally, the *Resource Perspective* (e.g. Wang 2007) points to the need of considering all the resources available to the individual that could help in transitioning successfully towards retirement. Such resources include good health, motivation, education, finances, a supportive social network, and aspects of personality. These may vary through time both before and after retirement, thus requiring not only agency to actively pursue changes in a desired direction, but also a degree of flexibility and capacity to adjust, as illustrated in *Stage Theory* (e.g. Wang et al. 2011).

Clearly, all those approaches to the transition into retirement are proactive, requiring commitment, introspection, and decision-making from the part of the retiree, and they also require a degree of involvement in social life. Therefore, they run against the views of disengagement theory that we mentioned in Chap. 2 (see Cumming 1975). But in Chap. 2 we also mentioned that a degree of "disengagement", as seen in Tornstam's (1989) concept of *gerotranscendence,* could be productively used by old people, especially those with a more contemplative personality, to experience fulfilment in retirement.

Once the individual is fully into retirement, then confronting the need to establish a plan of engagement with life becomes inescapable. We have seen above how a group of USA and New Zealand academics were planning to spend their retirement years (Winston and Barnes 2007). The forms of engagement can be manifested into life-fulfilling activities that will vary according to physical capacities, social context, material needs, and resources, a desire for a degree of continuity with past interests (e.g. professional interests), and motivation to change to something very different from past occupations.

There is a current interest in understanding the role of diverse activities in retirement and how they can help the individual cope with the changes that may occur with ageing, such as health deterioration, loss of partner, altered relationships within the family, and the experience of various stresses. Such activities may contribute to the individual well-being in retirement, giving meaning to his/her life and providing challenges that the individual may strive to overcome (Havighurst 1963). But what kind of activity is both possible and meaningful for each individual will be variable and a result of personal preferences and constraints (Nimrod et al. 2009; see also the *Selective Optimisation with Compensation* (SOC) model of Baltes and Baltes (1990) that we mentioned in Chap. 1 and the *Socioemotional Selectivity* theory of Carstensen 1992). Activities that represent either *change* or *continuity* with the pre-retirement life can be equally valid, as the individual will adjust his/her behaviour according to the chosen path to fulfilling ageing.

Getting the activity right, whatever it might be—a cultural activity, outdoor activities, spiritual activities or any hobby, as suggested by Nimrod (2006)—can in fact be an important key to fulfilling ageing across cultures. The wrong activity could potentially decrease well-being and life satisfaction (Nimrod et al. 2008). The alternative to disengage from the many usual aspects of life and enter a more contemplative state altogether (Cumming 1975; Tornstam 1989) is also a possibility that may appeal to some retirees, as already mentioned. In addition, preferred activities in retirement may also vary cross-culturally which is an issue to consider especially in multicultural societies that have to deal with the care of elders coming from a variety of ethnic backgrounds (e.g. Nimrod et al. 2008).

The diversity of activities expected in retirement has been the focus of some attempts at categorisation. For instance, focusing on old people in Israel, Galit Nimrod (2007) described four types of post-retirement behaviour:

(a) *Expanders*: people who increase the number of activities they engage in during retirement whilst doing so at a higher frequency than they used to before retirement.
(b) *Reducers*: people who participate in the same or in fewer activities, doing so at a frequency that is either the same or lower than it was pre-retirement.
(c) *Concentrators*: these people engage in the same number or in fewer activities, but they do so at a higher frequency than before retirement.
(d) *Diffusers*: people who expand the number of activities they participate in, performing them at the same or at a lower frequency as compared to their behaviour before retirement.

Interestingly, the *expanders* and the *concentrators* expressed higher degrees of life satisfaction than the *reducers* in Nimrod's (2007) work, in support of both the continuity and innovation aspects of retirement. Innovation in retirement, as expressed by the *expanders*, stresses the Jungian framework that we discussed in Chap. 2 and that Nimrod and Kleiber (2007) refer to as *Innovation theory*; whereas *concentrators* may experience great states of flow by instead focusing on a smaller number of preferred activities in retirement that are however performed more intensely.

In their work carried out in the USA, Nimrod et al. (2009) studied people who had transitioned into retirement. Data were collected when respondents were still working and then again when they were retired. About 67% of participants were Caucasian and 70% were women. Participation in the following activities was assessed: talking to friends and family, visiting friends and family, involvement in clubs/organisations, religious activities, walking, gardening, physical exercise/sports, and volunteering. The distribution of individuals across the four types of behaviours described by Nimrod (2007) was unequal, with 51% of respondents fitting the description of *reducers*, followed in frequency by the *expanders* (21%), whereas *concentrators* and *diffusers* accounted for 16 and 12% of respondents respectively. Post-retirement *reducers* decreased their activities in the areas of phone conversations, attending clubs, walking, exercising, gardening, religious activities, and visiting friends and family. *Expanders*, on the other hand, increased their visiting of friends and family. In addition, both *concentrators* and *diffusers* did more volunteering than the *expanders*. Importantly, everybody seemed to be satisfied with life in general and were not highly affected by depression, although the *concentrators* were relatively more satisfied than the rest. This suggests that each individual simply tries to optimise his/her own satisfaction in retirement, according to personal characteristics and preferences. There is not just one, but many ways of experiencing personal fulfilment in retirement.

Leisure in retirement is regarded as a way of increasing well-being and meaning in life across cultures. In a comparison between Israel and the USA, Nimrod et al. (2008) focused on leisure activities of retirees who were still engaged in some remunerative work. The USA sample tended to show continuity through time in the type of leisure activity they participated in in retirement. In Israel there was an increased emphasis on volunteering and physical activities. Subjective well-being, as measured by life satisfaction, was predicted in Israel by an increased involvement in visiting siblings and physical activity. The USA sample showed no association of life satisfaction with changes in leisure activity through retirement, this may explain why this sample is more inclined to show continuity.

Whether they are responding to direct financial needs or following a desire for continuity (Stebbins 2000), some retirees look for an opportunity to engage in paid work as one of their activities in retirement, beyond any period of transition represented by bridge employment. Such retirees, however, tend to confront a wall of difficulties that range from low availability of appropriate work for them to straightforward prejudice. This issue has been tackled in Japan through the development of a network of *Silver Human Resource Centres* (SHRCs) that allows retirees to apply for flexible jobs, according to their preferences and needs (Weiss et al. 2005). Thanks to this initiative, older Japanese can not only complement their income but also nurture a greater sense of well-being and life meaning, as encapsulated in the Japanese concept of *ikigai*, which translates as having a sense of "life worth living" (Sone et al. 2008).

The use of time by older people in retirement was also studied cross-culturally by Gauthier and Smeeding (2003) comparing participants from various countries: The Netherlands, Finland, Germany, Austria, Italy, Canada, and the USA. Each participant in the study was asked to keep a diary of activities and the entries were then compared across nationalities. Here we comment on some of the results obtained for time spent in housework, active pursuits (e.g. hobbies, volunteer work, education, sport, and social and out-of-home leisure), passive activities (e.g. listening to radio or tapes, relaxing, watching TV), and personal activities (self-care activities such as sleeping, eating, bathing), with emphasis on 75 years old and older participants (old-old and oldest-old).

Regarding housework, there is a striking difference between men and women, with old women devoting relatively more time to housework than old men across all ethnic groups, the difference between sexes being higher in Italy and smaller in Canada. The reverse is true for active pursuits that attracted more time among men than women, with the largest difference being seen among the Italians and the smallest among Canadians and the Dutch. Passive activities were performed relatively more often by men than women. The Dutch sample was an exception as women devoted slightly more time to those activities than men. The smallest difference between the sexes in this area was seen in the Finnish sample, whereas the largest difference was detected in the Italian sample. Finally, time devoted to personal activities tended to be slightly higher among women than men in most countries except for Finland and Italy, with the largest difference between the sexes being found in Canada, whereas the smallest difference was detected in the Italian, Austrian, and Dutch samples. In short, these qualitative trends suggest that the patterns of various activities tend to follow some gendered directions that are broadly consistent across these Western countries, with 75+ old men devoting less time doing housework than women, but more time in active and passive pursuits, whilst women tended to devote a bit more time to personal activities. In terms of the ethnic patterns in gender differences for those activities, the Italian sample shows a relatively higher gender difference than other ethnicities; with the exception of the time devoted to personal activities.

Many of the activities performed in retirement that we have mentioned above fall into the category of *leisure* (Havighurst 1961), which can be regarded as a modern version of the ancient *otium honestum*. Leisure activities are personally satisfying actions that are performed by choice (see the various categories of *hobbyist* retirees listed by Stebbins (2000), and some of the *active* and *passive* activities mentioned by Gauthier and Smeeding (2003)) and have been associated with personal growth and well-being in retirement. *Serious leisure* has some of the characteristics of actual work (e.g. commitment) but without the stress component. Stebbins (1992) described six characteristics that define serious leisure: (1) perseverance in the activity, (2) personal and social benefits accrued from it, (3) a significant effort devoted to the activity, (4) the leisure activity is an expression of self and identity, (5) career development, and (6) it is an expression of a unique ethos. It is expected that people engaged in serious leisure may be more likely to perform the various activities in a state of flow.

Whereas in antiquity leisure in old age was the prerogative of the wealthier classes, since the mid-twentieth century it has become more accessible to more old people. Adams et al. (2011) list three major types of social and leisure activities that older people engage in according to their cultural and individual preferences: *informal* (such as social interactions with a diversity of people), *formal* (interactions in formal groups), and *solitary* (e.g. non-social hobbies). Leisure activities tend to be positive for well-being and a sense of fulfilment in life when they involve physical exercise, helping the individual develop meaning, nurturing self-identity, and fostering a sense of accomplishment through the mastery of specific skills; and especially for the informal and formal categories of leisure activities, when they involve social support.

Adams et al. (2011) reviewed the published literature on the relationship between leisure activities and well-being in old people drawing on examples from various countries around the world: Spain, Germany, The Netherlands, Finland, Sweden, Japan, China, Taiwan, Israel, Great Britain, and the USA. Most studies report a positive relationship between participating in a leisure activity and psychosocial well-being, health, or survival across countries. Some important gender differences were found in terms of the effect of leisure activities on well-being. For instance, following the well-known pattern of greater sociability in women than men, solitary activities reduced mortality risks for men but not women and the social environment was more influential on women's well-being (including survival) than on men's well-being.

Ethnic differences but also similarities can be found in the performance of specific leisure activities in retirement. Dergance et al. (2003) compared attitudes towards barriers and benefits of *leisure-time physical activity* (LTPA) between sedentary elderly Mexican Americans and European Americans in Texas. They found that important barriers to LTPA in both ethnic groups included: Self-consciousness and lack of self-discipline, interest in the activity, company to perform it with, enjoyment of it, and knowledge. Respondents did acknowledge both the physical and psychological benefits of LTPA—with the acknowledgment being more prevalent among Mexican Americans—but they did confront some barriers to actively engage in LTPA. Given that this was a sample of sedentary elders, a question naturally arises: If they know the benefits of LTPA, why don't they do it? Dergance et al. (2003) venture the hypothesis that in part the prevailing culture is to blame, where the elders are seen either as naturally sedentary (which only reinforces sedentariness) or, alternatively, they are expected to engage in highly demanding physical exercises fit for a twenty-year old, which may discourage some elders. But their work also hints to additional barriers, both individual and social. Ethnic differences also appear in the contradiction between knowledge and action, Mexican Americans in this study were perfectly aware of the benefits of LTPA, even more so than their European American counterparts, yet they engaged less in LTPA.

Leisure activities have been also studied among the elderly in India (Prakash 1999, and references therein). In this case watching television is a major activity for old people, but especially men, old women tend to continue performing their traditional role as housekeepers, for as long as they physically can, in addition to occupying their time sewing, reading or knitting. Prakash (1999) interviewed 60 old women in the Bangalore area of southern India: 30 from Bangalore City and 30 from rural villages in the Bangalore district; a group of 60 old men was also included (30 for each location) for comparison. Watching TV was the most common leisure activity for all participants across the board, this was followed by listening to the radio. Urban men were the most likely to engage in reading, presumably consequent to their better education, whereas urban women reported spending leisure time listening to music or singing. Rural men also engaged in social activities such as visiting friends and public places like cafes and restaurants, where they would chat with friends and acquaintances whilst drinking tea. Travelling outside town was an activity performed by both urban and rural elders, usually associated with family matters. What it is clear from this work is that the activities that all participants were involved in did not seem to be part of a specific and conscious plan, rather they were some form of extension of the pass-times and other daily routines they had before retirement. This situation is a reminder of similar "slow-life" patterns that were also prevalent in urban and rural areas in western countries in the past and that some people are striving to recover (e.g. Heitmann et al. 2011).

Contrasting with the "slow-life" scenario is the leisure time in retirement as understood by the highly mobile and active *Grey Nomads*, *RVers* and *Snowbirds* mentioned in Chap. 2.

The North American *Snowbirds* have a counterpart in Europe among the migratory elder Northern Europeans (British and Germans in particular) who have been spending part of the year in coastal towns in Spain in ever greater numbers since the 1980s, being attracted by the mild Mediterranean climate, easy-going lifestyle and also affordable costs of living (O'Reilly 2007). O'Reilly (2000) identified four distinct patterns of residency among the British visiting the *Costa del Sol* in the province of Málaga, Andalucía: (1) *full residents*, (2) *returning residents*, (3) *seasonal migrants*, and (4) *peripatetic migrants*, "who move back and forth between countries, often having a home in more than one place" (O'Reilly 2007, p. 145). These lifestyles have been variably named *residential tourism*, *amenity migration*, *later-life migration*, *sunshine migration,* and *gerontomigration* (Haas 2013). O'Reilly describes *residential tourists* as people who are affluent enough to be able "to turn tourism, to some extent, into a way of life, and to construct fluid, leisured lifestyles betwixt and between places, and in which even when they ostensibly try to settle they still remain in some ways outside or above the community they have moved to" (O'Reilly 2007, p. 145).

Residential tourism is undoubtedly a by-product of not only improved and more affordable technologies, motility and modes of communication that have promoted globalisation, but in this case also of greater unification within the European Union (although at the time of writing this text, the British residential tourists in Spain may be in danger of suffering some changes as a result of both the *Brexit* vote of June 2016 and the 2019 general election in the UK; https://www.bbc.com/news/world-europe-50777995).

Older English people spending time in Spain do not only enjoy what the local environment and people can offer, but they also contribute to the local communities with traditions that they bring from England. Haas (2013) mentions the case of volunteering and charity, activities that are well established not only in England but also in other English-speaking countries.

In his work carried out in the *Costa Blanca* (Valencia province) Heiko Haas focused on the community of English residents in this part of Spain, stressing their strong devotion to place and people. Haas (2013, p. 1380) quotes a very touching example that perfectly illustrates the multidimensional character of migration, leading to the development of what we may call *culturally syncretic identities*: the mental synthesis, within an individual, of experiences lived in different cultures throughout a lifetime. The quote comes from a tombstone inscription in one of the local cemeteries: "Son of Poland, Citizen of England, Lover of Spain". We regard a *culturally syncretic identity* as a personal reality that is far more coherent in terms of meaning than the sometimes disconnected *cultural bifocality* mentioned by Steve Vertovec (2007): "According to Vertovec, bifocality means 'having a life world oriented to, or grounded in, more than one locality', a 'dual orientation' of cultural orientations and affiliations, that is 'most clearly discernible in social practices and conveyed in individual narratives'" (Haas 2013, p. 1391). Bifocality as such splits, whereas syncretism unites. In Chap. 8 we will introduce and embrace the concept of *multiculturalism*, which implies both the respect for cultural diversity (cultural bi/multifocality) and learning from what others have to offer (cultural syncretism).

Haas (2013) explains that the local English community organised a charity that was devoted to the assistance of English residents and residential tourists. Such assistance helped the English residents to better integrate and navigate their way through the new cultural environment. This creates a social network of acquaintances, friends, and family residents that not only helps newcomers, but it becomes a further attractor for more immigrants. If specific institutional support is added to this social network, then the process of immigration can become self-sustaining (Massey et al. 1987, 1993). In addition, charitable work is also directed to the Spanish community as exemplified by cases of fundraising on behalf of a local hospital, for instance (Haas 2013).

Various areas in Mexico are common destinations for retirement migration from the USA. Sunil et al. (2007) studied migrant retirees to the area of *Lake Chapala* in the state of Jalisco. Just as it is the case for English retirees going to Spain, a warmer climate and lower costs of living are major pull factors attracting North American retirees to Mexico. More specifically, respondents in Sunil et al.'s work listed the following major reasons to retire in Mexico: low costs of living in general (88%), climate (69%), the "natural beauty of the area" (66%), the fact that foreigners are welcomed (55%), the presence of an already established and large community of immigrants from the USA (53%), and the access to quality health care at an affordable cost (47%). A slower-pace lifestyle is another important attractor that is also relevant to older retirees moving from Northern Europe (the UK in particular) to Southern European countries, perhaps looking to escape the high-stress/high-competition environment in their native country (e.g. Florio 2004).

But of course, there is no need to adopt a nomadic or seasonally migratory lifestyle in order to enjoy a sense of leisure, even serious leisure, in retirement. Elders around the world can organise themselves to increase their well-being through various leisure-time activities. In the USA *Senior Games* are organised as annual state-based multi-events for people 50 years old and older. The *Games* involve competitive activities (e.g. athletics) that require a degree of preparation from the part of the participants (Heo 2010). In Australia there is an equivalent, biennial initiative known as the *Australian Masters Games* which are open to old people but not only, as they have a minimum age of participation of 30 years. Participation in such serious leisure activities can have positive effects on both optimism and the reinforcement of self-identity (Heo 2010).

Chinese *Baby Boomers* are an interesting case of retirees when it comes to studying various aspects of old age, including leisure activities. This is so because many *Baby Boomers* in China started their reproductive life under the family planning environment of the one-child policy, which was introduced in China in 1978–1980 and it began to be officially dismantled only in 2015. This means that Chinese *Baby Boomers* are living their old age with only the support of one adult child at the most. This could potentially put such elderly people in relatively greater financial stress than the previous generation, whenever they require material support from family. In a study carried out in the city of Xi'an, Northwest China, Liang (2011) interviewed people who were either retired or preparing for retirement. Among the major themes that emerged in the interviews respondents mentioned their orientation towards family and the expression of traditional filial piety under the constraints of the one-child policy. The decrease in the size of the family social network in this cohort of elders affected aspects of meaning of life in retirement, as they found themselves with much time at their disposal without enough activities to fill it in, leading to feelings of "emptiness" and loneliness as expressed by a participant in the study, Mrs. Cai.

Others, however, expressed an opinion about leisure time in retirement that is consistent with current trends towards active retirement that we have seen emerging in Western countries. Mr. Qi, for instance, stated that "I think I'll find something to do after retirement, just for fun rather than out of economic considerations. I have pensions and health insurance, so I have nothing to worry about...after retirement, I want to go travelling, hiking and exercise more. I can't idle around though—much ado about nothing. I should find something to do to kill the time" (Liang 2011, p. 292).

Various hobbies were also mentioned, some continuing from pre-retirement life, such as dancing. In this case dancing was also an activity that opened the possibility of extending the social network beyond the family. Maintaining and expanding the social network was also expressed as an objective by those who were interested in post-retirement activities that were a continuation of their previous work life. A common objective, however, was to perform activities that are good for physical and mental health, including enjoying "a peaceful and contemplative life", especially for those who had a stressful working life. Mr. Ma expressed this wish so:

> I'm too busy. I can't calm down my mind while working. I want to read more. I feel like I've already stepped into old age. I'm gardening and raising birds. If I'm retired, I want to take a seat, grab a tea and read. I don't want to attend senior universities though. I'm longing for freedom. I think self-learning can better help me apprehend the truth by referring to my own experiences...all I want is to set my mind free...I like nature. I will go out for a walk as often as possible after I retire. I like taking the dirt road to walk along, so I can get access to the soil vapor (Di Qi) (Liang 2011, p. 294).

As we have seen *qi* is a concept that derives from Taoist philosophy approximating the meaning of "life force".

We conclude this section by mentioning that leisure in retirement is not just a form of activity expressing our fulfilment in life, or an experience, or a personal quest (for challenges, meaning and/or company). These days it is also a business. Social researchers have been unleashed to understand the consumption preferences of cohorts of elders, in order to better target commercial products to potential buyers (e.g. Szmigin and Carrigan 2001). Although retirees do consume, and offering them goods and services to satisfy their needs is perfectly legitimate, seeing them just as targets for products to sell is reductionistic in the extreme, and at the very least a system of consumer protection should be in place to specifically shield old retirees against scams and exploitation (Martin 2009).

In sum, given the trend towards a longer lifespan seen around the world, retirement is going to represent a significant part of the life of very many old people now and into the future. Living such final part of our life in a manner that is personally fulfilling requires good planning, if we want to access good medical support, financial security, and a lifestyle devoted to the pursue of what individual retirees find most interesting and motivating, according to their cultural background. Initiatives springing from individual and social acts of self-organisation will certainly play an important role in achieving such goals, but the requirement for government support cannot be ignored.

3.10 Old Age and End of Life Across Cultures

Although some of us may wish that leisure and fun may last forever, we are all aware that the lifespan has a limited duration. To conclude this chapter, we now turn to some cross-cultural aspects of the end of life, as they specifically affect older people.

Just as it is the case with almost any other aspect of our human condition, the study of death has also its own academic discipline, that in this case is known as *thanatology* (from the Greek *thanatos*, meaning "death"). Thanatology aims "to construct a scientific comprehension of death, its rites, and its meanings" (Mascarenhas Fonseca and Testoni 2011–2012, p. 157) and it can be described as "the study of life with death left in" (Kastenbaum 1993, p. 76).

A detailed review of the cross-cultural diversity of beliefs and rituals associated with death is beyond the scope of this section, but it is important to remark that such rituals are highly determined by the religious beliefs prevalent in each society and they have an important psychological function in providing emotional support to the bereaved (e.g. Irish et al. 1993; Robben 2004; Parkes et al. 2015) and to reinforce social bonds, as it is the case during the performance of *rites of aggregation* (Van Gennep 1960). Rites of aggregation strengthen emotional ties among the survivors, promoting stability and continuity through a transition in life. In rites of aggregation there is an expression of grief that provides cathartic relief and then a breaking of the emotional ties with the deceased one that helps in returning to life as usual, after a period of bereavement (Kagawa-Singer 1995). This process unfolds through what J. William Worden (1982) described as the four tasks of mourning:

1. accepting that the loss is a reality;
2. experiencing the emotional pain produced by grief;
3. adjusting to the new reality where the loved one is no longer present;
4. achieve emotional detachment from the lost loved one.

Rites of aggregation are at the distal point of a *rite of passage* process that starts with separation and the rites associated with it (*rites of separation*) soon after death of a dear one, that then give way to *rites of liminality* or the crossing of a threshold that is marked by ambiguity (Van Gennep 1960; Turner 1969), where the deceased one "passes through a cultural realm that has few or none of the attributes of the past or coming state" (Turner 1969, p. 94).

Some cultures are more accepting of death than others. In some cultural traditions the dead are even incorporated into the daily life of the survivors as ancestors, whereas others, like our modern and increasingly secular Western cultures, tend to somewhat ignore death and do what they can to delay it as much as possible.

Beliefs and rituals about death and dying are then taken by ethnic groups with them as they migrate to a different country (Lobar et al. 2006), leading to potential misunderstandings and miscommunication if there is no widespread awareness of such beliefs in the host community. In multi-ethnic Australia, where a majority of people still declare themselves to be Christian, many funeral services companies do provide special plans tailored to the growing Chinese community, for instance, offering Buddhist and Taoist ceremonies in addition to the more prevalent rites officiated in accordance to various Christian traditions.

But even in modern Western cultures not all old people keep thoughts about death out of their mind. In fact, it is not unusual especially for the oldest-old to be relaxed about openly expressing their thoughts on death and dying, without showing any sign of distress, let alone depression. In fact, there is often a negative correlation between death anxiety and age (Hallberg 2004).

Using the *Death Attitude Profile* (DAP) instrument Gesser et al. (1987) have described a curvilinear association between fear of death/dying and age, with maximum values seen in middle ages, minimum values in older ages, and young ages displaying intermediate values (see Fig. 3.9). Although this suggests that older people may be less fearful of death than younger ones, Hallberg also notes that in interviews, old people who are anxious about death may simply not address the topic unless they are asked explicitly. The idea of death can also become more existentially acceptable to old people who have greater satisfaction with their life achievements and feel more content and fulfilled, and they are also old enough for their body to suffer from serious handicaps of old age, impeding the full enjoyment of life. In addition, having probably already lost very many dear people (a spouse or partner, perhaps some children, good friends) the oldest-old may feel that approaching death is more like the inevitable closure of not just one's existence, but also of a "historical chapter", to be followed by new "chapters" authored by others. This is known as the *life span perspective*. To quote Hallberg (2004, p. 88):

> The life span perspective is perhaps best understood within the framework of narrative gerontology suggesting, among other things, that the meaning and nature of time is connected to our lives as a story, and that the story is open to be changed, i.e., our lives are

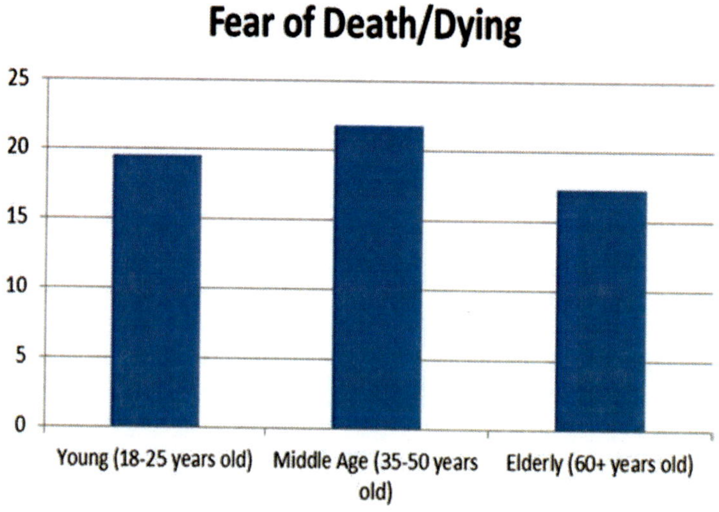

Fig. 3.9 Mean score of fear of death/dying at different ages recorded after the administration of the *Death Attitude Profile* (DAP) questionnaire; adapted from Gesser et al. (1987)

continuously re-storied. Supposing that death and dying is reviewed from a life span perspective, agony may differ in those who reach old age with a certain bitterness from those who look back at their life with satisfaction, but open to be re-storied.

Such lifespan perspective is likely to be a major differentiating point between the end of life at old or very old ages and the end of life at younger ages. At young ages (especially young adult ages) the feeling of life as an "unfinished business" is still strong, thus making the process of confronting death one of greater frustration, perhaps anger and often depression, with final acceptance coming only after some psychological effort. Old people can still have regrets, of course, but it is often the case that they may have done enough to look back to their life and find at least something that can bring a degree of satisfaction. That is why a life review exercise may be useful for the old and terminally ill patients (see Chochinov 2006 and Chap. 4 of Cordella and Poiani 2014). On the other hand, expecting to reach the end-of-life feeling to have achieved a stage of "finished business" is rather idealistic. After all, when can we ever say that any "business" is really "finished"? Life fulfilment is more easily experienced through a sense of having initiated or contributed to a "business", understood as any activity (research, contributing to a public or private institution in any capacity, children and family, the environment, a charity, a commercial company, a trade, a profession, teaching, and so forth) and then leave others to continue with it. What is completed at the end of our life is not necessarily a task, whatever it might be, but just our personal contribution to it in life.

Nevertheless, although death anxiety (or fear of death) is well known to decrease in old age across cultures as we have already stated (see also Cicirelli 2002; Depaola et al. 2003; Neimeyer et al. 2004 and references therein), it is not necessarily eliminated by the advancing years in all old people. The reason for this is simply that there are many factors that can affect death anxiety apart from age. We agree with Hallberg (2004, p. 100) that:

> Death/dying anxiety or fear of death/dying appears in a complex interaction, especially of aspects most probably related to factors such as personality aspects (hardiness, ego integrity, low neuroticism, external locus of control), social network, physical problems in the end-of life phase … and beliefs, including religious beliefs.

Hallberg also notes that in some cultures, including some Western cultures, feeling that one is a burden to others—family in particular—may be a source of additional anxiety to older people as they ponder about the process of dying. Other potential sources of death anxiety include psychological distress from whatever circumstance the individual may be confronting, low satisfaction in life, resilience, and place of residence such as institutions (e.g. a retirement village, etc.) (Neimeyer et al. 2004; see also Fortner et al. 2000). It is still unclear whether gender makes any difference in the display of death anxiety among older people (Fortner and Neimeyer 1999), although a recent study carried out in Malaysia suggests that death anxiety may be higher among women than men (age range 17–70 years) (Chin and Choo 2010); see also Chap. 8 for more examples of greater death anxiety in women than men. Oldest-old people tend to fear more the process of dying than the inevitable fact of death as such (Cicirelli 2002).

As an emotion, fear of death is likely to have deep roots in the biology of our mind, being obviously adaptive, but it can be modulated cognitively in different ways, and it can vary across ethnic groups and across cultures. A major theoretical framework to understand fear of death is provided by *Terror Management Theory* (Becker 1973) that we will analyse in more detail in Chap. 8. In the words of Neimeyer et al. (2004, pp. 327–328):

> human beings would be immobilized by dread if they lived constantly with the awareness of their mortality and the frailty of human life. Consequently, societies develop cultural worldviews with their associated cosmologies concerning the meaning of life and death as a way of ameliorating the uniquely human predicament of being able to anticipate our own eventual demise. In this account, the symbolic adherence to these worldviews is largely driven by a defensive function namely, the attempt to minimize the anxiety caused by the awareness of death.

We can decrease the anxiety caused by fear of death by suppressing our thoughts about it or by somehow rationalising death thus making it more acceptable (Greenberg et al. 1990). Acceptance of death can be achieved in various ways: by seeing death in a neutral manner as an inevitable fact of life (a highly-cognitive, low-emotional approach), by approaching death as a passage to a better afterlife or by seeing it as a relief of (an escape from) unbearable living conditions (Wong et al. 1994).

The combined cognitive-emotional nature of our confrontation with the inevitable fact of death has been clearly expressed by Neimeyer et al. (2004, pp. 329–330) in this quote:

> death acceptance is the conscious rational acknowledgement of the prospect of one's own inevitable death (confrontation with death, cognitive component) and at the same time the positive emotional appraisal of this realization (integration of death, affective component).

Many strategies have been developed across cultures to confront, integrate, and adaptively cope with the death of a dear one, to grieve and to mourn (Clements et al. 2003). Expressions of grief can be open such as crying and wailing, although among some ethnic groups—such as the Latinos—wailing loudly is a behaviour restricted to women. Men are expected to exert a higher degree of self-control, being considered the leaders of the family and therefore those who have the main duty of steering it through hard times. *Machismo* undoubtedly plays a role in the emotional self-control exerted by men. Comfort is sought not only in the company of close family and friends but also in physical contact with them, by hugging for instance (e.g. Clements et al. 2003). Among Catholics, such as a majority of Latin Americans and Southern Europeans, the family continues to maintain an emotional relationship with the deceased one through a period of mourning and beyond, not just in memory but also by regularly visiting his/her resting place, praying and, not unusually, talking to him/her. The relationship is often also maintained active through the mental recalling of the deceased in dreams, sometimes interpreted as "appearances" that in some traditions are believed to be motivated by the "willingness" of the deceased one to communicate.

Family and close friends are essential in bringing comfort to each other at a time of loss of a dear one across cultures. This is frequently helped by religious beliefs and the expectation to meet the deceased one in the afterlife. This linear view of life and death that can help cope with the anguish of loss and the general aspects of death anxiety is common to all major monotheistic religions (Christianity, Judaism, and Islam).

Among the *Navajo* Native Americans fear of death is also diminished by the belief in the linear continuation of existence into an afterlife. But, unlike the Christian tradition, remembrance and talking about the deceased one is not allowed, as it is believed that such recalling may interfere with the continuation of the journey of the deceased one. In addition, speaking of the deceased one could affect (physically or mentally) either him/her or the living relatives (Clements et al. 2003).

Fear of death is resolved in Hinduism with the belief in reincarnation, in which continuity is not linear but recursive. Good actions (*karmas*) accumulated through the various reincarnations, eventually lead to a state of self-knowledge and liberation (*jivanmukta*).

Following a religion can result in decreased death anxiety (e.g. through a belief in an afterlife or reincarnation) or it could increase it (e.g. through a belief in eternal punishment) (Jong and Halberstadt 2016), but death anxiety can be also controlled independently of religion. Chin and Choo (2010) carried out a study of death anxiety in Malaysia interviewing ethnic Malays, Chinese, Indians, and others spanning across four major religions: Islam, Buddhism, Christianity, and Hinduism. Importantly, the sample also included 50% of participants with no religion. No effect of religion (including intrinsic/extrinsic religiosity) was detected on death anxiety in this study, although Muslims seem to display a somewhat higher degree of death anxiety than Christians. Other studies, however, have shown that *intrinsic religiosity* (a more committed form of religiosity) is associated with decreased levels of death anxiety compared with *extrinsic religiosity* (which is a more superficial adherence to religious beliefs). In fact, some studies suggest that extrinsic religiosity could actively increase death anxiety (Neimeyer et al. 2004 and references therein).

Fear of death and death anxiety are likely to decrease at very old ages following a significant drop in life satisfaction, although this may be variable among individuals. In turn, life satisfaction decreases the closer we are to death. Gerstorf et al. (2008) investigated this issue in Germany using data from a longitudinal study: the German *Socio-Economic Panel Study* (SOEP) that includes 22 years of data obtained from 1637, 70 to 100 years old participants. Gerstorf et al. express their expectations regarding the relationship between old age and life satisfaction in clear, testable terms, reflecting what is known as the *Terminal Decline hypothesis* (Sliwinski et al. 2006):

> The underlying idea is that individuals typically have enough resources to maintain a sense of well-being, even as they face increased risks for losses in the social and health domains (i.e., normative age-related decline). As death approaches, however, individuals are faced with additional mortality-related burdens that make it increasingly difficult to maintain well-being—and the system collapses. If this speculation is true, intraindividual changes in

well-being occurring in the last years of life would be more closely associated to distance to death rather than distance from birth (i.e., chronological age) (p. 1149).

Thus, life satisfaction is expected to sharply decline in old people as they approach death. In fact, Gerstorf et al.'s results indicate that it is distance to death rather than age itself that is associated with a steep decline in life satisfaction. This is also in accordance with the findings of previous studies. They also found a transition point of about 4 years before death when the decline in life satisfaction significantly steepened. Previous studies had reported a similar transition point of between 2 and 6–8 years before death.

Physical and mental decline in old age may reach a point where intensive care may be required. Current views about how to approach such intensive care are strongly supportive of both patient- and family-centred approaches (e.g. Kaufman 1998). In addition, legalistic issues concerning the end of life have also come to prominence such as the requirement for *advance directives* (e.g. Blank 2011) which, however, may be affected by culture. There has been a recent push to globalise the practice of advance directives, in a process that Harrison et al. (2002) refer to as *ideational convergence*. However, critics have argued that local cultural traditions should be taken into account when designing such policies; in particular, concepts such as "individual autonomy" and "truth telling", that have been gaining prominence in some Western countries as end-of-life policies, are not universally accepted as best cultural practices around the world.

3.11 Conclusions

In this cross-cultural chapter we have seen how elders across cultures can remain engaged with life in different ways in the pursue of their life fulfilment and, in doing so, they can provide a contribution to society by transmitting their knowledge and life experience to the younger generations. They can also offer many useful services, especially to their family, through childcaring, and in some traditional societies they may hold religious roles. On the negative side, they can be the target of prejudice, negative discrimination and sometimes violence. Over time old people lose their ability to actively contribute, thus becoming a burden to their family. Much emphasis has been put in recent times around the world on the conditions that can help extend the active life of the elders and the opportunities to experience fulfilling ageing. Although experiencing fulfilling ageing can be helped by the provision of material and financial support for the elderly, other aspects of support such as showing respect and offering loving care are equally important. Respect for the elders and filial piety are still evident in Asian countries, where elders remain important for the maintenance of cultural continuity, despite recent changes coming with modernisation. With advancing age, the mental condition of the elders may deteriorate, which may increase the burden of care on their family caregivers. But for old people in good health, and in possession of enough wealth, new opportunities to enjoy life in retirement may open up, extending the possibilities for new forms of leisure and fulfilment. Irrespective of culture, life inevitably comes to an end. Across the world fear of death and death anxiety can be decreased in various ways: through some aspects of religiosity, other existential approaches to the meaning of life and through legacy.

References

Adams, K. B., Leibbrandt, S., & Moon, H. (2011). A critical review of the literature on social and leisure activity and wellbeing in later life. *Ageing and Society, 31*, 683–712.

Akiyama, H., Antonucci, T. C., & Campbell, R. (2009). Exchange and reciprocity among two generations of Japanese and American women. In J. Sokolovsky (Ed.), *The cultural context of aging: Worldwide perspectives* (p. 173). Westport, CT: Praeger. Available at www.stpt.usf.edu/~jsokolov/webbook/akiyama.htm

Albert, M., Jones, K., Savage, C., Berkman, L., Seeman, T., Blazer, D., et al. (1995). Predictors of cognitive change in older persons: Mac-Author studies of successful aging. *Psychology and Aging, 10*, 578–589.

Anetzberger, G. J., Korbin, J. E., & Tomita, S. K. (1996). Defining elder mistreatment in four ethnic groups across two generations. *Journal of Cross-Cultural Gerontology, 11*, 187–212.

Antonucci, T. C. (1985). Personal characteristics, social support, and social behavior. In R. H. Binstock & E. Shanas (Eds.), *Handbook of aging and the social sciences* (pp. 94–128). New York: Van Nostrand Reinhold Company.

Arnhoff, F. N., Leon, H. V., & Lorge, I. (1964). Cross-cultural acceptance of stereotypes towards aging. *The Journal of Social Psychology, 63*, 41–58.

Atchley, R. C. (1976). *The sociology of retirement*. New York: Halsted Press.

Atchley, R. C. (1989). A continuity theory of normal aging. *The Gerontologist, 29*, 183–190.

Avis, N. (2005). A universal menopausal syndrome? In: *NIH State-of-the-Science Conference on Management of Menopause-Related Symptoms*, March 21–23, 2005, pp. 43–46. William H. Natcher Conference Center, National Institutes of Health, Bethesda, MD.

Ayers, B., Forshaw, M., & Hunter, M. S. (2010). The impact of attitudes towards the menopause on women's symptom experience: A systematic review. *Maturitas, 65*, 28–36.

Baltes, P. B., & Baltes, M. M. (1990). Psychological perspectives on successful aging: The model of selective optimization with compensation. In P. B. Baltes & M. M. Baltes (Eds.), *Successful aging: Perspectives from the behavioral sciences* (pp. 1–34). Cambridge: Cambridge University Press.

Bardage, C., Pluijm, S. M. F., Pedersen, N. L., Deeg, D. J. H., Jylhä, M., Noale, M., et al. (2005). Self-rated health among older adults: A cross-national comparison. *European Journal of Ageing, 2*, 149–158.

Barker, J. C. (2009). Between humans and ghosts: The decrepit elderly in a Polynesian society. In J. Sokolovsky (Ed.), *The cultural context of aging: Worldwide perspectives* (pp. 606–621). Westport, CT: Praeger.

Barker, L. M. (1982). Building memories of food. In L. M. Barker (Ed.), *The Psychobiology of human food selection* (pp. 85–99). Westport, CT: AVI Publishing.

Barnes, D. E., Cauley, J. A., Lui, L.-Y., Fink, H. A., McCulloch, C., Stone, K. L., et al. (2007). Women who maintain optimal cognitive function into old age. *Journal of the American Geriatrics Society, 55*, 259–264.

Becker, E. (1973). *The denial of death*. New York: Macmillan.

Becker, G., Beyene, Y., Newsom, E., & Mayen, N. (2003). Creating continuity through mutual assistance: Intergenerational reciprocity in four ethnic groups. *Journal of Gerontology, 58B*, S151–S159.

Benatti, P., Peluso, G., Nicolai, R., & Calvani, M. (2004). Polyunsaturated fatty acids: Biochemical, nutritional and epigenetic properties. *Journal of the American College of Nutrition, 23*, 281–302.

Benedetti, R., Cohen, L., & Taylor, M. (2013). "There's really no other option": Italian Australians' experiences of caring for a family member with dementia. *Journal of Women and Aging, 25*, 138–164.

Benyamini, Y., Idler, E. L., Levental, H., & Levental, E. A. (2000). Positive affect and function as influences on self-assessments of health: Expanding our view beyond illness and disability. *Journal of Gerontology, 55B*, 107–116.

Bermúdez, O. I., Falcón, L. M., & Tucker, K. L. (2000). Intake and food sources of macronutrients among older Hispanic adults: Association with ethnicity, acculturation, and length of residence in the United States. *Journal of the American Dietetic Association, 100*, 665–673.

Bermudez, O. I., & Tucker, K. L. (2004). Cultural aspects of food choices in various communities of elders. *Generations, 28*, 22–27.

Berndt, R. M., & Berndt, C. H. (1992). The world of the first Australians. In *Aboriginal traditional life: Past and present*. Canberra, ACT: Aboriginal Studies Press.

Beyene, Y. (2009). Menopause: A biocultural event. In J. Sokolovsky (Ed.), *The cultural context of aging: Worldwide perspectives* (pp. 93–103). Westport, CT: Praeger.

Blank, R. H. (2011). End-of-life decision making across cultures. *Journal of Law, Medicine and Ethics Summer, 39*(2), 201–214.

Boduroglu, A., Yoon, C., Luo, T., & Park, D. C. (2006). Age-related stereotypes: A comparison of American and Chinese cultures. *Gerontology, 52*, 324–333.

Bombak, A. E. (2013). Self-rated health and public health: A critical perspective. *Frontiersin Public Health, 1*, 15. https://doi.org/10.3389/fpubh.2013.00015

Boorse, C. (1977). Health as a theoretical concept. *Philosophy of Science, 44*, 542–573.

Branscombe, N. R., Schmitt, M. T., & Harvey, R. D. (1999). Perceiving pervasive discrimination among African Americans: Implications for group identification and well-being. *Journal of Personality and Social Psychology, 77*, 135–149.

Brewer, L. (2001). Gender socialization and the cultural construction of elder caregivers. *Journal of Aging Studies, 15*, 217–235.

Bromberger, J. T., Meyer, P. M., Kravitz, H. M., Sommer, B., Cordal, A., Powell, L., et al. (2001). Psychologic distress and natural menopause: A multiethnic community study. *American Journal of Public Health, 91*, 1435–1442.

Buchowski, M. S., & Sun, M. (1996). Nutrition in minority elders: Current problems and future directions. *Journal of Health Care for the Poor and Underserved, 7*, 184–209.

Burbank, V. K. (2011). *An ethnography of stress: The social determinants of health in Aboriginal Australia*. New York: Palgrave Macmillan.

Burgess, E. W. (1960). *Aging in Western societies*. Chicago: University Press.

Carel, H., & Cooper, R. (2012). Introduction. In H. Carel & R. Cooper (Eds.), *Health, illness and disease: Philosophical essays* (pp. 1–22). Newcastle: Acumen.

Carstensen, L. L. (1992). Social and emotional patterns in adulthood: Support for socioemotional selectivity theory. *Psychology and Aging, 7*, 331–338.

Cattell, M. G. (2009). Global perspectives on widowhood and aging. In J. Sokolovsky (Ed.), *The cultural context of aging: Worldwide perspectives* (pp. 155–172). Westport, CT: Praeger.

Cattell, M. G., & Albert, S. M. (2009). Elders, ancients, ancestors and the modern life course. In J. Sokolovsky (Ed.), *The cultural context of aging: Worldwide perspectives* (pp. 115–133). Westport, CT: Praeger.

Chang, J., & Moon, A. (1997). Korean American elderly's knowledge and perceptions of elder abuse. *Journal of Multicultural Social Work, 6*, 139–154.

Cheang, M. (2002). Older adults' frequent visits to a fast-food restaurant: Nonobligatory social interaction and the significance of play in a "third place". *Journal of Aging Studies, 16*, 303–321.

Cheng, S.-T. (2009). Generativity in later life: Perceived respect from younger generations as a determinant of goal disengagement and psychological well-being. *The Journals of Gerontology B, 64B*, 45–54.

Cheng, S.-T., & Chan, A. C. M. (2006). Filial piety and psychological well-being in well older Chinese. *The Journals of Gerontology B, 61*, P262–P269.

Chin, C. L., & Choo, Y. C. (2010). Age, gender, and religiosity as related to death anxiety. *Sunway Academic Journal, 6*, 1–15.

Chiu, W. C. K., Chan, A. W., Snape, E., & Redman, T. (2001). Age stereotypes and discriminatory attitudes towards older workers: An East–West comparison. *Human Relations, 54*, 629–661.

Chochinov, H. M. (2006). Dying, dignity, and new horizons in palliative end-of-life care. *CA: A Cancer Journal for Clinicians, 56*, 84–103.

Cicirelli, V. G. (2002). Fear of death in older adults: Predictions from terror management theory. *Journal of Gerontology, 57B*, P358–P366.

Clapham, K., & Duncan, C. (2017). Indigenous Australians and ageing: Responding to diversity in policy and practice. In K. O'Loughlin, C. Browning, & H. Kendig (Eds.), *Ageing in Australia: Challenges and opportunities* (pp. 103–125). New York: Springer.

Clements, P. T., Vigil, G. J., Manno, M. S., Henry, G. C., Wilks, J., Das, S., et al. (2003). Cultural perspectives of death, grief, and bereavement. *Journal of Psychosocial Nursing and Mental Health Services, 41*, 18–26.

Cohen, G. D. (1994a). Journalistic elder abuse: It's time to get rid of fictions, get down to facts. *The Gerontologist, 34*, 399–401.

Cohen, L. (1994b). Old age: Cultural and critical perspectives. *Annual Reviews of Anthropology, 23*, 137–158.

Confucius. (1989). *The analects of confucius*. New York: Vintage Books.

Connell, C. M., & Gibson, G. D. (1997). Racial, ethnic, and cultural differences in dementia care-giving: Review and analysis. *The Gerontologist, 37*, 355–364.

Cooper, C., Selwood, A., & Livingston, G. (2008). The prevalence of elder abuse and neglect: A systematic review. *Age and Ageing, 37*, 151–160.

Cordella, M., & Poiani, A. (2014). *Behavioural oncology: Psychological , communicative, and social dimensions*. New York: Springer.

Costa, P. T., & McCrae, R. R. (1992). *Revised NEO Personality Inventory (NEO-PI-R) and NEO Five-Factor Inventory (NEO-FFI) professional manual*. Odessa, FL: Psychological Assessment Resources.

Cowgill, D. O. (1986). *Aging around the world*. Belmont, CA: Wadsworth.

Crimmins, E. M., Kim, J. K., Alley, D. E., Karlamangla, A., & Seeman, T. (2007). Hispanic paradox in biological risk profiles. *American Journal of Public Health, 97*, 1305–1310.

Cuddy, A. J. C., Norton, M. I., & Fiske, S. T. (2005). This old stereotype: The pervasiveness and persistence of the elderly stereotype. *Journal of Social Issues, 61*, 265–283.

Cumming, E. (1975). Engagement with an Old Theory. *International Journal of Aging and Human Development, 6*, 187–191.

De Groot, C. P. G. M., Schlettwein-Gsell, D., Schroll-Bjørnsbo, K., & van Staveren, W. A. (1998). Meal patterns and food selection of elderly people from six European towns. *Food Quality and Preference, 9*, 479–486.

Depaola, S. J., Griffin, M., Young, J. R., & Neimeyer, R. A. (2003). Death anxiety and attitudes toward the elderly among older adults: The role of gender and ethnicity. *Death Studies, 27*, 335–354.

Dergance, J. M., Calmbach, W. L., Dhanda, R., Miles, T. P., Hazuda, H. P., & Mouton, C. P. (2003). Barriers to and benefits of leisure time physical activity in the elderly: Differences across cultures. *Journal of the American Geriatrics Society, 51*, 863–868.

Dilworth-Anderson, P., & Gibson, B. E. (2002). The cultural influence of values, norms, meanings, and perceptions in understanding dementia in ethnic minorities. *Alzheimer Disease and Associated Disorders, 16*(Suppl. 2), S56–S63.

Dowd, J. (1983). Social exchange, class, and old people. In J. Sokolovsky (Ed.), *Growing old in different societies: Cross-culture perspectives* (pp. 29–42). Belmont, CA: Wadsworth.
Downs, M. (2000). Dementia in a socio-cultural context: An idea whose time has come. *Ageing and Society, 20*, 369–375.
Duménil, G., & Lévy, D. (2013). *The crisis of neoliberalism*. Cambridge, MA: Harvard University Press.
Ebrahim, S. (1996). Ethnic elders. *British Medical Journal, 313*, 610–613.
Edwards, H., & Noller, P. (1993). Perceptions of overaccommodation used by nurses in communication with the elderly. *Journal of Language and Social Psychology, 12*, 207–223.
Eriksson, B. G., Hessler, R. M., Sundh, V., & Steen, B. (1999). Cross-cultural analysis of longevity among Swedish and American elders: The role of social networks in the Gothenburg and Missouri longitudinal studies compared. *Archives of Gerontology and Geriatrics, 28*, 131–148.
Federation of Ethnic Communities' Councils of Australia. (2015). *Review of Australian research on older people from culturally and linguistically diverse backgrounds*. Deakin, ACT: FECCA.
Fernández-Ballesteros, R., Garcia, L. F., Abarca, D., Blanc, E., Efklides, A., Moraitou, D., et al. (2010). The concept of 'ageing well' in ten Latin American and European countries. *Ageing and Society, 30*, 41–56.
Florio, M. (2004). *The great divestiture: Evaluating the welfare impact of the British privatizations 1979–1997*. Cambridge, MA: The MIT Press.
Fortes, M. (1961). Pietas in ancestor worship. *The Journal of the Royal Anthropological Institute of Great Britain and Ireland, 91*, 166–191.
Fortner, B. V., & Neimeyer, R. A. (1999). Death anxiety in older adults: A quantitative review. *Death Studies, 23*, 387–411.
Fortner, B. V., Neimeyer, R. A., & Rybarczyk, B. (2000). Correlates of death anxiety in older adults: A comprehensive review. In A. Tomer (Ed.), *Death attitudes and the older adult* (pp. 95–108). New York: Brunner Routledge.
Gallois, C., Giles, H., Ota, H., Pierson, H. D., Ng, S. H., Lim, T.-S., et al. (1999). Intergenerational communication across the Pacific Rim: The impact of filial piety. In J.-C. Lasry, J. Adair, & K. Dion (Eds.), *Latest contributions to cross-cultural psychology* (pp. 192–211). Lisse: Swets and Zeitlinger.
Gauthier, A. H., & Smeeding, T. M. (2003). Time use at older Ages: Cross national differences. *Research on Aging, 25*, 247–274.
Gawande, A. (2015). *Being mortal: Illness, medicine, and what matters in the end*. London: Wellcome Collection.
George, L. K. (1993). Sociological perspectives on life transitions. *Annual Review of Sociology, 19*, 353–373.
Gerstorf, D., Ram, N., Estabrook, R., Schupp, J., Wagner, G. G., & Lindenberger, U. (2008). Life satisfaction shows terminal decline in old age: Longitudinal evidence from the German Socioeconomic Panel Study. *Developmental Psychology, 44*, 1148–1159.
Gesser, G., Wong, P. T. P., & Reker, G. T. (1987). Death attitudes across the life-span: The development and validation of the death attitude profile. *Omega, 18*, 113–128.
Giles, H., Noels, K. A., Williams, A., Ota, H., Lim, T.-S., Ng, S. H., et al. (2003). Intergenerational communication across cultures: Young people's perceptions of conversations with family elders, non-family elders and same-age peers. *Journal of Cross-Cultural Gerontology, 18*, 1–32.
Glascock, A. (2009). Is killing necessarily murder? Moral questions surrounding assisted suicide and death. In J. Sokolovsky (Ed.), *The cultural context of aging: Worldwide perspectives* (pp. 77–92). Westport, CT: Praeger.
Goffman, E. (1963). *Stigma*. Englewood Cliffs, NJ: Prentice-Hall.
Goffman, E. (1974). *Frame analysis: An essay on the organization of experience*. New York: Harper and Row.
Goodenough, W. H. (1981). *Culture, language, and society*. Menlo Park, CA: Benjamin/Cummings.
Greenberg, J., Pyszczynski, T., Solomon, S., Rosenblatt, A., Veeder, M., Kirkland, S., et al. (1990). Evidence for terror management II: The effects of mortality salience on reactions to those who

threaten or bolster the cultural worldview. *Journal of Personality and Social Psychology, 58*, 308–318.
Gusmano, M. K. (2009). Growing older in world cities: Benefits and burdens. In J. Sokolovsky (Ed.), *The cultural context of aging: Worldwide perspectives* (pp. 395–417). Westport, CT: Praeger.
Gutmann, D. (2002). Bringers of Allah: The Druse elders. In R. S. Weiss & S. A. Bass (Eds.), *Challenges of the third age: Meaning and purpose in later life* (pp. 115–136). Oxford: Oxford University Press.
Haas, H. (2013). Volunteering in retirement migration: Meanings and functions of charitable activities for older British residents in Spain. *Ageing and Society, 33*, 1374–1400.
Hallberg, I. R. (2004). Death and dying from old people's point of view. A literature review. *Aging Clinical and Experimental Research, 16*, 87–103.
Hamilton, G. G. (1990). Patriarchy, patrimonialism, and filial piety: A comparison of China and Western Europe. *The British Journal of Sociology, 41*, 77–104.
Harrison, S., Moran, M., & Wood, B. (2002). Policy emergence and policy convergence: The case of 'scientific-bureaucratic medicine' in the United States and United Kingdom. *British Journal of Politics and International Relations, 4*, 1–24.
Haski-Leventhal, D. (2009). Elderly volunteering and well-being: A cross-European comparison based on SHARE data. *VOLUNTAS: International Journal of Voluntary and Nonprofit Organizations, 20*, 388–404.
Hatfield, G. (2004). *Encyclopedia of folk medicine: Old World and New World traditions*. Santa Barbara, CA: ABC-CLIO.
Havighurst, R. J. (1961). Successful aging. *The Gerontologist, 1*, 8–13.
Havighurst, R. J. (1963). Successful aging. In R. H. Williams, C. Tibbitts, & W. Donahue (Eds.), *Processes of aging* (pp. 299–320). New York: Atherton Press.
Hayslip, B., Jr. (2009). Ethnic and cross-cultural perspectives on custodial grandparenting. In J. Sokolovsky (Ed.), *The cultural context of aging: Worldwide perspectives* (pp. 346–356). Westport, CT: Praeger.
Hegland, M. E., & Associates. (2009). Losing, using and crafting spaces for aging: Muslim Iranian American seniors in California's Santa Clara Valley. In J. Sokolovsky (Ed.), *The cultural context of aging: Worldwide perspectives* (pp. 302–324). Westport, CT: Praeger.
Heitmann, S., Robinson, P., & Povey, G. (2011). Slow food, slow cities and slow tourism. In P. Robinson, S. Heitmann, & P. Dieke (Eds.), *Research themes for tourism* (pp. 114–127). Wallingford: CAB International.
Henderson, J. N., & Henderson, L. C. (2002). Cultural construction of disease: A "supernormal" construct of dementia in an American Indian tribe. *Journal of Cross-Cultural Gerontology, 17*, 197–212.
Henderson, J. N., & Traphagan, J. W. (2005). Cultural factors in dementia: Perspectives from the anthropology of aging. *Alzheimer Disease and Associated Disorders, 19*, 272–274.
Henderson, N. (2009). Dementia in cultural context: Development and decline of a caregivers support group in a Latin population. In J. Sokolovsky (Ed.), *The cultural context of aging: Worldwide perspectives* (pp. 1–15). Westport, CT: Praeger. Available at: www.stpt.usf.edu/~jsokolov/webbook/henderson.htm
Heo, J. (2010). Serious leisure, health perception, dispositional optimism, and life satisfaction among senior games participants. *Educational Gerontology, 36*, 112–126.
Herne, S. (1995). Research on food choice and nutritional status in elderly people: A review. *British Food Journal, 97*, 12–29.
Hershey, D. A., & Henkens, K. (2014). Impact of different types of retirement transitions on perceived satisfaction with life. *The Gerontologist, 54*, 232–244.
Hershey, D. A., Henkens, K., & van Dalen, H. P. (2007). Mapping the minds of retirement planners: A cross-cultural perspective. *Journal of Cross-Cultural Psychology, 38*, 361–382.
Hill, K. R., & Hurtado, A. M. (1996). *Aché life history: The ecology and demography of a foraging people*. New York: Aldine de Gruyter.

Hummert, M. L., Garstka, T. A., Shaner, J. L., & Strahm, S. (1994a). Stereotypes of the elderly held by young, middle-aged and elder adults. *Journal of Gerontology, 49*, 240–249.
Hummert, M. L., Mazloff, D., & Henry, C. (1994b). *Responses of elderly adults to patronizing and affirming speech styles*. Paper presented at the Second International Conference on Communication, Aging and Health, Hamilton, ON.
Hung, K., & Crompton, J. L. (2006). Benefits and constraints associated with the use of an urban park reported by a sample of elderly in Hong Kong. *Leisure Studies, 25*, 291–311.
Indrizal, E., Kreager, P., & Schröder-Butterfill, E. (2009). The structural vulnerability of older people in a matrilineal society: The Minangkabau of West Sumatra, Indonesia. In J. Sokolovsky (Ed.), *The cultural context of aging: Worldwide perspectives* (pp. 383–394). Westport, CT: Praeger.
Ingersoll-Dayton, B., & Saengtienchai, C. (1999). Respect for the elderly in Asia: Stability and change. *International Journal of Aging and Human Development, 48*, 113–130.
Ip, D., Lui, C. W., & Chui, W. H. (2007). Veiled entrapment: A study of social isolation of older Chinese migrants in Brisbane. *Ageing and Society, 27*, 719–738.
Iris, M. (2009). African Americans growing older in Chicago: Living in a time and place of change. In J. Sokolovsky (Ed.), *The cultural context of aging: Worldwide perspectives* (pp. 357–367). Westport, CT: Praeger.
Irish, D. P., Lunquist, K. F., & Nelsen, V. J. (Eds.). (1993). *Ethnic variations in dying, death, and grief: Diversity in universality*. New York: Routledge.
Janevic, M. R., & Connell, C. M. (2001). Racial, ethnic, and cultural differences in the dementia caregiving experience: Recent findings. *The Gerontologist, 41*, 334–347.
Jenike, B. R., & Traphagan, J. W. (2009). Transforming the cultural scripts for aging and elder care in Japan. In J. Sokolovsky (Ed.), *The cultural context of aging: Worldwide perspectives* (pp. 240–258). Westport, CT: Praeger.
Jong, J., & Halberstadt, J. (2016). *Death anxiety and religious belief: An existential psychology of religion*. London: Bloomsbury.
Jost, J. T., & Banaji, M. R. (1994). The role of stereotyping in system-justification and the production of false consciousness. *British Journal of Social Psychology, 33*, 1–27.
Jylhä, M., Guralnik, J. M., Balfour, J., & Fried, L. P. (2001). Walking difficulty, walking speed, and age as predictors of self-rated health: The women's health and aging study. *Journal of Gerontology, 56A*, M609–M617.
Jylhä, M., Guralnik, J. M., Ferrucci, L., Jokela, J., & Heikkinen, E. (1998). Is self-rated health comparable across cultures and genders? *Journal of Gerontology, 53B*, S144–S152.
Kagawa-Singer, M. (1995). Diverse cultural beliefs and practices about death and dying in the elderly. *Gerontology and Geriatrics Education, 15*, 101–116.
Karp, A., Paillard-Borg, S., Wang, H. X., Silverstein, M., Winblad, B., & Fratiglioni, L. (2006). Mental, physical and social components in leisure activities equally contribute to decrease dementia risk. *Dementia and Geriatric Cognition Disorders, 21*, 65–73.
Kastenbaum, R. J. (1993). Reconstructing death in postmodern society. *Omega, 27*, 75–89.
Katz, R. (2009). Intergenerational family relations and subjective well-being in old age: A cross-national study. *European Journal of Ageing, 6*, 79–90.
Kaufman, S. R. (1998). Intensive care, old age, and the problem of death in America. *The Gerontologist, 38*, 715–725.
Kayser-Jones, J. (2001). Starved for attention. *Reflections on Nursing Leadership, 27*, 10–14.
Kim, S., & Feldman, D. C. (2000). Working in retirement: The antecedents of bridge employment and its consequences for quality of life in retirement. *Academy of Management Journal, 43*, 1195–1210.
Knight, B. G., & Sayegh, P. (2010). Cultural values and caregiving: The updated sociocultural stress and coping model. *Journal of Gerontology, 65B*, 5–13.
Krause, N., & Shaw, B. A. (2000). Giving social support to others, socioeconomic status, and changes in self-esteem in late life. *Journal of Gerontology, 55B*, S323–S333.

Kreuter, M. W., Lukwago, S. N., Bucholtz, D. C., Clark, E. M., & Sanders-Thompson, V. (2002). Achieving cultural appropriateness in health promotion programs: Targeted and tailored approaches. *Health Education and Behavior, 30*, 133–146.
Krondl, M., & Lau, D. (1982). Social determinants in human food selection. In L. M. Barker (Ed.), *The psychobiology of human food selection* (pp. 139–152). Westport, CT: AVI Publishing.
La Fontaine, J., Ahuja, J., Bradbury, N. M., Phillips, S., & Oyebode, J. R. (2007). Understanding dementia amongst people in minority ethnic and cultural groups. *Journal of Advanced Nursing, 60*, 605–614.
Lachman, M. E., Weaver, S. L., Bandura, M., Elliott, E., & Lewkowicz, C. J. (1992). Improving memory and control beliefs through cognitive restructuring and self- generated strategies. *Journal of Gerontology, 47*, 293–299.
Laidlaw, K., Wang, D., Coelho, C., & Power, M. (2010). Attitudes to ageing and expectations for filial piety across Chinese and British cultures: A pilot exploratory evaluation. *Aging and Mental Health, 14*, 283–292.
Lamb, S. (2009). Elder residences and outsourced sons: Remaking aging in cosmopolitan India. In J. Sokolovsky (Ed.), *The cultural context of aging: Worldwide perspectives* (pp. 418–440). Westport, CT: Praeger.
Lamb, S. (2020). Assemblages of care and personhood: "Successful ageing"across India and North America. In C. Brosius & R. Mandoki (Eds.), *Caring for old age. Perspectives from South Asia* (pp. 321–338). Heidelberg: Heidelberg University Publishing.
Lan, P.-C. (2002). Subcontracting filial piety: Elder care in ethnic Chinese immigrant families in California. *Journal of Family Issues, 23*, 812–835.
Legare, C. H., Evans, E. M., Rosengren, K. S., & Harris, P. L. (2012). The coexistence of natural and supernatural explanations across cultures and development. *Child Development, 83*, 779–793.
Lewis, J. P. (2011). Successful aging through the eyes of Alaska native elders. What it means to be an elder in Bristol Bay, AK. *The Gerontologist, 51*, 540–549.
Liang, J. (2011). Components of a meaningful retirement life - A phenomenological study of the 1950s birth cohort in urban China. *Journal of Cross-Cultural Gerontology, 26*, 279–298.
Lin, H., Bermudez, O. I., & Tucker, K. L. (2003). Dietary patterns of Hispanic elders are associated with acculturation and obesity. *Journal of Nutrition, 133*, 3651–3657.
Lindert, J., Luna, J., Torres-Gonzales, F., Barros, H., Ioannidi-Kopolou, E., Melchiorre, M. G., et al. (2013). Abuse and neglect of older persons in seven cities in seven countries in Europe: A cross-sectional community study. *International Journal of Public Health, 58*, 121–132.
Liu, D., Hinton, L., Tran, C., Hinton, D., & Barker, J. C. (2008). Re-examining the relationships among dementia, stigma, and aging in immigrant Chinese and Vietnamese family caregivers. *Journal of Cross Cultural Gerontology, 23*, 283–299.
Liu, J. H., Ng, S. H., Loong, C., Gee, S., & Weatherall, A. (2003). Cultural stereotypes and social representations of elders from Chinese and European perspectives. *Journal of Cross-Cultural Gerontology, 18*, 149–168.
Lobar, S. L., Youngblut, J. M., & Brooten, D. (2006). Cross-cultural beliefs, ceremonies, and rituals surrounding death of a loved one. *Pediatric Nursing, 32*, 44–50.
Lock, M. (2002). Symptom reporting at menopause: A review of cross-cultural findings. *Journal of the British Menopause Society, 18*, 132–136.
Lock, M., & Kaufert, P. (2001). Menopause, local biologies, and cultures of aging. *American Journal of Human Biology, 13*, 494–504.
Löckenhoff, C. E., De Fruyt, F., Terracciano, A., McCrae, R. R., De Bolle, M., Costa, P. T., Jr., et al. (2009a). Perceptions of aging across 26 cultures and their culture-level associates. *Psychology and Aging, 24*, 941–954.
Losada, A., Knight, B. G., Márquez-González, M., Montorio, I., Etxeberría, I., & Peñacoba, C. (2008). Confirmatory factor analysis of the familism scale in a sample of dementia caregivers. *Aging and Mental Health, 12*, 504–508.

Losada, A., Robinson Shurgot, G., Knight, B. G., Márquez, M., Montorio, I., Izal, M., et al. (2006). Cross-cultural study comparing the association of familism with burden and depressive symptoms in two samples of Hispanic dementia caregivers. *Aging and Mental Health, 10*, 69–76.
Luborsky, M. R., & Leblanc, I. M. (2003). Cross-cultural perspectives on the concept of retirement: An analytic redefinition. *Journal of Cross-Cultural Gerontology, 18*, 251–271.
Mahoney, D. F., Cloutterbuck, J., Neary, S., & Zhan, L. (2005). African American, Chinese, and Latino family caregivers' impressions of the onset and diagnosis of dementia: Cross-cultural similarities and differences. *The Gerontologist, 45*, 783–792.
Malley-Morrison, K., Nolido, N. E.-V., & Chawla, S. (2006). International perspectives on elder abuse: Five case studies. *Educational Gerontology, 32*, 1–11.
Markides, K. S., & Eschbach, K. (2005). Aging, migration, and mortality: Current status of research on the Hispanic paradox. *The Journals of Gerontology B, 60*, S68–S75.
Martin, N. (2009). Consumer scams and the elderly: Preserving independence through shifting default rules. *Elder Law Journal, 17*, 1–30.
Martinez, I. L. (2009). Aging in exile: Family support and emotional well-being among older Cuban immigrants in the United States. In J. Sokolovsky (Ed.), *The cultural context of aging: Worldwide perspectives* (pp. 325–345). Westport, CT: Praeger.
Massey, D. S., Alarcón, R., Durand, J., & González, H. (1987). *Return to Aztlan: The social processes of international migration from Western Mexico*. Los Angeles, CA: University of California Press.
Massey, D. S., Arango, J., Hugo, G., Kouaouci, A., Pellegrino, A., & Taylor, J. E. (1993). Theories of international migration: A review and appraisal. *Population and Development Review, 19*, 431–466.
Matsubayashi, K., Ishine, M., Wada, T., & Okumiya, K. (2006). Older adults' view of 'successful ageing': Comparison of older Japanese and Americans. *Journal of American Geriatrics Society, 54*, 184–187.
McCann, R. M., Cargile, A. C., Giles, H., & Bui, C. T. (2004). Communication ambivalence toward elders: Data from North Vietnam, South Vietnam, and the U.S.A. *Journal of Cross-Cultural Gerontology, 19*, 275–297.
McCrae, R. R., Costa, P. T., Jr., Pedroso de Lima, M., Simões, A., Ostendorf, F., Angleitner, A., et al. (1999). Age differences in personality across the adult lifespan: Parallels in five cultures. *Developmental Psychology, 35*, 466–477.
Meek, B. A. (2007). Respecting the language of elders: Ideological shift and linguistic discontinuity in a Northern Athapascan community. *Journal of Linguistic Anthropology, 17*, 23–43.
Mehta, K. (1999). Singaporeans' perceptions and preparations for retirement: A crosscultural inquiry. *Ageing International, 25*, 31–46.
Melby, M. K., Lock, M., & Kaufert, P. (2005). Culture and symptom reporting at menopause. *Human Reproduction Update, 11*, 495–512.
Mellor, D., Merino, M. E., Saiz, J. L., & Quilaqueo, D. (2009). Emotional reactions, coping and long-term consequences of perceived discrimination among the Mapuche people of Chile. *Journal of Community and Applied Social Psychology, 19*, 473–491.
Menec, V. H., Shooshtari, S., & Lambert, P. (2007). Ethnic differences in self-rated health among older adults: A cross-sectional and longitudinal analysis. *Journal of Aging and Health, 19*, 62–86.
Mintz, S. W., & Wolf, E. R. (1950). An analysis of ritual co-parenthood (compadrazgo). *Southwestern Journal of Anthropology, 6*, 341–368.
Moon, A., & Williams, O. (1993). Perceptions of elder abuse and help-seeking patterns among African-American, Caucasian, and Korean-American elderly women. *The Gerontologist, 33*, 386–395.
Morley, J. E. (2001). Decreased food intake with aging. *The Journals of Gerontology A, 56*, 81–88.
Moscovici, S. (1981). On social representations. In J. P. Forgas (Ed.), *Social cognition: Perspectives on everyday understanding* (pp. 181–209). London: Academic Press.

Mossakowski, K. N. (2003). Coping with perceived discrimination: Does ethnic identity protect mental health? *Journal of Health and Social Behavior, 44*, 318–331.
Murphy, W. P. (1980). Secret knowledge as property and power in Kpelle society: Elders versus youth. *Africa, 50*, 193–207.
Neimeyer, R. A., Wittkowski, J., & Moser, R. P. (2004). Psychological research on death attitudes: An overview and evaluation. *Death Studies, 28*, 309–340.
Ng, S. H., Loong, C. S. F., Liu, J. H., & Weatherall, A. (2000). Will the young support the old? An individual – And family – Level study of filial obligations in two New Zealand cultures. *Asian Journal of Social Psychology, 3*, 163–182.
Nimrod, G. (2006). Retirees' leisure: Activities, benefits and their contribution to life satisfaction. *Leisure Studies, 26*, 65–80.
Nimrod, G. (2007). Expanding, reducing, concentrating and diffusing: Post retirement leisure behavior and life satisfaction. *Leisure Sciences, 29*, 91–111.
Nimrod, G., Janke, M. C., & Kleiber, D. A. (2008). Retirement, activity, and subjective well-being in Israel and the Unites States. *World Leisure, 50*, 18–32.
Nimrod, G., Janke, M. C., & Kleiber, D. A. (2009). Expanding, reducing, concentrating and diffusing: Activity patterns of recent retirees in the United States. *Leisure Sciences, 31*, 37–52.
Nimrod, G., & Kleiber, D. A. (2007). Reconsidering change and continuity in later life: Toward an Innovation Theory of successful aging. *International Journal of Aging and Human Development, 65*, 1–22.
Numhauser-Henning, A. (2018). Ageism, age discrimination and employment law in the EU. In I. Doron & N. Georganzi (Eds.), *Ageing, ageism and the law: European perspectives on the rights of older persons* (pp. 98–115). Cheltenham: Edward Elgar Publishing.
O'Reilly, K. (2000). *The British on the Costa del Sol*. London: Routledge.
O'Reilly, K. (2007). Emerging tourism futures: Residential tourism and its implications. In C. Geoffroy & R. Sibley (Eds.), *Going abroad. Travel, tourism and migration* (pp. 144–157). Newcastle: Cambridge Scholars Publishing.
Obermeyer, C. M. (2000). Menopause across cultures: A review of the evidence. *Menopause, 7*, 184–192.
Olson, J. C. (1981). The importance of cognitive processes and existing knowledge structures for understanding food acceptance. In J. Solms & R. H. Hall (Eds.), *Criteria of food acceptance* (pp. 69–81). Zurich: Forster Publishing Ltd..
Ormel, J., Kernpen, G. I. J. M., Deeg, D. J. H., Brilman, E. I., van Sonderen, E., & Relyveld, J. (1998). Functioning, well-being, and health perception in late middle-aged and older people: Comparing the effects of depressive symptoms and chronic medical conditions. *Journal of the American Geriatrics Society, 46*, 39–48.
Parkes, C. M., Laungani, P., & Young, B. (Eds.). (2015). *Death and bereavement across cultures* (2nd ed.). East Sussex: Routledge.
Patterson, M., & Malley-Morrison, K. (2006). A cognitive-ecological approach to elder abuse in five cultures: Human rights and education. *Educational Gerontology, 32*, 73–82.
Phelan, E. A., Anderson, L. A., Lacroix, A.-Z., & Larson, E. B. (2004). Older adults' views of 'successful ageing': How do they compare with researchers' definitions? *Journal of American Geriatric Society, 52*, 211–216.
Pillai, J. A., & Verghese, J. (2009). Social networks and their role in preventing dementia. *Indian Journal of Psychiatry, 51*(Suppl. S1), 22–28.
Prakash, I. J. (1999). Senior women's perception of leisure in India. *Tourism Recreation Research, 24*, 82–85.
Radermacher, H., Feldman, S., & Browning, C. (2009). Mainstream versus ethno-specific community aged care services: It's not an 'either or'. *Australasian Journal on Ageing, 28*, 58–63.
Robben, A. C. G. M. (Ed.). (2004). *Death, mourning, and burial: A cross-cultural reader*. Malden, MA: Blackwell Publishing.
Roberts, S. B., & Rosenberg, I. (2006). Nutrition and aging: Changes in the regulation of energy metabolism with aging. *Physiological Reviews, 86*, 651–667.

Rohrer, J. (2008). Escaping the cult, recuperating victims. *Theory and Event, 11*, 2.
Rosenberg, H. G. (2009). Complaint discourse, aging and caregiving among the Ju/'hoansi of Botswana. In J. Sokolovsky (Ed.), *The cultural context of aging: Worldwide perspectives* (pp. 30–52). Westport, CT: Praeger.
Ryan, E. B., Kwong-See, S., Meneer, W. B., & Trovato, D. (1992). Age-based perceptions of language performance among younger and older adults. *Communication Research, 19*, 311–331.
Sampson, R. J., Raudenbush, S. W., & Earls, F. (1997). Neighborhoods and violent crime: A multilevel study of collective efficacy. *Science, 277*, 918–924.
Savishinsky, J. (2004). The volunteer and the sannyasin: Archetypes of retirement in America and India. *International Journal of Aging and Human Development, 59*, 25–41.
Schans, J. M. D., & Komter, A. E. (2010). Ethic differences in intergenerational solidarity in the Netherlands. *Journal of Aging Studies, 24*, 194–203.
Sharps, W. J., Price-Sharps, J. L., & Hanson, J. (1998). Attitudes of young adults toward older adults: Evidence from the United States and Thailand. *Educational Gerontology, 24*, 655–660.
Shirom, A., Toker, S., Berliner, S., Shapira, I., & Melamed, S. (2008). The effects of physical fitness and feeling vigorous on self-rated health. *Health Psychology, 27*, 567–575.
Silverman, P., & Maxwell, R. (1978). How do I respect thee? Let me count the ways: Deference towards elderly men and women. *Behavior Science Research, 13*, 91–108.
Silverstein, M., Burholt, V., Wenger, G. C., & Bengtson, V. L. (1998). Parent-child relations among very old parents in Wales and the United States: A test of Modernization Theory. *Journal of Aging Studies, 12*, 387–409.
Simmons, L. W. (1945). *The role of the aged in primitive society*. Hamden, CT: Archon Books.
Simmons, L. W. (1946). Attitudes toward aging and the aged: Primitive-societies. *Journal of Gerontology, 1*, 72–95.
Sliwinski, M. J., Stawski, R. S., Hall, R. B., Katz, M., Verghese, J., & Lipton, R. B. (2006). On the importance of distinguishing pre-terminal and terminal cognitive decline. *European Psychologist, 11*, 172–181.
Sokolovsky, J. (1985). Ethnicity, culture and aging: Do differences really make a difference? *Journal of Applied Gerontology, 4*, 6–17.
Sokolovsky, J. (2009a). Introduction: Human maturity and global aging in cultural context. In J. Sokolovsky (Ed.), *The cultural context of aging: Worldwide perspectives* (pp. xv–xxxv). Westport, CT: Praeger.
Sokolovsky, J. (2009b). A global vision of aging, culture and context. In J. Sokolovsky (Ed.), *The cultural context of aging: Worldwide perspectives* (pp. 1–12). Westport, CT: Praeger.
Sokolovsky, J. (2009c). The life course and intergenerational ties in cultural and global context. In J. Sokolovsky (Ed.), *The cultural context of aging: Worldwide perspectives* (pp. 105–114). Westport, CT: Praeger.
Sokolovsky, J. (2009d). Aging, globalization and societal transformation. In J. Sokolovsky (Ed.), *The cultural context of aging: Worldwide perspectives* (pp. 175–183). Westport, CT: Praeger.
Sokolovsky, J. (2009e). Aging proletariats in a twenty-first-century indigenous Mexican community. In J. Sokolovsky (Ed.), *The cultural context of aging: Worldwide perspectives* (pp. 216–239). Westport, CT: Praeger.
Sokolovsky, J. (2009f). The ethnic dimension in aging: Culture, context and creativity. In J. Sokolovsky (Ed.), *The cultural context of aging: Worldwide perspectives* (pp. 277–287). Westport, CT: Praeger.
Sokolovsky, J. (2009g). Ethnic elders and the limits of family support in a globalizing world. In J. Sokolovsky (Ed.), *The cultural context of aging: Worldwide perspectives* (pp. 289–301). Westport, CT: Praeger.
Sokolovsky, J. (2009h). Families, communities and elderscapes: Transforming cultural spaces for aging. In J. Sokolovsky (Ed.), *The cultural context of aging: Worldwide perspectives* (pp. 371–382). Westport, CT: Praeger.

Sommer, B., Avis, N., Meyer, P., Ory, M., Madden, T., Kagawa-Singer, M., et al. (1999). Attitudes toward menopause and aging across ethnic/racial groups. *Psychosomatic Medicine, 61*, 868–875.

Sone, T., Nakaya, N., Ohmori, K., Shimazu, T., Higashiguchi, M., Kakizaki, M., et al. (2008). Sense of life worth living (*ikigai*) and mortality in Japan: Ohsaki Study. *Psychosomatic Medicine, 70*, 709–715.

Stack, C., & Burton, L. (1993). Kinscripts: Reflections of family, generation and culture. *Journal of Comparative Family Studies, 24*, 157–170.

Stark, A. (2013). The matrilineal system of the Minangkabau and its persistence throughout history: A structural perspective. *Southeast Asia: A Multidisciplinary Journal, 13*, 1–13.

Stebbins, R. A. (1992). *Amateurs, professionals, and serious leisure*. Montreal, QC: McGill-Queen's University Press.

Stebbins, R. A. (2000). The extraprofessional life: Leisure, retirement and unemployment. *Current Sociology, 48*, 1–18.

Sung, K.-t. (1995). Measures and dimensions of filial piety in Korea. *The Gerontologist, 35*, 240–247.

Sung, K.-t. (2004). Elder respect among young adults: A cross-cultural study of Americans and Koreans. *Journal of Aging Studies, 18*, 215–230.

Sunil, T. S., Rojas, V., & Bradley, D. E. (2007). United States' international retirement migration: the reasons for retiring to the environs of Lake Chapala, Mexico. *Ageing and Society, 27*, 489–510.

Surbone, A., Kagawa-Singer, M., Terret, C., & Baider, L. (2007). The illness trajectory of elderly cancer patients across cultures: SIOG position paper. *Annals of Oncology, 18*, 633–638.

Szmigin, I., & Carrigan, M. (2001). Leisure and tourism services and the older innovator. *The Service Industries Journal, 21*, 113–129.

Tannahill, R. (1988). *Food in history*. London: Penguin Books Ltd.

Topping, D. (2007). Cereal complex carbohydrates and their contribution to human health. *Journal of Cereal Science, 46*, 220–229.

Tornstam, L. (1989). Gero-transcendence; a reformulation of the disengagement theory. *Aging: Clinical and Experimental Research, 1*, 55–63.

Torsch, V. L., & Ma, G. X. (2000). Cross-cultural comparison of health perceptions, concerns, and coping strategies among Asian and Pacific Islander American elders. *Qualitative Health Research, 10*, 471–489.

Traphagan, J. W. (2009). Brain failure, late life and culture in Japan. In J. Sokolovsky (Ed.), *The cultural context of aging: Worldwide perspectives* (pp. 568–575). Westport, CT: Praeger.

Tulli, R. (2016). L'ultima madre. In C. Casi (Ed.), *La donna nell'antichità: Archeologia e storia della condizione femminile dalla Preistoria al Medioevo*. Pitigliano, GR: Editrice Laurum.

Turner, V. (1969). *The ritual process: Structure and anti-structure*. London: Routledge & Kegan Paul.

Van Gennep, A. (1960). *Rites of passage*. Chicago: University of Chicago Press.

Veale, K. (2004). Discussing our family trees: A longitudinal analysis of online, community-based communication in genealogical newsgroups. *Interface: The Journal of Education, Community and Values, 4*(4). Available http://bcis.pacificu.edu/journal/ 2004/04/veale.php

Verhoeven, J. (1985). Goffman's frame analysis and modern micro-sociological paradigms. In H. J. Helle & S. N. Eisenstadt (Eds.), *Micro-sociological theory: Perspectives on sociological theory* (Vol. 2, pp. 71–100). London: Sage.

Vertovec, S. (2007). Migrant transnationalism and the modes of transformation. In A. Portes (Ed.), *Rethinking migration: New theoretical and empirical perspectives* (pp. 149–179). London: Berghan.

Wang, M. (2007). Profiling retirees in the retirement transition and adjustment process: Examining the longitudinal change patterns of retirees' psychological well-being. *Journal of Applied Psychology, 92*, 455–474.

Wang, M., Henkens, K., & van Solinge, H. (2011). Retirement adjustment: A review of theoretical and empirical advancements. *American Psychologist, 66*, 204–213.

Weibel-Orlando, J. (2009). *La cura dei nostri cari anziani*: Family and community elder care roles in contemporary Italy. In J. Sokolovsky (Ed.), *The cultural context of aging: Worldwide perspectives* (pp. 536–549). Westport, CT: Praeger.

Weiss, R. S., Bass, S. A., Heimovitz, H. K., & Oka, M. (2005). Japan's silver human resource centers and participant well-being. *Journal of Cross-Cultural Gerontology, 20*, 47–66.

Wellman, B. (1981). Applying network analysis to the study of support. In B. H. Gottlieb (Ed.), *Social networks and social support* (pp. 171–200). Beverly Hills, CA: Sage.

WHO. (2015). *The growing need for home health care for the elderly: Home health care for the elderly as an integral part of primary health care services*. Cairo: WHO, Regional Office for the Eastern Mediterranean.

Wilcox, S., Sharkey, J. R., Mathews, A. E., Laditka, J. N., Laditka, S. B., Logsdon, R. G., et al. (2009). Perceptions and beliefs about the role of physical activity and nutrition on brain health in older adults. *The Gerontologist, 49*, S61–S71.

Winston, N. A., & Barnes, J. (2007). Anticipation of retirement among baby boomers. *Journal of Women and Aging, 19*, 137–159.

Wong, J. Y., & Earl, J. K. (2009). Towards an integrated model of individual, psychosocial, and organizational predictors of retirement adjustment. *Journal of Vocational Behavior, 75*, 1–13.

Wong, P. T., Reker, G. T., & Gesser, G. (1994). Death attitude profile revised. In R. A. Neimeyer (Ed.), *Death anxiety handbook* (pp. 121–148). New York: Taylor and Francis.

Worden, J. W. (1982). *Grief counselling and grief therapy: A handbook for the mental health practitioner*. New York: Springer.

Wu, S., & Barker, J. C. (2008). Hot tea and juk: The institutional meaning of food for Chinese elders in an American nursing home. *Journal of Gerontological Nursing, 34*, 46–54.

Yeh, K. H. (2003). The beneficial and harmful effects of filial piety: An integrative analysis. In K. S. Yang, K. K. Hwang, P. S. Pederson, & I. Daibo (Eds.), *Progress in Asian social psychology: Conceptual and empirical contributions* (pp. 67–82). Westport, CT: Greenwood.

Yoon, C., Hasher, L., Feinberg, F., Rahhal, T. A., & Winocur, G. (2000). Cross-cultural differences in memory: The role of culture-based stereotypes about aging. *Psychology and Aging, 15*, 694–704.

Yovetich, N., & Rusbult, C. E. (1994). Accommodative behavior in close relationships: Exploring transformation of motivation. *Journal of Experimental Social Psychology, 30*, 138–164.

Yum, J. O. (1988). The impact of Confucianism on interpersonal relationships and communication patterns in East Asia. *Communication Monograph, 55*, 374–388.

Yun, R. J., & Lachman, M. E. (2006). Perceptions of aging in two cultures: Korean and American views on old age. *Journal of Cross Cultural Gerontology, 21*, 55–70.

Zeilig, H. (2014). Dementia as a cultural metaphor. *The Gerontologist, 54*, 258–267.

Zhang, H. (2009). The new realities of aging in contemporary China: Coping with the decline in family care. In J. Sokolovsky (Ed.), *The cultural context of aging: Worldwide perspectives* (pp. 196–215). Westport, CT: Praeger.

Chapter 4
Ageing in Better Mental Health

Abstract This develops the topic of *Ageing in better health*. Here we return to the biology of ageing, that was first introduced in Chap. 1, but with a special emphasis on brain plasticity, a very important topic that is the focus of a fast-developing research program, and we also review psychological health in old age. Initially mentally healthy persons may be at risk of experiencing serious deterioration of their mental capacities as they age; therefore, we devote a section in this chapter to review mental pathologies in the elderly. Special emphasis is given to the analysis of the various forms of dementia, including Alzheimer's disease. From there we proceed to address those aspects of ageing that affect sexual behaviour. The challenges experienced by people with a disability (mental or physical) who are becoming older are also the topic of a section in this chapter. We conclude the chapter with a section devoted to older people who reach very advanced ages: the centenarians, semi-supercentenarians, and supercentenarians.

The ageing process implies a gradual deterioration of both physical and mental capacities. However, ageing can be experienced quite differently by different individuals, as there is variability in biology and life experiences in the population of elders, and there is also a diversity of limitations in health in general, and mental health in particular: from serious impairment to a reasonable degree of healthy ageing. In this chapter, we describe some of the most common body and mental changes associated with ageing, with special reference to the brain, brain plasticity, and behaviour. Greater brain plasticity can help the mind of older people remain active and functional for longer, but ageing may also come with the development of mental disorders, which require both medical intervention and a considerable level of support from family and institutions. We then explore specific issues of gender in old age and how older people live their sexual life, to then focus on the challenges of old age for people with a disability. We end the chapter with an analysis of life in exceptional longevity: centenarians, semi-supercentenarians, and supercentenarians.

M. Cordella, A. Poiani, *Fulfilling Ageing*, International Perspectives on Aging 30,
https://doi.org/10.1007/978-3-030-60071-6_4

4.1 Body Changes with Age

Many of us, men in particular, tend to believe that the conditions enjoyed after we have reached some sort of adult "steady state" in our life will continue, unchanged, for a very long time, this is what has been termed *the end of history illusion* (Quoidbach et al. 2013); but in reality many biological functions tend to decline with age, thus negatively affecting the health of old people and shattering the illusion. Indeed, ageing compromises our entire biology, starting from the deep molecular level, as we have already seen in Chap. 1. Over time many molecules in our body experience what is known as *bystander damage*, which occurs when the rate of molecular damage exceeds the rate of repair carried out by enzymes or by processes of "rejuvenation" grounded on the synthesis of new molecules and turnover of old ones (Finch 2007). The immune, cardiovascular, and other systems are particularly affected by bystander damage with ageing.

Focusing on the immune system, Chung et al. (2001) proposed the *molecular inflammation* hypothesis to point to a major molecular aspect of ageing: chronic inflammation, which contributes to cause various aspects of biological deterioration in old age, including diseases such as arthritis, cancer, diabetes, osteoporosis, dementia, vascular diseases (Chung et al. 2009). The *molecular inflammation* hypothesis states that ageing leads to dysregulation of the immune system and altered redox balance (i.e. the balance between oxidants—or pro-oxidants—and antioxidants). The increased redox imbalance due to ageing is a result of weakened anti-oxidative defences and an over-production of *reactive oxygen species* (ROS), such as superoxide (O_2^-), hydroxyl radical ($^{\bullet}OH$), hydrogen peroxide (H2O2), reactive nitric oxide (NO), peroxynitrite ($ONOO^-$), and reactive lipid aldehydes. ROS over-production and dysregulation during ageing can activate various components of the immune system such as over-reactive macrophages in inflammatory processes. If chronic inflammation remains unresolved during ageing, it may cause age-related degenerative diseases over time (Chung et al. 2009).

ROS have positive physiological roles in the organism at low concentrations, such as maintenance of various functions like cell proliferation, immune defence, signal transduction, gene expression; but at high concentrations they may cause pathological conditions in various organs, including the brain.

Innate immune activity tends to increase with age, such activity is expressed as increased numbers of macrophages and local inflammation; but adaptive immunity shows a reduction of repertoire with ageing (e.g. decreased number of naïve T-cells, see Fig. 1.4). In old age there is also an increased concentration of enzymes (e.g. oxidases) that produce free radicals such as the ROS mentioned above, and there are changes in cell adhesion molecules and cytokines. In addition, inflammatory processes are a key to understanding changes in bones and joints occurring in the old that lead to osteoporosis. Inflammatory processes increase with age due to an increased dysregulation of the immune system as we become older, and dysregulation is also seen at the level of entire organs, which become functionally less effective with age (e.g. the lung function decreases with age).

Therefore interventions aimed at better regulating the immune system and keeping inflammatory reactions under control should help slow down the process of ageing. Both calorie restriction and exercise could help in controlling inflammatory reactions and the redox balance (e.g. Chung et al. 2009).

A better regulation of the immune system can be also helped by controlling the changes in the gut bacteria that occur with age. The human gut microbiota (GM) consists of more than 1000 species of microbes belonging to five phyla: *Firmicutes, Bacteroidetes, Actinobacteria*, *Verrucomicrobia*, and *Proteobacteria* (Santoro et al. 2018). Both immunosenescence and *inflamm-ageing*—i.e. the chronic, low-grade inflammatory status that is typical in the elderly—have an effect on GM, and changes in the GM can alter the health condition of the elder. For instance, an inflamed gut is a better habitat for aerobic pathobionts (e.g. *Staphylococcaceae*) that can outcompete probiotic bacteria and other mutualistic symbionts, thus further enhancing inflammation in a positive feedback mechanism that can worsen the health condition of the elder. More frail elders have a less diverse GM and centenarians tend to have a different species composition in their GM, partly because their immune system changes and partly because the relationship within the GM changes due to factors such as alterations in diet and physiology (Santoro et al. 2018). Changes in molecules of the immune system that are caused by GM may also affect the activity of the brain, sometimes causing pathologies (Sheedy et al. 2009; Berer et al. 2011; Hornig 2013).

Ageing is also evident at other systemic levels beyond the immune system. Table 4.1 summarises modifications observed in the skin, the digestive system, muscles and bones, kidney and bladder, and the cardiovascular and respiratory systems as we age. Broadly speaking, such changes indicate a gradual loss of functionality and capacity to sustain the basic biological activities required to be in good health, as we get older (Byles 2017). Weight loss in old age in particular is a strong predictor of mortality, even in old people who are not suffering from any specific disease (Newman et al. 2005). Some endocrine changes with ageing were mentioned in Chap. 1 (Table 1.3). Decline in body functioning with age also leads to increased susceptibility to diseases such as cancer, cardiovascular diseases, respiratory diseases, musculoskeletal conditions, neuropsychiatric disorders (including dementias), and diseases of the sense organs (Byles 2017).

As we age, our arteries become more rigid and accumulate lipids in their wall, this loss of arterial elasticity increases blood pressure in the old leading, in the worst cases, to blood clots or thromboses, stroke, and potentially death (Finch 2007).

Ageing is also clearly visible externally through a gradual decrease in muscle mass, the skin becomes thinner and less elastic, it also becomes drier and wrinkled. Although brown spots may appear on the skin, production of melanin decreases overall and therefore the skin becomes generally paler and the hair tends to whiten (e.g. Baumann 2007).

Sensory decline has also been well documented in old age affecting hearing, vision (Saunders and Echt 2007), taste, smell (Kaneda et al. 2000), sensitivity to touch (Wickremaratchi and Llewelyn 2006), temperature perception (Guergova and Dufour 2011), whereas the pain threshold changes with age in a stimulus-specific manner (e.g. Lautenbacher et al. 2005). Sensory decline with age may also go hand in hand with cognitive decline (Lindenberger and Baltes 1997), in part due to a common set of causes (Lindenberger and Ghisletta 2009).

Table 4.1 Some age-associated changes to body structure and function

Skin	Digestive system
• Thinner and more fragile	• Less efficient absorption of nutrients (especially vitamins)
• Slower healing	• Changes in smooth muscle—increased constipation
• Less efficient production of Vitamin D	• Smaller, less efficient liver—increasing the risk of drug toxicity
• Increased vulnerability to extremes of temperature	
• Reduced sensitivity to touch	
Muscle and bone	Kidney and bladder
• Loss of muscle mass	• Decreased renal function
• Loss of bone mass and demineralisation (osteoporosis) balance	• Difficulty maintaining fluid and electrolyte
• Tendon shortening	• Reduced bladder capacity
• Stiffer cartilages	• Weak and unstable bladder muscles (incontinence)
	• Urethral fragility
	• Enlarged prostate leading to bladder obstruction
Cardiovascular system	
• Stiffer arteries	
• Slowed electrical activity	
• Postural hypotension	
• Reduced right ventricular function	
Respiratory system	
• Stiffer lungs	
• Larger alveoli	
• Collapsed bronchioles	
• Weaker chest muscles	
• Reduced respiratory capacity	
• Reduced cough reflex	

From Byles (2017)

With ageing the brain reduces its schedule of activities, processing becomes more noisy, neuromodulatory control weakens, and learning becomes more challenging (Mahncke et al. 2006). Cognitive decline in old age, however, is variable among individuals.

Cognitive decline in normal ageing does not seem to be significantly associated with neuronal losses, even though such losses, especially shrinkage in neuronal size, are part of the normal development of the brain (Finch 2007). Instead, the ageing of the brain is more closely reflected by the changes in dendritic branching and spine density occurring in specific areas (Burke and Barnes 2006). That is, what decreases in the old is the number of neuronal interconnections (synapses) in the brain and such synaptic loss occurs at a rate of 1% per year after the age of 20 (Finch 2007). The overall brain size and the integrity of the white matter also decrease with ageing (Park and Reuter-Lorenz 2009). That is, ageing affects more significantly the integrity of the communication networks within the brain and between the brain and the rest of the body.

Cognitive decline with ageing includes lower capacities in memory, processing speed, visuospatial skills, executive functions, speech comprehension (Mahncke et al. 2006). Such changes are not pathological, instead they are part of the normal process of ageing (Aldwin and Gilmer 2004). Other changes, however, may be a preamble to a potential pathology. For instance, *mild cognitive impairment* (MCI) may, in 80% of cases, develop into dementia within a period of 5–8 years (Craft et al. 2003).

Brain areas that are especially affected by ageing include the hippocampus and the prefrontal and entorhinal cortices, the caudate, and the cerebellum, thus functions that rely on the activity of those brain regions also tend to decline with age. Such declining functions include spatial memory, episodic memory, working memory, spatial navigation, processing speed, executive function, but there are no great changes in vocabulary (Rajah and D'Esposito 2005; Raz et al. 2005; Burke and Barnes 2006; Salthouse 2010). Cognition, emotions, and memory aspects of normal ageing will be reviewed more specifically in Chap. 5.

Within the brain, ageing can lead to vascular complications (hyaline atherosclerotic changes) and also greater oxidative damage to proteins and lipids, increased glial cells activation, irreversible protein glycation (a mechanism that is associated with dementia, Kassaar et al. 2017), and also damage to DNA (Mrak et al. 1997).

Many of those biological changes occurring with ageing are coordinated by specific molecules such as osteocalcin, which in old age can concomitantly affect memory, bone fragility, and sarcopenia, for instance (Karsenty 2017). Levels of osteocalcin can be improved with exercise.

The idea that cellular decay increases with age in the brain was first proposed by C. F. Hodge in an article published in 1894 (Hodge 1894). Increase of free radicals in the brain that can cause oxidative damage to neurons is a mechanism that has been postulated to explain such decline of neuronal functioning with age. However, if we keep engaging neurons in mental activities then their functional decay with age may be slowed down, this could be explained by the production of molecules that are scavengers of free radicals and antioxidants such as superoxide dismutase, glutathione peroxidase, catalase (Swaab 1991).

The broad views about ageing of the brain can be encapsulated in the expressions: "wear and tear" and "use it or lose it"; see Aldwin and Gilmer (2004) for the *wear and tear* hypothesis, and Hultsch et al. (1999) for the *use it or lose it* hypothesis, the latter has been also referred to by Salthouse (2006) as the *differential-preservation* hypothesis.

In 1991, the Dutch neurobiologist Dick Swaab explained those two, non-contradictory views of the brain so: as a result of ageing, oxidative stress damages neurons to the extent that their functionality decreases, this *wear and tear* process would explain decreased brain faculties in old age. On the other hand, the brain also shows plastic capacities to keep functioning despite experiencing the stresses of age, the *brain may not be completely lost if we keep using it* (Swaab 1991). The concept of brain use as a strategy to help in the preservation of our brain faculties into old age harks back to at least the first century b.c.e., when Cicero wrote in his

De Senectute (On Ageing) that: "Much greater care should be given to the intellect and the mental faculties. For they too, like lamps, grow dim with time, unless we keep them supplied with oil" (Nussbaum and Levmore 2017, p. 77).

Therefore, individuals who experience greater mental stimulation throughout their life should keep performing well into old age, or at least they should perform better in old age than people who have experienced lower levels of mental stimulation, everything else being the same. In this case, an "extrinsic" factor, experience, can explain the difference between brain capabilities in old age. In addition, differences in mental capabilities in old age may be also intrinsic, simply reflecting brain differences that have been evident since the early ages of the individuals and have been preserved through old age. Such differences may have a heritable base or they may be due to the effects of early learning experiences (Shors et al. 2012), or both. The difference between these two hypotheses is nicely explained by Salthouse (2006, p. 70–71) with a metaphor that highlights the respective extrinsic and intrinsic character of mental preservation with age:

> The distinction can be elaborated by considering how each perspective would explain a finding that people who play bridge tend to have somewhat higher levels of cognitive functioning than people who do not play bridge ... The *differential-preservation hypothesis* would suggest that playing bridge builds mental muscle that prevents atrophy of mental ability, whereas the *preserved-differentiation hypothesis* would suggest that a minimum level of mental strength is needed for individuals of any age to be capable of playing bridge (italics ours).

In both cases, however, the brain of some individuals is capable of performing better than the brain of others well into old ages, in association with a mentally active life. This would suggest that intervention programs may be implemented to help individuals prepare themselves for old age since their younger years and to help people who are already old. Salthouse (2006) reviewed such programs much to his disappointment, as various methodological deficiencies impede reaching firm conclusions. However, we strongly agree with his view that:

> people should continue to engage in mentally stimulating activities because even if there is not yet evidence that it has beneficial effects in slowing the rate of age-related decline in cognitive functioning, there is no evidence that it has any harmful effects, the activities are often enjoyable and thus may contribute to a higher quality of life, and engagement in cognitively demanding activities serves as an existence proof—*if you can still do it, then you know that you have not yet lost it*. (Salthouse 2006, p. 84–85; *italics ours*)

The *"use it or lose it"* hypothesis relies on the ability of the nervous system to express and then retain a degree of plasticity with age. In the next section we explore brain plasticity, which is a major property of our brain that can underpin the level of resilience of our mental capacities to the processes of ageing and their detrimental effects.

4.2 Plasticity of the Ageing Brain

> The same riot and profusion which I mentioned in connection with Nature evinces itself in the brain. Suddenly, this part of one which had been in abeyance, which one hardly knew was there, begins to exfoliate in all directions. The mind becomes a steaming jungle of thoughts.
>
> Henry Miller (1957, p. 94)

(T)he human brain has apparently underestimated itself.
Norman Doidge (2010, p. xvi)

The term *plasticity* was introduced in the neurosciences by the Italian psychiatrist Ernesto Lugaro (the surname is often misspelt in the literature as "Lugano") in 1906, following up on previous works by Eugenio Tanzi who postulated that practice improves neuronal growth and shortens the synaptic cleft, the small space between functionally associated neurons, thus facilitating their interactions. Lugaro's concept of plasticity meant to point to the pattern "that throughout life the anatomo-functional relations between neurons can change in an adaptive fashion to enable psychic maturation, learning, and even functional recovery after brain damage" (Berlucchi 2002, p. 305). Important pioneering experimental work on neuroplasticity was subsequently carried out by Donald Hebb (Johansson 2004).

Plasticity can therefore be defined as the capacity to change, and when it comes to biological entities such change may be adaptive. The *adaptiveness* of organisms relies on both the *robustness* of their biological mechanisms and function and also their capacity to change adaptively during their lifetime: *adaptive plasticity* (Paillard 1976; Bateson and Gluckman 2011).

Neuroplasticity in particular is "the ability of the nervous system to respond to intrinsic and extrinsic stimuli by reorganizing its structure, function and connections" (Cramer et al. 2011, p. 1591; see also Duffau 2015). The brain is confronted with the challenges of guiding the organism through an environment characterised by both recurrence of events and variability, whilst maintaining coherence in bodily functions. This leads to a dynamic compromise between robustness—or stability—and plasticity (Pascual-Leone et al. 2011). Plasticity does not necessarily require the organism to return to a previous state once it has been confronted with an environmental challenge, a process known as *elasticity* (e.g. Poiani 2010). Plasticity also involves the capacity of the organism to transition from one state to a different one using a diversity of mechanisms, some of which may be functionally redundant, providing alternative pathways to achieve similarly adaptive results (Denes 2016).

The plasticity of the developing brain is well known and even lay observers can indirectly infer it from the changing behaviour seen in the child, teenager, and young adult and the processes of learning they go through. But the adult brain is also plastic and it is so in more than one way. On the one hand *structural plasticity* in the adult brain involves morphological changes that are affected by neurogenesis, for instance, that is the production of new neurons and their integration into neuronal circuits, whereas *functional plasticity* involves the reorganisation of neural activity that is supported by the modifiability of synaptic transmission between neurons (Patten et al. 2015; Denes 2016). In other words, the brain is plastic in both structure and function.

Plasticity of the nervous system is allowed by inter-neuronal connections, following the principle that "neurons that fire together wire together". The concept was originally proposed by Sigmund Freud, but it is currently known as *Hebb's law* (Doidge 2010). At the level of inter-neuronal connections the adult brain displays various forms of plasticity: (a) *Map expansion* (recruitment of adjacent neurons); (b) *sensory reassignment* (processing of environmental information using other senses when one is not working); (c) *compensatory masquerade* (*compensation*): activating redundant pathways to achieve the same goal once the main one has been disturbed, the concepts of *vicariance* (Stojanovic-Radic and DeLuca 2015) and *redundancy* (Hampstead and Sathian 2015) point to a similar mechanism; and (d) *extra-region recruitment* involving distant brain sites that take over a lost function (Cabeza et al. 2002; Greenwood 2007; Stojanovic-Radic and DeLuca 2015; Venkatesan et al. 2015); a specific example of extra-region recruitment is *mirror region takeover*: activation of the mirror area in the opposite hemisphere (Erickson et al. 2007; Doidge 2010; Chen and D'Esposito 2015), leading to "hemispheric asymmetry reduction in old age" (HAROLD) or, more generally, "compensatory-related utilization of neuronal circuits" (CRUNCH) (Hampstead and Sathian 2015).

In response to the deterioration of some brain functions, older individuals have been observed to engage in both brain map expansion and extra-region recruitment, including shifts in brain hemisphere usage, not just from one hemisphere used preferentially at young ages to the other at older ages (*mirror region takeover*), but even *bilateral activation* of brain areas (Gutchess 2014; see Fig. 4.1).

The recruitment of other brain areas that keep cognitive abilities going has been encapsulated in Park and Reuter-Lorenz's (2009) *Scaffolding Theory of Ageing and Cognition* (STAC) that we mentioned in Chap. 1. STAC posits that "behavior is maintained at a relatively high level with age, despite neural challenges and functional deterioration, due to the continuous engagement of compensatory scaffolding—the recruitment of additional circuitry that shores up declining structures whose functioning has become noisy, inefficient, or both" (Park and Reuter-Lorenz's 2009, p. 10). Studies carried out using *positron emission tomography* (PET) and *functional magnetic resonance imaging* (fMRI) indicate that old people do activate more regions of the brain than younger ones, even for a simple motor task, and they can do so with no detectable difference in accuracy, but with an increased reaction time, a greater level of activation of the areas involved, and with recruitment of additional cortical and sub-cortical areas (see Johansson 2004, for more cases of bilateral activation of brain regions in old people as compared with the young during the performance of a diversity of tasks).

On the other hand, the same plastic mechanisms that underlie variability in neurons connections can be also responsible for increased rigidity of neural networks and therefore behaviour. For instance, the force of habit may transform a plastic/variable neuronal circuit into a more rigid one, what Doidge (2010) calls the "plastic paradox". Indeed, it is even possible for plasticity to produce maladaptive outcomes (*maladaptive plasticity*, Edwards et al. 2015; Stojanovic-Radic and DeLuca 2015), as in the case of *focal hand dystonia*, a loss of muscular control in highly practised movements (Johansson 2004).

Plasticity expressed through activation or inhibition of current neuronal pathways is achieved by the release of internal signals (e.g. hormones, neurotransmitters), whereas plasticity in the architecture of cell interconnections (synapses) in specific areas, a process known as *synaptogenesis*, can be regulated by hormones (e.g. oestrogens), neurotransmitters (e.g. dopamine), glial cells (e.g. astrocytes and oligodendrocytes), and cholesterol (Naftolin et al. 1996; Nedergaard et al. 2003; Slezak and Pfrieger 2003). It is worth noting that the peak expression of synaptogenesis in humans is achieved at around 5 years, and extending up to 10 years later

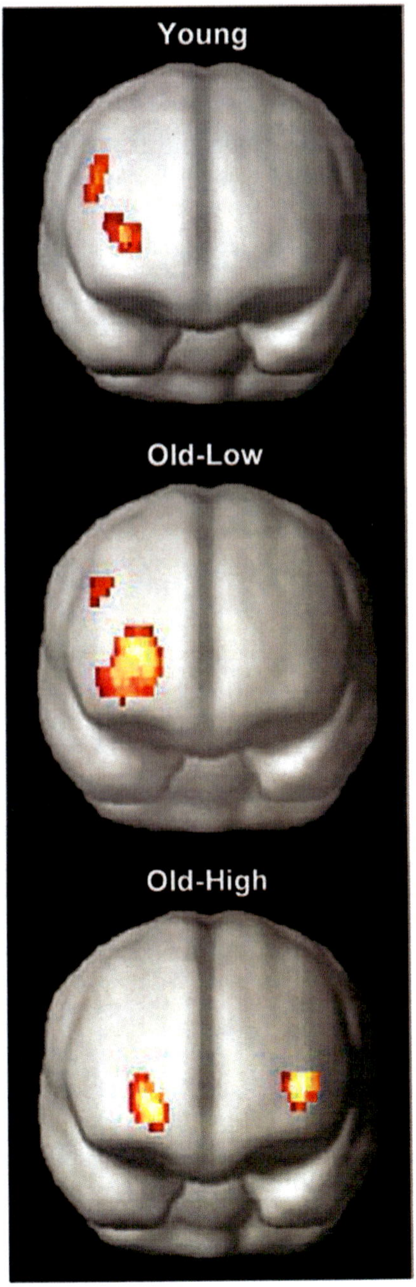

Fig. 4.1 Prefrontal cortex activity during a source memory task recalling recently studied words. Activity was right-lateralised in Young and Old-Low performing participants—although the latter activated a more expanded area than the former—but bilateral in Old-High performing participants. Reproduced with permission from Cabeza et al. (2002)

than it is the case in other primates where synaptogenesis peaks just a few months after birth (Denes 2016), a phenomenon that is consistent with the neotenic characteristics of the human species (Poiani 2010, see Chap. 9).

Brain plasticity can be achieved through increased synaptogenesis following two major mechanisms: (a) changes in the number of synapses without altering interconnectivity among neurons (i.e. no new neurons are connected through the new synapses) or (b) increased inter-neuronal connectivity (i.e. new neurons are connected) (Stepanyants et al. 2002). The latter could be achieved by an increase in the number of dendritic spines or by an increase in the topology of dendrites or of the axonal branches (Stepanyants et al. 2002; see Fig. 4.2).

The finding that the adult brain retains a degree of synaptogenic capability clearly opens the possibility not only for behavioural changes in adult individuals, but also changes in the structure of the brain itself. Moreover, behaviour can in turn morphologically and functionally reshape neuronal circuitry in the brain (Kleim et al. 1996; Zito and Svoboda 2002; Johansson 2004; Waites et al. 2005). From this, however, it cannot be concluded that the plasticity of the adult brain is unlimited, it is obviously not.

In their review of the relationship between the experience of adult individuals and brain plasticity, Markham and Greenough (2004) indicate how exposure to stimulus-enriched environments can affect the morphology of brain regions such as the auditory cortex, hippocampus, visual cortex, primary somatosensory cortex, entorhinal cortex, cerebellar cortex, amygdala, and basal ganglia. They point to the effects of a stimulus-enriched environment on increasing dendritic arborisation, density of dendritic spines, and number of synapses per neuron. The hippocampus

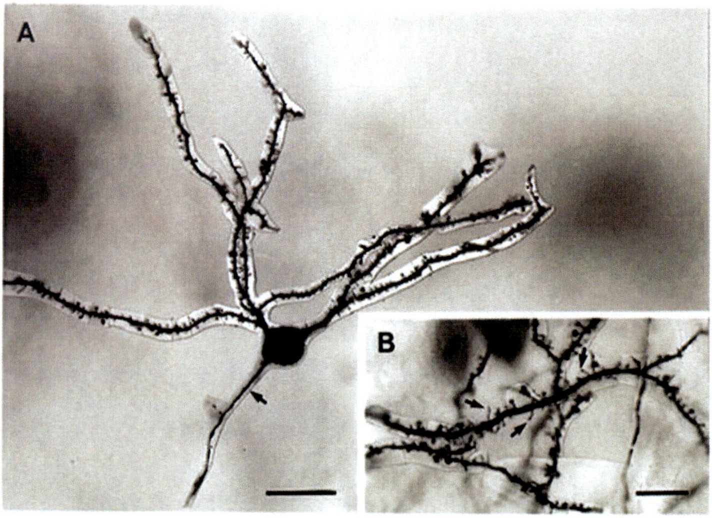

Fig. 4.2 Hippocampal neuron showing: (**a**) the soma (round black structure), dendrites (emerging from the soma), the axon (indicated by an arrow), and spines clearly visibly emerging from dendrites; (**b**) closer view of spines (indicated by arrows). From Frotscher et al. (2000), reproduced with permission

in particular is an important brain area for some aspects of learning and memory and its functions can suffer deterioration with age, this change is associated with a decrease in number of axo-dendritic synapses over the years (Jessberger and Gage 2008).

Specific areas of the adult brain can not only modulate their functions through activation/inhibition of established neuronal connections, or the formation of new connections among existing cells as we mentioned above, but also through an increase in the number of neurons (*neurogenesis*) or a decrease in the number of neurons following cell death (*apoptosis*) (Buss et al. 2006; Bredesen et al. 2006).

Neurons that connect to target cells (e.g. other neurons) can survive only if the target cell provides trophic factors called *neurotrophins* (e.g. nerve growth factor) (Black 1986; Sastry and Subba Rao 2000; Buss et al. 2006). Neurons that are not participant to a network and that therefore have not established active connections (e.g. synapses) with other cells are starved of neurotrophins and therefore die (apoptosis). Such apoptotic process can be prevented by sexual hormones, for instance (see Meier et al. 2000 for a review). The architecture of the brain is therefore strongly influenced by the tendency of neurons to undergo apoptosis, whereas apoptosis is modulated by survival signals received by the cell from its surrounding environment (Meier et al. 2000). Apoptotic processes occurring in the adult brain can provide additional plasticity if they are limited to specific areas of the brain. For instance, adaptive apoptotic changes in brain architecture can be associated with improved efficiency of local brain circuits and therefore improved mental processes and behaviours (e.g. Chambers et al. 2004).

Apoptosis does not only affect neurons but also glial cells (Buss et al. 2006). The death of neurons can also occur through changes in the *hemodynamic response*: flow of blood in the brain, oxygen metabolism, and *blood oxygenation level dependent* (BOLD) response (Denes 2016).

In the adult brain neurons may not only die, but new ones may be also produced through neurogenesis. That specific regions of the adult mammal brain are capable of undergoing neurogenesis is a fact that has been recognised only recently (see Markham and Greenough 2004; Ming and Song 2005 for reviews). Knowledge about neurogenesis in the adult avian brain was already well established in the 1980s by the work of Fernando Nottenbohm at Rockefeller University (Nottenbohm 1985, see Gross 2000 for a review). Neurogenesis in the mammalian adult brain has been described in the lateral ventricle and the subgranular zone of the dentate gyrus in the hippocampus (Ming and Song 2005; Hampstead and Sathian 2015) and the olfactory bulb (Buss et al. 2006), but whether it also occurs in other areas remains to be seen. Current evidence indicates that the new neurons in the adult mammal brain develop from astrocytes, a remarkable finding that suggests that some glial cells in fact may act as stem cells in the adult brain for the production of new neurons (Ming and Song 2005). Neurogenesis, just like apoptosis, can be also influenced by circulating hormones. For instance, it is inhibited by circulating corticosterone, whereas it is promoted by oestrogen and prolactin (Ming and Song 2005). Similarly to apoptosis, not only neurons but also glial cells can undergo population growth in specific areas of the adult brain (Markham and Greenough 2004).

Therefore, the brain displays a capacity to be adaptively plastic through its ability to: (a) activate or inhibit various brain areas, (b) regulate synaptogenesis, and promote (c) neurogenesis and (d) apoptosis.

The plastic capabilities of the adult brain can be harnessed in order to recover or at least compensate for the loss of some brain functions after injury, which may affect not only directly compromised areas of the brain but also some adjacent areas, a process known as *diaschisis* (Johansson 2004; Heilman 2015). After a trauma, brain plasticity may be able to return the injured brain to normality (*recovery* or *restitution*), or the original function may be only partially recovered, through recruiting other areas of the brain (*compensation* and *extra-region recruitment*) (Mayes 2015; Mogensen 2015; Denes 2016). Compensation may also involve a process of *reorganisation* (or permanent "rewiring") of specific areas in the brain (Stojanovic-Radic and DeLuca 2015). Compensation and extra-region recruitment, for instance, can explain the maintenance of language comprehension in Parkinson's disease patients via the increased activation of brain areas not compromised by the disease (Copland and Angwin 2015).

Recovery or compensation may be spontaneous, in which case they involve processes of *reactive neuroplasticity* (e.g. angiogenesis, neurogenesis, axonal growth, remyelination, reduction in necrotic and apoptotic cell death, neuronal migration, expansion of dendritic spines, synaptogenesis) and/or *dedifferentiation* (i.e. a regression of specialised cells to a stem cell-like state that can then contribute to tissue repair, Cai et al. 2007); or they may result from specific interventions: *experience-dependent neuroplasticity* (Nadeau 2015). Learning new tasks, a new language, for instance, can help the brain accumulate *cognitive reserve* (Katzman et al. 1988; Stojanovic-Radic and DeLuca 2015) that improves brain plasticity and memory in particular (Doidge 2010).

A common effect of normal ageing is the reduction in the level of brain plasticity (de Magalhães and Sandberg 2005; Denes 2016). For instance, the plasticity of episodic learning, memory, and executive functions—i.e. skills necessary for goal-directed behaviour (Stern et al. 2014)—decreases with age (Jones et al. 2006; Hampstead and Sathian 2015), but there are also additional factors, apart from age, that can contribute to either speeding up or slowing down the decrease of brain plasticity in older ages. Depression, for instance, further decreases brain plasticity with ageing, especially in the prefrontal cortex (e.g. McEwen and Morrison 2013), as it does Alzheimer's disease, especially with regard to memory (e.g. Mirmiran et al. 1996).

Some environmental factors can also affect brain plasticity in old age (Kramer et al. 2004). Diet and exercise can help improve brain plasticity (Murphy et al. 2014), whereas sustained stress decreases brain plasticity (Mora 2013). Sustained stress is detrimental to brain plasticity through its effects on dendritic atrophy and loss of excitatory synapses in the hippocampus and prefrontal cortex, lower growth of dendrites and spines in the amygdala, and also decreased levels of neurogenesis in the dentate gyrus (Radley and Morrison 2005; see also McEwen and Morrison 2013). In addition, stress decreases cognitive flexibility and functional connectivity, as seen in fMRI imaging studies (McEwen and Morrison 2013), and we have seen above how circulating corticosterone, a hormone released under stress, can inhibit neurogenesis. Therefore, interventions aimed at reducing stress are likely to also help improve brain plasticity. It has been shown that programs of *mindfulness-based stress reduction* (MBSR) can modify the volume of the medial prefrontal cortex in individuals suffering from chronic fatigue, and it can also change the volume of the amygdala in people showing reduced chronic anxiety (McEwen and Morrison 2013 and the references therein).

Training and adopting strategies to maintain brain plasticity into older ages—to the extent that it is possible—should clearly be an objective of fulfilling ageing (e.g. Baltes and Singer 2001). Building cognitive reserve throughout life is an important step to help the brain remain plastic for longer (Oberman et al. 2015), maintaining physical activity into old ages also helps in retaining a degree of cognitive plasticity (Kramer et al. 2006; Crosson and McGregor 2015; Voss et al. 2010 and the references therein) and neurogenesis, as it does living in a diverse environment (Kempermann et al. 2010). Mental training techniques, such as the use of virtual reality, can also help improve brain plasticity in older individuals or individuals who suffer from some brain impairment (Johansson 2004).

In a study carried out in the USA, that compared the effect of aerobic (walking) and non-aerobic (stretching and toning) fitness training programs on the brain activity of both young adults (18–35 years old) and old (55–80 years old), Voss et al. (2010) subjected the participants to 1-year training in each of the two modalities (aerobic vs non-aerobic) and assessed the activity of the *Default Mode Network* (DMN), a *fronto-executive network* (FE), and a *fronto-parietal network* (FP). Those are three broad brain networks that are involved in communication between the frontal cortex and the rest of the brain and that are negatively affected by ageing. In particular, the DMN is involved in memory consolidation, mind-wandering, autobiographical memory, and self-referential thought, along with executive control. The FE is associated with sustained task-set maintenance and maintaining the associations between actions and their outcomes, among other functions. The FP is known to be associated with the control of various cognitive functions (e.g. Zanto and Gazzaley 2013).

The activity of those brain regions was assessed pre- and post-intervention using fMRI. One year of walking improved the functional connectivity within the DMN and a FE, although a shorter period of training (6 months) had no significant effect. The stretching and toning regime produced an increased functional connectivity in two areas: the DMN after 6 months and the FP after 12 months. This increased functional connectivity was associated with an improvement in executive function. Exercise can also increase temporal lobe functional connectivity (Voss et al. 2013).

In an experimental study of both brain and movement plasticity carried out in Germany, Boyke et al. (2008) trained one group of 44 inexperienced individuals (54.5% women; 50 to 67 years old) in three-ball cascade juggling and compared their performance after the training period with that of a 25-individual control group (68.0% women; 55–67 years old) who were equally inexperienced at juggling before the beginning of the experiment. The intervention group had 3 months of training during the experiment. After the training and acquiring a degree of juggling ability, there was a transient increase in the size of the middle temporal area of the visual cortex, and juggling was also associated with a transient increase in grey matter in the hippocampus on the left hemisphere and in the nucleus accumbens bilaterally.

Mortimer et al. (2012) investigated the effects of a 40-week program on various neuropsychological variables in four randomised groups of, overall, 120 elderly Chinese in Shanghai: one group was exposed to a Tai Chi program (a non-aerobic exercise), another group followed a walking regime, a third group was engaged in social interactions, and there was also a fourth, control group, that was exposed to no intervention. Brain volume, as measured through MRI, increased in both the Tai Chi and the social intervention groups as compared to the control group. The Tai Chi group also performed better on a measurement of dementia (*Mattis dementia rating scale*), on a *trail making test* (A and B), the *auditory verbal learning test,* and a test of verbal fluency. The social interaction group also did well, but not as well as the Tai Chi group. This suggests that a combination of physical exercise and social activity can better help maintain cognitive plasticity in the elderly.

Training programs can also help improve, to an extent at least, the negative effects of ageing on memory (Li et al. 2008a); the level of success of such programs relies both on individual characteristics and also age (Singer et al. 2003a), but some programs have achieved a degree of memory improvement even among the old-old (Zinke et al. 2012).

Listening to music can also promote brain plasticity, to the point of even increasing cognitive recovery after an acute middle cerebral artery stroke (Särkämö et al. 2014); but playing a musical instrument can be even more effective, as it requires not only sensory but also motor skills that stimulate activity in several regions of the brain (Wan and Schlaug 2010). Musicians have been observed to have greater volume of grey matter in the left inferior frontal gyrus compared with non-musicians, they also show less reduction in total brain volume, dorsolateral prefrontal cortex, and the left inferior frontal gyrus with age.

Lustig et al. (2009) classified cognitive enhancing programs for old people into four major categories:

a. Strategy training: device strategies to increase performance of elderly people in tasks where they do not perform well.
b. Multimodal approaches: guiding towards lifestyle changes that include both individual cognitive and social components.
c. Cardiovascular training: aerobic exercise training.
d. Process training: focus on specific cognitive aspects without following a fixed strategy.

Aspects of strategy training, multimodal approaches, and process training were included by Mahncke et al. (2006) in a program aimed at increasing brain plasticity in old people. Such program is based on four basic principles:

a. Strongly engage the brain: through demanding exercises and daily, challenging activities.

b. Re-normalise noisy processing: improving the ability of the brain, through exercises, to enhance the representational fidelity of cognitive systems.
c. Enhance the neuromodulatory function: use exercises of arousal, attention, reward, and novelty to strengthen the basic function of neuromodulatory components that are essential for the regulation of learning and memory.
d. Strengthen critical life skills: guide the individual to abandon learned behaviours that can be detrimental to the health of the brain.

Preliminary applications of the program have shown encouraging results in terms of memory enhancement (Mahncke et al. 2006; see also Shah et al. (2017) for a review of computerised programs to enhance brain plasticity and cognition in older people).

In sum, the adaptiveness of neuroplasticity is strongly seen in the young brain, but to a degree it can be also extended to the brain of old people. Factors that can enhance brain plasticity in the elders include: an enhanced environment and a diversity of experiences, accumulation of cognitive reserve throughout life and also learning new skills in old age (e.g. a new language), reduce the level of stress and remain socially connected, increase physical activity, and use mental training techniques. Brain plasticity can be also used in specific therapies to help the elderly compensate, at least to an extent, for the effects of brain injuries, traumas, and diseases.

Given the decline in body functions and some mental capacities in the elders, the issue of mental health is obviously high on the agenda if fulfilling ageing is to be experienced. In the next section we will focus on mental health in the elderly.

4.3 Mental Health

In Chap. 1 we defined *subjective well-being* (or *existential well-being*) as "a subjective sense of happiness and vigour in life". It is difficult to imagine fulfilling ageing without also experiencing subjective well-being and, in turn, the more subjective well-being we experience the better our mental health should be (Diener and Chan 2011). Subjective well-being can be achieved in many ways that are variable according to opportunity and personality, but a sense of meaning in life seems to be central. We will focus on life meaning and the existential dimension of old age in more detail in Chap. 8 *Journey Towards the End of Life*, here we just address the relationship between mental health and well-being in a broad sense, the issue of mental disorders in old age will be the focus of the next section.

Mental health and well-being can be seriously compromised when the elder is subject to significant and chronic stresses. Stress can have direct negative effects on mental health through the release of cortisol in circulation, and previous studies have described how old, healthy humans have higher diurnal levels of cortisol than younger individuals. Sustained elevated levels of plasma glucocorticoids over the years in older adults negatively correlate with hippocampal volume and memory. In addition, high levels of glucocorticoids could negatively affect survival and/or function of neurons in the prefrontal cortex, thus affecting complex cognitive functions (Lupien et al. 2009).

Stress management is therefore central to improving our mental health and it can be achieved in a diversity of ways, including receiving social support and performing various personally pleasurable activities that induce feelings of well-being and fulfilment, leading to the expression of positive affect and the development of personal resilience. Specific psychological stress-management strategies that should be made available to elders, with outside help if necessary, include meditation, relaxation, spirituality, mindfulness, the arts, exercise, experiences of "loving-kindness"/compassion, and nurturing closeness or connectedness to others (Esch and Stefano 2005).

Sources of stress can be individual (e.g. psychological, financial) and/or social (e.g. social rejection). For instance, the work of Peggy Thoits on *Identity Theory* and Stress (Thoits 1991) combines both the individual and the social dimension of stress in old age. The fundamentals of *Identity theory* and its association with stress are clearly explained by Krause (2004, p. S287):

> Identity theory rests on two core constructs: social roles and role identities. Social roles are defined structurally as positions in a group (e.g., husband, father, provider), whereas identities refer to the kinds of information that people emphasize when thinking about themselves and presenting themselves to others. This information is grounded in the roles they occupy (e.g., 'I am a father'). Associated with each social role are clusters of normative expectations that tell role occupants what to do and what to expect. As a result, they provide a sense of direction and purpose in life as well as a basis for evaluating the adequacy of one's own role performance. These normative expectations are important because, as Thoits (1991) points out, they are a major source of meaning in life. According to Thoits (1991), *stressors arising in certain roles are likely to be especially noxious because they have the capacity to undermine the identity associated with the role* (italics ours).

Therefore, although stress can be caused by a loss of social roles (e.g. post-retirement) undermining identity in the elders, Thoits also points to an additional mechanism whereby it is some times the performance of specific normative but unwanted social roles in old age that may cause stress in the old person. That is, stress may further increase when there is inconsistency between the socially expected role in old age and the desire for a specific identity development in the elder. These sources of tension and stress, if not resolved, may potentially lead to a deterioration of mental health. On the other hand, when the social environment is supportive of a process of free meaning-making and identity creation by older people, the general consequences for the health of the elder can be positive (Krause 2004 and the references therein).

Thus, greater meaning in life and a more fulfilling ageing that decrease the levels of stress in old age can enhance mental health and well-being. Lars Tornstam, the ideator of the concept of *gerotranscendence* (see Chap. 1), has argued that meaning-making, increased purpose in life, and transcendence all contribute to greater well-being in old age through fostering both personal resilience in the face of life adversities and a sense of life coherence and purpose; that is, meaning and purpose can help buffer the individual from the effects of stress.

Nygren et al. (2005) tested the relationship between various measurements of psychological well-being in old people: *resilience* (Wagnild and Young 1990; Windle 2011), *sense of coherence* (Antonovsky 1979; Antonovsky and Sagy 1986), *meaning in life* (Frankl 1963) and *self-transcendence* (Reed 1991, 1991a); and mental and physical health. *Resilience* can be defined as "the process of effectively negotiating, adapting to, or managing significant sources of stress or trauma" (Windle 2011, p. 152). Resilience relies on the assets and resources available to individuals in their life and environment that facilitate their capacity for adaptation and their ability to "bounce back" in the face of adversity. On the other hand, *sense of coherence* has been defined by Antonovsky (1979, p. 132) as "a global orientation that expresses the extent to which one has a pervasive, enduring, though dynamic, feeling of confidence that one's internal and external environments are predictable and that there is a high probability that things will work out as well as can reasonably be expected". *Meaning in life* is an eminently personal construct, as what gives meaning to one person's life may not be equally relevant to others (Baumeister 1991); it is also a diverse and complex concept (Wong 2012, see also Chap. 8 in this book), but here we quote Battista and Almond's (1973, p. 409) definition of life meaning, with an addition to prevent gender bias: "an individual's belief that he [/she] is fulfilling his [/her] positively valued life-framework or life-goal". Finally, *self-transcendence* can be seen as: "the expansion of one's conceptual boundaries inwardly through introspective activities, outwardly through concerns about others' welfare, and temporally by integrating perceptions of one's past and future to enhance the present" (Reed 1991a, p. 5).

Variables of psychological well-being were measured by Nygren et al. using the *Resilience Scale* (RS), *Sense of Coherence Scale* (SOC), *Purpose in Life Test* (PIL), and the *Self-Transcendence Scale* (STS), whereas mental and physical health were measured through the *SF-36 Health Survey* (SF-36). The investigation was carried out in Sweden as part of the *Umeå 85+ study* and participants included 125 elders (69% women), with their age distribution being 21% in the group of 95 years old or older, 37% were 90 years old, and 42% were 85 years old. For all sexes combined there was a positive correlation between mental health and sense of coherence, purpose in life, resilience, and self-transcendence; but only purpose in life was positively correlated with physical health. However, when sex was controlled, the mental health correlations applied especially to women, even though men had higher scores than women on the *Purpose in Life Test*.

Another stress-reduction strategy that can improve mental health in old people is physical activity. Although some elders may not like to engage in regular physical activity and planned exercise, due to fear of falling, for instance (Bruce et al. 2002), exercising is an important source not only of physical but also mental health in old people. Physical activity increases mental well-being, vitality, social functioning, quality of life; it alleviates symptoms of depression and it decreases anxiety and cognitive decline in older people (Rejeski and Mihalko 2001; Reijneveld et al. 2003; Acree et al. 2006; Stathopoulou et al. 2006; Blake et al. 2009; Ströhle 2009; Sofi et al. 2011). There is also an association between physical exercise and reduced risk of developing Alzheimer's disease and Parkinson's disease (e.g. Podewils et al. 2005; Deslandes et al. 2009).

As we have already mentioned, the sources of health and well-being in old age are not just individual but also social. Sociality has a well-established role in the maintenance of mental health (see Cacioppo et al. 2007 for a review), and social support can improve mental resilience to the stresses of life beyond the current capacities of the individual (Davydov et al. 2010). It is well known that elders who are involved in group activities—such as group performances of music or dance—experience better mental and physical health (e.g. Cohen et al. 2006).

Social support may come from institutional sources (e.g. government agencies), cultural traditions (e.g. community organisations, social habits) but also specific levels of attachment between individuals. Attachment, acting through biological and psychological mechanisms, can lead to positive health outcomes for the individuals involved. The positive health effects of attachment have been somewhat poetically described by Esch and Stefano (2005) as the "benefits of love", and they are mediated by the operation of neuroendocrine mechanisms controlling emotions, attention, motivation, and memory, which are all functions that are dependent on the activity of brain limbic structures. There is also the additional involvement of the autonomic nervous system, especially in stress reduction. More closely attached individuals are also expected to be keener to help each other in case of need, including through the provision of emotional and financial support.

Financial troubles can be a major source of stress in old age. Neal Krause offered an intriguing analysis of the relationship between elders' financial stress and social support (Krause 1997). To understand both Krause's theoretical framework and analysis we have to remember that his work was carried out in the USA; that is, within a cultural framework that emphasises individualism and self-reliance without, traditionally, completely dismissing the value of social support, especially at the level of family and the local community. At the onset, Krause distinguished between received and anticipated support, establishing both the downside aspects of *received support* and the benefits of *anticipated support*. That anticipated support (i.e. the expectation of being helped if required) would bring some psychological benefits should be self-evident, but why would *received support* have downsides? According to Krause, the downsides of received support can emerge if such received support is seen as an admission of personal failure ("received support actually serves as a marker of failed or ineffective individual coping efforts" p. 285), which can make sense in a society where individual success and self-reliance are supposed to be important objectives in life. To complicate matters further, Krause also states that it is often the case that elders in financial need may be embedded within a social environment that is also characterised by financial limitations, in which case members of the direct social milieu may not be in the position of offering any or much financial help, and they may respond negatively to any request for help, thus leading to tensions and conflict. On the other hand, anticipated support does not put immediate pressure on the community for as long as the elder decides not to act on such expectation, but it can serve as an individual psychological spur to put more effort into coping with current hardships, and it helps in maintaining hope and a positive attitude towards the future. Eventually, when help is finally requested, the immediate family is the most likely source of financial aid that is expected to respond with the least resistance, as Krause also noted.

From Krause's perspective, therefore (p. 286):

a. "Anticipated support will buffer the noxious effects of economic stressful events on depressive symptoms" and
b. "received support will exacerbate the impact of financial problems on psychological distress".

In our view, although (a) seems to be a general pattern that is likely to be valid across cultures, (b) may not be equally general, but affected by the specific cultural environment. Receiving support when in need is not always perceived as a source of shame, thus causing distress, instead it can be perceived as a fact of life that will hopefully alleviate a dire financial situation, or it could even be seen as a right if there is a cultural expectation of mutual support (e.g. Cylwik 2002).

In his empirical work, Krause (1997) interviewed a sample of non-institutionalised, English-speaking, 65 years old or older retirees and found that past financial hardships were associated with both depressive and negative somatic symptoms in the elders. Such negative symptoms are especially evident in elders who report not having anyone to recur to for help in the future, whereas if the elder believes that help may be available if requested then the negative psychological effects of financial stress decrease, to eventually disappear altogether in the case of elders who hold strong beliefs of receiving help when required. Therefore, anticipated help buffers the individual against the stress of current financial hardships. Krause's data also support his prediction that received help is negatively related to individual well-being. But this could be caused by a combination of factors, a cultural one ("shame") as suggested above, but also older people requesting support when their psychological condition is already deteriorating. Krause tries to reconcile the results obtained for both anticipated and received support through an interpretation that puts anticipated support at the centre of the analysis:

> The findings reviewed up to this point suggest that received support may exacerbate the effects of recent economic stressors, while anticipated support may have important stress-buffering effects. Taken at face value, this may create the impression that received support is undesirable, and that it is better to merely believe that help will be available in the future if needed. However, this simple interpretation may not do justice to the complex processes that are at work. One possibility is that received support may affect well-being indirectly by reinforcing the belief that assistance will be forthcoming in the future should the need arise. (p. 290).

Although this interpretation that *received support* is acting as reinforcer of the belief that help will be available in the future is consistent with additional analyses carried out by Krause in this work, it seems to us that there may also be a more parsimonious alternative scenario. At some point in time the shame factor coming with a request for help will be out-weighted by the increasing needs of the elder, and therefore help will be explicitly requested much to the relief of the elder receiving it: a *swallow your pride effect*, so to speak.

Therefore the practical, beneficial effects of social support on older people's mental health do not just come from psychological factors such as attachment and anticipated support, but also from logistic help such as financial aid (received support). Many older people, women in particular, can find themselves under financial strain and when relief becomes available through social support the stress experienced by the individual is reduced, thus decreasing anguish, anxiety, demoralisation, depression, and increasing self-esteem (Krause 1987; Krause et al. 1991; Keith 1993; Angel et al. 2003). This is especially important when the elder has already experienced serious financial strain when she or he was younger, as long-term financial stress may have cumulative effects on health (Kahn and Pearlin 2006).

The positive effects of sociality on the well-being of older people are not just seen in the support they may receive but also in the acts of social support they perform for the benefit of others. Participating in volunteering activities, for instance, has positive effects on the well-being of elders (Morrow-Howell et al. 2003).

Although the benefits of sociality on the health of elders are clearly seen when they do receive (or provide) help and are socially well integrated, they can be also inferred from the negative effects of social isolation. Risks to own health (e.g. mortality, morbidity, immune dysregulation, depression, cognitive decline) tend to increase when elders live alone, have a small social network, participate infrequently in social activities (i.e. they are *socially disconnected*), and experience feelings of loneliness (i.e. they perceive themselves to be *socially isolated*) (Cornwell and Waite 2009; Miyawaki 2015; Courtin and Knapp 2017). From their own analysis of data coming from the *National Social Life, Health, and Aging Project* (NSHAP) of the USA, Cornwell and Waite (2009) concluded that social disconnectedness in older people is associated with poor physical health independently of feelings of loneliness or perceived lack of social support. Moreover, in the case of both socially connected and disconnected elders the perception of lacking social resources may have negative effects on physical health. This suggests that although elders who isolate themselves voluntarily may not suffer psychologically; they may still experience some negative physical effects of isolation. Therefore, both actual and perceived social disconnectedness have negative effects on the physical health of elderly people. In the case of mental health, however, it is perceived social isolation that truly matters.

The negative effects on mental health and well-being caused by social rejection and isolation can be especially serious when they affect elders who have made a considerable sacrifice for their country, and therefore for the community at large. In this case social isolation is compounded by a sense of unfairness that is likely to increase stress and frustration. For instance, Elder and Clipp (1989) describe the impact of social rejection on returnee soldiers who fought in unpopular wars (e.g. German soldiers in WWI or American soldiers in the Vietnam war). Although even extreme experiences such as war can be integrated into the development of an individual in some positive manner, as a learning experience (e.g. through coping with adversity, cooperation, self-discipline, friendship, and so forth), the traumas of combat are undeniable and they can be enduring, affecting the mental health of former soldiers into their old age. Problems such as sleep disturbance, depression, irritability, anxiousness, difficulty concentrating, and anger are common, along with drinking and drug dependence behaviour, existential anguish, and various other mental health problems that can interfere with affective relationships and the ability of finding and holding a job. Social rejection can only make such bad situation much worse (Elder and Clipp 1989).

Another population of elders that can be seriously affected by social disconnectedness and its negative effects on health and well-being is that of elders belonging to minority groups, such as immigrants (e.g. Mui and Kang 2006).

However, as already noted, not all cases of social disconnectedness are associated with deterioration of health (physical and/or mental) and well-being. Some elders may simply have an asocial, detached, and introverted personality, being well adapted to it even in old age. An extreme case is represented by the *Diogenes syndrome* which is characterised by self-neglect, aloofness, detachment, shrewdness, suspiciousness, and poor integration in society in spite of some old people diagnosed with the syndrome having high intelligence, a successful professional past, and falling within a normal range of personalities (Clark et al. 1975).

In the end, older people are more likely to achieve better mental health through a confluence of factors that can be individual (physical, mental), social, and financial, the latter being affected by both individual (personal wealth) and social factors (support from other individuals, groups, and institutions such as government) (Borg et al. 2006). Deterioration of health, physical condition, and fulfilment in life, increased strains in the social network, decline of financial status, or any combination of those may reach a point where some older people may contemplate hastening the end of their life—even through suicide if legal euthanasia is not accessible—and some may act on it (Kraaij et al. 2002; Duberstein et al. 2004). We will return to this issue in Chap. 8.

Two social and individual factors that have received much recent attention in research on elders' mental health and well-being are spirituality and religiosity.

In general, people tend to become more spiritual and religious the older they are, this is especially true for women (McFadden 1995), although this trajectory is obviously variable between individuals and presumably between cohorts of elders as well (e.g. Schwadel 2011). As suggested by Moberg (2005, p. 16): "there are a variety of patterns of religious transitions during the life course. For some people there is a high degree of stability (no change), but for others religiosity may increase, decrease, or show a curvilinear trajectory". Several explanations have been offered for these patterns, including a cohort effect (some generations may be more religious/spiritual than others); social disengagement at older ages (e.g. retirement from work) leading the individual to nurture the more religious/spiritual dimensions of the self; psychological compensation (i.e. the health and social challenges of old age lead to seeking relief in the spiritual dimension of life). Religiosity/spirituality may be also seen as direct promoters of good health through specific behavioural prescriptions (e.g. food, hygiene) that are followed in the expectation of potentially extending the lifespan (Van Ness and Larson 2002).

Spirituality and religiosity are two specific avenues used by many old people to improve their mental health, thanks to the better existential well-being and a more fulfilling ageing that they are supposed to afford (Corrigan et al. 2003, see McFadden 1995; Moreira-Almeida et al. 2006 for reviews). We addressed the broader relationship between spirituality/religiosity and health in our previous book *Behavioural Oncology* and here we start by summarising the basic concepts as they apply to the elderly.

The human interest in linking health with spiritual and religious beliefs is probably almost as old as our own species, and once organised religion emerged, churches also became a form of communities of mutual help and psycho-therapeutic practice (Allport 1963). On the negative side, religion, and especially institutionalised religion, can also be a source of fears, guilt, superstition, and social/psychological control that could jeopardise mental well-being.

The modern scientific interest in the link between spirituality/religiosity and health has been steadily increasing since at least the early 1990s. Such interest has been sparked by the role of religion and spirituality in the process of coping with the stresses of life (e.g. Stefanek et al. 2005; Koenig 2009).

Although we may have a personal understanding of "religiosity" and "spirituality", it is also important to provide a more precise definition of both concepts. As we already mentioned in Chap. 1, Brady et al. (1999, p. 418), quoting Elkins et al. (1988) and others, state that: "Religiosity has been defined as 'participation in the particular beliefs, rituals and activities of traditional religion'… It can serve as a 'nurturer and channel of expression' for spirituality …". Religiosity has been characterised as either *intrinsic* (referring to both religious practices and the more individualised spiritual aspects of religion) or *extrinsic* (more specifically associated with the rituals and practices of institutionalised religions) (see Allport and Ross 1967; Nelson et al. 2002). On the other hand, Brady et al. (1999, p. 418) note that spirituality is "'more basic' than religiosity … being a 'subjective experience that exists both within and outside of traditional religious systems' … Spirituality has been defined as 'the way in which people understand and live their lives in view of their ultimate meaning and value' … and is thought to include 'a present state of peace and harmony' … In a similar vein, the spiritual dimension has been said to relate to 'the need for finding satisfactory answers to … ultimate questions about the meaning of life, illness, and death' …". Cole and Pargament (1999, p. 397) also define spirituality as "one's experience of being in relationship to a transcendent reality" and interpret it as a strategy to recover some degree of control and meaning in life in situations where both control and existential harmony have been lost. Control brings more security and with it, decreased distress. Instrumental to such recovery from distress is the role of spirituality and religion in enhancing supportive social interactions.

A recent controversy about the use of the term "spirituality" in health studies has seen Pär Salander (2006) supporting alternative terms to replace spirituality, as distinct to religiosity, such as "meaning", "purpose", or "existential". The suggestion has been criticised by William Breitbart (2007) who defends the specific usefulness of the concept of "spirituality". In this regard he quotes Carl Sagan who defined spirituality as "the search each human being undertakes to find a sense of peace in one's relationship to the universe" (p. 106). It is clear to us that Salander and Breitbart are just emphasising two different, but complementary aspects of spirituality. On the one hand, Salander seems to be stressing the cognitive aspects of spirituality, whereas Breitbart, following Sagan, seems to be mainly concerned with the emotional aspects of spirituality. We agree with Breitbart that spirituality remains a useful concept, but only in so far as it synthesises both emotional and cognitive aspects of the life experience, especially when the individual is confronted with health issues.

Therefore, paraphrasing Sagan/Breitbart with a "Salander twist" and also taking into account others' views of spirituality (such as Brady et al. 1999; Cole and Pargament 1999), *spirituality* could be defined as *the search a human being undertakes to find a sense of peace, satisfaction, and meaning in life, it is an existential journey that engages the whole of our self.* It is clear from this definition that spirituality can be an important aspect of living a fulfilling life.

Thus, spirituality and religiosity provide meaning through a system of beliefs and emotionally reassuring support (especially support from equally minded people) that is particularly useful when external circumstances threaten our control on our own life. A third aspect of spirituality/religiosity that is relevant to mental health is the adoption of a healthy lifestyle as prescribed by the belief system.

Spirituality can acquire greater relevance to our life when we experience bodily or mental decay as we become older. Chao et al. (2002) identified the following main positive aspects of spirituality that can be especially relevant to old people: communion with self, communion with others, communion with nature, and communion with a higher being. Through spirituality and a sense of life coherence the deteriorations of ageing may be transformed from a state of "hopelessness" (Chochinov 2006) to a more meaningful experience, thus preventing the elder from plunging into a state of *existential anxiety* (Hackney and Sanders 2003).

But spirituality may also have a "dark" side in the form of *spiritual suffering* which, following Chochinov (2006, p. 89), is manifested in people "who are desperate to escape their situation". Spiritual suffering in the elders may be also enhanced by the feeling that they are a burden to others and that those others are suffering as a result. Such individuals find themselves in a situation of existential uncertainty or disharmony that may affect their mental health. Harmony may be returned if the elders' concerns are resolved through *acceptance*, for instance, thus achieving better *existential well-being*. In fact, existential well-being can be achieved in a variety of ways: e.g. through acceptance of specific realities that cannot be easily changed, but also through adjustments in the broader lifestyle. Seeking existential well-being through a specific system of religious beliefs may work for some elders but not necessarily for others (Edmondson et al. 2008).

Although our stance in this book is respectful of the sincere religious and spiritual beliefs of old people and of the metaphysical implications that such beliefs have for them, our analytical approach is scientific and therefore we also acknowledge the use of more general concepts such as Beadle et al.'s (2004) "positive illusory beliefs" that is beliefs in the improvement of a currently negative situation, which come together with a strong will to live. In our view, positive illusory beliefs include both those expressed in a structured manner through a religious or a spiritual system and those that are expressed in an intuitive manner: "I just feel that I will get better", with no additional external referent.

In religious discourse, the spiritual concept of *healing* refers not only to a process of recovery of health but also to a process leading to "inner peace; peace and reconciliation with God, self, and others; acceptance; letting go; and, finally, a peaceful death" (Puchalski et al. 2004, p. 700). Countries around the world vary in the degree of importance of religiosity and spirituality in the life of old people, and in their view on the capacity of religiosity and spirituality to *heal*; but all cultures have some set of beliefs organised into some form of religious or philosophical system, stories, myths, and traditions aimed at healing the "wounds of our life".

Religiosity can help improve various aspects of well-being in the elderly in many ways, including: healthy behaviours and lifestyles, social support, providing a belief system and also a cognitive and emotional framework that can help decrease stress (e.g. Koenig et al. 1988; Koenig 2006). Many studies have found a positive association between religiosity and specific aspects of well-being in old people: optimism and hope, self-esteem, meaningfulness and sense of purpose in life, internal locus of control, social support that can be material and/or psychological, coping with disease and disability, a healthy lifestyle prescribed by religious practices, somewhat lower levels of depression, anxiety, fears, frustration, anger, inferiority feelings, despondency and isolation, lower probability of committing suicide, and also higher probability of being married and having higher marital satisfaction (Levin and Chatters 1998; Van Ness and Larson 2002; Moreira-Almeida et al. 2006). Spirituality can provide greater meaning in life and cathartic avenues to relieve stress (Moreira-Almeida et al. 2006), and when religious elders are interviewed, many express the opinion that their well-being is anchored in their religious beliefs (Mackenzie et al. 2000). In addition, participation into a church community may also give elders access to material support that can improve their well-being: e.g. counselling, meals, home repair assistance, transportation, home visits and company, recreation, education, some forms of medical and nursing care (McFadden 1995). Long-term positive effects of participating in religious services may also include lower levels of cognitive dysfunction in old people (Van Ness and Kasl 2003). Although extrinsic (i.e. more institutionalised) religiosity may lead to higher neuroticism, emotional stability can follow from open-mature (intrinsic) religiosity and spirituality (Saroglou 2002).

Neal Krause has investigated how social support received from a religious group can become a promoter of better mental health in the elderly. Krause's model of religion-mediated mental health starts by stressing the perception of social cohesiveness within a church from the perspective of its members, older members in particular. Worshipping in a more cohesive church group may translate into greater spiritual and emotional support. Such support would then strengthen the religious relationship between the elder and his/her God. Closer relationship with God may then lead to greater optimism in life and, finally, more optimistic elders are expected to enjoy better mental health (Krause 2002, 2004b). Therefore, old people can increase their well-being through religion in a diversity of manners: Coping with the stresses and the unpredictability of life, forgiveness from God (existential reassurance), prayers that also provide reassurance and hope, and church-based social support/social cohesion (Krause 2004a).

In a study carried out to investigate the association between some of the aspects of religiosity mentioned by Krause with well-being in old people, Ardelt (2003) administered questionnaires to 103 Caucasians and African Americans in Florida, USA. Respondents were between 58 and 87 years old and 70% were women. A total of 93% of respondents were Christians of one denomination or another (38% Baptists, 23% Methodists, 24% belonging to other Protestant groups, 8% Catholics), whereas the remaining 7% followed other religions. Given that intrinsic religiosity points to a more personal and profound commitment to religious beliefs, it is not surprising that intrinsic religious orientation positively correlated with purpose in life, subjective well-being, and acceptance of death in this study, whereas intrinsic and extrinsic religiosity were negatively correlated to each other. Greater purpose in life and subjective health were associated with greater subjective well-being, the same positive association with subjective well-being was found for shared spiritual activities and belonging to a religious group. Age, however, was negatively related to subjective well-being, which is expected from the general decay observed with ageing, and purpose in life was negatively related to fear of death (Ardelt 2003).

Therefore, some elders who are confronted with deterioration in their subjective well-being may obtain specific benefits from intrinsic religiosity and from participating in the activities of their religious community.

Powell et al. (2003) reviewed the evidence linking religiosity and spirituality with health and although there is some empirical evidence for the positive health effects of religiosity and spirituality, more research is required before we can reach a firmer conclusion. Table 4.2 summarises Powell et al.'s main findings.

Note that point 9 in Table 4.2 tends to be valid only when the ill person is aware that others are praying for his prompt recovery (Cordella and Poiani 2014).

We have mentioned some gender effects in the relationship between religiosity/spirituality and well-being among old people, with old women being, in general, more reactive to the positive effects of religiosity/spirituality. McFarland (2010) investigated this issue in more detail in the USA using data from the *Religion,*

Table 4.2 Strength of evidence coming from tests of hypotheses for the relationship between religiosity/spirituality and health

Hypotheses	Evidence
1. Church/service attendance protects against death	Persuasive
2. Religion or spirituality protects against cardiovascular disease	Some
3. Religion or spirituality protects against cancer mortality	Inadequate
4. Deeply religious people are protected against death	Consistent failures
5. Religion or spirituality protects against disability	Inadequate/failures
6. Religion or spirituality slows the progression of cancer	Consistent failures
7. People who use religion to cope with difficulties live longer	Inadequate
8. Religion or spirituality improves recovery from acute illness	Consistent failures
– Religion or spirituality impedes recovery from acute illness	Some
9. Being prayed for improves physical recovery from acute illness	Some

From: Powell et al. (2003)

Aging, and Health Survey which include information from 66 to 95 years old people. McFarland's analyses show some interesting patterns. Overall, women are more religious than men, whether their religiosity is church based (organisational or extrinsic) or more personal and spontaneous (non-organisational or intrinsic). Men responded to high organisational religiosity with less depressive symptoms and greater levels of optimism and self-esteem. Organisational religiosity had no statistical effect on women. With regard to death anxiety, men responded to moderate levels of organisational religiosity with lower levels of death anxiety, whereas women responded with lower levels of death anxiety when engaged in either high and moderate levels of non-organisational religiosity or high levels of organisational religiosity. Therefore, in this study old men mainly benefitted from organisational religiosity, whereas women reaped benefits from all forms of religiosity.

To recap, a fulfilling life in old age can help decrease stress, improve mental health, and achieve better well-being. Control of stress and its health consequences can be also helped in old age by social initiatives. From a personal perspective, greater meaning in life, self-coherence, and purpose—as expressed through spirituality and religiosity for instance—but also physical activity and financial independence may help control the stresses and uncertainties of old age, thus favouring good mental health. Social support that helps foster mental health and well-being can come from family, friends, or the community at large.

Mental health may sometimes deteriorate with age to such an extent that a mental disorder becomes apparent. We devote the next two sections to an analysis of mental disorders in old age.

4.4 Mental Disorders in the Elderly

Although many people around the world can live to old age in good mental health, mental disorders are not uncommon in the elderly, but they are often overlooked being brushed off as effects of "normal ageing" (Jeste et al. 1999). Referring to the situation of mental health in old age in the USA, Bartels and Naslund (2013) stated that there are approximately 5.6–8 million Americans who are 65 years old or older and have been diagnosed with a mental health or substance-use disorder; moreover, the *Institute of Medicine* (IOM) estimates that their numbers will rise to 10.1–14.4 million by 2030. The predicted increase in the *number* of old people diagnosed with mental health problems is directly expected from the increasing number of people who can reach old ages.

Currently, there are an estimated 20.4% of 65 years old and older people who meet the criteria for a mental disorder in the USA (Karel et al. 2012). Around the world, approximately 15% of people aged 60 and over have been diagnosed with a mental disorder (https://www.who.int/news-room/fact-sheets/detail/mental-health-of-older-adults Downloaded on 21 April 2020). Although a prediction that the *percentage* (not just the number) of old people with mental disorders may increase in the future is currently disputed (Karel et al. 2012), there is an argument based on cohort characteristics suggesting that the ageing of the *Baby Boomer* generation may "herald an increase in substance use and mood disorders in late life" (Draper 2014). Present younger generations may also expect a future of greater stress in retirement, if they have been unable to accumulate enough personal financial resources and government support for retirees decreases, thus potentially leading to greater mental health issues in old age.

Future trends notwithstanding the current number of old people requiring psychiatric/psychological attention are already large enough to put considerable pressure on the health care system of many countries, especially those that are less economically advanced (Draper 2014). Draper (2014) lists a series of initiatives that should be enacted in order to improve geriatric mental health services around the world:

a. Establish clear national geriatric mental health policies and provide appropriate funding to implement them.
b. Focus on more mental disorders beyond dementia.
c. Fight the stigma and prejudice against geriatric psychiatry, including the effects of ageism.
d. Provide integrated, comprehensive geriatric mental health services that are also affordable to the elders and their family.
e. Better training for more specialist and primary care professionals.
f. Better remuneration and technical support for health care professionals.
g. Empower older persons.

Mental disorders vary in their prevalence across the elderly population. Table 4.3 shows a summary of various mental disorders and their prevalence values recorded in the elderly population in the USA. With regard to dementia (currently referred to as a *major neurocognitive disorder*), there is an increasing prevalence with age, whereas for mental disorders other than dementia there is a peak of prevalence in the 60s age bracket for elders living in the community, the prevalence decreases in the

Table 4.3 Prevalence of dementia (major neurocognitive disorder) and other mental disorders in the elderly population in the USA

(1)	(2) (%)	(3)	(4) (%)	(5) (%)	(6) (%)
65–69	1.4	65–74	9.6		
70–74	2.9				
75–79	7.6	75–84	6.8		
80–84	18.6				
85–89	32.7	85+	8.1		
90+	37.4				
Total 65+	10.0	Total 65+	8.5	31.5	20.4

(1) Age group (in years); (2) Prevalence of dementia; (3) Age group (in years); (4) Prevalence of mental disorders other than dementia in the community; (5) Prevalence of mental disorders other than dementia in residential care; (6) Total prevalence of mental disorders, including dementia, for older adults in the community and in residential care.
From Karel et al. (2012).

70s age bracket to then increase again in the 80s. The highest prevalence of mental disorders is seen among elders living in residential care (31.5%). That is, although older adults (75+) are more likely to develop various forms of dementia compared with younger adults, they display lower prevalence values of anxiety, impulse-control, mood, and substance-use disorders (e.g. Gum et al. 2009; Byers et al. 2010).

A more detailed frequency distribution for various mood, anxiety, co-morbid, neurocognitive, and other disorders found in people 55 years old and older is shown in Table 4.4. Mood, anxiety, and co-morbid mental disorders tend to decrease in prevalence with age—in contrast to neurocognitive disorders—but there is a slight increase in the oldest group compared to the previous age bracket: people older than 85 vs 75–84 (Byers et al. 2010), as it is also seen in Table 4.3. Anxiety and neurocognitive disorders tend to be the most prevalent types of mental disorders in old people, followed by mood disorders. Psychoses tend to occur at a comparatively lower frequency.

In Australia, Trollor et al. (2007) reported 15.8% prevalence of psychiatric disorders in the elderly that mainly included cognitive impairment (7.4%), personality disorder (2.9%), major depression (2.4%), and generalised anxiety disorder (2.4%), with women displaying higher prevalence values of affective disorder, major depression, generalised anxiety disorder, and drug use disorders than men, but alcohol use disorders are more prevalent in old men (Riedel-Heller et al. 2006). Cognitive impairment and personality disorder were found with similar prevalence among old women and men in Trollor et al.'s (2007) work. However, they also found a reduction in the prevalence of affective disorder, and major depression in particular, after the age of 75.

In general, mental disorders are more frequent in old women than old men (Jeste et al. 1999; Riedel-Heller et al. 2006; Trollor et al. 2007), which reflects a trend for higher prevalence values of mental disorders in the female than in the male population across all adult ages, with the exception of alcohol-related disorders which are more prevalent in old men, as already noted above by Riedel-Heller et al. (2006) (see Table 4.5).

We will devote the rest of this section to review in more detail depression, a major mood disorder in the elderly, and then anxiety which is also an important issue in old age, especially for people in their 60s (see Table 4.4); that is, people who are on the verge of retiring or have just retired.

Depressed old people feel sad and discouraged, displaying low energy and capacity to concentrate, sleep problems, low appetite, and weight loss (Riedel-Heller et al. 2006).

The DSM-5 (American Psychiatric Association 2013, p. 160–161) defines *Major Depressive Disorder* as a condition where:

> Five (or more) of the following symptoms have been present during the same 2-week period and represent a change from previous functioning; at least one of the symptoms is either (1) depressed mood or (2) loss of interest or pleasure.

1. Depressed mood most of the day, nearly every day, as indicated by either subjective report (e.g., feels sad, empty, hopeless) or observation made by others (e.g., appears tearful).

Table 4.4 Prevalence rates of some mental disorders in old individuals

Disorder	Total %	% by age 55–64 y	65–74 y	75–84 y	>85 y
Mood disorders					
Any mood	4.9	7.6	3.6	1.8	2.4
Major depressive disorder[a]	4.0	6.2	3.1	1.1	1.8
Dysthymia	0.8	1.2	0.5	0.5	0.5
Bipolar I or II disorder	0.9	1.4	0.5	0.7	
Mania/hypomania[b]	0.5				
Anxiety disorders					
Any anxiety	11.6	16.6	8.9	6.0	8.1
Panic disorder	1.3	2.0	0.6	1.0	0.5
Agoraphobia without panic	0.8	1.4	0.5	0.4	
Specific phobia	6.5	8.8	4.9	4.4	5.1
Social phobia	3.5	5.1	3.1	1.5	
Generalised anxiety disorder	2.0	3.2	1.4	0.8	1.8
Post-traumatic stress disorder	2.1	4.7	0.6	0.1	0.7
Adjustment disorder[b]	1.1				
Obsessive compulsive disorder[b]	0.9				
Co-morbid mood-anxiety disorder	2.8	4.8	1.7	0.9	
Neurocognitive disorders (NCD)					
Mild NCD[c]	3–19				
Major NCD[d]	5–7	1.6	3.4	10.1	32.4
		1.1	2.6	9.1	34.6
Psychotic disorders[b,e]	1.7	2.9[f]		3.5[f]	
Substance use disorders[b]					
Alcohol use disorders	1.0				
Drug use disorders	0.3				
Somatoform disorders[g]	1.5–13				

Most data are from Byers et al.'s (2010), which are complemented with additional information from other sources

[a]Range for major depression reported by Luppa et al. (2012): 4.6–9.3%. Twenty per cent of older people have any depressive disorder (Draper 2014), with the range being 4.5–37.4% according to Luppa et al. (2012)

[b]From Draper (2014)

[c]Gauthier et al. (2006)

[d]Prevalence of major NCD for people 60 years old and older, with a higher prevalence recorded in Latin America (8.5%) and lower prevalence values obtained in sub-Saharan African countries (2–4%) (Prince et al. 2013). First row of prevalence values is for Western Europe, the second row are values for the USA. When the age interval in this table included more than one age interval in Prince et al. (2013) Table 4.2, the arithmetic mean was calculated and reported here

[e]There are also reports of prevalence values of 10% for psychotic symptoms in non-demented elders (Riedel-Heller et al. 2006)

[f]Perälä et al. (2007). The prevalence value of 3.5% is for ages ≥65 years old

[g]Conditions characterised by medically unexplained symptoms. Data are from Hilderink et al. (2013)

Table 4.5 Prevalence (%) of mental disorders in six European countries for 18 years old and older males and females

	Lifetime prevalence		12-month prevalence	
	Male	Female	Male	Female
Any mental disorder	21.6	28.1	7.1	12.0
Any mood disorder	9.5	18.2	2.8	5.6
Any anxiety disorder	9.5	17.5	3.8	8.7
Any alcohol disorder	9.3	1.4	1.7	0.3
Major depression	8.9	16.5	2.6	5.0
Dysthymia	2.6	5.6	0.8	1.5
Generalised anxiety disorder	2.0	3.6	0.5	1.3
Social phobia	1.9	2.9	0.9	1.4
Specific phobia	4.9	10.3	1.9	5.0
Post-traumatic stress disorder	0.9	2.9	0.4	1.3
Agoraphobia	0.6	1.1	0.2	0.6
Panic disorder	1.6	2.5	0.6	1.0

From: Alonso et al. (2004)

2. Markedly diminished interest or pleasure in all, or almost all, activities most of the day, nearly every day (as indicated by either subjective account or observation).
3. Significant weight loss when not dieting or weight gain (e.g., a change of more than 5% of body weight in a month), or decrease or increase in appetite nearly every day.
4. Insomnia or hypersomnia nearly every day.
5. Psychomotor agitation or retardation nearly every day (observable by others, not merely subjective feelings of restlessness or being slowed down).
6. Fatigue or loss of energy nearly every day.
7. Feelings of worthlessness or excessive or inappropriate guilt (which may be delusional) nearly every day (not merely self-reproach or guilt about being sick).
8. Diminished ability to think or concentrate, or indecisiveness, nearly every day (either by subjective account or as observed by others).
9. Recurrent thoughts of death (not just fear of dying), recurrent suicidal ideation without a specific plan, or a suicide attempt or a specific plan for committing suicide.

The DSM-5 also recognises other depressive disorders that may affect old people, such as disruptive mood dysregulation disorder, dysthymia (persistent depressive disorder), substance/medication-induced depressive disorder, depressive disorder due to another medical condition, other specified depressive disorder, and unspecified depressive disorder.

Following up on previous work published by Dan Blazer (Blazer 1997), a pioneer in the study of depression in old people, Haigh et al. (2018) have recently summarised the available evidence for five of what Blazer called "myths and misconceptions" about depression in the elderly. The myths identified by Blazer are that, compared to younger ages, depression in older adulthood is:

1. symptomatically different;
2. more common;
3. more chronic;
4. more difficult to treat; and

5. more often caused by psychological factors.

Based on current evidence, Haigh et al. (2018) concluded that it is unclear whether depression in the old is different from depression at younger ages, hence myth 1 is not strongly supported, whereas accumulated evidence indicate that myth 2 is just that, a myth, as there is strong evidence that depression is less prevalent in the old than in young adults and the middle-aged. On the other hand, myth 3 is better supported, as depression in the old seems to take a more chronic trajectory than at younger ages. In principle, depression in late life is not more difficult to treat than at younger ages (against myth 4) but older adults may have a higher risk of relapse. Finally, myth 5 is not supported, as depression in old age is caused by a confluence of biological, psychological, and social factors that interact with each other, not just psychological factors.

Biologically, depression in the elders is associated with age-related dysregulation of the immune system that can also cause atherosclerosis and other inflammatory conditions in old age. Endocrine dysregulation is also associated with depression in old age, especially changes in the stress-related glucocorticoid feedback mechanisms leading to the development of hypercortisolism (Sapolsky et al. 1987; Tiemeier 2003). The endocrine and immune alterations caused by age-related dysregulation can eventually affect the functioning of brain areas such as the fronto-striatal region, the amygdala, and hippocampus that lead to the memory, cognitive, and mood changes characteristic of depression (Alexopoulos 2005; Dantzer et al. 2008).

Although depression can have a heritable component, major immune dysregulation in the old leading to depression can be accelerated by stress that, in turn, is increased under specific conditions such as: disability, social isolation, financial impoverishment, involuntary relocation (e.g. to an institution), engaging in caregiving, and bereavement (Anderson 2001; Kraaij et al. 2002; Alexopoulos 2005). Personality factors such as neuroticism can also predispose old individuals to develop depression even when they had no previous experiences with clinical episodes of depression when they were younger (Duberstein et al. 2008). Other risk factors for the development of depression in old age include: being a woman, being widower or single, low education level, suffering from a physical illness, use of multiple drugs, changes in the brain white matter, psychological stressors (Sözeri-Varma 2012). Susceptibility to stress in older people can also have roots into their earlier development (e.g. dysfunctional past family relationships, Brodaty et al. 2001).

A specific link through immunity such as chronic inflammation could also explain the relationship between depression and onset of dementia (Hermida et al. 2012), and depression can further complicate dementia when they co-occur (see Blazer 2003 and the references therein). A positive correlation between depression and broader cognitive impairment in the elders has been also described (Vinkers et al. 2004). In addition to immune dysregulation expressed in inflammatory changes, other processes that are caused by depression may eventually lead to the onset of dementia; they include: dysfunction of the vascular system and

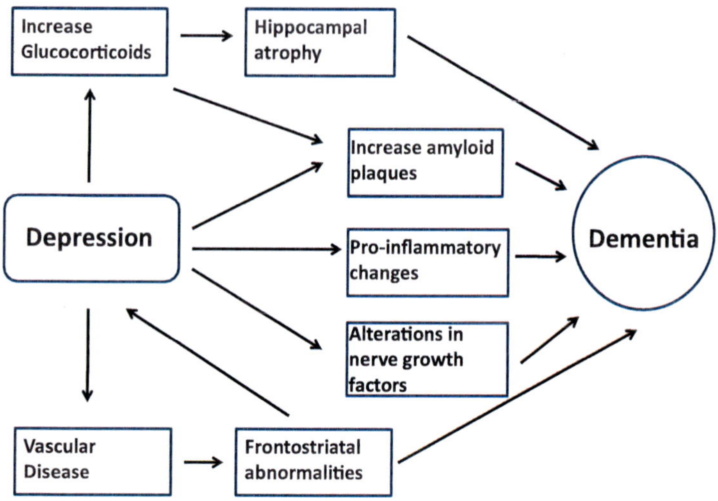

Fig. 4.3 Predominant pathways linking depression as a risk factor for the onset of dementia, as proposed by Byers et al. (2011)

glucocorticoids activity, hippocampal atrophy, increased deposition of β-amyloid plaques, and deficits of nerve growth factors (Byers and Yaffe 2011). Such mechanisms linking depression and dementia are summarised in Fig. 4.3.

Dementia in the elderly will be reviewed in more detail in Sect. 4.5 in this chapter.

One would assume that various social factors could also significantly affect the development of depression in old age, as we have already suggested, and indeed there is a considerable amount of evidence indicating that greater social support, participation in social activities, and an overall decreased level of social stress are linked to decreased depression in the elderly (Pahkala 1990; Russell and Cutrona 1991; Cervilla and Prince 1997; Croezen et al. 2015). A review by Cole and Dendukuri (2003), however, did not detect strong evidence for such effect. What they did find was that out of 13 risk factors for depression in the elderly that they reviewed, the following ones are significant: bereavement, sleep disturbance, disability, prior depression, and female gender. Whereas increasing older age, lower education level, being not married, and poor social support did not seem to be significant risk factors. Uncertain risk factors included poor health, cognitive impairment, living alone, and being diagnosed with a new medical illness. Such results notwithstanding specific individuals and sub-populations may be more susceptible to the negative effects of poor social support, and different factors may also interact. For instance, in the USA, older Latinos display "up to double the rates of clinically significant depression compared with both whites and blacks in similar population-based studies" (Aranda 2013, p. 2). Such high prevalence of depression among old Latinos could be explained by the interaction between old age and low social support, high stress, chronic financial strain, and low acculturation that characterise populations of immigrants, who may also be subject to a degree of discrimination in society. Some methodological issues may also affect the results of studies of social factors affecting depression in old age. In a review carried out by Schwarzbach et al. (2014) they concluded that, in qualitative analyses, greater social support, higher quality of relations, and greater presence of confidants are significantly associated with less depression. Quantitatively, however, whereas depressive symptoms significantly decrease with greater social integration of the elders and with satisfaction with their social support in studies carried out longitudinally, cross-sectional studies are more ambiguous. Such methodological issues may have affected the analyses carried out by Cole and Dendukuri (2003) mentioned above and therefore their conclusions.

Therefore, given the effects of stress on depression and the role of social support on stress relief, supportive social conditions remain a valid pathway to relieving depression in old age, at least to some extent (Lin et al. 1999). The limit of such effect is determined by the quality and adequacy of social support, and to what extent the biological mechanisms of depression are impervious to supportive social effects in any specific case.

Depression in the elderly can also occur with various co-morbidities, including excess weight, cardiovascular disease, decrease in bone mineral density, and it is in general associated with poor self-rated health over time, eventually leading to increased mortality (Schoevers et al. 2000; Blazer 2003; Mitchell and Subramaniam 2005). The effect of depression on increasing mortality is more pronounced in old men than old women (Holwerda et al. 2007).

As shown in Table 4.4 about 4% of the elderly population may suffer from major depressive disorder, with the prevalence decreasing from the mid-fifties to the early eighties, but then increasing again at older ages. The range of prevalence values for major depressive disorder provided by Luppa et al. (2012), however, is higher: 4.6% to 9.3%. Minor depression is more prevalent (4–13%), whereas dysthymic disorder is diagnosed in about 2% of the elderly population. Such mood disorders are especially prevalent in elders who are institutionalised (Alexopoulos 2005, see also Table 4.3). For instance, in the case of elders who are living in long-term care institutions, the prevalence values of minor depression or clinically significant symptoms of depression span between 17% and 35% (Aranda 2013), whereas Anstey et al. (2007) reported a prevalence rate of depressive disorders of 32.0% in nursing homes compared to 14.4% for community-dwelling elders. If we consider all depressive disorders combined, 20% of older people have a depressive disorder (Draper 2014), within a range of 4.5–37.4% (Luppa et al. 2012).

Old women are more frequently diagnosed with depression than men (Blazer 2003) and depression steadily increases with age more reliably in women than men (Glaesmer et al. 2011). For major depression, prevalence for 75 years old and older women ranges between 4.0% and 10.3%, whereas for men it is lower: from 2.8% to 6.9%. The trend between sexes is maintained for old people 85 years old and older, with prevalence values of depression in women ranging from 3.8% to 11.1% and for men it ranges from 2.1% to 5.1% (Luppa et al. 2012). In the case of minor depression, Luppa et al. (2012) provide prevalence values of between 1.2% and 3.1% for women and between 0.9% and 4.2% for men.

In nursing home residents 8.4% to 61.4% women are diagnosed with depressive disorders compared to 6.4% to 38.1% of men (Luppa et al. 2012).

More depressed old people have been found to be more frail (Vaughan et al. 2015) and to have poorer generic health-related quality of life than non-depressed elders (Sivertsen et al. 2015); therefore, it is not surprising that depression in the old is also associated with greater risk of natural mortality and also suicide (Anderson 2001). The association of depression with decreased quality of life and increased mortality may be also explained by the effect that depression can have on decreasing the size of older people's social network, thus leading to greater perceived loneliness and potentially also lower material social support, especially in men (Houtjes et al. 2014).

Given the negative effects that depression can have on old people's health, several strategies have been adopted in order to treat depression. Here we just focus on behavioural approaches.

In a review of the effectiveness of coping strategies in decreasing depression in the elderly, Bjørkløf et al. (2013) concluded that active strategies such as increasing control on own life, and also positive religious coping, were associated with fewer symptoms of depression in both community living elders and in those studied in clinical settings. Therefore organised psychological interventions should adopt proactive approaches, fostering active participation, involvement, and self-motivation in the elders. Many such organised psychological interventions have been used to alleviate depression in old people and they include *cognitive behavioural therapy* (CBT), *interpersonal psychotherapy* (IPT), *behavioural activation therapy* (BAT), *problem-solving therapy* (PST), *life-review therapy*, *psychodynamic psychotherapy*, *non-directive counselling*, *group therapy*, *family therapy* (Anderson 2001; Sözeri-Varma 2012; Cuijpers et al. 2014), and some of them have been proven successful to improve symptoms of depression in the elderly, CBT in particular (Liu and Aftab 2020). Such professionally guided therapies may be needed mainly in clinical cases, whereas other elders may decrease depressive symptoms using more spontaneous activities such as engaging in a personal program of light exercise, which can decrease depression in older people provided that the activity is continuous rather than sporadic (Lindwall et al. 2006) and/or participating in group activities within their local community. Some non-psychological interventions are also available (Taylor 2014).

Anxiety disorders are also common among the elderly and they can be co-morbid with depressive disorders (Beekman et al. 2000). The DSM-5 (p. 189) defines anxiety disorders so:

> Anxiety disorders include disorders that share features of excessive fear and anxiety and related behavioral disturbances. *Fear* is the emotional response to real or perceived imminent threat, whereas *anxiety* is anticipation of future threat. Obviously, these two states overlap, but they also differ, with fear more often associated with surges of autonomic arousal necessary for fight or flight, thoughts of immediate danger, and escape behaviors, and anxiety more often associated with muscle tension and vigilance in preparation for future danger and cautious or avoidant behaviors. Sometimes the level of fear or anxiety is reduced by pervasive avoidance behaviors. *Panic attacks* feature prominently within the anxiety disorders as a particular type of fear response... Anxiety disorders differ from developmentally normative fear or anxiety by being excessive or persisting beyond developmen-

tally appropriate periods… Many of the anxiety disorders develop in childhood and tend to persist if not treated. Most occur more frequently in females than in males (approximately 2:1 ratio).

The variety of anxiety disorders that can apply also to older individuals include: Generalised Anxiety Disorder, Separation Anxiety Disorder, Specific Phobia, Social Anxiety Disorder (Social Phobia), Panic Disorder, Panic Attack Specifier, Agoraphobia, Substance/Medication-Induced Anxiety Disorder, Anxiety Disorder Due to Another Medical Condition (American Psychiatric Association 2013).

Among the various anxiety disorders, *generalised anxiety disorder* (GAD) has been the focus of several theoretical and empirical studies. According to the DSM-5 the specific diagnostic criteria for GAD include excessive anxiety and worry that occur frequently over a period of at least 6 months and that the individual finds difficult to control. Such excessive anxiety and worry are associated with at least three of the following symptoms: restlessness or feeling keyed up or on edge, being easily fatigued, difficulty concentrating or mind going blank, irritability, muscle tension, and sleep disturbance. Such conditions cause significant distress or impairment in social, occupational, or other relevant areas of functioning.

Current research on the causes of GAD suggests that emotional dysregulation may be a major causative factor of the disorder (Mennin et al. 2005; Behar et al. 2009). The emotional dysregulation model proposes that people with GAD are subject to emotional hyperarousal, especially in response to negative experiences, and they have a negative attitude towards their emotions, thus responding poorly to their emotive experiences, being incapable of regulating and managing them in an adaptive manner. In this model, worry is an ineffective strategy to cope with emotions. Additional models have been built around this central theme of emotional dysregulation, emphasising "avoidance of worry" (Behar et al. 2009) or "intolerance of uncertainty" (Dugas et al. 1998); see Mochcovitch et al. (2014) for a review. Neurologically, GAD patients have been observed to fail to appropriately activate both the prefrontal cortex and the anterior cingulated cortex during tasks requiring emotional regulation. This has been suggested by Ball et al. (2013) to be a result of over-responsiveness of the limbic system interfering with cortical mechanisms of emotional regulation.

Several risk factors make the development of anxiety disorders in the elderly more likely. Vink et al. (2008; see also Zhang et al. 2015, Zhang et al. 2015a) concluded that, on the ground of current evidence, the following risk factors are associated with an increased probability of developing anxiety disorders in old age:

a. *Biological*: cognitive impairment, high blood pressure, poor self-perceived health, vision or hearing loss, lower body mass index, taking a high number of medications, chronic physical health disorders (respiratory disorders, arrhythmia and heart failure, dyslipidemia or abnormal levels of lipids in the blood).

b. *Psychological*: greater neuroticism, more dysfunctional coping, lack of self-efficacy, a psychiatric history especially of depression and phobias, cluster C personality disorder (anxious and fearful personality), recent adverse life events.

c. *Social*: lower income, lack of social support, over-protection by parents early in life, low affective support during childhood, history of mental problems in parents, attachment status, bereavement, recent negative life events, traumatic events.

For instance, based on data from the *Longitudinal Aging Study Amsterdam* (LASA), Beekman et al. (1998, 2000) list being female, having suffered extreme experiences (such as war), having lower levels of education and external locus of control as the dominant risk factors for anxiety disorders, whilst stresses such as recent losses in the family and chronic physical illness were also seen to play a part. A smaller size of the social network was also associated with the development of anxiety disorders. After analysing data from the same study, Schuurmans et al. (2005) also found that long-term anxiety in old people is associated with high scores on the personality trait of neuroticism at baseline (i.e. being moody, worrisome, fearful, prone to display anger and frustration).

Lenze et al. (2001) include physical illness and disability as clear risk factors for anxiety disorders in the elderly, but it is still unclear whether cognitive impairment decreases or increases the probability of developing anxiety disorders in old age. Loss of memory due to normal ageing processes, however, may be associated with the development of anxiety disorders that may subsequently conduce to depression first and then further cognitive decline (Sinoff and Werner 2003). Anxiety disorders can also manifest themselves in older people following unresolved childhood experiences and traumas (Raposo et al. 2014), as it is also the case for depression.

For overall anxiety disorders, Hendriks et al. (2008) give a range of prevalence values of between 10% and 15% for elderly living independently, whereas in their review of anxiety disorders in the old, Wolitzky-Taylor et al. (2010) provide a range of between 3.2% and 14.2%, with prevalence values decreasing in the population of over 80 years old. Prevalence is also lower in older adults in general as compared with pre-retirement adults (50 years old and younger) (Wolitzky-Taylor et al. 2010).

The above results are consistent with those obtained by Byers et al. (2010) who studied the prevalence of anxiety disorders in a sample of 2575, 55 years old or older participants in the USA. The prevalence of "any anxiety" was 11.6%, and it tended to decline with age, except for a surge at ages older than 85 (see Table 4.4). Married/cohabiting elders displayed less anxiety than divorced/separated/widowed and never married elders. Table 4.4 also lists prevalence values for specific anxiety disorders studied by Byers et al. (2010): panic disorders (1.3%), agoraphobia without panic (0.8%), specific phobia (6.5%), social phobia (3.5%), generalised anxiety disorder (2%), post-traumatic stress disorder (2.1%), adjustment disorder (1.1%), and obsessive-compulsive disorder (0.9%).

In a sample of elders studied in The Netherlands, Beekman et al. (1998) provided an overall prevalence value for anxiety disorders of 10.2%, which is consistent with the value obtained in the more recent works mentioned above. Generalised anxiety disorder was the most frequent disorder (7.3%), followed by phobic disorders (3.1%). Prevalence values for panic disorder (1.0%) and obsessive-compulsive disorder (0.6%) were low. Manela et al. (1996) reported a prevalence of 15% for anxiety disorders in a sample of elders from a socially deprived area in London. Phobic disorder was diagnosed in 12.0% of the sample along with generalised anxiety (4.7%), agoraphobia (7.9%), specific phobia (5.9%), social phobia (0.6%), and panic disorder (0.1%).

Flint (2005) gives prevalence values between 0.7% and 9% for generalised anxiety disorder among older people living in the community. Results that fall within the same range were obtained by Schaub and Linden (2000) as part of the *Berlin Aging Study*, with a prevalence value for anxiety of 4.5%, which was more reflective of the prevalence in the "younger" group of old people (70–84 years) in their population (4.3% diagnosed with anxiety overall), whereas the prevalence decreases in the older old (85–103 years), being 2.3%. Zhang et al. (2015) provide a range of prevalence values for GAD of between 4.6% and 11%.

Therefore, overall prevalence values for anxiety disorders in the elderly tend to vary between 10% and 15% and although it is possible that anxiety disorders may decline in old age to then experience a resurgence after the age of 80, some studies do not support this trend.

Anxiety is more prevalent in old women than old men. In fact, it was 2.8 times higher in women in Holwerda et al.'s (2007) study of a Dutch population of old people and 2.2 times more frequent in women in a study carried out in France (Zhang et al. 2015). According to McLean et al. (2011) the lifetime male:female prevalence ratio of any anxiety disorder is 1:1.7, whereas for 12 months it is 1:1.79. Schaub and Linden (2000) reported prevalence values of 2.9% for old men and 4.7% for old women in Germany. Byers et al. (2010) also indicate that anxiety is more prevalent among women (14.7%) than men (7.6%) in their sample from the USA.

Co-morbidity, especially between anxiety and depression disorders can also occur (Wolitzky-Taylor et al. 2010; Zhang et al. 2015), leading to more severe psychopathologies than non-anxious depression (Lenze et al. 2001), although the effect of co-morbidity is not always strong, especially in the older old. For instance, Van der Weele et al. (2009) found that co-occurrence of depression and anxiety did not add additional negative effects over those recorded for depression alone, except for a further increase in loneliness, after studying a sample of 90 years old Dutch people.

Unlike the link suggested between depression and dementia, anxiety disorders do not seem to be associated with the risk of developing dementia (de Bruijn 2015) nor do they increase mortality in old age (Holwerda et al. 2007). Nevertheless, they do decrease quality of life and psychological interventions have been devised to treat such disorders.

As reviewed by Sheikh and Cassidy (2000), psychological therapies to treat anxiety disorders can be categorised into three groups: those using *cognitive restructuring* (replacing thoughts that cause anxiety with others that do not), those focusing on *relaxation training*, and those based on *exposure* (mainly used to treat phobias). Such therapies include supportive psychotherapy and cognitive behavioural therapy.

Cognitive behavioural therapy (CBT) is the most frequently used cognitive restructuring treatment for generalised anxiety disorder and indeed it is the most frequent psychological intervention to treat anxiety disorders in the elderly. CBT aims at replacing thoughts that produce negative emotions in the subject with thoughts that produce more "balanced perceptions" (Wilkinson 2013). This is achieved through discussions, diaries, and behavioural experimentation (i.e. exploring new ways of acting in specific situations). CBT for anxiety disorders aims to overcome obsessive self-monitoring of symptoms that are the cause of anxiety. Stanley et al. (2003), for instance, have shown how, after CBT treatment, 60 years old and older participants showed a significant decrease in anxiety, worry, depression, fears, and an increased level of quality of life compared with minimal-contact controls.

Although studies have detected a consistently positive effect of CBT on controlling anxiety disorders in old people, recovery after a course of CBT seems to be modest (Flint 2005; see Hendriks et al. 2008 for a review).

In sum, some mental disorders increase in old age (e.g. major neurocognitive disorders), others decrease, or they decrease first to then increase at older ages (e.g. many mood and anxiety disorders). Most mental disorders in old age are more prevalent in women than men, reflecting gender differences that are already apparent at younger ages. Depression tends to decrease with age, but when it is persistent, it could be a causative factor for the development of neurocognitive disorders later in life in some individuals. Depression is significantly more likely to develop in institutionalised elders and in older women, with higher levels of depression being associated with an increased probability of death. Depressive disorders can be co-morbid with anxiety disorders, but the latter have their own specific aetiology and presentation. Generalised anxiety disorder (GAD) has been the target of intense study, being a result of emotional dysregulation. GAD can develop in the elderly due to biological, psychological, and/or social factors, and just like depression it is more prevalent among old women than old men. Cognitive behavioural therapy is one of the most frequently used behavioural interventions to help elders with depression and anxiety; but specific initiatives based on personal changes in lifestyle can also help, before any professional intervention is required.

In the next section we focus more specifically on neurocognitive disorders, which are more likely to develop in old age and are becoming a serious concern for health authorities around the world.

4.5 Neurocognitive Disorders: Causes, Prevention, and Psychological Interventions

Old people who are developing neurocognitive disorders tend to go through a series of changes that start with gradually becoming aware that they are having some cognitive problems such as memory issues, perhaps initially dismissing them as part of their normal ageing, or they may explain them away as the effect of stress. The elders may subsequently grow more concerned as the symptoms become more severe, but they may still excuse (or not) the changes depending on personality, for instance, until they receive a dramatic shake up when they are confronted with drastic events, such as being banned from driving. At this point denial becomes untenable. Steeman et al. (2006, p. 730) provide a detailed and graphic description of the complex process an older person who is becoming aware of developing a neurocognitive disorder goes through:

> Becoming aware that something is wrong produces frustration, uncertainty, and fear, which in turn puts stress on interpersonal relationships ... These feelings stem from inability to understand the perceived changes and their unpredictability, lack of self-assurance and feelings of being out of control ... Being aware that something is wrong also causes individuals with dementia to form strategies of watching and analysing themselves and others for signs of deterioration that eventually lead them to the stage of searching for meaning... Meanwhile, to maintain control and conceal growing memory problems, they develop strategies of vigilance and avoidance ... Attempts to keep the problems hidden and to maintain the veneer of normality become increasingly difficult and cause psychological strain… Eventually, these strategies fail and close friends and family begin to notice the cognitive decline and may express their concern…. In many cases, family members initiate the stage of sharing awareness (i.e. acknowledging the memory impairment), whereas in others the person with dementia does this... The individual's desire to acknowledge the problem may be triggered by several needs: the need for an explanation …, the need to relieve the pressure of maintaining a normal appearance, and the need to feel supported ... Sharing awareness may be limited to close family and friends ... However, sharing awareness cannot be taken for granted. While disclosure of memory problems may relieve an individual with dementia, family members may not always be willing to acknowledge that their loved one has a problem ... If this occurs, the individual may become frustrated, being uncertain as to what to do next ... Alternatively, people with dementia may not necessarily communicate the problem to their families ... This does not necessarily mean that they are unaware of having memory problems ... Alternately denying and acknowledging memory loss, or acknowledging memory loss without being able to describe it or without expressing concern or understanding may occur ... Awareness may eventually lead to the stage of seeking professional help.

Once the diagnosis is official, the reaction of the elder can be variable ranging from denial, fighting spirit, or quiet acceptance, which may lead to either focusing on the challenges ahead or giving up. Throughout the whole process, however, social support from family, friends, and well-trained medical and nursing personnel is essential.

We have seen in the previous section how the prevalence of neurocognitive disorders increases with age in older people. Following the DSM-5, neurocognitive disorders currently include *delirium*, *major neurocognitive disorder* (*major* NCD), and *mild* NCD. Here we list the subtypes of NCDs as recognised in the DSM-5, with information about prevalence in the elder population and relative distribution between the sexes for the most common subtypes:

- *NCD due to Alzheimer's disease*, that we will just refer to as *Alzheimer's disease* (AD), has a variable prevalence in the elders population of between 1.9% and 6.5% (>60 years old) (Rizzi et al. 2014), with the condition being more prevalent among old women than men (Lobo et al. 2000; Riedel-Heller et al. 2006). AD will be also the subject of a more in-depth analysis in this chapter.
- *Vascular NCD* is a neurocognitive disorder caused by cerebrovascular injury (Jefferson et al. 2014) which has a prevalence of 0.9–3.3% (>60 years old) (Rizzi et al. 2014), being more prevalent in men than women until the 70s, but then

becoming more prevalent in women than men from about the age of 85 (Lobo et al. 2000).

- *NCD with Lewy bodies* is a disorder that shares some characteristics with Parkinsonism (Fong and Press 2014), it has a prevalence of 0–5% (>65 years old) and it is more common among old men than women (Zaccai et al. 2005).
- The prevalence of *NCD due to Parkinson's disease* is 0.2–0.5% (≥65 years old) (Aarsland et al. 2005), with old men having a shorter time period without NCD due to Parkinson's than old women (Buter et al. 2008).
- *Frontotemporal NCD* is a condition characterised by a severe level of brain atrophy (Boxer 2014). The disease varies in prevalence, with Onyike and Diehl-Schmid (2013) reporting a range of between 7.7 and 8.9/100,000 (60–69 years old), and Neary et al. (2005) giving a range of 3.8–9.4/100,000 (60–79 years old). Prevalence of the disease is also variable between the sexes across studies, but overall it is similar between old women and men (Onyike and Diehl-Schmid (2013)).

Other NCDs include NCD due to traumatic brain injury, NCD due to HIV infection, substance/medication-induced NCD, NCD due to Huntington's disease, NCD due to prion disease, NCD due to another medical condition, NCD due to multiple aetiologies, and unspecified NCD. See also Morin (2014) and McKee and Gavett (2014) for additional forms of dementia.

In general, the NCD condition is characterised by a primary clinical deficit in cognitive function that has not been present in the individual since birth or early on in life. Such deficits include forgetfulness, language deterioration, impaired judgement; loss of initiative and mood changes can also occur (Mandell and Green 2014). Table 4.6 compares the function of different forms of memory: *episodic* (personally experienced episodes/events and details related to them), *semantic* (facts about language, functions of objects, appropriate behaviour), *procedural* (specific practical

Table 4.6 Selective memory system disruptions in common clinical disorders

Disease	Episodic memory	Semantic memory	Procedural memory	Working memory
Alzheimer's disease	+++	++	–	++
Frontotemporal dementia	++	++	–	+++
Semantic dementia	+	+++	?	–
Lewy body dementia	++	?	?	++
Vascular dementia	+	+	+	++
Parkinson's disease	+	+	+++	++
Head trauma	+	+	+/–	++
Depression	+	+/–	++	+/–
Anxiety	+	–	–	+/–

+++ = Early and severe impairment, ++ = moderate impairment, + = mild impairment; +/– = occasional impairment or impairment in some studies but not others, – = no significant impairment, ? = unknown (from Budson 2014)

skills), and *working* (short-term storage of information) in people with various mental disorders and other conditions affecting brain functioning, such as head trauma. Note how in the case of NCDs all forms of memory are significantly affected, with the exception of procedural memory where there are some areas of uncertainty regarding the changes in patients diagnosed with semantic or Lewy bodies dementia.

NCDs as defined by the DSM-5 are not the only conditions that can lead to serious cognitive deficit in old people. Fahr's syndrome, for instance, that is characterised by calcification of the basal ganglia and cerebral cortex, could lead to serious difficulties with concentration, memory, changes in personality, and ultimately psychosis and dementia (Saleem et al. 2013).

Although *dementia* is now formally subsumed into the concept of *major* NCD, the word is still commonly used, and it is also accepted by the DSM-5. Given the familiarity of most people (professionals and laypersons alike) with the term dementia, we will use it in this book, and we will also refer to *major* NCD when appropriate.

Dementia can be defined as a "progressive decline in mental function beyond what is considered a normal part of ageing" (Fuller et al. 2013, p. 2). The term "dementia" itself means "without mind" and it was introduced by the Roman medical writer Cornelius Celsus (c. 25 c.e.), although memory impairment in old age was already mentioned in the writings of Solon (seventh- to sixth-century b.c.e.) and Plato (fifth- to fourth-century b.c.e.) (Kurz 2013).

Major NCDs increase in prevalence with age (Table 4.3), this is the case of Alzheimer's disease (AD), the most common form of dementia, which is followed in order of importance by vascular dementia and NCD with Lewy bodies. In European data, the overall prevalence of dementia increases by one order (in some cases two orders) of magnitude from the age bracket of 65–69 years old to 90+. The general trend is consistent across countries (Table 4.7). Prevalence of *mild* NCD, however, does not always follow the same predictable increase with age (Riedel-Heller et al. 2006).

The strong association between increased risk of dementia and old age is also seen in the current prevalence distribution of dementia around the world. Table 4.8 shows how prevalence of dementia is higher in the population older than 60 in regions and subregions where people live longer (see also Table 1.1 in Chap. 1); e.g. Australasia and the Asia Pacific region—these regions include data for economically developed countries—Western Europe, North America, and Southern Latin America. It is also worth noting that dementia is more prevalent in women than men in Asia, Western Europe, and Latin America in the population older than 60. On the other hand, in the high-income regions of the Asia Pacific and in the USA dementia is more prevalent in men between the age of 60 and 79 years old, to subsequently become more prevalent in women from the age of 80 (Prince et al. 2013).

However, what changes around the world is not just the prevalence of dementia but also the cultural attitudes towards dementia. As we have seen in Chap. 3, cultural views about dementia may either hinder the provision of support required by the elder (e.g. by blaming dementia on some moral failure of the elder, stigmatising the elder) or they may help in the task of providing care for the elder (e.g. through a concept of duty such as *familism*, for instance).

Table 4.7 Prevalence of dementia in old age in the European Union

	Prevalence (%)					
	Age bracket					
Country	65–69	70–74	75–79	80–84	85–89	90+
Belgium	0.6	5.1	7.6	16.2	33.6	33.6
Denmark	3.7	5.5	9.8	13.8	–	51
Finland	–	–	–	13.1	26.7	–
France	0.9	1.1	3.5	7.6	9.6	26.1
Germany	–	0.0	5.9[a]	11.9[a]	25.6[a]	41.6[a]
United Kingdom	1.3[a]	2.1[a]	5.1[a]	10.9[a]	16.6	25.2
Italy	1.5[a]	2.6[a]	9.3[a]	20.8[a]	29.9[a]	44.4[a]
The Netherlands	1.0[a]	2.7[a]	5.6[a]	16.6[a]	36.3[a]	47.2[a]
Spain	0.9[a]	7.1[a]	9.6[a]	19.1[a]	32.3[a]	52.5[a]
Sweden	–	–	9.6[a]	17.1[a]	33.8[a]	32.0

Adapted in a simplified form from Riedel-Heller et al. (2006)
[a]Mean value of more than one study/diagnostic criteria, the calculation only includes data for that specific age interval

Given the current world trend towards an increased prevalence of people older than 65 (see Table 1.1) and the link between dementia and old age, there is a growing concern about what the future may hold and how to care for the projected large population of old people with cognitive disorders. This population is estimated to increase 225% between 2010 and 2050 according to Prince et al. (2013); see the last column of Table 4.8.

Some intriguing recent studies, however, suggest that the future may not be necessarily as bleak as predicted. Satizabal et al. (2016), for instance, report on the results of the *Framingham Heart Study* which is a community-based, longitudinal cohort study that started in 1948 in Framingham, Massachusetts (USA). Their analyses indicate that the risk of developing dementia in this population has in fact decreased since the late 1970s and early 1980s but such decrease was only observed among residents who had at least a high school diploma. The decline in dementia was also associated with an improvement in some indicators of cardiovascular health and an increased use of antihypertensive medications. Whether it is possible to decrease the current prevalence of dementia in ageing populations by decreasing cardiovascular risk factors, through medical interventions, improved education, and lifestyle changes is something that requires further study.

Mild NCD (also referred to as *mild cognitive impairment*, MCI) has been suggested to be a prodrome or early symptom of AD since the work of Petersen et al. (1999). *Mild* NCD is characterised by some loss of memory, about recent events in particular; a symptom that often elicits disbelief and denial in the affected elders (Fuller et al. 2013).

In dementias such as AD, however, loss of memory and deterioration in other aspects of cognition become serious and noticeable. The common risk factors for men and women for progressing from *Mild* NCD to dementia include carrying the

Table 4.8 Current (2010) and projected (2030, 2050) levels of dementia around the world (Global Burden of Disease regions) in the population of >60 years old

Region	Over 60 population (millions, 2010)	Crude estimated prevalence (%, 2010)	Number of people with dementia (millions) 2010	2030[a]	2050[a]	Proportionate increases (%) 2010–2030[a]	2010–2050[a]
Asia	406.55	3.9	15.94	33.04	60.92	107	282
Australasia	4.82	6.4	0.31	0.53	0.79	71	157
Asia Pacific	46.63	6.1	2.83	5.36	7.03	89	148
Oceania	0.49	4.0	0.02	0.04	0.10	100	400
Central Asia	7.16	4.6	0.33	0.56	1.19	70	261
East Asia	171.61	3.2	5.49	11.93	22.54	117	311
South Asia	124.61	3.6	4.48	9.31	18.12	108	304
South East Asia	51.22	4.8	2.48	5.30	11.13	114	349
Europe	160.18	6.2	9.95	13.95	18.65	40	87
Western Europe	97.27	7.2	6.98	10.03	13.44	44	93
Central Europe	23.61	4.7	1.10	1.57	2.10	43	91
Eastern Europe	39.30	4.8	1.87	2.36	3.10	26	66
The Americas	120.74	6.5	7.82	14.78	27.08	89	246
North America	63.67	6.9	4.38	7.13	11.01	63	151
Caribbean	5.06	6.5	0.33	0.62	1.04	88	215
Andean LA	4.51	5.6	0.25	0.59	1.29	136	416
Central LA	19.54	6.1	1.19	2.79	6.37	134	435
Southern LA	8.74	7.0	0.61	1.08	1.83	77	200
Tropical LA	19.23	5.5	1.05	2.58	5.54	146	428
Africa	71.07	2.6	1.86	3.92	8.74	111	370
North Africa/East Middle	31.11	3.7	1.15	2.59	6.19	125	438
Central SSA	3.93	1.8	0.07	0.12	0.24	71	243
East SSA	16.03	2.3	0.36	0.69	1.38	92	283
Southern SSA	4.66	2.1	0.10	0.17	0.20	70	100
West SSA	15.33	1.2	0.18	0.35	0.72	94	300
World	758.54	4.7	35.56	65.69	115.38	85	225

LA Latin America, SSA Sub-Saharan Africa
Reproduced with permission from Prince et al. (2013)
[a]Projected. Prevalence values are estimated from 2010 data, projections to the years 2030 and 2050 are a result of modeling

ApoE4 allele (see below in this section), lower education, loss in the performance of instrumental activities of daily living (IADL), and old age (Artero et al. 2008).

Alzheimer's disease was first described by Aloysius Alzheimer, a German psychiatrist (also referred to in the literature as Alois Alzheimer), after examining a

51-year-old patient, Auguste D. in 1901: "Her husband had noted a relatively sudden change in her behavior, dominated by panic, terror and suspicions of his having an affair with a neighbor. She neglected her housework, hid objects and fumbled in the kitchen. Over the next several months, she became increasingly restless and a disturbance to their neighbors" (Mandell and Green 2014, p. 3). Eventually she was admitted to hospital, where she met Alzheimer, who saw her condition deteriorate until she died in April 1906. Alzheimer published a report on Auguste D. in 1907, which does not contain figures of brain tissue, and there is even the suspicion that the patient may have suffered from atherosclerosis of the brain rather than what subsequently became known as Alzheimer's disease, a name given to it by Alzheimer's colleague Emil Kraepelin in 1910. A full report, however, is available for a second patient, Johann F., a 56-year-old demented man who had been admitted to the psychiatric clinic on the 12th of November 1907. The report indicated that the patient did indeed suffer from AD. The case of Johann F. has been the subject of a more recent investigation and genetic analysis (Graeber et al. 1997). After a period of very little research on AD, interest started to pick up, thanks to Sir Martin Roth who described in 1966 that neuritic plaques occurred in brains of the elderly, and that their number approximately correlated with the degree of severity of dementia. In 1976, Robert Katzman published a seminal article on the epidemiology of AD stressing that pre-senile and senile dementia were pathologically similar (Growdon 2014).

AD accounts for 55–80% of adult-onset dementia in the industrialised world (Thies and Bleiler 2013; Mandell and Green 2014), and it tends to be more prevalent in women than men (Fuller and Martins 2013). The prevalence of AD in the population increases exponentially from the age of 55 to older than 85 (see Table 4.4).

The major mechanisms that have been suggested to explain cognitive decline in old age and AD include cardiovascular risks and oxi-inflammatory loads that can negatively affect both neuroplasticity (e.g. through decreasing interconnections between neurons) and indeed the very survival of neurons themselves. Both processes can also co-occur, thus further increasing the probability of cognitive decline (Harrison et al. 2017).

The cognitive decline characteristic of AD includes memory loss as already mentioned, needing more time to accomplish tasks, abnormal reasoning, and problem-solving difficulties. Deterioration is also observed in language (losing track of a conversation and problems with expressing oneself coherently) and there is an inability to learn new information, loss of practical skills, decision-making difficulties, disorientation in space and time (not knowing time in the day or date, for instance). Elders affected by AD also have difficulty in using objects or devices, they may lose objects without knowing where they might be, experience mood changes and loss of judgement (Steeman et al. 2006; Fuller and Martins 2013). Such decline is also part of normal ageing, but people suffering from AD decline more rapidly and to a greater extent.

Table 4.9 compares normal and AD (early stages) levels of forgetfulness. Clearly, the level of forgetfulness that most people will experience in normal old age is markedly exaggerated in the case of an old person developing AD.

Table 4.9 Forgetfulness in normal ageing and in Alzheimer's disease

Normal forgetfulness	Early signs of dementia, especially AD
1. Occasional confusion about the date	Confusion about the month or season
2. Minor lapses of memory or losses of attention, such as misplacing keys or a wallet	Storing items in strange places, problems in making routine decisions and performing routine tasks
3. Walking into a room for a specific purpose and occasionally forgetting that purpose	Problems carrying out simple tasks such as getting dressed properly
4. Sometimes forgetting names of people shortly after being introduced	Forgetting the names of commonly used items, repeating questions or phrases several times in the same conversation
5. Having troubles following directions or a map in an unfamiliar place	Getting disoriented or lost in familiar places
6. Getting a bit slower at reading and understanding bills, dealing with payments and new regulations	Trouble doing simple calculations, confusion, and problems dealing with money
7. Occasionally feeling sad or moody or feeling tired at the thought of social interaction	Behavioural changes such as apathy, frustration, paranoia, agitation, anxiety; withdrawing from new social and mental challenges and situations and avoiding visiting friends

Simplified from Fuller et al. (2013)

As the gravity of AD increases, the cognitive problems associated with the disease become more pronounced, until at a later stage severe AD is manifested in greater levels of confusion and disorientation, often (but not always) high agitation and sometimes aggression, wandering (e.g. at night due to sleeplessness), abnormal body movements, little or no recognition of family and friends, needing help with virtually all daily activities, speech problems, eventually leading to lack of conversation, delusions and hallucinations, incontinence, weight loss, and muscle atrophy (Fuller et al. 2013). Table 4.10 shows the progressive deterioration in some cognitive and perceptual capacities as the disease transitions from *mild cognitive impairment* (MCI or *mild* NCD) to ever more severe forms of AD, note the strong deterioration in the various aspects of memory, as it was also stressed in Table 4.9. Dementia is therefore a specific form of acceleration of ageing processes that leads to greater deterioration of the brain and its functions. This concept is schematically illustrated in Fig. 4.4.

Memory consolidation is positively affected by good sleep, but AD patients experience significant degrees of sleep disturbance which negatively affects memory, in addition to both the direct effects of the disease on memory and the normal effects of ageing on memory (Harper 2014).

The strongest and most consistent risk factor for AD is old age (Bates and Martins 2013). With old age, several processes that damage cellular DNA occur that may lead to neurodegeneration. For instance, *segmental progeroid syndromes* are characterised by faster rates of change in such ageing processes (Yankner et al. 2008).

Table 4.10 Clinical characteristics of dementia of the Alzheimer's type

	MCI	Mild	Moderate	Severe
Memory				
Working	–	–/+	++	+++
Anterograde episodic	++	+++	+++	+++
Remote	–/+	–/+	++	+++
Semantic	–/+	+	+++	+++
Attention and executive	–/+	++	++	+++
Language	–	–/+	+	++
Visuospatial and perceptual[a]	–	–/+	++	++
Apraxia[b]	–	–	++	++

Number of + denotes the degree of impairment. MCI = mild cognitive impairment (from Mandell and Green 2014)
[a]Cronin-Golomb (2014)
[b]This is a group of cognitive motor disorders characterised by a loss or impaired ability to perform purposeful skilled movements

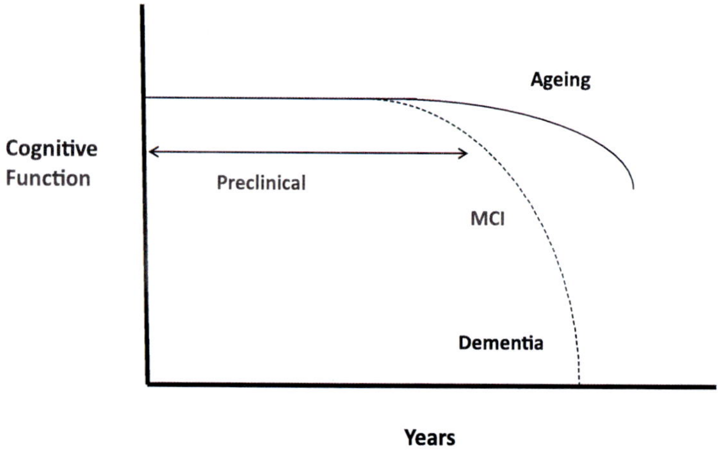

Fig. 4.4 Cognitive decline in normal ageing compared with the trajectory expected from mild cognitive impairment (MCI) to eventually dementia (redrawn from Lock 2013)

Some progeroid molecular changes in the brain of old people include *double-strand breaks* in DNA, which can be increased by the activity of reactive oxygen species (ROS). ROS cause damaging oxidative stress when they are present in excess, and accumulated molecular damage to DNA over time can lead to reduced transcription of genes that are involved in important brain functions such as synaptic activity, protein transportation, and mitochondrial function.

Another risk factor is family history, presumably associated with some genetic predisposition such as carrying specific alleles: e.g. *SORL1*, *ApoE4*, plus some mutations occurring in chromosomes 1, 14, and 21. But there are additional risk factors: systemic inflammatory activity, traumatic brain injury, muscle weakness in old age, low bone density in women, low serum docosahexaenoic acid (DHA, which is an essential fatty acid obtained from the diet that is important for normal functioning of the central nervous system, Schaefer et al. 2006). High plasma levels of homocysteine are associated with both brain atrophy and neurofibrillary tangles; metabolic syndrome is also a risk factor as are low folate intake, low thyroid stimulating hormone levels, hypercholesterolemia, obesity, smoking and alcohol consumption, weight loss in women and the elderly (Mandell and Green 2014; de Bruijn 2015).

Cardiovascular risk factors for AD include hypertension, smoking, hyperhomocysteinemia, low cerebral perfusion, and arterial stiffness (Bates and Martins 2013; Mandell and Green 2014; de Bruijn 2015). Some of those cardiovascular risks can be increased by diabetes mellitus (de Bruijn 2015). Higher circulating blood glucose in diabetes mellitus can lead to microvascular damage in the brain and neurotoxicity, thus increasing the risk of AD, whereas high insulin levels in the blood may decrease clearance of amyloid β in the brain. High serum levels of cholesterol and low-density lipoproteins may also increase the risk of *mild* NCD and dementia. The above vascular risk factors for dementia can also act in synergy (Middleton and Yaffe 2009).

The metabolic syndrome that is characterised by low levels of high-density lipoprotein, abdominal obesity, hypertriglyceridemia, hypertension, and/or hyperglycaemia has been linked to an increased risk of cognitive decline, especially in individuals presenting with high levels of inflammation (Middleton and Yaffe 2009).

Chronic depression (but not anxiety, de Bruijn 2015), being female, and also low educational achievement, poor verbal performance, visual memory, and other aspects of learning are additional risk factors alongside apathy and chronic psychological stress among others (Bates et al. 2013; Mandell and Green 2014). Wilson et al. (2006) tested the hypothesis that proneness to distress is associated with increased risk of developing AD in the USA. They followed up 600 old persons who did not have dementia at baseline for an average period of 3 years during which 55 participants were diagnosed with AD. After controlling for age, sex, and education they found that older persons with high levels of proneness to distress were 2.7 times more likely to develop symptoms of dementia than participants who were not prone to distress.

Depression is associated with high levels of circulating cortisol, a steroid hormone that may damage the hippocampus and increase the risk of dementia. Moreover, some studies have suggested that depressed people have enhanced deposition of β-amyloid (Aβ) plaques (Middleton and Yaffe 2009). We will explain the role of Aβ in AD below. Conversely, it would be also expected that people who have been diagnosed with AD may subsequently experience depression.

Increased daytime sleepiness is a feature of normal ageing, with an estimated 20–30% of older people experiencing sleeping bouts during the day, but excessive daytime sleepiness is associated with cortical thinning and β-amyloid accumulation in the brain and it increases the risk of dementia (Carvalho et al. 2018).

On the other hand, there are also some potential protective factors against AD such as: maintaining an active mind especially since middle age and into older ages (e.g. through learning new skills such as a second language), living a fulfilling life in old age without major chronic psychological stresses, keeping up physical exercise, and following a healthy diet that includes antioxidants as in the Mediterranean diet (Mandell and Green 2014; de Bruijn 2015).

The identified risk factors (or at least some of them) may be involved, directly or indirectly, in the molecular and cellular causative mechanisms of AD. What are such causative mechanisms?

At the cellular and molecular levels, AD is currently diagnosed through neuronal loss especially in the cerebral cortex and in the hippocampus, presence of extracellular neuritic plaques caused by β-amyloid deposits (Aβ or amyloid plaques), and also threshold densities of intra-neuronal neurofibrillary tangles (Finch 2007), even though both tangles and amyloid deposits may not be always sufficient for the development of AD (Yankner et al. 2008). Examples of neuritic plaques and neurofibrillary tangles are shown in Fig. 4.5.

Amyloid plaques in brain tissue are caused by the accumulation of 40 or 42 amino acids long Aβ peptides that are derived from the *amyloid precursor protein* (APP). Aβ also accumulates in perivascular regions producing *cerebral amyloid angiopathy* (CAA). Apart from producing plaques, soluble Aβ is associated with inflammation processes and it may normally accumulate throughout ageing. Over time Aβ oligomers can injure synapses (e.g. in the mediotemporal lobe in particular, where the hippocampus is) leading to neurodegeneration and therefore cognitive deficits (Fjell et al. 2014).

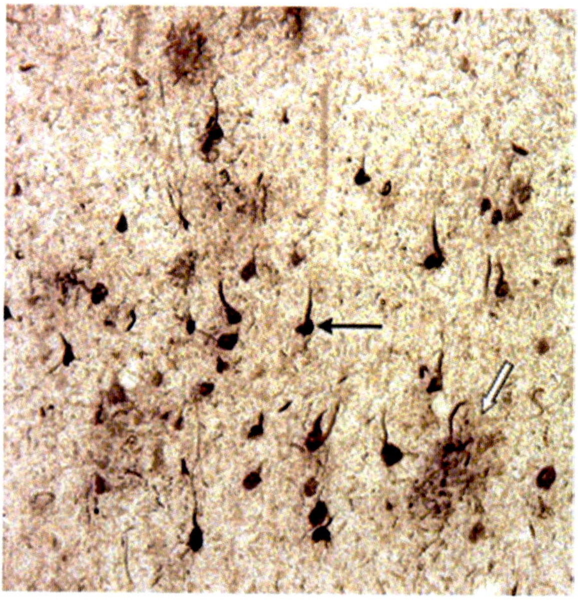

Fig. 4.5 Neurofibrillary tangles (black arrow) and neuritic plaques (white arrow) in AD brain. From Binder et al. (2005), reproduced with permission

These mechanisms configure what is known as the *Amyloid Cascade Hypothesis* (e.g. Kanekiyo et al. 2014). In 1992 John Hardy and Gerald Higgins published an influential article in the journal *Science* introducing what they called the *Amyloid Cascade Hypothesis* to explain the development of AD. They suggested that:

> deposition of amyloid β protein (AβP), the main component of the plaques, is the causative agent of Alzheimer's pathology and that the neurofibrillary tangles, cell loss, vascular damage, and dementia follow as a direct result of this deposition (Hardy and Higgins 1992, p. 184).

Figure 4.6 shows a diagrammatic representation of the hypothesis for the genetic dominant (usually expressed at younger ages or early-onset) and the non-dominant (usually expressed at older ages or late-onset) forms of AD (Abraham 2014). The accumulation of Aβ peptides in the brain (due to a dominant inherited condition or a metabolic failure occurring with ageing) affects efficiency of neuronal communication at the level of synapses, leading also to deposition of plaques over time. This stimulates inflammatory processes, resulting in oxidative damage and the formation of tangles, hence causing neuronal dysfunction, including some neuronal losses, and therefore dementia.

The hypothesis has been the subject of considerable attention and empirical research since the publication of Hardy and Higgins' article, receiving support but also criticisms and undergoing some adjustments. Recent results, however, suggest that the hypothesis remains viable (He et al. 2018), although the field is in continuous flux and more mechanisms are being considered, such as functional changes in

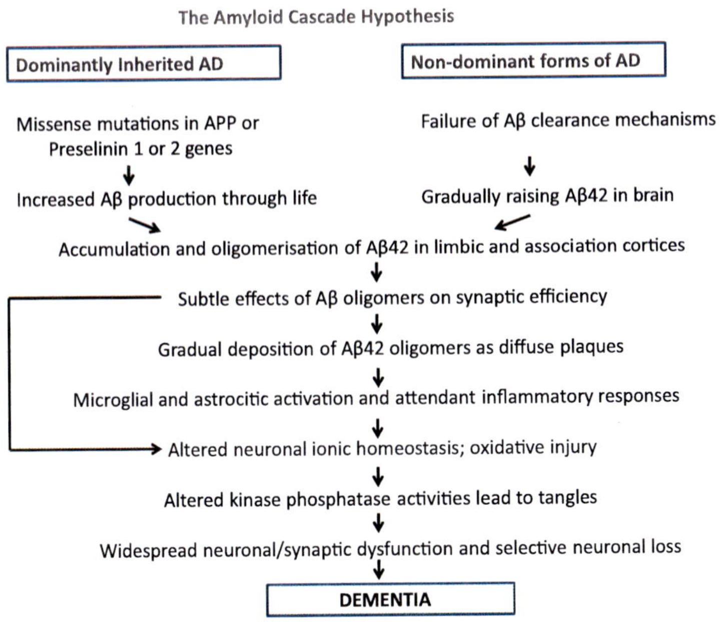

Fig. 4.6 The *amyloid cascade hypothesis*, slightly modified from Abraham (2014)

a sub-domain of the endoplasmic reticulum that is known as the mitochondria-associated ER membranes (MAMs) (Area-Gomez and Schon 2017) that are not included in the *Amyloid Cascade Hypothesis*.

The Aβ accumulation in the brain that is observed in many AD patients is associated with the presence of the ε4 allele of the *ApoE* gene, encoding for the lipid/cholesterol carrier apolipoprotein E (ApoE) (Abraham 2014). ApoE has been also associated with heart disease and longevity. The protein has a physiological role in transporting cholesterol, and it can therefore regulate cholesterol levels in the blood, as well as transporting cholesterol that can be used in the synthesis of steroids and to modulate neurite outgrowth. The ApoE4 variant of the protein is specifically associated with a shorter lifespan and it is virtually absent in centenarians (Finch 2007), it decreases the level of neuroprotection and increases neurotoxicity in the brain (e.g. Theendakara et al. 2013). However, not all homozygous for the *ApoE4* (ε4) allele develops dementia.

Another lipoprotein gene that can influence longevity is the gene encoding for the cholesteryl ester transferase protein (CETP) which is a lipid binding protein (Milman et al. 2014). A CETP promoter allele tends to be found in combination with the ε4 allele in AD patients (Rodríguez et al. 2006).

The frequency of the ε4 allele is about 15% in the general population, but it is approximately 40% in the population of AD patients (Farrer et al. 1997; 27% vs 59% in Snowden et al.'s 2007 work). The prevalence values for both ε4 carriers and ε4 homozygous individuals in the AD population vary around the world, with lowest prevalence values found in Southern Europe (ε4 carriers: 40.5%, ε4 homozygous: 4.6%) and the highest values are recorded for Northern Europe (ε4 carriers: 61.3%, ε4 homozygous: 14%) (Ward et al. 2012). The ε4 allele, however, is neither strictly necessary nor sufficient to cause AD and there are other genes that are implicated in late-onset AD, some of them are found on chromosomes 9 and 10. In fact, Hollingworth and Williams (2014) mention a total of 11 chromosomes that show evidence of linkage to AD. Evidence for genetic mechanisms of both vascular and frontotemporal dementia is also available (Hollingworth and Williams 2014).

Approximately 90% of people who are ε4 homozygotes develop AD and they do so at an earlier age (about 68 years old) compared to 47% who are heterozygotes and develop AD later in life (76 years old), whereas 20% of non-carriers can also develop AD but they do so at even older ages (about 84 years old) (Corder et al. 1993, see also Zuo et al. 2006; Liu et al. 2015). Therefore, although not all AD patients are carriers of ε4, most ε4 homozygotes do develop AD, but not all of them do so.

Other alleles of the *ApoE* gene are less of a genetic risk factor for AD (e.g. ε3) and still others, such as ε2, can be protective (Kanekiyo et al. 2014 and the references therein).

In sum, most cases of early-onset AD (30–60 years old) are due to inherited conditions (Fuller and Martins 2013; Verdile et al. 2013), with heritability values of the disease ranging between 92% and 100% and with between 35% and 60% of early-onset patients having at least one AD-affected first-degree relative. A range of 10%–15% of those early-onset AD patients display a mode of AD genetic transmission that is autosomal dominant. Late-onset AD is of more complex aetiology, displaying a relatively lower heritability that ranges between 70% and 80% (see Cacace et al. 2016, for a review).

Genes involved in the development of AD include the *ApoE* which we have already mentioned, with the ApoE peptide facilitating the deposition of the Aβ peptide in the brain, and both peptides can bind together to form amyloid plaques in AD patients. ApoE4 facilitates a greater speed of deposition of Aβ than ApoE3 (Hori et al. 2015). Other genes involved in AD are the *APP* (*amyloid precursor protein* (APP) gene) that encodes for the amyloid precursor protein. *APP* was one of the first genetic candidates for the development of AD and it is located on chromosome 21, which was singled out because people with trisomy 21 (Down syndrome), who have three copies of the gene, are more likely to develop AD, and the onset of the disorder is at an earlier age (Yanker et al. 2008). Of importance are also the alleles *PSEN1* and *PSEN2* of the *PSEN* gene (Thies and Bleiler 2013) that encodes for the proteolytic enzyme *presenilin* (Wolfe 2007).

Apart from amyloid plaques, AD can also present with intracellular neurofibrillary tangles (NFTs) in the brain that have high content of hyper-phosphorylated tau, a protein that is associated with the cell microtubules, and that severely compromises neuronal function (Abraham 2014). NFTs are typical of normal ageing processes, being especially frequent in neurons of the hippocampus, neocortex, and the limbic system. Post-translational modifications of the protein can lead to its aggregation into fibrillar bodies that can alter the transportation of molecules through the axon of the neuron and into the synapses. Tau-pathology can be seen in AD but especially other forms of dementia such as frontotemporal dementia with Parkinsonism that is associated with chromosome 17 (Yanker et al. 2008; Abraham 2014). The tau protein is encoded in the *microtubule-associated protein tau* gene (*MAPT*), and mutations in both such gene and the granulin gene (*GRN*) have been linked to dementia (Ghetti et al. 2015).

AD is also associated with various vascular abnormalities such as the increase in microvessels that occur in close association with plaques and that may contribute to the genesis of such plaques (Finch 2007). The mechanisms leading to vascular NCD are more specific, and they include hypertension which increases vascular tortuosities in the brain. The blood flow to the brain during ageing goes through a normal process of decline over time: 25% decline from 20 to 80 years old, with *obstructive sleep apnoea* adding to poor oxygenation of the brain leading to additional cognitive impairment (Mandell and Green 2014). But the conditions for the development of AD involve more significant decline in brain blood flow (hypoperfusion, microemboli), as a consequence of cardiovascular diseases, for instance (e.g. stroke, coronary heart disease, atrial fibrillation, the latter causing embolisms) (de Bruijn 2015).

A link with immune mechanisms could also explain some convergence between infectious diseases and the probability of developing dementia (Aliev et al. 2019). This has been seen in individuals diagnosed with HIV-AIDS, for instance (Finch 2007). Chronic infections are also associated with amyloid deposits in tissues.

Given that the most consistent risk factor for AD is old age, we should further explore the root causes of AD through the analysis of some physiological and

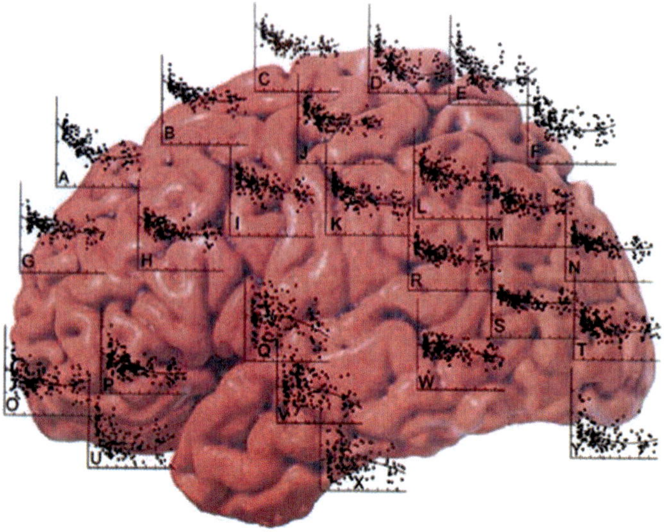

Fig. 4.7 Cortical change across the lifespan. The plots represent grey matter density (y-axis) versus age (from 7 to 87 years; *x*-axis). The trend is towards a variable and non-linear decline in grey matter in various cortical regions. Reproduced, with permission, from Fox and Schott (2004)

cellular mechanisms of ageing in the brain. As part of the process of normal ageing, the human brain loses both grey and white matter over time; that is, it loses both brain cells and also axonal/dendrite interconnections between them. Figure 4.7 illustrates the rate of loss of grey matter—which is overall formed by neurons and their neuropil (dendrites and myelinated as well as un-myelinated axons), glial cells (astrocytes and oligodendrocytes), and synapses—during normal ageing from 7 to 87 years old in cortical areas of the brain. The rate of decline of grey matter density differs according to the region in the brain, but the general trend is negative across the entire organ.

The loss of cells and interconnections between cells seen in normal ageing is further accelerated in AD, being the result of two processes: apoptosis and necrosis (Boziki et al. 2014). This has led some researchers to suggest that AD derives directly from the processes of normal ageing (Fjell et al. 2014). As suggested by Caleb Finch (2007, p. 8): "most age-associated diseases interact throughout with 'normal aging processes'". However, AD also shows some specificities.

Normal ageing in the brain begins with changes in areas such as the dentate gyrus subregion of the hippocampus, an area active in neurogenesis, whereas AD initially compromises the entorhinal cortex first and then spreads to the neocortex and the hippocampus. The entorhinal cortex, like the hippocampus, is also involved in memory functions (Hampstead and Sathian 2015). The hippocampal and entorhinal cortices, in turn, are not only interconnected with each other but also with the dorsolateral prefrontal cortex and the amygdala. The amygdala plays an important role in emotions production and regulation; hence, it is not surprising that people suffering from dementia display emotional dysregulation (Wright 2014).

The AD brain also differs from a normal brain by showing an annual rate of global atrophy of about 2–3%, which is one order of magnitude higher than the yearly atrophy rate of between 0.2% and 0.5% observed in the brain of healthy individuals (Fox and Schott 2004). In normal ageing brains the rate of synaptic loss is about 1% per year, for an overall decrease of 15–30% in the number of synapses in old people (Finch 2007), whereas in AD individuals loss of cortical synapses ranges from 25% to 35% (Selkoe 2002). The normal annual atrophy rates in the ageing brain also vary according to the brain region. For instance, the annual rate of decrease in volume is 0.79–2.0% for the hippocampus and 0.3–2.4% for the entorhinal cortex in healthy individuals (Fjell et al. 2014), whereas in AD individuals the annual rate of decrease in volume is 3.5% in the hippocampus (Jack et al. 2000) and 6.5% in the entorhinal cortex (Du et al. 2003; mean value for both hemispheres). The frontal and the temporal lobes show the highest magnitude of relative change, and important changes also occur in the medial parietal area (precuneus and the retrosplenial and posterior cingulate cortices).

Changes in brain structures with AD are also associated with changes in brain function. Figure 4.8 shows results of PET (positron emission tomography) scans that measure functional changes in the brain. More specifically, the scan detects alterations in cerebral metabolic rate of glucose, which is an index of synaptic function and density, and hence an indication of neurodegeneration when such function is low. In particular, the figure compares results for FDG-PET scans, which use [18F]-fluoro-2-deoxyglucose (FDG) to measure cerebral metabolic rate of glucose, and Pittsburgh Compound-B-PET scans (PIB-PET scans), which are used to detect amyloid deposition, a potential evidence of developing dementia.

Note from Fig. 4.8 how in the three brain areas of interest (centrum-semiovale, basal ganglia, and medial temporal lobe), old individuals diagnosed with AD show both higher amyloid deposition, as seen from PIB-PET scans, and also lower cerebral metabolic rate of glucose, as measured by FDG-PET scans, than normal brains. However, note also that a degree of both amyloid deposition and lower metabolic rate of glucose do also occur in normal ageing brains (Fjell et al. 2014).

The main changes in brains affected by AD such as the disruption of episodic memory function, brain atrophy, and accumulation of amyloid protein are also recorded in healthy ageing brains, where they are associated with a reduction in cognitive abilities such as mental speed, executive function, and episodic memory over time, whereas verbal abilities are typically maintained in healthy ageing brains (Fjell et al. 2014). Cognitive abilities are better maintained through normal ageing in women, in those who exercise regularly and in those who live with someone, whereas being a man and being unemployed increase the chances of cognitive decline with age. In AD individuals, the deterioration of brain functions observed in healthy ageing individuals is accelerated, it is more widespread, and it includes verbal capacities.

Which areas of the brain are affected by AD? The *retrogenesis* ("last in, first out") hypothesis suggests that later developing regions are the first ones affected by age deterioration of the brain (e.g. Tamnes et al. 2012). Many (but not all) such areas are also evolutionarily more recent. The sequence of brain regions undergoing atrophy in AD is therefore structured and it tends to start in the entorhinal cortex and the hippocampus, to then spread into the medial parietal, lateral temporal, and frontal

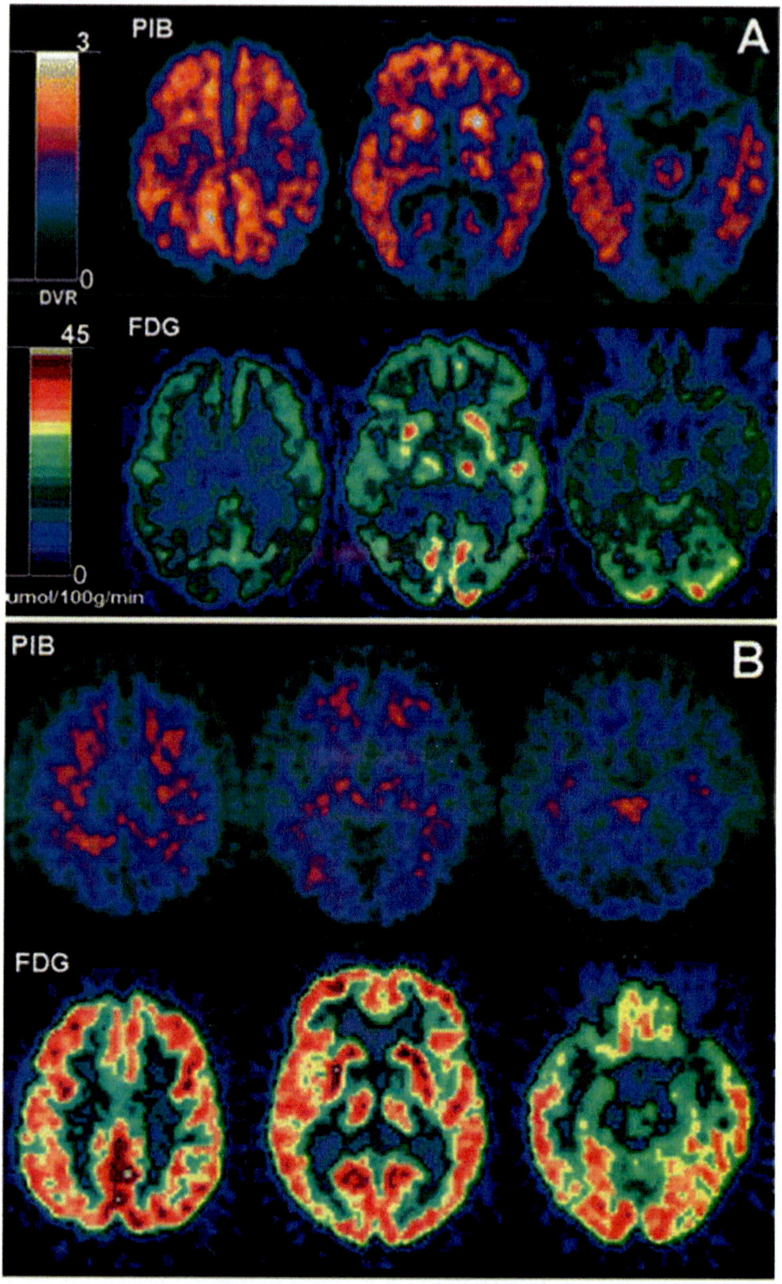

Fig. 4.8 PIB and FDG-PET scans from a 71-year-old male AD subject (**A**) and a 65-year-old male normal subject (**B**). Top row: PIB-PET images and bottom row: co-registered FDG-PET images. PET scans are displayed at the level of the centrum-semiovale (left), basal ganglia (centre), and medial temporal lobe (right). Reproduced, with permission, from Li et al. (2008b)

regions, and subsequently to the rest of the cortex (Fjell et al. 2014). However, atrophy of specific brain regions alone may not be a certain indication of cognitive deterioration, especially when the individual has great cognitive reserve and, in response to decreased neuroactivity in regions affected by AD, other regions of the brain are recruited to carry out mental functions.

Although some of the brain areas that are compromised by normal ageing (e.g. the fronto-striatal network controlling executive functions) are different from brain areas compromised by AD (e.g. entorhinal cortex, medial temporo-parietal memory network) (Fjell et al. 2014), memory may be a common function linking all those different areas to ageing and AD, as executive functioning controlled by the fronto-striatal cortex is strongly correlated with working memory (McCabe et al. 2010) and the entorhinal cortex is also associated with memory (episodic memory, Di Paola et al. 2007). In particular, the default mode network (DMN) is virtually all confined within age-declining regions of the brain that are also affected by AD, and the "DMN maps onto a network of core brain areas involved in episodic memory and imagination" (Fjell et al. 2014, p. 10). Other areas involved in episodic memory include: the inferior parietal cortex, middle temporal cortex, medial orbitofrontal cortex, precuneus, and the hippocampus, all areas that are also affected by AD (Baron et al. 2001; Khan et al. 2014).

What can we do to prevent the onset of AD? And once the disease has been diagnosed, how can we intervene on the patient to help decrease the effects of AD on the brain and perhaps slow down its progression? Although AD is the subject of much pharmacological research (see Kowall 2014; Bachurin et al. 2018; Cummings et al. 2019; Gupta et al. 2020, for reviews of current pharmacological interventions) here we only focus on behavioural strategies for prevention and intervention.

Some preventive measures to better defend the brain from developing AD include building cognitive reserve through formal education and various spontaneous forms of acquisition of new knowledge, diet, exercise, social engagement, and participating in fulfilling leisure activities. In what follows we expand on each one of those measures in turn.

Building cognitive reserve to better face the challenges of old age is a process that should start early on in life. In old age some cognitive abilities are likely to decrease but, in spite of still scanty neurological evidence supporting it, the *use it or lose it* approach is a good strategy to adopt in order to keep our mental capabilities sound for as long as possible.

It is known that maintaining specific regions of the brain active as well as promoting functional connectivity among brain regions is positively related to cognitive reserve (Arenaza-Urquijo et al. 2013). Brain functioning may be also made more efficient by multitasking (Just and Buchweitz 2017).

Building cognitive reserve is a lifetime process that is aided by life experiences, education, creativity, and overall intellectual engagement with life, such cognitive reserve may not only help in the prevention or delay of AD, but it may also help in decreasing the behavioural progression of the disease. This remains true in old age as well, where cognitive reserve can be built through enhancement of executive functions (e.g. planning and organising, multitasking) and by broadly increasing brain activity in the prefrontal cortex through formal education, reading books, magazines, newspapers, playing strategic games of cards or chess, and doing crosswords and other puzzles, for instancc (Middleton and Yaffe 2009; Vidovich et al. 2013).

In the words of Fjell et al. (2014: 21):

> Higher levels of cognitive reserve, as indicated by education level and socio-economic status, are protective of AD diagnosis ..., as individuals with greater reserve can maintain cognitive function in the face of higher levels of brain amyloid ... The weak correlations between amyloid level and cognitive function suggest that other mechanisms, such as functional compensation, impact cognitive ability. The view that cognitive function can be maintained in aging by compensatory cognitive processes, and that decline is seen when a person is no longer able to compensate for reduced workings of primary brain structures and circuits, has many supporters in contemporary research.

In fact, there is evidence for the relationship between capacity to perform complex mental activities and probability of developing dementia in later life. Valenzuela (2008) carried out a meta-analysis showing that there is an overall 46% risk reduction of dementia for individuals engaged in high mental activity compared with a low mental activity group. He notes that some important factors contributing to such difference are education, occupational complexity, and type of cognitive lifestyle.

There is also an additional issue to keep in mind regarding cognitive reserve. For instance, the more cognitive reserve a person has, that is the more a person is capable of maintaining good cognitive functioning in spite of brain incapacitation, the more difficult it will be to diagnose AD in that person at the early stages of the disease, hence cognitive reserve may delay diagnosis until the disease is well advanced. But this is only a problem when there is an effective treatment for the disease, when there is not, as it is the case for AD so far (Patnaik 2015), then greater cognitive reserve can be seen more as a benefit.

Bilingualism is a specific case of cognitive reserve that has received some recent attention in the context of AD. Language production and also comprehension are supported by cognitive processes that are highly interactive, such as production of sounds (phonology), semantics—the meanings of words—naming or production of lexicon, syntax—the grammatic rules that determine the structure of the language-—and narrative discourse which is produced through a cogent story that links events sequentially in a meaningful and cohesive manner.

In bilinguals such processes become a bit more complicated due to the potential interference between two languages. Specific brain regions are devoted to the task of *bilingual language control* (Abutalebi and Green 2007), they include a dorsal anterior cingulate region, the prefrontal cortex, the left caudate (basal ganglia), the left putamen, the thalamus, cerebellum, and the inferior parietal lobules (Calabria et al. 2018). Bilinguals are confronted with the task of maintaining proficiency in the use of a given language whilst retaining the ability to fully switch to the second language when required and avoiding interference between the two languages all along. Because of their need to use executive control brain functions in order to update, switch, and inhibit attention, bilinguals can develop an insurance against the deterioration of executive control mechanisms with ageing (Bialystok et al. 2016).

The suggested relationship between bilingual language control and executive functions of the brain—the latter monitor behaviour and control attention, cognitive flexibility, working memory, planning, and fluid intelligence (reasoning, problem solving)—has been the subject of much debate focusing on the alternative views of whether bilingualism is an advantage or a disadvantage compared to monolingualism. However, the performance of bilinguals has been found to be superior in tasks that rely on conflict monitoring and resolution, suggesting an advantage in terms of executive control functions in bilinguals (Del Maschio et al. 2018 and the references therein).

Bilingualism also requires the activation of more regions of the brain than monolingualism due to the complexity of juggling the use of more than one language. Current evidence indicates that the anterior cingulate cortex (ACC), which is associated with cognitive control, is differentially activated in bilinguals and monolinguals, the former display additional involvement of the ACC when they are operating bilingually. This extra-engagement could lead to structural changes in the ACC, as observed with respect to grey matter density. The left prefrontal cortex (PFC), on the other hand, is responsible for selecting the correct language to use, whereas the right PFC inhibits the language that is not being used by bilinguals. The inferior parietal lobules are involved in attentional tasks and in biasing the selection of language in bilinguals away from the language that is not in use (see Calabria et al. 2018, for a review).

Does such additional activation of the brain in bilinguals confer an advantage in terms of resilience against the processes of ageing? Current evidence suggests that bilingualism may have a long-term protective effect against age-related cognitive decline, resulting from greater integrity of neural networks and white matter, that is most apparent in older individuals, whereas the difference between bilinguals and monolinguals among younger adults is less prominent (Schroeder and Marian 2017 and the references therein). Greater connectivity of the brain through better white matter integrity can help maintain cognitive reserve in the face of functional losses in parts of the ageing brain.

Brain resilience to ageing may be increased in at least two ways by bilingualism: (a) through greater *neural reserve* (i.e. more grey and/or white matter or better white matter integrity compared to monolinguals) or (b) by means of *neural compensation* (better compensation for the loss of neural and therefore brain structures with age compared to monolinguals, through recruiting alternative brain areas).

Evidence for bilingualism-driven brain resilience in old age, however, is variable. Some studies have supported the neural reserve or compensation mechanisms, others have not. Still, the overall trend seems to lean more towards supporting a positive role for bilingualism in maintaining brain and mental functioning into old age (Bialystok et al. 2016). Del Maschio et al. (2018) provide recent evidence that bilingual speakers can suffer a significant level of cerebral atrophy without actually experiencing the loss of cognitive functioning that is typically observed in healthy ageing monolinguals. Greater neural reserve in bilingual older people compared to monolinguals has been also shown in the works of Luk et al. (2011), Abutalebi et al. (2015a, b), and Perani and Abutalebi (2015). In another recent work, Borsa et al.

(2018) studied 20 bilinguals (presumably speakers of Italian and German), 47–74 years old (60% females) from South Tyrol (Trentino-Alto Adige) in Italy, who acquired their second language (L2) early on when they were about 6 years old, and compared them with 20 Italian monolinguals (49–75 years old, 55% females). They tested monolinguals and bilinguals on a cognitive task, the *Attentional Network Task* (ANT) which requires the participants to respond to the direction of a central arrow (left or right) flanked on either side by two additional arrows pointing in the congruent (→→→→→) or incongruent direction (←← → ← ←) or by two straight lines (– – → – –) (neutral flanker). What the researchers measured was the accuracy of response and the reaction time. They then calculated the interference effect: the performance (accuracy, time) on incongruent trials minus the performance on neutral trials and the conflict effect: the performance on incongruent trials minus the performance on congruent trials; both indices are regarded as proxy measures of nonverbal cognitive control. Age was a significant predictor of mean interference and conflict effects in monolingual seniors but not in bilinguals, thus suggesting that bilingualism may help in slowing down the typical effects of mental ageing on this cognitive task. This role of bilingualism is strengthened by daily exposure to both languages: "our results support the view that the routine maintenance of the mechanisms used for cognitive control specifically within the linguistic domain can mitigate the typical effects of healthy aging" (Borsa et al. 201, p. 60).

Can bilingualism also mitigate the cognitive effects of AD? And given the association of AD with loss of language functions, could development of language capabilities as seen in bilingualism be a factor in the prevention of AD in the first place?

AD patients typically face language and communication challenges starting early on in the development of the disease, such as naming and discourse comprehension (e.g. Reilly et al. 2014). As the disease progresses the AD patient experiences *narrative dissolution*: "Difficulties in maintaining global connectedness necessary for a cohesive storyline, whereas others demonstrate impairment at the level of semantic propositional knowledge" (Reilly et al. 2014, p. 346) that can eventually develop into full mutism.

In terms of prevention, several independent studies have converged on a consensus that bilingualism confers an advantage that is translated into a 4–5 years delay in the onset of AD (Bialystok et al. 2007; Craik et al. 2010). More recently, Alladi et al. (2013) carried out a study of dementia and bilingualism in Hyderabad, India, where most people speak Telugu, the majority language, in informal contexts, whereas Dakkhini is spoken by the local Muslim community. In addition, most people speak Hindi—the official national language of India—and some speak English in formal situations. Alladi and collaborators showed that bilingual individuals developed dementia 4.5 years later than monolingual ones, independently of the effect of education, sex, occupation, and urban/rural residence, whereas for those who developed dementia, its severity was higher in monolinguals than bilinguals.

Some studies in this area have been criticised because the bilingual population included immigrants who were compared with monolingual natives. But this problem did not affect Alladi et al.'s (2013) work. Woumans et al. (2015) also controlled for this potential effect by comparing monolingual and bilingual members of Belgian ethnic groups. In Belgium two major languages are spoken: Dutch in Flanders and French in Wallonia, whereas the capital Brussels is overall Dutch–French bilingual. The authors included bilingual participants who lived either in Flanders or in Brussels and they had mainly acquired their L2 through their family (i.e. one French- and one Dutch-speaking parent) or by going to an L2 school. They reported that bilingualism was associated with a 4.6 years delay in the onset of AD symptoms, after controlling for occupation, gender, and education. Interestingly, they also found that more demanding occupations were associated with an earlier onset of AD, they suggested that this result could be potentially explained by the accumulated effects of stress.

Therefore, the available evidence suggests that bilingualism may confer an advantage at least in terms of delaying the development of AD for a few years. Can bilingualism also mitigate the effects of AD on mental capacities once the disease has been already diagnosed? This seems to be indeed the case. Table 4.11 shows significant effects of bilingualism in increasing visuospatial short-term memory, verbal short-term memory, and verbal long-term memory in bilingual AD patients compared with monolingual patients, in spite of both groups being statistically indistinguishable in terms of language production and in spite of monolinguals having more years of education (Perani et al. 2017).

If bilingualism can help not only in delaying dementia but also in improving the condition of dementia patients, then programs that encourage current bilinguals, and also increase the learning of a second language by current monolinguals, should be supported by governments. Alas, not all politicians understand the benefits of speaking more than one language.

In the USA, the former Republican congressman Newt Gingrich has led a push towards reinforcing English monolingualism in that country (http://www.aei.org/publication/make-english-our-official-language/), but more recently he has also been concerned with the public health issue of dementia. In a deliciously ironic quote, Bialystok et al. (2016: 7) note that:

> In a recent article in the New York Times, Newt Gingrich (2015) expressed concern for the problem of rising numbers of dementia patients in our aging population and argued that the National Institutes of Health should double its budget devoted to research on dementia in an attempt to hasten a solution. He wrote, 'Delaying the average onset of the disease by just

Table 4.11 Means and SDs of years of education, language production, and memory scores in bilingual and monolingual Alzheimer's disease patients and significance of t-test in the between-groups comparisons

	Bilinguals ($N = 45$)	Monolinguals ($N = 40$)	P
Education, *y*	8.26 ± 4.55	10.5 ± 4.07	0.019
Language production	2.06 ± 1.12	2.03 ± 1.51	0.94 *NS*
Visuospatial short-term memory	1.60 ± 1.48	0.58 ± 1.09	0.0035
Verbal short-term memory	1.46 ± 1.28	0.39 ± 0.62	0.000044
Verbal long-term memory	0.71 ± 0.89	0.08 ± 0.35	0.00041

Data are from Perani et al. (2017)

> 5 years would reduce the number of Americans with Alzheimer's in 2050 by 42%, and cut costs by a third.' These dramatic claims are consistent with scientific evidence: Brookmeyer et al. (2007) used epidemiological data from 2006 to predict the prevalence of Alzheimer's disease in 2050. Their model projected public health savings that would accrue from only a 1 year delay in symptom onset and showed a staggering 9.2 million fewer cases of the disease. In this context, Gingrich's projection for public savings based on a 5-year delay is completely reasonable, *and a five-year delay is almost exactly what has been found for bilingualism. A solution may be hiding in plain view*. (italics ours)

Indeed, governments have a solution that may be hiding in plain view, but it requires open-mindedness and the overcoming of prejudice in order to see it and then adopt it. Learning a second language and encouraging current bilinguals can be of benefit to the individual, society at large and even the "budget bottom line".

If bilingualism can provide some protection against AD and some help to AD patients, could trilingualism be even better, or are all the benefits exhausted with just a second language?

Schroeder and Marian (2017, p. 8) hypothesised that:

> trilingualism increases demands on executive processes, due to the need to inhibit, monitor, and switch among more languages, and/or increases demands on memory processes, due to the need to encode, store, and retrieve more linguistic information; these demands would lead to an increase in cognitive reserve, which, in turn, would allow memory to be maintained despite normal and pathological aging.

There are indeed some benefits of trilingualism above those of bilingualism and they mainly seem to apply in old age. Trilingual AD patients seem to be diagnosed somewhat later in life than bilingual patients (Chertkow et al. 2010), and trilinguals are significantly less likely than bilinguals to develop mild cognitive impairment (Perquin et al. 2013).

Such benefits of multilingualism (see also Kavé et al. 2008, for the beneficial effects of multilingualism on cognition) have been recently criticised by Mukadam et al. (2017), who have published a meta-analysis questioning the benefits of bilingualism, compared to monolingualism, with regard to the risks of dementia. This work was in turn criticised by Woumans et al. (2017) who argued that Mukadam et al. left out of their meta-analysis exactly those articles that show a protective effect of bilingualism against AD:

> If the entire relevant literature is considered, it becomes clear that there is considerable empirical support for a bilingual effect on dementia and cognitive decline, in different samples and contexts, controlling for a wide variety of variables. Importantly, the effects described in these studies are of a size to which no pharmacologic intervention can yet aspire (Woumans et al. 2017, p. 1238–1239).

Cognitive reserve can therefore be of help against AD, what about diet? After puberty dietary restrictions can decrease chronic diseases, increase lifespan, and they can also decrease the rate of ageing in various tissues, including decreasing the probability of developing AD, although experimental evidence is not clear cut. On the other hand, restricting food intake to the point of malnutrition may negatively affect the ability of the immune system to fight infections.

Various specific strategies based on diet have been suggested to help in preventing the onset of AD. For instance, oxygen free radicals produce oxidative stress that, in the long term, can damage cells and hence promote AD. This can be prevented by dietary antioxidants which include vitamins A, C, D, E, carotenoids, alpha-lipoic acid (ALPA), L-acetylcarnitine (LAC), phosphatidylserine (PS), pycnogenol, selenium (Creegan and McManus 2013; Boziki et al. 2014). Vitamins of the B group (B1, B2, B6, B9, B12) could also have protective effects against AD. Creegan and McManus (2013) list some food items that are good sources of antioxidants: berries, grapes, pomegranates, oranges, green tea, cabbage, Brussels, broccoli, potatoes, spinach, tomatoes, spices such as turmeric, rosemary, and animals such as fish or animal products such as kidney and liver. Dark chocolate also contains antioxidants. Given that dysregulation of the inflammatory response may favour the development of AD, anti-inflammatory compounds could help protect the integrity of brain cells against disease (e.g. essential fatty acids, vitamins A, D).

Polyunsaturated fatty acids are known to reduce the risk of cardiovascular disease and have been suggested to also protect against AD. However, although evidence for the former effect is good, for the latter it remains inconclusive. Still, Middleton and Yaffe (2009, p. 5) suggest that old people should be encouraged to adopt a healthy diet simply because it is "likely to have positive health outcomes regardless of the effect on cognitive functioning and without any adverse effects".

An adequate diet can also help control some risk factors for dementia such as diabetes, obesity, and hypertension.

Exercise can also have long-term benefits on ageing, through its actions on the immune system, improved balance and strength that reduce the risks of falls, reduction of stress, better blood flow to the brain, improved cognitive speed, executive function, attention, and memory, the latter is achieved through improved neurogenesis and neuroplasticity in the hippocampus (Finch 2007; Brown et al. 2013). However, exercise is more useful as a preventive measure in individuals who have not been diagnosed with AD; that is, physical activity is a protective factor against AD (Middleton and Yaffe 2009). For those already diagnosed with dementia, exercise may improve physical fitness, but it does not necessarily decrease the severity of the cognitive limitations (Lamb et al. 2018).

Several longitudinal studies have been carried out to determine the long-term effect of exercise and physical fitness on the probability of developing dementia in later life. The results are consistent in pointing out that better physical fitness in mid-life is associated with lower probability of developing AD in old age (Defina et al. 2013; Kulmala et al. 2014; Nyberg et al. 2014). In a 44-years-long longitudinal study carried out in Sweden, Hörder et al. (2018) established the initial level of cardiovascular fitness in 191 women (38–60 years old, first examined in 1968) using a stepwise-increased maximal ergometer cycling test. The group was regularly followed up with neuropsychiatric examinations to determine signs of dementia over the years. The cumulative incidence of dementia was 32% for low, 25% for medium, and 5% for high fitness. Overall, high fitness delayed the onset of dementia by 9.5 years. The mechanisms that could explain the reduction in the risk of dementia following exercise and better fitness include improvements in the control of blood pressure, cholesterol, and body mass, prevention of diabetes, and brain effects that could involve the better preservation of both form and function of neurons, synthesis of neurotransmitters and growth factors.

In addition, physical activity increases the levels of brain neurotrophic factors, which are involved in neurological repair. Physical activity also decreases vascular risks that can be associated with both increased risk of vascular dementia and AD. In rats, high levels of voluntary physical activity are associated with less β-amyloid plaque formation (Middleton and Yaffe 2009).

Social engagement may also be protective against dementia, a relationship that is probably mediated by stress control and cognitive reserve mechanisms (Middleton and Yaffe 2009). It is well known that availability of social networks has beneficial effects on health (e.g. coronary heart disease and stroke), cognitive abilities, and survival (see Fratiglioni et al. 2004 for a review).

Although, as it is often the case, not all studies in this field are carried out following a sound methodology that controls for various potential confounding variables, social networks have been shown to positively affect the health of old people in various ways: by providing opportunities to engage the mind, making social support available to the elder, exerting social influence; by fostering social engagement, improving person-to-person contact and access to various resources, both services and material goods (Fratiglioni et al. 2004). Therefore, social networks can decrease stress and improve the material condition of the elder. Social networks can also increase the *cognitive reserve* of old people—through education, for instance—thus slowing down, to an extent at least, the processes of mental ageing and perhaps the development of dementia (Katzman et al. 1988; Pillai and Verghese 2009). In addition, being able to sustain the level of effort required to maintain social contact is in itself evidence of better health, and cultural knowledge can be transmitted through social contact to influence the elder for the better (e.g. by improving lifestyle choices).

In a study of older Swedes (75 years old and older) carried out within the long-term *Kungsholmen Project*, Wang et al. (2002) reported how elders who participated in social activities had a lower risk of developing dementia than elders who did not. The same protective effect against dementia was reported for mental and productive activities. The authors stress that the above effects are independent of potential confounding factors such as sex, age, education, cognitive functioning, depressive symptoms, co-morbidity, and physical functioning.

The benefits of social interactions can be significantly improved and expanded in the elderly population through new information technology (IT) and Internet facilities. As the physical capacity of old people deteriorates, becoming a limiting factor for the maintenance of social contact, the Internet can, to some extent, improve the social connectedness of older people. This can be achieved through spontaneous initiatives of the individual user and through organised programs that are available around the world (see Morris et al. 2004, for examples). In addition, IT can help older people by providing memory aids, such as automatic reminders, and practicing name and face recognition, warning when people are available to engage in a conversation or not, ensuring privacy and so forth.

Preventative initiatives can also involve leisure activities. In a long-term prospective study spanning over 21 years that was carried out in the USA, Verghese et al. (2003) tested the hypothesis that involvement in leisure activities decreases the long-term probability of developing dementia in old age. Their results indicate that over a median follow-up period of 5.1 years the following leisure activities were associated with a lower decline in memory and lower risk of dementia (AD and vascular dementia): reading, playing board games, playing musical instruments, and dancing. Similar results were obtained by Wang et al. (2002) in Sweden, where they found that engaging in leisure activities of various kinds that involved mental effort, social interactions, and a productive objective was inversely related to the incidence of dementia.

Some of the preventative activities mentioned above such as choice of diet, exercising, social support, building up cognitive reserve can be also useful post-diagnosis and they may be adopted in spontaneous interventions with the help of family members. In addition, there are also more specific therapies in use for dementia patients that tend to be delivered by professionals.

One of the earlier interventions for dementia patients has been to enrich the patient's living environment, both in terms of décor and of interactions with people, animals, or plants. A complex environment stimulates mental activity, which may favour brain plasticity through the development of alternative neuronal pathways to process environmental inputs and to produce thoughts (Valenzuela 2008). Long-term initiatives of cognitive training such as the *Advanced Cognitive Training for Independent and Vital Elderly* (ACTIVE) program can also produce good results with the patient (Willis et al. 2006; Gross et al. 2011), but we agree with Valenzuela (2008) that the effectiveness of such interventions is likely to increase the more personally tailored the program is to the specific characteristics and needs of the elder.

Livingston et al. (2005), and more recently O'Neill et al. (2011), Nehen and Hermann (2015), McDermott et al. (2019), and MacDonald and Summers (2020) have published broad reviews of psychological interventions for people diagnosed with dementia. Here we provide a very basic summary of each intervention along with relevant references that offer a more extensive analysis. Interventions are categorised according to four major types: Cognitive/Emotion-oriented Interventions, Sensory Stimulation Interventions, Activity Interventions, and Environmental Manipulation Interventions.

Cognitive/Emotion-oriented Interventions:

- *Reminiscence therapy*: it makes use of materials such as old newspapers and personal or general household items to elicit memories and allow elders to share and value their personal experiences (Cotelli et al. 2012).
- *Validation therapy*: inspired by humanistic psychology it is aimed at giving an opportunity to resolve unresolved conflicts by encouraging and validating expressions of feelings (Neal and Barton Wright 2003).

- *Simulated presence therapy*: in this intervention video or audiotape recordings of family members (or a surrogate) are played to a person with dementia, with the objective of stimulating positive autobiographical memories (Abraha et al. 2017).
- *Reality orientation therapy*: it is based on the idea that increasing orientation information in dementia patients, e.g. through enhancing their ability to recall day, date, weather, time, and names, will improve daily functioning of the patient (Spector et al. 2000).
- *Cognitive stimulation therapy*: this involves a diversity of activities that have the objective to stimulate thinking, concentration, and memory, including discussing past and present events and topics of interest, playing word games, solving puzzles, and performing practical activities such as baking or indoor gardening. Typically, this is carried out in a social setting of four or five people with dementia for around 45 min, at least twice a week. Family caregivers may also be trained to provide cognitive stimulation to their relative on a one-to-one basis (Woods et al. 2012).
- *Individualised special instruction therapy*: the individual focuses his/her attention by participating in an appropriate activity for 30 min (Mitchell and Maercklein 1996).
- *Self-maintenance therapy*: this is intended to help the patient maintain a sense of personal identity, continuity, and also coherence. The intervention incorporates techniques from other therapeutic approaches such as psychotherapy, reminiscence, and validation (Romero et al. 2001).

Sensory stimulation interventions:

- *Acupuncture*: this intervention derives from an ancient Chinese medical practice and it consists of inserting a needle through one or more acupoints which activates specific pathways in the peripheral and central nervous systems. Electroacupuncture may increase the efficacy of classical acupuncture. However, the usefulness of acupuncture in the treatment of cognitive deficits in dementia patients remains unproven (Lee et al. 2009).
- *Aromatherapy*: it uses essential oils to affect the interaction between the brain and the rest of the body via olfactory stimulation. Lavender, for instance, has been postulated to have direct action on tryptophan, the precursor of serotonin, which helps in modulating activity, regulating mood, motivation, emotions, and sickness behaviour. It has been suggested that aromatherapy may help improve the behavioural symptoms of dementia and some cognitive functions, thus increasing quality of life. However, there may also be some adverse effects (Fung et al. 2012).
- *Light therapy*: it is based on the knowledge that rest-activity and sleep-wake cycles are controlled by the endogenous circadian activity of the suprachiasmatic nuclei (SCN) of the hypothalamus. Dementia may be associated with degenerative changes in the SCN, thus inducing circadian rhythm disturbances, which might be ameliorated through stimulation of the SCN with light. Available evidence, however, remain inadequate to reach a firm conclusion about the effec-

tiveness of this therapy on improving cognitive, sleep, functional, behavioural, and/or psychiatric disturbances in dementia patients (Forbes et al. 2009).

- *Massage/touch therapy*: these interventions tend to be used in order to decrease anxiety, disorientation, and mental confusion in dementia patients by enhancing relaxation through touch and massage. Current evidence suggests that massage can indeed decrease agitation in dementia patients to an extent (Moyle et al. 2013).
- *Music therapy*: The *World Federation of Music Therapy* defines music therapy as "the professional use of music and its elements as an intervention in medical, educational, and everyday environments with individuals, groups, families, or communities who seek to optimize their quality of life and improve their physical, social, communicative, emotional, intellectual, and spiritual health and wellbeing. Research, practice, education, and clinical training in music therapy are based on professional standards according to cultural, social, and political contexts" (http://www.wfmt.info/2011/05/01/announcing-wfmts-new-definition-of-music-therapy/) (Ueda et al. 2013). Early reports on the effects of music therapy on dementia patients were rather enthusiastic (e.g. Koger et al. 1999), but subsequent reviews have criticised the design and sample sizes used by researchers in this field (Vink et al. 2003), a concern shared by Sherratt et al. (2004). More recently, however, it was concluded that music interventions have a medium effect on decreasing agitation in dementia patients (Pedersen et al. 2017).
- *Snoezelen therapy/multisensory stimulation*: this therapeutical approach is grounded on the observation that sensory deprivation produces neuropsychiatric symptoms. Therefore multisensory stimulation is used to target sight, hearing, touch, taste, and smell (e.g. lights, surfaces for tactile stimulation, music, scents) in order to decrease neuropsychiatric symptoms in dementia patients. Sensory stimulation can be combined with relaxation (Chung and Lai 2002; Strøm et al. 2016).

Activity Interventions:

- *Therapeutic activity programs*: they involve various activities carried out individually or in small groups with the assistance of a helper (family, volunteer, professional) (Livingston et al. 2005).
- *Animal-assisted therapy*: these are activities conducted with single individuals or in groups using either real animals (usually social species that are domestic pets such as dogs, but other small animals have also been used) or robotic pets. These interventions are capable of reducing agitation and improving social interactions in dementia patients. They can also have a positive effect on coping and communication but there is no reliable evidence supporting an improvement in cognition (Bernabei et al. 2013).
- *Exercise*: regular physical exercise tends to have some positive effects on cognition, especially executive function, attention, and delayed recall in older people

diagnosed with mild neurocognitive disorder (mild cognitive impairment). However, there are contrasting results regarding the success of physical exercise in improving cognition in older people diagnosed with dementia. Some authors suggest that most studies that were performed on older people with dementia showed no significant effect of exercise on cognition (Öhman et al. 2014); but a more recent review is more optimistic, pointing to a positive effect of physical exercise on cognitive functions in AD patients (Du et al. 2018).

- *Montessori activities*: they use rehabilitation principles that are culturally relevant and are based on tasks breakdown, guided repetition, self-correcting, and modifiable tasks that are progressively more difficult. This approach emphasises the use of procedural memory (i.e. unconscious memory of skills and performance of specific tasks), which also happens to be better preserved with age than declarative memory (i.e. conscious memory of facts and events). Specific activities used in the intervention are consistent with the interests, abilities, and skills of the elder and they strongly emphasise the promotion of active engagement. There is good evidence for the benefits of Montessori-based activities on eating behaviours but weak evidence for benefits on cognition in dementia patients (Sheppard et al. 2016).
- *Social interactions*: improving social interactions of dementia patients is important in the context of therapy from both emotional and cognitive perspectives. Cognitively, sociality keeps verbal skills of patients in action and communication with others flowing. Interventions to improve communication can be structured or more spontaneous involving the participation of the elder's caregivers. Such interventions can have some positive effects on the elder, but care should be taken whenever contact with others is producing anxiety and fear in the elder (Vasse et al. 2010). Some social interventions are aimed at training caregivers to better support the dementia patient in his/her social interactions and for the caregiver to better cope with the caregiving effort (Livingston et al. 2005; Ayalon et al. 2006).
- *Decreased sensory stimulation*: this is aimed at reducing anxiety and agitation in dementia patients by reducing environmental disturbances that are upsetting them (Livingston et al. 2005).

Environmental Manipulation Interventions:

- *Visually complex environments*: living environments can be enhanced not only to increase safety but also to provide specific stimulation and sensory/cognitive aid/challenges to the elder diagnosed with dementia. Some added complexity in the environment provides sensory challenges that may enhance the sensory experience. A more stimulating sensory experience is ultimately expected to help in information processing and cognition, although care should be taken to prevent over-stimulation that could trigger anxiety; mirrors, for instance, could produce anxiety. The effectiveness of this kind of interventions is currently mixed but it could potentially be improved by designing individualised programs (Padilla 2011).

- *Sign-posting*: signs may be used to aid memory, object name recognition (or other cognitive tasks), and sensory stimulation. Such aids may help decrease anxiety but evidence is not strong (Livingston et al. 2005).
- *Unlocking doors*: this has been observed to lower neuropsychiatric symptoms and wandering, but it can have the obvious side effect of increasing the risk of accidents, if the door leads to an environment that can facilitate such accidents (Livingston et al. 2005).

Direct care for a dementia patient is one of the major interventions involving social interactions. Such care is usually provided by a combination of the patient's family network and professionals. Even before AD is medically diagnosed, the family may play a role in the management of the mental deterioration of their elder relative. They may note some unusual changes in behaviour that may lead them to suspect that something serious is happening. In this case, their role would be positive, leading up to an early diagnosis and best chances for professional intervention, even though it is often the case that the elder may initially respond to the concerns of the family with denial (Hall 2013). An early diagnosis also allows more time for adjustment, such as family members obtaining a power of attorney and giving an opportunity to family caregivers to establish a relationship with health and social care providers and to navigate through bureaucratic hurdles. But if members of the family are also in denial and are ready to interpret the changes as just evidence of normal ageing, then a diagnosis may be delayed and occur when the disease is already advanced, thus complicating the entire process.

The first challenge for the primary caregivers of an elderly person who has been recently diagnosed with AD is the role they play in helping the elder understand the diagnosis, accept the reality of the disease, and then plan the future action (e.g. Goozee and Dias 2013). The second challenge is to also organise some support for themselves.

As AD eventually takes hold of an old person and the level of mental deterioration becomes more critical, making the elder less independent in the performance of his/her basic activities of daily living, the challenges of caring for the elder significantly increase. The slower the development of the disease, the more chances the family will have to adjust, but at some point in time they will reach a stage where external aid may be required and, eventually, time will come to start considering and planning for institutional care for the elder.

Dementia Australia (https://www.dementia.org.au/support-and-services/families-and-friends/residential-care/deciding-on-residential-care) offers the following decision-making checklist to guide the choice of residential care for a dementia patient:

- What is the attitude of managers and workers—do they listen to you and ask for information?
- Does it feel like a friendly, welcoming place?

- Is there somewhere to sit privately?
- Is it OK for you to come to help the person with dementia eat and shower?
- Can the person have his or her own doctor?
- Are you satisfied with medical services and specialist services?
- What is the medication policy?
- Are there procedures in place in case of fire?
- Is there at least one person on duty at all times?
- Has the fee structure been fully explained?
- Are there extra costs?
- What are the individual resident's rights—own belongings, pets, mail, religious beliefs?
- Are you satisfied with the services and range of activities and choices for recreation?
- Are you satisfied with visiting times and access for family members?
- Can outings, overnight stays, and holidays with family members be easily arranged?
- Do other residents appear well cared for?
- Does anyone speak the language of the person with dementia?
- Are there additional services such as hairdressing and massage provided?
- Is there assistance for you and the person with dementia in preparing for the move?
- Will you be asked for suggestions and comments?
- Is there a policy for having your concerns looked into?
- Are surroundings such as buildings, grounds, and individual rooms suitable?
- What training have staff had?
- Can you join a residents' and relatives' committee?
- Can increased needs be catered for?

Goozee and Dias (2013) also provide some additional tips for the family caregiver who is responsible to decide about residential care:

1. Get to know the staff—they need to be friends, not enemies!
2. Provide a one-page "Who am I?" story to introduce the resident to the staff.
3. Add personal belongings to the room so that it looks as much like home as possible.
4. Consider using a communication book (in their room), which all visitors/family can write in when visiting, to enable staff to remind the elder that he/she had a visitor and who that visitor was. It can also be a great way to share an idea for improving care or to leave a personal note.

A degree of commitment and organisation will be also required from the part of the caregivers in order to assist on legal and official medical matters, protect the elder from undue external interference (e.g. scams, fraud), and organise a network of personal support for both the elder and themselves.

In the case of verbally aggressive AD patients, Hall (2013, p. 138) recommends to the caregivers to just "(t)ake a slow, long breath and look again, and you can see someone you love who's terrified or lost or tormented—probably all three—and who just doesn't know what to do except howl". Indeed, it is often frustration that unsettles the patient, making him/her aggressive. For physically violent patients, professional medical help will be required.

Being aware of what to expect from the elder as his/her condition deteriorates will also help the caregiver prepare, understand, and adapt. For instance, it is not uncommon for dementia patients to confabulate, making up their own reality, imagining, for instance, that somebody is trying to steal from them or plotting to cause harm. They may also say that they have taken a medicine when they have not. Caregivers should be aware of this and prepare themselves not to take false accusations seriously and to organise the daily schedules of medication themselves.

As the patients lose control of their inner and outer world, they will try to find protection in the caregiver, by clinging on to him/her and not letting the caregiver out of sight. This can be particularly exhausting to the caregiver, potentially leading to depression, which is common in caregivers of dementia patients, with prevalence ranging from 30% to 80%, affecting more frequently women and older caregivers (Schoenmakers et al. 2010). Only additional caregiving help can provide appropriate relief. Such relief can also give respite to the caregiver in order to get much needed sleep, especially when the elder and the caregiver had an agitated night. Circadian rhythms disrupted by lack of sleep can be reset in the caregivers through greater exposure to daylight (Hall 2013).

If any family member prefers not to be part of the caregiving network in a direct manner, she or he should be at least asked to cooperate financially if needed, as other members of the family commit their time and energy to the caregiving task (Hall 2013).

To recap, neurocognitive disorders (NCDs) are characterised by a clinical deficit in cognitive capacities, memory in particular. Among the NCDs, dementia (or *major* NCD) increases in prevalence with age from around 1% in sexagenarians to about 40% in nonagenarians and older individuals. Alzheimer's disease (AD) is the most common form of dementia, with the strongest risk factor for its onset being old age. Earlier-onset and late-onset forms of AD are associated with different and specific genetic predispositions (Cacace et al. 2016). Cardiovascular problems and chronic depression/distress are also strong risk factors at all ages. Preventive measures for AD include: building cognitive reserve throughout life, a diet that includes food items high in antioxidants, exercise, social engagement, being involved in fulfilling leisure activities. Although much can be done by the individual in prevention, this does not guarantee protection from dementia, although it can delay the age at onset by a few years. Eventually, medical assistance will be required as soon as the disease is diagnosed, and a point may be reached when institutionalised care will be needed. Nevertheless, the role of family as caregivers will remain important throughout.

In the next section we shift our focus to gender issues in the elders, including aspects of sexual life.

4.6 Ageing, Gender, and Sexual Life

We have already addressed some gender issues in old age in previous chapters. For instance, we know from Chap. 1 that longevity is currently higher for women than men, a trend that became demographically apparent only in the Renaissance, leading to a current female-biased sex ratio in the elderly population. Before the Renaissance the sex ratio was male biased, most likely due to high mortality of women during child birth (Chap. 2). The current difference in the lifespan of women and men mirrors the differences observed between females and males in other long-living vertebrates, being explained in part by the higher mortality rate observed in males but also by sex differences in the rates of ageing (Clutton-Brock and Isvaran 2007). However, on average men tend to reach older ages in better physical shape than women, with older women displaying more pronounced ADL disabilities and more obesity. Men also still reach old age in better financial shape than women—although this trend may be currently changing—but old women tend to benefit from better social support than men from a variety of sources, starting from the family.

In general, older men have historically enjoyed a relative advantage over women, through their role as official repositories of wisdom and knowledge and through controlling social power, although it seems that in the pre-Hellenic world the male-bias was less accentuated (Chap. 2). Throughout history, old men and women have been given diverse attributes both positive (e.g. wisdom) and negative (e.g. physical decay), but women have been usually treated far worse than men, in part as a consequence of body changes rendering them less sexually attractive to men, whereas men, who tended to control power and property, could still attract the interest of young women into old age. In addition, there is the biological fact that women stop conceiving after menopause, whereas men retain their biological reproductive capacity for longer.

Such historical social devaluation of old women may be at the root of the greater level of anxiety observed in old women than men across cultures (Chap. 3), with women having less access to property and basic resources, earning less, and being more likely poor than men; they are also a more frequent target of abuse. This produces greater dependency of women on close kin, children in particular which, however, also comes with an expectation of retribution. Women in general, including old women, are also the primary carers for relatives requiring assistance (e.g. those who are older and infirm), following the cultural expectations of *familism*. Such caregiving generates a degree of burden both physical and emotional.

In this chapter we have also mentioned how mental disorders seem to be, in general, more prevalent among old women than men.

In this section we start with a brief review of gender differences and similarities in behaviour, including body image. We then analyse aspects of gender differences in personality traits and cognition, followed by a review of the stress response and depression in old men and women. Coping with stress is then analysed from the perspective of gender and how gender affects aspects of subjective well-being, especially in the context of widowhood. We conclude with an analysis of sexual life in old age.

In a recent review, Janet Hyde focused on some major theoretical frameworks that have been used to understand gender differences in behaviour: cognitive social learning, sociocultural, evolutionary and expectancy-value theories, to which she also added a theoretical analysis of gender similarities (Hyde 2014). In brief, *Cognitive Social Learning* theory (Bussey and Bandura 1999) states that reinforcement and punishments shape children's and adults' behaviour which, added to the human tendency to imitate role models, leads to behavioural gender differentiation. Children can then internalise the gender-specific behaviours initially reinforced through punishment and reward, thus spontaneously adopting them thereafter and in so doing they strengthen their self-efficacy, or the belief in their ability to perform (or not) certain tasks. This process eventually leads to behavioural gender differentiation in adults. *Sociocultural* theory (Eagly and Wood 1999) posits that it is gender division of labour in society that drives all other behavioural gender differences. By accommodating to the diversity of gender-assigned tasks in society the individual strengthens the gender-specific characteristics of his/her behaviour. Some gender differences, however, may be grounded in biological differences. The acknowledgement of some specific biological predispositions in males and females has led to the proposal of *Evolutionary* theories of gender differences (e.g. Buss and Schmitt 1993). Sexual selection and parental investment are suggested to be the two major evolutionary processes that contribute to behavioural differentiation between the sexes. They would explain the tendency for males to be comparatively more aggressive than females and females to be comparatively more nurturing and sociable than males. This, of course, does not mean that all women are not aggressive, in fact they can be (Burbank 1987), or that all men are not nurturing/sociable, they can also be (Hanlon 2012). Finally, *expectancy-value* theory (Eccles 1994) states that decision-making about being involved in a task is guided by *expectancies* (i.e. expectations of success in the task) and *task values* (i.e. interest in and usefulness of the task). Various factors can also influence both expectancies and values such as self-perception (how good am I at the task), past achievements in the task, socialisers' beliefs and behaviours (e.g. the opinions of others about your abilities at the task), and the broader sociocultural milieu. Expectancies, values, and the specific influence of various factors can affect differentially men and women according to the sociocultural environment.

There are also theoretical frameworks that have been put forward to understand gender behavioural similarities, not just differences (Hyde 2014). From an evolutionary perspective, natural selection may determine some species-specific behavioural constraints that are somewhat similar between women and men (e.g. need to provide parental care; Geary and Flinn 2001), and recent cultural changes have allowed both men and women to converge, to a degree, in their behaviours (e.g. the drive to access a paid job, interest in obtaining an appropriate level of education). For instance, Hyde (2014) pointed out that genders do show a degree of similarity, especially in gregariousness, conscientiousness, sensitivity to rewards and other traits.

Broadly speaking, some gender similarities and differences have been found in various cognitive domains: men tend to outperform women on mathematical tests (but recent trends indicate that the gap is closing in some countries, Hyde 2014) and spatial tests (but if women are given more time to solve the task their rate of success increases), whereas women outperform men on verbal tests. In the case of verbal tests the difference is overall small and the direction of gender advantage depends on the specific verbal ability: verbal fluency seems to be stronger in women, but both genders are similar in terms of vocabulary, reading comprehension, and essay writing.

As an antecedent to personality, temperament factors such as attention, effortful control, persistence, emotionality, distress/fear, activity, and impulsivity differ only modestly between men and women. We will analyse gender differences in personality below. As far as impulsivity is concerned, women tend to be more sensitive to punishment and men are more sensation-seeking. Differences in emotions between men and women seem to be much smaller than suggested by current stereotypes that expect women to be more emotional than men. A significant difference between men and women, however, is found in the area of "interests", with men being more interested in "things", whereas women are more interested in people (Hyde 2014). Men are moderately more aggressive than women, as already mentioned, and both sexes also tend to differ in the behavioural mode they use to express their aggression: more physical in men, more verbal in women. Women also display a small advantage over men in communicative abilities. Interestingly, helping behaviour can be sex-biased but in a variable direction depending on the situation that requires a helping intervention: men help more in situations of danger; women help more in situations of a person in distress.

Self-esteem has been traditionally attributed more to men than women, but more recent analyses indicate that the situation is more complex, with men scoring higher than women on aspects of self-esteem such as physical appearance, athletic, and self-satisfaction, whereas women tend to score higher on behavioural conduct and moral-ethical self-esteem (Hyde 2014).

There is no significant difference in capacities for leadership between men and women, but the distribution of psychopathologies differs, despite areas of overlap between the sexes. For instance, women tend to be diagnosed more often with depression and anorexia than men, whereas boys are more likely than girls to be diagnosed with *attention deficit hyperactivity disorder* (ADHD). Among older people, we have seen in the previous section that AD is more prevalent among women than men, vascular NCD is more prevalent in men until they are 70 years old, to subsequently become more prevalent in women than men from about the age of 85. NCD with Lewy bodies is more prevalent in old men than old women, whereas, overall, frontotemporal NCD is similarly prevalent in the two sexes in old ages.

With regard to sexual behaviours, Petersen and Hyde (2010) carried out a meta-analysis in men and women and found that men tended to report more petting, intercourse incidence, frequent intercourse, younger age at first intercourse, more sexual partners, oral sex, anal sex, having more extramarital affairs, using condoms, engaging in cybersex, engaging more in casual sex, masturbation, and pornography use than women, whereas women participating in the study were more likely to report same-sex sexual behaviour than men. The magnitude of those differences, however, was small. Importantly, across variables gender differences decreased with age.

Therefore, although men and women do differ in many psychological variables, such differences seem to be less significant in magnitude than we may think at first sight (Hyde 2005). Moreover, the differences further decrease with ageing.

Gender roles, that is, masculinity and femininity, are a case in point to illustrate the degree of behavioural overlap between men and women as they age. As argued by Poiani (2010)—and first suggested by Bem (1974)—masculine and feminine gender roles are independent factors that co-occur in both men and women, with each individual displaying a dynamic mixture of feminine and masculine psychological traits spanning from a more masculine combination to a more feminine combination and all potential alternatives in between (androgyny). With ageing, women and men tend to become more similar in terms of their gender roles (see Strough et al. 2007 and the references therein), in part because old men become more feminine (Hyde et al. 1991), but there is also a degree of convergence towards greater androgyny by both old men and women. An alternative view suggests that old women, more than androgynous, may become gender undifferentiated ("degendering" of identities; Shimonaka et al. 1994; Silver 2003). Greater level of androgyny in old age has been associated with better health (e.g. Vafaei et al. 2014).

Russell (2007), however, argues that some specific gender differences can endure into old age, presumably following the different "lived experiences" that men and women have gone through in their lives. Such differences and their endurance into old age can be variable cross-culturally.

The broad gender role convergence between men and women observed with ageing can be partially mediated by changes in body image. Concerns about physical appearance tend to decrease at more advanced ages as the inevitable changes in the body are better accepted by the individual ("cognitive control" of body image, Tiggemann 2004), and body image has a smaller effect on self-esteem. Although some variability remains, as expressed in the ways people dress, use make-up or opt for plastic surgery, for instance, even at older ages ("secondary control" of body image, Thompson et al. 1998).

As some biological aspects of the female body deteriorate with ageing, they may affect body image. This may alter women's perception of their femininity (Winterich 2007), leading to a shift of emphasis towards more relevant traits for older women such as health. A similar process is observed in men who may relax their expectations of how their body should look as they age, against the perception of masculinity at younger ages, shifting their emphasis towards health instead (Liechty et al. 2014).

From a social/cultural perspective—exceptions notwithstanding—the ageing woman's body is seen as less attractive than the ageing man's body, with the image of both bodies being regarded similar only at very old ages (Tiggemann 2004 and the references therein). Aspects of the body that impoverish the image include decreased muscularity, changes in body shape and accumulation of fat, wrinkles, loss of skin elasticity, grey hair. The relatively greater effect of such changes on the feminine body image of women than the masculine body image of men is known as the "double standard of ageing" (Sontag 1972). The "double standard of ageing" is influenced by entrenched cultural stereotypes and expectations.

Therefore *body image* in old age is somewhat trapped in a conflict of interests between the desire of the individual to move on and focus more on both health and *body function* and the pressure from the social milieu that may continue to stress the relevance of image (e.g. Reboussin et al. 2000). Hence some aspects of body image such as grooming and clothing, that can be easily controlled, remain important in old age, especially for women. Women also increase dieting with ageing, which may have negative effects on their health if dieting is motivated by body image concerns alone, but it can be helpful to healthy ageing if it is driven by physiological needs, such as increasing the intake of vitamins, calcium, and other important nutrients and decreasing the intake of fat and sugar.

Another source of variability between genders in old age is personality. What are the personality characteristics of old men and women? Do they simply reflect their personality differences and similarities at younger ages?

In an early meta-analysis of personality traits in men and women, Feingold (1994, p. 42) concluded that:

> Males were found to be more assertive and had slightly higher self-esteem than females. Females were higher than males in extraversion, anxiety, trust, and, especially, tender-mindedness (e.g., nurturance). There were no noteworthy sex differences in social anxiety, impulsiveness, activity, ideas (e.g., reflectiveness), locus of control, and orderliness. Gender differences in personality traits were generally constant across ages, years of data collection, educational levels, and nations.

Feingold's conclusions have been broadly corroborated by subsequent research carried out within the framework of the *Five factor* model of personality of Costa and McCrae (Costa and McCrae 1992). The five personality factors are commonly referred to as the *Big Five*, and we briefly introduced them in Sect. 3.1 of Chap. 3: *Neuroticism*, the propensity to experience negative emotions such as anxiety, anger, self-consciousness, depression, and emotional lability; *Agreeableness* is characterised by altruism, empathy, and kindness, thus a tendency to cooperate and maintain social harmony; *Conscientiousness* is associated to self-discipline, organisation, and self-control in following rules or maintaining specific goals; *Extraversion* expresses sociability, assertiveness, and also positive emotionality, which are associated with sensitivity to rewards; and *Openness/intellect* reflects imagination, creativity, intellectual curiosity, and aesthetical appreciation.

Costa et al. (2001) initially suggested that women tend to score higher than men in Neuroticism, Extraversion, Agreeableness, and Conscientiousness, whereas men score higher than women in Openness to ideas.

According to Weisberg et al. (2011), the overall trend from more recent studies is that women score higher than men on Neuroticism and Agreeableness. There is also a trend for women to score higher than men on some aspects of Conscientiousness such as order and self-discipline, but this is variable across cultures. Gender differences in Extraversion are generally small, with a trend for women to score higher than men. The small gender differences in Extraversion could be due to one gender scoring higher in some aspects of the trait and the other scoring higher in others. For instance, women tend to score higher than men on warmth, positive emotions, and gregariousness, but men score higher on excitement-seeking and assertiveness. Overall, there are no significant gender differences in Openness/intellect.

The *Big Five* were also used by Chapman et al. (2007) to study personality differences and similarities between old men and women (65–98 years old). In accordance with studies done on younger people, Chapman et al. also reported higher scores for Neuroticism and Agreeableness in older women than older men, which suggests that some personality traits may be both different between genders and such gender difference may be enduring for a considerable amount of time. This has been further studied by Soto et al. (2011) who have shown that Neuroticism tends to decrease since about the age of 30, but it is consistently higher in women than men. Agreeableness tends to steadily increase with age from about the age of 30 years old, but then it tends to decrease since about 62 years old, being also consistently higher in women than men throughout.

Conscientiousness in Soto et al.'s study steadily increases with age from teenager to 60+ but at a slower rate since about the age of 45, and it is qualitatively higher in women than men. Extraversion is relatively stable throughout adult life, with women being more extraverted than men, but the gender difference tends to decrease from the age of 55 onward due to an increase in male extraversion. Openness tends to steadily increase from the age of 25 on, with men being qualitatively more open than women, but the difference between genders decreases at older ages and there is a drop in openness in both genders between the age of 60 and 65.

Lehmann et al. (2013) have also reported that women tend to display higher levels of Neuroticism, Extraversion, and Agreeableness than men. Men, on the other hand, are more open to experience, whereas in this work Conscientiousness was similar between genders.

The higher levels of Neuroticism and Agreeableness that have been consistently found in women than men in previous studies were also confirmed in a Canadian study of 17–85 years old (Weisberg et al. 2011). Women are also more Extraverted than men in this study, but Conscientiousness is similar between genders and Openness is higher in women than men.

Therefore, the two most consistent differences in personality traits found between genders are that women tend to score higher on Neuroticism and Agreeableness and such difference is maintained in old age; even though with ageing Neuroticism tends to decrease, whereas Agreeableness increases, at least until the age of 60.

One of the major concerns that old people have about ageing is the decay in their mind in general and their cognitive faculties in particular, not only because of fears of developing dementia, as we have already seen in this chapter, but also because of fears of being unable to properly take care of personal needs, administer money or property, being the target of scams without realising it and so forth. How do cognitive faculties change with age according to gender?

Despite broad similarities in cognitive capacity, men and women do tend to differ in some specific cognitive abilities such as episodic memory tasks and verbal production tasks where women outperform men, whereas men outperform women in the area of visuospatial abilities (De Frias et al. 2006). Some of those differences are maintained in old age.

Wahlin et al. (2006) compared samples of community-dwelling adults (61–95 years old) as part of the Canadian *Victoria Longitudinal Study*. They described a general cognitive decline with age that was moderated by the level of education. Old women tended to outperform old men on perceptual speed and episodic memory tasks, whereas men outperformed women on semantic memory. Relative gender advantages, however, could be decreased by individual differences in health. In a study carried out in Finland, Portin et al. (1995) assessed the cognitive performance of 389 individuals (both men and women) who were 62 years old when they were tested. Women outperformed men in tasks such as Object Memory, Paired Word Associates, Digit Symbol, and Months Backwards, whereas men did better than women in tasks such as Trail Making and untimed Block Design. Therefore old women outperformed old men in visuomotor and verbal speed and also in verbal and object memory, whereas old men outperformed old women in visuospatial speed and also visuoconstructive performance.

In the USA, Barrett-Connor and Kritz-Silverstein (1999) carried out cognitive tests on 800 women and 551 men, who were aged between 65 and 95 years old, in three different periods from the early 1970s to 1991. The performance on cognitive tests decreased with age and the rate of decrease was steeper for men than women in various memory tests, including category fluency. Also in the USA and using data from the *Baltimore Longitudinal Study of Aging*, McCarrey et al. (2016) analysed the cognitive performance of a group of 50.0–95.8 years olds (at baseline). Women outperformed men in mental status as assessed through the *Mini-Mental State Examination* (MMSE), women also outperformed men in verbal learning and memory as shown in *California Verbal Learning Test* (CVLT) results. Women were also more fluent in language production and performed better than men in a *Digit Symbol* task. Attention, concentration, visuomotor scanning, perceptuomotor speed, working memory, and set-shifting were assessed through the *Trail Making test* Parts A (Trails A) and B (Trails B) and women tended to be faster than men on Trails A but there was no difference on Trails B. The *Benton Visual Retention Test* (BVRT) is designed to measure visual constructional skills and short-term figural memory, men made fewer errors than women at baseline, but their rate of errors increased with age and it was higher than women's. Visuospatial abilities as measured with the *Card Rotations Test* were better for men at baseline, but men showed a faster rate of decline with age than women. However, no gender differences were found in the *Boston Naming Test*. Therefore in this study old women generally performed better than men on tests of verbal learning, memory, and language production. Interestingly, the cognitive decline with age was faster for men than women in all skills tested. In another study carried out in the USA, Pillemer et al. (2003) found that, among 68–79 years old, memory styles, as expressed in narratives, were more specific (or episodic) in women than men, with women putting purposeful reminiscence of people and events higher in their priorities than men.

After analysing data from 625 participants (35–80 years old) in the Swedish *Betula project*, De Frias et al. (2006) reported that men performed better than women on the block design cognitive test, whereas women performed better than men on episodic recall, face recognition, and verbal recognition. The differences did not vary with age. Hence women show an advantage over men in verbal and social skills, whereas men have an advantage in tasks assessing visuospatial ability.

Gerstorf et al. (2006) analysed 13-year longitudinal data from the *Berlin Ageing Study* that included 70–100 years old participants at baseline. Cognitive capacities of men and women decreased with age but the differences between genders run in parallel, with women outperforming men on digit letter, paired associates, memory for text, word beginnings, and spot a word, after controlling for the level of education. Pauls et al. (2013) compared memory abilities in women and men in a German sample of 696 participants (330 men and 366 women) between 16 and 69 years old. Their results indicate that older women outperform older men in verbal fluency and auditory episodic memory, whereas older men outperform older women in spatial perception, visual episodic memory, and visual working memory.

van Hooren et al. (2007) carried out a study in The Netherlands where they investigated gender effects on cognitive speed, verbal memory, executive functioning, and verbal fluency in a group of 578 healthy, 64–81 old individuals, within the framework of the *Maastricht Aging Study*. The ability to perform all cognitive tasks declined with age in both men and women, but the speed of decline could be diminished by better education. In general, women performed better than men on verbal memory tasks (e.g. immediate and delayed recall of the verbal learning test).

In an Internet study carried out in the UK, Maylor et al. (2007) also showed a decline in cognitive performance with age in both men and women, with older women outperforming older men in a category fluency test (verbal test) and an object location memory test, whereas men outperformed women on a mental rotation test and a correct line angle test.

Gender differences in cognitive functions in old age follow a somewhat different pattern in China, as compared to the trends we have seen in the West. In China, women have tended to be significantly more disadvantaged in cognitive functioning in old age than men, mainly due to historical and cultural factors related to low education, simpler occupational achievements, fewer opportunities to participate in leisure activities, and also a smaller social network available to women (Zhang 2006). Lei et al. (2012) carried out a study in the East Coast Chinese province of Zhejiang, a high-growth and industrialised area, and the West province of Gansu which is mainly agricultural and poor, as part of the *China Health and Retirement Longitudinal Study* (CHARLS). On measures of both episodic memory and intact mental status men scored much higher than women and the difference increases at older ages. In this case, the difference between genders in their level of schooling (higher for men, especially in the older generations) and access to sources of new knowledge in urban areas may contribute to explain this result. Zhang (2006) analysed gender differences in cognitive impairment from the *Chinese Longitudinal Healthy Longevity Survey* (CLHLS), focusing on participants who were 80–105 years old. The odds of cognitive impairment were 1.3 times higher in oldest-old women than oldest-old men after controlling for a number of variables including age, rural residence, ADL disability, social network, and leisure activities.

Therefore, based on studies carried out in Western countries, older women score higher than men in most cognitive tests of object location, although in some others (e.g. incomplete letters) older men score higher; but in verbal memory tests, verbal speed/fluency, auditory episodic memory, and overall language production older women score consistently better than older men, whereas older men score higher than women in cognitive tests of spatial rotation and navigation, that is visuospatial tasks (see also Li and Singh 2014). Such gender differences are consistent with evolutionary theories of human cognitive abilities, with men having evolved as mainly hunters (requiring high navigation skills) and women as mainly gatherers (requiring high object location skills) (Gur et al. 2007). Women's superiority in language production is also likely to have adaptive roots in the social life of women that promoted communication with other women and children during food gathering and child-rearing, without the constraints of silence imposed on men by their hunting activities (Joseph 2000). Such adaptive mechanisms, however, require that specific genes be specifically expressed in males and females, thus being under gender-specific natural selection. The fact that the X-chromosome contains many loci associated with cognitive abilities may suggest that gender-specific selection in cognition is possible (Zechner et al. 2001). There is also an overall steeper decrease in cognitive abilities with age in men than women. Proximally, the relative advantage of old women over old men in terms of some cognitive functions—some aspects of memory in particular—may be in part a result of positive effects of oestrogen on the long-term maintenance of brain functionality (e.g. Genazzani et al. 2007), whereas education, health, social connectedness, and an active lifestyle also play a role in explaining gender differences in cognition and speed of cognitive decline with age.

Given the impact of stressors on health (Schneiderman et al. 2005), it is worth asking whether, in old age, there are differences in the stress response between genders and in the levels of stress they experience in life.

It is often believed that life becomes less stressful at older ages, that is, older people are thought to be exposed to fewer eventful or discrete stressors. However, we also know that at older ages eventful stresses affecting our health have greater chance to occur, as are stresses associated with the death of a loved one, especially family members, friends, and acquaintances who are also old (Pearlin and McKean Skaff 1996). In addition, various chronic stresses are likely to emerge in older ages, such as chronic health conditions, apprehension about safety both at home and outside, concerns about physical and mental deterioration, financial constraints, tensions within the family, caregiving duties towards other old persons such as a spouse. The effects of such stressors on old people can be somewhat decreased by adopting individual coping strategies such as displaying *mastery*—or a broad sense of control over life events, and by receiving social support.

From a physiological perspective, the stress response involves the activation of the hypothalamus-pituitary-adrenal (HPA) axis that triggers the release of molecules such as cortisol in circulation in response to various stressors, including psychological ones (e.g. O'Connor et al. 2000), with potential negative psychological, immunological, and other health effects if the stress becomes chronic (Bauer 2008).

Experimental studies have shown that psychological/social stressors affect young and old, men and women differently. For instance, Kudielka et al. (2004) exposed males and females of different ages, including 30 older adults (mean age of 67.3 years), to the *Trier Social Stress Test* (TSST), where individuals are confronted with a psychological challenge such as speaking and performing a mental arithmetic task in public, that can be stressful. The stress response was measured through changes in the circulating concentration of adrenocorticotropic hormone (ACTH), a hormone secreted by the anterior pituitary gland in the brain in response to stress. ACTH is released through the action of the corticotropin releasing hormone (CRH) produced in the hypothalamus, and the ACTH, in turn, stimulates the release of cortisol from the adrenal glands. Kudielka et al. (see also Kudielka and Kirschbaum 2005, for a review) showed that younger adults displayed a higher ACTH response than older adults, this difference was especially due to a difference between men of different ages rather than women. Older women, however, had higher levels of plasma cortisol than younger women, and their cortisol levels were also higher than men of all ages. Older men, on the other hand, showed a higher free salivary cortisol response than older women. In general, studies have shown that women may experience greater increase in HPA axis response at older ages, whereas at younger ages there appears to be greater cortisol response among men or no significant difference between genders depending on the specific study (see Seeman et al. 2001 and the references therein).

Seeman et al. (2001) studied HPA axis reactivity in 9 young men and 17 young women (22–36 years old) and in 7 old men and 7 old women (67–88 years old) in the USA. A stress response was elicited through a battery of cognitive tests challenges. Older subjects had more elevated levels of salivary cortisol at baseline, with older men having more elevated salivary cortisol than older women. Cortisol reactivity to the cognitive challenge (stressor) differed between genders among the younger participants, with younger men showing a greater relative percentage increase in cortisol over the baseline than women. The opposite was detected in the older group: older women showed greater relative increase of cortisol compared with older men.

In a recent meta-analysis, Otte et al. (2015) confirmed that healthy elders tend to show a higher cortisol response to stressful challenges than younger people, and such response is stronger for older women than men. Are older women also more distressed than older men?

Cairney and Krause (2005) surveyed 5719 adults including both 50–59 years old and oldest-old (80 and older) in Canada to study factors affecting *distress* (a psychological response to strong levels of stress). Their results indicate that distress is higher in both the youngest and the oldest age groups and lowest in between, whereas old women are more likely to be more distressed than old men, in part due to differences in physical health. Interestingly, the authors also point out that for old women "social support" also comes with a source of distress, but not for old men.

This may be explained by the duties involved in maintaining close relationships within a social network, which may take a greater toll on old women. In another study also carried out in Canada, Nurullah (2010) found that older people tend to be less distressed than younger ones and that women are generally more distressed than men, with other factors apart from gender also significantly affecting the level of distress: education, marital status, income, tensions at work, life events, childhood and adulthood trauma, self-esteem, mastery, and sense of coherence.

After analysing data from the *ESRC/MRC HSRC Quality of Life Survey*, which targeted 65+ years old Great Britain residents, Paul et al. (2006) found that prevalence of psychologically distressed individuals increased with age: 15% in the age group 65–69, 22% in the 70–79 interval, and 24% among the 80+ years old. The prevalence of distressed individuals was also higher in women (22%) than men (18%), as it was the prevalence of those feeling lonely: 8% among women and 5% among men.

In a study carried out in Finland, Talala et al. (2008) investigated psychological distress in a sample of 5425 adults (54.3% women). Retirement was not associated with distress in either men or women, but lack of partner did increase the level of distress.

Hatch and Dohrenwend (2007) reviewed changes in the experience of traumatic and stressful events with age. Although traumatic events are more common at younger than older ages especially in men, the experience of stressful events is variable across ages and genders.

Therefore, we can say that among the elders studied in the past 20 years, it is women who, in general, seem to be more stressed than men, both physiologically and psychologically.

Distress can potentially turn into full blown depression. Do older men and women differ in the likelihood of developing depression and if so, do the patterns mirror the female bias we have just described for distress?

Although about 1–4% of the elderly population has been diagnosed with major depression, roughly twice (1.5–3 times, Culbertson 1997; Kessler 2003) as many old women present with the condition as old men around the world. A female bias in the prevalence of depression in the elderly was also reported in the works of Blazer (2003), Alonso et al. (2004), and Luppa et al. (2012) that we quoted above in this chapter in Sect. 4.4. Depression is also more heritable in old women than old men (Jansson et al. 2004). Such female bias in the current generations of old people diagnosed with depression may be also explained, at least in part, by greater physical disability and lower socio-economic status of older women compared with men (Blazer 2003). The gender gap in depression, however, narrows in the oldest-old (Fiske et al. 2009).

More depression in old women is also associated with greater levels of insomnia compared with men (Brabbins et al. 1993). Depression also presents somewhat differently between older men and women. For instance, Fiske et al. (2009) mention how depressed older women display more disturbance of appetite than men, whereas older men report being more agitated.

One context in which old men respond with more depression than old women is the loss of a spouse. Following widowhood, men are more likely to develop depression and they tend to remain depressed for a longer period of time than women (Fiske et al. 2009).

The association of depression with mortality may also vary with gender across countries. For instance, Blazer (2003) quotes studies carried out on Japanese Americans where depression was a cause of mortality in old women but not old men, whereas in another study mild depression was negatively related to mortality in women but it was not associated with mortality in men. In a study carried out in The Netherlands, however, Schoevers et al. (2000) found that psychotic depression was associated with a significant excess mortality in both old men and women, but neurotic depression was linked to a 1.67-fold higher mortality risk in men only.

The female gender bias in depression observed in the old starts early on in life, between 11 and 15 years old (Kessler 2003). Among adolescents the likelihood of depression ranges between 25 and 40% for girls and 20 and 35% for boys (Hankin and Abramson 2001). Some of the theories put forward to explain gender differences in depression include the *cognitive vulnerability-stress* model that suggests that females may be more vulnerable to depression as they may be more likely to confront a negative event in a more negative manner (Hankin and Abramson 2001). This could emerge from a special sensitivity to social approval, for instance, or a broader sensitivity and cognitive vulnerability to events that are perceived as negative and stressful. It should be noted that the events towards which individuals are psychologically vulnerable can change with age:

> Whereas adolescents confront the task of establishing an identity, elderly individuals question whether their lives had meaning. This view suggests that the domains most relevant for cognitive vulnerability may change over the life span. For example, health may be a more relevant domain for elderly people than for children (Hankin and Abramson 2001, p. 780)

Is the broad female bias in distress and depression an effect of gender differences in coping with the stresses of life? The answer is probably yes, but with some caveats.

Whenever we are confronted with a stressor, it is usually the case that we respond by enacting coping strategies that serve the purpose of decreasing the negative effects of such stressor. Two major strategies have been identified: one that could be described as "mature defences" (e.g. humour, altruism, absence of impulsivity and aggressive reactions, cognitive reassessment, acceptance), and another one that has been referred to as "immature/neurotic defences" (e.g. distortion of reality, denial). In general, it is expected that ageing brings a more frequent use of "mature defences" (e.g. less use of denial, more use of altruism), but this is variable among individuals (Diehl et al. 1996). Moreover, individuals may use both coping strategies simultaneously blending them according to circumstances and following their own personal cognitive and emotional development. One important difference between genders in their coping styles is that males tend to cope by externalising through aggressive behaviours, whereas females tend to cope by internalising, that is by turning against the self, seeking social support, and using escape-avoidance. That is, women could be restricted in their strategies to cope with stress if they lack social support, in which case women could become psychologically distressed and develop depression. This suggests that distress could be prevented in women by offering social support (e.g. Ha et al. 2008).

The differences in coping styles between genders seem to decrease as individuals become older, leading to a pattern of increasing "masculine coping" in old women and increasing "feminine coping" in old men (Diehl et al. 1996).

A greater ability of coping with stress should eventually lead to greater well-being. Given the above information, one would tend to expect that older women may, on average, feel unhappier than older men. Is this so, and what are the factors that modulate gender trends in well-being?

The psychological construct of *subjective well-being* (SWB) is expressed through various measurable variables and, broadly speaking, it seems that SWB is lower in women than men (e.g. Pinquart and Sörensen 2001). For instance, women, including old women, tend to report more negative emotions than men (including old men). Variable results, however, have been reported for general life satisfaction and positive affect (Tesch-Römer et al. 2008). Social and cultural factors have been suggested to be important causes of this gender difference in SWB. For instance, social disadvantage among women, leading to financial hardship, isolation, being the target of violence, negatively affects their SWB. Gender inequality in opportunities in life also negatively affects the well-being of women at all ages in many societies. Such negative effect of inequality on well-being results from preventing women to reach their goals in life and fulfilling their personal ambitions, thus limiting their ability to be happy. Preventing access to material resources (e.g. a well-paid job, ownership of property) also contributes to women's dissatisfaction in life, insecurity, and lower well-being.

Tesch-Römer et al. (2008) added an additional complication that may help to better understand the female negative bias in SWB. Cultural factors can affect well-being in women depending on whether gender inequality is widely accepted or not in society. If gender inequality is accepted, then women who have been able to achieve equality (e.g. good paid jobs) may experience lower well-being than men, presumably because the effort needed by women to overcome prejudice may negatively affect some aspects of their mental and physical health. In societies where gender equality is a cultural norm, women who can achieve equality experience equal well-being as men. Equality of opportunities at work, however, may also drag women retirees into the same conundrum that men are mostly facing now, that is: how to retain a good level of subjective well-being in retirement once the old role that was defined by paid work is lost (Kim and Moen 2002; Arber et al. 2003a; Fairhurst 2003).

Another construct that is aimed at describing well-being is *Health-Related Quality of Life* (HRQL). HRQL measures a number of variables related to physical, mental, emotional, and social functioning. Again, old women tend to show lower levels of HRQL than old men (e.g. Borglin et al. 2005a), especially given the higher prevalence of disability and chronic medical conditions that affect more significantly women than men in old age. Similar results were obtained by Kirchengast and

Haslinger (2008) who described the HRQL of women older than 70 to be relatively worse than that of men, although the difference was not statistically significant.

Well-being and life satisfaction can be also affected in the elderly by transitions in the marital status, with the effect being different for men and women. In old age life satisfaction declines with time in married women—in part perhaps due to declining social support from partner (Acitelli and Antonucci 1994)—but it does not seem to change in married men. Loss of a spouse can change this situation with both men and women experiencing a decline in life satisfaction, but the decline is more serious for men than women. Old men can revert the situation by marrying again, for old women, however, it is not so simple (Chipperfield and Havens 2001), but they can compensate through accessing a wider net of social support from family and friends (Antonucci and Akiyama 1987).

Old men seem to enjoy greater emotional stability when their wife/partner is emotionally independent, whereas the wife is more emotionally stable when she feels that she is important to her husband (Tower and Kasl 1996). Cultural and educational effects aside, we also see here a more emotional reliance of old women, and indeed women of all ages, on strong social ties.

Old men greatly benefit from women's tendency to be sociable and cooperative. This is seen in the women bias in caring for older people (Calasanti 2003). It is reasonable therefore for old women to expect a degree of reciprocation from old men in this regard. Although such reciprocation tends to focus on material transactions, a degree of empathy and emotional closeness is also expected.

The experiencing of well-being can be hindered or facilitated not just by social relationships but also by individual views on self-identity, life meaning, and the broader existential stance on life. We will focus on existential aspects of old age in great details in Chap. 8 *Journey towards the End of Life*, here we only touch on gender differences and similarities in broad existential issues of identity, life, and death.

Post-retirement, old people are confronted with the need to construct narratives of identity in an environment where old roles and social networks may have changed: direct family (children and grandchildren) may disperse, relationships with former colleagues may weaken. In particular, the home environment acquires a more central role for most retirees. Focusing on the concept of "home" as an instance for the construction of gendered identities, Russell (2007) studied a sample of 66 older people living in Australia, pointing out to common views that "home" is broadly seen as a "woman's place" by the generation of older people interviewed. After retirement, the view of "home" emerges in women and men from the direct interaction of the individual with the physical space, which may differ between genders: e.g. cleaning, cooking, and/or decorating for women, home repairs for men, and perhaps gardening for both. It is such greater intimate level of interaction that builds a rapport between a person's identity and his/her habitat. In the end there is a confluence of views about the importance of, and attachment to, home as living space and as property, providing material and emotional safety to both genders. A gender difference in old age, however, may affect the reaction or even propensity of old men and women to change residence and leave their home, especially the home where they raised their children. Women tend to be more reluctant to leave home than men, in spite of verbally expressing similar level of attachment to place. This could be a result of gender differences in the rolc of home in the construction of identity.

Additional gender-specific preferences are also retained in old age in the Australian elders interviewed by Russell (2007). Men, for instance, typically mentioned spending time in retirement doing sports, fishing, going to the races, playing snooker, carpentry, and the like.

Such endurance of gender-specific activities into old age may result, at least in part, from the strong gendered nature of social activities at younger ages that contribute to building a gendered identity. In some English-speaking societies (e.g. England, Australia) old men find strong social referents in men's clubs that may derive from previous gender-segregated experiences: school, workplace, armed services, team sports (Russell 2007 and the references therein).

Given the similarities and differences in their capacity to build a satisfactory identity in old age: Do women and men differ in their *will to live* as they become older and more frail? Will to live in old age is undoubtedly linked to the degree of well-being (physical, emotional, cognitive) experienced by the old person: greater well-being leads to greater will to live and vice versa. When life becomes a burden that is difficult to carry, satisfaction with life may start to crumble, thus leading to a decrease in the will to live. The process can be worsened by lack of material and social support. Religion and spiritual beliefs can further modulate the will to live. In some religious traditions will to live is supposed to be strong, in others death could be embraced as the moment when the transition to eternal life will finally occur, thus modulating down the will to live.

Sara Carmel (Carmel 2001) hypothesised that old women would show a lower will to live than men on the grounds that, in spite of living longer than men, they suffer greater physical morbidity and functional disability than men, leading them to use more medication and being more burdened with the financial costs of medical treatments. Old women also tend to be less wealthy because it is more likely that they have been employed in lower paid jobs compared to men before retirement, which exposes them to the stress of uncertainty about whether they will be able to satisfy their material needs in old age (housing, food, transport, medicines, and so forth). Greater stress in old women makes them also more likely to develop anxiety and depression, thus leading to lower will to live.

Indeed, in a study carried out in Israel, Carmel (2001) described a lower will to live in old women than men, most likely associated with lower self-esteem in women. The deterioration of health with age was also a strong determinant of decreased will to live. Will to live is a predictor of survival in old women in this population (Carmel et al. 2007). In addition, Carmel and collaborators have shown that it is a decay in will to live that leads to the expression of depressive symptoms in old people (75 years old and older) and not the other way around (Carmel et al. 2018). Similar gender-specific patterns are repeated across different countries, although with variability due to differences in various cultural, social, and economic factors (Carmel 2019).

No matter how high our will to live is, the end is obviously inevitable. The death of an old member of the family can affect the survivors in various ways, with gender being one of the variables that can modulate the response. Morbidity and mortality tend to increase in both old men and women after the loss of their partner, but the effect is stronger in men (widowers) than women (widows), especially immediately after the loss, during the *acute grieving period* (Stroebe et al. 2001). Availability of social support during bereavement, that may be more easily available to old women than old men, can compensate for the negative effects of the loss of a dear one, as social support has the capacity to buffer the individual from stress.

Over the longer term, widowers are supposed to cope less well than widows, given the better access of widows to external social support (e.g. Lopata (1973) mentions supportive "societies of widows", contrasted with the less common "societies of widowers"), but this can be compensated by a potentially better capacity of old men to find a new, perhaps younger, partner. Thus, in fact, the negative effects of widowhood experienced in the immediate aftermath of the death of a life-long partner may be better compensated by men than women over the longer term, especially when men can still attract a new partner or have a well-established identity that maintains them engaged with society (e.g. professionally), and they also possess the financial capacity to pay for help with household chores (Umberson et al. 1992).

Therefore, we can draw some general conclusions about gender differences and similarities in old age from the above brief review. Although gender differences tend to decrease in older ages there are some gender-specific trends that are identifiable. Older women tend to be more sociable than older men, whereas men are relatively more aggressive. Cognitively, older men display better visuospatial capacities, whereas women display better verbal and communicative abilities. Older men also decline more rapidly in their cognitive faculties with age than older women. From a personality perspective, older women score higher than men in Neuroticism and Agreeableness and they display a stronger stress response, being also more distressed and more often diagnosed with depression. Coping styles to confront such life stresses differ between genders: men tend to externalise through aggression, whereas women respond by internalising and seeking social support. If social support is missing, older women tend to display lower subjective well-being leading to a lower will to live. Social support is a major factor that can modulate the differences between genders.

We conclude this section with an analysis of some aspects of sexual life in old age.

Old age has been traditionally characterised by what Gott and Hinchliff (2003) call "sexual retirement"; this is not universal for either men or women, but it is general enough:

> For older men, experiencing health problems and/or erectile dysfunction determined the adoption of a position of 'sexual retirement'—they viewed this as their 'luck running out'. For older women, widowhood was the key determinant of sex no longer assuming any importance to them – they 'didn't want anyone else' (Gott and Hinchliff 2003, p. 65)

Note that Gott and Hinchliff also pointed out that widows had already lost much of their interest in sexual intercourse before losing their husband, but they continued to have sex out of "marital duty", a concept that was still *en vogue* when these old couples married. The sexual revolution that started in the 1960s–1970s (e.g. Robinson et al. 1991) subsequently changed this situation, giving more importance to mutual individual satisfaction and sexual compatibility. In fact, divorced old women may currently feel liberated to return to a more fluid and active sexual life, which provides an avenue to strengthening intimate relationship and attachment. One female respondent in Gott and Hinchliff's (2003a, p. 1625) study explained:

> I've got some lovely friends (but)...I've still felt lonely cause you tend to not feel like a woman and I've not been well over the last few years and not felt right great about myself, so it is nice that someone is interested in me as a woman and I find that to be the best bit (3358, divorced woman aged 52).

Interest in sex may be retained before turning 60, less interest is shown in the 70s and 80s, centenarians on the other hand predictably show no interest in sex (Taylor and Gosney 2011 and the references therein). The decline in sexual activity with age affects all domains of sexual function: desire, orgasm and, in men, erection and ejaculatory functions (Helgason et al. 1996). Variability among individuals can be determined by biological factors and psychological issues such as depression, but also by diverse priorities in life. Men in particular are confronted with the physiological challenges of erectile dysfunctions, which affect about 75% of men older than 70 (Camacho and Reyes-Ortiz 2005). Such dysfunctions include slower and less firm erections and longer refractory periods (DeLamater 2012), whereas old women may have complaints about vaginal soreness, dryness, and pain during intercourse, the latter is known as *dyspareunia* (Barlow et al. 1997; Taylor and Gosney 2011). The specific effect of menopause on sexual intercourse in women is variable: some studies show a decline in sexual activities after menopause others show no difference (DeLamater 2012).

Further health complications that can decrease sexual function in old age include cardiovascular disease, diabetes, obesity, and lower urinary tract problems (Camacho and Reyes-Ortiz 2005).

The above issues that may already interfere with sexual life in old age are compounded by some societal expectations that see older people as "asexual", a social attitude that could be internalised by some elders (Gott 2006; Taylor and Gosney 2011). Moreover, social expectations of elders' asexuality may also interfere with well-being in retirement institutions, where staff may have a negative attitude towards sexual relationships between the residents (e.g. Ehrenfeld et al. 1999). Recent cultural developments, however, have weakened the "asexual elder" stereotype and instead it is asexuality in older ages that has become a kind of "new deviancy" (Gott 2006). This shift towards a better acceptance of sexuality in older ages has been helped by the introduction of some medical aids to maintain sexual activity, alongside specific cultural changes. As a result, sexual life currently remains moderately or very important—and even extremely important—in older people, including some people older than 70 (Gott and Hinchliff 2003a, see Fig. 4.9).

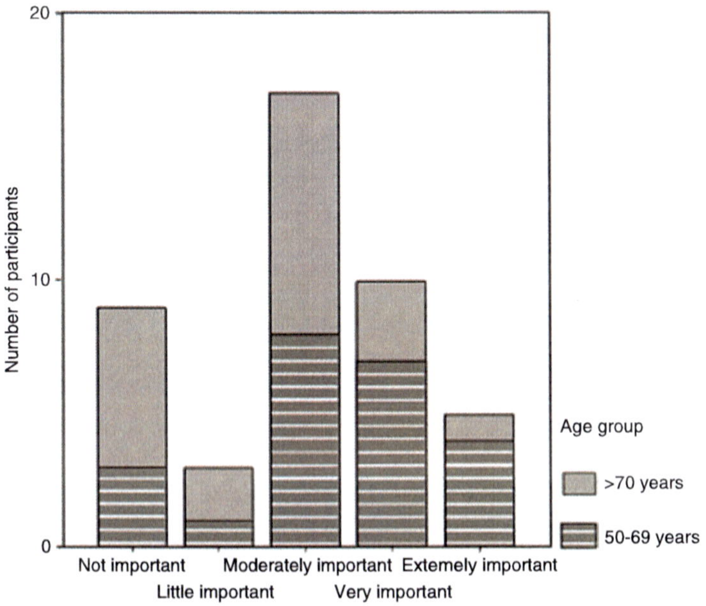

Fig. 4.9 Number of old people of two different age groups who regard sex life to various degrees of importance. From Gott and Hinchliff (2003)

In general, old women tend to lose interest in sex earlier than old men (e.g. Camacho and Reyes-Ortiz 2005), a difference that has recently been strengthened by the introduction of drugs such as Viagra© (sildenafil citrate) that help correct erectile dysfunction and thus maintain sexual function in men for longer, at apparently no significant long-term risks to their health (e.g. Gresser and Gleiter 2002), although some increase in the risk of headaches, dyspepsia, and visual disturbances has been reported (Tsertsvadze et al. 2009).

Erectile dysfunctions at older ages affect a variable percentage of between 20% and 40% of 60–69 years old men and 50–100% of men in their 70s and 80s (Gandaglia et al. 2014). Given that sexuality is at the centre of the traditional concept of "masculinity", erectile dysfunctions may jeopardise the self-identity of older men, whilst younger men suffering erectile dysfunctions feel "older" (Gott and Hinchliff 2003). Hence a correction of erectile dysfunction can have dramatic effects not only on old men's sexual activity, but also self-esteem and self-identity. In some studies, however, some respondents have indicated the use of Viagra© as an "unnatural" mean of achieving erection and penetration (Gott and Hinchliff 2003).

Loss of interest in sex in some old women may be also influenced by *pronatalist* views, or the emphasis on the aim of sexual intercourse being reproduction, thus with menopause and therefore the inability of reproducing, some women may be under a social pressure to become less interested in sexual intercourse (DeLamater 2012).

Although sex has been traditionally seen as more important for the expression of masculinity than femininity, the situation may be changing in current generations of old people. Women are currently more inclined to focus on their own sexual needs, and to abandon their sense of sexual duty towards their husband, leading to a stronger focus on the individual achievement of the beneficial effects of sex, including through masturbation (Gott and Hinchliff 2003). Table 4.12 indicates that masturbating alone is quite frequent in both old men and women even at advanced ages. Masturbating with the help of a partner is less frequent, but still ranging between 12.9 and 27.9% in men older than 50 and between 5.3 and 17.7% in women.

Laumann et al. (2006, p. 145) define *subjective sexual well-being* as "cognitive and emotional evaluation of an individual's sexuality", whereas DeLamater (2012, p. 127) define *sexual functioning* as "one's ability to engage in sexual expression and sexual relationships that are rewarding, and the state of one's physical, mental, and social well-being in relation to his or her sexuality". Despite the decreased level of interest in sex with age and some physical constraints, sexual functioning and well-being can be maintained by using a diversity of approaches to sexual stimulation (Table 4.12).

Similar results indicating that some old people remain sexually active well into their 70s, as it is shown in Table 4.12 which displays data from the USA, were obtained in other countries such as Finland and Australia (DeLamater 2012), with the main factors explaining such extension of sexual life being good health and having a sexual partner.

Maintenance of sexual activity during old age is certainly helped by good mental and physical health, positive attitudes about sex, being in a mutually satisfactory

Table 4.12 Sexual behaviours in the past year by gender and age

	Male (age)			Female (age)		
Variable	50–59[a]	60–69[b]	70+[c]	50–59[d]	60–69[e]	70+[f]
Masturbated alone (%)	72.1	61.2	46.4	54.1	46.5	32.8
Masturbated with partner (%)	27.9	17.0	12.9	17.7	13.1	5.3
Received oral sex from female (%)	48.5	37.5	19.2	0.9	0.6	1.5
Received oral sex from male (%)	8.4	2.6	2.4	34.2	24.8	7.8
Gave oral sex to female (%)	44.1	34.3	24.3	0.9	0.9	1.5
Gave oral sex to male (%)	8.0	2.6	3.0	36.2	23.4	6.8
Vaginal intercourse (%)	57.9	53.5	42.9	51.4	42.2	21.6
Inserted penis into anus (%)	11.3	5.8	1.7	–	–	–
Received penis in anus (%)	4.6	6.0	1.7	5.6	4.0	1.0

Data are taken from the *National Survey of Sexual Health and Behavior* (USA)
From DeLamater (2012)
[a]$n = 454$
[b]$n = 317$
[c]$n = 179$
[d]$n = 435$
[e]$n = 331$
[f]$n = 192$

partnership, and being well informed about the issues of sexuality in old age (DeLamater 2012). In addition, Gott and Hinchliff (2003a, p. 1625) point out that sex is valued by older people as a way of "expressing love for a partner, helping maintain their relationship, as well as for giving them pleasure and improving self-confidence, including for some women, body image".

Laumann et al. (2006) assessed four aspects of *subjective sexual well-being*: physical pleasure, emotional pleasure, satisfaction with sexual function, and importance of sex, in a sample of 13,882 women and 13,618 men (40–80 years old), in 29 countries. Figure 4.10 shows a dendrogram of a cluster analysis for the four aspects of subjective sexual well-being across countries.

Three major regions or clusters emerge from Fig. 4.10: A left cluster (cluster 1) that comprises countries characterised by a "gender-equal sexual regime" and where the levels of sexual satisfaction are high (e.g. Western European countries such as Germany, Sweden, and European-culturally linked Western countries such as USA, Australia). On the other hand, the cluster on the right of the dendrogram includes countries characterised by a "male-centred sexual regime" (cluster 2), where the levels of sexual satisfaction are intermediate (e.g. Islamic and some Asian and European countries). Finally, there is a central cluster (cluster 3) typified by East Asian countries (e.g. Japan, China, Indonesia) that display low levels of sexual satisfaction.

Figure 4.11 compares the levels of satisfaction in men and women across the three clusters and for the four aspects of *subjective sexual well-being*. The first obvious pattern to notice is that across the three clusters and for all aspects of subjective sexual well-being, a greater percentage of men are satisfied than women. Another important conclusion to be drawn from the results is that sexual well-being in both

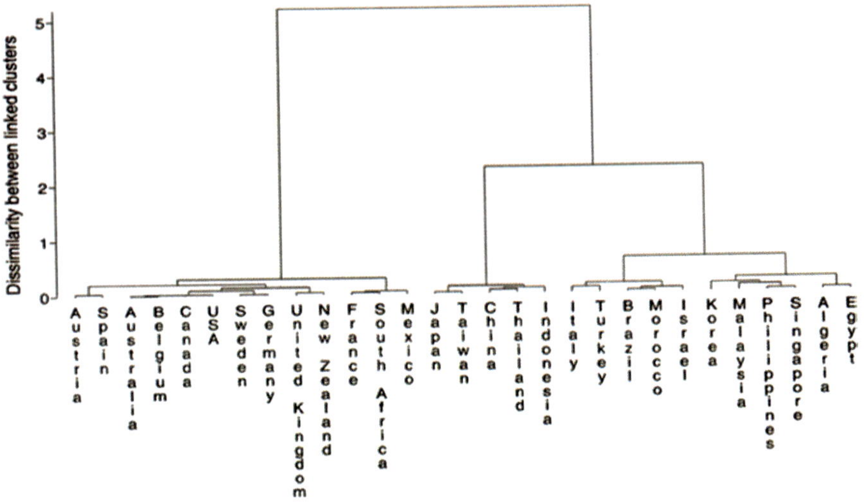

Fig. 4.10 Dendrogram of a cluster analysis associating countries into groups on the ground of positive responses to four aspects of subjective sexual well-being. From Laumann et al. (2006)

Fig. 4.11 Four aspects of subjective sexual well-being by gender and cluster. (**a**) Cluster 1: "gender-equal sexual regime" cluster; (**b**) cluster 2: "male-centred sexual regime" cluster; (**c**) cluster 3: "low levels of sexual satisfaction" cluster. From Laumann et al. (2006)

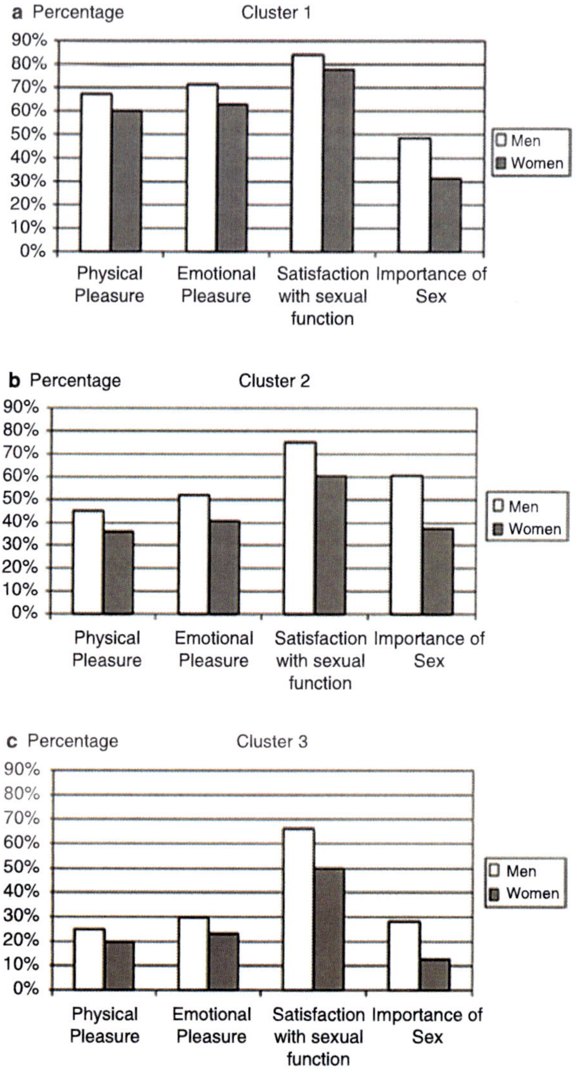

genders and at least for physical pleasure, emotional pleasure, and satisfaction with sexual function decreases from cluster 1 to 2 to 3. That is several aspects of subjective sexual well-being tend to be higher in societies where a "gender-equal sexual regime" prevails and are lower in some Asian countries. Interestingly, importance of sex is relatively higher for both genders in cluster 2, comprised of countries with a "male-centred sexual regime".

Sexual well-being in Laumann et al. (2006) study was also positively associated with physical activity. In particular, men engaging in low levels physical activity also experienced low sexual well-being. The association was weaker in women.

Relational satisfaction was decreased by depression, but satisfaction with sexual functioning and the importance of sex were unaffected. Sexual functioning was also negatively affected by pain during intercourse and lubrication problems in women and difficulties in maintaining an erection among men. Interestingly, in this study non-marital relationships (e.g. cohabitation and dating) were associated with higher subjective sexual well-being than marriage, for men in particular. However, men who were engaged in multiple partnerships were less satisfied with such relationships, but they were more likely to attribute importance to sex.

Laumann et al. (2006) also reported on the relationship between subjective sexual well-being and happiness as assessed by respondents. All four aspects of sexual well-being—physical and emotional satisfaction in particular—were significantly related to happiness.

In sum, interest in sex and engaging in sexual intercourse gradually decline with age in both genders and across cultures, although the rate of decline is variable individually and inter-culturally, and the importance of sex in old age remains higher for longer in men than women. Current availability of medical aids, better physical condition, and challenges to traditional views on sexual intercourse in old age have increased the interest in sex among older individuals in recent times.

In the next section we analyse the issue of ageing with a disability. What kind of challenges do people with disabilities confront as they become older? How can they overcome such challenges?

4.7 Ageing with a Disability

In this section, elderly people with a disability are those who were disabled before becoming old. Therefore, the question is: How does a disabled person who survives to old age experience the ageing process? In our analysis, we will address two major types of disability in turn: physical and intellectual.

In the early studies of social integration of people with disabilities, they were described as victims of prejudice and discrimination, in danger of becoming pariahs within their own community. They were also seen just as an economic liability for society and, even on the more positive side, acceptance did not go beyond a basic level of tolerance, involving limited participation (Hanks and Hanks 1948). The condition of such old people has generally improved but stigma remains, which in extreme cases may go as far as taking the form of *oppression* (Abberley 1987).

A disability therefore goes beyond the objective limitations of an old person as they can be described medically (intra-personal perspective), but it branches out into the realm of social attitudes and prejudice (interactionist perspective). Prejudice and stigma can lead to discrimination, thus calling for the consideration of the specific rights of elders with disabilities (Neufeldt and Mathieson 1995).

People can develop a life-long disability before the age of 60 in various ways. Some may experience a *developmental disability* (e.g. of genetic origin) that impairs their functioning and potentially lowers their quality of life for the rest of their life, often since birth, as compared to "normals". Others may have an *acquired disability* which they developed at some point in their life due to accidents or disease.

The term "normal" is used here *sensu* Goffman (1963, p. 5): "those who do not depart negatively from the particular expectations at issue". Consistent with our concept of fulfilling ageing, "normal" will be used only for comparative purposes, not as an "ideal", or "optimum", that people with a disability should aspire to (Walker and Walker 1998).

People ageing with a developmental disability have been variably referred to as a "little known group" or, more dramatically, the "invisible group" (e.g. Prakash 2003), in spite of individuals with such disabilities being currently more likely to reach old age than in the past due to improved medical treatments and health care practices.

Doka and Lavin (2003, p. 135) describe a paradox of ageing for people with a developmental disability: "As persons with developmental disabilities age, they are likely to experience cognitive and physical deficits that increase their need for services and support. Yet their family-based support systems (who are also ageing) may be less available and social services may be ill suited to assist". Decline in physical and mental capacities in ageing people with a developmental disability tends to be more accelerated than the decline observed in ageing people without such disabilities, thus putting additional stress on a healthcare system that is likely to be already over-stretched. In addition, their network of support may also thin over time more rapidly, as they are less likely to have had children.

Specific developmental disabilities also present with specific and predictable limitations in old age. Down syndrome individuals, for instance, develop cardiac, respiratory, and bowel problems, as they become older (Doka and Lavin 2003). In general, available research suggests that the population of elders with developmental disabilities tend to display cognitive declines that start at earlier ages and that are more extensive than in the population of "normal" elders. Moreover, they have fewer personal abilities to cope with ageing, thus requiring greater institutional support.

Acquired disabilities, on the other hand, include those that result from disease, war, accidents at work, or any other cause that can turn a potentially "normal" developmental path into one that is marked by difficulties in carrying out activities of daily living. Such disabilities may be acquired at any time after birth: in childhood, adolescence, or as an adult; and they may lead to developing specific problems in old age. Old people with a spinal cord injury, for instance, tend to experience osteoporotic changes and higher levels of depression potentially leading to higher rates of suicide, whereas old people with cerebral palsy tend to experience joint dislocations, and those who suffered polio at younger ages show neuromotor changes when they are old. In addition, some problems are common to a diversity of disabilities: increased fatigue and pain, for instance (Kemp and Krause 1999). Complications that develop in old people throughout "normal" ageing are often observed at younger ages in elders with acquired disabilities.

Those challenges notwithstanding life satisfaction may increase in people with acquired disabilities to the extent that they are able to carry out valued activities that give meaning to their life (Kemp and Krause 1999).

Another example of ageing with acquired disabilities is represented by old military veterans who may often experience significant physical and mental health problems in old age resulting from experiences in combat, or from having been prisoner of war, or from other service-related incidents (e.g. Wilmoth and London 2011). The war veteran population is well known to suffer high life-time prevalence of post-traumatic stress disorder (PTSD), the sequelae of which can be carried into old age. Such old former servicemen and women may still require proper professional attention delivered in specialised institutions, or they could receive assistance in their preferred living environment when needed. In the case of the USA, Wilmoth and London (2011) mention the *U.S. Department of Veteran Affairs* giving high priority to veterans with combat-related disabilities for access to various services.

In general, people who enter their old age with a disability, whether it is developmental or acquired, tend to experience a worsening of the disabling condition and its side effects over time (Zola 1989). Trieschmann (1987) proposed two hypotheses to understand how ageing can interact with disability. On the one hand, if the disability is stretching the capacity of the person to the limit, in order to cope with the demands of daily life, old age may just push the stress over the top, as suggested by Zola (1989). That is, the challenges of both disability and old age compound each other to make old age a more negative experience to the ageing disabled person. A second, more optimistic possibility is that a person with a disability may be better able to adjust to the changes coming with ageing, because their coping capacities may be already well developed due to their disability. Which one of those two possible scenarios applies in any specific case is likely to be affected by the personal characteristics of the elder and by how much social support she or he receives.

The rest of the section will be devoted to analysing in more detail the experience of old people with physical disabilities first and then those with mental disabilities.

4.8 Physical Disabilities

Old people with a physical disability are confronted with specific challenges in their life that often call for specialised forms of care, but society may not always be organised in a way that offers them the material and psychological support they need.

The attitude of society towards old people with a physical disability can seriously affect their welfare, starting from their mental health. We already mentioned in Sect. 4.4 that although depression tends to decrease with age in the elderly from the age of 55 to the mid-80s, there is an increase in prevalence in the oldest-old. How does the prevalence of depression change in old age in the case of old people with a physical disability?

In old people, including those who have a disability, just about any negative life event may contribute to develop depression (Kraaij et al. 2002), but as we have seen in the previous section, it is possible for the disability to compound the effects of those negative life events, thus making the development of depression even more likely. In fact, old people with a physical disability have an elevated risk of developing depression, and this is true for both men and women (Turner and Noh 1988). The reason for this increased risk can be found in their difficulties to master the basic needs of life due to both their age and disability, which is made much worse when they lack social support. Decreased self-confidence due to issues with body image can also add to the problem (Taleporos and McCabe 2002), especially in the young-old.

In an early study of depression in old people with a *spinal cord injury* (SCI), Decker and Schulz (1985) suggested that disabled 40 to 73 year old were not at higher risk of depression than non-disabled people. This may result from the effect of compensating factors such as better education, employment, and higher self-efficacy in performing activities of daily living. More recently, however, Charlifue et al. (2010) have reviewed the research on depression in people ageing with SCI, indicating that depression is indeed common in such people, being more prevalent among the older age groups, especially those who have been injured for a longer span of their life. At least among some such elders, SCI can have negative effects on life satisfaction.

Older people who became disabled from paralytic poliomyelitis may also experience depression whenever they see their disability negatively and have issues with poor family functioning (Kemp et al. 1997; Harrison et al. 2011).

Elders with physical disabilities who are depressed may be helped by providing psychological support to them, within a caring social environment. Morgan et al. (1984) stressed the importance of providing such supportive social environment to disabled people, whether the members of the social network are relatives or not.

Apart from the issue of depression that we have just mentioned, specific physical disabilities also confront old people with other challenges. Old people with SCI suffer from alterations in gastrointestinal function (e.g. constipation), haemorrhoids, genitourinary complications such as diminished bladder capacity and increased likelihood of urinary tract infections (Charlifue et al. 2010). Pressure ulcers can also develop in old people with SCI, due to their immobility, along with problems of the musculoskeletal system such as pain in their extremities and osteoporosis, the latter leading to greater risk of lower extremity fracture. Neuropathies have also been reported in this population of elders (Charlifue et al. 2010), and lack of adequate exercise may increase the chances of cardiovascular diseases. Tetraplegic elders, in particular, are also at higher risk of suffering from respiratory complications.

In a long-term longitudinal study carried out in the USA on people with SCI who were 55.7–67.6 years old at a 30-year follow-up point, Krause and Coker (2006) reported changes in sitting tolerance that increased during an initial period to then decrease afterwards, social and sexual life also decreased over time. This went hand in hand with a continuous increase in the number of non-routine visits to the doctor. Hence, one would expect lower levels of life satisfaction in such SCI elders. This was observed by Decker and Schulz (1985) in their study of a group of 40–73 years old people with SCI, but well-being and life satisfaction can improve if such elders feel in control of their lives, have good social support and judge their overall health to be good.

Paralytic poliomyelitis is a disease that affects mainly children and that can leave them incapacitated to move, breathe, and/or swallow normally. Harrison et al. (2011) carried out a study of 25 women (55–75 years old) who were diagnosed with polio at a variable age from 3 months to 13 years old. Participants suffered from various conditions that included hypertension, diabetes, arthritis, sleep apnoea, osteoporosis, scoliosis, muscle atrophy, pain, weakness, fatigue, and high cholesterol among others. Behaviourally, they felt a degree of tension in social interactions due to the contrast between their condition and gendered expectations.

Some disabilities—such as hearing impairment developed at young ages—have the social disadvantage of being associated with the normal decline in sensory acuity with age, thus making the hearing-impaired person automatically old or older than she or he may be. Hindhede (2011) interviewed 41 hearing-impaired people (23–70 years old) in Copenhagen, Denmark. The participants had been hearing-impaired for a period of 5–30 years and they expressed their "distaste about how ageing was brought to bear on their sense of self, and how the hearing aid linked them to the undesirable characteristic of 'being old'" (Hindhede 2011, p. 178). Unfortunately, the stigma associated with being hearing-impaired may be difficult to shake off, as elders with such condition may be treated as "deaf" due to their sensory limitations before they get a hearing aid, and they may be also treated as "deaf" afterwards because of the visible hearing aid. Hindhede (2011) also notes that such stigma is easier to overcome by more socioeconomically privileged people.

Women, who tend to be disadvantaged even when they are healthy, are particularly disadvantaged when they have a physical disability (e.g. Pentland et al. 1999); with old age, they may be confronted with a *triple jeopardy*: being woman, old, and disabled.

Given the challenges they face due to their limitations, elders with physical disabilities have specific needs that should be attended to if quality of life is to reach acceptable levels. Even routine activities of daily living (ADL) can become a considerable challenge to elders with a physical disability. To solve such problems it is often necessary to provide specialised assistance, as the individual capacity of the elder to cope—and also that of family members—is often not enough. For elders with physical disabilities living at home, support should be provided to them there, or if they have a different accommodation, they may be assisted in other residential settings such as retirement villages, collective and sheltered housing, or even hospital units (e.g. Goodall 1988).

Technology can be also put to good use in assisting older people with a physical disability. For instance, specific *assistive technologies* can provide solutions for seating, mobility, and access. Communication can be also facilitated by various technologies, as it can environmental control that improves the conditions of daily living. Help may be also provided by designing the living space in an appropriate manner that can compensate for the limitations of the disability (Oliver et al. 2001). Modifications may include thc building of ramps, grab bars, lifts, and others (Hammel et al. 2002).

We mentioned in the previous section that, on the optimistic side, people with a disability may be more capable of adjusting to the challenges of old age, having already developed some coping capacities given their disability. Are there such cases regarding physical disabilities? Indeed, Rybarczyk et al. (2012) pointed out that some people who age with a disability may show more resilience in the face of life challenges and more capacity to recover their psychological well-being. Coping capacities after such long-term physical disabilities may derive from personal competencies (such as reframing of the condition in a more positive light, optimism, and hope) and social support.

4.9 Intellectual Disabilities

People with intellectual disabilities have to deal with their own specific challenges and needs in old age, especially when the intellectual disability is profound. The most prevalent of those intellectual disabilities derive from Down syndrome, cerebral palsy, and epilepsy, and they are classified as mild, moderate, severe, or profound (Janicki et al. 2002).

Not surprisingly, intellectual disabilities can add additional negative effects to the mental health of old people on top of the mental health issues already expected in old age. For instance, earlier reports indicate that the prevalence of major psychiatric disorders is around 30–40% in people with an intellectual disability who are 40 years old and older, the prevalence being slightly lower (20%) in people older than 65. Minor psychiatric issues have been reported in about 25–50% such elders. This contrasts with a prevalence of 16.9% for major psychiatric disorders and 24% for minor psychiatric symptoms in the general population of old people (Day and Jancar 1994 and the references therein). Behavioural disorders (e.g. drug and alcohol use) are also more common in populations of intellectually disabled elders than "normals", but psychoses are less prevalent.

Younger people with an intellectual disability have higher mortality rates than same-age people who are "normal", but the difference decreases with age (Patja et al. 2001). The percentage of people with an intellectual disability who can reach old age has increased in recent times mainly due to better medical care. For instance, in the late 1920s the life expectancy at birth for children with Down syndrome was 9 years, this increased to 12–15 years by 1947, and then to 18 years in 1961. By the 1980s–1990s their mean life expectancy at birth had reached 50 years (Haveman 2004) and it is currently approaching 58 years (de Graaf et al. 2017). For old people with a mild intellectual disability, they are currently living beyond 70 years of age (Haveman et al. 2011). Reasons for this trend include regular health checks and vaccinations, better overall medical and nursing care, safer environments, a better diet, and a decrease in smoking, drinking alcohol, and use of illegal drugs (Haveman et al. 2011).

Elders with intellectual disabilities also develop several physical problems with ageing, such as various musculoskeletal and sensory issues (e.g. visual, hearing), cardiovascular and respiratory complications, mobility problems, fractures, incontinence, diabetes mellitus, emphysema, obesity, and in some cases epilepsy (Day and Jancar 1994; Evenhuis et al. 2000).

Table 4.13 shows prevalence values of various medical conditions in a population of 40–89 years old people with an intellectual disability in the USA, whereas more recent prevalence values for a Dutch sample of ≥50 years old are available in Table 4.14.

Both Tables 4.13 and 4.14 combined indicate that some of the major health issues for old people with an intellectual disability include cardiovascular and dermatological problems, sensory impairment, greater susceptibility to infectious diseases, osteoporosis, and problems with the digestive system.

More specifically, prevalence of hearing and visual impairments among older people with intellectual disabilities increases with age and it is similar or greater in value than in the general population (Evenhuis 1995; Evenhuis 1995a; Janicki and Dalton 1998). Better medical outcomes could be achieved with improved health care and monitoring, whenever such sensory problems of old age are due to preventable causes that are often diagnosed too late due to the intellectual disability of the elder.

Other health issues affecting old people with intellectual disabilities include hepatitis B, thyroid problems, obesity, and more (McCallion and McCarron 2004; Haveman et al. 2011, see Tables 4.13 and 4.14). On the positive side, prevalence of hypertension seems to be lower in elders with intellectual disabilities than in the broader population (McCallion and McCarron 2004).

In general, the prevalence of dementia is similar among people with intellectual disabilities as compared to the rest of the population, being also variable across studies: from 3.1% to 30.4% depending on the specific age group studied (Shooshtari et al. 2011). However, dementia is more prevalent in the population with Down syndrome, as we already mentioned in Sect. 4.5 of this chapter (see also Day and Jancar 1994; Coppus et al. 2006).

In an English sample of 23 people (65 years old and older) with an intellectual disability (13 mild and nine more severe; one participant could not be scored), 74% had one or more psychiatric symptoms, with one-third having been diagnosed with dementia. Three quarters of the sample had physical health problems: 74% poor sight, 22% hearing loss, and 30% mobility problems (Strydom et al. 2005).

Other psychiatric conditions that can be common among the elderly with an intellectual disability include anxiety and affective disorders, schizophrenia/delusional disorders, behaviour disorders, and autism spectrum disorders (Cooper 1999).

Depression has been also investigated in this group of elders and results indicate a very large variability in prevalence among studies, ranging from 2.2% to 43.9% (Whitaker and Read 2006; Shooshtari et al. 2011 and the references therein).

Table 4.13 Percentage of subjects with an intellectual disability by age group and by gender according to some medical diagnoses

	Age group (years)							
	40–49		50–59		60–69		70–89	
Diagnostic category	F (245[a])	M (312)	F (187)	M (186)	F (88)	M (106)	F (76)	M (77)
Cancer	3	1	5	3	16	6	17	13
Cardiovascular	14	14	17	23	40	28	38	43
Dermatological	53	49	55	52	45	53	51	58
Endocrine	19	16	27	17	33	10	32	18
Gastrointestinal	7	13	12	23	23	17	14	30
Gynaecological	18	–	25	–	26	–	21	–
Hearing impairment	18	13	16	27	34	26	57	52
Haematological	6	5	5	7	11	8	12	10
Infectious diseases	42	29	57	53	72	56	72	79
Musculoskeletal	13	5	14	12	34	9	50	19
Neurological	33	26	32	35	26	32	18	30
Psychiatric	44	51	49	47	36	48	29	27
Respiratory	7	4	9	9	19	14	16	29
Visual impairment	37	37	43	43	64	49	70	65

[a]Sample size. Data are from New York State, USA. From Janicki et al. (2002)

Table 4.14 Occurrence rates of specific conditions in a sample of elders with intellectual disability in The Netherlands

	≥50 years old	
Condition	*n*	Occurrence rate (%)
Dysphagia	929	52.1
Chronic constipation	806	43.3
Osteoporosis	771	42.9
Severe challenging behaviour	1027	32.4
Hearing impairment	879	30.3
Visual impairment	885	24.7
Epilepsy	898	21.8
Peripheral arterial disease	629	20.7
Gastro-esophageal reflux disease	898	19.9
Thyroid dysfunction	812	18.3
Autism	979	18.1
Other cardiovascular diseases	896	17.6
Depression	976	15.2
Anxiety	976	13.6
Asthma/COPD	807	13.6
Motor impairment	989	11.5
Diabetes mellitus I and II	812	10.0
Dementia	910	8.6
Cerebrovascular accident	902	5.8
Cancer	903	2.9

Simplified from Hermans and Evenhuis (2014)

Loneliness and lack of social integration are important contributors to the development of depression in intellectually disabled elders (Day and Jancar 1994).

In a work carried out in Manitoba, Canada, Shooshtari et al. (2011) studied people with intellectual developmental disabilities, that is people with intellectual disabilities who "have had such difficulties from a very early age". The sample comprised 6048 people of various ages (35.8% females) including participants younger than 20 years old to older than 55. Depression was diagnosed in 38.5% of the older participants, being higher than the prevalence observed in the general population of the same age. Thus this result falls within the upper limit of the range of prevalence of depression in the population of old people with an intellectual disability mentioned above.

Dagnan and Sandhu (1999) pointed out that similar factors that promote depression in the general population, such as lack of social support, are also important in the development of depression in people with an intellectual disability. However, people with an intellectual disability may have more difficulties in finding ways to improve their social standing and therefore their self-esteem, especially when they are conscious of stigma and prejudice. Better social support is therefore expected to exert a positive effect on their self-esteem and it is also protective against depression. In their study carried out in the UK, Dagnan and Sandhu (1999) interviewed 43 participants (41.9% women) who presented with mild or moderate intellectual disability, describing how positive self-esteem and some aspects of social comparison (e.g. achievement) are positively associated, and both social comparison and self-esteem are negatively associated with depression, as it is also expected in the general population.

Hermans and Evenhuis (2012) studied depression in 988 people (48.4% women; mean age 61 years old) with intellectual disabilities within the *Healthy Ageing and Intellectual Disabilities* (HA-ID) project based in The Netherlands. Not surprisingly, participants who had experienced more negative life events also displayed more depressive symptoms. The same pattern was observed for anxiety symptoms.

Finally, it is unclear whether personality disorders are different or similar in elders with an intellectual disability as compared with the general population (Whitaker and Read 2006).

Both physical and mental problems may also co-occur in multi-morbidity syndromes in elders with intellectual disabilities. In their work carried out in The Netherlands, Hermans and Evenhuis (2014) showed how multi-morbidity was present in 79.8% of the elders they studied (n = 1047, ≥50 years old, 48.7% women, mainly borderline/mild and moderate intellectual disability (72.5%)). Known intellectual disabilities included Down syndrome, fragile X syndrome, and other syndromes such as Rett and Angelman, and 46.8% of them had four or more conditions that were chronic. Across various studies carried out on Dutch elders with intellectual disabilities (≥55 years old), multi-morbidity (≥2 conditions) ranged between 36.6% and 82%. Multi-morbidity is positively associated with age and severe or profound intellectual disability, in Down syndrome in particular.

Addressing the above problems in the population of elders with intellectual disabilities is essential if they are to experience healthy and fulfilling ageing and a more extended lifespan (Evenhuis et al. 2000). Compared with the general population, people with mild intellectual disability tend not to differ in their life expectancy, whereas in people with a profound intellectual disability more than 20% of the expected lifespan is lost (Patja et al. 2000). Multiple disabilities (e.g. cerebral palsy, epilepsy, severe motor handicap) and Down syndrome also reduce life expectancy (Hogg et al. 2000). Most common causes of death include cardiovascular diseases, respiratory diseases, and neoplasms (Janicki et al. 1999; Patja et al. 2001; see also Haveman 2004 for a review), although full-blown cancer is not more common in this population than in the population at large (e.g. Evenhuis 1997).

Improving the living conditions of elders with intellectual disabilities is necessary but it may be compromised by stereotypes that may be held in society about those elders and by the stereotypes such elders themselves may hold about old age. On the other hand, old people with an intellectual disability may simply refer to their old age in a direct and descriptive manner rather than stereotypically. For instance, in a study carried out in Israel on mildly and moderately intellectually disabled people, Lifshitz (2002) found that 33% of the elderly group ascribed negative characteristics to old people, but they did so descriptively and non-judgementally: "walks with a cane", "has white hair", "lives in an old-age home", "in the cemetery", "dead". When they were asked: "Do you want to be an old person?" all of them also replied in a matter-of-fact fashion: "I'm already old". This, however, did not impede them from seeking the company of younger adults, as 66% of them expressed such wish in their response to the question: "With whom do you prefer to be and to spend time?" The company of younger adults may help them alleviate their fears about old age, which they saw as "threatening and frightening".

Providing better residential and hospital care for intellectually disabled elders requires the coordination of caregivers and agencies, and the use of best practices in both residential care for the elderly and in caring for the disabled (Day 1987; Bigby 2002). Appropriate government departments should develop the capacity to productively and efficiently coordinate the various stakeholders, in the provision of welfare for the disabled elders.

Haveman et al. (2011) interviewed 1235 adults: 16–90 years old, 49% females, from various European countries (Austria, Belgium, Finland, France, Germany, Ireland, Italy, Lithuania, the Netherlands, Norway, Romania, Slovenia, Spain, and the UK). Participants had variable degrees of intellectual disability: mild, moderate, severe, profound, and unknown. Most of them were living either in residential care (47%) or in the family home (34%), but older people tended to live in psychiatric hospitals, nursing homes, and other institutional care facilities. Social isolation increased with age in this group, but older people often engaged in unpaid work to stay active and to keep in touch with others, which presumably made their ageing more fulfilling. However, society at-large should not leave such obvious and often simple solutions to increasing fulfilling ageing in the elderly with a disability, to the vagaries and limitations of individual initiatives. Appropriate services should be made available, that fill the gap in the individual resources, when such gap is evident.

In the end, for both kinds of disabilities: physical and intellectual, social support is essential if prevention of mental and health problems, and if fulfilling ageing are going to be experienced, even though people with a physical disability usually tend to enjoy a better degree of social support than people with an intellectual disability (Lippold and Burns 2009). Social support in the form of community integration (Gulcur et al. 2007) may not only be of help to the elder but also to their family.

To recap, whether they are developmental or acquired, physical or intellectual, disabilities carried into old age simply tend to worsen the physical and psychological conditions that old people may already experience in virtue of their old age. Although individual resources could be put to good use in order to live a more fulfilling life under the circumstances, specific social and government support is often required.

We conclude this chapter with a review of health issues in very old ages.

4.10 Reaching Very Old Ages

Elders who live up to and beyond the age of 100 years are described as having *exceptional longevity* (Santos-Lozano et al. 2016). Such group of elders can be subdivided into three major age categories: the *centenarians* (i.e. 100–104 years old), the *semi-supercentenarians* (i.e. 105–109 years old), and the *supercentenarians* (or *ultracentenarians*, 110 years old and older).

Although centenarians occur in the population with a frequency of only 10–20 individuals per 100,000 in most industrialised countries (Willcox et al. 2006), many more people can live beyond the age of 100 years today than it was the case in past decades and centuries (see Fig. 1.1), as mortality rates among the oldest-old have been declining over time. It is still very unusual, however, for a person to live to 110 years, the probability being only about 2 women in 100,000, and for a man the probability is ten times lower (Rootzén and Zholud 2017).

The estimated *maximum reported age at death* (MRAD) was close to 108 years in the 1860s (with most cases falling in the range of 106–111 years old) but it increased to around 115–116 years in the 1990s (Wilmoth and Robine 2003). Since then, however, further increases have been recorded only sporadically (Vijg and Le Bourg 2017), whether such increases can become more frequent or not in the future is a contentious issue.

The verified oldest person ever recorded is Jeanne Calment, a French woman who died in 1997 at the age of 122 years and 164 days. Although Calment's record has been questioned, recent analyses have confirmed its validity (Robine et al. 2019; Robin-Champigneul 2020). The issue of verification of the date of birth, and therefore the confidence about the elder's correct age, is still casting some doubt on reports of cases of very long longevity in some remote areas, especially those surrounded by myths of being healing/healthy places where people are bound to live a long life (see López-Carr and Ervin 2012, for an example in the region of the Caucasus). Only verified ages should be used in scientific studies. The *International Database on Longevity* (IDL) based at the University of Montpellier, France, only includes properly validated cases (Robine and Vaupel 2002). Considering all validated cases of longevity, over a period of 20 years, the maximum age at death increased by roughly 10 years in the twentieth century: from 112 to 122 years (Robine and Vaupel 2002).

Is there a limit to the maximum human lifespan despite a trend towards increasing life expectancy (i.e. the average number of years an individual is still expected to live at any given age) observed in recent decades? The issue is being currently hotly debated (e.g. Le Bourg 2012) with life expectancy seemingly increasing at a lower rate in the 1960–2000s than it did in the 1880s–1960s (Vallin and Meslé 2009). Although the maximum lifespan is more likely to be limited by intrinsic biological constraints, and therefore future mutations and reshuffling of the human genome through interbreeding among different populations could potentially produce new genotypes and phenotypes associated with a more extended lifespan; the adoption of more healthy ageing lifestyles could also contribute to living a longer life. On the other hand, others have suggested that "there is very, very little heterogeneity within the human population in the 'individual Gompertz slope'—the rate at which a given individual's risk of death increases with age" (de Grey 2015, p. 1), which would make selection of higher longevity weak. This is consistent with the "compression of mortality" view (Fries 1980), where as "the adult modal age at death increases towards higher ages, the standard deviation in adult ages at death should decrease" (Robine and Cubaynes 2017, p. 66). Still, as de Grey noted, the more we study supercentenarians around the world, the more we seem to find cases that break records, although Jeanne Calment's record has remained unbroken for 23 years, at the time we are writing in 2020.

According to Vaupel (1997) and Robine and Vaupel (2002) in developed countries the population of centenarians is increasing at the very rapid rate of about 8% per year on average, especially because mortality after the age of 80 has declined, and the mortality rate (or risk of death per year) of supercentenarians is close to constant at 50% for the age interval between 110 and 115, but it starts to increase thereafter (de Grey 2015). Such longer lifespan is observed in both women and men, although women have lived longer than men throughout this period of expansion of the maximum age (Fig. 4.12).

Better medical treatments (including vaccinations, Caselli et al. 2018), hygiene, housing, diet, technology (here people born in developed countries have an advantage), having healthy habits (not smoking, responsible drinking habits, healthy diet) affect the chances of living a long life. Genetic factors (having some favourable genes, such as *ApoE2*) can also make a difference (Vaupel 1997). In addition, greater conscience about the need to exercise and remain mentally active, a positive view on life and social connectivity can all contribute to increasing longevity (Vaupel 2010; López-Carr and Ervin 2012; Solé-Ribalta and Borge-Holthoefer 2018).

When those various factors act in synergy the positive effect on longevity can be even more significant. This has provided the human population with increasing numbers of centenarians, semi-supercentenarians, and supercentenarians (Fig. 4.13). In other words, such growing number of elders living to and beyond the age of 100 around the world has been, to a great extent, an "artefact of civilisation" (Vaupel 2000).

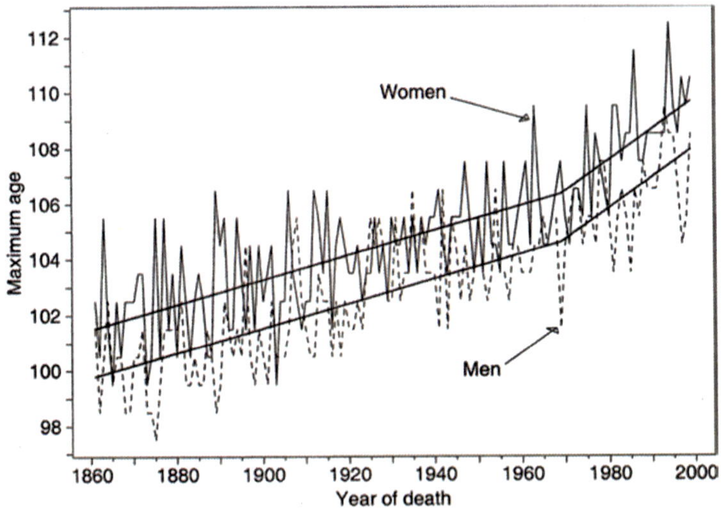

Fig. 4.12 Maximum reported age at death for men and women in Sweden over a period of 140 years. The breakpoint for the slope in the regression lines is in 1969, adding the breaking point maximises the goodness-of-fit. From Wilmoth (2000)

Regarding gender effects, women are more likely to survive to very old ages than men. Centenarians are mostly women, and although the female-to-male ratio among centenarians shows great degree of variation between populations, ranging from 2.5:1 in Sardinia, Italy, to 11.5:1 in South Korea (Arosio et al. 2017 and the references therein); and the ratio of female-to-male supercentenarians is 8:1, with also high variability between countries, from 22:1 in England and Wales to 4:1 in Japan (Robine and Vaupel 2002), women remain on top with regard to exceptional longevity. This means that if they were married, most centenarians and supercentenarians are widows and living alone or institutionalised. They tend to become hospitalised towards the end of their life, and the majority dies alone in hospital or in a retirement institution (Suzman and Riley 1985; Baltes and Smith 2003).

We will also see below that the ability to live to very old ages seems to be relatively less heritable in women than men, although there is variability between studies. Greater heritability of survival to very old age in men may be the result of greater selective pressure on male survival at younger ages, given that fewer of them do survive to very advanced ages (e.g. Smith and Baltes 1998). Men also tend to reach such advanced ages in better physical condition than women, which may be also an effect of selective processes that are in action at younger ages. What are the differences between men and women who survive to very old ages?

After studying participants in the *Berlin Aging Study*, including centenarians, Smith and Baltes (1998) described significant gender differences in social relationships, patterns of everyday activity and reported well-being. Women had a life characterised by more strains and difficulties—including illnesses, mobility issues, and functional impairment—than men, and they were also more likely to be widowed

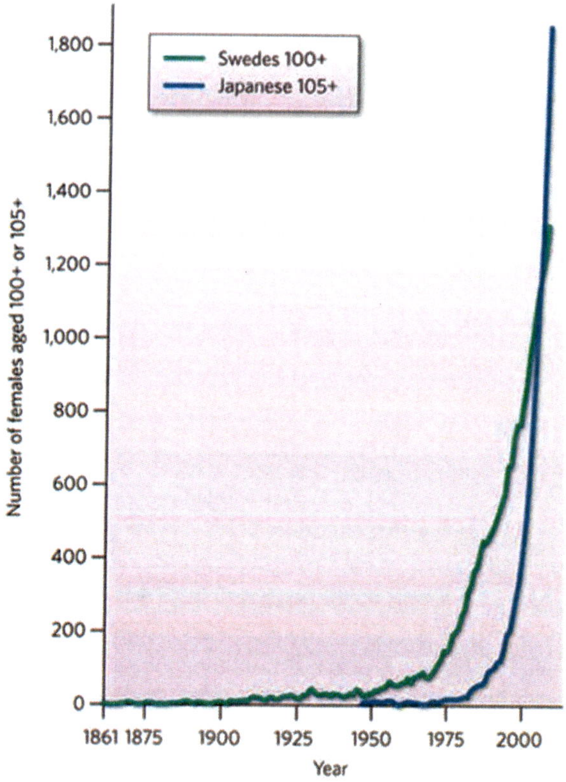

Fig. 4.13 Number of women aged 100+ in Sweden from 1861 to 2008 and those aged 105+ in Japan from 1947 to 2007. The number of centenarians and supercentenarias started to increase, quite rapidly, in the second half of the twentieth century. From Vaupel (2010)

and living alone. Women's social and educational status was lower than men's and their mental health was poorer. Such difference in educational status between men and women among the oldest-old is especially accentuated in China, leading to stronger financial dependence of old widows on their children (Yi et al. 2003).

Franceschi et al. (2000) studied gender differences in centenarians from the island of Sardinia and the province of Mantova in Italy. The two samples differed substantially in their gender ratio (female/male): 2:1 in Sardinia and 7:1 in the Mantova province, compared with 4:1 across Italy. Again, men were found to live to such oldest ages in a better physical and health condition than women.

Despite the various factors that can contribute to higher longevity, the controversy on the limits of human lifespan rages on. Dong et al. (2016) have recently noted the trend towards a decline in the rate of change of the maximum age in the twenty-first century, compared with a steady increase in the twentieth century (from the 1950s in Fig. 4.14). This suggests that the maximum age may be constrained in our species, the limit being around 122 years.

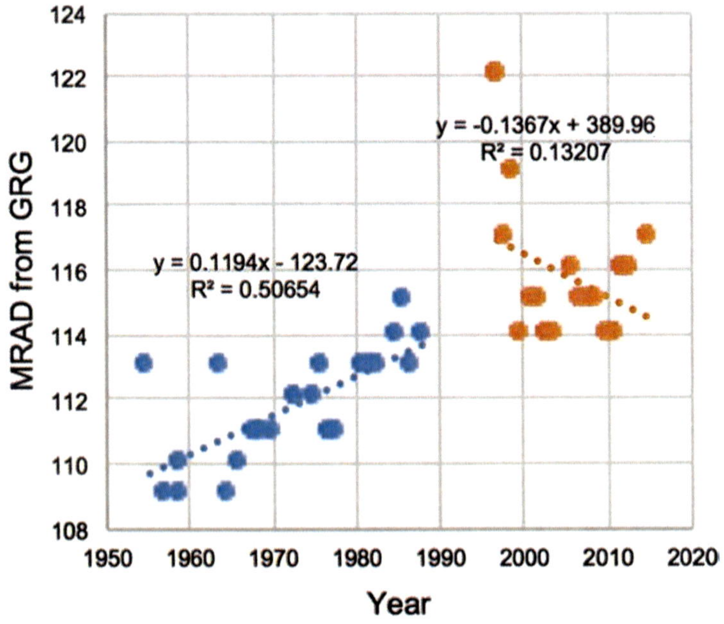

Fig. 4.14 Change of the yearly maximum reported age at death (MRAD) over time. Data are from the *Gerontology Research Group* (GRG; http://www.grg.org/) database. From Dong et al. (2016)

Interestingly, the data in Fig. 4.14 could be seen in two ways: (a) all as equally legitimate datapoints, in which case there would be either a trend to declining maximum age in the twenty-first century, as seen in the red regression line; or, at the very least, a plateau in the maximum age, with the upper boundary represented by Jeanne Calment. Alternatively, (b) one to two points in the late 1990s could be considered as outliers and disregarded, which would produce an overall continuous straight line, hence suggesting an expanding limit for the maximum human lifespan. Such potential ambiguity in Dong et al. (2016) work has been the focus of a recent criticism by Rootzén and Zholud (2017). Clearly, the issue can be only resolved over time as more validated cases of exceptional longevity are studied. A recent work by Gavrilova and Gavrilov (2019), however, supports the notion that human lifespan has indeed a current limit. Figure 4.15 suggests that for most supercentenarians the maximum lifespan has remained well below Jeanne Calment's age record. This has happened despite the number of supercentenarians increasing exponentially since the second half of the nineteenth century (Fig. 4.16). Therefore, it seems that most humans who are past the age of 113 years old fall off some kind of "ageing cliff", presumably determined by our biological limitations (Fig. 4.17).

Some social factors have also been suggested that may contribute to set a current limit to human lifespan. For instance, worsening socio-economic conditions and raising inequality, as they have emerged during the neoliberal phase of socio-economic development around the world (see Chaps. 2 and 3), tend to lead to a

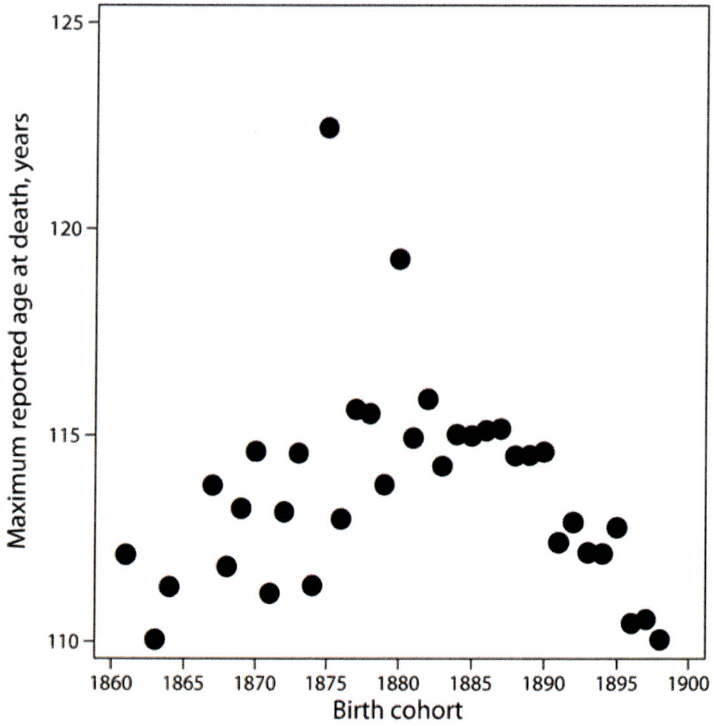

Fig. 4.15 Maximum reported age at death, by year of birth, data are from the International Database on Longevity. Reproduced, with permission, from Gavrilova and Gavrilov (2019)

deterioration of the recorded maximum lifespan (Solé-Ribalta and Borge-Holthoefer 2018). Note from Fig. 4.15 how the descent in the maximum reported age at death starts with the cohort born in the mid-1880s, these were elders who were about 115 years old in the year 2000, from then on most of the world, including areas of traditional high longevity such as Southern Europe and Japan, was shaken by the economic and social effects of the *Global Financial Crisis* of 2008.

Centenarians and supercentenarians themselves have their own views about the factors that can explain their longevity. Freeman et al. (2013) studied transcripts of interviews or videos of 19 English-speaking centenarians and supercentenarians (100–115 years old; 26.3% women) and described four central themes in the explanation of their longevity as identified by the interviewees: (a) lifestyle choices, (b) community and environment, (c) attitude towards life, and (d) goal setting and attainment; with the importance of social interactions and engagement with life being unifying themes. Interestingly, the vast majority of the elders expressed a positive view on life:

> I just kept on plugging away. People say how can you enjoy old age? But I still do. Frankly, I don't feel old at all. I don't drink or smoke and I train every day of the week except

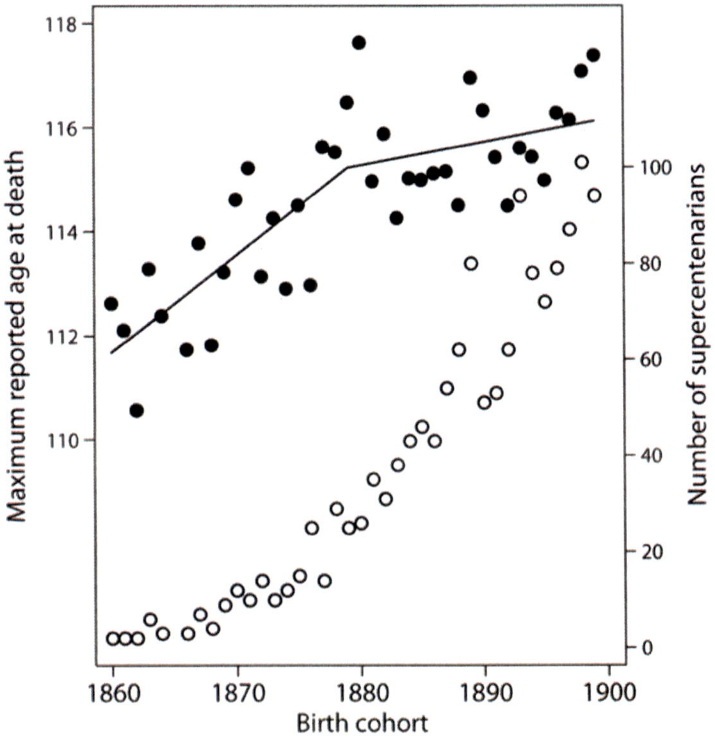

Fig. 4.16 Maximum reported age at death (black circles, left axis) and number of supercentenarians (white circles, right axis), by year of birth. Lines correspond to a linear fit before and after the 1879 birth year. Reproduced, with permission, from Gavrilova and Gavrilov (2019)

> Sunday. The secret to a long, healthy life is simple. I call it GEDS; genes, exercise, diet, and maintaining a good spirit. It works beautifully. [P9, age 100] (p. 724)

But there was an exception:

> When asked at her 106th birthday, how she felt, Participant 1 (P1) bluntly responded: Terrible. Everybody regards you as a freak and I guess in a way you are. [P1, age 106] (p. 721)

In a study of Polish centenarians carried out in Krakow (Mackowicz and Wnek-Gozdek 2017) religion/God were also mentioned by the interviewees as factors explaining their longevity:

> I know no onset. Probably, there isn't any. People pray and yet they die even when they're young. I never thought I could live for such a long time. I can only thank God that he spared me until now. [Anna, 105] (p. 105)

In this case, in Anna's view the mercy of God had granted her a long life so far. Others simply thought that an active life and mind can help:

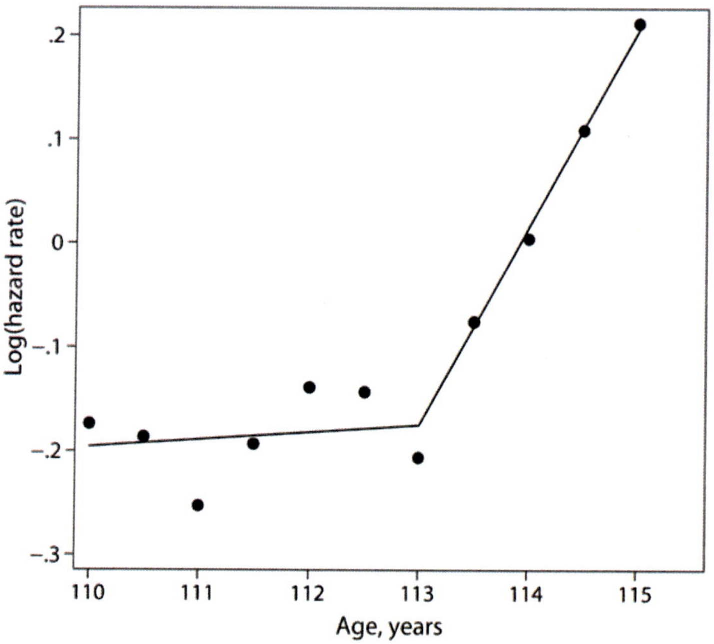

Fig. 4.17 Mortality of the 1885–1898 birth cohort as a function of age. Hazard rates are estimated for semi-annual age units. Lines correspond to a linear fit of the logarithm of hazard rate before and after 113 years of age. Reproduced, with permission, from Gavrilova and Gavrilov (2019)

> No scientist can tell you that, it's in your head, it's about what you do and how active you are. Well, I didn't feel old when I was 90. It was not until 10 years ago when I experienced the first symptoms of old age, this is relatively late, but some of my friends, even those in their 50s were already old and grumpy. [Jan, 100] (p. 106)

> Try to lead a good life or try to be active. [Regina, 103] (p. 106)

Social support and optimism were also mentioned:

> I, for example, live alone, but I never feel alone, because when I look at my neighbors, my friends, they are kind to me, I rather focus on the good things. [Klara, 102] (p. 108)

In this Polish work social relations, faith, and an optimistic attitude in life are the distinguishing features in the discourse of centenarians.

In spite of any expression of optimism, at some point people living to very old ages will experience a variable degree of deterioration of their physical and mental capacities. Baltes and Smith (2003) point to increased multi-morbidity affecting vision, hearing, strength, functional capacity, cognition, and more; systemic breakdown (physical and psychological), losses in the positive side of life (happiness, social contacts, independence, willingness to live). However, many centenarians and older people only seem to decline in their health condition in the last 3–5 years of their lives, previously having been able to function independently, thus many appear to delay disability until very old ages (Perls 2005).

Perls (2005) studied the age at onset of ten age-associated diseases: hypertension, heart disease, diabetes, stroke, non-skin and skin cancer, osteoporosis, thyroid condition, Parkinson's disease, chronic obstructive pulmonary disease, and cataracts, in 424 centenarians (323 men and 101 women) within the *New England Centenarian Study* (NECS) in the USA. Three morbidity profiles were described among the centenarians: *survivors*, *delayers,* and *escapers*. The survivors were diagnosed with an age-related illness before they turned 80 and they accounted for 24% of men and 43% of women in the sample. In the delayers the onset of age-related diseases was delayed to at least when they were 80, with this group accounting for 44% of men and 42% of women. Finally, the escapers reached the age of 100 without being diagnosed with an age-related disease and they accounted for 32% of men and 15% of women. It can be seen from those results that some centenarians are able to survive the health issues of old age, even after they are diagnosed with some serious life-threatening disease. However, only a small proportion of them are escapers and therefore fully capable of avoiding age-related morbidity. Note also that most escapers are men, thus supporting the idea that although women live longer than men, men reach advanced ages in better physical conditions than women.

Although centenarians are globally uncommon, there has been a recent increase in the population of centenarians and supercentenarians around the world that has been detected in many developed countries using validated cases: Denmark, France, Japan, Sweden, Switzerland (Robine and Cubaynes 2017); Belgium, UK, Canada, Spain, Germany, Italy, USA (Gampe 2010); Australia, Finland, Guadeloupe, The Netherlands, Portugal (Young et al. 2008); Poland (Mackowicz and Wnek-Gozdek 2017). A more complete list, including countries with only one record of validated supercentenarians, can be found in Santos-Lozano et al. (2015). In all countries with a sample size of $n \geq 10$, the vast majority of supercentenarians are women (80–95.4%).

Some specific areas of the world have attracted the attention of researchers for their unusual high density of supercentenarians. These include Ogliastra in the island of Sardinia (Italy), Okinawa in Japan, the Nicoya peninsula in Costa Rica, and the island of Ikaria in Greece (Poulain et al. 2013). Those areas have been called *Blue Zones*, a term that was initially introduced by Dan Buettner (2005) in a *National Geographic* magazine article: *The Secrets of a Long Life*. Some commonalities of *Blue Zones* include geographical isolation (e.g. islands and mountainous regions) which, among other factors, may affect the genetics and epidemiology of the population; maintenance of a traditional lifestyle involving intense physical activity extending beyond the age of 80; reduced stress; strong family and community support; and consumption of locally produced food (Poulain et al. 2013).

The population of supercentenarians in Okinawa in particular has been the subject of detailed scientific studies. In Okinawa the frequency of supercentenarians can be as high as 40–50/100,000 people (Willcox et al. 2006). Willcox et al. (2009, p. 507) quote the words appearing on a welcome stone marker in the village of

Ogimi that read: "At seventy you are but a child, at eighty you are merely a youth, and at ninety if the ancestors invite you into heaven, ask them to wait until you are one hundred… and then you might consider it". The wish seems to have been granted to nearly 1000 residents of Okinawa, who are at least 100 years old (Willcox et al. 2017). The renowned Japanese *ukiyo-e* painter and printmaker Katsushika Hokusai (1760–1849) had a similar view of the life course, extending beyond 100 years, with an added emphasis that is consistent with fulfilling ageing:

> I have been in love with painting ever since I became conscious of it at the age of six. I drew some pictures I thought fairly good when I was fifty, but really nothing I did before the age of seventy was of any value at all. At seventy-three I have at last caught every aspect of nature–birds, fish, animals, insects, trees, grasses, all. When I am eighty I shall have developed still further and I will really master the secrets of art at ninety. When I reach a hundred my work will be truly sublime and my final goal will be attained around the age of one hundred and ten, when every line and dot I draw will be imbued with life. (quoted in Miller 1957, p. i)

Willcox et al. (2009) characterise Okinawan centenarians as "healthy-agers", with slow decline in their functional capacities over time. In fact, about one-third were still functionally independent, one-third needed some major assistance with their activities of daily living, and the other one-third were very ill and disabled. Some behavioural factors that could contribute to such longevity in some natives of Okinawa include dietary preferences: caloric restriction, low salt in food, consumption of antioxidants with food, and low stress (Willcox et al. 2009).

The validity of the birth records and therefore the exact age status of Okinawan centenarians have been recently questioned by Poulain (2011) but see Willcox et al. (2008) for a validation.

The capacity to survive to very old ages requires a degree of good health that can be broadly defined as "a state of adequate physical and mental independence" in activities of daily living (ADL), which also implies an ability to be "free from serious disease" (Luyten et al. 2016, p. 772). Such good health can be achieved through "good luck" (good genes, chance avoidance of negative environmental effects such as life-threatening diseases, and so forth) or *adaptive behaviour* (adopting a healthy lifestyle: good alimentation, exercise, stress avoidance; accessing good quality health care, nurturing supportive social bonds among others; e.g. Montesanto et al. 2017). In real life, the capacity to reach very old age in good health is more likely than not a result of a combination of both "good luck" and lifetime adaptive behaviours. Adaptive behaviours can be encapsulated in the concept of *health capital* or the knowledge and skills an individual has acquired throughout life that promote good health (e.g. Bergland and Slettebø 2015).

The process of ageing comes with a degree of deterioration of our health condition and centenarians certainly do not get any better with time (e.g. Martin et al. 1996). No matter how healthy and functional the very old look like, we always add in our description—even for the most impressive cases of good health—the qualificative "for his/her age". von Heideken Wågert et al. (2006), for instance, described how the majority of the 85–95+ years old participants in the *Umeå 85+ Study* rated their general health and morale as good, but still displayed a very high prevalence of hypertension, hip fractures, and depression, with women having a higher rate of medical diagnoses and symptoms than men.

Still, some elders can achieve exceptional longevity in reasonably good physical and mental conditions. Almost half of the old participants in Grundy and Bowling's (1999) study obtained good scores on quality of life. Cevenini et al. (2014) showed that the probability of survival of a cohort of 90+ years old Italians increased in those with good self-reported health and an optimistic attitude (Cevenini et al. 2014). Self-reported health is positively associated with supportive living arrangements (i.e. those living with family report better health), better levels of education and occupational history, but it is poorer in individuals with lower social status (Liu and Zhang 2004). Self-reported good health is also negatively associated with depression (Lau et al. 2016).

In a study of 489 centenarians carried out in Greece (77% women, the sample included 2.4% supercentenarians), Stathakos et al. (2005) found that 98% of them lived at home, mainly with their children (86%), and most of them had worked in a farm for most of their life (75% of women and 66% of men). Interestingly, many of them (55%) "mentioned that they have experienced special hardships at some point of their lives, extending from poverty and starvation (mostly during war) to participation in battles, captivity, or exile" (p. 514), suggesting that they did not escape serious stresses in their life by sheer luck. Moreover, many of them had also suffered serious diseases in the past and survived: cardiovascular, infectious (malaria, tuberculosis, Spanish flu, diphtheria), and cancer; but others had been healthier throughout their life. Health issues that they were experiencing included cardiovascular disorders (71%), osteoporosis (48%), urinary incontinence (47%), and faecal incontinence (23%). Women tended to be in poorer health condition than men, and women were also more often diagnosed with dementia that affected 19% of participants overall. As one would expect, their diet had been traditional Mediterranean, based on a daily consumption of olive oil, dairy products, and vegetables; poultry, fish, and legumes two or three times a week and they only consumed red meat very occasionally, if ever. A total of 25% could carry out their activities of daily living autonomously, and 23% led an active social life. Therefore, it seems that this population of centenarians included at least two broad kinds of individuals: those who reached a very old age despite having suffered health problems in the past (survivors), and those who had been healthy throughout their entire life (escapers).

The above evidence suggests that a surprisingly large proportion of centenarians and supercentenarians are still in good physical condition and can carry out their activities of daily living with only little assistance and sometimes no assistance at all (Schoenhofen et al. 2006), although declining muscular strength occurs at advanced ages which is correlated with increased morbidity and eventually mortality (Ling et al. 2010). Other centenarians and supercentenarians may confront more challenges but have been able to live with recurrent illnesses until an advanced old age, as we have just seen in the case of Stathakos et al.'s (2005) study. In general, most of them experience morbidity and rapid physical decline mainly towards the latest stages of their life (Andersen et al. 2012).

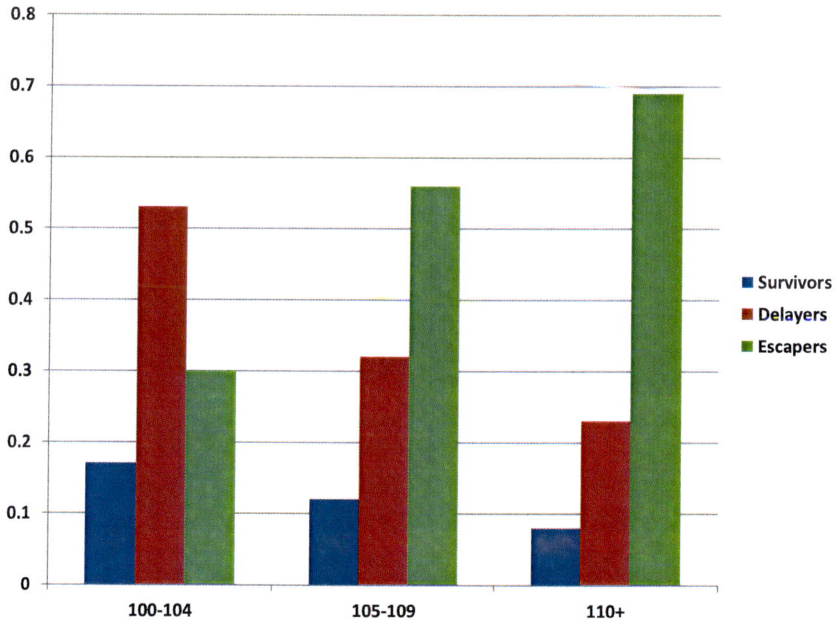

Fig. 4.18 Relative frequency of survivors (onset of at least one disease prior to 80 years old), delayers (onset of at least one disease between the ages of 80 and 99), and escapers (onset of at least one disease after the age of 100) among three age groups: centenarians, semi-supercentenarians, and supercentenarians. Redrawn from Andersen et al. (2012) with minor modifications

We mentioned the work that Perls (2005) carried out within the context of the NECS, where three major types of centenarians were described according to their morbidity profiles: *survivors*, *delayers,* and *escapers*. Andersen et al.'s (2012) work was part of the same NECS project. In this case they included 104 supercentenarians (87% women), 430 semi-supercentenarians (80% women), 884 centenarians (75% women), and 343 nonagenarians (87–99, 62% women). There was also a younger reference group of 436 subjects aged 47–86 years old. The authors observed that the relative frequency of both survivors and delayers decreased with age from 100 years old to 110+, whereas the frequency of escapers increased during the same interval (Fig. 4.18). That is escapers seem to be better endowed with the capacity to survive to very old ages, whereas the other groups face more difficulties to survive to the same very old ages. Reasons for this can be multiple and not mutually exclusive. Past diseases may take a toll that has a cumulative effect on survival over time, added perhaps to specific biological (e.g. genetic) and/or environmental differences that protect the escapers, thus allowing them to delay to very old age the usual diseases observed in the old: cancer, cardiovascular disease, chronic obstructive pulmonary disease, dementia, diabetes, and stroke.

A shift in the onset of disease to older ages was observed in Andersen et al. (2012) work across the different categories of old people. Whereas the median age of onset of morbidity in the younger reference group was 71 years, and in the nonagenarians it was 90 years; in centenarians it was 95 years, 100 years in semi-supercentenarians, and 109 years in supercentenarians. Morbidity was also compressed more in older-surviving participants. In fact, 10% of supercentenarians were disease free until the last 3 months of their lives, whereas only 4% of semi-supercentenarians were so, and 3% of centenarians.

In a survey of 8805 elders aged 80–105 carried out in China (*Chinese Healthy Longevity Survey*), it was found that rural oldest-old are significantly more engaged in their activities of daily living than urban oldest-old. Suggesting that the higher levels of physical exercise rural people tend to experience throughout their lives may help them to live a healthier old age (Yi and Vaupel 2002). In addition, the physical environment in rural areas may be less polluted than in the cities and therefore better suited to protect health (Yi et al. 2001). Yet, the gender difference in functional capacity is also retained in this sample: women are more disadvantaged in functional capacities and self-reported health as compared to men. Although functional capacity decreases with age, the proportion of elders satisfied with their current life is almost constant between the ages of 80 years old and 94, slightly declining afterwards (Yi and Vaupel 2002).

In a study of centenarians carried out in Turkey, Tufan et al. (2017) described high levels of dependency of the elders in their activities of daily living, which may have contributed to a majority of them (69%) having lost the will to live. Quality of life is also lower if the elders are in pain; and when pain is compounded by other constraints such as financial ones, quality of life may decrease even further. This has been shown to be the case especially for women in a study of Swedish oldest-old (Jakobsson et al. 2004).

Achieving better health in centenarians can be also aided by the living environment. There is an emotional advantage for older people to live in their home environment which may help in their capacity to be independent and to increase their health and well-being (Oswald et al. 2007). But older people residing in nursing homes can be also in the position of maintaining a good quality of life if conditions of care are adequate, in spite of their potential worse health compared with elders living in the community, the difference in health being usually the reason why they moved to a nursing home in the first place (Urciuoli et al. 1998). This suggests that older people retain an ability to adapt to a new, but caring and safe environment, presumably as a result of their tendency to simplify expectations and objectives in life. On the other hand, Urciuoli et al. (1998) also note that complaints may be lower in dependent elders for fear of reprisal and withdrawal of help if they give the impression of being ungrateful.

Panagiotakos et al. (2011) carried out a study of quality of life and healthy living among the oldest-old in one of the *Blue Zones* we mentioned in this section: the Greek island of Ikaria, describing the majority of participants as physically active in their daily life, having healthy eating habits (i.e. a vegetables-based diet), non-smoking, enjoying a rich social life and also mid-day naps, and they also displayed low rates of depression.

Therefore some old people are capable of living into advanced old ages in a reasonable health condition, and for those in this category who survive to the age of 90 the rate of deterioration afterwards slows down (e.g. Christensen et al. 2008). That is, as suggested by Luyten et al. (2016), some old people have a longer *healthspan*. Having a long healthspan is not the same as having a long lifespan. Some elders who suffer deterioration of their health may still live a long life, but their quality of life may not be so good. This can affect fulfilling ageing, as the will to live may decrease when life becomes unbearable.

We mentioned above that the capacity to live to 100 years and beyond can be affected by our genetic make-up. In which way? What are the genes involved and what do they exactly do that promotes a very long life?

Those aspects of ageing that are more directly dependent on the activity of genes are collectively known as *primary ageing*, whereas those aspects that involve the deterioration of biological functions due to "wear and tear" are referred to as *secondary ageing* (Santos-Lozano et al. 2016). Primary ageing is therefore more programmed, whereas secondary ageing can be better controlled by the individual through lifestyle choices, social interactions, and so forth. Primary and secondary factors of ageing can interact, and they are studied within the field of *epigenomics of ageing* (Santos-Lozano et al. 2016). In fact, to understand the biology of ageing a focus on genotype–epigenetic–phenotype relationships is required (Govindaraju et al. 2015).

The genetic component of human longevity is not insignificant, as we already mentioned in Chap. 1, with very many loci being implicated in the biology of ageing that are part of the nuclear and mitochondrial genomes, the number is estimated to be over 750 (Govindaraju et al. 2015). Heritability estimates for longevity range between 0.20 and 0.30 (Santos-Lozano et al. 2016 and the references therein), increasing after the age of 65 to 0.36, and to 0.40 after 85 years old, meaning that the ability of living to very old ages has a stronger genetic basis in centenarians, semi-supercentenarians, and supercentenarians (Cho and Suh 2014).

Although women are, on average, more likely to live to 100 years and beyond (see Fig. 4.12), heritability of living to not less than 100 years tends to be slightly higher in men than women: 0.33 in women and 0.48 in men (Sebastiani and Perls 2012), in some studies. In Okinawa female siblings of centenarians have a 2.58-fold likelihood of reaching the age of 90 years, whereas male siblings have a 5.43-fold likelihood, compared to their birth cohorts, thus suggesting a strong familial effect on longevity that is more favourable to males (Willcox et al. 2006). Korpelainen (2000) also showed greater heritability of longevity in sons than daughters. These results notwithstanding the gender bias in heritability of longevity are variable among studies, with some authors indicating greater heritability of longevity in old women than old men (e.g. Gavrilova et al. 1998; Cournil et al. 2000). The sources of such variability and their specific effects are still to be fully investigated (van den Berg et al. 2017). For instance, family effects could be a result of genetic factors, environmental factors, or a combination of both. As we have already mentioned, in Okinawa, for instance, a low-calorie and fish-vegetables based traditional diet could also contribute to longevity (Willcox et al. 2006, 2017), perhaps affecting men and women differently. In Korpelainen's (2000) Finnish study, heritability of longevity was higher in a sample of aristocrats—that showed a gender effect—than a rural sample—that did not show a gender effect—possibly pointing to environmental factors affecting longevity.

Various loci have been identified that can influence exceptional longevity (Brooks-Wilson 2013). So far, candidate genes include those that are involved in regulatory pathways such as genome maintenance, insulin signalling, stress resistance, lipid metabolism, immunoinflammatory response, cognitive pathways, and cardiovascular function, among others (Willcox et al. 2006; Govindaraju et al. 2015; Giuliani et al. 2017; Montesanto et al. 2017). The gene encoding for apolipoprotein E (*ApoE*) that is located on chromosome 19 is among them. We have seen in Sect. 4.5 of this chapter that apolipoprotein E is a major cholesterol carrier that promotes lipid transport and repair of brain injuries, along with being involved in protein metabolism and brain plasticity. In this case the allele *ApoE2* is positively associated with longevity, whereas *ApoE4* is negatively associated with longevity. *ApoE4* is also a risk factor for Alzheimer's disease. Genes encoding for polypeptides that have a function in the insulin/insulin-like growth factor system (IIS) are also implicated in human longevity. *FOXO3A* (also known as *FKHR1*) in particular has a regulatory function in the IIS, being located on chromosome 6. The protein encoded by *FOXO3A* is involved in various cell processes such as proliferation, apoptosis, and longevity, contributing to healthy human ageing and exceptional longevity, through regulating the IIS. *FOXO1A*, however, is either not associated with exceptional longevity or it is negatively correlated to it. The *TP53* gene that is found on chromosome 17 has also a positive association with longevity through controlling the stress response and apoptosis.

Given that human longevity is associated with an appropriate working of the immune system, including inflammatory responses, it may come to no surprise that genes encoding for anti-inflammatory cytokines may be implicated in exceptional longevity. The interleukin-10 gene (*IL-10*) found on chromosome 1 encodes for interleukin-10 which limits and also terminates inflammatory responses through inhibiting the action of T cells, macrophages, and monocytes. The gene is polymorphic and some alleles are associated with exceptional longevity. Low-level chronic inflammation has been associated with longevity in centenarians and semi-supercentenarians in a study carried out in Japan (Arai et al. 2015). The ability to achieve exceptional old ages has been also associated with loci on chromosome 4 (Perls et al. 2002).

A negative relationship with exceptional longevity has been described not just for *ApoE4* and potentially for *FOXO1A*, but also for the case of the *PON1* locus found on chromosome 7 that encodes for Paraoxonase 1, an antioxidant enzyme that is associated with high-density lipoproteins. The negative relationship was described in women (Christiansen et al. 2004). The gene that encodes for the glutathione peroxidase 1 (*GPX1*) enzyme is found on chromosome 3 and it has been also associated with reduced longevity, coronary disease, myocardial infarction before the age of 50 and also premature death.

Accumulation of mitochondrial DNA (mtDNA) mutations with age is an additional factor contributing to the broader process of ageing, therefore mechanisms protecting mitochondria from oxidative damage would favour a longer life (e.g. Costa et al. 2009). Protection of telomeres is also important. We have seen in Chap. 1 that telomeres are repetitive sequences of DNA found at the end of eukaryotic chromosomes that serve the function of stabilising the chromosome. Loss of telomere length may contribute to ageing pathologies and telomere length is better maintained in centenarians and semi-supercentenarians, such capacity can be transmitted to offspring through the paternal line (Arai et al. 2015).

Santos-Lozano et al. (2016) provide a comprehensive review of the genetic aspects of exceptional longevity. In a statement of an international panel of experts who met in Athens (Greece) to consider the relevance of genetic and non-genetic markers of ageing, longevity, and exceptional longevity, Avery et al. (2014, p. 660) pointed to the complexity of genetic and epigenetic mechanisms likely to be involved in the processes of ageing and they recommended, as one would expect, that more research to be done in this field. More precisely, they stated that:

> There is evidence identifying some genes related to longevity and ageing. Such genes are included in a variety of signalling pathways, i.e. insulin/insulin-like growth factor (IGF-1), nutrient-sensing (mTOR), oxidative stress and anti-oxidants, control of immune-inflammatory responses and lipid metabolism as well as in mitochondrial DNA (mtDNA). However, more evidence is needed. In addition, it is becoming clear that epigenetic changes linked to diet or to other environmental/life style factors (physical activity, emotional stress) play a role in longevity attainment.

As mentioned by Avery et al. (2014), epigenetic factors can be also at play in exceptional longevity. These are hereditary changes that can modify gene expression without changing the DNA sequence (e.g. DNA methylation, miRNAs, siRNA, the latter two are non-coding RNAs which have important gene regulation roles). The DNA of centenarians, for instance, is less methylated than the DNA of younger individuals (Govindaraju et al. 2015). The level of DNA methylation has provided a mechanism to estimate chronological age of cells, including cases where ageing is accelerated due to biological processes such as obesity, infections, and Down syndrome, this is known as the *epigenetic clock* (Lowe et al. 2016) and the age estimated by the epigenetic clock is known as "DNA methylation age" (DNAm age) (Horvath et al. 2015). The DNAm age of semi-supercentenarians is lower than their chronological age (Horvath et al. 2015).

Although genetic and epigenetic mechanisms may contribute to exceptional longevity, most people would be concerned not just with living longer but also reaching advanced old ages in good physical and mental condition. Centenarians and older people can experience impairments in their ability to perform activities of daily living, with physical capability continuing to decline as the individual becomes older (Andersen-Ranberg et al. 1999; Parker et al. 2005). This is made worse by an increasing perception of pain with age (Zarit et al. 2004), especially in women (Życzkowska et al. 2007). Social support, technological/medical aids, the maintenance of a good mental condition, and remaining physically active can slow down the physical deterioration of advanced age, but they cannot stop it (e.g. Femia et al. 2001).

Over time, some centenarians and older individuals may also experience mental health problems such as depression (Stek et al. 2006; Spira et al. 2012) which is affected by how the elders perceive their health, degree of loneliness, impairment of mobility, living in an institution, use of medications, overall well-being (Haynie et al. 2001; Stek et al. 2004; Bergdahl et al. 2005), and cognitive decline (Green et al. 2000), the latter potentially developing into dementia in some individuals (Kawas and Corrada 2006; Corrada et al. 2010; Jellinger and Attems 2010; Spira et al. 2012). Dementia also comes with multiple pathologies (James et al. 2012; Kawas et al. 2015), whereas delirium is regarded as a risk factor for dementia in this group of elders (Davis et al. 2012).

Such mental health problems may contribute to decrease the will to live (Erlangsen et al. 2004) and hasten death (Bergdahl et al. 2005).

In spite of the difficulties and challenges inherent to very old ages, many centenarians, semi-supercentenarians, and supercentenarians do have the capacity to show resilience and coping and to maintain a degree of subjective well-being and a sense of fulfilling ageing, at least to some extent (Baltes 1998; Blazer 2000; Smith et al. 2002; Enkvist et al. 2012; Nosraty et al. 2015), especially if they enjoy supportive social relationships (e.g. von Faber et al. 2001; Jopp et al. 2016).

The current view is that the capacity to cope reaches a limit over time, but there are cases where resilience is extended to almost the very end of life. Jopp and Rott (2006), for instance, analysed data from the *Heidelberg Centenarian Study* indicating high levels of happiness in centenarians, which was mainly explained by factors such as job training, a good cognitive condition, a good health and social network, and an extraverted personality. Self-efficacy and an optimistic attitude in life added to the factors that promoted happiness. Optimism among centenarians, regarding their perceived level of health, for instance, can be even maintained in spite of objective evidence that health is, in fact, deteriorating. This contradiction between an optimistically self-perceived health and a declining objective health status is known as the *Paradox of Ageing* (Wettstein et al. 2016). As suggested by Arosio et al. (2017), the paradox of ageing can be regarded as an adaptive response to deteriorating health, implicating a downward adjustment of what the individual perceives as "good health" in advanced age.

Johnson and Barer (1993) noted that coping can be aided in very old age by simplifying life and the living environment down to just those aspects that are both necessary for daily living requirements and personal fulfilment. Redefining what is "optimal" in terms of health can also help, as it will make life objectives more realistic and easier to achieve, thus leading to greater satisfaction. The coherence and meaningfulness of life can be maintained until a very advanced age to the extent that inevitable changes are accepted and incorporated, routines allow better control of the daily living, and social networks can be stabilised—in spite of the loss of close friends—through family and other sources of social support.

Cho et al. (2015) identified several proximal and also more indirect or distal factors that can influence fulfilling ageing in both octogenarians and centenarians, after studying participants in the USA *Georgia Centenarian Study*. Among the proximal factors considered by Cho et al. (2015) were current individual, social, and economic resources, whereas the distal factors included past personal experiences accumulated throughout life (e.g. education) and historical events. Taking positive affect as a measure of fulfilling ageing, Cho et al. described a complex network of factors that affect positive affect in a direct, indirect, positive, or negative manner. Positive affect directly decreases with an increased level of physical-health impairment, whereas health impairment is in turn decreased by better cognitive function, with the latter relying on better education and life experience. Positive affect is also directly affected by social resources, with more social resources leading to higher values of positive affect, and it is indirectly affected by the effect of cognitive functioning on social resources (i.e. better cognitive functioning leads to better social resources). Therefore, this model stresses that fulfilling ageing is more likely to be experienced by centenarians who are in better physical and mental condition, who enjoy better education, and who receive more social support.

The same conclusions were reached in a study carried out in Hong Kong by Chou and Chi (2002): better ageing was experienced in the oldest-old by having a better functional, cognitive, and affective condition, and also a more productive involvement in society, whereas it was constrained by factors such as age, financial strain, and sensory impairment (e.g. hearing), old women were also less likely to age well than men.

Although social engagement will necessarily start to wind down with advanced age and it may be reduced to a trickle beyond the age of 100 (Bowling and Browne 1991; Cherry et al. 2013), social support is one of the key factors that can help a centenarian experience fulfilling ageing. For centenarians and older people, social support remains central to their well-being until the very end of their life (e.g. Li et al. 2009), both emotionally and also in terms of instrumental help (Bowling and Browne 1991; Bondevik and Skogstad 1998). Social support and active engagement in society can have positive effects on the mental health of the oldest-old (Cherry et al. 2013) by decreasing loneliness (Valora Long and Martin 2000), stimulating cognition and maintaining functional ability (Margrett et al. 2011). Both the need for and capacity to attract social support are also dependent on personality factors (Margrett et al. 2011) and the physical and emotional ability of elders to maintain social relationships, which decline in advanced ages (Johnson and Barer 1992). Some elders are more dependent on social support than others, and some are less capable of attracting and retaining social support than others, even when they are in need. When support from family and friends is missing, some form of institutional social support may replace it.

Kahn and Antonucci (1980) stressed the importance of social relationships in the elders in their *social convoy model*. In this view, social support is an important determinant of well-being in old people. The convoy is just the personal social space of family, friends, and other people, providing the network of relationships through which social support is given and also received. Social support may promote well-being not only in the elders receiving help but also in those who provide help, through the feelings of satisfaction that may be experienced when supporting others. Through social support the effects of life stresses could be diminished and less stress can positively affect mental health. But social support also comes with a cost to family and friends, as the needs of a centenarian tend to increase over time, whereas older members of the social network confront their own constraints as they age (Jopp et al. 2016). In the end, centenarians are better helped when the helpers themselves are not left alone.

Johnson and Barer (1992, p. 362) identified four major patterns of adjustment among the oldest-old in their social interactions:

> First, with physical disability and depleted social networks, it is common for individuals to redefine their optimal level of social integration and report contentment with a narrower, more constricted social world. Second, they are likely to reject those norms which place expectations on them which are incongruent with their capacities. Third, respondents tend to redefine their time orientation from the future to the present and become content to live day to day. Fourth, as they detach themselves from social involvement, increasing introspection occurs.

The patterns identified by Johnson and Barer suggest that at advanced ages individuals minimise the waste of energy and the potential stress involved in maintaining social interactions with many individuals whilst redefining what makes them more content in terms of their relationships with others. This, however, does not mean that such elders do not require help from others, it just means that they will be unable to reciprocate the help received, thus restricting their interactions to fewer people and to those who provide better help. This can be seen in the following example from Chen and Short (2008).

Chen and Short (2008) analysed data from the *Chinese Longitudinal Healthy Longevity Survey* to understand the effects that living arrangements may have on oldest-old subjective well-being. Subjective well-being was lower in elders living alone and it was better in those living in co-residence with immediate family (a spouse or children). Emotional health, however, was better when the elder was living with a daughter than with a son, despite a cultural tendency towards preferring co-residence with a son. This can be explained by a greater level of care provided by daughters to their old parents than by either son or daughter in law or at least by a different emotional perception of care relative to who is providing it.

Fulfilling ageing and happiness in centenarians can be also promoted by a better cognitive condition (Jopp and Rott 2006). The prevalence of cognitive impairment in centenarians is variable across countries and studies, ranging from 7%, 27%, or 34% on the lower end of the spectrum to 67.7%, 74.2%, 77.5% on the upper end (Kliegel et al. 2004; Arosio et al. 2017). Cognitive abilities among centenarians are affected not only by the biological processes of advanced age but also social factors such as their cohort, ethnicity, education, and the social environment more broadly (e.g. urban vs rural) (Arosio et al. 2017).

Cognitive decline has been observed to proceed more rapidly in female centenarians than males (Arosio et al. 2017). This may well be a result of some biological differences between the sexes, but consideration should be also given to cohort effects, as when current centenarians were younger women had less access to formal education than men and therefore they might have not been in the position to build significant cognitive reserve over their lifetime. The situation has changed in recent decades. However, hormonal effects such as the fall in circulating oestrogen in women may also play a role (Arosio et al. 2017) as oestrogen can have a positive effect on cognition (Daniel 2006).

Despite the inevitable cognitive losses observed among centenarians and older people, good cognitive condition in very old ages can still be achieved to an extent. A variable percentage of between 15% and 25% of centenarians have been observed to preserve their cognitive functioning (Arosio et al. 2017). However, a good level of cognitive ability beyond the age of 100 is not something that can be fully built at such a late stage in life, the ground must be prepared in earlier years either through formal education and/or by engaging in cognitively stimulating activities in a practical and spontaneous manner (e.g. travelling and learning new languages). The learning experience accumulated throughout life can build *cognitive reserve*, thus establishing a base for cognitive resilience in older ages. This process has been encapsulated by Salthouse and Mitchell (1990) in their *disuse hypothesis*, also referred to in Sect. 4.1 of this chapter as the "*use it or lose it*" hypothesis or, following Salthouse (2006), the *differential-preservation* hypothesis.

Cognitive reserve clearly relies on our brain reserve—brain neuronal reserve in particular—which is a consequence of both genetic make-up, lifestyle and life-long learning and cognitive stimulation (Mortimer et al. 1981; Katzman 1993; Stern 2002). But it is more complex than that, as what matters in cognitive reserve is the actual activation of the brain reserve, the ability to dynamically recruit neuronal circuitry for the production of thought and action (even from alternative areas of the brain, as seen in Sect. 4.2). It is such ability that ultimately allows centenarians to continue functioning cognitively despite increasing deterioration over time.

We have seen in Sect. 4.5 that bilingualism and indeed multilingualism can build cognitive reserve, and studies on bilingual, trilingual, and multilingual oldest-old have shown the positive effects that learning languages has on cognition, independently of other educational factors (Kavé et al. 2008).

Kliegel et al. (2004) studied the effect of cognitive reserve on cognitive impairment in 90 centenarians (88.8% women) within the *Heidelberg Centenarian Study*. Participants were divided into a "moderate to strong cognitive impairment" group and a "no to mild cognitive impairment" group as determined from *Mini-Mental State Exam* (MMSE) scores. The objective was to determine the effect of variables that are associated with cognitive reserve on cognition at an advanced age. Such variables were: early education (more education, more cognitive reserve), occupational status (more occupational complexity, more cognitive reserve), and terminated intellectual activities (the more activities were terminated earlier in life the less cognitive reserve). As one might have expected, more education was associated with better cognitive status in the centenarians. More specifically, the probability of being in the "not or only mildly cognitively impaired" (the lowest level of cognitive impairment in this study) group of centenarians increased by a factor of 2.27 with every additional year of formal education. With regard to the effect of "terminated intellectual activities", the probability of being in the lowest group of cognitive impairment as a centenarian decreased by a factor of 0.45 with every activity given up by the age of 80. On the other hand, occupational status did not affect cognitive status in this group of centenarians. Therefore, better education and greater number of intellectual activities provide more cognitive reserve to centenarians.

Good cognitive condition can also promote a positive attitude towards life. This was the case in a sample of centenarians, most of whom appeared cognitively intact (82.4%), studied by Kato et al. (2013). A positive attitude to life, in turn, is associated with both fewer depressive symptoms (Kato et al. 2016) and personality traits such as extraversion, conscientiousness, emotional stability, and dispositional optimism (Kato et al. 2012). Those personality traits seem to be promoters of better health, cognitive preservation, and ultimately longevity (Weiss and Costa 2005: Masui et al. 2006; Terracciano et al. 2008; Kern and Friedman 2008). Cognitive preservation has been also associated with greater longevity in a sample of centenarians who participated in the Dutch *100-plus Study* (Holstege et al. 2018).

The preservation of cognitive abilities in some oldest-old can be extended to cognitive plasticity which, although declining with age, it can do so more gradually in elders enjoying better cognitive functioning, as demonstrated in Yang et al.'s (2006) work carried out within the *Berlin Aging Study*.

It is expected that across genders better physical and mental health will improve morale at advanced ages. This issue was studied by von Heideken Wågert et al. (2005) who analysed interviews of 85 years old and older people, including centenarians, in the context of the Swedish *Umeå 85 study*. Most of the interviewees displayed middle to high morale and the factors that explained such high morale included absence of depressive symptoms, living in ordinary housing (i.e. living in a "house or apartment with or without access to home care") rather than in institutional care, not feeling lonely and low number of medical symptoms.

Morale also improves the more meaningful the life of centenarians is (e.g. Tornstam 1997). Meaningfulness may help in compensating for objective declines in physical capacities, as the elder may retain meaning in life by transcending specific physical limitations (e.g. Nygren et al. 2005).

Finding meaning in life is essential for centenarians in order to retain their will to live. This takes us to the existential dimension of exceptional longevity. As we mentioned in Sect. 4.3 in this chapter in quoting the work of Nygren et al. (2005), various concepts have been put forward to encapsulate the existential dimension of old age and well-being that could sustain the will to live in centenarians and supercentenarians: *resilience*, *sense of coherence*, *meaning in life,* and *self-transcendence*.

Meaning in life provides the scaffold upon which a sense of coherence can be built in order to achieve self-transcendence, and self-transcendence helps in improving the resilience of the individual as she or he goes through the inevitable changes of very old age.

After studying a population of elders in Sweden, Nygren et al. (2005) found that resilience, sense of coherence, purpose in life, and self-transcendence were all strongly correlated with each other and they were associated with perceived mental health in women, but not in men in this study. Reed (1991a) also described a positive effect of self-transcendence on the mental health and well-being of the older old.

Undoubtedly there is a limit, as we will see in Chap. 8, beyond which the will to live collapses and transcendence acquires an ultimate "post-life dimension" that can differ among individuals following the differences in their beliefs system (e.g. legacy and generativity, nihilism, religiosity, spirituality).

In sum, achieving exceptional longevity is still extremely uncommon in the human population, with the currently recorded upper limit being 122 years. Whether humans can live to more advanced ages or not remains to be seen. However, more centenarians, semi-supercentenarians, and supercentenarians are recorded at present than ever before, mainly due to improved living conditions and medical treatments that can add to any genetic predisposition to achieving exceptional longevity. Other important factors that also play a role in crossing the 100-year-old threshold include low stress in life, physical exercise, a healthy diet, intellectual stimulation, social support, and full engagement with life. More women than men are able to achieve exceptional longevity, although they tend to do so in worse medical condition than men.

4.11 Conclusions

Ageing implies a gradual deterioration of our bodily functions, including aspects of our mind. However, we are not necessarily hopeless in the face of such deterioration, and to some extent we can be, to paraphrase William Ernest Henley, the masters of our fate, the captains of our soul. By using it—our body, our brain—we may decrease the chances of losing it prematurely through ageing. The limits to our brain plasticity can be only tested if we push them through our active and healthy life. This may not fully protect us from either the normal ageing processes or the more pathological aspects of ageing, but it may delay them through building cognitive and physical reserve. Fulfilling ageing can then be experienced by both old women and men in spite of their specific health challenges in old age, and this can be true also for people ageing with a disability and those who reach exceptional longevity.

References

Aarsland, D., Zaccai, J., & Brayne, C. (2005). A systematic review of prevalence studies of dementia in Parkinson's disease. *Movement Disorders, 20*, 1255–1263.

Abberley, P. (1987). The concept of oppression and the development of a social theory of disability. *Disability, Handicap and Society, 2*, 5–19.

Abraha, I., Rimland, J. M., Lozano-Montoya, I., Dell'Aquila, G., Vélez-Díaz-Pallarés, M., Trotta, F. M., et al. (2017). Simulated presence therapy for dementia. *Cochrane Database of Systematic Reviews*, (4), CD011882. https://doi.org/10.1002/14651858.CD011882.pub2

Abraham, C. R. (2014). Amyloid beta peptide and the Amyloid Cascade Hypothesis. In A. E. Budson & N. W. Kowall (Eds.), *The handbook of Alzheimer's disease and other dementias* (pp. 262–276). Chichester, West Sussex: Wiley.

Abutalebi, J., Canini, M., Della Rosa, P. A., Green, D. W., & Weekes, B. S. (2015a). The neuroprotective effects of bilingualism upon the inferior parietal lobule: A structural neuroimaging study in aging Chinese bilinguals. *Journal of Neurolinguistics, 33*, 3–13.

Abutalebi, J., & Green, D. W. (2007). Bilingual language production: The neurocognition of language representation and control. *Journal of Neurolinguistics, 20*, 242–275.

Abutalebi, J., Guidi, L., Borsa, V., Canini, M., Della Rosa, P. A., Parris, B. A., et al. (2015b). Bilingualism provides a neural reserve for aging populations. *Neuropsychologia, 69*, 201–210.

Acitelli, L. K., & Antonucci, T. C. (1994). Gender differences in the link between marital support and satisfaction in older couples. *Journal of Personality and Social Psychology, 67*, 688–698.

Acree, L. S., Longfors, J., Fjelstad, A. S., Fjelstad, C., Schank, B., Nickel, K. J., et al. (2006). Physical activity is related to quality of life in older adults. *Health and Quality of Life Outcomes, 4*, 37. https://doi.org/10.1186/1477-7525-4-37

Aldwin, C. M., & Gilmer, D. F. (2004). *Health, illness, and optimal aging*. Thousand Oaks, CA: Sage.

Alexopoulos, G. S. (2005). Depression in the elderly. *Lancet, 365*, 1961–1970.

Aliev, G., Bachurin, S. O., Mikhaylenko, E. V., Bragin, V., Avila-Rodriguez, M., Somasundaram, S. G., et al. (2019, November 29) Infectious burden and Alzheimer disease: Is there a link? *Psychiatric Times*https://www.psychiatrictimes.com/neuropsychiatry/infectious-burden-and-alzheimer- disease-there-link

Alladi, S., Bak, T. H., Duggirala, V., Surampudi, B., Shailaja, M., Shukla, A. K., et al. (2013). Bilingualism delays age at onset of dementia, independent of education and immigration status. *Neurology, 81*, 1938–1944.

Allport, G. W. (1963). Behavioral science, religion, and mental health. *Journal of Religion and Health, 2*, 187–197.

Allport, G. W., & Ross, J. M. (1967). Personal religious orientation and prejudice. *Journal of Personality and Social Psychology, 5*, 432–443.

Alonso, J., Angermeyer, M. C., Bernert, S., Bruffaerts, R., Brugha, T. S., Bryson, H., et al. (2004). Prevalence of mental disorders in Europe: Results from the European Study of the Epidemiology of Mental Disorders (ESEMeD) project. *Acta Psychiatrica Scandinavica, 109*(Suppl. 420), 21–27.

American Psychiatric Association. (2013). *Diagnostic and statistical manual of mental disorders* (5th ed.). Arlington, VA: American Psychiatric Association.

Andersen, S. L., Sebastiani, P., Dworkis, D. A., Feldman, L., & Perls, T. T. (2012). Health span approximates life span among many supercentenarians: Compression of morbidity at the approximate limit of life span. *Journal of Gerontology, 67A*, 395–405.

Andersen-Ranberg, K., Christensen, K., Jeune, B., Skytthe, A., Vasegaard, L., & Vaupel, J. W. (1999). Declining physical abilities with age: A cross-sectional study of older twins and centenarians in Denmark. *Age and Ageing, 28*, 373–377.

Anderson, D. N. (2001). Treating depression in old age: The reasons to be positive. *Age and Ageing, 30*, 13–17.

Angel, R. J., Frisco, M., Angel, J. L., & Chiriboga, D. A. (2003). Financial strain and health among elderly Mexican-origin individuals. *Journal of Health and Social Behavior, 44*, 536–551.

Anstey, K. J., von Sanden, C., Sargent-Cox, K., & Luszcz, M. A. (2007). Prevalence and risk factors for depression in a longitudinal, population-based study including individuals in the community and residential care. *American Journal of Geriatric Psychiatry, 15*, 497–505.

Antonovsky, A. (1979). *Health, stress and coping*. San Francisco: Jossey-Bass.

Antonovsky, A., & Sagy, S. (1986). The development of a sense of coherence and its impact on responses to stress situations. *Journal of Social Psychology, 126*, 213–225.

Antonucci, T. C., & Akiyama, H. (1987). An examination of sex differences in social support among older men and women. *Sex Roles, 17*, 737–749.

Arai, Y., Martin-Ruiz, C. M., Takayama, M., Abe, Y., Takebayashi, T., Koyasu, S., et al. (2015). Inflammation, but not telomere length, predicts successful ageing at extreme old age: A longitudinal study of semi-supercentenarians. *eBioMedicine, 2*, 1549–1558.

Aranda, M. P. (2013). Depression-related disparities among older, low-acculturated US Latinos. *Psychiatric Times*. Downloaded on October 29, 2016, from http://www.psychiatrictimes.com/major-depressive- disorder/depression-related-disparities-among-older-low-acculturated-us-latinos

Arber, S., Davidson, K., & Ginn, J. (2003a). Changing approaches to gender and later life. In S. Arber, K. Davidson, & J. Ginn (Eds.), *Gender and ageing: Changing roles and relationships* (pp. 1–14). Maidenhead: Open University Press.

Ardelt, M. (2003). Effects of religion and purpose in life on elders' subjective well-being and attitudes toward death. *Journal of Religious Gerontology, 14*, 55–77.

Area-Gomez, E., & Schon, E. A. (2017). On the pathogenesis of Alzheimer's disease: The MAM hypothesis. *FASEB Journal, 31*, 864–867.

Arenaza-Urquijo, E. M., Landeau, B., La Joie, R., Mevel, K., Mezenge, F., Perrotin, A., et al. (2013). Relationships between years of education and gray matter volume, metabolism and functional connectivity in healthy elders. *NeuroImage, 83*, 450–457.

Arosio, B., Ostan, R., Mari, D., Damanti, S., Ronchetti, F., Arcudi, S., et al. (2017). Cognitive status in the oldest old and centenarians: A condition crucial for quality of life methodologically difficult to assess. *Mechanisms of Ageing and Development, 165*, 185–194.

Artero, S., Ancelin, M.-L., Portet, F., Dupuy, A.-M., Berr, C., Dartigues, J.-F., et al. (2008). Risk profiles for mild cognitive impairment and progression to dementia are gender specific. *Journal of Neurology, Neurosurgery and Psychiatry, 79*, 979–984.

Avery, P., Barzilai, N., Benetos, A., Bilianou, H., Capri, M., Caruso, C., et al. (2014). Ageing, longevity, exceptional longevity and related genetic and non genetics markers: Panel statement. *Current Vascular Pharmacology, 12*, 659–661.

Ayalon, L., Gum, A. M., Feliciano, L., & Areán, P. S. (2006). Effectiveness of nonpharmacological interventions for the management of neuropsychiatric symptoms in patients with dementia: A systematic review. *Archives of Internal Medicine, 166*, 2182–2188.

Bachurin, S. O., Gavrilova, S. I., Samsonova, A., Barreto, G. E., & Aliev, G. (2018). Mild cognitive impairment due to Alzheimer disease: Contemporary approaches to diagnostics and pharmacological intervention. *Pharmacological Research, 129*, 216–226.

Ball, T. M., Ramsawh, H. J., Campbell-Sills, L., Paulus, M. P., & Stein, M. B. (2013). Prefrontal dysfunction during emotion regulation in generalized anxiety and panic disorders. *Psychological Medicine, 43*, 1475–1486.

Baltes, M. M. (1998). The psychology of the oldest-old: The fourth age. *Current Opinion in Psychiatry, 11*, 411–415.

Baltes, P. B., Lindenberger, U., & Staudinger, U. M. (1998). Life-span theory in developmental psychology. In W. Damon & R. M. Lerner (Eds.), *Handbook of child psychology. Vol. 1—Theoretical models of human development* (pp. 1029–1143). New York: Wiley.

Baltes, P. B., & Singer, T. (2001). Plasticity and the ageing mind: An exemplar of the bio- cultural orchestration of brain and behavior. *European Review, 9*, 59–76.

Baltes, P. B., & Smith, J. (2003). New frontiers in the future of aging: From successful aging of the young old to the dilemmas of the fourth age. *Gerontology, 49*, 123–135.

Barlow, D. H., Cardozo, L. D., Francis, R. M., Griffin, M., Hart, D. M., Stephens, E., et al. (1997). Urogenital ageing and its effect on sexual health in older British women. *British Journal of Obstetrics and Gynaecology, 104*, 87–91.

Baron, J. C., Chételat, G., Desgranges, B., Perchey, G., Landeau, B., de la Sayette, V., et al. (2001). *In vivo* mapping of gray matter loss with voxel-based morphometry in mild Alzheimer's disease. *NeuroImage, 14*, 298–309.

Barrett-Connor, E., & Kritz-Silverstein, D. (1999). Gender differences in cognitive function with age: The Rancho Bernardo study. *Journal of the American Geriatrics Society, 47*, 159–164.

Bartels, S. J., & Naslund, J. A. (2013). The underside of the silver tsunami - Older adults and mental health care. *New England Journal of Medicine, 368*, 493–496.

Bates, K. A., & Martins, R. N. (2013). Modifiable risk factors for Alzheimer's disease. In R. Martins (Ed.), *Understanding Alzheimer's* (pp. 87–102). Sydney, NSW: Macmillan.

Bateson, P., & Gluckman, P. (2011). *Plasticity, robustness, development and evolution*. Cambridge: Cambridge University Press.

Battista, J., & Almond, R. (1973). The development of meaning in life. *Psychiatry, 36*, 409–427.

Bauer, M. E. (2008). Chronic stress and immunosenescence: A review. *Neuroimmunomodulation, 15*, 241–250.

Baumann, L. (2007). Skin ageing and its treatment. *Journal of Pathology, 211*, 241–251.

Baumeister, R. F. (1991). *Meanings of life*. New York: Guilford Press.

Beadle, G. F., Yates, P. M., Najman, J. M., Clavarino, A., Thomson, A. D., Williams, G., et al. (2004). Illusions in advanced cancer: The effect of belief systems and attitudes on quality of life. *Psycho-Oncology, 13*, 26–36.

Beekman, A. T. F., Bremmer, M. A., Deeg, D. J. H., Van Balkom, A. J. L. M., Smit, J. H., De Beurs, E., et al. (1998). Anxiety disorders in later life: A report from the longitudinal aging study Amsterdam. *International Journal of Geriatric Psychiatry, 13*, 717–726.

Beekman, A. T. F., de Beurs, E., van Balkom, A. J. L. M., Deeg, D. J. H., van Dyck, R., & van Tilburg, W. (2000). Anxiety and depression in later life: Co-occurrence and communality of risk factors. *American Journal of Psychiatry, 157*, 89–95.

Behar, E., DiMarco, I. D., Hekler, E. B., Mohlman, J., & Staples, A. M. (2009). Current theoretical models of generalized anxiety disorder (GAD): Conceptual review and treatment implications. *Journal of Anxiety Disorders, 23*, 1011–1023.

Bem, S. L. (1974). The measurement of psychological androgyny. *Journal of Consulting and Clinical Psychology, 42*, 155–162.

Berer, K., Mues, M., Koutrolos, M., Rasbi, Z. A., Boziki, M., Johner, C., et al. (2011). Commensal microbiota and myelin autoantigen cooperate to trigger autoimmune demyelination. *Nature, 479*, 538–541.

Bergdahl, E., Gustavsson, J. M. C., Kallin, K., von Heideken Wågert, P., Lundman, B., Bucht, G., et al. (2005). Depression among the oldest old: The Umeå 85+ study. *International Psychogeriatrics, 17*, 557–575.

Bergland, A., & Slettebø, Å. (2015). Health capital in everyday life of the oldest old living in their own homes. *Ageing and Society, 35*, 2156–2175.

Berlucchi, G. (2002). The origin of the term plasticity in the neurosciences: Ernesto Lugaro and chemical synaptic transmission. *Journal of the History of Neurosciences, 11*, 305–309.

Bernabei, V., De Ronchi, D., La Ferla, T., Moretti, F., Tonelli, L., Ferrari, B., et al. (2013). Animal-assisted interventions for elderly patients affected by dementia or psychiatric disorders: A review. *Journal of Psychiatric Research, 47*, 762e773.

Bialystok, E., Abutalebi, J., Bak, T. H., Burke, D. M., & Kroll, J. F. (2016). Aging in two languages: Implications for public health. *Ageing Research Reviews, 27*, 56–60.

Bialystok, E., Craik, F. I. M., & Freedman, M. (2007). Bilingualism as a protection against the onset of symptoms of dementia. *Neuropsychologia, 45*, 459–464.

Bigby, C. (2002). Ageing people with a lifelong disability: Challenges for the aged care and disability sectors. *Journal of Intellectual and Developmental Disability, 27*, 231–241.

Binder, L. I., Guillozet-Bongaarts, A. L., Garcia-Sierra, F., & Berry, R. W. (2005). Tau, tangles, and Alzheimer's disease. *Biochimica et Biophysica Acta (BBA) - Molecular Basis of Disease, 1739*, 216–223.

Bjørkløf, G. H., Engedal, K., Selbæk, G., Kouwenhoven, S. E., & Helvik, A.-S. (2013). Coping and depression in old age: A literature review. *Dementia and Geriatric Cognitive Disorders, 35*, 121–154.

Black, I. B. (1986). Trophic molecules and evolution of the nervous system. *Proceedings of the National Academy of Sciences of the United States of America, 83*, 8249–8252.

Blake, H., Mo, P., Malik, S., & Thomas, S. (2009). How effective are physical activity interventions for alleviating depressive symptoms in older people? A systematic review. *Clinical Rehabilitation, 23*, 873–887.

Blazer, D. G. (1997). Depression in the elderly: Myths and misconceptions. *Psychiatric Clinics, 20*, 111–119.

Blazer, D. G. (2000). Psychiatry and the oldest old. *American Journal of Psychiatry, 157*, 1915–1924.

Blazer, D. G. (2003). Depression in late life: Review and commentary. *Journal of Gerontology, 58A*, 249–265.

Bondevik, M., & Skogstad, A. (1998). The oldest old, ADL, social network, and loneliness. *Western Journal of Nursing Research, 20*, 325–343.

Borg, C., Hallberg, I. R., & Blomqvist, K. (2006). Life satisfaction among older people (65+) with reduced self-care capacity: The relationship to social, health and financial aspects. *Journal of Clinical Nursing, 15*, 607–618.

Borglin, G., Jakobsson, U., Edberg, A.-K., & Rahm Hallberg, I. (2005). Self-reported health complaints and their prediction of overall and health-related quality of life among elderly people. *International Journal of Nursing Studies, 42*, 147–158.

Borsa, V. M., Perani, D., Della Rosa, P. A., Videsott, G., Guidi, L., Weekes, B. S., et al. (2018). Bilingualism and healthy aging: Aging effects and neural maintenance. *Neuropsychologia, 111*, 51–61.

Bowling, A., & Browne, P. D. (1991). Social networks, health, and emotional well-being among the oldest old in London. *Journal of Gerontology, 46*, S20–S32.

Boxer, A. L. (2014). Frontotemporal dementia. In A. E. Budson & N. W. Kowall (Eds.), *The handbook of Alzheimer's disease and other dementias* (pp. 145–178). Chichester, West Sussex: Wiley.

Boyke, J., Driemeyer, J., Gaser, C., Büchel, C., & May, A. (2008). Training-induced brain structure changes in the elderly. *The Journal of Neuroscience, 28*, 7031–7035.

Boziki, M., Papaliagkas, V., & Tsolaki, M. (2014). Other mechanisms of neurodegeneration: Neuronal cell death and neuroprotection in Alzheimer's disease. In A. E. Budson & N. W. Kowall (Eds.), *The handbook of Alzheimer's disease and other dementias* (pp. 277–300). Chichester, West Sussex: Wiley.

Brabbins, C. J., Dewey, M. E., Copeland, J. R. M., Davidson, I. A., McWilliam, C., Saunders, P., et al. (1993). Insomnia in the elderly: Prevalence, gender differences and relationships with morbidity and mortality. *International Journal of Geriatric Psychiatry, 8*, 473–480.

Brady, M. J., Peterman, A. H., Fitchett, G., Mo, M., & Cella, D. (1999). A case for including spirituality in quality of life measurement in oncology. *Psycho-Oncology, 8*, 417–428.

Bredesen, D. E., Rao, R. V., & Mehlen, P. (2006). Cell death in the nervous system. *Nature, 443*, 796–802.

Breitbart, W. (2007). Who needs the concept of spirituality? Human beings seem to! *Palliative and Supportive Care, 5*, 105–106.

Brodaty, H., Luscombe, G., Parker, G., Wilhelm, K., Hickie, I., Austin, M.-P., et al. (2001). Early and late onset depression in old age: Different aetiologies, same phenomenology. *Journal of Affective Disorders, 66*, 225–236.

Brooks-Wilson, A. R. (2013). Genetics of healthy aging and longevity. *Human Genetics, 132*, 1323–1338.

Brown, B. M., Taylor, P., & Martins, R. N. (2013). Physical activity and Alzheimer's disease. In R. Martins (Ed.), *Understanding Alzheimer's* (pp. 152–168). Sydney, NSW: Macmillan.

Bruce, D. G., Devine, A., & Prince, R. L. (2002). Recreational physical activity levels in healthy older women: The importance of fear of falling. *Journal of the American Geriatrics Society, 50*, 84–89.

Budson, A. E. (2014). Memory dysfunction in dementia. In A. E. Budson & N. W. Kowall (Eds.), *The handbook of Alzheimer's disease and other dementias* (pp. 315–335). Chichester, West Sussex: Wiley.

Buettner, D. (2005). The secrets of long life. *National Geographic, 2005*, 2–27.

Burbank, V. K. (1987). Female aggression in cross-cultural perspective. *Cross-Cultural Research, 21*, 70–100.

Burke, S. N., & Barnes, C. A. (2006). Neural plasticity in the ageing brain. *Nature Reviews Neuroscience, 7*, 30–40.

Buss, D. M., & Schmitt, D. P. (1993). Sexual strategies theory: An evolutionary perspective on human mating. *Psychological Review, 100*, 204–232.

Buss, R. R., Sun, W., & Oppenheim, R. W. (2006). Adaptive roles of programmed cell death during nervous system development. *Annual Reviews of Neuroscience, 29*, 1–35.

Bussey, K., & Bandura, A. (1999). Social cognitive theory of gender development and differentiation. *Psychological Review, 106*, 676–713.

Buter, T. C., van den Hout, A., Matthews, F. E., Larsen, J. P., Brayne, C., & Aarsland, D. (2008). Dementia and survival in Parkinson Disease: A 12-year population study. *Neurology, 70*, 1017–1022.

Byers, A. L., & Yaffe, K. (2011). Depression and risk of developing dementia. *Nature Reviews Neurology, 7*, 323–331. Downloaded on June 11, 2018, from https://doi.org/10.1038/nrneurol.2011.60

Byers, A. L., Yaffe, K., Covinsky, K. E., Friedman, M. B., & Bruce, M. L. (2010). High occurrence of mood and anxiety disorders among older adults. *Archives of General Psychiatry, 67*, 489–496.

Byles, J. (2017). Health and ageing. In K. O'Loughlin, C. Browning, & H. Kendig (Eds.), *Ageing in Australia: Challenges and opportunities* (pp. 63–81). New York: Springer.

Cabeza, R., Anderson, N. D., Locantore, J. K., & McIntosh, A. R. (2002). Aging gracefully: Compensatory brain activity in high-performing older adults. *NeuroImage, 17*, 1394–1402.

Cacace, R., Sleegers, K., & Van Broeckhoven, C. (2016). Molecular genetics of early-onset Alzheimer's disease revisited. *Alzheimer's and Dementia, 12*, 733–748.

Cacioppo, J. T., Amaral, D. G., Blanchard, J. J., Cameron, J. L., Carter, C. S., Crews, D., et al. (2007). Social neuroscience: Progress and implications for mental health. *Social Neuroscience, 2*, 99–123.

Cai, S., Fu, X., & Sheng, Z. (2007). Dedifferentiation: A new approach in stem cell research. *Bioscience, 57*, 655–662.

Cairney, J., & Krause, N. (2005). The social distribution of psychological distress and depression in older adults. *Journal of Aging and Health, 17*, 807–835.

Calabria, M., Costa, A., Green, D. W., & Abutalebi, J. (2018). Neural basis of bilingual language control. *Annals of the New York Academy of Sciences, 1426*, 221–235.

Calasanti, T. (2003). Masculinities and care work in old ages. In S. Arber, K. Davidson, & J. Ginn (Eds.), *Gender and ageing: Changing roles and relationships* (pp. 15–30). Maidenhead: Open University Press.

Camacho, M. E., & Reyes-Ortiz, C. A. (2005). Sexual dysfunction in the elderly: Age or disease? *International Journal of Impotence Research, 17*, S52–S56.

Carmel, S. (2001). The will to live: Gender differences among elderly persons. *Social Science and Medicine, 52*, 949–958.

Carmel, S. (2019). Health and well-being in late life: Gender differences worldwide. *Frontiers in Medicine (Lausanne), 6*, 218. https://doi.org/10.3389/fmed.2019.00218

Carmel, S., Baron-Epel, O., & Shemy, G. (2007). The will-to-live and survival at old age: Gender differences. *Social Science and Medicine, 65*, 518–523.

Carmel, S., Tovel, H., Raveis, V. H., & O'Rourke, N. (2018). Is a decline in will-to-live a consequence or predictor of depression in late life? *Journal of the American Geriatrics Society, 66*, 1290–1295.

Carvalho, D. Z., St Louis, E. K., Knopman, D. S., Boeve, B. F., Lowe, V. J., Roberts, R. O., et al. (2018). Association of excessive daytime sleepiness with longitudinal β-amyloid accumulation in elderly persons without dementia. *JAMA Neurology*. https://doi.org/10.1001/jamaneurol.2018.0049

Caselli, G., Battaglini, M., & Capacci, G. (2018). Beyond one hundred: A cohort analysis of Italian centenarians and semisupercentenarians. *The Journals of Gerontology. Series B, Psychological Sciences and Social Sciences, 75*(3), 591–600. https://doi.org/10.1093/geronb/gby033

Cervilla, J. A., & Prince, M. J. (1997). Cognitive impairment and social distress as different pathways to depression in the elderly: A cross-sectional study. *International Journal of Geriatric Psychiatry, 12*, 995–1000.

Cevenini, E., Cotichini, R., Stazi, M. A., Toccaceli, V., Palmas, M. G., Capri, M., et al. (2014). Health status and 6 years survival of 552 90+ Italian sib-ships recruited within the EU project GEHA (Genetics of healthy ageing). *Age (Dordrecht, Netherlands), 36*, 949–966.

Chambers, R. A., Potenza, M. N., Hoffman, R. E., & Miranker, W. (2004). Simulated apoptosis/neurogenesis regulates learning and memory capabilities of adaptive neural networks. *Neuropsychopharmacology, 29*, 747–758.

Chao, C. C., Chen, C., & Yen, M. (2002). The essence of spirituality of terminally ill patients. *Journal of Nursing Research, 10*, 237–244.

Chapman, B. P., Duberstein, P. R., Sörensen, S., & Lyness, J. M. (2007). Gender differences in five factor model personality traits in an elderly cohort: Extension of robust and surprising findings to an older generation. *Personality and Individual Differences, 43*, 1594–1603.

Charlifue, S., Jha, A., & Lammertse, D. (2010). Aging with spinal cord injury. *Physical Medicine and Rehabilitation Clinics of North America, 21*, 383–402.

Chen, A. J. W., & D'Esposito, M. (2015). Plasticity in prefrontal cortical networks after brain injury: Finding the optimal paths. In J. I. Tracy, B. M. Hampstead, & K. Sathian (Eds.), *Cognitive plasticity in neurologic disorders* (pp. 227–257). Oxford: Oxford University Press.

Chen, F., & Short, S. E. (2008). Household context and subjective well-being among the oldest old in China. *Journal of Family Issues, 29*, 1379–1403.

Cherry, K. E., Walker, E. J., Silva Brown, J., Volaufova, J., LaMotte, L. R., Welsh, D. A., et al. (2013). Social engagement and health in younger, older, and oldest-old adults in the Louisiana Healthy Aging Study (LHAS). *Journal of Applied Gerontology, 32*, 51–75.

Chertkow, H., Whitehead, V., Phillips, N., Wolfson, C., Atherton, J., & Bergman, H. (2010). Multilingualism (but not always bilingualism) delays the onset of Alzheimer disease: Evidence from a bilingual community. *Alzheimer Disease and Associated Disorders, 24*, 118–125.

Chipperfield, J. G., & Havens, B. (2001). Gender differences in the relationship between marital status transitions and life satisfaction in later life. *Journal of Gerontology, 56B*, P176–P186.

Cho, J., Martin, P., Poon, L. W., & for the Georgia Centenarian Study. (2015). Successful aging and subjective well-being among oldest-old adults. *The Gerontologist, 55*, 132–143.

Cho, M., & Suh, Y. (2014). Genome maintenance and human longevity. *Current Opinion in Genetics and Development, 26*, 105–115.

Chochinov, H. M. (2006). Dying, dignity, and new horizons in palliative end-of-life care. *CA: A Cancer Journal for Clinicians, 56*, 84–103.

Chou, K.-L., & Chi, I. (2002). Successful aging among the young-old, old-old, and oldest- old Chinese. *International Journal of Aging and Human Development, 54*, 1–14.

Christensen, K., McGue, M., Petersen, I., Jeune, B., & Vaupel, J. W. (2008). Exceptional longevity does not result in excessive levels of disability. *Proceedings of the National Academy of Sciences of the United States of America, 105*, 13274–13279.

Christiansen, L., Bathum, L., Frederiksen, H., & Christensen, K. (2004). Paraoxonase 1 polymorphisms and survival. *European Journal of Human Genetics, 12*, 843–847.

Chung, H. Y., Cesari, M., Anton, S., Marzetti, E., Giovannini, S., Seo, A. Y., et al. (2009). Molecular inflammation: Underpinnings of aging and age-related diseases. *Ageing Research Reviews, 8*, 18–30.

Chung, H. Y., Kim, H. J., Kim, J. W., & Yu, B. P. (2001). The inflammation hypothesis of aging: Molecular modulation by calorie restriction. *Annals of the New York Academy of Sciences, 928*, 327–335.

Chung, J. C. C., & Lai, C. K. Y. (2002). Snoezelen for dementia. *Cochrane Database of Systematic Reviews*, (4), CD003152. https://doi.org/10.1002/14651858.CD003152

Clark, A. N. G., Mankikar, G. D., & Gray, I. (1975). Diogenes syndrome: A clinical study of gross neglect in old age. *The Lancet, 305*, 366–368.

Clutton-Brock, T. H., & Isvaran, K. (2007). Sex differences in ageing in natural populations of vertebrates. *Proceedings of the Royal Society B, 274*, 3097–3104.

Cohen, G. D., Perlstein, S., Chapline, J., Kelly, J., Firth, K. M., & Simmens, S. (2006). The impact of professionally conducted cultural programs on the physical health, mental health, and social functioning of older adults. *The Gerontologist, 46*, 726–734.

Cole, B., & Pargament, K. (1999). Re-creating your life: A spiritual psychotherapeutic intervention for people diagnosed with cancer. *Psycho-Oncology, 8*, 395–407.

Cole, M. G., & Dendukuri, N. (2003). Risk factors for depression among elderly community subjects: A systematic review and meta-analysis. *American Journal of Psychiatry, 160*, 1147–1156.

Cooper, S.-A. (1999). The relationship between psychiatric and physical health in elderly people with intellectual disability. *Journal of Intellectual Disability Research, 43*, 54–60.

Copland, D. A., & Angwin, A. (2015). Cognitive plasticity in Parkinson's disease. In J. I. Tracy, B. M. Hampstead, & K. Sathian (Eds.), *Cognitive plasticity in neurologic disorders* (pp. 85–106). Oxford: Oxford University Press.

Coppus, A., Evenhuis, H., Verberne, G.-J., Visser, F., van Gool, P., Eikelenboom, P., et al. (2006). Dementia and mortality in persons with Down's syndrome. *Journal of Intellectual Disability Research, 50*, 768–777.
Cordella, M., & Poiani, A. (2014). *Behavioural oncology: Psychological , communicative, and social dimensions*. New York: Springer.
Corder, E. H., Saunders, A. M., Strittmatter, W. J., Schmechel, D. E., Gaskell, P. C., Small, G. W., et al. (1993). Gene dose of apolipoprotein E type 4 allele and the risk of Alzheimer's disease in late onset families. *Science, 261*, 921–923.
Cornwell, E. Y., & Waite, L. J. (2009). Social disconnectedness, perceived isolation, and health among older adults. *Journal of Health and Social Behavior, 50*, 31–48.
Corrada, M. M., Brookmeyer, R., Paganini-Hill, A., Berlau, D., & Kawas, C. H. (2010). Dementia incidence continues to increase with age in the oldest old The 90+ Study. *Annals of Neurology, 67*, 114–121.
Corrigan, P., McCorkle, B., Schell, B., & Kidder, K. (2003). Religion and spirituality in the lives of people with serious mental illness. *Community Mental Health Journal, 39*, 487–499.
Costa, M. D., Cherni, L., Fernandes, V., Freitas, F., Ben Ammar el Gaaied, A., & Pereira, L. (2009). Data from complete mtDNA sequencing of Tunisian centenarians: Testing haplogroup association and the "golden mean" to longevity. *Mechanisms of Ageing and Development, 130*, 222–226.
Costa, P. T., & McCrae, R. R. (1992). *Revised NEO Personality Inventory (NEO-PI-R) and NEO Five-Factor Inventory (NEO-FFI) professional manual*. Odessa, FL: Psychological Assessment Resources.
Costa, P. T., Terracciano, A., & McCrae, R. R. (2001). Gender differences in personality traits across cultures: Robust and surprising findings. *Journal of Personality and Social Psychology, 81*, 322–331.
Cotelli, M., Manenti, R., & Zanetti, O. (2012). Reminiscence therapy in dementia: A review. *Maturitas, 72*, 203–205.
Cournil, A., Legay, J.-M., & Schächter, F. (2000). Evidence of sex-linked effects on the inheritance of human longevity: A population-based study in the Valserine valley (French Jura), 18–20th centuries. *Proceedings of the Royal Society of London B, 267*, 1021–1025.
Courtin, E., & Knapp, M. (2017). Social isolation, loneliness and health in old age: A scoping review. *Health and Social Care in the Community, 25*, 799–812.
Craft, S., Cholerton, B., & Reger, M. (2003). Aging and cognition: What is normal? In W. R. Hazzard, J. P. Blass, J. B. Halter, J. G. Ouslander, & M. Tinetti (Eds.), *Principles of geriatric medicine and gerontology* (5th ed., pp. 1355–1372). New York: McGraw-Hill.
Craik, F. I. M., Bialystok, E., & Freedman, M. (2010). Delaying the onset of Alzheimer disease. *Bilingualism as a form of cognitive reserve. Neurology, 75*, 1726–1729.
Cramer, S. C., Sur, M., Dobkin, B. H., O'Brien, C., Sanger, T. D., Trojanowski, J. Q., et al. (2011). Harnessing neuroplasticity for clinical applications. *Brain, 134*, 1591–1609.
Creegan, R., & McManus, A. (2013). Diet and Alzheimer's disease. In R. Martins (Ed.), *Understanding Alzheimer's* (pp. 103–151). Sydney, NSW: Macmillan.
Croezen, S., Avendano, M., Burdorf, A., & van Lenthe, F. J. (2015). Social participation and depression in old age: A fixed-effects analysis in 10 European countries. *American Journal of Epidemiology, 182*, 168–176.
Cronin-Golomb, A. (2014). Visuospatial function in Alzheimer's disease and related disorders. In A. E. Budson & N. W. Kowall (Eds.), *The handbook of Alzheimer's disease and other dementias* (pp. 457–482). Chichester, West Sussex: Wiley.
Crosson, B., & McGregor, K. M. (2015). Aging-related changes in neural substrates of motor and cognitive systems. In J. I. Tracy, B. M. Hampstead, & K. Sathian (Eds.), *Cognitive plasticity in neurologic disorders* (pp. 282–298). Oxford: Oxford University Press.
Cuijpers, P., Karyotaki, E., Pot, A. M., Park, M., & Reynolds, C. F. (2014). Managing depression in older age: Psychological interventions. *Maturitas, 79*, 160–169.
Culbertson, F. M. (1997). Depression and gender: An international review. *American Psychologist, 52*, 25–31.

Cummings, J., Lee, G., Ritter, A., Sabbagh, M., & Zhong, K. (2019). Alzheimer's disease drug development pipeline: 2019. *Alzheimer's and Dementia: Translational Research and Clinical Interventions, 5*, 272–293.

Cylwik, H. (2002). Expectations of inter-generational reciprocity among older Greek Cypriot migrants in London. *Ageing and Society, 22*, 599–613.

Dagnan, D., & Sandhu, S. (1999). Social comparison, self-esteem and depression in people with intellectual disability. *Journal of Intellectual Disability Research, 43*, 372–379.

Daniel, J. M. (2006). Effects of oestrogen on cognition: What have we learned from basic research? *Journal of Neuroendocrinology, 18*, 787–795.

Dantzer, R., O'Connor, J. C., Freund, G. G., Johnson, R. W., & Kelley, K. W. (2008). From inflammation to sickness and depression: When the immune system subjugates the brain. *Nature Reviews, Neuroscience, 9*, 46–56.

Davis, D. H. J., Muniz Terrera, G., Keage, H., Rahkonen, T., Oinas, M., Matthews, F. E., et al. (2012). Delirium is a strong risk factor for dementia in the oldest-old: A population-based cohort study. *Brain, 135*, 2809–2816.

Davydov, D., Stewart, R., Ritchie, K., & Chaudieu, I. (2010). Resilience and mental health. *Clinical Psychology Review, 30*, 479–495.

Day, K., & Jancar, J. (1994). Mental and physical health and ageing in mental handicap: A review. *Journal of Intellectual Disability Research, 38*, 241–256.

Day, K. A. (1987). The elderly mentally handicapped in hospital: A clinical study. *Journal of Mental Deficiency Research, 31*, 131–146.

de Bruijn, R. F. A. G. (2015). *Emerging determinants of dementia*. PhD Thesis, Erasmus University, Rotterdam, The Netherlands.

De Frias, C. M., Nilsson, L.-G., & Herlitz, A. (2006). Sex differences in cognition are stable over a 10-year period in adulthood and old age. *Aging, Neuropsychology, and Cognition, 13*, 574–587.

de Graaf, G., Buckley, F., & Skotko, B. G. (2017). Estimation of the number of people with Down syndrome in the United States. *Genetics in Medicine, 19*, 439–447.

de Grey, A. D. N. J. (2015). Can super-duper centenarians possibly matter? Increasingly it seems they may. *Rejuvenation Research, 18*, 1–2.

de Magalhães, J. P., & Sandberg, A. (2005). Cognitive aging as an extension of brain development: A model linking learning, brain plasticity, and neurodegeneration. *Mechanisms of Ageing and Development, 126*, 1026–1033.

Decker, S. D., & Schulz, R. (1985). Correlates of life satisfaction and depression in middle-aged and elderly spinal cord-injured persons. *The American Journal of Occupational Therapy, 39*, 740–745.

Defina, L. F., Willis, B. L., Radford, N. B., Gao, A., Leonard, D., Haskell, W. L., et al. (2013). The association between midlife cardiorespiratory fitness levels and later-life dementia: A cohort study. *Annals of Internal Medicine, 158*, 162–168.

Del Maschio, N., Sulpizio, S., Gallo, F., Fedeli, D., Weekes, B. S., & Abutalebi, J. (2018). Neuroplasticity across the lifespan and aging effects in bilinguals and monolinguals. *Brain and Cognition, 125*, 118–126.

DeLamater, J. (2012). Sexual expression in later life: A review and synthesis. *Journal of Sex Research, 49*, 125–141.

Denes, G. (2016). *Neural plasticity across the lifespan: How the brain can change*. London: Routledge.

Deslandes, A., Moraes, H., Ferreira, C., Veiga, H., Silveira, H., Mouta, R., et al. (2009). Exercise and mental health: Many reasons to move. *Neuropsychobiology, 59*, 191–198.

Di Paola, M., Macaluso, E., Carlesimo, G. A., Tomaiuolo, F., Worsley, K. J., Fadda, L., et al. (2007). Episodic memory impairment in patients with Alzheimer's disease is correlated with entorhinal cortex atrophy: A voxel-based morphometry study. *Journal of Neurology, 254*, 774–781.

Diehl, M., Coyle, N., & Labouvie-Vief, G. (1996). Age and sex differences in strategies of coping and defense across the life span. *Psychology and Aging, 11*, 127–139.

Diener, E., & Chan, M. Y. (2011). Happy people live longer: Subjective well-being contributes to health and longevity. *Applied Psychology: Health and Well-Being, 3*, 1–43.

Doidge, N. (2010). *The brain that changes itself.* Melbourne, Vic: Scribe.

Doka, K. J., & Lavin, C. (2003). The paradox of ageing with developmental disabilities: Increasing needs, declining resources. *Ageing International, 28*, 135–154.

Dong, X., Milholland, B., & Vijg, J. (2016). Evidence for a limit to human lifespan. *Nature, 538*, 257–259.

Draper, B. (2014, February 14). Preparing for the graying of the world: Meeting the needs of older adults. *Psychiatric Times*. Downloaded on March 20, 2014, from http://www.psychiatrictimes.com/geriatric-psychiatry/preparing-graying-world- meeting-needs-older-adults/page/0/1

Du, A. T., Schuff, N., Zhu, X. P., Jagust, W. J., Miller, B. L., Reed, B. R., et al. (2003). Atrophy rates of entorhinal cortex in AD and normal aging. *Neurology, 60*, 481–486.

Du, Z., Li, Y., Li, J., Zhou, C., Li, F., & Yang, X. (2018). Physical activity can improve cognition in patients with Alzheimer's disease: A systematic review and meta-analysis of randomized controlled trials. *Clinical Interventions in Aging, 13*, 1593–1603.

Duberstein, P. R., Conwell, Y., Conner, K. R., Eberly, S., & Caine, E. D. (2004). Suicide at 50 years of age and older: Perceived physical illness, family discord and financial strain. *Psychological Medicine, 34*, 137–146.

Duberstein, P. R., Pálsson, S. P., Waern, M., & Skoog, I. (2008). Personality and risk for depression in a birth cohort of 70-year-olds followed for 15 years. *Psychological Medicine, 38*, 663–671.

Duffau, H. (2015). Plasticity of cognition in brain gliomas. In J. I. Tracy, B. M. Hampstead, & K. Sathian (Eds.), *Cognitive plasticity in neurologic disorders* (pp. 125–151). Oxford: Oxford University Press.

Dugas, M. J., Gagnon, F., Ladouceur, R., & Freeston, M. H. (1998). Generalized anxiety disorder: A preliminary test of a conceptual model. *Behaviour Research and Therapy, 36*, 215–226.

Eagly, A. H., & Wood, W. (1999). The origins of sex differences in human behavior: Evolved dispositions versus social roles. *American Psychologist, 54*, 408–423.

Eccles, J. S. (1994). Understanding women's educational and occupational choices: Applying the Eccles et al. model of achievement-related choices. *Psychology of Women Quarterly, 18*, 585–610.

Edmondson, D., Park, C. L., Blank, T. O., Fenster, J. R., & Mills, M. A. (2008). Deconstructing spiritual well-being: Existential well-being and HRQOL in cancer survivors. *Psycho-Oncology, 17*, 161–169.

Edwards, D. J., Medeiros, A. H., Cortes, M., & Pascual-Leone, A. (2015). Assessment and enhancement of human brain plasticity using electromagnetic stimulation. In J. I. Tracy, B. M. Hampstead, & K. Sathian (Eds.), *Cognitive plasticity in neurologic disorders* (pp. 363–392). Oxford: Oxford University Press.

Ehrenfeld, M., Bronner, G., Tabak, N., Alpert, R., & Bergman, R. (1999). Sexuality among institutionalised elderly patients with dementia. *Nursing Ethics, 6*, 144–149.

Elder Jr., G. H., & Clipp, E. C. (1989). Combat experience and emotional health: Impairment and resilience in later life. *Journal of Personality, 57*, 311–341.

Elkins, D. N., Hedstrom, L. J., Hughes, L. L., Leaf, J. A., & Saunders, C. (1988). Toward a humanistic-phenomenological spirituality: Definition, description, and measurement. *Journal of Humanistic Psychology, 28*, 5–18.

Enkvist, Å., Ekström, H., & Elmståhl, S. (2012). Life satisfaction (LS) and symptoms among the oldest-old: Results from the longitudinal population study called Good Aging in Skåne (GÅS). *Archives of Gerontology and Geriatrics, 54*, 146–150.

Erickson, K. I., Colcombe, S. J., Wadhwa, R., Bherer, L., Peterson, M. S., Scalf, P. E., et al. (2007). Training-induced plasticity in older adults: Effects of training on hemispheric asymmetry. *Neurobiology of Aging, 28*, 272–283.

Erlangsen, A., Jeune, B., Bille-Brahe, U., & Vaupel, J. W. (2004). Loss of partner and suicide risks among oldest old: A population-based register study. *Age and Ageing, 33*, 378–383.

Esch, T., & Stefano, G. B. (2005). Love promotes health. *Neuroendocrinology Letters, 26*, 264–267.

Evenhuis, H., Henderson, C. M., Beange, H., Lennox, N., & Chicoine, B. (2000). *Healthy Ageing - Adults with intellectual disabilities: Physical health issues*. Geneva: World Health Organization.

Evenhuis, H. M. (1995). Medical aspects of ageing in a population with intellectual disability: II. Hearing impairment. *Journal of Intellectual Disability Research, 39*, 27–33.

Evenhuis, H. M. (1997). Medical aspects of ageing in a population with intellectual disabilities: III. Mobility, internal conditions and cancer. *Journal of Intellectual Disability Research, 41*, 8–18.

Fairhurst, E. (2003). New identities in ageing: Perspectives on age, gender and life after work. In S. Arber, K. Davidson, & J. Ginn (Eds.), *Gender and ageing: Changing roles and relationships* (pp. 31–46). Maidenhead: Open University Press.

Farrer, L. A., Cupples, L. A., Haines, J. L., Hyman, B., Kukull, W. A., Mayeux, R., et al. (1997). Effects of age, sex, and ethnicity on the association between apolipoprotein E genotype and Alzheimer disease. A meta-analysis. APOE and Alzheimer Disease Meta Analysis Consortium. *JAMA., 278*, 1349–1356.

Feingold, A. (1994). Gender differences in personality: A meta-analysis. *Psychological Bulletin, 116*, 429–456.

Femia, E. E., Zarit, S. H., & Johansson, B. (2001). The disablement process in very late life: A study of the oldest-old in Sweden. *Journal of Gerontology, 56B*, P12–P23.

Finch, C. E. (2007). *The biology of human longevity: Inflammation, nutrition, and aging in the evolution of life spans*. Amsterdam: Elsevier.

Fiske, A., Loebach Wetherell, J., & Gatz, M. (2009). Depression in older adults. *Annual Review of Clinical Psychology, 5*, 363–389.

Fjell, A. M., McEvoy, L., Holland, D., Dale, A. M., Walhovd, K. B., & for the Alzheimer's Disease Neuroimaging Initiative. (2014). What is normal in normal aging? Effects of aging, amyloid and Alzheimer's disease on the cerebral cortex and the hippocampus. *Progress in Neurobiology, 117*, 20–40.

Flint, A. J. (2005). Generalised anxiety disorder in elderly patients: Epidemiology, diagnosis and treatment options. *Drugs & Aging, 22*, 101–114.

Fong, T. G., & Press, D. Z. (2014). Dementia with Lewy bodies. In A. E. Budson & N. W. Kowall (Eds.), *The handbook of Alzheimer's disease and other dementias* (pp. 131–144). Chichester, West Sussex: Wiley.

Forbes, D., Culum, I., Lischka, A. R., Morgan, D. G., Peacock, S., Forbes, J., et al. (2009). Light therapy for managing cognitive, sleep, functional, behavioural, or psychiatric disturbances in dementia. *Cochrane Database of Systematic Reviews*, (4), CD003946. https://doi.org/10.1002/14651858.CD003946.pub3

Fox, N. C., & Schott, J. M. (2004). Imaging cerebral atrophy: Normal ageing to Alzheimer's disease. *Lancet, 363*, 392–394.

Franceschi, C., Motta, L., Valensin, S., Rapisarda, R., Franzone, A., Berardelli, M., et al. (2000). Do men and women follow different trajectories to reach extreme longevity? *Aging Clinical and Experimental Research, 12*, 77–84.

Frankl, V. E. (1963). *Man's search for meaning: An introduction to logotherapy*. New York: Washington Square Press.

Fratiglioni, L., Paillard-Borg, S., & Winblad, B. (2004). An active and socially integrated lifestyle in late life might protect against dementia. *Lancet Neurology, 3*, 343–353.

Freeman, S., Garcia, J., & Marston, H. R. (2013). Centenarian self-perceptions of factors responsible for attainment of extended health and longevity. *Educational Gerontology, 39*, 717–728.

Fries, J. F. (1980). Aging, natural death, and the compression of morbidity. *New England Journal of Medicine, 303*, 130–135.

Frotscher, M., Drakew, A., & Heimrich, B. (2000). Role of afferent innervation and neuronal activity in dendritic development and spine maturation of fascia dentata granule cells. *Cerebral Cortex, 10*, 946–951.

Fuller, S. J., Goozee, K. G., & Martins, R. N. (2013). Diagnosing Alzheimer's disease. In R. Martins (Ed.), *Understanding Alzheimer's* (pp. 16–52). Sydney, NSW: Macmillan.

Fuller, S. J., & Martins, R. N. (2013). Introduction. In R. Martins (Ed.), *Understanding Alzheimer's* (pp. 1–15). Sydney, NSW: Macmillan.

Fung, J. K. K. M., Tsang, H. W. H., & Chung, R. C. K. (2012). A systematic review of the use of aromatherapy in treatment of behavioral problems in dementia. *Geriatrics and Gerontology International, 12*, 372–382.

Gampe, J. (2010). Human mortality beyond age 110. In H. Maier, J. Gampe, B. Jeune, J. M. Robine, & J. W. Vaupel (Eds.), *Supercentenarians, demographic research monographs* (pp. 219–230). Berlin: Springer.

Gandaglia, G., Briganti, A., Jackson, G., Kloner, R. A., Montorsi, F., Montorsi, P., et al. (2014). Systematic review of the association between erectile dysfunction and cardiovascular disease. *European Urology, 6*(5), 968–978.

Gauthier, S., Reisberg, B., Zaudig, M., Petersen, R. C., Ritchie, K., Broich, K., et al. (2006). Mild cognitive impairment. *The Lancet, 367*, 1262–1270.

Gavrilova, N. S., & Gavrilov, L. A. (2019). Are we approaching a biological limit to human longevity? *Journals of Gerontology Series A, Biological Sciences and Medical Sciences, 75*(6), 1061–1067. https://doi.org/10.1093/gerona/glz164

Gavrilova, N. S., Gavrilov, L. A., Evdokushkina, G. N., Semyonova, V. G., Gavrilova, A. L., Evdokushkina, N. N., et al. (1998). Evolution, mutations, and human longevity: European royal and noble families. *Human Biology, 70*, 799–804.

Geary, D. C., & Flinn, M. V. (2001). Evolution of human parental behavior and the human family. *Parenting, 1*, 5–61.

Genazzani, A. R., Pluchino, N., Luisi, S., & Luisi, M. (2007). Estrogen, cognition and female ageing. *Human Reproduction Update, 13*, 175–187.

Gerstorf, D., Herlitz, A., & Smith, J. (2006). Stability of sex differences in cognition in advanced old age: The role of education and attrition. *The Journals of Gerontology Series B, 61*, P245–P249.

Ghetti, B., Oblak, A. L., Boeve, B. F., Johnson, K. A., Dickerson, B. C., & Goedert, M. (2015). Frontotemporal dementia caused by *microtubule-associated protein tau* gene (*MAPT*) mutations: A chameleon for neuropathology and neuroimaging. *Applied Neurobiology, 41*, 24–46.

Giuliani, C., Pirazzini, C., Delledonne, M., Xumerle, L., Descombes, P., Marquis, J., et al. (2017). Centenarians as extreme phenotypes: An ecological perspective to getinsight into the relationship between the genetics of longevity andage-associated diseases. *Mechanisms of Ageing and Development, 165*, 195–201.

Glaesmer, H., Riedel-Heller, S., Braehler, E., Spangenberg, L., & Luppa, M. (2011). Age- and gender-specific prevalence and risk factors for depressive symptoms in the elderly: A population-based study. *International Psychogeriatrics, 23*(8), 1294–1300. https://doi.org/10.1017/S1041610211000780

Goffman, E. (1963). *Stigma*. Englewood Cliffs, NJ: Prentice-Hall.

Goodall, J. (1988). Living options for physically disabled adults: A review. *Disability, Handicap and Society, 3*, 173–193.

Goozee, K. G., & Dias, A. (2013). Caring for someone with dementia. In R. Martins (Ed.), *Understanding Alzheimer's* (pp. 247–291). Sydne, NSW: Macmillan.

Gott, M. (2006). Sexual health and the new ageing. *Age and Ageing, 35*, 106–107.

Gott, M., & Hinchliff, S. (2003). Sex and ageing: A gendered issue. In S. Arber, K. Davidson, & J. Ginn (Eds.), *Gender and ageing: Changing roles and relationships* (pp. 63–78). Maidenhead: Open University Press.

Gott, M., & Hinchliff, S. (2003a). How important is sex in later life? The views of older people. *Social Science and Medicine, 56*, 1617–1628.

Govindaraju, D., Atzmon, G., & Barzilai, N. (2015). Genetics, lifestyle and longevity: Lessons from centenarians. *Applied and Translational Genomics, 4*, 23–32.

Graeber, M. B., Kösel, S., Egensperger, R., Banati, R. B., Müller, U., Bise, K., et al. (1997). Rediscovery of the case described by Alois Alzheimer in 1911: Historical, histological and molecular genetic analysis. *Neurogenetics, 1*, 73–80.

Green, M. S., Kaye, J. A., & Ball, M. J. (2000). The Oregon Brain Aging Study: Neuropathology accompanying healthy aging in the oldest old. *Neurology, 54*, 105–113.

Greenwood, P. M. (2007). Functional plasticity in cognitive aging: Review and hypothesis. *Neuropsychology, 21*, 657–673.

Gresser, U., & Gleiter, C. H. (2002). Erectile dysfunction: Comparison of efficacy and side effects of the PDE-5 inhibitors sildenafil, vardenafil and tadalafil. Review of the literature. *European Journal of Medical Research, 7*, 435–446.

Gross, A. L., Rebok, G. W., Unverzagt, F. W., Willis, S. L., & Brandt, J. (2011). Cognitive predictors of everyday functioning in older adults: Results from the ACTIVE cognitive intervention trial. *The Journals of Gerontology: Series B, 66B*, 557–566.

Gross, C. G. (2000). Neurogenesis in the adult brain: Death of a dogma. *Nature Reviews, 1*, 67–73.

Growdon, J. H. (2014). Foreword. In A. E. Budson & N. W. Kowall (Eds.), *The handbook of Alzheimer's disease and other dementias* (pp. xii–xiv). Chichester, West Sussex: Wiley.

Grundy, E., & Bowling, A. (1999). Enhancing the quality of extended life years. Identification of the oldest old with a very good and very poor quality of life. *Aging and Mental Health, 3*, 199–212.

Guergova, S., & Dufour, A. (2011). Thermal sensitivity in the elderly: A review. *Ageing Research Reviews, 10*, 80–92.

Gulcur, L., Tsemberis, S., Stefancic, A., & Greenwood, R. M. (2007). Community integration of adults with psychiatric disabilities and histories of homelessness. *Community Mental Health Journal, 43*, 211–228.

Gum, A. M., King-Kallimanis, B., & Kohn, R. (2009). Prevalence of mood, anxiety, and substance-abuse disorders for older Americans in the National Comorbidity Survey - Replication. *American Journal of Geriatric Psychiatry, 17*, 769–781.

Gupta, R., Ambasta, R. K., & Kumar, P. (2020). Pharmacological intervention of histone deacetylase enzymes in the neurodegenerative disorders. *Life Sciences, 243*, 117278. https://doi.org/10.1016/j.lfs.2020.117278

Gur, R. C., Levin, S. L., Farzin, I., Panyavin, I. S., Thomson, J. W., & Platek, S. M. (2007). The evolution of sex differences in spatial cognition. In S. M. Platek, J. P. Keenan, & T. K. Shackelford (Eds.), *Evolutionary cognitive neuroscience* (pp. 383–402). Cambridge: MIT Press.

Gutchess, A. (2014). Plasticity of the aging brain: New directions in cognitive neuroscience. *Science, 346*, 579–582.

Ha, J.-H., Hong, J., Seltzer, M. M., & Greenberg, J. S. (2008). Age and gender differences in the well-being of midlife and aging parents with children with mental health or developmental problems: Report of a national study. *Journal of Health and Social Behavior, 49*, 301–316.

Hackney, C. H., & Sanders, G. S. (2003). Religiosity and mental health: A meta-analysis of recent studies. *Journal for the Scientific Study of Religion, 42*, 43–55.

Haigh, E. A. P., Bogucki, O. E., Sigmon, S. T., & Blazer, D. G. (2018). Depression among older adults: A 20-year update on five common myths and misconceptions. *American Journal of Geriatric Psychiatry, 26*, 107–122.

Hall, J. (2013). *Dementia essentials: How to guide a loved one through Alzheimer's or Dementia and provide the best care*. London: Vermilion.

Hammel, J., Lai, J.-S., & Helle, T. (2002). The impact of assistive technology and environmental interventions on function and living situation status with people who are ageing with developmental disabilities. *Disability and Rehabilitation, 24*, 93–105.

Hampstead, B. M., & Sathian, K. (2015). Cognitive plasticity in healthy older adults, mild cognitive impairment, and Alzheimer's disease: Contributory factors and treatment responses. In J. I. Tracy, B. M. Hampstead, & K. Sathian (Eds.), *Cognitive plasticity in neurologic disorders* (pp. 197–224). Oxford: Oxford University Press.

Hankin, B. L., & Abramson, L. Y. (2001). Development of gender differences in depression: An elaborated cognitive vulnerability-transactional stress theory. *Psychological Bulletin, 127*, 773–796.

Hanks, J. R., & Hanks Jr., L. M. (1948). The physically handicapped in certain non-occidental societies. *Journal of Social Issues, 4*, 11–20.

Hanlon, N. (2012). *Masculinities, care and equality: Identity and nurture in men's lives*. New York: Palgrave Macmillan.

Hardy, J. A., & Higgins, G. A. (1992). Alzheimer's disease: The amyloid cascade hypothesis. *Science, 256*, 184–185.

Harper, D. G. (2014). Sleep and circadian rhythms in dementia. In A. E. Budson & N. W. Kowall (Eds.), *The handbook of Alzheimer's disease and other dementias* (pp. 483–503). Chichester, West Sussex: Wiley.

Harrison, S. L., de Craen, A. J. M., Kerse, N., Teh, R., Granic, A., Davies, K., et al. (2017). Predicting risk of cognitive decline in very old adults using three models: The Framingham Stroke Risk Profile; the Cardiovascular Risk Factors, Aging, and Dementia Model; and Oxi-Inflammatory Biomarkers. *Journal of the American Geriatrics Society, 65*, 381–389.

Harrison, T., Stuifbergen, A., Walker, J., Scott, T., & Choban, R. (2011). The meaning of gender while aging with paralytic polio. *Journal of Gerontological Social Work, 54*, 138–158.

Hatch, S. L., & Dohrenwend, B. P. (2007). Distribution of traumatic and other stressful life events by race/ethnicity, gender, SES and age: A review of the research. *American Journal of Community Psychology, 40*, 313–332.

Haveman, M., Perry, J., Salvador-Carulla, L., Walsh, P. N., Kerr, M., Van Schrojenstein Lantman-De Valk, H., et al. (2011). Ageing and health status in adults with intellectual disabilities: Results of the European POMONA II study. *Journal of Intellectual and Developmental Disability, 36*, 49–60.

Haveman, M. J. (2004). Disease epidemiology and aging people with intellectual disabilities. *Journal of Policy and Practice in Intellectual Disabilities, 1*, 16–23.

Haynie, D. A., Berg, S., Johansson, B., Gatz, M., & Zarit, S. H. (2001). Symptoms of depression in the oldest old: A longitudinal study. *Journal of Gerontology, 56B*, P111–P118.

He, Z., Guo, J. L., McBride, J. D., Narasimhan, S., Kim, H., Changolkar, L., et al. (2018). Amyloid-β plaques enhance Alzheimer's brain tau-seeded pathologies by facilitating neuritic plaque tau aggregation. *Nature Medicine, 24*, 29–38.

Heilman, K. M. (2015). Foreword. In J. I. Tracy, B. M. Hampstead, & K. Sathian (Eds.), *Cognitive plasticity in neurologic disorders* (pp. ix–x). Oxford: Oxford University Press.

Helgason, A. R., Adolfsson, J., Dickman, P., Arver, S., Fredrikson, M., Gothberg, M., et al. (1996). Sexual desire, erection, orgasm and ejaculatory functions and their importance to elderly Swedish men: A population-based study. *Age and Ageing, 25*, 285–291.

Hendriks, G. J., Oude Voshaar, R. C., Keijsers, G. P. J., Hoogduin, C. A. L., & van Balkom, A. J. L. M. (2008). Cognitive-behavioural therapy for late-life anxiety disorders: A systematic review and meta-analysis. *Acta Psychiatrica Scandinavica, 117*, 403–411.

Hermans, H., & Evenhuis, H. M. (2012). Life events and their associations with depression and anxiety in older people with intellectual disabilities: Results of the HA-ID study. *Journal of Affective Disorders, 138*, 79–85.

Hermans, H., & Evenhuis, H. M. (2014). Multimorbidity in older adults with intellectual disabilities. *Research in Developmental Disabilities, 35*, 776–783.

Hermida, A. P., McDonald, W. M., Steenland, K., & Levey, A. (2012). The association between late-life depression, mild cognitive impairment and dementia: Is inflammation the missing link? *Expert Review of Neurotherapeutics, 12*(11), 1339–1350.

Hilderink, P. H., Collard, R., Rosmalen, J. G. M., & Oude Voshaar, R. C. (2013). Prevalence of somatoform disorders and medically unexplained symptoms in old age populations in comparison with younger age groups: A systematic review. *Ageing Research Reviews, 12*, 151–156.

Hindhede, A. L. (2011). Negotiating hearing disability and hearing disabled identities. *Health, 16*, 169–185.

Hodge, C. F. (1894). Changes in ganglion cells from birth to senile death. Observations on man and honey-bee. *Journal of Physiology, 17*, 128–134.

Hogg, J., Lucchino, R., Wang, K., Janicki, M. P., & Working Group. (2000). *Healthy ageing - Adults with intellectual disabilities: Ageing and Social Policy*. Geneva: World Health Organization.

Hollingworth, P., & Williams, J. (2014). Genetic risk factors for dementia. In A. E. Budson & N. W. Kowall (Eds.), *The handbook of Alzheimer's disease and other dementias* (pp. 197–234). Chichester, West Sussex: Wiley.

Holstege, H., Beker, N., Dijkstra, T., Pieterse, K., Wemmenhove, E., Schouten, K., et al. (2018). The 100-plus Study of Dutch cognitively healthy centenarians: Rationale, design and cohort description. *European Journal of Epidemiology*. https://doi.org/10.1101/295287

Holwerda, T. J., Schoevers, R. A., Dekker, J., Deeg, D. J. H., Jonker, C., & Beekman, A. T. F. (2007). The relationship between generalized anxiety disorder, depression and mortality in old age. *International Journal of Geriatric Psychiatry, 22*, 241–249.

Hörder, H., Johansson, L., Guo, X. X., Grimby, G., Kern, S., Östling, S., et al. (2018). Midlife cardiovascular fitness and dementia: A 44-year longitudinal population study in women. *Neurology, 2018*, e1–e8. https://doi.org/10.1212/WNL.0000000000005290

Hori, Y., Hashimoto, T., Nomoto, H., Hyman, B. T., & Iwatsubo, T. (2015). Role of apolipoprotein E in β-amyloidogenesis: Isoform-specific effects on protofibril to fibril conversion of Aβ in vitro and brain Aβ deposition in vivo. *The Journal of Biochemistry, 290*, 15163–15174.

Hornig, M. (2013). The role of microbes and autoimmunity in the pathogenesis of neuropsychiatric illness. *Current Opinions in Rheumatology, 25*, 488–795.

Horvath, S., Pirazzini, C., Bacalini, M. G., Gentilini, D., Di Blasio, A. M., Delledonne, M., et al. (2015). Decreased epigenetic age of PBMCs from Italian semisupercentenarians and their offspring. *Aging, 7*, 1159–1170.

Houtjes, W., van Meijel, B., van de Ven, P. M., Deeg, D., van Tilburg, T., & Beekman, A. (2014). The impact of an unfavorable depression course on network size and loneliness in older people: A longitudinal study in the community. *International Journal of Geriatric Psychiatry, 29*, 1010–1017.

Hultsch, D. F., Hertzog, C., Small, B. J., & Dixon, R. A. (1999). Use it or lose it: Engaged lifestyle as a buffer of cognitive decline in aging? *Psychology and Aging, 14*, 245–263.

Hyde, J. S. (2005). The gender similarities hypothesis. *American Psychologist, 60*, 581–592.

Hyde, J. S. (2014). Gender similarities and differences. *Annual Review of Psychology, 65*, 373–398.

Hyde, J. S., Krajnik, M., & Skuldt-Niederberger, K. (1991). Androgyny across the life span: A replication and longitudinal followup. *Developmental Psychology, 27*, 516–519.

Jack, C. R., Petersen, R. C., Xu, Y., O'Brien, P. C., Smith, G. E., Ivnik, R. J., et al. (2000). Rates of hippocampal atrophy correlate with change in clinical status in aging and AD. *Neurology, 55*, 484–490.

Jakobsson, U., Hallberg, I. R., & Westergren, A. (2004). Overall and health related quality of life among the oldest old in pain. *Quality of Life Research, 13*, 125–136.

James, B. D., Bennett, D. A., Boyle, P. A., Leurgans, S., & Schneider, J. A. (2012). Dementia from Alzheimer disease and mixed pathologies in the oldest old. *Journal of the American Medical Association, 307*, 1798–1800.

Janicki, M. P., & Dalton, A. J. (1998). Sensory impairments among older adults with intellectual disability. *Journal of Intellectual and Developmental Disability, 23*, 3–11.

Janicki, M. P., Dalton, A. J., Henderson, C. M., & Davidson, P. W. (1999). Mortality and morbidity among older adults with intellectual disability: Health services considerations. *Disability and Rehabilitation, 21*, 284–294.

Janicki, M. P., Davidson, P. W., Henderson, C. M., McCallion, P., Taets, J. D., Force, L. T., et al. (2002). Health characteristics and health services utilization in older adults with intellectual disability living in community residences. *Journal of Intellectual Disability Research, 46*, 287–298.

Jansson, M., Gatz, M., Berg, S., Johansson, B., Malmberg, B., McClearn, G. E., et al. (2004). Gender differences in heritability of depressive symptoms in the elderly. *Psychological Medicine, 34*, 471–479.

Jefferson, A. L., Gentile, A. M., & Kahlon, R. (2014). Vascular dementia. In A. E. Budson & N. W. Kowall (Eds.), *The handbook of Alzheimer's disease and other dementias* (pp. 92–130). Chichester, West Sussex: Wiley.

Jellinger, K. A., & Attems, J. (2010). Prevalence of dementia disorders in the oldest-old: An autopsy study. *Acta Neuropathologica, 119*, 421–433.

Jessberger, S., & Gage, F. H. (2008). Stem cell-associated structural and functional plasticity in the aging hippocampus. *Psychology of Aging, 23*, 684–691.

Jeste, D. V., Alexopoulos, G. S., Bartels, S. J., Cummings, J. L., Gallo, J. L., Gottlieb, G. L., et al. (1999). Consensus statement on the upcoming crisis in geriatric mental health: Research agenda for the next 2 decades. *Archives of General Psychiatry, 56*, 848–853.

Johansson, B. B. (2004). Brain plasticity in health and disease. *Keio Journal of Medicine, 53*, 231–246.

Johnson, C. L., & Barer, B. M. (1992). Patterns of engagement and disengagement among the oldest old. *Journal of Aging Studies, 6*, 351–364.

Johnson, C. L., & Barer, B. M. (1993). Coping and a sense of control among the oldest old: An exploratory analysis. *Journal of Aging Studies, 7*, 67–80.

Jones, S., Nyberg, L., Sandblom, J., Stigsdotter Neely, A., Ingvar, M., Magnus Petersson, K., et al. (2006). Cognitive and neural plasticity in aging: General and task- specific limitations. *Neuroscience and Biobehavioral Reviews, 30*, 864–871.

Jopp, D., & Rott, C. (2006). Adaptation in very old age: Exploring the role of resources, beliefs, and attitudes for centenarians' happiness. *Psychology and Aging, 21*, 266–280.

Jopp, D. S., Boerner, K., Ribeiro, O., & Rott, C. (2016). Life at age 100: An international research agenda for centenarian studies. *Journal of Aging and Social Policy, 28*, 133–147.

Joseph, R. (2000). The evolution of sex differences in language, sexuality, and visual–spatial skills. *Archives of Sexual Behavior, 29*, 35–66.

Just, M. A., & Buchweitz, A. (2017). What brain imaging reveals about the nature of multitasking. In S. E. F. Chipman (Ed.), *The Oxford handbook of cognitive science* (pp. 265–282). Oxford: Oxford University Press.

Kahn, J. R., & Pearlin, L. I. (2006). Financial strain over the life course and health among older adults. *Journal of Health and Social Behavior, 47*, 17–31.

Kahn, R. L., & Antonucci, T. C. (1980). Convoys over the life course: Attachment, roles, and social support. In P. B. Baltes & O. C. Brim (Eds.), *Life-span, development, and behavior* (pp. 254–283). New York: Academic Press.

Kaneda, H., Maeshima, K., Goto, N., Kobayakawa, T., Ayabe-Kanamura, S., & Saito, S. (2000). Decline in taste and odor discrimination abilities with age, and relationship between gestation and olfaction. *Chemical Senses, 25*, 331–337.

Kanekiyo, T., Xu, H., & Bu, G. (2014). ApoE and Aβ in Alzheimer's disease: Accidental encounters or partners? *Neuron, 81*, 740–754.

Karel, M. J., Gatz, M., & Smyer, M. A. (2012). Aging and mental health in the decade ahead: What psychologists need to know. *American Psychologist, 67*, 184–198.

Karsenty, G. (2017). Update on the biology of osteocalcin. *Endocrine Practice, 23*, 1270–1274.

Kassaar, O., Pereira Morais, M., Xu, S., Adam, E. L., Chamberlain, R. C., Jenkins, B., et al. (2017). Macrophage Migration Inhibitory Factor is subjected to glucose modification and oxidation in Alzheimer's Disease. *Nature Scientific Reports, 7*, 42874. https://doi.org/10.1038/srep42874

Kato, K., Zweig, R., Barzilai, N., & Atzmon, G. (2012). Positive attitude towards life and emotional expression as personality phenotypes for centenarians. *Aging, 4*, 359–367.

Kato, K., Zweig, R., Schechter, C. B., Barzilai, N., & Atzmon, G. (2016). Positive attitude toward life, emotional expression, self-rated health, and depressive symptoms among centenarians and near-centenarians. *Aging and Mental Health, 20*, 930–939.

Kato, K., Zweig, R., Schechter, C. B., Verghese, J., Barzilai, N., & Atzmon, G. (2013). Personality, self-rated health, and cognition in centenarians: Do personality and self-rated health relate to cognitive function in advanced age? *Aging, 5*, 183–191.

Katzman, R. (1993). Education and the prevalence of dementia and Alzheimer's disease. *Neurology, 43*, 13–20.

Katzman, R., Terry, R., DeTeresa, R., Brown, T., Davies, P., Fuld, P., et al. (1988). Clinical, pathological, and neurochemical changes in dementia: A subgroup with preserved mental status and numerous neocortical plaques. *Annals of Neurology, 23*, 138–144.

Kavé, G., Eyal, N., Shorek, A., & Cohen-Mansfield, J. (2008). Multilingualism and cognitive state in the oldest old. *Psychology and Aging, 23*, 70–78.

Kawas, C. H., & Corrada, M. M. (2006). Alzheimer's and dementia in the oldest-old: A century of challenges. *Current Alzheimer Research, 3*, 411–419.

Kawas, C. H., Kim, R. C., Sonnen, J. A., Bullain, S. S., Trieu, T., & Corrada, M. M. (2015). Multiple pathologies are common and related to dementia in the oldest-old: The 90+ Study. *Neurology, 85*, 535–542.

Keith, V. M. (1993). Gender, financial strain, and psychological distress among older adults. *Research on Aging, 15*, 123–147.
Kemp, B. J., Adams, B. M., & Campbell, M. L. (1997). Depression and life satisfaction in aging polio survivors versus age-matched controls: Relation to post-polio syndrome, family functioning and attitude towards disability. *Archives of Physical Medicine and Rehabilitation, 78*, 187–192.
Kemp, B. J., & Krause, J. S. (1999). Depression and life satisfaction among people ageing with post-polio and spinal cord injury. *Disability and Rehabilitation, 21*, 241–249.
Kempermann, G., Fabel, K., Ehninger, D., Babu, H., Leal-Galicia, P., Garthe, A., et al. (2010). Why and how physical activity promotes experience-induced brain plasticity. *Frontiers in Neuroscience, 4*, 189. https://doi.org/10.3389/fnins.2010.00189
Kern, M. L., & Friedman, H. S. (2008). Do conscientious individuals live longer? A quantitative review. *Health Psychology, 27*, 505–512.
Kessler, R. C. (2003). Epidemiology of women and depression. *Journal of Affective Disorders, 74*, 5–13.
Khan, U. A., Liu, L., Provenzano, F. A., Berman, D. E., Profaci, C. P., Sloan, R., et al. (2014). Molecular drivers and cortical spread of lateral entorhinal cortex dysfunction in preclinical Alzheimer's disease. *Nature Neurosciences, 17*, 304–311.
Kim, J. E., & Moen, P. (2002). Retirement transitions, gender, and psychological well- being: A life-course, ecological model. *Journal of Gerontology, 57B*, P212–P222.
Kirchengast, S., & Haslinger, B. (2008). Gender differences in health-related quality of life among healthy aged and old-aged Austrians: Cross-sectional analysis. *Gender Medicine, 5*, 270–278.
Kleim, J. A., Lussnig, E., Schwarz, E. R., Comery, T. A., & Greenough, W. T. (1996). Synaptogenesis and FOS expression in the motor cortex of the adult rat after motor skill learning. *Journal of Neuroscience, 16*, 4529–4535.
Kliegel, M., Zimprich, D., & Rott, C. (2004). Life-long intellectual activities mediate the predictive effect of early education on cognitive impairment in centenarians: A retrospective study. *Aging and Mental Health, 8*, 430–437.
Koenig, H. G. (2006). Religion, spirituality and aging. *Aging and Mental Health, 10*, 1–3.
Koenig, H. G. (2009). Research on religion, spirituality, and mental health: A review. *Canadian Journal of Psychiatry, 54*, 283–291.
Koenig, H. G., Smiley, M., & Gonzales, J. (1988). *Religion, health, and aging*. Westport, CT: Greenwood Press.
Koger, S. M., Chapin, K., & Brotons, M. (1999). Is music therapy and effective intervention for dementia? A meta-analytical review of literature. *Journal of Music Therapy, 36*, 2–15.
Korpelainen, H. (2000). Variation in the heritability and evolvability of human lifespan. *Naturwissenschaften, 87*, 566–568.
Kowall, N. W. (2014). Rational therapeutics for Alzheimer's disease and other dementias. In A. E. Budson & N. W. Kowall (Eds.), *The handbook of Alzheimer's disease and other dementias* (pp. 3001–3311). Chichester, West Sussex: Wiley.
Kraaij, V., Arensman, E., & Spinhoven, P. (2002). Negative life events and depression in elderly persons: A meta-analysis. *Journals of Gerontology: Psychological Science, 57B*, P87–P94.
Kramer, A. F., Bherer, L., Colcombe, S. J., Dong, W., & Greenough, W. T. (2004). Environmental influences on cognitive and brain plasticity during aging. *Journal of Gerontology, 59A*, 940–957.
Kramer, A. F., Erickson, K. I., & Colcombe, S. J. (2006). Exercise, cognition, and the aging brain. *Journal of Applied Physiology, 101*, 1237–1242.
Krause, J. S., & Coker, J. L. (2006). Aging after spinal cord injury: A 30-year longitudinal study. *Journal of Spinal Cord Medicine, 29*, 371–376.
Krause, N. (1987). Chronic financial strain, social support, and depressive symptoms among older adults. *Psychology and Aging, 2*, 185–192.
Krause, N. (1997). Anticipated support, received support, and economic stress among older adults. *Journal of Gerontology, 52B*, P284–P293.

Krause, N. (2002). Church-based social support and health in old age: Exploring variations by race. *Journal of Gerontology, 57B*, S332–S347.

Krause, N. (2004). Stressors arising in highly valued roles, meaning in life, and the physical health status of older adults. *Journal of Gerontology, 59B*, S287–S297.

Krause, N. (2004a). Religion, aging, and health: Exploring new frontiers in medical care. *Southern Medical Journal, 97*, 1215–1222.

Krause, N. (2004b). Common facets of religion, unique facets of religion, and life satisfaction among older African Americans. *Journal of Gerontology, 59B*, S109–S117.

Krause, N., Jay, G., & Liang, J. (1991). Financial strain and psychological well-being among the American and Japanese elderly. *Psychology and Aging, 6*, 170–181.

Kudielka, B. M., Buske-Kirschbaum, A., Hellhammer, D. H., & Kirschbaum, C. (2004). HPA axis responses to laboratory psychosocial stress in healthy elderly adults, younger adults, and children: Impact of age and gender. *Psychoneuroendocrinology, 29*, 83–98.

Kudielka, B. M., & Kirschbaum, C. (2005). Sex differences in HPA axis responses to stress: A review. *Biological Psychology, 69*, 113–132.

Kulmala, J., Solomon, A., Kåreholt, I., Ngandu, T., Rantanen, T., Laatikainen, T., et al. (2014). Association between mid- to late life physical fitness and dementia: Evidence from the CAIDE study. *Journal of Internal Medicine, 276*, 296–307.

Kurz, A. (2013). A brief history of Alzheimer's disease. In R. Martins (Ed.), *Understanding Alzheimer's* (pp. 53–66). Sydney, NSW: MacMillan.

Lamb, S. E., Sheehan, B., Atherton, N., Nichols, V., Collins, H., Mistry, D., et al. (2018). Dementia And Physical Activity (DAPA) trial of moderate to high intensity exercise training for people with dementia: Randomised controlled trial. *BMJ, 361*, k1675. https://doi.org/10.1136/bmj.k1675

Lau, B. H., Kwan, J. S., Cheung, K. S., & Martin, P. (2016). Depression moderates the frailty-subjective health link among Chinese near centenarians and centenarians. *American Journal of Geriatric Psychiatry, 24*, 753–761.

Laumann, E. O., Paik, A., Glasser, D. B., Kang, J.-H., Wang, T., Levinson, B., et al. (2006). A cross-national study of subjective sexual well-being among older women and men: Findings from the Global Study of Sexual Attitudes and Behaviors. *Archives of Sexual Behavior, 35*, 145–161.

Lautenbacher, S., Kunz, M., Strate, P., Nielsen, J., & Arendt-Nielsen, L. (2005). Age effects on pain thresholds, temporal summation and spatial summation of heat and pressure pain. *Pain, 115*, 410–418.

Le Bourg, É. (2012). Forecasting continuously increasing life expectancy: What implications? *Ageing Research Reviews, 11*, 325–328.

Lee, M. S., Shin, B.-C., & Ernst, E. (2009). Acupuncture for Alzheimer's disease: A systematic review. *International Journal of Clinical Practice, 63*, 874–879.

Lehmann, R., Denissen, J. J. A., Allemand, M., & Penke, L. (2013). Age and gender differences in motivational manifestations of the Big Five from age 16 to 60. *Developmental Psychology, 49*, 365–383.

Lei, X., Hu, Y., McArdle, J. J., Smith, J. P., & Zhao, Y. (2012). Gender differences in cognition among older adults in China. *Journal of Human Resources, 47*, 951–971.

Lenze, E. J., Mulsant, B. H., Shear, M. K., Alexopoulos, G. S., Frank, E., & Reynolds, C. F., III. (2001). Comorbidity of depression and anxiety disorders in later life. *Depression and Anxiety, 14*, 86–93.

Levin, J. S., & Chatters, L. M. (1998). Religion, health, and psychological well-being in older adults. *Journal of Aging and Health, 10*, 504–531.

Li, S.-C., Schmiedek, F., Huxhold, O., Röcke, C., Smith, J., & Lindenberger, U. (2008a). Working memory plasticity in old age: Practice gain, transfer, and maintenance. *Psychology and Aging, 23*, 731–742.

Li, Y., Rinne, J. O., Mosconi, L., Pirraglia, E., Rusinek, H., DeSanti, S., et al. (2008b). Regional analysis of FDG and PIB-PET images in normal aging, mild cognitive impairment, and

Alzheimer's disease. *European Journal of Nuclear Medicine and Molecular Imaging, 35*, 2169–2181.

Li, L. W., Zhang, J., & Liang, J. (2009). Health among the oldest-old in China: Which living arrangements make a difference? *Social Science and Medicine, 68*, 220–227.

Li, R., & Singh, M. (2014). Sex differences in cognitive impairment and Alzheimer's disease. *Frontiers in Neuroendocrinology, 35*, 385–403.

Liechty, T., Ribeiro, N. F., Sveinson, K., & Dahlstrom, L. (2014). "It's about what I can do with my body": Body image and embodied experiences of aging among older Canadian men. *International Journal of Men's Health, 13*, 3–21.

Lifshitz, H. (2002). Attitudes toward aging in adult and elderly people with intellectual disability. *Educational Gerontology, 28*, 745–759.

Lin, N., Ye, X., & Ensel, W. M. (1999). Social support and depressed mood: A structural analysis. *Journal of Health and Social Behavior, 40*, 344–359.

Lindenberger, U., & Baltes, P. B. (1997). Intellectual functioning in old and very old age: Cross-sectional results from the Berlin Aging Study. *Psychology and Aging, 12*, 410–432.

Lindenberger, U., & Ghisletta, P. (2009). Cognitive and sensory declines in old age: Gauging the evidence for a common cause. *Psychology and Aging, 24*, 1–16.

Lindwall, M., Rennemark, M., Halling, A., Berglund, J., & Hassmén, P. (2006). Depression and exercise in elderly men and women: Findings from the Swedish National Study on Aging and Care. *Journal of Aging and Physical Activity, 15*, 41–55.

Ling, C. H. Y., Taekema, D., de Craen, A. J. M., Gussekloo, J., Westendorp, R. G. J., & Maier, A. B. (2010). Handgrip strength and mortality in the oldest old population: The Leiden 85-plus study. *Canadian Medical Association Journal, 182*, 429–435.

Lippold, T., & Burns, J. (2009). Social support and intellectual disabilities: A comparison between social networks of adults with intellectual disability and those with physical disability. *Journal of Intellectual Disability Research, 53*, 463–473.

Liu, F., & Aftab, A. (2020). Management of treatment resistant depression in the elderly. *Psychiatric Times, 37*(1). Downloaded February 10, 2020, from https://www.psychiatrictimes.com/cme/management- treatment-resistant-depression-elderly

Liu, G., & Zhang, Z. (2004). Sociodemographic differentials of the self-rated health of the oldest-old Chinese. *Population Research and Policy Review, 23*, 117–133.

Liu, M., Bian, C., Zhang, J., & Wen, F. (2015). Apolipoprotein E gene polymorphism and Alzheimer's disease in Chinese population: A meta-analysis. *Scientific Reports, 4*, 4383. https://doi.org/10.1038/srep04383

Livingston, G., Johnston, K., Katona, C., Paton, J., Lyketsos, C. G., & Old Age Task Force of the World Federation of Biological Psychiatry. (2005). Systematic review of psychological approaches to the management of neuropsychiatric symptoms of dementia. *American Journal of Psychiatry, 162*, 1996–2021.

Lobo, A., Launer, L. J., Fratiglioni, L., Andersen, K., Di Carlo, A., Breteler, M. M. B., et al. (2000). Prevalence of dementia and major subtypes in Europe: A collaborative study of population-based cohorts. *Neurology, 54*, S4–S9.

Lock, M. (2013). *The Alzheimer conundrum: Entanglements of dementia and aging*. Princeton, NJ: Princeton University Press.

Long, M. V., & Martin, P. (2000). Personality, relationship closeness, and loneliness of oldest old adults and their children. *Journal of Gerontology, 55B*, P311–P319.

Lopata, H. Z. (1973). *Widowhood in an American city*. Cambridge, MA: Schenkman.

López-Carr, D., & Ervin, D. (2012). A geriatric fountain of youth in the Caucasus or spurious census data: Spooning through the yogurt myth. *Rural Studies, 27*, 135–147.

Lowe, D., Horvath, S., & Raj, K. (2016). Epigenetic clock analyses of cellular senescence and ageing. *Oncotarget, 7*, 8524–8531.

Luk, G., Bialystok, E., Craik, F. I., & Grady, C. L. (2011). Lifelong bilingualism maintains white matter integrity in older adults. *Journal of Neuroscience, 31*, 16808–16813.

Lupien, S. J., McEwen, B. S., Gunnar, M. R., & Heim, C. (2009). Effects of stress throughout the lifespan on the brain, behavior and cognition. *Nature Reviews Neuroscience, 10*, 434–445.

Luppa, M., Sikorski, C., Luck, T., Ehreke, L., Konnopka, A., Wiese, B., et al. (2012). Age- and gender-specific prevalence of depression in latest-life – Systematic review and meta-analysis. *Journal of Affective Disorders, 136*, 212–221.

Lustig, C., Shah, P., Seidler, R., & Reuter-Lorenz, P. A. (2009). Aging, training, and the brain: A review and future directions. *Neuropsychology Review, 19*, 504–522.

Luyten, W., Antal, P., Braeckman, B. P., Bundy, J., Cirulli, F., Fang-Yen, C., et al. (2016). Ageing with elegans: A research proposal to map healthspan Pathways. *Biogerontology, 17*, 771–782.

MacDonald, S., & Summers, S. J. (2020). Psychosocial interventions for people with intellectual disabilities and dementia: A systematic review. *Journal of Applied Research in Intellectual Disabilities: JARID, 33*(5), 839–855. https://doi.org/10.1111/jar.12722

Mackenzie, E. R., Rajagopal, D. E., Meibohm, M., & Lavizzo-Mourey, R. (2000). Spiritual support and psychological well-being: Older adults' perceptions of the religion and health connection. *Alternative Therapies in Health and Medicine, 6*, 37–45.

Mackowicz, J., & Wnek-Gozdek, J. (2017). The process of (non) ageing in the perception of Polish centenarians. *Educational Gerontology, 43*, 101–111.

Mahncke, H. W., Bronstone, A., & Merzenich, M. M. (2006). Brain plasticity and functional losses in the aged: Scientific bases for a novel intervention. *Reprogramming the Brain, 157*, 81–109.

Mandell, A. M., & Green, R. C. (2014). Alzheimer's disease. In A. E. Budson & N. W. Kowall (Eds.), *The handbook of Alzheimer's disease and other dementias* (pp. 3–91). Chichester, West Sussex: Wiley.

Manela, M., Katona, C., & Livingston, G. (1996). How common are the anxiety disorders in old age? *International Journal of Geriatric Psychiatry, 11*, 65–70.

Margrett, J. A., Daugherty, K., Martin, P., MacDonald, M., Davey, A., Woodard, J. L., et al. (2011). Affect and loneliness among centenarians and the oldest old: The role of individual and social resources. *Aging and Mental Health, 15*, 385–396.

Markham, J. A., & Greenough, W. T. (2004). Experience-driven brain plasticity: Beyond the synapse. *Neuron Glia Biology, 1*, 351–363.

Martin, P., Poon, L. W., Kim, E., & Johnson, M. A. (1996). Social and psychological resources in the oldest old. *Experimental Aging Research, 22*, 121–139.

Masui, Y., Gondo, Y., Inagaki, H., & Hirose, N. (2006). Do personality characteristics predict longevity? Findings from Tokyo Centenarian Study. *Age, 28*, 353–361.

Mayes, A. (2015). What factors determine how much it recovers over time? In J. I. Tracy, B. M. Hampstead, & K. Sathian (Eds.), *Cognitive plasticity in neurologic disorders* (pp. 258–281). Oxford: Oxford University Press.

Maylor, E. A., Reimers, S., Choi, J., Collaer, M. L., Peters, M., & Silverman, I. (2007). Gender and sexual orientation differences in cognition across adulthood: Age is kinder to women than to men regardless of sexual orientation. *Archives of Sexual Behavior, 36*, 235–249.

McCabe, D. P., Roediger, H. L., III, McDaniel, M. A., Balota, D. A., & Hambrick, D. Z. (2010). The relationship between working memory capacity and executive functioning: Evidence for a common executive attention construct. *Neuropsychology, 24*, 222–243.

McCallion, P., & McCarron, M. (2004). Ageing and intellectual disabilities: A review of recent literature. *Current Opinion in Psychiatry, 17*, 349–352.

McCarrey, A. C., An, Y., Kitner-Triolo, M. H., Ferrucci, L., & Resnick, S. M. (2016). Sex differences in cognitive trajectories in clinically normal older adults. *Psychology and Aging, 31*, 166–175.

McDermott, O., Charlesworth, G., Hogervorst, E., Stoner, C., Moniz-Cook, E., Spector, A., et al. (2019). Psychosocial interventions for people with dementia: A synthesis of systematic reviews. *Aging and Mental Health, 23*, 393–403.

McEwen, B. S., & Morrison, J. H. (2013). The brain on stress: Vulnerability and plasticity of the prefrontal cortex over the life course. *Neuron, 79*, 16–29.

McFadden, S. H. (1995). Religion and well-being in aging persons in an aging society. *Journal of Social Issues, 51*, 161–175.

McFarland, M. J. (2010). Religion and mental health among older adults: Do the effects of religious involvement vary by gender? *Journal of Gerontology, 65B*, 621–630.

McKee, A. C., & Gavett, B. E. (2014). The neuropathology of the dementing disorders. In A. E. Budson & N. W. Kowall (Eds.), *The handbook of Alzheimer's disease and other dementias* (pp. 235–261). Chichester, West Sussex: Wiley.
McLean, C. P., Asnaani, A., Litz, B. T., & Hofmann, S. G. (2011). Gender differences in anxiety disorders: Prevalence, course of illness, comorbidity and burden of illness. *Journal of Psychiatric Research, 45*, 1027–1035.
Meier, P., Finch, A., & Evan, G. (2000). Apoptosis in development. *Nature, 407*, 796–801.
Mennin, D. S., Heimberg, R. G., Turk, C. L., & Fresco, D. M. (2005). Preliminary evidence for an emotion dysregulation model of generalized anxiety disorder. *Behaviour Research and Therapy, 43*, 1281–1310.
Middleton, L., & Yaffe, K. (2009). Promising strategies for the prevention of dementia. *Archives of Neurology, 66*, 1210–1215.
Miller, H. (1957). *Big Sur and the oranges of Hieronymus Bosch*. New York: New Directions Publishing Corporation.
Milman, S., Atzmon, G., Crandall, J., & Barzilai, N. (2014). Phenotypes and genotypes of high density lipoprotein cholesterol in exceptional longevity. *Current Vascular Pharmacology, 12*, 690–697.
Ming, G., & Song, H. (2005). Adult neurogenesis in the mammalian central nervous system. *Annual Reviews of Neuroscience, 28*, 223–250.
Mirmiran, M., van Someren, E. J., & Swaab, D. F. (1996). Is brain plasticity preserved during aging and in Alzheimer's disease? *Behavioural Brain Research, 78*, 43–48.
Mitchell, A. J., & Subramaniam, H. (2005). Prognosis of depression in old age compared to middle age: A systematic review of comparative studies. *American Journal of Psychiatry, 162*, 1588–1601.
Mitchell, L. A., & Maercklein, G. (1996). The effect of individualized special instruction on the behaviors of nursing home residents diagnosed with dementia. *American Journal of Alzheimer's Disease and Other Dementias, 11*, 23–31.
Miyawaki, C. E. (2015). Association of social isolation and health across different racial and ethnic groups of older Americans. *Ageing and Society, 35*, 2201–2228.
Moberg, D. O. (2005). Research in spirituality, religion, and aging. *Journal of Gerontological Social Work, 45*, 11–40.
Mochcovitch, M. D., da Rocha Freire, R. C., Ferreira Garcia, R., & Nardi, A. E. (2014). A systematic review of fMRI studies in generalized anxiety disorder: Evaluating its neural and cognitive basis. *Journal of Affective Disorders, 167*, 336–342.
Mogensen, J. (2015). Recovery, compensation, and reorganization in neuropathology. In J. I. Tracy, B. M. Hampstead, & K. Sathian (Eds.), *Cognitive plasticity in neurologic disorders* (pp. 3–28). Oxford: Oxford University Press.
Montesanto, A., De Rango, F., Pirazzini, C., Guidarelli, G., Domma, F., Franceschi, C., et al. (2017). Demographic, genetic and phenotypic characteristics of centenarians in Italy: Focus on gender differences. *Mechanisms of Ageing and Development, 165*, 68–74.
Mora, F. (2013). Successful brain aging: Plasticity, environmental enrichment, and lifestyle. *Dialogues in Clinical Neuroscience, 15*, 45–52.
Moreira-Almeida, A., Lotufo Neto, F., & Koenig, H. G. (2006). Religiousness and mental health: A review. *Revista Brasileira de Psiquiatria, 28*, 242–250.
Morgan, M., Patrick, D. L., & Charlton, J. R. (1984). Social networks and psychological support among disabled people. *Social Science and Medicine, 19*, 489–497.
Morin, P. (2014). Other dementias. In A. E. Budson & N. W. Kowall (Eds.), *The handbook of Alzheimer's disease and other dementias* (pp. 179–193). Chichester, West Sussex: Wiley.
Morris, M., Lundell, J., & Dishman, E. (2004). Catalyzing social interaction with ubiquitous computing: A needs assessment of elders coping with cognitive decline. In *Proceedings of ACM CHI 2004*, Vienna, Austraia (Vol. 2, pp. 1151–1154).
Morrow-Howell, N., Hinterlong, J., Rozario, P. A., & Tang, F. (2003). Effects of volunteering on the well-being of older adults. *Journal of Gerontology, 58B*, S137–S145.

Mortimer, J. A., Ding, D., Borenstein, A. R., DeCarli, C., Guo, Q., Wa, Y., et al. (2012). Changes in brain volume and cognition in a randomized trial of exercise and social interaction in a community-based sample of non-demented Chinese elders. *Journal of Alzheimer's Disease, 30*, 757–766.

Mortimer, J. A., Schuman, L., & French, L. (1981). Epidemiology of dementing illness. In J. A. Mortimer & L. M. Schuman (Eds.), *The epidemiology of dementia: Monographs in epidemiology and biostatistics* (pp. 323–333). New York: Oxford University Press.

Moyle, W., Murfield, J. E., O'Dwyer, S., & Van Wyk, S. (2013). The effect of massage on agitated behaviours in older people with dementia: A literature review. *Journal of Clinical Nursing, 22*, 60–610.

Mrak, R. E., Griffin, W. S. T., & Graham, D. I. (1997). Aging-associated changes in human brain. *Journal of Neuropathology and Experimental Neurology, 56*, 1269–1275.

Mui, A. C., & Kang, S.-Y. (2006). Acculturation stress and depression among Asian immigrant elders. *Social Work, 51*, 243–255.

Mukadam, N., Sommerlad, A., & Livingston, G. (2017). The relationship of bilingualism compared to monolingualism to the risk of cognitive decline or dementia: A systematic review and meta-analysis. *Journal of Alzheimer's Disease, 58*, 45–54.

Murphy, T., Pereira Dias, G., & Thuret, S. (2014). Effects of diet on brain plasticity in animal and human studies: Mind the gap. *Neural Plasticity, 2014*, 563160, 32 pages. https://doi.org/10.1155/2014/563160

Nadeau, S. E. (2015). Neuroplastic mechanisms of language recovery after stroke. In J. I. Tracy, B. M. Hampstead, & K. Sathian (Eds.), *Cognitive plasticity in neurologic disorders* (pp. 61–84). Oxford: Oxford University Press.

Naftolin, F., Leranth, C., Horvath, T. L., & Garcia-Segura, L. M. (1996). Potential neuronal mechanisms of estrogen actions in synaptogenesis and synaptic plasticity. *Cellular and Molecular Neurobiology, 16*, 213–223.

Neal, M., & Barton Wright, P. (2003). Validation therapy for dementia. *Cochrane Database of Systematic Reviews*, (3), CD001394. https://doi.org/10.1002/14651858.CD001394

Neary, D., Snowden, J., & Mann, D. (2005). Frontotemporal dementia. *Lancet Neurology, 4*, 771–780.

Nedergaard, M., Ransom, B., & Goldman, S. A. (2003). New roles for astrocytes: Redefining the functional architecture of the brain. *Trends in Neurosciences, 26*, 523–530.

Nehen, H. G., & Hermann, D. M. (2015). Supporting dementia patients and their caregivers in daily life challenges: Review of physical, cognitive and psychosocial intervention studies. *European Journal of Neurology, 22*, 246–e20.

Nelson, C. J., Rosenfeld, B., Breitbart, W., & Galietta, M. (2002). Spirituality, religion, and depression in the terminally ill. *Psychosomatics, 43*, 213–220.

Neufeldt, A. H., & Mathieson, R. (1995). Empirical dimensions of discrimination against disabled people. *Health and Human Rights, 1*, 174–189.

Newman, A. B., Lee, J. S., Visser, M., Goodpaster, B. H., Kritchevsky, S. B., Tylavsky, F. A., et al. (2005). Weight change and the conservation of lean mass in old age: The Health, Aging and Body Composition Study. *American Journal of Clinical Nutrition, 82*, 872–878.

Nosraty, L., Jylhä, M., Raittila, T., & Lumme-Sandt, K. (2015). Perceptions by the oldest old of successful aging, Vitality 90+ Study. *Journal of Aging Studies, 32*, 50–58.

Nottenbohm, F. (1985). Neuronal replacement in adulthood. *Annals of the New York Academy of Science, 457*, 143–161.

Nurullah, A. S. (2010). Gender differences in distress: The mediating influence of life stressors and psychological resources. *Asian Social Science, 6*, 27–35.

Nussbaum, M. C., & Levmore, S. (2017). *Aging thoughfully: Conversations about retirement, romance, wrinkles, & regret*. New York: Oxford University Press.

Nyberg, J., Aberg, M. A., Schioler, L., Nilsson, M., Wallin, A., Toren, K., et al. (2014). Cardiovascular and cognitive fitness at age 18 and risk of early-onset dementia. *Brain, 137*, 1514–1523.

Nygren, B., Aléx, L., Jonsén, E., Gustafson, Y., Norberg, A., & Lundman, B. (2005). Resilience, sense of coherence, purpose in life and self-transcendence in relation to perceived physical and mental health among the oldest old. *Aging and Mental Health, 9*, 354–362.

O'Connor, T. M., O'Halloran, D. J., & Shanahan, F. (2000). The stress response and the hypothalamic-pituitary-adrenal axis: From molecule to melancholia. *QJM: An International Journal of Medicine, 93*, 323–333.

O'Neil, M., Freeman, M., Christensen, V., Telerant, A., Addleman, A., & Kansagara, D. (2011). Non-pharmacological interventions for behavioral symptoms of dementia: A systematic review of the evidence. VA-ESP Project #05–225; 2011. http://www.hsrd.research.va.gov/publications/esp/Dementia-Nonpharm.pdf.

Oberman, L. M., Rotenberg, A., & Pascual-Leone, A. (2015). Aberrant brain plasticity in autism spectrum disorders. In J. I. Tracy, B. M. Hampstead, & K. Sathian (Eds.), *Cognitive plasticity in neurologic disorders* (pp. 176–196). Oxford: Oxford University Press.

Öhman, H., Savikko, N., Strandberg, T. E., & Pitkälä, K. H. (2014). Effect of physical exercise on cognitive performance in older adults with mild cognitive impairment or dementia: A systematic review. *Dementia and Geriatric Cognitive Disorders, 38*, 347–365.

Oliver, R., Gyi, D., Porter, M., Marshall, R., & Case, K. (2001). A survey of the design needs of older and disabled people. In M. A. Hanson (Ed.), *Contemporary Ergonomics 2001, the Proceedings of the Annual Conference of the Ergonomics Society* (pp. 365–370). Cirencester: Royal Agricultural College, Taylor & Francis.

Onyike, C. U., & Diehl-Schmid, J. (2013). The epidemiology of frontotemporal dementia. *International Review of Psychiatry, 25*(2), 130–137.

Oswald, F., Wahl, H.-W., Schilling, O., Nygren, C., Fänge, A., Sixsmith, A., et al. (2007). Relationships between housing and healthy living in very old age. *The Gerontologist, 47*, 96–107.

Otte, C., Hart, S., Neylan, T. C., Marmar, C. R., Yaffe, K., & Mohr, D. C. (2015). A meta-analysis of cortisol response to challenge in human aging: Importance of gender. *Psychoneuroendocrinology, 30*, 80–91.

Padilla, R. (2011). Effectiveness of environment-based interventions for people with Alzheimer's disease and related dementias. *American Journal of Occupational Therapy, 65*, 514–522.

Pahkala, K. (1990). Social and environmental factors and depression in old age. *International Journal of Geriatric Psychiatry, 5*, 99–113.

Paillard, J. (1976). Réflexions sur l'usage du concept du plasticité en Neurobiologie. *Journal de Psychologie Normale et Pathologique, 1*, 33–47.

Panagiotakos, D. B., Chrysohoou, C., Siasos, G., Zisimos, K., Skoumas, J., Pitsavos, C., et al. (2011). Sociodemographic and lifestyle statistics of oldest old people (>80 Years) living in Ikaria Island: The Ikaria Study. *Cardiology Research and Practice, 2011*, 679187, 7 pages. https://doi.org/10.4061/2011/679187

Park, D. C., & Reuter-Lorenz, P. (2009). The adaptive brain: Aging and neurocognitive scaffolding. *Annual Review of Psychology, 60*, 173–196.

Parker, M. G., Ahacic, K., & Thorslund, M. (2005). Health changes among Swedish oldest old: Prevalence rates from 1992 and 2002 show increasing health problems. *Journal of Gerontology, 60A*, 1351–1355.

Pascual-Leone, A., Freitas, C., Oberman, L., Horvath, J. C., Halko, M., Eldaief, M., et al. (2011). Characterizing brain cortical plasticity and network dynamics across the age-span in health and disease with TMS-EEG and TMS-fMRI. *Brain Topography, 24*, 302–315.

Patja, K., Iivanainen, M., Vesala, H., Oksanen, H., & Ruoppila, I. (2000). Life expectancy of people with intellectual disability: A 35-year follow-up study. *Journal of Intellectual Disability Research, 44*, 591–599.

Patja, K., Mölsa, P., & Ilvanalnen, M. (2001). Cause-specific mortality of people with intellectual disability in a population-based, 35-year follow-up study. *Journal of Intellectual Disability Research, 45*, 30–40.

Patnaik, N. (2015). Cure for Alzheimer's disease. *World Journal of Neuroscience, 5*, 328–330.

Patten, A. R., Yau, S. Y., Fontaine, C. J., Meconi, A., Wortman, R. C., & Christie, B. R. (2015). The benefits of exercise on structural and functional plasticity in the rodent hippocampus of different disease models. *Brain Plasticity, 1*, 97–127.

Paul, C., Ayis, S., & Ebrahim, S. (2006). Psychological distress, loneliness and disability in old age. *Psychology, Health and Medicine, 11*, 221–232.

Pauls, F., Petermann, F., & Lepach, A. C. (2013). Gender differences in episodic memory and visual working memory including the effects of age. *Memory, 21*, 857–874.

Pearlin, L. I., & McKean Skaff, M. (1996). Stress and the life course: A paradigmatic alliance. *The Gerontologist, 36*, 239–247.

Pedersen, S. K. A., Andersen, P. N., Lugo, R. G., Andreassen, M., & Sütterlin, S. (2017). Effects of music on agitation in dementia: A meta-analysis. *Frontiers in Psychology, 8*, 742. https://doi.org/10.3389/fpsyg.2017.00742

Pentland, W., Tremblay, M., Spring, K., & Rosenthal, C. (1999). Women with physical disabilities: Occupational impacts of ageing. *Journal of Occupational Science, 6*, 111–123.

Perani, D., & Abutalebi, J. (2015). Bilingualism, dementia, cognitive and neural reserve. *Current Opinion in Neurology, 28*, 618–625.

Perani, D., Farsad, M., Ballarini, T., Lubian, F., Malpetti, M., Fracchetti, A., et al. (2017). The impact of bilingualism on brain reserve and metabolic connectivity in Alzheimer's dementia. *Proceedings of the National Academy of Sciences of the United States of America, 114*, 1690–1695.

Perälä, J., Suvisaari, J., Saarni, S. I., Kuoppasalmi, K., Isometsä, E., Pirkola, S., et al. (2007). Lifetime prevalence of psychotic and bipolar I disorders in a general population. *Archives of General Psychiatry, 64*, 19–28.

Perls, T. (2005). The different paths to age one hundred. *Annals of the New York Academy of Science, 1055*, 13–25.

Perls, T., Kunkel, L. M., & Puca, A. A. (2002). The genetics of exceptional human longevity. *Journal of the American Geriatrics Society, 50*, 359–368.

Perquin, M., Vaillant, M., Schuller, A. M., Pastore, J., Dartigues, J. F., Lair, M. L., et al. (2013). Lifelong exposure to multilingualism: New evidence to support cognitive reserve hypothesis. *PLoS One, 8*(4), e62030.

Petersen, J. L., & Hyde, J. S. (2010). A meta-analytic review of research on gender differences in sexuality, 1993–2007. *Psychological Bulletin, 136*, 21–38.

Petersen, R. C., Smith, G. E., Waring, S. C., Ivnik, R. J., Tangalos, E. G., & Kokmen, E. (1999). Mild cognitive impairment: Clinical characterization and outcome. *Archives of Neurology, 56*, 303–308.

Pillai, J. A., & Verghese, J. (2009). Social networks and their role in preventing dementia. *Indian Journal of Psychiatry, 51*(Suppl. S1), 22–28.

Pillemer, D., Wink, P., DiDonato, T., & Sanborn, R. (2003). Gender differences in autobiographical memory styles of older adults. *Memory, 11*, 525–532.

Pinquart, M., & Sörensen, S. (2001). Influences on loneliness in older adults: A meta- analysis. *Basic and Applied Social Psychology, 23*, 245–266.

Podewils, L. J., Guallar, E., Kuller, L. H., Fried, L. P., Lopez, O. L., Carlson, M., et al. (2005). Physical activity, APOE genotype, and dementia risk: Findings from the Cardiovascular Health Cognition Study. *American Journal of Epidemiology, 161*, 639–651.

Poiani, A. (2010). *Animal homosexuality: A biosocial perspective*. Cambridge: Cambridge University Press.

Portin, R., Saarijarvi, S., Joukamaa, M., & Salokangas, R. K. R. (1995). Education, gender and cognitive performance in a 62-year-old normal population: Results from the Turva Project. *Psychological Medicine, 25*, 1295–1298.

Poulain, M. (2011). Exceptional longevity in Okinawa: A plea for in-depth validation. *Demographic Research, 25*, 245–284.

Poulain, M., Herm, A., & Pes, G. (2013). The Blue Zones: Areas of exceptional longevity around the world. *Vienna Yearbook of Population Research, 11*, 87–108.

Powell, L. H., Shahabi, L., & Thoresen, C. E. (2003). Religion and spirituality: Linkages to physical health. *American Psychologist, 58*, 36–52.

Prakash, I. J. (2003). Aging, disability, and disabled older people in India. *Journal of Aging and Social Policy, 15*, 85–108.

Prince, M., Bryce, R., Albanese, E., Wimo, A., Ribeiro, W., & Ferri, C. P. (2013). The global prevalence of dementia: A systematic review and metaanalysis. *Alzheimer's and Dementia, 9*, 63–75.

Puchalski, C. M., Dorff, R. E., & Hendi, I. Y. (2004). Spirituality, religion, and healing in palliative care. *Clinics in Geriatric Medicine, 20*, 689–714.

Quoidbach, J., Gilbert, D. T., & Wilson, T. D. (2013). The end of history illusion. *Science, 339*, 96–98.

Radley, J. J., & Morrison, J. H. (2005). Repeated stress and structural plasticity in the brain. *Ageing Research Reviews, 4*, 271–287.

Rajah, M. N., & D'Esposito, M. (2005). Region-specific changes in prefrontal function with age: A review of PET and fMRI studies on working and episodic memory. *Brain, 128*, 1964–1983.

Raposo, S. M., Mackenzie, C. S., Henriksen, C. A., & Afifi, T. O. (2014). Time does not heal all wounds: Older adults who experienced childhood adversities have higher odds of mood, anxiety, and personality disorders. *American Journal of Geriatric Psychiatry, 22*, 1241–1250.

Raz, N., Lindenberger, U., Rodrigue, K. M., Kennedy, K. M., Head, D., Williamson, A., et al. (2005). Regional brain changes in aging healthy adults: General trends, individual differences and modifiers. *Cerebral Cortex, 15*, 1676–1689.

Reboussin, B. A., Rejeski, W. J., Martin, K. A., Callahan, K., Dunn, A. L., King, A. C., et al. (2000). Correlates of satisfaction with body function and body appearance in middle- and older aged adults: The activity counselling trial (ACT). *Psychology and Health, 15*, 239–254.

Reed, P. G. (1991). Toward a nursing theory of selftranscendence: Deductive reformulation using developmental theories. *Advances in Nursing Science, 13*, 64–77.

Reed, P. G. (1991a). Self-transcendence and mental health in oldest-old adults. *Nursing Research, 40*, 5–11.

Reijneveld, S. A., Westhoff, M. H., & Hopman-Rock, M. (2003). Promotion of health and physical activity improves the mental health of elderly immigrants: Results of a group randomised controlled trial among Turkish immigrants in the Netherlands aged 45 and over. *Community Health, 57*, 405–411.

Reilly, J., Troche, J., & Grossman, M. (2014). Language processing in dementia. In A. E. Budson & N. W. Kowall (Eds.), *The handbook of Alzheimer's disease and other dementias* (pp. 336–368). Chichester, West Sussex: Wiley.

Rejeski, W. J., & Mihalko, S. L. (2001). Physical activity and quality of life in older adults. *Journals of Gerontology, 56A*, 23–35.

Riedel-Heller, S. G., Busse, A., & Angermeyer, M. C. (2006). The state of mental health in old-age across the old European Union - A systematic review. *Acta Psychiatrica Scandinavica, 113*, 388–401.

Rizzi, L., Rosset, I., & Roriz-Cruz, M. (2014). Global epidemiology of dementia: Alzheimer's and vascular types. *BioMed Research International, 2014*, 908915, 8 pages. https://doi.org/10.1155/2014/908915

Robin-Champigneul, F. (2020). Jeanne Calment's unique 122-year life span: Facts and factors; longevity history in her genealogical tree. *Rejuvenation Research, 23*, 19–47.

Robine, J.-M., Allard, M., Herrmann, F. R., & Jeune, B. (2019). The real facts supporting Jeanne Calment as the oldest ever human. *Journals of Gerontology A Biological Sciences Medical Sciences, 74*, S13–S20.

Robine, J.-M., & Cubaynes, S. (2017). Worldwide demography of centenarians. *Mechanisms of Ageing and Development, 165*(Part B), 59–67.

Robine, J.-M., & Vaupel, J. W. (2002). Emergence of supercentenarians in low-mortality countries. *North America Actuarial Journal, 6*, 54–63.

Robinson, I., Ziss, K., Ganza, B., & Katz, S. (1991). Twenty years of the sexual revolution, 1965-1985: An update. *Journal of Marriage and Family, 53*, 216–220.

Rodríguez, E., Mateo, I., Infante, J., Llorca, J., Berciano, J., & Combarros, O. (2006). Cholesteryl ester transfer protein (CETP) polymorphism modifies the Alzheimer's disease risk associated with APOE ε4 allele. *Journal of Neurology, 253*, 181–185.

Rootzén, H., & Zholud, D. (2017). Human life is unlimited - But short. *Extremes, 20*, 713–728.

Russell, C. (2007). What do older women and men want? Gender differences in the 'lived experience' of ageing. *Current Sociology, 55*, 173–192.

Russell, D. W., & Cutrona, C. E. (1991). Social support, stress, and depressive symptoms among the elderly: Test of a process model. *Psychology and Aging, 6*, 190–201.

Rybarczyk, B., Emery, E. E., Guequierre, L. L., Shamaskin, A., & Behel, J. (2012). The role of resilience in chronic Illness and disability in older adults. *Annual Review of Gerontology and Geriatrics, 32*, 173–187.

Salander, P. (2006). Who needs the concept of "spirituality"? *Psycho-Oncology, 15*, 647–649.

Saleem, S., Aslam, H. M., Anwar, M., Anwar, S., Saleem, M., Saleem, A., et al. (2013). Fahr's syndrome: Literature review of current evidence. *Orphanet Journal of Rare Diseases, 8*, 156. Page 2 of 9 http://www.ojrd.com/content/8/1/156

Salthouse, T. A. (2006). Mental exercise and mental aging: Evaluating the validity of the "use it or lose it" hypothesis. *Perspectives on Psychological Science, 1*, 68–87.

Salthouse, T. A. (2010). Selective review of cognitive aging. *Journal of the International Neuropsychological Society, 16*, 754–760.

Salthouse, T. A., & Mitchell, D. (1990). Effects of age and naturally occurring experience on spatial visualization performance. *Developmental Psychology, 26*, 845–854.

Santoro, A., Ostan, R., Candela, M., Biagi, E., Brigidi, P., Capri, M., et al. (2018). Gut microbiota changes in the extreme decades of human life: A focus on centenarians. *Cellular and Molecular Life Sciences, 75*, 129–148.

Santos-Lozano, A., Sanchis-Gomar, F., Pareja-Galeano, H., Fiuza-Luces, C., Emanuele, E., Lucia, A., et al. (2015). Where are supercentenarians located? A worldwide demographic study. *Rejuvenation Research, 18*, 14–19.

Santos-Lozano, A., Santamarina, A., Pareja-Galeano, H., Sanchis-Gomar, F., Fiuza-Luces, C., Cristi-Montero, C., et al. (2016). The genetics of exceptional longevity: Insights from centenarians. *Maturitas, 90*, 49–57.

Sapolsky, R., Armanini, M., Packan, D., & Tombaugh, G. (1987). Stress and glucocorticoids in aging. *Endocrinology and Metabolism Clinics of North America, 16*, 965–980.

Särkämö, T., Ripollés, P., Vepsäläinen, H., Autti, T., Silvennoinen, H. M., Salli, E., et al. (2014). Structural changes induced by daily music listening in the recovering brain after middle cerebral artery stroke: A voxel-based morphometry study. *Frontiers in Human Neuroscience, 8*, 245. https://doi.org/10.3389/fnhum.2014.00245

Saroglou, V. (2002). Religion and the five factors of personality: A meta-analytical review. *Personality and Individual Differences, 32*, 15–25.

Sastry, P. S., & Subba Rao, K. (2000). Apoptosis and the nervous system. *Journal of Neurochemistry, 74*, 1–20.

Satizabal, C. L., Beiser, A. S., Chouraki, V., Chêne, G., Dufouil, C., & Seshadri, S. (2016). Incidence of dementia over three decades in the Framingham Heart Study. *New England Journal of Medicine, 374*, 523–532.

Saunders, G. H., & Echt, K. V. (2007). An overview of dual sensory impairment in older adults: Perspectives for rehabilitation. *Trends in Hearing, 11*, 243–258.

Schaefer, E. J., Bongard, V., Beiser, A. S., Lamon-Fava, S., Robins, S. J., Au, R., et al. (2006). Plasma phosphatidylcholine docosahexaenoic acid content and risk of dementia and Alzheimer disease: The Framingham heart study. *Archives of Neurology, 63*, 1545–1550.

Schaub, R. T., & Linden, M. (2000). Anxiety and anxiety disorders in the old and very old results from the Berlin Aging Study (BASE). *Comprehensive Psychiatry, 41*, 48–54.

Schneiderman, N., Ironson, G., & Siegel, S. D. (2005). Stress and health: Psychological, behavioral, and biological determinants. *Annual Review of Clinical Psychology, 1*, 607–628.

Schoenhofen, E. A., Wyszynski, D. F., Andersen, S., Pennington, J., Young, R., Terry, D. F., et al. (2006). Characteristics of 32 supercentenarians. *Journal of the American Geriatrics Society, 54*, 1237–1240.

Schoenmakers, B., Buntinx, F., & Delepeleire, J. (2010). Factors determining the impact of care-giving on caregivers of elderly patients with dementia. A systematic literature review. *Maturitas, 66*, 191–200.

Schoevers, R., Geerlings, M., Beekman, A., Penninx, B., Deeg, D., Jonker, C., et al. (2000). Association of depression and gender with mortality in old age: Results from the Amsterdam Study of the Elderly (AMSTEL). *British Journal of Psychiatry, 177*, 336–342.

Schroeder, S. R., & Marian, V. (2017). Cognitive consequences of trilingualism. *International Journal of Bilingualism, 21*, 754–773.

Schuurmans, J., Comijs, H. C., Beekman, A. T. F., de Beurs, E., Deeg, D. J. H., Emmelkamp, P. M. G., et al. (2005). The outcome of anxiety disorders in older people at 6-year follow-up: Results from the Longitudinal Aging Study Amsterdam. *Acta Psychiatrica Scandinavica, 111*, 420–428.

Schwadel, P. (2011). Age, period, and cohort effects on religious activities and beliefs. *Social Science Research, 40*, 181–192.

Schwarzbach, M., Luppa, M., Forstmeier, S., König, H.-H., & Riedel-Heller, S. G. (2014). Social relations and depression in late life - A systematic review. *International Journal of Geriatric Psychiatry, 29*, 1–21.

Sebastiani, P., & Perls, T. T. (2012). The genetics of extreme longevity: Lessons from the New England centenarian study. *Frontiers in Genetics, 3*, 277.

Seeman, T. E., Singer, B., Wilkinson, C. W., & McEwen, B. (2001). Gender differences in age-related changes in HPA axis reactivity. *Psychoneuroendocrinology, 26*, 225–240.

Selkoe, D. J. (2002). Alzheimer's disease is a synaptic failure. *Science, 298*, 789–791.

Shah, T. M., Weinborn, M., Verdile, G., Sohrabi, H. R., & Martins, R. N. (2017). Enhancing cognitive functioning in healthy older adults: A systematic review of the clinical significance of commercially available computerized cognitive training in preventing cognitive decline. *Neuropsychology Review, 27*, 62–80.

Sheedy, J. R., Wettenhall, R. E. H., Scanlon, D., Gooley, P. R., Lewis, D. P., McGregor, N., et al. (2009). Increased d-lactic acid intestinal bacteria in patients with chronic fatigue syndrome. *In Vivo, 23*, 621–628.

Sheikh, J. I., & Cassidy, E. L. (2000). Treatment of anxiety disorders in the elderly: Issues and strategies. *Journal of Anxiety Disorders, 14*, 173–190.

Sheppard, C. L., McArthur, C., & Hitzig, S. L. (2016). A systematic review of Montessori- based activities for persons with dementia. *Journal of the American Medical Directors Association, 17*, 117–122.

Sherratt, K., Thornton, A., & Hatton, C. (2004). Music interventions for people with dementia: A review of the literature. *Aging and Mental Health, 8*, 3–12.

Shimonaka, Y., Nakazato, K., & Marushima, R. (1994). Androgyny and psychological well-being among older and younger Japanese adults. *Aging Clinical and Experimental Research, 6*, 43–48.

Shooshtari, S., Martens, P. J., Burchill, C. A., Dik, N., & Naghipur, S. (2011). Prevalence of depression and dementia among adults with developmental disabilities in Manitoba, Canada. *International Journal of Family Medicine, 2011*, 319574, 9 pages. https://doi.org/10.1155/2011/319574

Shors, T. J., Anderson, L. M., Curlik, D. M., II, & Nokia, S. M. (2012). Use it or lose it: How neurogenesis keeps the brain fit for learning. *Behavioural Brain Research, 227*, 450–458.

Silver, C. B. (2003). Gendered identities in old age: Toward (de)gendering? *Journal of Aging Studies, 17*, 379–397.

Singer, T., Lindenberger, U., & Baltes, P. B. (2003a). Plasticity of memory for new learning in very old age: A story of major loss? *Psychology and Aging, 18*, 306–317.

Sinoff, G., & Werner, P. (2003). Anxiety disorder and accompanying subjective memory loss in the elderly as a predictor of future cognitive decline. *International Journal of Geriatric Psychiatry, 18*, 951–959.

Sivertsen, H., Bjørkløf, G. H., Engedal, K., Selbæk, G., & Helvik, A.-S. (2015). Depression and quality of life in older persons: A review. *Dementia and Geriatric Cognitive Disorders, 40*, 311–339.

Slezak, M., & Pfrieger, F. W. (2003). New roles for astrocytes: Regulation of CNS synaptogenesis. *Trends in Neurosciences, 26*, 531–535.

Smith, J., & Baltes, M. M. (1998). The role of gender in very old age: Profiles of functioning and everyday life patterns. *Psychology and Aging, 13*, 676–695.

Smith, J., Borchelt, M., Maier, H., & Jopp, D. (2002). Health and well-being in the young old and oldest old. *Journal of Social Issues, 58*, 715–732.

Snowden, J. S., Stopford, C. L., Julien, C. L., Thompson, J. C., Davidson, Y., Gibbons, L., et al. (2007). Cognitive phenotypes in Alzheimer's disease and genetic risk. *Cortex, 43*, 835–845.

Sofi, F., Valecchi, D., Bacci, D., Abbate, R., Gensini, G. F., Casini, A., et al. (2011). Physical activity and risk of cognitive decline: A meta-analysis of prospective studies. *Journal of Internal Medicine, 269*, 107–117.

Solé-Ribalta, A., & Borge-Holthoefer, J. (2018). *Socio-economic constraints to maximum human lifespan*. arXiv:1806.11376 [q-bio.PE].

Sontag, S. (1972). The double standard of aging. *Saturday Review of Literature, 39*, 29–38.

Soto, C. J., John, O. P., Gosling, S. D., & Potter, J. (2011). Age differences in personality traits from 10 to 65: Big Five domains and facets in a large cross-sectional sample. *Journal of Personality and Social Psychology, 100*, 330–348.

Sözeri-Varma, G. (2012). Depression in the elderly: Clinical features and risk factors. *Aging and Disease, 3*, 465–471.

Spector, A., Davies, S., Woods, B., & Orrell, M. (2000). Reality orientation for dementia: A systematic review of the evidence of effectiveness from randiomized controlled trials. *The Gerontologist, 40*, 206–212.

Spira, A. P., Rebok, G. W., Stone, K. L., Kramer, J. H., & Yaffe, K. (2012). Depressive symptoms in oldest-old women: Risk of mild cognitive impairment and dementia. *American Journal of Geriatric Psychiatry, 20*, 1006–1015.

Stanley, M. A., Beck, J. G., Novy, D. M., Averill, P. M., Swann, A. C., Diefenbach, G. J., et al. (2003). Cognitive–behavioral treatment of late-life generalized anxiety disorder. *Journal of Consulting and Clinical Psychology, 71*, 309–319.

Stathakos, D., Pratsinis, H., Zachos, I., Vlahaki, I., Gianakopoulou, A., Zianni, D., et al. (2005). Greek centenarians: Assessment of functional health status and life-style characteristics. *Experimental Gerontology, 40*, 512–518.

Stathopoulou, G., Powers, M. B., Berry, A. C., Smits, J. A. J., & Otto, M. W. (2006). Exercise interventions for mental health: A quantitative and qualitative review. *Clinical Psychology: Science and Practice, 13*, 179–193.

Steeman, E., Dierckx de Casterlé, B., Godderis, J., & Grypdonck, M. (2006). Living with early-stage dementia: A review of qualitative studies. *Journal of Advanced Nursing, 54*, 722–738.

Stefanek, M., Green Mcdonald, P., & Hess, S. A. (2005). Religion, spirituality and cancer: Current status and methodological challenges. *Psycho-Oncology, 14*, 450–463.

Stek, M. L., Gussekloo, J., Beekman, A. T. F., van Tilburg, W., & Westendorp, R. G. J. (2004). Prevalence, correlates and recognition of depression in the oldest old: The Leiden 85-plus study. *Journal of Affective Disorders, 78*, 193–200.

Stek, M. L., Vinkers, D. J., Gussekloo, J., van der Mast, R., Beekman, A. T. F., & Westendorp, R. G. J. (2006). Natural history of depression in the oldest old. *Bristish Journal of Psychiatry, 188*, 65–69.

Stepanyants, A., Hof, P. R., & Chklovskii, D. B. (2002). Geometry and structural plasticity of synaptic connectivity. *Neuron, 34*, 275–288.

Stern, R. A., Andersen, S. L., & Gavett, B. E. (2014). Executive functoning. In A. E. Budson & N. W. Kowall (Eds.), *The handbook of Alzheimer's disease and other dementias* (pp. 369–415). Chichester, West Sussex: Wiley.

Stern, Y. (2002). What is cognitive reserve? Theory and research application of the reserve concept. *Journal of the International Neuropsychological Society, 8*, 448–460.

Stojanovic-Radic, J., & DeLuca, J. (2015). Neuroplasticity in multiple sclerosis. In J. I. Tracy, B. M. Hampstead, & K. Sathian (Eds.), *Cognitive plasticity in neurologic disorders* (pp. 107–124). Oxford: Oxford University Press.

Stroebe, M., Stroebe, W., & Schut, H. (2001). Gender differences in adjustment to bereavement: An empirical and theoretical review. *Review of General Psychology, 5*, 62–83.
Ströhlc, A. (2009). Physical activity, exercise, depression and anxiety disorders. *Journal of Neural Transmission, 116*, 777–784.
Strøm, B. S., Ytrehus, S., & Grov, E.-K. (2016). Sensory stimulation for persons with dementia: A review of the literature. *Journal of Clinical Nursing, 25*, 1805–1834.
Strough, J., Leszczynski, J. P., Neely, T. L., Flinn, J. A., & Margrett, J. (2007). From adolescence to later adulthood: Femininity, masculinity, and androgyny in six age groups. *Sex Roles, 57*, 385–396.
Strydom, A., Hassiotis, A., & Livingston, G. (2005). Mental health and social care needs of older people with intellectual disabilities. *Journal of Applied Research in Intellectual Disabilities, 18*, 229–235.
Suzman, R., & Riley, M. W. (1985). Introducing the "oldest old". *Milbank Memorial Fund Quarterly/Health and Society, 63*, 177–186.
Swaab, D. F. (1991). Brain aging and Alzheimer's disease, "wear and tear" versus "use it or lose it". *Neurobiology of Aging, 12*, 317–324.
Talala, K., Huurre, T., Aro, H., Martelin, T., & Prättälä, R. (2008). Socio-demographic differences in self-reported psychological distress among 25- to 64-year-old Finns. *Social Indicators Research, 86*, 323–335.
Taleporos, G., & McCabe, M. P. (2002). Body image and physical disability - Personal perspectives. *Social Science and Medicine, 54*, 971–980.
Tamnes, C. K., Walhovd, K. B., Dale, A. M., Ostby, Y., Grydeland, H., Richardson, G., et al. (2012). Brain development and aging: Overlapping and unique patterns of change. *NeuroImage, 68C*, 63–74.
Taylor, A., & Gosney, M. A. (2011). Sexuality in older age: Essential considerations for healthcare professionals. *Age and Ageing, 40*, 538–543.
Taylor, W. D. (2014). Depression in the elderly. *New England Journal of Medicine, 371*, 1228–1236.
Terracciano, A., Lockenhoff, C. E., Zonderman, A. B., Ferrucci, L., & Costa, P. T., Jr. (2008). Personality predictors of longevity: Activity, emotional stability, and conscientiousness. *Psychosomatic Medicine, 70*, 621–627.
Tesch-Römer, C., Motel-Klingebiel, A., & Tomasik, M. J. (2008). Gender differences in subjective well-being: Comparing societies with respect to gender equality. *Social Indicators Research, 85*, 329–349.
Theendakara, V., Patent, A., Peters Libeu, C. A., Philpot, B., Flores, S., Descamps, O., et al. (2013). Neuroprotective Sirtuin ratio reversed by ApoE4. *Proceedings of the National Academy of Sciences of the United States of America, 110*, 18303–18308.
Thies, W., & Bleiler, L. (2013). Alzheimer's disease facts and figures. *Alzheimer's and Dementia, 9*, 208–245.
Thoits, P. A. (1991). Merging identity theory and stress research. *Social Psychology Quarterly, 54*, 101–112.
Thompson, S. C., Thomas, C., Rickabaugh, C. A., Tantamjarik, P., Otsuki, T., Pan, D., et al. (1998). Primary and secondary control over age-related changes in physical appearance. *Journal of Personality, 66*, 583–605.
Tiemeier, H. (2003). Biological risk factors for late life depression. *European Journal of Epidemiology, 18*, 745–750.
Tiggemann, M. (2004). Body image across the adult life span: Stability and change. *Body Image, 1*, 29–41.
Tornstam, L. (1997). Gerotrancendence: The contemplative dimension of aging. *Journal of Aging Studies, 11*, 143–154.
Tower, R. B., & Kasl, S. V. (1996). Gender, marital closeness, and depressive symptoms in elderly couples. *Journal of Gerontology, 51B*, P115–P129.
Trieschmann, R. (1987). *Ageing with a disability*. New York: Demos Publications.
Trollor, J. N., Anderson, T. M., Sachdev, P. S., Brodaty, H., & Andrews, G. (2007). Prevalence of mental disorders in the elderly: The Australian national mental health and well- being survey. *American Journal of Geriatric Psychiatry, 15*, 455–466.

Tsertsvadze, F. Y., Fink, H. A., MacDonald, R., Wilt, T. J., Bella, A. J., Ansari, M. T., et al. (2009). Oral sildenafil citrate (Viagra) for erectile dysfunction: A systematic review and meta-analysis of harms. *Urology, 74*, 831–836.e8.

Tufan, I., Arslaner, C., Seedsman, T., Durak, M., & Seeberger, B. (2017). A baseline investigation of independence among 100 community-dwelling centenarians in Turkey. *International Journal of Humanities and Social Science, 7*, 25–34.

Turner, R. J., & Noh, S. (1988). Physical disability and depression: A longitudinal analysis. *Journal of Health and Social Behavior, 29*, 23–37.

Ueda, T., Suzukamo, Y., Sato, M., & Izumi, S.-I. (2013). Effects of music therapy on behavioral and psychological symptoms of dementia: A systematic review and meta-analysis. *Ageing Research Reviews, 12*, 628–641.

Umberson, D., Wortman, C. B., & Kessler, R. C. (1992). Widowhood and depression: Explaining long-term gender differences in vulnerability. *Journal of Health and Social Behavior, 33*, 10–24.

Urciuoli, O., Dello Buono, M., Padovani, W., & De Leo, D. (1998). Assessment of quality of life in the oldest-olds living in nursing homes and at home. *Archives of Gerontology and Geriatrics, 27*(Suppl. 6), 507–514.

Vafaei, A., Alvarado, B., Tomás, C., Muro, C., Martinez, B., & Zunzunegui, M. V. (2014). The validity of the 12-item Bem Sex Role Inventory in older Spanish population: An examination of the androgyny model. *Archives of Gerontology and Geriatrics, 59*, 257–263.

Valenzuela, M. J. (2008). Brain reserve and the prevention of dementia. *Current Opinion in Psychiatry, 21*, 296–302.

Vallin, J., & Meslé, F. (2009). The segmented trend of highest life expectancy. *Population and Development Review, 35*, 159–187.

van den Berg, N., Beekman, M., Smith, K. R., Janssens, A., & Slagboom, P. E. (2017). Historical demography and longevity genetics: Back to the future. *Ageing Research Reviews, 38*, 28–39.

Van der Weele, G. M., Gussekloo, J., De Waal, M. W. M., De Craen, A. J. M., & Van der Mast, R. C. (2009). Co-occurrence of depression and anxiety in elderly subjects aged 90 years and its relationship with functional status, quality of life and mortality. *International Journal of Geriatric Psychiatry, 24*, 595–601.

van Hooren, S. A. H., Valentijn, A. M., Bosma, H., Ponds, R. W. H. M., van Boxtel, M. P. J., & Jolles, J. (2007). Cognitive functioning in healthy older adults aged 64–81: A cohort study into the effects of age, sex, and education. *Aging, Neuropsychology, and Cognition, 14*, 40–54.

Van Ness, P. H., & Kasl, S. V. (2003). Religion and cognitive dysfunction in an elderly cohort. *Journal of Gerontology, 58B*, S21–S29.

Van Ness, P. H., & Larson, D. B. (2002). Religion, senescence, and mental health: The end of life is not the end of hope. *American Journal of Geriatric Psychiatry, 10*, 386–397.

Vasse, E., Vernooij-Dassen, M., Spijker, A., Rikkert, M. O., & Koopmans, R. (2010). A systematic review of communication strategies for people with dementia in residential and nursing homes. *International Psychogeriatrics, 22*, 189–200.

Vaughan, L., Corbin, A. L., & Goveas, J. S. (2015). Depression and frailty in later life: A systematic review. *Clinical Interventions in Aging, 10*, 1947–1958.

Vaupel, J. W. (1997). The remarkable improvements in survival at older ages. *Philosophical Transactions of the Royal Society of London B, 352*, 1799–1804.

Vaupel, J. W. (2000). Longevity as an artefact of civilization. In R. N. Butler & C. Jasmin (Eds.), *Longevity and quality of life* (pp. 37–46). New York: Kluwer Academic/Plenum Publishers.

Vaupel, J. W. (2010). Biodemography of human ageing. *Nature, 464*, 536–542.

Venkatesan, U. M., Rajtmajer, S. M., & Hillary, F. G. (2015). Connectivity modeling and neuroplasticity after traumatic brain injury. In J. I. Tracy, B. M. Hampstead, & K. Sathian (Eds.), *Cognitive plasticity in neurologic disorders* (pp. 152–175). Oxford: Oxford University Press.

Verdile, G., Laws, S. M., & Martins, R. N. (2013). Genes and Alzheimer's disease. In R. Martins (Ed.), *Understanding Alzheimer's* (pp. 67–86). Sydney, NSW: Macmillan.

Verghese, J., Lipton, R. B., Katz, M. J., Hall, C. B., Derby, C. A., Kuslansky, G., et al. (2003). Leisure activities and the risk of dementia in the elderly. *New England Journal of Medicine, 348*, 2508–2516.

Vidovich, M., Lautenschlager, N. T., Shah, T., Weinborn, M., & Almeida, O. P. (2013). Use it or lose it: Mental exercise. In R. Martins (Ed.), *Understanding Alzheimer's* (pp. 169–193). Sydney, NSW: Macmillan.

Vijg, J., & Le Bourg, E. (2017). Aging and the inevitable limit to human life span. *Gerontology, 63*, 432–434.

Vink, A. C., Bruinsma, M. S., & Scholten, R. J. P. M. (2003). Music therapy for people with dementia. *Cochrane Database of Systematic Reviews*, (4), CD003477. https://doi.org/10.1002/14651858.CD003477.pub2

Vink, D., Aartsen, M. J., & Schoevers, R. A. (2008). Risk factors for anxiety and depression in the elderly: A review. *Journal of Affective Disorders, 106*, 29–44.

Vinkers, D. J., Gussekloo, J., Stek, M. L., Westendorp, R. G. J., & van der Mast, R. C. (2004). Temporal relation between depression and cognitive impairment in old age: Prospective population based study. *BMJ, 329*(7471), 881. https://doi.org/10.1136/bmj.38216.604664.DE

von Faber, M., Bootsma-van der Wiel, A., van Exel, E., Gussekloo, J., Lagaay, A. M., van Dongen, E., et al. (2001). Successful aging in the oldest old: Who can be characterized as successfully aged? *Archives of Internal Medicine, 161*, 2694–2700.

von Heideken Wågert, P., Gustavsson, J. M. C., Lundin-Olsson, L., Kallin, K., Nygren, B., Lundman, B., et al. (2006). Health status in the oldest old. Age and sex differences in the Umeå 85+ Study. *Aging Clinical and Experimental Research, 18*, 116–126.

von Heideken Wågert, P., Rönnmark, B., Rosendahl, E., Lundin-Olsson, L., Gustavsson, J. M. C., Nygren, B., et al. (2005). Morale in the oldest old: The Umeå 85 study. *Age and Ageing, 34*, 249–255.

Voss, M. W., Erickson, K. I., Prakash, R. S., Chaddock, L., Kim, J. S., Alves, H., et al. (2013). Neurobiological markers of exercise-related brain plasticity in older adults. *Brain Behavior and Immunity, 28*, 90–99.

Voss, M. W., Prakash, R. S., Erickson, K. I., Basak, C., Chaddock, L., Kim, J. S., et al. (2010). Plasticity of brain networks in a randomized intervention trial of exercise training in older adults. *Frontiers in Neuroscience, 2*, 32. https://doi.org/10.3389/fnagi.2010.00032

Wagnild, G., & Young, H. M. (1990). Resilience among older women. *Image: Journal of Nursing Scholarship, 22*, 252–255.

Wahlin, Å., MacDonald, S. W. S., deFrias, C. M., Nilsson, L. G., & Dixon, R. A. (2006). How do health and biological age influence chronological age and sex differences in cognitive aging: Moderating, mediating, or both? *Psychology and Aging, 21*, 318–332.

Waites, C. L., Craig, A. M., & Garner, C. C. (2005). Mechanisms of vertebrate synaptogenesis. *Annual Reviews of Neuroscience, 28*, 251–274.

Walker, A., & Walker, C. (1998). Normalisation and 'normal' ageing: The social construction of dependency among older people with learning difficulties. *Disability and Society, 13*, 125–142.

Wan, C. Y., & Schlaug, G. (2010). Music making as a tool for promoting brain plasticity across the life span. *The Neuroscientist, 16*, 566–577.

Wang, H.-X., Karp, A., Winblad, B., & Fratiglioni, L. (2002). Late-life engagement in social and leisure activities is associated with a decreased risk of dementia: A longitudinal study from the Kungsholmen project. *American Journal of Epidemiology, 155*, 1081–1087.

Ward, A., Crean, S., Mercaldi, C. J., Collins, J. M., Boyd, D., Cook, M. N., et al. (2012). Prevalence of apolipoprotein E4 genotype and homozygotes (APOE e4/4) among patients diagnosed with Alzheimer's disease: A systematic review and meta- analysis. *Neuroepidemiology, 38*, 1–17.

Weisberg, Y. J., DeYoung, C. G., & Hirsh, J. B. (2011). Gender differences in personality across the ten aspects of the Big Five. *Frontiers in Psychology, 2*, 178.

Weiss, A., & Costa Jr., P. T. (2005). Domain and facet personality predictors of all-cause mortality among Medicare patients aged 65 to 100. *Psychosomatic Medicine, 67*, 724–627.

Wettstein, M., Schilling, O. K., & Wahl, H. W. (2016). Still feeling healthy after all these years: The paradox of subjective stability versus objective decline in very old adults' health and functioning across five years. *Psychology of Aging, 31*, 815–830.

Whitaker, S., & Read, S. (2006). The prevalence of psychiatric disorders among people with intellectual disabilities: An analysis of the literature. *Journal of Applied Research in Intellectual Disabilities, 19*, 330–345.

Wickremaratchi, M. M., & Llewelyn, J. G. (2006). Effects of ageing on touch. *Postgraduate Medical Journal, 82*, 301–304.
Wilkinson, P. (2013). Cognitive behavioural therapy with older people. *Maturitas, 76*, 5–9.
Willcox, B. J., Willcox, D. C., & Suzuki, M. (2017). Demographic, phenotypic, and genetic characteristics of centenariansin Okinawa and Japan: Part 1- centenarians in Okinawa. *Mechanisms of Ageing and Development, 165*, 75–79.
Willcox, D. C., Willcox, B. J., He, Q., Wang, N.-c., & Suzuki, M. (2008). They really are that old: A validation study of centenarian prevalence in Okinawa. *The Journals of Gerontology: Series A, 63*, 338–349.
Willcox, D. C., Willcox, B. J., Hsueh, W.-C., & Suzuki, M. (2006). Genetic determinants of exceptional human longevity: Insights from the Okinawa Centenarian Study. *Age, 28*, 313–332.
Willcox, D. C., Willcox, B. J., Rosenbaum, M., Sokolovsky, J., & Suzuki, M. (2009). Exceptional longevity and the quest for healthy aging: Insights from the Okinawa centenarian study. In J. Sokolovsky (Ed.), *The cultural context of aging: Worldwide perspectives* (pp. 505–532). Westport, CT: Praeger.
Willis, S. L., Tennstedt, S. L., Marsiske, M., Ball, K., Elias, J., Koepke, K. M., et al. (2006). Long-term effects of cognitive training on everyday functional outcomes in older adults. *JAMA, 296*, 2805–2814.
Wilmoth, J. M., & London, A. S. (2011). Aging veterans: Needs and provisions. In R. A. Settersten Jr. & J. L. Angel (Eds.), *Handbook of sociology of aging* (pp. 445–561). New York: Springer.
Wilmoth, J. R., Deegan, L. J., Lundström, H., & Horiuchi, S. (2000). Increase of maximum life-span in Sweden, 1861–1999. *Science, 289*, 2366–2368.
Wilmoth, J. R., & Robine, J.-M. (2003). The world trend in maximum life span. In J. R. Carey & S. Tuljapurkar (Eds.), *Life span: Evolutionary, ecological, and demographic perspectives* (pp. 239–257). New York: Population Council.
Wilson, R. S., Arnold, S. E., Schneider, J. A., Kelly, J. F., Tang, Y., & Bennett, D. A. (2006). Chronic psychological distress and risk of Alzheimer's disease in old age. *Neuroepidemiology, 27*, 143–153.
Windle, G. (2011). What is resilience? A review and concept analysis. *Reviews in Clinical Gerontology, 21*, 152–169.
Winterich, J. A. (2007). Aging, femininity, and the body: What appearance changes mean to women with age. *Gender Issues, 24*, 51–69.
Wolfe, M. S. (2007). When loss is gain: Reduced presenilin proteolytic function leads to increased Aβ42/Aβ40: Talking Point on the role of presenilin mutations in Alzheimer disease. *EMBO Reports, 8*, 136–140.
Wolitzky-Taylor, K. B., Castriotta, N., Lenze, E. J., Stanley, M. A., & Craske, M. G. (2010). Anxiety disorders in older adults: A comprehensive review. *Depression and Anxiety, 27*, 190–211.
Wong, P. T. P. (2012). Introduction: A roadmap for meaning research and applications. In P. T. P. Wong (Ed.), *The human quest for meaning: Theories, research, and applications* (pp. xxix–xlvi). London: Routledge.
Woods, B., Aguirre, E., Spector, A. E., & Orrell, M. (2012). Cognitive stimulation to improve cognitive functioning in people with dementia. *Cochrane Database of Systematic Reviews*, (2), CD005562. https://doi.org/10.1002/14651858.CD005562.pub2
Woumans, E., Santens, P., Sieben, A., Versijpt, J., Stevens, M., & Duyck, W. (2015). Bilingualism delays clinical manifestation of Alzheimer's disease. *Bilingualism: Language and Cognition, 18*, 568–574.
Woumans, E., Versijpt, J., Sieben, A., Santens, P., & Duyck, W. (2017). Bilingualism and cognitive decline: A story of pride and prejudice. *Journal of Alzheimer's Disease, 60*, 1237–1239.
Wright, C. I. (2014). Emotion and behavior in Alzheimer's disease and other dementias. In A. E. Budson & N. W. Kowall (Eds.), *The handbook of Alzheimer's disease and other dementias* (pp. 416–456). Chichester, West Sussex: Wiley.
Yang, L., Krampe, R. T., & Baltes, P. B. (2006). Basic forms of cognitive plasticity extended into the oldest-old: Retest learning, age, and cognitive functioning. *Psychology and Aging, 21*, 372–378.

Yankner, B. A., Lu, T., & Loerch, P. (2008). The aging brain. *Annual Review of Pathology: Mechanisms of Disease, 3*, 41–66.
Yi, Z., & Vaupel, J. W. (2002). Functional capacity and self-evaluation of health and life of oldest old in China. *Journal of Social Issues, 58*, 733–748.
Yi, Z., Yuzhi, L., & George, L. K. (2003). Gender differentials of the oldest old in China. *Research on Aging, 25*, 65–80.
Young, R. D., Epstein, L., & Coles, L. S. (2008). Supercentenarian counts by nation of birth. *Rejuvenation Research, 11*, 981.
Zaccai, J., McCracken, C., & Brayne, C. (2005). A systematic review of prevalence and incidence studies of dementia with Lewy bodies. *Age and Ageing, 34*, 561–566.
Zanto, T. P., & Gazzaley, A. (2013). Fronto-parietal network: Flexible hub of cognitive control. *Trends in Cognitive Sciences, 17*, 602–603.
Zarit, S. H., Griffiths, P. C., & Berg, S. (2004). Pain perceptions of the oldest old: A longitudinal study. *The Gerontologist, 44*, 459–468.
Zechner, U., Wilda, M., Kehrer-Sawatzki, H., Vogel, W., Fundele, R., & Hameister, H. (2001). A high density of X-linked genes for general cognitive ability: A run-away process shaping human evolution? *Trends in Genetics, 17*, 697–701.
Zhang, J.-G., Ishikawa-Takata, K., Yamazaki, H., Morita, T., & Ohta, T. (2006). The effects of Tai Chi Chuan on physiological function and fear of falling in the less robust elderly: An intervention study for preventing falls. *Archives of Gerontology and Geriatrics, 42*, 107–116.
Zhang, X., Norton, J., Carrière, I., Ritchie, K., Chaudieu, I., & Ancelin, M.-L. (2015). Generalized anxiety in community-dwelling elderly: Prevalence and clinical characteristics. *Journal of Affective Disorders, 172*, 24–29.
Zhang, X., Norton, J., Carrière, I., Ritchie, K., Chaudieu, I., & Ancelin, M.-L. (2015a). Risk factors for late-onset generalized anxiety disorder: Results from a 12-year prospective cohort (The ESPRIT study). *Translational Psychiatry, 5*, e536. https://doi.org/10.1038/tp.2015.31
Zhang, Z. (2006). Gender differentials in cognitive impairment and decline of the oldest old in China. *Journal of Gerontology, 61B*, S107–S115.
Zinke, K., Zeintl, M., Eschen, A., Herzog, C., & Kliegel, M. (2012). Potentials and limits of plasticity induced by working memory training in old-old age. *Gerontology, 58*, 79–87.
Zito, K., & Svoboda, K. (2002). Activity-dependent synaptogenesis in the adult mammalian cortex. *Neuron, 35*, 1015–1017.
Zola, I. K. (1989). Aging and disability: Toward a unified agenda. *Journal of Rehabilitation, 55*, 6–8.
Zuo, L., van Dyck, C. H., Luo, X., Kranzler, H. R., Yang, B.-z., & Gelernter, J. (2006). Variation at APOE and STH loci and Alzheimer's disease. *Behavioral and Brain Functions, 2*(13). https://doi.org/10.1186/1744-9081-2-13
Życzkowska, J., Szczerbińska, K., Jantzi, M. R., & Hirdes, J. P. (2007). Pain among the oldest old in community and institutional settings. *Pain, 129*, 167–176.

Chapter 5
The Psychology of Older Ages

Abstract This chapter starts with an analysis of self-identity and then reviews emotions and cognition in the elderly. Aspects of memory and personality and how they can affect the psychology of older people are addressed next, to then conclude with a section devoted to an analysis of psychological stress in the elderly, followed by one on retirement and how the experience of retired people can be improved through an active life.

In Chap. 4 we focused on health issues in old age in general and mental health in particular. We pointed out that although a decline in mental health is expected with age, not all old people are the same and factors such as genetics, various aspects of lifestyle, social support, neurodegenerative diseases, and more can make a difference in the way we experience our old age. In particular, a lifelong cultivation of mental plasticity and good physical condition can be of great help to experience fulfilling ageing.

In this chapter we focus in more detail on specific psychological aspects of old age, starting from self-identity to then analyse some major emotional and cognitive issues, including memory. The interaction between personality and ageing will be analysed next to then focus on the ways in which psychological stress is affected in the old. We conclude this chapter with an analysis of the psychological effects of retirement and how an active life may be of help to old retirees.

5.1 Psychological Ageing

The psychological scaffolding needed to build fulfilling ageing relies on our capacity to maintain a degree of psychological functioning, of accepting some limitations or losses, and of making best use of what we have and the experience we have accumulated over time, in order to experience fulfilment in life. Gains and losses in our

M. Cordella, A. Poiani, *Fulfilling Ageing*, International Perspectives on Aging 30,
https://doi.org/10.1007/978-3-030-60071-6_5

mental capacities occur throughout life as we develop, and the common perception is that the balance between gains and losses favours gains at younger ages (e.g. growing experience) and losses at older ages (e.g. failing memory). However, both processes co-occur in old age: old people can still seek opportunities and gain experiences that are useful for personal growth.

Baltes et al. (1998) distinguished three central adaptive tasks that can be enacted throughout the lifespan:

- *growth* (behaviours that aim at reaching higher levels of functioning and adaptation to changing life circumstances),
- *maintenance/recovery* (resilience in the face of life changes), and
- *regulation of loss* (behaviours that maintain a degree of adequate functioning in response to effective loss).

Baltes et al. note that growth more closely characterises younger ages whereas resilience and regulation are more typical of older ages, but they also point out that, in fact, all functions can occur in old age.

An older person in good mental condition has the advantage of a lifetime experience that can be used to better understand the complexities of self, others, and life in general. In other words, elders can compensate for some limitations such as lower speed of mental processing, greater memory constraints or difficulties in keeping up with specific new technological developments, with greater wisdom and greater ability to incorporate complexity in their understanding of the broader world (e.g. Happé et al. 1998). This is what Elkhonon Goldberg refers to as the *Wisdom Paradox*: old age is the age of greater wisdom in spite of being the age of greater comparative brain decay (*neuroerosion*, is the term coined by Goldberg (2007)) and, more generally, body decay. An even more optimistic view is expressed by Ashley Montagu in his book *Growing young* (Montagu 1989, p. 175):

> In the matter of maturity it is not maturity, but *maturation*, which should be seen as lifelong, unending, process. It is foolish, though customary, to believe that one reaches a stage of development which represents the limit of wisdom and ripeness, following which it is 'downhill all the way'. Such a view of development negates the very nature of development, which implies increase in amplitude and complexity, insight and experience, in short, wisdom. (italics original)

Wisdom can be defined in various ways. Ardelt (1997, p. 15), for instance, mentions: "expertise", "advanced cognitive functioning", the "art of questioning", "awareness of ignorance" to then settle on a concept of wisdom that integrates various dimensions of mental activity: "integration of cognitive, reflective, and affective elements" (see also Clayton and Birren 1980; Bergsma and Ardelt 2012), which is closer to our own view of wisdom: *an integrative capacity that can be increased by accumulated experiences throughout life and reflection on such experiences*. There is a mutual interaction between wisdom and experiencing fulfilment in life: more wisdom, more capacity to integrate our life experience into a coherent whole, and therefore more capacity to reach fulfilment; more fulfilment, more ability to further develop wisdom by making sense of the life experiences that truly matter.

Paul Baltes offers an even more technical view of wisdom which is defined as *expert-level knowledge in the fundamental pragmatics of life* (Baltes and Smith 1990, p. 95). The

> domain of the fundamental pragmatics of life comprises insights into the quintessential aspects of the human condition and human life, including its biological finitude, cultural conditioning, and inter-individual variations. At the centre of this body of knowledge and its application are questions concerning the conduct, interpretation, and meaning of life. *Expert levels of performance in the fundamental pragmatics of life domain are categorized as wisdom* (Baltes et al. 1995, p. 155; italics ours)

In empirical studies the *fundamental pragmatics of life domain* is evaluated through establishing the level of adherence of individual behaviour to five criteria: rich factual knowledge, rich procedural knowledge, lifespan contextualism, value relativism, and recognition and management of uncertainty (see also Baltes et al. 2002). Thus, greater wisdom in Baltes' and collaborators' view is developed through the accumulation of knowledge, that is used to put our life in a coherent context, taking into consideration our own values; this is expected to improve our capacity to manage the unpredictability of life.

The perception of the role of wisdom in old age changes not only between individuals, but also between categories of individuals, such as men and women. For instance, Ardelt (1997) mentions that although wisdom is a relevant concept for both old men and women, the tendency for women to reach old ages in relatively worse health condition than men makes them more interested in focusing on wisdom to achieve greater life satisfaction. Whether the health issues are the sole or even the main factors explaining greater wisdom in old women than men is of course debatable.

Through wisdom older people can better adapt to their changing life circumstances, even the adverse ones. In particular, they may make use of accumulated experiences and also a better capacity to control negative emotions (emotional regulation) to overcome the challenges of life and increase agency, mastery, intimacy, communion, and a sense of belonging (Marcoen et al. 2007). *Emotional regulation* can be defined as "the extrinsic and intrinsic processes responsible for monitoring, evaluating, and modifying emotional reactions, especially their intensive and temporal features" (Thompson 1991, p. 271).

Wisdom can help regulate our emotions by reducing false or unrealistic expectations, by helping achieve "serene self-esteem" and living a less fear-motivated life. Wisdom can guide personal transformation, especially through the hard experiences of life, and it can help in the experiencing of fulfilment in life, including through the use of humour: "Humour can protect us from becoming stuck in resentments, and it thus opens us to living with gratitude" (Bianchi 2005, p. 96).

Wisdom is therefore an integrative concept, it synthesises our accumulated experience over a lifetime in all aspects of being: perceptual, emotional, cognitive, spiritual, and the achievement of some form of balance among them (Sternberg 1998). Aleman (2014, p. 140–142) lays bare the full complexity of wisdom in the following quote:

> wisdom is the ability to understand complex situations and thereby promote optimal behavior, so that the outcome satisfies as many people as possible and takes everyone's wellbeing into account..... [it is the] ability to understand complex issues and relationships, knowledge and life experience, self-reflection and self-criticism, acceptance of others' perspectives and values, empathy and love for humanity, and an orientation toward goodness.... [two more qualities can be added to the list:] emotional stability, and the ability to make decisions in ambiguous situations.... Though it isn't generally regarded as an essential component of wisdom, a sense of humour must surely be part of the self-knowledge

required for true wisdom.... It is also about achieving a balance between polar opposites: between weakness and strength, doubt and certainty, dependence and autonomy, transience and infinity. We consider people wise if they can give good advice in difficult circumstances and their judgments are balanced.... Wisdom [is also] the sum of life experiences, the ability to deal with emotions, self-control, love of God, compassion, humility, and self-sacrifice.

Neurobiologically and cognitively, Goldberg (2007) roots wisdom into our brain mechanisms that recognise patterns, the so-called *attractors* (see also Rolls 2010, 2020):

a concise constellation of neurons... with strong connections among them. A unique property of an attractor is that a very broad range of inputs will activate the same neural constellation, the attractor, automatically and easily... I believe that those of us who have been able to form a large number of such cognitive templates, each capturing the essence of a large number of pertinent experiences, have acquired 'wisdom', or at least a certain crucial ingredient thereof (p. 20–21).

Through the use of attractors the ageing brain can decide on matters of relevance to the life of the individual in spite of ever reduced neuronal resources. Neural attractors that have been reinforced throughout life through use and the development of competence (i.e. expert knowledge and skills) should be more resilient to the "neuroerosion" associated with ageing.

Although Goldberg mainly focuses on a cognitive understanding of wisdom, he certainly realises that brain centres that are associated with the development of wisdom, such as the frontal lobes, are also strongly involved in the integration of emotions, and the association of cognition and emotion that, in a broad evolutionary context, is adaptive from the perspective of social life. Therefore wisdom also involves the integration of cognition and emotion not only in an individual but also a social context: "Empathy, insight into the minds of others, and the capacity for moral reasoning are among the most important ingredients of wisdom by any definition, on a par with the capacity for effective problem-solving" (Goldberg 2007, p. 173).

The view of wisdom that we have expounded above is reflective of what is known as *practical wisdom*. Another approach to wisdom is known as *transcendent wisdom*:

a type of wisdom that is about transforming consciousness, or shifting a perspective ... Whereas practical wisdom entails accretion and the acquisition of knowledge and experiences that enable one to live a good life, adherents of transcendent wisdom see such accumulation of experience and knowledge as potentially a hindrance to insight and understanding. Transcendent wisdom is more about shedding beliefs, attachments and other psychological encumbrances that can impede one's vision of reality, not about acquiring cognitive complexity and experiences. Transcendent wisdom does not necessarily emphasise the importance of cognition and action, as with practical wisdom, but instead centres on quieting of the mind so as to gain direct intuitive insight. Others have also considered transcendent wisdom as a developmental process involving self-knowledge, detachment, integration and ultimately self-transcendence (Le 2008, p. 385–386)

Transcendent wisdom is more typically seen in practitioners of spiritual and religious disciplines, although such practitioners can also display practical wisdom. Conversely, practical persons may also benefit from a "quieting of the mind" on occasions, in order to appreciate the whole beyond the details.

Therefore, wisdom is the integration of the self through the knowledge, emotions, and experiences that the self has gone through throughout life, and thus wisdom necessarily contributes to the emergence of self-identity through self-examination, self-awareness, and self-insight (Bergsma and Ardelt 2012). Wisdom grows the more we strive to develop our self-identity in response to our experiences with the world outside the self (society, nature). In the next section we focus on self-identity in old age.

5.2 Self-Identity

Since the 1970s the self has experienced a resurgence as a subject of inquiry and has made strong inroads into psychological thinking and research. The concept of the self is perhaps best described as: "a theory that a person holds about himself as an experiencing, functioning being in interaction with the world" (Gecas 1982, p. 3, following Epstein 1973). Our own perception of ourself or *self-identity* is composed of both the more contingent aspects of the self (*self-image)* and the more enduring aspects of the self (*self-concept*) (Turner 1968), and it emerges from a greater consciousness of our identity and reality, which can also have a social referent. Such social referent of the self can be supportive or not of our own self-identity and, in fact, it can also affect its development (e.g. Holstein and Minkler 2003).

Gecas (1982) also noted the emergence of additional concepts alongside self-identity, self-image and self-concept such as: *self-awareness* (capacity to separate the self from its surroundings), *self-esteem* (or the evaluative and emotional dimensions of self-identity), *self-ideology* (conception of our self aimed at affirming and protecting such self), *self-evaluation* (analysis of our self), *self-attribution* (determination of the antecedents and consequences of own behaviours), *self-affirmation* and *self-protection* (support and strengthening the perceived integrity of the self), *self-efficacy* (our own beliefs in our ability to succeed in specific circumstances or accomplish an objective), *self-worth* (sense of our own value). The self is both subject and object in the world (Johnson 1985) and his/her identity is both private and public (McMullen and Luborsky 2006) both in young and old ages.

James (1890, p. 331) defined self-identity as "consciousness of personal sameness", but such consciousness does not preclude change over time. That is, self-identity is also consciousness of personal sameness through change: the child we were is both the same and different from the adult we are. Constructing a self-identity implies the integration of various aspects of the self across a diversity of domains and over time, throughout one's lifespan (McLean 2008). Moreover, there are as many self-identities as there are people, so much so that it is rather difficult to understand the process of ageing unless we have a grasp of the extent of individual diversity and how it comes about: "We need to keep searching for stories of ageing in order to expand our grasp" (Nussbaum and Levmore 2017, p. 16).

We have also mentioned in previous chapters how the self has been hijacked socially and politically by neoliberalism in recent decades, leading to an extreme individualistic view of self-identity, where "ego"—as imagined by Ayn Rand in *Anthem*, for instance—is in permanent competition with others in a quest to affirm itself and prevail. But the reality of the interaction between the self and others is far more complex: as we develop our identity, we learn from others and we give to others too. Over time we may reach a level of what Erickson (1963, 1968) called *ego integrity* (see Chap. 2) when we are able to look back on life and can be at peace with both our achievements and failures.

Self-identity is a complex combination of who we are, how we see ourselves, how we wish to be, how we are expected to be according to some social or moral framework, and how others see us and therefore interact with us and influence us following their own conceptions (e.g. Rosenberg 1979; Demo 1992; McMullen and Luborsky 2006). For instance, it is well known that many people tend to feel younger (and hope that others will agree with them) than what their actual age is (e.g. Kaufman and Elder 2002; Kirkpatrick Johnson et al. 2007; Barak 2009). In particular, both women and men are of the opinion that women start "old age" at a younger age than men, although at any given age women tend to feel younger than men or at least they so express it (Pinquart and Sörensen 2001).

In a study carried out in the USA as part of the panel on grandparents of the *Iowa Youth and Families Project* (IYFP), Kaufman and Elder (2002) described various sources of variability in the actual and perceived age of their interviewees. The grandparents sample had a mean actual age of 69 years, however, their subjective age was lower (61 years), and their desired age was even lower (47 years). They perceived "old age" as being 73 years old and on average they hoped to live to 83 years old. The pattern seems to be repeated across cultures: desired age is younger than subjective age, which is younger than actual age (Barak 2009).

When subjective old age becomes a combination of both current feelings and desires it may lead to perceptions of young age even at ages beyond 70. In fact, Logan et al. (1992) reported how among the 70–79 years old in their sample, 4% of men and 8% of women regarded themselves as young, health being an important factor influencing their perception. They also tended to regard themselves as being happy. Interestingly, an opposite pattern has been observed among Cambodian refugees in the USA, many of whom regarded the beginning of old age to be before 65 years old. In fact 16 participants thought that old age begins at 50, one participant was of the opinion that old age begins at 40 years old, whereas three went as far as to say that old age begins in the 20s. Several factors in combination could explain this shift of old age perception to younger chronological ages: cultural traditions, the physically hard working life they experienced in Cambodia, lack of healthcare support, and an early age of reproduction leading to becoming grandparent in their 40s; grandparenthood may increase the perception of old age (Dubus 2014).

Cultural and personal factors can therefore affect perception of old age. In contrast to the Cambodian example that we have just mentioned, Kaufman and Elder (2003), working in the USA, described that older people who enjoy being grandparents also feel younger and they hope to live longer than those who do not enjoy grandparenting. In this work, older people who were content with their own identity seemed to have favoured the positive integration of specific social roles into their life, thus somewhat alleviating the burden of caring for grandchildren.

The establishment and nurturing of self-identity (e.g. *identity assimilation*, Sneed 2005) is central to fulfilling ageing, and such self-identity may be stable over time or change according to circumstances. The self may grow, be maintained, recovered and regulated in response to individual experiences and changes, whilst keeping a sense of historical continuity over time (Troll and McKean Skaff 1997). Self-identity also benefits from the input from others, which can contribute to modify or reinforce such identity. In some cases, however, social influences may promote "no-identity zones" which "fail to sustain an authentic framework for supporting experiences in old age" (Phillipson and Biggs 1998, p. 11).

Erik Erikson combined both the individual and the social aspects of identity in a synthetic definition:

> The term 'identity' expresses such a mutual relation in that it connotes both a persistent sameness within oneself (selfsameness) and a persistent sharing of some kind of essential character with others (Erikson 1980, p. 109).

This sense of continuity in identity probably reaches its most extreme expression in the life of people who join a religious order or adopt a solid philosophical stance or become committed to specific views of the world (e.g. environmentalism) early on in their adult life, to then keep reinforcing and developing those basic principles that give meaning to their existence, within the same broad paradigm. A commitment to transfer such knowledge and principles to new generations (generativity) gives transcendence to their life (e.g. McAdams 2006).

Continuity may also have an additional adaptive function in such individuals in the form of sheltering them from the negative psychological effects of uncontrollable changes in their environment (Atchley 1989; see also Melia 1999).

Yet, identity can be also dynamic, changing to a variable extent as the individual grows and ages and incorporates new life experiences and thoughts into his/her own worldview (e.g. Troll and McKean Skaff 1997). Erikson (1959, 1968) also pointed to the changing character of self-identity, referring to it as an "evolving configuration", where modifications in the biological, psychological, and social spheres are continuously integrated, possibly into a coherent whole, that can itself change over time, after perhaps occasionally undergoing periods of crisis.

The persistence of the self through life, from birth to old age, does involve a sense of unity and continuity, but such unity and continuity are dynamic, with some people even changing their own name (e.g. through marriage, change of gender), external characteristics (e.g. through plastic surgery, behavioural masquerades, Biggs 1997), not to speak of views on life, political ideas, opinions, religious beliefs, career, and so forth, over their lifetime.

Ego integrity may require the adoption of new meanings and sometimes even substantial changes in our life as both external (e.g. social, cultural) and internal (e.g. biological ageing) circumstances change. Fulfilling ageing may call for change in adaptive responses over time whenever the old rules do not hold anymore (see also Dittman-Kohli 1990; Thompson 1992).

Through social interactions, our identity can change not only according to the views and perceptions of others (e.g. by means of self-stereotyping, for instance, Levy (2003); or, more positively, by learning from others), but also through our own experiences in various social roles: parent, partner, grandparent, sibling, offspring, colleague, friend, just to mention a few. In the words of George Mead (1934, p. 178): "The organization of the social act has been imported into the organism and becomes then the mind of the individual".

We do recognize ourself as a unit in space and time, but biologically, psychologically, and socially we are dynamic. What we finally tend to develop is not so much a "multiple identity" (e.g. Rosenberg 1997) but rather a "complex identity". Unity in complexity or, as Sökefeld (1999, p. 424) put it:

> This ability to manage different identities - to manage difference - is an important aspect of the self. Put the other way round, to conceive of a plurality of identities that can simultaneously and/or subsequently be embraced and enacted by the 'same' person we need something that somehow remains the 'same' - in spite of the various differences entailed by different identities. What remains the same is a reflexive sense of a basic distinction between the self and everything else.

Although a complex self-identity could switch from one state to another (and then back) in response to specific circumstances, some identity transitions may be more permanent or easier to achieve than others. We use the anthropological concept of *liminality* introduced by Victor Turner to refer to the transitional process from one identity to another during ageing. Turner (1969, p. 95; see also Turner 1967) stated that:

> The attributes of liminality or of liminal *personae* ('threshold people') are necessarily ambiguous, since this condition and these persons elude or slip through the network of classifications that normally locate states and positions in cultural space. Liminal entities are neither here nor there; they are betwixt and between the positions assigned and arrayed by law, custom, convention, and ceremonial.

The liminal person is under pressure to quickly transition and eventually define a specific new identity (no matter how complex or simple it might be) and move out of the liminal state, which may risk to become protracted (e.g. Beech 2011) if it is left unresolved. Helping the elder transition through the liminal phase between work and retirement, for instance, will require not only a personal effort but often also external support from family or community/government organisations. Such social support may help the elder find and stabilise a new identity, thus overcoming the liminal ambiguity (West et al. 2017).

The social effects on self-identity will continue beyond the liminal phase and can be further affected by our assumptions about what others think about us and what they actually think that we might think, which in itself can spark a recursive process that may further complicate the dynamics of self-identity throughout our life (Goffman 1963). The recursive process may occur in any social environment that is a source of identity development for the elderly: volunteering activities, leisure, grandparenting, intimate relationships with other adults, and so forth (Thompson 1992). In response to this possibility, we can proactively select to interact with people who tend to reaffirm the perception of our own identity, thus perhaps slowing down the dynamic of change or at least limiting the effect of those social influences that could spur self-ambiguity (e.g. Demo 1992).

The mutual interactions between the social sphere and self-identity can affect *agency* or the capacity and will to act in the pursue of a goal. More precisely, Albert Bandura defines agency as "the human capability to influence one's functioning and the course of events by one's actions" (http://professoralbertbandura.com/albert-bandura-agency.html). This requires the acquisition of some tools that will be used in appropriate circumstances, but also a degree of optimism that the effort will bear fruits. Agency relies on self-identity and therefore a list of mental and physical tools that the individual uses for action, but also on social relationships and the ability to recruit support in the pursue of such action (e.g. Emirbayer and Mische 1998). Bandura lists four characteristics of agency: (a) intentionality (plans and strategies to guide action), (b) forethought (an anticipatory plan for future action), (c) self-reactiveness (self-regulation), and (d) self-reflectiveness (assessment of personal efficacy and adoption of corrective adjustments when necessary). The better defined the self-identity of an older person is, the more likely it is that she or he can be an effective agent in his/her environment.

Hitlin and Elder (2007) described various forms of agency:

a. *Existential agency*, which is a form of self-initiated and directed action that relies on the belief that one is capable of such action (self-efficacy). Self-efficacy leads to the development of "personal empowerment" (Little et al. 2002).
b. *Pragmatic agency*, this is capacity to act that takes into consideration some contingent aspects of a situation, that may be also affected by social factors. The focus on contingency contrasts with action performed out of habit.
c. *Identity agency* involves action following established patterns that are part of the repertoire of personal identity behaviours. Such identity behaviours may have also been constrained by social interactions and cultural expectations.
d. *Life course agency* is action that attempts to shape our future life trajectory. This requires self-reflection about both our current capacity for action and our capacity for future action in maintaining a certain life course. Decreasing health and advanced age may impair to a degree the extent of our life course agency, but if the future life course is planned taking the constraints of ageing into account, then a degree of agency can be retained. Life course agency is put to the test especially during major life transitions, such as retirement.

In a study carried out in the UK, Thompson (1992) reported a strong message from the elders he interviewed, a message of pursuing personal goals through their agency such as a sense of coherence, meaning in life, and personal fulfilment that gave them strength in the face of adversity:

> … the message which comes most strongly from these accounts is of resilience in the face of the twists of fate; of adaptability; and in some of these lives, of a powerfully continuing ability to seize or create changes for fulfilment, whether in work, leisure, or love (p. 45)

When personal fulfilment is experienced in old age, the perception of old age itself may change and the individual may become just "as old as you feel" (Logan et al. 1992).

We mentioned above that self-identity has a dimension into the future, as in the case of forethought—a component of Bandura's concept of agency—or in relation to Hitlin and Elder's concept of *life course agency*. In fact, what can extend the self into the future is not just a positive concept of identity but also negative prejudice which in the case of older people can take the shape of ageistic stereotypes, even self-stereotypes (Nelson 2005). This will be the object of a more detailed analysis in Sect. 6.6 of Chap. 6.

Self-identity in the old may also reach extremes such as that observed in the so-called *Diogenes syndrome*, where the elder neglects aspects of socially established rules of behaviour (e.g. "hygienic neglect, squalid living conditions and general refusal to receive help", Band-Winterstein et al. 2012, p. 110) thus prioritising his or her own will and desires.

Self-identity leading to social exclusion, especially if exclusion is caused by the action of others as in prejudice and stigma—perhaps less so if it is caused by an individual decision, as it could be the case of some elders displaying *Diogenes syndrome*—may have negative consequences on the elder's well-being (e.g. Jose and Cherayi 2016). This negative aspect of self-identity is also confronted by immigrants who must juggle their desire to perhaps keep some aspects of their native culture (e.g. language, religion) and the need to integrate. For instance, in some Western countries it has become a challenge for Arabic-speaking Muslim immigrants to be able to retain some aspects of their identity. Still, with the correct policies and enough social support, immigrant elders could be better integrated (Ward 2000) and also allowed space to maintain important aspects of their identity. If we are arguing for the need to blend both continuity and change in the development of self-identity, then we should be open to the possibility that immigrants may develop their own identity, blending the old (from their original roots) with the new (from their adopted country).

The increasing complexity of identities resulting from immigration and other mass movements of people around the world, such as tourism, and improved communication and globalisation more generally (e.g. Hammack 2008), has recently led to the emergence of a phenomenon of *identity insecurity* (Giddens 1991), which is being exploited politically by nationalistic parties, especially in economically developed countries. In this regard, successful multilingual and multicultural societies (such as New Zealand, Canada) can provide good case studies for the development of self-identity in social coherence within a diverse cultural environment.

Gender is another factor affecting self-identity that is relevant throughout life and it continues to have some relevance into old age. For instance, old women tend to be more concerned about maintaining a youthful identity than men (Barrett 2005) even though, as we have seen in Sect. 4.6 of Chap. 4, overall gender differences tend to decrease as people become older, and gender itself becomes less central as a factor giving specific meaning to the life of the old (e.g. Silver 2003).

In addition, aspects of the living environment of the elder, the physical surroundings, for instance, the landscape, can also help shape self-identity. Rowles (1983) points to the development of emotional attachment to a specific locale which can be an important source of elders' well-being. This is valid not just for elders who have difficulties mobilising, but also for those who are fully ambulant; and it does not just apply to landscapes that can attract the admiration of the average observer but also to places that many of us may like to avoid if we can. As Rowles points out, even "dilapidated homes, boarded up stores, and bulldozed lots" can be a landscape contributor to self-identity for older residents who have grown in place. An *autobiographical affinity* to the environment (*insideness*, Buttimer 1980; or *environmental identity*, *environmental self*, Clayton and Opotow 2003) inevitably shapes self-identity, especially for those who have always lived in such environment and are already old, but often also for those who may have left it at a younger age. Rowles argues that in old age, insideness sustains self-identity, and it can provide emotional support and life meaning through reminiscence.

Clayton (2003, p. 45–46) has defined *environmental identity* as "a sense of connection to some part of the non-human natural environment, based on history, emotional attachment, and/or similarity, that affects the ways in which we perceive and act toward the world". Such ability to develop attachment to the environment is profound and deeply engrained in our evolution, although the specifics may be dynamic and learned (Schultz and Tabanico 2007).

Is the concept of self universal? Do cultures differ in their conception of self-identity, especially among the elders?

Although the basics of self-consciousness and the ability to develop a concept of the "self" are universal in the human species, the specifics can vary across cultures (Han and Northoff 2009). Cross-cultural studies suggest that a broad divide exists in the perception of the self along the lines of *Individualistic/Collectivistic* cultures (Katzko et al. 1998). Collectivistic cultures characterised by *familism,* for instance, to an extent restrain self-identity within a cultural mould of social expectations. Individualistic cultures also display a degree of social control on the individual, as some forms of self-identity tend to be better tolerated than others, but not to the same extent as collectivist cultures.

Across cultures, self-identity develops through narratives. Narratives of the self and about the self are embedded not only in the personal experience of the narrator but also in the cultural norms and traditions the individual has experienced throughout his/her life (Gone et al. 1999; Hammack 2008). Through narratives we story our life experiences, perhaps into a coherent whole, from which self-identity and meaning may emerge:

> It is through the process of telling one's own life story that individuals are able to make meaning of their lives, construct and better understand the self, and maintain a sense of personal identity (Dorfman et al. 2004, p. 188),

and in the words of Hammack (2008, p. 232):

> If the content of identity assumes an ideological quality - a cognition of self in relation to discourse - it is through the development of a personal narrative that such cognition is rendered comprehensible and meaningful to an individual and to the group or groups to which he or she belongs. If ideology provides the basic cognitive content of identity, it is in narrative that ideological identifications assume a coherent structure... It is through the construction of personal narrative that the life course achieves its coherence, its continuity in social, cultural, and historical time.

Dorfman et al. (2004) studied narratives of community-dwelling and institutionalised elders in a Midwestern rural community in the USA, with the objective of understanding their sources of personal identity. By asking elders to produce a narrative of their life we are actually also asking them to reflect on such life, and through the process of self-reflection self-identity may be put into a sharper focus. Life narratives are clearly affected not only by the current personal experiences of the individual, but also by past experiences and their memory (and memory failures), creative reconstruction of events and what Dorfman et al. (2004) call *generational imprinting*, which is equivalent to the cohort effects that we have already mentioned in previous chapters.

In Dorfman et al.'s work elders living in rural communities tended to produce narratives of self-identity that emphasised the importance of family, religion, hard work, self-reliance but also a sense of community within their town, the impact of specific historical events (e.g. WWII, the Great Depression) and a close relationship with the land.

The process of identity formation in old age through narratives, especially autobiographical narratives, can be complex and it relies on the ability of the individual for introspection and capacity to synthesise, perhaps in an optimistic manner, various experiences from the past, whilst using such process of identity-making to project the self towards what remains of the future.

McLean (2008) interviewed a group of young and old people in Canada asking them to write aspects of their life that were identity-defining (*self-defining memories*). Table 5.1 summarises the kind of narratives recorded by McLean according to type, with examples that show a diversity of ways of incorporating memories in the construction of self-identity: *Self-event connections* refer to the connections established between the self and experience; *event-event connections*, on the other hand, are associations established between past events, rather than links between events and the self, event-event connections help extend the life story; *change connections* are recollections of change events, they tend to be more common among younger participants; whereas *explanatory connections* are more representative of stability of self-identity and were more typical of older participants. Examples of the *processing of connections* are also shown in Table 5.1. Here processing points to the degree in which an individual reports "thinking about, or generally reflecting upon self-event connections".

Individuals differ in the level of processing of their past experiences and their capacity to use them for identity formation: some may use low level processing, others high level. Some have better capacity to integrate past experiences into a

Table 5.1 Types of narrative and examples recorded in Canada by McLean (2008)

Narrative type	Example
Self-event connection	"I learned a lot in that period because it was um, uh, I learned that I could enjoy life again"
Event-event connection	"Sometimes you blame someone for accidents and I think I blamed my husband ... Eventually, we were divorced". In this case the connected events are accident with son and divorce with husband
Change connection	"It sort of helped me become a more open person to new ideas and ... more willing to try new things"
Explanatory connection	"I don't think I understand it any differently ... At my age I think we know who we are and what we are"
High level of processing	"You live and learn ... I learned a lot from that experience ... I, learned a lot in that, I don't know if this is what you mean by defining, but I learned a lot in that period because it was um, uh, I learned"
Low level of processing	"I don't know. I don't know if it really means anything to my identity. Umm, hmm, I don't know"

coherent self (*thematic coherence*, or the "general theme of one's life story") others may struggle to make sense of it all. As expected, explanatory connections were more frequently produced by older participants, as were event-event connections and *thematic coherence*. Thus old age confers an advantage in terms of capacity to integrate identity into a more coherent self. On the other hand, change connections were more frequent among younger participants and also among women (the latter effect was especially evident among the old: older women produced more change connections in their narratives than older men). Although change at young ages can be a straightforward effect of the process of building an identity, which may be still in flux earlier in life, change among older women may point to a different phenomenon: a greater capacity of women to change and adapt compared to old men, at least in the cohort and cultural context considered in McLean's work.

Narratives that help create and sustain self-identity are also an instrument for the production of what Assmann (1994) refers to as *wholesome knowledge*, that is knowledge and cognitive activity that aim at preserving human beings (see also Dittmann-Kohli and Jopp 2007). In this case, narratives can help preserve the integrity of the self in older people in both a spontaneous manner and also in a more planned, therapeutical manner. We will return to the issue of narratives in Sect. 7.5 of Chap. 7.

In conclusion, experiencing fulfilling ageing requires an individually-tailored combination of continuity and change in self-identity, with the mixture itself being in dynamic flux. The flux reflects the reality of both the plastic capacity of our brain and its limitations in the construction of self-identity throughout the ageing process. In addition, the circumstances of our life can also change in a dynamic way, and such change may lead to, at the personal level, a degree of reaffirmation and strengthening of our principles, but also our questioning of such principles. At the social level, change of self-identity can result from interactions with others. With age, greater integration of the self and coherence in self-identity may become more likely, thus enhancing our ability to develop wisdom.

In the next section we focus on the emotional aspects of old age.

5.3 The Emotional Dimension

What are emotions? Given how ubiquitous the experiencing of emotions is among most of us, one would expect that a definition of emotions would be readily available. Instead, the concept has proven to be difficult to pin down with words. Some have tried to define emotions as feelings that can affect our behaviour and also our physiology. But then, what are those "feelings"? It is not our objective to dig deep into the complexities and difficulties of the definition of emotions (see, for instance, Mulligan and Scherer 2012) but it is important to mention that emotions have deep roots into our evolutionary history and therefore they are not only shared, to an extent, with some other animal species (e.g. Paul and Mendl 2018) but they also display many adaptive features especially related to individual experiences and social relationships; although maladaptive (e.g. pathological) outcomes of emotions may also results from specific social and developmental circumstances (see Buss 2008; Workman and Reader 2014 for an introduction to the evolutionary aspects of emotions). For instance, *interest* is an emotion that promotes information-gathering and that can have adaptive value in the solution of problems affecting survival, whereas *guilt* is more readily associated with an adaptation to a social environment. Thus we may see emotions as internal states that orient the organism towards displaying a specific subset of behaviours that are adaptive in a given circumstance.

Izard (2010) offered the following synthesis of the views on emotions that she derived from the replies to a survey of 35 specialists in the areas of behavioural and cognitive neuroscience, computational cognitive science, and clinical, cognitive, developmental, and social psychology:

> Emotion consists of neural circuits (that are at least partially dedicated), response systems, and a feeling state/process that motivates and organizes cognition and action. Emotion also provides information to the person experiencing it, and may include antecedent cognitive appraisals and ongoing cognition including an interpretation of its feeling state, expressions or social-communicative signals, and may motivate approach or avoidant behavior, exercise control/regulation of responses, and be social or relational in nature (p. 367).

It should be noted how the above definition contains not only the view of emotions as systems of behavioural regulation, but also as properties of our brain that interact with cognition.

Although the terms emotion and mood have been used interchangeably in the psychological literature, there is a general agreement that they refer to two different psychological phenomena (e.g. see Beedie et al. 2005). Broadly speaking, mood is a proactive state, a state of mind adopted by the individual irrespective of external circumstances. Whereas an emotion is seen as a reactive state that manifests itself in response to certain external circumstances (see, for instance, Damasio 1999). McKenzie and Crouch (2004) have suggested subsuming *mood* and *emotion* into the common umbrella term of *affect*.

The experiencing of emotions involves the adaptive coordination of a number of bodily systems and sub-systems that range from sensory to motor, and also various mental faculties, recruited to address a specific contingency (e.g. a cause of anger, fear, surprise, and so forth).

From a developmental perspective, various emotions emerge throughout the initial stages of post-natal life in humans. The earliest emotions to develop are joy, interest, sadness, anger, fear, and disgust. These are followed later in childhood by shame, guilt, shyness, and contempt (Izard and Ackerman 2000). That is, the more complex emotions associated with social interactions develop later in our childhood. Changes in emotional states, especially if they occur during the initial years of development, and also when they are intense in magnitude and protracted over long periods of time, may have cascading effects on a series of aspects of the individual's biology, many of which may either compromise or promote health, depending on the specific emotions involved. Emotions can affect health in both direct and indirect ways. Directly, they may affect health through the links between the brain and the immune, peripheral nervous and endocrine systems, and indirectly by influencing our ability to take care of—or perhaps neglect—our health under the influence of specific emotions (Salovey et al. 2000; Fredrickson 2003 and references therein).

Carroll Izard proposed a *Differential-emotions theory* for the various roles emotions have in humans (see, for instance, Izard and Ackerman 2000). Emotions determine various motivational states of the individual, by organising aspects of perception, cognition, and action. The individual responds to specific situations according to her/his particular emotional states, and such behavioural responses may vary according to sex, age, and various degrees of previous learning experiences. Personality provides a vast mental umbrella that encapsulates, among others, specific combinations of emotions and moods that are expressed more or less reliably by the individual. As we have already noted, emotions may also interact with cognition in the production of thought and manifest behaviour. Izard and Ackerman refer to this emotion/cognition interaction as an "affective-cognitive structure", regarding it as "the most common type of mental structure". The expression of emotions is also likely to vary across cultures as a result of this cognition/emotion interaction.

Emotions as adaptations are mainly a result of evolution by natural selection, which requires surviving individuals to pass on to their offspring those genetic characteristics that contributed to making them survivors. Emotions can be adaptive in the promotion of cooperation among kin, for instance (e.g. Carstensen and Löckenhoff 2003), and by promoting experiences of personal satisfaction, even pleasure, leading to greater well-being.

Izard and Buechler (1980) distinguished the following basic emotions that underpin the broader emotional variability in our species: *joy*, *surprise*, *sadness*, *anger*, *guilt*, *disgust*, *contempt*, *fear*, *interest*, and *shame/shyness*. The adaptive value of such emotions is readily clear from our daily experiences, especially in the context of survival, reproduction, social communication, and promoting the acquisition of knowledge.

Charles Darwin (1872) laid the groundwork for the evolutionary study of emotions in his book *The Expression of the Emotions in Man and Animals*. More recently, Izard (1992, see also Ekman 1994) has suggested a series of basic adaptive functions for the various emotions. The following summary is taken from Cordella and Poiani (2014):

Interest (but also *curiosity* and *wonder*) is adaptive in motivating play and other activities aimed at gathering information from the environment and producing creative solutions; *joy* (and also *happiness*) fosters social interactions and openness to new experiences, including social affiliative experiences; *sadness* serves the function of recruiting assistance from members of the social group, thus strengthening social bonds, and it also slows down cognitive functions, which is a potential adaptation for a better evaluation of the current situation before the individual embarks into any specific course of action; *disgust* prevents poisoning and also alerts witnesses about this potential danger, its use in social contexts not associated with food consumption is probably a derived state. *Anger* is a defensive strategy that signals the potential for physical aggression if a conflict is escalated, and it may also be used as a conveyor of disappointment with a specific situation or outcome; *fear* protects the individual from the consequences of potential danger by triggering hesitation or retreat. An individual expresses *shame* as a way to avoid the social consequences of a wrong action, shame is aimed at decreasing somebody else's anger. In this context, shame has the effect of promoting social conformity. *Surprise* may have originally evolved from the freezing behaviour released in the presence of sudden danger, and it was subsequently retained as a reaction to other kinds of novelties, such as a new object observed, or a new thought. *Guilt* may have initially evolved as a social signal of submission, whereas *contempt* could have had an initial adaptive value as a signal of social disapproval, at the bottom of a potential escalating process leading to full blown anger.

Note that although each one of those emotions may have been originally selected in a very specific adaptive context (see, for instance, *disgust* and poisoning prevention), a process that would define them as *adaptations*; the same emotions may have been subsequently co-opted into different functions (in the case of *disgust*, it could be its use as a display of social disapproval in some cultures). In the evolutionary literature the latter is known as an *exaptation* (Gould and Vrba 1982).

In general, old age tends to bring more emotional stability and well-being compared to younger ages, especially after the age of 65 (Consedine and Magai 2003; Yankner et al. 2008; Scheibe and Carstensen 2010), with older people displaying less autonomic response to various tasks than younger people (Levenson et al. 1991). Indeed, as stated by Levenson et al. (1991, p. 28):

> Old age has been described by several theorists as a time of pensive self-focus and dampened emotional intensity.

Moreover, in old age emotions may shift from negative to positive, not just from physically active to pensive, following a diversity of simultaneous processes that occur with ageing: greater wisdom in old age but also greater decay of biological functions. In this regard, Ross and Mirowsky (2008, p. 2391) stated that:

> The shift from negative to positive is consistent with the age as maturity perspective. The shift from active to passive supports the age as decline perspective. If these generalities are correct, then they should apply to positive emotions as well as negative emotions. We should see a shift in positive emotions from active (excitement) to passive (serenity), as well as in the negative emotions (from the agitation of anxiety and anger to the lethargy of depression). In order to accurately portray the shifts in emotional tone, *age may best be considered as simultaneously indicating maturity and decline.* (italics ours)

In their study carried out in the USA using data from the *1996 U.S. General Social Survey*, Ross and Mirowsky (2008) did indeed observe a shift from negative to positive emotions and from active to passive emotions in old age. This is expected whenever old people reach a stage where they feel that they have their life under control and have learned how to reduce the tension of daily challenges (e.g. Carstensen et al. 2011).

Consedine and Magai (2003) point to a general trend where older persons seem to be better at regulating negative affect than younger adults, in part because they select their environment in order to avoid conflict, but they also display less confrontation in their coping, whilst trying to increase their positive emotional experience. They also frequently use humour and affection to de-escalate situations where they find themselves in disagreement with somebody else.

The change from negative to positive emotions in old age can also affect memories, with old people tending to remember the past more positively than young people. This has been termed the *positivity effect* (e.g. Carstensen et al. 2011, see also Kunzmann et al. 2014). Such positivity is in part due to old people giving more prevalence to *positive affect*, which in this case is expressed through a more selective memory of those events that were happier and more satisfactory (Brümmer 2009). The adaptive value of positive affect is seen in that well-being is better achieved in old age through a greater prevalence of positive than negative emotions (e.g. Meeks et al. 2012).

The old also display better emotions regulation than younger adults (Orth et al. 2010). More emotional stability in old age, however, does not mean that older people are emotionally detached. Memories can still elicit emotions in old people, as can facial expressions and other social interactions, along with engaging in activities such as reading, observing a landscape, listening to music, watching a movie, and so forth (Schulz 1982). But older people are better able to exert emotional self-control during social interactions following specific social conventions (*display rules*) (Ekman and Friesen 1975). In this regard, it should also be noted that older men tend to be under greater social pressure to control, even hide, their emotions than women (Malatesta and Kalnok 1984).

We have already mentioned that emotions have a specific adaptive role in social interactions (e.g. Keltner and Haidt 1999). They can help initiate and strengthen social bonds (attachment) and foster social cohesion and cooperation, establish and reinforce social positions and roles, assist in finding a sexual partner and/or defend the self and others against threats. In addition, emotions can be manipulated in order to control others in the pursue of specific individual or group objectives (e.g. in the area of politics).

Different forms of attachment can be distinguished according to the predominant emotions mediating them. For instance, Consedine and Magai (2003) list *secure attachment* and *insecure/dismissive/avoidant* relationships within a continuum of social interactions. Secure attachment is associated with positive affect or the experiencing of positive emotions (e.g. joy, hope), whereas *insecure attachment* is associated with negative emotions (e.g. hate, anger). In older adults, secure attachment leads to greater joy and lower levels of anxiety, anger, and sadness (Magai et al. 2000).

Consedine and Magai (2003) investigated the association between some emotions and forms of attachment in a sample of 65–86 years old in the USA. Their results are summarised in Table 5.2. Forms of attachment such as secure attachment (*security* in Table 5.2) and insecure attachment (*fearful avoidance* and *dismissingness* in Table 5.2) are variably correlated with emotions. Secure attachment was positively associated with guilt, joy, disgust, sadness, surprise, shame, interest, fear, and anger. Fearful avoidance was also positively associated with most of those emotions but also with contempt, although it was not associated with guilt and joy. Dismissingness was the form of attachment least associated with emotions, being only positively correlated with disgust and interest; it was also positively correlated with fearful avoidance. In principle, it seems that dismissingness would be associated with asocial (even antisocial) tendencies or at least low interest in social interactions (Beech and Mitchell 2009), but deep inside some dismissing individuals may be looking for more social integration and acceptance (Carvallo and Gabriel 2006). Old women tend to display more interest but also more sadness and fear than old men, and they also tend to be more dismissing.

Emotions can significantly affect our sense of psychological well-being, independently on how well (or not so well) we are doing materially in retirement. Better regulation of emotions in old age can reduce stress, including the stress of maintaining specific social commitments, an issue that is central to Laura Carstensen's *Socioemotional selectivity theory* (SST) (e.g. Carstensen 1987, 1991; see Chap. 1). In fact, increased emotional stability in the old comes hand in hand with a relative reduction in their social network. However, a smaller but selected social network in old age can still have a positive effect on the emotional life of the elder. In addition, the realisation that we are approaching the end of our life leads us to prioritise immediate emotionally-gratifying experiences, as opposed to focusing on future rewards (Scheibe and Carstensen 2010).

In a study carried out in the USA, Carstensen et al. (2011) concluded that older ages are associated with greater emotional well-being, more emotional stability and also increased emotional complexity (i.e. co-occurrence of both positive and negative emotions). Experiencing more positive emotions later in life also helps decrease the rate of mortality.

Table 5.2 Correlations among variables measuring emotions and attachment

Variables	1	2	3	4	5	6	7	8	9	10	11	12	13	14	15
1. Age	–	0.12*	−0.05	0.03	0.02	0.04	0.01	0.01	−0.08*	0.00	0.02	−0.04	0.02	0.03	0.02
2. Sex		–	−0.06	0.06	0.04	0.13*	0.07	−0.04	0.03	0.12*	0.15*	0.05	0.02	0.05	0.10*
3. Guilt			–	0.13*	0.46*	0.53*	0.31*	0.25*	0.48*	0.16*	0.45*	0.50*	0.16*	0.29	0.02
4. Joy				–	−0.12*	−0.37*	−0.13*	−0.03	−0.16*	0.30*	−0.20*	−0.13*	0.08*	0.01	0.03
5. Disgust					–	0.52*	0.13*	0.32*	0.41*	0.17*	0.45*	0.54*	0.15*	0.40*	0.09*
6. Sadness						–	0.20*	0.22*	0.45*	0.08*	0.60*	0.54*	0.22*	0.29*	0.04
7. Surprise							–	0.17*	0.24*	0.34*	0.26*	0.29*	0.14*	0.22*	0.05
8. Contempt								–	0.21*	0.09*	0.20*	0.42*	0.01	0.18*	0.02
9. Shame									–	0.08*	0.49*	0.35*	0.11*	0.30*	0.05
10. Interest										–	0.08*	0.24*	0.32*	0.12*	0.21*
11. Fear											–	0.44*	0.22*	0.28*	−0.03
12. Anger												–	0.26*	0.27*	0.06
13. Security[a]													–	0.06	−0.01
14. Fearful avoidance[a]														–	0.26*
15. Dismissingness[a]															–

[a]Variables measuring attachment; all the others measure emotions. Sex was dummy coded such that 1 = female. *$P < 0.01$. Simplified from Consedine and Magai (2003)

There are other theories beyond the SST that focus on emotions in old age (see Scheibe and Carstensen 2010 for a review). For instance, *Dynamic integration* theory associates diminishing cognitive capacities in old age with an increased difficulty to integrate and accept negative feelings, thus leading to a narrowing of the emotional experience in old people (Labouvie-Vief 2003). On the other hand, *The life-span theory of control* suggests that older individuals are less capable of controlling their environment in order to achieve their "developmental goals", thus older adults may use more emotion regulation (a secondary control strategy) that focuses on adapting the self to the changing circumstances rather than trying to change such circumstances (Heckhausen et al. 2010).

Another view put forward to explain the trend to greater emotional stability in old age is Mendes' (2010) *Maturational dualism*, which links decreased emotionality in the old to their decreased peripheral perception and physiological reactivity. On the other hand, the *Neurological change* hypothesis simply abscribes the decreased capacity in old people to personally experience and also recognise emotions in others to age-related neurological changes in the brain (e.g. in the amygdala) (Brümmer 2009).

Scheibe and Carstensen (2010, p. 136) conclude that older people can adopt a combination of strategies to adapt to and regulate their changing emotional experiences throughout ageing:

"selective changes in emotional preferences, compensatory efforts to adapt to declining cognitive and control capacities, and the optimization of emotional behavior through lifelong learning and practice can all be expected to drive changes in emotional experience and regulation across adulthood".

Therefore older people not only tend to display more positive emotions but also more emotional regulation and stability. Greater emotional regulation can increase psychological stability and decrease anxiety and depression, and it can also improve sleeping patterns and increase the general well-being of the elders (Brümmer 2009). Emotional regulation can be also helped by the selection of activities (*situation selection*) that are favourable to the regulation of emotions or by the active modification of a situation (*situation modification*) in such a way that regulation of emotions is better achieved (Brümmer 2009).

On the other hand, the capacity for *emotion recognition* has been suggested to decline with age, at least in experimental settings where participants are usually confronted with situations of single emotions assessed through a single sensory channel (e.g. visual, vocal). However, when experiencing real life situations with all their complexity of stimuli, rather than reacting to simplified experimental conditions, emotion recognition seems to be better in the old than young adults (Sze et al. 2012). The adaptive value of emotion recognition is self-evident in any social context, and it may be argued that older individuals may be at greater risk of suffering negative social consequences if they make communicative mistakes. Reducing social conflicts, on the other hand, could be highly beneficial to the old and frail who rely on the support of others.

A phenomenon that is observed in some, although not all, older people is increased verbosity, expressed as lengthy, unfocused and not always coherent speech. This form of communication can have negative consequences on the elder, as social interactions are usually negatively affected by verbosity. Verbosity may be a consequence of emotional disinhibition and/or, as suggested by Ruffman et al. (2010), it may result from a decreased ability to decipher emotional cues coming from a listener. Ruffman et al. investigated verbosity in interviews with old men and women (60–85 years old), comparing them with younger men and women (18–35 years old). Both older men and women were more talkative and spent more time off-topic than younger participants, and especially men seemed to act in a verbose manner because they failed to pick up emotional cues from their listener, whenever the listener indicated that they were talking too much.

We conclude this section with a brief review of some key emotions and how they manifest themselves in the old.

Anger: Anger tends to decrease with ageing (Steptoe et al. 2015), however, old people do retain an ability to express anger and act accordingly. Anger in older people may lead to violence, domestic violence in particular, usually perpetrated against a partner. We tend to see older people as frail and vulnerable and therefore as victims of abuse, which indeed they can be (e.g. Dong 2015, see also Chap. 6), but in reality older people can be also perpetrators of abuse and violence, especially against their partner, most frequently female partners (Zink et al. 2006). Such instances of old-to-old abuse also come with a high risk of mortality. Most such cases of partner abuse in the elderly are instances of "domestic violence grown old", that is they are abusive relationships that have roots deep into the younger years of the couple involved (Wijeratne and Reutens 2016). Lazenbatt and Devaney (2014) report a mean duration of domestic violence of 39 years. As the physical abilities of the individuals decline with age, anger may be manifested more in terms of verbal violence and emotional abuse. Moreover, if the perpetrator of violence becomes impaired with old age (mentally, physically or both), it is possible for the roles to reverse and the victim to become the abuser (Wijeratne and Reutens 2016). Deterioration of mental health with age, especially in cases of dementia, may further exacerbate anger and violence among partners (e.g. Patel and Hope 1993). This results from various factors such as frustration in the patient due to his/her inhability to comprehend and communicate, hallucinations, agitated behaviour; but there may be also depression, frustration, and irritability in the caregiver.

Contempt: Contempt is an emotion (but it has also been described as a *sentiment*: a functional network of attitudes and emotions, Gervais and Fessler 2017) that is usually expressed in order to derogate others whilst at the same time staying calm (Fischer and Giner-Sorolla 2016, see also Matsumoto and Ekman 2004); and yet contempt is closely associated with anger (Scherer et al. 2013). A difference between the two emotions is that contempt specifically denotes a stance of superiority, "the appraisal that the other is inferior, maybe not even worth one's energy or attention, because he or she has a bad character" (Fischer and Giner-Sorolla 2016, p. 347). It involves an absence of respect for the other, which leads to a less openly aggressive reaction. Through contempt we may go as far as to dehumanise others (Haslam

2006). Contempt can be expressed in humans at any age and indeed the young may show contempt for the elders and vice versa. Young and old adults do not seem to differ significantly in their expressions of contempt, at least in experimental settings (Kunzmann and Grühn 2005). Contempt for the elderly has been also referred to as "negative deference" (Krassen-Maxwell and Maxwell 1980). In a cross-cultural study, Krassen-Maxwell and Maxwell (1980) have shown that across 95 societies contempt for the elderly is motivated mainly by: physical deterioration, deterioration of appearance, mental deterioration, assumption of negative traits (e.g. involvement in witchcraft or sorcery), lack or loss of children, lack or loss of skills, hoarding of wealth, lack or loss of wealth, in that order. Expressions of contempt for the elderly can vary and they may include actions such as feeding them scraps up to more extreme forms of contempt such as death hastening behaviours (see also Chaps. 3 and 8).

Disgust: This is an emotion that is closely linked to our senses of taste and smell, leading to negative reactions towards unpalatable food items. The adaptiveness of such behaviour is obvious in protecting us against potentially toxic or infected food (Tybur et al. 2013; Rottman 2014). But the rejection can be also extended to individuals we regard as socially unpaltable (e.g. Olatunji and Sawchuk 2005), and again there is an underlying potential adaptive function in avoiding close contact with people who may cause harm (e.g. through aggression, transmission of diseases) (e.g. Tybur et al. 2013). Some specific acts may also be regarded as disgusting, the disgust being reinforced by adaptive functions (e.g. defecating in inappropriate places). But the association between disgust, social unpalatability, and harm is not always warranted and the reaction of disgust towards others may be simply a consequence of "embodied moral judgment" (Schnall et al. 2008) expressed in the form of prejudice, stigma, and negative discrimination (see Chap. 6). We have seen in Chap. 2 how the old have been regularly perceived with a sense of disgust as ugly, mean, and tragic since ancient times (Gilleard 2007), although this has varied according to the social status of the elder. In old age, the adaptive functions of disgust mentioned above remain as valid as they are in the young (Fessler and Navarrete 2005), but attention should be paid to the discriminatory aspects of disgust felt against the elders.

Fear: There is a list of fears that although are not unique to older people they tend to be more accentuated in this age group due to their changing body. For instance, the fear of falling is especially widespread among old people, those who are physically frail in particular (e.g. Zhang et al. 2006; Zijlstra et al. 2007; Lee et al. 2008) and it increases with age, along with activities that aim at decreasing the risk of falling (Zijlstra et al. 2007a) (see also Table 5.3). Falling may have serious physical consequences that can render the elder less independent and significantly affect his/her quality of life. Fear of falling can be decreased in the elderly by a better lower-body performance, improved general body condition and strategies to increase the elder's self-confidence (Reyes-Ortiz et al. 2006). Frailty may also increase the fear of driving (Taylor et al. 2011) and promote a more general avoidance of various

Table 5.3 Fear of falling and avoidance of activities that may increase the chances of falling

Age (years)	(%)	Fear of falling (%)	Avoidance of activity (%)
70–74	39.8	45.8	29.9
75–79	33.7	55.2	37.7
≥80	26.5	65.6	50.5

Note that both fear of falling and avoidance of activity continuously increase with age between 70 and ≥80 years old. Simplified from Zijlstra et al. (2007)

activities (Delbaere et al. 2004; Hadjistavropoulos et al. 2007; Curcio et al. 2009). Another common fear observed in the elderly—especially those who are still independent, live on their own and go out and about also alone (e.g. shopping, visiting family and friends)—is fear of criminal victimisation (Clememte and Kleiman 1976; Eve and Eve 1984; Miethe and Lee 1984; Warr 1993; De Donder et al. 2005). Such fear may be high in the elders even when the crime rate in the area they live in is not high, and fear is further increased if they, in fact, become the victim of a crime. Old people may also feel fear and anxiety about medical problems such as dental fear: fear of losing teeth, but also fear of having dental issues requiring an expensive medical intervention. Such fear is especially elevated in older people who are not subject to regular dental check ups (Locker et al. 1991). Fear of developing dementia may also occur (Cantegreil-Kallen and Pin 2012), along with fear of ageing itself, which may negatively affect well-being (Klemmack and Lee Roff 1984; Nelson 2005). Older people are also more fearful of taking risks (Warr 1993), including the risks coming with social change (Eve and Eve 1984). The ultimate fear, of course, is the fear of death that can spur our effort to stay alive, often in spite of growing challenges to do so (Cicirelli 2002). We will return to the issue of fear of death in Sect. 8.3 of Chap. 8.

Guilt: This is an emotion that most likely evolved within the context of our sociality, as guilt can be seen as adaptive to avoid perpetrating unnecessary and/or continuous harm to others (Gilbert 2003). More proximally, guilt may be triggered by transgressions of an established code (e.g. a moral or legal code) (Rohan et al. 2014) and it is commonly observed in the context of religion (e.g. London et al. 1964; Studzinski 1987). Orth et al. (2010) observed that guilt steadily increases from the teenage years to old age in a diminishing-returns fashion, reaching a maximum between 60 and 70 years old, to then slightly decline afterwards. Women tend to feel more guilt than men (Orth et al. 2010).

Sadness: Sadness is less frequent in 65 years old and older people than in younger people (Gallo et al. 1999) although, of course, older people are not immune from sadness (Black 2009). In fact, Seider et al. (2011) showed experimentally that older people display greater reactivity to sadness than younger people, in response to a film that was expected to elicit sadness. Studies carried out using other stimuli, including autobiographical memories, support the heightened reactivity to sadness in old age. Thus there seems to be a difference in old age between the experiencing of sadness (which tends to be low) and the reactivity to sad situations (which is high). Ultimately, deeply saddened old people are more likely to become depressed, which may contribute to hastening their death (Cooper et al. 2002).

Shame: Shame is an emotion that is commonly expressed in the context of a transgression of established social norms (e.g. Hacker 2017) and it can have the same adaptive function as guilt. In the words of Tangney et al. (1996, p. 806; see also Aknouche and Noor 2014; Fussi 2015; Subandi and Good 2018): "Shame is a painful, ugly feeling that involves a global negative evaluation of the entire self. When people feel shame, they feel devalued as a person. Their sense of self — and self-efficacy—is impaired. And their awareness of others' negative evaluations (real or imagined) is highlighted". Shamed individuals may react by withdrawing from social interaction or by becoming aggressive and blaming others for their predicament. But shame could also spark a more constructive reaction leading to self-improvement (Aknouche and Noor 2014). Although shame is overall negatively correlated with age (Consedine and Magai 2003, see Table 5.2), it tends to increase in old age after decreasing from teens to middle age (Orth et al. 2010). Women tend to feel more shame than men (Orth et al. 2010).

The above could be seen as mainly negative emotions, but of course the old can also experience positive emotions.

Pride: Although pride has been seen in various cultures as negative (e.g. a sin in the Christian tradition), there is also a view of pride that conveys a positive meaning. Tracy et al. (2010) mention how Aristotle referred to pride positively as the "crown of the virtues", pointing to the just satisfaction we should feel about our achievements. Proper pride is regarded positively by Aristotle, whereas *excessive pride* and *devaluated pride* are seen as negative. Orth et al. (2010) studied pride across the lifespan distinguishing between the positive "authentic pride" (e.g. a sense of accomplishment and productivity) and the negative "hubristic pride" (e.g. characterised by arrogance and egotism). Authentic pride increases from adolescence to old age, whereas hubristic pride decreases from adolescence to middle adulthood, to reach a minimum at about 65 years old, to then increase into older ages.

Happiness: Being happy implies the experiencing of good quality of life and well-being (Bergsma and Ardelt 2012) and, in our view, the experiencing of fulfilment in life. Happiness, as measured through variables such as subjective well-being, can undergo different trajectories throughout life according to social, cultural, economic, and contingent historical factors (cohort effects). A concave pattern may occur, with high values of subjective well-being at young and old ages, but lower values in middle age; a convex curve of higher well-being in middle age and lower well-being towards the extremes is also possible, as it is a linear trend of subjective well-being throughout life, with the trend being constant around a specific value over time, or linearly increasing with age or decreasing with age (López Ulloa et al. 2013). Such patterns can even been found in the same population if we focus on specific levels of happiness. Based on data from the English *National Survey of Psychiatric Morbidity*, Cooper et al. (2011) reported variable patterns of change in happiness across four older age classes: <60 years old, 60–69, 70–79, and 80+ (see Table 5.4).

Table 5.4 Degrees of happiness among old people of different ages and gender

Factor studied	*n* (%)	Happiness scale		
		Not too happy (%)	Fairly happy (%)	Very happy (%)
Age				
<60	4789 (73.4)	8.1	52.7	39.3
60–69	1194 (12.9)	7.3	50.8	41.9
70–79	948 (9.1)	8.2	50.3	41.5
80+	530 (4.7)	9.2	57.3	33.5
Gender				
Female	4204 (51.4)	8.9	53.8	37.3
Male	3195 (48.6)	7.1	50.9	42.0

From Cooper et al. (2011)

The category "not too happy" shows an almost continuous trend towards increasing with age, the category "fairly happy" shows a concave pattern of change with age, whereas the category "very happy" has a convex pattern of change, presumably because after retirement high levels of happiness tend to be more frequent, but then the frequency tends to decrease at older ages as bodily functions deteriorate. In this study men were happier than women. Elders themselves can contribute to making their life happier, independently of current circumstances. An optimistic stance in life can certainly help, as it can a positive view of past life (*positive reminiscence*) (Pasupathi and Carstensen 2003; Bryant et al. 2005; Shmotkin and Shrira 2012). Other variables that have been associated with happiness in old age include: social engagement and altruism, marital/partner satisfaction, level of education, better economic resources, humour, less abuse of alcohol, following a healthy lifestyle that includes engaging in physical exercise, and fostering personal authenticity and self-identity (Post 2005; Berg 2008; Mathieu 2008; Tse et al. 2010; Cooper et al. 2011; Hsu and Chang 2015; Khazaee-Pool et al. 2015).

Humour, for instance, can be even used therapeutically to increase happiness and decrease pain in older people (see Fig. 5.1), whereas living a happy life can contribute to increasing longevity (Diener and Chan 2011).

Henricksen and Stephens (2010) have categorised the kind of activities conducing to happiness in old age into a scheme that we have complemented with some additions:

Other-Focused activities
- Time with significant others
- Meeting others
- Helping others

Personal Recreation and Interests
- Pursuing hobbies
- Entertainment and relaxation
- External engagement

Personal care
- Health-promoting activities

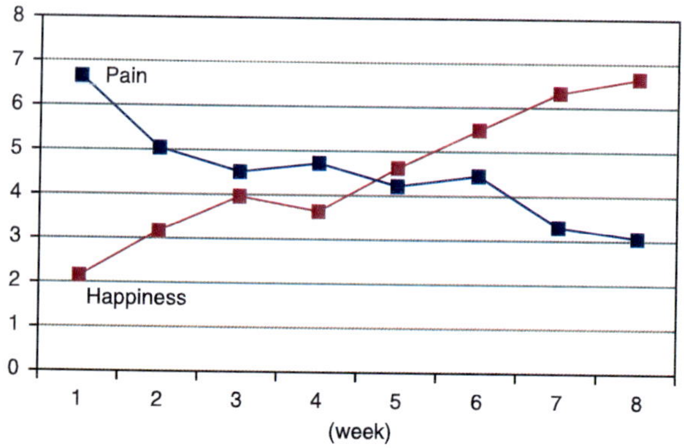

Fig. 5.1 Pain and happiness scores over an 8-week humour therapy programme. From Tse et al. (2010)

Financial security (variable according to personal material needs and expectations)

Thoughts and Attitudes

Transcendence

Giving thanks

Constructive/creative thinking

Achievement-Related activities

Small achievements

Longer-term goal progress

Curiosity: Curiosity has been traditionally seen as a mechanism of information-gathering to support better and more efficient decision-making, but in reality it can also spur individuals to obtain more information than they actually need at any given time, and therefore curiosity has also a function that transcends the immediate practical benefits of mastering useful knowledge (Lowenstein 1994). For instance, the curiosity of the scientific theoretician can open up new ways of understanding the world, that only later may lead to practical applications. In this context, curiosity can be associated with positive emotions "corresponding with seeking unknown information that an individual anticipates will stimulate interest" (Grossnickle 2016, p. 28), and learning more about topics of interest can be also enjoyable. Although curiosity tends to decline to some extent in old age (Sakaki et al. 2018), motivation towards greater curiosity can be improved through an increased engagement of old people with life. In a study of curiosity in older adults, Brink (2007) described a positive correlation between curiosity and both life satisfaction and personal growth initiatives. Curiosity may also help consolidate the memory of interesting events in the old (McGillivray et al. 2015) and it may even contribute to buffer old people against the negative emotional effects of social rejection (Kawamoto et al. 2017).

Surprise: Surprise describes our emotional reaction to the unexpected (Meyer et al. 1997). This initial emotional reaction subsequently leads to focusing our attention on the cause of our surprise (Horstmann 2015), to an evaluation of the situation and then a behavioural response. The ability to experience surprise does not seem to change perceptibly with ageing (Consedine and Magai 2003).

In sum, emotions can have an adaptive function within both individual and social contexts that can remain valid into old age. Broadly speaking, old age tends to be associated with greater emotional stability. As people enter their old ages they tend to shift from negative to positive emotions and from physical activities to a more pensive stance in life. With regard to the specific emotions reviewed here, authentic pride, a positive emotion, increases in old age, whereas hubristic pride tends to decrease up to the age of 65 years to then increase afterwards. The level of happiness in old age, on the other hand, is very variable, responding to the specific life experience of each elder. However, many old people currently tend to be fairly or very happy in old age, at least in some industrialised countries. Although curiosity is not readily associated with old age, elders retaining their curiosity in life are better emotionally. Surprise, on the other hand, does not change in old age. Negative emotions display a degree of variability. Some, such as anger and sadness, tend to decrease with old age. Others remain stable: contempt, disgust; whereas still others may increase, such as specific fears, guilt (which, however, stabilises from the age of 70), and shame.

From emotions we shift in the next section to cognition in old age.

5.4 The Cognitive Dimension

> Everyone wants to live for a long time, but no one wants to be old
>
> André Aleman (2014, p. 5)

The *cognitive function* could be defined as "intellectual functioning and the ability to process information" (Baumgartner 2004, p. 284), and it is expressed through attention, learning, memory, visual-spatial ability, psychomotor efficiency, manual dexterity, and general intelligence. Together, such abilities allow us to build a cognitive representation of the world based on inputs from the outside, our internal biological mechanisms—starting from sensory perception and then continuing with more complex production of thoughts—and the interaction between the two. Thoughts are therefore our representation of the world, and they are organised in our mind into synthetic *schemata* following an increasing level of abstraction (e.g. Craik and Bialystok 2006).

We have already mentioned in previous chapters (especially Chaps. 1 and 4) how ageing implies changes in our cognition. Most of those changes involve a deterioration of our cognitive abilities compared to our younger self (e.g. Lindenberger and Ghisletta 2009), but the rate and extent of such decline are quite variable among individuals. Deterioration can be slowed down the more substantial our brain reserve capacity is, with such reserve being built throughout life. Brain reserve capacity underlies our cognitive reserve which can protect us, to some extent, from the decay in brain functions observed in old age, as expressed in Park and Reuter-Lorenz's (2009) *Scaffolding Theory of Ageing and Cognition* (STAC) mentioned in

Chap. 1. The plasticity of the brain and its ability to adapt to the changes of old age can be also helped by a sense of optimism and emotion regulation.

Cognitive decline with ageing is mainly explained by the loss of neural connectivity within the brain, as explained in Chap. 4, being evident especially in the areas affecting memory, processing speed, visuospatial skills, executive functions, and speech comprehension. Such loss of connectivity results from oxidative stress that decreases the number of synapses; a process that is usually referred to as the *wear and tear* of the brain. But the brain also shows plasticity, which allows it to retain a degree of functionality in the face of the accumulated stresses of age. Brain plasticity is helped by us maintaining an active mental life. That is, the brain may deteriorate to a lesser extent if we keep using it (*use it or lose it*) (Swaab 1991). Cognitive decline with age, however, can take a turn for the worst in the case of the development of a pathology, such as dementia, as reviewed in Chap. 4.

Therefore, from previous chapters we can see that the cognitive dimension of old age offers three major areas of interest: cognitive changes in normal ageing, cognitive changes under the influence of neuropathology, and strategies to preserve cognition in spite of those changes.

The lifetime decline in cognitive capacities can follow different trajectories according to: (a) the specific cognitive variable considered, (b) the effect of additional variables (apart from age) on the sample of individuals studied (including generational effects), and (c) the individual circumstances of each person (Hedden and Gabrieli 2004). Diseases experienced in early life, for instance, may negatively affect cognitive functions in old age (Case and Paxson 2009).

In addition, studies of cognition and ageing based on cross-sectional data (i.e. data obtained from an age-diverse population sampled at a specific point in time) may yield somewhat different results to studies based on longitudinal data (i.e. a cohort studied for a length of time throughout their ageing process), but the most general conclusions seem to apply across methods.

In general, some cognitive decline can be observed after the age of 55, whereas significant cognitive decline in healthy individuals is mainly detected after the age of 70 (see Fig. 5.2b, which shows longitudinal data). Onset of decline at ages earlier than 55 years old is primarily seen in people suffering from various health problems. Figure 5.2 shows the rate of decline with age of six cognitive capacities: *inductive reasoning* (a reasoning procedure in which the premises supply some evidence for the truth of the conclusion), *spatial orientation*, *perceptual speed*, *numeric ability*, *verbal ability*, and *verbal memory*.

Note in Fig. 5.2 how verbal ability shows the slowest rate of decline with age in normally ageing elders. The relative rates of decline across cognitive capacities also vary according to the type of study: cross-sectional (Fig. 5.2a) vs longitudinal (Fig. 5.2b). Importantly, there is a contrast between the two types of studies in the variation of numerical ability over time, with a greater decline being observed in the longitudinal study. This may suggest, for instance, that although older generations in this population do decline in numerical ability over time, younger generations are currently not doing well in their development of numerical ability, hence the almost

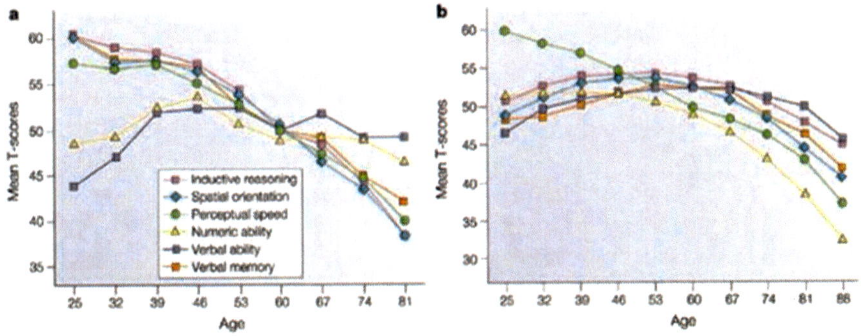

Fig. 5.2 (**a**) Cross-sectional data from the *Seattle Longitudinal Study* showing declines in cognitive capacities in all domains, with some domains starting to decline at an older age than others, and some declining faster over time than others. (**b**) Seven-year longitudinal data from the same study. Declines are evident in all domains after the age of 55. From Hedden and Gabrieli (2004)

flat trend for this cognitive skill in Fig. 5.2a. A consistent decline with age is observed for perceptual speed across the two methodological approaches.

As expected from the *use it or lose it* hypothesis, the cognitive abilities that show the lowest rate of decline with age, such as verbal ability, are the ones that have been used the most by the individual throughout his/her life (Hedden and Gabrieli 2004).

Cognitive abilities that are accumulated throughout life through learning are known as *crystallised intelligence*, whereas *fluid intelligence* is the capacity to think logically and solve problems in novel situations (Craik and Bialystok 2006; Aleman 2014). Crystallised intelligence declines more slowly with ageing than fluid intelligence (Verhaeghen and Salthouse 1997; Li et al. 2004; see Fig. 5.3c). Moreover, as different cognitive abilities may develop and then deteriorate with age at different rates, it is expected that they can bundle together earlier in life but then separate in old age as some become less efficient. This is shown in Fig. 5.3a (see also Fig. 5.2b), where all the cognitive capacities considered develop more or less closely until the age of 40 and then start to separate, with verbal faculties declining more slowly into old age than reasoning, mental mapping and memory. Based on data from the *Berlin Aging Study* (BASE), Singer et al. (2003) also found a differential decline of cognitive capacities with old age; with perceptual speed, memory, and also fluency showing a similar decline, whereas knowledge was better preserved and stable until about 90 years old (see also Park and Reuter-Lorenz 2009; and Fig. 1.4 in Chap. 1).

In old age it is not only specific cognitive faculties that are preserved differently, but individuals also vary in their overall cognition. The *Preservation differentiation* hypothesis (Salthouse et al. 1990; Krampe and McInnes 2007) propounds that cognitively competent older people have always been more competent than others (perhaps due to genetic or early developmental effects) and they better preserve the difference in old age even though they, just like everybody else, also decline in cognitive ability over time. On the other hand, the *Differential preservation* hypothesis suggests that even though people may start from similar initial cognitive capabilities, those who develop greater cognitive competence through life also tend to

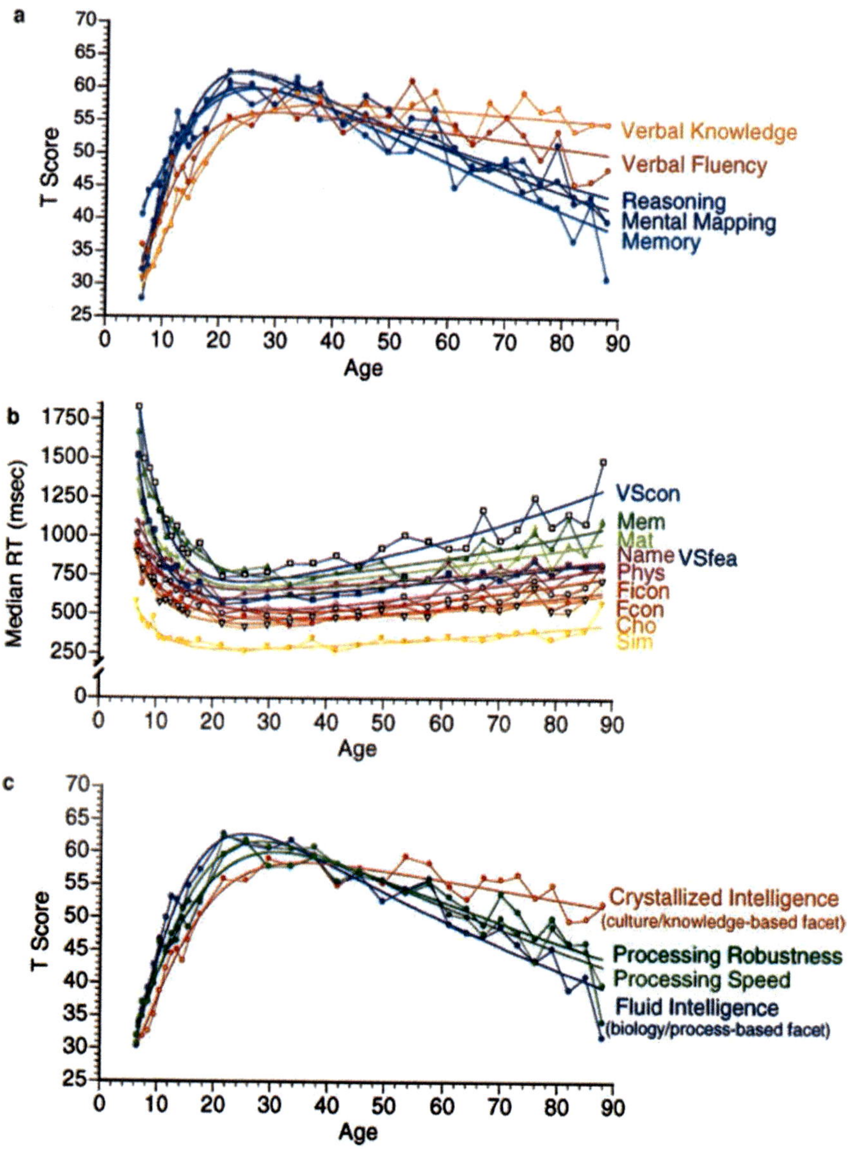

Fig. 5.3 Variations with age of intellectual abilities and cognitive processes. (**a**) Group means and estimated changes through ageing of five intellectual abilities; (**b**) group means and estimated age variations in reaction times (RTs) for 10 basic experimental cognitive tasks; and (**c**) age variations in fluid intelligence, crystallised intelligence, processing speed, and processing robustness. *VScon* visual conjunction search, *Mem* memory-matching task, *Mat* pattern-matching task, *Name* Posner task, name-identity condition, *VSfea* visual feature search, *Phys* Posner task, physical-identity condition, *Ficon* flanker task, incompatible condition, *Fcon* flanker task, compatible condition, *Cho* choice RT task, *Sim* simple RT task. From Li et al. (2004), reproduced with permission

retain such differentiation into old age, compared with people who experienced less cognitive development before becoming old (e.g. due to education, type of work, and so forth). Such differentiation affects all domains of cognition. Finally, the *Expertise preservation* hypothesis states that old people only retain a degree of competence in old age in those specific skills or cognitive domains that they developed the most earlier in life. Evidence suggests that older experts do retain some distinct capabilities in their field of expertise in spite of a general cognitive decline due to ageing. Such retention of specialised skills is also associated with ongoing practice (Krampe and McInnes 2007). The three hypotheses point to different broad mechanisms that may underlie cognitive preservation into old age: genetic/epigenetic and behavioural/environmental, it is clear that differences among individuals in their cognitive capacities in old age can be explained by any combination of those mechanisms.

Cognitive abilities accumulated throughout life constitute our *cognitive reserve*. We have mentioned in Chap. 4 that greater cognitive reserve improves brain plasticity, and memory in particular, in old age; it may help slow down cognitive deterioration due to dementia by increasing brain connectivity and it is helped by better white matter integrity (e.g. Scarmeas and Stern 2003). The building up of cognitive reserve throughout life can be promoted by education, by stimulating creativity and greater intellectual and social engagement in life. In particular, bilingualism and multilingualism are important contributors to cognitive reserve. A better cognitive reserve can have long-lasting effects on the individual, making a difference even in the cognition of centenarians.

Yaakov Stern reviewed the concept of cognitive reserve (Stern 2002) pointing out that two major views prevail with regard to cognitive reserve: one sees it as a *passive process* where reserve is measured as the amount of tissue damage the brain can sustain before any noticeable effects on behaviour are evident. The other regards cognitive reserve as an *active process*, pointing to the ability of the brain to actively compensate for tissue damage (e.g. through neuroplasticity, as seen in Chap. 4) in order to restore or at least prevent further decrease in cognitive capacity. Both processes are not mutually exclusive and can co-occur in the brain.

Among the passive processes, the *Threshold model* (e.g. Satz 1993) states that resilience of the brain to injuries or ageing varies according to the brain reserve capacity represented by the complexity of current interconnectivity of the brain tissues normally used to perform a task. Losing the functionality of some brain tissue may not translate into observable changes in behaviour until a threshold is exceeded, beyond which changes are observed.

Active processes, on the other hand, imply cognitive reserve that relies on the preferential use of brain circuits that are less susceptible to disruption (i.e. they are more resilient), and if this fails, then *compensation* may be activated in the form of other brain regions being used to accomplish a task, that are normally not activated for that specific task.

Passive and active processes may operate in tandem as we age, with the brain cognitive reserve capacity being used through passive processes for as long as the behaviour produced is functional, to then switch to compensatory active responses that recruit other regions of the brain onces damage becomes more extensive. The switch can be aided by therapeutical interventions (e.g. Pascual-Leone et al. 2005).

The plasticity of the brain, as seen in Chap. 4, underlies our ability to express fluid intelligence and build cognitive reserve, which in turn contribute to sustaining our cognitive plasticity, even into old ages. In addition, Baltes et al. (1980) suggested that, with age, greater *cognitive integration* may occur, which would explain the age-dependent development of wisdom; however, different specific cognitive skills do not necessarily become more integrated with each other in old age, as shown empirically by Tucker-Drob and Salthouse (2008). The decrease in efficiency of many cognitive abilities with age seems to be a result of a general underlying deterioration of mental faculties (e.g. Bates and Wolbers 2014), not of a coordinated interaction among those faculties.

In 1988, Paul Baltes and Ulman Lindenberger introduced the *Selective optimisation with compensation* (SOC) model, that we already mentioned in Chap. 1, in order to not only provide a theoretical framework to better understand cognitive plasticity but also a blueprint for the design of practical interventions.

The SOC model takes into consideration the cognitive losses that occur with ageing, which reduce cognitive reserve, and also the potential ability of the organism to compensate for those losses through plasticity (Baltes and Carstensen 1996). The ability to compensate for expected late-life cognitive losses should be built in earlier life, with the learning experience obtained at younger ages providing a degree of *mastery* in finding the solutions to problems, whereas plasticity provides the capacity to *adapt* to new circumstances.

SOC suggests that older people experiencing cognitive losses should start by *selecting* the kind of tasks that can be reasonably achieved given the circumstances. They are subsequently expected to engage in those tasks by using as much additional aid as possible in order to *compensate* for any current limitation (physical or mental). For instance, use of computer technology that can supplant memory deficiencies or to improve the speed of processing in complex mental activities (e.g. writing a book); and finally the current mental faculties could be *optimised* by helping the brain with better alimentation, exercise, and so forth. In the words of Baltes and Carstensen (1996, p. 412):

> Optimisation ... refers to the enrichment and augmentation of reserves or resources and, thus, the enhancement of functioning and adaptive fitness in selected life domains.

Thus cognitive plasticity can be protected and nurtured throughout life in such a way that it may still be expressed, at least to some extent, into old age (e.g. Greenwood 2007). Do we reach a limit to cognitive plasticity as we become older? Evidence suggests that we do. Baltes and Lindenberger (1988), for instance, demonstrated that younger individuals can maintain a high rate of learning in spite of increasing difficulty in the task to be learned, whereas older individuals reach a limit of efficiency earlier on in the scale of difficulty.

Executive functions are cognitive capacities that are defined as "basic skills people need to perform a task: the ability to organize, plan, initiate actions; and to remain on track, control impulses, regulate emotions, adapt, and recover" (Aleman 2014, p. 21). The core executive functions are *inhibition*, which includes *response inhibition*—that is, self-control or the ability to resist temptations and acting

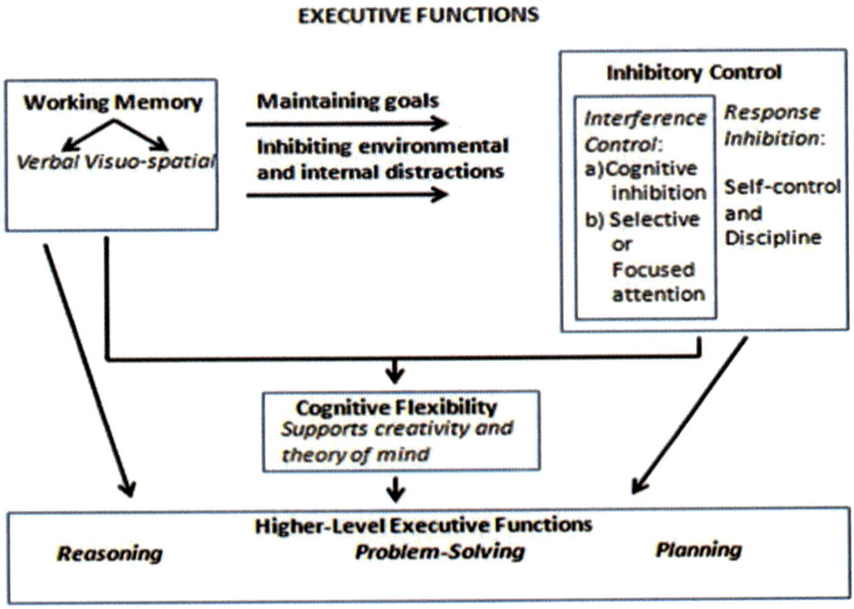

Fig. 5.4 Executive functions. Inhibitory control is also involved in self-regulation. Reasoning and problem-solving comprise what is also known as *fluid intelligence*. Redrawn in simplified form from Diamond (2013)

impulsively—and *interference control*, which involves selective attention and cognitive inhibition. *Working memory* is another core executive function as it is *cognitive flexibility*, the latter includes *creative thinking* and the capacity to adapt to changing circumstances (Diamond 2013). The broad functional relationships among executive functions are summarised in Fig. 5.4. Inhibitory control, cognitive flexibility, and working memory all contribute to the dynamics of higher level executive functions such as: reasoning, problem-solving, and planning which, in order to be better accomplished, require the use of past experiences stored in our memory, the ability to focus on the task ahead and finally the use of our creativity to come up with solutions or answers to our problems. Such executive functions decline with age (Aleman 2014).

A cognitive ability closely associated with higher level cognitive functions such as reasoning, problem-solving, and planning is *cognitive processing speed* or the ability to process information rapidly (Ebaid et al. 2017). Processing speed also declines with age (Aleman 2014), but it should be noted that with age and greater experience in life, more variables may be taken into consideration in the search for a solution to a problem, thus increasing *information overload* and therefore decreasing the speed of thoughts-processing.

The usually greater experience in life of older people, and presumably also their higher levels of accumulated knowledge, are reflected in their better performance in tests of general knowledge and vocabulary (Aleman 2014).

Environmental complexity is also associated with the ability to retain cognitive capacities into old age (Schooler 1984), this includes continued engagement in social activities (Lövdén et al. 2005), a positive effect of a complex and stimulating physical environment (Cassarino and Setti 2015) and both the complexity of the occupations the elders were engaged in during their younger adult years, that built cognitive reserve (Andel et al. 2007), and the complexity of activities carried out in old age (Schooler and Mulatu 2001).

Cognitive ageing may be also affected by self-perceptions (*subjective ageing*) (Schafer and Shippee 2010). Subjective views about performance in cognitive tasks may interfere with actual performance in either positive or negative ways. In turn, subjective cognitive ageing may be affected by the views of others also in a positive or negative manner. In an empirical study carried out in the USA, Schafer and Shippee (2010) detected a tendency, especially among women, to report age identities that tended to be younger than their actual chronological age. More pessimism about cognitive ageing was detected among women displaying negative affect towards cognitive ageing, only compensated by better education and a greater sense of control. Interestingly, among men, those in better health were more negative about cognitive ageing, perhaps because they may have been more aware of what they could lose if cognition deteriorated with age, in spite of their good physical condition.

Given that Schafer and Shippee (2010) detected some differences in self-perceptions of cognition among old men and women: Do men and women also objectively differ in their cognitive abilities? We have seen in Sect. 4.6 of Chap. 4 that there are some cognitive similarities and differences between men and women: men tend to perform better than women on mathematical tests—with the gap apparently closing in some countries—and some spatial tests; and women outperform men on verbal tests, a result that also varies according to the specific verbal skill tested. Kimura (2002) lists some more specific cognitive differences between the sexes: women tend to outperform men in object location memory, perceptual speed, verbal memory, numerical calculation, and dexterity; whereas men tend to outperform women in spatial orientation, visualisation, line orientation, mathematical reasoning, and throwing accuracy (see also Miller and Halpern 2014). Such cognitive differences between men and women could be attributed to genetic and developmental factors, endocrine effects, differences in brain structure and physiology, cultural influences including gender stereotypes and gender-biased education, and interactions among those variables (Miller and Halpern 2014).

Steen et al. (1998) studied two cohorts of 70 year olds in Göteborg, Sweden, measuring in each case various cognitive capacities in women and men. Their results are shown in Table 5.5.

Old men tended to perform better on tasks such as synonyms, figure grouping, block design and digits forward than women in both cohorts, whereas women performed better than men on identical forms and in the Thurstone memory test, thus broadly supporting the results of gender differences observed at earlier ages. van Hooren et al. (2007) described a better performance of 64–81-year-old women than men on a visual verbal learning task, as part of the *Maastricht Aging Study* of The Netherlands. However, Tilvis et al. (2004) administered the *Mini-Mental State Examination* (MMSE) to measure cognitive capacity in a sample of 75–85 years old

Table 5.5 Some cognitive test results in two 70-year-old cohorts of women and men

	Women						Men					
	Born in 1906–1907			Born in 1922			Born in 1906–1907			Born in 1922		
	N	*M*	*SD*	*N*	*M*	*SD*	*N*	*M*	*SD*	*N*	*M*	*SD*
Synonyms	220	18.1	6.9	172	20.9	5.5**	167	19.1	6.5	104	21.1	6.1**
Figure Grouping	220	13.8	5.1	170	15.2	4.4**	172	14.1	4.5	106	16.5	4.1***
Block Design	225	14.6	7.6	177	17.7	7.1**	173	16.6	6.6	106	18.3	6.1*
Identical Forms	219	19.6	7.7	168	23.2	7.3***	172	18.5	6.2	105	20.7	6.9**
Digits Forward	224	5.4	1.2	178	5.8	1.3**	174	5.5	1.2	108	6.1	1.1**
Digits Backward	223	4.1	1.2	178	4.3	1.1*	174	4.2	1.3	108	4.3	1.2 NS
Thurstone Memory Test	218	18.9	5.1	176	21.3	4.5***	173	18.3	4.8	105	20.1	4.5**

N number of probands, *M* mean, and *SD* standard deviation. Significance levels: $*p < 0.05$, $**p < 0.01$, $***p < 0.001$. From Steen et al. (1998)

men and women in Finland over a period of 10 years, and found no difference between genders after controlling for age.

Interestingly, in Steen et al.'s (1998) work there was a broad cohort effect across the cognitive skills tested that was detected in both genders, where the cohort born in 1906–1907 tended not to perform as well as that born in 1922. This is in spite of the 1922 cohort being generally in worse physical condition than the 1906–1907 cohort. The difference could be explained by the level of education, which was significantly greater in the 1922 cohort than in the 1906–1907 cohort, and the living conditions also improved from the 1906–1907 cohort to the 1922 cohort.

More recently, Barnes et al. (2007) reported on a longitudinal study carried out in the USA, where a group of old women was followed for 15 years, from an average age of 72 at baseline to and average of 85 at the end of the study. Nine percent of them maintained optimal cognitive functioning, 58% experienced a minor degree of decline, and 33% experienced major decline. This points to the importance of focusing on the specific situation of each individual as he or she ages. The cognitive decline with ageing is real, but not all elders decline cognitively to the same extent.

Cognitive decline associated with ageing can be slowed down not only by using and developing cognitive skills throughout life, such as better education (Wilson et al. 2009; but see Zahodne et al. (2011) for a critical view) and various forms of mental exercise suitable to each individual (Salthouse 2006); but also by nurturing an overall healthy body (*mens sana in corpore sano*) through physical exercise (Schäfer et al. 2006), healthy lifestyle choices (Middleton et al. 2010), and diet (e.g. Klimova et al. 2017), including calories restriction (e.g. Greenwood and Parasuraman 2012).

Specific interventions for cognitive training can also contribute to slowing down cognitive deterioration in the old, especially after retirement if the elder tends to remain passive. For instance, the *Advanced Cognitive Training for Independent and Vital Elderly* (ACTIVE) study of the USA has focused on 65 years old and older healthy people who went through training sessions targeting memory, reasoning or processing speed to be used in the solution of everyday problems (e.g. remembering grocery lists, a bus schedule). Training sessions improved cognitive abilities which resulted in better performance of activities of daily living (Ball et al. 2002).

Bilingualism and multilingualism can also help maintain cognitive capacities into old age, as already mentioned in Chap. 4, by providing greater cognitive reserve. Bilingualism helps cognitive development (Hakuta and Diaz 1985) thanks to the need to learn and master more than one language, and the complexity of manipulating different linguistic structures without making mistakes, whilst avoiding interference.

Kavé et al. (2008) carried out a study of multilingualism and cognition in a group of 75–94 years old in Israel, as part of the *Cross-Sectional and Longitudinal Aging Study* (CALAS). Participants were initially interviewed between 1989 and 1992, and there was a 12-year follow-up. The best cognitive condition was observed among multilinguals (speakers of four languages) compared to bilinguals and trilinguals; and multilingualism added cognitive capacity beyond the effect of additional demographic variables. Remarkably, the difference observed remained significant throughout the period of follow-up, even when participants were older than 90.

On the other hand, there are factors that can further increase the rate of cognitive decline with ageing beyond what one would expect in normal circumstances, such as diseases and mental disorders (see Chap. 4), use of medication (Starr et al. 2004) and an unhealthy lifestyle.

Greater cognitive decline is associated with higher probability of death. In a longitudinal study carried out in the mid-Monongahela valley of southwestern Pennsylvania (USA), Lavery et al. (2009) followed up 989 community-dwelling adults for up to 12 years, starting from an average baseline age of 76.3 years old, as part of the *The Monongahela Valley Independent Elders Survey* (MoVIES project). In this case the authors focused on the effect of cognitive decline on mortality within a period of old age. After correcting for the effect of dementia, a combined decline in general cognition, processing speed, language and executive function was predictor of mortality.

Lavery et al.'s (2009) result could be interpreted as elders who have already started a dying trajectory also showing a decline in cognitive capacities. However, long-term lifetime effects have been also detected associating cognitive development and mortality, which may indicate a causative path directly linking better cognition with longevity. Whalley and Deary (2001), for instance, have shown how childhood mental ability as measured through an intelligence test, was positively related to survival to the age of 76 years old in both women and men. They suggested various mechanisms that can explain a negative association between higher lifetime cognitive development and mortality in old age: (a) low cognition may mediate a relationship between negative experience in childhood and increased adult mortality; (b) poor cognitive ability may be a predictor of a less safe adult environment; (c) higher cognitive ability, on the other hand, may be a predictor of healthy behaviours and hence lower mortality; and (d) cognitive ability may be an indicator of broader "system integrity".

Nevertheless, not always better cognitive capacities are associated with low mortality. In a Swedish study that focussed on males, Hemmingsson et al. (2006) found that higher cognitive abilities were positively associated with mortality. In part this

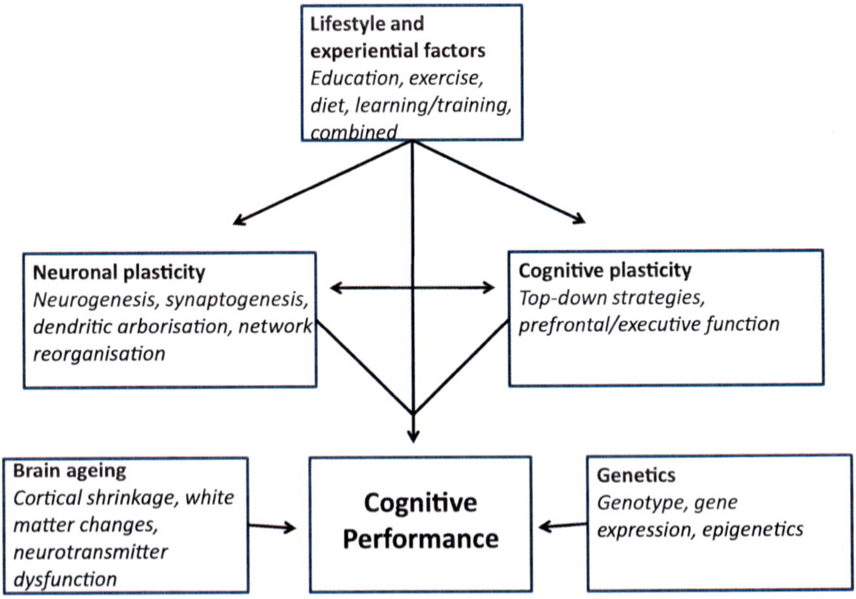

Fig. 5.5 Greenwood and Parasuraman's (2012) framework to understand how environmental and experiential factors can interact to improve cognitive ageing, in spite of genetic and ageing effects on cognitive performance. Redrawn with slight modifications from Greenwood and Parasuraman (2012)

could be explained by a positive association between cognition and the probability of committing suicide.

Greenwood and Parasuraman (2012) provided a broad scheme for a process of fulfilling ageing that can better nurture our cognitive capacities, taking into consideration the variables affecting cognition in old age, as already mentioned in this section. Lifestyle choices may help protect and perhaps even improve neural and cognitive plasticity throughout ageing, which may sustain good cognitive performance in spite of the physiological constraints of ageing and genetic factors (see Fig. 5.5).

One psychological risk of retirement is the potential decrease in cognitive stimulation that may decrease mental capacities and functioning. Engaging in cognitively stimulating activities in retirement helps our neuronal plasticity and cognitive performance. Greenwood and Parasuraman's (2012) provide some basic practical suggestions for the maintenance of cognitive abilities into old age:

1. Exercise.
2. Substitute saturated fats in the diet (e.g. red meat, butter) for sources of monosaturated fats (e.g. olive oil) and polyunsaturated fats (e.g. oily fish, walnuts).
3. Avoid excess intake of calories.
4. Learn new skills, keep expanding knowledge.

To recap, cognitive abilities decline with normal ageing but such decline varies according to the specific ability. Executive functions decline with age, as it does cognitive processing speed. Verbal skills, on the other hand, tend to decline more slowly than other cognitive capacities, as it does knowledge. Crystallised intelligence shows a relatively lower rate of decline with age than fluid intelligence. The rate of loss of cognitive functions with ageing is crucially affected by their relative use throughout life. The accumulation of cognitive skills over time constitutes our cognitive reserve which, along with plastic mechanisms of the brain can help compensate for cognitive losses incurred throughout the ageing process. Performance of cognitive skills can, however, be interfered with by subjective ageing in a positive or negative manner. Cognitive capacity can be better preserved in old age through developing and maintaining skills, better education including learning a second or third language, a healthy body, physical exercise, a healthy lifestyle and a better diet (calories restriction in particular). Specific therapeutical interventions can also help slow down the rate of cognitive loss expected with ageing.

In the next section we focus on a specific cognitive ability that can noticeably deteriorate in old age: memory.

5.5 Memory

> Older adults typically perform more poorly than younger adults on direct or explicit tests of memory. In recall, older adults omit more of the originally occurring information than do younger adults, include more never-presented items (intrusions), and repeat more previously recalled items. On recognition tests, older adults are more likely than younger adults to accept as old never-presented items (called foils or lures), especially if those lures share a conceptual, schematic, or perceptual resemblance to the presented items
>
> Zacks et al. (2000, p. 311).

Memory may be defined as the quantity and accuracy of information that can be recalled or recognised after a specific retention interval (Pansky et al. 2009). Memory can play tricks at any age. How many times have we tried to keep some piece of information in mind, but upon distracting ourselves in other activities we just forgot and missed the appointment, did not buy what we had to, or left something behind having forgotten about it? And the problem does not get any better as we become older, as Zacks et al.'s (2000) quote above indicates by referring to *intrusions* and *foils* occurring in old age.

We have seen in Chaps. 1 and 4 that memory declines with age (e.g. Denes 2016) following the ageing of brain centres controlling memory, such as the hippocampus. This decline is part of the normal process of ageing. More serious declines in memory do occur as a result of neuropathologies such as dementia (see Chap. 4). In general, information learned through automatic processes tends to be retained in memory for a longer time, including into old ages. This may be the reason why upon encountering an acquaintance in a non-familiar place we may recognise his/her face as familiar but may not remember the name (e.g. Grady and Craik 2000), whereas professionals (e.g. musicians) tend to retain some of their learned and frequently practiced abilities into old age. Cognitive and sensory systems may also interact with ageing whereby the loss of some sensory abilities (e.g. hearing) may impair memory capabilities in the old (Murphy et al. 2000).

Early on in the modern scientific study of memory, Ribot (1881) observed that brain-injured patients show memory problems that are affected in reverse order from the time of the injury, that is, the most recent memories are affected more seriously than the older ones. This *Ribot's law* is now known as *temporally graded retrograde amnesia* (Wixted 2004; Frankland and Bontempi 2005; a similar law was also proposed by Adolph Jost in 1897 (Jost 1897)). Ribot's observations indicate that memories, to be retained over the longer term, need to be consolidated. *Memory consolidation* is a concept first introduced by Müller and Pilzecker (1900), greater consolidation makes memories more resilient to accidental injuries but also to the processes of ageing.

Later in the twentieth century, Endel Tulving suggested a three-system structure for *long-term memory*: *procedural memory*, the capacity to retain learned relationships between stimuli and responses; *semantic memory*, the ability to recall representations of the world; and *episodic memory*, or the capacity to store and retrieve information about the relationship in space and time between objects and the self (Tulving 1985). Subsequently, Tulving and Schacter (1990) added a *perceptual representational system* (PRS) that improves the perceptual identification of objects and which is suggested to support *priming*, or the subsequent facilitation of performance due to a previous exposure to a stimulus, even when the previous experience with the stimulus is not consciously remembered (Hultsch et al. 1998). In addition to long-term memory Tulving also recognised *short-term memory*. Evolutionarily, procedural memory is the oldest memory system (Nilsson et al. 2002).

Memories can be established and also retrieved through various mechanisms: consciously—which is more effortful—such is the case with episodic, explicit, and declarative memory; and with low-consciousness, or low-effort establishment/retrieval as in the case of semantic, implicit, and procedural memory (Connor 2001). Low-effort memories are usually better preserved in old age, and memory remains functional for as long as we can *encode*, *store,* and *retrieve* information (Connor 2001). *Crystallised abilities* tend to be retrieved more effortlessly from memory and therefore they can remain functional for longer into old age, not so *fluid abilities* (see also Sect. 5.4 in this chapter).

Crystallised and fluid memory point to the issue of memory plasticity. Such memory plasticity broadly decreases with ageing, as do other aspects of brain plasticity, as we have seen in Sect. 4.2 of Chap. 4, although it is not completely lost in healthy elders and it may be preserved, although in a much reduced form, even in centenarians (Singer et al. 2003a). Working memory has been shown to be plastic and subject to some improvements through training (Li et al. 2008), whereas both episodic and working memory improve in the old with bilateral activation of the brain (Angel et al. 2011).

To become crystallised, and therefore enduring in long-term memory, information must experience a process of consolidation. Such memory consolidation tends to occur during periods of quite wakefulness and sleep (Brown and Maylor 2017). The problem with older people is that ageing is associated with more frequent wakening and disordered sleep (Wolkove et al. 2007), hence their difficulties in consolidating memories. As we have also seen from Tulving's work mentioned above, memory can be stored over the long term but ageing tends to decrease our ability to store information in this manner.

To the difficulty of storing information we also have to add the difficulty of retrieving stored information. A common phenomenon affecting the retrieval of information stored in memory is that of *tip-of-the-tongue experience* (TOT), which involves the inability to retrieve a word we are looking for from our memory, but we feel sure that we know it and that it is about to come back to us at any moment (Brown 1991). The TOT experience is also more common in old age (Rendell et al. 2005).

Primary memory stores information for the short term and such information can then be either retained more permanently or simply forgotten. Short-term information can be used in conjunction with working memory to execute specific tasks, such as the ability to carry out calculations—which is more permanently remembered—by using contingent information (specific numbers) that can be more easily forgotten (Connor 2001).

In addition to Tulving's initial classification, further types of memory have been also described. Here we provide a more comprehensive list with a brief definition of each type of memory:

- *Working memory*: it provides temporary storage and manipulation of the information necessary for complex cognitive tasks such as language comprehension, learning, and reasoning (Baddeley 1992). Working memory is multi-component (Baddeley 2010) operating on different sensory modalities such as phonological/auditory (Pallesen et al. 2010) or spatial (Budson 2014), visual (Awh et al. 2007), tactile (Harris et al. 2001), presumably also taste and smell (Baddeley 2010, 2012).
- *Episodic memory*: memory of personally experienced episodes/events, the details related to them and the level of consciousness associated with the recollection of a specific event, which includes the re-experiencing of details and the awareness of the continuity of self in time (i.e. *autonoetic awareness*) (Levine et al. 2002; Tulving 2002). Episodic memory could become *semanticised* in old age, that is generalised by becoming detached from specific temporal and spatial contexts (Piolino et al. 2002).
- *Semantic memory*: capacity to remember words, their meanings and relationships, these include facts about language, functions of objects, appropriate behaviour, names (e.g. president/prime minister of own country, the difference between two objects of common use) (e.g. Bowles and Poon 1985).
- *Autobiographical memory*: explicit memory of events that occurred in a specific time and place in one's past (Nelson and Fivush 2004). This involves memory of lifetime periods, general events, and event-specific knowledge (Tung and Ritchie 2011).
- *Procedural memory*: learning and remembering to perform specific tasks, such as driving a car (Tulving 1985).
- *Declarative memory*: conscious memory of facts and events (Squire 2004).
- *Implicit memory*: unconscious influence of previous information on subsequent performance (Hedden and Gabrieli 2004). Implicit memory is helped by *priming* (Hultsch et al. 1998).
- *Explicit memory*: recall of information following a specific request to do so (Graf 1990).

- *Short-term memory*: also known as *primary memory*, involves the retention of events in memory for only a short period of time (e.g. Baddeley 2012, Lewis-Peacock et al. 2012).
- *Long-term memory*: also known as *remote memory*, implies memory retention of events for longer periods of time (Ranganath and Blumenfeld 2005).
- *Prospective memory*: remembering to carry out a task in the future (Zeintl et al. 2007).
- *False memory*: implies remembering events that never happened (Dennis et al. 2007).

How are those types of memory expressed in old age?

- *Working memory*: declines with age (Hultsch et al. 1998; Bopp and Verhaeghen 2005; Denes 2016) and it can be interfered with by intrusion of other thoughts and not relevant information (e.g. Connor 2001).
- *Episodic memory*: this type of memory is negatively affected by ageing (Hultsch et al. 1998; Lövdén et al. 2004; Denes 2016). It can decline since middle age and more significantly after 70 years old, with some specific memory tasks declining faster than others. In a study carried out in Sweden, episodic memory was stable up to the age of 60 years followed by a significant and accelerating decline (Rönnlund et al. 2005). The decline in episodic memory with ageing could be explained by a reduced ability to encode and retrieve associated or linked units of information as we become older (Old and Naveh-Benjamin 2008). Failure of episodic memory in old age can result in anterograde amnesia and retrograde amnesia. Anterograde amnesia disrupts memory of new information (Aggleton and Brown 2006) whereas retrograde amnesia affects memories formed before the onset of mnemonic difficulties (Reed and Squire 1998).
- *Semantic memory*: is relatively well preserved through ageing (Levine et al. 2002; Lövdén et al. 2004), although Hultsch et al. (1998) found evidence of decline and it is usually the case that memory of names may specifically decline with age (Rendell et al. 2005; Luo and Craik 2008). Moreover, older people tend to experience an elevated frequency of TOT situations especially related to trying to remember names, compared with younger people (Rendell et al. 2005).
- *Autobiographical memory*: Conway (1996) divided autobiographical memory into recalling lifetime periods, general events, and sensory/perceptual event-specific knowledge. Autobiographical memory tends to be well preserved in old age (Hedden and Gabrieli 2004), but it could display *reminiscence bumps* especially in older individuals (Tung and Ritchie 2011) who may remember events that occurred during specific periods in life better than others. There is, however, a broad pattern of *infantile amnesia*, where memories from ages younger than 6 years old are fewer than those from later ages, and there is virtually no memory from ages younger than 3 years old (Piolino et al. 2002, and references therein). Autobiographical memory is characteristically biased in the old towards positive events, which may be indicative of greater emotion regulation in old age. In this case emotion regulation tends to dismiss or minimise unsettling memories and select uplifting ones (Mather and Carstensen 2005). Figure 5.6 shows the ability

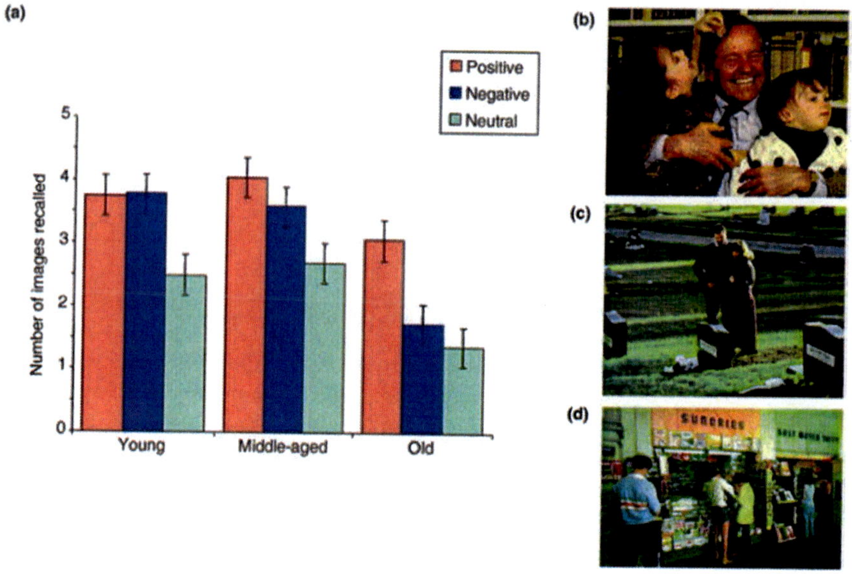

Fig. 5.6 (**a**) Total number of pictures recalled by younger (18–29 years old), middle aged (41–53 years old), and older (65–80 years old) adults; examples of (**b**) positive, (**c**) negative, and (**d**) neutral pictures used in the experiment. From Mather and Carstensen (2005)

to recall memorised pictures among three adult age classes: younger (18–29 years old), middle aged (41–53 years old), and older (65–80 years old). The three pictures have different emotional valence: positive (a grandfather enjoying playing with his two grandchildren), negative (a couple visiting a grave), and "neutral" (people shopping). The total number of images recalled differed among the three age classes, with the older participants recalling fewer pictures. The ability to recall, however, is also affected by the emotional valence of the picture, especially for the old: they are far more likely to remember the picture with positive valence than the other pictures, in contrast to young and middle-aged participants. An interesting example of such biased recalling in old age, possibly modulated by emotional effects and semantic memory, is reported by Mondini and Semenza (2006) who described the case of V.Z., a 66-year-old Italian woman diagnosed with progressive bilateral atrophy of the mesial temporal lobes and who showed symptoms of dementia. She had virtually no ability to recognise even the faces of her closest relatives or common objects and animals, however, she consistently recognised and named the face of Silvio Berlusconi, the mass media tycoon and at the time Italian Prime Minister. In this case it is possible that the pervasive media propaganda (a persistent stimulus with positive emotional valence) that has accompanied Mr Berlusconi political and business career likely crystallised the image of Berlusconi in V.Z.'s memory. In a study of the interaction between autobiographical, episodic, and semantic memory, Levine et al. (2002; see also Piolino et al. 2002) found that older adults were also better at remembering semantic details not related to a specific time and place.

- *Procedural memory*: is usually well preserved in old people (Smith et al. 2005).
- *Declarative memory*: this type of memory declines in capacity with ageing (Murty et al. 2009).
- *Implicit memory*: is often stable with ageing or just shows slight age-related changes (Graf 1990; La Voie and Light 1994; Hultsch et al. 1998; Grady and Craik 2000; Fleischman et al. 2004).
- *Explicit memory*: becomes more compromised in old age (Graf 1990; Fleischman et al. 2004).
- *Short-term memory*: this type of memory is well preserved in older people with regard to remembering single items, but it declines in old age for the case of remembering the binding between items and the context in which they are encountered (Fandakova et al. 2014).
- *Long-term memory*: the old tend to be less efficient at preserving this type of memory (Connor 2001).
- *Prospective memory*: it can decline significantly with age (Zeintl et al. 2007) but the decline may not be as strong as suggested by tests carried out in controlled experimental environments, as in natural settings old people display a better capacity for prospective memory (e.g. Grady and Craik 2000).
- *False memory*: production of false memories increases in old age (Dennis et al. 2007). Older adults experience false memories as a consequence of lower memory capacities for context and more reliance on general information and familiarity rather than details. The capacity to produce false memories can even vary throughout the day in old people (Intons-Peterson et al. 1999). False memories are affected by the interference of emotions with memory. In this case, negatively emotionally valenced memories are more distorted than the positively valenced ones (Brainerd et al. 2010). In addition, memory complaints can be more reflective of emotional states and personality than actual abilities to remember (Verhaeghen et al. 2000).

Therefore, none of the types of memory listed above show an increase in performance in old age with the exception of false memories, that are more common in old age but this may in part indicate an increased disruption of functionality, and also interference by emotions, rather than an improvement in memory capacities. The other kinds of memory either decline (working memory, episodic memory, declarative memory, explicit memory, long-term memory, and prospective memory) or remain the same (procedural memory); or mainly remain the same with the exception of some specific contexts: semantic memory (where there is a decline in remembering names), autobiographical memory (which can experience reminiscence bumps), implicit memory (which can sometimes decline slightly), and short-term memory (which is preserved for single items but declines for linked item-context).

There is inter-individual variability in memory even within a specific age cohort, and such variability increases with age (Meier et al. 2002 and references therein). Some factors that contribute to such memory variability in old age include: genetic makeup, health, physical condition, gender, personality, education, sociality, and lifestyle. In what follows we will briefly review each one of those factors.

Genetics Individual differences in memory can be in part explained by genetic makeup. For instance, Nilsson et al. (1996) studied the association between six genetic markers (serum protein polymorphisms) to study their contribution to the heritability of episodic memory. The authors were especially interested in polymorphisms of the complement C3 and the acute-phase reactant haptoglobin, simply because they are molecules involved in the immune response and it is known that memory is affected by immune activity (Pugh et al. 2001; Ziv and Schwartz 2008; Yirmiya and Goshen 2011). Differences were found between high and low performers in episodic memory tests relative to sex and genetic markers: high- and low-performing women differed significantly in four of the six genetic markers (complement C3, haptoglobin, transferrin C, and group-specific component); whereas men only differed in the haptoglobin system. This suggests that although immune response factors may be relevant in maintaining episodic memory in both men and women, the exact molecular mechanism may be somewhat different between the sexes.

A better information on the genetic basis of memory comes from the study of dementia, which can significantly compromise cognitive abilities, memory in particular (Nilsson et al. 2002). We have seen in Chap. 4 that Alzheimer's disease (AD) is characterised by accumulation of Aβ in the brain in association with the presence of the ε4 allele of the *ApoE* gene, which encodes for the lipid/cholesterol carrier apolipoprotein E (ApoE). The ApoE4 variant of the protein increases the risk of developing AD; other alleles are more weakly associated with AD (ε3) or can be protective (ε2). Roughly 90% of people who are ε4 homozygotes develop AD, therefore although not all AD patients are carriers of ε4, most ε4 homozygotes do develop AD.

The ε4 allele tends to be also associated with poor memory performance in individuals who have not been diagnosed with dementia, although this association is not always observed (Nilsson et al. 2002, and references therein). Longitudinal data from the *Betula Study* of Sweden suggest that the ε4 allele specifically decreases episodic memory performance in non-demented individuals, and such effect is stronger in homozygous than heterozygous people. Cross-sectional data show a different trend, with only small differences observed in episodic and semantic memory performance across ε2, ε3, and ε4 alleles, although one of the relatively strongest differences in ε4 was observed for the oldest age class (65–80 years old) in the case of episodic memory: low-memory-performance individuals tend to be slightly more likely to carry the allele than high-memory-performance individuals (see Table 5.6). The beneficial effects of ε2 are generally more evident.

Genetic aspects of episodic memory were also studied by Egan et al. (2003). They demonstrated the association of a methionine substitution in the 5′ pro-region of the human brain-derived neurotrophic factor (BDNF) protein with poorer episodic memory and abnormal hippocampal activity. The BDNF modulates plasticity in the hippocampus and hippocampal-dependent memory.

Genetic aspects of working memory have also been studied. Luciano et al. (2001) compared monozygotic and dizygotic twin pairs and showed evidence for genetic

Table 5.6 Relative frequencies of participants in the *Betula Study* of Sweden with *ApoE* alleles ε2, ε3, and ε4 by age group and by high and low performance in episodic memory and semantic memory tasks

	35–45 years		50–60 years		65–80 years	
	High	Low	High	Low	High	Low
Episodic memory						
ε 2	0.10	0.03	0.05	0.10	0.10	0.05
ε 3	0.78	0.87	0.73	0.73	0.83	0.82
ε 4	0.12	0.10	0.23	0.17	0.07	0.13
Semantic memory						
ε 2	0.10	0.00	0.13	0.10	0.13	0.09
ε 3	0.73	0.84	0.70	0.68	0.74	0.82
ε 4	0.17	0.16	0.17	0.22	0.13	0.09

From Nilsson et al. (2002)

influences on working memory. Ando et al. (2001) also compared monozygotic and dizygotic twins and reported heritability estimates for working memory of 43–49%.

Genetic effects on memory tend to be magnified in old age. For instance, Nagel et al. (2008) demonstrated that old individuals carrying specific polymorphisms affecting the enzyme catechol-o-methyltransferase (COMT) and the brain-derived neurotrophic factor (BDNF) protein, who were tested in a spatial working memory task, had even poorer memory task performance compared to young individuals, according to their genetic makeup.

Health Health is another important factor affecting memory. We have seen in Sect. 4.5 of Chap. 4 how memory problems characterise older people with neuro-cognitive disorders, Alzheimer's disease and other forms of dementia in particular. In addition, memory is also negatively affected by diabetes (Zelinski et al. 1998), blood pressure and other circulatory problems (Launer et al. 1995), depression (Verhaeghen et al. 2000 and references therein), and anxiety (Commissaris et al. 1993).

Physical Condition Physical exercise can improve our mnemonic capacities. In a Swedish study carried out within the *National Study on Aging and Care in Kungsholmen* (SNACK), Ferencz et al. (2014) followed up 2480, 60–100 years old individuals for up to 6 years measuring the level of activity they were engaged in and episodic memory. They reported reduced episodic memory performance in people at higher genetic risk of developing cognitive problems, but more physical activity diminished the negative effect of genetic makeup on episodic memory. The positive effect of physical activity on memory can be explained by physical activity directly affecting dendritic complexity, number of dendritic spines, hippocampal neurogenesis and volume, and synaptic plasticity in the brain (Makizako et al. 2015; Duzel et al. 2016, and references therein). Moreover, Smith et al. (2011) provided evidence suggesting that physical activity may selectively increase memory-related brain activation in cognitively intact elders who are at risk of developing Alzheimer's disease, being carriers of the *ApoE4* allele.

Gender Women display a better performance in episodic memory tasks than men (Herlitz et al. 1997), and old women also display an autobiographical memory that tends to be more specific than old men's, perhaps because women may be more motivated to remember more details of their life (Pillemer et al. 2003). In general, gender differences in memory tasks (a verbal advantage for women over men and a visuospatial advantage for men over women) tend to decrease in old age, but this is not always the case. For instance, Pauls et al. (2013) found that older women continue to outperform older men in verbal fluency and auditory episodic memory, whereas older men outperform older women in spatial perception, visual episodic memory, and visual working memory. On the other hand, Kramer et al. (2003) showed that men and women tend to display similar levels of decline in verbal episodic memory at older ages, and Hassing et al. (1998) did not find any gender difference in episodic memory functioning among healthy 90–100 years olds.

Personality Memory abilities in old age can be affected by personality traits (e.g. Hooper et al. 1986). In general, it is expected that in old age personality traits such as Extraversion could help individuals preserve their memory capacities through maintaining an active mind and providing sensory stimulation throughout the period of retirement. Neuroticism, on the other hand, is expected to be negatively correlated with memory as older people may be more sensitive to worries and various health and other complaints (Meier et al. 2002, and references therein). Meier et al. (2002) studied the relationship between those personality factors and memory in old age using data from 287 participants (31% women), 68–95 years old, from the *Basel Prospective Study*, Switzerland. Over a period of 24 years, participants displayed stability in Extraversion but a significant decline in their level of Neuroticism. Episodic memory was positively correlated with Extraversion and negatively correlated with Neuroticism. Thus, it looks like older people may react to the potentially negative effects of Neuroticism on memory by adopting coping strategies that decrease their levels of Neuroticism. On the other hand, personality may also affect the subjective perception of memory abilities (Hänninen et al. 1994).

Education Intellectually stimulating activities are associated with an increase in memory performance (Hultsch et al. 1999). Education in particular can positively affect memory in the old, starting from early ages, as the number of schooling years improves memory performance in old age (Glymour et al. 2008). Old people with poor education, however, can compensate and maintain better episodic memory capabilities into old age, by engaging in frequent cognitive activities such as writing (e.g. letters, stories), reading books, magazines or newspapers; doing word games such as crosswords, enrolling in educational lectures or courses and so forth (Lachman et al. 2010). This effect is illustrated in Fig. 5.7 where low-education individuals performing high cognitive activities can have better episodic memory than high education individuals performing low cognitive activities. As expected, however, the best performing group are those with both high education and high cognitive activity.

Sociality Engaging in social activities involving cognitive and physical stimulation is positively associated with memory performance (Christensen et al. 1996). In an analysis of data from the USA *Health and Retirement Study*, Ertel et al. (2008)

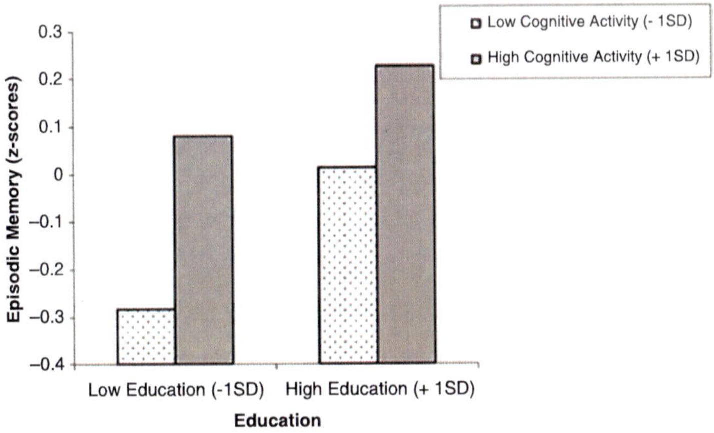

Fig. 5.7 Values of episodic memory for lower (−1 *SD*) and higher (+1 *SD*) cognitive activity frequency and lower (−1 *SD*) and higher (+1 *SD*) levels of education. From Lachman et al. (2010)

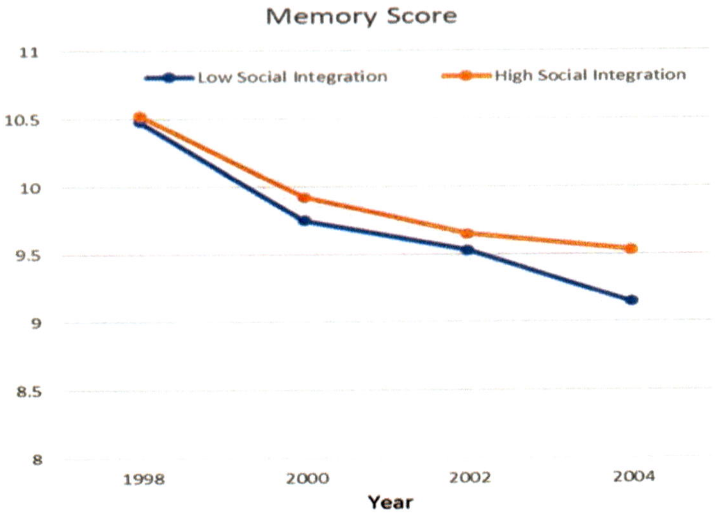

Fig. 5.8 Change in memory scores over 6 years of follow-up according to the level of social integration at baseline in people 50 years old and older. Redrawn from Ertel et al. (2008)

investigated the relationship between social integration and memory change (immediate and delayed recall) over a period of 6 years in a group of people 50 years old and older. The group was divided between those who were more socially integrated and those who were less so, with social integration being measured through frequency of contact with various people (children, parents, neighbours), marital status and participation in volunteering. More social integration at baseline was associated with a slower rate of memory decline with ageing, as shown in Fig. 5.8. However, not all social interactions have a positive effect on elders' memory. Negative social stereotypes about ageing, that may become internalised by the elder, have been observed to negatively affect memory in old people (Levy et al. 2011). Stereotypes, on the other hand, can be also positive, and positive stereotyping about ageing can in fact improve memory performance (Levy 1996; Hess et al. 2004).

Lifestyle Some lifestyles may be better than others to address the memory deficits of old age. We have seen above that maintaining a good health, engaging in physical exercise, keeping our intellectual curiosity and nurturing social positive relationships can all help in slowing down the decay in memory performance with ageing. Flöel et al. (2008) constructed a composite lifestyle index encompassing a diversity of variables: physical exercise, dietary habits, body mass index (BMI), smoking and alcohol consumption; and used it to study the association between lifestyle and memory in a group of 198 healthy 65–84 years olds. They defined the "healthiest behaviour" as one where BMI was less than 22, the diet was high in fruits, vegetables, wholemeal/low-fat foods, and unsaturated fatty acids, the energy expenditure through physical activity was higher than 13,000 kcal/week; having never been a smoker, and a level of alcohol consumption of 4–10 drinks/week. Their results indicate that a higher lifestyle index score was associated with a better memory score in healthy elders. Interestingly, single factors had a weaker association with memory than the composite index, suggesting a cumulative effect of a diversity of lifestyle variables on memory.

We briefly expand on the relationship between diet and memory improvement in old age, as controlling the diet may be a simple intervention to help the elders. Perrig et al. (1997) carried out a longitudinal study of 65–94 years olds in Switzerland, associating measurements of priming, working-memory, free recall, and semantic memory with plasma antioxidant levels obtained from diet: α-tocopherol (vitamin E), ascorbic acid (vitamin C), and β-carotene (a precursor of vitamin A). Their results indicate that higher plasma levels of ascorbic acid and β-carotene are associated with better memory performance. Witte et al. (2009) studied the effects of caloric restriction on memory in a group of 60.5 years old women. They divided the group into a sub-group that was subject to caloric restriction (30% reduction), a second sub-group was on a diet of relative increased intake of unsaturated fatty acids (20% increase, with unchanged total fat), and there was also a control group. The authors found a significant increase in verbal memory scores (a mean of 20%) following a calorie-restriction diet.

Given that some factors have been identified that can improve memory performance in old people, we conclude this section with a brief review of some interventions that have been suggested to achieve better memory outcomes for the elders.

Interventions to improve memory in old age have been in place since ancient times. The Roman Cicero (first century b.c.e.), for instance, suggested a method of "loci" where pieces of mnemonic information are associated with specific landmarks along a familiar walk. Once the need to retrieve the information arises then one only has to imagine performing the walk, encountering each landmark and the information it represents (Connor 2001).

Many methods of memory training have been proposed, including individual or group sessions, using tasks that range from paper-and-pencil approaches to computerised interventions (Stigsdotter Neely and Backman 1995; Zehnder et al. 2009) that teach individuals to use specific mnemonic strategies (Buschkuehl et al. 2008). An early meta-analysis of the effectiveness of episodic memory training for old people (60 years old and above) indicated that memory training can improve the mnemonic capabilities of elders, with the magnitude of gains being constrained by age and duration of training (Verhaeghen et al. 1992). In particular, Stigsdotter Neely and Backman (1995) demonstrated that a multi-factorial memory training programme can improve the ability to recall concrete words, and the improvement was maintained for 6 months. Buschkuehl et al. (2008) carried out a study of the effects of a working memory programme for 80 year olds based on a training intervention delivered twice a week for 45 min, over a 3 months period; a control group just engaged in physical activity with an eccentric bicycle ergometer following the same time schedule. Effects on memory were recorded upon completion of training and also one year later. Training had a short term but not a long-term effect on improving working memory. Training also had positive effects on working memory in a programme for 60–70 years olds described by Brehmer et al. (2011).

In their meta-analysis, Zehnder et al. (2009) concluded that memory training in healthy old people tend to have some positive but limited effect, mainly improving outcomes in face-name immediate recall and visuospatial memory tasks. More positive effects are achieved with memory training on elders diagnosed with *Mild Cognitive Impairment* (MCI).

Elders can be also helped in coping with their decreasing memory in various ways. *Instrumental* or *problem-focused coping* (Lazarus 1991) aims at alleviating the memory deficit through adopting some mnemonic strategies that may be spontaneous or the result of some more formal training, as we mentioned above. *Emotion-focused* or *cognitive coping* (Lazarus 1991), on the other hand, aims at decreasing the emotional distress caused by poor memory. This can be helped through social comparison, for instance, which allows the elder to realise that she or he is not alone in experiencing age-related memory problems. It is often the case that in groups of elders, as soon as such memory problems are shared by all, they may spread a sense of relief and even release a collective humourous response that lightens up the situation (Verhaeghen et al. 2000). In fact, by decreasing stress, this approach may also help the memory function. The relationship of stress with memory is complex, but chronic stress can reduce memory through the mediation of molecules such as cortisol (Kirschbaum et al. 1996; Newcomer et al. 1999), and therefore alleviating chronic stress through humour or other strategies can be of great help. On the other hand, acute stress may improve memory (Henckens et al. 2009). The different effects of chronic and acute stress on memory can be understood adaptively in terms of acute stresses resulting from negative experiences that are worth remembering in order to avoid them in the future; whereas chronic stresses, if they persist, can be psychologically and physiologically damaging to the broader mental capacities. Psychological therapies based on mindfulness aim at decreasing stress by reducing rumination (Chiesa and Serretti 2009), thus helping the memory function (Levi and

Rosenstreich 2019); although mindfulness meditation may also lead to the production of false memories (Wilson et al. 2015).

In sum, memory capabilities decline in old age but not all types of memory decline to the same extent. In general, low-effort forms of memory (semantic, implicit, procedural) are better preserved into old age. Plasticity of memory also decreases with ageing, which tends to result in a better level of preservation of crystallised memories over fluid ones. Various types of memories have been identified that store information valuable for survival (e.g. working, episodic, semantic memory) but other types of memories can store false information (e.g. false memory). Factors that affect our capacity to remember in old age include genetics, health, physical condition, gender, personality, education, sociality, and lifestyle. In spite of the decline in memory observed in old age, some spontaneous activities or formal interventions can be used in order to at least slow down the decline.

In the next section we shift to the effects of ageing on personality and how individuals with different personalities age.

5.6 Personality and Ageing

The *American Psychological Association* defines personality as the:

> (I)ndividual differences in characteristic patterns of thinking, feeling and behaving. The study of personality focuses on two broad areas: One is understanding individual differences in particular personality characteristics, such as sociability or irritability. The other is understanding how the various parts of a person come together as a whole. (https://www.apa.org/topics/personality/)

Personality can be also seen as an "organized, dynamic, agentic system functioning in the social world" (Mischel 2004, p. 2). To an extent, personality develops over time from early-life *temperament* (McAdams and Olson 2010). For instance, it has been suggested by Caspi et al. (2005) that affective and positive temperament in childhood may predispose to Extraversion and positive emotionality later in life, whereas temperament aspects of anxious/fearful distress and irritable distress may lead to Neuroticism and negative emotionality in adult ages. Conscientiousness and some aspects of Agreeableness may develop from a capacity for focused attention and effortful control in childhood. However, the childhood experience is not the only factor affecting personality characteristics in the old.

Early on in the modern psychological study of personality, James (1890) suggested that personality becomes "set like plaster" by age 30 or at least it is expected not to change substantially from then on; a proposition that Costa and McCrae (1994; see also Srivastava et al. 2003) later named the *Plaster hypothesis*. The *Plaster hypothesis*, however, has been challenged by subsequent research (Marsh et al. 2013), as personality traits can indeed change over time and into old age. This continuous change has been encapsulated in the *Maturity principle* (Caspi et al. 2005): personality keeps changing in older individuals leading to greater levels of maturity, especially in the ability to contribute to society (Helson et al. 2002; Roberts and Mroczek 2008; Roberts et al. 2008; Orth et al. 2010). The *Maturity Principle* implies that at least up to middle age but presumably beyond too, people are expected to become more agreeable, conscientious, and emotionally stable (Caspi et al. 2005), whilst decreasing in Neuroticism (McAdams and Olson 2010). The decrease in Neuroticism allows older people to better help members of their social network, such as their family, and also recruit help from them.

Marsh et al. (2013) proposed a view of lifetime personality development that combines elements of stasis and change. They named their model *La dolce vita effect*. As people become older and in spite of their inevitable physical deterioration, they tend to become more content, happier, less self-centred, more relaxed about whom they are and what they have, and also more accepting of their own self, leading to higher levels of self-esteem. This is better achieved the more capable the mind of the elder is of reaching a level of equilibrium in self-identity and the retention of a degree of plasticity and adaptability to be used as required, such as compensating for any losses (e.g. cognitive, sensory) and re-adjusting life goals, to make life objectives more achievable given the constraints of old age.

We agree with McAdams and Olson (2010, p. 536) that "personality across the human life course is a complex and multilayered affair… [Personality] develops as a dynamic constellation of dispositional traits (the person as actor), characteristic goals and motives (the person as agent), and integrative life stories (the person as author)". Some aspects of personality show more resilience across the lifespan, others are more dynamic, although the overall variability decreases in old age. McFadden (1999) refers to the relationship of personality and ageing as a condition of "constancy and change". Indeed, adulthood could be described as "a time of systematic and meaningful changes embedded in a larger framework of stability" (Nesselroade 1992, p. 236).

Change and stasis in personality traits may be particularly affected by the social environment, changing historical events and personal life experiences. Social, historical, and individual factors may determine the prevalence of some specific personality characteristics and variability within and between cohorts of old people. Personality changes that may occur throughout life may be influenced by what is regarded as *normative* in society, and what is normative is affected by biological processes, historical periods, social factors, culture (Roberts et al. 2006).

In an early review of personality traits and psychological changes in old age, Kuhlen (1945) described the old as more sedentary, relatively less engaged socially than they were in their youth, although they were also described as engaging more often in educational and musical activities. The level of social engagement seems to be determined not only by personality traits but especially by life circumstances, such as health and the availability of a social network. In fact Kuhlen observed that the desire to participate in various activities declined more slowly with age than actual participation. Old men also decline in their masculinity traits with age and the old in general increase their levels of introversion, which is seen as an adaptive psychological shift, rather than an indication of "emotional disturbance". The decline in Neuroticism in old age was also observed in the early reports. Menopause in women can also lead to some personality changes both during the period of transition and afterwards.

We mentioned in Chap. 1 how personality helps the individual establish a sense of continuity through time (Allemand et al. 2010), and yet old people are also confronted with the need to respond to changing life circumstances, especially after retirement, that may affect their ability to experience fulfilling ageing. In Chap. 3 we also introduced what is arguably one of the most important models of personality: Costa and McCrae's (1992) *Big Five* (Neuroticism, Extraversion, Openness, Agreeableness, and Conscientiousness) and we saw how the model is capable of describing personality characteristics across ethnic groups. We mentioned in that chapter the cross-cultural work of McCrae et al. (1999) describing personality traits in people 50 years old and older, where the authors broadly concluded that old people tend to be low in Neuroticism (that is, they are less anxious) compared to young adults; they also tend to be lower in Extraversion (are less active), and are lower in Openness (are less intellectually dynamic). On the other hand, old people tend to score higher on Agreeableness (they are more cooperative) and are also higher on Conscientiousness (more self-disciplined, dutiful). Although those personality characteristics of old age may reflect increasing constraints: being less active may be just a result of declining physical capabilities; other changes may be more intentional such as their increasing cooperativeness, which can have the double function of both attracting reciprocation from the recipients of their help and also helping the next generations (generativity). Decreasing Neuroticism in old age may also help attenuate the fear of death (death anxiety). We also mentioned in Chap. 4 that an introverted personality may lead older people to become less socially connected.

Personality may be also the locus of some disorders in old age, with Neuroticism predisposing old individuals to develop depression, whereas Extraversion and Conscientiousness, along with emotional stability and disposition to optimism, can be promoters of better health. Previously in this Chap. 5 we have also pointed out the role of personality as a mental umbrella synthesising aspects of emotionality and cognition, to the extent that personality can contribute to variability in memory capabilities for instance. Extraversion may contribute to a better preservation of memory in old age, whereas Neuroticism is less conducive to good memory in old age. Thus, the decrease in the level of Neuroticism in old age that has been observed may be also adaptive in terms of memory preservation.

Therefore it is clear that some aspects of personality can be preserved with ageing and others can change, often in an adaptive manner. Changes in personality factors throughout life can be punctuated by *turning points*. *Self-identified psychological turning points* (SPTPs), defined as "perceived, long-lasting *major changes* in how people view themselves and their lives, and learn new things about themselves and others" (Allemand et al. 2010, p. 148, italics original), can lead to personal growth.

In what follows we further explore some of the central issues concerning personality in old age within a lifespan perspective. In particular, we will devote special attention to the issues of continuity and change in personality throughout the ageing process.

Following from its widespread use in the study of personality, we start by further exploring the applications of the *Big Five* model to old individuals. As we already mentioned, McCrae et al. (2005) confirmed the broad validity of the *Big Five* model to describe and study personality traits across ages and cultures. However, McCrae and collaborators have tended to emphasise a lifelong maintenance of some basic personality traits, especially from mid-adulthood onwards (see also Allemand et al. 2010), whereas we have also seen that there is development of personality into adulthood, including old age.

Conscientiousness and Agreeableness tend to increase over the years from adolescence to late mid-life, whereas Neuroticism tends to decrease (McAdams and Olson 2010). Although Conscientiousness may then decrease in very old ages (Donnellan and Lucas 2008), empathy tends to be stable with age and narcissism tends to decrease (Orth et al. 2010, and references therein). This has been interpreted by Roberts et al. (2006) as an adaptation to an increasing need for social connections in old age, whereas Costa and McCrae (2006) have interpreted the same pattern of personality change as a result of genetically programmed biological maturation.

Marsh et al. (2013) studied the *Big Five* factors using data from the Wave 15 of the *British Household Panel Study* (BHPS): $n = 14{,}021$ participants; 54% women; ages 15–99 years old. They found that women scored higher for all *Big Five* factors, except for Openness, with such gender differences being consistent across the entire lifespan. Older participants were characterised as "happier (more agreeable and less neurotic), more self-content and self-centred (less extroverted and open), more laid back and satisfied with what they have (less conscientious, open, outgoing and extroverted), and less preoccupied with productivity" (Marsh et al. 2013, p. 1194), in spite of increasing physical difficulties due to ageing. Such results fit within the expectations of *La dolce vita effect*. Adapting to the growing limitations of ageing whilst enjoying life as much as possible are also fundamental characteristics of fulfilling ageing. *La dolce vita* has, to some extent, deep ancient roots into the old Roman ideal of *otium honestum* (see Chap. 2).

Roberts et al. (2006) carried out a meta-analysis of longitudinal studies to understand the patterns of personality change with age. They concluded that social vitality (a facet of Extraversion) shows a small decrease into old age, just like Openness; whereas social dominance (another facet of Extraversion) tends to increase up to the age of 40, beyond which it stabilises. Agreeableness tends to increase with age, especially in the period 50–60 years old, then it tends to decrease into older ages. Conscientiousness increases throughout life, including into old age, a similar trend was found for emotional stability. Therefore these personality traits do not display consistent stability into old age, let alone across the lifespan. Changes are mainly concentrated during young adulthood (20–40 years old) and although change decreases into older ages, it does not necessarily stop (Fig. 5.9). On the other hand, in a review of personality studies, Roberts and DelVecchio (2000) found that Extraversion is one of the most consistently maintained traits over time, including into old age. Agreeableness is also maintained but, as indicated by Roberts et al. (2006), this trait may in in fact be more stable at young and old ages than in middle age.

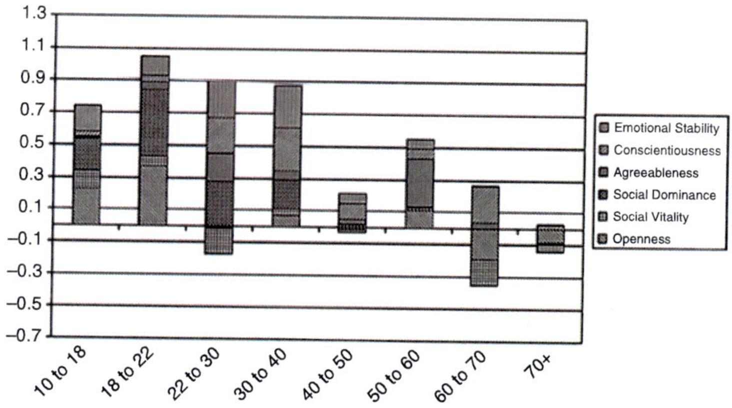

Fig. 5.9 Aggregate mean-level changes in six personality traits across the lifespan. Smaller changes indicate greater level of stability of the personality trait. Social vitality and social dominance are two facets of extraversion. From Roberts et al. (2006)

The overall picture emerging from various studies suggests that adults tend to increase in their socially desirable personality traits (e.g. Agreeableness and Conscientiousness) as they age, whilst decreasing in Neuroticism. A diversity of results has been obtained for Extraversion and Openness to experience. Openness tends to increase in young adults to then decrease somewhat in old age. Changes in Extraversion are even more variable (Allemand et al. 2010).

Although the *Big Five* personality traits can vary over time in each person, they are still ultimately grounded in the biology of the individual (DeYoung et al. 2010). It has been suggested that Extraversion (associated with positive emotions) and Neuroticism (associated with negative emotions) are manifestations of sensitivity to rewards and punishment, respectively (DeYoung et al. 2010 and references therein). Hence the lower Neuroticism observed in old age, when it would be expected that the individual becomes more inclined to avoid stress.

Loss of functionality with ageing may also affect personality traits. For instance, the expected lower levels of Neuroticism in old age may not occur if significant functional decline is experienced throughout the ageing process. In this case Neuroticism may increase as an effect of increased distress due to greater functional decline. Wahl et al. (2010) have suggested that an association of functional decline with increased Neuroticism may result from such functional losses being seen as a threat, leading to negative arousal and psychological distress. In this regard, Mordekar and Spence (2008) have also noted that an increase in obsessive-compulsive traits in the old, that are associated with Neuroticism (Schirmbeck et al. 2015), may be a response to decreased abilities to operate in life due to functional losses.

The association of personality traits with ageing is a result of the age-specific activation of specific brain regions (DeYoung et al. 2010). Therefore changes occurring in old age that affect those brain regions may also produce changes in personality traits. For instance, early stages in the development of Alzheimer's disease (AD) may affect some aspects of personality. Balsis et al. (2005) have shown, after studying a group of 62–100 years old participants (both men and women), that 47% of those who developed dementia after they had entered the study had already manifested some change in personality before they received the diagnosis. Personality changes observed in this group included increased rigidity (25%), growing apathy (24%), increased egocentricity (21%), and impaired emotional control (18%); and less frequently, diminished emotional responsiveness (9%) and purposeless hyperactivity (4%). With regard to the *Big Five* personality traits in particular, AD patients show pre-morbid personality characteristics such as increased Neuroticism, decreased Extraversion, and changes in Agreeableness (Smith-Gamble et al. 2002). After AD has been diagnosed, Neuroticism continues to increase, Extraversion decreases (Duchek et al. 2007) along with Conscientiousness, whereas Openness and Agreeableness are either stable or slightly decrease (Robins Wahlin and Byrne 2011; see Cipriani et al. 2015, for a review).

Other studies have investigated personality in old age from perspectives other than the *Big Five*, reaching similar conclusions. For instance, Ojha and Pramanick (2010) studied 1800 individuals in India spanning in age from 30 years to 80 years old. They studied the association of life satisfaction, ego strength, emotional stability, altruism, and alienation with ageing. Life satisfaction continuously declines with ageing, whereas ego strength and emotional stability display an inverted-U-shaped pattern of change with age, where both variables increase in value with age up to late middle age, to subsequently decline. Altruism and alienation increase linearly throughout life. Therefore, from this study old people are lower in both ego strength and emotional stability, which may contribute to increasing their sense of alienation and decreasing their overall life satisfaction. The potential negative effects of such changes on the psychology and health of old people may be somewhat compensated by their increasing sense of altruism with age, which may help them develop a new meaning in life (e.g. through generativity or helping in the domestic economy of adult children and their family) but also enjoy support from family as a form of reciprocation.

How general is the dynamic of change and consistency of personality traits over time? Roberts and DelVecchio (2000) reviewed the level of rank-order consistency—consistency in the relative positioning of individuals within a group—of personality traits over the lifespan. Personality traits may remain consistent over time at the individual level (*intra-individual consistency* and *ipsative stability*) or the population level (*mean-level consistency* and *rank-order consistency*), or they may change. As we have seen, consistency (or lack thereof) in personality traits over time may be explained by environmental (e.g. social), genetic, health, psychological factors (e.g. self-identity), and the interactions among them (Roberts and DelVecchio 2000). For instance, we have just seen how poor mental health may affect personality and therefore less mentally healthy elders may display lower personality consistency over time. Löckenhoff et al. (2009a) have shown that participants in their study who reported extremely adverse life events displayed an increased level of Neuroticism, a decrease in some aspects of Agreeableness and also a decrease in Openness to experience.

In their review, Roberts and DelVecchio (2000) concluded that personality trait consistency increases over the years up to the age of 50–59, to then plateau in the age interval 60–73 years old, with the pattern being similar between men and women. Specht et al. (2011) investigated personality trait stability in 14,718 Germans across their adult years to also conclude that Emotional Stability, Extraversion, Openness, and Agreeableness increase in rank-order consistency up to the age of 40–60 to then decrease. Conscientiousness, on the other hand continuously increases in rank-order consistency over the adult years. Gender patterns differed in this study, with women being more consistent in Emotional Stability and in terms of Openness than men and were less consistent over time in terms of Conscientiousness.

Allemand et al. (2010) measured stability of the *Big Five* over a 10-year period in a study carried out in the USA that included 40–60-year-old participants. Overall, personality traits displayed a mixture of stability and change that was variable across individuals and traits (see Table 5.7). Stability tended to prevail and when change occurred the trend was consistently towards decreasing the value of the trait. Conscientiousness showed the greatest level of stability, whereas Agreeableness displayed the lowest degree of stability. Neuroticism, Extraversion, and Openness displayed somewhat similar patterns of intermediate stability.

In a recent longitudinal study of personality stability and change carried out in the USA, Damian et al. (2019) analysed data from 1795 participants in the *Project Talent* who were followed up for over 50 years, since they were high school students. The sample included 52.5% females at baseline and 52% females at follow-up, with Caucasians representing 93% of participants at baseline and 95.3% at follow-up. Age range spanned on average from 16 to 67 years old over the entire period. Personality traits were measured through the *Project Talent Personality Inventory* (PTPI). Their results suggest that there is a degree of consistency over time in some traits (e.g. leadership, culture, vigour, sociability) but there is also predictable change over time which is consistent with the *Maturity principle*: higher Conscientiousness, Agreeableness, emotional stability, and dominance facet of Extraversion at older ages. At the individual level they found that on any given trait, about 40% of participants showed reliable change, whereas 60% did not, which is qualitatively consistent with Table 5.7: more individuals show consistency in personality trait than change. The degree of change accumulates the longer the individual lives.

Table 5.7 Individual differences in personality trait change over a 10-year period

	Decreased (%)	No change (%)	Increased (%)	X^2
Neuroticism	9.1	88.5	2.4	158.48***
Extraversion	7.5	88.7	3.8	99.49***
Openness	8.6	87.5	3.9	146.77***
Agreeableness	10.9	80.2	8.9	414.77***
Conscientiousness	2.1	96.3	1.6	3.74n.s.

The chi-square tests whether the observed distribution of changers and non-changers would differ from the expected distribution if change is random. *** $P < 0.001$; n.s. = not significant. From Allemand et al. (2010)

Personality traits may become stabilised or they may change adaptively as a strategy to cope with life circumstances. In a study carried out in the USA as part of the *Second Duke Longitudinal Study*, Koenig et al. (1990) interviewed 100 older adults in order to determine how they coped with stressful periods in their life through religion or through other means. Religious copers tended to be less aggressive-hostile, more humble and more submissive compared with non-religious copers. Religious copers were also more responsible, conscientious, and moralistic. For as long as the specific use of religiosity increases well-being, it tends to be positively associated with Extraversion and self-esteem and it is negatively associated with Neuroticism (McFadden 1999, and references therein). But coping can be achieved in different ways. Koenig (1994) studied older men (65–102 years old) and described a level of stability in their coping behaviour both in those who did use religious coping and in those who did not.

Social support is also an important source of coping for the elders. Does social support vary with personality? Krause et al. (1990) investigated the interaction between social support and personality in the UK based on interviews with 2349 participants in the 1985 *Health and Lifestyle Survey* who were 60 years old and older (53% women). A total of 1504 participants also returned a questionnaire. The results indicate that extraverts reported greater contact with friends than introverts, perhaps because extraverts may not only be better able to maintain social relationships but they may also require more social support as a result of experiencing more psychological distress. Extravert older people also tend to make greater use of the Internet (Berner et al. 2012).

Personality-dependent coping styles may lead to different levels of well-being in old age. In a study carried out in Spain, González Herero and Extremera (2010) investigated the relationship between personality variables (self-esteem and optimism) and subjective well-being in a group of 50–82 years old, describing a positive association between self-esteem and social, creative and outside-home activities, whereas optimism was positively correlated with social activities and games. Participation in such activities, social activities in particular, only partially mediates the association between personality traits and well-being, whereas there was a direct positive association between both self-esteem and optimism with well-being.

Moor et al. (2006) investigated the potential mediation of self-perceptions of own ageing (e.g. "Things keep getting worse as I get older") in the relationship between personality factors and subjective health in a group of German elders (men and women) aged 60–64. Their results indicate that subjective health is negatively associated with Neuroticism and such relationship is mediated by ageing self-perceptions (after controlling for objective health status, depression, and gender): "That is, the fact that individuals high in neuroticism report worse subjective health can be explained by the finding that those individuals also have more negative self-perceptions of aging, which in turn are associated with poorer subjective health" (Moor et al. 2006, p. 252).

Neuroticism has also been associated with both subjective cognitive complaints and higher risk of developing dementia, as already mentioned above (Smith-Gamble et al. 2002). As stated by Snitz et al. (2015), poorer self-appraisal and a proness to psychological distress reflect aspects of negative affect, which may be associated with higher risk of developing dementia.

Personality is also associated with the probability of developing a disability in old age. Krueger et al. (2006) carried out a longitudinal study in the USA, with participants being 813 older Catholic nuns, priests, and brothers who had not been diagnosed with dementia or had any disability at the beginning of the study. Risk of developing a disability was lower for participants scoring high on Extraversion and Conscientiousness, whereas higher Neuroticism was marginally associated with higher disability risk.

Older individuals may also develop disorders that affect their personality. Although personality disorders are difficult to diagnose in older people (e.g. Segal et al. 1996) they may occur with a prevalence of about 10% in the population (Mordekar and Spence 2008). Such personality disorders may include avoidant, anxious, obsessive-compulsive, dependent and borderline types and they may be comorbid with depression, phobias, acute stress disorder, somatisation, substance misuse, extreme self-neglect typical of the *Diogenes syndrome* and, as we have seen, dementia. Personality disorders in old age are also a risk factor for abuse. Management of personality disorders in the old include various approaches ranging from medication to a variety of psychological therapies (see Mordekar and Spence 2008 for a review).

The association of some personality traits with health may in turn have consequences on longevity in a variety of ways. In 1921 Lewis Terman started a long-term study of 1500 children with high abilities who were followed up from childhood to death (Terman 2018, see also Friedman and Martin 2011). The study found that conscientious people staid healthier and lived longer, whereas "cheerful and optimistic" children were less likely to reach old age than their more sober counterparts, presumably because some optimistic people may also engage in unhealthy behaviours trusting in that everything will finally work out alright for them, no matter what they do. In other words, such people may express *illusory optimism* which may lead to overlooking real threats. Not surprisingly, illusory optimism may become a constraint, a source of frustration and maladaptive behaviours, rather than an aid to fulfilling ageing. Real threats are better identified rather than overlooked, and actively overcome or avoided. On the other hand, we may also see the negatives of a construct such as *illusory pessimism*, which may become more common in some elders perhaps following an accumulated lifetime of negative experiences and disappointments. Again, illusory pessimism may also become a constraint to fulfilling ageing as it may limit our abilities to live a happy life into old age and it may produce a level of disenchantment that curtails our capacity for legacy and generativity. Friedman and Martin (2011) also mention the issue of *catastrophising*, which is a characteristic of pessimistic personalities. In the Terman's study, catastrophisers died younger than conscientious participants.

Personality traits may alter health-related behaviours, the frequency of engaging in social interactions, the tendency to experience negative emotions (Iwasa et al. 2008) thus affecting mortality. As expected from the studies reviewed in this section, Neuroticism is associated with higher risk of mortality in older men and women (Wilson et al. 2004; Mroczek and Spiro 2007), whereas Extraversion is associated with a lower risk of mortality in old people (Wilson et al. 2005) as it is Concientiousness (Wilson et al. 2004). Based on an analysis of a sample of 856 men and 672 women from the USA, who were studied in the context of the *Terman Life-Cycle Study* for a period extending from childhood in the 1920s to 1986, Friedman et al. (1993) also concluded that Conscientiousness is associated with greater longevity, whereas cheerfulness and optimism are linked to reduced longevity. The association of Conscientiousness with higher longevity may result from people displaying the trait being more concerned about self-care and engaging in less risk-taking behaviours, thus becoming less likely of contracting preventable diseases or adopting unhealthy habits. On the other hand, cheerful/optimistic people may not take the risks of their behaviours seriously enough, in spite of humour and optimism being useful coping mechanisms to decrease stress in specific situations. As stated by Friedman et al. (1993, p. 181): "The same optimistic thoughts that may promote recovery from surgery ('I'm going to be just fine') may prove deadly to a cigarette smoker, a hypertensive, or an overeater ('I'm going to be just fine')". Wilson et al. (2004) concluded that there is a personality-dependent variable risk of death in old people, ranging from lower risk associated with Extraversion/Agreeableness to a higher risk for Neuroticism (see Table 5.8).

In a study carried out in Japan, Iwasa et al. (2008) followed up a cohort of community-dwelling old people who were 65 years old at baseline, for a period of 5 years. All-cause mortality in this sample was inversely associated with Conscientiousness, Extraversion, and Openness, whereas Neuroticism and Agreeableness were not associated with mortality.

To recap, although some aspects of personality may remain consistent throughout life, compared with young adults old people tend to score lower in Neuroticism (i.e. they are less anxious), provided that they do not experience serious loss of functionality, in which case Neuroticism may increase in the old. Lower Extraversion (less activity) also characterises old age, but different aspects of extraversion show different patterns of change in old age: whereas social vitality decreases, social dominance stabilises after the age of 40 years old. The old tend to also be lower in Openness, that is they are less intellectually dynamic, but they score higher in Agreeableness (i.e. they are more cooperative), although Agreeableness may decrease in very old age. Conscientiousness (self-discipline) is also higher among

Table 5.8 Relative risk of death associated with personality traits, after controlling for age, gender, and education

Personality trait	Relative risk[a]	95% confidence interval
Neuroticism	1.049	1.022, 1.077
Extraversion	0.964	0.937, 0.991
Openness	1.005	0.973, 1.038
Agreeableness	0.964	0.927, 1.002
Conscientiousness	0.968	0.940, 0.996

[a]Relative risk is for about 1-point increase in each personality trait measure. From Wilson et al. (2004)

the old. Overall, as people grow older they display more emotional stability, but some personality traits may also remain consistent through ageing due to early social experiences and potentially genetic factors. On the other hand, those personality traits that change over time may do so not only due to the effects of ageing per se, but also due to a diversity of life experiences: e.g. the accumulation of very adverse life events may increase Neuroticism. In addition, some changes may be adaptive, such as the higher degree of agreeableness in old age or the increase in the level of optimism, that may be associated with greater levels of cooperation and attraction of social support. Finally, personality factors can affect the health and well-being of old people. For instance, Neuroticism is negatively associated with subjective health, it is also associated with a higher probability of developing dementia, higher risk of disability and, ultimately, a higher risk of mortality, hence the broad tendency for old people to score lower on Neuroticism for as long as they are in good health. Contientiousness, on the other hand, is linked to lower mortality risk.

Some personalities are more susceptible to the effects of psychological stress than others. In the next section we review the relationship between ageing and psychological stress.

5.7 Ageing and Psychological Stress

Psychological stress can be defined as a "negative emotional experience accompanied by predictable biochemical, physiological, cognitive and behavioral changes that are directed either towards altering the stressful event or accommodating to its effects" (Dimitroglou et al. 2003, p. 39, following Baum 1990).

We have already addressed some aspects of psychological stress in old age in Chap. 4 and in previous sections of this chapter. For instance, we have seen in Chap. 4 that sources of psychological stress can be individual or social and that sustained psychological stress can impair brain plasticity—and therefore cognitive flexibility—and increase the probability of developing depression, thus negatively affecting mental health and well-being in the old. The post-retirement period can be an especially problematic time for older people due to the potential loss of social role, decrease in income, the potential need to construct a new identity, and instability of the supportive social network, all of which can significantly increase psychological stress.

The negative effects of psychological stress call for initiatives to manage it in order to improve mental health, such initiatives may include nurturing a supportive social network, finding meaning in life, maintaining physical capacities, adopting a healthy lifestyle.

In previous sections of this chapter we have also pointed out that stress can reduce memory, and in turn stress may be reduced in old age by a better regulation of emotions. Personality can also make the individual more sensitive to psychological stress, or better protect him/her from the negative effects of such stress.

In this section we review some major aspects of psychological stress in old age, its frequency, major causes, types of stress and specific effects. We conclude the section with an analysis of the coping mechanisms that can be used by old people to decrease the negative consequences of psychological stress.

Although psychological stress can have long-term effects on the development of our brain (e.g. decreased hippocampal volume; McEwen and Gianaros 2011) and our psychology, that may start at young ages (Amir and Lev-Wiesel 2003; Lupien et al. 2009; Miller et al. 2011), or in middle age (Andel et al. 2011), and that can cause significant differences between individuals in old age, here we will mainly focus on the direct effects of stress in old age. In fact, the negative effects of stress on hippocampal volume and therefore memory, that are observed earlier in life, can be also seen in old age (Gianaros et al. 2007).

A stressful situation releases two main molecules from the hypothalamus in the brain: the neuropeptides corticotropin-releasing hormone (CRH) and vasopressin. CRH eventually stimulates the release of adenocorticotropic hormone from the pituitary, which finally stimulates the production of glucocorticoids in the adrenal glands.

Excessive and chronic release of glucocorticoids due to stress in old age can promote cognitive decline (memory decline in particular), immune dysregulation and susceptibility to infectious diseases, muscular weakness, depression, abdominal obesity, cardiovascular problems, upper respiratory diseases and debilitation of the bones (Chiriboga et al. 2002; McEwen 2002; de Kloet et al. 2005; Schneiderman et al. 2005; Stawski et al. 2006; Hasan et al. 2012).

Chronically stressed elders who may become depressed may also suffer from dysregulation of their immune system (e.g. exaggerated inflammatory response) (Gouin et al. 2008) as a consequence of depression, and such dysregulation of the immune system may increase the mortality rate (Kiecolt-Glaser and Glaser 2002).

Glucocorticoids secreted under chronic stress have been associated with *immunosenescence* (Bauer 2008; Bauer et al. 2009). Immunosenescence involves the decreased ability of the immune system to function effectively in old age. Such immune problems include defects in the haematopoietic bone marrow, peripheral lymphocyte migration, maturation and function; decreased functional capacity of the lymph nodes and atrophy of the thymus gland (Gruver et al. 2007). In our book *Behavioural oncology* (Cordella and Poiani 2014) we also reviewed the association of chronic stress with the development of cancer, haematological cancers in particular.

Stress can also negatively affect quality of life and subjective well-being, through compromising self-conception, self-worth, self-esteem, and self-efficacy (Tran et al. 1991).

Psychological stress is experienced to various degrees by 34–48% of elders, with some differences being observed between those living at home, who are usually less stressed, and those living in institutions. Some other factors affecting psychological stress in the old include: social isolation or lack of social support, and low cognitive, functional and socio-economic status (Martin et al. 2001; Angel et al. 2003; Cairney and Krause 2005). Those factors may also interact. For instance, it is possible for elders to display high levels of psychological stress when they are still living at home compared with an institution (Préville et al. 2002) if they do not enjoy good social support (Cruces et al. 2014). Elders' abuse in particular is a concerning cause of widespread stress for some old people at home, the abuse can be physical, sexual, emotional, and it may also take the form of financial exploitation and neglect (Dong et al. 2013). We will return to the issue of elders' abuse in more detail in the Sect. 6.3 of Chap. 6.

Additional sources of elders stress that are emerging, especially in recent decades, include the stress of caring for young children born late in the life of one or both parents, following from a trend towards increasing parental age (Bray et al. 2006); or old people still finding themselves in the need of supporting adult children (Suitor and Pillemer 1987; Pillemer and Suitor 1991). The frequency of grandparents who raise their grandchildren as *custodial grandparents* has also been increasing (Hayslip and Kaminski 2005).

Some studies have investigated more unusual stressors and their effects on the elderly population, such as natural disasters. Older individuals are more affected by the stresses of natural disasters, such as an earthquake or a hurricane, than are younger members of the population (Jia et al. 2010). Jia et al. (2010), for instance, reported prevalence rates of 22.5% of post-traumatic stress disorder (PTSD) symptoms among elderly Chinese people who survived the 2008 earthquake in the Sichuan province, and a prevalence of only 8% in the case of younger adults.

Stress further increases in elders when they suffer from medical conditions that make their daily living harder; e.g. arthritis, poor mental health (Roberts et al. 1994; Tak 2006).

The individual perception of psychological stress depends on our *cognitive appraisal* of it. In the words of Lazarus and DeLongis (1983, p. 249):

> Appraisal refers to the way a person construes the significance of an encounter for his or her well-being, that is, as irrelevant, benign, harmful, threatening, or challenging, the latter three being forms of stress appraisal.

Once a situation is indeed experienced as stressful by the elder, some consequences are expected such as lower cognitive performance, as observed in the study of Beluche et al. (2010). Such reduced cognitive performance under stress can be short term, as seen in responses to *acute stress*, or long term as in the case of *chronic stress*. Acute stress may lead to reduced attention—that can interfere with information processing (Stawski et al. 2006)—and to memory failures (Hidalgo et al. 2014) for instance. In the long term, chronic stress may accelerate cognitive decline, along with an increased probability of developing dementia (Johansson et al. 2010; Scott et al. 2015). With regard to dementia, the risk of developing an Alzheimer's Disease-like form of dementia increased approximately by 5.6% for each point on a distress proneness scale in a sample of elders studied by Wilson et al. (2006). Wilson et al. (2006) also described negative effects of distress on various cognitive abilities in the old such as semantic memory, working memory, perceptual speed, and visuospatial ability.

As stated by Stawski et al. (2006, p. 536):

> Thus, the psychological effects of stress on cognition are largely a result of a resource competition, such that stress acts as a cognitive load, creating a dual-task situation whereby attention is divided between task demands and coping with environmental demands.

Stawski et al. (2006) showed how stress interferes with attention and it decreases the performance of working memory in old people. It also interferes cognitively with processing speed and episodic memory. In addition, declarative memory was also shown to decrease under stress in old people (Lupien et al. 1997).

Negative effects of stress on cognition can be further magnified by cognition itself, through negative thoughts. *Unconstructive repetitive thinking* (URT) that includes aspects of rumination, worry, perseverative cognition, and cognitive interference, can prolong the physiological and psychological responses to daily and chronic stressors which, eventually, can negatively affect the cognitive function over the short and long term, as suggested by Scott et al. (2015).

Stress can also affect more specific aspects of the biology of old people that can have repercussions on psychology. We are all aware of the effects of ageing on our skin (wrinkles, sagging, spots, reduced elasticity, and increased dryness), in fact those skin changes are central to our visual judgment about ageing on ourselves and others. The better the skin condition is the more we look "younger than our age", the worse it is and the more we look "older than our age". Psychological stress has been suggested to be a factor in the ageing of our skin. Although direct evidence for the effect is limited, a mechanism has been proposed in which chronic stress may cause cumulative skin damage through the promotion of inflammation, oxidative stress and DNA damage (Dunn and Koo 2013).

The ageing skin is the locus of much of the ageing stereotyping prevalent in society, alongside mental changes and bodily functions. How does stress affect stereotypes towards old people? Here we will address this question from the perspective of the old people themselves and how their own stereotyping of old age can affect psychological stress: Negative ageing stereotyping can increase stress in the old, whereas positive stereotyping can reduce stress (e.g. Levy et al. 2000; Levy et al. 2016). Levy's (2009) theory of *stereotype embodiment* suggests that stereotypes (whether they are positive or negative) may become internalised in the target of the stereotype after being assimilated from the cultural environment. Levy et al. (2015) analysed the effects of stressful events (hospitalisation, death of a family member) on negative ageing stereotypes in a group of 70 years old and older individuals who were living independently in the community. The authors described that for people holding negative ageing stereotypes such stereotypes were maintained over a 10-year period among participants who experienced hospitalisations and bereavements, thus suggesting that negative stereotypes are difficult to change, especially when they are reinforced by stressful events. Interestingly, Levy et al. (2015) also expected the maintenance of stereotypes in the face of stressful experiences among old people who held positive stereotypes about ageing. In this case, the old person may be expected to adopt an optimistic stance and perceive "the stressful events as exceptions". In other words, some old people may tend not to change their age stereotypes—whether they are positive or negative—no matter what.

Among old immigrants, the stress of adapting to the local culture (*acculturative stress*) should be also added to the usual stresses that we have mentioned so far and that come with ageing. Fitting within the local culture, that may be quite alien, is not an easy task for an immigrant who immigrated in old age. Jang and Chiriboga (2010) divided acculturative stress into *task-oriented stress* ("difficulties due to language barriers and lack of knowledge about the new culture") and *emotion-oriented*

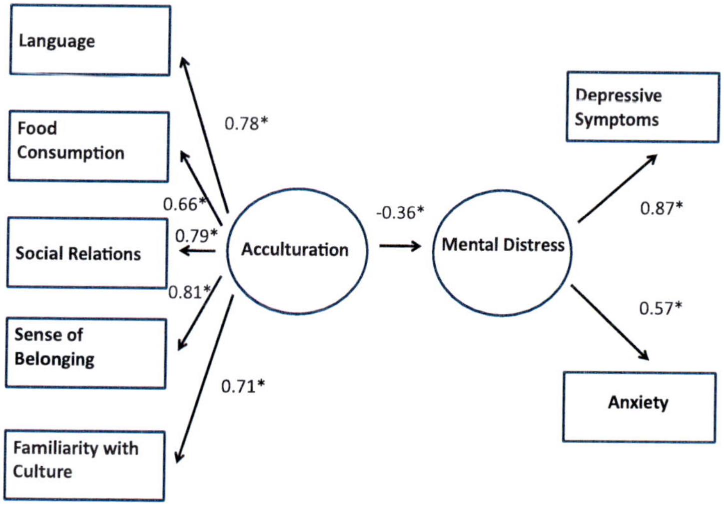

Fig. 5.10 Model of the relationship between acculturation, mental distress and other behavioural, psychological, and social variables. *$P < 0.05$. Redrawn from Jang and Chiriboga (2010)

stress ("feelings of alienation and homesickness"). In their study of Korean American elders who had immigrated to the USA after 1965 (60–94 years old, 57.9% females), Jang and Chiriboga (2010) described mental distress to be positively correlated with both anxiety and depressive symptoms. Greater acculturation decreases mental distress whilst increasing language competency, food consumption, social relations, sense of belonging and familiarity with the local culture (see Fig. 5.10).

The experiencing of psychological stress under a set of circumstances calls for mechanisms to cope with such stress, lest the elder may suffer the negative long-term health effects of chronic stress.

Coping with stress in old age fundamentally relies on life satisfaction which, in turn, is affected by many factors that can improve the life and well-being of old people: from health to social support, existential well-being, legacy, self-esteem, material comfort and security, and so forth (e.g. Lazarus 1999).

Lazarus and DeLongis (1983, p. 250) distinguished between two broader styles of coping: *problem-focused coping* and *emotion-focused coping*:

> Problem-focused coping changes the troubled person-environment relationship, whereas emotion-focused coping changes, through realistic or defensive reappraisal, the way an encounter is construed or attended to and therefore the emotional reaction to it. Emotion-focused coping operates through a variety of cognitive acts, such as attentional avoidance, intellectualised detachment, denial, reinterpretation of the past, humour, magical or wishful thinking, and religious faith. If one resolves the trouble through problem-focused efforts, there is no longer any reason to be threatened; if one engages in emotion-focused coping, the objective situation remains the same, although a more benign emotional reaction is still created.

Following up from Golden (1982), Lazarus and DeLongis (1983, p. 249) described, in more detail, three specific styles of coping with stress that are observed as we age:

- Confrontational style: this involves a major effort to cope with anger, guilt, and sadness by turning to intimate confidants, using strategies that magnify or extend the experience of positive emotions; and an active effort to bring stressful experiences to a closure.
- Denial style: repression of negative emotions, frequent use of what in psychoanalysis is known as "undoing", or the performance of a behaviour that is contrary to a performed action that is regarded as negative (e.g. being nice to somebody whom you treated badly shortly before), and a persistent seeking of elation, used perhaps as a buffer to distress.
- Avoidant style: suppression (rather than denial) of negative emotions, this may lead to experiencing frequent somatic symptoms such as headaches and indigestion, pervasive concern with blame, and infrequent psychological closure in stressful experiences.

As people age they may cope with the new stresses of life such as losing status after retirement, changes in the social networks, widowhood, and change in their financial status, by finding new ways of living their life that can maintain dignity and provide meaning and fulfilment.

Lazarus and DeLongis (1983) carried out a study to investigate stress-coping strategies in a group of 100 adults (52% women) that included 25, 60–64 years old. Data on stress, coping, and adaptation to stress came from interviews. They found that when health stressors are expected to increase, people tend to use any positive experience in life in order to uplift their morale. They also use more than one coping strategy. In particular, social resources can be an important aid in coping with stress as they can help alleviate the negative effects of distressful circumstances in life, thus increasing well-being (Martin et al. 2001).

Some stressors are unpredictable and age-independent, and they usually occur at low frequency (e.g. serious accidents). Preparation to cope with such stressors is not always easy to achieve. Other stressors, however, are more predictable (e.g. menopause, retirement) and therefore people have at least the possibility to prepare themselves at younger ages so they can better cope with the stressor once it actually occurs (Lazarus and DeLongis 1983).

Humour is a common way of coping with both predictable and unpredictable stresses and such function of humour is seen regularly in old people jokes. For instance, Lazarus (1999, p. 171) mentions a joke of two old men playing golf:

> One is about to make his drive from the tee, and says to his partner, 'I don't see so well, so please watch where the ball goes.' He hits the ball and turns to his partner, saying, 'Did you see where it went?' The partner responds ingenuously, 'Yes, I saw it, but I can't for the life of me remember where it is'.

It should be noted, however, that the coping function of a joke becomes evident only when it is the elders themselves who make the joke and laugh at it. Older people do retain the capacity to laugh at themselves, but the jokes should be better left to their own initiative, and support them whenever they are in the mood of joking about old age.

When the use of humour occurs socially, it may be considered a strategy to both cope with stress directly and also, indirectly, as a way of promoting social support, given that loneliness in the elders is associated with increased psychological stress (Paul et al. 2006). Social support helps in decreasing stress in old people through the provision of both material and psychological aid (Krause and Liang 1993). Social support may be promoted by living close to relatives or in communities of cooperating neighbours (Brown et al. 2009).

Positive stereotypes about the elders can also provide coping mechanisms to stressed old people (Levy et al. 2000). Such positive stereotypes build resilience against stressors.

Specific interventions may also involve the participation of old people in stress-reduction programmes based on various techniques; e.g. *Mindfulness-Based Stress Reduction* (MBSR) programmes (Kabat-Zinn 1990, Moss et al. 2015), *music therapy* (Lee et al. 2010), *gratitude interventions* (Killen and MacAskill 2015), *reminiscence interviews* (Rybarczyk and Auerbach 1990), *dance movement therapy* (Bräuninger 2014), *Tai Chi* (Wang et al. 2010), *yoga* (Field 2016), *humour and laughing therapy* (Bennett et al. 2003), just to mention a few.

In sum, psychological stress tends to increase in old age with negative consequences for health and well-being. In response to psychological stress, coping strategies can be used to help old people decrease the negative effects of such stress.

In the last part of this chapter we review the psychological aspects of retirement and the benefits expected from living a mentally and physically active life.

5.8 Retirement and Fulfilling Life

We have already addressed the issue of retirement in previous chapters from a variety of perspectives. For instance, in Chap. 1 we pointed out how retirement is a transition from formal working life to a potentially very long period of a possibly different life, which may involve adjustment to new ways of living. Therefore, retirement requires a degree of preparation that may help smooth the transition over. Retirement has the potential to either challenge our identity or liberate us from the shackles of the duties imposed by work, thus allowing us to develop a more satisfactory identity. In Chap. 2 we also analysed the important issue of retirement pensions and old age pensions from a historical perspective, and pointed out how greater activism and organisation allowed workers to enjoy the benefits of a reasonable safety net from the end of World War II to at least the 1970s, the so-called social democratic period of the evolution of capitalism. The subsequent period was marked by a neoliberal approach that has increased income inequality, decreased help from government through public services, and made retirement a more complicated and uncertain matter for many elders, both men and women. In Chap. 3 we devoted an entire section to a cross-cultural analysis of retirement: *Life after retirement*, and therefore we will not stress cross-cultural comparisons here.

The specific focus of this section will be the psychology of retirement. We analyse the psychological aspects of retirement looking at retirement as a process that has antecedents in the younger years of the individual. At some stage, the individual will experience a change that will lead him/her from perhaps an intense working life, packed with commitments, to full retirement. Such transition to retirement requires a period of preparation, leading up to a post-transition adjustment into the retirement years.

In an article published in 2006 Andrew Achenbaum asked: "What is retirement for?" (Achenbaum 2006). The question seems to interpret retirement just in a utilitarian perspective, which may be too reductive. However, Achenbaum does have a point, because retirement from a lucrative activity—such as an employment or a business—does release free time into the life of a retiree, and so the question comes naturally: What am I going to do with such free time? What is my retirement for?

Answering the above questions is easier said than done, and what is regarded as an acceptable answer is ultimately a matter of individual preference; but we will do our best throughout this section to at least unravel the diverse psychological aspects of the process leading up to retirement and of life in retirement.

In many cases retirement is simply a form of escape from the ever growing stress of most contemporary workplaces (e.g. Midanik et al. 1995), hence the recent trends towards early voluntary retirement (Shultz and Wang 2011). Escape from stress leads to the classic modern concept of retirement, with its promise of a relaxed lifestyle of socialising with family and friends, fishing, reading, writing, gardening, exercising, meditating, travelling and perhaps involvement in some charity work. Changes in the modern workplace have also challenged the old view that retirement occurs after a few decades spent working in a full-time and secure job with a single employer. The insecurity of a career marked by periods of unemployment or sub-employment may add to the accumulated stresses of the work life, leading older people to desire a change as soon as possible. On the other hand, many retirees suddenly realise that they may not have enough assets and funds in their superannuation account to sustain a good life after retirement, and government pensions and social security are currently not really good enough to fill the gap in any meaningful way. This is undoubtedly a cause of much distress and uncertainty: wanting to retire but lacking the material capacity to do so. The distress can be partially reduced with a change in lifestyle and a reduction of our expectations; in the longer term, an overhaul of the current neoliberal system of superannuation and pensions will be required. Table 5.9 summarises some major historical aspects of retirement that illustrate the increasing level of uncertainty facing retirees at present.

Life satisfaction in retirement seems to pivot around three major axes: health, material security, and self-fulfilment (the latter may include a social dimension through family, recognition, legacy, and generativity). The lack of any of them, let alone the three of them, can be a cause of much distress in old age, but we regard self-fulfilment as central. Ideally, when individuals enjoy the benefits of good health, some material/financial security and personal satisfaction in life, then transition to retirement and life in old age can be more fulfilling (e.g. Fors et al. 2012; Parker et al. 2013), but fulfilment in life can be also found by retirees living modestly and even by elders living with a disability. It is in the pursue of ticking all those

Table 5.9 History of retirement

Historical period	View of retirement
Prior to 1900	Retirement nonexistent for most workers as most adults simply worked until they were no longer able to do so
1900–1950	Retirement becomes a legitimate possibility with social security and employer pensions
1950–1980	Male workers continue to strive to retire earlier and earlier, while women in the workforce delay their retirement. There is no meaningful retirement for housewives
1980–2000	Economic and social conditions following the onset of neoliberalism halt the decline in age of retirement for men
2000 to present	The landscape of retirement is in flux and riddled with uncertainty, with trends emerging for a delay in the pensionable age

Modified from Shultz and Wang (2011)

boxes in retirement that elders unleash their creativity, reinventing themselves in old age, and exploring new lifestyles (Byles et al. 2013).

Shultz and Wang (2011) offer a good introduction to the psychological implications of retirement in current times. Today's nature of retirement is fluid: old people can retire from one job, then "un-retire" by returning to an employment (perhaps a different one), and then again "re-retire". They can also engage in "bridge employment" or a part-time/casual job that covers the period between full-time work and full retirement (Beehr and Bennett 2015). Such fluidity is dictated by multiple factors, which may be financial but also psychological such as boredom, feeling that something is missing in life, finding a new life meaning, anxiety, fear, and even depression. Some retirees may feel the burden of the loss of a well rehearsed activity, finding themselves idle and useless after retirement.

In response to the psychological challenges of retirement, older people may react with fluidity. Fluidity and diversity in retirement pathways can be determined by personal priorities, characteristics, and interests that are likely to differ among individuals. Fluidity allows the retiree to consider a smorgasbord of possibilities in the task of deciding about the future in retirement. However, as soon as a retiree decides to focus on a specific activity or set of activities, she or he may then embark in such activity with intensity and commitment. Here, Mihaly Csikszentmihályi's (1990) concept of *flow* becomes central (see Chap. 2). When the level of engagement with specific activities post-retirement is intense, it may conduce older people to experience states of flow during such experiences (e.g. Standridge et al. 2019). Experiencing flow can help us sustain the continued generation of our own individual self in the period post-retirement, with the commitment expressed in a state of flow being more likely to transform the activity into a path to fulfilment rather than just a way of "killing time".

Shultz and Wang (2011) conceptualise their psychological view of retirement into three major aspects: retirement as a *decision-making process*, retirement as an *adjustment process*, and retirement as a *stage in career development*.

Retirement as a Decision-Making Process This assumes that retirees seek and handle the information available to them in order to ground their decision-making process on as much knowledge as possible. Workers are expected to weigh the different factors and finally make an informed decision on whether to retire and when.

Retirement as an Adjustment Process This is especially important to the path of retirement for individuals who voluntarily decide to retire, but it is ultimately relevant to all retirees. It involves a continuous adjustment of lifestyle in preparation for retirement. A well planned transition is expected to better prepare the individual psychologically to live a fulfilling life in retirement.

Stage in Career Development This approach regards retirement as a continuation in the development of a person's career. In this way retirement becomes centred on the retiree's "agency and efficacy in keeping and pursuing their growth and renewal needs" (Shultz and Wang 2011, p. 175). In this context, "career" is not understood necessarily as a "professional career", but simply as a continued development of the self.

A satisfactory adjustment to life in retirement, however, may or may not occur depending on the individual and his/her circumstances. Atchley (1976) proposed a conceptual framework for the psychological phases of retirement. According to Atchley, the immediate aftermath of retirement is characterised by an euphoric and busy honeymoon, which may be followed by a disenchantment phase and, after that, a reorientation phase where the individual adapts to the new reality of retirement, to be followed by a stability phase when the retiree may achieve a satisfying lifestyle. This is a more detailed description of the *adjustment process* mentioned by Shultz and Wang (2011). Ekerdt et al. (1985) tested Atchley's model empirically and broadly confirmed the immediate post-retirement optimistic phase followed by letdown and reassessment (see also Gall et al. 1997). A similar result was obtained by Kim and Moen (2002) among male retirees: initial relief and increased well-being in the immediate post-retirement period, followed by distress and depression later on. Westerlund et al. (2009) provide evidence for the same process from the perspective of self-rated general health, which increases soon after retirement, alongside a decrease in mental fatigue, physical fatigue, and depressive symptoms in particular (Westerlund et al. 2010). Midanik et al. (1995) also showed that soon after retirement stress tended to decrease in their sample of retirees.

The model has remained productive but variability among individuals (e.g. in education and access to financial resources, for instance) may make retirement more satisfying to some retirees than others (e.g. Moen 1996; Wetzel et al. 2016). In particular, Dingemans and Henkens (2015) described how people forced to retire experienced a decline in life satisfaction in the immediate post-retirement period, whilst there was no detectable change in life satisfaction in the case of voluntary retirees.

Pinquart and Schindler (2007) found a diversity of patterns of satisfaction through retirement in their sample of 1,456 German retirees. Some of the participants experienced a large increase in satisfaction at retirement but then satisfaction declined as suggested by Atchley (1976) (see the Class 2 pattern in Fig. 5.11). In another group, satisfaction declined at retirement to then start increasing (Class 1 pattern in

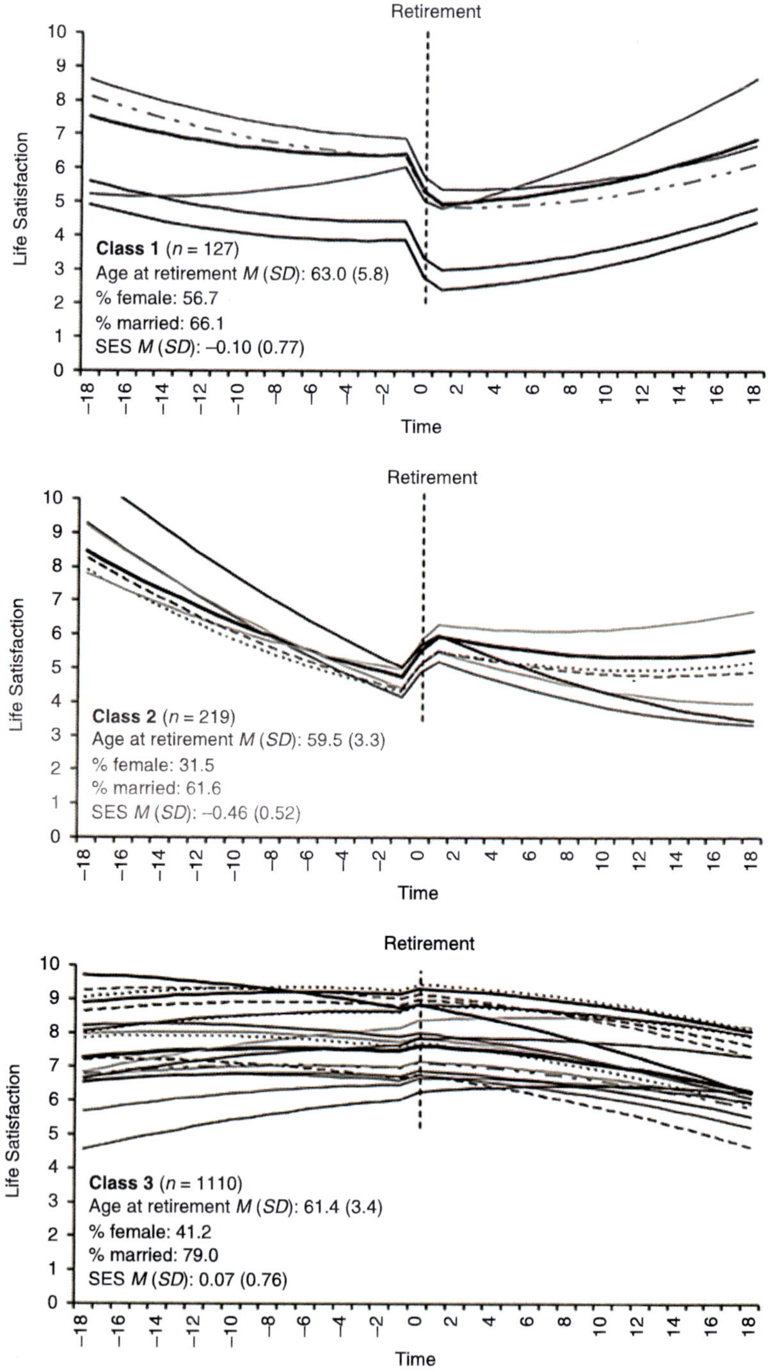

Fig. 5.11 Patterns of change in satisfaction pre- and post-retirement: class 1, class 2, and class 3. From Pinquart and Schindler (2007)

Fig. 5.11). Finally, in a third group satisfaction only showed a small and temporary increase at retirement to then slowly decline (Class 3 pattern in Fig. 5.11). This diversity of responses to retirement was also seen in the retirees interviewed by Nuttman-Shwartz (2004) in a work carried out in Israel. Perceptions of retirement ranged from 44.6% of participants who regarded retirement as a "period of uncertainty and crisis", but 28.6% saw it as an event that allows hope for change, whereas 26.8% described retirement as a continuation of their previous life course. In addition, 5.4% of the interviewees viewed retirement as a developmental transition (note that across the entire sample some of the respondents attributed more than one meaning to retirement). In this work, there were examples of both pessimistic and optimistic views about retirement. Some examples of pessimistic views included:

> I'm confused
> I'm scared not to have anything to do
> I feel superfluous
> If I don't have work, I'm alone. I don't have anyone
> I think that retiring is like dying
> I hope my life ends quickly so that I don't become a burden to anyone.

On the other hand, optimistic views about retirement included:

> There's life after retirement
> I'll finally realize my dream
> Going toward 'something new, something different'
> The future holds all my wishes and desires
> More for spirit and less for matter
> Enjoying my leisure.

Therefore, following Pinquart and Schindler (2007) retirement can involve a psychological improvement, deterioration and perhaps even a degree of stasis, in comparison to the situation before retirement (see also Oksanen et al. 2011).

Some other models of retirement are also worth commenting on:

- *Role theory*: Here the emphasis is on the loss of a specific role following retirement from a paid work, which is expected to lead to anxiety and depression and therefore low life satisfaction if the lost role was life-fulfilling; whereas if it was cause of stress then retirement may lead to greater life satisfaction (e.g. Merton 1957). Great life satisfaction in retirement is also expected to be maintained if a satisfying role does not change significantly after retirement. Consequently, the worst case scenario is to shift from a life-fulfilling role during working life to an unsatisfactory role in life after retirement, a change that can be frustrating. Moving from an unsatisfactory role before retirement to an equally unsatisfactory role after retirement may be also frustrating, leading to questioning the worthiness of retirement and of life more broadly.
- *Continuity theory*: In this theory retirement is not seen as a stressful transition between two different modes of life, but as a smooth accommodation of the life post-retirement to the new reality (e.g. Richardson and Kilty 1991).
- *Life Course perspective*: Here transitions in life are seen as occurring in the context of ongoing life trajectories. Therefore, retirement may be shaped by the resources (mental, financial, cultural, social) available to the retirees that have been accumulated during their working life in preparation for life after paid work (Elder 1995).

It should be noted that the above models assume a continuous trajectory from full-time work to retirement, which until recent decades mainly characterised men, whereas many women fulfilled a role of housewife that remained stable upon the retirement of the working husband. Thus the transition from work to retirement may have strained the relationship if the husband was unable to develop a new purpose in life, especially one that accommodated the expectations of the wife in her older age. Quick and Moen (1998) further elaborated this gender perspective on retirement indicating that at least until the 1970s women tended to have a more positive attitude towards retirement than men, but the greater participation of women in the workforce that has occurred since then has gradually shifted the view of retirement of men and women towards a convergence. Moreover, some studies have even suggested that both the transition to and the life after retirement may be currently somewhat more stressful for working women than men (Quick and Moen 1998, and references therein). In part this can be explained by the uneven career path that is still followed by many working women, due to interruptions from pregnancies and child rearing, for instance, but also by women usually working in lower paid jobs that generate less financial security in retirement (Alkadry and Tower 2006). To that we should also add workplace gender discrimination that may slow down the career progression of women (e.g. Heilman and Caleo 2018). We will return to gender issues in retirement below.

We mentioned above how Shultz and Wang (2011) conceptualised retirement as a decision-making process. Feldman and Beehr (2011, p. 193) subdivided the aspects of retirement as a *decision-making process* into three major phases:

- Phase 1: "imagining the possibility of retirement",
- Phase 2: "assessing when it is time to let go of long-held jobs", and
- Phase 3: "putting concrete plans for retirement into action at present".

Again, just as it was the case for the models we mentioned above, Feldman and Beehr's view of retirement also seems to focus on a traditional model of continuity of employment, perhaps with the same employer, until the age of retirement is reached. This pattern was broadly relevant until the 1970s in industrialised countries, but it is no longer valid for very many workers in those countries. Since the onset of neoliberal policies introduced in capitalist countries in the late 1970s and until present, the job experience of many individuals, both men and women, has been marked by instability, low wages, frequent bouts of unemployment and precariat, thus making the approaching retirement a period of uncertainty, insecurity, and concern. As noted by Shultz and Wang (2011, p. 170), retirement is a decision-making process which is "quickly evolving and shifting as the social, organizational, and societal contexts in which retirement takes place change".

We should keep the above caveat in mind as we consider Feldman and Beehr's framework that provides the theoretical underpinnings for those three phases, and that incorporates previous theoretical work.

Phase 1 (Imagining the Possibility of Retirement):

- Individual Differences: retirement can become a clear possibility for some people who have poor health and therefore feel that they cannot continue to work, or they are wealthy enough to feel that they can afford to retire, or think that they are old enough to start considering the possibility of retirement. Individual psychological aspects of self-esteem, affectivity and avoidance of uncertainty may also play a role on whether retirement is worth considering or not.

- Image theory: old people are expected to be motivated to maintain a positive self-image over the years. If the conditions for the maintenance of a positive self-image are not met yet, then people are less likely to retire and will try to stay on their job for as long as they can.
- Continuity theory: old people will be more willing to retire for as long as they can see that the basics of their current life at work (positive image, personal satisfaction, income) will not change dramatically after retirement.
- Social Identity theory: an important aspect of self-image that can affect the consideration of retirement is the social influence on identity. For as long that the expected social environment post-retirement could be seen as favouring the positive self-image of the elder, then the decision to retire is more likely to be made.
- Self-Categorisation theory: this is associated with *Social identity theory* in that a person is more likely to consider retirement seriously to the extent that the individual possesses a cognitive identity that self-categorises him/her as retiree. He/she is also expected to be affectively identified with that category in general and a specific group of retirees in particular (e.g. Michinov et al. 2008).

Phase 2 (Deciding the Time to Let Go):

- Life and Career Stage theories: retirement is seen as a normal stage in life.
- Social-Normative theories: the decision to retire or not, and when to do so, is guided by perceptions and norms that are affected by the social environment and culture.
- Disengagement theory: if disengagement with life as usual and also disengagement with social interactions is the result of retirement, retirement may be seen either negatively if the retiree becomes marginalised and more exposed to the stresses of life, which would delay the decision to retire; or positively if disengagement simply means pursuing transcendence through a "graceful withdrawal from society", as suggested by Cumming et al. (1960; quoted in Chap. 1), which would encourage retirement.
- Person-Environment Fit theory: whenever older workers feel that their skills and motivation do not really match the current requirements of their job, their interest in staying in the workforce may decrease over time. The decline in interest may be also enhanced by decreasing cognitive abilities and physical capacities coming with ageing. The more older workers perceive that they are becoming less fit for the work, the more likely it is that they may contemplate retiring.

Phase 3 (Putting Plans into Action to Transition into Retirement):

- Economic theory: The years into retirement require financial resources that will come from a combination of accumulated superannuation, government support through age full or part-pension, and additional personal savings/investments (e.g. Cagetti 2003). Older workers who cannot afford to retire due to financial constraints will try to delay retirement for as long as they can. Alternatively, prospect retirees may engage in bridge employment that can serve as a transition towards retirement, by maintaining a source of income that can boost savings.

- Motivation theories: to retire or not to retire is a decision that requires taking a number of factors into consideration. The financial factors considered in *Economic theory* being just some of them. A person may contemplate retirement whenever the motivations to stay in the workforce are fewer and less pressing than those to leave it. Some issues that may be considered in the decision-making process include: potential loss of status, what to do with the free time that will become suddenly available, whether the individual feels that she or he has accomplished enough at work/in the career, the level of social support expected in retirement, satisfaction in feeling a sense of "mission accomplished" at work. On the other hand, motivation to retire may be enhanced by increased opportunities to enjoy life with friends and family, greater time to pursue other interests in life without any pressure of time and performance, all of which can make life enjoyable and less stressful. How all those factors play in the decision to retire or not is variable individually.

Whatever the process that any single person goes through in order to evaluate his/her options and finally decide whether to retire or not, for most people the older they are the more likely it is that they will retire. Even people earning their living from an art or a specific craft or profession, who may potentially go on doing it until an advanced age (unlike most employees), tend to slow down as they become older due to the physical and mental constraints of old age. Therefore, asking what is the optimal age for retirement is a moot question, we can only argue, sensibly, about the major variables that are likely to play a role in deciding about retiring, such as old age as already mentioned, health, real opportunities to find work, financial status, motivation (e.g. Bloom et al. 2014). Establishing the optimal age of retirement is a personal decision, but in those cases where the obligation to retire is imposed externally there is no choice.

In the next three sub-sections, we will further elaborate on Feldman and Beehr's (2011) phases of retirement, including a phase of post-transition adjustment to retirement.

5.8.1 *Preparing for Retirement*

Preparation for retirement should start earlier in our working life than we may think, and as far as our health is concerned, it should start in our youth, even before we join the workforce (Damman et al. 2015). One would expect that a curious person with plenty of interests outside work, with a personally satisfactory social network, who was prudent enough to remain healthy and save and invest to secure a materially comfortable retirement, or who is lucky enough—and active enough with his/her vote—to live in a country that takes government support for retirees seriously, should be well prepared to retire and ready to live a fulfilling life into old age. Alas, not everybody is equally prepared to retire, and for many employees retirement becomes compulsory at a set age, whether they like it or not, whether they are prepared for it or not.

The ways in which we live our life can make all the difference in preparing ourselves for retirement. One of the advantages of employment during our working life is that work usually demands a degree of cognitive involvement that keeps our mind active. This mental exercise has a protective effect on our brain. Unless a similar degree of behavioural and environmental complexity is maintained in retirement, the retiree risks suffering cognitive deterioration (Finkel et al. 2009, and references therein). Therefore, preparation for retirement should take into consideration the planning of regular activities that engage the elder in intellectually demanding tasks, that can help in the maintenance of cognitive functions as suggested by the *use it or lose it hypothesis* mentioned in Chap. 4.

Without a clear idea of what each individual expects from the retirement years, it is difficult to make any preparation towards retirement, beyond the basics of keeping good health and wealth (although sometimes not even the basics are well planned for; e.g. Adams and Rau 2011). Hence preparation for retirement does require a degree of introspection, planning, and decision-making regarding the future, this being a deeply personal endeavour. Inputs from the outside may help in the process of planning for retirement: What to do? How to use my time? How much money do I need? But no retirement experience can be ultimately fulfilling unless the answers to those questions are satisfactory to the retiree.

Gee (1999) reported on the expectations for retirement from a group of English and Australian people aged 40 and older. Their most important expectation was that retirement would represent a new beginning for them, in contrast to the findings of other researchers who also reported a desire for continuity into retirement (Pushkar et al. 2010). A new beginning may be desired by people working in jobs that become less satisfactory over time, whereas those still excited and motivated with their job may wish to somehow continue with a similar activity after they have to formally retire. Seeing retirement has an opportunity to enhance leisure experiences in life may also make some old people more inclined to retire.

Although we believe that preparation for retirement is a personal initiative that relies on an active pursue of fulfilment through maturation and self-analysis, agency, forethought, intentionality, activity, and self-regulation (Adams and Rau 2011); some people preparing for retirement may benefit from participating in special pre-retirement programmes that may be available in their community (e.g. Rosenkoetter and Garris 2001). Such programmes may help prospect retirees in the area of financial planning but also psychologically. Rosenkoetter and Garris (2001) report how participating in pre-retirement programmes may help future retirees in terms of improving social interactions in retirement and also stimulate greater mental activity and making travelling an enjoyable experience that also facilitates personal growth; all of which can promote self-worth and social inclusion. In other words, better planning helps retirees to better adjust to their life in retirement, making the most of it.

Some individuals may postpone the planning of retirement simply because they may think that the whole exercise is distressing. In reality, planning decreases distress and depression by giving the retiree greater confidence in the future (Adams and Rau 2011).

Planning for retirement may involve finding substitutes for those aspects of paid work that promote self-fulfilment. For instance, life in retirement should be such that (a) the needs for survival are met, as are (b) the needs for relatedness, (c) self-determination, (d) meaning, and (e) generativity (e.g. Adams and Rau 2011). A more fulfilling life in retirement may come from volunteering, leisure activities, visiting friends and family, involvement in clubs, sports or hobbies; learning and developing artistic skills, becoming involved politically or in social welfare initiatives, protecting the environment, and so forth (e.g. Erlinghagen 2010; Kahana et al. 2013).

The process of planning for retirement eventually leads to the transition to retirement; although in reality planning is an ongoing process, as circumstances may change over time that will require adaptations and modifications of the initial retirement plan.

5.8.2 Transition to Retirement

The transition to retirement may involve experiencing a state of *liminality*–or, not knowing exactly where you stand—that will be hopefully resolved in one way or another. One potential transition to retirement that helps resolve the issue of liminality is working in a bridge employment (Beehr and Bennett 2015). *Bridge employment* can be defined as: "employment that takes place after a person's retirement from a full-time position but before the person's permanent withdrawal from the workforce" (Kim and Feldman 2000, p. 1195). Bridge employment can provide a degree of continuity through a phase of transition, thus potentially making the process leading to full retirement less stressful (Dingemans and Henkens 2015). For some employees bridge employment is a source of satisfaction and reassurance in the lead up to full retirement. Other employees, however, may not need bridge employment to readily adapt to the new life in retirement, most likely because they have already planned their retirement well and can become immediately involved in a series of alternative activities that are personally fulfilling. Still other employees may conclude that bridge employment is not worthy of consideration, if it comes with a substantially decreased income but similar duties and pressure at work.

Bridge employment is also affected by age. von Bonsdorf et al. (2009) showed that bridge employment is favoured by people younger than 60, whether bridge employment is in the same or a different field/career (Table 5.10). Whereas more people who are 60 years old and older prefer full retirement over bridge employment. Bridge employment is also preferred by men in comparison to women, especially bridge employment in a different field. This may perhaps explain, at least in part, the relatively better morale among men than women observed during the transition to retirement by Kim and Moen (2002).

During the neoliberal period, transitions to retirement are becoming "longer and fuzzier" (Moen et al. 2001). This process can undermine marital relationships, whenever the transition challenges established roles that had until then stabilised such relationships. For as long as the transition period is a stressful one, the shorter it is, the better (Holcomb 2010).

Table 5.10 Percentage of individuals by retirement and bridge employment intentions

Variables	Full retirement (n = 189)	Bridge employment in a different field (n = 175)	Career bridge employment (n = 74)	X^2
Age				
40–49 years	5.8	13.1	18.7	51.6***
50–54 years	29.5	34.3	57.3	
55–59 years	29.5	34.9	20.0	
60–64 years	27.4	13.7	2.7	
65 years	7.9	4.0	1.3	
Gender				
Men	58.7	77.7	50.7	22.4***
Women	41.3	22.3	49.3	

***$P < 0.01$. Simplified from von Bonsdorf et al. (2009)

5.8.3 *Post-Transition Adjustment*

Perhaps for the first time in the history of humankind, the old Roman ideal of *otium honestum* (see Chap. 2), that in ancient times was only the prerogative of the wealthier social classes, is now becoming available to very many retirees (with some clouds emerging that overshadow the future of young workers living through the neoliberal period). *Otium honestum* did not mean idleness, or just to be relieved from the stresses of specific duties at work, but to be free to pursue activities and interests that one finds personally fulfilling. The adjustment to life as a retiree is ultimately affected not only by individual characteristics, interests, needs, and resources, but also by family, actions and reactions from the rest of the community, and the support structure of government. Poor adjustment to the new life in retirement may lead to distress, inability to fully enjoy the freedom from the obligations of employment, and even physical illness (Nuttman-Shwartz 2004; see review in Wang et al. 2011).

Long-term adjustment becomes more difficult if life satisfaction in retirement does not increase after an initial period of transition (see the Class 3 pattern of retirement in Fig. 5.11). Given the prevailing cultural view that work defines not only our purpose in life but to a great extent also our identity, unless we are able to strengthen our identity and purpose during retirement, adjustment is likely to be difficult. In the words of Fred Darnley (1975, p. 217): “adjust or be miserable”.

To the extent that the leisure time available in retirement is utilised in a way that is meaningful and reflective of the personal purpose in life of the elder, adjustment may proceed satisfactorily for the retiree, in which case a positive message about the self would be transmitted to the rest of society (Darnley 1975). Depending on personality, such utilisation of leisure time may be undertaken with a more active or a more contemplative attitude, and there is no reason to believe that the former is in principle superior to the latter or vice versa. In fact, the so-called busy ethics in retirement (e.g. Ekerdt 1986) may become a double-edged sword when it leads to stress. The concept of "busy ethics", however, does have a more general beneficial function in the adjustment to retirement in terms of spurring the individual to continuously use her/his mind and also keeping muscles and other organs active, all of which have proven health benefits (see Chap. 4). Nonetheless, as with just about all things in life, retirement is also an exercise in life balance.

Darnley (1975) makes the important point that adjustment to life in retirement is more difficult to achieve unless one has already developed an attitude to adjust to changing circumstances. This requires the building up of mental plasticity throughout life, as a crystallised mind will find it difficult to adaptively adjust to an environment that has changed significantly; but also develop the skills and build the social network that can endure into retirement, according to personal needs.

On the other hand, some retirees may not achieve a good adjustment in retirement no matter what, thus experiencing poorer health, greater levels of loneliness and depression, and a lower degree of life satisfaction and happiness compared to the situation pre-retirement, leading to disenchantment about retirement (e.g. Gall et al. 1997).

Many issues have been identified that potentially make the post-retirement period fraught with challenges and difficulties. As partners have more time to spend together they have to find a way to avoid potential tensions that during their working life were hidden by them spending a significant amount of time apart and busy worrying about their job. Tensions may also arise if income falls in retirement dragging the social status of individuals down, thus affecting morale and potentially health (Rosenkoetter and Garris 1998). On the other hand, life in retirement may be positive and even exciting (e.g. Earl et al. 2015).

Based on results from interviews of a group of retirees in the USA, Rosenkoetter and Garris (1998) concluded that such retirees were adjusting reasonably well to their life in retirement, being mainly concerned about their health and the future, social roles, relationships, self-esteem, use of time, support groups and changes in life structure. Earl et al. (2015) suggest that direct predictors of adjustment to retirement include finances, health, mastery, and self-efficacy.

Although we would expect that in the retirement years health and mental functions (e.g. memory) are likely to deteriorate over the long run, due to the broad process of ageing, there are specific effects of retirement that could cause either improvement or deterioration of mental health and life satisfaction soon after an individual retires from paid work (see Class 2 and Class 1 patterns in Fig. 5.11).

The adoption of a healthy lifestyle can swing the balance towards improvement of mental health in retirement (Zantinge et al. 2013), and participating in leisure activities after retirement contributes positively to life satisfaction and psychological well-being (Ragheb and Griffith 1982; Nimrod 2007).

Retiring early has no significant effect on health and lifespan when confounding variables are controlled (Litwin 2007; although Rohwedder and Willis (2010) described negative effects of early retirement on cognition), the effect may be mediated by how voluntary early retirement is: voluntary early retirement is less likely to be followed by depression than forced retirement (Charles 2004). A degree of social support after retirement is also beneficial to mental health and life satisfaction (e.g. Diener and Diener 1995), especially when the support that a worker might have received in the workplace during his/her working life is compensated by support received from others in retirement (Bosse et al. 1993). When it is the retiree to give support (e.g. to family and friends) he/she can actively contribute to strengthening social bonds (van den Bogaard et al. 2014), which may be cognitively and emotionally beneficial. Social bonds within a couple in particular, can be reinforced post-retirement by sharing duties at home, participating in some activities together, communicating, keep growing as individuals and couple, and sharing at least some additional interests beyond the basics of health, home, family, and money (e.g. Szinovacz 2000; Leopold and Skopek 2015). The benefits of social relationships may also affect the decision where to live after retirement: e.g. in the community or a retirement village (Gardner et al. 2005).

Moen (1996) suggested a model for the relationship between retirement, health, and well-being that starts by acknowledging the specific historical circumstances of the retiree's life (Fig. 5.12). Such circumstances can affect the individual history of employment and conditions during the working life, and also the *structural exigencies* (e.g. corporate policies and government legislation) and *situational exigencies* (i.e. personal circumstances of the individual's life) prevalent at any given time, that eventually contribute to determining the timing and the overall decision-making process leading to retirement. After retirement, good health and subjective well-being may be experienced to the extent that the activities undertaken post-retirement

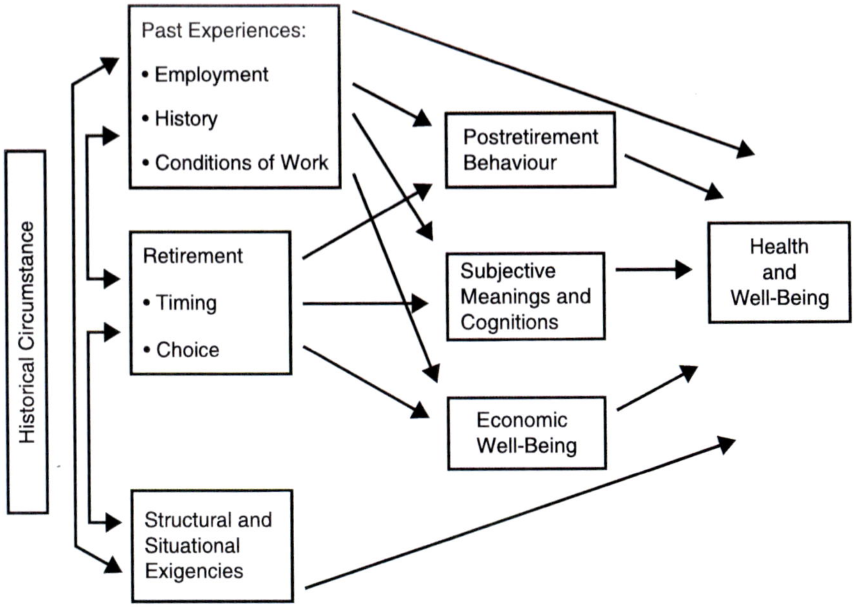

Fig. 5.12 Moen's (1996) life course model that links retirement with health and well-being

are conducive to healthy living, they are meaningful, fulfilling and they also safeguard the retiree's economic well-being.

Experiencing a fulfilling life in retirement may be helped by a degree of *hardiness*. Hardiness is a personality style of stress resistance that includes aspects of challenge, commitment, and control, and it enables individuals to withstand the threats of change, role transitions, and the stresses coming with declining health (Kobasa 1982). Hardy individuals are expected to adapt more easily to retirement, especially when retirement is involuntary. Hardiness also leads to a more positive appraisal of life events, thus promoting better satisfaction in life (e.g. Reis and Pushkar Gold 1993). In addition to hardiness, a more fulfilling life in retirement is helped by a disposition to enjoy life. Distress may be potentially decreased in retirement, and life satisfaction increased by the optimistic aspects of personality traits such as Extraversion, Openness, and Agreeableness.

On the negative side, a personality displaying "goal instability" faces more challenges of adjustment in retirement (Reis and Pushkar Gold 1993). Neuroticism may also jeopardise adaptation to retirement through its negative effects on health, as it may Type A personality (Reis and Pushkar Gold 1993). Type A personality is characterised by impatience, hostility, competitiveness, drive, difficulty expressing emotions, perfectionism, and unhealthy dependence on external rewards such as wealth, status, and power; that is, traits that are not very useful when traditional social power may decrease in retirement and the need for help from others increases.

It is not unusual that, whenever mental and physical health allows it, some of the current retirees may be seeking to reenter the workforce, after spending a few years in retirement. The reasons for becoming "un-retired" (e.g. Maestas 2010) may be different from individual to individual, ranging from boredom, seeking a clearer meaning in life and personal fulfilment, being recalled to help or, more simply, because the elder needs an additional source of income (e.g. Parry and Taylor 2007). Of course, for those seeking paid employment, getting one is not necessarily easy as they may confront competition from the younger generations. This is probably one of the reasons—among others, such as lack of interest or physical/mental constraints—why un-retired people tend to be a minority: e.g. 15% of male retirees and 13% of female retirees in Cahill et al. (2011) work.

If retirement is a stressful experience and no compensatory strategies are used to decrease such stress, chances are that retirement may contribute to an early death of the elder (e.g. Haynes et al. 1978; Marmot and Shipley 1996; Maier and Smith 1999). Hence, preparing for retirement in order to make it a fulfilling and enjoyable part of our life may also help extend such life.

5.8.4 *Gender and Retirement*

Kim and Moen (2001, p. 85) have summarised the main gender aspects of retirement for working women compared to men in this statement:

> Women are more likely than men to experience discontinuity, moving in and out of the labor force as their family responsibilities shift. Correspondingly, women tend to have more

> negative attitudes toward retirement, plan for it less, adjust to it more poorly, and are more likely to experience depression following retirement ... Moreover, women's adjustment to retirement is more adversely affected by poor health and inadequate incomes than men's.

Of course, not all women or men necessarily fit such description, and past trends may not be necessarily maintained in the future. For instance, the transition to "retirement age" may involve little change to women who have been fulfilling a traditional role of housewife for most of their adult life, whereas men who have been the main breadwinners for the same period may be confronted with a more structured path to retirement, and they may perceive retirement itself as a significant shift in their life circumstances. The same perception of a shift would apply in current situations where gender roles are reversed, with the wife being the main breadwinner.

From the perspective of gender, well-being in retirement relies on the independent experiencing of fulfilment by each individual: man and woman, even if they live as a co-habiting couple. For simplicity, in referring to couples here we will focus on heterosexual relationships, but for as long as each individual is expected to feel fulfilled in retirement, many conclusions are also valid for other forms of relationships.

Today it is not uncommon for man and woman to retire approximately at the same time (Kim and Moen 2001) in order to start a life in retirement which may emphasise shared leisure activities and sometimes a common project (e.g. a charity), alongside a more equal sharing of household chores, even though a division of labour may be retained (e.g. Loretto and Vickerstaff 2012, and references therein).

After two persons involved in a relationship retire, however, there may also be some divergence of paths between them, with women tending to put family higher in their priorities during retirement than men. Family members such as adult children—or other third persons—may become confidantes of one or the other old parent, this may further contribute to divergence (or even segregation) between partners, especially if the third person interferes with good communication within the couple that would be otherwise expected from a reciprocal style of intimacy (Kulik 1999).

In a study carried out in the UK, Loretto and Vickerstaff (2012) described a diversity of patterns in the relationship between man and woman in retirement. In many cases the partners decided to pursue different activities not just as an attractive opportunity offered by retirement but also as a "survival strategy" for their relationship. The perception of retirement for those who retired from paid work was that it represented the earned reward for years of hard work, but interestingly when work had been asymmetrical (traditionally the husband had worked full-time, whereas the wife mainly took care of the house and children) the wife also argued that she deserved a reward in retirement for the hard work carried out at home over the years. Retirement represents freedom, but it is unlikely that freedom will be allowed to be monopolised by any gender within a relationship. In the case of couples who had experienced long-term unemployment, retirement did not represent a significant change from life as usual.

Vo et al. (2015) carried out a large-scale study on retirement, gender, and mental health in New South Wales, Australia, based on a sample of 202,584 participants aged 45 years old and older, who replied to the questions of a postal survey. As expected, the proportion of participants who were fully retired increased with age for both men and women. However, in the oldest age range considered in this study (75–79 years old) the proportion of retired women was higher than men. The proportion of participants with high levels of psychological distress decreased at older ages of retirement, being similar between men and women for the fully retired participants. For fully retired people, distress was high among young retirees (both women and men). Distress was highest in the sick and disabled women and men of all ages, and it was lowest for men and women of all ages who were either partly retired or were working full/part-time or self-employed. The unemployed of all ages, both men and women, tended to be more distressed than the fully retired ones. That is, both men and women decrease their distress in retirement the older they are and the more active they remain.

Table 5.11 shows the reasons for retirement given by participants in Vo et al.'s study according to gender and age at retirement. Reaching usual retirement age was a reason given similarly frequently by men and women. Caring for family/friend, however, was given more frequently by women than men, although in both cases the frequency tended to decrease with age at retirement. Being made redundant at work is a reason given more frequently by men than women, with the reason reaching a maximum between 60 and 69 years old in both genders.

Table 5.11 Reasons for retirement according to age and gender, restricted to fully retired and partially retired participants

	Age (years)					
	45–54	55–59	60–64	65–69	70–74	75–79
Men						
Reached usual retirement age	0.6	8.1	16.6	28.7	32.0	37.9
To care for family/friend	7.6	6.4	5.0	3.9	3.4	3.1
Made redundant	10.8	16.4	18.7	19.1	17.3	14.4
Could not find a job	3.7	3.4	4.1	5.1	4.2	3.6
Lifestyle	30.4	43.2	40.1	28.6	25.6	21.7
Ill health	47.0	28.9	21.7	20.0	17.7	15.5
Other	12.4	11.2	10.5	10.0	10.7	11.4
Not answered	5.5	2.4	2.3	2.8	3.3	3.9
Women						
Reached usual retirement age	0.3	12.9	18.3	30.6	34.8	35.8
To care for family/friend	17.8	16.1	14.3	12.8	12.0	12.5
Made redundant	6.8	7.6	8.5	8.2	6.0	4.1
Could not find a job	2.4	3.1	3.7	3.3	2.3	1.4
Lifestyle	41.1	44.8	38.3	28.3	25.5	23.0
Ill health	26.3	17.4	15.9	13.9	9.7	7.7
Other	14.2	13.3	14.5	13.5	12.8	13.4
Not answered	7.8	3.8	4.3	5.5	8.3	11.8

Values are percentages. Reasons for retirement are not mutually exclusive. From Vo et al. (2015)

"Could not find a job" was the lowest reason for retirement in both men and women across most age groups, whereas "lifestyle" was among the most frequently mentioned reasons, especially among the men and women retirees in the younger age classes. "Ill health" was also frequently mentioned, especially among the younger retirees of both genders.

In sum, experiencing a fulfilling life in retirement requires some planning and preparation, which help retirees adapt to life after work through a transition period, and then start a path, perhaps a new path, for the continued development of the self into the future. However, the specific experience of retirement differs among people. Some retirees may see retirement as an opportunity to improve their life (positive change), others may see it as a deterioration of their lifestyle (negative change), whereas for others there may be continuity between their life before and after retirement. Retirees are constrained in their decisions about retirement by the legal framework operating in their country, that sets limits to retirement age. In addition, the decision when to retire is also conditioned by health, financial situation, opportunities, and motivation. The transition to retirement involves a period of adaptation that will hopefully overcome any state of liminality that the new retirees may experience. Such transition is currently affected by the instability of employment that has characterised the neoliberal period. Employment instability increases life uncertainty in retirement, especially due to financial insecurity, which is made worse by low government support. For others, retirement comes with increased freedom. The freedom represented by retirement is unlikely to be monopolised by any gender. Both men and women retirees rightly feel entitled to pursue their own happiness and fulfilment in life, whether they decide to do so alone or in a partnership. Personal fulfilment in retirement may not only help extend life, but above all our satisfaction with it.

5.9 Conclusions

In old age we have the opportunity to produce a synthesis of our life-so-far, and also to continue developing our self-identity and wisdom into the future. A better coherence of the self allows us not only to live a more fulfilling life but to also enhance our agency, thus promoting a better control of our life experiences and the stresses that they may cause.

Psychological ageing challenges us to understand our emotions and their expression in various contexts. Emotions tend to shift towards the more positive ones in old age, as we also become more emotionally stable. Cognitively, we tend to broadly decline in our capabilities as we age, especially in our memory, but the decline can be slowed down to an extent by building cognitive reserve throughout our life and living a cognitively rich life in old age. As far as personality traits are concerned, ageing tends to lead to decreased Neuroticism, increased Agreeableness, and increased or stable Conscientiousness. The decrease in Neuroticism seems to be an adaptation to increased psychological stress in old age. Psychological stress may challenge old people's health, potentially leading to disease and early death, unless effective coping strategies are adopted. Low Neuroticism dampens such negative effects of psychological stressors.

Most old people eventually retire from paid work in order to pursue a fulfilling life through other activities. The process of retirement is one requiring preparation and then adjustment in order to make the best of it: decrease stress, secure a good health and sufficient wealth, and increase the quality and meaning of our life as we enter the last part of it.

References

Achenbaum, W. A. (2006). What is retirement for? *Wilson Quarterly Spring, 30*(2), 50–56.

Adams, G. A., & Rau, B. L. (2011). Putting off tomorrow to do what you want today: Planning for retirement. *American Psychologist, 66*, 180–192.

Aggleton, J. P., & Brown, M. W. (2006). Interleaving brain systems for episodic and recognition memory. *Trends in Cognitive Sciences, 10*, 455–463.

Aknouche, N., & Noor, N. M. (2014). The primacy of the self in shame: Can shame be benevolent? *American International Journal of Social Science, 3, 59–79.*

Aleman, A. (2014). *Our ageing brain.* Melbourne, SA: Scribe.

Alkadry, M. G., & Tower, L. E. (2006). Unequal pay: The role of gender. *Public Administration Review, 66*, 888–898.

Allemand, M., Gomez, V., & Jackson, J. J. (2010). Personality trait development in midlife: Exploring the impact of psychological turning points. *European Journal of Ageing, 7*, 147–155.

Amir, M., & Lev-Wiesel, R. (2003). Time does not heal all wounds: Quality of life and psychological distress of people who survived the Holocaust as children 55 years later. *Journal of Traumatic Stress, 16*, 295–299.

Andel, R., Crowe, M., Kåreholt, I., Wastesson, J., & Parker, M. G. (2011). Indicators of job strain at midlife and cognitive functioning in advanced old age. *The Journals of Gerontology, Series B, 66*, 287–291.

Andel, R., Kåreholt, I., Parker, M. G., Thorslund, M., & Gatz, M. (2007). Complexity of primary lifetime occupation and cognition in advanced old age. *Journal of Aging and Health, 19*, 397–415.

Ando, J., Ono, Y., & Wright, M. J. (2001). Genetic structure of spatial and verbal working memory. *Behavior Genetics, 31*, 615–624.

Angel, L., Fay, S., Bouazzaoui, B., & Isingrini, M. (2011). Two hemispheres for better memory in old age: Role of executive functioning. *Journal of Cognitive Neuroscience, 23*, 3767–3777.

Angel, R. J., Frisco, M., Angel, J. L., & Chiriboga, D. A. (2003). Financial strain and health among elderly Mexican-origin individuals. *Journal of Health and Social Behavior, 44*, 536–551.

Ardelt, M. (1997). Wisdom and life satisfaction in old age. *Journal of Gerontology, 52B*, P15–P27.

Assmann, A. (1994). Wholesome knowledge: Concepts of wisdom in a historical and cross- cultural perspective. In D. L. Featherman, R. M. Lerner, & M. Perlmutter (Eds.), *Life-span development and behaviour* (Vol. 12, pp. 187–224). Hillsdale, NJ: Lawrence Erlbaum.

Atchley, R. C. (1976). *The sociology of retirement.* New York: Halsted Press.

Atchley, R. C. (1989). A continuity theory of normal aging. *The Gerontologist, 29*, 183–190.

Awh, E., Barton, B., & Vogel, E. K. (2007). Visual working memory represents a fixed number of items regardless of complexity. *Psychological Science, 18*, 622–628.

Baddeley, A. (1992). Working memory. *Science, 255*, 556–559.

Baddeley, A. (2010). Working memory. *Current Biology, 20*, R136–R140.

Baddeley, A. (2012). Working memory: Theories, models, and controversies. *Annual Reviews of Psychology, 63*, 1–29.

Ball, K., Berch, D. B., Helmers, K. F., Jobe, J. B., Leveck, M. D., Marsiske, M., et al. (2002). Effects of cognitive training interventions with older adults: A randomized controlled trial. *JAMA, 288*, 2271–2281.

Balsis, S., Carpenter, B. D., & Storandt, M. (2005). Personality change precedes clinical diagnosis of dementia of the Alzheimer type. *The Journals of Gerontology: Series B, 60*, P98–P101.

Baltes, M. M., & Carstensen, L. L. (1996). The process of successful aging. *Ageing and Society, 16*, 397–422.

Baltes, P. B., Cornelius, S. W., Spiro, A., Nesselroade, J. R., & Willis, S. L. (1980). Integration vs. differentiation of fluid-crystallized intelligence in old age. *Developmental Psychology, 16*, 625–635.

Baltes, P. B., Glück, J., & Kunzmann, U. (2002). Wisdom: Its structure and function in regulating successful life span development. In C. R. Snyder & S. J. Lopez (Eds.), *Handbook of positive psychology* (pp. 327–347). Oxford: Oxford University Press.

Baltes, P. B., & Lindenberger, U. (1988). On the range of cognitive plasticity in old age as a function of experience: 15 years of intervention research. *Behavior Therapy, 19*, 283–300.

Baltes, P. B., Lindenberger, U., & Staudinger, U. M. (1998). Life-span theory in developmental psychology. In W. Damon & R. M. Lerner (Eds.), *Handbook of child psychology. Vol. 1—Theoretical models of human development* (pp. 1029–1143). New York: Wiley.

Baltes, P. B., & Smith, J. (1990). The psychology of wisdom and its ontogenesis. In R. J. Steinberg (Ed.), *Wisdom: Its nature, origins, and development* (pp. 87–120). Cambridge: Cambridge University Press.

Baltes, P. B., Staudinger, U. M., Maercker, A., & Smith, J. (1995). People nominated as wise: A comparative study of wisdom-related knowledge. *Psychology and Aging, 10*, 155–166.

Band-Winterstein, T., Doron, I., & Naim, S. (2012). Elder self neglect: A geriatric syndrome or a life course story? *Journal of Aging Studies, 26*, 109–118.

Barak, B. (2009). Age identity: A cross-cultural global approach. *International Journal of Behavioral Development, 33*, 2–11.

Barnes, D. E., Cauley, J. A., Lui, L.-Y., Fink, H. A., McCulloch, C., Stone, K. L., et al. (2007). Women who maintain optimal cognitive function into old age. *Journal of the American Geriatrics Society, 55*, 259–264.

Barrett, A. E. (2005). Gendered experiences in midlife: Implications for age identity. *Journal of Aging Studies, 19*, 163–183.

Bates, S. L., & Wolbers, T. (2014). How cognitive aging affects multisensory integration of navigational cues. *Neurobiology of Aging, 35*, 2761–2769.

Bauer, M. E. (2008). Chronic stress and immunosenescence: A review. *Neuroimmunomodulation, 15*, 241–250.

Bauer, M. E., Moriguchi Jeckel, C. M., & Luz, C. (2009). The role of stress factors during aging of the immune system. *Annals of the New York Academy of Science, 1153*, 139–152.

Baum, A. (1990). Stress, intrusive imagery and chronic distress. *Health Psychology, 9*, 653–675.

Baumgartner, K. (2004). Neurocognitive changes in cancer patients. *Seminars in Oncology Nursing, 20*, 284–290.

Beech, A. R., & Mitchell, I. J. (2009). Attachment difficulties. In M. McMurran & R. Howard (Eds.), *Personality, personality disorder and risk of violence* (pp. 213–228). Chichester: Wiley.

Beech, N. (2011). Liminality and the practices of identity reconstruction. *Human Relations, 64*, 285–302.

Beedie, C. J., Terry, P. C., & Lane, A. M. (2005). Distinctions between emotion and mood. *Cognition and Emotion, 19*, 847–878.

Beehr, T. A., & Bennett, M. M. (2015). Working after retirement: Features of bridge employment and research directions. *Work, Aging and Retirement, 1*, 112–128.

Beluche, I., Carrière, I., Ritchie, K., & Ancelin, M. L. (2010). A prospective study of diurnal cortisol and cognitive function in community-dwelling elderly people. *Psychological Medicine, 40*, 1039–1049.

Bennett, M. P., Zeller, J. M., Rosenberg, L., & McCann, J. (2003). The effect of mirthful laughter on stress and natural killer cell activity. *Alternative Therapies in Health and Medicine, 9*, 38–45.

Berg, M. C. (2008). New age advice: Ticket to happiness? *Journal of Happiness Studies, 9*, 361–377.

Berner, J., Rennemark, M., Jogréus, C., & Berglund, J. (2012). Distribution of personality, individual characteristics and internet usage in Swedish older adults. *Aging and Mental Health, 16*, 119–126.

Bergsma, A., & Ardelt, M. (2012). Self-reported wisdom and happiness: An empirical investigation. *Journal of Happiness Studies, 13*, 481–499.

Bianchi, E. C. (2005). Living with elder wisdom. *Journal of Gerontological Social Work, 45*, 319–329.

Biggs, S. (1997). Choosing not to be old? Masks, bodies and identity management in later life. *Ageing and Society, 17*, 553–570.

Black, H. K. (2009). Pictures of suffering in elders' narratives. *Journal of Aging Studies, 23*, 82–89.

Bloom, D. E., Canning, D., & Moore, M. (2014). Optimal retirement with increasing longevity. *Scandinavian Journal of Economics, 116*, 838–858.

Bopp, K. L., & Verhaeghen, P. (2005). Aging and verbal memory span: A meta-analysis. *Journal of Gerontology, 60B*, P223–P233.

Bosse, R., Aldwin, C. M., Levenson, M. R., Spiro, A., III, & Mroczek, D. K. (1993). Change in social support after retirement: Longitudinal findings from the Normative Aging Study. *Journal of Gerontology, 48*, P210–P217.

Bowles, N. L., & Poon, L. W. (1985). Aging and retrieval of words in semantic memory. *Journal of Gerontology, 40*, 71–77.

Brainerd, C. J., Holliday, R. E., Reyna, V. F., Yang, Y., & Toglia, M. P. (2010). Developmental reversals in false memory: Effects of emotional valence and arousal. *Journal of Experimental Child Psychology, 107*, 137–154.

Bräuninger, I. (2014). Dance movement therapy with the elderly: An international Internet-based survey undertaken with practitioners. *Body, Movement and Dance in Psychotherapy, 9*, 138–153.

Bray, I., Gunnell, D., & Smith, G. D. (2006). Advanced paternal age: How old is too old? *Journal of Epidemiology and Community Health, 60*, 851–853.

Brehmer, Y., Rieckmann, A., Bellander, M., Westerberg, H., Fischer, H., & Bäckman, L. (2011). Neural correlates of training-related working-memory gains in old age. *NeuroImage, 58*, 1110–1120.

Brink, A. (2007). *Curiosity, personal growth initiative, and life satisfaction in older adults*. Doctor of Psychology dissertation, Faculty of The Chicago School of Professional Psychology, Chicago.

Brown, A. S. (1991). A review of the tip-of-the-tongue experience. *Psychological Bulletin, 109*, 204–223.

Brown, H., & Maylor, E. A. (2017). Memory consolidation effects on memory stabilization and item integration in older adults. *Psychonomic Bulletin and Review, 24*, 1032–1039.

Brown, S. C., Mason, C. A., Lombard, J. L., Martinez, F., Plater-Zyberk, E., Spokane, A. R., et al. (2009). The relationship of built environment to perceived social support and psychological distress in Hispanic elders: The role of " Eyes on the Street". *Journal of Gerontology, 64B*, 234–246.

Brümmer, L. (2009). *Ageing and emotion regulation* (151 pp). PhD Thesis, Faculty of Medicine, Health and Life Sciences, School of Psychology, University of Southampton, UK.

Bryant, F. B., Smart, C. M., & King, S. P. (2005). Using the past to enhance the present: Boosting happiness through positive reminiscence. *Journal of Happiness Studies, 6*, 227–260.

Budson, A. E. (2014). Memory dysfunction in dementia. In A. E. Budson & N. W. Kowall (Eds.), *The handbook of Alzheimer's disease and other dementias* (pp. 315–335). Chichester, West Sussex: Wiley.

Buschkuehl, M., Jaeggi, S. M., Hutchison, S., Perrig-Chiello, P., Däpp, C., Müller, M., et al. (2008). Impact of working memory training on memory performance in old–old adults. *Psychology and Aging, 23*, 743–753.

Buss, D. M. (2008). *Evolutionary psychology: The new science of the mind*. Boston: Pearson.

Buttimer, A. (1980). Home, reach, and the sense of llace. In A. Buttimer & D. Seamon (Eds.), *The human experience of space and place* (pp. 166–187). London: Croom Helm.

Byles, J., Tavener, M., Robinson, I., Parkinson, L., Warner Smith, P., Stevenson, D., et al. (2013). Transforming retirement: New definitions of life after work. *Journal of Women and Aging, 25*, 24–44.

Cagetti, M. (2003). Wealth accumulation over the life cycle and precautionary savings. *Journal of Business and Economic Statistics, 21*, 339–353.

Cahill, K. E., Giandrea, M. D., & Quinn, J. F. (2011). Reentering the labor force after retirement. *Monthly Labor Review, 134*, 34–42.

Cairney, J., & Krause, N. (2005). The social distribution of psychological distress and depression in older adults. *Journal of Aging and Health, 17*, 807–835.

Cantegreil-Kallen, I., & Pin, S. (2012). Fear of Alzheimer's disease in the French population: Impact of age and proximity to the disease. *International Psychogeriatrics, 24*, 108–116.

Carstensen, L. L. (1987). Age-related changes in social activity. In L. L. Carstensen & B. A. Edelstein (Eds.), *Handbook of clinical gerontology* (pp. 222–237). New York: Pergamon Press.

Carstensen, L. L. (1991). Socioemotional activity theory: Social activity in life-span context. *Annual Review of Gerontology and Geriatrics, 11*, 195–217.

Carstensen, L. L., & Löckenhoff, C. E. (2003). Aging, emotion, and evolution: The big picture. *Annals of the New York Academy of Science, 1000*, 152–179.

Carstensen, L. L., Turan, B., Scheibe, S., Ram, N., Ersner-Hershfield, H., Samanez-Larkin, G. R., et al. (2011). Emotional experience improves with age: Evidence based on over 10 years of experience sampling. *Psychology and Aging, 26*, 21–33.

Carvallo, M., & Gabriel, S. (2006). No man is an island: The need to belong and dismissing avoidant attachment style. *Personality and Social Psychology Bulletin, 32*, 697–709.

Case, A., & Paxson, C. (2009). Early life health and cognitive function in old age. *American Economic Review: Papers and Proceedings, 99*, 104–109.

Caspi, A., Roberts, B. W., & Shiner, R. L. (2005). Personality development: Stability and change. *Annual Review of Psychology, 56*, 453–484.

Cassarino, M., & Setti, A. (2015). Environment as 'Brain Training': A review of geographical and physical environmental influences on cognitive ageing. *Ageing Research Reviews, 23*, 167–182.

Charles, K. K. (2004). Is retirement depressing?: Labor force inactivity and psychological well being in later life. *Research in Labor Economics, 23*, 269–299.

Chiesa, A., & Serretti, A. (2009). Mindfulness-based stress reduction for stress management in healthy people: A review and meta-analysis. *The Journal of Alternative and Complementary Medicine, 15*, 593–600.

Chiriboga, D. A., Black, S. A., Aranda, M., & Markides, K. (2002). Stress and depressive symptoms among Mexican American elders. *Journal of Gerontology, 57B*, P559–P568.

Christensen, H., Korten, A., Jorm, A. F., Henderson, A. S., Scott, R., & Mackinnon, A. J. (1996). Activity levels and cognitive functioning in an elderly community sample. *Age and Ageing, 25*, 72–80.

Cicirelli, V. G. (2002). Fear of death in older adults: Predictions from terror management theory. *Journal of Gerontology, 57B*, P358–P366.

Cipriani, G., Borin, G., Del Debbio, A., & Di Fiorino, M. (2015). Personality and dementia. *The Journal of Nervous and Mental Disease, 203*, 210–214.

Clayton, S. (2003). Environmental identity: A conceptual and an operational definition. In S. Clayton & S. Opotow (Eds.), *Identity and the natural environment* (pp. 45–65). Cambridge, MA: MIT Press.

Clayton, S., & Opotow, S. (2003). *Identity and the natural environment*. Cambridge, MA: MIT Press.

Clayton, V. P., & Birren, J. E. (1980). The development of wisdom across the life-span: A reexamination of an ancient topic. In P. B. Baltes & O. G. Brim Jr. (Eds.), *Life- span development and behavior* (Vol. 3, pp. 103–135). New York: Academic Press.

Clememte, F., & Kleiman, M. B. (1976). Fear of crime among the aged. *The Gerontologist, 13*, 207–210.
Commissaris, C. J. A. M., Jolles, J., Verhey, F. R. J., Ponds, R. W. H. M., Damoiseaux, V., & Kok, G. J. (1993). Vergeetachtig of dement? Wie maakt zich zorgen en waarom? *Tijdschrift voor Gerontologie en Geriatrie, 24*, 144–149.
Connor, L. T. (2001). Memory in old age: Patterns of decline and preservation. *Seminars in Speech and Language, 22*, 117–125.
Consedine, N. S., & Magai, C. (2003). Attachment and emotion experience in later life: The view from emotions theory. *Attachment and Human Development, 5*, 165–187.
Conway, M. A. (1996). Autobiographical knowledge and autobiographical memories. In D. C. Rubin (Ed.), *Remembering our past: Studies in autobiographical memory* (pp. 67–93). Cambridge: Cambridge Universioty Press.
Cooper, C., Bebbington, P., King, M., Jenkins, R., Farrell, M., Brugha, T., et al. (2011). Happiness across age groups: Results from the 2007 National Psychiatric Morbidity Survey. *International Journal of Geriatric Psychiatry, 26*, 608–614.
Cooper, J., Harris, Y., & McGready, J. (2002). Sadness predicts death in older people. *Journal of Aging and Health, 14*, 509–526.
Cordella, M., & Poiani, A. (2014). *Behavioural oncology: Psychological , communicative, and social dimensions*. New York: Springer.
Costa, P. T., & McCrae, R. R. (1992). *Revised NEO Personality Inventory (NEO-PI-R) and NEO Five-Factor Inventory (NEO-FFI) professional manual*. Odessa, FL: Psychological Assessment Resources.
Costa Jr., P. T., & McCrae, R. R. (1994). Set like plaster: Evidence for the stability of adult personality. In T. F. Heatherton & J. L. Weinberger (Eds.), *Can personality change?* (pp. 21–40). Washington, DC: American Psychological Association.
Costa, P. T., & McCrae, R. R. (2006). Age changes in personality and their origins: Comment on Roberts, Walton, and Viechtbauer (2006). *Psychological Bulletin, 132*, 26–28.
Craik, F. I. M., & Bialystok, E. (2006). Cognition through the lifespan: Mechanisms of change. *Trends in Cognitive Sciences, 10*, 131–138.
Cruces, J., Venero, C., Pereda-Pérez, I., & De la Fuente, M. (2014). The effect of psychological stress and social isolation on neuroimmunoendocrine communication. *Current Pharmaceutical Design, 20*, 2014–2020.
Csikszentmihályi, M. (1990). *Flow: The psychology of optimal experience*. New York: Harper & Row.
Cumming, E., Dean, L. R., Newell, D. S., & McCaffrey, I. (1960). Disengagement - a tentative theory of aging. *Sociometry, 1*, 23–35.
Curcio, C.-L., Gomez, F., & Reyes-Ortiz, C. A. (2009). Activity restriction related to fear of falling among older people in the Colombian Andes mountains: Are functional or psychosocial risk factors more important? *Journal of Aging and Health, 21*, 460–479.
Damasio, A. R. (1999). *The feeling of what happens*. London: Heinemann.
Damian, R. I., Spengler, M., Sutu, A., & Roberts, B. W. (2019). Sixteen going on sixty-six: A longitudinal study of personality stability and change across 50 years. *Journal of Personality and Social Psychology, 117*, 674–695.
Damman, M., Henkens, K., & Kalmijn, M. (2015). Missing work after retirement: The role of life histories in the retirement adjustment process. *Gerontologist, 55*, 802–813.
Darnley Jr., F. (1975). Adjustment to retirement: Integrity or despair. *The Family Coordinator, 24*, 217–226.
Darwin, C. (1872). *The expression of the emotions in man and animals*. London: John Murray.
De Donder, L., Verté, D., & Messelis, E. (2005). Fear of crime and elderly people: Key factors that determine fear of crime in West Flanders. *Ageing International, 30*, 363–376.
de Kloet, E. R., Joëls, M., & Holsboer, F. (2005). Stress and the brain: From adaptation to disease. *Nature Reviews Neuroscience, 6*, 463–475.
Delbaere, K., Crombez, G., Vanderstraeten, G., Willems, T., & Cambier, D. (2004). Fear-related avoidance of activities, falls and physical frailty. A prospective community-based cohort study. *Age and Ageing, 33*, 368–373.

Demo, D. H. (1992). The self-concept over time: Research issues and directions. *Annual Review of Sociology, 18*, 303–326.
Denes, G. (2016). *Neural plasticity across the lifespan: How the brain can change*. London: Routledge.
Dennis, N. A., Kim, H. K., & Cabeza, R. (2007). Effects of aging on the neural correlates of true and false memory formation. *Neuropsychologia, 45*, 3157–3166.
DeYoung, C. G., Hirsh, J. B., Shane, M. S., Papademetris, X., Rajeevan, N., & Gray, J. R. (2010). Testing predictions from personality neuroscience: Brain structure and the Big Five. *Psychological Science, 21*, 820–828.
Diamond, A. (2013). Executive functions. *Annual Review of Psychology, 64*, 135–168.
Diener, E., & Chan, M. Y. (2011). Happy people live longer: Subjective well-being contributes to health and longevity. *Applied Psychology: Health and Well-Being, 3*, 1–43.
Diener, E., & Diener, M. (1995). Cross-cultural correlates of life satisfaction and self- esteem. *Journal of Personality and Social Psychology, 68*, 653–663.
Dimitroglou, E., Zafiropoulou, M., Messini-Nikolaki, N., Doudounakis, S., Tsilimigaki, S., & Piperakis, S. M. (2003). DNA damage in a human population affected by chronic psychogenic stress. *International Journal of Hygiene and Environmental Health, 206*, 39–44.
Dingemans, E., & Henkens, K. (2015). How do retirement dynamics influence mental well-being in later life? A 10-year panel study. *Scandinavian Journal of Work, Environment and Health, 41*, 16–23.
Dittmann-Kohli, F. (1990). The construction of meaning in old age: Possibilities and constraints. *Ageing and Society, 10*, 279–294.
Dittmann-Kohli, F., & Jopp, D. (2007). Self and life management: Wholesome knowledge for the third age. In J. Bond, S. Peace, F. Dittmann-Kohli, & G. Westerhof (Eds.), *Ageing in society* (pp. 268–295). London: Sage.
Dong, X., Chen, R., Chang, E.-S., & Simon, M. (2013). Elder abuse and psychological well-being: A systematic review and implications for research and policy—A mini review. *Gerontology, 59*, 132–142.
Dong, X. Q. (2015). Elder abuse: Systematic review and implications for practice. *Journal of the American Geriatrics Society, 63*, 1214–1238.
Donnellan, M. B., & Lucas, R. E. (2008). Age differences in the Big Five across the life span: Evidence from two national samples. *Psychology and Aging, 23*, 558–566.
Dorfman, L. T., Murty, S. A., Evans, R. J., Ingram, J. G., & Power, J. R. (2004). History and identity in the narratives of rural elders. *Journal of Aging Studies, 18*, 187–203.
Dubus, N. M. (2014). Self-perception of when old age begins for Cambodian elders living in the United States. *Journal of Cross-Cultural Gerontology, 29*, 185–199.
Duchek, J. M., Balota, D. A., Storandt, M., & Larsen, R. (2007). The power of personality in discriminating between healthy aging and early-stage Alzheimer's disease. *Journal of Gerontology B, 62*, P353–P361.
Dunn, J. H., & Koo, J. (2013). Psychological stress and skin aging: A review of possible mechanisms and potential therapies. *Dermatology Online Journal, 19*(6), 18561.
Duzel, E., van Praag, H., & Sendtner, M. (2016). Can physical exercise in old age improve memory and hippocampal function? *Brain, 139*, 662–673.
Earl, J. K., Gerrans, P., & Halim, V. A. (2015). Active and adjusted: Investigating the contribution of leisure, health and psychosocial factors to retirement adjustment. *Leisure Sciences, 37*, 354–372.
Ebaid, D., Crewther, S. G., MacCalman, K., Brown, A., & Crewther, D. P. (2017). Cognitive processing speed across the lifespan: Beyond the influence of motor speed. *Frontiers in Aging Neuroscience, 9*, 62. https://doi.org/10.3389/fnagi.2017.00062
Egan, M. F., Kojima, M., Callicott, J. H., Goldberg, T. E., Kolachana, B. S., Bertolino, A., et al. (2003). The BDNF val66met polymorphism affects activity-dependent secretion of BDNF and human memory and hippocampal function. *Cell, 112*, 257–269.
Ekerdt, D. J. (1986). The busy ethic: Moral continuity between work and retirement. *The Gerontologist, 26*, 239–244.

Ekerdt, D. J., Bosse, R., & Levkoff, S. (1985). An empirical test for phases of retirement: Findings from the Normative Aging Study. *Journal of Gerontology, 40*, 95–101.
Ekman, P. (1994). All emotions are basic. In P. Ekman & R. Davidson (Eds.), *The nature of emotion* (pp. 15–19). Oxford: Oxford University Press.
Ekman, P., & Friesen, W. (1975). *Unmasking the face*. Englewood Cliffs, NJ: Prentice Hall.
Elder, G.H. Jr. (1995). The Life Course Paradigm: Social change and individual development. In: E. Moen, G.H. Elder, Jr. and K.L. Uscher (Eds.), Examining lives in contexts: Perspectives on the ecology of human development. Pp. 101–139. American Psychological Association, Washington, DC.
Emirbayer, M., & Mische, A. (1998). What is agency? *American Journal of Sociology, 103*, 962–1023.
Epstein, S. (1973). The self-concept revisited or a theory of a theory. *American Psychologist, 28*, 404–416.
Erikson, E. H. (1959). *Identity and the life cycle*. New York: Norton.
Erikson, E. H. (1963). *Childhood and society* (2nd ed.). New York: Norton.
Erikson, E. H. (1968). *Identity: Youth and crisis*. New York: W.W. Norton and Company.
Erikson, E. H. (1980). *Identity and life-cycle*. New York: Norton.
Erlinghagen, M. (2010). Volunteering after retirement. *European Societies, 12*, 603–625.
Ertel, K. E., Glymour, M. M., & Berkman, L. F. (2008). Effects of social integration on preserving memory function in a nationally representative US elderly population. *American Journal of Public Health, 98*, 1215–1220.
Eve, R. A., & Eve, S. B. (1984). The effects of powerlessness, fear of social change, and social integration on fear of crime among the elderly. *Victimology, 2*, 290–295.
Fandakova, Y., Sander, M. C., Werkle-Bergner, M., & Shing, Y. L. (2014). Age differences in short-term memory binding are related to working memory performance across the lifespan. *Psychology and Aging, 29*, 140–149.
Feldman, D. C., & Beehr, T. A. (2011). A three-phase model of retirement decision making. *American Psychologist, 66*, 193–203.
Ferencz, B., Laukka, E. J., Welmer, A.-K., Kalpouzos, G., Angleman, S., Keller, L., et al. (2014). The benefits of staying active in old age: Physical activity counteracts the negative influence of *PICALM*, *BIN1*, and *CLU* risk alleles on episodic memory functioning. *Psychology and Aging, 29*, 440–449.
Fessler, D. M. T., & Navarrete, C. D. (2005). The effect of age on death disgust: Challenges to terror management perspectives. *Evolutionary Psychology, 3*, 279–296.
Field, T. (2016). Yoga research review. *Complementary Therapies in Clinical Practice, 24*, 145–161.
Finkel, D., Andel, R., Gatz, M., & Pedersen, N. L. (2009). The role of occupational complexity in trajectories of cognitive aging before and after retirement. *Psychology and Aging, 24*, 563–573.
Fischer, A., & Giner-Sorolla, R. (2016). Contempt: Derogating others while keeping calm. *Emotion Review, 8*, 346–357.
Fleischman, S. A., Wilson, R. S., Gabrieli, J. D. E., Bienias, J. L., & Bennett, D. A. (2004). A longitudinal study of implicit and explicit memory in old persons. *Psychology and Aging, 19*, 617–625.
Flöel, A., Witte, A. V., Lohmann, H., Wersching, H., Ringelstein, E. B., Berger, K., et al. (2008). Lifestyle and memory in the elderly. *Neuroepidemiology, 31*, 39–47.
Fors, S., Modin, B., Koupil, I., & Vågerö, D. (2012). Socioeconomic inequalities in circulatory and all-cause mortality after retirement: The impact of mid-life income and old-age pension. Evidence from the Uppsala Birth Cohort Study. *Journal of Epidemiology and Community Health, 66*, e16. https://doi.org/10.1136/jech.2010.131177
Frankland, P. W., & Bontempi, B. (2005). The organization of recent and remote memories. *Nature Reviews Neuroscience, 6*, 119–130.
Fredrickson, B. L. (2003). The value of positive emotions: The emerging science of positive psychology is coming to understand why it's good to feel good. *American Scientist, 91*, 330–335.

Friedman, H. S., & Martin, L. R. (2011). *The Longevity Project: Surprising discoveries for health and long-life from the landmark eight-decade study*. New York: Plume.
Friedman, H. S., Tucker, J. S., Tomlinson-Keasey, C., Schwartz, J. E., Wingard, D. L., & Criqui, M. H. (1993). Does childhood personality predict longevity? *Journal ot Personality and Social Psychology, 65*, 176–185.
Fussi, A. (2015). Aristotle on shame. *Ancient Philosophy, 35*, 113–135.
Gall, T. L., Evans, D. R., & Howard, J. (1997). The retirement adjustment process: Changes in the well-being of male retirees across time. *Journal of Gerontology, 52B*, P110–P117.
Gallo, J. J., Rabins, P. V., & Anthony, J. C. (1999). Sadness in older persons: 13-year follow- up of a community sample in Baltimore, Maryland. *Psychological Medicine, 29*, 341–350.
Gardner, I. L., Browning, C., & Kendig, H. (2005). Accommodation options in later life: Retirement village or community living? *Australasian Journal of Ageing, 24*, 188–195.
Gecas, V. (1982). The self-concept. *Annual Reviews of Sociology, 8*, 1–33.
Gee, S. (1999). Happily ever after? An exploration of retirement expectations. *Educational Gerontology, 25*, 109–128.
Gervais, M. M., & Fessler, D. M. T. (2017). On the deep structure of social affect: Attitudes, emotions, sentiments, and the case of "contempt". *Behavioral and Brain Sciences, e225*, 1–63. https://doi.org/10.1017/S0140525X16000352
Gianaros, P. J., Jennings, J. R., Sheu, L. K., Greer, P. J., Kuller, L. H., & Matthews, K. A. (2007). Prospective reports of chronic life stress predict decreased grey matter volume in the hippocampus. *NeuroImage, 35*, 795–803.
Giddens, A. (1991). *Modernity and self-identity: Self and society in the late modern age*. Stanford, CA: Stanford University Press.
Gilbert, P. (2003). Evolution, social roles, and the differences in shame and guilt. *Social Research, 70*, 1205–1230.
Gilleard, C. (2007). Old age in ancient Greece: Narratives of desire, narratives of disgust. *Journal of Aging Studies, 21*, 81–92.
Glymour, M. M., Kawachi, I., Jencks, C. S., & Berkman, L. F. (2008). Does childhood schooling affect old age memory or mental status? Using state schooling laws as natural experiments. *Journal of Epidemiol and Community Health, 62*, 532–537.
Goffman, E. (1963). *Stigma*. Englewood Cliffs, NJ: Prentice-Hall.
Goldberg, E. (2007). *The wisdom paradox: How your mind can grow stronger as your brain grows older*. London: Pocket Books.
Golden, G. (1982). *Coping with aging: Denial and avoidance in middleaged care-givers*. Unpublished doctoral dissertation, University of California, Berkeley, CA.
Gone, J. P., Miller, P. J., & Rappaport, J. (1999). Conceptual self as normatively oriented: The suitability of past personal narrative for the study of cultural identity. *Culture and Psychology, 5*, 371–398.
González Herero, V., & Extremera, N. (2010). Daily life activities as mediators of the relationship between personality variables and subjective well-being among older adults. *Personality and Individual Differences, 49*, 124–129.
Gouin, J.-P., Hantsoo, L., & Kiecolt-Glaser, J. K. (2008). Immune dysregulation and chronic stress among older adults: A review. *Neuroimmunomodulation, 15*, 251–259.
Gould, S. J., & Vrba, E. S. (1982). Exaptation - A missing term in the science of form. *Paleobiology, 8*, 4–15.
Grady, C. L., & Craik, F. I. M. (2000). Changes in memory processing with age. *Current Opinion in Neurobiology, 10*, 224–231.
Graf, P. (1990). Life-span changes in implicit and explicit memory. *Bulletin of the Psychonomic Society, 28*, 353–358.
Greenwood, P. M. (2007). Functional plasticity in cognitive aging: Review and hypothesis. *Neuropsychology, 21*, 657–673.
Greenwood, P. M., & Parasuraman, R. (2012). *Nurturing the older brain and mind*. Cambridge, MA: MIT Press.

Grossnickle, E. M. (2016). Disentangling curiosity: Dimensionality, definitions, and distinctions from interest in educational contexts. *Educational Psychology Review, 28*, 23–60.

Gruver, A. L., Hudson, L. L., & Sempowski, G. D. (2007). Immunosenescence of ageing. *Journal of Pathology, 211*, 144–156.

Hacker, P. (2017). Shame, embarrassment, and guilt. *Midwest Studies in Philosophy, 41*, 202–224.

Hadjistavropoulos, T., Martin, R. R., Sharpe, D., Lints, A. C., McCreary, D. R., & Asmundson, G. J. G. (2007). A longitudinal investigation of fear of falling, fear of pain, and activity avoidance in community-dwelling older adults. *Journal of Aging and Health, 19*, 965–984.

Hakuta, K., & Diaz, R. M. (1985). The relationship between degree of bilingualism and cognitive ability: A critical discussion and some new longitudinal data. In K. E. Nelson (Ed.), *Children's language* (Vol. 5, pp. 319–344). Hillsdale, NJ: Erlbaum.

Hammack, P. L. (2008). Narrative and the cultural psychology of identity. *Personality and Social Psychology Review, 12*, 222–247.

Han, S., & Northoff, G. (2009). Understanding the self: A cultural neuroscience approach. *Progress in Brain Research, 178*, 203–212.

Hänninen, T., Reinikainen, K. J., Helkala, E.-L., Koivisto, K., Mykkänen, L., Laakso, M., et al. (1994). Subjective memory complaints and personality traits in normal elderly subjects. *Journal of the American Geriatrics Society, 42*, 1–4.

Happé, F. G. E., Winner, E., & Bromwell, H. (1998). The getting of wisdom: Theory of mind in old age. *Developmental Psychology, 14*, 358–362.

Harris, J. A., Harris, I. M., & Diamond, M. E. (2001). The topography of tactile learning in humans. *Journal of Neuroscience, 21*, 1056–1061.

Hasan, K. M. M., Rahman, M. S., Arif, K. M. T., & Sobhani, M. E. (2012). Psychological stress and aging: Role of glucocorticoids (GCs). *Age, 34*, 1421–1433.

Haslam, N. (2006). Dehumanization: An integrative review. *Personality and Social Psychology Review, 10*, 252–264.

Hassing, L., Wahlin, Å., & Bäckman, L. (1998). Minimal influence of age, education, and gender on episodic memory functioning in very old age: A population-based study of nonagenarians. *Archives of Gerontology and Geriatrics, 27*, 75–87.

Haynes, S. G., McMichael, A. J., & Tyroler, H. A. (1978). Survival after early and normal retirement. *Journal of Gerontology, 33*, 269–278.

Hayslip Jr., B., & Kaminski, P. L. (2005). Grandparents raising their grandchildren: A review of the literature and suggestions for practice. *The Gerontologist, 45*, 262–269.

Heckhausen, J., Wrosch, C., & Schulz, R. (2010). A motivational theory of life-span development. *Psychological Review, 117, 32*(1). https://doi.org/10.1037/a0017668

Hedden, T., & Gabrieli, J. D. E. (2004). Insights into the ageing mind: A view from cognitive neuroscience. *Nature Reviews Neuroscience, 5*, 87–96.

Heilman, M. E., & Caleo, S. (2018). Gender discrimination in the workplace. In A. J. Colella & E. B. King (Eds.), *Oxford library of psychology. The Oxford handbook of workplace discrimination* (pp. 73–88). Oxford: Oxford University Press.

Helson, R., Kwan, V. S. Y., John, O. P., & Jones, C. (2002). The growing evidence for personality change in adulthood: Findings from research with personality inventories. *Journal of Research in Personality, 36*, 287–306.

Hemmingsson, T., Melin, B., Allebeck, P., & Lundberg, I. (2006). The association between cognitive ability measured at ages 18–20 and mortality during 30 years of follow-up-a prospective observational study among Swedish males born 1949-51. *International Journal of Epidemiology, 35*, 665–670.

Henckens, M. J. A. G., Hermans, E. J., Pu, Z., Joëls, M., & Fernández, G. (2009). Stressed memories: How acute stress affects memory formation in humans. *The Journal of Neuroscience, 29*, 10111–10119.

Henricksen, A., & Stephens, C. (2010). An exploration of the happiness-enhancing activities engaged in by older adults. *Ageing International, 35*, 311–326.

Herlitz, A., Nilsson, L. G., & Bäckman, L. (1997). Gender differences in episodic memory. *Memory and Cognition, 25*, 801–811.

Hess, T. M., Hinson, J. T., & Statham, J. A. (2004). Explicit and implicit stereotype activation effects on memory: Do age and awareness moderate the impact of priming? *Psychology and Aging, 19*, 495–505.
Hidalgo, V., Almela, M., Villada, C., & Salvador, A. (2014). Acute stress impairs recall after interference in older people, but not in young people. *Hormones and Behavior, 65*, 264–272.
Hitlin, S., & Elder Jr., G. H. (2007). Time, self, and the curiously abstract concept of agency. *Sociological Theory, 25*, 170–191.
Holcomb, T. F. (2010). Transitioning into retirement as a stressful life event. In T. W. Miller (Ed.), *Handbook of stressful transitions across the lifespan* (pp. 133–148). New York: Springer.
Holstein, M. B., & Minkler, M. (2003). Self, society, and the "new gerontology". *The Gerontologist, 43*, 787–796.
Hooper, J. O., Hooper, F. H., Colbert, K., & McMahan, R. (1986). Cognition, memory, and personality in elderly students. *Educational Gerontology, 12*, 219–229.
Horstmann, G. (2015). The surprise–attention link: A review. *Competitive Visual Processing Across Space and Time, 1339*, 106–115.
Hsu, H.-C., & Chang, W.-C. (2015). Social connections and happiness among the elder population of Taiwan. *Aging and Mental Health, 19*, 1131–1137.
Hultsch, D. F., Hertzog, C., Dixon, R. A., & Small, B. J. (1998). *Memory change in the aged.* Cambridge, UK: Cambridge University Press.
Hultsch, D. F., Hertzog, C., Small, B. J., & Dixon, R. A. (1999). Use it or lose it: Engaged lifestyle as a buffer of cognitive decline in aging? *Psychology and Aging, 14*, 245–263.
Intons-Peterson, M. J., Rocchi, P., West, T., McLellan, K., & Hackney, A. (1999). Age, testing at preferred or nonpreferred times (testing optimality), and false memory. *Journal of Experimental Psychology: Learning, Memory, and Cognition, 25*, 23–40.
Iwasa, H., Masui, Y., Gondo, Y., Inagaki, H., Kawaai, C., & Suzuki, T. (2008). Personality and all-cause mortality among older adults dwelling in a Japanese community: A five-year population-based prospective cohort study. *American Journal of Geriatric Psychiatry, 16*, 399–405.
Izard, C., & Buechler, S. (1980). Aspects of consciousness and personality in terms of differential emotions theory. In R. Plutchik & H. Kellerman (Eds.), *Emotion: Theory, research, and experience. Vol. 1: Theories of emotion* (pp. 165–187). New York: Academic Press.
Izard, C. E. (1992). Basic emotions, relations among emotions, and emotion-cognition relations. *Psychological Review, 99*, 561–565.
Izard, C. E. (2010). The many meanings/aspects of emotion: Definitions, functions, activation, and regulation. *Emotion Review, 2*, 363–370.
Izard, C. E., & Ackerman, B. P. (2000). Motivational, organizational, and regulatory functions of discrete emotions. In M. Lewis & J. H. Haviland-Jones (Eds.), *Handbook of emotions* (2nd ed., pp. 253–264). New York: Guilford.
James, W. (1890). *The principles of psychology*. New York: Dover.
Jang, Y., & Chiriboga, D. A. (2010). Living in a different world: Acculturative stress among Korean American elders. *Journal of Gerontology, 65B*, 14–21.
Jia, Z., Tian, W., Liu, W., Cao, Y., Yan, J., & Shun, Z. (2010). Are the elderly more vulnerable to psychological impact of natural disaster? A population-based survey of adult survivors of the 2008 Sichuan earthquake. *BMC Public Health, 10*, 172.
Johnson, F. (1985). The Western concept of self. In A. J. Marsella, G. DeVos, & E. L. K. Hsu (Eds.), *Culture and self: Asian and Western perspectives* (pp. 91–138). London: Tavistock.
Johansson, L., Guo, X., Waern, M., Östling, S., Gustafson, D., Bengtsson, C., et al. (2010). Midlife psychological stress and risk of dementia: A 35-year longitudinal population study. *Brain, 133*, 2217–2224.
Jose, J. P., & Cherayi, S. (2016). Age identity, social exclusion and wellbeing: Implications for elder inclusive societies. *Indian Journal of Gerontology, 30*, 143–167.
Jost, A. (1897). Die Assoziationsfestigkeit in ihrer Abhängigkeit von der Verteilung der Wiederholungen. *Zeitschrift für Psychologie und Physiologie der Sinnesorgane, 16*, 436–472.
Kabat-Zinn, J. (1990). *Full catastrophe living.* New York: Dell.

Kahana, E., Bhatta, T., Lovegreen, L. D., Kahana, B., & Midlarsky, E. (2013). Altruism, helping, and volunteering: Pathways to well-being in late life. *Journal of Aging and Health, 25*, 159–187.

Katzko, M. W., Steverink, N., Dittmann-Kohli, F., & Rubio Herrera, R. (1998). The self- concept of the elderly: A cross-cultural comparison. *International Journal of Aging and Human Development, 46*, 171–187.

Kaufman, G., & Elder Jr., G. H. (2002). Revisiting age identity: A research note. *Journal of Aging Studies, 16*, 169–176.

Kaufman, G., & Elder Jr., G. H. (2003). Grandparenting and age identity. *Journal of Aging Studies, 17*, 269–282.

Kavé, G., Eyal, N., Shorek, A., & Cohen-Mansfield, J. (2008). Multilingualism and cognitive state in the oldest old. *Psychology and Aging, 23*, 70–78.

Kawamoto, T., Ura, M., & Hiraki, K. (2017). Curious people are less affected by social rejection. *Personality and Individual Differences, 105*, 264–267.

Keltner, D., & Haidt, J. (1999). Social functions of emotions at four levels of analysis. *Cognition and Emotion, 13*, 505–521.

Khazaee-Pool, M., Sadeghi, R., Majlessi, F., & Rahimi Foroushani, A. (2015). Effects of physical exercise programme on happiness among older people. *Journal of Psychiatric and Mental Health Nursing, 22*, 47–57.

Kiecolt-Glaser, J. K., & Glaser, R. (2002). Depression and immune function: Central pathways to morbidity and mortality. *Journal of Psychosomatic Research, 53*, 873–876.

Killen, A., & MacAskill, A. (2015). Using a gratitude intervention to enhance wellbeing in older adults. *Journal of Happiness Studies, 16*, 947–964.

Kim, J. E., & Moen, P. (2001). Is retirement good or bad for subjective well-being? *Current Directions in Psychological Science, 10*, 83–86.

Kim, J. E., & Moen, P. (2002). Retirement transitions, gender, and psychological well- being: A life-course, ecological model. *Journal of Gerontology, 57B*, P212–P222.

Kim, S., & Feldman, D. C. (2000). Working in retirement: The antecedents of bridge employment and its consequences for quality of life in retirement. *Academy of Management Journal, 43*, 1195–1210.

Kimura, D. (2002). Sex hormones influence human cognitive pattern. *Neuroendocrinology Letters, 23*(Suppl. 4), 67–77.

Kirkpatrick Johnson, M., Berg, J. A., & Sirotzki, T. (2007). Differentiation in self- perceived adulthood: Extending the confluence model of subjective age identity. *Social Psychology Quarterly, 70*, 243–261.

Kirschbaum, C., Wolf, O. T., May, M., Wippich, W., & Hellhammer, D. H. (1996). Stress- and treatment-induced elevations of cortisol levels associated with impaired declarative memory in healthy adults. *Life Sciences, 58*, 1475–1483.

Klemmack, D. L., & Lee Roff, L. (1984). Fear of personal aging and subjective well- being in later life. *Journal of Gerontology, 39*, 756–758.

Klimova, B., Valis, M., & Kuca, K. (2017). Cognitive decline in normal aging and its prevention: A review on non-pharmacological lifestyle strategies. *Clinical Interventions in Aging, 12*, 903–910.

Kobasa, S. C. (1982). The hardy personality: Toward a social psychology of stress and health. In G. S. Sanders & J. Suls (Eds.), *Social psychology of health and illness* (pp. 3–32). Hillsdale, NJ: Lawrence Erlbaum Associates.

Koenig, H. G. (1994). *Aging and God: Spiritual pathways to mental health in midlife and later years*. New York: Haworth.

Koenig, H. G., Siegler, I. C., Meador, K. G., & George, L. K. (1990). Religious coping and personality in later life. *International Journal of Geriatric Psychiatry, 5*, 123–131.

Kramer, J. H., Yaffe, K., Lengenfelder, J., & Delis, D. C. (2003). Age and gender interactions on verbal memory performance. *Journal of the International Neuropsychological Society, 9*, 97–102.

Krampe, R. T., & McInnes, L. (2007). Competence and cognition. In J. Bond, S. Peace, F. Dittmann-Kohli, & G. Westerhof (Eds.), *Ageing in society* (pp. 255–267). London: Sage.

Krassen-Maxwell, E., & Maxwell, R. J. (1980). Contempt for the elderly: A cross-cultural analysis. *Current Anthropology, 21*, 569–570.

Krause, N., & Liang, J. (1993). Stress, social support, and psychological distress among the Chinese elderly. *Journal of Gerontology, 48*, P282–P291.

Krause, N., Liang, J., & Keith, V. (1990). Personality, social support, and psychological distress in later life. *Psychology and Aging, 5*, 315–326.

Krueger, K. R., Wilson, R. S., Shah, R. C., Tang, Y., & Bennett, D. A. (2006). Personality and incident disability in older persons. *Age and Ageing, 35*, 428–433.

Kuhlen, R. G. (1945). Age differences in personality during adult years. *Psychological Bulletin, 42*, 333–358.

Kulik, L. (1999). Continuity and discontinuity in marital life after retirement: Life orientations, gender role ideology, intimacy, and satisfaction. *Families in Society, 80*, 286–294.

Kunzmann, U., & Grühn, D. (2005). Age differences in emotional reactivity: The sample case of sadness. *Psychology and Aging, 20*, 47–59.

Kunzmann, U., Kappes, C., & Wrosch, C. (2014). Emotional aging: A discrete emotions perspective. *Frontiers in Psychology, 5*, 380.

La Voie, D., & Light, L. L. (1994). Adult age differences in repetition priming: A meta- analysis. *Psychology and Aging, 9*, 539–553.

Labouvie-Vief, G. (2003). Dynamic integration: Affect, cognition, and the self in adulthood. *Current Directions in Psychological Science, 12*, 201–206.

Lachman, M. E., Agrigoroaei, S., Murphy, C., & Tun, P. A. (2010). Frequent cognitive activity compensates for education differences in episodic memory. *American Journal of Geriatric Psychiatry, 18*, 4–10.

Launer, L. J., Masaki, K., Petrovitch, H., Foley, D., & Havlik, R. J. (1995). The association between midlife blood pressure levels and late-life cognitive function: The Honolulu- Asia aging study. *JAMA, 274*, 1846–1851.

Lavery, L. L., Dodge, H. H., Snitz, B., & Ganguli, M. (2009). Cognitive decline and mortality in a community-based cohort: The Movies Project. *Journal of the American Geriatrics Society, 57*, 94–100.

Lazarus, R. S. (1991). *Emotion and adaptation*. New York: Oxford University Press.

Lazarus, R. S. (1999). *Stress and emotion: A new synthesis*. New York: Springer.

Lazarus, R. S., & DeLongis, A. (1983). Psychological stress and coping in aging. *American Psychologist, 38*, 245–254.

Lazenbatt, A., & Devaney, J. (2014). Older women living with domestic violence: Coping resources and mental health and wellbeing. *Current Nursing Journal, 1*, 10–22.

Le, T. N. (2008). Age differences in spirituality, mystical experiences and wisdom. *Ageing and Society, 28*, 383–411.

Lee, F., Mackenzie, L., & James, C. (2008). Perceptions of older people living in the community about their fear of falling. *Disability and Rehabilitation, 30*, 1803–1811.

Lee, Y. Y., Chan, M. F., & Mok, E. (2010). Effectiveness of music intervention on the quality of life of older people. *Journal of Advanced Nursing, 66*, 2677–2687.

Leopold, T., & Skopek, J. (2015). Convergence or continuity? The gender gap in household labor after retirement. *Journal of Marriage and Family, 77*, 819–832.

Levenson, R. W., Carstensen, L. L., Friesen, W. V., & Ekman, P. (1991). Emotion, physiology, and expression in old age. *Psychology and Aging, 6*, 28–35.

Levi, U., & Rosenstreich, E. (2019). Mindfulness and memory: A review of findings and a potential model. *Journal of Cognitive Enhancement, 3*, 302–314.

Levine, B., Svoboda, E., Hay, J. F., Winocur, G., & Moscovitch, M. (2002). Aging and autobiographical memory: Dissociating episodic from semantic retrieval. *Psychology and Aging, 17*, 677–689.

Levy, B. (1996). Improving memory in old age through implicit self-stereotyping. *Journal of Personality and Social Psychology, 71*, 1092–1107.

Levy, B. (2009). Stereotype embodiment: A psychosocial approach to aging. *Current Directions in Psychological Sciences., 18*, 332–336.

Levy, B. R. (2003). Mind matters: Cognitive and physical effects of aging self-stereotypes. *Journal of Gerontology, 58B*, P203–P211.

Levy, B. R., Hausdorff, J. M., Hencke, R., & Wei, J. Y. (2000). Reducing cardiovascular stress with positive self-stereotypes of aging. *Journal of Gerontology: Psychological Sciences, 55*, 205–213.

Levy, B. R., Moffat, S., Resnick, S. M., Slade, M. D., & Ferrucci, L. (2016). Buffer against cumulative stress: Positive age self-stereotypes predict lower cortisol across 30 years. *GeroPsych: The Journal of Gerontopsychology and Geriatric Psychiatry, 29*, 141–146.

Levy, B. R., Slade, M. D., Chung, P. H., & Gill, T. M. (2015). Resiliency over time of elders' age stereotypes after encountering stressful events. *Journals of Gerontology B, 70*, 886–890.

Levy, B. R., Zonderman, A. B., Slade, M. D., & Ferrucci, L. (2011). Memory shaped by age stereotypes over time. *The Journals of Gerontology, Series B: Psychological Sciences and Social Sciences, 67*, 432–436.

Lewis-Peacock, J. A., Drysdale, A. T., Oberauer, K., & Postle, B. R. (2012). Neural evidence for a distinction between short-term memory and the focus of attention. *Journal of Cognitive Neuroscience, 24*, 61–79.

Li, S.-C., Lindenberger, U., Hommel, B., Aschersleben, G., Prinz, W., & Baltes, P. B. (2004). Transformations in the couplings among intellectual abilities and constituent cognitive processes across the life span. *Psychological Science, 15*, 155–163.

Li, S.-C., Schmiedek, F., Huxhold, O., Röcke, C., Smith, J., & Lindenberger, U. (2008). Working memory plasticity in old age: Practice gain, transfer, and maintenance. *Psychology and Aging, 23*, 731–742.

Lindenberger, U., & Ghisletta, P. (2009). Cognitive and sensory declines in old age: Gauging the evidence for a common cause. *Psychology and Aging, 24*, 1–16.

Little, T. D., Hawley, P. H., Henrich, C. C., & Marsland, K. W. (2002). Three views of the agentic self: A developmental synthesis. In E. L. Deci & R. M. Ryan (Eds.), *Handbook of self-determination research*. New York: University of Rochester Press.

Litwin, H. (2007). Does early retirement lead to longer life? *Ageing and Society, 27*, 739–754.

Löckenhoff, C. E., De Fruyt, F., Terracciano, A., McCrae, R. R., De Bolle, M., Costa, P. T., Jr., et al. (2009a). Perceptions of aging across 26 cultures and their culture-level associates. *Psychology and Aging, 24*, 941–954.

Locker, D., Liddell, A., & Burman, D. (1991). Dental fear and anxiety in and older adult population. *Community Dentistry and Oral Epidemiology, 19*, 120–124.

Logan, J. R., Ward, R., & Spitze, G. (1992). As old as you feel: Age identity in middle and later life. *Social Forces, 71*, 451–467.

London, P., Schulman, R. E., & Black, M. S. (1964). Religion, guilt, and ethical standards. *The Journal of Social Psychology, 63*, 145–159.

López Ulloa, B. F., Møller, V., & Sousa-Poza, A. (2013). *How does subjective well-being evolve with age? A literature review* (FZID Discussion Paper, No. 72-2013). Universität Hohenheim, Forschungszentrum Innovation und Dienstleistung (FZID), Stuttgart, http://nbn-resolving.de/urn:nbn:de:bsz:100-opus-8412

Loretto, W., & Vickerstaff, S. (2012). The domestic and gendered context for retirement. *Human Relations, 66*, 65–86.

Lövdén, M., Ghisletta, P., & Lindenberger, U. (2005). Social participation attenuates decline in perceptual speed in old and very old age. *Psychology and Aging, 20*, 423–434.

Lövdén, M., Rönnlund, M., Wahlin, Å., Bäckman, L., Nyberg, L., & Nilsson, L.-G. (2004). The extent of stability and change in episodic and semantic memory in old age: Demographic predictors of level and change. *Journal of Gerontology, 59B*, P130–P134.

Lowenstein, G. (1994). The psychology of curiosity: A review and reinterpretation. *Psychological Bulletin, 116*, 75–98.

Luciano, M., Wright, M. J., Smith, G. A., Geffen, G. M., Geffen, L. B., & Martin, N. G. (2001). Genetic covariance among measures of information processing speed, working memory, and IQ. *Behavior Genetics, 31*, 581–592.

Luo, L., & Craik, F. I. M. (2008). Aging and memory: A cognitive approach. *Canadian Journal of Psychiatry, 53*, 346–353.
Lupien, S. J., Gaudreau, S., Tchiteya, B. M., Maheu, F., Sharma, S., Nair, N. P. V., et al. (1997). Stress-induced declarative memory impairment in healthy elderly subjects: Relationship to cortisol reactivity. *Journal of Clinical Endocrinology and Metabolism, 82*, 2070–2075.
Lupien, S. J., McEwen, B. S., Gunnar, M. R., & Heim, C. (2009). Effects of stress throughout the lifespan on the brain, behavior and cognition. *Nature Reviews Neuroscience, 10*, 434–445.
Maestas, N. (2010). Back to work: Expectations and realizations of work after retirement. *Journal of Human Resources, 45*, 718–748.
Magai, C., Hunziker, J., Mesias, W., & Culver, L. C. (2000). Adult attachment styles and emotional biases. *International Journal of Behavioral Development, 24*, 301–309.
Maier, H., & Smith, J. (1999). Psychological predictors of mortality in old age. *The Journals of Gerontology, 54B*, P44–P54.
Makizako, H., Liu-Ambrose, T., Shimada, H., Doi, T., Park, H., Tsutsumimoto, K., et al. (2015). Moderate-intensity physical activity, hippocampal volume, and memory in older adults with mild cognitive impairment. *The Journals of Gerontology: Series A, 70*, 480–486.
Malatesta, C. Z., & Kalnok, M. (1984). Emotional experience in younger and older adults. *Journal of Gerontology, 39*, 301–308.
Marcoen, A., Coleman, P. G., & O'Hanlon, A. (2007). Psychological ageing. In J. Bond, S. Peace, F. Dittmann-Kohli, & G. Westerhof (Eds.), *Ageing in society* (pp. 38–67). London: Sage.
Marmot, M. G., & Shipley, M. J. (1996). Do socioeconomic differences in mortality persist after retirement? 25 Year follow up of civil servants from the first Whitehall study. *BMJ, 313*, 1177–1180.
Marsh, H. W., Nagengast, B., & Morin, A. J. S. (2013). Measurement invariance of big-five factors over the life span: ESEM tests of gender, age, plasticity, maturity, and la dolce vita effects. *Developmental Psychology, 49*, 1194–1218.
Martin, M., Grünendahl, M., & Martin, P. (2001). Age differences in stress, social resources, and well-being in middle and older age. *Journal of Gerontology, 56B*, P214–P222.
Mather, M., & Carstensen, L. L. (2005). Aging and motivated cognition: The positivity effect in attention and memory. *Trends in Cognitive Sciences, 9*, 496–502.
Mathieu, S. I. (2008). Happiness and humor group promotes life satisfaction for senior center participants. *Activities, Adaptation and Aging, 32*, 134–148.
Matsumoto, D., & Ekman, P. (2004). The relationship among expressions, labels, and descriptions of contempt. *Journal of Personality and Social Psychology, 87*, 529–540.
McAdams, D. P. (2006). The redemptive self: Generativity and the stories Americans live by. *Research in Human Development, 3*, 81–100.
McAdams, D. P., & Olson, B. D. (2010). Personality development: Continuity and change over the life course. *Annual Review of Psychology, 61*, 517–542.
McCrae, R. R., Costa, P. T., Jr., Pedroso de Lima, M., Simões, A., Ostendorf, F., Angleitner, A., et al. (1999). Age differences in personality across the adult lifespan: Parallels in five cultures. *Developmental Psychology, 35*, 466–477.
McCrae, R. R., Terracciano, A., & 78 Members of the Personality Profiles of Cultures Project. (2005). Universal features of personality traits from the observer's perspective: Data from 50 cultures. *Journal of Personality and Social Psychology, 88*, 547–561.
McEwen, B. S. (2002). Sex, stress and the hippocampus: Allostasis, allostatic load and the aging process. *Neurobiology of Aging, 23*, 921–939.
McEwen, B. S., & Gianaros, P. J. (2011). Stress- and allostasis-induced brain plasticity. *Annual Review of Medicine, 62*, 431–445.
McFadden, S. H. (1999). Religion, personality, and aging: A life span perspective. *Journal of Personality, 67*, 1081–1104.
McGillivray, S., Murayama, K., & Caste, A. D. (2015). Thirst for knowledge: The effects of curiosity and interest on memory in younger and older adults. *Psychology and Aging, 30*, 835–841.
McKenzie, H., & Crouch, M. (2004). Discordant feelings in the lifeworld of cancer survivors. *Health, 8*, 139–157.

McLean, K. C. (2008). Stories of the young and the old: Personal continuity and narrative identity. *Developmental Psychology, 44*, 254–264.
McMullen, C. K., & Luborsky, M. R. (2006). Self-rated health appraisal as cultural and identity process: African American elders' health and evaluative rationales. *The Gerontologist, 46*, 431–438.
Mead, G. H. (1934). *Mind, self, and society*. Chicago: University of Chicago Press.
Meeks, S., Van Haitsma, K., Kostiwa, I., & Murrell, S. A. (2012). Positivity and well- being among community-residing elders and nursing home residents: What is the optimal affect balance? *The Journals of Gerontology B, 67*, 460–467.
Meier, B., Perrig-Chiello, P., & Perrig, W. (2002). Personality and memory in old age. *Aging, Neuropsychology and Cognition, 9*, 135–144.
Melia, S. P. (1999). Continuity in the lives of elder Catholic women religious. *International Journal of Aging and Human Development, 48*, 175–189.
Mendes, W. B. (2010). Weakened links between mind and body in older age: The case for maturational dualism in the experience of emotion. *Emotion Review, 2*, 240–244.
Merton, R. K. (1957). *Social theory and social structure*. Glencoe, IL: Free Press.
Meyer, W.-U., Reisenzein, R., & Schützwohl, A. (1997). Toward a process analysis of emotions: The case of surprise. *Motivation and Emotion, 21*, 251–274.
Michinov, E., Fouquereau, E., & Fernandez, A. (2008). Retirees' social identity and satisfaction with retirement. *International Journal of Aging and Human Development, 66*, 175–194.
Midanik, L. T., Soghikian, K., Ransom, L. J., & Tekawa, I. S. (1995). The effect of retirement on mental health and health behaviors: The Kaiser Permanente Retirement Study. *Journal of Gerontology, 50B*, S59–S61.
Middleton, L. E., Barnes, D. E., Lui, L.-Y., & Yaffe, K. (2010). Physical activity over the life course and its association with cognitive performance and impairment in old age. *Journal of the American Geriatrics Society, 58*, 1322–1326.
Miethe, T. D., & Lee, G. R. (1984). Fear of crime among older people: A reassessment of the predictive power of crime-related factors. *The Sociological Quarterly, 25*, 397–415.
Miller, D. I., & Halpern, D. F. (2014). The new science of cognitive sex differences. *Trends in Cognitive Sciences, 18*, 37–45.
Miller, G. E., Chen, E., & Parker, K. J. (2011). Psychological stress in childhood and susceptibility to the chronic diseases of aging: Moving towards a model of behavioral and biological mechanisms. *Psychological Bulletin, 137*, 959–997.
Mischel, W. (2004). Toward an integrative science of the person. *Annual Review of Psychology, 55*, 1–22.
Moen, P. (1996). A life course perspective on retirement, gender, and well-being. *Journal of Occupational Health Psychology, 1*, 131–144.
Moen, P., Kim, J. E., & Hofmeister, H. (2001). Couples' work/retirement transitions, gender, and marital quality. *Social Psychology Quarterly, 64*, 55–71.
Mondini, S., & Semenza, C. (2006). How Berlusconi keeps his face: A neuropsychological study in a case of semantic dementia. *Cortex, 42*, 332–335.
Montagu, A. (1989). *Growing young*. Massachusetts: Bergin and Garvey Publishers, Inc..
Moor, C., Zimprich, D., Schmitt, M., & Kliegel, M. (2006). Personality, aging self- perceptions, and subjective health: A mediation model. *International Journal of Aging and Human Development, 63*, 241–257.
Mordekar, A., & Spence, S. A. (2008). Personality disorder in older people: How common is it and what can be done? *Advances in Psychiatric Treatment, 14*, 71–77.
Moss, A. S., Reibel, D. K., Greeson, J. M., Thapar, A., Bubb, R., Salmon, J., et al. (2015). An adapted mindfulness-based stress reduction program for elders in a continuing care retirement community: Quantitative and qualitative results from a pilot randomized controlled trial. *Journal of Applied Gerontology, 34*, 518–538.
Mroczek, D. K., & Spiro III, A. (2007). Personality change influences mortality in older men. *Psychological Science 18*: 371–376.

Müller, G. E., & Pilzecker, A. (1900). Experimentelle Bieträge zur Lehre vom Gedächtnis. *Zeitschrift für Psychologie Ergänzungsband, 1*, 1–300.

Mulligan, K., & Scherer, K. R. (2012). Toward a working definition of emotion. *Emotion Review, 4*, 345–357.

Murphy, D. R., Craik, F. I. M., Li, K. Z. H., & Schneider, B. A. (2000). Comparing the effects of aging and background noise on short-term memory performance. *Psychology and Aging, 15*, 323–334.

Murty, V. P., Sambataro, F., Das, S., & Tan, H.-Y. (2009). Age-related alterations in simple declarative memory and the effect of negative stimulus valence. *Journal of Cognitive Neuroscience, 21*, 1920–1933.

Nagel, I. E., Chicherio, C., Li, S.-C., von Oertzen, T., Sander, T., Villringer, A., et al. (2008). Human aging magnifies genetic effects on executive functioning and working memory. *Frontiers in Human Neuroscience, 2*, 1. https://doi.org/10.3389/neuro.09.001.2008

Nelson, K., & Fivush, R. (2004). The emergence of autobiographical memory: A social cultural developmental theory. *Psychological Review, 111*, 486–511.

Nelson, T. D. (2005). Ageism: Prejudice against our feared future self. *Journal of Social Issues, 61*, 207–221.

Nesselroade, J. R. (1992). Adult personality development: Issues in assessing constancy and change. In R. A. Zucker, A. I. Rabin, J. Aronoff, & S. J. Frank (Eds.), *Personality structure in the life course* (pp. 221–275). New York: Springer.

Newcomer, J. W., Selke, G., Melson, A. K., Hershey, T., Craft, S., Richards, K., et al. (1999). Decreased memory performance in healthy humans induced by stress-level cortisol treatment. *Archives of General Psychiatry, 56*, 527–533.

Nilsson, L.-G., Nyberg, L., & Bäckman, L. (2002). Genetic variation in memory functioning. *Neuroscience and Biobehavioral Reviews, 26*, 841–848.

Nilsson, L.-G., Sikström, C., Adolfsson, R., Erngrund, K., Nylander, P. O., & Beckman, L. (1996). Genetic markers associated with high versus low performance on episodic memory tasks. *Behavior Genetics, 26*, 555–562.

Nimrod, G. (2007). Expanding, reducing, concentrating and diffusing: Post retirement leisure behavior and life satisfaction. *Leisure Sciences, 29*, 91–111.

Nussbaum, M. C., & Levmore, S. (2017). *Aging thoughfully: Conversations about retirement, romance, wrinkles, & regret*. New York: Oxford University Press.

Nuttman-Shwartz, O. (2004). Like a high wave: Adjustment to retirement. *The Gerontologist, 44*, 229–236.

Ojha, H., & Pramanick, M. (2010). Do personality characteristics change with advancement of age? *Journal of the Indian Academy of Applied Psychology, 36*, 55–68.

Oksanen, T., Vahtera, J., Westerlund, H., Pentti, J., Sjösten, N., Virtanen, M., et al. (2011). Is retirement beneficial for mental health? Antidepressant use before and after retirement. *Epidemiology, 22*, 553–559.

Olatunji, B. O., & Sawchuk, C. N. (2005). Disgust: Characteristic features, social manifestations, and clinical implications. *Journal of Social and Clinical Psychology, 24*, 932–962.

Old, S. R., & Naveh-Benjamin, M. (2008). Differential effects of age on item and associative measures of memory: A meta-analysis. *Psychology and Aging, 23*, 104–118.

Orth, U., Robins, R. W., & Soto, C. J. (2010). Tracking the trajectory of shame, guilt, and pride across the life span. *Journal of Personality and Social Psychology, 99*, 1061–1071.

Pallesen, K. J., Brattico, E., Bailey, C. J., Korvenoja, A., Koivisto, J., Gjedde, A., et al. (2010). Cognitive control in auditory working memory is enhanced in musicians. *PLoS One, 5*, e11120.

Pansky, A., Goldsmith, M., Koriat, A., & Pearlman-Avnion, S. (2009). Memory accuracy in old age: Cognitive, metacognitive, and neurocognitive determinants. *European Journal of Cognitive Psychology, 21*, 303–329.

Park, D. C., & Reuter-Lorenz, P. (2009). The adaptive brain: Aging and neurocognitive scaffolding. *Annual Review of Psychology, 60*, 173–196.

Parker, V., Andel, R., Nilsen, C., & Kåreholt, I. (2013). The association between mid-life socioeconomic position and health after retirement – Exploring the role of working conditions. *Journal of Aging and Health, 25*, 863–881.

Parry, J., & Taylor, R. F. (2007). Orientation, opportunity and autonomy: Why people work after state pension age in three areas of England. *Ageing and Society, 27*, 579–598.
Pascual-Leone, A., Amedi, A., Fregni, F., & Merabet, L. B. (2005). The plastic human brain cortex. *Annual Review of Neuroscience, 28*, 377–401.
Pasupathi, M., & Carstensen, L. L. (2003). Age and emotional experience during mutual reminiscing. *Psychology and Aging, 18*, 430–442.
Patel, V., & Hope, T. (1993). Aggressive behaviour in elderly people with dementia: A review. *International Journal of Geriatric Psychiatry, 8*, 457–472.
Paul, C., Ayis, S., & Ebrahim, S. (2006). Psychological distress, loneliness and disability in old age. *Psychology, Health and Medicine, 11*, 221–232.
Paul, E. S., & Mendl, M. T. (2018). Animal emotion: Descriptive and prescriptive definitions and their implications for a comparative perspective. *Applied Animal Behaviour Science, 205*, 202–209.
Pauls, F., Petermann, F., & Lepach, A. C. (2013). Gender differences in episodic memory and visual working memory including the effects of age. *Memory, 21*, 857–874.
Perrig, W. J., Perrig, P., & Stähelin, H. B. (1997). The relation between antioxidants and memory performance in the old and very old. *Journal of the American Geriatrics Society, 45*, 718–724.
Phillipson, C., & Biggs, S. (1998). Modernity and identity: Themes and perspectives in the study of older adults. *Journal of Aging and Identity, 3*, 11–23.
Pillemer, D., Wink, P., DiDonato, T., & Sanborn, R. (2003). Gender differences in autobiographical memory styles of older adults. *Memory, 11*, 525–532.
Pillemer, K., & Suitor, J. J. (1991). "Will I ever escape my child's problems?" Effects of adult children's problems on elderly parents. *Journal of Marriage and the Family, 53*, 585–594.
Pinquart, M., & Schindler, I. (2007). Changes of life satisfaction in the transition to retirement: A latent-class approach. *Psychology and Aging, 22*, 442–455.
Pinquart, M., & Sörensen, S. (2001). Influences on loneliness in older adults: A meta- analysis. *Basic and Applied Social Psychology, 23*, 245–266.
Piolino, P., Desgranges, B., Benali, K., & Eustache, F. (2002). Episodic and semantic remote autobiographical memory in ageing. *Memory, 10*, 239–257.
Post, S. G. (2005). Altruism, happiness, and health: It's good to be good. *International Journal of Behavioral Medicine, 12*, 66–77.
Préville, M., Hébert, R., Bravo, G., & Boyer, R. (2002). Predisposing and facilitating factors of severe psychological distress among frail elderly adults. *Canadian Journal of Aging, 21*, 195–204.
Pugh, C. R., Fleshner, M., Watkins, L. R., Maier, S. F., & Rudy, J. W. (2001). The immune system and memory consolidation: A role for the cytokine IL-1β. *Neuroscience and Biobehavioral Reviews, 25*, 29–41.
Pushkar, D., Chaikelson, J., Conway, M., Etezadi, J., Giannopoulus, C., Li, K., et al. (2010). Testing continuity and activity variables as predictors of positive and negative affect in retirement. *Journal of Gerontology, 65B*, 42–49.
Quick, H. E., & Moen, P. (1998). Gender, employment, and retirement quality: A life course approach to the differential experiences of men and women. *Journal of Occupational Health Psychology, 3*, 44–64.
Ragheb, M. G., & Griffith, C. A. (1982). The contribution of leisure participation and leisure satisfaction to life satisfaction of older persons. *Journal of Leisure Research, 14*, 295–306.
Ranganath, C., & Blumenfeld, R. S. (2005). Doubts about double dissociations between short- and long-term memory. *Trends in Cognitive Sciences, 9*, 374–380.
Reed, J. M., & Squire, L. R. (1998). Retrograde amnesia for facts and events: Findings from four new cases. *The Journal of Neuroscience, 18*, 3943–3954.
Reis, M., & Pushkar Gold, D. (1993). Retirement, personality, and life satisfaction: A review and two models. *Journal of Applied Gerontology, 12*, 261–282.
Rendell, P. G., Castel, A. D., & Craik, F. I. M. (2005). Memory for proper names in old age: A disproportionate impairment? *The Quarterly Journal of Experimental Psychology, 58A*, 54–71.

Reyes-Ortiz, C. A., Ayele, H., Mulligan, T., Espino, D. V., Berges, I. M., & Markides, K. S. (2006). Higher church attendance predicts lower fear of falling in older Mexican-Americans. *Aging and Mental Health, 10*, 13–18.

Ribot, T.-A. (1881). *Les maladies de la memoire*. Paris: Editions L'Harmattan.

Richardson, V., & Kilty, K. (1991). Adjustment to retirement: Continuity versus discontinuity. *International Journal of Aging and Human Development, 32*, 151–169.

Roberts, B. L., Dunkle, R., & Haug, M. (1994). Physical, psychological, and social resources as moderators of the relationship of stress to mental health of the very old. *Journal of Gerontology, 49*, S35–S43.

Roberts, B. W., & DelVecchio, W. F. (2000). The rank-order consistency of personality traits from childhood to old age: A quantitative review of longitudinal studies. *Psychological Bulletin, 126*, 3–25.

Roberts, B. W., & Mroczek, D. (2008). Personality trait change in adulthood. *Current Directions in Psychological Science, 17*, 31–35.

Roberts, B. W., Walton, K. E., & Viechtbauer, W. (2006). Patterns of mean-level change in personality traits across the life course: A meta-analysis of longitudinal studies. *Psychological Bulletin, 132*, 1–25.

Roberts, B. W., Wood, D., & Caspi, A. (2008). The development of personality traits in adulthood. In O. P. John, R. W. Robins, & L. A. Pervin (Eds.), *Handbook of personality: Theory and research* (pp. 375–398). New York: Guilford Press.

Robins Wahlin, T. B., & Byrne, G. J. (2011). Personality changes in Alzheimer's disease: A systematic review. *International Journal of Geriatric Psychiatry, 26*, 1019–1029.

Rohan, J., Newman, S., & El Alaoui, J. (2014). Cultural reflections on shame and guilt: An intergenerational dialogue. *Journal of Intergenerational Relationships, 12*, 448–453.

Rohwedder, S., & Willis, R. J. (2010). Mental retirement. *Journal of Economic Perspectives, 24*, 119–138.

Rolls, E. T. (2010). Attractor networks. *WIREs Cognitive Science, 1*, 119–134.

Rolls, E. T. (2020). Neural computations underlying phenomenal consciousness: A higher order syntactic thought theory. *Frontiers in Psychology, 11*, 655. https://doi.org/10.3389/fpsyg.2020.00655

Rönnlund, M., Nyberg, L., Bäckman, L., & Nilsson, L.-G. (2005). Stability, growth, and decline in adult life span development of declarative memory: Cross-sectional and longitudinal data from a population-based study. *Psychology and Aging, 20*, 3–18.

Rosenberg, M. (1979). *Conceiving the self*. New York: Basic Books.

Rosenberg, S. (1997). Multiplicity of selves. In D. Ashmore & L. Jussim (Eds.), *Self and identity: Fundamental issues* (pp. 23–45). New York: Oxford University Press.

Rosenkoetter, M. M., & Garris, J. M. (1998). Psychosocial changes following retirement. *Journal of Advanced Nursing, 27*, 966–976.

Rosenkoetter, M. M., & Garris, J. M. (2001). Retirement planning, use of time, and psychosocial adjustment. *Issues in Mental Health Nursing, 22*, 703–722.

Ross, C. E., & Mirowsky, J. (2008). Age and the balance of emotions. *Social Science and Medicine, 66*, 2391–2400.

Rottman, J. (2014). Evolution, development, and the emergence of disgust. *Evolutionary Psychology, 12*, 417–433.

Rowles, G. D. (1983). Place and personal identity in old age: Observations from Appalachia. *Journal of Environmental Psychology, 3*, 299–313.

Ruffman, T., Murray, J., Halberstadt, J., & Taumoepeau, M. (2010). Verbosity and emotion recognition in older adults. *Psychology and Aging, 25*, 492–497.

Rybarczyk, B. D., & Auerbach, S. M. (1990). Reminiscence interviews as stress management interventions for older patients undergoing surgery. *The Gerontologist, 30*, 522–528.

Sakaki, M., Yagi, A., & Murayama, K. (2018). Curiosity in old age: A possible key to achieving adaptive aging. *Neuroscience and Biobehavioral Reviews, 88*, 106–116.

Salovey, P., Rothman, A. J., Detweiler, J. B., & Steward, W. T. (2000). Emotional states and physical health. *American Psychologist, 55*, 110–121.

Salthouse, T. A. (2006). Mental exercise and mental aging: Evaluating the validity of the "use it or lose it" hypothesis. *Perspectives on Psychological Science, 1*, 68–87.
Salthouse, T. A., Babcock, R. L., Skovronek, E., Mitchell, D. R. D., & Palmon, R. (1990). Age and experience effects in spatial visualization. *Developmental Psychology, 26*, 128–136.
Satz, P. (1993). Brain reserve capacity on symptom onset after brain injury: A formulation and review of evidence for threshold theory. *Neuropsychology, 3*, 273–295.
Scarmeas, N., & Stern, Y. (2003). Cognitive reserve and lifestyle. *Journal of Clinical and Experimental Neuropsychology, 25*, 625–633.
Schafer, M. H., & Shippee, T. P. (2010). Age identity, gender, and perceptions of decline: Does feeling older lead to pessimistic dispositions about cognitive aging? *Journal of Gerontology: Social Sciences, 65B*, 91–96.
Schäfer, S., Huxhold, O., & Lindenberger, U. (2006). Healthy mind in healthy body? A review of sensorimotor - Cognitive interdependencies in old age. *European Review of Aging and Physical Activity, 3*, 45–54.
Scheibe, S., & Carstensen, L. L. (2010). Emotional aging: Recent findings and future trends. *Journal of Gerontology, 65B*, 135–144.
Scherer, K. R., Shuman, V., Fontaine, J. R. J., & Soriano, C. (2013). The GRID meets the wheel: Assessing emotional feeling via self-report. In J. R. J. Fontaine, K. R. Scherer, & C. Soriano (Eds.), *Components of emotional meaning: A sourcebook* (pp. 281–299). Oxford: Oxford University Press.
Schirmbeck, F., Boyette, L.-L., Valk, R. V. D., Meijer, C., Dingemans, P., Van, R., et al. (2015). Relevance of five-factor model personality traits for obsessive–compulsive symptoms in patients with psychotic disorders and their un-affected siblings. *Psychiatry Research, 225*, 464–470.
Schnall, S., Haidt, J., Clore, G. L., & Jordan, A. H. (2008). Disgust as embodied moral judgment. *Personality and Social Psychology Bulletin, 34*, 1096–1109.
Schneiderman, N., Ironson, G., & Siegel, S. D. (2005). Stress and health: Psychological, behavioral, and biological determinants. *Annual Review of Clinical Psychology, 1*, 607–628.
Schooler, C. (1984). Psychological effects of complex environments during the life span: A review and theory. *Intelligence, 8*, 259–281.
Schooler, C., & Mulatu, M. S. (2001). The reciprocal effects of leisure time activities and intellectual functioning in older people: A longitudinal analysis. *Psychology and Aging, 16*, 466–482.
Schultz, P. W., & Tabanico, J. (2007). Self, identity, and the natural environment. *Journal of Applied Social Psychology, 37*, 1219–1247.
Schulz, R. (1982). Emotionality and aging: A theoretical and empirical analysis. *Journal of Gerontology, 37*, 42–51.
Scott, S. B., Graham-Engeland, J. E., Engeland, C. G., Smyth, J. M., Almeida, D. M., Katz, M. J., et al. (2015). The effects of stress on cognitive aging, physiology and emotion (ESCAPE) Project. *BMC Psychiatry, 15*, 146.
Segal, D. L., Hersen, M., Van Hasselt, V. B., Smith Silberman, C., & Roth, L. (1996). Diagnosis and assessment of personality disorders in older adults: A critical review. *Journal of Personality Disorders, 10*, 384–399.
Seider, B. H., Shiota, M. N., Whalen, P., & Levenson, R. W. (2011). Greater sadness reactivity in late life. *SCAN, 6*, 186–194.
Shmotkin, D., & Shrira, A. (2012). Happiness and suffering in the life story: An inquiry into conflicting expectations concerning the association of perceived past with present subjective well-being in old age. *Journal of Happiness Studies, 13*, 389–409.
Shultz, K. S., & Wang, M. (2011). Psychological perspectives on the changing nature of retirement. *American Psychologist, 66*, 170–179.
Silver, C. B. (2003). Gendered identities in old age: Toward (de)gendering? *Journal of Aging Studies, 17*, 379–397.
Singer, T., Verhaeghen, P., Ghisletta, P., Lindenberger, U., & Baltes, P. B. (2003). The fate of cognition in very old age: Six-year longitudinal findings in the Berlin Aging Study (BASE). *Psychology and Aging, 18*, 318–331.

Singer, T., Lindenberger, U., & Baltes, P. B. (2003a). Plasticity of memory for new learning in very old age: A story of major loss? *Psychology and Aging, 18*, 306–317.

Smith, C. D., Walton, A., Loveland, A. D., Umberger, G. H., Kryscio, R. J., & Gash, D. M. (2005). Memories that last in old age: Motor skill learning and memory preservation. *Neurobiology of Aging, 26*, 883–890.

Smith, J. C., Nielson, K. A., Woodard, J. L., Seidenberg, M., Durgerian, S., Antuono, P., et al. (2011). Interactive effects of physical activity and APOE-ε4 on BOLD semantic memory activation in healthy elders. *NeuroImage, 54*, 635–644.

Smith-Gamble, V., Baiyewu, O., Perkins, A. J., Gureje, O., Hall, K. S., Ogunniyi, A., et al. (2002). Informant reports of changes in personality predict dementia in a population-based study of elderly African Americans and Yoruba. *American Journal of Geriatric Psychiatry, 10*, 724–732.

Sneed, J. R. (2005). Models of the aging self. *Journal of Social Issues, 61*, 375–388.

Snitz, B. E., Weissfeld, L. A., Cohen, A. D., Lopez, O. L., Nebes, R. D., Aizenstein, H. J., et al. (2015). Subjective cognitive complaints, personality and brain amyloid-beta in cognitively normal older adults. *American Journal of Geriatric Psychiatry, 23*, 985–993.

Sökefeld, M. (1999). Debating self, identity, and culture in anthropology. *Current Anthropology, 40*, 417–447.

Specht, J., Egloff, B., & Schmukle, S. C. (2011). *Stability and change of personality across the life course: The impact of age and major life events on mean-level and rank-order stability of the Big Five* (SOEPpapers on Multidisciplinary Panel Data Research, No. 377). Berlin: Deutsches Institut für Wirtschaftsforschung (DIW).

Squire, L. R. (2004). Memory systems of the brain: A brief history and current perspective. *Neurobiology of Learning and Memory, 82*, 171–177.

Srivastava, S., John, O. P., Gosling, S. D., & Potter, J. (2003). Development of personality in early and middle adulthood: Set like plaster or persistent change? *Journal of Personality and Social Psychology, 84*, 1041–1053.

Standridge, S. H., Dunlap, R., & Hamilton, G. (2019). Retirement and flow: Can the casual leisure pursuits of older adults in retirement create the experience of flow? *Activities, Adaptation and Aging*. https://doi.org/10.1080/01924788.2019.1651177

Starr, J. M., McGurn, B., Whiteman, M., Pattie, A., Whalley, L. J., & Deary, I. J. (2004). Life long changes in cognitive ability are associated with prescribed medications in old age. *International Journal of Geriatric Psychiatry, 19*, 327–332.

Stawski, R. S., Sliwinski, M. J., Smyth, J. M., & Syracuse University Department of Psychology and Center for Health and Behavior, Syracuse University. (2006). Stress-related cognitive interference predicts cognitive function in old age. *Psychology and Aging, 21*, 535–544.

Steen, G., Berg, S., & Steen, B. (1998). Cognitive function in 70-year-old men and women. A 16-year cohort difference population study. *Aging Clinical and Experimental Research, 10*, 120–126.

Steptoe, A., Deaton, A., & Stone, A. A. (2015). Psychological wellbeing, health and ageing. *Lancet, 385*, 640–648.

Stern, Y. (2002). What is cognitive reserve? Theory and research application of the reserve concept. *Journal of the International Neuropsychological Society, 8*, 448–460.

Sternberg, R. J. (1998). A balance theory of wisdom. *Review of General Psychology, 2*, 347–365.

Stigsdotter Neely, A., & Backman, L. (1995). Effects of multifactorial memory training in old age: Generalizability across tasks and individuals. *Journal of Gerontology, 50B*, P134–P140.

Studzinski, R. (1987). Impasse, finitude, and guilt: Some religious challenges of aging. *Bulletin of the Menninger Clinic, 51*, 436–446.

Subandi, M. A., & Good, B. J. (2018). Shame as a cultural index of illness and recovery from psychotic illness in Java. *Asian Journal of Psychiatry, 34*, 33–37.

Suitor, J. J., & Pillemer, K. (1987). The presence of adult children: A source of stress for elderly couples' marriages? *Journal of Marriage and Family, 49*, 717–725.

Swaab, D. F. (1991). Brain aging and Alzheimer's disease, "wear and tear" versus "use it or lose it". *Neurobiology of Aging, 12*, 317–324.

Sze, J. A., Goodkind, M. S., Gyurak, A., & Levenson, R. W. (2012). Aging and emotion recognition: Not just a losing matter. *Psychology and Aging, 27*, 940–950.

Szinovacz, M. E. (2000). Changes in housework after retirement: A panel analysis. *Journal of Marriage and Family, 62*, 78–92.

Tak, S. H. (2006). An insider perspective of daily stress and coping in elders with arthritis. *Orthopaedic Nursing, 25*, 127–132.

Tangney, J. P., Wagner, P. E., Hill-Barlow, D., Marschall, D. E., & Gramzow, R. (1996). Relation of shame and guilt to constructive versus destructive responses to anger across the lifespan. *Journal of Personality and Social Psychology, 70*, 797–809.

Taylor, J. E., Alpass, F., Stephens, C., & Towers, A. (2011). Driving anxiety and fear in young older adults in New Zealand. *Age and Ageing, 40*, 62–66.

Terman, L. M. (2018). *Terman life-cycle study of children with high ability, United States, 1922–1991*. Ann Arbor, MI: Inter-University Consortium for Political and Social Research.

Thompson, P. (1992). 'I don't feel old': Subjective ageing and the search for meaning in later life. *Ageing and Society, 12*, 23–47.

Thompson, R. A. (1991). Emotional regulation and emotional development. *Educational Psychology Review, 3*, 269–307.

Tilvis, R. S., Kähönen-Väre, M. H., Jolkkonen, J., Valvanne, J., Pitkala, K. H., & Strandberg, T. E. (2004). Predictors of cognitive decline and mortality of aged people over a 10-year period. *Journal of Gerontology, 59A*, 268–274.

Tracy, J. L., Shariff, A. F., & Cheng, J. T. (2010). A naturalist's view of pride. *Emotion Review, 2*(2), 163–177.

Tran, T. Y., Wright, R., Jr., & Chatters, L. (1991). Health, stress, psychological resources, and subjective well-being among older blacks. *Psychology and Aging, 6*, 100–108.

Troll, L. E., & McKean Skaff, M. (1997). Perceived continuity of self in very old age. *Psychology and Aging, 12*, 162–169.

Tse, M. M. Y., Lo, A. P. K., Cheng, T. L. Y., Chan, E. K. K., Chan, A. H. Y., & Chung, H. S. W. (2010). Humor therapy: Relieving chronic pain and enhancing happiness for older adults. *Journal of Aging Research, 2010*, 343574, 9 pages. https://doi.org/10.4061/2010/343574

Tucker-Drob, E. M., & Salthouse, T. A. (2008). Adult age trends in the relations among cognitive abilities. *Psychology and Aging, 23*, 453–460.

Tulving, E. (1985). How many memory systems are there? *American Psychologist, 40*, 385–398.

Tulving, E. (2002). Episodic memory: From mind to brain. *Annual Review of Psychology, 53*, 1–25.

Tulving, E., & Schacter, D. L. (1990). Priming and human memory systems. *Science, 247*, 301–306.

Tung, V. W. S., & Ritchie, J. R. B. (2011). Investigating the memorable experiences of the senior travel market: An examination of the reminiscence bump. *Journal of Travel and Tourism Marketing, 28*, 331–343.

Turner, V. (1967). *The forest of symbols: Aspects of Ndembu ritual*. Ithaca, NY: Cornell University Press.

Turner, R. H. (1968). The self-conception in social interaction. In C. Gordon & K. J. Gergen (Eds.), *The self in social interaction. Vol. I: Classic and contemporary perspectives* (pp. 93–106). New York: Wiley.

Turner, V. (1969). *The ritual process: Structure and anti-structure*. London: Routledge & Kegan Paul.

Tybur, J. M., Lieberman, D., Kurzban, R., & DeScioli, P. (2013). Disgust: Evolved function and structure. *Psychological Review, 120*, 65–84.

van den Bogaard, L., Henkens, K., & Kalmijn, M. (2014). So now what? Effects of retirement on civic engagement. *Ageing and Society, 34*, 1170–1192.

van Hooren, S. A. H., Valentijn, A. M., Bosma, H., Ponds, R. W. H. M., van Boxtel, M. P. J., & Jolles, J. (2007). Cognitive functioning in healthy older adults aged 64–81: A cohort study into the effects of age, sex, and education. *Aging, Neuropsychology, and Cognition, 14*, 40–54.

Verhaeghen, P., Geraerts, N., & Marcoen, A. (2000). Memory complaints, coping, and well-being in old age: A systemic approach. *The Gerontologist, 40*, 540–548.

Verhaeghen, P., Marcoen, A., & Goossens, L. (1992). Improving memory performance in the aged through mnemonic training: A meta-analytic study. *Psychology of Aging, 7*, 242–251.

Verhaegen, P., & Salthouse, T. A. (1997). Meta-analyses of age-cognition relations in adulthood: Estimates of linear and nonlinear age effects and structural models. *Psychological Bulletin, 122*, 231–249.

Vo, K., Forder, P. M., Tavener, M., Rodgers, B., Banks, E., Bauman, A., et al. (2015). Retirement, age, gender and mental health: Findings from the 45 and Up Study. *Aging and Mental Health, 19*, 647–657.

von Bonsdorf, M., Shultz, K. S., Leskinen, E., & Tansky, J. (2009). The choice between retirement and bridge employment: A continuity theory and life course perspective. *International Journal of Aging and Human Development, 69*, 79–100.

Wahl, H.-W., Schmitt, M., Danner, D., & Coppin, A. (2010). Is the emergence of functional ability decline in early old age related to change in speed of cognitive processing and also to change in personality? *Journal of Aging and Health, 22*, 691–712.

Wang, C., Bannuru, R., Ramel, J., Kupelnick, B., Scott, T., & Schmid, C. H. (2010). Tai Chi on psychological well-being: Systematic review and meta-analysis. *BMC Complementary and Alternative Medicine, 10*, 23. https://doi.org/10.1186/1472-6882-10-23

Wang, M., Henkens, K., & van Solinge, H. (2011). Retirement adjustment: A review of theoretical and empirical advancements. *American Psychologist, 66*, 204–213.

Ward, V. E. (2000). Immigrant elders: Are we missing them? An examination of leisure participation and identity among immigrant elders. *Journal of Aging and Identity, 5*, 187–195.

Warr, M. (1993). Fear of victimization. *The Public Perspective, 5*, 25–28.

West, K., Shaw, R., Hagger, B., & Holland, C. (2017). Enjoying the third age! Discourse, identity and liminality in extra-care communities. *Ageing and Society, 37*, 1874–1897.

Westerlund, H., Kivimäki, M., Singh-Manoux, A., Melchior, M., Ferrie, J. E., Pentti, J., et al. (2009). Self- rated health before and after retirement: Findings from the French GAZEL cohort study. *Lancet, 374*, 1889–1896.

Westerlund, H., Vahtera, J., Ferrie, J. E., Singh-Manoux, A., Pentti, J., Melchior, M., et al. (2010). Effect of retirement on major chronic conditions and fatigue: French GAZEL occupational cohort study. *BMJ, 341*, c6149. https://doi.org/10.1136/bmj.c6149

Wetzel, M., Huxhold, O., & Tesch-Römer, C. (2016). Transition into retirement affects life satisfaction: Short- and long-term development depends on last labor market status and education. *Social Indicators Research, 125*, 991–1009.

Whalley, L. J., & Deary, I. J. (2001). Longitudinal cohort study of childhood IQ and survival up to age 76. *British Medical Journal, 322*, 1–5.

Wijeratne, C., & Reutens, S. (2016). When an elder is the abuser: Improving the recognition and management of domestic violence among elderly couples. *Medical Journal of Australia, 205*, 246–247e1.

Wilson, B. M., Mickes, L., Stolarz-Fantino, S., Evrard, M., & Fantino, E. (2015). Increased false-memory susceptibility after mindfulness meditation. *Psychological Science, 26*, 1567–1573.

Wilson, R. S., Arnold, S. E., Schneider, J. A., Kelly, J. F., Tang, Y., & Bennett, D. A. (2006). Chronic psychological distress and risk of Alzheimer's disease in old age. *Neuroepidemiology, 27*, 143–153.

Wilson, R. S., Hebert, L. E., Scherr, P. A., Barnes, L. L., Mendes de Leon, C. F., & Evans, D. A. (2009). Educational attainment and cognitive decline in old age. *Neurology, 72*, 460–465.

Wilson, R. S., Krueger, K. R., Gu, L., Bienias, J. L., Mendes de Leon, C. F., & Evans, D. A. (2005). Neuroticism, extraversion, and mortality in a defined population of older persons. *Psychosomatic Medicine, 67*, 841–845.

Wilson, R. S., Mendes de Leon, C. F., Bienias, J. L., Evans, D. A., & Bennett, D. A. (2004). Personality and mortality in old age. *Journal of Gerontology, 59B*, P110–P116.

Witte, A. V., Fobker, M., Gellner, R., Knecht, S., & Flöel, A. (2009). Caloric restriction improves memory in elderly humans. *Proceedings of the National Academy of Sciences of the United States of America, 106*, 1255–1260.

Wixted, J. T. (2004). On common ground: Jost's (1897) Law of Forgetting and Ribot's (1881) Law of Retrograde Amnesia. *Psychological Review, 111*, 864–879.

Wolkove, N., Elkholy, O., Baltzan, M., & Palayew, M. (2007). Sleep and aging: 1. Sleep disorders commonly found in older people. *Canadian Medical Association Journal, 176*, 1299–1304.

Workman, L., & Reader, W. (2014). *Evolutionary psychology* (3rd ed.). Cambridge: Cambridge University Press.

Yankner, B. A., Lu, T., & Loerch, P. (2008). The aging brain. *Annual Review of Pathology: Mechanisms of Disease, 3*, 41–66.

Yirmiya, R., & Goshen, I. (2011). Immune modulation of learning, memory, neural plasticity and neurogenesis. *Brain, Behavior, and Immunity, 25*, 181–213.

Zacks, R. T., Hasher, L., & Li, K. Z. H. (2000). Human memory. In F. I. M. Craik & T. A. Salthouse (Eds.), *The handbook of aging and cognition* (pp. 293–357). Hillsdale, NJ: Erlbaum Associates.

Zahodne, L. B., Glymour, M. M., Sparks, C., Bontempo, D., Dixon, R. A., MacDonald, S. W. S., et al. (2011). Education does not slow cognitive decline with aging: 12-year evidence from the Victoria Longitudinal Study. *Journal of the International Neuropsychological Society, 17*, 1039–1046.

Zantinge, E. M., van den Berg, M., Smit, H. A., & Picavet, H. S. J. (2013). Retirement and a healthy lifestyle: Opportunity or pitfall? A narrative review of the literature. *European Journal of Public Health, 24*, 433–439.

Zehnder, F., Martin, M., Altgassen, M., & Clare, L. (2009). Memory training effects in old age as markers of plasticity: A meta-analysis. *Restorative Neurology and Neuroscience, 27*, 507–520.

Zeintl, M., Kliegel, M., & Hofer, S. M. (2007). The role of processing resources in age- related prospective and retrospective memory within old age. *Psychology and Aging, 22*, 826–834.

Zelinski, E. M., Crimmins, E., Reynolds, S., & Seeman, T. (1998). Do medical conditions affect cognition in older adults. *Health and Psychology, 17*, 504–512.

Zhang, J.-G., Ishikawa-Takata, K., Yamazaki, H., Morita, T., & Ohta, T. (2006). The effects of Tai Chi Chuan on physiological function and fear of falling in the less robust elderly: An intervention study for preventing falls. *Archives of Gerontology and Geriatrics, 42*, 107–116.

Zijlstra, G. A. R., van Haastregt, J. C. M., van Eijk, J. T. M., van Rossum, E., Stalenhoef, P. A., & Kempen, G. I. J. M. (2007). Prevalence and correlates of fear of falling, and associated avoidance of activity in the general population of community-living older people. *Age and Ageing, 36*, 304–309.

Zijlstra, G. A. R., van Haastregt, J. C. M., van Rossum, E., van Eijk, J. T. M., Yardley, L., & Kempen, G. I. J. M. (2007a). Interventions to reduce fear of falling in community-living older people: A systematic review. *Journal of the American Geriatrics Society, 55*, 603–615.

Zink, T., Jacobson, C. J. J., Pabst, S., Regan, S., & Fisher, B. S. (2006). A lifetime of intimate partner violence: Coping strategies of older women. *Journal of Interpersonal Violence, 21*, 634–651.

Ziv, Y., & Schwartz, M. (2008). Immune-based regulation of adult neurogenesis: Implications for learning and memory. *Brain, Behavior, and Immunity, 22*, 167–176.

Chapter 6
The Social Dimension of Older Ages

Abstract This chapter focuses on the many factors affecting the social integration of the elderly and the psychological complexities associated with loneliness. Partners and the family are first in line as potential social factors that can contribute to elders' well-being, but the community at large, including friends and acquaintances, can also become important, especially if direct family members are not available. Social interactions can be also negative and therefore we analyse the important issue of stigma affecting the elderly, and the various ethical issues, including those associated with the abuse of old people, that are relevant to old age. The oldest-old represent a special case as they face unique social challenges, not only due to their physical frailty, but also because at very advanced ages their peer network of social relationships weakens considerably. In multi-ethnic societies older immigrants experience specific problems of adaptation that cannot be ignored, and thus a section is also devoted to this topic. Other minorities are also important, and they may have unique needs associated with their ageing. Here we address the social issues relevant to older gay, lesbian, bisexual, transsexual/transgender (trans), and intersex people, and in a separate section we focus on the experience of ageing prison inmates. We conclude the chapter with an analysis of the challenges associated with caring for older people and the consequences of such care in terms of caregiver burden.

Humans are an eminently social species and our propensity to live in groups has a genetic basis (Donaldson and Young 2008). In addition, human sociality modulates, and it is in turn modulated by culture and individual choices. Sociality exposes individuals to a variety of other people, and over time such contact may give rise to a common culture. A common culture can then affect the ways in which individuals continue to interact into the future. However, in spite of some important cultural commonalities such as language and traditions, group members may also differ individually in many ways (Page 2007): cognitively (specific ways of thinking and processing ideas), various aspects of personal identity (age, sex, sexual orientation,

M. Cordella, A. Poiani, *Fulfilling Ageing*, International Perspectives on Aging 30,
https://doi.org/10.1007/978-3-030-60071-6_6

personality, religion, race, ethnicity, immigrant status), and they may also vary in their individual preferences (tastes, values). Such diversity brings various benefits to the members of a group, such as greater adaptability, flexibility, and creativity in both thinking and acting, and more human capital to solve problems of common interest (Jones et al. 2014); but it can also introduce a degree of social instability if the members of a diverse group are incapable of finding a common ground for a peaceful and productive coexistence.

Family, friends, and acquaintances, as well as local and national government and non-government organisations provide the social network that can facilitate (and sometimes difficult) the achievement of individual daily goals and the performance of various activities, along with the fulfilment of more long-term plans. This remains true at any age including, of course, old age.

Old people are usually embedded in a *social network*. We remind from Chap. 1 that a social network can be defined as:

> a structure of individuals with a designed relationship to the focal person, as well as an average frequency of contact and a specified geographical proximity to that person. Social networks can be thought of as a key resource over the life course, a form of social capital that potentially influences the exchange of supports (Ajrouch et al. 2001, p. S112).

Social networks can be somewhat stable, or they can be variable and fluid. Moreover, because the network can influence the individual both positively and negatively, sometimes a loose social network can be of more benefit to the individual than a tighter and more restrictive one (Granovetter 1973).

The dynamics of individual social networks vary, as people age, in response to their specific life course, the number of close friends and family members that are still alive, the individual mobility from one community to another, and so forth (Ajrouch et al. 2001). In this chapter we explore several issues that together illustrate the social dimension of old age. We start by analysing the challenges that older people encounter in their social integration, including issues of loneliness and abuse, to then focus on the more specific structure of social relationships as they relate to family and the community at large. The effects of stigma and the broader issues of the ethics of ageing are going to be discussed next. The rest of the chapter will focus on more detailed social aspects of old age starting with an analysis of the social issues confronting the oldest-old, to then address the challenges experienced by old immigrants, old people who are members of a gender minority, and also old prison inmates. We conclude the chapter with a review of the social and psychological aspects of caring for older people, including aspects of caregiver burden.

6.1 Social Integration of the Elderly

Since the second half of the twentieth century three major social approaches to ageing have emerged: ageing may be regarded as a matter of social interest affecting the individual's health and psychology, or as an economic/employment issue, or as a global issue.

As we have already noted in Chap. 1, from an individual perspective older people have been regarded as either proactive in society (*Activity theory*) or withdrawing from mainstream social activities and roles (*Disengagement theory*). The social integration of the elderly, understood as the quality and quantity of their social relationships, implicit in *Activity theory* is achieved through various kinds of behaviours ranging from the communicative (i.e. adherence to a system of shared meanings in communication), functional (i.e. the degree of interdependence or self-sufficiency in the exchange of goods and services), cultural (i.e. adherence to common cultural standards), and normative (i.e. degree in which the cultural standards are effective norms that regulate behaviour) (de Jong Gierveld and Hagestad 2006). That is, social integration emerges from a common set of values and norms shared through communicating in a common language. Greater social integration is expected to lead to greater well-being in the elders. On the other hand, disengagement from society could be associated with greater well-being if it is a personal decision (voluntary disengagement) leading to less stress and greater personal fulfilment (Phillipson and Baars 2007) or disengagement could be stressful and associated with unhealthy behaviours such as heavy drinking and smoking if it is forced or involuntary (Henkens et al. 2008). Voluntary disengagement could potentially also lead to self-neglect, as in the case of the so-called *Diogenes syndrome* (Clark et al. 1975). Obligate disengagement that can have positive consequences may also occur, such as compulsory retirement potentially opening new fulfilling possibilities in life for old people.

The recognition that the ageing process varies not only according to biological but also to social, cultural, and historical factors has been made explicit through various theoretical models such as *Age Stratification* theory: old people can be seen as belonging to specific cohorts that may be distinguishable through various stages of history, in spite of some internal diversity; and *Life Course* theory: old people in the present can be seen as a point in a life course process, which is an age-graded pattern of events that are embedded into social institutions and a historical context (Elder et al. 2003).

An even more integrative perspective on the social aspects of ageing can be found in George Engel's *Biopsychosocial* model, which simply points to the multi-causal character of some aspects of our reality as a social organism. Our mind has causal roots into our biology and our social relationships (Engel 1977), but we may add that our mind is affected by our relationships with other components of our natural environment as well (ecological dimension) and, more personally, also by our spiritual experience. This has led to a further extension of the model into a *Biopsychosocial-spiritual* model (e.g. Sulmasy 2002). The usefulness and limitations of these integrative models have been recently discussed in the journal *Psychiatric Times* (https://www.psychiatrictimes.com/couch-crisis/can-we-salvage-biopsychosocial-model), see also Hogan (2019).

Although there is variability among old individuals belonging to a social group, there is also a general trend towards nurturing a more restricted social network as we age (e.g. Fung et al. 2001). This has been variously interpreted as a tendency by older people to emotionally disengage from the rest of society, as expected from the *Disengagement theory* mentioned above (Cumming and Henry 1961) or as a necessary outcome of the impediments to cultivate social relationships as our physical abilities decline with age (as in *Activity theory*; Havighurst and Albrecht 1953).

More recently, restricted social networks in old age have been interpreted as an adaptive response to the realisation that the time left in life is decreasing, and therefore effort should be better focused on the self and a selected group of relevant others (*Socioemotional Selectivity theory*; Carstensen 1993).

Given the personal aspects affecting social integration in old age, it is also important to distinguish between *objective social integration* (e.g. number and types of interactions and social roles) and *subjective social integration* (e.g. intimacy in social relationships, feelings of loneliness or integration, happiness and satisfaction). Objective social integration may be jeopardised in older people by physical constraints such as cessation of driving (e.g. Mezuk and Rebok 2008); but even when there is no physical impediment to social interaction the lift in morale expected from integration is not guaranteed. For instance, Liang et al. (1980) demonstrated that objective social integration requires the intervening effect of subjective social integration in order to positively affect morale in older people. Therefore, the material availability of interactants can be of benefit to an older person but only for as long as the interaction does help the elder lift his/her morale. This relationship between the *objective* and the *personal/subjective* will be a common feature of our analysis of sociality in old age throughout this chapter.

From a socio-economic perspective, and in response to the current neoliberal period of social and economic change around the world, *Critical, Cultural and Humanistic Gerontology* (CCHG) has emphasised the social construction of ageing by stressing aspects of social inequality (e.g. in retirement), and the effects of the relationships among culture, social structure, and individual agency in the achievement of social change. CCHG also emphasises "issues of meaning and the nature of subjectivity in later life" (Phillipson and Baars 2007), which is an individual dimension, a point that is also shared with *Socioemotional Selectivity theory* and *Disengagement theory* (Cumming and Henry 1961).

Ageing as a global social issue brings to the fore important demographic, economic, health care, and social welfare concerns that urgently require a re-evaluation of the meaning of retirement in old age, but also taxation and the role of government in ensuring the development of a harmonious society in the new context of an ageing population (Kinsella and Phillips 2005).

In what follows we analyse in more detail various important factors affecting sociality in the elders: disengagement, social support, social capital, and social cohesion and integration.

We already introduced the concept of disengagement from life-as-usual in old age in response to signs of biological decline (see also Chap. 1). Such disengagement is intended to shield the individual from the frustrations of failing to achieve unachievable goals and to help the elder satisfactorily accommodate life goals to the constraints of old age, such as insufficient energy to sustain a large network of social relationships.

According to Hughes and Heycox (2010, p. 76) *Disengagement theory* (Cumming and Henry 1961) has four central tenets:

1. Disengagement is a gradual process and is a reaction to a series of events, such as the loss of parental roles, retirement, and the reduction of activity in community and social group.
2. Disengagement is viewed as inevitable for all those living in older age.

3. Disengagement is a mutually satisfying experience both for society and individual. Older people are happy to disengage from the demands, activities, and pressures of their social roles. Society benefits by those roles being taken over by younger individuals, allowing them to fulfil their objectives in life.
4. Disengagement is viewed as the norm for society. It should be facilitated by social policy and legislation where necessary—for example, via mandatory retirement policies—and the prospect of a fulfilling life in retirement.

For as long as the disengagement process is led by the individual and supported by society, it could be a fulfilling experience, providing best outcome for both the old person (e.g. decreasing stress and allowing more leisure time in retirement, supported by a good pension system) and the community (e.g. releasing job opportunities for the younger generations). It is such mutually beneficial interaction between individual and society that crucially determines the quality of life in old age (e.g. Mead 1934). But as we have already seen, disengagement may be enforced rather than voluntary, traumatic rather than smooth and fulfilling, thus leading to situations that are far from optimal from the perspectives of both the old person and society at large.

Contrasting the disengagement approach, the concept of *successful ageing* supports a more proactive stance in old age (Rowe and Kahn 1997). Proactivity or disengagement can be pursued by different people in their old age or a mixture of both could be a better option for others. Ultimately, what matters in our view is not so much to achieve "success" but to live a *fulfilling* old age in the way we choose.

The desire for fulfilling ageing experienced within a supportive social network is rooted in the evidence that positive social relationships are well known to help decrease both morbidity and mortality through socioemotional support (e.g. affection) and instrumental support (e.g. material help). In fact, when Bowling and Dieppe (2005) asked their respondents about the prominent aspects of what they referred to as "successful ageing", *social roles and activities* were listed by older people themselves among the top three, only behind *health* and *psychological factors*, and followed in fifth place by *social relationships*, after *finances* (see Fig. 6.1, Bowling and Dieppe 2005).

In a study carried out in the 1980s, that is before the widespread use of the Internet, Seeman and Berkman (1988) analysed how instrumental and emotional support vary according to the structural characteristics of social networks in the case of community-dwelling people aged 65 and over in the USA. After controlling for various factors, such as age, sex, race, and income, they concluded that not only the total network size, but also the number of face-to-face contacts and the number of proximal ties between interacting people are associated with greater levels of instrumental and emotional support. Such social support is deemed more adequate to the extent that it involves a greater degree of face-to-face contacts. Interestingly, it is the quality of the social relationship that matters, more than the type of interactant (e.g. a spouse or other). In addition, they also found what we may describe as a social "division of labour", where family members are more closely associated with instrumental support, and friends and confidantes with emotional support. The frequency of face-to-face contacts is associated with an increased availability of both emotional and instrumental support for the elderly, as seen in Table 6.1, although some support can be also available when the number of face-to-face contacts is zero.

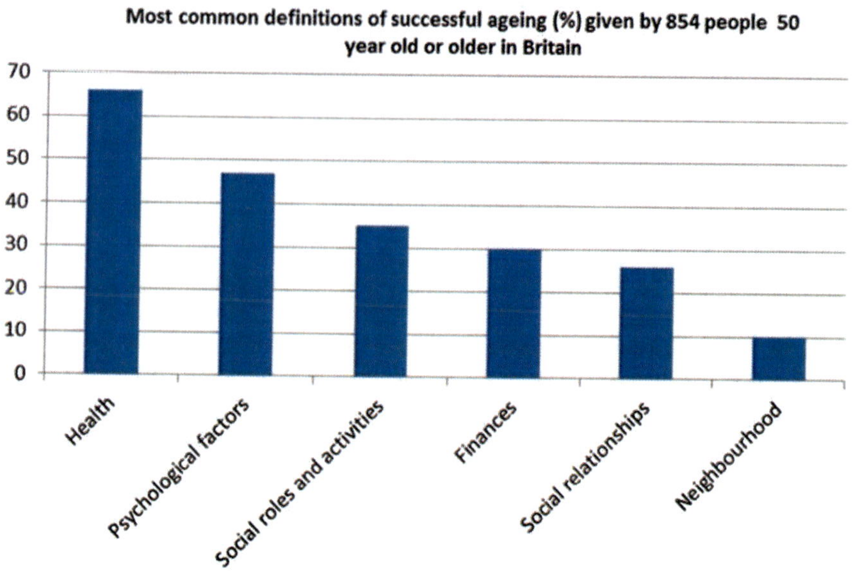

Fig. 6.1 Frequency distribution of major aspects of "successful ageing" according to a group of British older people (redrawn from Bowling and Dieppe 2005)

Table 6.1 Availability of support by total number of monthly face-to-face contacts with children and close friends and relatives

	Proportion reporting availability of support	
Face-to-face contacts	Instrumental support	Emotional support
0	65.6[a]	62.2
1–2	87.6	72.5
3–5	91.6	84.1
6–10	94.1	87.9
11+	97.2	90.9

[a]Percentage of contacts observed within each category where such type of support is available; from Seeman and Berkman (1988)

Positive social relationships may also help improve the level of *resilience* in the old (Rowe and Kahn 1997); that is, the capacity of old people to recover from stressful episodes in life and return to a more fulfilling life course is increased by social support.

The network of social support that helps achieve a better and more fulfilling experience in old age is commonly referred to as *social capital* (Helliwell and Putnam 2004). Social capital is measured through the strength of family and extra-family ties (e.g. neighbourhood, community; also known as the *informal support system*) that facilitate good physical health and psychological well-being. Despite their dynamic nature, such supportive social networks are broadly characterised by stabilising norms of reciprocity and high levels of trust among the members. Complementing the informal support system is the *formal public system of support* relying on government and private welfare agencies (Cantor 1980).

Some of the benefits of social networks as social capital for the members of a community also include improving the ability to cope, better enjoying life and experiencing greater hope (Jacobs 1960; Coleman 1990; Putnam 1995a; Cattell 2001). Social capital is also an insurance against the stresses of poverty (Cattell 2001). In principle, a broader network of supportive social relationships involving both close family ties and more diffused ties within an informal support system could be more effective and useful to the older person than a more restricted network (Litwin and Landau 2000); but in the end it is the quality of social ties that matter and for most elders this can be achieved with only a limited number of individuals.

More specifically, the structure of social capital can take the form of bonding, bridging, and linking capitals (Curry and Fisher 2013). The *bonding capital* refers to the ties existing between individuals within a specific social group; *bridging capital* refers to more distant ties between people across different groups; *linking capital* can be political or financial and it connects people to the individual or institutional power structure present in society.

In modern urban centres, social capital may decrease as a result of decreased levels of *social cohesion*, with such loss of social capital and cohesion being especially felt by older people. The concept of social cohesion emphasises:

> the need for a shared sense of morality and common purpose; aspects of social control and social order; the threat to social solidarity of income and wealth inequalities between people, groups and places; the level of social interaction within communities or families; and a sense of belonging to place. By implication, a society lacking cohesion would be one which displayed social disorder and conflict, disparate moral values, extreme social inequality, low levels of social interaction between and within communities and low levels of place attachment (Forrest and Kearns 2001, p. 2128).

Social cohesion may vary according to the level of analysis: whereas a neighbourhood may be cohesive, it could still be in conflict with another neighbourhood, due to some specific issues that cause friction (e.g. resource sharing, government investment in infrastructure, and so forth). In addition, the larger the community is, the more difficult it may be to achieve social cohesion due to increased anonymity and decreased quality of personal interactions in larger communities (e.g. Wilkinson 2008), this tends to negatively affect older people's social integration.

Forrest and Kearns (2001) have identified various domains of social cohesion, which we list in Table 6.2. Such domains of social cohesion point to the need to incorporate all the members of a social group including, crucially, the elderly, into a fair, just, participative and meaningful coexistence, a fulfilling life experience. Ideally, such harmonious coexistence should be already evident at the microlevel of the neighbourhood, where many older people tend to spend most of their time. An example of this is the case of the *Wagga Wagga Aboriginal Elders Group Incorporated* (WWAEGI) in Australia (Miliken et al. 2007). The WWAEGI was created in 1999 to provide an oral history program for the benefit of the local community. This resulted in the emergence of a level of *connectedness* not only among the elder women who were part of the program but also with their community at large. The social capital thus created was put into practical use through the writing of a book and the production of a video: *Making waves for a new dreaming*, which facilitated further work beneficial to the community. A similar program has been

Table 6.2 Domains of social cohesion

Domain	Description
Common values and a civic culture	Common aims and objectives, common moral principles and codes of behaviour, support for political institutions, and participation in politics
Social order and social control	Absence of general conflict and threats to peaceful coexistence, absence of incivility, effective informal social control, tolerance, respect for difference, intergroup cooperation
Social solidarity and reductions in wealth disparities	Harmonious economic and social development and common standards, redistribution of public finances and of opportunities, equal access to services and welfare benefits, ready acknowledgement of social obligations and willingness to assist others
Social networks and social capital	High degree of social interaction within communities and families, civic engagement and associational activity, easy resolution of collective action problems
Place attachment and identity	Strong attachment to place, intertwining of personal and place identity

From Forrest and Kearns (2001), with some modifications

implemented in Hawai`i (*Hā Kūpuna: National Resource Center for Native Hawaiian Elders*) (Mokuau et al. 2008).

Given the potential for social conflict that we have mentioned, one of the challenges to the concept of social cohesion is how to achieve a harmonious coexistence among the members of a community, whilst also respecting their individual and cultural diversity. Just as we expect cohesion within a specific linguistic-racial group that, however, may be diverse in terms of personalities, religion, habits, age structure, political views, and so forth; we are quite justified to expect the achievement of social cohesion within a social group that may be racially, linguistically, and culturally diverse. This is especially important when we consider the integration of elderly migrants into a community. Such social cohesion may be achieved in a multicultural context through the establishment of some basic common grounds, such as the respect of the law, the use of a common language—apart from other languages that may coexist in multilingual communities-, the interest in learning new ways of living, tolerance of religious/philosophical diversity, and a general interest to care for the common environment and the well-being of all members of the community. The alternative of building virtual or real walls that transform cities into a patchwork of "gated communities" (Forrest and Kearns 2001) is bound to decrease rather than increase social cohesion. Thus, social cohesion can benefit from a stance whereby those who are "different", such as the elders, are nevertheless integrated. Fostering *civic engagement* of the elders will likely improve social cohesion (e.g. Putnam 1995).

Therefore, the social integration of the elderly can prevent their isolation into a patchwork of communities and subcultures (e.g. Rose 1965; Longino et al. 1980). Integration relies on the absence of age barriers and the presence of inter-generational interactions. Such social integration of the elderly can be achieved through better relationships with various components of the community, and it can improve the sense of purpose and meaning in life in old people, thus improving their well-being (e.g. Durkheim 1897; Su and Ferraro 1997). Uhlenberg (2000, p. 261) defines an age-integrated social structure as one that "does not use chronological age as a criterion for entrance, exit, or participation. Conversely, when a structure makes use of age criteria as barriers, it is, to some extent, age segregated. Barriers to age integration may, in many ways, be similar to those that produced racial segregation".

Social integration of the elderly may be more difficult to achieve if social relationships are seen from the perspective of *social exchange*, where interactions among individuals are equated to economic transactions. In old age, some individuals may not have enough "capital" to sustain a "profitable economic exchange" with other members of society, thus potentially leading to their social segregation and abandonment in destitution. Su and Ferraro (1997), however, note that older people may be regarded as accruing a "social credit" from past efforts helping the younger generations (e.g. their children), thus leading to an expectation for a "repayment" in old age.

Social integration becomes obviously more likely to the extent that interacting people are more closely attached to each other. Cicirelli (2010) studied such attachment relationships in a group of 80 elders (67.5% women, 47.5% widowed), who were mostly in good health, in Indiana, USA. Respondents identified a number of attachment figures classified according to three attachment functions: *proximity seeking* (preferred person to be close to when in a dangerous or threatening situation), *secure base* (preferred person to be with when feeling insecure), *safe haven* (preferred person to be with when feeling emotionally upset and in distress and wanting to receive comfort in order to relax). The size of the supportive social network varied, with men having smaller networks than women, and widowed elders having smaller networks than the married ones. Table 6.3 shows, for each attachment figure (spouse, friend, and so forth), the percentage of the sample that identified such attachment figure with each of the three attachment functions mentioned above and the mean ranking for such function for each attachment figure (1 being most important). Note how for the three attachment functions most elders named members of the nuclear family (spouse, daughter, and son) first, followed by God as the most frequent attachment figures. Regarding the ranking of various attachment figures according to function, respondents considered both spouse and God as most important, followed by the doctor and nurse-caregiver. Thus, the most important figures of attachment in this sample are the nuclear family—the spouse in particular—and God, indicating the importance of both family and religion in this group of elders (see also Magilvy et al. 2000).

Although a major context for social integration in the elders is the family—the nuclear family in particular—integration within the community at large is also important, especially for elders who do not have family members living close by. In a study carried out in England, Davidson et al. (2003a) investigated older men's participation in 37 types of social groups and organisations (e.g. voluntary organisations, local authority, church, university of the third age (U3A), and others). Their findings suggest that, provided that older men keep in reasonably good health, they do not diminish their interest to participate in such organisations as they age, and they enjoy such participation provided that appropriate facilities and activities are available. Although such broad conditions may be general for both men and women, the sexes are also somewhat different in the way they interact with peers outside their family. For instance, Davidson et al. (2003b) list some major differences in the

Table 6.3 Attachment figures named by elderly respondents for proximity seeking, secure base, and safe haven, and percent of the sample naming each figure and mean ranking of figure ($n = 80$)

Attachment figure	Percent of sample			Mean ranking		
	Prox[a]	Sec	Safe	Prox	Sec	Safe
Living family						
Spouse	28	26	30	1.59	1.42	1.50
Daughter	30	44	34	2.46	2.18	2.12
Son	58	41	34	2.12	2.17	2.55
Sister	7	7	5	2.29	2.29	3.20
Brother	12	10	5	2.17	2.40	2.40
Parent	3	3	4	2.00	2.33	2.25
Other kin	4	4	3	3.75	3.25	3.00
In-law	8	4	5	2.38	2.00	2.00
Fictive kin	2	3	2	5.00	3.00	5.00
Deceased family						
Spouse	13	10	10	2.54	2.60	2.40
Parent	1	3	2	2.00	3.33	1.00
Other kin	3	3	5	3.33	2.00	2.20
Friends						
Female friend	12	20	25	2.67	2.65	2.56
Male friend	18	11	12	2.33	2.09	2.08
Religious figures						
God	20	25	19	1.25	1.68	1.42
Minister	3	5	11	3.33	2.40	1.64
Health care figures						
Doctor	1	2	2	1.00	2.50	1.50
Nurse, caregiver	2	2	3	1.50	1.50	2.33
Animals dog	4	0	0	2.50	0	0

[a]Prox = proximity seeking, Sec = secure base, Safe = safe haven. The mean ranking refers to the importance of an attachment figure, where 1 = most important, and it was calculated only for the portion of the total sample naming that figure. From Cicirelli (2010)

degree of intimacy women show in their friendships as compared to men: intimacy being higher in women than men (see Table 6.4).

Another instance of socialisation outside the family is provided by volunteering. We briefly mentioned the work of Morrow-Howell et al. (2003) in Chap. 1, who carried out a study in the USA on the effects of volunteering on the well-being of older adults. They found that volunteering improves well-being by affording older people the opportunity to play a specific constructive role in the broader society. A constructive role in society helps improve the capacity of older people to live a better and more fulfilling life (e.g. Sieber 1974). This is especially so when the level of commitment in volunteering activities is attuned to the capacities of the elders and does not cause stress (Goode 1960). Moreover, the specific benefits of volunteering can be variable across individuals due to personality, family circumstances, and other factors (Moen et al. 1992).

Table 6.4 Gendered characteristics of friendships

Women's friendships are broadly	Men's friendships are broadly
Characterised by connectedness	Characterised by separateness
Intimacy—face-to-face	Sociability—side-by-side
Mutual disclosure	Self-disclosure rare
Focus on talk	Focus on activity
Context: home	Context: pub/sports club, etc.

Slightly modified from Davidson et al. (2003b); based on cultural aspects of some Western societies in the second half of the twentieth century

Socialising with the rest of the community is also facilitated by a better urban design and participation in leisure activities in natural settings. For instance, life satisfaction may be increased in the elderly by providing parks and green outdoor spaces where elders can meet and interact with other members of the neighbourhood at their leisure (e.g. Kweon et al. 1998). In a study carried out in the USA, Steinkamp and Kelly (1987) conducted telephone interviews with 400 people (66% women) from 45 to 75+ years old. Older members of the community became better socially integrated in part through participation in leisure activities, which increased their life satisfaction.

Social integration can have not only positive psychological effects on the elderly but also positive effects on their health, even to the extent of decreasing mortality (e.g. Luo et al. 2012). In an early review of the published literature, House et al. (1988) found that across studies the mortality rate tends to decrease with social integration. The pattern is valid for both men and women, but the effect is especially strong in men, perhaps because men tend to be, on average, less socially integrated than women and therefore any improvement in their level of sociality can be especially beneficial. Pennix et al. (1997) studied a group of 55–85 year old who participated in the *Longitudinal Aging Study Amsterdam* between 1992 and 1995. Mortality was reduced in this sample among the participants who expressed fewer feelings of loneliness and among those who received moderate and high levels of emotional support. Individual characteristics also helped in improving survival rate in this sample, as greater feelings of *mastery*, which is a personal coping resource, decreased the risk of death. Berkman et al. (2004) also demonstrated an association between low social integration and high mortality among men and women in a large sample of 16,699 French employees of the *Electricity of France* and *Gas of France* companies. Fewer social ties were also associated with higher probability of mortality in a cohort of 1987 men and women studied in Germany (Baumann et al. 1998). In a study carried out in Spain, Zunzunegui et al. (2003) also found that poor social networks, social integration, and social engagement were associated with higher mortality in a group of 65 years old and older people.

More recently, Holt-Lunstad et al. (2010) have carried out a meta-analysis of the empirical evidence linking sociality and risks of mortality. After including a dataset of 148 studies, comprising 308,849 participants, they concluded that there is a 50% increase in the likelihood of survival in the case of participants who experience stronger social relationships. The trend was consistent across ages.

Broader positive effects of social integration on elders' health have been reported by many authors (e.g. Berkman 2000; Obisesan and Gillum 2009; Tanaca and Johnson 2010; Yang et al. 2013; Coll-Planas et al. 2017). In a study carried out in the USA, Faulkner et al. (2003) showed that social integration as assessed through family and friendship networks may protect against falls in older women living in the community. Interactions with the local community at large also afford elderly people the opportunity to engage in leisure activities that can help enhance their health-related quality of life. Better health is more likely achieved when such leisure activities are performed in group, therefore satisfying needs in both the personal and social domains (e.g. Cheung et al. 2009). In a cross-cultural study of the relationship between health and social integration/contributions, including data from Fiji, Korea, Malaysia, and the Philippines, Su and Ferraro (1997) analysed interviews of 3577 individuals to find that social integration into family and friends networks improved health in countries such as Malaysia, the Philippines, and Korea, but integration into voluntary associations displayed positive effects in Fiji and Korea only. Overall, the results point to the cross-cultural importance of social support for the well-being of older people.

Social support was also positively associated with health-related quality of life in a group of Korean elderly studied by Kim (2013), and a group of elderly Americans studied by Szanton et al. (2011). Social support is not only associated with higher life satisfaction in elders living independently in the community (Hall Gueldner et al. 2001; English 2013), but even among elders living in nursing homes social support, as shown in the number of visits received from specific individuals, is associated with increased life satisfaction (Hall Gueldner et al. 2001).

Through greater life satisfaction, integration into social networks could also positively affect mental health among older people. Fiori et al. (2006) studied 1669 adults aged 60 or older in the USA and described improved depressive symptomatology among participants who enjoyed positive support from their social network, irrespective of the specific composition of such network: family or friends. The authors interpreted this effect through the *Social convoy model* (Kahn and Antonucci 1980, see also Chap. 4) which states that individuals can be embedded into different types of social networks that can vary over time according to various factors both personal—such as age, gender, and personality-, and also situational (role expectations, resources, and demands), with women having closer contacts with their network members and also having larger and more diverse networks. The social connectedness displayed within the network is what it can presumably affect, usually positively, the level of mental health of older people. In a study carried out in Germany, Schwarzbach et al. (2013) also described lower levels of depression among 75 years old and older participants who were better socially integrated. Likewise, in a study carried out in Israel, Litwin (2001) showed that morale among the elderly is higher when they are embedded in diverse or friends networks.

Memory can be also affected in older people by the level of their social integration. Ertel et al. (2008) investigated the effect of social integration, measured as marital status, volunteering, and frequency of contact with children, parents, and neighbours on the change in memory over 6 years in a sample of older people in the USA. Higher social integration predicted a slower rate of memory decline. Serious loss of memory is also one of the distinctive characteristics of dementia, hence we may expect that the quality of the social network may be also associated with the development of dementia in a population of elders. Fratiglioni et al. (2000) studied exactly this issue in Sweden in a sample of 1203, 75–90+ year olds (74.5% women) who were initially not demented. The cohort was followed for an average of 3 years at the end of which 14.6% of the participants were diagnosed with dementia. The authors found that a social network that is either poor or limited is associated with a 60% increase in the risk of developing dementia. In a subsequent meta-analysis of longitudinal studies, Fratiglioni et al. (2004) concluded that an active and socially integrated lifestyle later in life protects against dementia, Alzheimer's disease in particular.

The association of cognitive decline in old age with lower levels of social integration and smaller or weaker social networks has been described in various works. In a longitudinal study of 964 community-dwelling 65 years old and older people carried out in Spain, Zunzunegui et al. (2003) measured cognitive decline in those elders over four years (from 1993 to 1997) in relation to social networks, social integration, and also social engagement with friends, children, and relatives. Low levels of social connections and infrequent engagement in social activities (e.g. going to church, social centre for elderly people, visit a park) are predictors of decline in cognitive abilities in the elderly. Greater contact with family and community was associated with lower cognitive decline in both men and women, whereas specific interactions with friends afforded more protection against cognitive decline in women than men.

Therefore, better mental health can be sustained in the elders who keep their morale, happiness and satisfaction with life high, and this can be helped by a better integration of older people into social life.

But some aspects of sociality may also affect the health of older people negatively. For instance, Hislop and Arber (2003) point to the change occurring in old age on the patterns of sleeping, which becomes lighter and more disrupted, leading to fatigue, a condition that affects more frequently older men than women. Snoring (e.g. caused by obstructive sleep apnoea) and periodic limb movements also affect more frequently older men than women. Disrupted sleeping is not only associated with various comorbidities (e.g. heart and respiratory conditions, dementia) but also with social distress between the partners sleeping together, with women usually finding themselves distressed by the disrupted sleeping caused by their partner's snoring. In this case, the couple may tend to decrease the level of social contact by sleeping in different rooms inside their shared habitation. Social contact can also increase the spread of pathogens that can be especially dangerous to old people. As we write this text we are in the middle of an international pandemic of coronavirus (COVID-19) that can spread by close contact between individuals and that has caused higher mortality rates in older individuals (>70 years old, especially men), with mortality rates increasing the more underlying conditions the individual had (Banerjee et al. 2020).

Nevertheless, given the many health and psychological benefits of social integration, whatever means there are that can be used to enhance social integration in the elderly (even through the use of modern computer-based technologies; e.g. Cabrera

Giráldez and Rodríguez Casal 2005; Teixeira et al. 2012), they should be tried and then adopted if successful. Montoro-Rodriguez and Pinazo (2005), for instance, describe a program offered by the University of Valencia in Spain inspired on the *University of the Third Age* (U3A). The U3A was originally implemented in France by Pierre Vellas (e.g. Groombridge 1995; see also the Open University model established in the United Kingdom in 1969, Weinbren 2015). Montoro-Rodriguez and Pinazo (2005) found that old students participating in the U. of Valencia program (average age about 62 years, mainly women) displayed a high degree of social integration and personal growth. Health benefits of social integration are well-established effects of participating in U3A programs around the world (see Bridgstock 2018; Formosa 2019, for reviews).

Although political organisations tend not to be currently regarded as a major source of social experiences and integration by old people, in some countries political participation is more significant than in others. Some old people may find participation in political organisations to have generative value, for instance. Serrat et al. (2017) studied this issue in Catalonia, Spain—a region that has a long historical tradition of people engaging in politics—after collecting data from 182 members of political parties, trade unions, and single-issue organisations. Indeed, older people involved in political organisations had higher levels of generative concerns and generative goals. This can help improve not only their social integration into some sectors of society, but also their well-being by increasing their sense of personal fulfilment and life meaning.

We conclude this section with a brief mention of the cultural context of social integration. Although one important aspect of social integration in the elderly is their relationship with family: both nuclear and extended (stepfamily, in-laws, adopted children), such relationship follows specific *cultural scripts*. Cultural scripts are "prescribed roles and norms for members of a social group… (that) explain the underlying organising principles and the strategies applied for attaining social integration" (Mehta 1997, p. 255). Cultural scripts are clearly not static, and they may vary, to different degrees, not only over time but also in space across society. As an illustration of the diversity in cultural scripts of old age, we quote Mehta (1997, p. 255) who described variability across the Chinese, Malay, and Indians:

> In the Chinese culture, scripts regarding old age emphasise propriety of behaviour, especially in relation to interpersonal relationships. The Confucian concept of *li* …, or 'propriety', is aimed at regulating ethical relations in society. The Malay culture stressed old age as a time of retribution for past sins, and for progress in religious knowledge …. A significant aspect of Malay Muslim cultural expectations is the behavioural dimension. The Malay cultural script stresses adherence to Islamic religion, and places high value on good, helpful behaviour in the daily context within family and community. In the Indian context, in the last stage of the life cycle, *Sanyasa*, there is a strong emphasis on spiritual pursuits and a gradual decrease of worldly involvement. The Indian cultural script stresses personal religiosity, the goal of spiritual realisation and withdrawal from worldly concerns …. The common setting for the enactment of this script is the family, where the older adult's needs are often met.

Thus, we can see how across cultures elders can approach their social lives according to the expectation of various models: *Activity* (e.g. Confucian, Muslim) or *Disengagement* (e.g. traditional Hindu), but in all cases there is a common cultural script emphasising good relationships within the family.

In a study carried out on older people in Singapore (a multi-ethnic society), Mehta (1997) described social integration—in contexts ranging from family to the community at large—to be underpinned by philosophical principles of reciprocity. Supporting each other was expressed by respondents of all ethnic backgrounds in terms of satisfying recipients' need for help. Reciprocity was more intense among closer kin (e.g. following filial piety) extending itself beyond death through the worship of ancestors in all ethnic groups. Ancestors were mainly remembered for their contribution in life, to the family: advising, helping at times of difficulties, child-rearing, cooking, cleaning, doing the laundry, accompanying grandchildren to and from school, and also providing care for a daughter-in-law during her post-natal period, helping in maintaining harmony in the family. Inability or unwillingness of the elderly to reciprocate tended to lead to low social integration, even though such unwillingness might have been justified in the mind of the elder in terms of an expectation of inter-generational reciprocation for past help they provided to their family.

The celebration of ethnic/religious festivals is also an important opportunity to nurture the social integration of the elderly within the family and the community at large across cultures (Mehta 1997). In the Western world, *New Year* is traditionally an opportunity to celebrate in group, as it is *Christmas* for Christians in particular. Malays celebrate festivals such as *Hari Raya Puasa*, the Muslim *Eid al-Fitr* which is a festival that marks the end of the month of fasting, or Ramadan, and it involves visiting friends and relatives, eating special foods and exchanging gifts and cards. The Indians celebrate *Deepavali* and for the Chinese there is the *Chinese New Year*.

Social integration of the elders relies on the maintenance of social harmony, especially within the family. This is achieved in various ways across cultures: preserving traditions such as various filial obligations, observance of family rituals and participation in festivities, maintaining non-confrontational and tolerant attitudes aimed at preserving family harmony, and an attitude of striving for financial self-reliance. In some cultures (e.g. among the Malay Muslims), financially destitute elderly can be socially integrated thanks to concepts such as "universal brotherhood" and *ihsan* (benevolent conduct), with the elder becoming an "adoptive kin", a practice known as *saudara angkat* (Mehta 1997).

To recap, in this opening section we have introduced a broad framework to understand the social integration of the elders. Two major models: *Disengagement* and *Activity*, were explained. The models provide the limits to a spectrum of potential social pathways to fulfilling ageing: from more individualistic to more socially integrated. Fulfilling ageing combines both the individual and the social aspects of the life experience. The social dimension of fulfilling ageing involves an active stance of the elders in society, but also a societal respect for individual priorities and the objective limitations of older people. Greater social capital facilitates fulfilling ageing through improved social cohesion and support. Social integration of the elders relies on a strong sense of attachment among individuals, which can eventually lead to improved general health and psychological well-being, ultimately contributing to an extended lifespan. Nevertheless, the social network of older people may also fail if there are not enough people to interact with, or some elders may prefer to withdraw from social interactions due to personal or cultural motivations.

When the social network fails to integrate old people, the elders may experience loneliness. Loneliness is a complex experience that can have multiple causes, and it is the topic of the next section.

6.2 Loneliness

As a social species, humans are usually not very fond of being alone, especially over the long term, although we can clearly see variability in the population due to personality, individual past experiences, culture, gender (men tend to be comparatively less social than women; e.g. de Jong Gierveld 2003) and age (Day et al. 2018).

Human sociality is rooted in our biology and the development of social behaviour in our species is not only affected by experience, but it is also under a degree of heritable genetic control, leading to a propensity to seek the company of others. This has been demonstrated not only in humans (Fowler et al. 2009) but also in other social species such as yellow-bellied marmots (*Marmota flaviventris*) (Lea et al. 2010; Blumstein 2013) and, among other primates, rhesus macaques (*Macaca mulatta*) (Brent et al. 2013).

The strength of the tendency to live socially in some species is clearly shown when social animals are confined in solitude. A. Poiani remembers an experience as a young undergraduate student doing a feeding experiment on a goat. The goat was kept in a pen, familiar to the animal, as she was offered different food items to choose from; but apparently eating was not high in her priorities, as she kept bleating, seeking contact with the rest of the herd, which was farther away. He finally decided to release the goat to put an end to her distress. The experiment obviously needed a change in design: providing the experimental goat with some company. Was the obvious behavioural distress shown by the goat evidence of her feeling "lonely"?

Being physically isolated from others may lead a person to feel alienated and lonely (Dean 1961), but some people may feel lonely even when they are in the company of others (Wenger et al. 1996; de Jong Gierveld 1998; de Jong Gierveld et al. 2006). Therefore, what is *loneliness*?

Rokach (2012, p. 2; see also Cacioppo et al. 2006) described loneliness in these adaptive terms: "just like physical pain, which, while it is unpleasant, has a protective function and serves as an alarm bell to guard against damage to the organism—loneliness is a social pain that is triggered by disconnection from others, from the group to which the human individual belongs and that can assure us of food, defense, social connection, and support…". Thus, being disconnected from others triggers in us what Moustakas (1961) referred to as *loneliness anxiety* (Perlman and Peplau 1981), which drives us to seek the company of others.

However, a person can also experience *existential loneliness* (Moustakas 1961), which is independent from the presence of others. Existential loneliness can be described as a "reality of being human, of being aware, and of facing ultimate experiences of upheaval, tragedy, and change, the intrinsic loneliness of being born, of living at the extremes, of dying" (Moustakas 1972, p. 20). In a sense, existential loneliness is a result of having lost contact with the self.

On the other hand, existential loneliness may also have some positive effects: "involving periods of self-confrontation and providing an avenue for self-growth. While it can be painful, it can also lead to 'triumphant creation'" (Perlman and Peplau 1981, p. 33). That is, overcoming existential loneliness can be achieved by reconnecting with the self, perhaps in a different way.

Thus, loneliness can produce anxiety that can be decreased by increasing social interactions or that it can be decreased by increasing the existential meaningfulness of our life.

Given the variability in the experiences of loneliness and our reactions to it, Perlman and Peplau (1981, p. 32) adopted a more relativistic and social-cognitive approach to loneliness based on the concept of *cognitive discrepancy* (Pinquart and Sörensen 2001). In their view, loneliness emerges from "a discrepancy between one's desired and achieved levels of social relations". Thus, a person who desires to be in the company of others but for some reason cannot, may feel lonely; whereas another person may desire to be alone for any personal reason (e.g. to concentrate and reflect) and therefore that person will not feel lonely.

Therefore, loneliness is a fundamentally personal experience that emerges from both social isolation and individual characteristics and priorities. This complexity is clearly exemplified in this quote from de Jong Gierveld (1998, p. 74; see also the review by Dury 2014):

> The relationship between objective social isolation and subjective experiences of loneliness is mediated by factors such as the characteristics of the relationships that are available and of the relationships that are not (or no longer) available; the saliency of the relationships that are missed; the time perspective, and the possibilities one sees to upgrade and enlarge one's network of relationships... The subjective evaluation of the network is the intermediating factor between the descriptive, objective characteristics of the network and loneliness. Additionally, the norms and values of a society concerning an optimal set of relationships may influence the risk of becoming lonely, as well as the personal capacities to adjust one's norms and values concerning an optimal set of relationships to changing circumstances. So, people can react to a situation of relational deficits in different ways, for example by resigning themselves to it or by attempting to change it.

de Jong Gierveld (1998, p. 73–74), building on Perlman and Peplau (1981), expanded the definition of loneliness into a multidimensional concept that synthesises the various "positive" and "negative" aspects of the experience of loneliness:

> a situation experienced by the individual as one where there is an unpleasant or inadmissible lack of (quality of) certain relationships. This includes situations in which the number of existing relationships is smaller than is considered desirable or admissible, as well as situations where the intimacy one wishes for has not been realized. Thus, loneliness is seen to involve the manner in which the person perceives, experiences, and evaluates his or her isolation and lack of communication with other people.

Such evaluation of social isolation may be positive (*positive loneliness* or *privacy*), as in the case of withdrawal in order to meditate, or negative (*negative loneliness*) which leads to anxiety (de Jong Gierveld et al. 2006). In Chap. 1 we also referred to *positive loneliness* as *aloneness*, following Ó Luanaigh and Lawlorz (2008), that may lead to a process of voluntary mutual disengagement between individuals (Mutran 1987).

Emotional isolation, just like existential loneliness, can lead to experiencing loneliness even when an individual may not be physically isolated from others (Weiss 1973). Prejudice and discrimination can cause emotional isolation, for instance. On the other hand, some individuals may self-isolate physically without experiencing the psychological pain and distress of loneliness, following specific aspects of personality. In spite of such subtleties and diversity, as a general rule the less socially embedded an individual is the more lonely she/he is likely to feel, simply because through the network of intimate relationships the individual obtains "cohesion, feelings of belongingness, and protection against loneliness" (de Jong Gierveld and van Tilburg 1995, p. 158). When the network fails, loneliness may ensue.

Emotional loneliness may emerge from emotional isolation when important emotional ties are broken and an attachment figure is lost, such as in some cases of divorce and widowhood; whereas *social loneliness* may develop when old social ties are broken (e.g. through migration) and new ones are not established yet in the new social environment. Thus, social loneliness stresses the emotional impact of lack of social integration; whereas emotional loneliness stresses the absence of an attachment figure (Heylen 2010).

In some cases, individuals who are lonely, but not personally inclined to loneliness, may reach a point where they confront their social isolation with a degree of *apathetic acceptance* (Perlman and Peplau 1981).

Rokach (2012, see also Moustakas 1961; Perlman and Peplau 1981) listed three major characteristics of the experience of loneliness:

a. Loneliness is a phenomenon found universally among humans and it is an intrinsic part of our human nature.
b. Although it is a general experience, loneliness is also a subjective experience that varies individually and between situations.
c. Despite its complexity, loneliness is more often than not a painful experience.

Factors that increase loneliness include: individual characteristics such as shyness, low self-esteem, poor social skills; degree of similarity/dissimilarity with other members of the social group, gender, marital status, income, and age, but also cultural factors and social norms, among others (Perlman and Peplau 1981). Loneliness can be also affected by interactions with and feedback from the social milieu that may even enhance the effect of personal characteristics. *Stereotypes* and *attributional factors* ("Is the person to blame for his/her loneliness?") for instance, are important social modulators of personal reactions to loneliness. Hence an already shy person who feels the pressure of social discrimination is likely to feel even more lonely than others and to attribute some of that loneliness to some personal "failure" or "inadequacy" derived from shyness.

Belongingness, embeddedness or *connectedness* are the opposite of loneliness (de Jong Gierveld and van Tilburg 1995; de Jong Gierveld et al. 2006; Stanley et al. 2010) and the lowest level of social contact that can prevent experiencing loneliness in any specific individual is known as the *loneliness threshold* (Johnson and Mullins 1987). Needless to say, the loneliness threshold varies from person to person and across cultures (e.g. Jylhä and Saarenheimo 2010).

Belongingness is "a subjective sense of … 'being a part of' in order to avoid feelings of loneliness and alienation" (Lee and Robbins 1995, p. 232) and it has been suggested by Lee and Robbins to be composed of three aspects: companionship, affiliation, and connectedness; whereas *social embeddedness* can be defined as "the composition and the functioning of an individual's network of social relationships" (de Jong Gierveld 2003). We would expect that loneliness should decrease the more socially embedded and connected elders are, and therefore the greater their sense of belongingness is. This is affected by the size of the network, but the quality of the relationships established within it is especially important.

In a study carried out in The Netherlands, de Jong Gierveld (2003) investigated the size of the social network of 1190 participants 55–89 year old, describing a larger network size for women than men (see also Aronson 1992). This was associated with lower scores of loneliness for women than men, with women being also more dependent on their social network. Thus, a larger social network and more dependency on it are associated with lower loneliness. But in fact, even relatively less intimate *living apart together* (LAT) relationships (e.g. Borell and Ghazanfareeon Karlsson 2003) can decrease the level of loneliness, suggesting that social embeddedness can take various shapes within a diversity of social networks, that can be useful in decreasing loneliness according to circumstances.

Older individuals are particularly at risk of feeling lonely, as their social network may reduce over time (Cornwell et al. 2008) thus decreasing their social embeddedness, without other alternatives being necessarily available; whereas health and financial resources may also deteriorate, which may compound any additional factor associated with personality and cognitive/emotional condition (Fees et al. 1999). However, older people may compensate for the decrease in the size of their social network by increasing the strength of connectedness with the members of the reduced network (Cornwell et al. 2008).

Prevalence values of loneliness in the elderly vary among studies but they could reach highs of up to 40% (Routasalo and Pitkala 2003; Hawkley and Cacioppo 2010), with cross-countries studies also indicating variability. Interestingly, Routasalo and Pitkala (2003; see also Jylhä and Jokela 1990; Fokkema et al. 2012) mention European data suggesting higher prevalence of loneliness in some southern European countries (e.g. Greece (36%) and Portugal (23%)) than in Scandinavian countries (Denmark and Sweden (4–6%)) (see also Table 6.5 showing a clear North-South gradient in loneliness among the elderly in Europe). The same result was obtained by Jylhä and Jokela (1990) after directly comparing older people in Tampere, Finland, and rural Greece. Prima facie this is counter-intuitive given the

Table 6.5 Older adults who often feel lonely in various European countries (%)

Denmark	<5
Finland, Germany, Netherlands, UK	5–9
Belgium, France, Ireland, Luxembourg, Spain	10–14
Italy, former Yugoslavia	15–19
Greece, Portugal	>19

From Dykstra (2009)

culturally entrenched concept of familism in southern Europe that is supposed to promote social support to older family members. But we have seen how modernisation has been breaking old cultural habits of co-residence and family cohesion, thus it is possible that older generations in northern Europe, with their stronger emphasis on self-reliance, but also with their usually better organised social welfare systems, higher economic standards and lower levels of socio-economic inequality are better endowed with the psychological, social and material tools to overcome loneliness in older ages than in southern Europe. This is also supported by data on suicide rates among the elderly which tend to be higher (in both men and women) in southern than northern European countries (Pritchard and Baldwin 2000). These results point to the importance of the *normative climate*, or set of culturally determined expectations, that can vary inter-culturally and that can regulate loneliness in each specific case (de Jong Gierveld et al. 2006).

Among the elderly, loneliness tends to be especially prevalent in rural areas (but with exceptions; e.g. Jones et al. 1985), at older ages, in those living alone or in a residential home, among the widowed, men (but sometimes it is old women who feel more lonely; e.g. Jones et al. 1985; de Jong Gierveld and van Tilburg 1995; Golden et al. 2009; Aartsen and Jylhä 2011), caregivers, those with low level of education, low income, poor health, poor functional capabilities (including loss of vision or hearing), and decreased social network (e.g. lack of friends) (Kivett 1979; Pinquart and Sörensen 2001; Routasalo and Pitkala 2003; Savikko et al. 2005; Hazer and Boylu 2010). Victor et al. (2000) list some correlates of loneliness (and of living alone and in isolation) in old age that are summarised in Table 6.6.

Although some factors such as older age, widowhood, having no children, entry into an aged care institution and bereavement are positively correlated with living alone, isolation and also loneliness; sometimes living alone (e.g. because the old person never married) is not associated with loneliness, whereas an old person may feel lonely even if she or he may have physical access to a social network.

Victor et al. (2000) (see Fig. 6.2) synthesised the relationship between isolation and loneliness as it is affected, directly or indirectly, by various factors such as age, sex, ethnicity, health, and other resources, including social resources, and various life events that are often beyond the control of the old person, as they were summarised in Table 6.6. We agree with Victor et al.'s model that loneliness is a complex multicausal phenomenon and that isolation and loneliness also affect each other: more feelings of loneliness may increase social isolation and, vice versa, more social isolation may further increase feeling lonely in many elders.

Table 6.6 Correlates of living alone, isolation and loneliness

	Living alone	Isolation	Loneliness
Personal circumstances			
↑Age	+	+	+
Female	+		
Ethnicity	?		
Widowhood	+	+	+
Never married	+		
No children	+	+	+
Resources			
↓Social class	+		
Income			
Physical ill health	+	+	
Mental ill health	+		
Restricted social network	+		
Life events			
Retirement migration	?	+	+
Entry into care	+	+	+
Bereavement	+	+	+

From Victor et al. (2000)

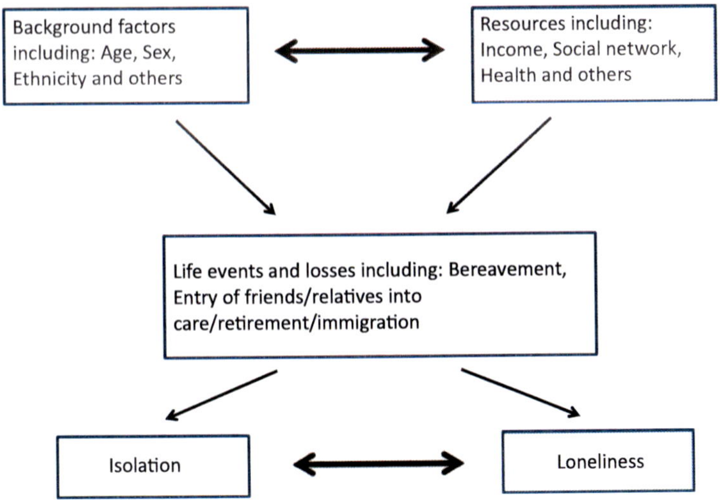

Fig. 6.2 Model of the relationship between life events, resources, and loneliness/isolation. Slightly modified from Victor et al. (2000)

Age effects on loneliness are not linear, but curvilinear (Fig. 6.3), with a rapid increase in loneliness among the oldest-old (Pinquart and Sörensen 2001; Dykstra 2009), an effect that can be explained by factors such as the loss of partner, reduced social activities, moving into aged care facilities, a fast deteriorating health, and increased feelings of low mood, uselessness, and nervousness (Aartsen and Jylhä

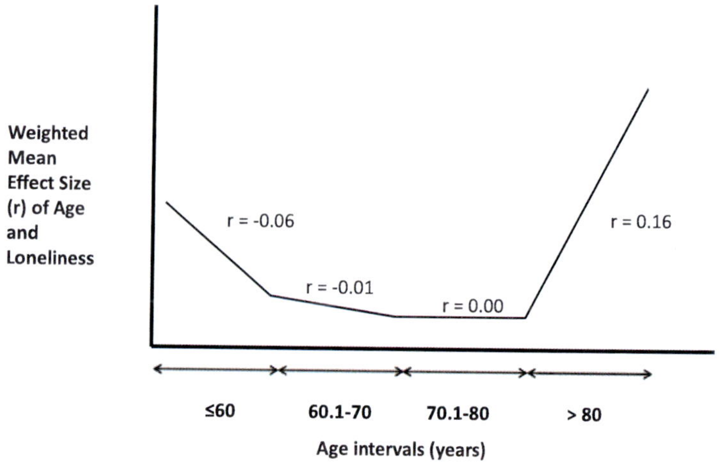

Fig. 6.3 Change in the weighted mean effect size (*r*) of age and loneliness in four age intervals (data are from Table 2 of the meta-analytical study of Pinquart and Sörensen 2001, similar results were reported by de Jong Gierveld and van Tilburg 1995; Dykstra 2009). Age and loneliness are negatively correlated for people younger than 60, the relationship starts to flatten between 60 and 70 years old to then flatten between 70 and 80, and it finally sharply raises above 80 years old. That is, the oldest people rapidly feel lonelier as they become older

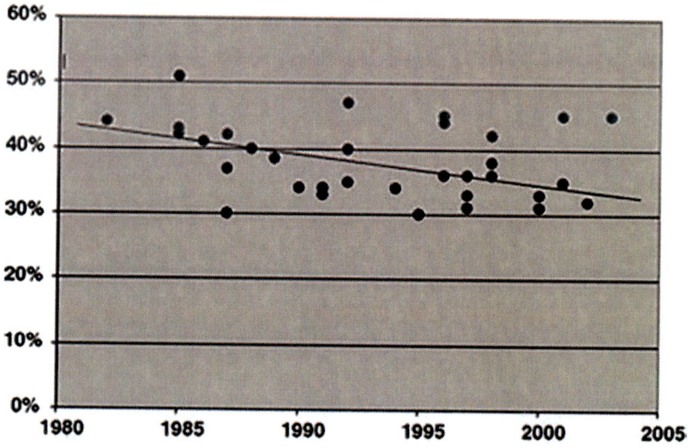

Fig. 6.4 Repeated assessments of loneliness in the Netherlands (% lonely) over a 20-year period (from Dykstra 2009; based on van Tilburg 2005)

2011). Regarding social contacts, Pinquart and Sörensen (2001) reported that it is the quality, even more than the quantity, of social contacts that can decrease loneliness, especially social contacts with friends. This may partially explain why the level of loneliness has been decreasing over time in the general population in some countries in recent years (Fig. 6.4), as people may have more chances of interacting with relevant others in various ways, including travelling and by using digital communication. It is a *well-integrated social network* that helps in decreasing loneliness, rather than just being physically surrounded by other people (Golden et al. 2009).

An accumulation of negative life events can predict the development of loneliness in old age, in interaction with personality traits such as neuroticism. In particular, decreased perception of competence as affected by negative life events increases loneliness (Hensley et al. 2012). But predictors of loneliness in old people are varied and they include both personal and social factors. Theeke (2009 and references therein) lists: self-report of poor health, including mental health, living alone and low social contact, motor impairment, domestic violence, unemployment and low income, low education, poor self-efficacy.

Loneliness can result in reduced quality of life (Victor et al. 2000; Ivbijaro 2013), with lonely elderly people being sometimes stigmatised in their community (Stanley et al. 2010). Loneliness can also increase the rate of physiological ageing, thus leading to poorer physical health and higher mortality rates. Mental faculties also tend to deteriorate with loneliness in some old individuals because of increased depression, hopelessness, impaired cognition, and higher risk of dementia (Golden et al. 2009; Hawkley and Cacioppo 2010; Mushtaq et al. 2014). Physical and mental deterioration among lonely older people may be in part mediated by dysregulation of the immune system (Hawkley and Cacioppo 2010 and references therein).

In a study carried out in the USA involving 755 participants, 60 years old and older, who were resident of New Mexico, Tomaka et al. (2006) found that feelings of belongingness, hence less loneliness, were positively associated with better outcomes for diabetes, hypertension, arthritis, and emphysema; although worse outcomes were detected for liver disease, a result that could be explained by higher consumption of alcohol in more socially integrated elders in this sample. Heart disease was associated with living alone in this study, and social support was negatively correlated with stroke. Increased probability of having a coronary condition was also detected in more lonely elders (58–90 years old) in a study carried out in the USA by Sorkin et al. (2002).

In rural China significant migratory movements towards the large cities have occurred in recent decades, especially among the younger generations, in order to seek work and a better material future. Such migratory flow has resulted in older parents in rural areas being potentially exposed to the loneliness derived from the *empty nest syndrome* (Fahrenberg 1986; Pillay 1988) and its potential effects on health. Empty nesters do not live with their children, who have become independent and left their family home. Liu and Guo (2007) compared empty-nest elders with not-empty-nest elders and found that empty-nest elders scored higher in loneliness and they also scored lower in both mental and physical health variables.

Kaasa (1998) reported the results of a study on loneliness in 202 elderly people (over 80 years old) carried out in Norway. Loneliness was significantly associated with low self-perceived health, low vision, poor hearing, low levels of performance of activities of daily living (ADL), and social variables such as loss of a spouse and a smaller social network. The author correctly stressed the need to more clearly understand the causative connection between loneliness and some aspects of poor health: is loneliness leading to the development of health problems or are older people already affected by health problems more prone to feel lonely? Potentially, both situations may occur in specific cases and indeed they could reinforce each other through feedback: loneliness may conduce to health deterioration, which may increase loneliness, that could lead to further health deterioration, and so forth. The same positive feedback loop may be triggered starting from an initial state of health deterioration that could increase loneliness.

Depression is among one of the most common mental effects of loneliness in older people (e.g. Prince et al. 1997; McInnis and White 2001; Luanaigh and Lawlorz 2008; Aylaz et al. 2012) which can be mitigated by social and material (e.g. financial) support. When they are left unaddressed, depression and loneliness may potentially lead to suicide. In fact, being lonely and living alone are associated with *suicide ideation* and *parasuicide* (a suicide attempt) (Stravynski and Boyer 2001). In a study carried out in the USA by Duberstein et al. (2004), they found that among 50 years old and older people, suicide was less likely in participants who were married, had children, and lived with family or, more generally, actively participated in their community. That is, suicide is less likely in people who enjoy better social support that can decrease loneliness and depression (Cacioppo et al. 2010).

Given the increasing concern about dementia in old age, it is also important to understand the relationship between loneliness and the development of dementia. In a four-year longitudinal study carried out in the USA, Wilson et al. (2007) described a more than double risk of developing Alzheimer's disease-like dementia in lonely as compared with not-lonely older people, although loneliness was not related to amyloid plaques, neurofibrillary tangles, or cerebral infarctions in this study. Lonely elders had already a lower level of cognition at baseline, which declined more rapidly throughout the study. The authors suggest that "loneliness might somehow compromise neural systems underlying cognition and memory, thereby making lonely individuals more vulnerable to the deleterious effects of age-related neuropathology (ie, decreasing neural reserve)" (p. 239). In a recent meta-analysis, Penninkilampi et al. (2018) concluded that dementia risk increases with poor social engagement and support whereas although the risk of dementia tends to increase with greater loneliness, the association is statistically not significant in their study, but only marginally so ($P = 0.063$). However, in a recent study that was not included in Penninkilampi et al.'s (2018) meta-analysis, Zhou et al. (2018) found that loneliness is positively associated with developing dementia, the same conclusion was reached by Lara et al. (2019) in a more recent meta-analysis.

Loneliness can also increase in old people after they have developed dementia. In a Swedish study carried out within the *Kungsholmen* longitudinal project, Holmén et al. (2000) compared subjective social and emotional loneliness, as measured through interviews, in demented and non-demented old people. Although the two groups did not differ in terms of their emotional experience of loneliness, the non-demented participants reported lower levels of loneliness, social loneliness in particular. More loneliness was also detected in mild AD patients as compared with healthy controls by El Haj et al. (2016).

Hence loneliness could potentially speed up the development of dementia among those who already show some propensity for it, and dementia itself could increase the level of loneliness in the old person once it has developed.

Social isolation as such and the psychological state of loneliness can have independent negative effects on health in old people. In a study carried out in the USA, York Cornwell and Waite (2009) studied the association between *social disconnectedness* (e.g. reduced social network, infrequent engagement in social activities) and *perceived isolation* (e.g. loneliness, perception of lack of social support) and physical and mental health in older adults (57–85 years old), concluding that both social disconnectedness and perceived isolation are associated with low levels of physical and mental health.

Luo et al. (2012, and references therein) summarise the results of various studies on the negative effects of loneliness on physiology and mental health. Such negative effects include increased depression, impaired sleep function, lower physical activity, impaired mental health and cognition, increased vascular resistance and systolic blood pressure, enhanced hypothalamic pituitary adrenocortical activity, and altered immunity and increased risk of developing dementia (Wilson et al. 2007), potentially contributing to increased probability of mortality.

Given all those physical and mental effects of loneliness in the elderly, special attention should be given to loneliness in older hospital patients who are already suffering from specific health problems and may also experience lack of mental concentration and engagement (*cathectic deprivation*) (Proffitt and Byrne 1993).

Loneliness can be managed in various ways including through adopting a positive attitude in life, engaging in meaningful activities (Stanley et al. 2010), and increasing the quality of social contacts (Perlman and Peplau 1981; Holmén and Furukawa 2002). More broadly, we have mentioned above how belongingness, embeddedness, and connectedness can help decrease loneliness.

Rokach and Brock (1998) listed six strategies to cope with loneliness: (1) acceptance, (2) self-development/understanding, (3) (re-)establishing social relations, (4) denying the problem, (5) seeking comfort in religion/faith, and (6) taking part in social activities (see also Patzelt 2017).

Various therapeutic approaches have been adopted in order to decrease loneliness among the elderly. One approach to decrease emotional loneliness that focuses on self-development/understanding is *reminiscence therapy* (Chiang et al. 2010). This is structured around recalling good and positive or, alternatively, sad and profound experiences, thus giving an opportunity to the elder to realise that he/she has lived a meaningful life. The effectiveness of the exercise increases when reminiscing occurs as a group experience.

Given that loneliness can develop in elders who lack social support, providing the elder with a supportive social network can be of significant help in managing loneliness in old age. In the case of elders who are widowed—or lack a partner, more broadly—social support may come from family members, friends or public and private institutions. Tomassini et al. (2003) summarised the sources of help for two samples of elders (men and women) in Italy and Great Britain (Table 6.7). The most important sources of help are family and private help, followed by friends and

Table 6.7 Percentage of unmarried men and women aged 65 and over who received care from different sources in Italy and Britain

	Help from family members	Help from friends	Public help	Private help	Any help	*N*
Italy						
Men	13.6	3.9	1.0	14.5	29	2576
Women	14.3	4.3	2.0	12.3	27	709
Total	14.1	4.2	1.8	12.7	27	3285
Great Britain						
Men	20.1	6.3	15.0	11.6	43	379
Women	32.2	8.7	17.9	13.1	55	980
Total	28.8	8.0	17.1	12.7	52	1359

From Tomassini et al. (2003) see also references therein

public help, which are variable between the two countries: British elders tended to receive more care from various sources than Italian elders. Women were more likely to receive help than men in the British sample, whereas help was more evenly received by men and women in the Italian sample.

However, as we have already seen, loneliness is not always perceived negatively by the elders. See, for instance, the positive aspects of existential loneliness noted by Perlman and Peplau (1981) and the positive aspects of loneliness-as-privacy mentioned by de Jong Gierveld et al. (2006) that we referred to at the beginning of this section. Sometimes old people may perceive some positive benefits of loneliness in terms of having an opportunity to grow personally, whilst focusing on the self (*positive loneliness* or *aloneness*). This can be especially beneficial when they can still access the company of others when needed. In this case elders can benefit from the enriching relationship with friends and family when required, as women often do (Rokach et al. 2007), whilst finding their own personal space to develop individual interests when in the mood to do so.

In a review of strategies to alleviate loneliness in older people, Cattan et al. (2005) listed various types of programs based on group or one-to-one activities, the provision of various kinds of services and community-based strategies. The most effective types of interventions involve group approaches with an educational focus, or programs based on targeted support activities. In general, one-to-one interventions conducted in the elder's home are not very effective. This clearly points to the crucial importance of engaging in social activities to help older people overcome issues of loneliness. In further support of this view, Moorer and Suurmeijer (2001) provided results from a study of 723 elderly people (66 years old and older) from the city of Gröningen, The Netherlands, indicating that loneliness is negatively associated with the size of the social network. The same result was obtained by van Tilburg et al. (1998) comparing groups of elderly Italians from Tuscany and Dutch (55–89 years old): the Dutch scored lower in loneliness due to them being better socially integrated within a larger network, that included their local community and it was not mainly restricted to family, as it was the case in the Tuscan sample.

These days the social network can be further extended through the Internet. Sum et al. (2008) have described lower levels of loneliness following the use of the Internet for communication purposes among Australians older than 55. In fact, more emotionally lonely people in this study tended to recur to the Internet, in a quest to find new people to interact with.

Providing social support to elders can be important to decrease loneliness especially when such elders need assistance with the routine performance of various activities of daily living. Appropriate others providing such assistance allow the elders not only the chance of overcoming their physical limitations, but also to benefit from social contact and emotional support, as shown by Drageset (2004) in a study of 113 residents (65–101 years old) carried out in various nursing homes in Bergen, Norway.

In sum, loneliness has both personal and social causes. Feeling lonely may be a result of lack of social contacts but also a characteristic of personality. Loneliness tends to cause anxiety and even depression, but some aspects of existential loneliness may respond to a personal need for more privacy and they may even conduce to personal growth. Therefore, loneliness in old age is a complex phenomenon. Embeddedness is the opposite of loneliness, and greater social embeddedness in the elderly tends to increase their well-being. Loneliness in the elderly varies across ethnic groups, socio-economic status, sex, and age, all of which interact in various ways with social isolation. Loneliness can have negative health effects on the elderly, being associated with a trend towards increased mortality rates. Coping with loneliness requires both individual and social resources.

We have seen how social interactions can have benefits for the elders in terms of decreasing loneliness, but being in contact with others also exposes the elders to potentially negative and abusive experiences, an issue that will be reviewed in the next section.

6.3 Abuse

The scientific study of elder abuse is relatively recent and the early pioneers had to overcome some initial skepticism about the frequency and relevance of such abuse in Western, modern and economically developed societies, ever since the phenomenon of *granny battering* was described in Great Britain in the mid-1970s (e.g. Baker 1975; Beck and Phillips 1983; Pillemer and Finkelhor 1988).

The *World Health Organisation* (WHO 2014, p. 8) defines elder abuse as: "A single or repeated act, or lack of appropriate action, that causes harm or distress to an older person". In this WHO definition, harm or distress include neglect, serious loss of dignity and respect, abandonment, physical, sexual, psychological, emotional, financial, and material abuse.

Over the years, there have been various approaches to understanding the causes and dynamics of elder abuse. Much of the theoretical effort has focused on the direct effects of interpersonal relationships on elder abuse and the role of social context, along with the effects of various additional factors (e.g. see Bergeron 2001). *Social Exchange* theory, for instance, posits that disruption of the balance in the exchange of material and non-material goods in the relationship may lead to abuse from the part of the party who feels shortchanged (Homans 1958). On the other hand, from the perspective of *Social Learning* theory elder abuse may simply derive from a process of learning, where an individual may have incorporated the use of violence as an acceptable way of resolving conflicts (Bandura 1978). The *Caregiver Stress* theory points to the effects of stress on caregivers leading to frustration, loss of emotional control and abuse of the elder who is in their care (Anetzberger 2000). In cases where the abuse is persistently directed to old people of a specific sex, women, for instance, some feminist theories point to the issue of sex-biased power relationships leading to violence and abuse (*Power and Control* theory, e.g. Brandl 2002).

Other theories have integrated various levels of analysis. For instance, *Positioning* theory links interpersonal relationships, institutional effects, and societal/cultural factors, such as ageism, in the explanation of elder abuse (van Langenhove and Harré 1999). Another multidimensional approach is provided by *Social Ecology* theory which tries to understand elder abuse in terms of complex interactions between individuals (micro-system), local community and institutions (meso-system), society at large (exo-system), and the more general cultural values prevailing in society (macro-system) (Bronfenbrenner 1979; Norris et al. 2013; Wango et al. 2014).

In other words, the theoretical effort directed to understanding elder abuse has strongly emphasised social dynamics of power, habits, discrimination, and culture.

In an early review of elderly abuse by both formal (e.g. institutional, as in nursing homes) and informal (e.g. family members, neighbours) caretakers, Kimsey et al. (1981) reported various types of abuse: physical, psychological, material, and fiscal. Physical abuse takes more commonly the form of physical neglect, but physical violence can also occur as in the case of *granny battering* mentioned above (Giurani and Hasan 2000), with such abuse often hiding behind excuses such as the elder "was going to die anyway". Psychological abuse may be seen in attitudes such as infantilising the elder, causing stress through threats and other forms of verbal abuse, fear, violence, isolating the elder, deprivation and causing feelings of shame and powerlessness (Johannesen and LoGiudice 2013). Material abuse can involve blatant theft, whereas fiscal abuse may include extracting money from the elder in various inappropriate or even illegal ways. More recently, Wyandt (2004) has pointed to some disturbing findings suggesting that young people tend not to recognise elderly abuse so easily, whereas for some older respondents in some studies even single acts of abuse can be of concern, with perception being variable culturally, suggesting that ethnicity must be taken into account when trying to understand abuse of old people.

More broadly, Gorbien and Eisenstein (2005, see also Brandl 2002, Anetzberger 2005) provided a classification of forms of abuse that include:

- Physical abuse
- Sexual abuse

- Emotional or psychological abuse
- Financial or material exploitation
- Abandonment
- Neglect
- Self-neglect (this may result from personal demoralisation or socially promoted demoralisation)

To this list, Schiamberg and Gans (2000, see also De Donder et al. 2011) add:

- Violation of personal rights (including human and civil rights such as violation of privacy, deprivation of autonomy and freedom).

Some forms of abuse, such as sexual abuse, may combine physical and psychological aspects (e.g. Benbow and Haddad 1993; Teaster and Roberto 2004). In addition, Cohen (1994) point to the phenomenon of "journalistic elder abuse" which is the use of pejorative labels to refer to the elders in the media (e.g. *greedy geezers*). Such abuse through the media can be particularly damaging when it can influence the perception of public opinion about the elders, reinforcing old negative stereotypes or even introducing new ones (e.g. van Dijk 2011). Abuse, however, is sometimes resisted by the elders in their attempt to defend themselves (Homer and Gilleard 1990).

Pillemer et al. (2016) report the following prevalence values for various forms of elder abuse from studies carried out around the world: Physical abuse (0.2–4.9%); sexual abuse (0.04–0.8%), financial abuse (1.0–9.2%); emotional/psychological abuse (0.7–27.3%); neglect (0.2–5.5%). When care recipients are interviewed, prevalence values of overall abuse vary from 21% to 25.7% (35.2% in the study of Dong et al. 2007); neglect in particular can reach heights of 18% (16% in a rural Spanish village studied by Garre-Olmo et al. 2009), verbal abuse 14% (51% in the UK, Cooney et al. 2006); 6–18% financial abuse, 4% physical or sexual abuse (Cooper et al. 2008). Focusing on responses from caregivers, Cooper et al. (2008) reported that 5% of family caregivers admitted physical abuse towards care recipients with dementia, whereas 16% of care home staff admitted significant levels of psychological abuse. Older people in institutions have been reported in the same review by Cooper et al. to suffer relatively high levels of abuse ranging from 11.9% to 53%, but the rate could be even higher as many instances of abuse in institutionalised elders are not reported. Cooper et al. (2008) indicate that more than 80% of members of staff in nursing homes have observed instances of abuse.

Risk factors for elders abuse include: advanced age, race, low socio-economic status, low level of education, social isolation and loneliness, substance abuse by elder or caregiver, caregiver stress, poor communication with caregiver, elder showing demanding and/or aggressive behaviour, functional disability and poor health, cognitive impairment (e.g. dementia), depression, belonging to a gender minority or other minority, previous history of family violence, history of psychological problems (Campbell Reay and Browne 2001; Wyandt 2004; Gorbien and Eisenstein 2005; Dong et al. 2007;Wiglesworth et al. 2010; Walsh et al. 2011; Johannesen and LoGiudice 2013; Pillemer et al. 2016). In addition, a *spoiled identity* ("a change for the worse in the perception of the identity of the elder", Beck and Phillips 1983) may decrease respect for the elder thus lowering the threshold for abuse. Both elderly men and women can suffer abuse.

Some risk factors for elder abuse are specific to some regions. In some parts of Africa, for instance, it is not uncommon for elderly people—women in particular—to be abused (even killed) after being accused of witchcraft (Ferreira 2005; in Europe elderly abuse related to witchcraft waned in the eighteenth century, Thurston 2001).

Giurani and Hasan (2000, p. 218, see also Campbell Reay and Browne 2001; Kurrle and Naughtin 2008) provide the following list of circumstances in which an individual is more likely to commit abuse against the elderly:

- Most abuse is perpetrated by adults who care directly for an older person. As most of these caregivers are close family members, abusers are usually either spouses (40%) or children and grandchildren (50%).
- The abuser has looked after the abused person for a considerable amount time.
- Alcoholism and drugs misuse are common among the abusers.
- History of mental illness and depression.
- Cognitive impairment or dementia.
- History of long standing interpersonal problems between the caregivers and their dependents.
- History of physical abuse in the early life of the caregiver.
- A caregiver who is under considerable stress because of inadequate support (but see Brandl 2002, for an alternative view).
- History of isolation, and some dependence of the abusers (e.g. financial and emotional) on the abused.
- Finally, it is important to remember that the abuse of older people has not only been increasingly recognised in the institutional setting, but that we may be seeing just the tip of the iceberg. Low staff-morale, negative attitudes towards the elderly, low pay, and low professional status of working with elderly residents are important factors increasing the probability of elder abuse.

As indicated by Giurani and Hasan (2000), elder abuse can occur in various settings, which include home, hospital, assisted living arrangements, and nursing homes (Wilson 1994; Wolf 1996; Lachs and Pillemer 2004; Fisher and Regan 2006). Abuse in aged care facilities can have a complex causation (e.g. Griffin and Aitken 1999), involving institutional organisation and culture, available infrastructure, individual characteristics of caretakers and interactions between caretakers and challenging elders (e.g. sufferers of dementia). In a study carried out in German nursing homes, Goergen (2001) reported 79% of staff indicating that they had abused or neglected a resident at least once during a given period, whereas 66% had witnessed acts of resident victimisation performed by colleagues. On the other hand, 59% of staff in the same study reported having been the target of physical or verbal aggression by residents. Various explanations have been put forward to explain elder abuse by staff in long-term care facilities (Goergen 2001), these include:

1. Caregiver stress and burnout.
2. Behaviours resulting from offenders' personality disorders and various unresolved biographical problems.
3. Group behaviours emerging from professional and institutional subcultures.
4. Differences in power between elder and institutional caregiver.
5. Lack of appropriate control of life in the nursing home.

Ramsey-Klawsnick (2000) proposed a classification of offenders in cases of elder abuse that includes five profiles:

1. the *overwhelmed* (the burden of care produces stress that leads the caretaker to breaking point and abusive reactions);
2. the *impaired* (caretakers may have some impediment to providing effective care and may cause distress or harm often without realising it);
3. the *narcissistic* (they are motivated by personal gains and may financially abuse the elder under their care);
4. the *domineering* or *bullying* (they abuse the elder but are ready to justify their behaviour and blame the elder for it); and
5. the *sadistic* (causing extremely dangerous levels of abuse).

Offenders may have also been victim of abuse in childhood (Yan and Tang 2003). In addition, a caregiver may be more predisposed to abuse under circumstances of depression, anxiety, experiencing low morale, feeling guilty and angry as a consequence of providing care for an older person (Giurani and Hasan 2000). Caregivers may also become abusive when they are guided by cultural views such as ageism and sexism that promote *power imbalances* and *oppression* between members of a social group (Walsh et al. 2011).

Elder abuse by caregivers is therefore rather common (Homer and Gilleard 1990; Cooney et al. 2006) and abusers may include family members such as children, spouse, siblings, and various other relatives (Steuer and Austin 1980; Bagshaw et al. 2009). The greater the dependence of the elder on the family caregiver the more likely it is that a family caregiver may commit abuse on the elder (Yan and Tang 2004), especially when the caregiver lacks sufficient support. Abuse by a caregiver derives from various co-occurring factors such as personal situation and characteristics of the caregiver, personal condition and characteristics of the elder, local and broader environmental/social conditions (Schiamberg and Gans 2000). In addition, there is also a temporal dimension that involves the accumulated experience of interactions between caregiver and elder over a lifetime. Such "ecological" complexity has been summarised by Schiamberg and Gans (2000) in their *Applied Ecological Bi-Focal Intergenerational Model.*

The temporal dimension we mentioned above, involving a cumulative experience of abuse over a lifetime, could make interventions on abused elders more difficult. In a study of focus groups carried out in Ontario and Alberta, Canada, Walsh et al. (2007) described four temporal pathways to elder abuse: (a) inter-generational cycles of abuse, (b) violence across the lifespan, (c) exposure to multiple subtypes of elder abuse, and (d) ongoing spouse abuse that shifted into elder abuse. In many instances, elders were suffering abuse in silence; some of the reasons that the elders gave to explain this attitude included: "fear of retaliation", "shame or embarrassment", "love of children", and "not wanting to put blame on their family". Therefore, cases of elder abuse within a family may be protracted over a long time and the ongoing process may not be stopped if the abuser is relentless, no external witness intervenes, and the elder him/herself cannot or does not want to accuse the abuser.

Assessing whether elder abuse is occurring is something that should be done carefully, thoroughly but also speedily (e.g. Lowenstein 2009). Recognising physical and psychological signs of elder abuse allows for the provision of appropriate medical, psychological, and legal assistance to the abused elder. Wyandt (2004, see also Giurani and Hasan 2000) lists various signs that could alert people interacting with elders about potential abuse: bruises, burns, scars, fractures, various other forms of injuries, effects of overmedication or undermedication and non-compliance with medical treatment, unexplained sexually transmitted diseases, dehydration, malnutrition, ulcers, poor personal hygiene, extreme withdrawal, depression, agitation, sleep disturbances, eating problems, recurring headaches, high levels of psychological distress, fearfulness, anger, passivity, embarrassment, suicidal ideation, and helplessness. As such these are signs that should raise concern, but also start an investigation when it is regarded appropriate and their origin is doubtful, to ascertain their actual cause.

Various strategies could be adopted to prevent elder abuse (Peri et al. 2009; Ash 2013; Pillemer et al. 2016):

a. Interventions on caregivers: relieve caregiver burden, education, support groups. It will be often necessary to help the caregiver remove the "cognitive mask" that prevents her/him seeing the abuse suffered by the elder. In the words of Ash (2013): "The cognitive mask does not hide what is underneath; rather, it obfuscates and obscures what is 'out there'".
b. Money management: safe help in managing money and paying bills provided by accredited money managers.
c. Helplines: telephone helplines attended by volunteers or paid professionals.
d. Emergency shelters: providing a safe accommodation to gain time and determine what to do next.
e. Multidisciplinary teams: they can provide a more coordinated response to the needs of the abused elder from various specialised perspectives (legal, medical, psychological, financial).
f. Legal provisions.

In addition, societal interventions are warranted, including targeting the media through pressure groups, to promote a more positive image of old people and ageing (Peri et al. 2009). This could help achieve cultural change that can lower the risks of abuse and decrease the stereotyping and discrimination coming with ageism. More broadly, the more effective social support is for the individual, the more likely it is that potential cases of abuse could be prevented; and once abuse occurs, social support could help in detecting, reporting, and ending it. The positive effect of social support in preventing elder abuse has been shown, for instance, in a study carried out in seven European communities: Ancona (Italy), Athens (Greece), Granada (Spain), Kaunas (Lithuania), Ludwigsburg (Germany), Porto (Portugal), and Stockholm (Sweden), by Melchiorre et al. (2013).

To recap, elders may suffer abuse in a wide range of settings: family, local community, institutions; and in various ways: physical, psychological, material, and fiscal abuse. The prevalence of elder abuse is variable in society following the distribution of various risk factors. Abuse within a family setting is often the result of caregiver burnout or the caregiver suffering from a psychological disorder. Signs of elders' potential abuse such as malnutrition, external evidence of potential violence, distress, and fearfulness should be investigated to determine their real cause. If abuse is confirmed, appropriate action should be taken.

We next shift our analysis of sociality and the elders to the specific context of the family.

6.4 Family

The family can be defined in many ways, whether it is from a biological, psychological, sociological, historical, or legal perspective. However, here we simply regard the *family* as a social unit of individuals who are linked by genetic and/or social relationships. The genetic and the social are two independent variables in the definition of the family. Thus, a family may include father, mother and their biological children; or just one parent and his/her biological children; or a homosexual or heterosexual partnership with the biological children of one partner, both as in *blended families*, or neither (adopted children). In particular, the *nuclear family* is composed of parents and children, whereas the *extended family* includes grandparents, aunts/uncles, cousins.

Although current means of communication and easy access to fast modes of transport allow the elders to establish and maintain social contacts with many people over a vast geographical scale, and the availability of different forms of social interaction (clubs, various organisations, institutions, charities, churches, and others) allows them to be more independent from the traditional social ties with their family, the family (both nuclear and extended) remains an important source of social relationships and support for many elders around the world. Himes et al. (1996), for instance, described how the living arrangements of most old Americans mainly involved living with spouse, kin or both, irrespective of ethnicity (see Table 6.8), whereas 70–80% of elders' care is still provided by members of their family (Crist 2005).

Social contact with family members can be modulated by multiple factors, starting from the quality of the relationship the elders have with their relatives (are they in good terms?), to continue with the ability to physically stay in contact by phone/Internet or visiting, with visiting in particular being conditioned by health and also wealth, as transport can be expensive (Arber et al. 2003b).

Table 6.8 The living arrangements of Americans aged 60 or older by ethnicity

Ethnic group	Living alone	Living with spouse only	Living with spouse, kin Present	Living with no spouse, kin present	Living With non-kin present	Institutionalised
Total	24.5	**45.7**	**12.3**	**11.8**	2.6	3.2
White	24.9	**48.8**	**10.8**	**9.8**	2.3	3.3
Black	26.7	**21.6**	**17.9**	**25.9**	4.8	3.1
Hispanic	16.2	**27.3**	**26.3**	**24.3**	4.3	1.6
Asian	10.0	**25.9**	**33.8**	**25.5**	3.6	1.2
Native American	23.9	**30.2**	**19.2**	**20.7**	3.6	2.3

Values shown are percentages. Figures in bold are percentages of old people living with spouse, kin, or both. In all groups, those arrangements represent more than 60% of cases. From Himes et al. (1996)

The family is a major source of *social capital* for many elders around the world (Crist 2005; Gray 2009), and social capital is a function of both the personal attitude of the elder and the characteristics of their social network. The family social capital is expected to deliver benefits to the elder in terms of various forms of social support, whether it is emotional or material. When the family fails to provide much needed social support (e.g. due to declining solidarity following the process of modernisation, for instance), the elder must seek it elsewhere, or suffer the negative effects of social isolation.

For most people, the nuclear family is not just the first social environment, but it also tends to be an enduring one throughout life, even though the family environment can be variable among individuals and at different ages. Orphans will have specific early social experiences that may differ from those of children living with their biological parents. Being raised by a single parent, or by grandparents, or in a blended family, or in a community; to migrating to a different country thus potentially losing contact with close relatives, are also alternatives to the traditional model of co-residential, stable, nuclear family-living. In addition, not all old people have had a partner in their life and/or children (Arber et al. 2003b), but in their case the family could be still relevant through their network of relatives or extended family.

Although the concept of family has been shaped by various myths over time, such as that of the *Rosy Family*: "a nuclear family with intergenerational extension, that provides strength, love, and sustenance to all its members" (Nydegger 1983, p. 26). The myth does contain both a core of truth and an ideal that many may strive to achieve. Elders not only receive support from, but they also provide support to the family (e.g. babysitting, helping with some house cores, lending money, and so forth). Yet, we have already seen in Chaps. 2 and 3 that the relationship between elder and the other members of the family has been also marred by tensions and conflicts—such as power struggles for control of resources and status—both through historical times and across cultures, and there is a point when a rapid decline of the elder's mental and physical conditions can become a significant burden on the family. In addition, in modernised societies, whenever the elder is in full control of his/her mental and physical capabilities, the relationship with family tends to be asymmetrical, with the elder often providing more help to family—children and grandchildren, for instance—than vice versa, and it is often the elders who tend to take the initiative to keep in touch with the rest of the family (Lindley et al. 2008).

The variability seen in modern families suggests that social interactions among family members may be described following the *solidarity-conflict paradigm* of Bengston and Roberts (1991; Roberts et al. 1991), in which both cooperation (*inter-generational solidarity*) and competition (*inter-generational conflicts*) can modulate the relationships between elders and the rest of the family. This dynamic can lead to some complexities where both solidarity and conflicts can coexist, an issue that has been stressed by the *Ambivalence model* of Lüscher (2004), whereby "rather than operating exclusively on the basis of affection, assistance and solidarity, or under threat of conflict or dissolution, the dynamics of intergenerational relations among adults revolve around sociological and psychological *contradictions or dilemmas*" (Pillemer et al. 2007, p. 777; italics original).

Indeed, the interrelation between competition and cooperation seen in interactions between generations within a family can lead to situations of ambivalence and ambiguity that will require a resolution to prevent the social dynamic from becoming unstable. Ideally, a net shift towards cooperation may occur, which is likely to resolve a given conflict (Lowenstein 2007), especially when the alternative to leave the family may not be easily available to the elders. But a competitive resolution is also possible, often leading to elders' destitution.

Askham et al. (2007, p. 186) list the following four characteristics that can be found in personal relationships within a family and that focus on cooperation. Cooperative family relationships are those in which:

1. intimate communication takes place or a confidante relationship can be established;
2. activities or goods are shared (held or done in common);
3. interactions are emotionally laden, with love or affection, trust, loyalty, caring about the other;
4. practical care, personal service to the other, or exchange of resources take place.

Such cooperation generates a broad environment that makes the family a favourite source of *emotional regulation* (Kahn and Antonucci 1980; Cicirelli 1985; Carstensen 1993; see also Chap. 5). Hence, a member of the family feeling sad may find comfort in the family, another member feeling angry and frustrated may relax in the supportive security of the family. The positive effect of cooperative interactions on emotional regulation remains valid into old age.

In old age the chances that parents will still be alive obviously decrease, but some siblings may be available, who may potentially provide a source of support. Such potential, however, may or may not be realised depending on the specific relationships between siblings. Gold (1989) describes the following five types of sibling relationships in old age:

1. Intimate siblings: displaying a very close and devoted relationship.
2. Congenial siblings: Close and caring, but they prioritise their own marriage and children.
3. Loyal siblings: supportive of each other in a crisis and keep in contact regularly.
4. Apathetic siblings: feel indifferent towards each other and are seldom in contact.
5. Hostile siblings: their relationship is based on anger and resentment.

Most siblings tend to be either congenial or loyal (Askham et al. 2007), suggesting a good degree of social support coming from siblings.

In addition, the elders may have the opportunity to establish a relationship with children and grandchildren. However, the styles of parenting and grandparenting are variable, following the diversity of personal characteristics of the elders and past relationships with their children and grandchildren. Such diversity will necessarily affect the relationship between individuals and determine the effects of the relationship on the elder. Herlyn and Lehmann (1998, cit. in Askham et al. 2007) list five broad styles of grandparenting—that could be also extended to relationships with adult children—with the *duteous grandparent* representing the most supportive style:

1. Duteous grandparents: provide regular care to grandchildren following a personal emphasis on "immortality through clan" or generativity. Generativity can also underpin relationships with adult children.
2. Autonomous and highly engaged grandparents: do provide care for grandchildren but when allowed to choose they prefer their own personal activities. Some parents may also establish a similar relationship with adult children.
3. Integrated grandparents: well-integrated within the family, but they show less commitment than (1) and (2).
4. Ambivalent grandparents: show conflicting needs of closeness and distance.
5. Family-independent grandparents: they show a detached relationship with grandchildren, often because they have little in common. The same may occur in their relationship with their adult children.

Therefore, the family is both the donor of help to the elders and the receiver of help from the elders. Magilvy et al. (2000) provide an example of family acting as donors of help to their elders. They studied the relationship between old Hispanics and their family in rural USA. In this case, the family was extended including both close and more distant relatives and in-laws. Moreover, close friends were also regarded as "family", a situation not uncommon among Hispanics. Together, extended family and friends form an integrated network of supporters and potential supporters for the elders, who keep in touch by various means. Such varied network of mutual assistance is guided by a strong sense of familism (see Chap. 3), which facilitates the social integration of the elders and the provision of help such as caring for sick elders and providing reassurance.

Familism can even lead to situations where caregivers may risk a deterioration of their own health by refusing to seek external aid, thus stretching their endurance to the limit. On the other hand, familism can help alleviate the burden of caring for a family elder, whenever family members take turns in the performance of daily care duties (Magilvy et al. 2000); this is especially important for daughters who tend to provide more care to old parents than sons (Adams and Blieszner 1995; see also Losada et al. 2010). Modernisation is introducing changes in the concept of family and lower commitment to traditional duties towards family members, especially in the young, to such an extent that elders in the Hispanic community studied by Magilvy et al. did raise some serious concerns about their future.

In rural China older people tend to receive differential support from various family members, which has been negatively affected by the low levels of co-residence coming with modernisation. Assistance with household chores and personal care works better from the perspective of the elder when there is coordination of duties among sons, daughters-in-law, and daughters (Cong and Silverstein 2008). Such coordination is dictated by a cultural expectation that elders should receive specific types of care from specific kinds of family caregivers, with mismatches potentially leading to distress. For instance, lack of expected support from daughter-in-law may be especially distressing for the old mother-in-law, especially when they are co-resident; whereas father's expectations rely more on the support from daughter.

Care for old parents by adult children can be constrained by the work commitment of children, especially when the elder is impaired or chronically ill and thus requires intense levels of care. In this case some children may decrease help but only if others step in. In the USA, Stoller (1983) described how adult children modulate their help according to actual need and the availability of alternative sources of support.

In a work carried out in Southern India, Sudha et al. (2006) investigated family social support for 60-year-old and older men and women. In India old men and women are under pressure to keep some control on their wealth in order to retain respect from the rest of the family without feeling that they are a burden; with gender, caste, class, and family situation being variables modulating the respect that elders are expected to receive in the community. Still, many elders do find themselves in the situation of needing both material and emotional support. Support is mainly expected from sons and, eventually, daughters too, with family support being especially important for older women.

Assistance provided by the family can be especially critical whenever the older person does not feel well and requires hospitalisation. In this case, family members may be of great help through the provision of emotional support, helping with various activities of daily living (ADL), helping with transport, exchanging valuable information and addressing bureaucratic hurdles, collaborating with the health care team, and providing financial aid if needed (e.g. Li et al. 2000, see Bridges et al. 2010 for a review). Contact between a sick elder who is in hospital and his/her family may be also helped by modern communication technologies which allow not only long-distance vocal but also visual contact between individuals. Once the elder is discharged from hospital the family is usually confronted with ongoing duties of care and responsibilities, which require careful planning in order to decrease the burden on the family and achieve a more successful recovery of the elder's health (Bauer et al. 2009). Indeed, technology can be also useful to improve communication and the provision of care to frail elders at home, not just when they are in hospital (Magnusson et al. 2004).

Providing effective care for elders with handicaps poses specific challenges to the family. McIlvane and Reinhardt (2001) carried out a study in the USA of 130 women and 111 men (age range from 65 to 99 years old) who were visually impaired (macular degeneration, cataracts, glaucoma). Social support was important for this group of elders, with special emphasis on the quality rather than the quantity of social interactions. Some differences also emerged between men and women, with men being relatively more dependent on help from family members, whereas women could also rely on the provision of support from friends. Despite the emphasis on quality relationships, the size of the social network as such was also important for a better adaptation to loss of vision.

Religiosity and spirituality can also provide a general cultural/ethical framework to facilitate family support to older members (e.g. Kim et al. 2011), especially when the elder is in urgent need of such support, as in the case of the seriously impaired old people that we have just mentioned.

Whenever elders are treated with respect and are integrated into their family life, they usually also tend to reciprocate to the younger generations in different ways, that are variable according to cultural and personal factors (Magilvy et al. 2000). In which ways does the family benefit from the help provided by the elders?

In the USA, Eggebeen (1992) described various helping functions of old parents in support of their adult children: give advice, emotional support, assisting with money, childcare and household chores. Such help is provided more frequently by married old parents than widowed or divorced ones. Never married adult children who live independently also tend to be frequent recipients of support, whereas adult stepchildren are less beneficiaries of support from old stepparents. Likewise, Greek grandparents studied by Svensson-Dianellou et al. (2010) also provided various forms of childcare and emotional and financial support to their children's family, with some division of labour observed, where grandmothers provided more household help and childcare than grandfathers. Support through the performance of household chores and childcare is facilitated by co-residence between grandparents and the child's family, a situation that is more common in some countries than others (Jappens and Van Bavel 2012). Grandparental help can be especially important for the well-being of parents in the case when a grandchild has a developmental disability (Trute 2003).

Grandparents can also act as family stabilisers during periods of economic hardship by, for instance, mediating in family conflicts and decreasing the level of stress on parents with their helping (Botcheva and Feldman 2004). In the most extreme of cases grandparents may even act as *custodial grandparents*, taking full responsibility for the well-being and education of their grandchildren (Richards 2001; Hayslip and Kaminski 2005).

The role of grandparents is also being reshaped by divorce and re-marriage of offspring (e.g. Borell and Ghazanfareeon Karlsson 2003). Perceived duties of care of older parents towards adult children and their family may even influence the decision whether and when to retire (Szinovacz et al. 2001).

In other words, grandparents can fulfil a variety of helping roles within the family that can be beneficial to their children, grandchildren or, more often than not, both (Timonen 2018).

Modernisation, with its emphasis on consumerism, flexible personal identities, loss of the centrality of marriage, late age parenthood, small number of children, extended period of grandchild/grandparent relationship, and genetically mixed family due to changes of partners ("patchwork family") (Askham et al. 2007) has contributed to modifying the concept of family, thus complicating the dynamics of interactions between elders and the rest of the family members. For instance, adult children may not have the time or financial resources to provide much needed help to old parents, thus potentially leading to neglect, that may not be compensated by the elder if his/her own personal resources (money, health, including psychological health) are scarce (see Aboderin 2004a for a review). In this case the ethics of familism may be eroded by modernisation, through the imposition of a heavy workload and/or financial burden that leaves the family with little resources to be used to help the elders. The neoliberal emphasis on individualism, self-reliance, and competition can only make things worse, by weakening the traditional norms of filial obligations (Yue and Ng 1999; Aboderin 2004a). Aboderin (2004a) has summarised the complex effects of modernisation on the relationship between old people and family members, which can be seen in Fig. 6.5. Modernisation makes the elders both more obsolete and more in need to be self-reliant, whereas the family is less capable—but also less interested through individualism—of offering help to their elders.

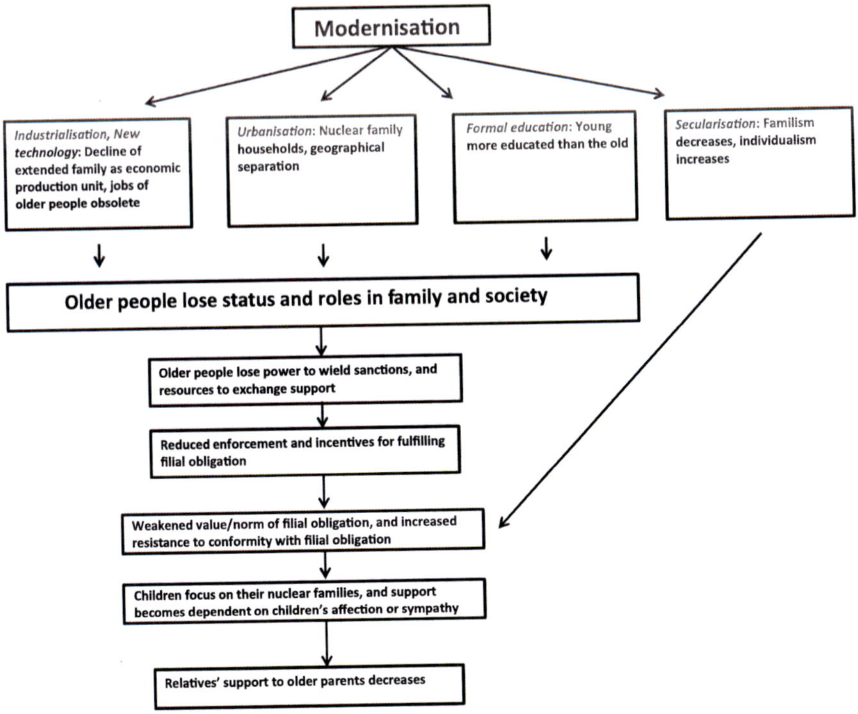

Fig. 6.5 Negative effects of modernisation on the supportive relationships between elders and their family members. Redrawn from Aboderin (2004a)

The negative effects of modernisation on the relationships between elders and the rest of their family are likely to become worse, as the population of older people is increasing and future generations of elders, who were workers during the neoliberal period, may have not sufficient personal resources to support themselves in old age (e.g. Yi and George 2000). In addition, the family may also lack the resources to support their elders (e.g. Aboderin 2004b), which can be a serious problem especially in countries that do not have an adequate government-based welfare system for the old. Apt (1996, p. 22) quotes a Ghanaian say: "When your elders take care of you while you cut your teeth, you must in turn take care of them while they are losing theirs", but such traditional inter-generational reciprocity has been challenged by neoliberalism and modernisation around the world (Standing 2016).

The disruption of traditional family structures by modernisation is clearly seen in the case of China that has undergone significant social and economic changes in the last 30-plus years started by Deng Xiaoping in the 1980s. Rural regions in China have been especially affected by such modernisation (e.g. Silverstein et al. 2006), which is expressed as declining multi-generational households and increased migration by working-age members of the family to larger urban centres. These changes challenge traditional values of family-oriented Confucianism, that are prevalent in the region and that are based on Confucian *filial piety* (e.g. Chow 1993), a concept that stands on three major pillars: respecting parents, bringing no dishonour to parents, and taking good care of parents (Sung 1990). Moreover, Li Chi (1879, p. 451) stressed that filial piety should be expressed in a natural and spontaneous manner: "Care for parents should not be a tiresome obligation; the son and his wife will do it with an appearance of pleasure to make their parents feel at ease". Guided by Confucian filial piety, traditional multi-generational households have been associated with well-being and life satisfaction for the elders in Chinese society (Silverstein et al. 2006), but also with useful help for adult children, especially in three-generation households where the elder can provide childcare support. Adult children who migrate seeking better job opportunities elsewhere cause a fracture in this traditional social system of mutual support, although they may still contribute to the well-being of the old parents left behind through remittances.

The concept and practice of Confucian filial piety seem to be more resilient to change in Korean society, where respect for both parents and elders is expected to be practiced through *reverence*. Reverence is in fact extended also to past generations through the maintenance of genealogical records (e.g. Sung 1990). The concept of filial piety is of such importance to Korean society that in 1973 a national *Filial Piety Prize* was established and maintained by the Ministry of Health and Social Affairs of the Korean government. Sung (1990) studied some of the cases behind recipients of such prize and identified the provision of specific care services to the elders, more frequently carried out by daughters, and emotional and financial support, more frequently provided by sons, as major expressions of filial piety.

Modernisation, but also other factors, may lead to an increased level of *inter-generational ambivalence* in the relationships between elders and their younger family members (Pillemer and Suitor 2002; Lüscher 2004; Pillemer 2004). As we already mentioned, Pillemer and Suitor (2002) describe the ambivalence in inter-generational relations as an experience of contradictory attitudes or emotions towards partners in a relationship or towards the relationship itself. Such ambivalence can derive from contrasting normative expectations, such as changing roles and demands; or changes in developmental stage (i.e. changes in the individuals and their relationship over time: what worked in the past may not work anymore in the present).

Inter-generational ambivalence has been affected by the greater mobility that has come with modernisation. For instance, in the case of *commuter marriages* we find arrangements where there is a priority for career development and partners may work in different locations during the week, whilst sharing a common residence on weekends. In such cases there may be family situations involving children that resemble "divorce". Actual divorce itself usually leads to the establishment of "binuclear" families (e.g. Borell and Ghazanfareeon Karlsson 2003), whereas commuter marriages may lead to greater ambivalence and instability. The same dynamic may affect the relationship of the family with grandparents when contacts between the two are only sporadic.

Although a family is characterised by the presence of children and, for grandparents, grandchildren, in old age the specific relationship between husband and wife also becomes important, given the increased time together that partners have after retirement. A stable partnership fulfils various functions such as that of fostering friendship, intimacy, and sexuality, but also love, compassion, emotional support, and mutual help (Koren 2011) and it can be seen in different ways: as an institution (e.g. marriage), a caring relationship, a disclosing intimacy, an identity project (Askham et al. 2007). But such views of partnerships may change dynamically according to circumstances in the periods before and after retirement. For instance, as part of the caring aspects of marriage a husband may increase his commitment to household duties if he retires before his wife does, but then decrease them when she also retires, this may occur especially when the relationship is dynamic and it is constantly renegotiated, rather than fixed by cultural norms. In more traditional cultural settings where the husband works and therefore retires from work at one stage, whereas the wife is permanently committed to a housewife role, tensions may arise when the wife is expected to maintain a high workload, regardless of age, whereas the husband can find more opportunities to rest after retirement. This may explain the reluctance of many older and recently widowed women to re-marry or, if they seek a new companion, their interest to establish with him a relationship that may differ from traditional marriage.

The social dynamics of modern separated/divorced or widowed elders has become more flexible and dynamic. Second couplehood in old age can take various forms such as: re-marriage, unmarried cohabitation, and *living apart together* (LAT, whereby two persons maintain an intimate relationship but live in separate residences, DeLamater 2012) and it may involve a degree of continuity or discontinuity with past experiences of couplehood (Koren 2011).

Continuity and discontinuity in the relationship between older partners can be interpreted in the context of the *Selection Optimization Compensation* (SOC) processes introduced in Chap. 1. SOC processes illustrate the dynamic and adaptive nature of human development that makes positive use of the opportunities and capacities available to the elder at any given point in time (Freund and Baltes 2002). LAT relationships, for instance, are becoming more common and they are a manifestation of such adaptive plasticity, combining a degree of continuity of attachment to a loved person with discontinuity in the specific arrangement of life-sharing, which in this case involves living in separate homes (e.g. Askham et al. 2007; Koren 2011). LAT relationships also tend to be favoured by women. In a study of LAT relationships carried out among 18 couples of highly educated and healthy 60–90 years old (57% women) in Sweden, Borell and Ghazanfareeon Karlsson (2003) found that the top three reasons for both men and women to prefer LAT rather than cohabiting relationships were: (a) "importance of having a home of one's own"; (b) "importance of practical reasons concerning living arrangements"; and (c) "importance of being freed from duties that may arise if one were married". In particular, women wanted to be free from commitments in order to visit grandchildren at their own leisure. Finances are also usually separated no matter how long the LAT relationship has been. The theme of individual independence also appeared when considering the provision of help to the partner in case of illness or any health problem. Partners were keen to help each other but not on a full-time basis. Therefore, a LAT relationship is characterised by a dynamic balance between commitment to the other and autonomy (Borell and Ghazanfareeon Karlsson 2003). This pattern is also in accordance with *Socioemotional Selectivity* theory, which predicts that in older ages there is an increasing trend to greater selectivity in social relationships according to their emotional significance (Carstensen 1992). Clearly, LAT relationships are consistent with such a tendency.

In a study carried out in Israel, Koren (2011) compared 20 couples of 66–92-year-old men and women who had become involved in a second couplehood in old age, such couples were re-married, or cohabiting, or LAT. Reasons for second couplehood were either widowhood (85% of participants) or divorce (15% of participants). That widowhood is associated with second couplehood is not surprising, as recently widowed old people may experience depression (Turvey et al. 1999), which could be alleviated by finding a new partner. Depression in old people can be also alleviated by more general family and other sources of social support (Sun 2004). Most participants perceived second couplehood as a discontinuity compared to their previous life in a married couple. Second couplehood also tended to be perceived as less integrative of the couple than first couplehood, with each partner retaining more independence, whereas cohesion was more likely in first couplehood. The weakening of couple's cohesion also came hand in hand with a strengthening of individuality and a relative decrease in the motivation to put effort into caring for each other. In contrast, at least in some cases, sexual life may benefit from the experience of second couplehood.

At some point in time, whether they have been living alone, with family or with somebody else, elderly people may decide to move from their current residence. The move may be motivated by various factors: health, finances, relationships, personal issues/interests, constraints of the current residence given the ageing process (e.g. stairs) or other circumstances, and it could mean a translocation to not only a different home but to a different social environment, some times in a new city, perhaps even a different country (e.g. Sergeant and Ekerdt 2008). The move may also involve entering a care facility for older people. It is often the case that the process of decision-making about moving may also involve a degree of input from the part of the rest of the family. Typically, this may occur when the family is concerned that the current residence is unsafe, or it is located too far from them, thus constraining their ability to provide assistance if needed. Family input may be taken negatively by the elder, sensing undue interference and pressure, or more positively, especially when the family just provides the relevant information to substantiate their concerns, but leaves the final decision entirely in the hands of their old relative (Sergeant and Ekerdt 2008).

Finally, what happens when the old person is childless and therefore the amount of support from family may be seriously jeopardised? According to the *principle of substitution* (Shanas 1979; Cicirelli 1982) when a person lacks the support of some expected others (such as close family members) she or he will seek support elsewhere as required, including from non-kin. Childless elderly would tend to seek support first and above all from a partner, and if a partner is not available, from close or even distant kin and non-related persons; failing that, the chances of institutionalisation may rapidly increase (Johnson and Catalano 1981). In some cases, childlessness may decrease the degree of socialisation of the elders, whether they are single or still married, a process that Johnson and Catalano (1981) describe as *social regression*. Such social regression may lead to the strengthening of interdependence between the members of a partnership, thus potentially leading to a "symbiotic" situation that could potentially difficult the establishment of social links with the "outside". The death of a partner in such a situation may have catastrophic effects on the surviving member of the couple (e.g. Erlangsen et al. 2004).

To recap, the nuclear/extended family remains a major source of support for the elders, despite various social changes that have occurred under the pressure of modernisation in recent decades. Modernisation tends to undermine the effectiveness of the family as a cooperative social unit, by putting additional stresses on the network of relationships among relatives. The family provides a complex social environment to the elders, offering a dynamic mixture of solidarity and conflicts. Of course, one expects the elders to benefit more from a cooperative family environment, but such environment ultimately benefits the entire family, as all members can also enjoy the help provided by the elders (e.g. child-caring, housework). When family support becomes more unstable and unpredictable, the elders seek such help elsewhere in the community. Within the family context, the specific relationship between old partners remains important, so much so that new forms of more dynamic partnerships in old age are emerging, such as living apart together (LAT).

Beyond the family there is the wider community, and the social interactions between older people and other members of their community are of course of great importance. We therefore turn now to analysing such interactions in more detail.

6.5 Community

Just as it is the case for the social relationships between family and elders, the community at large can be a source of either support or additional distress for old people (Adams and Blieszner 1995; Barker 2002). To better understand the overall social relationships of the elders within a community, it is necessary to consider the relevant aspects of the family and the extra-familial social environments within a more general perspective.

In 1991, Marjorie Cantor proposed a comprehensive *Social Care model* (SCM) (Cantor 1991), in an effort to integrate the family and community social dimensions of the life of old people. As we have already seen in this chapter, caring for the elders involves the satisfaction of some fundamental needs of old people, such as: socialisation, self-affirmation, self-actualisation, assistance with tasks of daily living, and other personal care needs, especially in the case when the elder suffers from severe disability. Such forms of support for the elderly can come from various sources: family, friends, neighbours, community organisations, social agencies, government. Figure 6.6 summarises the overall social interactive network of older people according to the SCM. Note how the interactive arrows are always bi-directional (i.e. the elder receives but also gives) and the various social levels can affect the elder both directly and indirectly, through the network of bi-directional interactions between levels.

The SCM is consistent with Adams and Blieszner's (1995) view that sees personal relationships as consisting of both structure and processes. In this case the structure is the way of linking individuals through informal, quasi-formal and

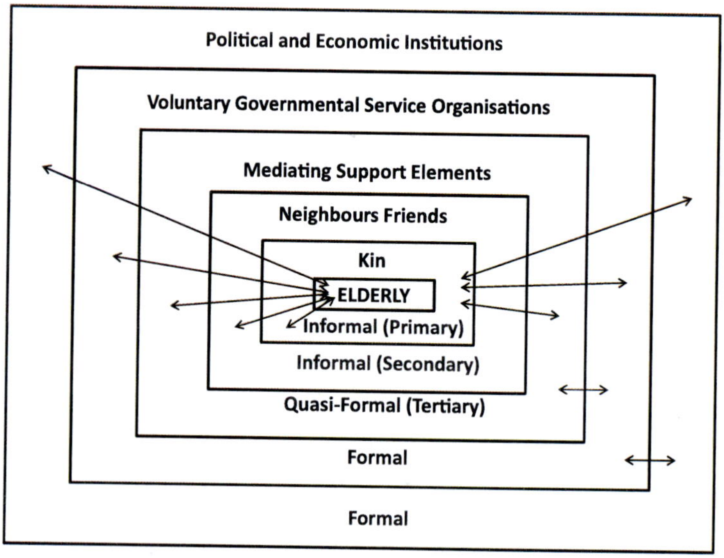

Fig. 6.6 Cantor's (1991) *Social Care model* for the elderly

formal relationships at various levels from kin to political and economic institutions, as shown in Fig. 6.6. To that we should also add the processes such as thoughts, feelings, and behaviours that give shape to the relationships over time.

Thus, for instance, parts of the formal support system such as voluntary organisations interact with broader political and economic institutions, and the outcome of such interactions (e.g. increased or decreased government funding) will eventually affect the well-being of the elders. The first level of support for the elderly comes from the family and the kin network as we have already seen, this can then extend to the broader social network constituted by non-kin (e.g. neighbours), to more complex private and government organisations and institutions. Private voluntary and government-controlled organisations are becoming more important for old people as modernisation tends to decrease the level of support from family (Chow 1993).

The balance between family and government support for the elders varies between countries, even within the same region (e.g. Europe), according to the specific economic and social characteristics of each country and its history. For instance, Fig. 6.7 shows greater support for the elderly coming from the welfare state in Norway, following a long social democratic tradition in this country and indeed the rest of Scandinavia (e.g. Block 2011), whereas in Spain the family remains very important as a consequence of the long-standing culture of familism still prevalent in this country and a relatively lower level of government support (Daatland and Lowenstein 2005; Losada et al. 2010). Figure 6.7 indicates Israel as the country with the largest proportion of "others" among those providing help. Whether "others" such as non-relatives and non-institutional caregivers would spend time and effort helping older people is likely the result of broad cultural values that favour cooperation. In this regard, a socially widespread positive image of the old may help in recruiting support from others; conversely, a negative image may limit the level of help offered (e.g. Arnold-Cathalifaud et al. 2008). Support from non-relatives and non-institutional caregivers is also expected to decrease where the welfare state is better developed, as in Norway (Fig. 6.7).

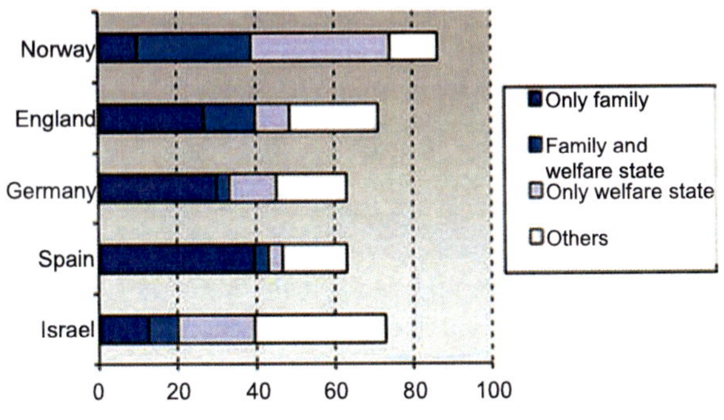

Fig. 6.7 Help rates (percentages) and help sources by country for those aged 75+. From Daatland and Lowenstein (2005)

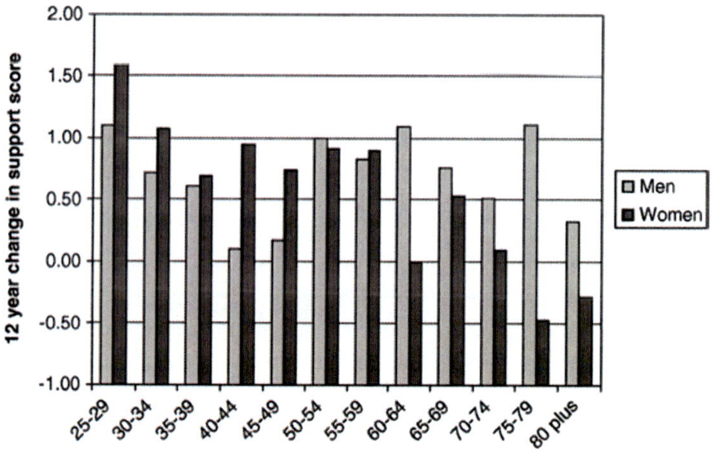

Fig. 6.8 Change in support scores for participants in the study over a period of twelve years, between 1991 and 2003. From Gray (2009)

A better established secondary informal and a formal social network can benefit older people not only materially (e.g. by providing financial assistance in case of need) but also emotionally (e.g. by decreasing loneliness).

Gray (2009) studied changes in social support in a group of older people in Great Britain over a 12-year period between 1991 and 2003. During this time, support decreased with age especially for women (Fig. 6.8). This was compensated by a greater participation of women than men in different community activities (Table 6.9, see also Arber et al. 2003a).

Note also from Table 6.9 how over the 12-year period both religious organisations and sports clubs experienced increased membership of both men and women. Activity in political parties decreased, but it rose somewhat in both environmental and tenants'/residents' groups. Men in this study were more active than women in trade unions, sports, and social clubs; but women were more active in religious organisations and in community and voluntary-service groups.

Activities that can help integrate older people into their local community are important, not least because most old people prefer to "age in place" rather than in a nursing home (Thomas and Blanchard 2009; Pfeifer 2016). *Ageing in place* does not mean just ageing at home, which may be a depressive experience if social links are severed. Ageing in place means to maintain the social network that was built over years of residence and interactions with members of a local community, and perhaps build a new one as a person becomes older and the community changes its composition. Links with the members of a local community represent part of the social capital of the elder (Oswald and Rowles 2006), which can be vital to live a better life in later years. In some circumstances, a group of elders who share similar interests, motivations, and goals in life may organise themselves and live together in *intentional communities* (Thomas and Blanchard 2009), such as the ashrams of India, the kibbutzim of Israel or co-housing communities in the USA and various countries in Europe and around the world. Alternatively, groups of elders living in

Table 6.9 Activism and membership of organisations in people aged 50 or more years in Great Britain in 1991 and 2003

	1991				2003				% of members active in 2003	
	Members		Active		Members		Active			
	Men (%)	Women (%)	Men (%)	Women (%)	Men (%)	Women (%)	Men (%)	Women (%)	Men	Women
Political party	n.a.	n.a.	4.2	2.3	4.6	3.3	1.7	1.9	37.0	57.6
Trades union	18.0	4.3	2.9	0.9	14.5	7.6	3.2	1.6	22.1	21.1
Environmental group	2.7	1.8	2.0	0.9	3.7	2.8	2.3	1.7	62.2	60.7
Parents' group	1.0	0.6	1.2	1.0	1.0	1.0	1.5	1.9	150.0	190.0
Tenants' or residents' group	8.4	8.9	3.9	4.6	8.7	9.3	4.8	5.2	55.2	55.9
Religious organisation	9.7	16.8	9.5	16.9	12.8	19.9	12.5	18.8	97.7	94.5
Voluntary service group	3.4	6.2	3.3	6.7	4.8	6.3	4.1	5.5	85.4	87.3
Other community group	4.7	4.4	4.1	4.6	3.6	3.6	3.1	3.6	86.1	100.0
Social club	19.0	8.2	13.4	7.0	17.7	6.2	14.0	5.5	79.1	88.7
Sports club	13.2	5.2	12.1	5.4	17.2	7.0	16.0	6.9	93.0	98.6
Women's Institute	0.0	7.3	0.0	7.0	0.0	4.9	0.0	4.7	0.0	95.9
Other women's organisation	0.0	1.7	0.0	1.7	0.0	3.1	0.0	2.6	0.0	83.9
Any other organisation	12.4	11.0	11.3	11.1	10.7	10.6	8.1	9.9	75.7	93.4
Professional organisation	n.a.	n.a.	n.a.	n.a.	10.7	4.1	5.5	1.6	51.4	39.0
Pensioners' organisation	n.a.	n.a.	n.a.	n.a.	4.2	4.6	2.7	5.0	64.3	108.7
Scouts/guides	n.a.	n.a.	n.a.	n.a.	1.2	1.1	1.1	1.1	91.7	92.4
Average number of memberships	1.10	0.98			1.16	0.94				
Average number[a] in which active			1.03	0.80			1.00	0.76		

In some organisations (e.g. parents' groups and pensioners' organisations) the percentage of activists is larger than 100% because in such cases there are more activists than members; i.e. some people may join in to help in specific activities without being a member

[a]Number of organisations. From Gray (2009)

the same neighbourhood or small town may organise into a *spontaneous community* (Thomas and Blanchard 2009, p. 16):

> Within small towns and suburban and urban areas in every state, one can find a community in which traditional ideas about caring for one's neighbors still prevails.

Thomas and Blanchard (2009) list the following beneficial effects of ageing in an intentional community that is organised around the objectives of cooperation, promoting health, and sustainability:

- Sense of inclusiveness: old people of all ages, races/ethnicities, and abilities are welcome.
- Commitment to sustainability: residents are committed to a lifestyle that is sustainable environmentally, economically, and socially (*triple bottom line* approach, Elkington 1994).
- Healthy lifestyle: the community encourages and supports wellness of the mind, body, and spirit and, to the same degree, plans and prepares programs and systems that support those dealing with disease, disability, and death.
- Accessibility: the setting provides easy access to the home and community. Multiple modes of transportation are encouraged.
- Interdependence: the community fosters reciprocity and mutual support among family, friends, neighbours, and across generations.
- Engagement: promotion of opportunities for community participation, social engagement, education, and creative expression.

Therefore, the community social network is important in supporting the ageing process of older people. But such social network is not stable, and it may change over time. Although it is expected that over time many older people may decrease the size of their overall social network, due to increased selectivity, lower capacity to nurture a large number of relationships—which requires time and energy—and simply due to the decrease in the number of same-age friends, as they may pass away (Stoller and Pugliesi 1991; Litwin 2001; Askham et al. 2007), there is also the potential for a dynamic compensation of losses through some gains. Indeed, over the years theoretical work on the relationship between social network size and age in the elders has been prolific, with the reality being, as it often is, rather complex. For instance, Eugene Litwak (1985) suggested that with age and increased frailty social support networks are expected to increase in size, a pattern also consistent with *Exchange theory* (Stoller and Pugliesi 1991), but such increase is also expected to come with a dilution of the per capita support effort from the members of the network, given the inability of the elder to reciprocate. The *Social Convoy model* of Kahn and Antonucci (1980), on the other hand, suggests a degree of stability in the social network of support with age, as friends and acquaintances may retain their level of support for the old person simply because the greatest burden of aid will fall onto the shoulders of the family. Whereas we have seen that *Disengagement theory* (Cumming and Henry 1961) predicts a decreased level of sociality with ageing.

In a study carried out in the USA, Stoller and Pugliesi (1991) showed that the reality is a bit more complex than what each available theory suggests, combining aspects of the three views mentioned above. In fact, over time, social network size increased for 45.5% of the sample of elders studied by Stoller and Pugliesi, it decreased for 38.3%, and remained the same for 16.2%.

Different elders can be embedded within different social network types. Based on a study carried out in Wales, Wenger (1996) described five types of social networks:

a. Family-dependent network: mainly relying on close family members.
b. Locally integrated network: it includes close interactions with family, neighbours, and friends.
c. Local self-contained network: a small network that is mainly neighbour-based.
d. Wider-community-focused network: this is a large and mainly friends-centred network.
e. Private-restricted network: which is characterised by the exclusion of local kin, and minimal interactions with neighbours.

This typology can be broadly observed across countries, with different emphasis on the various types of networks from one country to the other. Some additional general similarities in the social networks of elders have been also described across countries (Litwin 2001): being still married and maintaining good health is usually associated with larger social networks, whereas the social network decreases for the old-old and oldest-old.

In a study carried out in Israel, Litwin (2001) found that although the family was an essential part of the social network in the elders, networks were also dense with interactions with friends, neighbours, and members of the same religion. Five social network types were described that to an extent parallel Wenger's (1996) classification: diverse, friends, neighbours, family, and restricted. Elders belonging to a diverse network were typically men of European origin, they had a higher income, a higher level of education, and had low levels of disability. The friends network was similar to the diverse network but elders who belonged to it tended to be younger. Most of the elders in the neighbours network were lower income women. Elders in the family network mainly originated from Eastern Europe and had lower income, a lower level of education and greater degrees of disability. Finally, the restricted network included elders with low income, less education, greater disability, and they were also likely to be older. This suggests that belonging to and maintaining a large social network requires time and effort (physical, psychological, financial) from the part of the elder, hence those who cannot actively maintain their network are bound to decrease it and restrict it to family.

van Tilburg (1998) studied changes in social networks with age in a group of 55–84 years old in the Netherlands as part of the *Longitudinal Aging Study Amsterdam*. Oldest participants tended to have smaller network sizes than younger participants, but over a four-year period of longitudinal observations individuals did not significantly change their social network size. This seems to be a result of a dynamic process where relatives and non-relatives may replace each other over time. In this study, a decrease in the number of contacts with friends and other non-relatives was compensated by an increase of contacts with children, children-in-law, and siblings.

Old people who may be at risk of losing family social ties with age, can benefit from the social capital available in their community, both in terms of company and services. Indeed, better social capital can be of direct assistance to elders even in terms of their safety, health, and probability of survival in case of an environmental disaster. Cannuscio et al. (2003), for instance, described how a heat wave that struck Chicago in 1995 was more likely to cause death among socially isolated older people. As social capital is eroding in countries such as the USA (Putnam 1995), following the neoliberal focus on individualism, governments will be confronted with the need to assist isolated elders. Alternatively, modern technologies such as the Internet could potentially help restore and maintain social links. Self-organisation by members of the older generations into specific communities, or groups within broader communities, could also help maintain or recover a degree of social capital.

The social capital afforded to the elders by friends is also variable across ethnic groups. For instance, Askham et al. (2007) mention how English and Italians tend to have higher contact rates with friends than Swedes and Dutch. In Bangladesh social capital for the elders mainly derives from the support received from family and from the community at large, both sources of social capital significantly improve quality of life in the elders (Nilsson et al. (2006). Old people can experience greater subjective well-being across ethnic groups after enjoying the benefits of greater social capital (Pinquart and Sörensen 2009).

Neighbours represent a variable source of social capital. They are physically close and therefore conveniently located to provide and receive help, but the level of intimacy among neighbours may not always be high enough to support a close relationship, and the requirements for privacy may also tone down the level of contact. Eventually, if the relationship is regulated to the satisfaction of both neighbours—especially if they both live alone—then it can lead to increased well-being in the elders. In fact, some relationships between neighbours may be quite intimate and long-lasting, involving acts of kindness and care when required (e.g. Barker 2002), even to the point where elders involved in such relationships may use kin terms to refer to each other, thus becoming *fictive kin* (MacRae 1992).

In a study carried out in the USA, Barker (2002) investigated the various forms of caregiving relationships observed among non-kin aged 65 years old or older. Most non-kin caregivers were women and although the relationship tended not to last long, in some cases it extended over 5 years. A large proportion of caregivers (31%) were within 10 years of each other in age. Interestingly, in many instances, partners were different not only in gender, but also in ethnic and economic backgrounds, which may suggest that when the need for support is significant, people may become less discriminatory both in terms of whom to give and from whom to receive. Four styles of relationships were found by Barker in increasing order of intensity: *Casual*, *bounded*, *committed,* and *incorporative*. An incorporative relationship was characterised by a longer duration of the relationship and it mainly involved the performance of helping tasks such as shopping, house-keeping, and personal care; whereas the less intense casual relationship mainly involved socialising and helping with paying bills, completing paperwork, and also shopping.

Social capital can be particularly important to the elderly in small rural communities where access to various services may be more limited. In a study carried out in England and Wales, Curry and Fisher (2013) described various manners in which elderly people engage with their local rural community. Some of such forms of involvement can enhance social capital, although the full gamut of approaches to the community can range from anti-social, asocial to pro-social. According to Curry and Fisher (2013) elders in the community can be *antagonizing* ("actively seeking to oppose or disrupt"), *absenting* ("deliberate choice to 'opt out' of community activity, usually for the desire of privacy"), *being* ("just living in a place"), but they can also become positively involved through *belonging* ("having an identity with place"), *bestowing* ("actively assisting others or taking part in voluntary activity"), and *communing*, which is the most intense form of involvement with the local community.

Participating in social activities with people outside their family can be restorative to older people in terms of their cognitive abilities and general functioning in life. This can be better achieved by engaging in various forms of development of the self, whilst interacting with others: arts and creativity, altruism, contact with nature, nurturing social connections, being exposed to cognitive challenges, engaging in physical activities, reading, exploring spirituality and reflection, cultural initiatives, travelling, and more (Jansen and von Sadovszky 2004, see also Jansen 2008). Through active participation in the social activities of their local community, older people may also achieve what *Attention Restoration* theory (Kaplan 1995, see also Jansen 2006) defines as a better capacity to direct attention on specific matters that require effort in order to be sustained. Hence, participating in cooperative group projects can help in keeping focus on specific tasks, maintain an active mind, and experience states of *flow*.

According to Kaplan, *restorative activities* (i.e. activities that aid in improving mental and physical capacities, Resnick et al. 2013) help the individual feel more refreshed, at peace with him/herself, and mentally better able to face the challenges that may still lie ahead. Older people, however, may confront some barriers in their striving to participate in restorative activities, and some of such barriers are social, such as ageist attitudes, lack of companion, family separation (Jansen 2005). In addition, the elder may also experience other barriers such as: transportation difficulties, financial constraints, weather issues, limited opportunities, safety, but health limitations are the most important impediment to participating in restorative activities in the elderly (Jansen 2005).

Elders' participation in society can be improved if the social environment encourages and facilitates such participation. Above we have seen various cases of spontaneous support for the elderly coming from their local community, but support may be also more organised and institutional (see Figs. 6.6 and 6.7 and Table 6.9). Social services are particularly important to assist old people who live alone or are disabled or chronically ill, and they may involve handyman services (Osterkamp and Chapin 1995), health care, and assistance with various activities of daily living (ADL) (Kemper 1992).

Just as older people can receive support from various public services, they can also adopt a proactive approach and self-organise, by participating in the life of their community within the frame of local organisations of elders, clubs, branches of political parties, churches, and so forth. Volunteering, for instance, is a common activity carried out by old people in many countries that can increase the social capital available to the members of the community (e.g. Warburton and McLaughlin 2005). The elders enjoy the benefit of such volunteering simply by participating in them, but although they usually do not expect any official recognition, they do appreciate such recognition when it is bestowed on them. Warburton and McLaughlin (2005) studied volunteering in Australia in a group of 184 community-dwelling residents, 55–93 years old. Participants expressed their disconcert about their volunteering activities being often devalued or at least simply taken for granted, which just played into the negative bias observed in day-to-day discourse about ageing. One of the respondents said (p. 720):

> Older people are the basis of any community. They are the ones who give, give, give. I do feel that they are undervalued by government and I believe it's past the time when government should be stepping forward and saying 'hey, how can we help you?' not 'how can *we* help *them*'. I also think we've done a lot to debunk the myth that older people are a drain upon community because we are not, with a few notable exceptions! No, we do a good job and we need to all give ourselves a pat on the back because we have learnt throughout our lives. (italics original)

This is a loud cry for recognition, recognition of the value of actions performed by older people in support of their community. Participants in Warburton and McLaughlin's study also pointed to the many acts of "kindness" they regularly performed in the community on behalf of others, such as collecting neighbours' mail, providing support in times of ill health, talking to others who may feel lonely, and more. Old migrants, in particular, may provide support to other old migrants, as they could be more sensitive to their feelings, given their common experience of migration (Warburton and McLaughlin 2005).

After carrying out a meta-analysis, Gorey (1998) stressed the bi-directional nature of elders' volunteering: volunteering increases the sense of well-being in older volunteers and, in turn, it is also useful to other members of the community.

Some programs that support elders' participation in the community are more specialised, aiming at attracting elders with specific interests and, through participating in such specific activities those programs provide social support to the elders and promote their well-being. Ross (2007), for instance, describes the experience of the *Company of Elders*, which is a London based dance and performance group for people 60 years old and older. Through dancing, participants expect to increase creativity, boost their self-confidence, extend various skills, bring a sense of purpose into their life, improve fitness, and enjoy social interactions with people who share similar interests. The practice of Tai-Chi is also a common social activity that fosters well-being in the elderly (Chen et al. 2002), as are yoga (Chen et al. 2011) and laughter groups (Yim 2016).

Some programs are organised to offer participation in self-help groups to elderly people who are experiencing specific challenges. For instance, DeCoster and George (2005) described a *Diabetes Club* that provided self-care training to elders diagnosed with diabetes.

Community support for the elders, however, can be constrained by various obstacles, as in the case of elders who could benefit from the provision of some public services, such as home nursing care, but they do not know how to access such services (Alster and Keshian 1990). In other cases, the barriers to access supportive services may be self-imposed as a result of lack of interest and motivation (Dergance et al. 2003).

Technology such as the Internet can not only facilitate communication between elders and their families as we mentioned in the Sect. 6.4, but also with other members of the community, thus helping the elders remain engaged with their extended social network. This is especially true for isolated elders in rural and remote areas, but it is also valid in urban settings (Buckwalter et al. 2002).

In spite of its potential benefits in helping older people remain connected with the rest of society, the use of the Internet varies greatly among the elders between and within countries, with the full spectrum of possibilities ranging from the *silver surfers* who competently make use of Internet services, to *offline seniors* who have no interest in the use of this technology (Gardner et al. 2012). However, a constructive use of the Internet can help increase self-efficiency, mental stimulation, self-determination, knowledge, and social interactions (*social connectedness*) in the elderly (Hendrix 2000). Perhaps IT companies may assist with the reduction in the complexity of interfaces and other changes in the design of features and displays that could help older people (e.g. Fischer et al. 2014).

Finally, elders in culturally and linguistically diverse communities can provide expert knowledge especially to students, at all levels, including tertiary students, by means of transmitting their cultural traditions and language (Anonson et al. 2014; Cordella and Huang 2016).

We will return to the issue of elders actively contributing to their community, whilst at the same time enjoying the many benefits of social connectedness, in the last chapter *Fulfilling ageing*.

In sum, beyond the family, elders can enjoy the benefits of social support coming from the local community and various levels of government. Such sources of support can provide both emotional and material aid. The social capital provided by better integrated local communities allows the elders the opportunity to live a more fulfilling experience of ageing in place, according to their individual preferences. Ageing in place can be not only creative but also socially useful for both the elders and the community they are embedded into, as the elders can contribute with their accumulated experience and specific skills to the needs of others.

In the next sections we explore the complex interplay between ethics and the negative effects of social interactions on the elders.

6.6 The Ethics of Ageing: Stereotypes, Stigma, and Ageism

The ways we interact with old people are modulated by our ethical principles, which determine whether our relationships with the elders are positive or negative, being marked by cooperation and understanding or by negative stereotypes, stigma, and ageism. This section will be organised into four subsections: the first one deals with some ethical issues of old age, it will be followed by a section on stereotypes, and then one focused on stigma, to finish with a section on ageism.

6.6.1 Old Age and Ethics

We behave in an ethical manner for as long as we question the ways in which we conduct ourselves towards others, and for as long as we question what is "good" and what is "bad", what is "right" and what is "wrong" in such relationships (e.g. Doron 2007; McArdle 2012).

Our interactions with old people generate various ethical questions that require appropriate answers. We cannot reduce an old person to being just a "consumer", for instance, a body in need of various kinds of resources to keep functioning. This is a narrow economic approach that ignores central aspects of old people's well-being such as the conditions under which life is worth living. Inevitably, we are confronted with a much broader ethical dimension of old age.

The ethics of old age comprises a set of key issues that society is called to address and provide a resolution to. Some of the major ethical themes of old age include dignity, autonomy, moral obligation, and justice (Tinker 2003; Shultziner 2007).

Human *dignity* refers to the fundamental equal worth of any person (Habermas 2010), and it can be seen as "a justification for various rights and duties in legal instruments", a justification deriving from both an egalitarian worldview that encapsulates such rights and duties and also a sense of *honour* or "worthiness", within the context of specific social expectations (Shultziner 2007). Rights and duties define the confines of dignity. When rights are violated, we experience a "degrading attitude or humiliation of human worth" (Shultziner 2007). Dignity also relies on the capacity of the individual to operate with a degree of *autonomy*, which in society is dynamic and also subject to constraints that are variable over time: "A primary moral difficulty is the preservation of such a dignity at a time when there is a declining capacity for autonomous decision-taking" (Fenech 2003, p. 232). Thus, ageing confronts the individual and society with the moral obligation of preserving individual rights, worthiness, and autonomy, or in short preserving elder's human dignity, in response to specific transformations that are experienced by the elder and that can change the capacity of the individual to continue interacting as usual with others. Justice for the elders can be better achieved through the fulfilment of such moral obligation to preserve dignity.

Older people may see their dignity compromised even by their own family, whenever they may lose autonomy through unfair relationships, as in *filialism*: the limitation of "freedom and responsibility of older people by the well-meant actions of sons and daughters" (Fenech 2003, p. 232). On the other hand, the family can contribute to support the dignity of the elders through the positive aspects of *familism*. Autonomy could be also affected by the actions of the family, especially in critical circumstances as in the case when the elder may be considering end-of-life options such as euthanasia.

A dignified treatment of elders in society is also linked to the need to act with fairness. *Fairness* can be interpreted in various ways: as *equality* or, where returns are conditional to effort, *equity* (Jones et al. 2014). It is a response to the perception of unjustified inequalities. Fairness requires a proactive stance from the part of individuals in society (*social actors*, e.g. Whetten and Mackey 2002), not just absence of discrimination against older people, in order to position the elder in a situation where she or he can fulfil achievable objectives.

As members of society, we are all required to act on *moral obligations* if we want society to function in a stable and harmonious way. Moral obligations towards the elders spring from a recognition that they are vulnerable, that we can proactively protect them against the negative effects of such vulnerability, that such commitment must be non-discriminatory, and that it should lead the elders to experiencing a good quality of life (e.g. Nunes 2015). The need to fulfil moral obligations, however, may also expose individuals to *moral distress* "when one knows the right thing to do for the Other but institutional control or constraint make it nearly impossible to pursue the right course of action" (Lützén and Ewalds-Kvist 2013, p. 318). A supportive social environment may help the individual to better reflect and act on moral obligations, thus potentially decreasing moral distress.

Justice is another concept that is key to the ethics of old age. Justice is expected to achieve equity in the treatment of older people in society; e.g. if they are in need, they should receive help. But what about other age groups: should they also receive help if in need? And who should ultimately receive such help when resources are limited? Although we question the constraints of "limited resources" for social services in countries that are economically advanced—in those countries the money is there in droves, but under neoliberalism many governments have become reluctant to touch it, by establishing regimes of low effective taxation —it is not unreasonable to define just support as a combination of needs and objective ability to self-satisfy such need. It should be regarded as a moral obligation to help yourself if you can, so that the collective can help those who cannot overcome their challenges on their own.

The issue of resources allocation has become a controversial point at the root of decision-making in our relationship with the elders, especially the frail and infirm (Moody 1995). It is our ethical view on old age, not just our economic considerations, that will determine where and how government money should be spent on which types of interventions.

Achieving equity, however, is not easy for older people, even when they enjoy the protection of a fair system of laws. It is often the case that elders need to exert a degree of social power to receive just treatment from the rest of society. Social power may increase when elders do not act in isolation but as an organised group, being also able to attract the support of other sectors of society: elders seeking improvement to pensions may recruit the help not only of other elders, but also of churches and political parties. The ability to attract wider support should increase with greater levels of intersectionality in the community of elders.

Here we use the term *intersectionality* to express the complex reality of elders as a collective of identities (old, member of a church, member of a political party, etc.), criss-crossing various sectors of society. Greater intersectionality allows elders to come together, alongside other members of the communities they belong to, to better achieve a given objective.

A more specific set of ethical principles applies to the relationship between health care providers and the old people under their care. Tinker (2003, see also Marquis 2002) stresses the need to base the relationship between old people and caregivers on the ethical principles of *consent*, *confidentiality,* and *conduct*, the latter must be not only lawful but also respectful of the dignity of the elder. Health care services for the older generations are also increasingly confronted with ethical issues relating to the costs of providing such care (Dodds 2005) and how to allocate limited resources. Indeed, the question whether resources are in fact limited or not is not just economical but also ethical and of course political as we have suggested regarding neoliberalism.

The health problems that ageing people face and the various recent medical advances that address such problems have also shaped new fields of ethics. In particular, *bioethics* and *biogerontology* have questioned the concept of "normal ageing", as the quality of the ageing experience is variable among elders and becoming increasingly dependent on the unequal access to both health services and other resources that affect quality of life (Rees Jones and Higgs 2010). Thus, what is supposed to be "normal", may not be experienced by many elders due to such variability in the population.

Advances in medicine, changes in behaviours and access to material resources can significantly contribute to modifying our conception of what is achievable as we age and therefore our conception of "normality"; even providing room to wild speculations about human "immortality" (Turner 2004). This has opened new ethical challenges concerning expectations about what is attainable in old age. But being honest about the medical limits of biogerontology—and its field of technological applications: *biogerontechnology*—is a pressing ethical requirement for any medical practitioner (e.g. Turner 2004; Ehni and Spindler 2011). Ageing and, ultimately, mortality are not going to disappear any time soon, as gradual decay and final death remain defining aspects of the "normality" of ageing.

Although technology (e.g. *assistive technology*) cannot eliminate mortality, it can help older people live a better life and increase their *resourcefulness*, by compensating for some loss of capabilities or by, to some extent, slowing down such loss (e.g. Zwijsen et al. 2011; Giaccardi et al. 2016). Accessibility to such technology is an important ethical issue in modern societies.

Other kinds of technologies are not exempt from ethical considerations either. For instance, information technology (IT), that can support a better level of social connectivity in older people and greater dissemination of medical information through *E-health* (i.e. the use of information and communication technologies to treat, educate, monitor, and carry out research in public health) and *telemedicine* (a branch of *E-health* used to deliver medical care and information across long distances), also raises some ethical concerns. Thompson et al. (2012, see also Zwijsen et al. 2011) have reviewed this issue in the context of the European Union. Some ethical concerns about *E-health* include: *protection of privacy*, a concern that directly derives from the ethical principle of autonomy, a defining aspect of dignity; *informed consent*, which again derives from the autonomy of the elder; *equity* and *accessibility*, as access to IT is not easily available to all persons who may need it; and *risk of confinement*, that is, to what extent remote interactions through IT may isolate the older person from closer, face-to-face social contact with others. Some monitoring devices could be considered too *intrusive* (*obtrusive*) by the elder (Zwijsen et al. 2011), and they could also expose the elders to some potential effects of stigma. Providing educational programs of digital literacy, and public accessibility to electronic devices that are safe, affordable and that can be easily used by older people, should help address at least some of the above ethical concerns.

Therefore, treating the elders with dignity, fostering their autonomy, understanding and enacting society's moral obligations, and ensuring that there is justice in the achievement of equity in the elders' population are central ethical issues in our approach to old age.

In the next three sections we will review the social processes that can lead to stigma and discrimination based on age, and that can therefore undermine the ethical principles that should guide our cooperative relationship with the elders. Stigma and discrimination derive from stereotypes about old age, and they may threaten the dignity of old people and their capacity to be autonomous. When stereotypes are negative and become entrenched over time, they could lead to the development of stigma and age-based negative discrimination or ageism.

6.6.2 Stereotypes of Old Age

We have seen in Chap. 2 how myths such as the *Fountain of Youth* have expressed, throughout history, a deep desire to escape the ravages of old age. Old age is not only an indication of our relative closeness to the end of life, but it is also a reminder of our lost youth. The body changes as we age, unveiling what has been called a *mask of ageing* (Biggs 1997; Andrews 1999). The mask may reflect who we really are or perhaps it may hide a more youthful inner self. In some cases that inner self may keep craving and dreaming an experience of agelessness, thus motivating the individual to pass for a young person through changes in behaviour, attire and, these days, also plastic surgery. The youthful inner self, "trapped" behind an aged mask, may rebel against the stereotypes that expect it to behave as an old person and not a younger one, and it may respond by creating a defined *persona* that will hopefully live a more fulfilling life (Biggs 1997).

In a fast changing world, elders are constantly confronted with the social pressure of stereotyping—expressed as pigeon-holing, premature generalisations, and unwarranted simplifications (Barron 1953)—that may clash with the reality and desires of the individual who is the target of the stereotype.

Ehrlich (1973, p. 20) defined a *stereotype* as "a set of beliefs and disbeliefs about any group of people". Stereotypes ignore individual variability and subsume a group of people under a single category, assuming homogeneity (Hamilton and Trolier 1986). Ehrlich distinguished between stereotyping as a process, which can be generalised, and "stereotype assignment" which: "signal(s) the socially approved and accessible targets for the release of hostility and aggression" (p. 21). Such targets, of course, can vary greatly according to circumstances and time (Stangor 2009), and the stereotype itself may undergo changes over time (Biernat 2009; Dasgupta 2009; Devine and Sharp 2009).

A social target for stereotyping must necessarily be "visible". Visibility allows others to establish a degree of consensus about the stereotype. In some cases, the stereotype against an individual may be established early on in his/her life (e.g. on the ground of being a member of a specific minority) (Isaacs and Bearison 1986; Levy and Hughes 2009), and it is often the case that powerful groups in society are the ones stereotyping the less powerful ones (Vescio et al. 2009). Moreover, the stereotyping behaviour itself may start to develop at early ages (e.g. Burke 1982; Davidson et al. 1995). For instance, Isaacs and Bearison (1986) suggested that by elementary-school age, children have already started assimilating negative stereotypes about old people.

Whereas a stereotype relies on oversimplified information about a group of people, the elders in this case, a *prejudice* is a preconceived opinion about something, whereas *social prejudice* in particular is a pre-established attitude towards a specific group which may or may not be grounded on any objective information (e.g. Fiske 2000).

Stereotyping and prejudice are rooted in both biological/evolutionary processes and the more dynamic social and cultural domains that are affected by learning, all contributing in various degrees to the act of pigeon-holing others and its variability in space and time (Fiske 2000). A possible adaptive evolutionary root of stereotyping is that it may facilitate the differentiation between friends and foes (Cuddy et al. 2009).

Once established, stereotypes may perpetuate themselves *automatically*, although they can also undergo changes over time. Bargh (1994; see also Bodenhausen et al. 2009) listed some criteria to identify some automatic mental processes that could lead to stereotyping: spontaneity (stereotyping occurs in the absence of any intention); efficiency (the execution of a stereotyping act requires low levels of attentional resources); uncontrollability (stereotyping can operate in a speedy fashion and it is hard to stop once it has been triggered); and unconsciousness (the subjects are unaware of their stereotyping). In this way stereotypes become incorporated into what Biggs (1993) described as *commonsense reality*. In the words of Angus and Reeve (2006, p. 141):

> Commonsense reality signifies a set of unquestioned beliefs that enables people to make sense of everyday life in a complex world. This commonsense reality is a socially constructed space that operates in ways that legitimate behaviors and limits the possibility for imagining and acting on alternative realities. Participants in such judgments of behavior are often unaware of the basis for their assumptions that inform their commonsense reality.

Thus, through the construction of a "commonsense reality" people may use stereotypes against the elders whilst being unaware of what they are doing, and when challenged, they often believe that, in fact, they are expressing a fair and "unbiased" opinion, which makes the task of changing the stereotypes much harder (Scholl and Sabat 2008).

Stereotypes may sometimes contain a degree of truth, but such truth is usually exaggerated, overgeneralised, additional explanatory factors ignored, and variability across individuals disregarded (e.g. Jussim et al. 2009).

Although stereotypes can be either positive (e.g. the elder seen as a sage; old people are more stable, honest, trustworthy, loyal, committed to their job; Posthuma and Campion 2009; Iversen et al. 2009) or negative (e.g. Stangor 2009), stereotypical old age mainly elicits negative images of physical and mental limitations (poor memory in particular), ill health, hardened views and decreased flexibility and capacity to learn, along with passivity, withdrawal or disengagement from society (isolation). The old are also stereotypically seen as unproductive, ineffective, helpless, a burden to society and low in sexual motivation; not trying, irritating, self-oriented, living outside the mainstream, unattractive, uninteresting, frail, senile, silly, irrelevant, over the hill, narrow-minded, vulnerable, dowdy; along with complaining, bitter, and demanding (Minichiello et al. 2000; Ory et al. 2003; Posthuma and Campion 2009; Hughes and Heycox 2010; Esqueda 2011). In the workplace, older workers may be regarded more negatively than younger ones on the ground of their perceived lower trainability, adaptability, creativity, and interest in learning a new technology (Gringart et al. 2005).

Negative stereotypes against the old are often unconscious and coexisting with expressed views that may be supportive of the elders (Hense et al. 1995).

Rowe and Kahn (1998, see also Ory et al. 2003) subsumed many of those stereotypes into six common myths that can be easily proven to be at the very least facile or half-truths, and in many cases plainly wrong as a general characterisation of old people:

Myth 1: *To be old is to be sick*. But in reality many old people can perform their activities of daily living effectively even at advanced ages, and in spite of increasing probability of becoming ill.

Myth 2: *You cannot teach an old dog new tricks*. But in reality older people do retain a capacity to learn.

Myth 3: *The horse is out of the barn*. But in reality it is never too late to adopt behaviours that can improve physical and mental health.

Myth 4: *The secret to fulfilling ageing is to choose your parents wisely*. But in reality genetics does not determine the full extent of how we age.

Myth 5: *The lights may be on, but the voltage is low*. But in reality sexual activity may continue into old age given the appropriate circumstances, attitudes and these days medical aid.

Myth 6: *The elderly do not pull their own weight*. But in reality many older people do have the capacity to sustain themselves financially and many do also contribute to society through unpaid work and other charitable initiatives.

Such myths tend to be perpetuated through person-to-person social interactions and the media; and they rely on strong images and carefully chosen expressions to both penetrate deep into the mind of people and become refractory to change. Refractoriness to change is helped by using catchy metaphors like the ones used in some of the myths listed by Rowe and Kahn (1998). For instance, Myth 2 "You can't teach an old dog new tricks" is a strong and widespread one. Thornton (2002) reminds us of the powerful image of old age created by William Shakespeare: "sans teeth, sans eyes, sans taste, sans everything" that reinforces the view of old age as a process of decay, loss, and hopelessness as the body descends down the road to death.

The use of specific stereotypes about old people eventually regulates the attitude that the rest of society is going to adopt towards the elders. For instance, the *affective response* directed to the old will change according to the prevailing stereotype in society. Ehrlich (1973, p. 98–99) lists the following "norms of affective response" that characterise the emergence, spread, and change/stability of a stereotype:

- Distinction: Stereotypes are distinguishable on the ground of the belief statements associated with them.
- Diffusion: Belief statements about stereotypes are widely diffused in a society.
- Consensus: There is high consensus on the stereotype associated with old people.
- Stability: Belief statements associated with stereotypes are highly stable.
- Change: Belief statements associated with stereotypes change as the relations of the rest of the community with the elders also change.
- Visibility: Under conditions of stability, the greater the visibility of the elders, the greater the distinctiveness, diffusion, and consensus of stereotypes associated with them.

In turn, the affective response can be modulated by broad categorisations of the elders. Those categorisations generate from within society and they establish a place for the elders in that society. The individual may rebel against such pigeonholing, but she/he may not always be successful at asserting a specific personal identity. As stated by Ehrlich (1973, p. 100): "The less knowledge available about a social object, the greater the likelihood that it will be assigned the characteristics of its social category", thus becoming stereotyped.

Hence, through stereotyping, individuals lose their individuality and are regarded just as members of a specific category along with others, a category that can be precisely described (encoded). In the absence of additional knowledge about the individual, all old people become just the "same". This may lead to what is known as *social identity threat*: "a state of psychological discomfort that people experience when confronted by an unflattering group or individual reputation in situations when the reputation can be confirmed by one's behaviour" (Aronson and McGlone 2009, p. 154). Strongly negative attitudes towards the elders may eventually lead to the extreme of *gerontophobia* that can be the cause of serious threats to the social identity of old people. Gerontophobia can be defined as an "unreasonable fear and or irrational hatred of older people by society and by themselves" (Bunzel 1972, p. 116).

Negative and positive stereotypes contributing to the establishment and reinforcement of prejudice have been formally synthesised into a *Stereotype Content model* (SCM) by Fiske et al. (2002, see also Cuddy et al. 2009) (see Fig. 6.9). The SCM organises stereotypes into a two-dimensional space of Warmth and Competence, producing four broad areas of interest where high and low values of the two major variables combine: a *Paternalistic* stereotype is elicited by a perception of Low Competence/High Warmth in the target, *Envious* (High Competence/Low Warmth), *Contemptuous* (Low Competence/Low Warmth), and *Admiration* (High Competence/High Warmth). More diversity of categories can be produced if an additional "medium" level of both competence and warmth is added. Ultimately, however, both variables can be expressed as a continuum from which a subtle diversity of stereotypes can be derived.

Stereotype Content model

Competence

Warmth	**Low**	**High**
High	**Paternalistic** **Low status, not competitive** **Pity, sympathy** **(e.g., elderly people, disabled people, house-wives)**	**Admiration** **High status, not competitive** **Pride, admiration** **(e.g., in-group, close allies)**
Low	**Contemptuous** **Low status, competitive** **Contempt, disgust, anger, resentment** **(e.g. , welfare recipients, poor people)**	**Envious** **High status, competitive** **Envy, jealousy** **(e.g., Asians, Jews, rich people, feminists)**

Fig. 6.9 Model of stereotypes based on a Stereotype Content model, eliciting distinct emotions that are associated with various types of positive (e.g. admiration) or negative (e.g. contemptuous) stereotypes. Slightly modified from Fiske et al. (2002)

In a survey of respondents carried out in the USA by Fiske et al. (2002), the elderly were found in a Low Competence/High Warmth stereotype cluster along with the disabled; i.e. elders are regarded as people who experience difficulties operating in life, and deserve to be treated paternalistically with a degree of pity and sympathy, unlike the poor, welfare recipients and the homeless who clustered in the Low Competence/Low Warmth category, thus eliciting contempt. Who were the ones regarded as High Competence/High Warmth and deserving of admiration? Given that this work was carried out in the USA, perhaps not surprisingly this group included: Christians, middle class people, and Whites. Interestingly, women and students clustered in the same category, presumably indicating that the USA society has been undergoing some developments in the direction of gender equality. In fact, in recent decades there has been a modest increase in some aspects of gender equality around the world (Seguino 2016).

The Low Competence/High Warmth stereotype associated with the elderly is both pervasive and persistent within and across countries (Cuddy et al. 2005).

Given that such stereotypes are widespread and tend to be resilient to change, they may feedback on the old people themselves, becoming reinforced in their mind and behaviour, in a process that Erving Goffman (1963) described as "recursive". Thus, old people may be more likely to become passive if they incorporate societal expectations about their passivity, that is if they *internalise* the stereotype. Breaking the recursive process requires determination from the part of the elder, but the task can be facilitated if they are helped by the "wise", that is other members of society who reject the stereotype (Goffman's concept of the "wise" will be further explained in the next section).

The recursive process of stereotyping is illustrated in Fig. 6.10. Awareness of existing stereotypes gives social meaning to some individual characteristics that do conform to the stereotype. Consequently, the target of the stereotype may be under *identity threat* by others, with the stereotype challenging the ability of the individual to develop and sustain an independent identity. This can potentially lead to impaired performance which reinforces the initial characteristics, thus further supporting the stereotype. Therefore, old people who are continuously reminded that they are "senile" may help reinforce the stereotype after internalising stereotypical senile behaviours.

Becca Levy's (2009) theory of *stereotype embodiment* states that stereotypes become internalised after being assimilated from the cultural environment, by means of modifying the definition of self, which produces effects on behaviour and potentially also health. Internalisation occurs over time, the process is most likely unconscious, and a feeling of self-relevance strengthens the adherence to the stereotype through multiple perceptual and information processing mechanisms. Negative stereotypes about old people may also become self-fulfilling prophecies when the old person does detect some objective decline (e.g. in memory) that may reinforce the initially external stereotype in his/her mind. In this manner the stereotype may become embodied (Nelson 2009; North and Fiske 2012).

Internalising a stereotype can lead to self-stereotyping, a process that has been well documented in psychological research (see Levy 1996; Scholl and Sabat 2008).

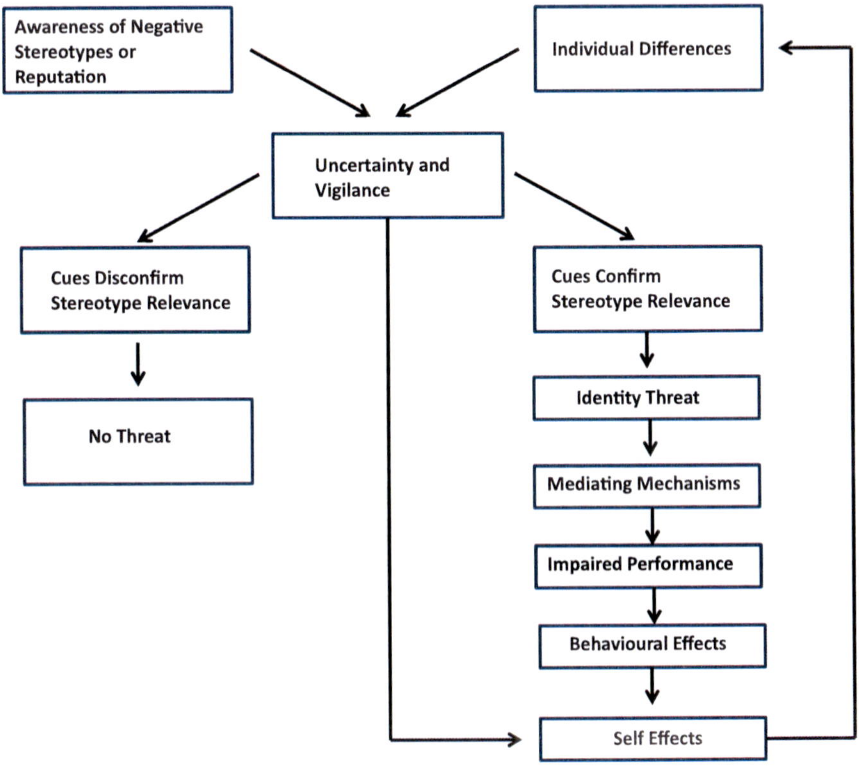

Fig. 6.10 Recursive process of stereotyping where stereotypes may impair the actual performance of the target, thus confirming first and subsequently reinforcing the stereotype. Redrawn from Aronson and McGlone (2009)

The roots of self-stereotyping in old age are found in early development, when the adoption of stereotypes against the elderly may initiate and then grow through reinforcement. Self-stereotyping is often unconscious, being under the influence of the social environment (family, friends, the media) (Scholl and Sabat 2008). Interestingly, Scholl and Sabat also suggest that negative stereotypes against old people may be so entrenched that they become refractory to positive experiences with elders, with non-stereotypical old people being simply dismissed as "exceptions to the rule" (see Levy and Banaji 2002).

Attitudes and behaviours adopted by the elder in conformity to stereotypes may have potential negative consequences at the physiological level. For instance, self-stereotyping may have effects on memory: believing that memory declines with age may lead to poor performance and low efforts to improve memory, thus contributing to memory decline (e.g. Scholl and Sabat 2008).

Stereotyping feeds prejudice that can then build frustration in the targets of the stereotype, sometimes resistance and, in more extreme cases, violence. The effects of prejudice on the targets can ultimately also affect physiological processes through emotional responses (e.g. Mendes et al. 2008).

We mentioned above how a social prejudice is a pre-established attitude towards a specific group, an attitude that may or may not be based on any objective information. Prejudice has three main components: (a) cognitive (beliefs, thoughts); affective (feelings, emotions); and (b) behavioural (avoidance, exclusion) which interact with each other (Jones et al. 2014), often in an unconscious manner. Prejudice against specific target groups leads to biases that can be expressed not only in personal behaviours and attitudes but also, institutionally, through laws, policies, and practices that perpetuate such prejudice. Through prejudice we establish and reinforce a consensus view about a group of individuals, attributing a set of characteristics (negative, positive) to them that, in our mind, are cognitively consistent and are accepted as part of the norm, but that may not be grounded on any objective evaluation of the target of prejudice.

Prejudice is reinforced and transmitted through socially accepted ideas such as that of a "Just World". The *Just World* hypothesis (Lerner 1980) states that: "people need to see the world as a just place in which we get what we deserve and deserve what we get … the illusion of a just world is comforting" (Jones et al. 2014, p. 122). The concept of a just world may be comforting but it is also one of the root causes of victim-blaming. In addition, prejudice possesses its own mechanisms of reinforcement that make it resistant to change. When prejudice is challenged, we may experience an aversive state of *cognitive dissonance*, whereas recovering the prejudice returns our mind to reassuring *consonance*.

Cognitive dissonance theory proposes that: "holding inconsistent thoughts or beliefs arouses psychological tension and discomfort, which people strive to reduce, often by changing their attitudes to be more consistent with their actions. Feeling tense, anxious, or uncomfortable motivates us to do something to make us feel better. In this case, inconsistency signals that we do not really understand the world, and we immediately become motivated to restore consistency and have the world make sense again" (Jones et al. 2014, p. 125). Through eliminating cognitive dissonance, we protect our prejudice against potential challenges. Other cognitive mechanisms that reinforce prejudice include *illusionary correlations*: "Like our attributions, our need for consistency can fuel our biases. We often see relationships that do not exist because it makes the world seem consistent" (Jones et al. 2014, p. 125); *confirmatory biases*: "Because people's first impressions, by definition, come first, … people will adjust their subsequent perceptions, attributions, and beliefs to be consistent with their initial attributions. That is, people have a confirmatory bias" (Jones et al. 2014, p. 126); and *self-fulfilling prophecies*: "The effects of the need for consistency reach much farther than affecting just our own behaviour, however. It also shapes others' reactions, leading them to behave in ways consistent with what we expected in the first place" (Jones et al. 2014, p. 127).When we meet a specific individual (e.g. an old person) who is expected to behave in a specific manner according to our prejudice, but that person does not conform to our expectations, we may be confronted with *stereotype disconfirmation*. The response to such disconfirmation, however, is often that of dismissing the case as just an "exception", which will inevitably confirm the stereotype: "the exception that confirms the rule" that we mentioned above (e.g. Levy and Banaji 2002).

Negative attitudes and prejudice against the elders may also derive from a sense of perceived threat from the elders as a demographic and social category (Stephan et al. 2009). For instance, elders could be perceived as competitors for jobs, users of government resources, and so forth. Ultimately, negative discrimination against the elders may lead to elder abuse. This has led to the establishment of the *International Network for the Prevention of Elder Abuse* (INPEA, http://www.inpea.net/) in 1997 (Lachs and Pillemer 2004).

Given the characteristics of stereotypes and prejudice affecting old people, could such stereotypes and prejudice be eventually eliminated? Yes, eliminating stereotypes and prejudice is indeed possible. Overcoming the self-stabilising mechanisms of prejudice requires offering a new paradigm that can satisfactorily and harmoniously replace the old one. Such change should not be expected to come easy, however. At the very least, groups that benefit from the maintenance of stereotypes and prejudice will try to resist any change (Major and Sawyer 2009) but, to a great extent, a degree of resistance against stereotypes and prejudice from the part of the elders may help in achieving change.

Rothbart (2003) mentions two broad strategies that have been suggested to modify stereotypes. One involves contact with "individual exemplars" of the stereotyped category of people who display attributes (e.g. behaviours) that do not conform to the stereotype. If such specific individuals who disconfirm the stereotype are regarded as representative of the norm, then the close experience with them can help in modifying the stereotype. On the one hand, if those individuals are disregarded as the "exception that confirms the rule", then the stereotype will be maintained. The other strategy is to expand the boundaries of the category that is stereotyped. Stereotyping narrows our view of people, therefore by expanding such view and producing a perception that takes diversity into consideration, we create a more complex and more realistic view of the world.

Therefore, factors that can help overcome stereotypes and prejudice include gathering additional factual information about the target (*individuation*, see Posthuma and Campion (2009) for a myth-debunking exercise about old age), overcoming potential bias (*correction*), act on the conscious elimination of stereotypes and prejudice (*suppression*, *disidentification*, Aronson and McGlone 2009), and intentionally inhibit and replace stereotypes (*self-regulation*) (Devine and Sharp 2009). To those we may also add an attitude of considering older members of society as potential partners in the achievement of common goals (Crocker and Garcia 2009).

Some form of resistance against prejudice and discrimination leading to change is also possible and it can be expressed in many ways by the targets. For instance, passive resistance could be adopted or more direct forms of active resistance, as it may occur in the case of old workers who may lodge a formal complaint because they are discriminated against in the workplace (Snape and Redman 2003). Over time, employers may come to realise that it is in their own self-interest to eliminate prejudice and discrimination in the workplace and keep all workers happy and performing, but anti-discrimination laws can certainly help.

To recap, stereotypes and prejudice about old people can become entrenched, as they keep reinforcing themselves, thus imposing upon older people a constraining image of ineptitude, physical limitations, pig-headedness, disengagement, frailty. Such image often constrains the ability of old people to operate in life. Prejudice against the old can reach extremes such as gerontophobia. Resistance against prejudice and stereotypes is possible and change can be achieved through the activism of old people themselves, but cooperation from the rest of society and their willingness to change are also required.

Entrenched negative stereotypes about the elders can eventually lead to older people becoming stigmatised. In the next section we address the issue of stigma affecting old people.

6.6.3 Stigma

We start this section with a general introduction to Goffman's analysis of stigma (see also Cordella and Poiani 2014), to then continue addressing further aspects of stigma, its impact on the life of old people and the strategies that could be adopted to overcome it.

Erving Goffman (1963) pioneered the study of stigma in the field of social psychology and his work has become very influential, providing the basic framework for many subsequent developments in the study of stigma in various social contexts (Link et al. 2004).

The word stigma comes from classic Greek and it originally designated external body features, or marks, that could function as signals to distinguish members of a specific group, such as a slave. A stigmatised person possesses a social identity that is given to him/her by the rest of society, an "attribute that is deeply discrediting"; such identity determines the position of the individual as "a tainted, discounted one" (Goffman 1963, p. 3). Following Jones et al. (2014, p. 208; see also Vázquez et al. 2011), the social status of *stigmatised* person is established by four conditions: "(a) individuals are differentiated and labelled, (b) cultural beliefs attach negative attributes to people who are labelled as different, (c) labelled individuals are placed in groups that are separated between 'us' and 'them', and (d) labelled individuals experience loss of status and are discriminated against, leading to social inequality." Through watching the actions of others and through gossip, stigma is produced and reproduced, transmitted and sustained within a local community (Gamliel and Hazan 2006). Presumably, the process can be even more effective when the community is small.

The rest of the members of society are called "normal" by Goffman. Stigmatised person and "normal" are clearly relative concepts as the stigmatised one in one context may become a "normal" in another. The immediate outcome of stigma is the denial to full membership into a group or, at least, the denial of membership within the terms that the stigmatised one would wish for. Therefore, the issue of social acceptance and full group membership is at the core of the concept of stigma.

A stigma can be felt by a person through fear of *expected* discrimination, given the possession of undesirable attributes, even in situations when others have not had opportunities to interact with the individual yet. Eventually the individual may experience *enacted* stigma as soon as others do act on their discrimination.

Stigmatised individuals may adopt various strategies to decrease the stresses and disadvantages of stigma. For instance, for as long as they remain within the group, they may reduce social interactions. The side effect of this strategy, however, is that social isolation may potentially lead to stress, anxiety, and depression.

A very different reaction to stigma is open rebellion. But then, a rebellious stigmatised person risks falling into Goffman's "recursive process" mentioned in the previous section (see Fig. 6.10), as openly rebelling against a negative view of self may simply reinforce such negative view in the mind of others: an elder labelled as "cranky" who rebels aggressively against the stigma, is likely to be regarded as even more "cranky". Moreover, Goffman also alludes to an effect of self-awareness on the person interacting with the stigmatised one. In fact, both interactants may experience such self-awareness affecting self-esteem. Garcia et al. (2005, p. 38) explain this issue succinctly:

> When people who are stigmatized and nonstigmatized interact with each other, both experience threats to self-esteem, but for different reasons. Individuals who are stigmatized may experience self-esteem decrements because they feel that their group is devalued in the eyes of others. Those who are nonstigmatized may fear that their actions will be perceived as biased, thereby threatening their self-image as an unprejudiced person. Individuals who are stigmatized and nonstigmatized act in ways that make their worst fears more than likely come true.

One potential effect of such an escalating process of self-awareness may be to decide to cut social relationships altogether, and in the face of uncertainty about how to behave in the presence of a carrier of stigma, some "normals" may suspend social contact with the stigmatised one.

Stigmatised persons can also look for a strategy to cope with stigma, not necessarily by ingratiating the "normals" but by establishing and strengthening social relationships with peers: other equally stigmatised people. Although this strategy avoids personal isolation, it may nevertheless lead to potential social segregation. In the case of age-based social segregation, the segregation can be spatial, cultural, and even institutional (e.g. Hagestad and Uhlenberg 2006). Goffman also identifies a third group of interactants: the "wise", that is "normals" who are especially sensitive and supportive of the stigmatised one. Being a "wise" is not necessarily cost-free, as she/he may run the risk of also being labelled with a "courtesy stigma".

Given that the identification of the stigmatised one is mediated by his/her possession of discernible symbols or marks, there is the tendency in some of them to hide such symbols (such as wrinkles, hair colour, in the case of old people) if possible. Such behaviour may lead to what Goffman calls "double living" where the carrier of stigma may behave in a specific manner in the presence of peers (or at least some peers) and the "wise" and in a different manner in the presence of "normals". However, although a stigma requires the presence of an identifying mark, possessing such mark is not sufficient to become stigmatised, it is the effect of that mark on the rest of the group that produces stigma. In the most extreme of cases socially imposed stigma may lead to "loss of personhood" or "social death".

The behaviour of the stigmatised one in the public domain may sometimes go as far as to display what Goffman (1963) terms *minstrelisation*; that is, the performance of ingratiating acts in the presence of "normals" such as displaying a "full dance of bad qualities imputed to his kind, thereby consolidating a life situation into a clownish role" (p. 110). Minstrelisation may decrease the stress of stigma by simply incorporating it into an accepted view of the self, and perhaps even reap some benefits from it. On the other hand, if minstrelisation corrodes self-identity without a full redefinition of it, in the long run it may threaten self-esteem.

Groups of people can be stigmatised following various criteria. Esqueda (2011, p. 188), based on Towler and Schneider's (2005) work, provides a simple list of such criteria (with examples) to which we have also added age, sex, and ethnic identity:

Age	Old, young, and indeed any age can be the target of stigma.
Physically disabled	The blind, epileptics, deaf, rape victims, cancer patients, multiple sclerosis sufferers, stutterers, Alzheimer's patients, car accident victims, amputees, the wheel-chaired, asthma sufferers, diabetics, paraplegics, burn victims, AIDS patients, dyslexics.
Mental	The depressed, psychiatric patients in general, the suicidal, people who have had a nervous breakdown, schizophrenics, obsessive people, neurotics.
Physical appearance	The obese, people with severe acne, people regarded as ugly, facially scarred people, people with body odour, the unkempt, the bald, eczema sufferers, strabismus sufferers.
Sexual orientation	Gay, lesbians, trans, bisexuals.
Sex	Males, females, intersex.
Racial identity	Blacks, Hispanics, Asians, whites, Native Americans, and so forth.
Ethnic identity	Middle Eastern, Indian, Chinese, Southern European, and so forth.
Social deviants	Murderers, reformed felons, drug addicts, smokers, alcoholics, sex offenders, people with tattoos, skinheads, people who body-pierce.
Economically disadvantaged	The homeless, welfare recipients, the unemployed, the precariat.

Although stigma has an obvious social referent, a not less important aspect of it is *self-stigma*—of which minstrelisation is a specific example. Public stigma is expressed in terms of stereotypes, prejudice, and discrimination that are imposed on the individual by others. But such stereotypes, prejudice, and discrimination may be also turned by the individual onto him/herself through the process of self-awareness that we mentioned above, thus producing self-stigma (e.g. a casual lapse in memory in an old person may elicit the comment from him that he has just had a "senior moment"). One potential consequence of self-stigma is the so-called why try effect which may become an impediment to action.

As a result of the awareness of stigma, the elder may turn stereotypes inwards, which may result into low self-esteem and low self-efficacy, and consequently failure to achieve life's goals. Agreement with others regarding such "failure" leads to the devaluation of self in the form of the “why try” effect: “I know I won’t be able to achieve it, so it’s not even worth trying”. Demoralisation often follows, that may lead to the avoidance of social contact and, potentially, depression. Interestingly, not all individuals react to the internalising of stigma by developing low self-esteem and low self-efficacy, others may react more proactively, by drawing a sense of “power” from the challenge of confronting and overcoming stigma. Such proactive approach may be helped by cooperative interactions in peer groups and by support received from the “wise”.

Self-stigma can have negative consequences on the individual. Vázquez et al. (2011), for instance, described how in a group of participants diagnosed with bipolar disorder, higher scores of self-perceived stigma were associated with lower scores of functioning. In a work carried out in Israel, Werner et al. (2009) found that old people scoring higher on the *Geriatric Depression Scale* (GDS) also scored higher on self-stigmatisation and displayed lower levels of self-esteem.

Although it must be also said that self-awareness can have a positive side, as in the case of people feeling younger despite their old age. In a study of subjective perception of age, Kleinspehn-Ammerlahn et al. (2008, p. 378) argued that positive self-perceptions of ageing (i.e. being content with own ageing) “serve to sustain levels of social activity and engagement, enhance self-esteem and well-being, and boost biophysiological functioning”. Presumably, self-stigma would achieve the opposite.

The stigmatised person may perpetuate the process of stigmatising through behaving as a “normal” towards those who carry an even more conspicuous stigma.

Following up on the work of Goffman, Jones et al. (1984) identified six dimensions of stigma:

1. Concealability: the capacity of the individual to hide the marks of stigma.
2. Course: reversibility or irreversibility of stigma, which affects the ability of the stigmatised one to reverse the negative attitudes of society against him/her.
3. Disruptiveness: the degree in which stigma negatively affects interpersonal relationships.
4. Aesthetics: whether the mark of stigma triggers a reflective negative emotional response from others (e.g. a reaction of disgust).
5. Origin: how the condition eliciting stigma originated and whether the individual can be held personally responsible for it.
6. Peril: an emotional response of danger or threat elicited by the mark of stigma on others.

Various aspects of both the body and the behaviour of elders may be interpreted by others as “marks” eliciting stigma. For instance, and abject or unclean body may be the target of stigma and discrimination (Gilleard and Higgs 2011). Hearing loss in older people, which can be already evident to the observer from the use of hearing aids, may be associated with a perception of disability (Wallhagen 2009). According to the prevailing cultural stereotype women and men may be expected to wear some specific kind of clothes rather than others; transgression of the stereotype may attract stigma, which may be resisted or not by the elder (e.g. Clarke et al. 2009).

Stigma can affect the life and well-being of older people not just in terms of direct discrimination but also because when a stigma is internalised it may constrain old persons to behave in ways that may be detrimental to their health. For instance, stigma, and the shame associated with it, may difficult the access to some public health services (e.g. see the case of incontinence in Horrocks et al. 2004).

If the stigma of old age was not enough, some elders may find themselves being the target of multiple stigmas. It is not uncommon that a double stigma may involve age and disability (e.g. Sheets 2005). Thomas and Shute (2006) reported a double stigma of age and mental illness among older Australians. See the double stigma of old age and dementia, in particular (Vernooij-Dassen et al. 2005, Milne 2010). The concern for the mental disorder-old age double stigma has also become a regional issue in Europe (de Mendonça Lima et al. 2003).

But multiple stigmas may be diverse and complex. In a study of mostly 50 and 60 year old who were attending a methadone maintenance clinic in the USA, Conner and Rosen (2008) identified 8 stigmas affecting the participants: drug addiction, ageing, taking psychotropic medications, depression, being on methadone maintenance (MM), poverty, race, and HIV status. Although drug addiction was a core stigma, most people felt being the target of the combined stigmas of age and drug addiction. However, 33% of participants reported being the target of three stigmas: (a) drug addiction, MM, and being on psychotropic medications; or (b) drug addiction, ageing, and poverty. Some respondents reported four stigmas, with the most common combination (66%) being drug addiction, ageing, psychotropic medications, and depression. Conner and Rosen also found that the two participants reporting the highest number of stigmas were both African American men.

Although Emlet (2006) did not find that older people living with HIV suffered from an age-medical condition double stigma (as compared to younger HIV participants), sufferers of HIV are indeed stigmatised (Emlet 2007).

We mentioned above that stigmatised elders may become socially segregated, but sometimes such social segregation may be self-motivated. In the specific case of old people who have retired and are still living in the community, they may age-segregate if they join specific clubs for the elderly. Spatial segregation may also result from institutionalisation in healthcare centres or living in retirement villages. Cultural age segregation may emerge when older people mainly interact with other old people who share the same cultural background and language (Hagestad and Uhlenberg 2006). In the above cases of segregation stigma may not be always a cause of it, but it could be an effect of it. For instance, cultural segregation may reinforce the prejudice that "foreigners do not integrate" leading to the stigma of "otherness" (e.g. Sharma 2014).

Although age segregation can have beneficial effects in terms of the elderly maintaining themselves active and engaged within a network of supportive social relationships, it does have several drawbacks. Hagestad and Uhlenberg (2006) list:

a. Producing and reproducing ageism (we will focus on ageism in the next section).
b. Threatening embeddedness and increasing the risk of isolation in later life.
c. Interfering with socialisation between young and old.
d. Impeding generativity or the elders' contribution to the younger generations through transmission of their experience (see Chap. 2, Erikson 1997).

In addition, the dynamics of socialisation in segregated elder communities may be quite different depending on cultural and socio-economic background. Gamliel and Hazan (2006) contrast the environment in a "sheltered housing project" for well-off elders in Israel, which was dominated by a more constructive and satisfying way of living, and where the identities of the residents were grounded in their professional past or their active roles within the residential community; with that of an old age home, where residents had a worse health condition, lower spirits and the social interactions were marked by gossip and name-calling. *Gossip* is a social mechanism that can produce *labelled identities*, which may then become internalised by the targets and subsequently consolidated (crystallised) within the group (Melucci 1996). If such identities take a positive tone, conferring to the individual a charismatic status, stigmatisation may not occur despite the labelling, but a negative label may lead to stigma. Gamliel and Hazan note how the labels identifying various individuals did achieve the establishment of an identity for such individuals, even when the identity might have been somewhat stigmatic. The label could be retained in the collective memory and then transmitted over time after the person had passed away. Gamliel and Hazan quote Becker (1973) pointing to this process as the establishment of some kind of "symbolic immortality". We can only wonder whether "immortality through stigma" is the kind of legacy an old person may look forward to pass on to posterity.

The benefits of social support for a stigmatised person do not necessarily require segregation into a group of equally minded people. A more positive attitude from family, friends, neighbours, and the community at large will decrease stigma and increase integration into the community. Apart from broader social support, other compensatory factors may help in the process of coping with stigma: being engaged in a fulfilling activity, possessing higher financial status, being in a stable relationship with a partner (Lyons et al. 2013). Satisfaction with past achievements and using such background to continue building a satisfactory identity into the future can also help overcome the stigma of ageing (Gamliel and Hazan 2006).

In a hospital setting, medical and nursing personnel can have an important role in helping overcome both the fear of stigma in the elder and stigma itself:

> This means creating a context in which people feel comfortable to take risks, believe they will not be judged, and give the goal of putting their issues on the table above the goal of avoiding rejection or disapproval from others (Garcia et al. 2005, p. 48).

Ultimately, if resistance strategies fail and the elder remains unsatisfied, there is always the option to move on and leave the stigmatising environment to find a more accepting community elsewhere (Phillipson 2007).

International organisations (such as WHO), professional associations (such as the *World Psychiatric Association*), and various non-governmental organisations around the world have produced statements over the years with the aim of providing guidelines to reduce the negative effects of stigma on old people. In general, the strategies to reduce stigma involve targeting attitudes and practices, increasing awareness about the risk of stigmatising old people, supporting empowerment in the elders, prevent the exclusion of elders, advocate on their behalf, educate the community, and allocate appropriate resources equitably so that specific programs for the elders can be financed and sustained over time (e.g. Graham et al. 2003).

Resistance to stigma, however, may sometimes produce unintended negative consequences. For instance, disregarding potential threats to own health by adopting attitudes usually associated with youthful exuberance in an attempt to counter the stigma of old age, but doing so without the body capable of performing accordingly, may cause health problems such as fractures in the case of athletic exploits (e.g. Day and Hitchings 2011).

To recap, elders who are labelled with age-specific negative attributes may become stigmatised in the community. Stigma then leads to discrimination that could become self-reinforced, developing into full blown ostracism, with potentially serious consequences for the well-being of the elder. Old people can resist being stigmatised by behaving in a way that is contrary to the stigma, but this may not be always sufficient, recruiting the active support from other members of the community may be also necessary.

The stigma of age can derive into ageism, which is the topic of our next section.

6.6.4 Ageism and Discrimination

The story of the origin of the term *ageism* is recounted by W. Andrew Achenbaum in an article published in the *American Society on Aging* website in 2015 entitled: *A history of ageism since 1969* (http://www.asaging.org/blog/history-ageism-1969):

> The term 'ageism' was coined in 1969 by Robert N. Butler, M.D., then a 42-year-old psychiatrist ... Butler used the term 'age-ism' during a Washington Post interview ... The Post story, 'Age and race fears seen in housing opposition,' described the apprehension of homeowners in Chevy Chase, Maryland, an affluent Washington, D.C., suburb, who were distressed by the NCHA's [National Capital Housing Authority] decision to turn an apartment complex into public housing (Bernstein 1969). The project was intended to offer residences for the elderly poor—including African Americans—and was opposed by residents who feared Chevy Chase would never be the same.
>
> 'People talk about aging gracefully, which is what they want to do of course. So, naturally, they don't want to look at people who may be palsied, can't eat well . . . who may sit on the curb and clutter up the neighborhood with canes,' Butler told Bernstein. 'Until our society builds [a] more balanced perspective about age groups, this lends to embittered withdrawal by old people'.

Although in this specific case a triple-stigma (old, African American, poor) might have contributed to the negative response of the wealthy residents to the idea of building public housing in the neighbourhood, age was most likely also part of it.

Butler (1969) then went on to explain in more detail his views on ageism. He defined ageism as "a form of bigotry", "prejudice by one age group toward other age groups". Therefore, in principle, the targets of ageism are not just the old, but potentially any age class. However, we will focus here on ageism against old people, just as Butler (1987, p. 22) subsequently also restricted ageism to discrimination against the old: "a process of systematic stereotyping and discrimination against people because they are old". Following Butler, Todd Nelson (2009) further distinguished between *malignant ageism*: which is negative in character and is characterised by extreme dislike of older people and a tendency to diminishing their worth; and *benign ageism*: a view of older people through the lenses of prejudice and stereotypes motivated by our own fear of ageing (*gerontophobia*). It is often the case that ageist attitudes may be implicit or expressed without a conscious understanding of their implications for the target; but it is not uncommon for ageist behaviours to be explicit or conscious and intentional (Levy and Banaji 2002).

Becca Levy (2001, p. 578) defined *implicit ageism* as: "the thoughts, feelings, and behaviors toward elderly people that exist and operate without conscious awareness or control, with the assumption that it forms the basis of most interactions with older individuals". The insidiousness of implicit ageism is its unconscious operation that leads the actor to negate being discriminatory or prejudicial against the evidence.

A somewhat mixed ageist stereotype that combines some negative aspects about the old (e.g. frailty) with positive ones (e.g. deserving) is encapsulated in the concept of *compassionate ageism* (Binstock 1983, 1985). Binstock (2010, p. 575) defines compassionate ageism as:

> the attribution of the same characteristics, status, and just deserts to a heterogeneous group of 'the aged' that tended to be stereotyped as poor, frail, dependent, objects of discrimination, and above all 'deserving'.

Although the young tend to hold more negative than positive stereotypes about the old, most people tend to see the elders with a combination of "fondness and pity" (Nelson 2009, p. 433). According to Nelson, in general, ageism and the stereotypes associated with it can be located along two dimensions: competence (capability) and warmth (friendliness). The more the competent and warm the elder is regarded, the more positive the discrimination could be, whereas more negative discrimination may affect older people who are regarded as both incompetent and unfriendly (see also Table 6.10).

Table 6.10 Twenty variants of ageism as combinations of four dimensions and five combinations of components/levels

	Micro-level			Meso-level	Macro-level
Components	Cognitive (stereotypes)	Affective (prejudice)	Behavioural (discrimination)	Discrimination in the social networks	Institutional and cultural discrimination
Explicit/negative	A	B	C	D	E
Explicit/positive	F	G	H	I	J
Implicit/negative	K	L	M	N	O
Implicit/positive	P	Q	R	S	T

Reproduced from Iversen et al. (2009), with slight modifications

The notion of deserving old people led to some positive outcomes such as the introduction of better social security and public health programs across the industrialised world, especially during the 1960s. This has changed since the late 1970s with the onset of neoliberalism (see Chaps. 2 and 3). Ageism lost a significant degree of its positive *compassionate ageism* lustre with neoliberalism, to be replaced with negative stereotypes such as that of the *greedy geezers*, and views about the old as:

> prosperous, hedonistic, politically powerful, and selfish… America's new elite—healthy, wealthy, powerful, and 'staging history's biggest retirement party'… A dominant theme in such accounts of older Americans was that their selfishness was ruining the nation (Binstock 2010, p. 576).

The negative ageistic stereotypes promoted by neoliberalism appealed to a sense of resentment from the rest of the community towards the elders, with the objective of eliciting a negative reaction against them as a "privileged class" on the one hand and, paradoxically, also a negative reaction against those elders who did not succeed, being regarded as personally culpable for their failures and therefore not deserving of social support through government programs (Kelchner 2000). The ultimate objective, of course, was to decrease government expenditures on social programs for the elders, paving the way for tax cuts for the wealthy.

In this environment of negative attitudes towards old people, concerns about inter-generational equity set the stage for the breaking out of what can be described as "inter-generational warfare" (Ory et al. 2003): a highly competitive environment defined by conflicts between the old and retired and the young and working, with the former being seen as "enjoying life" at the expense of the "hard-working young". Such inter-generational tensions are further fuelled by biases against the old disseminated through the media (Fealy et al. 2011) that can only worsen the levels of malignant ageism in society.

In a more detailed view on ageism, Bytheway and Johnson (1990, p. 36–37; see also Bytheway 2005) consider the following issues:

> (i) Ageism is a set of beliefs originating in the biological variation between people and relating to the ageing process. (ii) It is in the actions of corporate bodies, what is said and done by their representatives, and the resulting views that are held by ordinary ageing people, that ageism is made manifest. In consequence, it follows that:
> (a) Ageism generates and reinforces a fear and denigration of the ageing process, and stereotyping presumptions regarding competence and the need for protection. (b) In particular ageism 'legitimates' the use of chronological age to mark out classes of people who are systematically denied resources and opportunities that others enjoy, and who suffer the consequences of such denigration—ranging from well-meaning patronage to unambiguous vilification.

Iversen et al. (2009, p. 15) also provided their own synthetic perspective on ageism:

> Ageism is defined as negative or positive stereotypes, prejudice and/or discrimination against (or to the advantage of) elderly people on the basis of their chronological age or on the basis of a perception of them as being 'old' or 'elderly'… The concept includes the classic social psychological components in the form of; (1) cognitive (stereotypes), (2)

affective (prejudice), (3) and behavioral components (discrimination), in other words, how we on the basis of the chronological age or age categorization mistakenly; (1) think of, (2) feel for, (3) and act on the aging human being. Furthermore, ageism can operate both consciously (explicitly) and unconsciously (implicitly) and it can manifest itself on three different levels; the individual (micro-level), in social networks (meso-level) and on institutional and cultural level (macro-level).

This synthetic view is summarised in Table 6.10.

Thus, according to Iversen et al. (2009) the structure of ageism comprises several dimensions that are a function of both the personal views of the ageist individual (stereotypes, prejudice, discrimination) and of broader cultural and institutional forms of discrimination that can act in various combinations of implicit/explicit and positive/negative expressions. Such manifestations of ageism are subject to automatic, unconscious psychological responses, or biases, that can be further reinforced through learning (Perdue and Gurtman 1990).

Although ageism tends to involve either discrimination of younger people against old ones or the reverse, Giles and Reid (2005) note that ageism can be also seen within the same generation (*intra-generational ageism*), in this case discrimination may go from an elder who is in better mental and/or physical condition regarding him/herself as "young", to one who is in a worse condition, being treated as "old".

Eventually, ageism may become *institutionalised* that is, it becomes incorporated into the matter-of-fact perceptions about the elderly that are prevalent in society (Nelson 2011), as if such perceptions were truthful, objective, and self-evident descriptions of reality.

Internalisation (an individual process) and institutionalisation (a social process) of ageism transform it into a *commonsense reality* which perpetuates stereotypes. It is through this mechanism that ageism claims to be based on "objectivity", by appealing to scientific evidence and producing political, social, and cultural structures that keep recreating and reinforcing the prejudice (Angus and Reeve 2006).We receive challenges to our ageism with surprise and dismay, followed by our denial about any discriminatory intent, and a firm reassurance that we are merely describing raw facts.

In this approach to ageism we can see a link between descriptors that are based on visible changes occurring in the individual over time (e.g. sagging, wrinkling, greying; Calasanti 2005), and the social consequences imposed on those individuals on the ground of prevalent views about old people. Descriptors, however, may also refer to behaviours. Two important and opposite behaviours have been used to sustain ageist attitudes during the neoliberal period of the late twentieth and early twenty-first centuries. We already mentioned one above in this section: the lifestyle of the *greedy geezers* who are regarded as "well-off retirees"; the other is the *dependency* of the less well-off retirees (Angus and Reeve 2006). The neoliberal focus on individualism and competitiveness stresses the requirement for the elders to be self-reliant and, by doing so, to show their responsibility towards themselves and others by not becoming a burden. This leads to an ageist stereotype against dependent elders that is then transformed into social pressure exerted upon them. In a notion that harks back to at least the nineteenth century (see Chap. 2), and that divides older people into the deserving of social support and the undeserving ones, old people who are less self-reliant are also seen as less deserving. The more self-reliant ones are regarded as more deserving, in this case they are deserving of tax cuts, for instance, thus producing greater inequality (Stiglitz 2012).

Ageism can be expressed through many behaviours in day-to-day interactions. For instance, using a patronising language when interacting with old people (e.g. baby talk and infantilising more generally) is a common manifestation of ageism (Nelson 2005, 2009). In addition, ageism may lower the threshold for elder abuse, such as neglect by caregivers, violence, fraud, and broader forms of exploitation. Although ageism could be proximately sustained by social mechanisms of control, it may be also grounded on psychological mechanisms of fear towards ageing and getting old, leading to a negative attitude towards the elders (Snyder and Meine 1994). Fear of ageing ultimately derives from the knowledge of the increased proximity to death as we become older, thus a negative attitude towards the elders may just be a defensive mechanism to distance the self from death. This has been encapsulated by Greenberg et al. (1986) in their *Terror Management theory* (see also Solomon et al. 1991; Nelson 2011) which states that in our quest to manage our fears we impose order on a chaotic universe, therefore the fear of the old and the ageing process that they represent is attenuated by distancing ourselves from them (Nelson 2009). Ageism is thus a mechanism to restore order where order may be lost (e.g. at the thought of losing control on our life because of advanced age). After all, the old person is a mirror that reflects our own image into the future.

On the other hand, in cultures where death is seen as a "normal part of life", and it is culturally integrated through regular activities (e.g. religion, rituals, and festivities), one would expect lower levels of ageism in society. Nelson does recognise this issue, and he asks (Nelson 2009, p. 436): "Is the TMT [Terror Management Theory] explanation of ageism applicable in countries or cultures where people do not fear death?". The author offers an answer through a comparison between the more individualistic Western cultures and the more collectivistic Eastern cultures. In modern times death is feared and ageism is widespread in Westernised cultures, whereas in the more collectivistic Eastern cultures death and the reverence for the ancestors are better incorporated into the culture, and ageism is traditionally less evident. In addition, in traditional Mexican culture, with their concept of death and close cultural links with the dead (see the annual celebration of the *Día de los Muertos*; e.g. Brandes 2000), there is also a sense of respect for the elders, just as it is the tradition in Asia. However, we should also note that both in Asia and in countries such as Mexico ageism has been increasing in recent times under the pressure of modernisation (e.g. Zuniga 1997, Sokolovsky 2009).

Other theoretical approaches have been adopted to understand ageism, which are reviewed by North and Fiske (2012). For instance, *Terror Management theory* that we have mentioned is an *individual-level theory* where ageism has a protective function from the perspective of the ageist person. On the other hand, *Social Identity theory* (SIT) states that identification with own group produces the rejection of members of other groups, in this case the young discriminating against the old or vice versa. Such discrimination is expected to increase self-esteem in the discriminator. *Interpersonal theories* stress the negative perception towards older people on the ground of their appearance that may have negative connotations within the culture. Some *evolutionary theories* of ageism point to the adaptiveness of preferentially helping individuals who are more likely to reciprocate (e.g. the young) but this can be modified by inter-generational reciprocation: "Today I help my old parents, tomorrow my children will help me when I am old". *Socio-cultural theories*, on the other hand, state that major historical events influence the spread of ageism in society; for instance, writing makes vocal transmission of knowledge and storytelling, a traditional activity of the elders, less needed, thus devaluing the elders.

Given that ageism can have such negative consequences on old people, how can it be changed? Old people may resist ageism and perhaps even respond in kind to ageistic attitudes, by expressing stereotypical views against younger people (Giles and Reid 2005). This approach, however, just perpetuates ageism through a tit-for-tat process: if the old are regarded as incompetent because they are not up to date, the old retaliate by regarding the young also as incompetent because they do not have enough experience in life, leading to further retaliation by the young, and so forth.

A better approach should be used that can stop the vicious circle of tit-for-tat ageism. Giles and Reid (2005, p. 400–401) list the following strategies to address and hopefully change ageism:

- Educational courses to better manage death anxieties in the community, thus reducing negative age stereotypes.
- Replacing current egocentric life goals with ones that assist others and transcend a lifespan.
- Adopting personalised rather than categorised communication styles with older people.
- Promoting inter-generational contact, especially in family contexts.
- Encourage awareness about and capacity to recognise ageist images, in the media for instance.
- Discourage age-segregated communities and favour more integration with old people.

From their part, old people may actively help promote new and more positive images of old age (Minichiello et al. 2000) by adopting a positive self-perception (i.e. positive attitudes towards their own ageing experience) (North and Fiske 2012). For instance, the capacity for more complex thinking in old age is certainly a positive of being old and it should be further developed and then shared with others (Pennebaker and Stone 2003).

More involvement of women in educating the community on inter-generational matters would be also welcome, as women tend to have less discriminatory attitudes than men on average, and of course more contact with old people may help dispel some myths about them (Allan and Johnson 2009; Cordella and Huang 2016) through communication and self-disclosure. *Self-disclosure* is "the act of voluntarily providing information to another that is of an intimate or personal nature" (Tam et al. 2006, p. 415). Such disclosure is important in the development and maintenance of a relationship, by reducing a negative bias towards an outgroup, for instance, and in the establishment and maintenance of friendships. Through the promotion of individuation and familiarity, self-disclosure narrows the space of threatening behaviours in social interactions, especially when disclosure is mutual. The more positive the attitude is towards specific elders, due to greater communication, disclosure, and understanding, the more likely it is that the opinion about the elders in general may also change and become more positive in the community (Harwood et al. 2005). Tam et al. (2006) note that social contact and communication can better decrease prejudice when interactants perceive themselves as being of equal social status in any specific situation, they have common goals, are ready to cooperate, and benefit from the support of institutions and authorities in promoting positive social experiences.

Thus, through self-disclosure, elders may help change the ageist attitudes of society, by challenging the characteristics of the stereotypical old person. By sharing their experiences with others, they may provide evidence of their ability to live a fulfilling life (Minichiello et al. 2000). Showing a positive capacity to contribute to society may also help transform the currently negative ageism into a positive *sageism*: the elders as sages and transmitters of valuable life experience (Minichiello et al. 2000; see also Cordella and Huang 2016).

Hard work is necessary if change is to be achieved, as ageism runs deep into the mind of many people and it often manifests itself in subtle, rather than open and explicit ways. In a study carried out in Canada, Hurd (1999) describes various strategies used by old people to resist the negative discrimination of ageism. For instance, one attitude of resistance against ageism is to emphasise current abilities, rather than those activities the individual cannot perform anymore. Rather than using the word "old", participants in Hurd's work used words or expressions such as "youthful", "young at heart" or "young-old", to describe themselves. In fact, appearing youthful is a confirmation that they are actually "not old" and the contrast between chronological age and look can even become a badge of honour when the individual does look youthful. Still, this strategy to resist ageist discrimination may backfire, as old people who indeed have poor health may look "deviant" (North and Fiske 2012). Pretending to be young whilst obviously being old is not always a successful strategy to overcome discrimination, as the challenge to keep looking young with increasing age is uphill; unlike discrimination against the very young that can be potentially eliminated once those young grow up (Garstka et al. 2004). In addition, some botched attempts at looking "youthful" may only end up making the old person look "ridiculous". We suggest that the concept of *fulfilling ageing* may provide a positive scaffold to the old that may help them build resistance against discrimination, without imposing impossible goals on the elders. After all, life fulfilment can be experienced by anyone, no matter how they look, for as long as the elder has a positive story to tell.

Ageism can be also reduced at the individual level by providing more information about the old person beyond his/her age: Who is he/she? What has he/she done so far in life? Once more context that clarifies the life experience of the elder is available, a significant decrease in ageism may be achieved (Kite et al. 2005).

Although ageist stereotypes can be unconsciously activated and enacted, they can be also inhibited once an individual is made aware of them (Gilbert and Hixon 1991; von Hippel et al. 2000). Success in inhibiting ageism, however, is variable as individuals differ in such capacity. For instance, older people themselves tend to be less able to inhibit stereotypes (von Hippel et al. 2000).

Equal opportunity and anti-age discrimination legislation may help change discriminatory attitudes against older people, especially at work, thus improving the sense of dignity in the old (Duncan 2008). The challenge remains, however, especially if the case of discrimination is taken to a court, where proving a case beyond reasonable doubt may be difficult (Duncan 2003). Equal opportunity could better thrive in an environment of inter-generational solidarity, but for this to be achieved, the entrenched obsession about competition should be superseded. In addition, better education about old age and becoming conscious of stereotypes are necessary first steps to start tackling negative discrimination in society.

Gringart et al. (2008) also hold some hope for a potential awareness campaign carried out through the media to be effective in tackling ageism, but this remains to be tested. They suggest that interventions aimed at decreasing ageism could be based upon *cognitive dissonance*: "behaving in a way that is incongruent with one's self-concept creates unpleasant cognitive dissonance. In order to reduce this dissonance, the person adjusts his or her subsequent behavior to be more in line with his or her self-concept" (Gringart et al. 2008, p. 754); this requires bringing the subconscious biases of people behaving in an ageist manner to the fore and expose them, to then contrast them with their expressed ethical views, in order to highlight contradictions. Such contradictions can be threatening to the concept of self: if fairness is an important concept for you why is that you are behaving in such an unfair manner towards older people? Contradictions could finally be resolved through a change in attitude. The change can be also facilitated by providing additional information (e.g. about the actual capabilities of old people) that reinforces the appropriateness of such change.

In sum, ageism is systematic stereotyping and discrimination against old people. Sometimes such discrimination may be somewhat "positive" (e.g. compassionate ageism) but it is most often negative (e.g. malignant ageism). The complexity of ageism is clearly illustrated in Iversen et al.'s (2009) model that considers both the individual and the broader social factors affecting the expression of ageism. Ageism may become internalised and institutionalised and then incorporated into the "commonsense reality" of individuals. Such commonsense reality perpetuates ageist discrimination. However, strategies can be adopted in society to change ageist attitudes.

In the next section we shift our focus on the specific social issues that affect the lives of the oldest-old.

6.7 Oldest-Old in Society

In Chap. 1 we defined the age category of oldest-old as individuals who are 80 years old and older. This age category has been increasing over time, especially in industrialised countries. For instance, the number of centenarian women has increased at a very high rate since about the 1950s–1970s (e.g. Vaupel 2010). The increase in the number of centenarian women is much faster than that of centenarian men, although the latter have also increased in numbers since the 1970s (Robine and Cubaynes 2017, see Fig. 6.11). In general, such advanced age comes with a significant degree of physical frailty and mental decay, although some oldest-old can retain a fair level of mental capacity.

The conundrum of very advanced ages is beautifully illustrated by this quote from Baltes and Smith (2003, p. 124) referring to the ancient Greek saga of Tithonos (also known as Tithonus) and Eos:

> The saga is about Eos, the Greek goddess of the dawn, who fell in love with the mortal earthling Tithonos, the prince of Troy. True to her own immortality, Eos wanted to go on

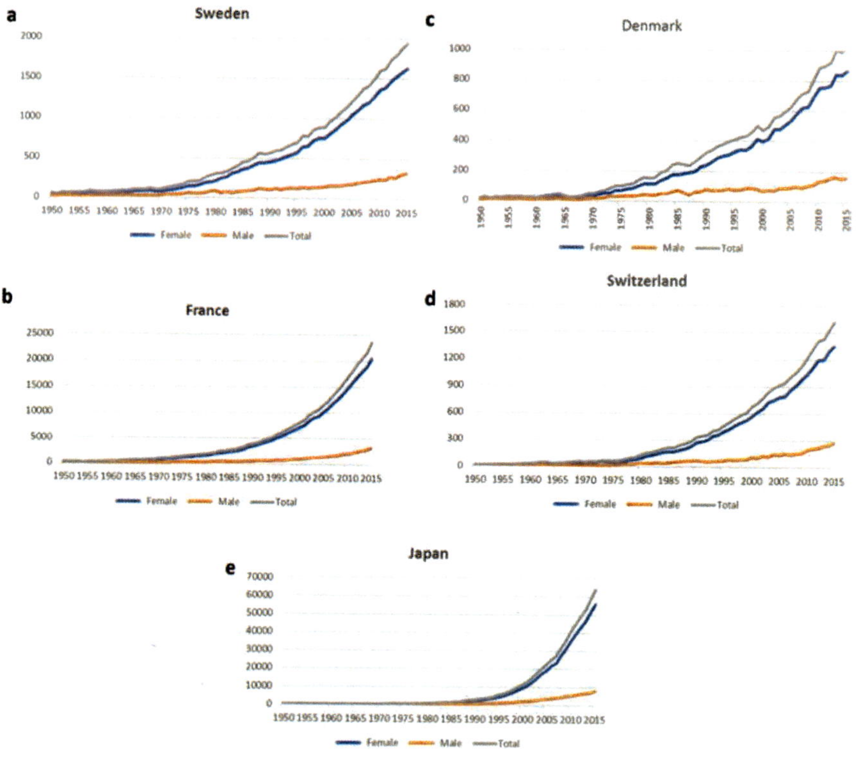

Fig. 6.11 Annual centenarian (100+) population estimates by sex in Sweden since 1751, France since 1816, Denmark since 1835, Switzerland since 1876, and Japan since 1947; from Robine and Cubaynes (2017)

living with and loving him forever. In this spirit, the goddess begged the master of all Greek gods, Zeus, to make her lover immortal. Zeus granted Eos this wish and bestowed eternal life on Tithonos. Not included in the gift of Zeus, however, was another condition that Greek gods enjoy, namely eternal youth and vitality. Despite immortality, Tithonos aged like a human: he became frailer and frailer and although his body remained alive, his mind died. With much pain in her heart, Eos decided to move her former lover into a separate chamber where, according to the Greek saga, he continued to live mindlessly.

Advanced age confronts the individual with the same conundrum Tithonos found himself in, that is living longer but in ever worsening physical and mental conditions, thus adding a question mark to the proposition whether life at very advanced ages is really worth living or not.

In the face of the inevitable physical decay, the oldest-old may nurture the desire to remain if not "youthful" at least autonomous. In many Western cultures being autonomous means to be independent (including economically independent), retain the capacity to think and operate in life, and controlling bodily functions (Becker 1994). In more collectivistic cultures the capacity to operate in life also includes, importantly, social support, especially from family.

The oldest-old are therefore confronted with the issue of losing autonomy at some stage as their frailty, or their limitations in the ability to carry out daily activities, increases with age (e.g. Chou and Chi 2002). But there is also variability among the oldest-old in their capacity to maintain autonomy, and in some cases a reasonable degree of capacity to operate in life can be retained until very advanced ages, especially in individuals with an optimistic personality and good health (Isaacowitz and Smith 2003; Jopp and Rott 2006). Although autonomy, health, and well-being may be improved in the oldest-old by non-social activities (e.g. Lennartsson and Silverstein 2001) social support is important to keep functioning at advanced ages (Field and Minkler 1988).

Lower autonomy in the oldest-old may attract social support as an act of solidarity, but lower autonomy is also an impediment to establishing and maintaining social relationships. In this case, one approach used by the oldest-old as they become less autonomous is to limit the size of the social network, whilst maintaining a rich pattern of relationships with the members of this reduced network, that are commensurate to their physical capacities, needs and priorities in life (e.g. Johnson and Barer 1992). In this way they may still enjoy the benefits of social support without paying too high a cost of maintaining social relationships. The restriction of social relationships in very advanced ages to fewer but stronger social bonds is also the result of a process of selection among potential interactants, whereby as the difficulties to interact with the elder increase due to very advanced age, only a few and highly committed individuals (e.g. offspring, best of friends) are likely to continue engaging with the elder.

Autonomy, therefore, is unlikely to be maintained up to the very end of life. Despite individual differences in the ageing experience (*differential ageing*), in the long-run deterioration is inevitable. It is especially at the point in life when autonomy cannot be sustained anymore, that the social network becomes an even more important resource to the oldest-old (Baltes 1998).

The decrease in the size of the social network experienced by the oldest-old, following widowhood, for instance, and death of close friends, but also due to their own intention to limit social contact to a selected few; should make it relatively easier for the oldest-old to retain some capacity of nurturing their social ties by remaining emotionally attached to the members of such smaller network until the very end of life (Lang 2001). In fact, Fung et al. (2001), in a study carried out in the USA, showed that an overall smaller social network in old ages is associated with greater selectivity for emotionally closer partners.

Although the oldest-old tend to rely heavily on the immediate family network (*kinship network*), other social networks can be also relevant; for instance, using data from the *Berlin Aging Study*, Fiori et al. (2007) derived a series of social network types after carrying out a cluster analysis on a sample of 516 participants (70–103 years old). Four major types of networks were identified: family-focused, restricted, diverse, and friend focused.

Friend-focused networks may include a special category of very close friends known as *fictive kin* (or *chosen kin, constructed kin*). Such special friends are particularly relevant when the network of kin is small (Perry and Johnson 1994). Fictive kin can be defined as "those people to whom one considers to be related but who are not related by blood or marriage" (Johnson 1999, p. S369).

In which ways do social networks benefit the oldest-old?

Social networks can help the oldest-old in many ways that include emotional support—leading to greater personal well-being and health—, and material aid, both financial and in the performance of activities of daily living. In a study carried out in the USA, Cherry et al. (2013) investigated a group of oldest-old (90–97 years old) within the *Louisiana Healthy Aging Study* concluding that social engagement is important in maintaining good health. Better quality of social relationships in the oldest-old is also associated with greater well-being (Lang 2001). In a study carried out in Sweden on a sample of 80–98 years old (66% women) from the *Swedish OCTO-Twin-study*, Berg et al. (2006) found that life satisfaction was positively affected not only by better physical and mental health but also by the quality of the social network. In a subsequent work, Berg et al. (2009) indicated that although life satisfaction decreased with age in their sample of Swedish oldest-old, the maintenance of social contacts (e.g. with a spouse) to an extent buffered them against the decline. In another Swedish study of the oldest-old, the *Umeå 85+ study*, Nyqvist et al. (2006) found that the combined effects of *attachment* (i.e. strength of ties outside and/or within the family), *social integration* (i.e. bonding relationships), and *social network* (i.e. family ties or strong ties to other people), which together define the *structural component of social capital*, were higher in women than men. That is, oldest-old Swedish women have greater social capital than men and greater social capital was associated with better mental health.

In Israel, Litwin (2001) described higher morale in old people embedded in a more diverse social network, including larger friends, neighbours, and family networks, although among the oldest-old the greatest social support was received from the network of kin (Litwin and Landau 2000). In the work carried out by Fiori et al. (2007) that we have already quoted in this section, a diverse social network provided greater instrumental and emotional support, being also associated with higher levels of well-being in old people. The oldest-old could also actively nurture their links with their social network through *collective*, *productive*, and *political participation* in their community, within their limitations, which helps in giving them more meaning in life and increasing their well-being (Bukov et al. 2002).

Greater social contact, whether it is by telephone or visiting friends and relatives, was also associated with more fulfilling ageing in a study carried out by Garfein and Herzog (1995). Similar results were obtained by Bondevik and Skogstad (1998) who found that emotional and social loneliness in the oldest-old decrease with higher frequency of social contacts with family, friends or neighbours.

In an analysis of data from the *Chinese Longitudinal Healthy Longevity Survey*, Chen and Short (2008) found that emotional well-being in the oldest-old was associated with the social context of their living arrangements: lower well-being was detected when living alone as compared to co-residing with immediate family. Moreover, co-residence with a daughter was better for emotional well-being compared with co-residence with a son.

After studying a group of 662 oldest-old in the UK, Bowling and Browne (1991) described how life satisfaction was positively correlated with social network size, number of relatives, number of confidants, and number of children who were alive. Moreover, number of confidants and household size were the social network variables which showed the strongest negative correlations with frequency of loneliness.

The oldest-old can nurture a broader community-based social network if cognitive and communicative skills remain adequate, or they can shift to a more restricted family network (if available) when cognitive-communicative skills may decline (Keller-Cohen et al. 2006). This may allow the oldest-old to maintain a relatively adequate degree of well-being despite changes in personal circumstances. Even in a nursing home comfort can be derived from "a community of caring, safety, closeness and inclusion" (Whitaker 2010, p. 99). Whenever the maintenance of social activities involves a net cost to the oldest-old (e.g. in terms of energy and effort) then we have seen that withdrawal into a more solitary life may be expected (e.g. Sun and Liu 2006). Would such withdrawal lead to loneliness? The answer is not straightforward. Loneliness is more likely to manifest itself in elders who desire some company, especially if they also require some practical assistance with their daily activities but are forced into a solitary life. The elder may not feel lonely if she/he desires to be alone and retains some independence in the performance of activities of daily living. In other words, three major variables may interact to explain loneliness or lack thereof in the oldest-old:

1. Personal desire: to be alone or to be with others.
2. Practical need: need of practical assistance or no such need.
3. Actual company: solitary living or social living.

The combination of those variables may be useful to understand each specific case, as it will be shown below in the results obtained by Long and Martin (2000).

In Sect. 6.2 in this chapter we have shown how lack of social support and low social embeddedness are predictors of loneliness in the elders in general. The same effect is observed in the oldest-old. In fact, we have seen in Fig. 6.3 that for the oldest-old loneliness increases at a fast rate with age as compared with younger old ages, the effect of reduced social support being a co-contributing factor, as shown by Long and Martin (2000).

Long and Martin studied the effects of both individual traits such as personality and attachment and social factors such as family solidarity, on the loneliness of the oldest-old in the USA. Figure 6.12 summarises their results in a path analysis. Here we would like to point out the positive correlation between attachment and *affective solidarity* (i.e. "the nature and extent of positive sentiment toward other members of the family" p. P313) that establishes the basic emotional ground from which the practical benefits of cooperation within the family can emerge; and the negative correlation between affective solidarity and loneliness. Thus, as the support from the social network decreases in the oldest-old, their loneliness increases. Note also how affective solidarity is negatively correlated with anxiety. Extraverted individuals, however, are less dependent on social relationships to decrease loneliness. Hence extraverted elders who intentionally decrease their social network in order to minimise the costs of maintaining social relationships may not be affected by such reduction, whereas less extraverted elders who reduce their social networks due to increasing physical constraints and incapacity to nurture many social relationships, may indeed suffer from lack of social support. Such elders may benefit if they become recipient of *normative solidarity*. Normative solidarity in the family context is higher the greater "the degree of perceived filial responsibility" is (Long and Martin (2000, p. P313).

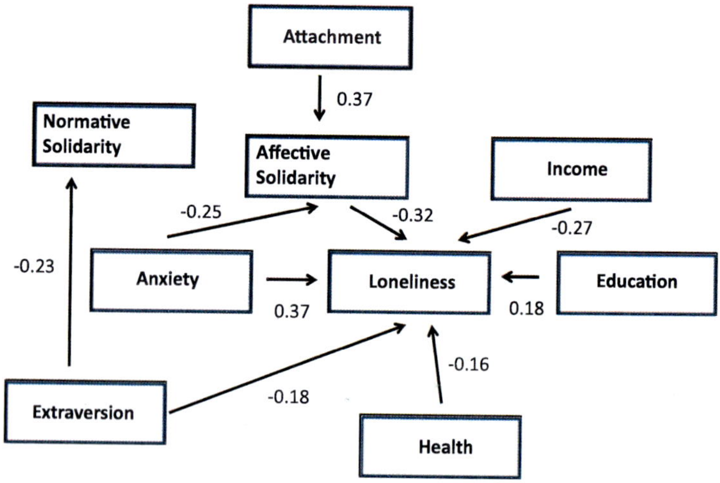

Fig. 6.12 Predictors of oldest-old loneliness. Shown are correlation coefficients from a path analysis, all the effects are statistically significant. Redrawn from Long and Martin (2000)

Therefore, as the oldest-old withdraw from various social activities due to their increasing constraints to maintain social networks, the family is expected to step in in order to provide the still needed social support that will help in preventing the onset of loneliness. Less commitment from family will be needed in the case of oldest-old who want to—and are capable of—living alone.

Loneliness can negatively affect mental health and physiology (see the Sect. 6.2 of this chapter), which means that social interactions that decrease loneliness can have beneficial effects on health, thus perhaps even contributing to extending the lifespan. In fact, poor social relationships are associated with increased mortality (e.g. Avlund et al. 1998, and references therein). Widowhood can specifically increase mortality in the oldest-old, especially in men (Erlangsen et al. 2004).

In a prospective study carried out in Denmark, Avlund et al. (1998) investigated the relationship between social interactions and mortality after a period of 11 years in a group of 366 men and 368 women who were 70 years old at baseline. During the period of this study 52% of men and 33% of women passed away. Interestingly, neither education nor income were predictors of mortality. However, some social factors did predict mortality: men living alone and those with no regular contact with children were more likely to die. Women with no social support were also more likely to die. Elders with no social activities outside the family were also more likely to die in the study period. The effect of low social contact on increasing mortality was maintained even for elders with better functional abilities.

The social network also affected the survival of 85-year-old and older participants in the long-term Swedish *Kungsholmen Project*. Rizzuto et al. (2012) showed how survival was higher in those participants who had a healthy lifestyle, participated in at least one leisure activity, and interacted with a rich or moderate social network.

On the other hand, as shown by Sun and Liu (2006) using data from the *Chinese Longitudinal Healthy Longevity Survey* that included people 80 years old and older, a degree of withdrawal from social activities in the older ages may be also positive in terms of survival if it involves a decreased level of effort and stress, as the social network significantly narrows down to just the most beneficial of relationships. Being embedded in a selected and loving social environment of family and friends may also help the oldest-old make the journey towards the end of life more bearable (Hedberg et al. 2013).

To recap, the deterioration observed in the oldest-old ages in terms of loss of functional capacities and autonomy, although inevitable, can be somewhat slowed down, or rendered more bearable, not only by individual initiatives but also social support. The social network, however, tends to decrease in size over time for the oldest-old, but it could retain its supportive effectiveness if cooperative members of the network—family in particular—are selectively retained. Some non-family members of the network can be also important. A supportive social network can help the oldest-old decrease the experience of loneliness and decrease mortality rate, although different individuals may have different levels of social needs.

In the next section we review the social aspects affecting the lives of older immigrants.

6.8 Older Immigrants

The flow of people around the world—alongside that of goods, services, and information—is not a recent phenomenon, but it has been significantly increasing since at least the *Industrial Revolution* in the second half of the nineteenth century (Ferrie and Hatton 2014). This has led to the current high level of globalisation which relies on accessibility to continuously improving means of transportation and communication (e.g. Crafts 2004). As a result of growing globalisation, many individuals have sought to resettle permanently elsewhere, with a recent *G20 Global Displacement and Migration Trends Report 2017* from the *Organisation for Economic Co-operation and Development* (OECD 2017) showing that net migration to G20 countries has been increasing since the 1950s, with only a slight recent decrease due to an outflow of EU citizens following the negative social and economic effects of the *Global Financial Crisis* that started in 2007–2008. On the other hand, asylum seekers applications to G20 countries have increased exponentially since about 2012.

Some of those economic migrants and asylum seekers leave their native country when they are already old, following on the path of younger members of the family. These migrants usually face important obstacles to their integration in the host country, such as learning the local language, overcoming social isolation, and accessing services. Others migrate young and age and become old in their new country of residence (*ageing in place*); in which case they may be more likely to be better integrated in their community in old age (Treas and Mazumdar 2002). Still others may migrate (temporarily or permanently) when they are old, but they do so on their own initiative, as "international retirement migrants" (IRM) for instance (Ciobanu et al. 2017). IRMs are usually in a good financial situation (see the Canadian *Snowbirds* mentioned in Chap. 2) or they try to maximise the purchasing power of their pension by taking advantage of convenient exchange rates and lower costs of living in a new country. There are also ***transnational grandparents***, i.e. those who may commute more or less regularly between a foreign country—where a child and his/her family reside—and their own country of residence (e.g. Plaza 2000; Li and Chong 2012; Sigad and Eisikovits 2013; Baldassar and Merla 2014). Transnational grandparents are members of the so-called *zero generation* (i.e. the parents of first-generation migrants) (Nedelcu 2009). However, such transnational grandparents become limited in their ability to travel from one country to another over the years.

King et al. (2017) have recently suggested the following typology for older people who are affected by migration, including those who remain in their native country and those who leave it: (a) older people left behind by migration of younger relatives, (b) older family-joining migrants, (c) affluent international retirement migrants, (d) older economic migrants, (e) older return migrants (migrants who return to their native country in old age), and (f) ageing in place migrants, these are people who migrate younger and age in the country of migration (e.g. Johansson et al. 2013).

What are the challenges confronting old immigrants in their new country? How do they cope and adapt? How does the population in the host country interact with old immigrants and either help or obstacle their ability to live a fulfilling life?

The USA provides an interesting example to understand the experience of older immigrants, as it is a country with a long tradition of immigration from around the world (Treas and Batalova 2009). Not surprisingly, the number of immigrants tends to increase in the USA during periods of economic expansion and it decreases regularly during periods of economic recession or periods of large world conflicts. This means that, in general, even if immigrants can be integrated into the workforce when they first arrive, at some point in time they may become redundant if the economic situation in the country deteriorates. Economic downturns may affect the ability of older immigrants to find or retain a job, following the double stigma of being both a "foreigner" and old.

The immigrant population resident in the USA is becoming proportionally older, starting at least in the mid-1990s, reversing a trend of smaller representation of immigrants in the population of old people observed between 1900 and 1995. In Canada, another country of large immigrant intake, the immigrant population is also becoming older, accounting for 28.4% of the population of old people (Durst 2005). The proportion of 65 years old and older people who are immigrants is projected to increase across OECD countries in the future (Withers 2002).

Immigrants in the USA are not only becoming older, but their ethnic composition has also changed, as reported by Treas and Batalova (2009): in the 1970s most immigrants came from Europe, whereas more recent immigrants have mainly come from other countries in the American continent (e.g. Mexico) and Asia.

Although Europe has been traditionally a source of migrants to other continents, it has also been a recipient of immigrants especially in the post-World War II period, and the immigrant population has also been ageing. In Germany, for instance, the percentage of over-60s in the total foreigner population trebled from 3.1% in 1970 to 9.7% in 2002 (White 2006). A rapid increase has also been observed in the Netherlands, Sweden, Switzerland (White 2006), and other European countries (Ciobanu et al. 2017).

Therefore, old immigrants are becoming an important section of the population in some countries, which potentially exposes them to various vulnerabilities such as the need for better social integration—poor social integration leads to loneliness and health problems—, financial issues, and access to services and social support. In addition to some of them being under social pressure as a "foreigner", they may be also at risk of becoming the target of ageism, potentially compounded by further prejudice against their specific ethnic group and race. No matter how important those vulnerabilities are, we must keep in mind that not all old immigrants are the same and their specific condition is variable, many are indeed quite capable of experiencing fulfilling ageing in their adopted country.

Upon arriving in their new country of residence, many immigrants may experience a degree of *acculturative stress* or *culture shock*: "a set of complex psychological consequences, usually distressful and maladaptive, to unfamiliar cultural environments" (Tsytsarev and Krichmar 2000, p. 36). Over time, however, the initial culture shock may give way to a process of acculturation. *Acculturation* can be broadly defined as: "a social process in which members of one cultural group adopt the beliefs and behaviors of another group" (Lopez-Class et al. 2011, p. 1556). This process goes through a series of adjustments over time, displaying a diversity of pathways and outcomes. In his classical *Fourfold theory of acculturation*, Berry (1997) described four major potential outcomes of the process of acculturation.

a. Marginalisation (low affiliation with both cultures).
b. Separation (high affiliation with culture of origin, low affiliation with new culture).
c. Assimilation (high affiliation with new culture, low affiliation with culture of origin).
d. Integration (high affiliation with both cultures).

Although Berry's theory was subsequently criticised (e.g. by Rudmin 2003) on the ground that it is an oversimplification, hiding the true diversity and dynamics of the acculturation process and outcomes, it remains a useful basic classificatory system for some of the major experiences of immigrants. The level of social embeddedness of immigrants increases as they progress from an initial stage of *marginalisation* to that of *integration*. Moreover, the more *multicultural* a society is (see Sect. 8.6 of Chap. 8), the easier the process of integration is expected to be, as multiculturalism integrates immigrants into society without imposing upon them the requirement to abandon their identity, past, memories, language and the aspects of their original culture that they want to retain and that do not clash with the law.

Culture shock is less intense in societies where integration policies prevail, as compared to societies under assimilation policies (e.g. Tsytsarev and Krichmar 2000, and references therein), but it may also be less intense in immigrants with more open-minded views of life and mental flexibility and adaptability. Integration allows the immigrant to enrich him/herself through the contact with the new culture, without losing those aspects of the culture of origin that remain meaningful and useful.

Integration is helped by social support. Migrants tend to establish supportive social networks within their ethnic community (Ciobanu et al. 2017), with preference shown for their family (Die and Seelbach 1988; Boyd 1991; Wilmoth et al. 1997; Treas and Mazumdar 2002). The family is usually important to help the migrant negotiate the difficulties of living in a foreign country, including navigating the intricacies of the public health and social services bureaucracy (e.g. Treas and Mazumdar 2002), decrease loneliness and social isolation, and often also buffer against cultural shock (Hossen 2012). The family becomes especially important to migrants who immigrate in old age, as they are more dependent on family support than those who migrate at younger adult ages (Angel et al. 1999). However, immigrants may confront a paradox as they age: they probably spent energy and effort helping their family integrate into the new country, but as they become older they may require more of their family support, whilst the family may grow some distance from their elders, as they are busy living their own lives, becoming more socially active and embedded into the new country. This may lead to issues of loneliness in the immigrant elders—with or without physical isolation—and neglect, that could even affect their health (Treas and Mazumdar 2002).

Palmberger (2017) studied social embeddedness in the ageing Turkish immigrant community in Vienna, Austria, and for such migrants, embeddedness mainly took the shape of integration within local Turkish associations, whether they are religious, cultural, or political. That is, embeddedness meant strengthening social ties with members of the same ethnic group. In part, this may be a side effect of discrimination experienced in their interaction with non-Turkish members of the community, which may reinforce the tendency to find support among equally discriminated people. If embeddedness is understood as "social relations that foster a sense of 'rootedness' and belonging, not in a geographically binding way but more in the form of socialties" (Palmberger 2017, p. 237), then older migrants may well be embedded within their local ethnic community, but this may not completely protect them against negative experiences in their interactions with the mainstream of society.

Elderly immigrants who have poor health and less mobility are also less capable of maintaining social contacts and therefore tend to receive less social support, leading to lower social embeddedness. Better cognitive capabilities and better general mental health, on the other hand, may help them enjoy a better social life (Wu and Hart 2002).

Therefore, social networks are important for old migrants. Amit and Hefer (2010) have studied this issue in elderly Russian Jews who migrated to Israel. During the late 1980s and mid-1990s a wave of Jewish immigrants moved from the old Soviet Union to Israel, which confronted them to various issues of acculturation such as learning the language and understanding the culture, but also adapting to the new climate (Ron 2001). Sooner or later such immigrants started feeling the loneliness of being in a foreign land, surrounded by foreign people, despite sharing some aspects of a common cultural heritage. Establishing and nurturing supportive social networks then became an important strategy to adjust to the situation.

Litwin (1995) described four major types of social networks observed in the old members of the Russian Jewish immigrant community: (a) family-intensive network, (b) kin network, (c) friend-focused network, and (d) a diffuse tie network. Each component of the system of networks can contribute in specific ways towards the emotional well-being of the old migrant, the instrumental support he or she receives, and other forms of support (e.g. advocacy). This general structure mirrors the broad organisation of social networks that also characterises non-immigrants, but there are differences and shifts in the relative use of networks consequent to migration. Litwin (1997) states that the shifts included an increase in reliance on kin networks which already involve strong social ties, and within such networks there was a shift from the broader-kin network to the more restricted family-intensive network after migration. As suggested by Litwin (1997), such differences are consequent to a change from "networks of choice" to social "networks of necessity" in response to the challenges of life in a new country. Stronger social ties usually imply greater reliability of receiving support whenever it is needed.

Help from a supportive social network of kin becomes especially important in the case of older migrants of low socio-economic background, and those who may be at risk of suffering abuse and neglect (Lee and Eaton 2009, see also Iecovich 2005). Whereas support from non-kin becomes more important when the kin network fails or the help provided is not enough to satisfy the needs of old people. In a study of 60+ year old, low-income Korean immigrants to the USA, Yoo and Zippay (2012) showed how this group of elderly people had to resort to non-kin support (against the Korean tradition) to compensate for insufficient help received from family members. Such non-kin support came from daycare centres and churches that were sensitive to the cultural needs of this group of immigrants. Kin provided insufficient support for various reasons that included distance and work commitment, but there was also an issue that touches more intimately on the condition of the immigrant: some of their younger family members did not speak Korean.

Communicating in their native language can be a spirit-lifting experience for older immigrants, but the inability to do so with their own kin may lead to disappointment. In addition, the sense of filial piety prevalent in traditional Korean society tends to be diluted in the younger family members of immigrants, consequent to acculturation and modernisation (Kim et al. 1991). A similar strain in the family relationships in old Korean immigrants was also described in the USA by Lee (2007). The more isolated the immigrants are the higher their sense of alienation is expected to be. Moon and Pearl (1991) showed that in the absence of integration and embeddedness into the broader society, including difficulties of integration with sections of the family, alienation among older Korean immigrants can be decreased by living in a large, cohesive, and also ethnically homogeneous community where old people can have plenty of opportunities to speak Korean and also to socialise with other Koreans.

Old immigrants may also help improve their social integration by using modern technologies, especially when it comes to keep in touch with family members who do not live close by, but also for other practical purposes. In another study of old immigrants from the former Soviet Union established in Israel, Khvorostianov et al. (2011) described the use of the Internet for the purposes of managing health, nurturing professional interests, appreciating the past, enjoying leisure, and maintaining and extending social networks. This provided such elders with an opportunity to improve their quality of life.

Whenever members of the family may not be physically able to provide help directly, they may "sub-contract" filial piety (e.g. Lan 2002). This has been described by Lan (2002) in middle-class Hong Kong and Taiwanese immigrants in California, USA, where older parents may enjoy the services of home-care workers who are treated as "fictive kin" (see Sect. 6.7 in this chapter). This somehow would fulfil children's duty of repayment, in adulthood, for the care they received from their parents when they were children (*bao-da* or payback), but with only a financial cost to them, rather than a cost in time.

But old members of the *zero generation* are not just recipients of care from family, they also return care and, in a situation of immigration, with life getting frantic and the work more demanding, such care may be highly valued by the family. This has been studied among Chinese immigrant grandparents (Xie and Xia 2011) and other nationalities in the USA (Treas and Mazumdar 2004). In Europe older immigrants display a diversity of patterns of variable support for their family members (Albertini et al. 2019).

In fact, care provided by the older generation in migrant families may even involve some form of *transnational caregiving*, as already mentioned at the beginning of this section. In a study of transnational caregiving in zero generation Moroccan migrants in Belgium, Zickgraf (2017) has recently shown that the zero generation continues to offer material and emotional support to their migrant families in the form of transnational care. This often requires some adaptations from the part of both family and elders. In this respect, changes in travel restrictions and visas from and to some countries may become an impediment to the viability of this kind of caregiving in some instances (e.g. Zhou 2013). Horn (2017) studied the case of transnational caregiving in the case of zero generation Peruvian elders (57–86 years old) whose family members had migrated to Spain. Horn mentions some of the motivations leading grandparents to engage in transnational caregiving such as extending the traditional custom of providing care for family members in the new situation of migration. This not only satisfies the need of those receiving care but also the sense of duty and emotional satisfaction of the elder providing care, especially when grandchildren are involved. In this context, the use of the Internet may provide a complementary venue for interactions, especially when there are visa or financial constraints to travelling.

Even if they are socially well supported, older immigrants inevitably experience health issues consequent to their ageing. What kind of specific health challenges do they have to confront given their condition as immigrants? How do they try to overcome them?

Some studies have suggested that the migrant population—males in particular—self-select on the ground of health, and that it is mainly healthy individuals who decide to leave their native country to establish themselves elsewhere (Swallen 1997, and references therein). In addition, there may be also an effect of selection from the part of the accepting country which will likely reject chronically ill applicants. The *health selection effect* is seen in lower mortality rates among immigrants than among the native population (e.g. Swallen 1996). Given that most immigrants migrate relatively young, is such difference in health between immigrant and native populations maintained into old age? Swallen (1997) addressed this issue in a study of adult immigrants in the USA and found that indeed immigrants live longer and reach older ages in a better health condition. Nevertheless, old immigrants do also confront various health issues that need to be addressed.

Individually, some old immigrants may respond to their increasing health challenges with stoicism. Among the older Chinese immigrants in Canada studied by Chow (2010), for instance, their tendency to display reserve and resignation to life made them accept their medical challenges with a sense of stoicism. They also tended to seek help from traditional Chinese medicine as well as modern medicine, but their health was also positively affected by support provided by their social network, alongside a good financial situation.

Mental health is an area of growing concern as far as communities of migrant elders are concerned. Some elderly Hispanic immigrants in the USA tend to develop physical and mental health problems, in part derived from their sense of alienation from the rest of the community, especially in areas where their ethnic group is under-represented (Ailinger 1989). In another example, although old Jewish immigrants to Israel tend to develop good levels of subjective well-being regardless of their ethnic origin, this is conditional on their economic status, social capital, and health. In the case of migrants from the former Soviet Union to Israel, they may be negatively affected in their well-being, caused by some specific problems of assimilation, including issues of language proficiency and finding a job, in part due to their relatively older ages (Amit and Hefer 2010, see also the work of Ron (2001) quoted previously in this section).

Depression is an important mental health problem confronting older immigrants. Older immigrants who may have troubles learning a new language and may therefore encounter greater difficulties to integrate into the social environment in the new country, are more likely to develop depression (Tsytsarev and Krichmar 2000). Among Korean immigrants in the USA their perception of depression intermingles with their folk understanding of illness, that includes the concepts of *han*, *hwabyung*, and *shinggyongshaeyak* (Pang 1994). *Han* encapsulates a psychological syndrome of unexpressed grudges, sorrows, and regrets, self-pity, and sadness. According to Pang (1994) *han* may express itself as "a lump in the epigastric area called *oong-uh-ri*, which may be painful" (p. 210). *Han* may be caused in elderly people by others behaving in an oppressive, disloyal, or unfaithful fashion towards them or by neglecting the elder thus causing unhappiness and a decrease in well-being. *Hwabyung* is a syndrome caused by suppressed anger involving aspects of depression, whereas *shinggyongshaeyak* is characterised by "affective instability, irritability, nervousness, tension, low moods, and low productivity" (p. 210). Pang (1990) studied *hwabyung* in more detail in immigrant elderly Korean women in the USA. Old Korean immigrants tried to overcome the psychological issues associated with *hwabyung* through psychological support from the family, spiritual support from religion an also traditional Korean medicine remedies, alongside biomedical interventions. Social support, especially from family, is also important among Chinese immigrants in the USA to overcome depression (Mui 1996).

Depression symptoms appear regularly in the discourse of migrants as a possible cause of ill health. In a study of immigrants in Sweden, Emami and Torres (2005) analysed the text of 60 interviews with elderly Iranians. Respondents described their sense of loneliness when their adult children first left Iran to migrate to Sweden. Loneliness also came with a sense of insecurity about the future, as in Iran old parents rely heavily on the support from adult children, as the country lacks an adequate government support system, and to a great extent, it was the insecurity of life in Iran that prompted them to leave and join their children in Sweden. Some respondents attributed their health deterioration to their sense of loss. Moreover, even after joining their children in Sweden, they realised that the family dynamic was not the same as in Iran, as their children were experiencing a process of acculturation into Swedish life and did not show the same respect for their elders as they were supposed to. Elders used the word *ghorbat* to describe their feelings, this is a Persian (Farsi) word meaning "home-sickness, exclusion, and alienation", and such feelings were regarded as causes of illness because they were distressing. In addition, the role reversal experienced in Sweden—whereby the elders become subordinate and dependent on their children rather than the reverse, which is traditionally expected in Iranian culture—added additional distress through loss of authority. In the words of Mr Ali:

> I have had to swallow my pride and dignity...all old men should possess that. If we had stayed in Iran the children would have turned to me for help and support. Here it's the opposite. I am the one who doesn't know anything and lacks all resources and has a great need for help and support. Somehow I feel that my swallowed dignity has gotten stuck in my throat. That's what gives me my ill health. I should have not given up my life in Iran. By doing that I have lost my role as an honorable, respectable man (Emami and Torres 2005, p. 159–160).

Emami and Torres correctly point out that this kind of role reversal experienced in immigration is especially frustrating to the elders, as it does not follow from specific losses in the capacity to operate in life due to the deterioration caused by ageing, but due to the sudden change in the social environment. Some participants indicated the loss of role and purpose in life in the new country as a cause of illnesses.

Migratory grief is an experience of loss among migrants, whether it is for a lost "loved person, object, or abstract thing" (Casado and Leung 2001, p. 7). The stresses of immigration and acculturation can produce migratory grief that could eventually develop into depression. Casado and Leung (2001) studied this issue in a group of 150 Chinese immigrants in the USA who were between 55 and 86 years old. Immigrants with more elevated migratory grief also had higher levels of depression.

Older refugees are a special case of migrants. Refugees' migration is often caused by fear for their life and that of their family. They may be escaping violence, political turmoil or environmental disasters that are causing famine in their native country. Refugees can be quite traumatised from their experience, unlike other immigrants who may decide to leave to pursue better economic opportunities. Becker and Beyene (1999) interviewed a group of 50–70 years old Cambodian refugees in the USA. Such refugees often suffer from the loneliness inherent in their uprootedness, which may involve living dissociated from the rest of the community, the family being a major social referent. As the young become acculturated within the receiving country cultural tensions with the elders may emerge, a phenomenon that of course is not peculiar to the refugee status but that it may be felt particularly strongly by older refugees due to their accumulated traumas. This makes elder refugees more susceptible to developing frustration and depression. In this context old women may be better pre-adapted to cope with the stresses of migrating as refugees, as they can focus on child-caring and home support activities, just as they did in their native country.

Therefore, older immigrants confront various health challenges that are in part explained by their age; but their health problems, such as cardiovascular and mental issues (Reijneveld et al. 2003) can be also worsened by the added stresses of migration. How do older immigrants address and overcome their health issues, including mental health problems? How can they be helped in this task by public health interventions?

A combination of greater levels of acculturation, better information, and social support can help improve the health of old immigrants in general and their mental health in particular, such as decreasing depressive symptoms (Lee et al. 1996; Mui 2000; Kim et al. 2012). In fact, many social activities can aid older immigrants in improving their mental health and general well-being (Althausen 1993), and in the case that an intervention is required involving healthcare professionals, such professionals should be aware of the cultural characteristics of the elder; here the adult children will be important in order to build a bridge between elder and healthcare providers.

Religious activities, for instance, involve a social component that can be especially relevant to immigrants from some ethnic groups, and we have seen in Sect. 4.3 of Chap. 4 that religiosity and spirituality can have some positive effects on health. Such religious activities tend to involve routine encounters with co-religionaries in the community. In a study carried out on 57+ year old Rumanian refugees in France and Switzerland, Ciobanu and Fokkema (2017) explained the good level of coping by this group of immigrants through their participation in religious activities. Religious activities decreased the levels of loneliness, not only because religious events involve social contact and the nurturing of a feeling of belonging, but also because God was seen as "best friend", thus providing a space of virtual social relationships that can help cope with loneliness. Religious/spiritual coping is also important to old Chinese immigrants in the USA, the effectiveness of such coping being aided by strong values, fait, and beliefs (Lee and Chan 2009).

More specific interventions can be also designed to help older immigrants. A program of health education and physical exercises was trialed by Reijneveld et al. (2003) to help 45 years old and older Turkish first-generation immigrants in the Netherlands. Participants did improve their mental health and well-being, although no improvement was detected in physical well-being. Physical exercise was also shown to be psychologically beneficial (e.g. improving life satisfaction, optimism) to old immigrants by Kim et al. (2016).

Emami et al. (2000) described a community program introduced by a Swedish municipality in a day care centre. The program is intended to reduce stress and social isolation and to help the integration process of elderly immigrants from Iran. All staff members in the centre are Iranian and speak the official language of that country, Farsi. The centre offers information about Sweden, Swedish language courses, Persian meals, various forms of social entertainment such as music and dance, opportunities to exercise and reading of poetry, which is very popular in the centre and it is also a traditional cultural activity in Iran. The centre has been seen to have various beneficial effects on the community of elders, which include developing social cohesion and a sense of belonging, a sense of purpose, communication, and cultural development, among others.

Participation in senior social centres can improve the general well-being of older immigrants, especially if the centre not only provides ethnically appropriate forms of socialising but also health support and other forms of practical aid to the elders. For instance, Chinese immigrants to Canada who participated in Lai's (2001) study used various services offered by the senior centre they attended, such as help in filling forms and other administrative services, going to picnics and social dinners.

Beyond organised centres, the neighbourhood as such could also become a focus of ethnic integration, rather than of ghettoisation. Activities organised by the "ethnic neighbourhood", including festivals, markets, commemorations, may not only help old immigrants live in a welcoming social environment, but they could also become a bridge connecting specific ethnic communities with others, including the mainstream of society (Valtonen 2002). For instance, Vietnamese immigrants in Finland who are in their 50s and 60s seek integration through their participation in the activities of their "ethnic neighbourhood" (Valtonen 2002). This suggests that ethnic social clustering, rather than ghettoising the migrants, may serve as functional communities that may facilitate the process of settlement. The ethnic neighbourhood provides advantages of residential proximity for older immigrants, facilitating social interactions within a stable social network. This favours the creation and maintenance of strong social ties. The purpose of the community is not isolationism but to become a safe link between the migrants and the mainstream of society. In many cities around the world several streets are characterised by a high concentration of shops and restaurants belonging to specific ethnic communities. Such streets showcase what a specific culture has to offer to the rest of society. For instance, in the Australian city of Melbourne the local Chinatown is located in Little Burke St., Lygon St. is a point of encounter for the Italian community, Lonsdale St. for the Greek community, Victoria St. also known as "Little Saigon" for the Vietnamese community, and so on. These streets are very popular and attract a great number and variety of visitors from around the city and beyond, thus acting as a cultural bridge.

Centres and social clubs can also help older immigrants fill some of the emotional vacuum that in some cases develops from longing for the *homeland*, even in immigrants who otherwise may enjoy good physical and economic conditions. Patzelt (2017) studied elderly German immigrants in Canada who migrated during the post-World War II period and who, overall, had become well integrated in the Canadian society. Such good integration notwithstanding, the ageing process did come with some nostalgia and a degree of loneliness, which they addressed by strengthening social relations especially within local German organisations, but also by travelling back to Germany when possible.

Although the concept of *homeland* is somewhat fluid. Quoting Patzelt (2017, p. 221–222; see references therein) we could say that homeland:

> is often associated with aspects such as 'language, nation, home, family, community, tradition, landscape, region, or place'... However, although the homeland may be perceived as a spatial place, it is mainly a 'mental image' which is caught in time: 'The true homeland exists in the mind of the migrant,' and, furthermore: '[w]hat is peculiar about the homeland is that it emerges as a mental image in a recollection once one is no longer there. [...] It represents a lack of something, a missing piece. It is nostalgia'.

This description encapsulates very well the idiosyncratic nature of homeland for a migrant, as a mental state sustained by memories that, however, may change over time and may be distorted by emotions and longing. In addition, the homeland itself, the culture, even some aspects of the language may change over time; hence the surprise of many old migrants when they return to visit their country of origin after having spent many years away, only to experience a mixed feeling of belongingness and alienation.

From a more institutionalised perspective, interventions involving social support organisations and social workers should ensure that enough resources and training are available to help old immigrants when needed (e.g. Angel 2003). For instance, 65–79-year-old migrants from Italy, Spain, and Portugal in Switzerland use public and non-for-profit services as much as locals, but they make less use of the more expensive private and semi-private services (Bolzman and Vagni 2017), therefore unless the public safety net for the elderly is well resourced some immigrant elders will miss out on adequate support.

Both the form of support and the attitude of the professionals delivering it should be compatible with the specific cultural requirements of immigrant elders (Lai and Chau 2007). Much has been written about some ethnic groups using the resources of the public health system beyond what they may really need, following culturally established habits (e.g. Aroian et al. 2001). Some education of the immigrant communities may help but, on the other hand, their real needs cannot be overlooked. Specific forms of support and care may be required even for older immigrants who otherwise share many cultural aspects with the local community. This is the case of Finnish immigrants in Sweden, for instance (Heikkilä and Ekman 2000) and Latin American immigrants in Spain who can be especially prone to suffer stress-related health problems (Revollo et al. 2011).

Taking the cultural context of old immigrants into account is essential, if they are going to be helped in experiencing fulfilling ageing. Emami et al. (2001) stress the importance of understanding the old immigrant *lifeworld*. The lifeworld is a concept introduced by Edmund Husserl (1970, orig. 1936) that encapsulates both the external influence of reality on the individual, including the social reality, and the individual internal capacity to generate his/her own thoughts and views. Together, such influences produce the lifeworld. Modulating interventions with a view on the old immigrant's lifeworld (e.g. his/her understanding of illness, social relationships, and so on) will allow for a better use of the resources offered by the relevant immigrant ethnic community, and also a better understanding of the constraints faced by health care practitioners or welfare officers in communicating with such elder. Emami et al. (2001) suggest the adoption of an attitude that recruits elements of the lifeworld of the elder to provide better and more effective care and support.

One example of a lifeworld approach in providing assistance to elderly immigrants is offered by the tendency of old immigrants to prefer paid care based at home than moving to an institution, where they fear to be marginalised and neglected (Karl et al. 2017). If they stay at home, elders should be assisted in successfully navigating the intricacies of bureaucracy to access those services that are available and that they are entitled to (e.g. Bolzman et al. 2004). This is not always easy. In Switzerland, for instance, Bolzman (2012) indicated that old immigrants (e.g. Italian, Spanish, and former Yugoslavians) may not always access the full benefits they are entitled to from the public system, simply because they are unaware of them (e.g. due to language difficulties) or perhaps they feel social pressure to find private solutions to their needs. This tends to lead them towards looking for help elsewhere, such as family members. In a study conducted in rural Victoria, Australia (Radermacher and Feldman 2017), a male Italian farmer in his 80s who had been living in the country for over 60 years would "refuse to accept the care or assistance provided by local government or service provider agencies, preferring to turn to his family members for support" (p. 97). This suggests that even in those cases where there may be knowledge about the available services, some elders may still prefer to rely on the family.

When the social conditions in the host country are not satisfactory enough for the elder immigrant, then there is always the possibility of returning to their native country. But how feasible is such alternative? What are the constraints of returning?

Again, Bolzman et al. (2006) have studied this issue for the case of old immigrants in Switzerland. Some factors that can swing the decision towards returning include: family reunion and the support the immigrant may receive from them once she or he returns, being able to spend the last years of life living in a familiar linguistic and cultural environment, being sure of receiving at least some degree of social welfare from the native country if required. On the negative side, children may have been born in the host country and they may decide to stay, hence returning may involve an emotional cost of separation from children and grandchildren. One possible solution to this dilemma could be to commute between the two countries—at least for as long as the elder is capable of travelling—which, in the case of Europe, is quite feasible due to the relatively short distances involved and easy accessibility by both land and air, not to speak of the open-borders facility that is accessible to citizens of the *Schengen Area* countries (De Capitani 2014). At one point, however, the commuting may be done more by children and grandchildren. Bolzman et al. (2006) illustrate this issue with data from Italian and Spanish immigrants in Switzerland. These groups of old immigrants behave in a somewhat different way: the Italians lean more towards staying in Switzerland (35%) than the Spaniards (21%); and the Italians also tend to prefer the commuting alternative of living partly in Switzerland and partly in their native country (38%) more than the Spaniards do (28%). The Spaniards, on the other hand, prefer the alternative of returning to their native country (40%) more than the Italians do (18%). In part, this may reflect the differences in distance between native countries and Switzerland, making the commuting trips easier for the Italians than for the Spaniards, but other factors may be also at play.

According to the Turkish *Ministry of Foreign Affairs* (http://www.mfa.gov.tr/the-expatriate-turkish-citizens.en.mfa, accessed on 12 August 2017) 5.5 million Turks still live in Western European countries, whereas about 3 million Turkish migrants have already returned to Turkey. Liversage and Mirdal (2017) have recently investigated the issue of return in the case of Turkish migrant women living in Denmark. In a long-term study of two Turkish women interviewed over a period of three decades, Liversage and Mirdal detected a pattern that progressed from a desire to return to Turkey when they had no children or young children, to a preference for staying once the children had grown up (men show a higher tendency to return independently of children). As they age, Turkish women immigrants may still hold thoughts about returning but usually without making them a reality.

In sum, older immigrants face various challenges that range from learning a new language (or a different version of the same language) to establishing a new network of social support. For older migrants their ethnicity can be seen as either a "resource" that can help them integrate into the wider community through specific contributions: festivals, food, music, language teaching, and so forth, this requires the rest of the community to be also ready to accept the contributions of migrants; or ethnicity may be experienced as a "vulnerability factor" (Palmberger 2017), thus triggering defensive mechanisms potentially leading to isolation and ghettoisation. The family is central to the integration of older immigrants into their new country, but social contact with non-family members is also relevant. A better degree of social embeddedness can be of great help to the older immigrants in order to overcome their grief and longing for their native country, and to live a more fulfilling life in better health. Achieving those goals can be helped by various community interventions.

In the next section we review the issues affecting older members of another minority group: the gay, lesbian, bisexual, trans, and intersex people.

6.9 Older Gay, Lesbian, Bisexual, Transsexual/Transgender(trans), and Intersex People

In addition to migrants, another group of elders who deserve more attention, if we want to understand their specific needs, is people belonging to gender minorities. That lesbian, gay, bisexual, transsexual/transgender (trans), and intersex (LGBTI) people have been the target of a considerable degree of discrimination over a long period of time is certainly well known and documented, and we refer the reader to the very large literature on the subject for more details (e.g. Friend 1989; Adelman 1991; Badgett and Frank 2007; Gordon and Meyer 2007; Fredriksen-Goldsen 2012; Powell et al. 2020, and references therein). Such discrimination has been an obstacle to the full integration of LGBTI people in society. Moreover, although discrimination is often blatant, it may sometimes be more subtle and unintentional, as in the case of many services that are geared up to cater for heterosexual users, thus creating an environment of *heteronormativity* (i.e. the assumption that users are and identify as either male or female and that they are heterosexual) to the detriment of the specific needs of LGBTI people (e.g. Sharek et al. 2015).

Here we focus on the social challenges experienced by older members of the LGBTI community and how to help them overcome such challenges. Incidentally, throughout this section we will refer to the LGBTI community as our default nomenclature, but some specific works were more restricted and for the sake of precision we will use the acronym that best reflects the community studied in each work (e.g. LGBT, LGB, and so forth).

We have mentioned in Chap. 1 how old people differ according to their cohort. The cohort marks the specific experience, historical and social circumstances that affected the life of that person over decades until reaching old age (see also Rosenfeld 1999; Shankle et al. 2003). The current generation of old LGBTI people is no different (De Vries and Herdt 2012) and they have been marked by rich and dynamic past experiences that van Wagenen et al. (2013, p. 2, see also Butler 2004) describe very succinctly in these terms:

> LGBT older adults share a unique historic location as witnesses to dramatic, rapid, and ongoing social changes in the construction of minority sexual and gender identity. The oldest among them have lived through the emergence of the modern construction of 'the homosexual,' the concomitant social exclusion and medicalization of homosexuality as a mental disorder, the rise of the gay liberation and lesbian feminist movements, the emergence and devastating impact of HIV/AIDS, the proliferation of sexual and gender minority identities (including bisexual, transgender, and queer), the 'normalization' of the movement and shift towards a politics of civil rights, and the increasing visibility and incorporation of LGBT issues into mainstream social and political discourse.

In part, the heavy discrimination that the current old LGBTI people were subject to in their youth can explain why many of them still prefer to remain "invisible", hidden in the "closet" (Brotman et al. 2003; Butler 2004; Jones and Pugh 2005). The HIV/AIDS epidemic has been an additional factor that has affected the capacity of some older LGBTI people to become more visible (e.g. Shankle et al. 2003). However, although older LGBTI people may still prefer to live under the protection of invisibility, as a defence against stigma, the younger generations are more open and more likely to out their homosexuality as an essential part of their identity (Rosenfeld 1999). Their LGBTI pride is openly demonstrated in parades and festivals in many countries around the world, and in living their normal life without feeling the need to hide from the rest of society. For instance, as we write this chapter in Brisbane, Australia, this country passed legislation to allow same-sex marriage on 7 December 2017.

How does their experience so far affect the life of older LGBTI people? How can their needs be better satisfied to experience fulfilling ageing? In which ways does the rest of society react to the pleas of LGBTI elders? How can we improve the level of integration of older LGBTI people within society at large? We will address these and other questions in the rest of this section.

There is a diversity of experiences and opinions among LGBTI people regarding their old age and their ageing process (Beeler et al. 1999). On the somewhat "optimistic" side some have argued that given the experience of discrimination in the past, old LGBTI people may be better equipped to face potential discrimination when they are old; others have suggested that suffering a double stigma of being "old and gay" makes life even worse (e.g. see Heaphy et al. (2004) and references therein). Discrimination, however, is just one of the many challenges that LGBTI people confront and should be hopefully able to overcome in old age. In fact, their challenges span across all dimensions of life from health to psychological issues and social interactions.

The major health, psychological, and social problems experienced by LGBTI elders could be summarised in the following list (see Shankle et al. 2003; Addis et al. 2009; Smith et al. 2010; Cartwright et al. 2012):

- Increasing problems finding a partner as they become older.
- Issues with sexually transmitted diseases.
- Issues with smoking and drinking.
- Recurrent problems of social marginalisation and oppression.
- More depression, distress, panic, and anxiety than heterosexual elders. Although older LGBTI people are less likely to suffer depression than younger ones.
- Issues of mental health and suicidality. But older LGBTI people have fewer suicidal thoughts than younger ones.
- A mental health counsellor who understands grief and loss from a LGBTI perspective is not always available.
- Problematic relationships with family (parents, siblings, and so forth).
- Discrimination in retirement care facilities leading to hiding their sexual identity.
- Desire for retirement homes for LGBTI people.
- They are of the opinion that adjustment to later life depends to a great extent on the acceptance of ageing, the maintenance of high level of life satisfaction and on being active in the LGBTI community.
- Their social network largely consists of partners and friends, with friends being an especially important element. But members of their family of origin are also included, along with children. Strengthening such network is important.
- Desire of a better access to legal advice regarding LGBTI-specific issues.
- Interest in finding LGBTI-friendly social events for people their age.
- Access to a LGBTI-friendly social worker to find services needed by older LGBTI people.
- Finding older LGBTI-friendly spiritual places and communities.
- Need for health care providers who are approachable and are knowledgeable about LGBTI health issues.

In a study carried out in the UK by Heaphy et al. (2004), LGB respondents mentioned the increasing difficulties that they were experiencing in order to find a partner as they aged, which led to more old people living on their own, although such trend was also caused by a degree of active preference to live alone, as it is also observed in some heterosexual old people. Relationships with family also tend to be currently more positive than is usually believed, although the family reaction to an individual sexual orientation continues to be negative in many cases (Rothman et al. 2012). However, it is the relationship with friends that is especially important to old LGBTI people. In the end, whenever old LGBTI people are in need of help, they turn to all the potential sources of support: partner, friends, family, local community, exactly as old heterosexuals would do, and just as expected from *Hierarchical Compensatory theory* (Cantor and Mayer 1978) which predicts that when older people require help, they turn first to close members of the family, but when this source of support fails, they will turn to more distant relatives, followed by friends, neighbours and eventually formal community-based support. In the case of old

LGBTI people, however, not all traditional sources of social support are as forthcoming as they are for old heterosexuals, despite some recent improvements.

Hierarchical compensation was also shown by Brennan-Ing et al. (2014) who studied the composition of the social support network of old LGBT people (average age was 60 years) in the USA. Friends are described in this work as the "backbone" of the social support network. If friends failed, then respondents relied on community-based services. In general, respondents needed support to fulfil socialisation needs, but also other needs such as housing, finances, and guidance in accessing entitlements from government services.

The supportive social network of old LGBTI people tends to be more diverse than that of heterosexual old people, most likely out of simple necessity, as family and partner are not always available and if they are present, they may not be always willing to help. Hughes and Kentlyn (2011) list the following members of old LGBTI people's social support network: partners, LGBTI friends, heterosexual friends, community agencies and, to some extent, also siblings, extended family members and in some cases adult children. Moreover, Hughes and Kentlyn also distinguish between a broad *social support network* and a more committed *care support network* (or *community of practice*: "groups or networks of people who regularly interact because of a mutual engagement in a joint enterprise", Hughes and Kentlyn 2011, p. 440) the latter is especially important in the specific case of old LGBTI people.

Social and care support networks can act as a substitute for family among LGBTI elders, becoming a *family of choice*. This has been investigated by Orel (2004) in a group of 26 LGB elders (65–84 years old) in the USA. Participants were from a diverse ethnic background: African Americans, European Americans, Asian Americans, Latino/Latinas. Through various focus groups, they all expressed a primary concern for their health care and legal needs. All participants preferred to age in place and most did not like the prospect of being in a nursing home, unless "nursing homes were gay/lesbian-only and owned and operated by gay/lesbian personnel" (Orel 2004, p. 67). All participants indicated an increasing interest in spirituality as they grew older but felt that the negative attitudes of some churches regarding homosexuality deterred them from being more active in this area, although they sought participation in gay-friendly religious organisations. Social relationships were mainly grounded in the LGBT community, that provided a *family of choice*, "comprising close friends and neighbors, in contrast to the biological family or 'family of origin' that is the foundation for most heterosexual older adults" (Brennan-Ing et al. 2014, p. 24), but with a longing for better and stronger support from the *family of origin*. Although the participants considered that they were overall socially and mentally well-adjusted, some had sought treatment for depression, anxiety, and substance abuse.

Therefore, social support is arguably one of the key factors in helping old LGBTI people experience fulfilling ageing, no matter where such social support comes from (e.g. Jacobson and Samdhal 1998).

Social support may also help old LGBTI people better overcome their physical and mental issues more specifically. In a study framed within the concept of "successful ageing" that was carried out in the USA, van Wagenen et al. (2013) interviewed twenty-two 60–80 years old LGBT participants to determine their level of "successful ageing" across four degrees of success: "traditional success" (absence of problems in four domains: physical health, mental health, emotional state, and social engagement), "surviving and thriving" (experiencing concerns or problems but they thrive and live a satisfying old age by coping with their problems), "working at it", "ailing" (not coping well). Physical health problems did not differ from those of other elders of their own age, whereas mental health was a common issue raised by respondents (see also McCann et al. 2013), especially depression and anxiety that they attributed to various factors including stigma and undergoing psychotherapy to try to change their sexual orientation. Social engagement included participation in community organisations and interactions with their broader network of social relationships. Respondents were distributed across all four categories of "success", but the most frequent one was *surviving and thriving* (46%), followed by *working at it* (27%), *ailing* (18%), and finally *traditionally successful* (9%). The low prevalence of *traditionally successful* respondents may be a combined effect of personal issues and lack of social support expressed as negative discrimination (stigma) and pressure to change their sexual orientation.

Although family and friends are important in helping LGBTI elders overcome their challenges, it is difficult to "train" them in the task to produce cultural change. Broad cultural change in society is possible, but it takes time. Healthcare providers, however, can and should be formally trained to be competent regarding the needs of LGBTI people, those of old LGBTI people in particular. This will require input from the LGBTI community itself. Fredriksen-Goldsen et al. (2014) provide a list of recommendations to develop such competencies among healthcare professionals that we summarise here, with only some changes in wording:

1. Critically analyse personal and professional attitudes towards sexual orientation, gender identity, and ageing, and understand how additional factors such as culture, religion, media, and health and human service systems influence attitudes and ethical decision-making.
2. Understand and articulate the ways in which broader social and cultural contexts may have negatively impacted LGBTI older adults as a historically disadvantaged population.
3. Distinguish similarities and differences within the subgroups of LGBTI older adults, as well as their intersecting identities (such as age, gender, race and ethnicity, socio-economic and health status) to develop tailored and responsive health strategies.
4. Apply the most up-to-date knowledge available to engage in culturally competent practice with LGBTI older adults.
5. When conducting a biopsychosocial assessment, attend to the ways that the larger social context and structural and environmental risks and resources may impact LGBTI older adults.
6. When using empathy and sensitive interviewing skills during assessment and intervention, ensure the use of appropriate language for working with LGBTI older adults to establish and build rapport.

7. Understand and articulate the ways in which agency, program, and service policies do or do not marginalise and discriminate against LGBTI older adults.
8. Understand and articulate the ways that local and broader national laws negatively and positively impact on LGBTI older adults, to advocate on their behalf.
9. Provide sensitive and appropriate outreach to LGBTI older adults, their families, caregivers, and other supports to identify and address service gaps, fragmentation, and barriers that impact LGBTI older adults.
10. Enhance the capacity of LGBTI older adults and their families, caregivers, and other supports to navigate ageing, social, and health services.

Regardless of the various issues of discrimination and ostracism, even from the part of some members of their own family, many older LGBTI people do provide care for family members and others when needed (e.g. Hughes and Kentlyn 2011). In fact, some studies have found that older LGBTI people may be even slightly more likely to provide care to other adults compared to heterosexuals (Fredriksen-Goldsen 2012). Such caregiving, however, is not cost free. For instance, Fredriksen-Goldsen (2012) mentions the following potential costs confronted by LGBTI caregivers: caregiver burden, family and employment conflicts, social isolation, depression, sleep disturbance, physical problems, relationship pressures, increased risk of physical and psychological strain, poor nutrition, and financial constraints.

In a study carried out in Australia, Hughes (2010) analysed data from a *Queensland Association for Healthy Communities* (QAHC) survey to understand aspects of social support among gay men and lesbians. Most respondents expected to receive emotional and social support from friends in the LGBTI community, or their partner or even heterosexual friends, with old lesbians expecting to receive more support than gay men. Lower in the priorities as potential sources of social support were family, siblings, children, and lowest of all were neighbours. A sizeable 15.4% had nobody to rely on for social support in older age, and 24.8% expected to live alone. Most respondents (62.3%) relied on their own capacities for financial support and at the most they considered the partner—or even government—just as an alternative source of such support. Not many respondents mentioned expecting financial support from siblings, extended family members, children, LGBT friends or heterosexual friends. Neighbours were not mentioned by any of the respondents in this context.

In another Australian study, this time carried out in the states of Victoria, Western Australia, and Queensland, Barrett et al. (2015) interviewed 11 gay men and lesbians, 65–79 years old. Respondents described their experience growing up in a social environment of culturally entrenched homophobia that compromised their social relationships with family and friends (e.g. they described enforced therapies, imprisonment, termination of employment and being disinherited by family, for instance). In some cases, they had to walk away from family in order to protect their own well-being. A same-sex partnership provided a refuge to the participants in this study against outside discrimination, but especially when the relationship was kept hidden. They felt that coming out would expose them to rejection and homophobia. Ageing was perceived to have a negative effect on their ability to maintain and nurture an active social network of friends, due to the increasing physical challenges and health constraints, with the added fear that in the case of need for aged care they may find themselves immersed in a homophobic social environment and would be pressed to hide their sexual orientation.

It is not uncommon for LGBTI people to live in an environment of *psychological ambivalence* in their relationships with their own family. Psychological ambivalence is a "simultaneous experience of positive and negative sentiments in particular situations; for example, the contradictory feelings of some parents who love their gay or lesbian child but hold negative views about their child's status as gay or lesbian" (Connidis 2003, p. 81). Such social environment of ambivalent relationships may continue into old age for some members of the LGBTI community. Negotiating such ambivalence can be helped by increasing self-acceptance and self-confidence, alongside introducing meaningful changes in the community at large regarding a better acceptance of diversity in sexual orientations, roles, and identities.

Therefore, the personal experience of old LGBTI people is diverse. Many have experienced severe homophobia and abuse throughout their lives, leading to isolation (e.g. Shankle et al. 2003), which may have caused them to develop depression if not suicidal thoughts at some stage (D'Augelli et al. 2001). Some may have even internalised such homophobia leading to the development of feelings of self-hatred. Others have accepted their identity, embraced it and become affirmative in their actions in favour of gender minorities (e.g. Harrison 2006), such LGBTI older people tend to be well integrated in society and are well adjusted to their ageing process.

Integration of an individual within a social group is possible provided that the individual is not stigmatised. Unfortunately, many members of the LGBTI community continue to confront the stigma of their sexual orientation and the homophobia that comes with it. Homophobia rides on the back of various stereotypes about LGBTI people. In the 1970s in the USA, typical stereotypes included old homosexuals being regarded as "alcoholics mourning their lost youth", or homosexuality being just for young people. Old homosexuals are supposed to be facing a life of loneliness and isolation, and be also sexless, poorly psychologically adjusted, fearful, and anxious, along with sad, depressed, and bitter, and old male homosexuals are also believed, according to the stereotype, to sexually depredate on young gay men (Wahler and Gabbay 1997; Kimmel et al. 2006; De Vries and Herdt 2012). Kelly (1977, see also Kimmel 1979–1980) was one of the first to challenge those stereotypes about old homosexuals. Kimmel et al. (2006, p. 2) provide this comment about Kelly: "Kelly presented his work at the 1972 meeting of the Gerontological Society of America and won the student award for the year's best dissertation. Ironically, when the society's president announced the award at the opening reception, he said the paper was on 'an unmentionable topic'".

The early works pointed to what has been later supported by further studies: that older homosexuals do not have more problems than any other old person provided that the aspects of discrimination specifically attributed to their sexual orientation are controlled. Helping in the process of social change have been the various organisations of LGBTI people like the *Senior Action in a Gay Environment* (SAGE) that was created in the USA in 1977 with the mission of supporting elderly gay and lesbian people, and various other organisations such as the *American Society on*

Aging's LGBT Aging Issues Network (LAIN) or the *Gays and Lesbians Older and Wiser* (GLOW) organisation dedicated to the mutual support of older gay men and lesbians; not to speak of historical movements whose activism has slowly improved the stance of LGBTI people in at least some Western countries (e.g. Slusher et al. 1996; Butler 2004; Genke 2004; Haber 2009).

Fighting the stigma of belonging to the LGBTI community requires first and above all developing a positive identity as a LGBTI person. A positive identity is more likely to help such person to better respond to the challenges of ageing as a member of a minority group (e.g. Friend 1991). The process must start from the individual, of course, but it cannot be completed without the help of other sectors of society, such as equally stigmatised people united in solidarity, and even more importantly the "wise" and the "normal".

In what follows we will briefly review some more specific studies carried out, in turn, on elderly gay men, lesbians, bisexuals, trans, and intersex people.

Given the emphasis on youth found especially in gay culture (e.g. Shankle et al. 2003; Slevin and Linneman 2010), old gay men are at risk of suffering ageist discrimination even within their own gay community, in addition to the overall discrimination against old people found in the community at large, where it compounds discrimination against homosexuals.

Homophobia and ageism are indeed a recurrent concern for old gay men, alongside their worries about social support in old age and access to appropriate care facilities (Kushner et al. 2013). In addition, internalised ageism further increases the "invisibility" of old gay men, even at ages (e.g. their forties) when heterosexual men hardly feel themselves as being old (Genke 2004; Slevin and Linneman 2010). Young (1990) described this phenomenon as *cultural imperialism* (see also Genke 2004): at the same time as negative stereotypes that prevail in the community at large may be in part internalised by gay people, they may also find themselves striving to rise above the stereotype. Discrimination may be further aggravated in older members of gender minorities as a result of the accumulated effects of multiple stigmas: old, gay, perhaps HIV-positive, and in some cases also poor, member of an ethnic minority or belonging to a stigmatised race, and so forth (Emlet 2006).

How can old gay men find some respite against the psychological and social pressures discrimination and stigma? Where can they obtain support? Clearly, the network of friendships within the gay community is a major source of support against discrimination for older gay men. Attachment to gay friends may also help older gay men overcome symptoms of depression, by providing a sense of belongingness in society. In this way old gay men can become more resilient. On the other hand, living alone exposes the old gay man to the brunt of loneliness and to the risk of developing depression (McLaren 2016).

In an attempt to overcome stigma and become better accepted into a social matrix of prevalent heterosexuality, some gay men may adopt an attitude that could be described as the opposite to minstrelisation, and that is to behave in an exaggerated masculine manner. It is doubtful whether gay men are really improving their social integration through either minstrelisation, which in this case could be expressed through exaggerated feminine behaviours, or through mocking the hegemonic masculine stereotype. Alternatively, gay men may develop more authentic and honest forms of gender role, such as *non-hegemonic patterns of masculinity* (Connell and Messerschmidt 2005), *androgyny* (Zoccali et al. 2008), and genuine expressions of *femininity* (VanderLaan et al. 2016).

Although in an early review of ageing in male homosexuals Wahler and Gabbay (1997) pointed out that many studies found that old gay men are indeed well adapted to their ageing, against commonly held views about their troubles in old age (e.g. Friend 1989; Adelman 1991), care should be taken to identify exactly what kinds of people were interviewed in such studies. If samples are biased towards well-educated, financially secure, Caucasian, urban homosexuals, living in currently reasonably tolerant societies, there is little surprise if many of them, especially in recent years, are in fact well-adjusted and quite happy with their life (e.g. McCann et al. 2013). For instance, it is rather difficult to imagine why an open gay person such as the successful singer Sir Elton John, who is in his 70s, would not feel currently adjusted to his sexuality. However, there is diversity in personal situations within the gay community and the root causes of such diversity must be properly understood.

With regard to old lesbians, previous studies have suggested that they may have at least some advantages compared to old heterosexual women: youth, for instance, is less important in lesbian relationships, lesbians tend to prefer same-age partners, physical attributes also tend to be relatively less important and the lesbian community tends to be quite supportive of its members (Gabbay and Wahler 2002).

We mentioned above how old gay men may suffer a double stigma, being both old and gay. Old lesbians may be confronted with a triple stigma: old, lesbian, and woman (Gabbay and Wahler 2002; Nystrom and Jones 2012). The negative effects of stigma may be compensated by social support, which is important to maintain well-being in old age. Such social support for lesbians usually comes from friends and *chosen family* (De Vries and Herdt 2012; Nystrom and Jones 2012), the latter usually comprises a group of especially close friends whom the individual regards as family. Old lesbians experience greater sense of freedom than they had at younger ages and a strong sense of personal fulfilment in the social context of a good network of friends that tends to be more easily built and maintained than is the case with gay men (Nystrom and Jones 2012).

One important characteristic of lesbians in general, including older ones, is their sexual fluidity, which is indeed a distinguishing quality of women in general; and to be even more precise, it is a broad trait found in female social mammals (Poiani 2010). Sexual fluidity in women is such that it can include periods of heterosexual and homosexual relationships. Fluidity in romantic relationships in lesbians has been supported by evidence from various studies. Averett et al. (2012), for instance, carried out an online survey of 456 older lesbians (51–86 years old; 86.9% Caucasian) in the USA showing fluidity alongside a greater focus of the respondents on social relationships than sexuality as such. Interestingly, Averett et al. note that despite such fluidity, respondents kept identifying themselves as lesbian (rather than bisexual, for instance). The authors suggest various reasons for this, including the possibility that the reality of sexual fluidity had not been consciously incorporated in the mind of those respondents, or that the definition of sexuality was based on the present circumstances and did not take into consideration the whole pattern experienced in the long term. But there may be also an effect of negativity against bisexuality that polarises the identity spectrum between two "extremes": homosexual (lesbian in this case) and heterosexual, a point also made by Poiani (2010). Such polarisation can be further stabilised by feminist political movements; Averett et al. mention "radical lesbianism", for instance.

An interesting comparison between gay men and lesbians regarding their views on old age was provided by Schope (2005) with data from the *Baby Boomers* generation in the USA. Table 6.11 shows the results of responses to a questionnaire. The first question asks the respondents to provide their view on the perception of ageing prevalent in the homosexual society. The responses of gay men and lesbians are quite different. In the view of gay men, the society of male homosexuals has a relatively negative view of ageing (terrible-tolerable), whereas lesbians think that the view of old age in the lesbian society is acceptable-good. Do these results match the views of the respondents about their own ageing? The match is good for lesbians, not so for gay men. Male homosexuals have a far more positive view about their own ageing (tolerable-acceptable-good) than what they believe is the common view in the gay society (terrible-tolerable). This suggests that lesbians and gay men are roughly equally positive about their own ageing (perhaps with relatively more lesbians regarding their ageing as good-fantastic), but they differ in their perception of the common view in others: lesbians being more of the view that their opinion is shared with other lesbians, whereas gay men believe that their personal opinion contrasts with the general view in the gay community. This is not surprising, given the well-known difference in sociability between males and females (e.g. Poiani 2010).

Table 6.11 Gender differences in ageing

	Terrible	Tolerable	Acceptable	Good	Fantastic
How does the gay society view ageing according to the interviewee?					
Total males ($n = 74$)	45.9	37.8	14.9	0.0	1.4
Younger males ($n = 52$)	50.0	38.5	9.6	0.0	1.9
Older males ($n = 22$)	36.4	36.4	27.3	0.0	0.0
Total females ($n = 104$)	3.8	18.1	37.1	32.4	8.6
Younger females ($n = 51$)	5.9	21.6	37.3	27.5	7.8
Older females ($n = 53$)	1.9	18.3	36.5	32.7	8.7
How does a gay person view own ageing?					
Total males ($n = 74$)	5.4	28.4	27.0	33.8	5.4
Younger males ($n = 52$)	5.8	28.8	26.9	34.6	3.8
Older males ($n = 22$)	4.5	27.3	27.3	31.8	9.1
Total females ($n = 104$)	0.9	11.0	32.2	38.5	18.3
Younger females ($n = 51$)	0.0	11.3	34.0	30.2	24.5
Older females ($n = 53$)	1.8	10.9	29.1	45.5	12.7

The table is reproduced from Schope (2005) with slight modifications. Values reported are percentages

Two major theories of ageing in homosexuals have been produced, with special reference to men: the theory of *Accelerated Ageing* which states that homosexual men see themselves as being old at relatively younger ages than heterosexual men do (lesbians, on the other hand, seem to be well adjusted to their ageing as shown in Table 6.11), and the theory of *Crisis competence*, purporting that gay men cope with ageing better than heterosexual men do: "successful management of one stigmatized identity early in the life course … creates skills that transfer to the successful management of a later stigmatized identity" (Fredriksen-Goldsen and Muraco 2010, p. 376). On the ground of the variable evidence available, both theories have their supporters and detractors, suggesting that neither of them is capable of accounting, on its own, for the full diversity of behaviours observed (Schope 2005). Fredriksen-Goldsen and Muraco (2010) also list additional theoretical approaches that have been used to understand ageing among homosexuals.

Although most of the traditional interest in the members of the LGBTI community has been focused on the "L" and the "G", in part because both lesbians and gay men have been especially visible and politically active, bisexuals are also beginning to raise their voice, and their concerns in old age are being heard (e.g. Rodríguez Rust 2012). Bisexuals have been historically squeezed out of the public consciousness by an increasing vocal clash between homosexuals and heterosexuals demanding a clear distinction between the two sides. But in fact, there is a diversity of sexual orientations and identities that runs across a continuum, with bisexuals occupying a considerable region between strict homosexuals and strict heterosexuals. Bisexuality, or sexual attraction for both males and females, encompasses a varied spectrum of preferences and combinations that can vary over time, which simply derives from our human behavioural sexual plasticity (e.g. Poiani 2010). Such plasticity makes the process of coming out more complicated for bisexuals than for homosexuals, as bisexuals may change the direction of their sexual preferences (towards women or towards men and back) in different periods of their lives.

On the positive side, bisexuals are more likely to have children than homosexuals. Therefore, as they age, they are also more likely to find support in their family, unless, of course, children are intolerant of their parent bisexuality (e.g. Rodríguez Rust 2012).

Transsexual people, on the other hand, are unhappy with their biological sex, feeling that they are a member of the other sex. Such feelings may lead some transsexuals to transition from female to male (FTM) or from male to female (MTF) by undergoing gender-reassignment surgery and hormone treatment, although not all transsexual people undergo transition, preferring instead to define their sexuality as fluid, without being restricted by the male-female binary (e.g. Cook-Daniels 2006; Witten and Eyler 2012). On the other hand, *transgender* "is a general term with an evolving meaning… 'transgender people are individuals who transgress societally constructed gender norms in one manner or another' and includes persons such as full-time cross-dressers, drag queens, drag kings, gender-blended people, and gender queers, as well as myriad other members of the 'gender community'" (Witten and Eyler 2012, p. 190). Both transsexual and transgender are currently encapsulated in the word *trans* (Witten and Whittle 2004), which saves us from growing the

LGBTI acronym to a "frightening" extent (e.g. LGBTQQUI: lesbian, gay, bisexual, transsexual, queer, questioning, unsure, intersex; see Cronin and King 2009), whilst retaining the full diversity of sexual orientations and identities found in society. *Trans* particularly stresses the fluidity found in human gender identities, sexual orientations (bisexuality in particular), and gender roles (Cronin and King 2009; Poiani 2010).

Many transsexuals transition later in life, when social restrictions and the potential for stigma decrease, as they have retired from work, their children are adult and have become independent, their parents may have passed away, whilst they feel that they are getting older and they either transition now or never. However, transitioning in older ages also implies some challenges such as health concerns regarding surgery and hormone therapy, fear of disrupting their social network, difficulties in finding a new partner, legal issues such as social security, and also employment issues if the person is still expecting to work (Cook-Daniels 2006).

Just as it is the case for other sexual orientations and identities, old trans people also experience cohort effects. A fluid gender identity is relatively more acceptable now than it was in the past, hence transitioning today is becoming relatively easier for a trans person.

Despite recent improvements in social acceptance, trans people have been and continue to be the target of *transphobia* (Witten 2009, 2014), expressed as discrimination, stigmatisation, and abuse, sometime resulting in hate crimes (e.g. Witten and Whittle 2004; Witten 2009). This has led to a significant psychological burden felt by trans people resulting in mental health problems: "The legacy of stigmatization, prejudice, and victimization can have a profound negative impact on the well-being of transgender older adults, including physical and mental health, social success, and financial stability" (Witten and Eyler 2012, p. 209). The burden can be alleviated through social support and better integration into society, especially if acceptance of diversity and respect for individuals' self-determination become more widespread (Harrison 2005), but many old trans people face loneliness and social isolation in later life (Witten 2009).

Trans people can also regard themselves as trans-gay or trans-lesbian, according to the sex of their preferred partner. Witten (2015) studied older trans-lesbians (mostly from the USA, 89.4%; mostly Caucasian, 93.8%; and mainly 51 years old and older, 61.1%) with a focus on issues of late life, preparations for such late life and end-of-life arrangements. Witten observes that although many of the respondents indicated to be generally fulfilled in their ageing, they did express some fears about the future and its insecurities especially with regard to access to health care (21.2% were unsure how they would pay for health care in later life, and end-of-life care in particular) and later-in-life legal and social support (e.g. 23.1% stated that they were unsure who would care for them in case of a major illness for instance). Other late-life concerns included fear of being discriminated against by caregivers, fear of being the target of cruelty and abuse, fear of becoming homeless, and fear of being impeded to live their final years as a true trans person.

A disturbing case of discrimination is recounted by Tina Donovan (2001, p. 21), a 61-year-old MTF trans woman:

> One incident in particular that took place in a hospital emergency room was extremely traumatic for me. I was in the emergency room seeking treatment, which under the best of conditions is already a stressful experience. One of the nursing staff members saw on my chart the name 'Thomas Donovan.' The name and the face did not go together. He asked me some questions, and did not like my answers. He proceeded, in a very nasty and belittling manner, to try to lift my skirt to show the other nurses and patients that I was not what I appeared to be. He should have treated me first as a patient and as a person; my gender identity should not have been an issue for ridicule. Certainly, the nurse should have shown far more sensitivity than he did. Instead, his behavior was malicious and abusive. I was tied down and sedated. I spent two weeks under sedation before I was released. It took me a long time to get over this episode. But, this is just one of the many negative experiences in the lives of transgender people.

Social isolation in old trans people could be alleviated through the use of the Internet, but accessing such service can be costly to some, especially those living close to the poverty line (Witten 2009). This financial issue also affects their access to appropriate housing and professional care.

In a review of ageing in trans persons, Finkenauer et al. (2012) identified several major issues that require addressing in order to offer trans people a better chance in life: end the violence and abuse trans people are subject to, removal of discriminatory policies and practices in health and mental health care, improve HIV/AIDS education, prevention, and treatment strategies specifically targeted to trans older adults; remove the obstacles in accessing education, employment, government services, and housing, and help with social support networks. From the perspective of health, Fredriksen-Goldsen et al. (2013) have stressed how trans people are at higher risk of suffering from poor physical health, showing symptoms of depression and displaying the effects of stress compared with non-trans people. They are also disadvantaged through internalised stigma and victimisation, fear of seeking help from health services, and lack of social support.

Disorders of sex development (DSD) are "conditions in which an individual's genital or reproductive anatomy, or chromosomal constellation, are defined as 'ambiguous', or atypical, or in which there is a specific type of reproductive endocrinological deficiency or imbalance". On the other hand, *intersex* describes "a group of people who identify as intersex because the experience of being born with, and treated for, a DSD has influenced personal experiences and self-perceptions" (Talley and Casper 2012, p. 273). Unfortunately, most of the research carried out on intersex people has been focused on young individuals and very little is known about old intersex individuals. Indeed, Talley and Casper's book chapter ends with a list of very many questions that are still waiting for an answer.

What are the options for old LGBTI people who want to retire?

Various studies have indicated that old LGBTI people regard traditional nursing homes and other forms of residential care as undesirable (Heaphy et al. 2004). Instead, they prefer to live independently in retirement (*ageing in place*) or, alternatively, retire in communities of similarly-minded elders (e.g. Neville and Henrickson 2010; Hughes and Kentlyn 2011); either in *Continuing Care Retirement Communities* (CCRC) where independent adults reside and where they can access various health care services or in *Naturally Occurring Retirement Communities* (NORC), which are neighbourhoods with well organised social networks of elders. If the elder decides to retire in a community, then LGTBI-friendly retirement communities are obviously preferred for their acceptance, understanding of specific needs and provision of adequate services (Gabbay and Wahler 2002).

Retirement communities for old gay men have been created in the USA, for instance. In Santa Fe, New Mexico, *Rainbow Vision* has become "the first retirement community specifically created for LGBT older adults" (De Vries and Herdt 2012, p. 117). Other such retirement communities in the USA include: *Palms of Manasota* in Florida that was established in 1998; *Birds of a Feather* in New Mexico; *Stonewall Communities* in Massachusetts; *Gay and Lesbian Association of Retiring Persons* (GLARP) retirement communities in California (Shankle et al. 2003).

Alternatively, *nursing homes* and *assisted living facilities* may be available that provide ongoing specialised assistance to LGBTI elders who require special care (Shankle et al. 2003).

An analysis of the limitations experienced by support facilities, such as *Area Agencies on Aging* (AAA) in the USA, when assisting old LGBTI people is available in Knochel et al. (2011, see also Knochel et al. 2012). In Australia, deficiencies have been described in the capacity of current aged care facilities to properly cater for LGBTI elders (e.g. Phillips and Marks 2008), hence some initiatives have been taken to provide specific accommodation services for old LGBTI people (Hughes 2007), although not all such initiatives have been successful in the long run.

In nursing homes where both heterosexual and LGBTI old people co-reside, both care staff and managers should be specifically trained to be able to provide appropriate care and quality service that is satisfactory to all residents (Willis et al. 2016).

We conclude this section with a brief mention of the issues confronted by old LGBTI people at the end of their life and the challenges of bereavement for the survivors. Although the end of life is a process that will be inevitably confronted by both LGBTI and heterosexuals alike, their personal experiences are likely to differ, simply because their life experiences are also likely not to have been the same, with many differences being specifically associated to their sexual orientation.

First, it is the accumulated discrimination that old LGBTI people have suffered that will likely affect the kind of social support they may need at the end of their life, even from their biological family. Sensitivity to this extends to the medical and nursing staff who will be caring for older terminal LGBTI patients, and therefore they should be properly trained to provide the best and most effective care (Rawlings 2012).

Given the devastating effects that the HIV-AIDS epidemic had especially (but not only) on the LGBTI community, survivors have to cope with their own ongoing worries, whereas partners of those who passed away have to face a process of bereavement that in this case may be aggravated by social discrimination against LGBTI people. This can be expressed in the phenomenon of *disenfranchised grief*, in which "the grief of a bereaved person is not acknowledged or perceived as legitimate" (Cartwright et al. 2012, p. 539), especially when the family of choice of the deceased LGBTI person is not recognised by the biological family. Moreover, social isolation among LGBTI elders (e.g. Almack et al. 2010) may even lead them to seek treatment for serious medical conditions when it is too late, thus increasing not only the medical complications but also the intensity of the crisis and the difficulties of its management and, ultimately, the probability of an unfavourable outcome.

From a legal perspective, members of the LGBTI community are advised to check the local legislation to determine the exact steps to be taken to write legally binding advance health care directives for their palliative care. A similar legal action can be taken to help prevent cases of discrimination, when it comes to the execution of the will of the deceased (e.g. Cartwright et al. 2012; Rawlings 2012; Witten 2014).

Spiritual and religious support may be also required by some members of the LGBTI community to help address the common problem of loneliness confronted by people towards the end of their life (Witten 2014). Thus, such religious and spiritual support should be facilitated.

To recap, LGBTI people face some recurrent challenges in their old age that require awareness and support from the rest of society. Stigma and discrimination need to be confronted and overcome. Providing retirement options and medical services that are specifically tailored to the needs of LGBTI elders is particularly important.

6.10 Ageing Prison Inmates

Another group of especially neglected old people are prison inmates: those who have been jailed for committing a crime in old age, the recidivists or "career criminals" who regularly come and go through the prison system, and also those who are ageing in prison as a result of receiving long-term sentences (e.g. Beckett et al. 2003).

In 2015, inmates 55 years old and older accounted for 10.6% of all prisoners in the USA jails system (Carson and Anderson 2016), as part of a trend that has seen an increase in the proportion of older inmates in recent years, and not just in the USA (Lemieux et al. 2002; Arndt et al. 2002; Fazel 2008). This trend is due to a combination of factors: more people commit crimes at an older age (the "new elderly") and more inmates age in jail (the "old timers" and also the "career" or "chronic" offenders, that is recidivists who age in and out of prison) (Aday 2003). In addition, Yorston and Taylor (2006) suggest the additional factor that in the past older people committing a crime were treated with some leniency by the criminal courts, the accused being able to plea bargain smaller sentences than the younger offenders who committed the same type of crime, but this is no longer the case, especially since the 1980s (Maschi et al. 2012). Mandatory minimum sentencing, reduced chances for early release, and increased frequency of crimes that attract longer jail terms are also contributing factors to an ageing population in prison (Baidawi et al. 2011). The increase is also seen in the Australian prison system (Table 6.12) and other countries as well (Baidawi et al. 2011, and references therein).

In the early 1980s, the vast majority of elderly inmates in USA prisons were men (Goetting 1983), the situation has not changed much in the following 30 years or so,

Table 6.12 Number of prisoners in Australia by age and year

Age (years)	Australian prisoners in 2000	Australian prisoners in 2010	Increase, 2000–2010 (%)
50–54	848	1445	70.4
55–59	459	825	79.7
60–64	281	529	81.8
65+	218	527	141.7
Total prison population	21,714	29,696	36.8

From Baidawi et al. (2011)

with the elderly population of inmates being 95.2% men in 2015 (Carson and Anderson 2016).

Goetting (1983) points out that a large proportion of elderly offenders are incarcerated for violent crimes. The trend has remained true in the early 2000s with older offenders being most likely to have committed crimes such as aggravated assault, homicide, and sex crimes; with these crimes being often committed against members of the family in a domestic environment (Lemieux et al. 2002; Fazel 2008; Reimer 2008), including "mercy killing" of a sick wife (Aday 2003). It has been suggested that this trend may be a result of "midlife crises" and psychological effects of retirement, often mediated by abuse of substances such as alcohol (Reimer 2008).

Other, non-violent crimes are also common: fraud, embezzlement, and larceny-theft in particular. This trend is valid for both old men and women (e.g. Fazel 2008; Wahidin 2011).

The prison environment has been seen through a variety of perspectives as far as the effects on old people are concerned. On the one hand the old can be subject to a regime of violence and oppression from the part of younger and more aggressive inmates, leading to fear and attitudes of bitterness (e.g. Aday 1994a; Kerbs and Jolley 2007; Baidawi et al. 2011; Trotter and Baidawi 2015). Also on the negative side, long-term prisoners in particular may experience a level of *depersonalisation*, expressed as boredom, resentment, loss of identity, sensory deprivation, and isolation (Aday 2003). If they have family, being in prison may lead to loneliness, anxiety, depression, and a feeling of longing for the company of their loved ones (Lemieux et al. 2002). In addition, the ageing process seems to be accelerated in a prison environment (Reimer 2008). For instance, Handtke and Wangmo (2014) report that a prison inmate who is 50 or 55 years old displays a health status equivalent to a 60–65 years old from the general population. The reasons for this difference include unhealthy lifestyles before entering prison, lower socio-economic status, and also the prison environment itself.

On the other hand, old and poor people who may be living in extreme conditions of risk to their safety in the community, may find a level of security and material well-being in prison, that they would not be able to access outside (Goetting 1983; Foucault 1988). Dependency on the prison environment (*institutional dependency*) may develop in older inmates especially if they are unmarried, are chronic offenders and have fewer alternative sources of both material and emotional support outside prison (Goetting 1983; Aday 2003): "Coming from a lifestyle filled with poverty and lacking access to health care, older inmates are frequently unaccustomed to the regular food, medicine, housing, and clothing now readily available" (Aday 2003, p. 117).

Although older inmates seem to be better adjusted to the prison environment than the younger ones, behaving in a less disruptive manner, there are exceptions (e.g. McShane and Williams 1990) and they do also report suffering from mental health problems such as depression (e.g. Aday 1994b; Fazel et al. 2001a; Caverley 2006; Baidawi et al. 2011), in part caused by *victimisation* (psychological, property, physical, sexual; Kerbs and Jolley 2007). Curbing victimisation is one of the arguments used to support age segregation in prison (Kerbs and Jolley 2009). An argument against age segregation in prison is provided next.

In contrast to the contemporary trend in society where older people are no longer seen as repositories of valuable knowledge due to the rapid changes society is going through as a result of modernisation, in prison older long-term inmates do command respect for their knowledge of the prison system and their demonstrated ability to survive in such environment (Goetting 1983, 1985). Young prisoners may be more knowledgeable about various aspects of modern life than old prisoners, but there is little in such knowledge that prepares the young for the specific challenges they confront in prison. Yorston and Taylor (2006, p. 334) suggest that such characteristic of older inmates may put them in the situation of exerting a "stabilizing or deflating force in the high-tension institutional environment where violence and riot are ever-present threats". The potential for this stabilising effect has been suggested as a reason to discourage age segregation in the prison system.

Life in prison has also many challenges for older inmates that may affect their mental health and that compound the psychological burden they already carry from their life outside prison, such as a history of stressful events experienced since childhood. In a study of an ethnically diverse sample of older inmates carried out in the USA, Haugebrook et al. (2010) noted that inmate participants in their study reported having experienced not just stressful events in their life such as parental death or abandonment and substance abuse by parents, but also traumatic experiences such as physical, sexual, and emotional abuse. Psychologically negative effects of such experiences were noted by the participants. Maschi et al. (2011b) carried out a study of lifetime traumas and stressors experienced by young (18–24 years old) and older (55 and over years old) inmates serving time in New Jersey (USA) prisons. Both groups had suffered traumas and stresses throughout their lives. Being victims to violence and witnessing violence were common experiences to all age groups, with physical assault that occurred during childhood being reported by 43.1% of all participants, followed by physical assault in adolescence (35.1%) and also adulthood (30.9%). Some of the stressors differed between young and old inmates. For instance, young prisoners reported witnessing someone beaten up, or punched, or kicked more frequently than older prisoners did; whereas the latter reported witnessing someone having been sexually assaulted more frequently. Older prisoners did not report living in a violent neighbourhood as frequently as young prisoners did.

In a literature review of the prevalence of mental illness in old prison inmates (50 years old and older), Kakoullis et al. (2010, additional data are from Maschi et al. 2012) provided the following range of prevalence values for various mental conditions: personality disorder (30–65%), antisocial personality disorder (8–47%), depressive disorder (7.7–42%), psychosis (3.3–13.3%), anxiety disorders (2.1–11.5%), substance misuse/dependence (0–4.9%), schizophrenia (1.1–2.5%), bipolar disorder (1.1–2.5%), dementia (0.8–13% and up to 44%), and organic mood disorder (1.1%). Mental illness seems to be as prevalent in men as it is in women prisoners (Fazel 2008), but it is more prevalent among prison inmates than in the non-incarcerated community (Diamond et al. 2001).

In a survey of personnel working within the prisons system in the USA, Williams et al. (2012, 2012a) concluded that there is a general belief that the prevalence of cognitive impairment, including dementia, among older prisoners is relatively high. Mental problems suffered by older prison inmates may also put them at risk of self-harm and of committing suicide (Kuhlmann and Ruddell 2005).

Apart from mental problems, both old male and female inmates also experience additional health challenges that are mostly typical of their age but that can be increased by the prison experience (Collins and Bird 2007). For instance, Williams et al. (2006) studied 120 older female inmates in California state prisons and found that fall rates among them ranged between 33% and 63%, other forms of functional impairment were also reported that were attributed to the effects of the prison environment. In a study carried out by Reviere and Young (2004) in the USA, only 30% of 50–64-year-old women prisoners were free from chronic conditions, disability, or other forms of functional limitations, especially among African Americans.

Older male prisoners have been reported to suffer from one or more major illnesses, including cardiovascular, musculoskeletal, respiratory, and sexually transmitted diseases (e.g. Fazel et al. 2001, 2004; Yorston and Taylor 2006; Deaton et al. 2009).

The medical treatment that prisoners receive is usually adequate but not always (Fazel et al. 2004). Solving the health problems of prisoners is possible but it requires resources, commitment, and proper understanding of the issues involved. Williams et al. (2009), for instance, point out that lack of proper training in some correctional officers may stand on the way of identifying first and then addressing health problems that affect older prisoners.

Given the challenges of confinement, how do old prisoners cope with the restrictions of life in prison?

Increasing *self-efficacy* can improve the health condition of old inmates. In a study carried out in the USA, Loeb and Steffensmeier (2006) investigated the level of self-efficacy in a group of old inmates. Following Albert Bandura (Bandura 1986, 2004), *self-efficacy* is defined as an "individual's confidence in his or her ability to mobilize the motivation, cognitive resources, and action plans necessary to exert control over events that occur in the person's life. Someone who has high self-efficacy is likely to attempt a new behavior, whereas one with low self-efficacy is more likely to try to avoid one" (Loeb and Steffensmeier 2006, p. 271). Old inmates with higher self-efficacy in the management of their own health rated their health status to be better, and they were also ready to be involved in health-enhancing behaviours.

Greater self-efficacy increases the coping capacity of the old inmates. Maschi et al. (2015, see also Maschi et al. 2011a) suggest that the effects of lifetime traumas that characterise many old inmates can be specifically ameliorated, and a better well-being achieved through the development of coping mechanisms that take advantage of physical exercise, positive thinking, expression of emotions, social engagement, and personal enrichment through spirituality (Fig. 6.13).

In Maschi et al.'s (2015) model, older inmates may recur to spirituality to better cope with life in prison. Allen et al. (2008) investigated the role of religiousness and spirituality in protecting old male inmates' mental health in the USA. Inmates were incarcerated mainly for murder and sexual crimes and they were older than 50. Results indicate that depressed inmates may try to cope with their negative emotional state by seeking help in religion. The authors also remark that inmates who felt that they had been abandoned by God also displayed more symptoms of depression and a greater propensity for hastened death. In another study also carried out in the USA, Koenig (1995) reported the opinion of 50–72 years old prisoners. Most of them were religious as they came from the USA so-called Bible belt, 86.4% had Christian religious beliefs and they did participate in regular chapel services and other religious activities. Religion was an important coping factor for many of them and religiousness was negatively associated with depression. Other forms of coping used by inmates, however, can be detrimental to their health, such as various forms of substance abuse (Arndt et al. 2002).

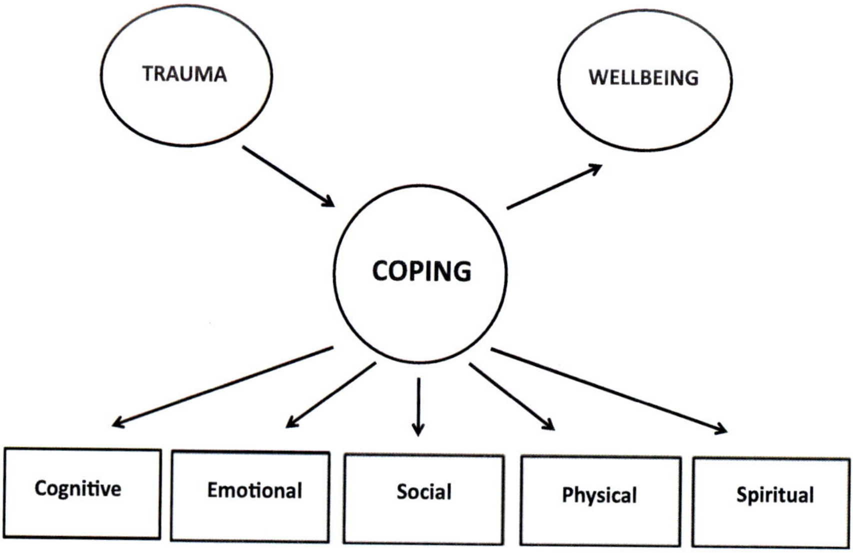

Fig. 6.13 Maschi et al.'s (2015) model of coping with the stresses of life in prison for prisoners 50 years old and older. The model is reproduced here in a simplified version

Coping can be better achieved when inmates have access to a variety of services in prison, but jail systems are often constrained in the number and quality of services that they provide (Bretschneider and Elger 2014).

In the case of older inmates serving long sentences the concern may arise about the possibility of dying in prison. Death in prison can come from various causes including suicide, violence, accidents, and illnesses (Handtke and Wangmo 2014). In a study carried out in Switzerland, Handtke and Wangmo (2014) recorded 35 interviews with 51–71 years old inmates in 12 prisons with the objective of investigating their stance about the possibility of dying in prison and their concept of "dying with dignity". Dying with dignity involves relieving pain and suffering, help in achieving readiness to die, and keep the individual in control of the process, allowing as much autonomy as possible. In particular, the authors stress Allmark's (2002) concept of "death without indignities"; an approach based on measures that reinforce the degree of autonomy and the removal of barriers to dignity (Handtke and Wangmo 2014, p. 375). In a prison environment, inmates' autonomy is necessarily restricted by their limited opportunities to engage in social interactions, a reduced access to end-of-life services—which include doctor-assisted suicide where it is legal—and the broader handling of inmates' deaths by the prison administration; lack of opportunities for bereavement, an overall negative experience of death, and a limited choice about the place of death. More autonomy could be allowed in prison by facilitating visits to a terminally ill inmate by family, acquaintances or support groups, and better access to palliative care services (Handtke and Wangmo 2014).

In general, it does not seem that older prisoners experience different levels of death anxiety compared to elders of the same age living freely. What concerns them is not so much death per se but the time and circumstances in which death will occur (e.g. Aday 2006). This concern may be a result of their mandated confinement that may strip them of the possibility of dying close to their loved ones. Aday (2006) reports results of interviews with 102 inmates between 50 and 84 years old in a prison in the state of Mississippi, USA. Most had been incarcerated for crimes such as homicide and manslaughter, criminal sexual conduct, and drugs. In this group death anxiety decreased inversely with age, self-reported mental health, and life satisfaction; but it increased with the number of medical problems (see also Deaton et al. 2009), feeling unsafe in the prison environment and loss of social support. Social support was sought from other inmates, especially in cases where they did not have support from family:

> Because of the isolation from the outside world, friendships in prison may be even more highly valued than kin relationships. Close personal relationships with friends can help cushion the shock of physical deterioration, the loss of loved ones, and other sources of stress in the prison environment. (Aday 2006, p. 207).

Aday (2006) also reports that old prisoners feared dying in prison for the stigma that such an event may have on the family and for the existential sadness of dying "abandoned, humiliated, and lonely". Aday also mentions religious beliefs as a common coping strategy to withstand the thoughts of dying in prison. In other cases, however, death may be regarded as the door through which they can walk towards their final freedom.

Among a group of 50–78-year-old women serving time in a USA prison (Arkansas, Georgia, Kentucky, Mississippi, and Tennessee), death anxiety was negatively correlated with self-reported health and self-reported mental health, but it was positively correlated with: age, involvement in social activities, number of health problems and anxiety in general.

But some old inmates may not die in prison. Instead they may regain their freedom after serving their sentence, being released into the community. Even when such older former prisoners may have a family that is willing to help them, support from the community will be most likely also required. Maschi et al. (2013) propose an integrative approach to the recovery of older former prisoners that is based on the re-acquisition of the lost personal agency leading to greater individual empowerment, capacity to decide autonomously, greater well-being and a more successful integration into the community. These objectives are difficult to achieve unless the individual enjoys the support of an "enabling social environment" (Maschi et al. 2013) and of aid coming from the structures of government.

Several rehabilitation programs specifically targeted to older prisoners who have regained their liberty are available around the world, such as: the *Senior Ex-Offender Program* (SEOP) in the USA, the *Resettlement and Care for Older Ex-Offenders* (Recoop) and the *RESTORE 50+ Program* in England, the *Reintegration Effort for Long-term Infirm and Elderly Federal Offenders* (RELIEF) in Canada.

The family can be important not only in making life in prison more bearable for the old inmate, but also after his/her release into the community. On the other hand, we may also consider the situation of old relatives (parents, grandparents) of people who are serving time in jail. What happens, for instance, with grandparents who have taken custody of their grandchildren because a daughter or a son is in prison? Dressel and Barnhill (1994) investigated this issue in the USA collecting data from 8 grandmothers and 21 grandchildren and reported the economic challenges grandmothers were facing and also the difficulties of dealing with the growing child without additional help. The availability of support groups can provide some help to grandmothers, but not in all cases; some of them may remain overwhelmed by ongoing concerns over their grandchildren's unruly behaviours, severe economic problems that they may be experiencing, and conflict with the imprisoned daughter (or son) (Dressel and Barnhill 1994).

We wish to conclude this section with an extreme example of grandmothers concerned about their grandchildren whose mother was taken into custody. This is the case of the *grandmothers of Plaza de Mayo* from Argentina. Unlike the cases we have seen so far, this situation does not involve common criminals doing time, but political dissidents who were taken by military or security services personnel and who were killed. In some cases their children were also taken and given up for adoption.

On the 24th of March 1976 the Argentinian democratic government of Isabel Perón—who, as vice-president, had succeeded her late husband president Juan

Domingo Perón—was ousted in a military coup led by Lt. General Jorge Videla. The military dictatorship that ensued only ended seven years later, in 1983, following the Argentinian defeat in the Falklands/Malvinas war. During the period of military rule the regime was involved in the brutal repression of political dissidents and it is estimated that about 30,000 people disappeared. Around 30% of the disappeared were women and 3% of them were pregnant (Argentine National Commission on the Disappeared 1986), but before they were killed they were allowed to give birth and the children given up for adoption to especially selected families regarded as "good" by the military regime, with the idea of educating the children in such a way that they would not grow into "subversives" (Arditti and Brinton Lykes 1992).

In 1977 the *Association of the Grandmothers of the Plaza de Mayo* was formed with the objective of seeking the restitution of more than 400 children who had been either kidnapped or were born in captivity. Note how the *Grandmothers* organised their association when the military were still in power, this means that they were facing grave peril to their lives as they secretly met in various places with the objective of organising the search of their grandchildren and return them to their legitimate families. Arditti and Brinton Lykes (1992, p. 464) describe how the *Grandmothers* "learned, on their own, to prepare writs of *habeas corpus* to present to the judicial system, and in January 1978 they made their first public international appearance by sending a letter to Pope Paul VI, asking him to intercede on their behalf to find out the fate of their missing relatives. The letter, which they sent by regular mail, was never answered".

The *Grandmothers* are a remarkable example of not only love and devotion to their grandchildren but of active social involvement in a situation highly risky to their life, of fighting for justice and against impunity, challenging the code of silence dictated by fear, and rejecting what Arditti (2002) refers to as the "numbing" notion of forgetting. In doing so they are also challenging many stereotypes about the elderly. We mentioned in this chapter some of the common stereotypes that allegedly characterise older people: mental limitations, decreased flexibility and capacity to learn, passivity, withdrawal or disengagement from society, unproductive, ineffective, helpless, a burden to society, self-oriented, irrelevant, vulnerable. We can only wonder what the *Grandmothers of the Plaza de Mayo* would make of all that.

The effort of the *Grandmothers* has produced some successes, with about 100 children having recovered their original identity. In 2005 one of those children, Horacio Pietragalla, published the following text in a Buenos Aires newspaper, telling the story of the recovery of his lost identity:

> Today I lead my life without doubting, and for this I can only thank my biological parents, my only and genuine parents, and the Abuelas de Plaza de Mayo [...]. The Abuelas [is] today an association that not only searches for grandchildren, but also carries out education in the right to identity (Gatti 2012, p. 353).

But not all grandchildren have been found. In 2009, the old and frail "Chicha" Mariani sent a touching email through the Internet, a last desperate attempt to transmit her legacy to her granddaughter Clara Anahí who disappeared 3 months after her birth, in 1976:

> I want to tell you that your paternal grandfather dedicated himself to music while my calling was the visual arts; that your maternal grandparents dedicated themselves to the sciences; that your mother loved literature and that your father was a graduate in economics [...] Some of all that will be present in just your own inclinations, because, in spite of your being brought up in a different home, one carries the genes of one's forebears inside oneself. (Gatti 2012, p. 353).

In sum, the percentage of older people in the prison inmate population has been on the increase, but the population has constantly remained heavily male biased. There are many negative aspects of ageing in prison, including issues of mental health and an accelerated ageing process. Remarkably, there are also some positive sides, especially in the case of poor old people who lived in conditions of destitution outside and who therefore may find a level of relative security in custody. In addition, older prisoners may be treated with some respect by younger ones if they are recognised as knowledgeable in the art of surviving inside the prison environment. Still, old inmates do face many health problems. The life of old inmates can be improved with various programs, often involving a focus on spirituality, at least in the USA system. Old grandparents taking care of the children of daughters or sons serving time in prison face severe hardships due to the burden of care of raising a child at an old age, but such burden probably pales in comparison to the angst of looking for a grandchild born in prison and lost under tragic political circumstances.

We conclude this chapter with a review of the challenges of caring for older people and the caregiver burden associated with them.

6.11 Caring for Older People and Caregiver Burden

As we become older and our personal capacity to take care of ourselves continuously diminishes, we reach a point in our life when we may require a degree of care. Such care may be minimal (e.g. somebody calling or visiting from time to time just to make sure that everything is fine) or intense (fully institutionalised care). We start this last part of the chapter with some considerations about the challenges of caring for old people, to end with a review of caregiver burden and the strategies that can be adopted to alleviate it.

6.11.1 Caring for Older People

There is a point in their life when older people are likely to need a significant degree of care, whether it is at home or in an institution. Assistance is required whenever the individual assets or capacities (physical and mental) to live an independent life are overrun by the deficiencies (Rockwood et al. 1994). Such elderly usually require assistance with their activities of daily living (ADL), having become too frail to be able to be completely independent.

Who are the caregivers of such elders? Most care provided to older people comes from family members (Sanborn and Bould 1991). When care cannot be provided by family anymore, due to the worsening conditions of the elder, ageing of the family caregivers or competing demands on caregivers' time, then institutionalised care will be required (Wolff and Kasper 2006; Nolan et al. 2009a).

Ageing in place usually relies on care commitment from the part of the direct family (e.g. spouse, children), which tends to be the most important source of caregiving for most older people. In fact, with an ageing population the role of the family is becoming even more important.

More female family members than males tend to be involved in caregiving, wife and daughters in particular, although the number of males has been increasing in recent years (Sanborn and Bould 1991; Schumacher et al. 2006; Family Caregiver Alliance 2017), and the older the caregiver becomes the more likely it is that other members of the family will step in to lend a hand (Wolff and Kasper 2006). Interestingly, the situation seems to be reversed among LGBTI people, with male caregivers providing more help than females (Family Caregiver Alliance 2017).

The *Family Caregiver Alliance* (2017) have recently published a list of characteristics of caregivers for elderly people. They define an *informal caregiver* (of which family caregivers are a specific example) as an "unpaid individual (for example, a spouse, partner, family member, friend, or neighbor) involved in assisting others with activities of daily living and/or medical tasks." On the other hand, *formal caregivers* are "paid care providers providing care in one's home or in a care setting (day care, residential facility, long-term care facility)". Carretero et al. (2009) described the characteristics of the most frequent informal caregivers as: family members, usually females (wife or a daughter) between 45 and 65 years old, unemployed or working part-time, with a low educational and socio-economic level.

Care therefore tends to be mainly provided by a person who is emotionally close to the elder, which is represented in Fig. 6.14 as a darker inner oval. Such elder-caregiver dyad is embedded within a larger care network which also provides aid, but usually not as regularly as the principal caregiver does.

The diversity of helping patterns reflecting a gradation of commitment from the part of various types of caregivers that we have seen in Fig. 6.14, can be further expressed in the intensity of care, as it is shown in the five types of help suggested by Matthews and Rosner (1988):

- *Routine*: regular assistance with activities of daily living.
- *Backup*: predictable help that is expected to be available on specific occasions.
- *Circumscribed*: predictable help that is bounded (e.g. helping with financial activities).
- *Sporadic*: help provided at the convenience of the helper.
- *Dissociation*: the individual is unavailable to help.

In Fig. 6.14 the level of routine help increases with the darkness of the colour, whereas the relative degree of dissociation increases with its lightness.

Nolan et al. (2009, p. 139) use a different classificatory approach to the care provided by the family, focusing on qualitative differences in the forms of help,

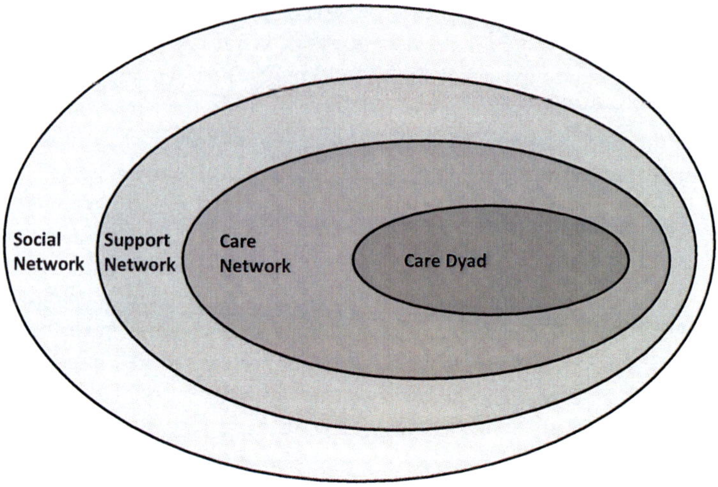

Fig. 6.14 Varying degrees of embeddedness in the social network of care for older people: the darker the colour the higher the level of personal commitment to care (redrawn from Hughes and Heycox 2010)

rather than a gradation in quantitative commitment. In this classification care can be described as:

- *Anticipatory*: speculations about what the family would do in the case an old member would require help.
- *Preventative*: subtle support provided with the intention of preventing major difficulties from arising.
- *Supervisory*: more specific support that is kept hidden from recipient.
- *Instrumental*: typical support associated with the caring role.
- *Protective*: care aimed at preventing the recipient from becoming aware of his/her frailty.

After the elder may start living in a nursing home, care from family becomes mainly *Preservative*, which includes preserving the personal identity of the elder.

Although family and other voluntary caregivers do not receive a monetary reward for their effort, the work they provide does have an economic value. On average, caregivers may spend 13 days of the month performing helping tasks such as shopping, preparing food, various forms of housekeeping, doing the laundry, transporting the elder, and taking care of the medication they need. It is also calculated that they spend 6 days per month on feeding, dressing, grooming, walking, bathing, and assistance toileting the elder under their care; and 13 h/month doing research on care services or gathering information on diseases, coordinating visits to the doctor and managing the finances of the elder. The *Family Caregiver Alliance* (2017) calculated the value of such work at U.S. $470 billion in 2013. This money is obviously saved by the system of public care institutions.

Pinquart and Sörensen (2011) carried out a meta-analysis to compare the caregiving patterns of spouses, adult children and children-in-law. Children seem to step in with caregiving duties when the spouse is too old to withstand the stresses of caregiving. Caregiving spouses were less likely to be female in this study as compared with caregiving children, whereas there were more females among the caregiving children-in-law than among the caregiving children.

In some cultures where the concept of familism is still strong, children and children-in-law are the members of the family who are expected to provide most of the care for the elders. We introduced the concept of familism in Chap. 3 and it can be viewed as a strong sense of loyalty and solidarity with the members of the family, leading to the provision of care for relatives in need of assistance. Rosenthal Gelman (2014) has studied familism as it relates to the care of elders among Latinos in the USA. The challenges faced by older Latinos are diverse and they include greater needs for social services as they tend to be poorer than other members of society, this also leads to poorer health conditions, ethnic discrimination and facing communication barriers due to their inadequate knowledge of English. Overcoming those problems often requires the active help of better socially integrated and younger members of the family. Even within a broad tradition of familism, however, there are cultural differences. For instance, although across cultures care is mainly carried out by spouse or children (wife and daughter in particular) in some Asian cultures a great degree of care is expected from the daughter-in-law (e.g. Kumamoto et al. 2006).

Cordella and Rojas-Lizana (2019) studied the effects of familism on caregiving expectations in 19 Spanish-speaking older people (65–84 years old; 52.6% women) who were independently living in Brisbane, Australia. The interviewees had lived in Australia for a variable period of between 22 and 40 years (mean = 31.7 years). They had decided to leave their Central or South American native country mainly to escape political and/or financial problems. Familism was one of the four major themes emerging from the interviews, the others were "feelings of vulnerability and fear of ageing in a foreign land", "a pervasive fear to 'end up' in an aged care facility", and "the importance of religion as a coping mechanism". The interviewees pointed to inter-generational tensions regarding familism, as younger generations who grew up in Australia behaved more individualistically than expected from familism, leading to a loosening of the social bonds among the members of the extended family. Showing concerns for the elders was expected to be not so much an externally imposed "obligation", but an internal "moral duty" that could be expressed by as simply as keeping in touch, even if it was by phone. Receiving demonstrations of love and affection from their children was important for the elders. Such show of affection was needed by older people especially in virtue of their many vulnerabilities, both material and emotional. Loss of contact with family, for instance, increased the fears of "ending up" in a nursing home. The fear that in the future they may not receive care and support from their family was cause of much distress in these immigrant elders. They tried to cope with such distress by using both active and passive coping strategies. Active coping strategies included increasing connection with family members and seeking community support, whereas passive strategies included justifying family behaviour and resigning themselves by seeking religious comfort (Rojas-Lizana and Cordella 2020).

We mentioned above that caregiving tends to be performed more by females than males, with some recent trends showing a more levelled performance of caregiving between the sexes. We also mentioned that males provide more caregiving than females among LGBTI people. In Williams and Dilworth-Anderson's (2002) study carried out in the USA among African Americans, caregivers were typically daughters (49%) who were caring for their mother, and therefore the vast majority of primary caregivers were women (82%). In a study of caregiving also carried out in the USA, Navaie-Waliser et al. (2002) found that in their sample women caregivers were older than men caregivers, they were also more likely to be African American, married, better educated but unemployed. Table 6.13 shows their results for women and men caregivers. Overall, caregiving is carried out more by people younger than 65, but the proportion of women is higher than that of men in the older category (≥65). Older care recipients are more frequent and are equally distributed across both sexes.

Overall, women caregivers perform various activities (both ADL and IADL) more frequently than men, with the difference being statistically significant or highly significant in most cases (Table 6.14). During the performance of caregiving, women caregivers also experience more challenges than men caregivers, such as providing more intensive care, not receiving help from institutional or formal caregivers, facing difficulties in the provision of care and experiencing unmet needs during the provision of care.

Apart from the direct provision of support, the family can also be of great help in terms of detecting early signs that something is not going well with an elderly relative. For instance, family members may be able to notice early symptoms of Alzheimer's disease or other forms of dementia and refer the patient to a specialist for evaluation (e.g. Mahoney et al. 2005). After the diagnosis has been produced, the elder could be initially cared for in the home environment. In a review of home

Table 6.13 Characteristics of informal caregivers and care recipients in a sample from the USA

Characteristic	Women *n* (%)	Men *n* (%)
Informal caregivers		
Age (years)		
<65	482 (85)	438 (90)
≥65	83 (15)	48 (10)
Care recipients		
Age (years)		
<65	202 (36)	163 (34)
≥65	359 (64)	321 (66)
Sex		
Male	188 (33)	181 (37)
Female	383 (67)	311 (63)

Data are from Navaie-Waliser et al. (2002)

Table 6.14 Activities and challenges of caregivers

Variable	Women *n* (%)	Men *n* (%)
Caregiver activities		
Type of care provided		
Activities of daily living (ADL)		
Bathing	183 (35)	68 (15)**
Dressing	265 (50)	143 (32)**
Feeding	103 (20)	61 (13)**
Incontinence	110 (21)	59 (13)**
Transferring	219 (42)	177 (39)
Ambulating	204 (39)	133 (29)**
Instrumental activities of daily living (IADL)		
Shopping	503 (88)	398 (81)**
Housework	425 (74)	331 (67)
Preparing meals	385 (68)	244 (50)**
Transportation	436 (76)	368 (75)
Telephone calls	409 (72)	221 (45)**
Managing finances	318 (56)	193 (39)**
Government program assistance	191 (34)	121 (25) 1.29
Complex care		
Dressing changes	128 (25)	57 (12)**
Medical equipment use	93 (18)	50 (11)*
Assist with ≥2 medications	463 (90)	337 (76)*
Provides ≥20 h/week of care	248 (45)	165 (35)
Has been providing care for ≥1 years	450 (79)	377 (77)
Caregiver challenges		
Provides more intensive care	244 (47)	141 (31)**
Has no help from formal caregivers	440 (77)	406 (83)
Has difficulty providing care	296 (52)	155 (32)**
Has unmet needs with care provision	124 (22)	66 (13)**

*$P < 0.05$, **$P < 0.01$. Data are from Navaie-Waliser et al. (2002)

care for dementia patients, Verbeek et al. (2009) suggested that small, home-like environments could indeed be beneficial to old people with dementia.

Beyond the family, care for the elders may be also recruited from other sections of the community, such as friends and neighbours (Wells et al. 2009), but elders may eventually require a more professional style of care. Institutional and community workers' help could be provided even when the elders stay at home, where their needs could be better satisfied in a familiar environment and where they can be more independent (Hughes and Heycox 2010). However, the more the elder becomes frail and in need of continuous support and attention, the higher the stress will be on family caregivers (Zarit 2002). At one point the burden may be such that the possibility of institutionalising the elder will be considered seriously by the family (Gaugler et al. 2003; Wolff and Kasper 2006).

Although institutionalised elders are supposed to enjoy professional care in a safe and stimulating environment, they are not always happy with their life and the treatment they receive from staff. Complaints about negative attitudes and incapacity to fully understand the needs of the elders are common (e.g. Clarke et al. 2003). This has led to the introduction of *person-centred care* which aims at addressing the needs of the elders as directly expressed by them. For instance, elders may be encouraged to engage in biographical conversations with both their caregivers and other elders, with the aim of allowing staff to achieve a better understanding of the elders' life, their current reality and needs (e.g. Clarke et al. 2003).

Within an institution the adoption of person-centred care is pivotal for the well-being of the elders. Nay et al. (2009, p. 109–110; see also Davis et al. 2009; Koch et al. 2009) suggested a list of approaches to be adopted in order to better achieve person-centred care, that we report here with only slight modifications:

- Acknowledging the individual as an experiencing person in spite of physical and mental limitations.
- Offering and respecting individual choices and preferences.
- Using the person's history and biography in care.
- Focusing on the abilities rather than the disabilities.
- Supporting individual rights, values, and beliefs.
- Providing unconditional positive regard.
- Interpreting all behaviour as meaningful.
- Maximising potential and providing shared decision-making.

Therefore, whether it is institutional or informal, by respecting individual choices, focusing on the abilities, supporting the elders' values and beliefs, fostering meaningfulness and maximizing potential, care should aim at providing the elder with an opportunity to experience fulfilling ageing.

Although ageing people tend to be seen as recipients of care, they are often also the providers of care, especially to family members who may be even older and more frail than they are (e.g. oldest-old parents). Or the recipients of care from old people could be younger members of the family (e.g. offspring) who may suffer from serious disabilities (Greenberg et al. 1993). Outside the family, old people may provide volunteering care for elders requiring special attention, whether at home or in an institutionalised setting. However, there is also the option for old people being paid for their care services. Hwalek et al. (2008) analysed this issue in the USA. They carried out a survey of 696 low-income, 55 years old and older participants and their results indicate that there was interest in such paid work assisting other elders, although respondents preferred to work in home-care settings rather than in nursing homes. They were confident that their health was good enough to withstand the stresses of such work, but they felt that they had to confront the resistance and prejudice of many employers who believe that older workers may carry additional costs and be less willing to learn to use new technology.

In the next section we turn to the issue of caregiver burden and how to alleviate it for the benefit of both elders and their caregivers.

6.11.2 Caregiver Burden

Although caring for a loved one could be a happy and rewarding experience, the reality in many cases is that providing care for older people can be costly in more ways than one, with caregivers likely to experience a degree of burden, even exhaustion, that can be physical, mental and often also financial (e.g. Haley 2003). In the analysis of the provision of care for elderly people, two central concepts should be distinguished: *caregiver burden* and *caregiver burnout*. Following Kasuya et al. (2000, p. 1), caregiver burden can be defined as:

> the strain or load borne by a person who cares for an elderly, chronically ill, or disabled family member or other person. It is a multidimensional response to physical, psychological, emotional, social, and financial stressors associated with the caregiving experience. The caregiver's perception of the burden, rather than the perception of other family members or healthcare providers, determines the impact on his or her life.

On the other hand, caregiver burnout "is the progression of caregiver burden to the point where the experience is no longer a viable or healthy option for either the caregiver or the person receiving care" (see also Carretero et al. 2009, for a review). Thus, the experience of caregiving may go through a process of escalating stress that from burden may lead to burnout.

Caregiver burden has been traditionally described in either subjective (e.g. feelings, concerns) or objective terms (e.g. financial costs, symptoms of stress and depression) (Brodaty and Donkin 2009). But as the above definition indicates, burden is a multidimensional phenomenon where subjective experiences, lived in the course of providing care for frail elders, also come hand in hand with the objective consequences of such care (e.g. Caserta et al. 1996; Nolan et al. 1996; Jervis et al. 2010).

A diversity of factors affect the degree of caregiver burden and these include the degree of impairment suffered by the elder receiving care (e.g. Pallant and Reid 2014), the caregiver perception of burden, and the actual impact of such burden on the caregiver's health (Poulshock and Deimling 1984); the financial capacity of the caregiver, the amount and quality of social support available to the caregiver, but also gender effects, personality characteristics (e.g. self-mastery, optimism, self-esteem), the coping strategies used, and also the quality of the emotional relationship between the caregivers and the elder who is in their care (Schulz and Martire 2004).

Thus, caregiver burden is simultaneously determined by both internal/subjective factors and external factors, which in fact interact. For instance, external social support may help decrease the burden of care; the same can be achieved through an internal sense of optimism, but optimism may be further enhanced by knowing that one counts with some help as a caregiver. Therefore, effective interventions to support caregivers should target not only the external sources of stress but also the psychological capacity of the caregiver to cope with such stress (e.g. Campbell et al. 2008). Campbell et al. (2008) mention the caregiver's sense of "role captivity" as an important causation of burden and also burnout. Role captivity could be alleviated with a combination of external social support bringing relief to the caregiver, and by strengthening the caregiver's psychological resilience.

Several instruments have been designed to measure caregiver burden, the instruments most frequently used include the: *Caregiver Strain Index* (CSI), *Relatives Stress Scale* (RSS), *Caregiver Burden Scale* (CBS), *Sense of Competence Questionnaire* (SCQ), *Caregiver Burden Inventory* (CBI), *Caregivers Reaction Assessment* (CRA), *Zarit Burden Interview* (ZBI), *Self-Rated Burden Scale* (SRB), *Cost of Care Index* (CCI), *Carer Quality of Life* (CarerQol), and *Bakas Caregiving Outcomes Scale* (BCO) (Chou et al. 2003; Brouwer et al. 2006; Bachner and O'Rourke 2007; Rigby et al. 2009). However, works evaluating some of such instruments suggest that the CSI and the SRB may be especially useful (van Exel et al. 2004), as are multidimensional instruments such as the CCI and CBI (Caserta et al. 1996; Chou et al. 2003).

We have already seen that the immediate family is usually the initial and often major provider of care for the elderly. Why would family members provide care for an elderly relative in the knowledge that it involves some cost or burden to them? We could simply answer this question by saying that love for family, and principled, religious, and legal considerations of duty of care play a role, but the reality is a bit more complex than that.

Let us start with some theoretical considerations. Given that old relatives are not reproductive anymore, caring for them cannot be explained by the evolutionary theory of *kin selection* (Hamilton 1963), that involves the indirect transmission of copies of our genes through helping the reproductive effort of a close relative. Instead such costly care may be explained by reciprocation (e.g. *reciprocal altruism*, Trivers 1971). That is, I help you now and pay a cost for it, you help me next and you pay a cost for it too. In the sociological and anthropological literature such reciprocation is understood within the framework of *Social Exchange theory* (e.g. Levi-Strauss 1964). Following Molm and Cook (1995), the four central assumptions of *Social Exchange theory* include: people depend on each other to achieve desirable objectives (*interdependence*); people's behaviour is such that it is aimed at increasing valued outcomes and decreasing undesirable outcomes; people engage in cooperative exchanges with specific partners over time; and outcomes follow a principle of satiation, or diminishing marginal returns with increasing costs. But what kind of benefit can the elder provide to the caregiver in return for the care received, to make reciprocation a viable option? Elders who are still in good physical and mental condition can reciprocate to their caregivers in many ways: babysitting and helping with house cores, lending money, and so forth. This assumes a degree of symmetry in the exchange: that is, that what is received at one time is approximately as valuable as what is given at a different time. However, power asymmetry between interacting people may skew the extraction of benefits preferentially in one direction, despite increasing costs for the donor of help. In this case, although the burden of caring for an old person is likely to be experienced more strongly when perception of costs (physical, psychological) overwhelms the perception of benefits in the donor, the donor may withstand the burden, at least for a while, following a response to power asymmetries. This is seen in many cultures in the relationship between old parent and adult child (e.g. Trommsdorff 2006). But a threshold, although shifted, will remain.

Social Exchange Theory assumes that people cooperating with each other are rational players who weigh the costs and benefits of their actions. The benefits may be immediate, or they may be delayed in time. When benefits are delayed one of the players is incurring a "debt" that is expected to be paid at some point in time into the future (Thiede Call et al. 1999). It is often the case that reciprocity (simultaneous or delayed) is mandated by social norms that may be legal, or traditional, or dictated by religious beliefs; some of such norms may weaken the more distant the family relationship is between elder and caregiver (lower degree of kinship).

Hence, when older people find themselves in the inability to reciprocate, they may feel discomfort unless the expectations for reciprocation decrease as the elder's general condition deteriorates with age. The same occurs with family caregivers. When the condition of the elder deteriorates to the point where she or he needs formal institutional care, family caregivers may feel a sense of guilt if they are incapable of providing such care (Thiede Call et al. 1999).

Although we mentioned that social norms dictating reciprocation may weaken as the genetic distance between relatives increases (lower degree of kinship) there are some exceptions affected by culture. In some cultures, the daughter-in-law is expected to provide a substantial amount of care to elderly parents-in-law but, presumably, this interaction is stabilised by the duties of care that the husband must also show towards his wife.

Therefore, when inter-generational reciprocity operates, elders tend to return the favours received from younger members of the family through helping in various ways. When elders cannot immediately reciprocate, they may feel guilty (Dwyer et al. 1994), but some elders may reciprocate in a delayed manner through allocation of inheritance.

At the same time, the younger generations have already accrued a debt with their parents from the care the parents provided when they were children, and hence some provision of help from their part is expected (e.g. Jervis et al. 2010). A final possibility in this spectrum of reciprocations is a phenomenon that could be termed the *Arrow of Time reciprocation*, this involves adult children helping their old parents without expecting any immediate reciprocation from them, in the expectation that the third generation, the grandchildren, will follow the same behaviour and help their parents in turn once the latter are old and in need of care.

Over time, however, the burden of caring for an older parent experiencing a rapidly deteriorating body and/or mind may be significant, potentially exposing the offspring to caregiver burnout (Dwyer et al. 1994).

The negative effects of burden on the caregiver include the development of depression and anxiety, exhaustion, emotional mood disorders, deterioration of physical and cognitive capacities, increased levels of psychosomatic and immunological disorders, cardiovascular issues, and increased dependency in the performance of activities of daily living. The levels of chronic stress caused by the burden of caregiving may at some point lower the threshold for the development of some diseases through the immune-dysregulating effects of such stress (Vitaliano et al.

2003; Schulz and Sherwood 2008). This overall clinical picture may ultimately lead to greater risk of premature death in the caregiver (Carretero et al. 2009).

In addition, caregivers may also risk seeing their employment situation deteriorate, whether they have a job and want to retain it or they are seeking employment. Lack of a job will obviously affect the financial situation of the caregiver thus further increasing the burden of care (Carretero et al. 2009).

The negative effects of caregiving for the caregiver have been recently summarised by Adelman et al. (2014, see also Schulz and Sherwood 2008), as seen here in Table 6.15. Higher burden tends to be experienced by women caregivers, those

Table 6.15 The epidemiology of caregiver burden

Domain/feature	Risk factor and/or outcome status of caregiving demographic
Female sex	Risk factor for becoming a caregiver and for caregiver burden (>twofold rate of high caregiver burden compared with men)
Low education	Associated with the highest levels of caregiver burden in various medical conditions
Cohabitation with care recipient	Associated with caregiver burden
Clinical outcomes of caregiver burden	
Mortality	Caregiver burden identified as an independent predictor of caregiver mortality with a 63% increased risk of death
Weight loss	Caregiver burden associated with caregiver weight loss
Poor self-care	Higher incidence of low self-care behaviours and of ignoring self-health
Sleep deprivation	Especially common in caregivers of dementia patients who experience disruption to the sleep-wake cycle and may also be a risk factor for caregiver burden
Psychosocial	
Depression and depressive symptoms	Identified as risk factor for and outcome of caregiver burden
Coping strategies	Using a smaller number of coping strategies (e.g. seeking advice, exercising) is associated with caregiver burden
Perceived patient distress	Greater likelihood of experiencing caregiver burden in caregivers who perceive distress in the care recipient
Social isolation and decreased social activity	Risk factors for caregiver burden and outcomes of caregiver burden
Anxiety	A possible effect of caregiving in caregivers
Suicide	A potential outcome of caregiver burden
Caregiving context	
Caregiving time and effort	Duration in the caregiving role and hours spent caregiving associated with caregiver burden
Financial stress	A risk factor for caregiver burden and an outcome of caregiver burden
Lack of choice	Lack of choice in becoming a caregiver is associated with caregiver burden
Inability to continue regular employment	Caregiving may preclude regular employment; when more than one potential caregiver exists, the family member who is not regularly employed is more likely to assume the caregiving role

Modified from Adelman et al. (2014)

with lower education, those cohabiting with the elder being cared for, those who are trapped in the role and have no choice, and those who are under financial or other stresses. Burden may lead to various medical conditions in the caregiver, to the point of decreasing life expectancy as already mentioned.

Different caregivers may experience different levels of burden according to their personal characteristics and the level of effort put into the task of providing care for the elder. For instance, in a study carried out by Pinquart and Sörensen (2011) spouse caregivers reported suffering from higher physical and financial burden, and higher levels of depressive symptoms along with a greater degree of relationship strain than adult child caregivers. Adult children, however, reported more job strain and depressive symptoms than children-in-law.

Depression is a frequent effect of burden on caregivers (e.g. Butler et al. 2005; Schulz and Sherwood 2008; Adelman et al. 2014). Conversely, caregivers can experience considerable burden when they care for elders who suffer depression (e.g. Scazufca et al. 2002, and references therein).

Even more challenging than caring for depressed elders is caring for elders suffering from dementia, which involves a considerable level of stress for the caregiver, along with some psychiatric and physical risks, especially due to disruptive behaviours and inadequate social functioning of the elder, as well as the financial costs involved (e.g. Deimling and Bass 1986; Langa et al. 2001; Schulz and Martire 2004; see Etters et al. 2008, for a review).

Most of the older dementia sufferers mainly receive care from their spouse or their adult children when the spouse is not available, with adult daughters and also daughters-in-law more likely to provide regular care than adult sons and friends (Schulz and Martire 2004; Brodaty and Donkin 2009). Although caregiving is usually female-biased, the participation of males in the caregiver role has been increasing in recent times (Brodaty and Donkin 2009).

Caring for an elder diagnosed with dementia tends to be more stressful for the caregiver than caring for an old person who only has some physical impairment. Caregiving is further complicated by many old dementia patients also suffering from depression which further affects their cognitive capacities. This in turn leads to negative effects of the burden of care on the caregivers: depressive and anxiety disorders, low degree of life satisfaction, and increased health problems (see Schulz and Martire 2004, for a review). This process has been summarised by Schulz and Martire (2004) in their *Stress/Health model* of caregiving (Fig. 6.15), whereby the caregiver attending a dementia patient is under the effect of both direct stressors—determined by the condition of the elder—and secondary stressors that are associated with family conflicts and potential problems at work. Depending on the capacity of the caregiver to cope with such stresses a specific perception of the difficulties experienced in the situation will be produced, leading to a perceived degree of stress. The caregiver may respond to such perceived stress in several ways, but if the response is inadequate the stress condition may lead to increased morbidity and, in extreme cases, higher probability of mortality. The column on the right of Fig. 6.15 lists several potential interventions that could be carried out on both the caregiver

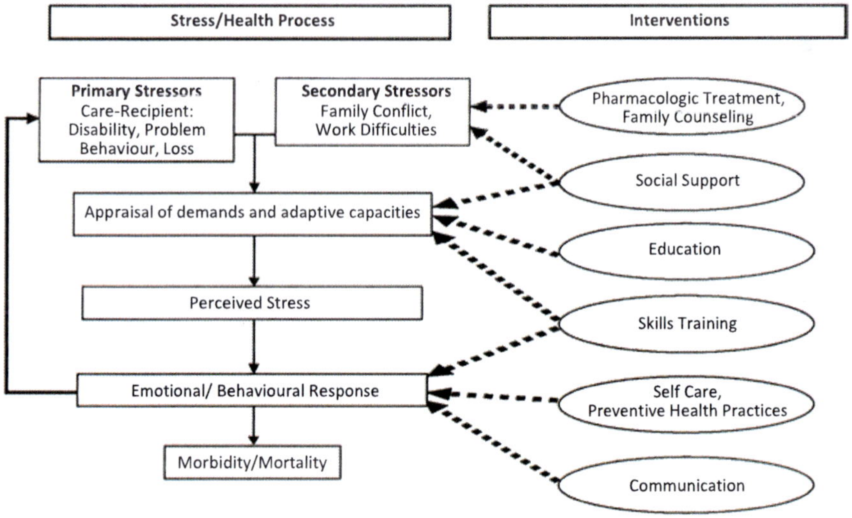

Fig. 6.15 Schulz and Martire's (2004) *Stress/Health model* as applied to caregiving. The model includes interventions to prevent the burden of care developing into health problems and potentially affecting the probability of morbidity and mortality in the caregiver

(e.g. skills training, communication) and her/his social milieu, such as family and workplace in particular (e.g. counselling).

Skills training is an important intervention to help caregivers. Training can include aspects of positive thinking about caregiving. Andrén and Elmståhl (2005, see also Brodaty and Donkin 2009) list the following positive aspects of the caregiving experience as indicated by caregivers themselves: "it is nice to see the person I care for clean, comfortable and well turned out", "it is nice when something I do gives the person I care for pleasure", "I get pleasure from seeing the person I care for happy" and "maintaining the dignity of the person I care for is important"; "caring makes me feel needed and wanted", "the person I care for is appreciative of what I do", "I feel that if the situation were reversed, the person I care for would do the same for me" and "caring is one way of expressing my love for the person I care for". Undoubtedly such positive thinking can help alleviate the burden of care, to some extent at least.

Although the caregiving experience could be potentially seen as positive, we have seen that depressive symptoms are common in caregivers of dementia patients and some may also develop negative health behaviours such as increased alcohol consumption, lack of exercise, poor sleeping patterns, smoking, and increased weight (Gallant and Connell 1997). Almberg et al. (1997) investigated the level of burden and burnout in 52 Swedish family caregivers of elderly people suffering from dementia. Burnout was more frequently reported in older wives and daughters, with some daughters-in-law also displaying a degree of burnout; whereas sons were less burnt out than daughters. Such burnout could be explained by the degree of effort put into caring for their old family member, a limited social life, health deterioration and poor prospects that the effort of caring would produce significant positive results in the elder. In short, *high-effort-with-frustration* seems to be a condition that facilitates burnout.

The burden of caring for dementia patients can also negatively affect the sense of coherence of caregivers and their perceived health through the mediation of feelings of isolation and disappointment, which can weaken their coping abilities (Andrén and Elmståhl 2008).

Black and Almeida (2004) reviewed the association between the behavioural and psychological symptoms of dementia and burden of care, concluding that indeed worsening behavioural and psychological aspects of dementia are associated with increased burden of care and also increased distress and depression in the caregiver, and higher probability of patient's institutionalisation. Nevertheless, some behavioural and psychological symptoms of dementia may be a greater cause of caregiver burden than others, and differences may also occur across cultures. For instance, in a study carried out in Japan, Matsumoto et al. (2007) found that caregivers living with dementia patients had the highest scores of distress in association with expressions of agitation and aggression by the patient. In addition, the stronger the symptoms of delusion, apathy/indifference, irritability/lability and aberrant motor behaviour in the elder, the greater the distress experienced by the caregiver.

Papasravrou et al. (2007) studied a sample of 172 family caregivers (mostly a daughter (48.3%), but also husband/wife (41.3%), and to a lesser extent son (5.8%) and others (4.1%)) and the elder dementia patients under their care in Cyprus. Results indicate that 68% of caregivers were highly burdened and 65% of them also manifested symptoms of depression.

Thommessen et al. (2002) investigated the burden of care in spouses of old people suffering from a mild degree of dementia, Parkinson's disease and also stroke in Norway. They identified the following major manifestations of burden: disruption of household routines, difficulties with the organisation of holidays, limitation of social life, and sleep disturbances. The level of burden increased the more cognitively limited the patient was, with wives suffering the brunt of caregiver burden.

In a study carried out in the USA, Chumbler et al. (2003) measured caregiver burden in 305 family caregivers of 70 years old and older memory-impaired individuals. More caregiver burden was experienced by adult children than by more distant relatives. In other studies, it is the spouse who experiences more caregiver burden (e.g. depression, fatigue) than children, but there is variability in this trend. Whereas yet other studies specifically point to daughters suffering high levels of caregiving stress (Chumbler et al. 2003 and references therein). Powell Lawton et al. (1991) investigated the level of burden and satisfaction in 285 spouses and 244 children caregivers of elderly people diagnosed with Alzheimer's disease in the USA. Their results are summarised in Table 6.16.

Regarding the spouse, Powell Lawton et al.'s (1991) results indicate that the more impaired the elder is the less help the spouse provides, presumably because in that case more help is recruited from elsewhere. By doing so, the spouse decreases the personal burden of care and low burden lowers the probability of developing

Table 6.16 Correlation coefficients among variables for spouses and adult child caregivers

	IPS	CH	HFO	HBC	CS	CB	PA	D
Impaired person symptoms	–	**0.10**	**−0.05**	**−0.17***	**0.05**	**0.27***	**0.03**	**0.19***
Caregiver health	0.25*	–	**−0.02**	**−0.07**	**0.07**	**0.36***	**0.30***	**0.46***
Help from others	0.01	−0.06	–	**0.14***	**0.08**	**−0.02**	**0.04**	**0.01**
Help by caregiver	0.27*	−0.19*	0.16*	–	**0.08**	**−0.20***	**0.09**	**−0.05**
Caregiver satisfaction	0.09	0.03	0.11	0.22*	–	**0.33***	**0.32***	**0.26***
Caregiver burden	0.34*	0.41*	0.03	−0.41*	0.11	–	**0.26***	**0.63***
Positive affect[a]	0.01	0.26*	0.04	0.17*	0.12*	0.36*	–	**0.42***
Depression	0.20*	0.51*	0.06	−0.18*	0.10	0.61*	0.43*	–

*$p < 0.05$

The correlations for spouses are above the diagonal and in bold, whereas those for adult children are below the diagonal

[a]Psychological well-being. Slightly modified from Powell Lawton et al. (1991)

depression. Spouses are willing to contribute more to caregiving the better their health and psychological well-being is and the more help they receive from others.

Help provided by children increases with the level of impairment of the elder and with the level of help provided by others, but it decreases when their own health is better. The greater their level of satisfaction with the caregiving tasks, the more they help. The burden of care increases with the level of impairment of the elder and with the caregiver own health, and they also help more the less burdened they feel. Their psychological well-being increases with: their health, the help they receive from others and their own levels of satisfaction. Children's depression increases with the level of impairment of the elder, the caregiver health and the level of caregiving burden.

Comparing spouse and children caregivers, the results from Powell Lawton et al.'s study indicate that they both increase their help as the condition of the elder deteriorates, and they both suffer from the effects of the burden of such care, expressed in the form of depression in particular. Help provided by others is not associated with a decrease in the help given by both spouse and children, rather it seems to spur more help from these close family members.

The burden of care is also mainly experienced by close family members in the care of adults with schizophrenia, just as it was the case for elders with dementia. Schizophrenia is a condition that can develop early on in life, around the age of 20 on average (Faraone et al. 1994), and thus the level of family-accumulated burden into the older ages of the schizophrenic can be considerable. Magaña et al. (2007) studied the effects of caring for a schizophrenia patient on family caregivers in a sample of Latinos in the USA. Most caregivers were mother (58.8% of all caregivers), followed by spouse (16.4%), sibling (14.1%), father (5.8%) or other relatives (4.7%). About 40% of caregivers displayed the symptoms of depression (as compared with 12–18% in the general population of Mexican Americans). More caregiver depression was observed in association with a younger age of the caregiver, lower levels of education and worse mental condition of the patient. Caregivers' depression was associated with their own perception of the burden and with stigma.

Caregiving burden can also vary according to the sex of the caregiver. The subjective reactions of women to their caregiving tend to be more negative than those observed in men. Women also report higher levels of caregiver burden and psychological distress, along with lower morale. Their depression levels are higher than those of men and they express more negative feelings. This effect may simply result from women providing more care than men, but women also tend to suffer greater levels of guilt for being unable to provide appropriate levels of care; and they tend to be more affected by emotional strain than men, leading to greater desire to institutionalise the older person under their care (Barusch and Spaid 1989, and references therein). Therefore, the greater level of caregiving provided by women may lead to greater burden, that is aggravated by an increased sense of guilt when conflicting duties decrease the ability of women to provide care.

In a study carried out in the USA, Barusch and Spaid (1989) described how women experienced higher subjective caregiver burden than men. Such burden was especially associated with memory problems in the elder under their care. Younger caregivers also experienced higher subjective burden of care than older ones. With regard to coping with the burden of caregiving, men were better able to cope with situations of feeling angry, having arguments with the spouse, and worrying about future financial challenges. On the other hand, men tended to have difficulties in coping with the spouse being particularly dependent, and with people making excessive demands on them.

In a study of husband/wife caregivers of a spouse diagnosed with Alzheimer's disease that was carried out in the USA, Zarit et al. (1986) found that the burden of care was higher for women than men caregivers, but such difference tended to level over time. This was mainly due to women decreasing their burden of care over time as they tended to place their husband in institutional care.

In general, female caregivers seem to be more likely to engage in personal care activities such as feeding or dressing the elder, and household tasks such as cleaning and cooking than male caregivers (Kramer and Kipnis 1995). Other caregiving activities like helping with financial support and with access to public services are more gender balanced. On the other hand, help with work for the home (gardening, plumbing, painting, etc.) is more likely to be carried out by men. As a result of such daily effort, women tend to suffer greater burden of caregiving than men whose house duties are performed more sporadically (Kramer and Kipnis 1995; see also Miller and Cafasso 1992, for an early review). In their study carried out in the USA, Kramer and Kipnis (1995) described greater levels of burden in female caregivers, especially when they were caring for parents or in-laws. In Chumbler et al.'s (2003) study a total of 69% of caregivers were women. In contrast to male caregivers, female caregivers experienced higher blood pressure and had problems with nervousness.

The general female bias in caregiver burden can be explained by a confluence of factors, including gender-role socialisation throughout development, which is affected by current cultural expectations, but also labour market segregation that can affect who carries out home-based activities, including caring for older members of the family (e.g. Pinquart and Sörensen 2006). Pinquart and Sörensen (2006) carried out a meta-analysis of 229 studies on gender differences in caregiver burden concluding that female caregivers have higher degrees of behavioural problems, they spend more hours in caregiving activities, they engage in more diverse caregiving tasks and they suffer more caregiver burden, depression, and lower physical health and subjective well-being.

The greater the burden experienced by the caregiver the greater the chances that some negative effects may spill over onto the elder under their care. Forced institutionalisation is of course one obvious outcome, but before that happens there is also the potential for the elder suffering mistreatment (Carretero et al. 2009).

The best way to decrease caregiver burden is, from start, for the elder to be able to reach an old age in as good as a mental and physical condition as possible. This will stretch the capacity of the elder to be autonomous (Spillman and Pezzin 2000), thus decreasing the burden on caregivers. But when assistance is required, then caregivers can alleviate their burden by adopting personal strategies to cope and/or through social support (family, institutional). Personal coping strategies include a more positive redefinition of the experience of care and focusing on the acquisition of greater capacity for problem-solving (Deeken et al. 2003; Carretero et al. 2009; Nolan et al. 2009a). Additionally, psychological interventions on the caregiver, such as counselling, psychological therapies and participation in mutual help groups, education and training for the improvement of coping skills may also be of some help.

Adopting an optimistic and constructive attitude where beneficial effects of caregiving are made explicit and prominent is one of the personal strategies that can be used to cope with caregiving. Some beneficial effects of caregiving for the caregiver could include increased expertise in caregiving activities, that could then be used in the future; and a personal feeling of satisfaction and life fulfilment (Haley 2003; Carretero et al. 2009). The quality of marital relationships and affectional bonds could be also increased when one member of a couple needs care and that care is provided by the other member (e.g. Haley 2003). A caregiving experience could also potentially enhance personal strength, the sense of life meaning and spirituality, along with personal growth, self-acceptance, autonomy and gratification for the achievement of good results under the circumstances. After being involved in the care of seriously impaired elders, caregivers may re-evaluate their existence and perhaps start enjoying those small things in life that they might have previously neglected. Family relationships could be mended if they had been tarnished by past conflicts. The caregiver may even acquire a better knowledge of the process of ageing and use such knowledge to better prepare for his/her own future (Hogstel et al. 2005/6).

Social support is also a very effective coping mechanism (Zarit et al. 1980; Thompson et al. 1993; Garlo et al. 2010; Jervis et al. 2010), although it may be interfered with by family conflicts. Receiving relief external to the family usually requires the intervention of government or charitable, community-based organisations that may provide respite services (Thompson et al. 1993; Kumamoto et al. 2006). But when all strategies to relieve caregiver burden fail, the family may decide to transfer the elder to an appropriate aged care facility (Gaugler et al. 2003).

In what follows we analyse in some additional detail the ways in which social support from family, community and institutions can be used as an intervention to decrease caregiver burden in family members caring for frail elders, and therefore to decrease the chances that they experience burnout.

In a study of coping with the burden of caring for older people by female primary caregivers (mainly spouses and children, 63%) carried out in the USA, Wood and Parham (1990) found that social support was important in alleviating the burden of care. In fact, the overall burden was negatively correlated with household size. Ethnicity could also explain differences in caregiver burden that reflect differences in social support, as African Americans displayed both lower burden and more social support (from both family and friends) than Caucasians. In fact, even religion—that could be regarded as a personal coping strategy—can be a source of social support received from co-religionaries.

Williams and Dilworth-Anderson (2002) investigated the specific social support of the informal (family, friends, and neighbours), church, and formal (professionals, paid helpers) types and their interactions in the USA. Informal social support is effective in alleviating the burden of care for as long as the social network is cohesive. *Social network cohesiveness* can be strengthened by cultural values that support mutual help. An important source of social support for caregivers that is not often mentioned in the literature is the one that could come from peers, that is other caregivers of old persons. In fact, Stoltz et al. (2004) mention that such need to link-up with peers is expressed as an explicit wish by family caregivers themselves, perhaps because they know that other caregivers could more easily empathise with them and also provide valuable feedback based on their direct caregiving experience.

Brodaty et al. (2003) reviewed individual and social interventions to alleviate the burden of care on informal caregivers. Interventions ranged from counselling and education, including family counselling, to patient involvement, support group/programs, and training and stress management. Such interventions tended to alleviate caregivers' psychological distress, increase their knowledge as well as improving the elder's mood, but did not greatly affect caregivers' burden. The success of the interventions increased if all participants were actively involved: both caregivers and the elders under their care.

Sörensen et al. (2002, see also Gallagher-Thompson and Coon 2007) carried out a meta-analysis of the diversity of interventions used on caregivers, including social interventions, to alleviate their burden of care. The authors classified such interventions into psycho-educational, supportive, respite/adult day care, psychotherapy, interventions to improve care receiver competence, and multicomponent interventions. Psycho-educational interventions provide information to caregivers about the condition of the elder (e.g. dementia) and the resources, services, and further training available to the caregivers (e.g. Haley 2003). Supportive interventions consist in either professionally led and/or peer-led unstructured support groups aimed at building rapport among the participants, thus creating an opportunity to discuss the various issues confronted by caregivers and provide solutions. Respite/adult day care interventions simply provide assistance with daily tasks in order to give some respite to the caregiver. They also offer activities to stimulate the elder. Psychotherapy interventions are based on the establishment of a therapeutic relationship between a trained professional and the caregiver. The therapist usually guides the caregiver towards challenging negative thoughts, develop self-monitoring, improve problem-solving capacities, and engage in positive experiences. Interventions to improve care receiver competence (e.g. in the performance of daily activities) help caregivers by improving the self-reliance ability of the elder, which should alleviate the effort required from the caregiver. Finally, multicomponent interventions combine aspects of two or more of the above approaches.

In general, interventions tend to give positive results in terms of alleviating, to a variable extent, caregiver burden and depression, whilst increasing subjective well-being, ability/knowledge and perceived caregiver satisfaction.

Both the costs/stresses of caregiving and some strategies to alleviate them have been synthesized in a model proposed by Pearlin et al. (1990). Caregivers perceive increasingly higher costs and stress of caregiving the more the condition of the elder deteriorates and therefore she or he becomes less likely to reciprocate. Pearlin et al. (1990) see the dynamics of caregiver stress as a process that incorporates four domains and various components (Fig. 6.16):

a. *The background and context of stress*: Various factors can affect the probability of a caregiver experiencing stress during the delivery of care. Socio-economic characteristics may determine the ability of the caregiver to receive some respite, the social network is also important in this respect.

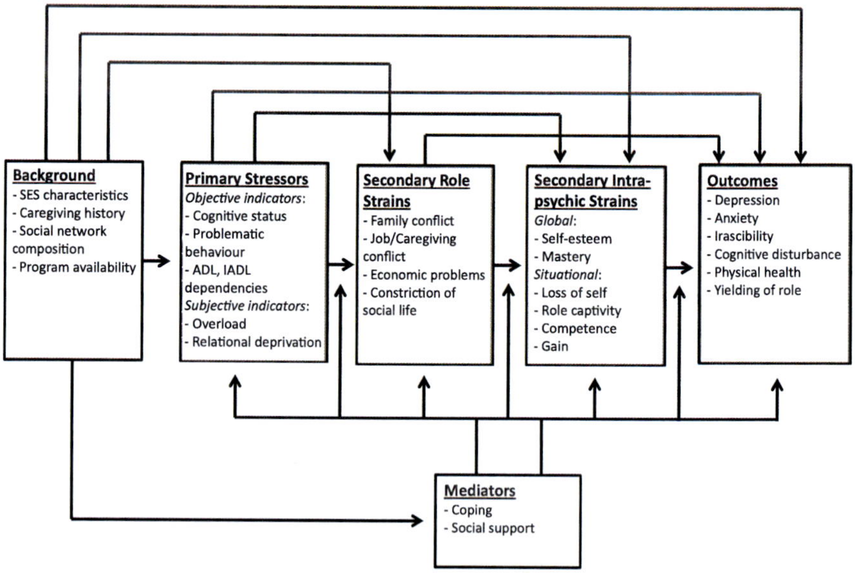

Fig. 6.16 Pearlin et al.'s (1990) conceptual model of caregiver stress. Four domains describe the stress process: the background and context of stress, the stressors (primary, secondary), the mediators of stress, and the outcomes or expressions of stress. *SES* socio-economic status, *ADL* activities of daily living, *IADL* instrumental activities of daily living. Slightly simplified from Pearlin et al. (1990)

b. *The stressors (primary and secondary)*: Sources of caregiver stress can be either "external" such as having to deal with problematic behaviours in the elder, a decline in his/her cognitive capacity, and dependence of the elder on the caregiver for the basic activities of daily living or they can be more "internal" to the caregiver (e.g. personality). Primary stressors directly affect the stress levels of the caregiver, which can then lead to further (secondary) sources of stress (e.g. increased number and intensity of conflicts within the family).
c. *The mediators of stress*: The effects of caregiving stress can be alleviated through personal coping strategies and social support.
d. *The outcomes or manifestations of stress*: If the burden of caregiving is not alleviated the stresses derived from it may ultimately lead to the caregiver experiencing burnout and suffering from various disorders such as depression, anxiety, diminished cognitive capacities, and poor physical health.

Although Parlin et al.'s model mainly focuses on the various sources of stress for the caregiver and the personal and social resources that the caregiver can use to cope with the burden of care; it also considers the possibility of some direct *gains* for the caregiver, gains that are derived from the experience of caring for the elders. Such gains should be seen as part of the coping mechanisms as well.

In sum, older people in need of care usually receive it from family members first and above all—women relatives and in-laws in particular -, but if the family fails to provide care others may step in, usually in the form of friends but also institutionalised caregivers. Institutionalised care requires greater focus on person-centred care that can deliver better results for the elders. The amount and continuity of caregiving that family members provide to their elders is modulated by many factors but of those, caregiver burden is particularly important. Caregiver burden may lead to burnout that can finally break the caring relationship between caregiver and elder, rendering institutionalisation of the elder inevitable. This outcome can be delayed by providing the caregiver with both the personal tools and above all the social support that can decrease the burden of care.

6.12 Conclusions

Although some old people may prefer to withdraw from the commitments and complexities of social life, the social dimension remains important in old age, especially as the elder becomes frail and in greater need of support. Other people can assist the elder in many ways, from helping in the development of greater meaning in life to providing material aid. Critical to such social dimension of old age are the family and close friends, but in many ways also the community at large and public institutions that exist with the intended purpose of helping older people. Social interactions, however, are not always positive and some aspects of stereotypes, stigma and ageism can in fact be detrimental to the well-being of old people. Caring for older people requires attention to their needs, including the specific needs of the oldest-old, older immigrants, older members of the LGBTI community and even those of older prison inmates. Providing care to the elders can involve a degree of burden for the caregivers—the family in particular—therefore services should be provided in order to decrease such burden on the family. Community and government organisations should offer support to caregivers who are in strife and are at risk of burnout, including support in terms of providing affordable and efficient care facilities, assisted living in retirement villages, and nursing homes for the elders.

References

Aartsen, M., & Jylhä, M. (2011). Onset of loneliness in older adults: Results of a 28 year prospective study. *European Journal of Ageing, 8*, 31–38.

Aboderin, I. (2004a). Modernisation and ageing theory revisited: Current explanations of recent developing world and historical Western shifts in material family support for older people. *Ageing and Society, 24*, 29–50.

Aboderin, I. (2004b). Decline in material family support for older people in urban Ghana, Africa: Understanding processes and causes of change. *Journal of Gerontology, 59B*, S128–S137.

Adams, R. G., & Blieszner, R. (1995). Aging well with friends and family. *American Behavioral Scientist, 39*, 209–224.

Aday, R. H. (1994a). Golden years behind bars special programs and facilities for elderly inmates. *Federal Probation, 58*, 47–54.

Aday, R. H. (1994b). Aging in prison: A case study of new elderly offenders. *International Journal of Offender Therapy and Comparative Criminology, 38*, 79–91.

Aday, R. H. (2003). *Aging prisoners: Crisis in American corrections*. Westport, CT: Praeger.

Aday, R. H. (2006). Aging prisoners' concerns toward dying in prison. *Omega, 52*, 199–216.

Addis, S., Davies, M., Greene, G., MacBride-Stewart, S., & Shepherd, M. (2009). The health, social care and housing needs of lesbian, gay, bisexual and transgender older people: A review of the literature. *Health and Social Care in the Community, 17*, 647–658.

Adelman, M. (1991). Stigma, gay lifestyles, and adjustment to aging. *Journal of Homosexuality, 20*, 7–32.

Adelman, R. D., Tmanova, L. L., Delgado, D., Dion, S., & Lachs, M. S. (2014). Caregiver burden: A clinical review. *Journal of the American Medical Association, 311*, 1052–1059.

Ailinger, R. L. (1989). Functional capacity of Hispanic elderly immigrants. *The Journal of Applied Gerontology, 8*, 97–109.

Ajrouch, K. J., Antonucci, T. C., & Janevic, M. R. (2001). Social networks among blacks and whites: The interaction between race and age. *Journal of Gerontology, 56B*, S112–S118.

Albertini, M., Mantovani, D., & Gasperoni, G. (2019). Intergenerational relations among immigrants in Europe: The role of ethnic differences, migration and acculturation. *Journal of Ethnic and Migration Studies, 45*, 1693–1706.

Allan, L. J., & Johnson, J. A. (2009). Undergraduate attitudes toward the elderly: The role of knowledge, contact and aging anxiety. *Educational Gerontology, 35*, 1–14.

Allen, R. S., Phillips, L. L., Roff, L. L., Cavanaugh, R., & Day, L. (2008). Religiousness/spirituality and mental health among older male inmates. *The Gerontologist, 48*, 692–697.

Allmark, P. (2002). Death with dignity. *Journal of Medical Ethics, 28*, 255–257.

Almack, K., Seymour, J., & Bellamy, G. (2010). Exploring the impact of sexual orientation on experiences and concerns about end of life care and on bereavement for lesbian, gay and bisexual older people. *Sociology, 44*, 908–924.

Almberg, B., Grafström, M., & Winblad, B. (1997). Caring for a demented elderly person—Burden and burnout among caregiving relatives. *Journal of Advanced Nursing, 25*, 109–116.

Alster, K. B., & Keshian, J. C. (1990). Falling between the cracks: A student experience in providing home care for underserved community residing elders. *Journal of Community Health Nursing, 7*, 15–24.

Althausen, L. (1993). Journey of separation: Elderly Russian immigrants and their adult children in the health care setting. *Social Work in Health Care, 19*, 61–75.
Amit, K., & Hefer, E. (2010). The subjective well-being of immigrants aged 50 and older in Israel. *Social Indicators Research, 98*, 89–104.
Andrén, S., & Elmståhl, S. (2005). Family caregivers' subjective experiences of satisfaction in dementia care: Aspects of burden, subjective health and sense of coherence. *Scandinavian Journal of Caring Sciences, 19*, 157–168.
Andrén, S., & Elmståhl, S. (2008). The relationship between caregiver burden, caregivers' perceived health and their sense of coherence in caring for elders with dementia. *Journal of Clinical Nursing, 17*, 790–799.
Andrews, M. (1999). The seduction of agelessness. *Ageing and Society, 19*, 301–318.
Anetzberger, G. J. (2000). Caregiving: Primary cause of elder abuse? *Generations, 24*, 46–51.
Anetzberger, G. J. (2005). The reality of elder abuse. *Clinical Gerontologist, 28*, 1–25.
Angel, J. L. (2003). Devolution and the social welfare of elderly immigrants: Who will bear the burden? *Public Administration Review, 63*, 79–89.
Angel, R. J., Angel, J. L., Lee, G.-Y., & Markides, K. S. (1999). Age at migration and family dependency among older Mexican immigrants: Recent evidence from the Mexican American EPESE. *The Gerontologist, 39*, 59–65.
Angus, J., & Reeve, P. (2006). Ageism: A threat to "aging well" in the 21st Century. *The Journal of Applied Gerontology, 25*, 137–152.
Anonson, J., Huard, S., Kristoff, T., Clarke-Arnault, V., Wilson, E. V., & Walker, M. E. (2014). The role of elders in post-secondary educational institutes. *The Canadian Journal of Native Studies, 34*, 1–18.
Apt, N. A. (1996). *Coping with old age in a changing Africa*. Aldershot: Avebury.
Arber, S., Davidson, K., & Ginn, J. (2003a). Changing approaches to gender and later life. In S. Arber, K. Davidson, & J. Ginn (Eds.), *Gender and ageing: Changing roles and relationships* (pp. 1–14). Maidenhead: Open University Press.
Arber, S., Price, D., Davidson, K., & Perren, K. (2003b). Re-examining gender and marital status: Material well-being and social involvement. In S. Arber, K. Davidson, & J. Ginn (Eds.), *Gender and ageing: Changing roles and relationships* (pp. 148–167). Maidenhead: Open University Press.
Arditti, R. (2002). The grandmothers of the Plaza De Mayo and the struggle against impunity in Argentina. *Meridians, 3*, 19–41.
Arditti, R., & Brinton Lykes, M. (1992). "Recovering identity": The work of the grandmothers of Plaza de Mayo. *Women's Studies International Forum, 15*, 461–471.
Argentine National Commission on the Disappeared. 1986. Nunca más: The report of the Argentine National Commission on the Disappeared. Farrar, Straus, and Giroux, New York.
Arndt, S., Turvey, C. L., & Flaum, M. (2002). Older offenders, substance abuse, and treatment. *American Journal of Geriatric Psychiatry, 10*, 733–739.
Arnold-Cathalifaud, M., Thumala, D., Urquiza, A., & Ojeda, A. (2008). Young people's images of old age in Chile: Exploratory research. *Educational Gerontology, 34*, 105–123.
Aroian, K. J., Khatutsky, G., Tran, T. V., & Balsam, A. L. (2001). Health and social service utilization among elderly immigrants from the former Soviet Union. *Journal of Nursing Scholarship., 33*, 265–271.
Aronson, J. (1992). Women's sense of responsibility for the care of old people: "*But who else is going to do it?*". *Gender and Society, 6*, 8–29.
Aronson, J., & McGlone, M. S. (2009). Stereotype and social identity threat. In T. D. Nelson (Ed.), *Handbook of prejudice, stereotyping, and discrimination* (pp. 153–178). New York: Psychology Press.
Ash, A. (2013). A cognitive mask? Camouflaging dilemmas in street-level policy implementation to safeguard older people from abuse. *British Journal of Social Work, 43*, 99–115.
Askham, J., Ferring, D., & Lamura, G. (2007). Personal relationships in later life. In J. Bond, S. Peace, F. Dittmann-Kohli, & G. Westerhof (Eds.), *Ageing in society* (pp. 186–208). London: Sage.

Averett, P., Yoon, I., & Jenkins, C. L. (2012). Older lesbian sexuality: Identity, sexual behavior, and the impact of aging. *The Journal of Sex Research, 49*, 495–507.
Avlund, K., Trab Damsgaard, M., & Holstein, B. E. (1998). Social relations and mortality. An eleven year follow-up study of 70-year-old men and women in Denmark. *Social Sciences and Medicine, 47*, 635–643.
Aylaz, R., Aktürk, Ü., Erci, B., Öztürk, H., & Aslan, H. (2012). Relationship between depression and loneliness in elderly and examination of influential factors. *Archives of Gerontology and Geriatrics, 55*, 548–554.
Bachner, Y. G., & O'Rourke, N. (2007). Reliability generalization of responses by care providers to the Zarit Burden Interview. *Aging and Mental Health, 11*, 678–685.
Badgett, M., & Frank, J. (2007). *Sexual orientation discrimination: An international perspective*. New York: Routledge.
Bagshaw, D., Wendt, S., & Zannettino, L. (2009). Preventing the abuse of older people by their family members. *Australian Domestic and Family Violence Clearinghouse Stakeholder Paper, 7*, 1–16.
Baidawi, S., Turner, S., Trotter, C., Browning, C., Collier, P., O'Connor, D., et al. (2011). Older prisoners - A challenge for Australian corrections. Trends and Issues in Crime and. *Criminal Justice, 426*, 1–8.
Baker, A. A. (1975). Granny battering. *Modern Geriatrics, 5*, 20–24.
Baldassar, L., & Merla, L. (2014). *Transnational families, migration and the circulation of care: Understanding mobility and absence in family life*. New York: Routledge Chapman and Hall.
Baltes, M. M. (1998). The psychology of the oldest-old: The fourth age. *Current Opinion in Psychiatry, 11*, 411–415.
Baltes, P. B., & Smith, J. (2003). New frontiers in the future of aging: From successful aging of the young old to the dilemmas of the fourth age. *Gerontology, 49*, 123–135.
Bandura, A. (1978). Social learning theory of aggression. *Journal of Communication, 28*, 12–29.
Bandura, A. (1986). *Social foundations of thought and action: A social cognitive theory*. Englewood Cliffs, NJ: Prentice Hall.
Bandura, A. (2004). Health promotion by social cognitive means. *Health Education and Behavior, 31*, 143–164.
Banerjee, A., Pasea, L., Harris, S., Gonzalez-Izquierdo, A., Torralbo, A., Shallcross, L., et al. (2020). Estimating excess 1-year mortality associated with the COVID-19 pandemic according to underlying conditions and age: A population-based cohort study. *Lancet, 395*, 1715–1725.
Barker, J. C. (2002). Neighbors, friends, and other nonkin caregivers of community-living dependent elders. *Journal of Gerontology, 57B*, S158–S167.
Barrett, C., Whyte, C., Comfort, J., Lyons, A., & Crameri, P. (2015). Social connection, relationships and older lesbian and gay people. *Sexual and Relationship Therapy, 30*, 131–142.
Bargh, J. A. (1994). Four horsemen of automaticity: Awareness, intention, efficiency, and control in social cognition. In R. S. Wyer Jr. & T. K. Srull (Eds.), *Handbook of social cognition* (Vol. 1, 2nd ed., pp. 1–40). Mahwah, NJ: Erlbaum.
Barron, M. L. (1953). Minority group characteristics of the aged in American society. *Journal of Gerontology, 8*, 477–482.
Barusch, A. S., & Spaid, W. M. (1989). Gender differences in caregiving: Why do wives report greater burden? *The Gerontologist, 29*, 667–676.
Bauer, M., Fitzgerald, L., Haesler, E., & Manfrin, M. (2009). Hospital discharge planning for frail older people and their family. Are we delivering best practice? A review of the evidence. *Journal of Clinical Nursing, 18*, 2539–2546.
Baumann, A., Filipiak, B., Stieber, J., & Löwel, H. (1998). Family status and social integration as predictors of mortality: A 5-year follow-up study of 55- to 74-year-old men and women in the Augsburg area. *Zeitschrift für Gerontologie und Geriatrie, 31*, 184–192.
Beck, C. M., & Phillips, L. R. (1983). Abuse of the elderly. *Journal of Gerontological Nursing, 9*, 96–101.
Becker, E. (1973). *The denial of death*. New York: Macmillan.
Becker, G. (1994). The oldest old: Autonomy in the face of frailty. *Journal of Aging Studies, 8*, 59–76.

Becker, G., & Beyene, Y. (1999). Narratives of age and uprootedness among older Cambodian refugees. *Journal of Aging Studies, 13*, 295–314.
Beckett, J., Peternelj-Taylor, C., & Johnson, R. L. (2003). Growing old in the correctional system. *Journal of Psychosocial Nursing and Mental Health Services, 41*, 12–18.
Beeler, J. A., Rawls, T. W., Herdt, G., & Cohler, B. J. (1999). The needs of older lesbians and gay men in Chicago. *Journal of Gay and Lesbian Social Services, 9*, 31–49.
Benbow, S. M., & Haddad, P. M. (1993). Sexual abuse ofthe elderly mentally ill. *Postgraduate Medical Journal, 69*, 803–807.
Berg, A. I., Hassing, L. B., McClearn, G. E., & Johansson, B. (2006). What matters for life satisfaction in the oldest-old? *Aging and Mental Health, 10*, 257–264.
Berg, A. I., Hoffman, L., Hassing, L. B., McClearn, G. M., & Johansson, B. (2009). What matters, and what matters most, for change in life satisfaction in the oldest-old? A study over 6 years among individuals 80+. *Aging and Mental Health, 13*, 191–201.
Bergeron, L. R. (2001). An elder abuse case study. *Journal of Gerontological Social Work, 34*, 47–63.
Berkman, L. F. (2000). Social support, social networks, social cohesion and health. *Social Work in Health Care, 31*, 3–14.
Berkman, L. F., Melchior, M., Chastang, J.-F., Niedhammer, I., Leclerc, A., & Goldberg, M. (2004). Social integration and mortality: A prospective study of French employees of Electricity of France–Gas of France: The GAZEL Cohort. *American Journal of Epidemiology, 159*, 167–174.
Berry, J. W. (1997). Immigration, acculturation, and adaptation. *Applied Psychology, 46*, 5–34.
Biernat, M. (2009). Stereotypes and shifting standards. In T. D. Nelson (Ed.), *Handbook of prejudice, stereotyping, and discrimination* (pp. 137–152). New York: Psychology Press.
Biggs, S. (1993). *Understanding ageing: Images,attitudes and professional practice*. Buckingham: Open University Press.
Biggs, S. (1997). Choosing not to be old? Masks, bodies and identity management in later life. *Ageing and Society, 17*, 553–570.
Binstock, R. H. (1983). The aged as scapegoat. *The Gerontologist, 23*, 136–143.
Binstock, R. H. (1985). The oldest old: A fresh perspective or compassionate ageism revisited? *The Milbank Memorial Fund Quarterly, 63*, 420–451.
Binstock, R. H. (2010). From compassionate ageism to intergenerational conflict? *The Gerontologist, 50*, 574–585.
Black, W., & Almeida, O. P. (2004). A systematic review of the association between the behavioral and psychological symptoms of dementia and burden of care. *International Psychogeriatrics, 16*, 295–315.
Block, F. (2011). Reinventing social democracy in the 21st century. *Journal of Australian Political Economy, 67*, 5–21.
Blumstein, D. T. (2013). Yellow-bellied marmots: Insights from an emergent view of sociality. *Philosophical Transactions of the Royal Society B, 368*, 20120349. https://doi.org/10.1098/rstb.2012.0349
Bodenhausen, G. V., Todd, A. R., & Richeson, J. A. (2009). Controlling prejudice and stereotyping: Antecedents, mechanisms, and contexts. In T. D. Nelson (Ed.), *Handbook of prejudice, stereotyping, and discrimination* (pp. 111–135). New York: Psychology Press.
Bolzman, C. (2012). Democratization of ageing: Also a reality for elderly immigrants? *European Journal of Social Work, 15*, 97–113.
Bolzman, C., Fibbi, R., & Vial, M. (2006). What to do after retirement? Elderly migrants and the question of return. *Journal of Ethnic and Migration Studies, 32*, 1359–1375.
Bolzman, C., Poncioni-Derigo, R., Vial, M., & Fibbi, R. (2004). Older labour migrants' well being in Europe: The case of Switzerland. *Ageing and Society, 24*, 411–429.
Bolzman, C., & Vagni, G. (2017). Forms of care among native Swiss and older migrants from Southern Europe: A comparison. *Journal of Ethnic and Migration Studies, 43*, 250–269.
Bondevik, M., & Skogstad, A. (1998). The oldest old, ADL, social network, and loneliness. *Western Journal of Nursing Research, 20*, 325–343.
Borell, K., & Ghazanfareeon Karlsson, S. (2003). Reconceptualizing intimacy and ageing: Living apart together. In S. Arber, K. Davidson, & J. Ginn (Eds.), *Gender and ageing: Changing roles and relationships* (pp. 47–62). Maidenhead: Open University Press.

Botcheva, L. B., & Feldman, S. S. (2004). Grandparents as family stabilizers during economic hardship in Bulgaria. *International Journal of Psychology, 39*, 157–168.
Bowling, A., & Browne, P. D. (1991). Social networks, health, and emotional well-being among the oldest old in London. *Journal of Gerontology, 46*, S20–S32.
Bowling, A., & Dieppe, P. (2005). What is successful ageing and who should define it? *BMJ, 331*, 1548–1551.
Boyd, M. (1991). Immigration and living arrangements: Elderly women in Canada. *The International Migration Review, 25*, 4–27.
Brandes, S. (2000). El Día de Muertos, el Halloween y la búsqueda de una identidad nacional mexicana. *Alteridades, 10*, 7–20.
Brandl, B. (2002). Power and control: Understanding domestic abuse in later life. *Generations, 24*, 39–45.
Brennan-Ing, M., Seidel, L., Larson, B., & Karpiak, S. E. (2014). Social care networks and older LGBT adults: Challenges for the future. *Journal of Homosexuality, 61*, 21–52.
Brent, L. J. N., Heilbronner, S. R., Horvath, J. E., Gonzalez-Martinez, J., Ruiz-Lambides, A., Robinson, A. G., et al. (2013). Genetic origins of social networks in rhesus macaques. *Nature Scientific Reports, 3*, 1042. https://doi.org/10.1038/srep01042
Bretschneider, W., & Elger, B. S. (2014). Expert perspectives on Western European prison health services: Do ageing prisoners receive equivalent care? *Bioethical Inquiry, 11*, 319–332.
Bridges, J., Flatley, M., & Meyer, J. (2010). Older people's and relatives' experiences in acute care settings: Systematic review and synthesis of qualitative studies. *International Journal of Nursing Studies, 47*, 89–107.
Bridgstock, M. (2018). *The impact of the Universities of the Third Age upon the health and welfare of their membership*. Retrieved June 11, 2020, from https://www.u3abrisbane.org.au/index.php/news
Brodaty, H., & Donkin, M. (2009). Family caregivers of people with dementia. *Dialogues in Clinical Neuroscience, 11*, 217–228.
Brodaty, H., Green, A., & Koschera, A. (2003). Meta-analysis of psychosocial interventions for caregivers of people with dementia. *Journal of the American Geriatrics Society, 51*, 657–664.
Bronfenbrenner, U. (1979). *The ecology of human development: Experiments by nature and design*. Cambridge, MA: Harvard University Press.
Brotman, S., Ryan, B., & Cormier, R. (2003). The health and social service needs of gay and lesbian elders and their families in Canada. *The Gerontologist, 43*, 192–202.
Brouwer, W. B. F., van Exel, N. J. A., van Gorp, B., & Redekop, W. K. (2006). The CarerQol instrument: A new instrument to measure care-related quality of life of informal caregivers for use in economic evaluations. *Quality of Life Research, 15*, 1005–1021.
Buckwalter, K. C., Davis, L. L., Wakefield, B. J., Kienzle, M. G., & Murray, M. A. (2002). Elders giving back to their communities. *Family and Community Health, 25*, 31–40.
Bukov, A., Maas, I., & Lampert, T. (2002). Social participation in very old age: Cross- sectional and longitudinal findings from BASE. *Journal of Gerontology, 57B*, P510–P517.
Bunzel, J. H. (1972). Note on the history of a concept: Gerontophobia. *The Gerontologist, 12*, 116. (cont. on p. 203).
Burke, J. L. (1982). Young children's attitudes and perceptions of older adults. *The International Journal of Aging and Human Development, 14*, 205–222.
Butler, R. N. (1969). Age-ism: Another form of bigotry. *Gerontologist, 9*, 243–246.
Butler, R. N. (1987). Ageism. In G. Maddox (Ed.), *The encyclopedia of aging* (pp. 22–23). New York: Springer.
Butler, S. S. (2004). Gay, lesbian, bisexual, and transgender (GLBT) elders. *Journal of Human Behavior in the Social Environment, 9*, 25–44.
Butler, S. S., Turner, W., Kaye, L. W., Ruffin, L., & Downey, R. (2005). Depression and caregiver burden among rural elder caregivers. *Journal of Gerontological Social Work, 46*, 47–63.
Bytheway, B., & Johnson, J. (1990). On defining ageism. *Critical Social Policy, 27*, 27–39.
Bytheway, B. (2005). Ageism and age categorization. *Journal of Social Issues, 61*, 361–374.
Cabrera Giráldez, M., & Rodríguez Casal, C. (2005). The role of ambient intelligence in the social integration of the elderly. In G. Riva, F. Vatalaro, F. Davide, & M. Alcañiz (Eds.), *Ambient*

intelligence: The evolution of technology communication and cognition towards the future of human-computer interaction (pp. 267–282). Amsterdam: IOS Press.
Cacioppo, J. T., Hawkley, L. C., & Thisted, R. A. (2010). Perceived social isolation makes me sad: 5-Year cross-lagged analyses of loneliness and depressive symptomatology in the Chicago Health, Aging, and Social Relations Study. *Psychology and Aging, 25*, 453–463.
Cacioppo, J. T., Hughes, M. E., Waite, L. J., Hawkley, L. C., & Thisted, R. A. (2006). Loneliness as a specific risk factor for depressive symptoms: Cross-sectional and longitudinal analyses. *Psychology and Aging, 21*, 140–151.
Calasanti, T. (2005). Ageism, gravity, and gender: Experiences of aging bodies. *Generations, 29*, 8–12.
Campbell, P., Wright, J., Oyebode, J., Job, D., Crome, P., Bentham, P., et al. (2008). Determinants of burden in those who care for someone with dementia. *International Journal of Geriatric Psychiatry, 23*, 1078–1085.
Campbell Reay, A. M., & Browne, K. D. (2001). Risk factor characteristics in carers who physically abuse or neglect their elderly dependants. *Aging and Mental Health, 5*, 56–62.
Cannuscio, C., Block, J., & Kawachi, I. (2003). Social capital and successful aging: The role of senior housing. *Annals of Internal Medicine, 139*, 395–399.
Cantor, M. H. (1980). The informal support system: Its relevance in the lives of the elderly. In E. F. Borgatta & N. G. McCluskey (Eds.), *Aging and society: Current research and policy perspectives* (pp. 131–144). Beverly Hills, CA: Sage.
Cantor, M. H. (1991). Family and community: Changing roles in an aging society. *The Gerontologist, 31*, 337–346.
Cantor, M. H., & Mayer, M. (1978). Factors in differential utilization of services by urban elderly. *Journal of Gerontological Social Work, 1*, 47–61.
Carretero, S., Garcés, J., Ródenas, F., & Sanjosé, V. (2009). The informal caregiver's burden of dependent people: Theory and empirical review. *Archives of Gerontology and Geriatrics, 49*, 74–79.
Carson, E. A., & Anderson, E. (2016). Prisoners in 2015. *Office of Justice Programs*. U.S. Department of Justice, USA.
Carstensen, L. L. (1992). Social and emotional patterns in adulthood: Support for socioemotional selectivity theory. *Psychology and Aging, 7*, 331–338.
Carstensen, L. L. (1993). Motivation for social contact across the life span: A theory of socioemotional selectivity. In J. E. Jacobs (Ed.), *Developmental perspectives in motivation [Nebraska Symposium on Motivation, 1992]* (pp. 209–254). Lincoln, NE: University of Nebraska Press.
Cartwright, C. M., Hughes, M., & Lienert, T. (2012). End-of-life care for gay, lesbian, bisexual and transgender people. *Culture, Health and Sexuality, 14*, 537–548.
Casado, B. L., & Leung, P. (2001). Migratory grief and depression among elderly Chinese American immigrants. *Journal of Gerontological Social Work, 36*, 5–26.
Caserta, M. S., Lund, D. A., & Wright, S. D. (1996). Exploring the caregiver burden inventory (CBI): Further evidence for a multidimensional view of burden. *International Journal of Aging and Human Development, 43*, 21–34.
Cattan, M., White, M., Bond, J., & Learmouth, A. (2005). Preventing social isolation and loneliness among older people: A systematic review of health promotion interventions. *Ageing and Society, 25*, 41–67.
Cattell, V. (2001). Poor people, poor places, and poor health: The mediating role of social networks and social capital. *Social Science and Medicine, 52*, 1501–1516.
Caverley, S. J. (2006). Older mentally ill inmates: A descriptive study. *Journal of Correctional Health Care, 12*, 262–268.
Chen, F., & Short, S. E. (2008). Household context and subjective well-being among the oldest old in China. *Journal of Family Issues, 29*, 1379–1403.
Chen, K.-M., Snyder, M., & Krichbaum, K. (2002). Tai chi and well-being of Taiwanese community-dwelling elders. *Clinical Gerontologist, 24*, 137–156.
Chen, K.-M., Wang, H.-H., Li, C.-H., & Chen, M.-H. (2011). Community vs. institutional elders' evaluations of and preferences for yoga exercises. *Journal of Clinical Nursing, 20*, 1000–1007.

Cherry, K. E., Walker, E. J., Silva Brown, J., Volaufova, J., LaMotte, L. R., Welsh, D. A., et al. (2013). Social engagement and health in younger, older, and oldest-old adults in the Louisiana Healthy Aging Study (LHAS). *Journal of Applied Gerontology, 32*, 51–75.
Cheung, M. C., Ting, W., Chan, L. Y., Ho, K. S., & Chan, W. M. (2009). Leisure participation and healthrelated quality of life of communitydwelling elders in Hong Kong. *Asian Journal of Gerontological Geriatrics, 4*, 15–23.
Chiang, K.-J., Chu, H., Chang, H.-J., Chung, M.-H., Chen, C.-H., Chiou, H.-Y., et al. (2010). The effects of reminiscence therapy on psychological well- being, depression, and loneliness among the institutionalized aged. *International Journal of Geriatric Psychiatry, 25*, 380–388.
Chou, K.-L., & Chi, I. (2002). Successful aging among the young-old, old-old, and oldest- old Chinese. *International Journal of Aging and Human Development, 54*, 1–14.
Chou, K.-R., Chu, H., Tseng, C.-L., & Lu, R.-B. (2003). The measurement of caregiver burden. *Journal of Medical Sciences, 23*, 73–82.
Chow, H. P. H. (2010). Growing old in Canada: Physical and psychological well-being among elderly Chinese immigrants. *Ethnicity and Health, 15*, 61–72.
Chow, N. (1993). The changing responsibilities of the state and family toward elders in Hong Kong. *Journal of Aging and Social Policy, 5*, 111–126.
Chumbler, N. R., Grimm, J. W., Cody, M., & Beck, C. (2003). Gender, kinship and caregiver burden: The case of community-dwelling memory impaired seniors. *International Journal of Geriatric Psychiatry, 18*, 722–732.
Cicirelli, V. G. (1982). Kin relationships of childless and one-child elderly in relation to social services. *Journal of Gerontological and Social Work, 4*, 19–33.
Cicirelli, V. G. (1985). Sibling relationships throughout the life cycle. In L. L. Abate (Ed.), *The handbook of family psychology and therapy* (pp. 177–214). Homewood, IL: Dorsey Press.
Cicirelli, V. G. (2010). Attachment relationships in old age. *Journal of Social and Personal Relationships, 27*, 191–199.
Ciobanu, R. O., & Fokkema, T. (2017). The role of religion in protecting older Romanian migrants from loneliness. *Journal of Ethnic and Migration Studies, 43*, 199–217.
Ciobanu, R. O., Fokkema, T., & Nedelcu, M. (2017). Ageing as a migrant: Vulnerabilities, agency and policy implications. *Journal of Ethnic and Migration Studies, 43*, 164–181.
Clark, A. N. G., Mankikar, G. D., & Gray, I. (1975). Diogenes syndrome: A clinical study of gross neglect in old age. *The Lancet, 305*, 366–368.
Clarke, A., Hanson, E. J., & Ross, H. (2003). Seeing the person behind the patient: Enhancing the care of older people using a biographical approach. *Journal of Clinical Nursing, 12*, 697–706.
Clarke, L. H., Griffin, M., & Maliha, K. (2009). Bat wings, bunions, and turkey wattles: Body transgressions and older women's strategic clothing choices. *Ageing and Society, 29*, 709–726.
Cohen, G. D. (1994). Journalistic elder abuse: It's time to get rid of fictions, get down to facts. *The Gerontologist, 34*, 399–401.
Cohen, G. D., Perlstein, S., Chapline, J., Kelly, J., Firth, K. M., & Simmens, S. (2006). The impact of professionally conducted cultural programs on the physical health, mental health, and social functioning of older adults. *The Gerontologist, 46*, 726–734.
Coleman, J. S. (1990). *Foundations of social theory*. Harvard: Harvard University Press.
Collins, D. R., & Bird, R. (2007). The penitentiary visit - A new role for geriatricians? *Age and Ageing, 36*, 11–13.
Coll-Planas, L., del Valle Gómez, G., Bonilla, P., Masat, T., Puig, T., & Monteserin, R. (2017). Promoting social capital to alleviate loneliness and improve health among older people in Spain. *Health and Social Care in the Community, 25*, 145–157.
Cong, Z., & Silverstein, M. (2008). Intergenerational support and depression among elders in rural China: Do daughters-in-law matter? *Journal of Marriage and Family, 70*, 599–612.
Connell, R. W., & Messerschmidt, J. W. (2005). Hegemonic masculinity: Rethinking the concept. *Gender and Society, 19*, 829–859.
Conner, K. O., & Rosen, D. (2008). "You're nothing but a junkie": Multiple experiences of stigma in an aging methadone maintenance population. *Journal of Social Work Practice in the Addictions, 8*, 244–264.

Connidis, I. A. (2003). Bringing outsiders in: Gay and lesbian family ties over the life course. In S. Aber, K. Davidson, & J. Ginn (Eds.), *Gender and ageing: Changing roles and relationships* (pp. 79–94). Maidenhead: Open University Press.

Cook-Daniels, L. (2006). Trans aging. In D. Kimmel, T. Rose, & S. David (Eds.), *Lesbian, gay, bisexual, and transgender aging: Research and clinical perspectives* (pp. 20–35). New York: Columbia U. Press.

Cooney, C., Howard, R., & Lawlor, B. (2006). Abuse of vulnerable people with dementia by their carers: Can we identify those most at risk? *International Journal of Geriatric Psychiatry, 21*, 564–571.

Cooper, C., Selwood, A., & Livingston, G. (2008). The prevalence of elder abuse and neglect: A systematic review. *Age and Ageing, 37*, 151–160.

Cordella, M., & Huang, H. (Eds.). (2016). *Rethinking second language learning: Using intergenerational community resources*. Bristol: Multilingual Matters.

Cordella, M., & Poiani, A. (2014). *Behavioural oncology: Psychological , communicative, and social dimensions*. New York: Springer.

Cordella, M., & Rojas-Lizana, S. (2019). Aging and migration: The value of familism for Spanish speakers. *Journal of Cross-Cultural Gerontology, 35*(2). https://doi.org/10.1007/s10823-019-09389-1

Cornwell, B., Laumann, E. O., & Schumm, P. (2008). The social connectedness of older adults: A national profile. *American Sociological Review, 73*, 185–203.

Cornwell, E. Y., & Waite, L. J. (2009). Social disconnectedness, perceived isolation, and health among older adults. *Journal of Health and Social Behavior, 50*, 31–48.

Crafts, N. (2004). Globalisation and economic growth: A historical perspective. *The World Economy, 27*, 45–58.

Crist, J. D. (2005). The meaning for elders of receiving family care. *Journal of Advanced Nursing, 49*, 485–493.

Crocker, J., & Garcia, J. A. (2009). Downward and upward spirals in intergroup interactions: The role of egosystem and ecosystem goals. In T. D. Nelson (Ed.), *Handbook of prejudice, stereotyping, and discrimination* (pp. 229–245). New York: Psychology Press.

Cronin, A., & King, A. (2009). Power, inequality and identification: Exploring diversity and intersectionality amongst older LGB adults. *Sociology, 44*, 876–892.

Cuddy, A. J. C., Fiske, S. T., Kwan, V. S. Y., Glick, P., Demoulin, S., Leyens, J.-P., et al. (2009). Stereotype content model across cultures: Towards universal similarities and some differences. *British Journal of Social Psychology, 48*, 1–33.

Cuddy, A. J. C., Norton, M. I., & Fiske, S. T. (2005). This old stereotype: The pervasiveness and persistence of the elderly stereotype. *Journal of Social Issues, 61*, 265–283.

Cumming, E., & Henry, W. E. (1961). *Growing old: The process of disengagement*. New York: Basic Books.

Curry, N., & Fisher, R. (2013). Being, belonging and bestowing: Differing degrees of community involvement amongst rural elders in England and Wales. *European Journal of Ageing, 10*, 325–333.

D'Augelli, A. R., Grossman, A. H., Hershberger, S. L., & O'Connell, T. S. (2001). Aspects of mental health among older lesbian, gay, and bisexual adults. *Aging and Mental Health, 5*, 149–158.

Daatland, S. O., & Lowenstein, A. (2005). Intergenerational solidarity and the family– welfare state balance. *European Journal of Ageing, 2*, 174–182.

Dasgupta, N. (2009). Mechanisms underlying the malleability of implicit prejudice and stereotypes: The role of automaticity and cognitive control. In T. D. Nelson (Ed.), *Handbook of prejudice, stereotyping, and discrimination* (pp. 267–284). New York: Psychology Press.

Davidson, D., Cameron, P., & Jergovic, D. (1995). The effects of children's stereotypes on their memory for elderly individuals. *Merrill-Palmer Quarterly, 41*, 70–90.

Davidson, K., Daly, T., & Arber, S. (2003a). Older men, social integration and organisational activities. *Social Policy and Society, 2*, 81–89.

Davidson, K., Daly, T., & Arber, S. (2003b). Exploring the social worlds of older men. In S. Arber, K. Davidson, & J. Ginn (Eds.), *Gender and ageing: Changing roles and relationships* (pp. 168–185). Maidenhead: Open University Press.

Davis, S., Dorevitch, M., & Garratt, S. (2009). Person-centred comprehensive geriatric care. In R. Nay & S. Garratt (Eds.), *Older people: Issues and innovations in care* (3rd ed., pp. 168–188). Chatswood, NSW: Churchill Livingstone.

Day, F. R., Ong, K. K., & Perry, J. R. B. (2018). Elucidating the genetic basis of social interaction and isolation. *Nature Communications, 9*, 2457.

Day, R., & Hitchings, R. (2011). 'Only old ladies would do that': Age stigma and older people's strategies for dealing with winter cold. *Health and Place, 17*, 885–894.

De Capitani, E. (2014). The Schengen system after Lisbon: From cooperation to integration. *ERA Forum, 15*, 101–118.

De Donder, L., Luoma, M.-L., Penhale, B., Lang, G., Santos, A. J., Tamutiene, I., et al. (2011). European map of prevalence rates of elder abuse and its impact for future research. *European Journal of Ageing, 8*, 129–143.

de Jong Gierveld, J. (1998). A review of loneliness: Concept and definitions, determinants and consequences. *Reviews in Clinical Gerontology, 8*, 73–80.

de Jong Gierveld, J. (2003). Social networks and social well-being of older men and women living alone. In S. Arber, K. Davidson, & J. Ginn (Eds.), *Gender and ageing: Chaging roles and relationships* (pp. 95–110). Maidenhead: Open University Press.

de Jong Gierveld, J., & Hagestad, G. O. (2006). Perspectives on the integration of older men and women. *Research on Aging, 28*, 627–637.

de Jong Gierveld, J., & van Tilburg, T. (1995). Social relationships, integration, and loneliness. In C. P. M. Knipscheer, J. de Jong Gierveld, T. G. van Tilburg, & P. A. Dykstra (Eds.), *Living arrangements and social networks of older adults* (pp. 155–172). Amsterdam: VU University Press.

de Jong Gierveld, J., van Tilburg, T. G., & Dykstra, P. A. (2006). Loneliness and social isolation. In A. Vangelisti & D. Perlman (Eds.), *Cambridge handbook of personal relationships* (pp. 485–500). Cambridge: Cambridge University Press.

de Mendonça Lima, C. A., Levav, I., Jacobsson, L., & Rutz, W. (2003). Stigma and discrimination against older people with mental disorders in Europe. *International Journal of Geriatric Psychiatry, 18*, 679–682.

De Vries, B., & Herdt, G. (2012). Aging in the gay community. In T. M. Witten & A. E. Eyler (Eds.), *Gay, lesbian, bisexual and transgender aging* (pp. 84–129). Baltimore: The Johns Hopkins University Press.

Dean, D. G. (1961). Alienation: Its meaning and measurement. *American Sociological Review, 26*, 753–758.

Deaton, D., Aday, R. H., & Wahidin, A. (2009). The effect of health and penal harm on aging female prisoners' views of dying in prison. *Omega, 60*, 51–70.

DeCoster, V. A., & George, L. (2005). An empowerment approach for elders living with diabetes: A pilot study of a community-based self-help group—The Diabetes Club. *Educational Gerontology, 31*, 699–713.

Deeken, J. F., Taylor, K. L., Mangan, P., Yabroff, K. R., & Ingham, J. M. (2003). Care for the caregivers: A review of self-report instruments developed to measure the burden, needs, and quality of life of informal caregivers. *Journal of Pain and Symptom Management, 26*, 922–953.

Deimling, G. T., & Bass, D. M. (1986). Symptoms of mental impairment among elderly adults and their effects on family caregivers. *Journal of Gerontology, 41*, 778–784.

DeLamater, J. (2012). Sexual expression in later life: A review and synthesis. *Journal of Sex Research, 49*, 125–141.

Dergance, J. M., Calmbach, W. L., Dhanda, R., Miles, T. P., Hazuda, H. P., & Mouton, C. P. (2003). Barriers to and benefits of leisure time physical activity in the elderly: Differences across cultures. *Journal of the American Geriatrics Society, 51*, 863–868.

Devine, P. G., & Sharp, L. B. (2009). Automaticity and control in stereotyping and prejudice. In T. D. Nelson (Ed.), *Handbook of prejudice, stereotyping, and discrimination* (pp. 61–87). New York: Psychology Press.

Diamond, P. M., Wang, E. W., Holzer, C. E., Thomas, C., & Cruser, A. (2001). The prevalence of mental illness in prison. *Administration and Policy in Mental Health, 29*, 21–40.
Die, A. H., & Seelbach, W. C. (1988). Problems, sources of assistance, and knowledge of services among elderly Vietnamese immigrants. *The Gerontologist, 28*, 448–452.
Dodds, S. (2005). Gender, ageing, and injustice: Social and political contexts of bioethics. *Journal of Medical Ethics, 31*, 295–298.
Donaldson, Z. R., & Young, L. J. (2008). Oxytocin, vasopressin, and the neurogenetics of sociality. *Science, 322*, 900–904.
Dong, X. Q., Simon, M. A., Gorbien, M., Percak, J., & Golden, R. (2007). Loneliness in older Chinese adults: A risk factor for elder mistreatment. *Journal of the American Geriatrics Society, 55*, 1831–1835.
Donovan, T. (2001). Being transgender and older. *Journal of Gay and Lesbian Social Services, 13*, 19–22.
Doron, I. (2007). Court of ethics: Teaching ethics and ageing by means of role-playing. *Educational Gerontology, 33*, 737–758.
Drageset, J. (2004). The importance of activities of daily living and social contact for loneliness: A survey among residents in nursing homes. *Scandinavian Journal of Caring Sciences, 18*, 65–71.
Dressel, P. L., & Barnhill, S. K. (1994). Reframing gerontological thought and practice: The case of grandmothers with daughters in prison. *The Gerontologist, 34*, 685–691.
Duberstein, P. R., Conwell, Y., Conner, K. R., Eberly, S., Evinger, J. S., & Caine, E. D. (2004). Poor social integration and suicide: Fact or artifact? A case-control study. *Psychological Medicine, 34*, 1331–1337.
Duncan, C. (2003). Assessing anti-ageism routes to older worker re-engagement. *Work, Employment and Society, 17*, 101–120.
Duncan, C. (2008). The dangers and limitations of equality agendas as means for tackling old age prejudice. *Ageing and Society, 28*, 1133–1158.
Durkheim, E. 1897 (1951). Suicide : a study in sociology. The Free Press, New York.
Durst, D. (2005). Aging amongst immigrants in Canada: Population drift. *Canadian Studies in Population, 32*, 257–270.
Dury, R. (2014). Social isolation and loneliness in the elderly: An exploration of some of the issues. *British Journal of Community Nursing, 19*, 125–128.
Dwyer, J. W., Lee, G. R., & Jankowski, T. B. (1994). Reciprocity, elder satisfaction, and caregiver stress and burden: The exchange of aid in the family caregiving relationship. *Journal of Marriage and the Family, 56*, 35–43.
Dykstra, P. A. (2009). Older adult loneliness: Myths and realities. *European Journal of Ageing, 6*, 91–100.
Eggebeen, D. J. (1992). Family structure and intergenerational exchanges. *Research on Aging, 14*, 427–447.
Ehni, H. J., & Spindler, M. (2011). Abolishing ageing? The importance of being realistic about the biomedical application of biogerontology and its ethical and social implications. *Bioethica Forum, 4*, 44–50.
Ehrlich, H. J. (1973). *The social psychology of prejudice: A systematic theoretical review and propositional inventory of the American social psychologicalstudy of prejudice*. New York: Wiley.
El Haj, M., Jardri, R., Larøi, F., & Antoine, P. (2016). Hallucinations, loneliness, and social isolation in Alzheimer's disease. *Cognitive Neuropsychiatry, 21*, 1–13.
Elder Jr., G. H., Johnson, M. K., & Cronsoe, R. (2003). The emergence and development of life course theory. In J. T. Mortimer & M. J. Shanahan (Eds.), *Handbook of the life course* (pp. 3–19). New York: Kluwer Academic/Plenum Publishers.
Elkington, J. (1994). Towards the sustainable corporation: Win–win–win business strategies for sustainable development. *California Management Review, 36*, 90–100.
Emami, A., Benner, P., & Ekman, S.-L. (2001). A sociocultural health model for late-in-life immigrants. *Journal of Transcultural Nursing, 12*, 15–24.

Emami, A., & Torres, S. (2005). Making sense of illness: Late-in-life migration as point of departure for elderly Iranian immigrants' explanatory models of illness. *Journal of Immigrant Health, 7*, 153–164.
Emami, A., Torres, S., Lipson, J. G., & Ekman, S.-L. (2000). An ethnographic study of a day care center for Iranian immigrant seniors. *Western Journal of Nursing Research, 22*, 169–188.
Emlet, C. A. (2006). A comparison of HIV stigma and disclosure patterns between older and younger adults living with HIV/AIDS. *AIDS Patient Care and STDs, 20*, 350–358.
Emlet, C. A. (2007). Experiences of stigma in older adults living with HIV/AIDS: A mixed- methods analysis. *AIDS Patient Care and STDs, 21*, 740–752.
Engel, G. (1977). The need for a new medical model: A challenge for biomedicine. *Science, 196*, 129–136.
English, E. (2013). *The effect of community participation on subjective well-being in community dwelling elders*. Honors Projects. Paper 161. Illinois Wesleyan University. Downloaded on March 22, 2017, from http://digitalcommons.iwu.edu/psych_honproj/161
Erikson, E. H. (1997). *The life cycle completed*. New York: Norton.
Erlangsen, A., Jeune, B., Bille-Brahe, U., & Vaupel, J. W. (2004). Loss of partner and suicide risks among oldest old: A population-based register study. *Age and Ageing, 33*, 378–383.
Ertel, K. E., Glymour, M. M., & Berkman, L. F. (2008). Effects of social integration on preserving memory function in a nationally representative US elderly population. *American Journal of Public Health, 98*, 1215–1220.
Esqueda, C. W. (2011). Cross-cultural perspectives on stigma. In R. L. Wiener & S. L. Willborn (Eds.), *Disability and aging discrimination* (pp. 185–204). New York: Springer.
Etters, L., Goodall, D., & Harrison, B. (2008). Caregiver burden among dementia patient caregivers: A review of the literature. *Journal of the American Academy of Nurse Practitioners, 20*, 423–428.
Fahrenberg, B. (1986). Coping with the empty nest situation as a developmental task for the aging female - An analysis of the literature. *Zeitschrift für Gerontologie, 79*, 323–335.
Family Caregiver Alliance. (2017). *Caregiver statistics: Demographics*. Downloaded on April 24, 2017, from https://www.caregiver.org/caregiver-statistics-demographics
Faraone, S. V., Chen, W. J., Goldstein, J. M., & Tsuang, M. T. (1994). Gender differences in age at onset of schizophrenia. *British Journal of Psychiatry, 164*, 625–629.
Faulkner, K. A., Cauley, J. A., Zmuda, J. M., Griffin, J. M., & Nevitt, M. C. (2003). Is social integration associated with the risk of falling in older community-dwelling women? *Journal of Gerontology, 58A*, 954–959.
Fazel, S., Hope, T., O'Donnell, I., Piper, M., & Jacoby, R. (2001). Health of elderly male prisoners: Worse than the general population, worse than younger prisoners. *Age and Ageing, 30*, 403–407.
Fazel, S., Hope, T., O'Donnell, I., & Jacoby, R. (2001a). Hidden psychiatric morbidity in elderly prisoners. *British Journal of Psychiatry, 179*, 535–539.
Fazel, S. (2008). Psychiatric aspects of crime and the elderly. In R. Jacoby, C. Oppenheimer, & T. Dening (Eds.), *Oxford textbook of old age psychiatry* (pp. 747–751). Oxford: Oxford University Press.
Fazel, S., Hope, T., O'Donnell, I., & Jacoby, R. (2004). Unmet treatment needs of older prisoners: A primary care survey. *Age and Ageing, 33*, 396–398.
Fealy, G., McNamara, M., Pearl Treacy, M., & Lyons, I. (2011). Constructing ageing and age identities: A case study of newspaper discourses. *Ageing and Society, 32*, 85–102.
Fees, B. S., Martin, P., & Poon, L. W. (1999). A model of loneliness in older adults. *Journal of Gerontology, 54B*, P231–P239.
Fenech, F. F. (2003). Ethical issues in ageing. *Clinical Medicine, 3*, 232–234.
Ferreira, M. (2005). Elder abuse in Africa: What policy and legal provisions are there to address the violence? *Journal of Elder Abuse and Neglect, 16*, 17–32.
Ferrie, J. P., & Hatton, T. J. (2014). Two centuries of international migration. In B. Chiswick & P. Miller (Eds.), *Handbook of the economics of international migration, Vol. 1A: The immigrants* (pp. 1–53). Amsterdam: North Holland.

Field, D., & Minkler, M. (1988). Continuity and change in social support between young-old and old-old or very-old age. *Journal of Gerontology, 43*, P100–P106.

Finkenauer, S., Sherratt, J., Marlow, J., & Brodey, A. (2012). When injustice gets old: A systematic review of trans aging. *Journal of Gay and Lesbian Social Services, 24*, 311–330.

Fiori, K. L., Antonucci, T. C., & Cortina, K. S. (2006). Social network typologies and mental health among older adults. *Journal of Gerontology, 61B*, P25–P32.

Fiori, K. L., Smith, J., & Antonucci, T. C. (2007). Social network types among older adults: A multidimensional approach. *Journal of Gerontology, 62B*, P322–P330.

Fischer, S. H., David, D., Crotty, B. H., Dierks, M., & Safran, C. (2014). Acceptance and use of health information technology by community-dwelling elders. *International Journal of Medical Informatics, 83*, 624–635.

Fisher, B. S., & Regan, S. L. (2006). The extent and frequency of abuse in the lives of older women and their relationship with health outcomes. *The Gerontologist, 46*, 200–209.

Fiske, S. T. (2000). Stereotyping, prejudice, and discrimination at the seam between the centuries: Evolution, culture, mind, and brain. *European Journal of Social Psychology, 30*, 299–322.

Fiske, S. T., Cuddy, A. J. C., Glick, P., & Xu, J. (2002). A model of (often mixed) stereotype content: Competence and warmth respectively follow from perceived status and competition. *Journal of Personality and Social Psychology, 82*, 878–902.

Fokkema, T., de Jong Gierveld, J., & Dykstra, P. A. (2012). Cross-national differences in older adult loneliness. *The Journal of Psychology, 146*, 201–228.

Formosa, M. (2019). Active ageing through lifelong learning: The university of the third age. In M. Formosa (Ed.), *The university of the third age and active ageing: European and Asian-Pacific perspectives* (pp. 1–18). Cham: Springer.

Forrest, R., & Kearns, A. (2001). Social cohesion, social capital and the neighbourhood. *Urban Studies, 38*, 2125–2143.

Foucault, M. (1988). *Madness and civilization*. New York: Vintage Books.

Fowler, J. H., Dawes, C. T., & Christakis, N. A. (2009). Model of genetic variation in human social networks. *Proceedings of the National Academy of Sciences of the United States of America, 106*, 1720–1724.

Fratiglioni, L., Wang, H.-X., Ericsson, K., Maytan, M., & Winblad, B. (2000). Influence of social network on occurrence of dementia: A community-based longitudinal study. *Lancet, 355*, 1315–1319.

Fratiglioni, L., Paillard-Borg, S., & Winblad, B. (2004). An active and socially integrated lifestyle in late life might protect against dementia. *Lancet Neurology, 3*, 343–353.

Fredriksen-Goldsen, K. I. (2012). Informal caregiving in the LGBT communities. In T. M. Witten & A. E. Eyler (Eds.), *Gay, lesbian, bisexual and transgender aging* (pp. 59–83). Baltimore: The Johns Hopkins University Press.

Fredriksen-Goldsen, K. I., Cook-Daniels, L., Kim, H.-J., Erosheva, E. A., Emlet, C. A., Hoy-Ellis, C. P., et al. (2013). Physical and mental health of transgender older adults: An at-risk and underserved population. *The Gerontologist, 54*, 488–500.

Fredriksen-Goldsen, K. I., Hoy-Ellis, C. P., Goldsen, J., Emlet, C. A., & Hooyman, N. R. (2014). Creating a vision for the future: Key competencies and strategies for culturally competent practice with lesbian, gay, bisexual, and transgender (LGBT) older adults in the health and human services. *Journal of Gerontological Social Work, 57*, 80–107.

Fredriksen-Goldsen, K. I., & Muraco, A. (2010). Aging and sexual orientation: A 25-year review of the literature. *Research on Aging, 32*, 372–413.

Freund, A. M., & Baltes, P. B. (2002). Life-management strategies of selection, optimization, and compensation: Measurement by self-report and construct validity. *Journal of Personality and Social Psychology, 82*, 642–662.

Friend, R. A. (1989). Older lesbian and gay people: Responding to homophobia. *Marriage and Family Review, 14*, 241–263.

Friend, R. A. (1991). Older lesbian and gay people: A theory of successful aging. *Journal of Homosexuality, 20*, 99–118.

Fung, H. H., Carstensen, L. L., & Lang, F. R. (2001). Age-related patterns in social networks among European Americans and African Americans: Implications for socioemotional selectivity across the life span. *International Journal of Aging and Human Development, 52*, 185–206.

Gabbay, S. G., & Wahler, J. J. (2002). Lesbian aging. *Journal of Gay and Lesbian Social Services, 14*, 1–21.

Gallant, M. P., & Connell, C. M. (1997). Predictors of decreased self-care among spouse caregivers of older adults with dementing illnesses. *Journal of Aging and Health, 9*, 373–395.

Gallagher-Thompson, D., & Coon, D. W. (2007). Evidence-based psychological treatments for distress in family caregivers of older adults. *Psychology and Aging, 22*, 37–51.

Gamliel, T., & Hazan, H. (2006). The meaning of stigma: Identity construction in two old-age institutions. *Ageing and Society, 26*, 355–371.

Garcia, J. A., Crocker, J., & Wyman, J. F. (2005). Breaking the cycle of stigmatization: Managing the stigma of incontinence in social interactions. *Journal of Wound, Ostomy and Continence Nursing, 32*, 38–52.

Gardner, P., Kamber, T., & Netherland, J. (2012). Getting turned on: Using ICT training to promote active ageing in New York City. *The Journal of Community Informatics 8*(1). Retrieved March 21, 2017, from http://ci-journal.net/index.php/ciej/article/view/809/907

Garfein, A. J., & Herzog, A. R. (1995). Robust aging among the young-old, old-old, and oldest-old. *Journal of Gerontology, 50B*, S77–S87.

Garlo, K., O'Leary, J. R., Van Ness, P. H., & Fried, T. R. (2010). Caregiver burden in caregivers of older adults with advanced illness. *Journal of the American Geriatrics Society, 58*, 2315–2322.

Garre-Olmo, J., Planas-Pujol, X., López-Pousa, S., Juvinyà, D., Vilà, A., Vilalta-Franch, J., et al. (2009). Prevalence and risk factors of suspected elder abuse subtypes in people aged 75 and older. *Journal of the American Geriatrics Society, 57*, 815–822.

Garstka, T. A., Schmitt, M. T., Branscombe, N. R., & Hummert, M. L. (2004). How young and older adults differ in their responses to perceived age discrimination. *Psychology and Aging, 19*, 326–335.

Gatti, G. (2012). Imposing identity against social catastrophes. The strategies of (re)generation of meaning of the abuelas de Plaza de Mayo (Argentina). *Bulletin of Latin American Research, 31*, 352–365.

Gaugler, J. E., Kane, R. L., Kane, R. A., Clay, T., & Newcomer, R. (2003). Caregiving and institutionalization of cognitively impaired older people: Utilizing dynamic predictors of change. *The Gerontologist, 43*, 219–229.

Genke, J. (2004). Resistance and resilience: The untold story of gay men aging with chronic illnesses. *Journal of Gay and Lesbian Social Services, 17*, 81–95.

Giaccardi, E., Kuijer, L., & Neven, L. (2016). Design for resourceful ageing : Intervening in the ethics of gerontechnology. In P. Lloyd & E. Bohemia (Eds.), *Proceedings of DRS 2016, Design + Research + Society Future-Future-Focused Thinking* (Vol. 1, pp. 1–14). 50th Anniversary International Conference, Brighton, UK, 27–30 June 2016.

Gilbert, D. T., & Hixon, J. G. (1991). The trouble of thinking: Activation and application of stereotypic beliefs. *Journal of Personality and Social Psychology, 60*, 509–517.

Giles, H., & Reid, S. A. (2005). Ageism across the lifespan: Towards a self-categorization model of ageing. *Journal of Social Issues, 61*, 389–404.

Gilleard, C., & Higgs, P. (2011). Ageing abjection and embodiment in the fourth age. *Journal of Aging Studies, 25*, 135–142.

Giurani, F., & Hasan, M. (2000). Abuse in elderly people: The Granny Battering revisited. *Archives of Gerontology and Geriatrics, 31*, 215–220.

Goergen, T. (2001). Stress, conflict, elder abuse and neglect in German nursing homes: A pilot study among professional caregivers. *Journal of Elder Abuse and Neglect, 13*, 1–26.

Goetting, A. (1983). The elderly in prison: Issues and perspectives. *Journal of Research in Crime and Delinquency, 1983*, 291–309.

Goetting, A. (1985). Racism, sexism, and ageism in the prison community. *Federal Probation, 49*, 10–22.

Goffman, E. (1963). *Stigma.* Englewood Cliffs, NJ: Prentice-Hall.

Gold, D. T. (1989). Sibling relationships in old age: A typology. *International Journal of Aging and Human Development, 28*, 37–51.
Golden, J., Conroy, R. M., Bruce, I., Denihan, A., Greene, E., Kirby, M., et al. (2009). Loneliness, social support networks, mood and wellbeing in community-dwelling elderly. *International Journal of Geriatric Psychiatry, 24*, 694–700.
Goode, W. J. (1960). A theory of role strain. *American Sociological Review, 25*, 483–496.
Gorbien, M. J., & Eisenstein, A. R. (2005). Elder abuse and neglect: An overview. *Clinics in Geriatric Medicine, 21*, 279–292.
Gordon, A. R., & Meyer, I. H. (2007). Gender nonconformity as a target of prejudice, discrimination, and violence against LGB individuals. *Journal of LGBT Health Research, 3*, 55–71.
Gorey, K. M. (1998). The beneficial effects of volunteering for older volunteers and the people they serve: A meta-analysis. *International Journal of Aging and Human Development, 47*, 69–79.
Graham, N., Lindesay, J., Katona, C., Bertolote, J. M., Camus, V., Copeland, J. R. M., et al. (2003). Reducing stigma and discrimination against older people with mental disorders: A technical consensus statement. *International Journal of Geriatric Psychiatry, 18*, 670–678.
Granovetter, M. S. (1973). The strength of weak ties. *American Journal of Sociology, 78*, 1360–1380.
Gray, A. (2009). The social capital of older people. *Ageing and Society, 29*, 5–31.
Greenberg, J., Pyszczynski, T., & Solomon, S. (1986). The causes and consequences of a need for self-esteem: A terror management theory. In R. F. Baumeister (Ed.), *Public self and private self* (pp. 189–212). New York: Springer.
Greenberg, J. S., Seltzer, M. M., & Greenley, J. R. (1993). Aging parents of adults with disabilities: The gratifications and frustrations of later-life caregiving. *The Gerontologist, 33*, 542–550.
Griffin, G., & Aitken, L. (1999). Visibility blues: Gender issues in elder abuse in institutional settings. *Journal of Elder Abuse and Neglect, 10*, 29–42.
Gringart, E., Helmes, E., & Speelman, C. (2008). Harnessing cognitive dissonance to promote positive attitudes toward older workers in Australia. *Journal of Applied Social Psychology, 38*, 751–778.
Gringart, E., Helmes, E., & Speelman, C. P. (2005). Exploring attitudes toward older workers among Australian employers: An empirical study. *Journal of Aging and Social Policy, 17*, 85–103.
Groombridge, B. (1995). Emergent challenges for Universities of the Third Age. In E. Heikkinen, J. Kuusinen, & I. Ruoppila (Eds.), *Preparing for aging* (pp. 27–38). New York: Plenum.
Haber, D. (2009). Gay aging. *Gerontology and Geriatrics Education, 30*, 267–280.
Habermas, J. (2010). The concept of human dignity and the realistic utopia of human rights. *Metaphilosophy, 41*, 464–480.
Hagestad, G. O., & Uhlenberg, P. (2006). Should we be concerned about age segregation? Some theoretical and empirical explorations. *Research on Aging, 28*, 638–653.
Haley, W. E. (2003). Family caregivers of elderly patients with cancer: Understanding and minimizing the burden of care. *Journal of Supportive Oncology, 1*(Suppl. 2), 25–29.
Hall Gueldner, S., Loeb, S., Morris, D., Penrod, J., Bramlett, M., Johnston, L., et al. (2001). A comparison of life satisfaction and mood in nursing home residents and community-dwelling elders. *Archives of Psychiatric Nursing, 15*, 232–240.
Hamilton, D. L., & Trolier, T. K. (1986). Stereotypes and stereotyping: An overview of the cognitive approach. In J. F. Dovidio & S. L. Gaertner (Eds.), *Prejudice, discrimination, and racism* (pp. 127–163). San Diego, CA: Academic.
Hamilton, W. D. (1963). The evolution of altruistic behaviour. *American Naturalist, 97*, 354–356.
Handtke, V., & Wangmo, T. (2014). Ageing prisoners' views on death and dying: Contemplating end-of-life in prison. *Bioethical Inquiry, 11*, 373–386.
Harrison, J. (2005). Pink, lavender and grey: Gay, lesbian, bisexual, transgender and intersex ageing in Australian gerontology. *Gay and Lesbian Issues and Psychology Review, 1*, 11–16.
Harrison, J. (2006). Coming out ready or not! Gay, lesbian, bisexual transgender and intersex ageing and aged care in Australia: Reflections, contemporary developments and the road ahead. *Gay and Lesbian Issues and Psychology Review, 2*, 44–53.

Harwood, J., Hewstone, M., Paolini, S., & Voci, A. (2005). Grandparent–grandchild contact and attitudes toward older adults: Moderator and mediator effects. *Personality and Social Psychology Bulletin, 30*, 770–786.

Haugebrook, S., Zgoba, K. M., Maschi, T., Morgen, K., & Brown, D. (2010). Trauma, stress, health, and mental health issues among ethnically diverse older adult prisoners. *Journal of Correctional Health Care, 16*, 220–229.

Havighurst, R. L., & Albrecht, R. (1953). *Older people*. Green, NY: Longmans.

Hawkley, L. C., & Cacioppo, J. T. (2010). Loneliness matters: A theoretical and empirical review of consequences and mechanisms. *Annals of Behavioral and Medicine, 40*, 218–227.

Hayslip Jr., B., & Kaminski, P. L. (2005). Grandparents raising their grandchildren: A review of the literature and suggestions for practice. *The Gerontologist, 45*, 262–269.

Hazer, O., & Boylu, A. A. (2010). The examination of the factors affecting the feeling of loneliness of the elderly. *Procedia Social and Behavioral Sciences, 9*, 2083–2089.

Heaphy, B., Yip, A. K. T., & Thompson, D. (2004). Ageing in a non-heterosexual context. *Ageing and Society, 24*, 881–902.

Hedberg, P., Gustafson, Y., Brulin, C., & Aléx, L. (2013). Purpose in life among very old men. *Advances in Aging Research, 2*, 100–105.

Heikkilä, K., & Ekman, S.-L. (2000). Health care experiences and beliefs of elderly Finnish immigrants in Sweden. *Journal of Transcultural Nursing, 11*, 281–289.

Helliwell, J. F., & Putnam, R. D. (2004). The social context of well-being. *Philosophical Transactions: Biological Sciences, 359*, 1435–1446.

Hendrix, C. C. (2000). Computer use among elderly people. *Computers in Nursing, 18*, 62–68. quiz 69-71.

Henkens, K., van Solinge, H., & Gallo, W. T. (2008). Effects of retirement voluntariness on changes in smoking, drinking and physical activity among Dutch older workers. *European Journal of Public Health, 18*, 644–649.

Hense, R. L., Penner, L. A., & Nelson, D. L. (1995). Implicit memory for age stereotypes. *Social Cognition, 13*, 399–415.

Hensley, B., Martin, P., Margrett, J. A., MacDonald, M., Siegler, I. C., & Poon, L. W. (2012). Life events and personality predicting loneliness among centenarians: Findings from the Georgia Centenarian Study. *The Journal of Psychology, 146*, 173–188.

Herlyn, I., & Lehmann, B. (1998). Grossmutterschaft im Mehrgenerationenenzusammenhang: Eine empirische Untersuchung aus der Perspektive von Grossmüttern. *Zeitschrift für Familienforschung, 10*, 27–45.

Heylen, L. (2010). The older, the lonelier? Risk factors for social loneliness in old age. *Ageing and Society, 30*, 1177–1196.

Himes, C. L., Hogan, D. P., & Eggebeen, D. J. (1996). Living arrangements of minority elders. *Journal of Gerontology, 51B*, S42–S48.

Hislop, J., & Arber, S. (2003). Sleep as a social act: A window on gender roles and relationships. In S. Arber, K. Davidson, & J. Ginn (Eds.), *Gender and ageing: Changing roles and relationships* (pp. 186–205). Maidenhead: Open University Press.

Hogan, A. J. (2019). Social and medical models of disability and mental health: Evolution and renewal. *Canadian Medical Association Journal, 191*, E16–E18.

Hogstel, M. O., Curry, L. C., & Walker, C. (2005/2006). Caring for older adults: The benefits of informal family caregiving. *Journal of Theory Construction and Testing, 9*, 55–60.

Holmén, K., Ericsson, K., & Winblad, B. (2000). Social and emotional loneliness among non-demented and demented elderly people. *Archives of Gerontology and Geriatrics, 31*, 177–192.

Holmén, K., & Furukawa, H. (2002). Loneliness, health and social network among elderly people - A follow-up study. *Archives of Gerontology and Geriatrics, 35*, 261–274.

Holt-Lunstad, J., Smith, T. B., & Layton, J. B. (2010). Social relationships and mortality risk: A meta-analytic review. *PLoS Medicine, 7*, e1000316. https://doi.org/10.1371/journal.pmed.1000316

Homans, G. C. (1958). Social behaviour as exchange. *American Journal of Sociology, 63*, 597–606.

Homer, A. C., & Gilleard, C. (1990). Abuse of elderly people by their carers. *British Medical Journal, 301*, 1359–1362.

Horn, V. (2017). Cross-border mobility and long-distance communication as modes of care circulation: Insights from the Peruvian 'zero generation'. *Journal of Ethnic and Migration Studies, 43*, 303–320.

Horrocks, S., Somerset, M., Stoddart, H., & Peters, T. J. (2004). What prevents older people from seeking treatment for urinary incontinence? A qualitative exploration of barriers to the use of community continence services. *Family Practice, 21*, 689–696.

Hossen, A. (2012). Social isolation and loneliness among elderly immigrants: The case of South Asian elderly living in Canada. *Journal of International Social Issues, 1*, 1–10.

House, J. S., Landis, K. R., & Umberson, D. (1988). Social relationships and health. *Science, 241*, 540–545.

Hughes, M. (2007). Older lesbians and gays accessing health and aged-care services. *Australian Social Work, 60*, 197–209.

Hughes, M. (2010). Expectations of later life support among lesbian and gay Queenslanders. *Australasian Journal on Ageing, 29*, 161–166.

Hughes, M., & Heycox, K. (2010). *Older people, ageing and social work: Knowledge for practice*. Crows Nest, NSW: Allen and Unwin.

Hughes, M., & Kentlyn, S. (2011). Older LGBT people's care networks and communities of practice: A brief note. *International Social Work, 54*, 436–444.

Hurd, L. C. (1999). "WE'RE NOT OLD!": Older women's negotiation of aging and oldness. *Journal of Aging Studies, 13*, 419–439.

Husserl, E. (1970). *The crisis of European sciences and transcendental phenomenology*. Evanston, IL: Northwestern University Press.

Hwalek, M., Straub, V., & Kosniewski, K. (2008). Older workers: An opportunity to expand the long-term care/direct care labor force. *The Gerontologist, 48*, 90–103.

Iecovich, E. (2005). Elder abuse and neglect in Israel: A comparison between the general elderly population and elderly new immigrants. *Family Relations, 54*, 436–447.

Isaacowitz, D. M., & Smith, J. (2003). Positive and negative affect in very old age. *Journal of Gerontology, 58B*, P143–P152.

Isaacs, L. W., & Bearison, D. J. (1986). The development of children's prejudice against the aged. *International Journal of Aging and Human Development, 23*, 175–194.

Ivbijaro, G. (2013). Loneliness and the elderly: Opportunities for health promotion. *Mental Health in Family Medicine, 10*, 1–2.

Iversen, T. N., Larsen, L., & Solem, P. E. (2009). A conceptual analysis of Ageism. *Nordic Psychology, 61*, 4–22.

Jacobs, J. (1960). *The death and life of great American cities*. New York: Vintage.

Jacobson, S., & Samdhal, D. M. (1998). Leisure in the lives of old lesbians: Experiences with and responses to discrimination. *Journal of Leisure Research, 30*, 233–255.

Jansen, D. A. (2005). Perceived barriers to participation in mentally restorative activities by community-dwelling elders. *Activities, Adaptation and Aging, 29*, 35–53.

Jansen, D. A. (2006). Attentional demands and daily functioning among community- dwelling elders. *Journal of Community and Health Nursing, 23*, 1–13.

Jansen, D. A. (2008). Mentally restorative activities and daily functioning among community-dwelling elders. *Activities, Adaptation and Aging, 32*, 181–197.

Jansen, D. A., & von Sadovszky, V. (2004). Restorative activities of community-dwelling elders. *Western Journal of Nursing Research, 26*, 381–399.

Jappens, M., & Van Bavel, J. (2012). Regional family norms and child care by grandparents in Europe. *Demographic Research, 27*, 85–120.

Jervis, L. L., Boland, M. E., & Fickenscher, A. (2010). American Indian family caregivers' experiences with helping elders. *Journal of Cross-Cultural Gerontology, 25*, 355–369.

Johannesen, M., & LoGiudice, D. (2013). Elder abuse: A systematic review of risk factors in community-dwelling elders. *Age and Ageing, 42*, 292–298.

Johansson, K., Rudman, D. L., Mondaca, M., Park, M., Luborsky, M., Josephsson, S., et al. (2013). Moving beyond 'aging in place' to understand migration and aging: Place making and the centrality of occupation. *Journal of Occupational Science, 20*, 108–119.

Johnson, C. L. (1999). Fictive kin among oldest old African Americans in the San Francisco bay area. *Journal of Gerontology, 54B*, S368–S375.

Johnson, C. L., & Barer, B. M. (1992). Patterns of engagement and disengagement among the oldest old. *Journal of Aging Studies, 6*, 351–364.

Johnson, C. L., & Catalano, D. J. (1981). Childless elderly and their family supports. *The Gerontologist, 21*, 610–618.

Johnson, D. P., & Mullins, L. C. (1987). Growing old and lonely in different societies: Toward a comparative perspective. *Journal of Cross-Cultural Gerontology, 2*, 257–275.

Jones, D. A., Victor, C. R., & Vetter, N. J. (1985). The problem of loneliness in the elderly in the community: Characteristics of those who are lonely and the factors related to loneliness. *Journal of the Royal College of General Practitioners, 35*, 136–139.

Jones, E. E., Farina, A., Hastorf, A. H., Marcus, H., Miller, D. T., & Scott, R. A. (1984). *Social stigma: The psychology of marked relationships*. New York: Freeman and Company.

Jones, J., & Pugh, S. (2005). Ageing gay men: Lessons from the sociology of embodiment. *Men and Masculinities, 7*, 248–260.

Jones, J. M., Dovidio, J. F., & Vietze, D. L. (2014). *The psychology of diversity: Beyond prejudice and racism*. West Sussex: Wiley.

Jopp, D., & Rott, C. (2006). Adaptation in very old age: Exploring the role of resources, beliefs, and attitudes for centenarians' happiness. *Psychology and Aging, 21*, 266–280.

Jussim, L., Cain, T. R., Crawford, J. T., Harber, K., & Cohen, F. (2009). The unbearable accuracy of stereotypes. In T. D. Nelson (Ed.), *Handbook of prejudice, stereotyping, and discrimination* (pp. 199–227). New York: Psychology Press.

Jylhä, M., & Jokela, J. (1990). Individual experiences as cultural - A cross-cultural study on loneliness among the elderly. *Ageing and Society, 10*, 295–315.

Jylhä, M., & Saarenheimo, M. (2010). Loneliness and aging: Comparative perspectives. In C. R. Phillipson & D. Dannefer (Eds.), *The SAGE handbook of social gerontology* (pp. 317–328). Thousand Oaks, CA: Sage.

Kaasa, K. (1998). Loneliness in old age: Psychosocial and health predictors. *Norwegian Journal of Epidemiology, 8*, 195–201.

Kahn, R. L., & Antonucci, T. C. (1980). Convoys over the life course: Attachment, roles, and social support. In P. B. Baltes & O. C. Brim (Eds.), *Life-span, development, and behavior* (pp. 254–283). New York: Academic Press.

Kakoullis, A., Le Mesurier, N., & Kingston, P. (2010). The mental health of older prisoners. *International Psychogeriatrics, 22*, 693–701.

Kaplan, S. (1995). The restorative benefits of nature: Toward an integrative framework. *Journal of Environmental Psychology, 15*, 169–182.

Karl, U., Ramos, A. C., & Kühn, B. (2017). Older migrants in Luxembourg – Care preferences for old age between family and professional services. *Journal of Ethnic and Migration Studies, 43*, 270–286.

Kasuya, R., Polgar-Bailey, P., & Takeuchi, R. (2000). Caregiver burden and burnout: A guide for primary care physicians. *Postgraduate Medicine, 108*, 1–7.

Kelchner, E. S. (2000). Ageism's impact and effect on society. *Journal of Gerontological Social Work, 32*, 85–100.

Kelly, J. (1977). The aging male homosexual: Myth and reality. *The Gerontologist, 17*, 328–332.

Kemper, P. (1992). The use of formal and informal home care by the disabled elderly. *Health Services Research, 27*, 421–451.

Kerbs, J. J., & Jolley, J. M. (2007). Inmate-on-inmate victimization among older male prisoners. *Crime and Delinquency, 53*, 187–218.

Kerbs, J. J., & Jolley, J. M. (2009). A commentary on age segregation for older prisoners philosophical and pragmatic considerations for correctional systems. *Criminal Justice Review, 34*, 119–139.

Khvorostianov, N., Elias, N., & Nimrod, G. (2011). 'Without it I am nothing': The internet in the lives of older immigrants. *New Media and Society*. Downloaded on April 8, 2017, from https://doi.org/10.1177/1461444811421599

Kim, B. J., Sangalang, C. C., & Kihl, T. (2012). Effects of acculturation and social network support on depression among elderly Korean immigrants. *Aging and Mental Health, 16*, 787–794.
Kim, J., Chun, S., Heo, J., Lee, S., & Han, A. (2016). Contribution of leisure-time physical activity on psychological benefits among elderly immigrants. *Applied Research Quality Life, 11*, 461–470.
Kim, J. I. (2013). Levels of health-related quality of life (EQ-5D) and its related factors among vulnerable elders receiving home visiting health care services in some rural areas. *Journal of the Korean Academy of Community Health Nursing, 24*, 99–109.
Kim, K.-C., Kim, S., & Hurh, W. M. (1991). Filial piety and intergenerational relationship in Korean immigrant families. *International Journal of Aging and Human Development, 33*, 233–245.
Kim, S. S., Reed, P. G., Hayward, R. D., Kang, Y., & Koenig, H. G. (2011). Spirituality and Psychological well-being: Testing a theory of family interdependence among family caregivers and their elders. *Research in Nursing and Health, 34*, 103–115.
Kimmel, D., Rose, T., Orel, N., & Greene, B. (2006). Historical context for research on lesbian, bisexual, and transgender aging. In D. Kimmel, T. Rose, & S. David (Eds.), *Lesbian, gay, bisexual, and transgender aging: Research and clinical perspectives* (pp. 1–19). New York: Columbia University Press.
Kimsey, L. R., Tarbox, A. R., & Bragg, D. F. (1981). Abuse of the elderly- the hidden agenda. I. The caretakers and the categories of abuse. *Journal of the American Geriatrics Society, 29*, 465–472.
King, R., Lulle, A., Sampaio, D., & Vullnetari, J. (2017). Unpacking the ageing – Migration nexus and challenging the vulnerability trope. *Journal of Ethnic and Migration Studies, 43*, 182–198.
Kinsella, K., & Phillips, D. R. (2005). Global aging: The challenge of success. *Population Bulletin, 60*, 5–42.
Kite, M. E., Stockdale, G. D., Whitley, B. E., Jr., & Johnson, B. T. (2005). Attitudes toward younger and older adults: An updated meta-analytic review. *Journal of Social Issues, 61*, 241–266.
Kivett, V. R. (1979). Discriminators of loneliness among the rural elderly: Implications for intervention. *The Gerontologist, 19*, 108–115.
Kleinspehn-Ammerlahn, A., Kotter-Grühn, D., & Smith, J. (2008). Self-perceptions of aging: Do subjective age and satisfaction with aging change during old age? *Journal of Gerontology, 63B*, P377–P385.
Knochel, K. A., Quam, J. K., & Croghan, C. F. (2011). Are old lesbian and gay people well served? Understanding the perceptions, preparation, and experiences of aging services providers. *Journal of Applied Gerontology, 30*, 370–389.
Knochel, K. A., Croghan, C. F., Moone, R. P., & Quam, J. K. (2012). Training, geography, and provision of aging services to lesbian, gay, bisexual, and transgender older adults. *Journal of Gerontological Social Work, 55*, 426–443.
Koch, S., Hunter, P., & Nair, K. (2009). Older people in acute care. In R. Nay & S. Garratt (Eds.), *Older people: Issues and innovations in care* (3rd ed., pp. 153–167). Sydney, NSW: Churchill Livingstone Elsevier.
Koenig, H. G. (1995). Religion and older men in prison. *International Journal of Geriatric Psychiatry, 10*, 219–230.
Koren, C. (2011). Continuity and discontinuity: The case of second couplehood in old age. *The Gerontologist, 51*, 687–698.
Kramer, B. J., & Kipnis, S. (1995). Eldercare and work-role conflict: Toward an understanding of gender differences in caregiver burden. *The Gerontologist, 35*, 340–348.
Kuhlmann, R., & Ruddell, R. (2005). Elderly jail inmates: Problems, prevalence and public health. *Californian Journal of Health Promotion, 3*, 49–60.
Kumamoto, K., Arai, Y., & Zarit, S. H. (2006). Use of home care services effectively reduces feelings of burden among family caregivers of disabled elderly in Japan: Preliminary results. *International Journal of Geriatric Psychiatry, 21*, 163–170.
Kurrle, S., & Naughtin, G. (2008). An overview of elder abuse and neglect in Australia. *Journal of Elder Abuse and Neglect, 20*, 108–125.

Kushner, B., Neville, S., & Adams, J. (2013). Perceptions of ageing as an older gay man: A qualitative study. *Journal of Clinical Nursing, 22*, 3388–3395.
Kweon, B.-S., Sullivan, W. C., & Wiley, A. R. (1998). Green common spaces and the social integration of inner-city older adults. *Environment and Behavior, 30*, 832–858.
Lachs, M. S., & Pillemer, K. (2004). Elder abuse. *Lancet, 364*, 1263–1272.
Lai, D. W. L. (2001). Use of senior center services of the elderly Chinese immigrants. *Journal of Gerontological Social Work, 35*, 59–79.
Lai, D. W. L., & Chau, S. B. Y. (2007). Predictors of health service barriers for older Chinese immigrants in Canada. *Health and Social Work, 32*, 57–65.
Lan, P.-C. (2002). Subcontracting filial piety: Elder care in ethnic Chinese immigrant families in California. *Journal of Family Issues, 23*, 812–835.
Lang, F. R. (2001). Regulation of social relationships in later adulthood. *Journal of Gerontology, 56B*, P321–P326.
Langa, K. M., Chernew, M. E., Kabeto, M. U., Herzog, A. R., Ofstedal, M. B., Willis, R. J., et al. (2001). National estimates of the quantity and cost of informal caregiving for the elderly with dementia. *Journal of General Internal Medicine, 16*, 770–778.
Lara, E., Martín-María, N., De la Torre-Luque, A., Koyanagi, A., Vancampfort, D., Izquierdo, A., et al. (2019). Does loneliness contribute to mild cognitive impairment and dementia? A systematic review and meta-analysis of longitudinal studies. *Ageing Research Reviews, 52*, 7–16.
Lea, A. J., Blumstein, D. T., Wey, T. W., & Martin, J. G. A. (2010). Heritable victimization and the benefits of agonistic relationships. *Proceedings of the National Academy of Sciences of the United States of America, 107*, 21587–21592.
Lee, E.-K. O., & Chan, K. (2009). Religious/spiritual and other adaptive coping strategies among Chinese American older immigrants. *Journal of Gerontological Social Work, 52*, 517–533.
Lee, H. Y., & Eaton, C. K. (2009). Financial abuse in elderly Korean immigrants: Mixed analysis of the role of culture on perception and help-seeking intention. *Journal of Gerontological Social Work, 52*, 463–488.
Lee, M. S., Crittenden, K. S., & Yu, E. (1996). Social support and depression among elderly Korean immigrants in the United States. *International Journal of Aging and Human Development, 42*, 313–327.
Lee, R. M., & Robbins, S. B. (1995). Measuring belongingness: The social connectedness and the social assurance scales. *Journal of Counseling Psychology, 42*, 232–241.
Lee, Y.-M. (2007). The immigration experience among elderly Korean immigrants. *Journal of Psychiatric and Mental Health Nursing, 14*, 403–410.
Lemieux, C. M., Dyeson, T. B., & Castiglione, B. (2002). Revisiting the literature on prisoners who are older: Are we wiser? *The Prison Journal, 82*, 440–458.
Lennartsson, C., & Silverstein, M. (2001). Does engagement with life enhance survival of elderly people in Sweden? The role of social and leisure activities. *Journal of Gerontology, 56B*, S335–S342.
Lerner, M. J. (1980). *The belief in a just world: A fundamental delusion*. New York: Plenum Press.
Levi-Strauss, C. (1964). Reciprocity, the essence of social life. In R. L. Coser (Ed.), *The family: Its structure and functions* (pp. 36–48). New York: St Martin's Press.
Levy, B. (1996). Improving memory in old age through implicit self-stereotyping. *Journal of Personality and Social Psychology, 71*, 1092–1107.
Levy, B. (2009). Stereotype embodiment: A psychosocial approach to aging. *Current Directions in Psychological Sciences., 18*, 332–336.
Levy, B. R. (2001). Eradication of ageism requires addressing the enemy within. *The Gerontologist, 41*, 578–579.
Levy, B. R., & Banaji, M. R. (2002). Implicit ageism. In T. D. Nelson (Ed.), *Ageism: Stereotyping and prejudice against older persons* (pp. 49–75). Cambridge, MA: The MIT Press.
Levy, S. R., & Hughes, J. M. (2009). Development of racial and ethnic prejudice among children. In T. D. Nelson (Ed.), *Handbook of prejudice, stereotyping, and discrimination* (pp. 23–42). New York: Psychology Press.
Li, C. (1879–1885). *Sacred books of the East. I & II* (J. Legge, Trans.). Oxford: Clarendon Press.

Li, H., Stewart, B. J., Imle, M. A., Archbold, P. G., & Felve, L. (2000). Families and Hospitalized elders: A typology of family care actions. *Research in Nursing and Health, 23*, 3–16.
Li, W. W., & Chong, M. D. (2012). Transnationalism, social wellbeing and older Chinese migrants. *Graduate Journal of Asia-Pacific Studies, 8*, 29–44.
Liang, J., Dvorkin, L., Kahana, E., & Mazian, F. (1980). Social integration and morale a re- examination. *Journal of Gerontology, 35*, 746–757.
Lindley, S. E., Harper, R., & Sellen, A. (2008). Designing for elders: Exploring the complexity of relationships in later life. In *Proceedings of the 22nd British HCI group annual conference on people and computers: Culture, creativity, interaction-Volume 1* (pp. 77–86). London: British Computer Society.
Link, B. Q., Yang, L. H., Phelan, J. C., & Collins, P. Y. (2004). Measuring mental illness stigma. *Schizophrenia Bulletin, 30*, 511–541.
Litwak, E. (1985). *Helping the elderly: The complementary roles of informal networks and formal systems*. New York: Guilford.
Litwin, H. (1995). The social networks of elderly immigrants: An analytic typology. *Journal of Aging Studies, 9*, 155–174.
Litwin, H. (1997). The network shifts of elderly immigrants: The case of Soviet Jews in Israel. *Journal of Cross-Cultural Gerontology, 12*, 45–60.
Litwin, H. (2001). Social network type and morale in old age. *The Gerontologist, 41*, 516–524.
Litwin, H., & Landau, R. (2000). Social network type and social support among the old-old. *Journal of Aging Studies, 14*, 213–228.
Liu, L.-J., & Guo, Q. (2007). Loneliness and health-related quality of life for the empty nest elderly in the rural area of a mountainous county in China. *Quality of Life Research, 16*, 1275–1280.
Liversage, A., & Mirdal, G. M. (2017). Growing old in exile – A longitudinal study of migrant women from Turkey. *Journal of Ethnic and Migration Studies, 43*, 287–302.
Loeb, S. J., & Steffensmeier, D. (2006). Older male prisoners: Health status, self-efficacy beliefs, and health-promoting behaviors. *Journal of Correctional Health Care, 12*, 269–278.
Long, M. V., & Martin, P. (2000). Personality, relationship closeness, and loneliness of oldest old adults and their children. *Journal of Gerontology, 55B*, P311–P319.
Longino Jr., C. F., McClelland, K. A., & Peterson, W. A. (1980). The Aged Subculture Hypothesis: Social integration, gerontophilia and self-conception. *Journal of Gerontology, 35*, 758–767.
Lopez-Class, M., González Castro, F., & Ramirez, A. G. (2011). Conceptions of acculturation: A review and statement of critical issues. *Social Science and Medicine, 72*, 1555–1562.
Losada, A., Marquez-Gonzalez, M., Knight, B. G., Yanguas, J., Sayegh, P., & Romero-Moreno, R. (2010). Psychosocial factors and caregivers' distress: Effects of familism and dysfunctional thoughts. *Aging and Mental Health, 14*, 193–202.
Lowenstein, A. (2007). Solidarity–conflict and ambivalence: Testing two conceptual frameworks and their impact on quality of life for older family members. *Journal of Gerontology, 62B*, S100–S107.
Lowenstein, A. (2009). Elder abuse and neglect—"Old phenomenon": New directions for research, legislation, and service developments. *Journal of Elder Abuse and Neglect, 21*, 278–287.
Luanaigh, O. C., & Lawlorz, B. A. (2008). Loneliness and the health of older people. *International Journal of Geriatric Psychiatry, 23*, 1213–1221.
Luo, Y., Hawkley, L. C., Waite, L. J., & Cacioppo, J. T. (2012). Loneliness, health, and mortality in old age: A national longitudinal study. *Social Sciences and Medicine, 74*, 907–914.
Lüscher, K. (2004). Conceptualising and uncovering intergenerational ambivalence. In K. Pillemer & K. Lüscher (Eds.), *ntergenerational ambivalences: New perspectives on parent-child relations in later life* (pp. 23–62). Oxford: Elsevier Science.
Lützén, K., & Ewalds-Kvist, B. (2013). Moral distress and its interconnection with moral sensitivity and moral resilience: Viewed from the philosophy of Viktor E. Frankl. *Bioethical Inquiry, 10*, 317–324.
Lyons, A., Pitts, M., & Grierson, J. (2013). Factors related to positive mental health in a stigmatized minority: An investigation of older gay men. *Journal of Aging and Health, 25*, 1159–1181.

MacRae, H. (1992). Fictive kin as a component of the social networks of older people. *Research on Aging, 14*, 226–247.
Magaña, S. M., Ramírez García, J. I., Hernández, M. G., & Cortez, R. (2007). Psychological distress among Latino family caregivers of adults with schizophrenia: The roles of burden and stigma. *Psychiatric Services, 58*, 378–384.
Magilvy, J. K., Congdon, J. G., Martinez, R. J., Davis, R., & Averill, J. (2000). Caring for our own: Health care experiences of rural Hispanic elders. *Journal of Aging Studies, 14*, 171–190.
Magnusson, L., Hanson, E., & Borg, M. (2004). A literature review study of Information and Communication Technology as a support for frail older people living at home and their family carers. *Technology and Disability, 16*, 223–235.
Mahoney, D. F., Cloutterbuck, J., Neary, S., & Zhan, L. (2005). African American, Chinese, and Latino family caregivers' impressions of the onset and diagnosis of dementia: Cross-cultural similarities and differences. *The Gerontologist, 45*, 783–792.
Major, B., & Sawyer, P. J. (2009). Attributions to discrimination: Antecedents and consequences. In T. D. Nelson (Ed.), *Handbook of prejudice, stereotyping, and discrimination* (pp. 89–110). New York: Psychology Press.
Marquis, R. (2002). Quality in aged care: A question of relational ethics? *Australasian Journal on Ageing, 21*, 25–29.
Maschi, T., Morgen, K., Zgoba, K., Courtney, D., & Ristow, J. (2011a). Age, cumulative trauma and stressful life events, and post-traumatic stress symptoms among older adults in prison: Do subjective impressions matter? *The Gerontologist, 51*, 675–686.
Maschi, T., Kwak, J., Ko, E., & Morrissey, M. B. (2012). Forget me not: Dementia in prison. *The Gerontologist, 52*, 441–451.
Maschi, T., Morrisey, M. B., & Leigey, M. (2013). The case for human agency, well- being, and community reintegration for people aging in prison: A statewide case analysis. *Journal of Correctional Health Care, 19*, 194–210.
Maschi, T., Viola, D., Morgen, K., & Koskinen, L. (2015). Trauma, stress, grief, loss, and separation among older adults in prison: The protective role of coping resources on physical and mental well-being. *Journal of Crime and Justice, 38*, 113–136.
Matsumoto, N., Ikeda, M., Fukuhara, R., Shinagawa, S., Ishikawa, T., Mori, T., et al. (2007). Caregiver burden associated with behavioral and psychological symptoms of dementia in elderly people in the local community. *Dementia and Geriatric Cognitive Disorders, 23*, 219–224.
Matthews, S. H., & Rosner, T. T. (1988). Shared filial responsibility: The family as the primary caregiver. *Journal of Marriage and the Family, 50*, 185–195.
McArdle, P. (2012). Ageing: The new ethical frontier. *Journal of Religion, Spirituality and Aging, 24*, 20–29.
McCann, E., Sharek, D., Higgins, A., Sheerin, F., & Glacken, M. (2013). Lesbian, gay, bisexual and transgender older people in Ireland: Mental health issues. *Aging and Mental Health, 17*, 358–365.
McIlvane, J. M., & Reinhardt, J. P. (2001). Interactive effect of support from family and friends in visually impaired elders. *Journal of Gerontology, 56B*, P374–P382.
McInnis, G. J., & White, J. H. (2001). A phenomenological exploration of loneliness in the older adult. *Archives of Psychiatric Nursing, 15*, 128–139.
McLaren, S. (2016). The relationship between living alone and depressive symptoms among older gay men: The moderating role of sense of belonging with gay friends. *International Psychogeriatrics, 28*, 1895–1901.
McShane, M. D., & Williams, F. P. I. I. I. (1990). Old and ornery: The disciplinary experiences of elderly prisoners. *International Journal of Offender Therapy and Comparative Criminology, 34*, 197–212.
Mead, G. H. (1934). *Mind, self, and society*. Chicago: University of Chicago Press.
Mehta, K. (1997). Cultural scripts and the social integration of older people. *Ageing and Society, 17*, 253–275.

Melchiorre, M. G., Chiatti, C., Lamura, G., Torres-Gonzales, F., Stankunas, M., Lindert, J., et al. (2013). Social support, socio-economic status, health and abuse among older people in seven European countries. *PLoS One, 8*(1), e54856. https://doi.org/10.1371/journal.pone.0054856

Melucci, A. (1996). *The playing self: Person and meaning in the planetary society*. Cambridge: Cambridge University Press.

Mendes, W. B., McCoy, S., Major, B., & Blascovich, J. (2008). How attributional ambiguity shapes physiological and emotional responses to social rejection and acceptance. *Journal of Personality and Social Psychology, 94*, 278–291.

Mezuk, B., & Rebok, G. W. (2008). Social integration and social support among older adults following driving cessation. *Journal of Gerontology, 63B*, S298–S303.

Miliken, N., Shea, S., & the Wagga Wagga Aboriginal Elders Group Incorporated. (2007). From oral history to leadership in the Aboriginal community: A five year journey with the Wagga Wagga Aboriginal Elders Group Incorporated. *Rural Society, 17*, 299–307.

Miller, B., & Cafasso, L. (1992). Gender differences in caregiving: Fact or artifact? *The Gerontologist, 32*, 498–507.

Milne, A. (2010). The 'D' word: Reflections on the relationship between stigma, discrimination and dementia. *Journal of Mental Health, 19*, 227–233.

Minichiello, V., Browne, J., & Kendig, H. (2000). Perceptions and consequences of ageism: Views of older people. *Ageing and Society, 20*, 253–270.

Moen, P., Dempster-McClain, D., & Williams, R. (1992). Successful aging: A life-course perspective on women's multiple roles and health. *American Journal of Sociology, 97*, 1612–1638.

Mokuau, N., Browne, C. V., Choy, L. B., & Braun, K. L. (2008). Using a community-based participatory approach to create a resource center for native Hawaiian elders. *Education for Health 20*(3). Downloaded on March 22, 2017, from http://www.educationforhealth.net/

Molm, L. D., & Cook, K. S. (1995). Social exchange and exchange networks. In K. S. Cook, G. A. Fine, & J. S. House (Eds.), *Sociological perspectives on social psychology* (pp. 209–235). Needham Heights, MA: Allyn and Bacon.

Montoro-Rodriguez, J., & Pinazo, S. (2005). Evaluating social integration and psychological outcomes for older adults enrolled at a university intergenerational program. *Journal of Intergenerational Relationships, 3*, 65–81.

Moody, H. R. (1995). Ageing, meaning and the allocation of resources. *Ageing and Society, 15*, 163–184.

Moon, J.-H., & Pearl, J. H. (1991). Alienation of elderly Korean immigrants as related to place of residence, gender, age, years of education, time in the U.S., living with or without children, and living with or without spouse. *International Journal of Aging and Human Development, 32*, 115–124.

Moorer, P., & Suurmeijer, T. P. B. M. (2001). The effects of neighbourhoods on size of social network of the elderly and loneliness: A multilevel approach. *Urban Studies, 38*, 105–118.

Morrow-Howell, N., Hinterlong, J., Rozario, P. A., & Tang, F. (2003). Effects of volunteering on the well-being of older adults. *Journal of Gerontology, 58B*, S137–S145.

Moustakas, C. (1972). *Loneliness and love*. Englewood Cliffs, NJ: Prentice-Hall.

Moustakas, C. E. (1961). *Loneliness*. Englewood Cliffs, NJ: Prentice-Hall.

Mui, A. C. (1996). Depression among elderly Chinese immigrants: An exploratory study. *Social Work, 41*, 633–645.

Mui, A. C. (2000). Stress, coping, and depression among elderly Korean immigrants. *Journal of Human Behavior in the Social Environment, 3*, 281–299.

Mushtaq, R., Shoib, S., Shah, T., & Mushtaq, S. (2014). Relationship between loneliness, psychiatric disorders and physical health? A review on the psychological aspects of loneliness. *Journal of Clinical and Diagnostic Research, 8*, WE01–WE04.

Mutran, E. (1987). Family, social ties and self-meaning in old age: The development of an affective identity. *Journal of Social and Personal Relationships, 4*, 463–480.

Navaie-Waliser, M., Spriggs, A., & Feldman, P. H. (2002). Informal caregiving: Differential experiences by gender. *Medical Care, 40*, 1249–1259.

Nay, R., Bird, M., Edvardsson, D., Fleming, R., & Hill, K. (2009). Person-centred care. In R. Nay & S. Garratt (Eds.), *Older people: Issues and innovations in care* (3rd ed., pp. 107–119). Australia: Elsevier.

Nedelcu, M. (2009). La 'génération zéro': Du sédentaire à l'acteur circulant. Effets de mobilité sur la génération des parents des migrants roumains hautement qualifiés à Toronto à l'ère du numérique. In G. Cortes & L. Faret (Eds.), *Les circulations transnationales. Lire les turbulences migratoires contemporaines* (pp. 187–198). Paris: Armand Colin.

Nelson, T. D. (2005). Ageism: Prejudice against our feared future self. *Journal of Social Issues, 61*, 207–221.

Nelson, T. D. (2009). Ageism. In T. D. Nelson (Ed.), *Handbook of prejudice, stereotyping, and discrimination* (pp. 431–440). New York: Psychology Press.

Nelson, T. D. (2011). Ageism: The strange case of prejudice against the older you. In R. L. Wiener & S. L. Willborn (Eds.), *Disability and aging discrimination* (pp. 37–47). New York: Springer.

Neville, S., & Henrickson, M. (2010). 'Lavender retirement': A questionnaire survey of lesbian, gay and bisexual people's accommodation plans for old age. *International Journal of Nursing Practice, 16*, 586–594.

Nilsson, J., Masud Rana, A. K. M., & Nahar Kabir, Z. (2006). Social capital and quality of life in old age: Results from a cross-sectional study in rural Bangladesh. *Journal of Aging and Health, 18*, 419–434.

Nolan, M. R., Grant, G., & Keady, J. (1996). *Understanding family care: A multidimensional model of caring and coping*. Buckingham: Open University Press.

Nolan, M., Bauer, M., & Nay, R. (2009). Supporting family carers: Implementing a relational and dynamic approach. In R. Nay & S. Garratt (Eds.), *Older people: Issues and innovations in care* (3rd ed., pp. 136–152). Australia: Elsevier.

Nolan, M., Edvardsson, D., Choowattanapakorn, T., & Oo, C. (2009a). Care and support for older people: Some international reflections. In R. Nay & S. Garratt (Eds.), *Older people: Issues and innovations in care* (3rd ed., pp. 3–8). Australia: Elsevier.

Norris, D., Fancey, P., Power, E., & Ross, P. (2013). The critical-ecological framework: Advancing knowledge, practice, and policy on older adult abuse. *Journal of Elder Abuse and Neglect, 25*, 40–45.

North, M. S., & Fiske, S. T. (2012). An inconvenienced youth? Ageism and its potential intergenerational roots. *Psychological Bulletin, 138*, 982–997.

Nunes, L. (2015). Bioethical perspectives for ageing. The documents produced by the Ethics Councils. *RIASE, 1*, 218–242.

Nydegger, C. N. (1983). Family ties of the aged in cross-cultural perspective. *The Gerontologist, 23*, 26–32.

Nyqvist, F., Gutavsson, J., & Gustafson, Y. (2006). Social capital and health in the oldest old: The Umeå 85+ study. *International Journal of Ageing and Later Life, 1*, 91–114.

Nystrom, N. M., & Jones, T. C. (2012). Aging in the lesbian community. In T. M. Witten & A. E. Eyler (Eds.), *Gay, lesbian, bisexual and transgender aging* (pp. 130–161). Baltimore: The Johns Hopkins University Press.

Obisesan, T. O., & Gillum, R. (2009). Cognitive function, social integration and mortality in a U.S. national cohort study of older adults. *BMC Geriatrics, 9*, 33. https://doi.org/10.1186/1471-2318-9-33

OECD. (2017). *G20 global displacement and migration trends report 2017*. Downloaded on August 7, 2017, from www.oecd.org/els/G20-OECD-migration.pdf

Orel, N. A. (2004). Gay, lesbian, and bisexual elders: Expressed needs and concerns across focus groups. *Journal of Gerontological Social Work, 43*, 57–77.

Ory, M., Hoffman, M. K., Hawkins, M., Sanner, B., & Mockenhaupt, R. (2003). Challenging aging stereotypes strategies for creating a more active society. *American Journal of Preventative Medicine, 25*, 164–171.

Osterkamp, L. B., & Chapin, R. K. (1995). Community-based volunteer home-repair and home-maintenance programs for elders. *Journal of Gerontological Social Work, 24*, 55–76.

Oswald, F., & Rowles, G. D. (2006). Beyond the relocation trauma in old age: New trends in today's elders' residential decisions. In H.-W. Wahl, C. Tesch-Römer, & A. Hoff (Eds.), *New dynamics in old age: Environmental and societal perspectives* (pp. 127–152). Amityville: Baywood.

Page, S. E. (2007). *The difference: How the power of diversity creates better groups, firms, schools and societies*. Princeton, NJ: Princeton University Press.

Pallant, J. F., & Reid, C. (2014). Measuring the positive and negative aspects of the caring role in community versus aged care setting. *Australasian Journal on Ageing, 33*, 244–249.

Palmberger, M. (2017). Social ties and embeddedness in old age: Older Turkish labour migrants in Vienna. *Journal of Ethnic and Migration Studies, 43*, 235–249.

Pang, K. Y. C. (1990). Hwabyung: The construction of a Korean popular illness among Korean elderly immigrant women in the United States. *Culture, Medicine and Psychiatry, 14*, 495–512.

Pang, K. Y. C. (1994). Understanding depression among elderly Korean immigrants through their folk illnesses. *Medical Anthropology Quarterly, 8*, 209–216.

Papasravrou, E., Kalokerinou, A., Papacostas, S. S., Tsangari, H., & Sourtzi, P. (2007). Caring for a relative with dementia: Family caregiver burden. *Journal of Advanced Nursing, 58*, 446–457.

Patzelt, A. (2017). 'A totally new world has been opening up for me' – Experiences of older German migrants who are actively involved in the German-speaking community in Ottawa, Canada. *Journal of Ethnic and Migration Studies, 43*, 218–234.

Pearlin, L., Mullan, J., Semple, S., & Skaff, M. (1990). Caregiving and the stress process: An overview of concepts and their measures. *The Gerontologist, 30*, 583–594.

Pennebaker, J. W., & Stone, L. D. (2003). Words of wisdom: Language use over life. *Journal of Personality and Social Psychology, 85*, 291–301.

Penninkilampi, R., Casey, A. N., Singh, M. F., & Brodaty, H. (2018). The association between social engagement, loneliness, and risk of dementia: A systematic review and meta- analysis. *Journal of Alzheimer's Disease, 66*, 1619–1633.

Perdue, C. W., & Gurtman, M. B. (1990). Evidence for the automaticity of ageism. *Journal of Experimental Social Psychology, 26*, 199–216.

Peri, K. A., Fanslow, J. L., Hand, J. E., & Parsons, J. G. M. (2009). Keeping older people safe by preventing elder abuse and neglect. *Social Policy Journal of New Zealand, 35*, 159–172.

Perlman, D., & Peplau, L. A. (1981). Toward a social psychology of loneliness. In S. W. Duck & R. Gilmour (Eds.), *Personal relationships in disorder* (pp. 31–56). London: Academic Press.

Perry, C. M., & Johnson, C. L. (1994). Families and support networks among African American oldest-old. *International Journal of Aging and Human Development, 38*, 41–50.

Pfeifer, G. M. (2016). Helping elders "age in place". *American Journal of Nursing, 116*, 20–21.

Phillips, J., & Marks, G. (2008). Ageing lesbians: Marginalising discourses and social exclusion in the aged care industry. *Journal of Gay and Lesbian Social Services, 20*, 187–202.

Phillipson, C. (2007). The 'elected' and the 'excluded': Sociological perspectives on the experience of place and community in old age. *Ageing and Society, 27*, 321–342.

Phillipson, C., & Baars, J. (2007). Social theory and social ageing. In J. Bond, S. Peace, F. Dittmann-Kohli, & G. Westerhof (Eds.), *Ageing in society* (pp. 68–84). London: Sage.

Pillay, A. L. (1988). Midlife depression and the "empty nest" syndrome in Indian women. *Psychologiacl Reports, 63*, 591–594.

Pillemer, K. (2004). Can't live with 'em, can't live without 'em: Mothers' ambivalence toward their adult children. In K. Pillemar & K. Lüscher (Eds.), *Intergenerational ambivalences: New perspectives on parent–child relations in later life* (pp. 115–132). Oxford: Elsevier Science.

Pillemer, K., Burnes, D., Riffin, C., & Lachs, M. S. (2016). Elder abuse: Global situation, risk factors, and prevention strategies. *The Gerontologist, 56*, S194–S205.

Pillemer, K., & Suitor, J. J. (2002). Explaining mothers' ambivalence toward their adult children. *Journal of Marriage and Family, 64*, 602–613.

Pillemer, K., Suitor, J. J., Mock, S. E., Sabir, M., Pardo, T. B., & Sechrist, J. (2007). Capturing the complexity of intergenerational relations: Exploring ambivalence within later-life families. *Journal of Social Issues, 63*, 775–791.

Pillemer, K. A., & Finkelhor, D. (1988). The prevalence of elder abuse. A random sample survey. *Gerontologist, 28*, 51–57.

Pinquart, M., & Sörensen, S. (2001). Influences on loneliness in older adults: A meta-analysis. *Basic and Applied Social Psychology, 23*, 245–266.
Pinquart, M., & Sörensen, S. (2006). Gender differences in caregiver stressors, social resources, and health: An updated meta-analysis. *Journal of Gerontology, 61B*, P33–P45.
Pinquart, M., & Sörensen, S. (2009). Influences of socioeconomic status, social network, and competence on subjective well-being in later life: A meta-analysis. *Psychology and Aging, 15*, 187–224.
Pinquart, M., & Sörensen, S. (2011). Spouses, adult children, and children-in-law as caregivers of older adults: A meta-analytic comparison. *Psychology and Aging, 26*, 1–14.
Plaza, D. (2000). Transnational grannies: The changing family responsibilities of elderly African Caribbean-born women resident in Britain. *Social Indicators Research, 51*, 75–105.
Poiani, A. (2010). *Animal homosexuality: A biosocial perspective*. Cambridge: Cambridge University Press.
Posthuma, R. A., & Campion, M. A. (2009). Age stereotypes in the workplace: Common stereotypes, moderators, and future research directions. *Journal of Management, 35*, 158–188.
Poulshock, S. W., & Deimling, G. T. (1984). Families caring for elders in residence: Issues in the measurement of burden. *Journal of Gerontology, 39*, 230–239.
Powell, A., Scott, A. J., & Henry, N. (2020). Digital harassment and abuse: Experiences of sexuality and gender minority adults. *European Journal of Criminology, 17*, 199–223.
Powell Lawton, M., Moss, M., Kleban, M. H., Glicksman, A., & Rovine, M. (1991). A two-factor model of caregiving appraisal and psychological well-being. *Journal of Gerontology, 46*, P181–P189.
Prince, M. J., Harwood, R. H., Blizard, R. A., Thomas, A., & Mann, A. H. (1997). Social support deficits, loneliness and life events as risk factors for depression in old age. The Gospel Oak Project VI. *Psychological Medicine, 27*, 323–332.
Pritchard, C., & Baldwin, D. (2000). Effects of age and gender on elderly suicide rates in Catholic and Orthodox countries: An inadvertent neglect? *International Journal of Geriatric Psychiatry, 15*, 904–910.
Proffitt, C., & Byrne, M. (1993). Predicting loneliness in the hospitalized elderly: What are the risk factors? *Geriatric Nursing, 14*, 311–314.
Putnam, R. D. (1995). 'Bowling alone': America's declining social capital. *Journal of Democracy, 6*, 65–78.
Putnam, R. D. (1995a). Tuning in, tuning out: The strange disappearance of social capital in America. *PS: Political Science and Politics, 28*, 664–683.
Radermacher, H., & Feldman, S. (2017). Cultural diversity, health and ageing. In K. O'Loughlin, C. Browning, & H. Kendig (Eds.), *Ageing in Australia: Challenges and opportunities* (pp. 83–101). New York: Springer.
Ramsey-Klawsnik, H. (2000). Elder-abuse offenders: A typology. *Abuse and Neglect of Older People Summer*, 17–22.
Rawlings, D. (2012). End-of-life care considerations for gay, lesbian, bisexual, and transgender individuals. *International Journal of Palliative Nursing, 18*, 29–34.
Rees Jones, I., & Higgs, P. F. (2010). The natural, the normal and the normative: Contested terrains in ageing and old age. *Social Science and Medicine, 71*, 1513–1519.
Reijneveld, S. A., Westhoff, M. H., & Hopman-Rock, M. (2003). Promotion of health and physical activity improves the mental health of elderly immigrants: Results of a group randomised controlled trial among Turkish immigrants in the Netherlands aged 45 and over. *Community Health, 57*, 405–411.
Reimer, G. (2008). The graying of the U.S. prisoner population. *Journal of Correctional Health Care, 14*, 202–208.
Resnick, B., Galik, E., & Boltz, M. (2013). Function focused care approaches: Literature review of progress and future possibilities. *Journal of the American Medical Directors Association, 14*, 313–318.
Reviere, R., & Young, V. D. (2004). Aging behind bars: Health care for older female inmates. *Journal of Women and Aging, 16*, 55–69.

Revollo, H.-W., Qureshi, A., Collazos, F., & Casas, M. (2011). P01-470 - The impact of social context on the acculturative process: Acculturative stress in Latin American immigrants. *European Psychiatry, 26*(Suppl.1), 474.
Richards, A. (2001). 'Second time around': A survey of grandparents raising their grandchildren. *Adoption and Fostering, 25*, 67–69.
Rigby, H., Gubitz, G., & Phillips, S. (2009). A systematic review of caregiver burden following stroke. *International Journal of Stroke, 4*, 285–292.
Rizzuto, D., Orsini, N., Qiu, C., Wang, H.-X., & Fratiglioni, L. (2012). Lifestyle, social factors, and survival after age 75: Population based study. *BMJ, 345*, e5568. https://doi.org/10.1136/bmj.e5568
Roberts, R. E. L., Richards, L. N., & Bengtson, V. L. (1991). Intergenerational solidarity in families: Untangling the ties that bind. *Marriage and Family Review, 16*, 11–46.
Robine, J.-M., & Cubaynes, S. (2017). Worldwide demography of centenarians. *Mechanisms of Ageing and Development, 165*(Part B), 59–67.
Rockwood, K., Fox, R. A., Stolee, P., Robertson, D., & Beattie, B. L. (1994). Frailty in elderly people: An evolving concept. *Canadian Medical Association Journal, 150*, 489–495.
Rodríguez Rust, P. C. (2012). Aging in the bisexual community. In T. M. Witten & A. E. Eyler (Eds.), *Gay, lesbian, bisexual and transgender aging* (pp. 162–186). Baltimore: The Johns Hopkins University Press.
Rojas-Lizana, S., & Cordella, M. (2020). Ageing in a foreign land: Stressors and coping strategies in the discourse of older adult Spanish speakers in Australia. *Transitions: Journal of Transient Migration, 4*, 5–23.
Rokach, A., & Brock, H. (1998). Coping with loneliness. *The Journal of Psychology, 132*, 107–127.
Rokach, A., Matalon, R., Rokach, B., & Safarov, A. (2007). The effects of gender and marital status on loneliness of the aged. *Social Behavior and Personality, 35*, 243–254.
Rokach, A. (2012). Loneliness updated: An introduction. *The Journal of Psychology, 146*, 1–6.
Ron, P. (2001). The process of acculturation in Israel among elderly immigrants from the former Soviet Union. *Illness, Crisis and Loss, 9*, 357–368.
Rose, A. M. (1965). The subculture of the aging: A framework for research in social gerontology. In A. M. Rose & W. A. Peterson (Eds.), *Older people and their social world*. Philadelphia, PA: Davis Company.
Rosenfeld, D. (1999). Identity work among lesbian and gay elderly. *Journal of Aging Studies, 13*, 121–144.
Rosenthal Gelman, C. (2014). *Familismo* and its impact on the family caregiving of Latinos with Alzheimer's disease: A complex narrative. *Research on Aging, 36*, 40–71.
Ross, F. (2007). Company of elders. *Working with Older People, 11*, 37–40.
Rothbart, M. (2003). Category dynamics and the modification of outgroup stereotypes. In R. Brown & S. Gaertner (Eds.), *Blackwell handbook of social psychology: Intergroup processes* (pp. 45–64). Malden, MA: Blackwell.
Rothman, E. F., Sullivan, M., Keyes, S., & Boehmer, U. (2012). Parents' supportive reactions to sexual orientation disclosure associated with better health: Results from a population-based survey of LGB adults in Massachusetts. *Journal of Homosexuality, 59*, 186–200.
Routasalo, P., & Pitkala, K. H. (2003). Loneliness among older people. *Reviews in Clinical Gerontology, 13*, 303–311.
Rowe, J. W., & Kahn, R. L. (1997). Successful aging. *The Gerontologist, 37*, 433–440.
Rowe, J. W., & Kahn, R. L. (1998). *Successful aging*. New York: Pantheon Books.
Rudmin, F. W. (2003). Critical history of the acculturation psychology of assimilation, separation, integration, and marginalization. *Review of General Psychology, 7*, 3–37.
Sanborn, B., & Bould, S. (1991). Intergenerational caregivers of the oldest old. *Marriage and Family Review, 16*, 125–142.
Savikko, N., Routasalo, P., Tilvis, R. S., Strandberg, T. E., & Pitkäla, K. H. (2005). Predictors and subjective causes of loneliness in an aged population. *Archives of Gerontology and Geriatrics, 41*, 223–233.

Scazufca, M., Menezes, P. R., & Almeida, O. P. (2002). Caregiver burden in an elderly population with depression in São Paulo, Brazil. *Social Psychiatry and Psychiatric Epidemiology, 37*, 416–422.
Schiamberg, L. B., & Gans, D. (2000). Elder abuse by adult children: An applied ecological framework for understanding contextual risk factors and the intergenerational character of quality of life. *International Journal of Aging and Human Development, 50*, 329–359.
Scholl, J. M., & Sabat, S. R. (2008). Stereotypes, stereotype threat and ageing: Implications for the understanding and treatment of people with Alzheimer's disease. *Ageing and Society, 28*, 103–130.
Schope, R. D. (2005). Who's afraid of growing old? Gay and lesbian perceptions of aging. *Journal of Gerontological Social Work, 45*, 23–39.
Schulz, R., & Martire, L. M. (2004). Family caregiving of persons with dementia: Prevalence, health effects, and support strategies. *American Journal of Geriatric Psychiatry, 12*, 240–249.
Schulz, R., & Sherwood, P. R. (2008). Physical and mental health effects of family caregiving. *American Journal of Nursing, 108*(Suppl), 23–27.
Schumacher, K., Beck, C. A., & Marren, J. M. (2006). Family caregivers: Caring for older adults, working with their families. *American Journal of Nursing, 106*, 40–49.
Schwarzbach, M., Luppa, M., Sikorski, C., Fuchs, A., Maier, W., van den Bussche, H., et al. (2013). The relationship between social integration and depression in non-demented primary care patients aged 75 years and older. *Journal of Affective Disorders, 145*, 172–178.
Seeman, T. E., & Berkman, L. F. (1988). Structural characteristics of social networks and their relationship with social support in the elderly: Who provides support. *Social Sciences and Medicine, 26*, 737–749.
Seguino, S. (2016). Global trends in gender equality. *Journal of African Development, 18*, 9–30.
Sergeant, J. F., & Ekerdt, D. J. (2008). Motives for residential mobility in later life: Post- move perspectives of elders and family members. *International Journal of Aging and Human Development, 66*, 131–154.
Serrat, R., Villar, F., Giuliani, M.-F., & Zacarés, J.-J. (2017). Older people's participation in political organizations: The role of generativity and its impact on well-being. *Educational Gerontology, 43*, 128–138.
Shanas, E. (1979). Social myth as hypothesis: The case of family relations of old people. *The Gerontologist, 19*, 3–9.
Shankle, M. D., Maxwell, C. A., Katzman, E. S., & Landers, S. (2003). An invisible population: Older lesbian, gay, bisexual, and transgender individuals. *Clinical Research and Regulatory Affairs, 20*, 159–182.
Sharek, D. B., McCann, E., Sheerin, F., Glacken, M., & Higgins, A. (2015). Older LGBT people's experiences and concerns with healthcare professionals and services in Ireland. *International Journal of Older People Nursing, 10*, 230–240.
Sharma, S. (2014). The conflict and challenge of integrating the "others" in Europe. *European Research Studies Journal, 17*(4), 67–92.
Sheets, D. J. (2005). Aging with disabilities: Ageism and more. *Generations, 29*, 37–41.
Shultziner, D. (2007). Human dignity - Functions and meanings. In J. Malpas & N. Lickiss (Eds.), *Perspectives on human dignity: A conversation* (pp. 73–92). Dordrecht: Springer.
Sieber, S. (1974). Toward a theory of role accumulation. *American Sociological Review, 39*, 567–578.
Sigad, L. I., & Eisikovits, R. A. (2013). Grandparenting across borders: American grandparents and their Israeli grandchildren in a transnational reality. *Journal of Aging Studies, 27*, 308–316.
Silverstein, M., Cong, Z., & Li, S. (2006). Intergenerational transfers and living arrangements of older people in rural China: Consequences for psychological well-being. *Journal of Gerontology, 61B*, S256–S266.
Slevin, K. F., & Linneman, T. J. (2010). Old gay men's bodies and masculinities. *Men and Masculinities, 12*, 483–507.
Slusher, M. P., Mayer, C. J., & Dunkle, R. E. (1996). Gays and Lesbians Older and Wiser (GLOW): A support group for older gay people. *The Gerontologist, 36*, 118–123.

Smith, L. A., McCaslin, R., Chang, J., Martinez, P., & McGrew, P. (2010). Assessing the needs of older gay, lesbian, bisexual, and transgender people: A service-learning and agency partnership approach. *Journal of Gerontological Social Work, 53*, 387–401.

Snape, E., & Redman, T. (2003). Too old or too young? The impact of perceived age discrimination. *Human Resource Management Journal, 13*, 78–89.

Snyder, M., & Meine, P. (1994). Stereotypingof the elderly: A functional approach. *British Journal of Social Psychology, 33*, 63–82.

Sokolovsky, J. (2009). Aging proletariats in a twenty-first-century indigenous Mexican community. In J. Sokolovsky (Ed.), *The cultural context of aging: Worldwide perspectives* (pp. 216–239). Westport, CT: Praeger.

Solomon, S., Greenberg, J., & Pyszczynski, T. (1991). Terror management theory of self-esteem. In C. R. Snyder & R. D. Forsyth (Eds.), *Handbook of social and clinical psychology: The health perspective* (pp. 21–40). New York: Pergamon Press.

Sörensen, S., Pinquart, M., & Duberstein, P. (2002). How effective are interventions with caregivers? An updated meta-analysis. *The Gerontologist, 42*, 356–372.

Sorkin, D., Rook, K. S., & Lu, J. L. (2002). Loneliness, lack of emotional support, lack of companionship, and the likelihood of having a heart condition in an elderly sample. *Annals of Behavioral Medicine, 24*, 290–298.

Spillman, B. C., & Pezzin, L. E. (2000). Potential and active family caregivers changing networks and the "sandwich generation". *The Milbank Quarterly, 78*, 347–374.

Standing, G. (2016). *The precariat: The new dangerous class*. London: Bloomsbury.

Stangor, C. (2009). The study of stereotyping, prejudice, and discrimination within social psychology: A quick history of theory and research. In T. D. Nelson (Ed.), *Handbook of prejudice, stereotyping, and discrimination* (pp. 1–22). New York: Psychology Press.

Stanley, M., Moyle, W., Ballantyne, A., Jaworski, K., Corlis, M., Oxlade, D., et al. (2010). 'Nowadays you don't even see your neighbours': Loneliness in the everyday lives of older Australians. *Health and Social Care in the Community, 18*, 407–414.

Steinkamp, M. W., & Kelly, J. R. (1987). Social integration, leisure activity, and life satisfaction in older adults: Activity theory revisited. *International Journal of Aging and Human Development, 25*, 293–307.

Stephan, W. G., Ybarra, O., & Rios Morrison, K. (2009). Intergroup threat theory. In T. D. Nelson (Ed.), *Handbook of prejudice, stereotyping, and discrimination* (pp. 43–59). New York: Psychology Press.

Steuer, J., & Austin, E. (1980). Family abuse of the elderly. *Journal of the American Geriatrics Society, 28*, 372–376.

Stiglitz, J. E. (2012). *The price of inequality*. New York: W.W. Norton & Company Inc.

Stoller, E. P. (1983). Parental caregiving by adult children. *Journal of Marriage and Family, 45*, 851–858.

Stoller, E. P., & Pugliesi, K. L. (1991). Size and effectiveness of informal helping networks: A panel study of older people in the community. *Journal of Health and Social Behavior, 32*, 180–191.

Stoltz, P., Udén, G., & Willman, A. (2004). Support for family carers who care for an elderly person at home – A systematic literature review. *Scandinavian Journal of Caring Sciences, 18*, 111–119.

Stravynski, A., & Boyer, R. (2001). Loneliness in relation to suicide ideation and parasuicide: A population-wide study. *Suicide and Life-threatening Behavior, 31*, 32–40.

Su, Y., & Ferraro, K. F. (1997). Social relations and health assessments among older people: Do the effects of integration and social contributions vary cross-culturally? *Journal of Gerontology, 52B*, S27–S36.

Sudha, S., Suchindran, C., Mutran, E. J., Rajan, S. I., & Sarma, P. S. (2006). Marital status, family ties, and self-rated health among elders in South India. *Journal of Cross Cultural Gerontology, 21*, 103–120.

Sulmasy, D. P. (2002). A biopsychosocial-spiritual model for the care of patients at the end of life. *Gerontologist, 42*, 24–33.

Sum, S., Mathews, R. M., Hughes, I., & Campbell, A. (2008). Internet use and loneliness in older adults. *CyberPsychology and Behavior, 11*, 208–211.

Sun, R. (2004). Worry about medical care, family support, and depression of the elders in urban China. *Research on Aging, 26*, 559–585.

Sun, R., & Liu, Y. (2006). Mortality of the oldest old in China. *Journal of Aging and Health, 18*, 37–55.

Sung, K. (1990). A new look at filial piety: Ideals and practices of family-centered parent care in Korea. *The Gerontologist, 30*, 610–617.

Svensson-Dianellou, A., Smith, P. K., & Mestheneos, E. (2010). Family help by Greek grandparents. *Journal of Intergenerational Relationships, 8*, 249–263.

Swallen, K. C. (1996). *Morbidity, mortality and nativity: The influence of early-life effects and health selection on health at old age*. Doctoral Dissertation, Department of Demography, University of California, Berkeley, CA.

Swallen, K. C. (1997). Do health selection effects last? A comparison of morbidity rates for elderly adult immigrants and US-born elderly persons. *Journal of Cross-Cultural Gerontology, 12*, 317–339.

Szanton, S. L., Thorpe, R. J., Boyd, C., Tanner, E. K., Leff, B., Agree, E., et al. (2011). Community Aging in Place, Advancing Better Living for Elders (CAPABLE): A bio-behavioral-environmental intervention to improve function and health- related quality of life in disabled, older adults. *Journal of the American Geriatrics Society, 59*, 2314–2320.

Szinovacz, M. E., DeViney, S., & Davey, A. (2001). Influences of family obligations and relationships on retirement: Variations by gender, race, and marital status. *Journal of Gerontology, 56B*, S20–S27.

Talley, H. L., & Casper, M. J. (2012). Intersex and aging: A (cautionary) research agenda. In T. M. Witten & A. E. Eyler (Eds.), *Gay, lesbian, bisexual and transgender aging* (pp. 270–289). Baltimore: The Johns Hopkins University Press.

Tam, T., Hewstone, M., Harwood, J., Voci, A., & Kenworthy, J. (2006). Intergroup contact and grandparent-grandchild communication: The effects of self-disclosure on implicit and explicit biases against older people. *Group Processes and Intergroup Relations, 9*, 413–429.

Tanaka, K., & Johnson, N. E. (2010). Social integration and healthy aging in Japan: How gender and rurality matter. *Journal of Cross Cultural Gerontology, 25*, 199–216.

Teaster, P. B., & Roberto, K. A. (2004). Sexual abuse of older adults: APS cases and outcomes. *The Gerontologist, 44*, 788–796.

Teixeira, V., Pires, C., Pinto, F., Freitas, J., Sales Dias, M., & Mendes Rodrigues, E. (2012). Towards elderly social integration using a multimodal human-computer interface. In *Proceedings of the 2nd International Living Usability Lab Workshop on AAL Latest Solutions, Trends and Applications - Volume 1: AAL* (pp. 3–13). Algarve: Vilamoura.

Theeke, L. A. (2009). Predictors of loneliness in U.S. adults over age sixty-five. *Archives of Psychiatric Nursing, 23*, 387–396.

Thiede Call, K., Finch, M. A., Huck, S. M., & Kane, R. A. (1999). Caregiver burden from a social exchange perspective: Caring for older people after hospital discharge. *Journal of Marriage and Family, 61*, 688–699.

Thomas, K., & Shute, R. (2006). The old and mentally ill in Australia: Doubly stigmatized. *Australian Psychologist, 41*, 186–192.

Thomas, W. H., & Blanchard, J. M. (2009). Moving beyond place: Aging in community. *Generations, 33*, 12–17.

Thommessen, B., Aarsland, D., Braekhus, A., Oksengaard, A. R., Engendal, K., & Laake, K. (2002). The psychological burden on spouses of the elderly with stroke, dementia and Parkinson's disease. *International Journal of Geriatric Psychiatry, 17*, 78–84.

Thompson, O., Koumanakos, G., & Hadjri, K. (2012). The ethical and policy implications of e-health and telemedicine: An ageing-focused review. *Clinical Ethics, 7*, 147–156.

Thompson Jr., E. H., Futterman, A. M., Gallagher-Thompson, D., Rose, J. M., & Lovett, S. B. (1993). Social support and caregiving burden in family caregivers of frail elders. *Journal of Gerontology, 48*, S245–S254.

Thornton, J. E. (2002). Myths of aging or ageist stereotypes. *Educational Gerontology, 28*, 301–312.
Thurston, R. W. (2001). *Witch, wicce, mother goose: The rise and fall of the witch hunts in Europe and North America.* New York: Pearson Education.
Timonen, V. (Ed.). (2018). *Grandparenting practices around the world.* Bristol: Policy Press.
Tinker, A. (2003). Older people and ethics. *Australasian Journal on Ageing, 22*, 206–210.
Tomaka, J., Thompson, S., & Palacios, R. (2006). The relation of social isolation, loneliness, and social support to disease outcomes among the elderly. *Journal of Aging and Health, 18*, 359–384.
Tomassini, C., Glaser, K., & Askham, J. (2003). Getting by without a spouse: Living arrangements and support of older people in Italy and Britain. In S. Arber, K. Davidson, & J. Ginn (Eds.), *Gender and ageing: Chaging roles and relationships* (pp. 111–126). Maidenhead: Open University Press.
Towler, A. J., & Schneider, D. J. (2005). Distinctions among stigmatized groups. *Journal of Applied Social Psychology, 35*, 1–14.
Treas, J., & Batalova, J. (2009). Immigrants and aging. In P. Uhlenberg (Ed.), *International handbook of population aging* (pp. 365–394). New York: Springer.
Treas, J., & Mazumdar, S. (2002). Older people in America's immigrant families: Dilemmas of dependence, integration, and isolation. *Journal of Aging Studies, 16*, 243–258.
Treas, J., & Mazumdar, S. (2004). Kinkeeping and caregiving: Contributions of older people in immigrant families. *Journal of Comparative Family Studies, 35*, 105–122.
Trivers, R. L. (1971). The evolution of reciprocal altruism. *Quarterly Review of Biology, 46*, 35–57.
Trommsdorff, G. (2006). Parent-child relations over the lifespan: A cross-cultural perspective. In K. H. Rubin & O. B. Chung (Eds.), *Parenting beliefs, behavior, and parent-child relations. A cross-cultural perspective* (pp. 143–183). New York: Psychology.
Trotter, C., & Baidawi, S. (2015). Older prisoners: Challenges for inmates and prison management. *Australian and New Zealand Journal of Criminology, 48*, 200–218.
Trute, B. (2003). Grandparents of children with developmental disabilities: Intergenerational support and family well-being. *Families in Society: The Journal of Contemporary Social Services, 84*, 119–126.
Tsytsarev, S., & Krichmar, L. (2000). Relationship of perceived culture shock, length of stay in the U.S., depression, and self-esteem in elderly Russian-speaking immigrants. *Journal of Social Distress and the Homeless, 9*, 35–49.
Turner, L. (2004). Biotechnology, bioethics and anti-aging interventions. *Trends in Biotechnology, 22*, 219–221.
Turvey, C. L., Carney, C., Arndt, S., Wallace, R. B., & Herzog, R. (1999). Conjugal loss and syndromal depression in a sample of elders aged 70 years or older. *American Journal of Psychiatry, 156*, 1596–1601.
Uhlenberg, P. (2000). Introduction: Why study age integration? *The Gerontologist, 40*, 261–308.
Valtonen, K. (2002). The ethnic neighbourhood: A locus of empowerment for elderly immigrants. *International Social Work, 45*, 315–323.
van Dijk, T. A. (2011). Discourse and ideology. In T. A. van Dijk (Ed.), *Discourse studies: A multidisciplinary introduction* (pp. 379–407). London: Sage.
van Exel, N. J., Scholte op Reimer, W. J., Brouwer, W. B., Berg, B., Koopmanschap, M. A., & van der Bos, G. A. (2004). Instruments for assessing the burden of informal caregiving for stroke patients in clinical practice. *Clinical Rehabilitation, 18*, 203–214.
van Langenhove, L., & Harré, R. (1999). Introducing positioning theory. In R. Harréand & L. van Langenhove (Eds.), *Positioning theory: Moral contexts of intentional actions* (pp. 14–31). Oxford: Blackwell Publishers.
van Tilburg, T. (1998). Losing and gaining in old age: Changes in personal network size and social support in a four-year longitudinal study. *Journal of Gerontology, 53B*, S313–S323.
van Tilburg, T. (2005). *Loneliness from 1978 to 2003: A historical time analysis among Dutch adults.* Paper prepared on the occasion of the farewell of Jenny Gierveld, Vrije Universiteit, Amsterdam.

van Tilburg, T., de Jong Gierveld, J., Lecchini, L., & Marsiglia, D. (1998). Social integration and loneliness: A comparative study among older adults in the Netherlands and Tuscany, Italy. *Journal of Social and Personal Relationships, 15*, 740–754.

van Wagenen, A., Driskell, J., & Bradford, J. (2013). "I'm still raring to go": Successful aging among lesbian, gay, bisexual, and transgender older adults. *Journal of Aging Studies, 27*, 1–14.

VanderLaan, D. P., Petterson, L. J., & Vasey, P. L. (2016). Femininity and kin-directed altruism in androphilic men: A test of an evolutionary developmental model. *Archives of Sexual Behavior, 45*, 619–633.

Vaupel, J. W. (2010). Biodemography of human ageing. *Nature, 464*, 536–542.

Vázquez, G. H., Kapczinski, F., Magalhaes, P. V., Córdoba, R., Lopez Jaramillo, C., Rosa, A. R., et al. (2011). Stigma and functioning in patients with bipolar disorder. *Journal of Affective Disorders, 130*, 323–327.

Verbeek, H., van Rossum, E., Zwakhalen, S. M. G., Kempen, G. I. J. M., & Hamers, J. P. H. (2009). Small, homelike care environments for older people with dementia: A literature review. *International Psychogeriatrics, 21*, 252–264.

Vernooij-Dassen, M. J. F. J., Moniz-Cook, E. D., Woods, R. T., De Lepeleire, J., Leuschner, A., Zanetti, O., et al. (2005). Factors affecting timely recognition and diagnosis of dementia across Europe: From awareness to stigma. *International Journal of Geriatric Psychiatry, 20*, 377–386.

Vescio, T. K., Gervais, S. J., Heiphetz, L., & Bloodhart, B. (2009). The stereotypic behaviors of the powerful and their effect on the relatively powerless. In T. D. Nelson (Ed.), *Handbook of prejudice, stereotyping, and discrimination* (pp. 247–265). New York: Psychology Press.

Victor, C., Scambler, S., Bond, J., & Bowling, A. (2000). Being alone in later life: Loneliness, social isolation and living alone. *Reviews in Clinical Gerontology, 10*, 407–417.

Vitaliano, P. P., Zhang, J., & Scanlan, J. M. (2003). Is caregiving hazardous to one's physical health? A meta-analysis. *Psychological Bulletin, 129*, 946–972.

von Hippel, W., Silver, L. A., & Lynch, M. E. (2000). Stereotyping against your will: The role of inhibitory ability in stereotyping and prejudice among the elderly. *Personality and Social Psychology Bulletin, 26*, 523–532.

Wahidin, A. (2011). Ageing behind bars, with particular reference to older women in prison. *Irish Probation Journal, 8*, 109–123.

Wahler, J., & Gabbay, S. G. (1997). Gay male aging. *Journal of Gay and Lesbian Social Services, 6*, 1–20.

Wallhagen, M. I. (2009). The stigma of hearing loss. *The Gerontologist, 50*, 66–75.

Walsh, C. A., Olson, J. L., Ploeg, J., Lohfeld, L., & MacMillan, H. L. (2011). Elder abuse and oppression: Voices of marginalized elders. *Journal of Elder Abuse and Neglect, 23*, 17–42.

Walsh, C. A., Ploeg, J., Lohfeld, L., Horne, J., MacMillan, H., & Lai, D. (2007). Violence across the lifespan: Interconnections among forms of abuse as described by marginalized Canadian elders and their care-givers. *British Journal of Social Work, 37*(3), 491–514. https://doi.org/10.1093/bjsw/bcm022

Wango, T., Teaster, P. B., Grace, J., Wong, W., Mendiondo, M. S., Blandford, C., et al. (2014). An ecological systems examination of elder abuse: A week in the life of adult protective services. *Journal of Elder Abuse and Neglect, 26*, 440–457.

Warburton, J., & McLaughlin, D. (2005). 'Lots of little kindnesses': Valuing the role of older Australians as informal volunteers in the community. *Ageing and Society, 25*, 715–730.

Weinbren, D. (2015). *The open university: A history*. Manchester: Manchester University Press.

Weiss, R. S. (1973). *Loneliness: The experience of emotional and social isolation*. Cambridge, MA: MIT Press.

Wells, Y., Foreman, P., & Ryburn, B. (2009). Community care for older Australians: Issues and future directions. In R. Nay & S. Garratt (Eds.), *Older people: Issues and innovations in care* (3rd ed., pp. 78–103). Sydney, NSW: Elsevier.

Wenger, G. C. (1996). Social networks and gerontology. *Reviews in Clinical Gerontology, 6*, 285–293.

Wenger, G. C., Davies, R., Shahtahmasebi, S., & Scott, A. (1996). Social isolation and loneliness in old age: Review and model refinement. *Ageing and Society, 16*, 333–358.

Werner, P., Stein-Shvachman, I., & Heinik, J. (2009). Perceptions of self-stigma and its correlates among older adults with depression: A preliminary study. *International Psychogeriatrics, 21*, 1180–1189.

Whetten, D. A., & Mackey, A. (2002). A social actor conception of organizational identity and its implications for the study of organizational reputation. *Business and Society, 41*, 393–414.

Whitaker, A. (2010). The body as existential midpoint—The aging and dying body of nursing home residents. *Journal of Aging Studies, 24*, 96–104.

White, P. (2006). Migrant populations approaching old age: Prospects in Europe. *Journal of Ethnic and Migration Studies, 32*, 1283–1300.

WHO. (2014). *Elder abuse*. Violence in the Western Pacific. Downloaded on March 15, 2017, from http://iris.wpro.who.int/handle/10665.1/12400

Wiglesworth, A., Mosqueda, L., Mulnard, R., Liao, S., Gibbs, L., & Fitzgerald, W. (2010). Screening for abuse and neglect of people with dementia. *Journal of the American Geriatrics Society, 58*, 493–500.

Wilkinson, D. (2008). Individual and community factors affecting psychological sense of community, attraction, and neighboring in rural communities. *Canadian Review of Sociology, 45*, 305–329.

Williams, B. A., Lindquist, K., Hill, T., Baillargeon, J., Mellow, J., Greifinger, R., et al. (2009). Caregiving behind bars: Correctional officer reports of disability in geriatric prisoners. *Journal of the American Geriatrics Society, 57*, 1286–1292.

Williams, B. A., Lindquist, K., Sudore, R. L., Strupp, H. M., Willmott, D. J., & Walter, L. C. (2006). Being old and doing time: Functional impairment and adverse experiences of geriatric female prisoners. *Journal of the American Geriatrics Society, 54*, 702–707.

Williams, S. W., & Dilworth-Anderson, P. (2002). Systems of social support in families who care for dependent African American elders. *The Gerontologist, 42*, 224–236.

Williams, B. A., Goodwin, J. S., Baillargeon, J., Ahalt, C., & Walter, L. C. (2012). Addressing the aging crisis in U.S. criminal justice healthcare. *Journal of the American Geriatrics Society, 60*, 1150–1156.

Williams, B. A., Stern, M. F., Mellow, J., Safer, M., & Greifinger, R. B. (2012a). Aging in correctional custody: Setting a policy agenda for older prisoner health care. *American Journal of Public Health, 102*, 1475–1481.

Willis, P., Maegusuku-Hewett, T., Raithby, M., & Miles, P. (2016). Swimming upstream: The provision of inclusive care to older lesbian, gay and bisexual (LGB) adults in residential and nursing environments in Wales. *Ageing and Society, 36*, 282–306.

Wilmoth, J. M., De Jong, G. F., & Himes, C. L. (1997). Immigrant and non-immigrant living arrangements among America's White, Hispanic, and Asian elderly population. *International Journal of Sociology and Social Policy, 17*, 57–82.

Wilson, G. (1994). Abuse of elderly men and women among clients of a community psychogeriatric service. *British Journal of Social Work, 24*, 681–700.

Wilson, R. S., Krueger, K. R., Arnold, S. E., Schneider, J. A., Kelly, J. F., Barnes, L. L., et al. (2007). Loneliness and risk of Alzheimer disease. *Archives of General Psychiatry, 64*, 234–240.

Withers, G. (2002). Population ageing and the role of immigration. *Australian Economic Review, 35*, 104–112.

Witten, T. M. (2009). Graceful exits: Intersection of aging, transgender identities, and the family/community. *Journal of GLBT Family Studies, 5*, 35–61.

Witten, T. M. (2014). End of life, chronic illness, and trans-identities. *Journal of Social Work in End-of-Life & Palliative Care, 10*, 1–26.

Witten, T. M. (2015). Elder transgender lesbians: Exploring the intersection of age, lesbian sexual identity, and transgender identity. *Journal of Lesbian Studies, 19*, 73–89.

Witten, T. M., & Eyler, A. E. (2012). Transgender and aging: Beings and becomings. In T. M. Witten & A. E. Eyler (Eds.), *Gay, lesbian, bisexual and transgender aging* (pp. 187–269). Baltimore: The Johns Hopkins University Press.

Witten, T. M., & Whittle, S. (2004). Transpanthers: The greying of transgender and the law. *Deakin Law Review, 9*, 503–522.

Wolf, R. S. (1996). Elder abuse and family violence. *Journal of Elder Abuse and Neglect, 8*, 81–96.
Wolff, J. L., & Kasper, J. D. (2006). Caregivers of frail elders: Updating a national profile. *The Gerontologist, 46*, 344–356.
Wood, J. B., & Parham, I. A. (1990). Coping with perceived burden: Ethnic and cultural issues in Alzheimer's family caregiving. *The Journal of Applied Gerontology, 9*, 325–339.
Wu, Z., & Hart, R. (2002). Social and health factors associated with support among elderly immigrants in Canada. *Research on Aging, 24*, 391–412.
Wyandt, M. A. (2004). A review of elder abuse literature: An age old problem brought to light. *Californian Journal of Health Promotion, 2*, 40–52.
Xie, X., & Xia, Y. (2011). *Grandparenting in Chinese immigrant families*. Faculty Publications, Department of Child, Youth, and Family Studies. 82. https://digitalcommons.unl.edu/famconfacpub/82
Yan, E., & Tang, C. S.-K. (2003). Proclivity to elder abuse: A community study on Hong Kong Chinese. *Journal of Interpersonal Violence, 18*, 999–1017.
Yan, E. C.-W., & Tang, C. S.-K. (2004). Elder abuse by caregivers: A study of prevalence and risk factors in Hong Kong Chinese families. *Journal of Family Violence, 19*, 269–277.
Yang, Y. C., Li, T., & Ji, Y. (2013). Impact of social integration on metabolic functions: Evidence from a nationally representative longitudinal study of US older adults. *BMC Public Health, 13*, 1210. https://doi.org/10.1186/1471-2458-13-1210
Yi, Z., & George, L. (2000). Family dynamics of 63 million (in 1990) to more than 330 million (in 2050) elders in China. *Demographic Research, 2*, 5. Downloaded on March 20, 2017, from www.demographic-research.org/Volumes/Vol2/5/
Yim, J. E. (2016). Therapeutic benefits of laughter in mental health: A theoretical review. *Tohoku Journal of Experimental Medicine, 239*, 243–249.
Yoo, J. A., & Zippay, A. (2012). Social networks among lower income Korean elderly immigrants in the U.S. *Journal of Aging Studies, 26*, 368–376.
Yorston, G. A., & Taylor, P. J. (2006). Older offenders - No place to go? *Journal of the American Academy of Psychiatry and the Law, 34*, 333–337.
Young, I. M. (1990). *Justice and the politics of difference*. Princeton: Princeton University Press.
Yue, X., & Ng, S. H. (1999). Filial obligations and expectations in China: Current views from young and old people in Beijing. *Asian Journal of Social Psychology, 2*, 215–226.
Zarit, S. H. (2002). Caregiver's burden. In S. Andrieu & J. P. Aquino (Eds.), *Family and professional carers: Findings lead to action* (pp. 20–24). Paris: Serdi Edition and Fondation Médéric Alzheimer.
Zarit, S. H., Reever, K. E., & Bach-Peterson, J. (1980). Relatives of the impaired elderly: Correlates of feelings of burden. *The Gerontologist, 20*, 649–655.
Zarit, S. H., Todd, P. A., & Zarit, J. M. (1986). Subjective burden of husbands and wives as caregivers: A longitudinal study. *The Gerontologist, 26*, 260–266.
Zhou, Y. R. (2013). Toward transnational care interdependence: Rethinking the relationships between care, immigration and social policy. *Global Social Policy, 13*, 280–298.
Zhou, Z., Wang, P., & Fang, Y. (2018). Loneliness and the risk of dementia among older Chinese adults: Gender differences. *Aging and Mental Health, 22*, 519–525.
Zickgraf, C. (2017). Transnational ageing and the 'zero generation': The role of Moroccan migrants' parents in care circulation. *Journal of Ethnic and Migration Studies, 43*, 321–337.
Zoccali, R., Muscatello, M. R., Bruno, A., Serrano, D., Campolo, D., Pandolfo, G., et al. (2008). Gender role identity in a sample of Italian male gays. *Journal of Homosexuality, 55*, 265–273.
Zuniga, M. E. (1997). Counseling Mexican American seniors: An overview. *Journal of Multicultural Counseling and Development, 25*, 142–155.
Zunzunegui, M.-V., Alvarado, B. E., Del Ser, T., & Otero, A. (2003). Social networks, social integration, and social engagement determine cognitive decline in community-dwelling Spanish older adults. *Journal of Gerontology, 58B*, S93–S100.
Zwijsen, S. A., Niemeijer, A. R., & Hertogh, C. M. P. M. (2011). Ethics of using assistive technology in the care for community-dwelling elderly people: An overview of the literature. *Aging and Mental Health, 15*, 419–427.

Chapter 7
Language and Communication

Abstract *Language and communication* starts with an analysis of intra-generational communication, or the ways in which old people communicate with each other. Intra-generational communication tends to differ from communicating with younger generations (inter-generational communication) which, in turn, poses specific challenges to both the old and the young. Communication is expressed through narratives that can unravel our priorities, concerns, emotions, beliefs, ways of processing life events and understanding of the relationships we establish with other people. The narratives of the elderly are to some extent a window into their mind. Such narratives can however remain more hidden in multicultural societies due to language barriers, in which case older people will require the services of bilingual family members or specifically trained interpreters and translators to communicate more effectively, especially in institutional and medical settings.

The ability to communicate effectively is central to older people to access assistance and to remain integrated with their family and community, including maintaining productive contact with healthcare institutions and social services agencies. But communication has even more functions than the pragmatic issues associated with care.

It is through language and communication that individuals portray who they are. Communication helps older people "maintain their sense of identity and place in the larger world" (Nussbaum et al. 2000, p. 4; see also Lambert and Cordella 2016) and inform others about their existential views and concerns. Verbal and non-verbal communication should be also flexible and varied enough to respond to the changing needs of the life of ageing individuals and their relationships with others.

There has been a growing interest in the communicative aspects of ageing within the field of gerontology in the past 20 years or so, with some major themes emerging: a relative decrease in the communicative ability of older people with age has been noted, which is variable across individuals; a decrease in the actual contact with others, especially inter-generational contacts, has been also observed; negative effects of

M. Cordella, A. Poiani, *Fulfilling Ageing*, International Perspectives on Aging 30,
https://doi.org/10.1007/978-3-030-60071-6_7

ageist language and use of stereotypes against older people in day-to-day communication are commonplace (Cordella and Huang 2016): "young adults construe older communicators as quite nonaccommodative by their talking, seemingly excessively, about their own situations (often painful), as not listening, and their being overly-stereotypic" (Giles and Dorjee 2004, p. 262). Stereotypes, however, can run both ways: from old to young and vice versa. Giles and Dorjee (2004) also point out that stereotypes about the ways in which older people behave and communicate are strengthened by caricatures portrayed in the media. This tends to depersonalise older people, making all of them members of a single, constructed category, without regard for their actual individual capacities, realities, and diversity. This can be very frustrating especially to those elders who retain a perfectly advanced ability to communicate, reason, display knowledge, capacity to learn, and be creative.

In this chapter we start with an overview of the broader issue of ageing and communication, including theoretical and cross-cultural aspects of it, to then focus on more specific social contexts of communication in old people, both intra- and inter-generational. We subsequently analyse the narratives of older people and how they are portrayed in the media, to end this chapter with sections on translating and interpreting for older immigrants, and the specific challenges observed in doctor-old patient communication.

7.1 Ageing and Communication: An Overview

Through language we shape our personal identity, make sense of our history and that of the world around us, and we do so in interaction with others within our community. In this way older people can create their own version of personal elderliness that can continue to be modified through their agency and the agency of the people they interact with, as their body changes over time along with their mental capacities. In this process, self-identity and the changing body may either proceed in harmony or they may split leading to denial of ageing. Denial of ageing is a feature not uncommon these days in the discourse of elderly people in Westernised societies (e.g. Bultena and Powers 1976), being heavily influenced by the ideal of "healthy ageing" as it is portrayed by marketing agencies and the media (Cardona 2008).

In their verbal communication with others, older people may address various issues that can be general or more personal (e.g. reminiscence), perhaps even intimately so. But elders can also communicate non-verbally, to transmit emotions, for instance. Through communication old people may seek help especially if they are lonely, and in the process, they may receive confirmation or disconfirmation of their beliefs from the people they interact with. Through maintaining a degree of communicative competence, they may also try to exert some control on their social interactions (Nussbaum et al. 2000).

Intimate aspects of reminiscence are frequent topics of conversation in the elders. Such act of reminiscing can revive past experiences just for the pleasure of it, or it can become a more serious process of evaluation and life review that older people share with others and that can help them cope with their ageing. Obsessive reminiscence can also occur, especially when elders are haunted by guilt, or are distressed by grief. As expressed by Nussbaum et al. (2000, p. 41), reminiscence among the elders "can be viewed as coping mechanism, as art, as identity management, as well as serving several other functions .... cope with the fear of death, with grief and depression, and with the loss of self-identity and self-esteem". In some instances reminiscence can also serve the function of "denial of ageing". In fact, reminiscing can be even used in psychological therapy, potentially leading to serenity and wisdom, better social cohesiveness, more intimacy, improved feelings of self-worth, and greater life satisfaction. Although it is also possible that reminiscence of negative events could lead to guilt and even depression or, potentially in extreme cases, a sense of panic.

Intimacy in communication, whether it is expressed through reminiscence or in any other form, implies a degree of mutual trust between interactants, and it requires that an initial threshold of understanding be achieved. Satisfaction with intimate relationships leads to psychological well-being, and better self-esteem and emotional health (Nussbaum et al. 2000).

But older people may decrease their communicative competence over time following a reduction in the size of their social network with age, not to speak of any process of cognitive deterioration that may be under way. The reduction in the social circle and cognitive abilities may lead older people to become more cautious in their communicative interactions, preferring more certainty in what they say before they say anything, and they may also have more difficulties understanding some aspects of the conversation due to their lower cognitive processing speed (e.g. Bashore et al. 2014), in which case they may take longer to reply. Alterations in communicative competence can negatively affect the level of intimacy that can be achieved between people communicating with each other, whereas intimacy could be restored through a better understanding of the situation and greater adaptation.

Given that communication is ultimately a social behaviour, *Social Identity theory* (SIT, Tajfel and Turner 1986) is a good starting point to begin unravelling the theoretical structure of intra- and inter-generational communication. This theory posits that people develop an in-group positivity bias that then affects their communicative relationships with people from within and outside their social group. In particular, the level of *accommodation* in discourse (i.e. the effort to adjust to the characteristics of the other interactant/s) can vary according to the composition of the communicating dyad, potentially producing a *staircase pattern*, where communication may be more successful with people of the same age and becoming gradually less successful as interactants differ more in their age (Ota et al. 2012).

Therefore, in-group/out-group views may affect social and communicative relationships between old people and others. Using a network approach, Marcum and Koehly (2015) analysed inter-generational social networks seen as a set of actors, young and old, and the relationships that exist among them. Within the network, actors engage in direct relations such as communication, exchanges (e.g. of favours), and transfers (e.g. of goods). Such relations can be *mutual* (or symmetric), *asymmetric*, or *null*, in the latter case communication is non-existent even when both actors may be physically close to each other.

The network of communicative relationships established between generations, whether such relationships are symmetric, asymmetric, null, or a combination of those according to circumstances, can alter the shape of social structures such as that of the *sandwich generation* (Miller 1981), where adults must cope with the double demand and burden of caring for their own children and their ageing parents. The sandwich or middle generation may often fulfil a liaison role between the other generations, which would require the maintenance of some form of communication and relationship, whether it is symmetric or not. Sometimes the mid-generation is not available, or it is disengaged (null relationship), and old and young generations interact directly (e.g. grandparents being legal custodians of grandchildren), this structure is referred to as the *skipped generation*. Marcum and Koehly (2015) identify two kinds of skipped generation: *downward*, when the older generation directly cares for the younger one; and *upward*, when it is the younger generation that cares for the older one. On the other hand, there is also the possibility of an inter-generational structure known as the *needy generation* "whereby resources are directed from at least two other generations toward a different, focal generation" (Marcum and Koehly 2015, p. 14). A structure where relationships are closer within- than between-generations is one in which *generational solidarity* prevails (Silverstein and Bengtson 1997). However, it is also possible to have a structure where intra-generational interactions are avoided—such as in situations of high intra-generational competition—, this may be referred to as *generational conflict*. Such social structures are also characterised by different patterns of communication between individuals, with social interactions that are grounded on greater levels of solidarity being expected to display more intimate forms of communication.

As soon as two strangers (e.g. one young and one old, or two old people) meet they may attempt to gather some information about each other with the objective of finding points in common and reducing uncertainty, which may lead to the establishment of a relationship (Berger and Calabrese 1975). According to the *Uncertainty Reduction theory* (URT), individuals who are engaged in a preliminary information-gathering process may not reveal negative information about themselves early on in the conversation. This, however, does not seem to be the case in young–elder encounters, where old people tend to be happy to open up and disclose information of personal nature (e.g. Nussbaum et al. 2000), although the young person may or may not feel comfortable with such open disclosure from the part of the elder.

Throughout the process of uncertainty reduction in the conversation older people may use *confirming statements* which acknowledge and also validate the interlocutor and his/her life experience, but they may also use *disconfirming statements* which do not provide such acknowledgment and validation (Nussbaum et al. 2000). *Imperviousness*, or lack of responsiveness to the specific circumstances of others, is a special case of disconfirmation that can in fact increase uncertainty about the direction of the conversation; e.g. "John communicates to Mary that he knows how she is feeling or what she has experienced better than she does. If Mary says, "I don't feel well", and John says, "Oh, yes you do—you feel just fine", Mary will feel disconfirmed by John's imperviousness" (Nussbaum et al. 2000, p. 55).

Disconfirmation can be also achieved with the use of *tangential remarks*:

Mary: "I don't feel well".
John: "I'm feeling pretty good today" (Nussbaum et al. 2000, p. 55).

Frequent use of various forms of disconfirmation may seriously damage communication with elders and affect their self-identity, as they may start doubting that what they say, think, and feel may not be right.

Ageing may affect communication between individuals in many ways, starting from physiological barriers to interaction with others that follow from physical changes in the ageing body. This *Deficit Paradigm* (Coupland et al. 1991) is expressed as a decrease in visual acuity, for instance, that may interfere with the ability of older people to perceive non-verbal cues during a conversation. Missing non-verbal cues may interfere with appropriate responses that are expected from communication rules.

Regarding verbal communication, loss of hearing ability with age, also known as *presbycusis*, can clearly be a problem as it can reduce the capacity of older people to discriminate some sound frequencies (Nussbaum et al. 2000; Scarinci et al. 2008). Hearing impairment may also interfere with the personal feedback loop that improves speech, thus eventually affecting how older people with hearing limitations produce speech, as well as the time they may take to encode and decode messages, which may affect the fluidity and smoothness of the conversation. The ability to react and respond on verbal cues is also affected by age, again impacting on the flow of the conversation.

On the other hand, older people with decreased hearing capacity may use other senses to try to keep up with the conversation, for instance by filling the gaps through their knowledge of the topic and the conversational style used, or even by lip reading. Medical interventions that could improve visual acuity and the availability of hearing aids may help solve some sensory problems affecting communication in the elders (Cacchione 2014).

Apart from communication problems caused by their hearing or visual difficulties, older people may also run the risk of breaching some social and cultural rules, such as keeping an appropriate interpersonal distance, if they get too close to their interlocutor in order to understand better what she or he is saying. Discomfort may lead a person to end the conversation prematurely. The conversation may be also ended by any of the interactants if the other adopts a less than genuine style of communication (Nussbaum et al. 2000).

Touch can be important for older people in social contexts, as being gently touched could be both relaxing and reassuring for some elders, thus contributing to their well-being. But touch may not be perceived positively by other individuals; therefore, it is vital to show a level of sensitivity and awareness in order to prevent misunderstandings.

On the cognitive side, older people may experience a decrease not only in their memory but also in their discourse comprehension, alongside their ability to infer meaning from a text (Bayles and Kaszniak 1987), whilst at the same time many increase their verbosity, syntactic errors, and they also tend to use a simpler syntax. On the other hand, older people tend to use narratives that are more integrative and interpretative in style and their knowledge of grammar remains good (Coupland et al. 1991).

Such broad decrease in sensory and some communicative abilities with age could negatively affect older people's confidence in inter-personal interactions, which may call for a degree of cooperation from younger speakers, who may find themselves in the need to talk in a more intelligible manner. This may lead younger people to use *baby talk* (i.e. talking more slowly, more loudly, and in a simpler style) which, in turn, may elicit feelings of uneasiness or even anxiety in the elder, thus making the conversation even more difficult, reinforcing a negative view about the elder. This self-reinforcing mechanism in communication is encapsulated in the *Communication Predicament of Ageing* model (CPA, Ryan et al. 1986; Harwood et al. 1993) which states that from baby talk used by a young interactant, communication may lead to feelings of incompetence in the elder, which can be perceived by the young person, thus reinforcing the use of simplified language and finally resulting in the premature end of the conversation and reluctance to engage with the same person in the future. More specifically, the CPA states that:

> where negative assumptions about age are made salient by young people by their particular forms of communication (e.g., patronizing talk) to older people, the latter may begin to self-stereotypically enact behaviors considered symptomatic of old age themselves, including 'learned helplessness' … In turn, this leads to negative outcomes such as a felt loss of personal control, low self esteem, lessened cognitive activity and social interactions, with concomitant decrements in psychological and physiological well-being for the older person (Giles and Dorjee 2004, p. 263).

That is, this model focuses on some negative aspects of inter-generational interaction in communication and the positive feedback mechanism that leads stereotypes to become internalised in the mind of the elders and entrenched in society (see also Williams and Garrett 2002). Internalisation of stereotypes could further affect those attributes (e.g. poor memory, lack of concentration, sensory decline) that provided an initial factual justification for the stereotype, thus reinforcing it (Savundranayagam and Ryan 2008, see also Chap. 6).

The point to make here is that in response to any objective limitations that old people may have in communication, others should not use a patronising speech such as baby talk, but simply adjust their speech in a natural way to make it more easily understandable, but not discriminatory. For instance, if one tells something to an old person and that old person requests the information to be repeated, the information should indeed be repeated, perhaps using different words but in a natural tone. Repeating the information using the same words, but with a higher pitch and in a staccato manner, would tell the elder that you are both annoyed and that you believe that he or she is dumb.

The stereotypes and prejudice that are at the core of the *Communication predicament of ageing* model can be also perpetuated through the media and through patronising ways of communicating with the elders, thus threatening their self-esteem and perpetuating ageist prejudices (Coupland et al. 1991). Communicatively, this can occur in various ways through: the *baby talk* that we have already mentioned (showing high care for the elder but also high control over the interaction), *directive talk* (showing low care and high control), *overly personal talk* (high care, but low control), and *superficial talk* (low care, low control). A variant on baby talk, at least in terms of the use of simplified speech expressed at a slow pace, is *elderspeak* (e.g. Savundranayagam and Ryan 2008). *Elderspeak* is a "simplified speech register with exaggerated pitch and intonation, simplified grammar, limited vocabulary, and slow rate of delivery" (Kemper et al. 1998, p. 56). On the positive side, *elderspeak* may help the speech comprehension ability of those elders who are indeed facing cognitive and/or sensory challenges (e.g. Gould et al. 2002), thus becoming an example of accommodative adaptation in communication (Coupland et al. 1991), but it may become an expression of prejudice and stereotyping when the elder has no such constraints. The elders themselves, on the other hand, may help reinforce negative ageist speech stereotypes by using "old age" as an excuse for communication problems: "I had a senior moment".

Giles et al. (2008) studied the effects of stereotypes in inter-generational communication as seen from the perspective of young and old Canadians, to conclude that young participants perceived older people as benevolent but with decreased personal vitality, alongside seeing them as more non-accommodating in their communication behaviour, which resulted in an intention to avoid old people.

Positive aspects of inter-generational communication are also possible, and this has been reflected in Ryan et al.'s (1995) *Communicative Enhancement of Ageing* model (CEA) which emphasises the accommodative process required between young and old interactants in order to enhance the effectiveness of their communication, and the achievement of positive outcomes. Effective communication leading to positive self-perception and self-esteem in the elders can decrease loneliness (Tian 2016) and even improve some aspects of the elders' health and well-being, to the extent of increasing longevity (e.g. Cai et al. 1998; Noels et al. 1999; Savundranayagam and Ryan 2008; Tian 2016). Positive self-perception in the elders can be enhanced by a process of *selective assertiveness* (Ryan et al. 2005) whereby one is expected to carefully choose "one's battles and responding (vs. reacting) in a clear, calm, confident manner" (Savundranayagam and Ryan 2008, p. 60).

Once the danger of the downward spiral in communication identified above in the CPA is understood, then people may prepare themselves and make changes to accommodate the difficulties that self and others may have, this could lead to forms of communication that can better achieve a satisfactory interaction as envisaged in the CEA.

The constructive approach of the CEA is clearly explained by Abrams and Giles (1999, p. 212) in the following quote:

> The Communication Enhancement of Aging (CEA) model was developed in an attempt to reduce the gap between stereotyped expectations and individual competence that may result from potential intergenerational miscommunications. Specifically, it focuses on the need for 'personalized' encounters between younger and older people. The CEA process begins with speakers having a better understanding of so-called 'normal' aging processes ... and being prepared to communicate on a more consultative and participatory basis than is typically the case. In this way, and after an educational intergenerational program, speakers are expected to have positive expectations and stereotypes of older people in general. This change in stereotypes will enable speakers to recognize more about an older person as an *individual* and not as an undifferentiated member of a socially stigmatized category. This assessment can enable youths to make more appropriate adjustments in communication style which could also increase the satisfaction of both partners, empowering the older person, reinforcing their competence, and enhancing their well-being and self-esteem.

Although in the above quote Abrams and Giles are referring to an organised training programme, the same constructive attitude in communication with old people can be achieved in spontaneous conversations through a simple grasp of the core issues: know the elder, understand his/her personal reality, do not impose stereotypes and prejudice on the interaction, then freely continue communicating with the elder in a mutually satisfactory fashion. If mutual satisfaction is not achieved, then communication will naturally break down as it is the case in all social interactions.

Ultimately, communication between old and young or between old and old is a dynamic process that can be facilitated or constrained by the specific actions of interactants. Barker and Giles (2003) proposed an integration of both *predicament* and *enhancement* as envisaged in the CPA and CEA, respectively, into an *Enhanced Predicament model* (EPM) of ageing. This model sees communication as a transactional process involving feedback effects coming and going between the interactants. The process is also affected by interpersonal similarities/differences that in turn may reflect intergroup and inter-cultural similarities/differences. Similarities are expected to facilitate communication and differences to difficult it. Prevalent stereotypes about the elders or about the young may affect the tendency to overaccommodate or underaccommodate the modes of communication. Eventually, an "optimal" form of accommodation may be achieved through the process of communication. Accommodation may be facilitated if interactants are more strongly associated through a common cultural identity, or ideological convergence.

Convergence between individuals can be facilitated when they are members of the same family, for instance. It is expected that elders may be more likely to engage in positive communication with relatives (e.g. parent–offspring), as suggested by the *Developmental Stake hypothesis* (Bengtson and Kuypers 1971), due to greater levels of intimacy between them. Although this expected trend may vary from one old generation to another according to historical-cultural factors, and from case to case according to personal history, we will see in Chap. 9 that the pattern is also expected from deeper evolutionary processes. Winkeler et al. (2000) proposed an alternative *Leniency hypothesis* that is consistent with expected life-history trends of increased altruism in old age, that extends elders' positivity beyond the family and towards others as well. In Winkeler et al.'s (2000, p. 175) words:

> aged subjects (as opposed to middle-aged or younger people) are more 'prosocial', in other words, more concerned with the fate of [a]... person in need and more generous in donating money to this person, even after controlling for economic resources. Such findings could lead to the—cautious—assumption that aged (as opposed to younger) people are more inclined to perceive their social world in a positively biased manner. Such a 'leniency' or 'positivity' effect may not just operate in helping or other kinds of altruistic behaviour; it may also determine the aged's perceptions of the quality of interpersonal relationships in a more general way, in other words, it is not restricted to perceptions of their relationships with their offspring.

Winkeler et al. (2000) tested the *Developmental Stake* and *Leniency* hypotheses in Germany in collaboration with 809 male and female 40 to 50 years old ($n = 467$) and 65 to 75 years old ($n = 342$) participants. The authors noted the reactions of the various participants to situations where there might have been intra- or inter-generational conflict. The *Stake* hypothesis predicts a more positive interpretation of inter-generational conflicts by older people, whereas the *Leniency* hypothesis predicts a positive interpretation in the case of both inter- and intra-generational conflicts. The results support the *Leniency* hypothesis as old individuals were more positive and less confrontational than middle-age individuals in assessing interactions with all types of individuals (same or different generation). This cohort of elders seems to be generally biased towards positivity in social relationships.

Communication Accommodation theory (CAT, Giles 1973; Coupland et al. 1988; Ryan 1994) specifically focuses on the accommodative effort expected from interactants if communication is going to be successful, especially when one or more of such interactants do face some specific impediments to a smooth communication (e.g. hearing loss):

> Essentially, CAT proposes that speakers come to interactions with an initial orientation, which is informed by past interpersonal and intergroup experiences, as well as the prevailing sociohistorical context. In interaction, speakers adjust their communicative behavior based on evaluations of their fellow interactants' communicative characteristics, as well as their own desire to establish and maintain a positive personal and social identity. Each speaker evaluates and makes attributions about the interaction, as well as about the other speaker, on the basis of their perceptions of that other speaker's, as well as their own, communication. These attributions and evaluations then affect the quality and nature of both the present interaction between these speakers and speakers' intent to engage in future interaction with each other (Dragojevic et al. 2016, pp. 2–3).

Figure 7.1 summarises the CAT. Through the accommodative effort of both individuals engaged in communication, such individuals may better adjust the mode of their interaction to hopefully improve its quality and the outcome of future encounters, that will eventually affect the overall quality of the relationship.

Throughout the conversation, interactants may achieve a degree of *convergence* by adjusting their language use and patterns (Gallois et al. 2005; Giles et al. 2012), which may lead to better mutual understanding. A greater level of understanding may conduce to a more satisfying mutual experience (Williams and Garrett 2002).

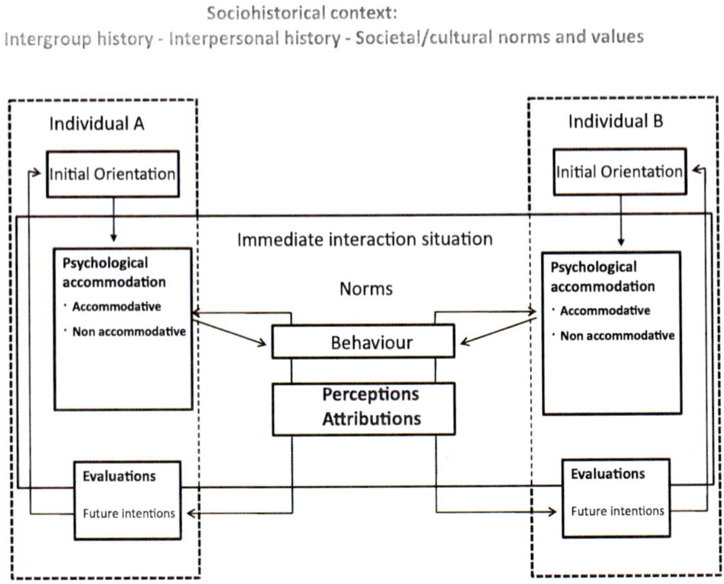

Fig. 7.1 Giles' *Communication Accommodation theory*, redrawn from Dragojevic et al. (2016) with modifications

But people interacting with each other may also experience *divergence* through differentiation in their communicative strategies; or *maintenance*, where each interactant simply retains his/her initial mode of communication. Divergence and maintenance result from an attempt to strengthen a positive individual identity in intergroup encounters, ensuring differentiation with the interlocutor (Williams and Garrett 2002), but such self-centred attitude may lead to a lower probability of building social cohesion.

Convergence and divergence in communicative styles throughout the conversation can be also *upward* (change to a more socially prestigious way of communicating) or *downward* (change to a less prestigious way, even a stigmatised one). Accommodation is *symmetrical* when both interactants move in the same direction, adjusting their speech to a closer consensus style, whereas it is *asymmetrical* when there is no reciprocation. In addition, both convergence and divergence can be *unimodal* or *multimodal*. They are unimodal when shifts occur only on one feature of language (e.g. accent), whereas multimodal convergence/divergence implies simultaneous shifts in various features of communication: accent, topics, posture (Dragojevic et al. 2016).

Therefore, in general, convergence tends to facilitate communication, whereas divergence and maintenance may hinder communication.

The above variability can produce a diversity of communicative interactions between elders and others, combining a variety of accommodative strategies where convergence, divergence, and maintenance may be symmetrical or asymmetrical (see Table 7.1 for a list of examples). For instance, *symmetrical convergence* in communication implies a similar effort from the part of both the elder and his/her partner to accommodate to the communicative characteristics of the other.

Communication can be also *overaccommodative* when it "overshoots recipients' desired level of adjustment in some way" or *underaccommodative* when it is "insufficiently adjusted relative to the recipient's needs or desires" (Dragojevic et al. 2016, p. 6).

Table 7.1 List of accommodative moves

Explanation	Elder Partner
Symmetrical convergence	A----> <----B
Asymmetrical convergence	A----> <--B
Convergence with maintenance	A----> B
Convergence with divergence	A----> B---->
Symmetrical maintenance	A B
Symmetrical divergence	<----A B---->
Asymmetrical divergence	<----A B-->
Divergence with maintenance	<----A B
Divergence with convergence	<----A <----B

The direction of the arrow indicates who is accommodating to whom, and the length indicates the intensity of the accommodative effort. Slightly modified from Dragojevic et al. (2016)

Different aspects of a conversation may even simultaneously combine overaccommodative and underaccommodative behaviours.

Interactants can adjust their communication style motivated by an affective desire to manage social distance and identity concerns, or a cognitive objective to facilitate comprehension and increase efficiency in communication (Dragojevic et al. 2016). Failure to accommodate may be perceived negatively or not to the extent that interactants conclude that such failure was intentional (which carries a negative connotation) or unintentional/attributed to the situation, which may be excused.

Dragojevic et al. (2016, p. 17) summarise the CAT according to six basic principles originally proposed by Giles et al. (2007, 2007a). Here we reproduce Dragojevic et al.'s account of CAT verbatim:

1. Communication accommodation is fundamental to interaction, and serves two major functions: first, to facilitate coherent interaction and, second, to manage social distance between individuals and individuals as members of groups.
2. Individuals have expectations about what constitutes appropriate and desirable accommodation in context, and these expectations are informed by the sociohistorical context of interaction, interpersonal and intergroup histories, and idiosyncratic preferences.
3. Speakers will increasingly accommodate to the communicative patterns they believe characteristic of their interactants, the more they wish to affiliate (i.e., decrease social distance) with their interactants on either an individual or group level, or make their message more easily understood.
4. As a function of the intentions and motives believed to underlie a speaker's communication, perceived accommodation increasingly and cumulatively decreases perceived social distance, enhances interactional satisfaction and positive evaluations of speakers, and facilitates mutual understanding.
5. Speakers will increasingly *non*accommodate to the communicative patterns they believe characteristic of their interactants, the more they wish to disaffiliate (i.e., increase social distance) with their interactants on either an individual or group level, or make their message more difficult to understand.
6. As a function of the intentions and motives believed to underlie a speaker's communication and the potential consequences of associated outcomes, perceived non-accommodation increasingly and cumulatively increases perceived social distance, diminishes interactional satisfaction and positive evaluations of speakers, and impedes mutual understanding. Such non-accommodation may take the form of either overaccommodation or underaccommodation relative to a recipient's needs, with underaccommodation typically resulting in more negative evaluative responses.

In a study of intra- and inter-generational communication carried out in the UK, Williams and Garrett (2002) interviewed 490 adults (20–59 years old; subdivided into different age classes: 20–29, 30–39, 40–49, and 50–59 years old) regarding their experiences communicating with non-family members aged 13–16 years old, or 65–85 years old, or their own age. The experiences were evaluated following a questionnaire. Communication was more frequent with peers than with members of other age groups and communication with older or younger people was similarly frequent overall. However, as people age, they tend to communicate more with older people despite seeing them as less accommodative, although teenagers are also seen as less accommodative. This suggests that communicative needs may be somewhat resilient to lower levels of accommodation by partners, for as long as the elder may feel a strong need to communicate with somebody they believe may be able to understand them. In fact, older people enjoyed the benefits of a greater level of accommodation in communication by younger people whenever such accommodation occurred in their interactions. But young people are not always keen to accommodate.

Some cultural traditions, such as filial piety, lead to the expectation that younger members of the family show respect towards their elders by displaying greater accommodation in the conversation, thus biasing the direction of power in communication towards the elders. This culturally expected bias in communicative accommodation tends to be challenged by current young generations, in response to the lack of reciprocation from the elders and a diversity of perceptions about old age: older people "expect respect and yet can seemingly taunt young people with perceived disrespect ...; receive help yet engender pity ...; and are seen to possess positive as well as negative interactional skills" (Giles and Dorjee 2004, p. 266).

Therefore, accommodation in communication is a dynamic process that can lead to a smooth and mutually satisfactory interaction between people of the same or a different age, but only for as long as communication is a fulfilling experience for all interactants.

Other theoretical approaches to communication are also available. van Dyke et al. (2009) studied intra- and inter-generational communication in the USA in the context of the *Situational Theory of Publics* (Grunig 1978) which states that a problem is recognised as such to the extent that it actually affects an individual, and such perception determines the level of involvement of the individual in solving the problem. Therefore, the theory predicts that "high problem recognition and level of involvement and low constraint recognition increase the likelihood that someone will respond to communication messages" (van Dyke et al. 2009, p. 20). Problem recognition could be affected by specific generational characteristics (both physical and cultural) that may lead to miscommunication when different generations interact. For instance, a *staircase pattern* in communication is expected to the extent that accommodation within generation is better than between generations and, between generations accommodation decreases as the age gap increases. This may result from a greater likelihood of recognising common problems within- than between-generations. Exceptions to this rule, however, are possible.

Van Dyke et al. (2009) carried out 60 interviews with members of four generations in the USA: *Seniors* (61 years old and older), *Baby Boomers* (41–60 years old), *Generation Xers* (26–40 years old), and *Generation Nexters* (18–25 years old). From the interviews, members of different generations demonstrated to have a diversity of characteristics. For instance, *Generation X* tended to focus on work, their careers, and the attainment of some financial goals; the *Baby Boomers* displayed a tendency to express more filial piety than younger generations, whereas *Seniors* self-described as conservative, fiscally responsible, and also slow-paced, religious, and family-focused. The members of *Generation Next* were mainly influenced by family and technology. Both intra- and inter-generational communication were regarded as being either satisfactory or neutral by most participants. Some barriers to communication that are believed to make such communication less satisfactory included: technological barriers (e.g. *Seniors* tended to use Internet-based communication technologies less frequently), differences in lifestyles, and resentment over being stereotyped. Although some topics of conversation were more age-specific (older participants tended to include issues about health; whereas younger adults focused on study, work issues, career, finances, current events, and also raising a family), the conversation could focus on a common topic when both individuals had similar experiences. Talking about topics of personal interest with people who share such interests reinforced a sense of identity across participants and facilitated a more satisfactory communication, whether this was within or across generations, as expected from the *Situational Theory of Publics*.

Communication satisfaction in either intra- or inter-generational encounters can be achieved if, throughout the interaction, the partners socially support each other "discouraging negative perceptions, discounting worries, validating one's social identity ... providing cognitive alternatives to life circumstances ... and so forth" (Williams and Giles 1996, p. 222). Such mutual social support in communication, signalling in-group solidarity, empathy, and affiliation, can be expressed in the process of *attuning*: "a range of communication management strategies along dimensions of discourse management (e.g., topic), interpretability (e.g., clarity), and interpersonal control (e.g., positive and negative face)" (Williams and Giles 1996, p. 224). Conversely, when we intend to signal out-group membership our speech patterns may diverge from those of our interlocutor in a process of *counter-attuning*. Attuning and counter-attuning are communicative strategies reflecting, respectively, a positive or a negative disposition to accommodate.

Williams and Giles (1996) studied satisfaction in inter-generational communication in a group of 129 Anglo-American university students (median age 18.5 years, 72% females) in California. Through questionnaires the authors evaluated the degree of inter-generational contact experienced by respondents and with whom that contact was, along with their level of satisfaction in conversations with older or younger people and topics discussed in such conversations. The young respondents spent most of their time with people younger than 35 years (85%), followed by middle-age people (13%), and they only spent 4.5% of their time with older people (more than 65 years old). Again, this is a *staircase pattern*. The time spent in inter-generational contacts was not only small in this sample, but the level of satisfaction with such interactions was also low: the young respondents felt that interactions with peers were more satisfying than inter-generational interactions. The topics of inter-generational conversations varied according to the relationship between old and young interactants. Among strangers, the conversation could be satisfying if it focused on "life in general", whereas among interactants already known to each other the topics of conversation were more diverse (e.g. family, relationships, school, young person's plans for the future). Interestingly, young participants reported a mixed bag of emotions when they engaged in inter-generational interactions, including initial trepidation potentially followed by relaxation as the conversation unfolded, and a combination of happiness and sadness. The most important perception of the younger persons regarding the style of communication of older people was that the latter were underaccommodating: "the older person was not 'with' them, either not listening, interrupting, inattentive, or unable to align with the younger person's communicative needs" (Williams and Giles 1996, p. 233). Young participants found themselves in some dilemmas regarding how to respond when older people behaved in an underaccomodating way (p. 234):

> My partner kept repeating himself, saying the same thing over and over again… rambling". "He does not hear my comments all the time… I felt like I could not communicate.

Such situations did elicit polite responses from the part of young participants, out of a sense of obligation to accommodate, but not without feelings of uneasiness. Uneasiness from the part of young participants was also elicited by the expression of "angry complaints" and "painful self-disclosure" from the part of the elders. The young became more defensive when older people expressed negative stereotypical opinions about the younger generations (e.g. "He makes false accusations at me, concerning those of a stereotypical teenager").

On the positive side, supportive communication between older and young occurred whenever they demonstrated a degree of interest and attentiveness towards the other or provided some emotional or instrumental support or various forms of encouragement, compliments, and welcome advice. Support from the older person was greeted by the younger one with responses of happiness, relaxation, increased self-esteem, and feelings of security.

Narratives and storytelling were also an important part of inter-generational communication in Williams and Giles' study; such narratives were mainly greeted with a sense of happiness from the part of the listener.

In sum, the understanding of communicative interactions involving old people has developed following a series of theoretical models, each emphasising a distinctive aspect of such interactions. Given that face-to-face communication is first and above all a social encounter, we started with an emphasis on preliminary interactions that function as information-gathering experiences between strangers. According to *Uncertainty Reduction theory* the aim of such preliminary encounters is to facilitate subsequent communication. Achievement of good communication through uncertainty reduction, however, is constrained by inter-generational and intra-generational factors. For instance, it is commonly observed that good communication can be better achieved the closer in generation the interactants are, following the *staircase pattern* of *Social Identity theory*. This can then lead to various patterns of intra- and inter-generational solidarity and conflicts. Belonging to the same generation can lead to a convergence and agreement on the identification of problems, and the best way to solve them (*Situational Theory of Publics*); whilst awareness of generational issues can increase the understanding of those sensorial and mental constraints experienced by the elders, that may difficult smooth communication. Through using better generational knowledge interactants can adopt a more positive attitude in the conversation (*Deficit Paradigm*). Two major approaches have characterised the theoretical work on elders' communication: one emphasises the negative aspects of ageing and the other emphasises some positive aspects. On the negative side, the *Communication Predicament of Ageing* model points to the objective (or assumed, that is, stereotypical) communicative limitations of the elderly, which may lead to patronising and baby-talking by younger interactants, thus potentially resulting in learned helplessness by the elders and a breakdown in communication. On the positive side, *Communication Accommodation theory* focuses on the cooperative effort expected from interactants if they are going to achieve a better level of communication. Similarly, the *Communicative Enhancement Model of Ageing* states that communication between old and young that accommodates to the limitations displayed by the elder, may lead to a better and more satisfactory interaction. Positivity in communicative outcomes can be also facilitated if older generations adopt a more constructive attitude towards younger generations, whether the interaction involves close relatives (e.g. grandparents and grandchildren) (*Developmental Stake* hypothesis) or non-related acquaintances (*Leniency* hypothesis). Finally, both positive and negative aspects of communication have been synthesised in the *Enhanced Predicament* model where communication is seen as a transactional process that can progress thanks to the mutual feedback and accommodative effort of interactants. A specific focus on the accommodative effort required by both interactants to achieve better communication is also found in the *Communication Accommodation* theory. A better degree of accommodation can facilitate communication especially with elders experiencing some sensory limitations such as hearing loss.

In the next section we focus on cross-cultural aspects of communication.

7.2 A Cross-Cultural Perspective on Communication in Old People

Studies of communication in old people have been also carried out in a cross-cultural perspective, indicating that some negative attitudes towards the old are widespread across cultures (e.g. Giles et al. 1992). A frequent comparison is between Western and Asian cultures which tends to show some current better attitudes of the young towards old people in the West than in Asia, against the traditional expectations of filial piety and respect for the elderly that are supposed to characterise most Asian cultures. Such trend seems to be the effect of increased modernisation observed in Asian countries in recent decades. In the West, elders are also perceived as displaying higher vitality and there is greater social equality between generations and emotional closeness with family elders (e.g. grandparents); whereas at present young Asians perceive their elders as being less accommodating to the younger generations (less encouraging, supportive, attentive) and the young are less satisfied with communication with their elders (Giles et al. 2003), even though the young in Asian countries are under greater social pressure to show respect for their elders following local traditions of *filial piety*.

Williams et al. (1997) carried out a study of old-young communication in a cross-cultural perspective. They studied the perceptions of young adult participants ($n = 1631$, 18–29 years old) from nine countries about their conversations with 65–85 years old people. The study focused on both elder non-accommodation as perceived by young participants: older people negatively stereotyping the young without responding to their communication needs; and elder accommodation. Elder accommodation was reported whenever older people were seen as "supportive, attentive and generally encouraging to young people". In addition, the study investigated respect/obligation and *age-irrelevant positivity*: "a situation where young people felt conversations with much older people were emotionally positive and satisfying, age did not matter". Participants were undergraduate university students and were drawn from four Western countries with a strong Anglo-Celtic cultural influence, that are broadly characterised as "individualistic": USA, New Zealand, Australia, and Canada; and five Eastern cultures that are traditionally characterised as "collectivistic": People's Republic of China, Hong Kong, Korea, Japan, and the Philippines. Participants' responses to a questionnaire (*Perceptions of Intergenerational Communication* (PIC)) were recorded, with the overall group of participants being female biased.

After carrying out a principal component analysis, the variance in the data was explained by four factors: *respect-obligation*, *elder non-accommodation*, *age-irrelevant positivity*, and *elder accommodation*, in that order of importance. Regarding *respect-obligation*, young participants considered themselves to be in the obligation to show politeness, they restrained themselves from expressing some of their feelings and thoughts and felt defensive in the conversation. Elders were broadly regarded as *non-accommodative* when they were perceived as angry and complaining, close-minded, out of touch, forcing attention, and negatively stereotyping young people. *Age-irrelevant positivity* was evident in those opinions indicating that age does not matter, older persons are emotionally positive, and interactions with older persons are satisfying. *Accommodative* elders were described as supportive, giving compliments and useful advice, they are attentive to the younger person, and tell interesting stories.

There was no effect of East–West cultures on elder non-accommodation and respect-obligation, whereas participants from Eastern cultures did not think that older people behaved in a communicatively positive manner with them, as expressed in their levels of support, attentiveness, and compliments, for instance. Significant differences were also observed in age-irrelevant positivity. In this case participants from Eastern countries were less likely to regard their interactions with older people as positive and satisfying; they were also less likely to agree that age did not matter in the interaction. Therefore, this study points to a significant degree of convergence between Eastern and Western young adults regarding their communication with old people, but also to a relatively higher degree of inter-generational tension in communication in some Asian countries.

Williams et al.'s (1997) work illustrates the pattern of erosion of respectful communication with elders expected from filial piety in Asian societies, where filial piety was culturally entrenched. Modernisation has somewhat flattened some cultural differences between West and East without, however, eliminating them completely. McCann et al. (2004) investigated the views of young adults on their intra- and inter-generational communication experiences comparing South Vietnam, North Vietnam, and the USA, for a total sample of 554 participants (66.7% females; around 20 years old). The PIC scale was used to measure personal experiences in inter-generational communication. In this study participants from the USA used more accommodative forms of communication than participants from both North and South Vietnam; moreover, women were overall more accommodative than men across groups. Both accommodation and non-accommodation were greater in the case of older targets of communication. Interestingly, respect-obligation in inter-generational communication was more common among the South Vietnamese (especially regarding older targets) than among the North Vietnamese and USA respondents. Such respect-obligation for older targets was especially accentuated among female participants. Avoidant communication was also more prevalent when the target was older, especially for the case of young female participants. Thus, counter to the traditional emphasis on filial piety in Asian countries, this study also points to greater levels of accommodation among Western participants, but with a convergence towards more accommodation in communication with elders observed in women across cultures.

In a cross-cultural study of inter-generational communication focusing on two Asian countries (Japan and Thailand), Ota et al. (2012) investigated young adults recalled past experiences of communication with people of various ages: young, middle age, and older. They studied a set of specific aspects of communication such as *approach/avoidance* (e.g. respectful accommodation, disengagement, self-restraint), *politeness* (i.e. respectful attitudes towards others), *deference norms* (i.e. modesty and non-assertiveness), and satisfaction with and enjoyment of the conversation through a questionnaire. Both samples of participants were female biased (63.7% females in the Japanese sample, 76.3% in the Thai sample). Their main results are summarised in Table 7.2.

With regard to respect, avoidance, deference, and politeness, all between-age differences were significant in pairwise comparisons and they were also incremental from young to elder, thus demonstrating a staircase pattern in both cultures: older people enjoyed greater respect, deference, and politeness, but also greater avoidance. Conversational satisfaction and enjoyment showed cultural differences. In Japan, young adults enjoy more and obtain more satisfaction in communication with younger than older people, showing an inverse staircase pattern as compared with their responses for respect, avoidance, deference, and politeness. Thus Japanese participants, although showing respect for their elders just as Thai do, are more satisfied and enjoy more when communicating with their peers. Thai participants were somewhat different, as they were equally satisfied with communication with peers and elders, and indeed they enjoyed communicating with their elders more than with any other age group (Table 7.2). This pattern can be a result of the

Table 7.2 Means and standard deviations for recalled past experiences of communicative behaviours, norms, and satisfaction in Japan and Thailand

	Age group					
	Young		Middle		Older	
Communication variable	Japan	Thailand	Japan	Thailand	Japan	Thailand
Respect	3.98 (1.09)	4.51 (1.05)	5.06 (0.86)	5.25 (0.84)	5.33 (0.82)	5.73 (0.80)
Avoidance	3.00 (1.07)	3.07 (1.09)	4.03 (1.16)	3.72 (1.05)	4.29 (1.23)	4.15 (1.04)
Deference	2.60 (0.86)	3.77 (1.07)	3.30 (0.97)	4.43 (0.93)	3.55 (1.01)	4.95 (0.99)
Politeness	3.63 (1.21)	5.17 (1.26)	5.36 (0.89)	5.92 (1.01)	5.74 (0.85)	6.38 (0.76)
Conversational satisfaction	5.27 (1.19)	4.82 (0.95)	4.18 (1.27)	4.62 (0.85)	3.99 (1.12)	4.82 (1.02)
Conversational enjoyment	5.21 (1.27)	4.94 (1.15)	4.24 (1.36)	4.86 (1.22)	4.06 (1.21)	5.25 (1.27)

From Ota et al. (2012)

combined effects of the Thai cultural norm of *kreng jai* (respect to others) alongside a strong sense of age-consciousness found in Thai culture.

In a comparative study of perceptions of inter-generational communication in the workplace (i.e. banks), McCann and Giles (2006, 2007) compared 168 Americans (23 years old on average, 69.0% females) and 180 Thai (29 years old on average, 65.5% females) participants. Communication beliefs were self-assessed by the participants through the *Global Perceptions of Intergenerational Communication* (GPIC) scale. The results indicate that older bank employees (bankers) were regarded as more non-accommodative than younger ones, and younger bankers thought that they should be respectfully avoidant in their communication with the older ones. Young Thai bankers were seen as more non-accommodative than American ones, which not only runs against the traditional respect that elders are expected to receive from younger generations in Asian societies, but also with the alleged exceptionality of Thai culture as suggested in Ota et al.'s (2012) work. It is possible that accommodation in communication with older people varies in modern Thai culture according to the specific context (e.g. day-to-day interactions vs. professional context).

Giles et al. (2003) asked 731 undergraduate students (16–30 years old, 56.1% females) from both Western (Canada, USA, and New Zealand) and Asian (The Philippines, South Korea, and Japan) universities to complete a questionnaire about their perceptions regarding family elders, non-family old people, and same-age peers. Consistent with a *staircase* pattern, both Western and Asian samples reported more positive interactions with individuals of the same age than with old family members, and they also had more conversations with family elders than they had with non-family elders. In addition, both groups had a similar degree of contact with same-age peers, although the Western group of participants had more conversations with both groups of older people compared with the Asian group. More contact with any specific age group was also associated with a higher perception of accommodation and low non-accommodation, and respondents felt greater levels of *age-irrelevant positivity* towards that group. Interestingly, whereas respondents from the different Western countries seemed to be homogeneous in their responses, among Asians there were differences, with Japanese participants displaying greater influence of modernity and greater departure from traditional values of familism.

In a survey aimed at comparing the communication experiences of 16–46 years old with elderly people among New Zealanders (NZ) of European or Chinese origin, Ng et al. (1997) analysed 50 questionnaires returned by NZ Europeans (52% females) and 50 returned by NZ Chinese (50% females). Positivity in communication, contact, and positive feelings—negative feelings showed the reverse pattern—increased from non-family members to family members and from old to same-age peers, they were also higher for females than males. Greater accommodation was reported by both genders in communication with family elders than with non-family elders; however, in the comparison between family elders and peers, females showed no difference in accommodation, whereas more accommodation with family elders was reported by males. Most measures showed no effect of culture, with the exception of negative feelings and frequency of contact, in this case NZ Europeans scored more positively. There was also a greater perception of elders' vitality in the NZ Chinese than the NZ Europeans. Positive stereotypes were more prevalent among women than among men; they were higher among NZ Chinese than among NZ Europeans and were higher for peers than for the elderly.

Many participants concurred in that communication with older people required some pragmatic adjustments such as talking slowly, loudly, and repeating phrases, as elderly people had problems hearing, and sometimes they may have forgotten what had just been said. Some of the participants pointed to the stubbornness and verbosity of older people and how demanding they could be. On the positive side, they remarked that old people could tell "interesting stories", give good advice, and give supportive compliments (Ng et al. 1997).

In another study also carried out on inter-generational interactions within a community of Chinese immigrants and European New Zealanders, Liu et al. (2000) specifically focused on filial piety. Participants included 286 people: 150, 35–55 years old parents and 136, 10–25 years old offspring; there was an equal number of men and women. Upon being asked about their filial piety responsibilities towards parents and grandparents, participants did not distinguish between the two categories of relatives regarding filial piety obligations, although they suggested that filial piety expectations by parents were higher than by grandparents. NZ Chinese had greater expectations of filial piety to be expressed by children and grandchildren than NZ Europeans. Interestingly, both Chinese and New Zealand identity were associated with a stronger sense of filial piety. In addition, a New Zealand identity was correlated with the obligation to show "respect" and also to "retain contact" with family elders, whereas a Chinese identity was strongly associated with "assist them financially" and "please them and make them happy". Obey the elders was not associated with any of the two ethnic identities. Finally, the more positive the subjective experience of inter-generational communication was the greater the level of filial piety also was, this seems to be more a result of internally felt obligations than external expectations. Relatives engaging in spontaneous conversations with elders, following a filial piety inclination, also experienced greater happiness as a result.

Lin and Zhang (2008) studied intra- and inter-generational communication in Taiwanese elders using the theoretical framework provided by *Social Identity theory* (SIT) and *Communication Accommodation theory* (CAT), also with an emphasis on filial piety as in the work of Liu et al. (2000), but now from the perspective of older people. They interviewed 16 female and 15 male 60–82 years old ethnic Taiwanese, Chinese mainlanders, and Hokka. The interviewees expressed great interest to accommodate to the conversational needs of young people by engaging in topics more closely matching the interests of the young (e.g. job, school, marriage, future career plans), whereas they were more reluctant to disclose personal experiences, especially the painful ones. The conversation, however, tended to be deeper when the old participants were communicating with a young family member than a non-family member, with the elders taking the initiative to offer their advice to the young based on their life experience, and expressing some concerns as to whether the younger generations would be able to fulfil the cultural expectations of filial piety towards their old parents (*underaccommodation*). Older participants comforted each other in cases of inter-generational disappointment or unfulfilled expectations, as when the elders perceived a level of self-centredness in the young that is not attuned to the cultural expectations of filial piety: e.g. the young displaying excessive materialism and low ability to endure hardships. Comfort was obtained by adopting a strategy of *downward comparison*: "at least they do not come and ask for money, right?", suggesting that things could be even worse, after all.

In Lin and Zhang's (2008) work, discourse on intra-generational relationships focused on *self-exception*, in the case of comparisons with elders who are worse off, *denial of personal disadvantage* (e.g. denying being the object of prejudice) and *self-inclusion*, a discourse where older people acted as in-group members in contrast to the young out-group. In all these cases older people simply tried to manage their self-identity in the best way they could.

Whereas Lin and Zhang (2008) studied Taiwanese elders, Lin et al. (2004) focused on inter-generational communication from the perspective of young Taiwanese. The authors especially targeted *inter-generational communication schemata* (ICS), which are cognitive representations of communication. Harwood (1998) originally described six ICS: *helping*, *learning*, *gerontophobic*, *gerontophilic*, *pity*, and *polite*.

In their work, Lin et al. (2004) interviewed 41 tertiary education students (68.3% females, mean age 20.3 years old). Participants were instructed to imagine various kinds of conversations with diverse categories of old people (65 years old or above): a family member vs. non-family member, a senile vs. healthy elder, a socially inept and bitter elder vs. one who is wise and kind; they were also requested to describe a positive vs. a negative experience, an ideal vs. a worst conversation, and a conversation with strangers on the bus or in the subway. Conversations were described in various ways. *Positive* conversations were characterised by the young person's general satisfaction with the experience. Participants also described conversations that could be *mutually satisfying*, whereas *helping* conversations were those in which young participants made an effort to make the elder feel happy. In *mixed feelings* conversations the young were ambivalent: whereas learning from the old was seen as positive, the constraints imposed by the rules of politeness were seen as negative. The conversation could also be *negative* for either one or both interactants (*mutually unpleasant* conversation). Finally, *small talk* conversation did not attract much interest from participants.

The *helping* schema used in conversation was the one most commonly mentioned in Lin et al.'s work, but the young participants tended to reject helping when it was delivered in a *laodao* fashion. *Laodao* is a repetitious form of advice provided by older adults which may be also associated with nagging or over-parenting: "When advice is communicated to young people in a '*laodao*' way, they tend to feel being criticized and imposed upon with unwanted advice, and associate traits such as 'outdated,' 'stubborn,' 'conservative' and 'living in the past' with older adults" (Lin et al. 2004, p. 335). Thus, inter-generational communication in this Taiwanese sample reflects a combination of respect towards older generations and also assertion of own identity and agency from the part of the young.

In another study of Taiwanese young adults Giles et al. (2001) reported that they tend to be more positive regarding their communication with peers than with older adults, in contrast to Euro-Americans young adults, who generally perceive inter-generational interactions more positively than their Taiwanese counterparts.

Young adults' perceptions of older people have been also compared between the USA and some African countries such as Ghana and South Africa. Giles et al. (2005) remark that throughout Africa elders still retain an aura of respect conferred to them by their social status but, just like in Asia, recent trends are eroding the traditional social power of older people in this continent, increasing the negative stereotypes that young people have of their elders, thus leading to frustration and dissatisfaction in the latter. In Giles et al.'s (2005) study the opinions of university students about their communicative experiences with non-family elders were assessed using a questionnaire and then compared among participants from the above three countries, with the USA sample being mainly Caucasian (65.7%). Men and women were similarly represented across the three samples. In all ethnic groups, older people were regarded as benevolent and they were expected to be treated in a polite, deferent, and respectful manner; whereas attributions of vitality and communication satisfaction decreased with increasing age, leading to a high degree of avoidance. Avoidance was associated with less enjoyment of the conversation with elders.

Giles et al. (2007) also compared perceptions of inter-generational communication between American and Indian young adults (mean age 20 years, 52.6% females) in the USA. Again, older ages were associated with attributions of benevolence, norms of politeness (with greater politeness shown towards old people in the American than the Indian sample), deference, respect but also communicative avoidance. Respect for older people was higher in the American than the Indian sample, which is consistent with the broad trend of erosion of respect towards old people observed in Asia in recent decades. Vitality (higher in the Indian than the

American sample) and communication satisfaction decreased with old age. Indian youth satisfaction in their communication with older people increased the more those youth perceived elders as benevolent and displaying greater vitality.

Nayar and Sandhu (2006) studied a Punjabi community resident in Western Canada focusing on three-generation households: children (born in Canada), parents (born in India), and grandparents (born in India and who had mainly migrated to Canada through family reunion). Communication across generations in this case may prove to be a challenge due to distinctive cultural differences among them. Focusing on communication styles alone, grandparents in this case had been mainly exposed to oral communication traditions in India, whereas parents and especially children had been more exposed to a literacy tradition and to the analytical tradition of Western culture. Nayar and Sandhu (2006) point to some barriers in inter-generational communication in this community such as perceptions that the older generation just lives in the past and cannot understand the reality of current young generations, the modern views about marriage, and the necessity or not of getting married. Some contradictions between cultural beliefs and actual behaviour dictated by the necessities of modern life also make younger generations wonder. For instance, a grandchild noted (p. 146):

> ... look at my mother, she works in the real world and takes the bus home at 11:30 at night. If respect is the most important thing, why doesn't my mother stay home and take care of my grandmother. If my father thinks safety is important for women, then why does he let my mother work late hours?

The grandchild argues on the grounds of analytics and knowledge, pointing to contradictions between the Indian cultural traditions of the elders and the actual behaviour of the parental generation, which operates under the constraints imposed by a modern Western society.

To recap, communication with elders in Asian countries is currently seen less positively by younger generations than in Western countries, this may be an effect of modernisation, a phenomenon that also affects other parts of the world, with similar consequences. There is a broad cross-cultural staircase pattern for respectful, deferent, and polite communication, but also avoidant and negative communication that characterise interactions with increasingly older people, but with some specific differences in some cultures. Finally, studies have shown that it is usually women who tend to be more accommodating in their communication with elders than men across cultures.

In the next two sections we focus more specifically on intra-generational communication first, or how old people communicate with their peers, and then on inter-generational communication.

7.3 Intra-Generational Communication

The theoretical framework introduced in the first section of this chapter, including those models that were originally devised to understand inter-generational communication, could be also applied to intra-generational communication in the older generations. For instance, *Communication Accommodation theory* (CAT) (need for mutual accommodation to achieve better communication), *Social Identity theory* (SIT) (both solidarity and conflict determine the outcome of social interactions), and the *Communication Predicament of Ageing model* (CPA) (role of stereotypes in communication) are all useful to understand the dynamics of either intra- or inter-generational communication. The synthesis provided by the *Enhanced Predicament model* (EPM) of ageing also applies to intra-generational communication, seen here as a transactional process unfolding between similarly aged individuals.

Despite the importance of accommodation in social interactions, old people are not always accommodative in their intra-generational communication (e.g. Ota et al. 2007), just as they not always accommodate to communicating with younger people. A lower level of accommodation in intra-generational communication seems to be more evident in Asian than Western cultures (e.g. Ota et al. 2007). However, when accommodation in intra-generational communication does occur in older adults, it can have positive effects on self-esteem. This pattern has been described cross-culturally (Ota et al. 2007).

We have already mentioned in the introductory section of this chapter how stereotypes can affect the course of communication between individuals. Stereotypes can be positive or negative and therefore they can affect positively or negatively the way in which we communicate with peers. Although most theoretical models and empirical research tend to focus on negative stereotypes, a more comprehensive approach would also consider the possibility of greater satisfaction in communication whenever stereotypes are positive. For instance, Hummert's (1994) *Stereotype-Sensitive* model focuses on the communicative mechanisms that may achieve greater satisfaction in social interactions. In this case, it is positive stereotypes (e.g. seeing other old people as "golden agers") that may enhance communication satisfaction among the elders.

In a work carried out in the USA, Chen and King (2002) found that communication was more satisfactory in the elders the more positive their age stereotypes were. Moreover, they also found that old adults displayed more positive stereotypes towards people of their own age than towards younger adults, without being completely negative towards the young. Old adults in this study simply appear to develop a greater sense of acceptance of others with age.

Better intra-generational communication among the elders could also positively affect the care of old people, when caregiving is provided by another old person, whether they live in the community or in a retirement village (Buys 2001). Mbanaso et al. (2006) studied intra-generational caregiving among elderly African Americans. A total of 129 participants (mostly women: 88% and 93% in two groups, one urban and one rural) were recruited who were taking care of a 65–104 years old spouse, or partner, relative, or a friend. Information gathered from focus groups suggest that old caregivers helped with *activities of daily living* (ADL): bathing, dressing, taking a medicine; and they also helped with *instrumental activities of daily living* (IADL): shopping, cooking meals, doing some housework, lifting, managing money. As far as communication is concerned, companionship was an important caregiving activity as it was accompanying the elder to medical visits.

Old people who can talk and communicate with other old people in face-to-face contact can find ways to share common problems and be better understood in their needs (Downs et al. 1988 and references therein). For instance, a good nursing home environment may not only provide more opportunities for communication with same-age people but also decrease the level of apprehension about social interactions, which may characterise some older people living alone at home. Communication apprehension is "an individual's level of fear or anxiety associated with either real or anticipated communication with another person or persons" (McCroskey 1984, p. 13). Communication among elders in a nursing home, however, may often require some encouragement from the outside: nursing home personnel, visitors, volunteers. On the other hand, living at home does not necessarily lead to isolation, as elders may be well integrated into the social life of their neighbourhood and they can also use modern computer-based technologies to stay in touch with friends near and far (e.g. Consolvo et al. 2004).

Intra-generational communication in the elderly can be especially challenging in the case of immigrants, whether such older immigrants communicate with old members of their own ethnic community or old members of the broader community. People migrating as adults or young adults have already acquired full competence in their native language (L1, e.g. a Romance language or an Asian language) but are confronted with the need to learn an additional language (L2, e.g. English). As they become older, they often show a degree of deterioration in their L2 and a reversion to L1, sometimes even involving a recall of L1 vocabulary not used in many years. This is known as *first language reversion* (De Bot and Clyne 1989). First language reversion poses some challenges to migrants' families if some or most of the members are not bilingual, and the challenge is especially acute in aged care facilities where old people of various ethnicities may live together. The issue could be addressed by becoming a member of an elder ethnic organisation, and by actively participating in their activities.

The relative deterioration of L2 in old age is known as *second language attrition*, involving a loss of vocabulary and grammatical rules of L2 that elders had learned when they were younger. Both reversion and attrition, however, are variable and their extent is dependent on the individual characteristics of the elder. According to Clyne (1981) reversion increases with lack of close social contact with native speakers of L2, it is also affected by how well L2 was learned and therefore how likely is it that it can be retrieved from memory in old age. More reversion is expected if brain centres controlling L2 deteriorate in old age, and reversion is also affected by the age at which L2 was learned: younger learners can better retain both L1 and L2 in old age. More frequent intra-generational communication with both speakers of L1 and L2 may help older immigrants retain their overall communicative competence.

7.4 Inter-Generational Communication

Although old people may interact with other old people in the broader community or in specific settings such as a retirement village, they are also likely to interact with younger people either occasionally or on a regular basis. Communication between the young and the old is affected by their individual characteristics (e.g. personality), experiences (e.g. education), and willingness to interact with each other. Some concepts, colloquial words, idioms, for example, that are used by old people may be unfamiliar to the young and vice versa, whereas some inclinations towards favouring specific social or communicative behaviours may not be taken positively by the young. All this can affect the likelihood that young and old can engage constructively in a conversation. For instance, Coupland et al. (1991) analysed the phenomenon of *painful self-disclosure* (PSD) which refers to the "revealing of personal and often intimate information about ill health, bereavement, immobility, loneliness, and so on" (p. 22) and which is often observed in elders' conversations. Women seem to self-disclose more than men, especially to other women (women also tend to be more sociable than men, Chap. 6), as do more lonely people. Self-disclosure is described in terms of its content, such as the kind of information being disclosed, and its implementation, or the way in which information is disclosed. In addition, the tendency of some people to self-disclose more than others is affected by personality, life circumstances, personal identity, norms of interaction, stereotypes, and preferred discourse strategies among others.

More tendency to self-disclose may be beneficial in terms of increasing mutual knowledge and fostering accommodation in the interaction, but there is also the potential cost of losing face. Disclosing personal, often painful experiences is a face-threatening act (FTA) (Brown and Levison 1978), that may affect the discloser and/or the recipient.

PSD was the focus of Coupland et al.'s (1991) work carried out in Cardiff, Wales. They analysed the contents of 40 videotaped interactions between 70–87 years old and 30–40 years old participants to find that self-disclosure was common and it frequently referred to frail health. In their discourse, elderly participants expressed a self-association with the past and its specific historical context but a self-dissociation from the present, this determined a point of potential rupture with a younger participant, who has obviously less past but is very anchored in the present and projecting into the future.

In spite of the tendency to dissociate from the present, the current self-identity of older participants was variable: it could be positive, negative, or even neutral (e.g. "that's how it is"). The self-identity disclosed by elders and younger participants in the conversation may help or interfere with a smooth communication depending on the level of *ideological matching*, or convergence of views, between young and old. The process of ideological matching may even lead to extremes such as *self-deprecation*—i.e., belittling or undervaluing oneself (self-handicapping) (Coupland et al. 1988)—as an expression of empathy whenever one of the interacting people is disclosing a painful experience.

Topics addressed in PSD conversations in Coupland et al.'s work included: bereavement, ill health, physical handicaps, disengagement, loneliness, and various forms of troubles (e.g. family, financial). Whatever the topic that was addressed, the self-disclosure event may have ended in various ways: the discloser may have taken the initiative to move on by changing perspective (e.g. through rationalisation, contextualisation, minimisation, resignation, face compensation, inversion) or by shifting the topic to one that is not painful. On the other hand, the recipient could take the initiative to end the disclosure event also by changing perspective (e.g. proposing an improbable solution to the situation being disclosed by the elder), shifting the topic to something less painful or through a more abrupt switching of the topic to a new, but non-painful one.

Given that PSD is a face-threatening act, why do some elders do it? What are the benefits? Self-disclosure may in fact have some positive-face benefits, if by disclosing personal experiences the elder may show to have some interesting information to offer. In addition, the disclosure of painful personal experiences may be therapeutic for the elder. Self-disclosure may also reinforce own identity, even when the disclosure is painful, but perhaps more importantly self-disclosing may attract empathy and social support. Moreover, disclosing painful experiences that the elder was able to overcome may put the elder in a position of "hero", which can be supportive of a positive self-identity. Therefore PSD can have an existential function in old people through reflection on own life and also gathering support and empathy from listeners.

Interestingly, Coupland et al. (1991) also reported frequent self-disclosure of own age, even among women. Old people in Western cultures are more likely to disclose their chronological age when they look younger than their age, but they are less likely to disclose it when they look as old or older than their age. Face-threatening is most likely at play here.

We have seen that when confronted with elders' painful self-disclosure, younger persons may either converge in an accommodative manner, by also expressing opinions about their own painful experiences, or diverge by subtly changing the topic of the conversation or avoid personal engagement by only listening to the elders' narrative but otherwise remaining passive (Giles 1984; Coupland et al. 1991). Throughout the conversation old and young may eventually *attune* their discourse to each other, that is they may become conversationally accommodating by getting closer to each other's views, or alternatively they could distance themselves. *Overaccommodating* in the conversation is also possible. This is a strategy that leads to miscommunication, "where at least one participant perceives a speaker (who may be herself or himself or another) to 'go beyond' a sociolinguistic style judged necessary for attuned talk on a particular occasion. Examples are speaking too loudly or too slowly, or being unnecessarily cautious and careful" (Coupland et al. 1991, p. 30). Stereotypes, for instance, may lead to overaccommodation in speech when talking to elderly people, potentially leading to talking down, baby talk, and patronising. On the other hand, *underaccommodation* is observed in intergenerational communication "when some style or quality of talk is underplayed (relative to needs or wishes)" (Coupland et al. 1991, p. 30).

Coupland et al. (1991) provided the following classification of young-to-elderly communicative overaccommodating or underaccommodating strategies based on previous empirical studies and their emphasis on *attunement*:

1. Sensory overaccommodation: a young speaker perceives (rightly or wrongly) an old listener to experience a sensory handicap and overaccommodates the speech by talking too loudly or too slowly.
2. Dependency-related overaccommodation: use of excessively directive and authoritarian talk to elderly people, a strategy that may be often used by carers.
3. Intergroup divergence: by increasing the speech difference with those of the elder a young person may increase distinctiveness in discourse that is intended to reflect age-group differences.
4. Intergroup overaccommodation: negative discourse elicited in the young by the simple fact of recognising the interlocutor as being a member of a social category that is subject to negative stereotypes.
5. Intergroup underaccommodation: in this case even though the interactional goals in the conversation may be positive the focus on the other person is decreased, as in the case of a young person being somewhat self-focused and therefore low attuning to the needs of the older person. This can often occur when the interactants have not had recent experience with people of a different age.

From the perspective of the old, it is also possible to identify some broad categories of patterns in their communication styles used with young people (Coupland et al. 1991):

1. Intergroup underaccommodation (as per in (5) above).
2. Self-protective underaccommodation: this involves the introduction of "safe" topics in a conversation that decrease the potential level of conflict.
3. Self-handicapping: a face-saving strategy whereby the elder excuses him/herself from some potential failures in communication on the grounds of suffering an illness or any other problem. With this the elder can decrease the level of potential criticism if the performance in communication is poor but also receive praise if the performance is better than anticipated.
4. Self-stereotyping: here the elders adopt a mode of communication that is stereotypically expected from them, a strategy that is more likely to perpetuate the stereotype than anything else.
5. Intergroup divergence: elders may express their disapproval in discourse through aggressive speech. This, however, may only confirm negative stereotypes that the young may have against the elders as "grumpy and irritable".

We mentioned that patronising may occur in inter-generational conversations. A patronising style of communication could be described as "inappropriate modifications based on age stereotypes of incompetence and dependence rather than on the actual communication needs of older persons" (Hummert and Ryan 1996, p. 149). Armstrong and McKechnie (2003, p. 16) list some non-verbal expressions of patronising behaviour which we reproduce here verbatim with only a minor modification:

1. Voice: high pitch, exaggerated intonation, loud, slow, exaggerated pronunciation.
2. Gaze: low eye contact, staring, roll eyes, wink.
3. Proxemics: stand too close, stand over a person seated or in bed.
4. Body expression: condescending facial expressions apart from gaze (e.g. smirking), shake head, shrug shoulders, hands on hips, cross arms, abrupt movements.
5. Touch: pat on head, pat on hand, arm, shoulder.

Patronising can be also manifested through the use of specific *language attitudes*. Verbal forms of patronising include directive talk, superficial talk, use of diminutives and baby talk, and they are usually seen as a disrespectful form of communication, especially when the speech is produced in a loud manner and with a more exaggerated intonation (Ryan et al. 1991). Some styles of speech directed to the elders, such as the use of *prestige dialects* (Giles and Coupland 1991), can be also used as a form of patronising.

Although elders usually see patronising negatively, especially when patronising occurs outside the family context and in institutionalised environments (hospitals, nursing homes) (Giles et al. 1993; Ryan et al. 1995), in a more intimate context where a loving relationship between young and old is well established, patronising forms such as *baby talk* may be expressed in a jocose manner that can be consensually taken with good humour. The same with proxemics, as in some cultures close body contact between young and old family members is a demonstration of affection rather than patronising.

Hummert et al. (1994a) described four responses of older people to patronising communication: appreciative, assertive, passive, and condescending. The assertive response is the strongest conveyor of a negative reaction of the elder against patronising, whereas the condescending response is useful as a face-saving strategy. The appreciative response may convey a mixed message: on the one hand, it seems to support the stance of the patroniser (when patronising is not face-threatening, as in the case mentioned in the previous paragraph), but on the other it allows the older person to retain a degree of authority. Finally, the passive response conveys a message of "submission" and it tends to contribute to further patronising.

In the end, through regular communication and interaction, old and young can develop a degree of *relational competence* that can help find a common ground in the relationship. Relational competence is a concept first introduced by Wienmann et al. (1990) which points to a process of joint construction and negotiation in discourse between interacting people, leading to mutual understanding. Relational competence can be achieved once both partners: the young and the old, have a better understanding of their own and the other's individual communicative strategies.

Thus, once both elder and young are aware of the dynamic characteristics of inter-generational communication, they may take initiatives to prevent miscommunication, as in the case when both young and old may refrain from using patronising discourse based on stereotypes.

In Coupland et al.'s (1991) work mentioned above, younger female partners in conversation were more engaging and they attuned better to the painful self-disclosure and the disclosure of age by the elders. Through their attunement, these younger participants can be described as acting as *facilitators* of self-disclosure by the elders.

Attuning to others' discourse is not always an easy task, whether the interacting people are of the same or a different generation, but in inter-generational interactions people are confronted with potential generational differences that, in extreme cases, may lead to *incompatibility* (contradictory viewpoints) or *incommensurability* (implying radically different world views) of discourses (Slife 2000). When world views are incommensurable, interactants may face the impossibility to reach a common ground, potentially leading to the end of communication.

People from different generations are likely to have been affected in their psychological development and the construction of their identity by different historical events and different societies, even if they grew up in the same country. This generational idiosyncrasy has been termed by Pilcher (1994) *generational consciousness*. Generational consciousness can even become a platform for political action on behalf of the members of a generation. In the words of Biggs et al. (2011, p. 1112; see also Mannheim 1952):

> Social actors moved from simply holding certain experiences in common, to a politicised generational consciousness, which Mannheim saw as contingent upon historical context at the time of emerging adulthood. Generational action, that is, acting with an awareness of one's own generational circumstances, is the strong suit, the ideal type for generational consciousness. A generation that becomes consciously aware of itself as a social or historical force can then become a motor for social change in its own right, especially during periods of social transition.

But generational consciousness could also lead to increased inter-generational friction, as in the case when different generations compete for the same scarce resources (e.g. jobs, government financial support). In this case generational consciousness may interfere with smooth inter-generational communication, through incompatibility and incommensurability.

The generational experience affects our identity through modifying our "inner world" (e.g. emotions, cognition) which then will guide our social relationships. During the period when the lives of young and old overlap, old people go through life already loaded with a baggage of previous experiences, whereas for the young their learning is influenced by a shorter time span of previous experiences. These differences are likely to come to the fore when young and old interact with each other.

To understand the complexities of generational factors in inter-generational communication, Biggs and Lowenstein (2011) advanced a *Generational Intelligence* model. Generational intelligence is defined as "an ability to reflect and act, which draws on an understanding of one's own and others' life-course, family and social history, laced within social and cultural context" (Biggs and Lowenstein 2011, p. 2). In the social and temporal space comprised of birth cohort, familial lineage, and personal maturation, generational intelligence describes the "experience of that space, the degree to which people are aware of it, how they react to it, the effect it has on the sense of who they are and how they behave towards others" (Biggs et al. 2011, p. 1109).

Thus, generational intelligence puts greater emphasis on self-awareness and personal relationships, whereas generational consciousness emphasises action, including political action, that springs from self-identity.

In their model, Biggs and Lowenstein propose that through a process of self-awareness and empathy interactants may become aware of their own generational identity and its role in modulating social relationships, this may then put them in the position of trying to understand that others may also be equally constrained by the cultural and historical events of their own generation. This may lead to a better comprehension of inter-generational differences—and perhaps some similarities as well—that can be achieved through communication. Once the frames within which each generation operates are understood, the possibility emerges for productive negotiations in discourse and for a convergence towards common solutions and better mutual understanding.

Within each generation, self-awareness can lead to specific forms of self-perception, the latter being especially affected by health (better health leads to a more youthful and positive self-perception), wealth, and own perception of fulfilment in life. Self-perception leads to the construction of *subjective age* (how individuals personally feel about themselves), which can be contrasted with an *ideal age* (how they would like to be) and *social age* (how others see the person). One would assume that better health and wealth, along with a coherent self-identity and a sense of personal fulfilment, may help in producing a degree of convergence between subjective, ideal and social age, thus facilitating communication between generations.

Therefore, generational intelligence can help build solidarity between generations whenever it favours inter-generational integration.

Inter-generational integration is better achieved when communication is more spontaneous and open rather than channelled through the filter of *social masks* (or *masquerades*; Woodward 1991), a concept akin to the Jungian *persona*. Social masks can be physical (e.g. cosmetic treatments) or behavioural (e.g. adopting youthful attitudes) and they may be of value to the elder as a form of coping and reinforcer of self-identity, feeling content with oneself and transmitting a social image of youthfulness. But a social mask may also convey a message of "insincerity", in which case it would make inter-generational communication more difficult.

If a person is perceived to be insincere by using a social mask, trust could be broken between interactants and solidarity would be more difficult to build. Social masks are more likely to be successful in brief communicative encounters, where the opportunities for an in-depth interaction are more limited, they are less likely to be so in longer-term relationships. However, obvious masks may be also ignored among close acquaintances (friends, some family members) whenever such masks are regarded as a "game". In that case, interactants may be happy to play the game for the sake of maintaining a good and valued relationship.

By fostering mutual understanding, generational intelligence can also help neutralise phenomena such as *telescoping*, that is the projection of conflicts relevant to a past generation to members of a current generation who are alien to those conflicts, or do not regard them as central to their life (Faimberg 2005); and it can also reinforce any good opinion that younger generations may have of the elders, especially in the family sphere (see the *grandmother* or *grandfather* effects, Biggs 1989).

By shifting social interactions towards more cooperation between generations, generational intelligence may strengthen various dynamics of reciprocity. For instance, Feldman et al. (2016) mention inter-generational *mutual reciprocity* (i.e. one elder helping one younger and then the younger reciprocating to the same elder) and also *transitional reciprocity* or what we referred to in Chap. 6 as *Arrow of Time reciprocation*: e.g. a grandparent helping a grandchild and then the grandchild, once s/he becomes older, helps his/her own grandchild. Note that the word *reciprocation* emphasises interactions between individuals who are in a more intimate relationship, whereas *reciprocity* is extended to broader relationships. Inter-generational cooperation may also involve the *pivotal generation*, which is the middle generation in a sequence of three. The younger generation may receive help from both the pivotal (parents) and older (grandparents) generations, especially in the form of financial support. Increased lifespan across generations tends to give an opportunity to the three generations to start converging in their cooperative exchanges, through communication, as they all become older (Riley and Riley 2000). Such convergence may give rise to *age integration* within the family and within society at large.

Age integration occurs in social structures that use age as a criterion for entrance and participation (Uhlenberg 2000a, b), and it is also seen in situations when the barriers for communication across generations are diminished (Loscocco 2000). This form of integration in society provides better social flexibility to elders, it decreases ageism, and it makes old age more socially productive and personally fulfilling. When age becomes a barrier to inclusion, the result is *age segregation* (Riley and Riley 1994). Age segregation involves restricting communication to people of the same or similar age.

Age integration favoured by better communication between young and old can help decrease inter-generational conflicts (Foner 2000) and strengthen young–old relationships at all levels in society, starting from the family (Uhlenberg 2000a); but it is jeopardised by a frenetic pace of life (e.g. Loscocco 2000) and intense inter-generational competition.

We have seen in Chap. 6 and in the section *Ageing and communication: an overview* in this chapter, how social interactions between the old and the rest of the community can be marred by stereotypes, stigma, and ageism. Negative prejudice (e.g. Atchley 1980) against old people can be felt not only in general attitudes, unsupportive and discriminatory behaviour, and non-verbal communication, but also verbally in day-to-day discourse and in articles and programmes available through the media. Stereotypes can determine the way in which we approach somebody else verbally and therefore the kind of rapport we may establish initially. A negative initial rapport may potentially evolve into a more positive one over time as people get to know each other better and may become more inter-dependent, but this is not always the case. Stereotypes can not only be expressed verbally, but they are also constructed verbally (Giles and Reid 2005). Language, as a product of our mind, does not simply "match" an external reality, but it actually constructs such reality (Whorf 1956).

Nussbaum et al. (2000; see also Allan and Burridge 2006) provide a sampler of common ageist words used in the English language: "senior', "mature", "golden ager", "old man/woman" including some derogatory terms referring to an old woman as "bag", "coot", "old maid", "hag", "witch". Old men can be also described using discriminatory words such as: "geriatric", "geezer", "codger", "goat". Such words reflect our perception of old people, but they also help reinforce such perception and further promote ageist negative attitudes, as they become commonplace through repetition. Discriminatory language may contribute to frame old people as an out-group in society (Ryan et al. 1986).

Hummert et al. (1998) studied the style of communication that 40 young and 40 adult participants would use upon approaching two kinds of older people, one fitting a negative stereotype of "despondent" (e.g. depressed, afraid, neglected) and one positive: "golden ager" (e.g. active, sociable, well-informed). Responses indicated the use of a more patronising talk with the despondent than the golden age stereotypical elder, although the response also changed with context: golden age elders were also addressed in a patronising manner in a hospital environment. This suggests that patronising may be a reaction to a message of "helplessness" coming from the elder, which elicits a response similar to the one triggered by a child in difficulties. According to Hummert et al. (1998), patronising talk can take different forms ranging from a "nurturing style" (use of simple language, diminutives, and baby-talk intonations) to a "highly controlling" and "directive" style, a less friendly form of communication, using "short imperative statements that are delivered in a cold, uncaring tone and sometimes include disparaging remarks about the receiver" (Hummert et al. 1998, p. 127) that can go as far as to become verbal abuse (Hummert and Ryan 1996).

Elderspeak, which as we mentioned previously in this chapter is regarded as a secondary form of baby talk (e.g. referring to an old person as "honey" or "sweetie"), is also commonly used to express prejudice against the elders (e.g. Ng 1997). When the use of elderspeak and baby talk influences older people to such an extent that they themselves start spontaneously using it, they may be undergoing a process that has been termed *instant ageing* (Giles et al. 1994).

Ng et al. (2015) have detected a trend in the English language towards increasing age-stereotype negativity in vocabulary over the past 200 years.

The level of patronising by younger people towards the elders may decrease to the extent that the elder is wealthy, or enjoys a level of social prestige of one kind or another. In such circumstances the social stigma of old age could be overcome (Nussbaum et al. 2000). Of course, it is also true that older people can patronise younger individuals, which is not regarded as a pleasant experience by the target (Nussbaum et al. 2000).

But stereotypes can be also positive. In a work carried out by McCann et al. (2005) they surveyed 137 undergraduate students (almost equal representation of

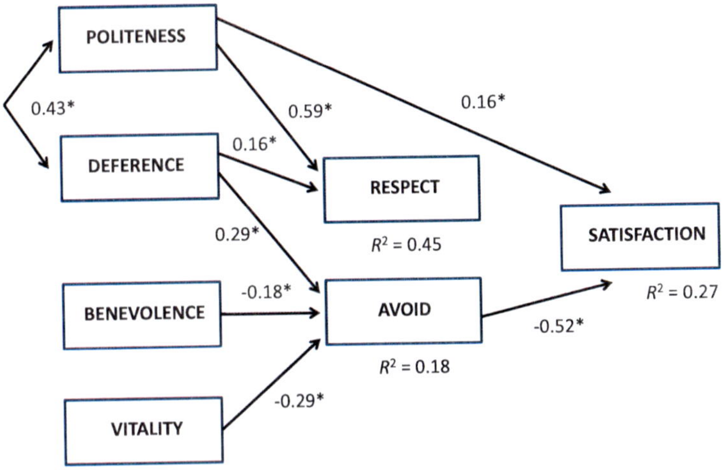

Fig. 7.2 Result of a path analysis for trait attribution of old people showing how *age stereotypes* and *respect norms* affect communication between young and old and how such communication in turn affects satisfaction with the conversation. Redrawn from McCann et al. (2005)

males and females, 65.7% Caucasians) and found that as the age of a target person increased, respondents also tended to attribute traits of benevolence, politeness, deference to the elders, and they also displayed communicative respect and satisfaction, even though traits such as personal vitality decreased with age. Figure 7.2 shows the results of a path analysis of McCann et al.'s (2005) data. The more positive the stereotypes were in the conversation (e.g. benevolence, vitality) the less avoidance there was; and the more politeness and deference there was in the conversation, the more respect was shown. More politeness and less avoidance were associated with greater satisfaction in the conversation. Therefore, satisfactory inter-generational interactions were those in which elders displaying positive attributes engaged with younger people in a polite conversation.

Cross-culturally, the stereotypes associated with communication with elders display some interesting patterns. Harwood et al. (2000) obtained questionnaire responses from a total of 615 adult volunteers who were older than 48 years (68.8 years old on average) from five countries: Australia, People's Republic of China (PRC), Hong Kong, the Philippines, and Thailand, with the sex ratio being male biased for the PRC and Hong Kong, and female biased for Australia, the Philippines, and Thailand. The questionnaire asked those respondents to provide their opinions, following a list of "trait-adjectival semantic scales", about people of three age classes: young (20–30 years old), middle-aged (45–55 years old), and older (65–85 years old) adults. There was some cross-cultural convergence in the results, whereby old people are seen as less attractive, physically weaker, less active, more conservative (but the pattern for conservatism is less strong in more westernised areas such as Australia and Hong Kong), less healthy and flexible than younger people. Across cultures, the elders are also seen as "kind", but they differ with regard to "wisdom" and "generosity", which are seen as worsening with age by respondents from the PRC and Hong Kong, a trend that we have observed repeatedly in various studies of contemporary Asian countries, as perception of older people is worsening against the expectations from traditional filial piety.

Inter-generational communication and age stereotypes were also studied cross-culturally by McCann and Keaton (2013). In this case the focus was the perception of younger workers towards workers who are older or of the same age. They compared one Western country, the USA, with one Asian country, Thailand, with the sample being female biased (76.4% overall). Previous research has shown that stereotypes usually associated with older workers include, on the negative side, that older workers have a more impaired memory and less ability to learn new tasks, they are less comfortable with new technology, are physically more frail, less productive and efficient, they are also seen as less flexible but, on the positive side, they are regarded as more loyal and showing greater levels of commitment to their work and employer. More specifically, McCann and Keaton (2013) analysed the responses of 267 participants (142 from the USA and 125 from Thailand) to the *Global Perceptions of Intergenerational Communication* (GPIC) instrument. In general, young Thais held more ageistic stereotypes against the old than their American counterparts. In addition, young Thai participants perceived both other young and older Thai workers as being more avoidant in communication than it is the case in the USA. Moreover, young Thais were also of the opinion that members of their own age group communicate in a more non-accommodating style than it is the case for young USA workers. Therefore, following also the conclusion that we reached earlier in this chapter in the section *A cross-cultural perspective on communication in old people*, it seems that inter-generational interactions are more conflictive within the Thai than within the USA contemporary society, and this is reflected in differences in prejudice and styles of communication.

Positive inter-generational communication can assist in building bridges between different generations, leading to better social integration. In particular, better communication can increase the level of affiliation between grandchildren and grandparents, thus also promoting pride in each other and decreasing social distance. Such greater levels of affiliation may result in social bonds between generations that are akin to friendship, defined as a "voluntary interdependence between two persons over time, that is intended to facilitate social-emotional goals of the participants, and may involve varying types and degrees of companionship, intimacy, affection, and mutual assistance" (Hays 1988, p. 395).

Satisfaction with such inter-generational communication can be affected by the level of solidarity between young and old. In the case of grandparent–grandchild relationships the specific behaviour of grandparents in the stories they share with their grandchildren can affect satisfaction in communication, as it does grandparents' own perception of the role of grandchildren in supporting their relationship (e.g. mutual self-disclosure) (Harwood and Lin 2000).

Harwood and Lin (2000) described four major themes emerging in conversations between grandchild and grandparent (mainly Caucasian grandmothers, 81%) in a study carried out in the USA: affiliation, pride (especially in the achievements of the grandchild), exchange (of information: grandparents giving advice, grandchildren informing about modern society), and distance (both geographical and generational).

Grandmothers tend to be more communicative than grandfathers (a broad feminine characteristic) and their topics of conversation also tend to differ: grandmothers tend to focus more on the family and its history, whereas grandfathers are more focused on talking about health issues and their own personal experiences when they were younger (e.g. Nussbaum and Bettini 1994). In general, older women develop more emotionally interconnected communicative networks than older men who, in turn, tend to communicate and interact within a more casual social network (Bettini and Norton 1991, and references therein). In a study carried out in a retirement centre in the USA, Bettini and Norton (1991) interviewed 60 elders (average age 84 years old, 80% females) to find out that they did maintain some social relationships and communication with young people, whom they considered as friends. Young people tended to establish social relationships with older people mainly with the purpose of seeking advice, and contact was facilitated by a young family member or the local church community.

When interactions with older people run into difficulties, miscommunication may occur, which may negatively affect the relationship between young and old. Miscommunication can arise in inter-generational conversations as a result of various factors: different expectations (schemata) between interactants, negative stereotypes, adoption of a communicative style that may elicit negative reactions (e.g. use of directives, patronising), lack of familiarity between interactants (that may increase the probability of misunderstandings), attitudes (non-accommodative), not recognising some sensory limitations of the elder (e.g. hearing), serious cognitive limitations of the elder (Edwards and Giles 1998). Miscommunication can then result in conflicts between young and old.

In a study carried out in the PRC, Zhang and Hummert (2001) investigated inter-generational communication in a group of 20 young adults (19–33 years old) and 13 older adults (62–72 years old) and described instances of miscommunication that were caused mainly by negativity, bossiness, and patronising styles used by elders in their communication, this was rendered even worse by the tendency of older people to talk unceasingly and repeating themselves (*laodao*). On the other hand, young adults could also use forms of communication that were perceived as negative by the elders: using the wrong forms of address, speaking in a condescending manner to the elder, not revealing personal information. Zhang (2004) subsequently studied written accounts of inter-generational communication in conflict situations in the PRC. In a conflictual social environment, the probability of miscommunication is likely to be high. Conflicts may arise between two interacting people if one (or both) expresses criticisms about the other, if any of them makes any illegitimate or unfair demand, if any of them rebuffs the other or fails to respond as it is socio-culturally expected, and if there is cumulative annoyance in the interaction (Peterson 1983). All the above forms of inter-generational conflicts were reported by Zhang (2004) after analysing the opinions of 441 university students (18–25 years old, 77.3% females). How are inter-generational conflicts managed? Zhang et al. (2005) investigated this issue also in a sample of old and young Chinese from the PRC.

Previous research identified four interpersonal conflict management styles: *competing* (i.e. a negative, confrontational, assertive, and uncooperative style), *avoiding* (i.e. a style that although is non-confrontational it is however under-responsive to a conflict), *accommodating* (i.e. this emphasises relational harmony through the recognition of a partner's needs, apologising, and also being non-assertive), and *problem-solving* (i.e. an assertive and cooperative style based on the display of empathy and the searching for a mutually acceptable solution) (Sillars 1980; van de Vliert and Hordijk 1989; Cai and Fink 2002; Zhang et al. 2005). Avoiding both conflict and miscommunication is necessary in order to achieve a smooth interaction that is not face-threatening and that can better convey the true intentions of the interactants. In fact, all participants in Zhang et al.'s study regarded the accommodating style of communication as the most respectful one, whereas the competing style was seen as the least respectful. Older participants in particular also regarded accommodation as the most appropriate style of communication. In terms of effectiveness, again old participants regarded accommodation as the most effective style, but the young ones saw problem-solving as most effective. Therefore, although there is some degree of convergence between old and young in the preferred styles to be used to resolve conflicts, there are also differences that could undermine smooth communication, with young generations being relatively more pragmatic than the older ones.

Inter-generational communication within gender minorities also warrants attention, given the growing interest on the issue of social minorities in general. Inter-generational communication within the homosexual community in particular is confronted with the same challenges that we have already seen in the heterosexual community: stereotypes, stigma, miscommunication, and so on, with the added difficulty that homosexuals have to maintain their identity in a broader social environment that may be specifically discriminatory against gender minorities (e.g. Tajfel and Turner 1986; Poiani 2012). Such inter-generational communication problems are affected by the emphasis on the body, physical activity, beauty, and youthfulness that tend to characterise the male homosexual culture. Those issues appear to be less relevant to lesbians (e.g. Poiani 2010). There is also a level of fear that interacting with older homosexuals may attract negative ageist discrimination—by association—within communities of homosexuals (e.g. Hajek and Giles 2002). Strategies to cope with an age-related threat to social identity among homosexuals and other LGBTI elders include individual mobility (leave), social creativity (transform, re-interpret), social competition (organise into new groups).

7.4.1 Improving Inter-Generational Communication: Intervention Programmes

Many social intervention programmes have been designed with the objective of improving inter-generational communication, and with the hope of achieving a better understanding and cooperation between young and old (Pinquart et al. 2000, for a review). It is known that such programmes can increase school attendance by children, they can improve the attitudes of young people towards the elderly, they can even reduce drug use among youth, and foster an overall increased sense of well-being (Abrams and Giles 1999). Ultimately, these programmes can contribute to achieve greater inter-generational social integration.

Inter-generational programmes are a way of restoring—to an extent at least—the social environment of three-generation families, that are currently becoming less common in industrialised countries due to the disruption of family structure, and despite greater longevity. In multi-generational households, opportunities for direct knowledge transfer between young and old are obviously higher (Newman and Hatton-Yeo 2008), when such households are no longer available, then opportunities for inter-generational interactions can be created in a different manner. For instance, inter-generational communication programmes that integrate local elders into a school community, by contributing their life experience and knowledge, have been available in the USA, Canada, and other countries since the 1990s (e.g. Smith and Yeager 1999; Strom and Strom 1999). Some of those programmes have been evaluated for their effects on either the young, the elders, or both.

Inter-generational Programmes (IPs) can be defined as "planned ongoing activities that purposefully bring together different generations to share experiences that are mutually beneficial. Typically, the programs involve interactions that promote social growth and learning between the young and the old" (Newman and Hatton-Yeo 2008, p. 32). Through such programmes, elders who would otherwise live their lives in social isolation can not only acquire a new meaning in life and a personal sense of connectedness and fulfilment, but they can also become a valuable *social capital* to the rest of the community (Newman and Hatton-Yeo 2008).

IPs can deliver benefits in terms of increased general learning, the feeling of being valued, accepted, and respected, increased specific knowledge and skills, and the development of a more meaningful and trusting inter-generational relationship, which can accrue to both young and old. Self-fulfilment and better understanding of younger generations are benefits specific to the old, whereas increased self-esteem and a better understanding of the old can be obtained in the younger generations. IPs usually aid in developing a degree of reciprocity and empowerment in both interacting generations (Newman and Hatton-Yeo 2008).

Many such IPs have been implemented around the world (see Newman and Hatton-Yeo 2008, for a review) and we will mention some of them below.

Strom and Strom (1999) suggest that inter-generational programmes should be characterised by flexibility and spontaneity. If the programme is implemented in a school, then both teachers and older volunteers should be well coordinated and specific tasks assigned that motivate the interests of both students and volunteers. Moreover, it is often the case that the volunteers may require some training if they have not had recent contacts with children, and these days they may also require to pass some form of *working with children checks* and *police checks*; the workload must be bearable for both children and elders and they should be asked for feedback, alongside providing due recognition to the elders for their effort.

In a small-scale programme carried out in the USA involving three older adult volunteers (72–82 years, one male and two females), 6 weeks duration, and 12 sessions; Angerbash and Jones-Forster (1999) organised encounters between the elders and 16 young children (50% females, 4–5 years old). Joint activities that were carried out were commensurate to the age of the children and the philosophy of the childcare centre. The experience resulted in high level engagement from all participants, after an initial period of hesitation, with many positive interactions being reported.

In the case of older children, such inter-generational interaction programmes are often carried out in the school environment. In one such programme implemented in the USA (Newman et al. 1999), old people acted as mentors/tutors to the students. Teachers reported that students became more responsive to teacher's help after undergoing an experience of mentoring with old-age volunteers. Teachers also observed an improvement in the homework activities of students and in their overall learning performance, and that the students seemed to be calmer, more attentive, and less disruptive in the classroom.

Inter-generational programmes have also been used to help young problematic students. A programme implemented in the USA (*Intergenerational Work/Study Program*, IWSP) targeting problematic high school students provided them with a "grandparent" figure who was expected to improve the students' levels of self-esteem, life-coping skills, motivation, attitude, and also planning skills (Brabazon 1999). Students showed lower levels of school absenteeism when exposed to a moderate programme of inter-generational activities, whereas a more intense programme of one-on-one inter-generational activities was followed by greater levels of absenteeism. This suggests that for this kind of students an approach that is less constraining may be more effective. Inter-generational mentoring can be also useful in programmes designed to reduce adolescents drug abuse (tobacco, alcohol, and other drugs). Taylor et al. (1999) evaluated one such programme (*Across Ages*) in the USA which used elders as mentors (55 years old and older). Mentors went through a process of screening and selection and then training, so they could understand the objectives of the programme and what was expected of them. They subsequently received follow-up supervision and feedback. With the parents also involved in the various activities, this becomes an intervention that integrates various members of the local community, which may better help in developing a sense of social integration in the youth. A controlled test of the effectiveness of this programme suggests that mentored students scored better than non-mentored students (controls) on various measures of well-being and attitudes toward the school, the future, elders, community service, and both their attitudes and reactions to drug use. Mentoring also improved school attendance.

In the *Landed Learning* programme based in an urban organic farm close to the University of British Columbia, Canada, elders (retired farmers and keen gardeners) and school children join together in agricultural activities carried out outside the usual learning environment of school: "These intergenerational groups work together throughout the year to plan, plant, and harvest food crops in raised beds that are grouped in a space at the farm designated the Land, Food, and Community

Fig. 7.3 Young and old working together in a community garden (from Mayer-Smith et al. 2007)

Garden." (Mayer-Smith et al. 2007, p. 79) (see Fig. 7.3). Teachers then integrate these outdoor experiences into a more formal learning in the classroom. Through such hands-on experiences, children improved their knowledge about the environment, their relationship with it becoming more "personal".

Some programmes have been also implemented in which inter-generational communication can have the additional utilitarian function of collecting memories and knowledge from older generations. One of such programmes, the *Memory Line Project*:

> aims at training groups of elderly and young citizens resident in the project's partner countries to collect records (stories, songs, poems, experiences, etc.) in order to ensure their conservation and dissemination with a view to creating a model of cooperation and inter-regional and intergenerational learning based on memory and communication and to promote innovative experiences within the area of lifelong training (Pieri and Diamantini 2010, p. 2423).

In their work carried out in Italy, Pieri and Diamantini (2010) organised focus groups with the participation of both old and young people where the main topic of discussion was the elders' memories and how to preserve them (e.g. with the use of information and communication technologies, ICTs).

Grandparents have also been successfully included in family therapies aimed at decreasing parental stress. Such family therapies support the traditional role of grandparents as helpers, as source of emotional support, and as brokers in inter-generational conflicts within three-generation families (e.g. Griff 1999).

In the case of ethnic minorities living in linguistically and culturally diverse communities, programmes that foster communication between elders and younger generations can become vital means of transferring and preserving language, culture, and memory in an environment where such culture may be in danger of disappearing (e.g. Cook 1999). In English-speaking countries, old immigrants who are native speakers of languages other than English can be of benefit to the general community, through helping younger native speakers of English who are studying a second language. In this manner old immigrants may become valuable assets to the community, contributing not only to knowledge transfer but also to strengthening the social bonds in a multicultural society (Cordella 2016a, b). According to Radermacher et al. (2016, p. 151): "Intergenerational programs can provide opportunities for older people to continue to engage with their communities, opportunities that might otherwise not be available". In addition, health and well-being of older people benefit from social contacts with younger generations, where they can learn from each other (reciprocity) and the elders can engage in generative activities (i.e. teach the young ones and transmit the wisdom and experience of their generation).

One of us, Marisa Cordella, has been involved in the design and implementation of one of such programmes in Australia, and the experience is reported in the book edited by Cordella and Huang (2016) *Rethinking second language learning: Using intergenerational community resources.*

According to the Australian Bureau of Statistics, about 26% of Australians are born overseas and overseas born people are older than the population born in Australia (http://www.abs.gov.au/ausstats/abs@.nsf/Latestproducts/2024.0Main%20Features22016). Many of such older immigrants do retain their first language (L1), which they may use to communicate with family members and/or with the members of their ethnic community. In Australia, where 400 community languages are spoken in the home environment, speakers of those languages can become a cultural resource and a source of knowledge transfer to younger generations in both institutional (e.g. schools) and everyday settings. The specific programme that was implemented by Cordella and her collaborators paired more than 160 upper secondary school students with 106 older participants from the local community. The students were learning Chinese (Mandarin), German, or Spanish at school, and were paired with a native speaker of the target language. Old participants were 60–95 years old (60.3% women). The elders and the young students were encouraged to talk freely on topics previously chosen in consultation with the class teacher. The conversations were recorded over a three-year period and were analysed following discourse analysis techniques (Huang and Cordella 2016).

This programme of inter-generational language and culture transfer is guided by the objective of promoting better inter-generational relationships through mentoring, generativity, and reciprocity. Here mentoring is delivered in the form of an old native speaker assisting in the second language (L2) learning of a young student. Language learning facilitated by elder native speakers is also a form of generativity, as in the process the elders transmit their native culture, experiences in life, and their view of the world to a younger generation. In addition, there is a degree of reciprocity, as the student can help the elder to better understand the reality of younger generations (Feldman et al. 2016).

It was observed that during the inter-cultural and inter-generational encounters the old person may take on various potential *stances* or roles such as that of *language instructor* (conveying knowledge of language), *sociocultural guide* (transmitting personal experiences associated with his/her native culture and the history of the native country), and *ethical-moral adviser* (offering advice on ethical-moral issues based on personal experience) (Cordella 2016a). Such roles were modulated by the degree of rapport that emerged between young and old. Inter-generational communication is expected to run more smoothly the stronger the rapport between older and younger interactants is. Rapport can be improved for as long as communication is effective, face is not threatened, and *Lakoff's rules of rapport* are followed: do not impose, give options, and maintain camaraderie (Lakoff 1973, p. 1975). This will facilitate the development of friendly relationships between older and younger persons. Cordella and Kokubu (2016, p. 117) have recently proposed a cyclical model of rapport "in which participants experience mutual attentiveness, positivity and coordination".

Direct communicative experiences with old native speakers improved the attitude of students towards learning a second language, and they also improved their self-efficacy, that is their ability and confidence in learning a second language (Huang 2016). This was achieved through the co-construction of a conversation between participants working together.

This inter-generational communication project led to the formulation of a model which encapsulates the benefits of encounters between an old native speaker of a foreign language and a student learning such second language. From a social perspective, these encounters help not only in improving second language acquisition but also cross-cultural understanding, appreciation of multicultural societies, development of inter-generational empathy, and engagement of older people in community life (Huang et al. 2016, see Fig. 7.4).

In a recent study carried out in a boys School in Brisbane, Australia, Cordella (2013) paired early senior students, who were between 13 and 14 years old, with old Spanish-speaking migrants, with the objective of enhancing the students' Spanish language skills. A set of questions around a topic learnt in class was given to the pairs to facilitate spontaneous inter-generational conversations. To maximise the chances of having a good communicative experience, the formation of the pairs was done organically, allowing individuals to interact freely among themselves during a morning tea. However, it took two additional contact experiences and the introduction of songs and games to make both young and old feel at ease. Young people reported having had very limited chances to interact with people older than their own parents (late 30s early 40s). One activity that played the role of icebreaker in communication was to publicly display each other's talents. It was not only a chance to discover how good dancers, singers, and cheeky the older people were, but also how appreciative they were of any abilities the younger students showed to them. As a result of a mix of admiration, shared laughter and support, inter-generational conversations started to flourish.

Although in many inter-generational programmes it is the elder who goes to a meeting place, a school, or childcare centre in order to interact with the young, the

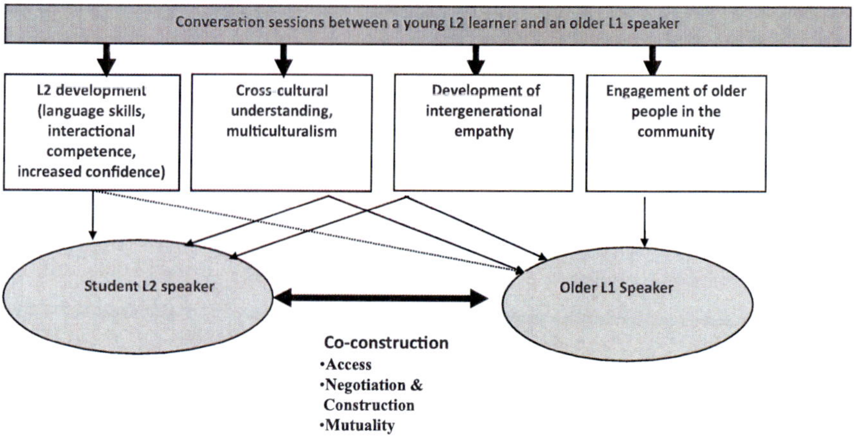

Fig. 7.4 An inter-generational and inter-cultural model of language learning and social inclusion (from Huang et al. 2016)

opposite can also occur, with the young going to nursing homes. Ivey (2001) evaluated one of such programmes in the USA to conclude that adolescents may improve their views of older people after they have had contact with them and listened to their stories. Similar positive results were obtained in a study carried out by Schwalbach and Kiernan (2002).

As we have already mentioned in this chapter, Ryan et al. (1995) proposed the *Communication Enhancement of Ageing model* to guide interventions to improve inter-generational communication especially in institutional settings:

> This framework ... emerged from a consideration of health promotion strategies that focus on the roles of self-care, mutual aid, and social support. The communication encounter with an older person is part of a positive feedback loop. This positive cycle can be achieved through a person-centered, as opposed to category-based, approach to communication with older individuals. This approach requires not only a consideration of the individual characteristics of an older conversational partner at the beginning of an interaction, but also a constant reassessment of the interaction as it progresses. If the partners engage in appropriate adaptations, the enhancement model sees positive outcomes for both parties in terms of empowerment, increased competence, satisfaction, health, and effective communication (Ryan 2010, p. 79).

The focus here is in designing a communicative intervention that is attuned to the specific individual requirements of the elder. This calls not only for an understanding of what the elder needs, but also for the capacity and resources to deliver the intervention suggested by such knowledge.

We have also pointed out in this chapter that communication in the elderly can be enhanced through *selective assertiveness*, whenever social interactions are not producing the desired effects for the elder. Specific programmes could be designed to enhance the level of selective assertiveness in elderly people that can help in their communicative competence. Ryan (2010) proposed a model of *Selective Assertiveness* of communication (see Fig. 7.5) that indicates that older people can

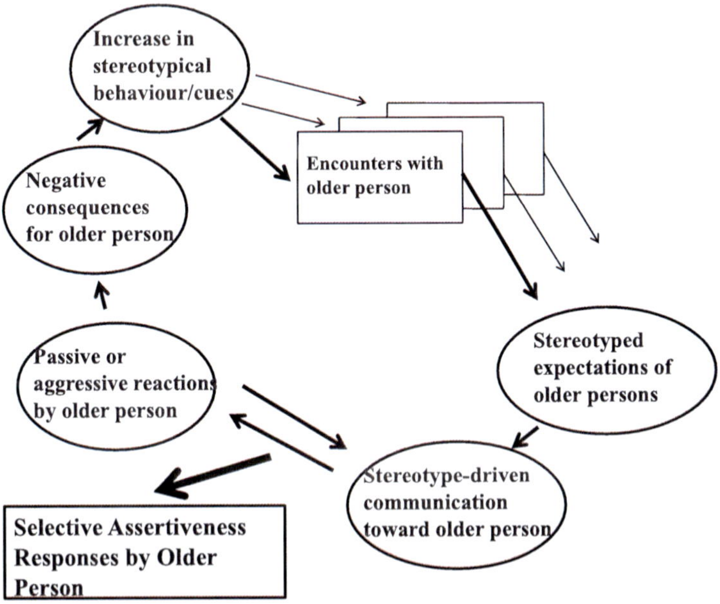

Fig. 7.5 *Selective Assertiveness* model of inter-generational communication. Selective assertiveness breaks the cycle of communication that would otherwise reinforce the stereotypical treatment of an older person in communicative encounters, which can lead to negative consequences for the elder (redrawn from Ryan 2010)

be proactive in breaking the cycle of stereotypically-driven communication by being more assertive in their responses to the people they are communicating with. Selective assertiveness avoids aggressive responses, which only serve to further reinforce negative stereotypes about the elderly. Instead, the elders are invited to respond to stereotypes in a calm manner, with self-confidence and clear language, in order to explain their requirements (such as when lodging a complaint): "The assertive speaker is tactful: aware of the social context and the other person's perspective, knows when to be direct or indirect, and acknowledges the communication partner's positive behaviors when appropriate" (Ryan 2010, p. 80).

Given that negative attitudes tend to seriously interfere with the success of inter-generational communication, and given that such negative attitudes are often grounded on prejudice and stereotypes that can thrive in an environment of low social contact between generations, better communication between generations can be achieved through initiatives that favour more inter-generational contact, such as the ones we mentioned above (Pinquart et al. 2000).

We conclude this section with a mention of technological aids used in inter-generational communication. Inter-generational connectedness can be helped by programmes and spontaneous initiatives that take advantage of modern Internet-based communication technologies, especially when people of different generations live far apart, sometimes even in different countries (e.g. Rodríguez et al. 2009).

Rodríguez et al. (2009) found that elders were keen to embrace modern communication technologies provided that their use was simple and straightforward. Websites such as YouTube that permit to share videos and also allow viewers to leave messages commenting on the video, are a platform that could be used for the purpose of improving communication between elders and others, thus enhancing their social connectivity (e.g. Harley and Fitzpatrick 2009).

Sayago et al. (2012) studied the use of YouTube by 32 people, 65–90 years old, in Scotland. Older people used YouTube to occasionally seek information but did not use the comments section to engage others, rather they commented on their experiences through their own social networks via email and face-to-face conversations. Participants were unwilling to upload their own videos on YouTube, although they knew how to do it, due to privacy concerns.

Social media such as Facebook can be also used by older people, although recent privacy concerns associated with the exploitation of personal data available on Facebook by Cambridge Analytica—a UK-based data analytics company that helps political campaign organisations reach voters online—has attracted criticisms (see, for instance, https://www.cnet.com/news/facebook-cambridge-analytica-data-mining-and-trump-what-you-need-to-know/). Younger generations (e.g. the *Millennials*, "the first generation to come of age in the third millennium" (Taylor and Keeter 2010, p. 4)), who are frequent users of social media, however, are not as strongly concerned about their privacy as are older people. Although Facebook has been the target of recent criticisms, social networking through Skype remains popular across generations, blogs are also becoming more popular, although older people still use email as a major information technology medium for communication. Web-based technologies allow older people to increase communication and their connectedness with family and friends (Siibak and Tamme 2013).

To recap, there are many benefits to be gained from improved intra- and intergenerational communication, such as increased elders' well-being and the achievement of a more fulfilling ageing. Greater accommodation between interacting people, whether of the same or different generations, facilitates communication, whereas negative stereotypes and painful self-disclosure from the part of the elders can interfere with smooth communication. Communicative improvement can be achieved through attunement that may require adjusting verbal and/or non-verbal strategies of communication, whereas patronising should be avoided. A process of self-awareness and empathy leading to developing greater levels of generational intelligence may help achieve better relational competence that can lead to finding common grounds in communication. Improved relational competence can eventually result in better social age integration through cooperation. Inter- and intra-generational communication programmes can help improve communication and promote a more fulfilling ageing in old people, but they should be tailored to the specific cultural and individual needs of the elders, taking into consideration the diversity found in contemporary societies. Such programmes could be helped by an appropriate use of modern communication technologies.

7.5 Narratives of the Elders

Two major forms of discourse can be identified in our day-to-day communication: *agentive discourse* and *narrative discourse*. Agentive discourse refers to the constructive dimension of language and the agency of the speaker, being expressed in the form of speech acts that are oriented to "taking and assigning, accepting and repudiating of responsibilities for actions" (Harré 1995, p. 123). Thus, through agentive discourse a speaker becomes a responsible performer of his/her own actions. *Narrative discourse*, on the other hand, is a more complex form of discourse that includes aspects of our imagination (Brockmeier 2009).

In constructing narratives (a process referred to as *narrativity*) people act as storytellers, that is:

> as authors and co-authors who creatively read and evaluate the texts of life and literatures. A narrative perspective focuses on existing institutions as providing 'plots' that are always in the process of re-creation rather than existing as settled scripts. Viewing human communication narratively stresses that people are full participants in the making of messages, whether they are agents (authors) or audience members (co-authors) (Fisher 1987, p. 18).

If we just rely on a dictionary definition, a narrative is a spoken/written account of connected events, a story. But more specifically, Kreuter et al. (2007, p. 222) define a narrative as a "representation of connected events and characters that has an identifiable structure, is bounded in space and time, and contains implicit or explicit messages about the topic being addressed". Importantly, narratives give "meaning, connectedness, and directionality to a series of otherwise isolated events" (Nouri and Helterline 1998, p. 38), and they are produced collaboratively by interacting actors using a variety of communicative resources (Goodwin 2015).

Narratives transmit our mental representation of the world and our understanding of it, through text. That is, narratives convey what Zwaan (2008) calls *situation models* or integrated mental representations of time, space, entity, causation, and intentionality. This is achieved within an interactional context that shapes and negotiates meaning among interactants (De Fina and Georgakopoulou 2015). From descriptions and meaning making, however, narratives can also become drivers of action and agency, especially *narratives of identity* (Hubble and Tew 2013).

A narrative of identity "is a person's internalized and evolving life story, integrating the reconstructed past and imagined future to provide life with some degree of unity and purpose" (McAdams and McLean 2013, p. 233). Narratives of identity can give not only unity to the diversity of life experiences that an individual has gone through, but they can also provide "redemptive meanings" to suffering, by strengthening personal agency. It has been shown that people who can construct a coherent narrative of identity also enjoy better mental health, well-being, and maturity (McAdams and McLean 2013).

Narratives may help shape identities also through a process of *habitus* that is by reinforcing specific patterns of wanting, feeling, thinking, doing, and interacting within a social context (Pickel 2005).

The process of identity formation through narratives is dynamic and ongoing (Hubble and Tew 2013) and although it can be grounded on some aspects of the reality of our external world, constructing our identity is ultimately a process shaped by individual characteristics in interaction with the individuality of others. Such individual characteristics define our subjectivity, they are affected not only by our biological reality and the reality of the external world that affects us, but also by our imagination and our communication with others (intersubjectivity) through which we co-construct our perceived reality, that we then express through narratives, and that finally gives shape to our identity.

$$\left[\text{Reality}\right] \rightarrow \text{Subjectivity} \rightarrow \text{Imagination} \rightarrow \text{Intersubjectivity} \rightarrow \\ \text{Co}-\text{constructed reality} \rightarrow \text{Narratives} \rightarrow \text{Identity}$$

Therefore, through narratives we "story" our experiences in life, giving structure and meaning to those experiences. In a narrative, we express our thoughts and feelings through the process of *storytelling*. Through storytelling we share meaning with others and convey the affective, reflective, and perceptual aspects of our reality, in the process of *intersubjectivity* (Hubble and Tew 2013). As we become older, we accumulate the stories of our life into a broader and more general story, following a process of *narrative accrual*. Narrative accrual "connects the various stories of a person's life into a whole, so it also addresses continuity and change, but as an element in a life's story" (Nouri and Helterline 1998, p. 57).

Life stories, to become public, require an attentive audience, without which the story remains personal. But the stories in our narratives are not just descriptions, they are more than that, in the words of Randall and Kenyon (2004, p. 333): "making stories is how we make meaning". The stories provide the substrate to give meaning to our life experience.

Therefore, storying our life is more than just an exercise in memory and reminiscence (Webster 1999), it is more than just providing an account of facts as experienced and perhaps modified through time, "(w)illingness or eagerness to 'story' one's life is tantamount to the desire to live—even when the story is flavoured with disaster" (Bruner 1999, p. 9). In the process of storying our life we embark in an exercise of *life review* and *meaning making* that can then be used as a springboard for future life experiences (Butler 1963), no matter how short or extended that future may be.

In antiquity myths of youth and old age permeated the stories about older people, as we have seen in Chap. 2. The stories about old age contributed—and still do—to the perceptions of older people in society and, vice versa, experiences with specific older people have affected and continue to affect the development of stories about them. Indeed, stories "colonize the life world" (Habermas 1989) of individuals, they modulate their perceptions and thoughts and in so doing affect social interactions.

Traditional folk narratives about the elders may take the form of tales leading to myths (e.g. the elders, women in particular, as witches) that can then be perpetuated inter-generationally. Bascom (1965) also uses the more general expression "prose narrative" (myths, legends, folktales). Such narratives of old age may display recurrent themes but also great diversity of issues and emphasis (Holstein and Minkler 2003).

We have seen in Chap. 1 that modern narratives about old age are permeated with concepts of "successful ageing", a "disability-free" life, a "culture of narcissism" and material well-being (e.g. Gilleard 2007). As correctly pointed out by Gilleard (2007, p. 82) we are living in a world where the prevailing narrative propounds that: "Successful old age is old age without old age". The concept of "successful ageing" has been the target of increasing criticism (e.g. Andrews 2009) due to its emphasis on competitiveness and the striving toward unrealistic goals. As we contend in this book, the alternative narrative of "fulfilling ageing" is not only more realistic but also less frustrating to the elders.

The narratives of the elders have acquired a specific academic status in the field known as *Narrative Gerontology* (Randall 2002, p. 56):

> A central conviction of narrative gerontology is that lives are biographical entities as much as they are biological ones... In contrast to the biomedical approach that has often characterised the study of ageing, 'biographicity'... is seen as integral to who we are. This implies a shift in the way that we think about ageing itself. In addition to being a physiological, psychological or social process, ageing is also a hermeneutical process. It is bound up with how, over time, we interpret the events and circumstance of our existence *in* time, through memory and imagination. It is linked to how we weave such interpretations into a complex, multi-layered, ever-extending text... or story that we assume when explaining who we are, whether to others or to ourselves.

Randall and Kenyon (2004) provide an analysis of narrative gerontology focusing on various central dimensions: the *temporal*, *poetical*, *spiritual* and also the issue of *wisdom*. The *temporal dimension* of narrative gerontology is self-evident from the recount of stories past, experiences of youth, lessons learned, memories of good/bad experiences that abound in the discourse of old people. As we become older, time is a dimension of existence that acquires a special status. We reach a point where we are quite aware of the asymmetry in our position within our life curve: time past looks longer than time future. The present may become more important and legacy is a concern that acquires a new urgency. For old people, the narrative in a time dimension often becomes an exercise in autobiography, and autobiographical time "is not that of clocks, ticking or flashing in a mechanical marking of instant after instant. It is time not as measured but as felt, not real time but human time ... *Human time*, we could say, is essentially narrative time, or *story*time. Where clock time epitomizes objective time, storytime epitomizes subjective time, the time of our lives" (Randall and Kenyon 2004, p. 334). From the perspective of our body the temporal narrative can be also a *narrative of decline* (Gullette 2004; Randall 2009).

Subjective time may take diverse priorities, according to each one of us. The past may become prominent, or perhaps the present may take centre place, or the future may become the object of our main concern (including spiritual/religious concerns regarding the afterlife). Interestingly, Randall and Kenyon (2004) point to the oft-observed exercise in *counterfactuality* ("what if…") that older people engage in when looking at their past life. An autobiographical story full of "what if" questions can become a counterfactual trap from which there is usually little possibility to escape. When the “what if” question points to a mistake that was made at some crucial point in life, the answer may be didactical but of not much use to change the past. Alternatively, one may in fact be happy with the decisions made in life and ask a counterfactual “what if” question simply to reinforce the feel-good impression of our own autobiography: What if rather than making this good decision at the age of 20, my decision was a different one (e.g. like the one made by a friend, relative, acquaintance), what a mess would have my life been? Counterfactuals may have a pedagogical function in informing younger generations about the circumstances in which one may make the wrong or correct decision in life.

Randall and Kenyon’s *poetical dimension* of elders’ narratives points to the creative exercise involved in producing a *lifestory* (McAdams 1996). We know that the story must make some sense, but gaps in memory may jeopardise the meaningfulness of past experiences, hence we may indulge in creatively filling in such gaps to maintain continuity (Nouri and Helterline 1998). Ultimately, over time some experiences may acquire the status of myth, thus becoming embellished with added colours and flavours that lift the story out of the more ordinary plain facts. Steele (1986) refers to such creative exercises as works of “faction”, a composite between fact and fiction. Ultimately, a *lifestory* is indeed a work of faction.

In the *spiritual dimension* of narrative gerontology, we produce stories that give a deeper meaning to our lives, stories that transcend the ordinary and the here and now. This may be associated with a specific set of religious beliefs or simply with a broader form of transcendence, such as a sense of generativity (Erickson 1997). In this spiritual dimension narratives could lead older people to greater meaning, a more coherent self-identity and wisdom; in whichever way wisdom may be understood by the individual: “enlightenment, good judgement, compassion, union with God” and so forth (Randall and Kenyon 2004, p. 339). This can be facilitated if the elder is embedded in what Randall and Kenyon (2002) call a *wisdom environment*, that is an environment that provides the elder with the opportunities “to discover their wisdom through a process of re-storying that *includes* meditation” (Randall and Kenyon 2004, p. 340).

To the individual dimensions of narratives about ageing we must also add social dimensions. In fact, the narratives on ageing are a result of a combined process of individual and social construction of discourse (Nouri and Helterline 1998). The social sphere provides a *narrative environment* (Randall 2009) where stories are born and develop, and they can be perpetuated over time. Following Foucault (1970), Westerhof and Tulle (2007, p. 236) write: “Discourse refers to sets of institutional and cultural systems that provide meaning, map out the space in which social practices are made possible and constitute ‘official’ knowledge or truth”. The social construction of discourse can potentially limit the possibilities of individual elders to imagine a different life.

Rees Jones and Higgs (2010) offer an analysis of discourses on old age that are prevalent in society, that is discourses that are regarded as *natural*, *normal*, and *normative* in a specific social context. Any society has expectations on what old age should be about in any given historical period. Such expectations may be shared by all, a majority or simply a section of society, in the latter case a diversity of expectations may coexist. Expectations establish the limits of what can be regarded as "natural" in old age. This implies that transgressing the boundaries of the "natural" may attract social criticism, leading to ostracism in the most extreme cases. Of course, the "natural" is especially affected by the biological boundaries of what is achievable in old age, but there is still a wide region of possibilities that is defined by culture. In particular, cultural limits are important to define what is regarded as "normal" in old age. Finally, the "normative" in old age is what society expects from old people and such expectations may be specific and narrow or they may be plural and wide, the latter being increasingly the case at present under the influence of modernisation.

We have seen that since the late 1970s modernisation has been marked by neoliberal policies and cultural changes that emphasise individualism and competition, and that are especially expressed through consumerism. The current narrative of "successful ageing" was developed under the influence of such neoliberal cultural environment. In the words of Rees Jones and Higgs (2010, p. 1516):

> For older citizens the paradigm of choice and consumerism is articulated through the ideas of healthy ageing, independent living and successful ageing as well as notions of 'agelessness'. The articulation of this paradigm is driven by a range of institutions, socio-technical devices and scientific/cultural intermediaries which also includes the anti-ageing movement.

Success becomes normative and elders not conforming to optimistic expectations must add to their old age the knowledge of being a "failure". A narrative of life fulfilment, on the other hand, opens a universe of possibilities to the elders, within which each one of us should be allowed to find his/her own way.

Given the inevitable decline of the physical and mental conditions in old age, *narratives of suffering* have also attracted the attention of scholars and lay people alike. *Suffering* can be regarded as psychological and/or physical distress that threatens the integrity of the self, thus limiting our ability to fulfil our objectives in life (see also Black and Rubinstein 2004; Black 2009).

The narrative of ageing as suffering and as a period of increasing limitations that should be overcome is further reinforced by the medicalisation of discourse on ageing that has led to the "construction of ageing as disorder, failure, abnormality, alteration, run down, defective, unstable, diminished, even dangerous and deleterious" (Westerhof and Tulle 2007, p. 237).

Suffering tends to produce a reaction that aims at alleviating it, and new autobiographical narratives may help in this regard. As stated by Black (2009, 82): "When an experience of suffering disrupts daily life, usual routines are suspended. Making sense of the experience may be accomplished by reporting it in the life story, which provides a way of incorporating the suffering experience into the self". We have already seen from Frankl (1988) how suffering can be constructively incorporated into our life to help our personal growth.

Depending on the culture, suffering may be repressed or expressed in verbal and non-verbal communication, but it is often the case that at the point where suffering becomes overwhelming, communicating about it may be stopped, as a way to avoid further pain and distress. Although suffering can be affected by negative or traumatic social interactions (e.g. racism, discrimination, ostracism, violence, poverty, loneliness), it is ultimately an individual experience. The level of resilience in the individual can alter the degree of suffering experienced by such individual under specific circumstances, but there is a point where resilience is exhausted and the desire to end suffering becomes prominent.

In a study carried out in the USA, Black and Rubinstein (2004) investigated suffering as expressed through narratives in a group of 40 people 70 years old and older. The authors identified three main themes of suffering as they appeared in discourse: suffering as lack of control, suffering as loss, and the personal value of suffering.

Suffering as lack of control is expressed in narratives that realise the loss of independence, increasing frailty and helplessness. The difficulties to correct mistakes made in the past, because in old age there is less time left and insufficient energy to do much about it, can be also a cause of suffering due to lack of control. On the other hand, suffering as a loss mainly appeared in narratives in the form of loss of close relatives, a child in particular; loss of valued roles in life, loss of former good health. Finally, the personal value of suffering was expressed in the narratives of those elders in various ways according to the specific circumstances of the elder (e.g. having to take many medicines without understanding why). Elders relieved suffering using a diversity of strategies, including religious beliefs (see also Coats et al. 2015), which can also play a protective role against loneliness (Ciobanu and Fokkema, 2017; Cordella and Rojas-Lizana 2019). Religious beliefs as a way to relieve suffering could be challenged by some individuals, by questioning why would God want them to suffer as they approach the end of their life, not to speak of the fear of potential greater suffering in the afterlife if, as in the Christian tradition, they dread their soul being judged to deserve eternal damnation.

But suffering can be also transformative. Many elders in Black and Rubinstein's (2004) study "proposed that suffering was an event or state of being that altered their self-view and worldview in a fundamental way" (p. S22). A 76-year-old European American woman was of the opinion that "(s)uffering is a constant search for solutions. It's positive. I mean even though it hurts a lot, you're always busy asking questions" (p. S22). Surviving experiences of suffering can be also seen positively, thanks to the personal "hardiness, optimism, or resilience" (Black 2009, p. 83) required to achieve it. Such transformative narratives aim at demonstrating the capacity of the elder to overcome adversity, repair any damage to the integrity of the self, and adapt by using the negative experience to support personal growth. In other words, transformative narratives help in the transition from suffering to healing (Coats et al. 2015).

Resilience in the face of adversity can help overcome suffering in old age, and narratives have a role in building such resilience. Shklarov (2009) studied the narratives of resilience of old Soviet Jews who survived the Holocaust as children. She analysed the narratives using the concept of *anchor scripts*: "personal scripts that defined the identity of the storyteller through meaningful realities other than trauma" (p. ii) and identified major common themes of power, agency, and choice. Following Shklarov, anchor scripts can be seen as acts of meaning making used in the construction of our identity and life history in the process of overcoming traumatic experiences.

The survivor narratives studied by Shklarov expressed the capacity of the speaker for resilience, both personal and as a group, providing some remarkable examples of the ability to overcome trauma. Liza, for instance, adopted the attitude of not allowing the traumatic memories of her childhood to define herself, by following the anchor script of *I did not take it into my soul*. Fira adopted a different anchor script: *We survived to tell others what we went through*. That is, she grew resilience through an educational stance:

> She assumes the obligation of *a keeper:* the person who must preserve and cherish the memories and lessons of the past, along with the knowledge about the communal values that gave her people the strength to survive and cope. The keeper saves these memories and meanings from oblivion and shares them with others, when they are ready to listen (Shklarov 2009, p. 112).

Abram, on the other hand, used an anchor script to build resilience that made him focus on himself, becoming more attached to his work: *My brain and my soul were always busy, and I had no time to think about those things*. Finally, Hanna used narratives that speak of "personal and intergenerational continuity, and her universal script is based on drawing her strength from traditional spiritual wisdom" (p. 141): *You know your roots are Jewish. You can believe it or not, you can want it or not, your roots have already grown—it's impossible to cut them out.*

Through their scripts the elders try to intentionally overcome suffering by showing power (e.g. capacity of survival), agency (e.g. taking an active stance in life), and choice (e.g. preventing traumatic events from fully defining themselves and their life). A common source of resilience in the interviewees was the confidence that all material goods could be taken away from them, but not the knowledge they had acquired through their education. They expressed this concept with the Latin motto *omnia mea mecum porto* ("all that is mine I carry with me"; a quote from Cicero who took it from the Greek sage Bias of Priene).

Shklarov (2009) also mentions an additional strategy to overcome suffering in old age: *disengagement*. By abstracting from disturbing realities and memories the elder may be able to avoid the stress coming from a sense of guilt or remorse, for instance. The narrative of disengagement is simply a void in the space of narratives about specific issues, that are perhaps too traumatic to address. Conversely, when narratives are used for the purpose of healing, they become explicit, overt and often collective. Sometimes they may even be coloured with humour.

Another example of the narrative of suffering in the old is associated with mental illness such as dementia. Here, eliciting narratives from patients can even have therapeutic value for the elder (e.g. Charon 2006; Randall 2009), especially when cooperation is offered by other individuals (including members of the family) in order to help the elder construct his/her narrative in spite of the lapses in memory about facts, circumstances, names, and basic words, depending on how advanced dementia is.

Thus narratives can be used in old age to express either the positive in life or suffering, and we have already mentioned in Chap. 1 how narratives can be also used by old persons to reinforce a current identity ("fixity") or shape a new one ("flux") (Biggs 2004); in other words, narratives can unfold but also extend the space of *possible selves* (Smith and Freund 2002), and through a process of synthesis a coherent narrative of identity may emerge that can promote life fulfilment.

In old age the *narrative of identity* is constantly finding its way between an ever-increasing past and an ever-shrinking future. *Possible selves* are still viable in the present and the future, but as far as the past is concerned, they can only be counterfactuals (What would my life be had I done this or that in the past?), as we already mentioned. The narrative of identity:

> aims to provide a person's life with some semblance of unity, purpose, and meaning… Narrative identity is the storied understanding that a person develops regarding how he or she came to be and where he or she is going in life. It is a narrative reconstruction of the autobiographical past and imagined rendering of the anticipated future, complete with demarcated chapters, key scenes (high points, low points, turning points), main characters, and intersecting plot lines … [it is] … an internalized and evolving life story that a person begins to develop in late adolescence to provide life with meaning and purpose" (McAdams and Olson 2010, p. 527).

The narrative of identity is enriched by *biographical reasoning*, or "the tendency to draw summary (semantic) conclusions about the self from autobiographical episodes" (McAdams and Olson 2010, p. 529).

Therefore, what we call the *self* may be seen, in the first instance, as a multiplicity of entities that can be expressed in time according to personal and social circumstances. The various identities may also relate to each other in a hierarchical manner: Some may be more pervasive whereas others may be more situation-specific; as they all strive to maintain a sense of unity and self-coherence. Variables affecting self-identity in old age include: personal resources (age, education, income, health), loss of role (retirement, widowhood, absence of children at home), development of new roles, and compensatory activities (leisure and group activities) (Mutran 1987).

In a study carried out in the USA, Mutran (1987) described how old women think about their identity in more affective terms than men, with the affective perspective increasing over the years, as health becomes poorer and women are more likely to become widows. In spite of this diversity, fulfilling ageing may still require a unity of the self, without which it may be difficult to construct meaning in life. This opens up a question about the concept of *possible selves*: Do we really juggle our way in life confronting a multiplicity of identities? Or do we simply experience a developmental process that is dynamic and fluid? Should not we be really referring to the *possibilities of the self*, rather than *possible selves*?

Indeed, as we have seen, the diverse life experiences of the self can be synthesised into a narrative of identity or *narrative emplotment* (Ricoeur 1984) that can:

> configure the diverse events and actions of one's life into a meaningful whole. One's self-concept or self-identity is fashioned by adaptation of plots from one's cultural stock of stories and myths. Stories of personal identity differ from literary productions in that they

> are constructed within an unfolding autobiography and incorporate the accidental events and unintended consequences of actions (Polkinghorne 1991, p. 135).

When the self-narrative is shaken by stressful circumstances, then anxiety, depression, and meaninglessness may ensue; but from such circumstances, personal growth and a new identity may also emerge. Life shocks may become transition points from one self-identity to another. Sometimes the transition is smooth, sometimes it looks more like a fracture, but the process also implies a temporal continuity retained in our memory.

There is also a role of the mass media in the construction of narratives of identity in old age, that may lead to a perception of old people from being "invisible" to the reinforcement of a diversity of stereotypes about old age. Although we have seen in Chap. 6 that many stereotypes of old age are negative, some of them may be positive (e.g. grandparents as repositories of valuable life experiences) and both positive and negative stereotypes could be transmitted by the media. Elders themselves can be proactive and resist the imposition of old-age stereotypes on themselves by adopting behaviours and attitudes that contradict media-reinforced negative stereotypes (e.g. Westerhof and Tulle 2007). However, there is no guarantee of achieving the neutralisation of stereotypes: an old person who is stereotypically expected to be passive and tired but who behaves in a more active and creative manner, may still be stereotypically judged as an "embarrassment" by others, a denier of old age. Nevertheless, resistance may at least reassure the elder that there is a way of reaffirming his/her own identity in spite of social constraints.

In an interesting work focusing on old people's identities, Townsend et al. (2006) interviewed 84 residents in two English towns, and they also recorded opinions of participants in focus groups. Participants were mainly working-class people and included 72.6% women, with the overall age distribution being 37.7% under 75 years of age and the rest older, including 21.4% who were older than 85. Narratives expressed the elders' daily struggles living with limited financial resources, which made social support from friends and family important, despite self-identity being marked by a strong sense of pride in being independent. Participants did appreciate some improvements provided by modern life; in fact, some went as far as suggesting that the access to more information from around the world, opportunities to travel and to broaden their experiences made them feel younger now than they were in the past. Ill health, however, constrained such optimism but they adopted a fighting-spirit attitude towards deteriorating health: "pushing yourself, not letting yourself go, not giving up" (p. 890). Their description of other old people also expressed a wide spectrum of perceived identities. Townsend et al. (2006) recorded three broad categories of older people as identified by peers in their narratives: "heroes/heroines", "villains", and "victims".

Heroes and heroines exemplify the most optimistic and positive views on old age and the chances of living a fulfilling life. In the discourse of these elders, almost all "heroes" were in fact heroines (e.g. mothers and sisters). Some characteristics of the identity of such heroines were their capacity to remain optimistic in the face of adversity and to never give up, their independence, intellectual flexibility, cheerfulness, adaptability, and altruism. "Villains", on the other hand, appear in the narratives of the elders displaying characteristics opposite to those of the heroines: "moaning, complaining, inactive, introverted and awkward" (p. 892). Villains tended not to be close relatives and they were frequently men. The pessimism and passivity of the villains was antithetic to the optimism and proactivity of the heroes/heroines, hence the latter tended to avoid the former. Finally, the "victims" are objectively impeded from being more active by their physical or mental limitations, and they were seen by the participants in the study as having had a "raw deal", being treated with sympathy and empathy.

Although grouping people according to broad categories may convey some aspects of reality (e.g. personality), it is also important to remember that most people's behaviour tends to be fluid and responsive to changing circumstances. At one stage or other in our life, we may have all been heroes, villains, or victims.

In Chap. 1—and in the *Inter-generational communication* section in this Chap. 7—we mentioned Biggs' (2004) concept of *masquerades* as a communicative device used to construct identity. Masquerades point to a social aspect of identity in which we "perform" one or more versions of our self, according to context and audience, thus potentially offering a complex and dynamic reality of the self (*fluidity*). But masquerades, not unlike the narratives behind the concepts of "heroines", "villains", and "victims" that we have just seen, can be not only descriptive of the dynamic reality of the self, but also constraining to the further development of the self, especially when a specific "mask" is expected socially and any departure from it could be risky to self-identity (*fixity*).

The interplay between *fluidity* and *fixity of identity* is central to the process of developing a narrative of self-identity during ageing. The body changes, the social situation may also change (e.g. retirement, children leaving home, friends passing away) nothing seems to be really fixed, not even culture and language, which may change over time with the addition of new concepts, social habits, words, expressions, and yet we see ourselves as a single biological entity: somebody who was born and has been living ever since. In our view the linchpin keeping the fluidity and fixity of our lives together is the narrative of identity that underpins life fulfilment. For as long as the current state of affair is fulfilling, then the individual will manoeuvre in order to retain such identity; when it is not, change will be needed. Conserving a state of fulfilment through ageing can be seen metaphorically as equivalent to conserving our biological unity as an individual, despite constant molecular and cellular changes. Fulfilment keeps the unity of the self together through change, by guiding the dynamic process of the possibilities of the self.

With the concept of masquerade, however, Biggs (2004) takes fluidity/fixity further. A masquerade may become an empty shell hiding the lack of a solid identity, a view that contrasts with our emphasis on life fulfilment, which is built on honestly expressing our self. A masquerade is fixed in its emptiness and yet fluid in its capacity to change according to the requirements of the circumstances: emptiness packed with superficial identities. Biggs (2004, p. 50) nicely expresses this situation in these words: "None of these disconnected images hold any intrinsic or authentic meaning and are still therefore essentially hollow". Such disconnectedness in the multiplicity of identities may go as far as being regarded as pathological, as in the case of *dissociative identity disorder* (formerly known as *multiple personality disorder*) which is defined in the DSM-5 as a:

> disruption of identity characterized by two or more distinct personality states, which may be described in some cultures as an experience of possession. The disruption in identity involves marked discontinuity in sense of self and sense of agency, accompanied by related alterations in affect, behavior, consciousness, memory, perception, cognition, and/or sensory-motor functioning (American Psychiatric Association, DSM-5, p. 292).

Beyond the masquerade, giving true meaning to life post-retirement may be a challenge for people who for decades defined themselves through their work. Fluidity can help in reinventing yourself, fixity can help in maintaining the path once the path has been found. Some elders may find a new meaning through participating in voluntary activities in the community (see section *Inter-generational communication* in this chapter), which may be especially fulfilling when those activities involve contact with younger generations, thus allowing the elder to engage in generativity. Lundgren (2010), for instance, describes a Swedish programme of that kind that allowed older participants to construct an identity of "grandparents" through the transmission of their life experiences to school children. Children may benefit from listening to eye-witness stories of the past, the elders may benefit through the reinforcement of their identity as "useful citizens":

> I feel that I am needed, and it feels good to give something to society. Too many [older persons] feel lonely and they get sick… Then it's better to do this, to have something to do, and to contribute to society (Lundgren 2010, p. 252).

Identity change that is achieved through the participation in inter-generational programmes is not uncommon, as such programmes can challenge older participants in some of their assumptions, especially those concerning modern life and younger generations, thus "resulting in self-reflection and changes in self-understanding" (McGowan and Blankenship 1994, p. 590).

We have seen at the beginning of this section how through narratives we shape the story of our life in a process of *storytelling*. Human communication has evolved mainly through the production of stories or "symbolic interpretations of aspects of the world occurring in time and shaped by history, culture and character" (Fisher 1987, p. xi) that possess a "narrative logic" characterised not only by *creativity* but also *coherence* and *fidelity* (i.e. the truth value of the story, Fisher 1985). Storytelling is a communicative process of building consensus through what Goffman called *transactional inter-individuation* (Goffman 1959). Such ability to build consensus through inter-individual exchanges of creative, coherent, and faithful texts that are expressed in narratives is innate, but what is learned is the final consensus reality that is built over time through our inter-subjective capacity for imagination (Hubble and Tew 2013).

The narratives of the elders as they unfold in inter-generational communication may involve autobiographical or more creative forms of storytelling, with the two styles being often intermingled, sometimes intentionally sometimes not. Within the family environment, grandparent-to-grandchild storytelling is common and through such storytelling grandparents try to engage in generativity by transmitting their experience in life and accumulated wisdom:

> Grandparents give grandchildren love and affection, support, trust, connections to the past, family heritage, moral values, and act as role models ... In return, grandchildren give grandparents joy, inspiration, love, and hope for the future (Ryan et al. 2004, p. 376).

Moreover, through storytelling grandparents may display aspects of their self-identity and indeed they may even engage in further identity construction, as we have already seen. Ryan et al. (2004) mentions some of the roles that grandparents may adopt in communication with their grandchildren through storytelling that can give added meaning to the elder's life: centrality of grandparenthood, valued elder as transmitter of tradition and roles, link with past generations, indulgent grandparent. In their study carried out in Canada, Ryan et al. invited grandparents to write a story, a letter, or a poem that was intended for their grandchild. The grandparents' stories focused on some major themes, such as general historical events that affected their life (war, the Great Depression) and the family history, with special emphasis on knowledge of own roots. They also gave advice through their stories as a form of inter-generational transmission of experience ("I can honestly say that I have lived a richer life by engaging with my mistakes, I have come to realize that the one who never makes a mistake is likely someone who never takes a risk either … My story of mistakes is meant to help you avoid making the same ones" p. 382). The historical theme also appeared in the form of life stories that recounted personal experiences that had been especially important for the grandparent. Here we may also recall the dramatic message sent by a *Grandmother of Plaza de Mayo* to her disappeared granddaughter, that we quoted in Chap. 6, a message that aimed at linking the child with her past generations, but without any certainty that the message would be ever received. Sending the message, however, must have put the grandmother, "Chicha" Mariani, somewhat at peace.

Dorfman et al. (2004) analysed the specific case of identity narratives of rural elders in the USA Midwest. In brief, the major themes in these elders' narratives included: World War I and II, the Great Depression as defining historical events in their lives, their life values (self-reliance, hard work, and religion), attachment to place and community, and the central importance of both family and work.

Both Ryan et al.'s (2004) and Dorfman et al.'s (2004) works point to the phenomenon of *generational imprinting* that we already mentioned in this book (e.g. Chap. 1). Historical events will mark each generation of old people as a broadly distinct group, in spite of individual differences and some commonalities seen across cohorts of old people due to their old age, and such experience of specific historical events will also mark the elders' narratives.

Elders as storytellers are especially relevant in ethnic groups still largely dependent on direct inter-generational transmission of cultural knowledge. Such ethnic groups may be under assimilation pressure from a dominant culture, their language may be threatened and their cultural heritage thinning as the old pass away without having had the opportunity to transmit their knowledge to the younger generations.

This is especially critical when elements of the culture and their explanation are not available in recorded form (visual, written, audio).

A considerable amount of work has been carried out in Canada involving native elders as transmitters of culture through storytelling. Atleo (2009) studied cultural transmission through storytelling in a small group of *Nuu-chah-nulth* elders from the province of British Columbia. The author investigated the story of *U-meek*, the first whaler "a person who can get (meek) anything (U) because he or she knows how to learn to get it" (p. 454). This is a story that conveys specific practical information about the traditional life of the *Nuu-chah-nulth* and a mythical explanation of the extension of their hunting knowledge from seals to also include whales. Through their stories, grandparents are conveyers of cultural knowledge relevant to various aspects of life: "The experiences of grandparents were expected to provide examples of successful living and models of activities, attitudes, and behaviors for their grandchildren" (p. 461). In order to achieve learning, elders developed a discipline of "spiritual bathing" or *oosumch* which aims at obtaining the level of humility and self-control that are necessary in order to learn: "Spiritual bathing can be seen as a ritual activity of purification that cleans the body, mind, and spirit so that the person melds with the environment" (p. 462). The discipline can be also transmitted to younger generations.

In another study of First Nations elders from Canada (Northwest Territories), James (2016) focused on narratives reflecting the qualities of a "capable person" and how such qualities are taught and transmitted to younger generations. The figure of a capable person develops in the narratives starting from an awareness of self and his/her identity. The self must develop a healthy body in order to become capable. The body becomes healthy through "natural foods, avoiding certain parts of wild meat, physical work activities like hunting, fishing, and trapping, limiting alcohol use, and being out on the land either collecting berries, digging roots, 'picking pukok,' or gathering medicines. It was all part of self-care in developing one's physical self and identity through guidance from parents, self awareness, and even making mistakes" (p. 225). The self must also develop a healthy mind, because it is by using the mind that one can solve the problems of survival. However, each individual can contribute to the well-being of the community within his/her own capacities: "even when family members did not possess a well-honed level of mental intelligence, which exhibited such advanced land-based intellectual capacities, they were not de-valued. In many cases, they became even more valued because they learned the skills that were necessary for thriving in family and camp life, such as cooking, cleaning, and camp setup. It took all levels of knowledge to function in traditional communities to create comfort, ease, and happiness" (p. 228). Becoming a capable person also requires a well-developed "heart" (emotionality and social connectedness) which helps in enjoying life. The capable person is held together and guided through personal development by a sense of spirituality and he or she passes cultural knowledge on to the next generation in his/her storytelling.

Graham and Davoren (2015) studied the narratives of *Métis* parents in British Columbia, Canada, and their role in inter-generational cultural transmission. *Métis* is French for a person of mixed blood, in this case European and Native American but they are considered distinct people, with their own culture and language (Michif). In the case of the *Métis*, social disadvantage, disruption of family connections, and instability impeded the smooth inter-generational transmission of culture, which was however more recurrent in those cases where children grew up in a stable and supportive social environment.

In native populations that suffered historical events of colonisation, such as the examples of Canadian First Nations we have just mentioned, the maintenance of their native culture crucially depends on the transmission of their language to younger generations, as language is not just a means of communication but a way to understand reality (Whorf 1956). But in such communities the maintenance of a native language is sometimes seen by the young as shameful and ostracising (e.g. Meek 2007), in just the same way as speakers of an official language may often refrain from using dialects in some social contexts for fear of discrimination (e.g. Macaulay 1975). In this respect, elders are best suited to become vehicles for the transmission of language, as they may convey both linguistic knowledge and cultural pride.

Through their narratives and knowledge of culture elders can become teachers of language and traditions not just to the younger generations in their community but to others as well, whether they do so in a spontaneous way or through organised programmes. For instance, in one such programme implemented in Canada, *Kaska* elders contributed to the teaching of a language course where they acted as cultural and spiritual guides, tradition-bearers, oral models (i.e., teachers of correct native language pronunciation), and as experiential teachers (transmitters of their own life experiences) (Moore 2003). Traditional stories were also collected and transmitted to others, including younger generations, in a programme involving *Noongar* elders in Western Australia using a diversity of media: oral history recordings, short film, photography, textiles, and painting (Quayle et al. 2016). Stories addressed issues such as dispossession and power relationships and how they negatively affected indigenous communities. On the other hand, the stories of the elders also demonstrated strength and resilience in adversity, resistance, survival and both cultural continuity and change.

Sometimes interference with other languages and cultures make transmission of native knowledge to younger generations unreliable (e.g. Viatori 2007), not to speak of cases involving blatantly fraudulent claims of being a knowledgeable elder within a specific ethnic group when the person is not (McIntosh 2009). However, many valuable programmes are available that use narratives in inter-generational transmission of cultural knowledge, meaning making, and identity construction.

Hubble and Tew (2013) described a narrative programme implemented in the UK, the *Fiction and Cultural Mediation of Ageing Project* (FCMAP) that aims at understanding the cultural representations of ageing, elders' social relationships, and the role of narratives in fostering elder's identity and agency through social interactions. The project is longitudinal, and participants are over 50 years old, with some of them having been followed up to over the age of 80, that is, continuously for 15 years. *Fiction narratives* are used to understand the process of meaning making in the elders through the personal worlds they construct in such narratives.

Meaning is achieved in the text through a confluence of lived experiences and imagination that are given shape in the narrative. The narrative may change and evolve over time, thus expressing a dynamic self under the influence of a variable and changing social environment. Such variability, however, may be regarded as an extreme in this case, as fictional change is clearly not constrained by negative feedback from the social environment and thus it can express itself in virtually unlimited form:

> As tropes of ageing are largely founded on narratives and symbols concerned with the self, fiction offers a capacity to allow another imaginary space of otherness (an empathic demand being part of its structure as well as requirement for self-identification) that initiates new ideas and responses, precisely because literature has the capacity to resist reductive, one dimensional narratives and images of ageing. Thus it is capable of conveying a heterogeneous view, centred on individual experience, physical and mental adjustments, and the process of memory, that continues to shape identity (Hubble and Tew 2013, p. 32).

Reading and discussion groups, plus writing diaries, are some of the tools that can be used—either as a personal initiative or in more organised programmes—to explore (and construct) identity in old age through narratives (Hubble and Tew 2013).

Therefore, narratives can help older people explore the possibilities of the self and experience a more fulfilling ageing by constructing meaningfulness in life, recovering memory, nurturing social links in the present (e.g. legacy) and the past (through memory), incorporate experiences of triumph and suffering, and shaping self-identity and *self-integration* in a dynamic manner involving both fluidity and fixity.

We have mentioned above that older people live in a space of multiple selves. In part, such multiplicity is explained by different social roles: partner, friend, father/mother, grandparent, colleague, and so forth, but there is also a component of multiple interests shaping a complex identity: professional interests, hobbies, political views, philosophy of life, and more. However, we have also stressed that in order to experience a fulfilling old age such multiplicity of selves must be somehow integrated into a coherent whole, a personal identity, as suggested at the beginning of this section. Narratives can help in this process of self-integration, an objective that is central to Erikson and Erikson's (1997) concept of *gerotranscendence* (see Chap. 1) and to others (see Chapman 2005).

Chapman (2005) notes that in recent times the concept of self-integration has been challenged by social and cultural trends that have been leading to the dispersal of the self into separated and independent entities, each one pursuing the satisfaction of its own objectives: the artist in us wants to be a free person, whilst the narcissist in us wants more power and control, whereas the socially conscious in us wants more fairness, whilst the capitalist wants to accumulate some wealth, and the environmentally sensitive within our self is concerned that the pursue of excessive profits may damage the environment, and so forth. We are not supposed to be able to integrate such diversity, simply because there are too many inherent contradictions. But in Chapman's (2005, p. 14) words: "gerontologists are describing aging well as the ongoing co-construction and reconstruction of multiple selves as an open-ended process of meaning-making amid later-life events and transitions". In fact, through meaning making the integration of the self is possible, both in terms of our multiplicity of interests at any given point in time and our dynamic development over time. Following from the above example, the artist in us wants to be a free person but you can only be free if you have the power to control the direction of your life, and you can support social fairness in spite of being a capitalist if you agree with the concept of redistribution of wealth through taxation that aims at spreading opportunities—and therefore freedom—to all, whilst profit can be obtained from commercial activities that are environmentally sustainable, and so forth.

A coherent self, once achieved, can still change, just like an organism can go through various developmental stages without stopping being a living organism. A person may be religious in youth and become an atheist in old age (or vice versa), there is both coherence at each stage and a historical link between those two states of the self that are only punctuated, presumably, by a period of crisis.

The underlying narrative of modernisation, that has produced a current perception of multiplicity and instability of the self, can be clearly seen in this quote from Chapman (2005, p. 14):

> Selves have long been understood as day-to-day constructions of interpretation, through introspective self-meaning and social interaction with others ... However, issues of plurality of life-course experiences, contradictory constructions of 'old age,' technological innovations, and changes in the meaning of time and space in society combine to influence individuals' efforts to keep up with these selves as they multitask and juggle paid and unpaid work.

Thus, the modern world is giving us instability and disconnectedness of the self, but as we have seen we can challenge it in pursue of an integrated and fulfilled self. Multiplicity of roles and interests, not to speak of past experiences can become both meaningful and integrated into a narrative of identity. In this view, major transitions, such as retirement, can become the life course equivalent to "metamorphosis": change with maintenance of integration. Or at least, that is the objective.

Chapman (2005, p. 14) states that "Rather than an introspective pursuit of self-integration, aging well may be an open-ended negotiation of the co-construction of multiple selves". But then she also admits that such negotiation involves making sense of the diversity and change, and here we ask: Would not we better able to achieve meaning through a process of self-integration and fulfilment? Diversity and change without meaning and integration are, in our view, more likely to lead to a life review in old age marked by dissatisfaction, a sense of directionless meandering through life, with no real purpose but the basic need to stay alive and attend to immediate material needs.

Fulfilling ageing offers a pathway for the integration of the self in an ever changing life experience as we age: as life proceeds into older ages, the individual is confronted with the need to experience fulfilment with ever diminishing resources (physical, mental, perhaps financial, and often social). However, at the same time as the ageing body and mind constrain some activities, the concept of fulfilment in life can remain adaptive. An old person who cannot drive anymore can use public transport, when even that is also precluded, she or he may require some home support from others (e.g. government services), until the frail elder may finally need more intense care in specialised facilities. For as long as the elder finds a degree of fulfilment throughout those experiences, and the support required is made available, the fact that individual resources are diminishing may be less of a trauma than one may think. The self can remain integrated through all those changes. But there is a limit, of course, and at some point in time deterioration may be such that the narrative of ageing may become "broken" (Andrews 2009), with no possibility of "repair". What may happen next will be the focus of Chap. 8.

Experiencing personal fulfilment in life provides the individual with *inner strength*, which can be seen as a "central human resource that promotes well-being and healing" (Nygren et al. 2007, p. 1061). Inner strength is expressed as resilience, sense of coherence, self-transcendence and purpose in life in old individuals, and a narrative that reflects them; therefore, life adversities can be better confronted when the elder has experienced greater personal fulfilment that gives her/him inner strength.

Nygren et al. (2007) investigated inner strength in a sample of 18 elders (61% women) who participated in the Swedish *Umeå 85+ project*. The text of interviews was analysed using a *phenomenological hermeneutical method* that follows three steps: naive reading, thematic structural analysis, and critical reading leading to a comprehensive understanding. Generally speaking, this sample of Swedish oldest-old defined inner strength as:

> ... looking back on one's life with pride, knowing that one has overcome most of the challenges one has met. It means relying on one's body, still meeting the challenges of life with faith in oneself, being determined but not too stubborn, and being aware that one can ask for and receive help if it is needed. It means the ability to accept and adjust to a situation, to forgive others, and also to see that something positive comes from the negative: One is learning from life. Being an oldest old person with inner strength means feeling needed, enjoying life, taking one day at a time but still 'looking forward,' and being aware of one's limitations and possible future difficulties. It means being connected with the people of today and also those of the past by thinking back (Nygren et al. 2007, p. 1063).

That is, for those oldest-old interviewees, inner strength came from both personal and social resources and an optimistic attitude in life, but one that is also realistic. It also came from acceptance of self that provides inner peace, and connectedness with others and the past.

An area that is little explored in the study of elders' narratives is that of the narratives of the old belonging to gender minorities. In a study carried out in Australia, Hughes (2008) focused on the narratives of 14 older gay and lesbian Australians (59–72 years old, 35.7% women). The narratives expressed concepts of identity as a lesbian or gay man and also attachment to community and friendship networks. Similar narratives were recorded by Woody (2015) after interviewing 15 African American self-identified lesbians, ranging from 57 to 72 years old. Identity appeared in the narratives especially in the context of old lesbians feeling "invisible" in their community, which mainly meant that they lacked social support, felt isolated, and threatened by discrimination. Loss of social support had consequences on their feeling unsafe, which often led to denial of their sexual orientation in order to increase their safety. Discrimination and lack of social support eventually affected their ability to live a fulfilling life at a very practical level; e.g. difficulties in accessing appropriate nursing homes. In spite of the obstacles, many lesbians in this study had been able to develop strategies of resistance that allowed them to live a more fulfilling life under the circumstances:

> A 60-year-old Black lesbian respondent recounted that in a Bible study class she attended with her wife, the minister started making comments about 'those funny women.' She told the minister 'You will not get another check from me to support ministry that is so mean-spirited!' (Woody 2015, p. 55).

Narratives of identity and of fulfilling ageing are also common among immigrants. Immigrants face the challenge of trying to construct their identity in a foreign land within a social environment that is not always supportive, and it is often hostile. They have to juggle the relationships between their past, present, and future in ways that non-immigrants may find difficult to imagine. The external pressures to redefine their identity can be strong and that may affect their chances of experiencing fulfilment in old age.

In countries where immigrants are seen as co-contributors to the construction of a national identity, their integration is likely to be better and the process less traumatic. Old immigrants can find spaces where their identity can retain the experiences of the past (e.g. their native language, some traditions) whilst constructively interacting with the rest of society during their working life and then through the younger generations in their family, that are born in the new country.

Becker (2003) carried out a long-term (10 years) study of the experiences of older immigrants ($n = 211$, 50 years old and older) in Northern California (USA). Ethnicities included Latinos, Filipino Americans, and Cambodian Americans. The analysis of narratives recorded in interviews suggests that "cultural traditions associated with [their homeland] continued to inform how they viewed the world" (p. 135) and depending on the size of the community, entire neighbourhoods could construct a cultural environment that is friendly to the maintenance of their culture, language, and traditions. In the USA this is common among Latinos.

Multi-generational households are not uncommon in those communities of immigrants, being in part a consequence of economic hardship, but also of traditional cultural habits that foster close relationships within the extended family (Becker 2003). The home as such, Becker notes, was a central pillar of their psychological security. The immigration experience is characterised by change and unpredictability that far from being "exciting" are more often than not sources of serious stress; hence a stable environment represented by a house and the family living in it becomes a sort of island of security in the middle of a dangerous ocean:

> Home was a central repository of meaning. People who were threatened with the loss of their homes, however modest, were often overwhelmed. The loss of their home was distressing and represented loss of security. Their situation was all the more frightening because of the difficulty of finding an affordable space in which to live (p. 142).

The relationship of old immigrants with both their native and adopted countries is varied and reflected in their complex narratives. Such narratives of old immigrants, as we have already seen for all narratives, are a mixture of memories based on personal lived experiences but also myths and further modifications, sometimes romanticised modifications, interspersed with voids caused by lapses in memory, that are often filled somehow:

> The narratives of the elders are not coherent wholes or necessarily wholly 'true'. Frequently ambiguous, they are filled with silences, illusive references and contradictions. Indeed, we would probably be misguided if we assumed that oral histories ever produce coherent or objective accounts of the past. Instead, it may be more useful to think of them as myths. This does not necessarily mean they are false, for the unconscious shaping of life stories does not jettison the truth, but rather, involves a recognition that memory is inherently revisionist; what is forgotten may be as important as what is remembered (Gardner 1999, p. 65).

To old immigrants their narratives can be an important component in the construction of their identity, combining experiences in two (or more) worlds. In a study of Bengali elders carried out in the UK, Gardner (1999) recorded narratives expressing difficulties with integration, finding employment, especially in the case of the more recent arrivals, and the constant preoccupation with money and their financial situation:

> I used to sleep, eat, work and sleep. Get up in the morning and work and work. I would just work and work, save and count my money (p. 67).

But, in the narratives, there was also determination to make a living for themselves and their families in their new country.

There are more specific narratives of the elders that can be described, apart from the narratives of old members of gender minorities and the migrants that we have just mentioned. Caring for frail older people is associated with specific caregiver narratives not only of joy and pride but also of burden and distress. In fact, narratives could be even used therapeutically to help caregivers switch from the stresses of burden to a more fulfilling perception of caregiving (e.g. Clark and Standard 1997), especially when external material support (financial, relief from duties) is made available to the caregiver.

Caregiver narratives can convey entrenched cultural beliefs that may be helpful in terms of alleviating caregiver burden. For instance, in many collectivistic cultures the family is central as a source of support, and social relationships within the family are framed in terms of the concept of *familism*, that we have already introduced and analysed in Chaps. 3 and 6. In a study of Latino elders carried out in the USA, Ruiz and Ransford (2012) noted how Latino elders relied on family support especially when they were suffering from serious medical issues. The narrative and practice of family solidarity can help in relieving primary carers of some of their burden when one of the members of the family is in need of care, although the narrative of familism and its expectations in terms of shared burden of care are being challenged by current trends of increasing individualism (Rosenthal Gelman 2014).

Caregivers are confronted with an especially intense burden when the elders they are caring for are losing or have already lost their identity, as in the case of patients with advanced Alzheimer's disease (AD). In this case little or no positive feedback is expected from the elder, that could somehow relieve the caregiver of some of her/his burden, and the caregivers' narratives are expected to reflect their frustration. Moreover, caregivers in this case are also burdened with the realisation that what they are witnessing in their old relative may be a potential reflection of themselves into the future (e.g. Hinton and Levkoff 1999).

In a study of the narratives of caregivers of elders suffering from AD, Hinton and Levkoff (1999, p. 454) mentioned the concerns of a 45 year old, Mrs. Jones, who was caring for her mother, about the extent to which her mother's identity was lost due to the disease:

> Reflecting on the words of another group member, she wonders if her own mother is still there, still alive. Who has she become? As she describes how the illness has changed her relationship with her mother, she expresses sadness, anger, and, in other parts of the interview, fear. The theme of lost identity runs throughout Mrs. Jones' interview as she describes the onset of her mother's illness, the family's search for a diagnosis, and the day-to-day experiences of caregiving as her mother became progressively more dependent and disabled.

The stories of lost identity and frustration reported by caregivers of AD patients are common due to the very nature of the disease: "Though the person is there physically, a discontinuity exists between his identity before and after the onset of the illness" (Hinton and Levkoff 1999, p. 461). Reconstructing a new identity with new narratives can be difficult for the patient if ageing and disease come with serious cognitive disabilities, but it is possible for the caregiver, especially when the caregiver can benefit from the use of social and other resources that can help in the task.

The burden of care could be somewhat alleviated, at least psychologically, in the cultural tradition of some Asian countries (e.g. China), where AD is just interpreted as a natural development of old age. By re-interpreting the disease, the Chinese attempt to minimise the added burden for the patient, of the disclosure that what they have is a serious health problem:

> while aware that their patients have Alzheimer's disease or another type of dementia, they see little benefit in 'burdening' elders and their families with a potentially stigmatizing diagnostic label that carries with it little hope of cure. Instead, they use explanations they view as more culturally accepted and reassuring (Hinton and Levkoff 1999, p. 463).

We conclude this section by mentioning the use of narratives in therapies for the elders. Narrative therapies can be especially helpful if older people are disconnected from their social surroundings, and they can be delivered either in institutionalised settings or in a home environment. Given that narratives are complex pieces of text that require not just memory but also ability to construct a coherent discourse, they can be specifically used to improve the cognitive capacity of elders (Tun 1989).

In the words of Kropf and Tandy (1998, pp. 6–7):

> Narrative therapy is an approach that focuses on client stories with the goal of challenging existing meaning systems and creating more functional ones. The therapeutic process begins with deconstruction of the current meaning system. In the initial work with a client, the problem or symptom is externalized [which] makes the situation more manageable ... The second phase of narrative therapy involves reconstruction of the client's stories. By careful listening and understanding, the practitioner is a co-collaborator in the transformation of the client's narrative.

Kropf and Tandy (1998) suggested using narratives to achieve *deconstruction* of text by challenging the meaning systems to which the elder subscribes; and creating an alternative meaning system through *reconstruction*. In this way, negative meanings can be transformed into positive ones, thus improving self-identity and the self-value of the elders and increasing their optimism in life.

The collaboration between elder and therapist in the construction of the elder's narrative mentioned by Kropf and Tandy, however, must be taken with some caution, as the therapist should only elicit action from the elder towards the self-construction of a new and personal meaning system, not guide the process in any specific direction of identity and meaning, be such direction political, religious, or other.

Young (2010) points to the use of narrative therapy to reconcile the narratives of their own life with the actual life experiences of elders. In this context the narrative simply externalises and gives meaning to life events, even if those life events were not all positive. External societal pressures that encase the life narrative within expected cultural moulds, such as a "happy life", a "successful life", are resisted in order to highlight those elements that can make the lived experience more fulfilling and, in any event, more truthful and coherent to the elder. For elders experiencing loss of memory, therapy can stress those elements of the past life that are still retained, reinforcing them with visual and audio stimuli.

Rood (2009) used narratives work to interact with five women and one man, 66–93 years old, suffering from various stages of dementia in a community care home. The narrative work was accomplished through several approaches: remembering conversations, use of letters, documents, and certificates; use of specific knowledge and the sharing of such knowledge between participants, linking people's lives to others' lives. Central to this exercise was producing a life review and taking steps towards planning for some future actions (e.g. advance care directives). The narrative work was able to give a clearer voice to the elders about themselves and their life experiences.

Elders diagnosed with bipolar disorder may also benefit from narrative therapy which can be used with the objective of increasing their "sense of power and importance, reconciling past negative feelings, and rediscovering interest in past activities or hobbies" (Burgin and Gibbons 2016). More generally, narrative therapy can be useful to old people who are dealing with some mental health issues (including those derived from substance abuse), especially when the therapy is delivered in group settings, as shown by Poole et al. (2009; see also Gardner and Poole 2009) in their work carried out in Canada.

Narrative therapy has been also used therapeutically in elders in conjunction with Robert Butler's (1963) life review (Caldwell 2005); and in a study carried out in South Korea, Ahn (2017) used narrative therapy to successfully decrease fear of death and increase ego-integrity in a group of 29 elders (75.8% women). Ego integrity can be strengthened through a positive evaluation of the individual's life in narrative therapy, which helps decrease death anxiety and it fosters the development of wisdom (Goodcase and Love 2017; Nozari et al. 2019).

In sum, the narratives of the elders can be a way of "storying" their experience in life, and they are an important conduit for their legacy. From the perspective of the individual, narratives—whether they are "uplifting" or focused on "suffering"—contribute to the development of self-identity, a better integrated self, and to growing resilience. Better self-identity and meaning making can help experience a more fulfilling life in old age. This can be helped by day-to-day interactions within an established community and, in some cases, by more targeted narrative therapeutical interventions. Socially, narratives may contribute to a better integration of the elders into their social milieu, including an improved inter-generational integration through generativity.

7.6 Elderly and the Media

Elderly people tend to spend a considerable amount of their time using various media: newspapers, television, radio and, more recently, the Internet (Nussbaum et al. 2000). Through the use of the media, older people obtain entertainment, information, and also topics that motivate discussions with friends and family. In addition, the mainstream media fulfil an important role in shaping the views that people in general have of the elders, whether such views are positive or negative (e.g. Bishop and Krause 1984; Gibb and Holroyd 1996; Blakeborough 2008; Wilińska and Cedersund 2010). However, because the media are in many instances private businesses that aim at maximising their profits, they also tend to reinforce already prevailing prejudices in society that can attract the interest of more users/customers. For instance, the media tend to portray older people of higher social status more favourably than older people of lower social status. For as long as the media are diverse (e.g. private vs public), there is the possibility that they may also provide a diversity of views on the elders; but when the media are a reflection of an ideological monopoly, then they could be operating "in ways that promote apathy, cynicism, and quiescence, rather than active citizenship and participation" (Gamson et al. 1992, p. 373).

Although old people make use of a diversity of media, according to Nussbaum et al. (2000) they tend to decrease the amount of time spent reading, listening to radio, and going to the movies as they continue to age. Participation in various activities, however, may change according to wealth: more expensive activities (e.g. going to the theatre) can more likely be maintained over time by wealthier elderly; and health, with healthier elders being more likely to remain active and engaged.

Watching TV is usually regarded as a "passive" activity that in some elder care institutions is discouraged. However, some studies (e.g. Östlund 2010) have indicated that TV viewing can have some positive effects in terms of access to information, opportunities for reflection, and as a contribution to social integration. Older people also tend to prefer TV programmes that include older characters (Nussbaum et al. 2000), being more supportive of a non-stereotypical and more realistic treatment of the elders (Healey and Ross 2002).

The Internet is also becoming an important medium for elderly people to access information and to communicate. Gatto and Tak (2008) list the following benefits that older people can obtain from the use of the Internet: sense of connectedness, satisfaction, utility, and positive learning experiences. But the authors also list some barriers to the use of the Internet such as frustration, physical and mental limitations, and mistrust. In spite of the potential benefits, older generations at the beginning of the twenty-first century, and even in technologically advanced societies, may find the use of information and communication technologies (ICTs) daunting, which has the consequence of increasing their social isolation (Lim and Tan 2003). This situation is very likely to change in the future, as the currently young and middle-age generations, who are regular users of ICTs, become old. Costs are also an important factor limiting old people's access to online services (Hill et al. 2008), but here the provision of free public Internet access—in local libraries and community centres, for instance—could help solve the problem.

7.7 Translating and Interpreting for Older Immigrants

In the modern world, that is characterised by intense movements of people across national borders, cultural and linguistic diversity has been increasing especially in those countries that are recipients of large numbers of migrants. Australia, for instance, is currently highly linguistically diverse and multicultural. With an ageing, multilingual population, such countries are confronted with the need to provide effective translation and interpreting services to help old members of immigrant communities (e.g. Okafor 2009; Hurley et al. 2013). As remarked by Radermacher and Feldman (2017) health and well-being in older immigrants are influenced, among other factors, by their communicative abilities, their proficiency in the official language of the new country of residence (e.g. English) and access to interpreting services.

Translation and interpreting involve rendering the meaning of a text that is available in a *source language* into a different, *target language*.

More often than not older immigrants, especially if they emigrated later in their life and had only limited contact with native speakers in the new country, may have only a basic knowledge of the language of immigration or L2. Such older people are referred to as LALP (limited autochthonous language proficiency) immigrants and they are the ones who will need most help to communicate with others.

In multilingual communities, translation aid is to be found first and above all at home and in the neighbourhood, where older immigrants can obtain help from their own children, grandchildren, or from neighbours (Flynn and van Doorslaer 2016). Children are often involved in translating or *language brokering* for members of the family who are not proficient in the language of immigration (De Fina 2012; Willoughby et al. 2018).

Technically we should distinguish between *translating*, of a usually written text, and *interpreting* which is the translation of the discourse of a person carried out as she or he is speaking. Translating is commonly used in research and in information leaflets for immigrants, for instance, whereas interpreting is more commonly used in face-to-face contacts in various settings (e.g. medical, government offices, and so on).

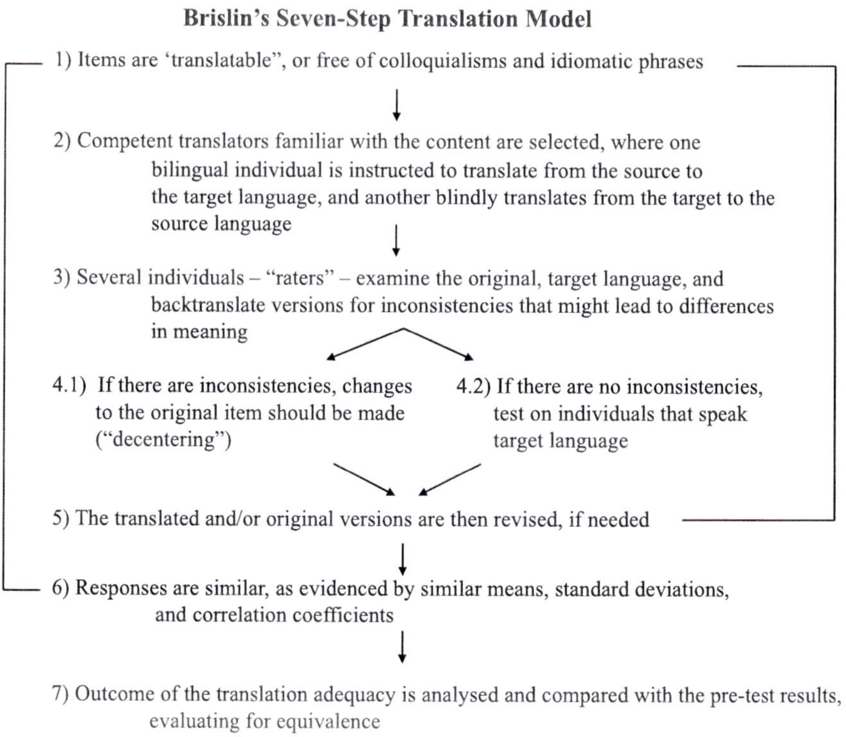

Fig. 7.6 Brislin's seven-step translation procedure. This is an adapted version of a flow chart originally produced by Lopez et al. (2008) based on Brislin's work and that we first published in Cordella and Poiani (2014)

The translation process aims at making the translated text not only intelligible and meaningful to the elder, but to also retaining the crucial information contained in the original text (Choi et al. 2012). In order to achieve maximum accuracy a *Seven-Step Translation Model* originally devised by Richard W. Brislin (1986) is usually recommended (see Fig. 7.6; see also Cordella and Poiani 2014).

Brislin's procedure starts by identifying those parts of the discourse that are actually "translatable", that is, that can be rendered meaningfully in a target language. The task of translation is then assigned to two selected competent translators who are not only fluent in both source and target languages but are also knowledgeable of the contents of the text (e.g. procedures to apply for a pension). One translator renders the text from the source language to the target language, whilst the other independently translates the text back to its source language. Both original translation and back-translation are then rated in terms of consistency by several independent individuals. This process, known as *decentering*, may lead to adjustments of the original translation when inconsistencies are found. Once an acceptable degree of consistency is achieved, the translation is then tested using native speakers of the target language. Their feedback will suggest whether further adjustments to the text are needed in order to convey the correct meaning. This will lead to the production of a final translation that can be further evaluated by showing it to bilingual subjects (see Brislin 1970, 1980). Brislin's method is indeed capable of delivering very good quality translations, but it is certainly laborious and time consuming and therefore more useful for the translation of documents intended for wide circulation. When the translation is more limited in purpose and it is subject to time and/or budgetary constraints, the steps can be simplified, but we strongly recommend retaining the concept of two independent translators reaching an agreement on the final text.

A successful translation necessarily requires not just proficiency in both the source and target languages but, importantly, a thorough understanding of culture (e.g. Daniel et al. 2011; Choi et al. 2012; Guan et al. 2016); and when it comes to translating the speech of older people, the understanding of the specific use of language by older generations will be also needed.

We mentioned above how bilingual members of immigrant families may act as language brokers for their older relatives. Among the many areas of concern in the public sphere with regard to *language brokering* (Tse 1995, also known as *natural translation*, Harris and Sherwood 1978) by family members, is the participation of children (variably referred to as *language brokers*, *natural translators*, *immigrant children mediators, informal interpreters*, *family interpreters*, *para-phrasers*, Reynolds and Orellana 2009) who, on occasions, may be the only members of the family with sufficient bilingual knowledge to act as language brokers (e.g. Castañeda 2007; Angelelli 2010; Esquivel 2012; Alvarez 2014). In the USA, young children may not be allowed to act as language brokers in some institutional settings (e.g. translating legal and medical information, Morales and Hanson 2005), as they may translate inaccurately and the experience may be traumatising to them (e.g. the old relative being diagnosed with a terminal illness), or it may affect child–relative relationships. The burden of translating and interpreting for older family members may also distract children from other important activities that are expected at their age, such as school duties or deserved leisure time spent with friends. However, there are also reports suggesting that some children may find language brokering a positive experience and they may acquire important cognitive skills compared with their monolingual peers (Morales and Hanson 2005 and references therein). In addition, they may also develop a better *transcultural perspective-taking* or "the ability to understand cultural values and practices different from one's own" (Guan et al. 2014, p. 335).

Figure 7.7 provides an example of the kind of translating activities carried out by immigrant children on behalf of their adult relatives. These include translating specific words, letters, and various types of documents among others. In general, female members of the family tend to act more often as language brokers for older relatives than males (e.g. Guan et al. 2014).

Beyond the family, older immigrants may also require translation and interpreting services from social workers (Engstrom et al. 2009) or specialised interpreters (Hadziabdic et al. 2015).

Medical consultations are especially sensitive to the provision of accurate and professional interpreting services due to the potentially negative consequences of a

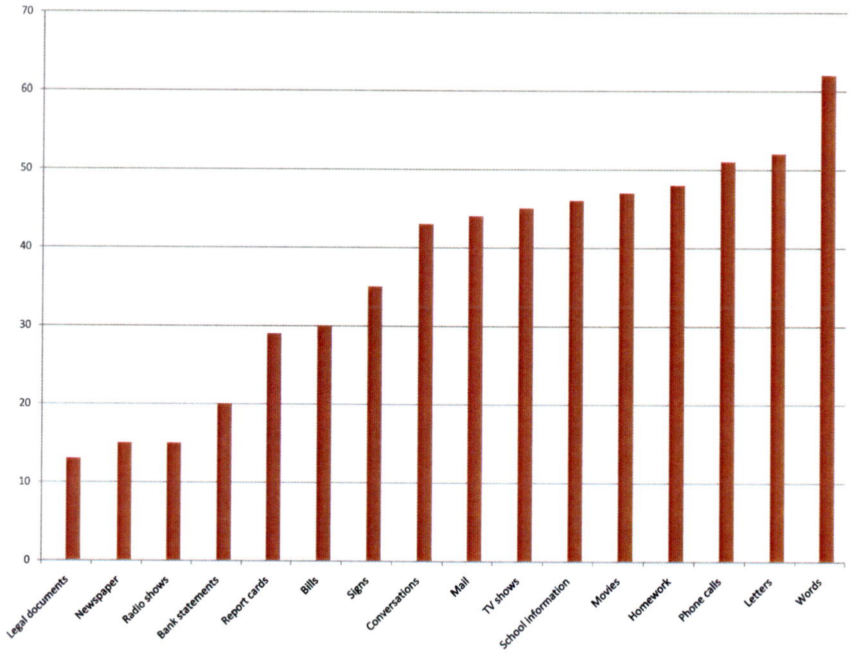

Fig. 7.7 Percentage of 236 Spanish-speaking, fifth- and sixth-graders who were involved in various experiences of translation for family members (redrawn from Orellana et al. 2003)

mistake. Hadziabdic et al. (2015) studied medical interpreting for elderly immigrants in Sweden in a healthcare setting where the objective is to provide *transcultural care*:

> Transcultural care emphasizes the need to provide care based on an individual's or group's cultural beliefs, practices and values in order to avoid ethnocentric assessments. It proceeds from an understanding that sees the main aim of caring as promoting or regaining health, and underlines that communication is central for caring... The transcultural approach focuses not only on the patients' cultural views but also stresses that we need to include the cultural understandings built into our medical institutions as well as the cultural background of the individual healthcare staff in order to provide culturally congruent healthcare (p. 2).

In spite of Sweden being a country well organised in the provision of services to their population, in the particular case of healthcare for LALP elderly not all runs necessarily smoothly and further improvement is required: "guidelines were lacking in the organizational content concerning the use of interpreters, the responsibility for arranging an interpreter service and the procedure for booking an interpreter" (Hadziabdic et al. 2015, p. 9). Ideally, medical interpreters should be also knowledgeable on the various technical aspects of health care, which usually means that bilingual doctors tend to be asked to help with immigrant patients. But this may add an additional burden on those professionals, contributing to their burnout.

The burden of care is not just felt by medical doctors asked to act as interpreters, both social workers acting as interpreters and even professionally trained interpreters may also feel over-burdened and fatigued by their duties depending on the intensity of their interpreting activities carried out during the day.

Medical interpreters are also confronted with the need to not only translate some traditional medical beliefs of the patients to the doctor, but to also translate back to the patient the reply of the doctor, which may be sometimes dismissive of the patients' beliefs. Here the interpreter may play an important role in keeping the medical consultation running smoothly and constructively, preventing a potential breakdown of communication between doctor and patient. The way of achieving this is not by "mis-translating" of course, but by prefacing the translation (both forward and backward) with an explanation in the language of the recipient (the patient or the doctor). Putting the text in cultural context may help the doctor adapt his/her approach and the patient understands what the doctor is trying to do.

Good communication between doctor and patient that is facilitated by a trusted interpreter, allows the patient to open up and perhaps disclose traditional medicine practices that she or he does use at home (e.g. Chinese elders may use coining (*cao gio*), cupping (*giac hoi*), herbal medicine, and other practices, Ngo-Metzger et al. 2003). This disclosure can help the doctor give best advice both in the case when traditional medical practices may interfere with Western medical treatments and in the case when there is no interference or there may even be complementarity.

Searight and Gafford (2005, p. 517) have provided some useful guidelines for medical interviews that include the participation of an interpreter, that aim at maximising the effectiveness of the encounter and the benefits for the patient. We list Searight and Gafford's suggestions here with only minor modifications:

1. Ideally, the interpreter should not be a family member.
2. Interpreters should be trained to respect patient confidentiality.
3. Physicians should orient the interpreter to the process of the medical encounter.
4. Physicians should request a literal, word-for-word translation.
5. Physicians should request the interpreter to ask the physician to restate or clarify unfamiliar terms.
6. After making a complete statement, the physician should pause for translation.
7. The physician should look directly at the patient, rather than at the interpreter, when either the physician or patient is speaking.
8. The physician should speak in the second person. For example, he or she might ask, "Where is your pain?" rather than "Can you ask him where he hurts?""

In addition, professional translators and interpreters are also bound by a code of ethics that in Australia includes to:

- maintain professional detachment, impartiality, objectivity, and confidentiality;
- strive for excellence through continuous regular professional development;
- decline work beyond their competence;
- promote working conditions, relationships, and an understanding of roles that facilitate collaboration and quality service delivery;
- adhere to dispute resolution procedures.

(Australian Institute of Interpreters and Translators, AUSIT; https://ausit.org/AUSIT/Documents/Code_Of_Ethics_Full.pdf)

We have ended this section with a brief analysis of interpreting in medical encounters. However, doctor–patient communication is important for old people whether the doctor and the patient speak the same language or not. Therefore, we conclude this chapter with a brief analysis of the specific challenges of doctor-old patient communication.

7.8 Doctor–Patient Communication in the Elderly

A visit to the doctor is a common experience for just about everybody, young and old, but inevitably older people are more likely to require some medical assistance at some point in time. In all cases, however, the medical consultation has greater chances to be satisfactory and effective the better the level of communication is between doctor and patient (e.g. Liang et al. 2002; Belcher et al. 2006, see also Chap. 7 of Cordella and Poiani 2014), and the more proactive the patient is (Kahana and Kahana 2003).

Nevertheless, interactions between medical doctors and old patients display some specific characteristics that are not always positive. For instance, doctors may spend less time with older patients, they may use a more condescending and less tolerant attitude, and an excessive technical language; they also tend to control the interview, ask older adults only few questions, and they may tend to use close-ended questions (e.g. those just requiring a yes/no answer). In addition, doctors may not include older patients in joint decision-making, and are less likely to recommend preventive strategies such as exercise (Savundranayagam and Ryan 2008). Some negative attitudes toward the elderly may also make communication more difficult (Adelman et al. 1991). These aspects of communication may become even more extreme in a hospital situation.

Some unsatisfactory outcomes of doctor–patient communication may be also an effect of ageing in the patient: sensory deficits, cognitive impairments, and functional limitations; or they may result from older patients' behaviour, such as not raising their most important concerns as soon as possible in the visit, meaning that if time is limited, the patient may miss the opportunity to talk about important issues (Nussbaum et al. 2000). Forgetfulness in older patients may obviously aggravate this situation, and forgetfulness could also affect compliance with treatment, such as not taking medications at the appropriate time in the day.

Adams et al. (2002) list three major challenges that medical practitioners mention when asked about their experiences with older patients:

- The complexity of older people's medical issues and the chronicity of illness; in particular, they refer to patients' vulnerability to adverse events.
- Personal and interpersonal constraints, including time pressure, communication problems, and ethical dilemmas.

- The administrative burden involved in dealing with the case, including more telephone calls and the requirement for specific documentation.

Older patients, on the other hand, may complain about ageist attitudes from the part of the doctor. Adelman et al. (1991) list these aspects of ageism that could be manifested in medical encounters:

1. Discounting or trivialising the medical problems of elderly patients, inappropriately attributing such problems to the "natural processes" of ageing.
2. Being less likely to suggest preventive regimens or treating medical or psychiatric problems aggressively when the patient is older.
3. Using derogatory terms when referring to older patients, for example, "crock" and "geezer".
4. Spending less time with older patients.
5. Considering elderly patients more "difficult" to deal with.

In spite of the above potential difficulties, Osborn et al. (2014) have recently reported a survey of 15,617 sixty-five or older individuals from 11 industrialised Western countries, indicating generally good levels of doctor–patient relationships, but some failures of communication were also noted in the areas of sharing information, and talking with patients about how they can promote good health or end-of-life care options.

An added complication in doctor—older patient communication appears when the medical visit is triadic. Most triadic visits involve the participation of the doctor, the patient, and a companion, usually a relative. In the field of geriatrics such triadic medical encounters are rather common, given the frailty of many older patients (e.g. Ishikawa et al. 2005).

We analysed triadic medical consultations in great detail in Chap. 7 of our book *Behavioural oncology* (Cordella and Poiani 2014), here it will suffice to say that medical consultations that include a companion of the patient have the potential to either facilitate or further complicate doctor–older patient communication. Communication is facilitated when the companion acts as mediator, smoothing the interaction between doctor and patient; but the companion may further complicate the situation when she or he is pushing her or his own agenda, sometimes against the will of both doctor and patient, or may be engaged in the formation of coalitions (with the patient against the doctor, or with the doctor against the patient). In addition, given that the consultation is limited in time, unless the companion makes a constructive contribution, she or he may detract time from the doctor or the patient to express their own opinions and concerns (Greene et al. 1994). In this case, however, it is up to the doctor to actively prevent the companion from becoming too intrusive and stealing too much time from the patient (Ekberg et al. 2015). It is usually the case that the doctor approximately maintains his/her level of participation, whereas the patient and companion share the rest of the time (e.g. Savundranayagam and Ryan 2008).

The potential roles of the companion in the consultation have been summarised by Adelman et al. (1987; see also Adelman 1991) into three major types: The *advocate* (a promoter and supporter of the patient's agenda), the *passive participant* (minimally involved in the consultation), and the *antagonist* (a promoter of a personal agenda against the objectives of patient and/or doctor). An ambiguous role of the companion, that can be seen as advocate in one perspective or as antagonist in another is that of becoming a *shield from bad news* (e.g. Fried et al. 2003). If the patient does not want to be fully informed about prognosis, for instance, then a companion who does not ask for prognosis information will be acting as an advocate. On the other hand, if the patient wants to be informed but the companion does not convey the information, then the behaviour could be better described as antagonist. Cordella (2011) identified the following specific aspects of the *advocate* role of the companion, after analysing triadic medical consultations involving cancer patients in Santiago, Chile:

- *Secretary*: making appointments, taking care of paperwork.
- *Carer*: checks medical details such as prescriptions.
- *Financial assistant*: mentions financial matters in the consultation.
- *Health advisor*: observes health changes in the patient and reports to the doctor.
- *Social communicator*: shares with the doctor aspects of the patient's social life.
- *Reporter*: directly answers questions from the doctor that refer to the patient.
- *Partnership builder*: reinforces the alliance with the patient during the consultation.

The most common roles recorded by Cordella were carer, health advisor, social communicator, and partnership builder.

In a study carried out in Japan, Ishikawa et al. (2005) reported that patients in triadic consultations were older, less educated, and tended to be in poorer health condition. Companions mainly performed the role of asking questions to the doctor, but they also provided information about the patient and acted as patient's advocates. In this study antagonism was uncommon. Indeed, companions may assist older people when their voices are not heard in medical visits. In a study conducted in Melbourne, Australia, involving an L1-Italian patient, a companion and a doctor, both companion—a family member—and the doctor made a concerted effort to speak in Italian throughout the consultation instead of switching to English, which allowed a smoother integration of the patient into the conversation (Willoughby et al. 2018).

How can communication between doctor and older patients be improved?

Programmes have been proposed to improve doctor–older patient communication that have a diversity of aims. Following Bethea and Balazs (1997, pp. 134–135; reported here verbatim, with only minor modifications) some of the main objectives of such programmes include:

1. Develop an increased awareness of ageist stereotyping in the doctor–older patient interaction.
2. View each interaction as a negotiation to reduce miscommunication.
3. Develop a unique relationship with each patient.
4. Promote the use of repetition and sensitive interrogation to help older adult patients understand technical jargon, diagnoses, and treatment options.

5. Utilise metaphors and examples salient to individual patients' interests to explain medical terms and procedures.
6. Enhance the relational interaction on an affective level so patients will be satisfied and remain loyal.
7. Develop a holistic understanding of each patient by listening to patient narratives/historical life reviews.
8. Understand patients in relation to their cohort membership and its effects on their expectations of the doctor–patient interaction.

Some interventions that are inspired by the *Communication Enhancement* model are especially directed to elders who are members of ethnic minorities (Ryan et al. 1995). In this case and in addition to Bethea and Balazs' (1997) aims listed above, we agree with Ryan et al. (1995, p. 75) to also add the aims of "guarding against ethnocentric behavior, developing sensitivity to cultural issues vis-a-vis health, avoiding ethnic myths and stereotypes, practicing empathy, and being aware of individual differences within groups", with the objective of promoting a respectful relationship between healthcare providers and elders who have a different ethnocultural background. The same approach is valid for the relationship between healthcare providers and elders belonging to any other kind of minority group (e.g. gender minorities).

Robinson et al. (2006) have offered some tips to the doctors to achieve those objectives of a better doctor–older patient communication:

1. Allow extra time for older patients.
2. Avoid distractions.
3. Sit face to face.
4. Maintain eye contact.
5. Listen.
6. Speak slowly, clearly, and loudly.
7. Use short, simple words and sentences.
8. Stick to one topic at a time.
9. Simplify and write down your instructions.
10. Use charts, models, and pictures.
11. Frequently summarise the most important points.
12. Give patients an opportunity to ask questions and express themselves.

    Besides Robinson et al. (2006) recommendations we should also add:

13. Adopt 6, 7, 9, 11 above following specific patient's needs, not as an automatic response to the age of the patient, as old people in good mental state may potentially find such communicative styles patronising and offensive.

Old patients who are at home may also receive care from a distance through modern communication technologies in the form of "televisits" (e.g. Mickus and Luz 2002, see also the concepts of "telehealth" and "telecare", Magnusson et al. 2004).

Recently, there have been also initiatives to provide some kind of "training" to the patients to improve their communicative abilities with medical doctors (e.g. Cegala et al. 2001).

7.9 Conclusions

Elders engage in communication with both same-age and other-age people, including young members of the community. Such communication is dynamic, displaying diverse patterns that have been summarised in various theoretical models. Conflicts, uncertainties, negative stereotypes, ageism and discrimination, patronising, deficits (physical, mental), distance in social relationship with the interlocutor, and other factors can contribute to poor communication, both intra- and inter-generational. On the other hand, the better predisposition there is to accommodate to each other, the more likely it is that communication will run smoothly and more productively. Accommodation is especially required for better communication when the elder engages in painful self-disclosure. Mutual accommodation in discourse can be achieved through a process of understanding of generational characteristics and attuning, and here women tend to attune better to a conversation with older people than men. Underaccommodation and overaccommodation are also possible and they may interfere with good communication with old people. Some intervention programmes have been introduced to improve communication with older people. Surprisingly, at present, it is in Asian countries, where traditional filial piety is expected to be culturally strong, that inter-generational communication seems to be more difficult. This is presumably the result of recent trends towards cultural modernisation in Asia. Communication with the elders unfolds through narratives, it is through narratives that we reminisce and tell the story of our life, which gives meaning and unity to the self in spite of the changes we have experienced over the years. More life meaning contributes to experiencing a more fulfilling ageing, even as we go through periods of suffering, the latter can be expressed in the elders' narratives either in terms of despair or resilience. Narratives can extend the possibilities of the self; they allow a synthesis to be produced, a coherent life story to emerge. Through narratives old people can modulate the fluidity and fixity of the self and find inner strength, but also transmit their personal and cultural knowledge to the younger generations (generativity). Some specific contexts require special consideration of elders' communicative abilities, such as translations and interpreting services and support in doctor–patient communication.

References

Abrams, J., & Giles, H. (1999). Intergenerational contact as intergroup communication. *Child and Youth Services, 20*, 203–217.

Adams, W. L., McIlvain, H. E., Lacy, N. L., Magsi, H., Crabtree, B. F., Yenny, S. K., et al. (2002). Primary care for elderly people: Why do doctors find it so hard? *The Gerontologist, 42*, 835–842.

Adelman, R. D., Greene, M. G., & Charon, R. (1987). The physician–elderly patient– companion triad in the medical encounter: The development of a conceptual framework and research agenda. *Gerontologist, 27*, 729–734.

Adelman, R. D., Greene, M. G., & Charon, R. (1991). Issues in physician-elderly patient interaction. *Ageing and Society, 11*, 127–148.
Ahn, Y. (2017). The impact of narrative therapy on fear of death anxiety and ego-integrity in elderly. *Information, 20*, 2945–2954.
Allan, K., & Burridge, K. (2006). *Forbidden words: Taboo and the censoring of language*. Cambridge: Cambridge University Press.
Alvarez, S. (2014). Translanguaging *tareas*: Emergent bilingual youth as language brokers for homework in immigrant families. *Language Arts, 91*, 326–339.
Andrews, M. (2009). The narrative complexity of successful ageing. *International Journal of Sociology and Social Policy, 29*, 73–83.
Angelelli, C. V. (2010). A glimpse into the socialization of bilingual youngsters as interpreters: The case of Latino bilinguals brokering communication for their families and immediate communities. *Monografías de Traducción e Interpretación, 2*, 81–96.
Angersbach, H. L., & Jones-Forster, S. (1999). Intergenerational interactions: A descriptive analysis of elder-child interactions in a campus-based child care center. *Child and Youth Services, 20*, 116–128.
Armstrong, L., & McKechnie, K. (2003). Intergenerational communication: Fundamental but under-exploited theory for speech and language therapy with older people. *International Journal of Language and Communication Disorders, 38*, 13–29.
Atchley, R. C. (1980). *The social forces in later life*. Belmont, CA: Wadsworth.
Atleo, M. R. (2009). Understanding aboriginal learning ideology through storywork with elders. *The Alberta Journal of Educational Research, 55*, 453–467.
Bascom, W. (1965). The forms of folklore: Prose narratives. *The Journal of American Folklore, 78*, 3–20.
Bashore, T. R., Jr., Wylie, S. A., Ridderinkhof, K. R., & Martinerie, J. M. (2014). Response- specific slowing in older age revealed through differential stimulus and response effects on P300 latency and reaction time. *Neuropsychology, Development, and Cognition, 21*, 633–673.
Bayles, K. A., & Kaszniak, A. W. (1987). *Communication and cognition in normal aging and dementia*. London: Taylor and Francis.
Becker, G. (2003). Meanings of place and displacement in three groups of older immigrants. *Journal of Aging Studies, 17*, 129–149.
Becker, G., Beyene, Y., Newsom, E., & Mayen, N. (2003). Creating continuity through mutual assistance: Intergenerational reciprocity in four ethnic groups. *Journal of Gerontology, 58B*, S151–S159.
Belcher, V. N., Fried, T. R., Agostini, J. V., & Tinetti, M. E. (2006). Views of older adults on patient participation in medication-related decision making. *Journal of General and Internal Medicine, 21*, 298–303.
Bengtson, V. L., & Kuypers, J. A. (1971). Generational differences and the developmental stake. *Aging and Human Development, 2*, 249–260.
Berger, C. R., & Calabrese, R. J. (1975). Some explorations in initial interaction and beyond: Toward a developmental theory of interpersonal communication. *Human Communication Research, 1*, 99–112.
Bethea, L. S., & Balazs, A. L. (1997). Improving intergenerational health care communication. *Journal of Health Communication, 2*, 129–137.
Bettini, L. M., & Norton, M. L. (1991). The pragmatics of intergenerational friendships. *Communication Reports, 4*, 64–72.
Biggs, S. (1989). Professional resistances to work with older people. *Ageing and Society, 9*, 43–60.
Biggs, S. (2004). Age, gender, narratives, and masquerades. *Journal of Aging Studies, 18*, 45–58.
Biggs, S., Haapala, I., & Lowenstein, A. (2011). Exploring generational intelligence as a model for examining the process of intergenerational relationships. *Ageing and Society, 31*, 1107–1124.
Biggs, S., & Lowenstein, A. (2011). *Generational intelligence: A critical approach to age relations*. London: Routledge.
Bishop, J. M., & Krause, D. R. (1984). Depictions of aging and old age on Saturday morning television. *The Gerontologist, 24*, 91–94.
Black, H. K. (2009). Pictures of suffering in elders' narratives. *Journal of Aging Studies, 23*, 82–89.

Black, H. K., & Rubinstein, R. L. (2004). Themes of suffering in later life. *Journal of Gerontology, 59B*, S17–S24.

Blakeborough, D. (2008). "Old people are useless": Representations of aging on The Simpsons. *Canadian Journal on Aging, 27*, 57–67.

Brabazon, K. (1999). Student improvement in the Intergenerational Work/Study Program. *Child and Youth Services, 20*, 50–61.

Brislin, R. (1970). Back-translation for cross-cultural research. *Journal of Cross-Cultural Psychology, 1*, 186–216.

Brislin, R. W. (1980). Translation and content analysis of oral and written materials. In H. C. Triandis & J. W. Berry (Eds.), *Handbook of cross-cultural psychology* (Vol. 2, pp. 389–444). Boston, MA: Allyn & Bacon.

Brislin, R. W. (1986). The wording and translation of research instruments. In W. J. Lonner (Ed.), *Field methods in cross-cultural research*. Beverly Hills, CA: Sage.

Brockmeier, J. (2009). Reaching for meaning: Human agency and the narrative imagination. *Theory and Psychology, 19*, 213–233.

Brown, P., & Levinson, S. (1978). *Politeness: Some universals in language usage*. Cambridge: Cambridge University Press.

Bruner, J. (1999). Narratives of aging. *Journal of Aging Studies, 13*, 7–9.

Bultena, J. L., & Powers, E. A. (1976). Denial of ageing: Age identification and reference group orientation. *Journal of Gerontology, 33*, 784–754.

Burgin, E. C., & Gibbons, M. M. (2016). "More life, not less": Using narrative therapy with older adults with bipolar disorder. *Adultspan, 15*, 49–61.

Butler, R. N. (1963). The life review: An interpretation of reminiscence in the aged. *Psychiatry, 26*, 65–76.

Buys, L. R. (2001). Life in a retirement village: Implications for contact with community and village friends. *Gerontology, 47*, 55–59.

Cacchione, P. Z. (2014). Sensory impairment: A new research imperative. *Journal of Gerontological Nursing, 40*, 3–5.

Cai, D., Giles, H., & Noels, K. (1998). Elderly perceptions of communication with older and younger adults in China: Implications for mental health. *Journal of Applied Communication Research, 26*, 32–51.

Cai, D. A., & Fink, E. L. (2002). Conflict style differences between individualists and collectivists. *Communication Monographs, 69*, 67–86.

Caldwell, R. L. (2005). At the confluence of memory and meaning - Life review with older adults and families: Using narrative therapy and the expressive arts to re-member and re-author stories of resilience. *The Family Journal: Counseling and Therapy for Couples and Families., 13*, 172–175.

Cardona, B. (2008). "Healthy ageing" policies and anti-ageing ideologies and practices: On the exercise of responsibility. *Medicine, Health Care and Philosophy, 11*, 475–483.

Castañeda, A. I. (2007). Language and other lethal weapons: Cultural politics and the rites of children as translators of culture. In J. Flores & R. Rosaldo (Eds.), *A companion to Latina/o studies* (pp. 126–137). Malden, MA: Blackwell.

Cegala, D. J., Post, D. M., & McClure, L. (2001). The effects of patient communication skills training on the discourse of older patients during a primary care interview. *Journal of the American Geriatrics Society, 49*, 1505–1511.

Chapman, S. A. (2005). Theorizing about aging well: Constructing a narrative. *Canadian Journal on Aging, 24*, 9–18.

Charon, R. (2006). *Narrative medicine*. New York: Oxford University Press.

Chen, Y., & King, B. E. (2002). Intra- and intergenerational communication satisfaction as a function of an individual's age and age stereotypes. *International Journal of Behavioral Development, 26*, 562–570.

Choi, J., Kushner, K. E., Mill, J., & Lai, D. W. L. (2012). Understanding the language, the culture, and the experience: Translation in cross-cultural research. *International Journal of Qualitative Methods, 11*, 652–665.

Ciobanu, R. O., & Fokkema, T. (2017). The role of religion in protecting older Romanian migrants from loneliness. *Journal of Ethnic and Migration Studies, 43*, 199–217.

Clark, M. C., & Standard, P. L. (1997). The caregiving story: How the narrative approach informs caregiving burden. *Issues in Mental Health Nursing, 18*, 87–97.

Clyne, M. (1981). *Deutsch als Muttersprache in Australien*. Wiesbaden: Steiner.

Coats, H., Crist, J. D., Berger, A., Sternberg, E., & Rosenfeld, A. G. (2015). African American elders' serious illness experiences: Narratives of "God did," "God will," and "Life is better". *Qualitative Health Research, 27*(5), 634–648. https://doi.org/10.1177/1049732315620153

Consolvo, S., Roessler, P., Shelton, B. E., LaMarca, A., Schilit, B., & Bly, S. (2004). Technology for care networks of elders. *Pervasive Computing, 3*(2), 22–29.

Cook, P. (1999). Capacity-building partnerships between indigenous youth and elders. *Child and Youth Services, 20*, 189–202.

Cordella, M. (2011). Overcoming language and cultural differences in medical encounters: The use of a language and culture training course (LACT) in educating IMGs in Australia. In B. J. Hoekje & S. M. Tipton (Eds.), *English language and the medical profession: Instructing and assessing the communication skills of international physicians* (pp. 175–210). Bingley: Emerald Group.

Cordella, M. (2013). *Connecting classroom 2 community (CC2C) secondary project: Report to ISQ. School of Languages and Comparative Cultural Studies*. St. Lucia, QLD: The University of Queensland.

Cordella, M. (2016a). The immigrant potential: Multiculturalism, language skills and community resources. In M. Cordella & H. Huang (Eds.), *Rethinking second language learning: Using intergenerational community resources* (pp. 3–25). Bristol: Multilingual Matters.

Cordella, M. (2016b). Taking a stance: Older native speakers with young language learners. In M. Cordella & H. Huang (Eds.), *Rethinking second language learning: Using intergenerational community resources* (pp. 51–65). Bristol: Multilingual Matters.

Cordella, M., & Huang, H. (Eds.). (2016). *Rethinking second language learning: Using intergenerational community resources*. Bristol: Multilingual Matters.

Cordella, M., & Kokubu, C. (2016). Creating, maintaining and challenging rapport across languages and age groups. In M. Cordella & H. Huang (Eds.), *Rethinking second language learning: Using intergenerational community resources* (pp. 110–148). Bristol: Multilingual Matters.

Cordella, M., & Poiani, A. (2014). *Behavioural oncology: Psychological , communicative, and social dimensions*. New York: Springer.

Cordella, M., & Rojas-Lizana, S. (2019). Aging and migration: The value of familism for Spanish speakers. *Journal of Cross-Cultural Gerontology, 35*(2). https://doi.org/10.1007/s10823-019-09389-1

Coupland, N., Coupland, J., Giles, H., & Henwood, K. (1988). Accommodating the elderly: Invoking and extending a theory. *Language in Society, 17*, 1–41.

Coupland, N., Coupland, J., & Giles, H. (1991). *Language, society and the elderly*. Oxford: Blackwell.

Daniel, M., Miller, A., & Wilbur, J. E. (2011). Multiple instrument translation for use with South Asian Indian immigrants. *Research in Nursing and Health, 34*, 419–432.

De Bot, K., & Clyne, M. (1989). Language reversion revisited. *Studies in Second Language Acquisition, 11*, 167–177.

De Fina, A. (2012). Family interaction and engagement with the heritage language: A case study. *Multilingua - Journal of Cross-Cultural and Interlanguage Communication, 31*, 349–379.

De Fina, A., & Georgakopoulou, A. (2015). *The handbook of narrative analysis*. Hoboken, NJ: Wiley.

Dorfman, L. T., Murty, S. A., Evans, R. J., Ingram, J. G., & Power, J. R. (2004). History and identity in the narratives of rural elders. *Journal of Aging Studies, 18*, 187–203.

Downs, V. C., Javidi, M., & Nussbaum, J. F. (1988). A comparative analysis of the relationship between communication apprehension and loneliness for elderly nursing home and non-nursing home residents. *Western Journal of Speech Communication, 52*, 308–320.

Dragojevic, M., Gasiorek, J., & Giles, H. (2016). Communication accommodation theory. In C. R. Berger & M. E. Roloff (Eds.), *The international encyclopedia of interpersonal communication* (pp. 1–20). Hoboken, NJ: Wiley.

Edwards, H., & Giles, H. (1998). Prologue on two dimensions: The risk and management of intergenerational miscommunication. *Journal of Applied Communication Research, 26*, 1–12.

Ekberg, K., Meyer, C., Scarinci, N., Grenness, C., & Hickson, L. (2015). Family member involvement in audiology appointments with older people with hearing impairment. *International Journal of Audiology, 54*, 70–76.

Engstrom, D. W., Piedra, L. M., & Min, J. W. (2009). Bilingual social workers: Language and service complexities. *Administration in Social Work, 33*, 167–185.

Erikson, E. H. (1997). *The life cycle completed.* New York: Norton.

Erikson, E. H., & Erikson, J. (1997). *The life cycle completed: Extended version.* New York: Norton.

Esquivel, A. (2012). *Language brokering a dynamic phenomenon: A qualitative study examining the experiences of Latina/o language brokers.* Scripps Senior Theses. Paper 52. Downloaded on February 2, 2018, from http://scholarship.claremont.edu/scripps_theses/52

Faimberg, H. (2005). *The telescoping of generations: Listening to the narcissistic links between generations.* London: Routledge.

Feldman, S., Radermacher, H., & Browning, C. (2016). Contemporary intergenerational relationships. In M. Cordella & H. Huang (Eds.), *Rethinking second language learning: Using intergenerational community resources* (pp. 25–36). Bristol: Multilingual Matters.

Fisher, W. R. (1985). The narrative paradigm: An elaboration. *Communication Monographs, 52*, 347–367.

Fisher, W. R. (1987). *Human communication as narration: Toward a philosophy of reason, value, and action.* Columbia, SC: University of South Carolina Press.

Flynn, P., & van Doorslaer, L. (2016). City and migration: A crossroads for non- institutionalized translation. *European Journal of Applied Linguistics, 4*, 73–92.

Foner, A. (2000). Age integration or age conflict as society ages? *The Gerontologist, 40*, 272–276.

Foucault, M. (1970). *L'orde du discourse.* Paris: Gallimard.

Frankl, V. E. (1988). *The will to meaning: Foundations and applications of logotherapy.* New York: A Plume Book.

Fried, T. R., Bradley, E. H., & O'Leary, J. (2003). Prognosis communication in serious illness: Perceptions of older patients, caregivers, and clinicians. *Journal of the American Geriatrics Society, 51*, 1398–1403.

Gallois, C., Ogay, T., & Giles, H. (2005). Communication accommodation theory. In W. B. Gudyunst (Ed.), *Theorizing intercultural communication* (pp. 121–148). Thousand Oaks, CA: Sage.

Gamson, W. A., Croteau, D., Hoynes, W., & Sasson, T. (1992). Media images and the social construction of reality. *Annual Reviews of Sociology, 18*, 373–393.

Gardner, K. (1999). Narrating location: Space, age and gender among Bengali elders in East London. *Oral History, 27*, 65–74.

Gardner, P. J., & Poole, J. M. (2009). One story at a time: Narrative therapy, older adults, and addictions. *Journal of Applied Gerontology, 28*, 600–620.

Gatto, S. L., & Tak, S. H. (2008). Computer, internet, and e-mail use among older adults: Benefits and barriers. *Educational Gerontology, 34*, 800–811.

Gibb, H., & Holroyd, E. (1996). Images of old age in the Hong Kong print media. *Ageing and Society, 16*, 151–175.

Giles, H. (1973). Accent mobility: A model and some data. *Anthropological Linguistics, 15*, 87–105.

Giles, H. (Ed.). (1984). The dynamics of speech accommodation. *International Journal of the Sociology of Language, 46*, Mouton, Amsterdam.

Giles, H., Bonilla, D., & Speer, R. B. (2012). Acculturating intergroup vitalities, accommodation and contact. In J. Jackson (Ed.), *Routledge handbook of language and intercultural communication* (pp. 244–259). Abingdon: Routledge.

Giles, H., & Coupland, N. (1991). *Language: Contexts and consequences.* Pacific Grove, CA: Brooks/Cole.

Giles, H., Coupland, N., Coupland, J., Williams, A., & Nussbaum, J. (1992). Intergenerational talk and communication with older people. *The International Journal of Aging and Human Development, 34*, 271–297.

Giles, H., & Dorjee, T. (2004). Communicative climates and prospects in cross-cultural gerontology. *Journal of Cross-Cultural Gerontology, 19*, 261–274.

Giles, H., Fox, S., Harwood, J., & Williams, A. (1994). Talking age and aging talk: Communicating through the life-span. In M. L. Hummert, J. M. Wiemann, & J. F. Nussbaum (Eds.), *Interpersonal communication in older adulthood: Interdisciplinary theory and research* (pp. 130–161). Newbury Park, CA: Sage.

Giles, H., Fox, S., & Smith, E. (1993). Patronizing the elderly: Intergenerational evaluations. *Research on Language and Social Interaction, 26*, 129–149.

Giles, H., Liang, B., Noels, K. A., & McCann, R. M. (2001). Communication across and within generations: Taiwanese, Chinese-Americans, and Euro-Americans perceptions' of communication. *Journal of Asian Pacific Communication, 11*, 161–179.

Giles, H., Makoni, S., & Dailey, R. M. (2005). Intergenerational communication beliefs across the lifespan: Comparative data from Ghana and South Africa. *Journal of Cross Cultural Gerontology, 20*, 191–211.

Giles, H., Noels, K. A., Williams, A., Ota, H., Lim, T.-S., Ng, S. H., et al. (2003). Intergenerational communication across cultures: Young people's perceptions of conversations with family elders, non-family elders and same-age peers. *Journal of Cross-Cultural Gerontology, 18*, 1–32.

Giles, H., & Reid, S. A. (2005). Ageism across the lifespan: Towards a self-categorization model of ageing. *Journal of Social Issues, 61*, 389–404.

Giles, H., Willemyns, M., Gallois, C., & Anderson, M. C. (2007). Accommodating a new frontier: The context of law enforcement. In K. Fiedler (Ed.), *Social communication* (pp. 129–162). New York: Psychology Press.

Giles, H., Dailey, R. M., Sarkar, J. M., & Makoni, S. (2007a). Intergenerational communication beliefs across the lifespan: Comparative data from India. *Communication Reports, 20*, 75–89.

Giles, H., Ryan, E. B., & Anas, A. P. (2008). Perceptions of intergenerational communication by young, middle-aged, and older Canadians. *Canadian Journal of Behavioural Science, 40*, 21–30.

Gilleard, C. (2007). Old age in ancient Greece: Narratives of desire, narratives of disgust. *Journal of Aging Studies, 21*, 81–92.

Goffman, E. (1959). *Presentation of self in everyday life*. Garden City, NY: Doubleday Anchor Books.

Goodcase, E. T., & Love, H. A. (2017). From despair to integrity: Using narrative therapy for older individuals in Erikson's last stage of identity development. *Clinical Social Work Journal, 45*, 354–363.

Goodwin, C. (2015). Narrative as talk-in-interaction. In A. De Fina & A. Georgakopoulou (Eds.), *The handbook of narrative analysis* (pp. 197–218). Hoboken, NJ: Wiley.

Gould, O. N., Saum, C., & Belter, J. (2002). Recall and subjective reactions to speaking styles: Does age matter? *Experimental Aging Research, 28*, 199–213.

Graham, C., & Davoren, T. (2015). *Sharing their stories: Narratives of young Métis parents and elders about parenting*. Prince George, BC: National Collaborating Centre for Aboriginal Health.

Greene, M. G., Majerovitz, S. D., Adelman, R. D., & Rizzo, C. (1994). The effects of the presence of a third person on the physician-older patient medical interview. *Journal of the American Geriatrics Society, 42*, 413–419.

Griff, M. D. (1999). Intergenerational play therapy. *Child and Youth Services, 20*, 63–76.

Grunig, J. E. (1978). Defining publics in public relations: The case of a suburban hospital. *Journalism Quarterly, 55*, 109–118.

Guan, S.-S. A., Greenfield, P. M., & Orellana, M. F. (2014). Translating into understanding: Language brokering and prosocial development in emerging adults from immigrant families. *Journal of Adolescent Research, 29*, 331–355.

Guan, S.-S. A., Nash, A., & Faulstich Orellana, M. (2016). Cultural and social processes of language brokering among Arab, Asian, and Latin immigrants. *Journal of Multilingual and Multicultural Development, 37*, 150–166.

Gullette, M. (2004). *Aged by culture*. Chicago: University of Chicago Press.

Habermas, J. G. (1989). *The theory of communicative action: Volume two. The critique of functionalist reason*. Cambridge: Polity Press.

Hadziabdic, E., Lundin, C., & Hjelm, K. (2015). Boundaries and conditions of interpretation in multilingual and multicultural elderly healthcare. *BMC Health Services Research, 15*, 458. https://doi.org/10.1186/s12913-015-1124-5

Hajek, C., & Giles, H. (2002). The old man out: An intergroup analysis of intergenerational communication among gay men. *Journal of Communication, 52*, 698–714.

Harley, D., & Fitzpatrick, G. (2009). YouTube and intergenerational communication: The case of Geriatric1927. *Universal Access in the Information Society, 8*, 5–20.

Harré, R. (1995). Agentic discourse. In R. Harré & P. Stearns (Eds.), *Discoursive psychology in practice* (pp. 120–136). London: Sage.

Harris, B., & Sherwood, B. (1978). Translating as an innate skill. In D. Gerver & H. W. Sinaiko (Eds.), *Language interpretation and communication* (pp. 155–170). New York: Plenum Press.

Harwood, J. (1998). Young adults' cognitive representations of intergenerational conversations. *Journal of Applied Communication Research, 26*, 13–31.

Harwood, J., Giles, H., Fox, S., Ryan, E. B., & Williams, A. (1993). Patronizing young and elderly adults: Response strategies in a community setting. *Journal of Applied Communication Research, 21*, 211–226.

Harwood, J., & Lin, M.-C. (2000). Affiliation, pride, exchange, and distance in grandparents' accounts of relationships with their college-aged grandchildren. *Journal of Communication, 50*, 31–47.

Harwood, J., McKee, J., & Lin, M.-C. (2000). Younger and older adults' schematic representations of intergenerational communication. *Communication Monographs, 67*, 20–41.

Hays, R. B. (1988). Friendship. In S. W. Duck (Ed.), *Handbook of personal relationships* (pp. 391–408). Chichester: Wiley.

Healey, T., & Ross, K. (2002). Growing old invisibly: Older viewers talk television. *Media, Culture and Society, 24*, 105–120.

Hill, R., Beynon-Davies, P., & Williams, M. D. (2008). Older people and internet engagement. Acknowledging social moderators of internet adoption, access and use. *Information Technology and People, 21*, 244–266.

Hinton, W. L., & Levkoff, S. (1999). Constructing Alzheimer's: Narratives of lost identities, confusion and loneliness in old age. *Culture, Medicine and Psychiatry, 23*, 453–475.

Holstein, M. B., & Minkler, M. (2003). Self, society, and the "new gerontology". *The Gerontologist, 43*, 787–796.

Huang, H. (2016). Gaining L2 self-confidence in conversations with native speakers. In M. Cordella & H. Huang (Eds.), *Rethinking second language learning: Using intergenerational community resources* (pp. 171–186). Bristol: Multilingual Matters.

Huang, H., & Cordella, M. (2016). Community resources on our doorstep: Language learning in action. In M. Cordella & H. Huang (Eds.), *Rethinking second language learning: Using intergenerational community resources* (pp. 37–48). Bristol: Multilingual Matters.

Huang, H., Cordella, M., Browning, C., & Baumgartner, R. (2016). An innovative model for second language learning and social inclusion. In M. Cordella & H. Huang (Eds.), *Rethinking second language learning: Using intergenerational community resources* (pp. 226–237). Bristol: Multilingual Matters.

Hubble, N., & Tew, P. (2013). *Ageing, narrative and identity: New qualitative social research*. Hampshire: Palgrave Macmillan.

Hughes, M. (2008). Imagined futures and communities: Older lesbian and gay people's narratives on health and aged care. *Journal of Gay and Lesbian Social Services, 20*, 167–186.

Hummert, M. L. (1994). Stereotypes of the elderly and patronizing speech. In M. L. Hummert, J. M. Wiemann, & J. F. Nussbaum (Eds.), *Interpersonal communication in old adulthood: Interdisciplinary research* (pp. 162–184). Thousand Oaks, CA: Sage.

Hummert, M. L., Garstka, T. A., Shaner, J. L., & Strahm, S. (1994a). Stereotypes of the elderly held by young, middle-aged and elder adults. *Journal of Gerontology, 49*, 240–249.

Hummert, M. L., & Ryan, E. B. (1996). Toward understanding variations in patronizing talk addressed to older adults: Psycholinguistic features of care and control. *International Journal of PsychoLinguistics, 12*, 149–169.

Hummert, M. L., Shaner, J. L., Garstka, T. A., & Henry, C. (1998). Communication with older adults: The influence of age stereotypes, context, and communicator age. *Human Communication Research, 25*, 124–151.

Hurley, C., Panagiotopoulos, G., Tsianikas, M., Newman, L., & Walker, R. (2013). Access and acceptability of community-based services for older Greek migrants in Australia: User and provider perspectives. *Health and Social Care in the Community, 21*, 140–149.

Ishikawa, H., Roter, D. L., Yamazaki, Y., & Takayama, T. (2005). Physician–elderly patient–companion communication and roles of companions in Japanese geriatric encounters. *Social Science and Medicine, 60*, 2307–2320.

Ivey, J. B. (2001). Somebody's grandma and grandpa: Children's responses to contacts with elders. *The American Journal of Maternal/Child Nursing, 26*, 23–27.

James, V. A. (2016). *The shaping influences of 'a capable person:' A narrative research of elders' stories of raising children to inform Aboriginal education in the Northwest Territories*. DEd Thesis, Faculty of Education, Simon Fraser University, Burnaby, BC.

Kahana, E., & Kahana, B. (2003). Patient proactivity enhancing doctor–patient–family communication in cancer prevention and care among the aged. *Patient Education and Counseling, 50*, 67–73.

Kemper, S., Finter-Urczyk, A., Ferrell, P., Harden, T., & Billington, C. (1998). Using elderspeak with older adults. *Discourse Processes, 25*, 55–73.

Kreuter, M. W., Green, M. C., Cappella, J. N., Slater, M. D., Wise, M. E., Storey, D., et al. (2007). Narrative communication in cancer prevention and control: A framework to guide research and application. *Narrative Cancer Communication, 33*, 221–235.

Kropf, N. P., & Tandy, C. (1998). Narrative therapy with older clients. *Clinical Gerontologist, 18*, 3–16.

Lakoff, R. (1973). The logic of politeness, or minding your P's and Q's. In C. Corum, T. C. Smith-Stark, & A. Weiser (Eds.), *Papers from the Ninth Regional Meeting of the Chicago Linguistics Society* (pp. 292–305). Chicago: Chicago Linguistics Society.

Lambert, B., & Cordella, M. (2016). The migration experience and the ethos of self. In M. Cordella & H. Huang (Eds.), *Rethinking second language learning: Using intergenerational community resources* (pp. 66–84). Bristol: Multilingual Matters.

Liang, W., Burnett, C. B., Rowland, J. H., Meropol, N. J., Eggert, L., Hwang, Y.-T., et al. (2002). Communication between physicians and older women with localized breast cancer: Implications for treatment and patient satisfaction. *Journal of Clinical Oncology, 20*, 1008–1016.

Lim, S. S., & Tan, Y. L. (2003). Old people and new media in wired societies: Exploring the socio-digital divide in Singapore. *Media Asia, 30*, 95–102.

Lin, M.-C., & Zhang, Y. B. (2008). Taiwanese older adults' perceptions of aging and communication with peers and young adults. *Journal of Asian Pacific Communication, 18*, 135–156.

Lin, M.-C., Zhang, Y. B., & Harwood, J. (2004). Taiwanese young adults' intergenerational communication schemas. *Journal of Cross-Cultural Gerontology, 19*, 321–342.

Liu, J. H., Ng, S. H., Weatherall, A., & Loong, C. (2000). Filial piety, acculturation, and intergenerational communication among New Zealand Chinese. *Basic and Applied Social Psychology, 22*, 213–223.

Lopez, G. I., Figueroa, M., Connor, S. E., & Maliski, S. L. (2008). Translation barriers in conducting qualitative research with Spanish speakers. *Qualitative Health Research, 18*, 1729–1737.

Loscocco, K. (2000). Age integration as a solution to work–family conflict. *The Gerontologist, 40*, 292–300.

Lundgren, A. S. (2010). "In the good old days": Insidious nostalgia and the constitution of old age identity. *Journal of Aging Studies, 24*, 248–256.

Macaulay, R. K. S. (1975). Linguistic insecurity. In J. D. McClure (Ed.), *The Scots language in education, ASLS Occasional Papers No 3* (pp. 35–43). Aberdeen: Waverley Press.
Magnusson, L., Hanson, E., & Borg, M. (2004). A literature review study of Information and Communication Technology as a support for frail older people living at home and their family carers. *Technology and Disability, 16*, 223–235.
Mannheim, K. (1952). The problem of generations. In K. Mannheim (Ed.), *Essays on the sociology of knowledge* (pp. 378–403). London: Routledge.
Marcum, C. S., & Koehly, L. M. (2015). Inter-generational contact from a network perspective. *Advances in Life Course Research, 24*, 10–20.
Mayer-Smith, J., Bartosh, O., & Peterat, L. (2007). Teaming children and elders to grow food and environmental consciousness. *Applied Environmental Education and Communication, 6*, 77–85.
Mbanaso, M. U., Shavelson, J., & Ukawuilulu, J. (2006). Elderly African Americans as intragenerational caregivers. *Journal of Gerontological Social Work, 47*, 3–15.
McAdams, D. (1996). Narrating the self in adulthood. In J. Birren, G. Kenyon, J.-E. Ruth, J. Schroots, & T. Svensson (Eds.), *Aging and biography: Explorations in adult development* (pp. 131–148). New York: Springer.
McAdams, D. P., & McLean, K. C. (2013). Narrative identity. *Current Directions in Psychological Science, 22*, 233–238.
McAdams, D. P., & Olson, B. D. (2010). Personality development: Continuity and change over the life course. *Annual Review of Psychology, 61*, 517–542.
McCann, R. M., Cargile, A. C., Giles, H., & Bui, C. T. (2004). Communication ambivalence toward elders: Data from North Vietnam, South Vietnam, and the U.S.A. *Journal of Cross-Cultural Gerontology, 19*, 275–297.
McCann, R. M., Dailey, R. M., Giles, H., & Ota, H. (2005). Beliefs about intergenerational communication across the lifespan: Middle age and the roles of age stereotyping and respect norms. *Communication Studies, 56*, 293–311.
McCann, R. M., & Giles, H. (2006). Communication with people of different ages in the workplace: Thai and American data. *Human Communication Research, 32*, 74–108.
McCann, R. M., & Giles, H. (2007). Age-differentiated communication in organizations: Perspectives from Thailand and the United States. *Communication Research Reports, 24*, 1–12.
McCann, R. M., & Keaton, S. A. (2013). A cross cultural investigation of age stereotypes and communication perceptions of older and younger workers in the USA and Thailand. *Educational Gerontology, 39*, 326–341.
McCroskey, J. C. (1984). The communication apprehension perspective. In J. A. Daly & J. C. McCroskey (Eds.), *Avoiding communication: Shyness, reticence, and communication apprehension* (pp. 13–38). Beverly Hills, CA: Sage.
McGowan, T. G., & Blankenship, S. (1994). Intergenerational experience and ontological change. *Educational Gerontology, 20*, 589–604.
McIntosh, J. (2009). Elders and "frauds": Commodified expertise and politicized authenticity among Mijikenda. *Africa, 79*, 35–52.
Meek, B. A. (2007). Respecting the language of elders: Ideological shift and linguistic discontinuity in a Northern Athapascan community. *Journal of Linguistic Anthropology, 17*, 23–43.
Mickus, M. A., & Luz, C. C. (2002). Televisits: Sustaining long distance family relationships among institutionalized elders through technology. *Aging and Mental Health, 6*, 387–396.
Miller, D. A. (1981). The 'sandwich' generation: Adult children of the aging. *Social Work, 13*, 419–423.
Moore, P. J. (2003). Lessons on the land: The role of Kaska Elders in a University language course. *Canadian Journal of Native Education, 27*, 127–139.
Morales, A., & Hanson, W. E. (2005). Language Brokering: An integrative review of the literature. *Hispanic Journal of Behavioral Sciences, 27*, 471–503.
Mutran, E. (1987). Family, social ties and self-meaning in old age: The development of an affective identity. *Journal of Social and Personal Relationships, 4*, 463–480.

Nayar, K. E., & Sandhu, J. S. (2006). Intergenerational communication in immigrant Punjabi families: Implications for helping professionals. *International Journal for the Advancement of Counselling, 28*, 139–152.

Newman, S., & Hatton-Yeo, A. (2008). Intergenerational learning and the contributions of older people. *Ageing Horizons, 8*, 31–39.

Newman, S., Morris, G. A., & Streetman, H. (1999). Elder-child interaction analysis. *Child and Youth Services, 20*, 129–145.

Ng, R., Allore, H. G., Trentalange, M., Monin, J. K., & Levy, B. R. (2015). Increasing negativity of age stereotypes across 200 years: Evidence from a database of 400 million words. *PLoS One, 10*(2), e0117086. https://doi.org/10.1371/journal.pone.0117086

Ng, S. H., Liu, J. H., Weatherall, A., & Loong, C. F. (1997). Younger adults' communication experiences and contact with elders and peers. *Human Communication Research, 24*, 82–108.

Ngo-Metzger, Q., Massagli, M. P., Clarridge, B. R., Manocchia, M., Davis, R. B., Iezzoni, L. I., et al. (2003). Linguistic and cultural barriers to care: Perspectives of Chinese and Vietnamese immigrants. *Journal of General Internal Medicine, 18*, 44–52.

Noels, K. A., Giles, H., Cai, D., & Turay, L. (1999). Perceptions of inter- and intra- generational communication in the United States of America and the People's Republic of China: Implications for self-esteem and life satisfaction. *South Pacific Journal of Psychology, 10*, 120–135.

Nouri, M., & Helterline, M. (1998). Narrative accrual and the life course. *Research on Aging, 20*, 36–64.

Nozari, Z., Mo'tamedi, A., Eskandari, H., & Ahmadivand, Z. (2019). Narrative group therapy to improve aging perceptions and reduce thanatophobia (death anxiety) in older adults. *Elderly Health Journal, 5*, 117–123.

Nussbaum, J. F., & Bettini, L. (1994). Shared stories of the grandparent-grandchild relationship. *International Journal of Aging and Human Development, 39*, 67–80.

Nussbaum, J. F., Pecchioni, L. L., Robinson, J. D., & Thompson, T. L. (2000). *Communication and aging* (2nd ed.). Mahwah, NJ: Lawrence Erlbaum Associates Publishers.

Nygren, B., Norberg, A., & Lundman, B. (2007). Inner strength as disclosed in narratives of the oldest old. *Qualitative Health Research, 17*, 1060–1073.

Okafor, M. C. (2009). Avenues and barriers to access of services for immigrant elders: State and local policies for OAA units on aging. *Journal of Gerontological Social Work, 52*, 555–564.

Orellana, M. F., Dorner, L., & Pulido, L. (2003). Accessing assets: Immigrant youth's work as family translators or "para-phrasers". *Social Problems, 50*, 505–524.

Osborn, R., Moulds, D., Squires, D., Doty, M. M., & Anderson, C. (2014). International survey of older adults finds shortcomings in access, coordination, and patient- centered care. *Health Affairs, 33*, 2247–2255.

Östlund, B. (2010). Watching television in later life: A deeper understanding of TV viewing in the homes of old people and in geriatric care contexts. *Scandinavian Journal of Caring Sciences, 24*, 233–243.

Ota, H., Giles, H., & Somera, L. P. (2007). Beliefs about Intra- and intergenerational communication in Japan, the Philippines, and the United States: Implication for older adults' subjective well-being. *Communication Studies, 58*, 173–188.

Ota, H., McCann, R. M., & Honeycutt, J. M. (2012). Inter-Asian variability in intergenerational communication. *Human Communication Research, 38*, 172–198.

Peterson, D. R. (1983). Conflict. In H. H. Kelley, E. Berscheid, A. Christensen, J. H. Harvey, T. L. Huston, G. Levinger, E. McCintock, L. A. Peplau, & D. R. Peterson (Eds.), *Close relationships* (pp. 360–396). New York: Freeman and Company.

Pickel, A. (2005). The habitus process: A biopsychosocial conception. *Journal for the Theory of Social Behaviour, 35*, 437–461.

Pieri, M., & Diamantini, D. (2010). Young people, elderly and ICT. *Procedia Social and Behavioral Sciences, 2*, 2422–2426.

Pilcher, J. (1994). Mannheim's sociology of generations: An undervalued legacy. *The British Journal of Sociology, 45*, 481–495.

Pinquart, M., Wenzel, S., & Sörensen, S. (2000). Changes in attitudes among children and elderly adults in intergenerational group work. *Educational Gerontology, 26*, 523–540.

Poiani, A. (2010). *Animal homosexuality: A biosocial perspective*. Cambridge: Cambridge University Press.

Poiani, A. (2012). Evolutionary arguments against the *de facto* re-pathologising of homosexuality. In A. Poiani (Ed.), *Pragmatic evolution: Applications of evolutionary theory* (pp. 261–280). Cambridge: Cambridge University Press.

Polkinghorne, D. E. (1991). Narrative and self-concept. *Journal of Narrative and Life History, 1*, 135–153.

Poole, J., Gardner, P., Flower, M. C., & Cooper, C. (2009). Narrative therapy, older adults, and group work?: Practice, research, and recommendations. *Social Work with Groups, 32*, 288–302.

Quayle, A., Sonn, C. C., & van den Eynde, J. (2016). Narrating the accumulation of dispossession: Stories of Aboriginal elders. *Community Psychology in Global Perspective, 2*, 79–96.

Radermacher, H., Browning, C., & Feldman, S. (2016). 'I feel very happy that I can contribute to siciety': Exploring the value of the project for older people. In M. Cordella & H. Huang (Eds.), *Rethinking second language learning: Using intergenerational community resources* (pp. 149–167). Bristol: Multilingual Matters.

Radermacher, H., & Feldman, S. (2017). Cultural diversity, health and ageing. In K. O'Loughlin, C. Browning, & H. Kendig (Eds.), *Ageing in Australia: Challenges and opportunities* (pp. 83–101). New York: Springer.

Randall, W. L. (2002). Teaching story: The pedagogical potential of narrative gerontology. *Education and Ageing, 17*, 55–71.

Randall, W. L. (2009). The anthropology of dementia: A narrative perspective. *International Journal of Geriatric Psychiatry, 24*, 322–324.

Randall, W. L., & Kenyon, G. M. (2002). Reminiscence as reading our lives: Toward a wisdom environment. In J. Webster & B. Haight (Eds.), *Critical advances in reminiscence: Theoretical, empirical, and clinical perspectives* (pp. 233–253). New York: Springer.

Randall, W. L., & Kenyon, G. M. (2004). Time, story, and wisdom: Emerging themes in narrative gerontology. *Canadian Journal on Aging, 23*, 333–346.

Rees Jones, I., & Higgs, P. F. (2010). The natural, the normal and the normative: Contested terrains in ageing and old age. *Social Science and Medicine, 71*, 1513–1519.

Reynolds, J. F., & Orellana, M. F. (2009). New immigrant youth interpreting in White public space. *American Anthropologist, 111*, 211–223.

Ricoeur, P. (1984). *Time and narrative* (Vol. 1). Chicago: University of Chicago Press.

Riley, M. W., & Riley, J. W., Jr. (1994). Age integration and the lives of older people. *The Gerontologist, 34*, 110–115.

Riley, M. W., & Riley, J. W., Jr. (2000). Age integration: Conceptual and historical background. *The Gerontologist, 40*, 266–270.

Robinson, T. E., II, White, G. L., Jr., & Houchins, J. C. (2006). Improving communication with older patients: Tips from the literature. *Family Practice Management, 13*(8), 73–78.

Rodríguez, M. D., Gonzalez, V. M., Favela, J., & Santana, P. C. (2009). Home-based communication system for older adults and their remote family. *Computers in Human Behavior, 25*, 609–618.

Rood, B. (2009). A time to talk: Re-membering conversations with elders. *The International Journal of Narrative Therapy and Community Work, 1*, 18–28.

Rosenthal Gelman, C. (2014). *Familismo* and its impact on the family caregiving of Latinos with Alzheimer's disease: A complex narrative. *Research on Aging, 36*, 40–71.

Ruiz, M. E., & Ransford, H. E. (2012). Latino elders reframing *familismo*: Implications for the health and caregiving support. *Journal of Cultural Diversity, 19*, 50–57.

Ryan, E. B. (1994). Intergenerational communication: Evaluations and analyses of talk exchanged between older adults and younger adults. *The International Journal of Aging and Human Development, 39*, 1–3.

Ryan, E. B. (2010). Overcoming communication predicaments in later life. In L. Hickson (Ed.), *Hearing care for adults 2009: Proceedings of the Second International Adult Conference* (pp. 77–86). Chicago.

Ryan, E. B., Bajorek, S., Beaman, A., & Anas, A. E. (2005). I just want you to know that "them" is me: Intergroup perspectives on communication and disability. In J. Harwood & H. Giles (Eds.), *Intergroup communication: Multiple perspectives* (pp. 117–137). New York: Peter Lang.

Ryan, E. B., Bourhis, R. Y., & Knops, U. (1991). Evaluative perceptions of patronizing speech addressed to elders. *Psychology and Aging, 6*, 442–450.

Ryan, E. B., Giles, H., Bartolucci, G., & Henwood, K. (1986). Psycholinguistic and social psychological components of communication by and with the elderly. *Language and Communication, 6*, 1–24.

Ryan, E. B., Meredith, S. D., MacLean, M. J., & Orange, J. B. (1995). Changing the way we talk with elders: Promoting health using the communication enhancement model. *International Journal of Aging and Human Development, 41*, 89–107.

Ryan, E. B., Pearce, K. A., & Anas, A. P. (2004). Writing a connection: Intergenerational communication through stories. In M. W. Pratt & B. E. Fiese (Eds.), *Family stories and the lifecourse: Across time and generations* (pp. 375–398). Mahwah, NJ: Erlbaum.

Savundranayagam, M. Y., & Ryan, E. B. (2008). Social psychological aspects of communication and aging. *Annual Review of Applied Linguistics, 28*, 51–72.

Sayago, S., Forbes, P., & Blat, J. (2012). Older people's social sharing practices in YouTube through an ethnographical lens. In *Proceedings of the BCS HCI 2012 People and Computers XXVI* (pp. 185–194). Birmingham: BISL.

Scarinci, N., Worrall, L., & Hickson, L. (2008). The effect of hearing impairment in older people on the spouse. *International Journal of Audiology, 47*, 141–151.

Schwalbach, E., & Kiernan, S. (2002). Effects of an intergenerational friendly visit program on the attitudes of fourth graders toward elders. *Educational Gerontology, 28*, 175–187.

Searight, H. R., & Gafford, J. (2005). Cultural diversity at the end of life: Issues and guidelines for family physicians. *American Family Physician, 71*, 515–522.

Shklarov, S. (2009). *Narratives of resilience in aging Soviet Jewish child survivors of the holocaust*. PhD Thesis, Interdisciplinary Graduate Program, University of Calgary, Alberta, Canada.

Siibak, A., & Tamme, V. (2013). 'Who introduced Granny to Facebook?': An exploration of everyday family interactions in web-based communication environments. *Northern Lights, 11*, 71–89.

Sillars, A. L. (1980). The sequential and distributional structure of conflict interactions as a function of attributions concerning the locus of responsibility and stability of conflicts. In D. Nimmo (Ed.), *Communication yearbook 4* (pp. 217–236). New Brunswick, NJ: Transaction Books.

Silverstein, M., & Bengtson, V. L. (1997). Intergenerational solidarity and the structure of adult child-parent relationships in American families. *American Journal of Sociology, 103*, 429–460.

Slife, B. D. (2000). Are discourse communities incommensurable in a fragmented psychology? The possibility of disciplinary coherence. *The Journal of Mind and Behavior, 21*, 261–271.

Smith, B. J., & Yeager, A. (1999). Intergenerational communities. *Child and Youth Services, 20*, 25–32.

Smith, J., & Freund, A. M. (2002). The dynamics of possible selves in old age. *The Journals of Gerontology, 57B*, P492–P500.

Steele, R. (1986). Deconstructing history: Toward a systematic criticism of psychological narratives. In T. Sarbin (Ed.), *Narrative psychology: The storied nature of human conduct* (pp. 256–275). Westport, CT: Praeger.

Strom, R., & Strom, S. (1999). Establishing school volunteer programs. *Child and Youth Services, 20*, 175–188.

Tajfel, H., & Turner, J. C. (1986). An integrative theory of intergroup conflict. In S. Worchel & W. Austin (Eds.), *Psychology of intergroup relations* (pp. 2–24). Chicago: Nelson-Hall.

Taylor, A. S., Losciuto, L., Fox, M., Hilbert, S. M., & Sonkowsky, M. (1999). The mentoring factor. *Child and Youth Services, 20*, 77–99.

Taylor, P., & Keeter, S. (2010). *Millennials. A portrait of generation next*. Washington, DC: Pew Research Center, Pew Research Social & Demographic Trends. Downloaded on March 23, 2018, from http://www.pewsocialtrends.org/files/2010/10/millennials-confident-connected-open-to-change.pdf

Tian, Q. (2016). Intergeneration social support affects the subjective well-being of the elderly: Mediator roles of self-esteem and loneliness. *Journal of Health Psychology, 21*, 1137–1144.
Townsend, J., Godfrey, M., & Denby, T. (2006). Heroines, villains and victims: Older people's perceptions of others. *Ageing and Society, 26*, 883–900.
Tse, L. (1995). Language brokering among Latino adolescents: Prevalence, attitudes, and school performance. *Hispanic Journal of Behavioral Sciences, 17*, 180–193.
Tun, P. A. (1989). Age differences in processing expository and narrative text. *Journal of Gerontology, 44*, 9–15.
Uhlenberg, P. (2000a). Introduction: Why study age integration? *The Gerontologist, 40*, 261–308.
Uhlenberg, P. (2000b). Integration of old and young. *The Gerontologist, 40*, 276–279.
van de Vliert, E., & Hordijk, J. W. (1989). A theoretical position of compromising among other styles of conflict management. *Journal of Social Psychology, 129*, 681–690.
van Dyke, M. A., Haynes, C., & Ferguson-Mitchel, J. (2009). Bridging the divide: A public relations perspective on intergenerational communication. *Public Relations Quarterly, 52*, 19–23.
Viatori, M. (2007). Zápara leaders and identity construction in Ecuador: The complexities of indigenous self-representation. *Journal of Latin American and Caribbean Anthropology, 12*, 104–133.
Webster, J. D. (1999). World views and narrative gerontology: Situating reminiscence behaviour within a lifespan perspective. *Journal of Aging Studies, 13*, 29–42.
Westerhof, G. J., & Tulle, E. (2007). Meanings of ageing and old age: Discursive contexts, social attitudes and personal identities. In J. Bond, S. Peace, F. Dittmann-Kohli, & G. Westerhof (Eds.), *Ageing in society* (pp. 235–254). London: Sage.
Whorf, B. L. (1956). *Language, thought, and reality*. Cambridge: MIT Press.
Wiemann, J., Gravell, R., & Wiemann, M. (1990). Communication with the elderly: Implications for health care and social support. In H. Giles, N. Coupland, & J. Wiemann (Eds.), *Communication, health and ageing (Proceedings of Fullbright International Colloquium 1988)* (pp. 229–242). Manchester: Manchester University Press.
Wilińska, M., & Cedersund, E. (2010). "Classic ageism" or "brutal economy"? - Old age and older people in the Polish media. *Journal of Aging Studies, 24*, 335–343.
Williams, A., & Garrett, P. (2002). Communication evaluations across the life span: From adolescent storm and stress to elder aches and pains. *Journal of Language and Social Psychology, 21*, 101–126.
Williams, A., & Giles, H. (1996). Intergenerational conversations young adults' retrospective accounts. *Human Communication Research, 23*, 220–250.
Williams, A., Ota, H., Giles, H., Pierson, H. D., Gallois, C., Ng, S. H., et al. (1997). Young people's beliefs about intergenerational communication: An initial cross-cultural comparison. *Communication Research, 24*, 370–393.
Willoughby, L., Cordella, M., Musgrave, S., & Bradshaw, J. (2018). Triadic medical interaction with a bilingual doctor. *Communication and Medicine, 15*, 222–232.
Winkeler, M., Filipp, S.-H., & Boll, T. (2000). Positivity in the aged's perceptions of intergenerational relationships: A "stake" or "leniency" effect? *International Journal of Behavioral Development, 24*, 173–182.
Woodward, K. (1991). *Aging and its discontents: Freud and other fictions*. Indianapolis, IN: Indiana University Press.
Woody, I. (2015). Lift every voice: Voices of African-American lesbian elders. *Journal of Lesbian Studies, 19*, 50–58.
Young, E. (2010). Narrative therapy and elders with memory loss. *Clinical Social Work Journal, 38*, 193–202.
Zhang, Y. B. (2004). Initiating factors of Chinese intergenerational conflict: Young adults' written accounts. *Journal of Cross-Cultural Gerontology, 19*, 299–319.
Zhang, Y. B., Harwood, J., & Hummert, M. L. (2005). Perceptions of conflict management styles in Chinese intergenerational dyads. *Communication Monographs, 72*, 71–91.
Zhang, Y. B., & Hummert, M. L. (2001). Harmonies and tensions in Chinese intergenerational communication: Younger and older adults' accounts. *Journal of Asian Pacific Communication, 11*, 203–230.
Zwaan, R. A. (2008). Time in language, situation models, and mental simulations. *Language Learning, 58*(Suppl. 1), 13–26.

Chapter 8
Journey Towards the End of Life

Abstract This deals with the existential aspects of old age, including changes in life meaning and concerns about legacy. Existential well-being often benefits from the development of spirituality which, in turn, can be expressed in several ways, including through specific religious beliefs and practices. Existential issues are brought into sharper focus in the life of an old person by the obvious fact that inevitably, as time passes, death nears. This brings to the fore questions about what constitutes a "good death", and thus the debate over euthanasia and suicide is the focus of a specific section. We conclude this chapter with an exploration of the challenges of the end of life in multicultural societies, and the issues that affect end-of-life care.

Everyone is alone at the heart of the earth,
pierced by a ray of the sun's light;
and it is suddenly night

Salvatore Quasimodo, *And it is suddenly night*[1]

Although in modern times life expectancy has tended to increase and therefore the direct experience of witnessing the death of family members and friends at young ages is relatively less frequent now—especially in economically developed countries—than it was in past historical times, the end of life remains an inescapable fact.

The inevitability of death coming with old age, Quasimodo's "sudden night", does not, as such, limit the specific ways in which we approach the final stage of our existence. The end of life is a deeply personal experience, and it is open to a vast

[1] *Ognuno sta solo sul cuor della terra*
trafitto da un raggio di sole:
ed è subito sera.
Salvatore Quasimodo, *Ed è subito sera*

M. Cordella, A. Poiani, *Fulfilling Ageing*, International Perspectives on Aging 30,
https://doi.org/10.1007/978-3-030-60071-6_8

array of possibilities that are affected not only by our existential view of life and dying, but also the physical, social and cultural circumstances that we experience throughout the entire process. For many, however, walking the path to the end of life is also a personal journey of synthesis.

The discourse on death has accumulated different meanings over the centuries and across cultures: death as a mystery, death as an inevitable outcome of life that must be embraced with courage, death as the necessary passage for the continuation of the self into the afterlife; and always death as loss and bereavement, to the more modern inclination of trying to ignore death for as long as we can, which is aided by the advances of medicine and the generally improved material quality of life.

Death is seen as undermining "our taken-for-granted consciousness of the reality of day-to-day life" (Mellor and Shilling 1993, p. 413), hence the effort we put individually and as a society to push the end of life as far as possible into the future whenever life is regarded as good and desirable, and even to avoid talking about death altogether until the need to do so is unavoidable. When the last stage of the end of life starts in earnest, modern society offers us a medicalisation of death (e.g. Carpentier and Van Brussel 2012) that may help relieve us from pain and the responsibility to think about dying and to act on it.

Current discourses on death have added to the medical dimension thus enriching the view of death and dying with concepts such as control, autonomy, dignity, awareness, and heroism, leading to the distinction between "good death" and "bad death" (Carpentier and Van Brussel 2012). *Good death* could be defined as one which is "free from avoidable distress and suffering for patients, families and caregivers; in general accord with patients' and families wishes; and reasonably consistent with clinical, cultural and ethical standards" (Emanuel and Emanuel 1998, p. 21). But a good death is not only death that is free from pain and suffering, it is also one that comes naturally at the end of a fulfilling life. A bad death is marked by pain, suffering, despair that can be not only physical but also existential and spiritual. Existential suffering may be relieved through greater life fulfilment, self-esteem, autonomy, integrity, meaning, and dignity, for instance (Whitaker 2010).

Nussbaum et al. (2000, p. 304, based on Kalish 1976) refer to *appropriate death*, which is distinguished by:

> caring relationships, an open awareness of impending death, and a belief system that provides some meaning to the death. … feelings that one will have a 'good' death eases the coping process… Most dying persons want to avoid unnecessary pain, have an opportunity to take leave of their loved ones, right their shortcomings with God, and maintain their dignity.

Gott et al. (2008; see also Lloyd et al. 2011) provide the following characteristics of what may constitute "good death" within a palliative care model:

a. Attention to the individually unique physical, psychological, social, and spiritual needs of the dying elder.
b. Control of pain and other distressing symptoms.
c. Death at home if requested, accompanied by family and friends.
d. Open awareness of the impending death for all involved, from the dying person to the family, and those providing care and treatment.

e. Choice and control by the dying patient over treatment decisions.
f. Support and comfort for the bereaved, both prior and subsequently to the death of their loved one.

Therefore, a good death is not only an individual experience that extends a sense of personal fulfilment to the very end of life, but also an experience that involves others, mostly our immediate social network of family and friends. Another important aspect of a "good death" mentioned by Gott et al. is the choice of the specific place to die, and most people prefer, when possible, to die at home, surrounded by the people they love. Those who cannot choose, tend to die in a hospital or aged care facility (Perrels et al. 2014).

A central issue in the process of a good death is dying with dignity. We have already mentioned in Sect. 6.1 of Chap. 6 the importance of maintaining human dignity in old age. Promoting end of life in dignity decreases existential stress by increasing meaning and fulfilment (Chochinov et al. 2012). Greater dignity towards the end of life also helps in controlling death anxiety, with the process being aided by social support (Garrow and Walker 2001).

We have also seen in Chap. 5 how old age can lead to greater self-identity and the development of a degree of wisdom. Wisdom can contribute to a perception of life that goes beyond the immediate experience to one that projects the self through time. Such projection of the meaning of our life into the future, our legacy, can increase satisfaction in the elders despite the approaching end.

Cultural factors also affect the search for meaning at the end of life. The ways of thinking about life and the human condition in Western cultures have been marked by two major influences that have been put in sharp contrast especially in the nineteenth and twentieth centuries: one emphasising the collective, the social; and the other focusing on the individual, his/her personal experience and development. We will illustrate this point through a brief description and critique of the work of Carl Gustav Jung, with the objective of stressing the combined importance of the individual and the social dimensions of a dying person.

The post-World War II period was characterised in the West by a rejection of strict state collectivism in response to the experience of both Nazism–Fascism and Communism, thus the attention was turned to the individual and the opportunities for personal development in a society that is respectful and even supportive of individual needs. This is seen, for instance, in the psychological work of Carl Gustav Jung (e.g. Jung 1958) who stressed the importance of individual fulfilment in life, whilst interpreting social influences not as components of an environment where such fulfilment could be more easily experienced, but as constraining factors to personal development.

As illustrated in his autobiography *Memories, dreams, reflections* (Jung 1963), Jung's views are strongly focused on the self and the development of the wholeness of the self throughout life. Needs for social contact seem to have been mainly satisfied in Jung by God, the knowledge of past people and their ideas as illustrated in books, and the characters inhabiting his fertile dreams. He regarded his experience with friends as more valuable when he personally felt firmer in his own convictions and ideas. This miniaturisation of the social dimension in Jung can be understood, in our view, not only through his own experience as a child and his personality, but also through his rejection of totalitarianism and all its consequences for the control and annihilation of individual freedom, leading to his belief that social interactions may be conducive to the loss of self-identity and the development of neuroses:

> Like the initiate of a secret society who has broken free from the undifferentiated collectivity, the individual on his lonely path needs a secret which for various reasons he may not or cannot reveal. Such a secret reinforces him in the isolation of his individual aims. A great many individuals cannot bear this isolation. They are the neurotics, who necessarily play hide-and-seek with others as well as with themselves, without being able to take the game really seriously. As a rule, they end by surrendering their individual goal to their craving for collective conformity—a procedure which all the opinions, beliefs, and ideals of their environment encourage. Moreover, no rational arguments prevail against the environment. (Jung 1963, p. 343)

Correctly, Jung regarded each "self" as a unique individual whose identity is a mixture of specific biologically and culturally inherited characteristics, that are deeply buried in the unconscious, the "archetypes", and unique aspects developed through our personal life experience. Correctly he also identifies the need for each individual to develop a clear self-identity. Incorrectly, he only seems to have seen the negatives of social interactions, missing the positives coming with cooperation, company, emotional attachment, and learning from others' experiences.

We have argued in Chap. 6 that there are aspects of social life that tend to threaten the ability of old people to live a fulfilling life (e.g. stereotypes, stigma, and ageism), but social interactions at all levels (with immediate family, neighbourhood, and even government) are also a source of much support for an ageing person, especially one who is approaching the end of life. In this regard we wish to recall the South African philosophy of *ubuntu* which stands for a celebration of community. *Ubuntu* asserts that society gives us our humanity and the word literally means "I am because we are":

> *Ubuntu* is the capacity in African culture to express compassion, reciprocity, dignity, harmony and humanity in the interests of building and maintaining community with justice and mutual caring (Nussbaum 2003, p. 2)

Ubuntu is a nice reminder that despite the eminently personal character of the end-of-life experience, we fundamentally remain a social species right to the end. In fact, the often remarked tendency to increase the focus on the self in old age is in part also a result of both the increased difficulties that old people have in maintaining and nurturing social relationships, and the tendency of modern society to push older people to the periphery (Längle 2001). Society should not abandon the individual at the end of his/her life.

Just to remain within the field of psychoanalysis, Jung's emphasis on the individual had a contemporary counterpoint in Alfred Adler's greater emphasis on the social (Adler 1927, p. 36):

> This social feeling remains throughout life, changed, colored, circumscribed in some cases, enlarged and broadened in others until it touches not only the members of his own family, but also his clan, his nation, and finally, the whole of humanity. It is possible that it may extend beyond these boundaries and express itself toward animals, plants, lifeless objects, or finally towards the whole cosmos. An understanding of the necessity for dealing with

man as a social being is the essential conclusion of our studies. Once we have grasped this, we have gained an important adjunct to the understanding of man's behavior.

A synthesis of the personal and social aspects of the end of life is seen in Lars Tornstam's concept of *gerotranscendence* that we first introduced in Chap. 1 and further discussed in Chap. 2. Gerotranscendence involves "a redefinition of the self and of relationships to others, as well as a new understanding of fundamental existential questions. It is not farfetched to associate this with a kind of transpersonal development" (Tornstam 2011, p. 166).

Wadensten (2005), based on Tornstam (1996), provides a list of signs that are apparent whenever Tornstam's *three levels of gerotranscendence* are experienced: the *cosmic*, the *self*, and the *social*. It should be noted that beyond their specific focus, each level contains both individual and social references:

The cosmic level

- *Time and space*. Transcendence of the definition of time and space occurs with blurring of the borders between past and present.
- *Connection to earlier generations*. Increasing attachment, a social aspect.
- *Life and death*. Loss of fear of death and a new understanding of life and death.
- *Mystery in life*. Life acquires a dimension that is beyond our grasp.
- *Subject of rejoicing*. Enjoying the experiencing of events at all scales: from macro to micro.

The self-level

- *Self-confrontation*. Discovery of hidden aspects of the self, both good and bad.
- *Decrease of self-centredness*. Removal of self from the centre of one's universe, which opens up to the possibility of seeing the self in relationship to others (a social dimension) and the rest of our surroundings.
- *Development of body transcendence*. Care of the body continues, but without obsessing about it.
- *Self-transcendence*. Shift from egoism to altruism. That is, a shift from self-centredness to valuing and contributing to the cooperative relationships with others.
- *Rediscovery of the child within*. Return to and transfiguration of childhood.
- *Ego integrity*. Achieving a synthesis of the scattered experiences of one's life.

Social and individual relations level

- *Changed meaning and importance of relations*. Less interest in superficial relations, more selectiveness in social interactions, increased need for moments of solitude. The latter may vary with gender and ethnic background.
- *Role-play*. New understanding of roles in life.
- *Emancipated innocence*. The addition of innocence to maturity.
- *Modern asceticism*. Material wealth loses importance and a new conception of freedom based on "asceticism" may develop.
- *Everyday wisdom*. Transcendence of the right-wrong duality following the realisation that the distinction between right and wrong is not always clear.

Therefore, through gerotranscendence we maintain a link with generations past and present and use it, alongside our personal experiences, to help achieve a synthesis within the self and a social expansion beyond the self through our legacy, cooperation, and even altruism. Gerotranscendence can develop through a confluence of individual capacities and motivations, and the help provided to and received from others in specific social interventions and experiences (e.g. Wadensten 2005).

Gerotranscendence solves the dilemma of older people in modern times, who dream of living a longer life but without experiencing any deterioration (Mascarenhas Fonseca and Testoni 2011–2012), a combination that is ultimately destined to remain largely a dream. Peace of mind may come despite the inability to achieve such a dream, when the individual sheds the obsession with the self-here-and-now and reaches beyond the self in order to develop a higher degree of transcendence through attachment, generativity, and altruism. Self-transcendence may help the individual go through the last phase of his/her life—when the personal condition deteriorates to such an extent that even a wish to die may arise—by feeling personally fulfilled and wishing for an end without suffering; in the conviction that there is little else that the individual can do in life anymore (e.g. Linden and Barnow 1997), but that his/her legacy may endure through others. Thus, gerotranscendence helps the individual in his/her preparation to the end of life.

Despite the inevitability of death and a preparation for death through a process such as gerotranscendence, the closeness to death may still bring a sense of anxiety. Death anxiety tends to decrease with age across cultures, in part because life satisfaction also tends to decrease with older ages and older people may wish for an end to their suffering, but also because older people are mentally better prepared to the fact that their life is coming to an end (Gerstorf et al. 2008, see also Hallberg 2004, Lacey et al. 2006). But death anxiety is not necessarily eliminated in all elders.

The more anxious old people feel about death the less willing they are to talk about it (Hallberg 2004). The cultural practices that help decrease death anxiety (e.g. spirituality, religion) have been the subject of study by *Terror Management* theorists (Becker 1973) and, as stated in Chap. 3, each culture has developed specific strategies to manage the anxiety caused by death, both in the dying old person and the members of his/her close social network of family and friends.

Factors that could cause anxiety towards the end of our life include the process of dying itself and the type of end-of-life care: the awareness of it, its control, the decision-making involved, the level of dependency, maintenance of dignity, feeling treated as a "whole person"; the kind of nutrition received, the desire of not prolonging life, maintaining meaning in life—including being meaningful to others—and the expectation that symptoms such as pain be controlled (Hallberg 2004). Social support may complement individual resources to help alleviate such death anxiety.

We have mentioned that the process of dying is a deeply personal experience, but this does not mean that it should be a lonely experience. According to their characteristics, some people may prefer to *die alone*, that is in conditions where they live on their own, as a personal choice (e.g. Kellehear 2009). Whereas others may prefer to die in proximity to loved ones. A *lonely death*, on the other hand, may be a result of isolation that can be social or existential and that is not the choice of the elder, but an effect of their current situation. Quasimodo's quote at the beginning of this chapter indicates the profoundly personal and existential experience of approaching death, which has nothing to do with abandonment or loneliness, but with the fact that it is us and only us who are dying. Thus, dying alone as a choice may be perfectly fulfilling, dying in unwanted loneliness may lead to helplessness and anguish.

Interest in death and disciplines of inquiry into death have very ancient roots in human history. Once consciousness evolved, routine events of life acquired meaning and a new cognitive and emotional depth. Death, as the ultimate disruptor of life and of physical attachment with relevant others, is obviously no exception and looking for its meaning springs naturally. It is beyond the scope of this book to carry out a thorough analysis of the concept of death and the cultural practices and rituals associated with the dead around the world and throughout history. For a brief review we refer the readers to Sect. 3.10 in Chap. 3, where we pointed out that old people who feel more satisfied with their life also tend to be less worried about death; and we will end this chapter with a brief mention of some mortuary practices.

Researching the final stages of the end of life is not easy given the deeply personal character of the experience and the seriousness of the matter; but all people involved, from professional caregivers to researchers, are strongly advised to first inform themselves on the specific personal and cultural perceptions of the end of life in the elders they are approaching. Without such knowledge it would be virtually impossible to be able to effectively—but also ethically—investigate the process of dying, with appropriate respect and with the provision of the required form of care (e.g. Lloyd et al. 2011).

In this chapter we review the various dimensions and complexities of the end-of-life process. We start by exploring the existential aspects of the end of life, to be followed by an exploration of life meaning and the role of legacy in giving the dying elder a sense of continuity through time, or gerotranscendence. Regardless of our best intentions to experience an end of life that is positive, fear of death and death anxiety are common, often requiring specific attention and interventions. In this regard religiosity and spirituality, but also other approaches to meaning making and transcendence, have an important role to play in decreasing death anxiety. When suffering becomes unbearable in the old, or when staying alive has simply lost its meaning, some individuals may consider euthanasia, or even suicide whenever legal euthanasia is not available. These are profoundly personal decisions of the elder at the end of life and society should do all what it can in order to help dying individuals decrease their suffering in a manner that is fulfilling to the self and preserving of their dignity. We conclude this chapter with two sections: one on the end of life in multicultural societies and finally one addressing the practical issues of end-of-life care.

8.1 The Existential Dimension

> Whether a life is fulfilled does not depend on how great one's radius of action is, but rather only on whether the circle is fully filled out … And the life of a person who has not completely filled out his (relatively) larger life circle, remains more unfulfilled than that of a person who truly meets the tasks he finds in his more drawn circle.
>
> Viktor Frankl (2019, p. 41)

As the end of life nears, we may ask some questions about ourselves that perhaps we have not asked before: What is the purpose of being born and having gone through many experiences in life, given that now it is all coming to an end? What is our legacy, and is leaving a legacy important at all? Are we happy? Are we fulfilled? If not, why, and what to do about it in the short time left? The answers to such existential questions vary across individuals, ranging from a nihilistic view of a life without meaning that abruptly ends at death to religious views of a life that will last forever, in eternal bliss after death, provided that we have lived according to specific precepts; and a variety of other alternatives that include finding meaning and transcendence through legacy and generativity. Giving meaning to life has the additional advantage of helping maintain good mental health in the final stages of our existence (e.g. Längle 2001), despite experiencing a continuously deteriorating body condition (Whitaker 2010).

We have already addressed some aspects of the existential dimension of old age in previous chapters. In Chap. 1, for instance, we have seen how fulfilling ageing is crucially dependent on finding meaning in life, and we quoted Viktor Frankl's work in the context that such meaning could be found even in the most challenging of situations, whereas lack of meaning in life can result in existential vacuum. As Frankl indicated in the above quote, any life can be fulfilling to the person who is living it, for as long as such life is full of meaning.

More meaning in life leads to a better quality of life and well-being (Fry 2001) and, we may also add a better quality of death. Developing greater life meaning gives to the dying individual an improved "sense of identity, integrity, satisfaction, beauty, love, respect, positive attitudes, inner peace and harmony" (Velasco-Gonzalez and Rioux 2013, p. 1125) which promotes greater spiritual well-being. A sense of life fulfilment and transcendence may help smooth the experiences of physical pain at the end of life and complement medical interventions to relieve pain, thus contributing to prevent the onset of demoralisation and depression.

Even as we are close to the end of our life, we may want to retain a degree of agency and ability to control some events, as we suggested in Chap. 5. Such control should contribute to maintaining existential well-being up to the end. Accumulated wisdom and a clear self-identity may further enhance our agency and our existential fulfilment throughout the inevitable process of dying.

In this section we explore some basic existential aspects of old age: how living an authentic life helps achieve fulfilment and how to ameliorate existential loneliness at the end of life.

A fundamental aspect of the existential approach to the end of life is that the guaranteed finitude of our existence compels us to focus on our life course: its meaning and its fruits. In the words of the existentialist philosopher Martin Heidegger: "As soon as a human comes to life, he is old enough to die" (Heidegger 1963, p. 245, translated from German by Hermans 1989, p. 11). That is, as we are confronted with the inevitability of death, we may choose between making the best of our life for as long as it lasts or simply go along with our basic biological necessities until the end inevitably comes.

Whatever our choice is, through our lived experience we become, to an extent, agents of our "destiny", and towards the end of our life we can evaluate the accumulated results of our *agency* (Bakan 1966), alongside the effects that other, less controllable factors had on our existence. Although we are responsible for many events of our life, we are not completely in charge: our agency has limitations.

Many regrets may surface towards the end of life, but analysis post-facto is always easier than decision-making in real time. In the end it is the broad direction of our life experience that matters. We take responsibility for our decisions but accept that, as human beings we are fallible, this can bring a sense of fulfilment and life satisfaction leading to existential well-being towards the end of our life in spite of occasional bumps and departures from our initial ideal (e.g. Kwee 2012).

Just a cursory analysis of anyone's life is likely to reveal a complicated mixture of uncontrolled events, purpose and striving to achieve goals, good and bad fortune, good and bad decisions, opportunities seized, and opportunity lost, with regret. Help received, help given, and help denied. In the end it is only us who can judge whether the whole path of our existence has been fulfilling or not.

Living a fulfilling life in old age requires attending to various central issues that are relevant to our existence as human beings. Laura Olson (1994) proposed the following list: securing the economic means that facilitate a decent survival, protecting one's physical and mental capacities, retiring at the appropriate age, strengthening social relationships to prevent loneliness, engaging in sports and physical activities more broadly, having meaningful work commitments, being able to use new technologies that help remain connected with the rest of society, satisfying intellectual curiosity, remain updated in the field of one's profession if so desired, satisfying sexual and emotional needs, and obtaining warmth and joy by helping others, rather than withdrawing into one self. To these, Guttmann (2001) also adds a broadly spiritual dimension, following Frankl (1963). Frankl understands spirituality as a "fundamental component of human nature", being a "man's capacity for a sense of awe, wonder, and mystery, even reverence, in one's assessing the meaning, value, and purpose of one's own personal life" (Morgan 2012, p. 100).

To some extent, we are free to consciously direct our life in just about any direction as we develop our life fulfilment, given the plasticity of our brain that we referred to in Chap. 4, whilst also being constrained by many factors (e.g. physical, psychological, social, cultural) at the time of making any decision. For instance, the existential concerns that may beset us towards the end of our life can be modulated by our personality. Personality traits such as Neuroticism may lead to excessively ruminating about the negatives of our past, whereas Extraversion may favour reminiscence with the objective of cultivating our memories. Thinking about life meaning as death nears is especially strong in people characterised by Openness (Cappeliez and O'Rourke 2002). Ultimately, however, it is our responsibility to decide and to find a specific meaning in life within the characteristics of our personality and other life constraints (Eliason et al. 2010).

Finding meaning in life does not just provide a clearer path to the lived experience of the individual, but it also gives strength to overcome the obstacles met along the way, as illustrated by the work of Viktor Frankl. Finding meaning in life, according to Frankl, involves focusing on living our best life in the present, despite whatever constraint we may be confronting, and taking personal responsibility for our actions and inactions and their consequences on our life. A similar focus on self-awareness and taking personal responsibility for our own actions is also found among *Gestalt* theorists such as Perls et al. (1958). The existential approach to psychology continues to be developed by others at present, including Irvin Yalom, who regards the concerns about death, freedom, existential isolation, and meaninglessness as the root causes of our existential conflicts (Yalom 1980); and Robert Neimeyer (2002) who focuses on the construction of life meaning after traumas, such as the loss of beloved ones (see also Eliason et al. 2010).

A major transition in our life, that may confront us with important existential concerns and that may lead to a redefinition of the meaning of life, is retirement. To some people retirement may produce a degree of existential anxiety, given the uncertainties about the future, what to do with the remaining time left in our life, and how to live in the social space of family and friends in the new role as retiree. Others may be better prepared, having laid the groundwork for their life in retirement in previous years.

The transition to retirement may start with a process of *life review* that can then establish the foundations to live an authentic and fulfilling life into the future (Malette and Oliver 2006). The final years of our life may be lived in a diversity of manners, but whatever we decide to do it may help living those years authentically by "being true to oneself" (Osborne 2012). Living an authentic life at this later stage of our existence is made easier if we feel that there is little that we stand to lose by being authentic. As stated by Macquarrie (1973, p. 206):

> Existence is authentic to the extent that the existent has taken possession of himself and shall we say, he has moulded himself in his own image. Inauthenticity, on the other hand, is moulded by external influences, whether these be circumstances, moral codes, political or ecclesiastical authority or whatever.

What authenticity means to each one of us is obviously something personal, idiosyncratic, and we are free to incorporate (or not) into our self—and in what way—any external influence that we are exposed to. We may recall here the concept of gerotranscendence that incorporates aspects of both the self and the social, as we proceed in our path to a fulfilling life.

Older people also become more frail and vulnerable with ageing, as their physical and psychological capacities become less effective, and such relative decay may lead them to experience *existential loneliness*, in which case they should be helped, whether it is through normal channels of social interactions with family, friends, other acquaintances, and so forth or through more structured psychological interventions. We defined existential loneliness in Chap. 6, following Moustakas (1972), as being aware of and facing the ultimate experiences of upheaval, tragedy, and change of living and dying. We also mentioned in the same chapter that although existential loneliness may have negative effects on the life of old people, it may also have some positive effects as described by Perlman and Peplau (1981), where periods of self-confrontation may provide an opportunity for self-growth, perhaps leading to painful but ultimately "triumphant creation".

Existential loneliness may become more pressing as old people feel that they are approaching the end of their life, especially if they are suffering from a terminal disease. Awareness of existential loneliness towards the end of life puts a burden on caregivers to provide effective care to the dying elder, such care is expected to help the elder feel reconciled with self, family and past life in general, whilst feeling supported and cared for all the way. The more the dying elder feels free to express his/her feelings, concerns, suffering, or ideas outwardly to somebody, the less loneliness he/she may experience (e.g. Sjöberg et al. 2018).

Sjöberg et al. (2018) carried out a study of existential loneliness in Sweden where they interviewed 23 participants, 75 years old and older (47.8% women) who were receiving various types of care: outpatient primary care centre ($n = 2$), homecare ($n = 4$), residential care ($n = 8$), hospital ($n = 2$), specialised palliative home care ($n = 6$), and specialised palliative ward ($n = 1$). Interviewees understood existential loneliness mainly as "being trapped in a frail and deteriorating body", "being met with indifference", "having nobody to share life with", and "lacking purpose and meaning". Indifference from others elicited feelings of worthlessness and abandonment, as shown in this excerpt:

> P: … It annoyed me … if there is something that is not done properly, I tell them …
> I: Don't they listen to you then?
> P: … No, not at all … They take it for granted that you are already 'lost' when you come here … (p. 1360)

Alternatively, there may be attention, but one that does not make the elder feel dignified. For instance, Ulla experienced shortness of breath, dizziness, and fatigue but felt that she was treated like an inanimate object:

> I don't think … that they can say 'you can get out of bed before we come and start in the bathroom' … I then have to stand there and wait … And they did not help me to shower, because it takes too long. I am so sad that I cry … You are not regarded as a human being, but as … a thing to 'make up' and then leave again. (p. 1360)

Loneliness comes with not being able to share your life and to have emotional and physical contact with others, thus leading to experiencing sadness, sorrow, emptiness, and abandonment. In the words of Inge, who was bedridden and almost blind:

> Yes … I feel really lonely sometimes … like being the only person in the world … having nobody … feeling … useless, so to say. Then, again, I think about leaving … leaving this present life … Oh, what a relief, I think. (p. 1360)

The same feelings were expressed by Mrs. Smith, a 96 years old who was interviewed by Kimble and Ellor (2001, p. 14):

> She has multiple chronic impairments including arthritis and a right brain stroke. She is functionally bed bound. Whenever anyone stops to talk to her, she frequently asks the question, 'Why won't God let me die?' Her explanation of this statement reflects her pain. 'I am 96 years old. My husband is dead, my parents are dead, my six brothers and sisters are all dead, my three children are all dead, even two of my twelve grandchildren are dead. All of my friends are dead and even the nurse who used to take care of me has taken a job somewhere else and is no longer here. I am of no earthly use to anyone … Why won't God let me die?'"

Existential loneliness is also expressed in the form of lacking purpose and meaning in life, which is reflected in views such as existing in a "vacuum" or being in a "waiting room" (Sjöberg et al. 2018). The fact that many friends, family members, and acquaintances may have passed away, whilst the physical capacities of the elder fade at an increasing rate, can only magnify the sense of loneliness and lack of meaning in staying alive.

In the words of one of the interviewees in Sjöberg et al.'s (2018) work, Ruth, who was mainly bedridden, just like Mrs Smith above:

> I don't want them to revive me if something happens … this is enough … I can't see any meaning in that … lying in a bed staring at the walls, what's the meaning? … No one comes out of here alive … it's the waiting-room for death. (p. 1361)

Some interventions, such as the presence of family and the support they may provide, may help improve the sense of meaning in dying individuals, but the inevitable trajectory towards death is set and the purpose of improving meaning at this stage is not so much that of extending life (although improving meaning and existential well-being are indeed associated with better health, Ryff 2012) but that of achieving a "good death".

Sjöberg et al. (2018) conclude that although existential loneliness is ultimately inevitable, given the strictly personal experience of dying, it can be ameliorated by increasing meaning and purpose (e.g. through legacy), connectedness with others (e.g. Suri 2010) and by decreasing physical pain.

Figure 8.1 summarises the negative aspects of existential loneliness at the end of life. The dying person experiences accelerated existential loneliness and disconnection with life the faster the body is deteriorating, the less supportive the social environment is and the less meaningful and purposeful it is to remain alive. However, although the end is inevitable and deeply personal, it does not need to be marked by loneliness and sadness: a fulfilling life can conclude with a fulfilling death.

It is more likely that we may experience a fulfilling death if we learn to live a fulfilling life in the knowledge that such life is finite, that each moment counts and that we become increasingly vulnerable as we age. Accepting the finitude of life and therefore the ultimate inevitability of death can also help decrease fear of death: "fear of death can be avoided because death is *un*avoidable" (Baars 2017, p. 3). Fear

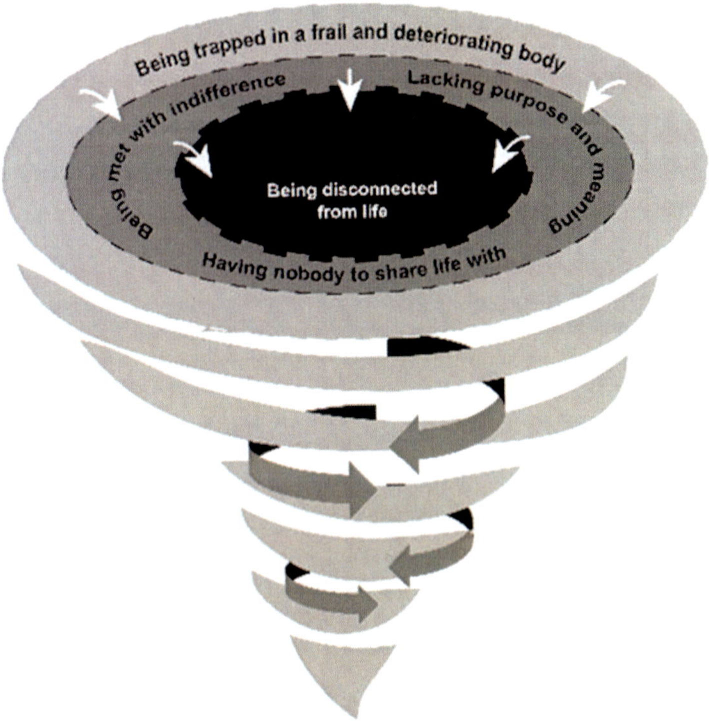

Fig. 8.1 The downward spiralling of existential loneliness (from Sjöberg et al. 2018)

of death can be also decreased if we abandon the myth of "success" (Baars 2017). How can we possibly live our life in contentedness if we are in constant fear of death?

A finite life gives time a special place in our existence, as we realise that the time of our life will come to an end upon death. The perception of time also changes towards the end of life, as time in old age is full of past and present, whereas the future is fast shrinking. But time operates at different scales, such as the scale of a single lifetime or the scale of generations, and in the case where old people seek self-transcendence through their beliefs in life beyond death or through generativity, time may be perceived by the individual as extending beyond death.

The knowledge of the finitude of life allows us to focus on our existence, our actions, and relationships. We can still plan, because we do not know for sure when the end will come, unless we purposefully bring it upon us, but the planning is free from the expectations of youth. In the experiencing of the finitude of life, Baars (2017) acknowledges not only the crucial importance of the self, but also the equally crucial influence of the social world, so much so that he prefers to talk not of "human condition" but of "interhuman condition". In the end:

> Finitude is a permanent condition of life and is deeply connected with the uniqueness of situations and human lives. Although a human being may seem nothing more than a short,

> egocentric breath in a huge cosmic wind, it is only in the short breath of human lives that everything of importance manifests itself and all questions arise. After all, it is a miracle that we exist at all. (Baars 2017, p. 7)

Baars reminds us that the finitude of life is what puts pressure on us to live our life to the full and in a meaningful way: "(the) preciousness of life is intrinsically related to its finitude" (Baars 2013, p. 11). Although living our life to the full may sometimes increase the risk of ending such life prematurely, it is us who will have to decide whether such risk is worth taking or not. It is a fine balance between spoiling life for the fear of death, as Jan Baars puts it with reference to Stoic thinking, or curtailing your future possibilities prematurely because you do not have enough fear of death. The conundrum should become easier to resolve as the end of life nears in old age. At that stage, our body and mind are likely to put some serious boundaries on what "living life to the full" means.

Although the end of life is a time of increasing vulnerability, given that the body becomes more frail (Whitaker 2010), it is also a time for recollection, synthesis, meaning making, generativity, and transcendence, a time when our individual uniqueness also embraces its social context. This process can continue deep into the late phases of our life, until we reach a point when there is nothing else that we can offer to life or that life can offer us. This is what has been referred to as a stage of *narrative foreclosure* (e.g. Laceulle and Baars 2014).

In sum, our existence should be hopefully fulfilling. In old age, we can look to the past to evaluate the present and decide our steps for the remaining future. But as the end nears, avoiding existential loneliness may be a desirable action, helped by ending our suffering and engaging in self-transcendence, be it spiritual, generative, or both. In this endeavour, support from others may be important. *Nihilism* offers an alternative—and simpler—existential path to the end of life, one that spares us from making sense of the often chaotic and nonsensical events of our life. In the words of Friedrich Nietzsche, nihilism is "the radical repudiation of value, meaning and desirability" (Nietzsche, 1968, p. 7). If life is meaningless, just relax and let it go. Although we do not personally find nihilism useful, given our view that humans have a natural propensity to find some meaning in their life and that fulfilment can be experienced by anyone, older people at the end of their life will decide what is best for them given their own reality and circumstances.

8.2 Life Meaning and Legacy

In this section we further expand on the concept of life meaning as it is expressed towards the end of life, and how meaning can be increased through legacy, a core issue in the process of generativity. The section will be organised into two major sub-themes. The first one will focus on *Life meaning* and in the second one we develop our view on *Legacy*.

8.2.1 Life Meaning

> Life's but a walking shadow, a poor player
> That struts and frets his hour upon the stage
> And then is heard no more. It is a tale
> Told by an idiot, full of sound and fury,
> Signifying nothing.
>
> William Shakespeare, *Macbeth*

The quote from Shakespeare's *Macbeth* is less than optimistic about life meaning. We are undeniably born without previous consultation, but is that enough reason to say that life has no meaning? After all, it is us who afterwards decide to keep going for as long as we want or can. If we stay alive, we surely have our own motivations to do so. What is our life during the period of our existence, is it but a "walking shadow", "signifying nothing"? Is it really an endless and meaningless Sisyphean endeavour to keep us busy? Or is it something else?

It is true that finding meaning in modern life has become an imperative driven by the fear of meaninglessness, a situation described by Moody (2002, p. 51) with these words:

> We have, alas, slipped into a one-sided stance in this attitude toward meaning and aging, and this one-sidedness manifests itself in insidious ways: sentimentality, hyperactivity, secret despair, and, above all, unrelenting efforts to escape from facing the possibility that what Jung called 'the forces of meaninglessness' will gain the upper hand.

But in fact, staying alive and finding meaning are closely linked aspects of our human nature. From a biological perspective staying alive is in-built in our genes that are the product of millions of years of evolution. Life now is the future of life past, and it is the past of future life. That is far more impressive than a mere walking shadow and, although basic, the perpetuation of life is a necessary condition for the emergence of any other aspects of life meaning.

In *The myth of Sisyphus*, Camus (1955/1975, p. 11) stated that "(t)here is but one truly philosophical problem and that is suicide", but the solution to that problem is more easily found in biology than philosophy. The emptiness and meaninglessness of *existential nihilism* can be filled with the pulsating life of our cells. Unbearable suffering signals the end of the capacity to live, thus potentially leading to actions—such as suicide—to end suffering, but before we reach such a point of no-return most of us have in-built resources that can be used to overcome the shaking clashes with life or the periods of nauseating boredom and existential emptiness, and if in addition there is help coming from others, the better.

Even in the most tragic of situations such as war or life in a concentration camp meaning can emerge against the pressures of existential nihilism, as shown in the experience of Viktor Frankl, for instance, and other survivors of the Holocaust (e.g. Wiesel 1958; Levi 1958; Frankl 1963; Dov Kulka 2013). In fact, life meaning can give us the strength to overcome the stresses, fears, and anguish of traumatic experiences.

Nevertheless, even in the most optimistic of cases there is a dimension of life meaning where the pessimistic view of the existentialists does acquire a pungent reality, this is clearly expressed in this quote from *A free man's worship* by Bertrand Russell (Russell 1967, p. 107):

> All the labors of the ages, all the devotion, all the inspiration, all the noonday brightness of human genius, are destined to extinction in the vast death of the solar system, and... the whole temple of man's achievement must inevitably be buried beneath the debris of a universe in ruins....

Extinction as a species does indeed mark the end of the possibility of life meaning but, for as long as we are not extinct, we may not want to kill meaning prematurely.

Life meaning and the search for meaning have roots into the organisation of the human brain and therefore emerge from our evolutionary history (e.g. Klinger 2012). Evolutionarily the "meaning of life" is simply to persist and perpetuate life: "sometimes the meaning in life is not found in striving but just being" (Ebersole and Quiring 1991, p. 115).

Adaptively, however, seeking meaning gives human life an existential purpose that may specifically help alleviate the burden of stressful situations. Loss of meaning may lead to negative emotions such as sadness, potentially developing into depression, whereas finding meaning in life stirs up positive emotions, such as joy. Ultimately, meaning derives from more basic aspects of the mind such as pursuing goals and fulfilling those goals, achieving a sense of coherence, and the emotional rewards coming with it (Baumeister 1991). Therefore, we may say that there is a cognition-emotion-meaning interactive system that we use to navigate our way through life.

Humans display a hierarchy of needs to be fulfilled. Once our basic biological needs are met, we can continue with the quest for self-actualisation and the development of more complex levels of meaning through spirituality, artistic creativity, knowledge, wisdom (e.g. Maslow 1943). As organisms, humans do strive for a degree of stability, with life meaning being one of the psychological tools to achieve such stability (Baumeister and Vohs 2002), but our environment can change. We can change too throughout our development as an individual, and we react to changing conditions by adjusting. Adjustment to altered conditions may require the strengthening of our old perception of life meaning or it may demand change, sometimes even radical change.

The sources of meaning in life are varied, and to those mentioned above we may add the effort to maintain our physical and emotional well-being (e.g. Peterson and Park 2012). For Seligman (2002, p. 278) meaning arises in life when "people feel connected to something external to and larger than themselves". Others see the goals in life as providing meaning to such life (e.g. Emmons 1986), or they see meaning as emerging from the demands imposed by the external world on the individual, demands that are regarded worthy of investment (Antonovsky 1979). Wong (1989) sees meaning as self-transcendence, a view also broadly shared by Damon et al. (2003). The stronger our life meaning is the more we engage with life and the more pleasure we derive from such engagement (Peterson et al. 2005).

How does meaning change throughout life? Does life meaning increase or decrease in old age?

Steger et al. (2009) collected data from a group of participants through a website with the objective of studying the developmental aspects of life meaning. The participants were subdivided by age according to Erikson (1968) classification of adult life stages into Young Adults (25–34 years), Middle-Age Adults (35–64 years), and Older Adults (over 65 years). In addition, they produced the Emerging Adults (18–24 years) age group following Arnett (2000), for a total of four life stages. A total of 8756 individuals (66.3% women), mostly from Western countries (97.7%) submitted responses. Greater *presence* of life meaning was associated with both older ages (see Fig. 8.2, upper graph) and greater wellbeing experienced throughout life. Moreover, *search* for meaning continues into old age, although to a lesser extent than at younger ages, presumably because the more meaningful life becomes at older ages the less need there is to search for further meaning (Fig. 8.2, lower graph). Older women tend to find more meaning in old age than men, and therefore they also show lower levels of searching for meaning.

Thus, the slowing down of the search for meaning in old age seems to simply derive from the greater life meaning usually achieved by older people. However, is it possible that lower search for meaning in old age may be also a result of a descent into a life of routines, a shrinking social network, and increased isolation (Weiss and Bass 2002)? That is certainly a possibility, but it does not need to be so for all old people. If elders feel satisfied with their life meaning, one may expect search to slow down.

Maintaining some sort of meaning in life, whether it is expressed in a complex set of philosophical, religious, or political beliefs or a more basic enjoyment of the good things that life offers: love of family, good food, the warm weather of summer or the cosy feelings of winter, the splendour of the colours of nature, and so forth, comes naturally from our nature as human beings. Meaning can be found in old age through those aspects of the self that are retained throughout life, the enduring aspects of our personal identity, and it can be also found in generativity and therefore social life, reminiscence, and religion-spirituality. In addition, there are historical aspects of life that affect life meaning in each generation of elders (e.g. independence movements, wars, civil rights movements, and so forth), alongside more contingent aspects such as retirement, physical relocation, bereavement, deteriorating physical capacities. Changes in those factors over time may lead to the restructuring of meaning as we become older (Weiss and Bass 2002, 2002a).

Historically, psychological research on life meaning has mainly focused on cognition. Emphasising the cognitive component of meaning, Steger (2013, p. 165) defined meaning in life as:

> the web of connections, understandings, and interpretations that help us comprehend our experience and formulate plans directing our energies to the achievement of our desired future. Meaning provides us with the sense that our lives matter, that they make sense, and that they are more than the sum of our seconds, days, and years. Comprehending our experience in this way builds the cognitive component of meaning in life. The cognitive component of meaning in life thus refers to the understandings that we develop of who we are, what the world is like, and how we fit in with and relate to the grand scheme of things.

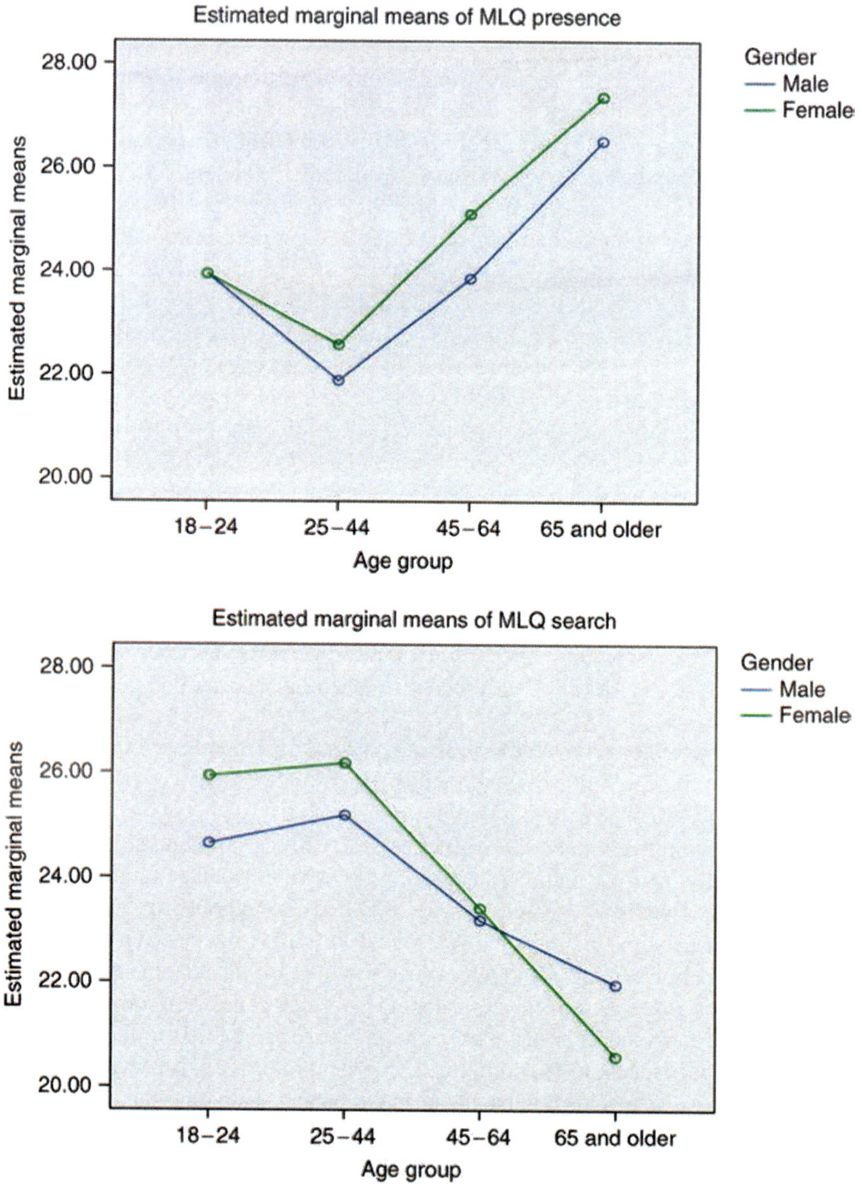

Fig. 8.2 Mean levels of the Meaning in Life Questionnaire (MLQ) *presence* and *search* sub-scales across life stages, by gender. Reproduced, with permission, from Steger et al. (2009)

Baumeister and Vohs (2002, p. 608–609) also stressed the cognitively connective nature of meaning: "The essence of meaning is connection. Meaning can link two things even if they are physically separate entities, such as if they belong to the same category, are owned by the same person, or are both used for a common goal. The connection between the two is not part of their physical makeup and thus can only be appreciated by a human mind (or some other mind capable of processing meaning)".

Thus, meaning provides sense to life by connecting the various threads of our life experience.

Reker (1997, p. 710) goes a bit further, defining meaning in life as "having a purpose in life, having a sense of direction, a sense of order, and a reason for existence, a clear sense of personal identity, and a greater social consciousness". The theme of meaning providing the individual with a sense of direction is also taken up by Klinger (2012, p. 25): "That is, by implication, meaning provides direction and hence intention"; and Steger et al. (2009, p. 43), for whom meaning is "the extent to which people comprehend, make sense of, or see significance in their lives, accompanied by the degree to which they perceive themselves to have a purpose, mission, or over-arching aim in life".

Therefore, meaning involves a synthetic view of our world, where components are cognitively interconnected into a whole—or in the words of Baumeister (1991, p. 15): "shared mental representations of possible relationships among things, events, and relationships. Thus, meaning connects things"—that provides a sense of direction towards some "worthy purpose", according to Frankl. This sense of purpose represents the motivational component of meaning (Steger 2013). Steger (2013) suggests that meaning binds together the cognitive (comprehension) and motivational (purpose) aspects of our personal approach to life.

Dittmann-Kohli's (1990, p. 281) view of meaning specifically reconciled those cognitive and motivational aspects:

> The personal meaning system is the pattern of valuations and concepts by which individuals represent to themselves, at a given phase of existence, what they (a) believe, and (b) want in relation to their own life and self.

Peterson (1999) went into even greater details proposing a conception of meaning at three levels of cognitive-motivational analysis:

a. Meaning in everyday living: serving the adaptive purposes of self-maintenance, self-regulation, and self-propagation.
b. Meaning used to overcome negative or threatening situations.
c. Meaning serving two major psychological purposes: (1) myth making, religion, and spirituality used to restore meaning after a crisis; (2) spur exploration, curiosity, fantasy, or meta-narratives with the intention of creating new meaning when all our basic needs are fully satisfied and boredom or atrophy may set in.

Peterson's analysis points to the importance of developing strength of character with the objective of increasing meaning in life. Likewise, Paul Wong (2012, p. xxxiii) states that the "top four character strengths most significantly related to meaning are religiousness, gratitude, hope, and zest for life".

Therefore, meaning springs out of a process of self-reflection (Hermans 1989) that is made possible by our consciousness and both cognitive and motivational capacities. Through self-reflection a sense of coherence may emerge that can lead to greater meaning in life (Antonovsky 1979, 1987). A sense of coherence helps make life manageable and comprehensible within the context of a "big picture".

In a further development of the cognitive-motivational aspects of meaning, Neal Krause (Krause 2005) identified four dimensions of life meaning that also include a temporal perspective:

a. Values: provide the basis for behavioural guidance.
b. A sense of purpose: the belief that our actions have a specific place in the larger order of things.
c. Goals: striving to fulfil some expectations for the future.
d. Reconciling the past: reconciling our initial goals in life with what it was finally achieved.

Krause's scheme, however, falls short in providing a useful structure for life meaning, as it seems to be too rooted into the concept of "successful ageing": we establish some goals that make sense to us, strive to achieve them and then review whether we succeeded or not. That is fine if we indeed succeed, what about those who did not? Did they "waste their life"? What kind of end of life are they going to experience if they convince themselves that they have been a "failure"? Values, a sense of purpose, even some specific goals should be seen as parts of a dynamic experience, a plastic path in life that we may re-invent as we go along, either because we just want to or because we are forced by unforeseen circumstances to do so. There is no point in limiting our life review to comparing what our own self dreamt of at the age of 18 and what our 65 years old self sees as she/he looks back at how life turned out, only to feel frustrated about any mismatch between initial dream and final reality, and then become depressed as a result. The initial dream was an idea, but our own self—living in the reality of a world that is part agency and part external events—might have drifted away in other directions afterwards. The issue is to find meaning and fulfilment in the life we have lived and that we will live in what remains of the future.

In our view, it was Wong (1989, p. 517) who provided what we believe is the best and most comprehensive approach to meaning, by identifying not just the basic cognitive and motivational aspects of it, but also the affective component of life meaning; with the only proviso, as stated above, that the pursue of goals in the *motivational component* must be seen as a dynamic process rather than a restrictive one:

> The *cognitive* component refers to one's beliefs and interpretations of the world as one sees it. It attempts to make sense of one's experiences and it primarily consists of ascriptions of meaning to events, activities and life as a whole. The *motivational* component refers to the pursuit as well as attainment of personal goals that are consistent (with) one's subjective values, needs and wants. The *affective* component refers to the feelings of fulfilment or satisfaction that come from the conviction that one's life is worth living.

From those components Wong derives the following definition of meaning (p. 517):

> In short, personal meaning may be defined as an individually constructed cognitive system, that is grounded in values and is capable of endowing life with personal significance and satisfaction.

We agree with Wong's comprehensive definition, which is the one that we will adopt. In other words, life meaning reflects the specific manner how we mentally organise the world, as we dynamically experience life fulfilment grounded in a system of values.

Wong (2012a) emphasised two major psychological perspectives to life meaning that expand the concept beyond cognition to also involve aspects of our motivational and emotional life. The *positive psychology perspective* to life meaning focuses on positive experiences in life and positive emotions. There is a reciprocal influence between positive emotions and life meaning: "The dispositional tendency to enjoy life, to find joy and excitement in what life has to offer, may promote the search for meaning: likewise, striving for meaning may cultivate traits of positive emotionality" (McAdams 2012, p. 110–111). On the other hand, the *existential perspective* involves learning to live with the negatives of life, with suffering, meaninglessness, loneliness, and death, whilst constructing meaning through life choices and actions. Wong (2012, p. xxx) defines the concept of *existential meaning* as "one's philosophy of life and worldviews to make sense of life", through such philosophy of life we provide a specific *situational meaning* to our day-to-day experiences: "why am I doing this rather than that right now?"

Together, the positive psychology and the existential perspectives offer tools to increase meaning in life that can overcome existential suffering, thus opening the door to greater happiness, life satisfaction, and emotional well-being.

Baird (1985, p. 118) asked whether life meaning is created or discovered:

> There is more to this contrast than just a way of speaking, for our way of talking reflects an attitude. To speak of discovering, or looking for, or searching for the meaning of life (phrases heard all the time) suggests that there is some kind of meaning 'out there' which is hidden from view, and if we are diligent enough or lucky enough we will uncover it or bump into it. Here the meaning of life is likened unto a beautiful piece of already published music which, if discovered, would fill one's days with harmony. On the other hand, to speak of creating a meaningful life suggests that we have a responsibility for its coming into being. This is a crucial shift in attitude. Now a meaningful life is likened unto composing a piece of music, a composing in which the harmony that emerges is a product of the creative activity of the self.

Baird leans towards the personal creation of life meaning, whereas Kimble (1991) is of the view that meaning is not created but discovered, following a spontaneous process of personal development. But perhaps meaning is more likely a combination of both creation and discovery. Each one of us possesses his/her own nature that comes with our biology and our life experience since childhood, that can display the propensity to give more meaning to some experiences than others (*implicit meaning*, Thompson and Janician 1988), and we also have a plastic dimension that allows us to create new meanings throughout our life (*created meaning*, Baird 1985). To the ancient Greek dictum *know thyself* that implies the discovery of the self, we must add that of *become yourself*, the creation of the self. Who we are today helps and contributes—but it is not sufficient—to become who we will be tomorrow. Throughout our life we cannot escape our responsibility to give meaning to our existence, but the way we do it and the meaning we finally adopt, will be affected by many contingencies, some of which are beyond our control.

Baird describes three main factors that he considers crucial to a meaningful life: (a) the quality of our social relationships, (b) the objectives in life that we pursue with intensity and in a state of flow, and (c) the capacity to produce a satisfactory narrative of our life. More precisely, the development of life meaning is helped by:

a. The quality of our social relationships and our personal contribution to the social sphere, along with the benefits we obtain in terms of personal growth from our interactions with others. This is equivalent to Frankl's (1967) longing for contact or union with the other.
b. Our objectives in life, our projects, the goals that can keep us in a state of *flow* also contribute to a meaningful life. We should only keep in mind that such goals can be fluid, there is nothing wrong with changing our priorities at some point in time, if that is what we have decided is the best course of action for us. But for as long as we pursue a specific goal, we may as well take the task seriously. In Chaps. 2 and 5 we mentioned Mihaly Csikszentmihályi's (1990) concept of *flow* or, paraphrasing Nakamura and Csikszentmihalyi (2002), the capacity to focus on a specific task and persisting on it single-mindedly, disregarding hunger, fatigue, and discomfort, yet rapidly losing interest in the task once it has been completed. A state of flow emerges whenever an activity becomes intrinsically motivated (*autotelic*) and self-rewarding. It could be expected that proximity to the end of life may put the individual under the urgent need to focus on specific tasks, to such an extent that a state of flow is more easily achieved. After all, time is fast running out and second opportunities may be no longer available. On the other hand, if the neurological mechanisms underlying the flow experience deteriorate with advanced age, experiencing flow may be less likely in later life. Collins et al. (2009) investigated flow in old age and found that older individuals experienced states of flow less frequently than younger participants in their study. The capacity of older people to experience states of flow, however, is also dependent on various other factors besides age, such as social context, individual differences, location (Heo et al. 2010), broader physical health (Hirao et al. 2012), and the characteristics of the task the elder is involved in (e.g. how demanding it is) combined with the degree of plasticity of the elder's brain (Payne et al. 2011). Being able to retain the capacity to experience states of flow is intrinsic to what Bakan (1966) describes as *striving for self-enhancement*, and some old people may have clearer objectives to strive for and capacity to do so than others.
c. The ability to create a redeeming story, a narrative, that can make sense of our life no matter how that life has been, can also contribute to life meaning. As we have seen in Chap. 7, the *narrative of identity* connects all meaningful aspects of life as they emerge from personality characteristics and individual experiences. The narrative of identity "is a story that distinguishes me from all others and yet shows how I am connected to others as well. It is a story that narrates the evolution of a particular self, but it is a self in cultural context" (McAdams 2012, p. 114). Through narratives of identity we produce a diversity of *autobiographical memories* such as various achievements, good personal relations, social

service, and a feeling of significance; freedom from hardship, creativity, legacy/self-transcendence/generativity, setting goals in life, seeking purpose in life, seeking enjoyment and personal growth, a sense of fulfilment (Battista and Almond 1973; Bar-Tur and Prager 1996; Emmons 2003; Beike and Crone 2012). Focusing on *autobiographical memory* as a source of meaning, Beike and Crone (2012) point to our capacity to construct personal life narratives that can interweave our memories of past experiences into a meaningful story representing our current view of ourselves. Some memories will be more central and therefore more meaningful than others in the context of contributing to such identity (e.g. *self-defining memories*). More emotionally charged memories may be also better retained. In the case of memories of past experiences that were emotionally unsettling, interventions may be necessary to bring such memories to a *psychological closure*. Of course, the incorporation of past experiences into an overarching meaning of our life through a narrative does require a degree of autobiographical memory, but accessing autobiographical memories is not enough to produce a meaningful view of our life. Beike and Crone (2012, p. 318) add other aspects of our psychology that are required to produce meaning from memory through a narrative, including coherence and a moral context:

> A satisfying and meaningful life story is therefore composed of a coherent narrative, which provides both temporal and causal explanations for events. Moreover, a satisfying life story should be grounded in a moral context that is understood by the society in which the story is being told. Coherent and morally grounded self-narratives provide a sense of stability. These well-defined narratives place the self contextually in space and time and integrate the person's life experience and self-concept, thereby providing a sense of a self with purpose… A well-functioning life narrative should therefore provide a stable sense of meaning and purpose.

Experiences that do not immediately fit the narrative are either incorporated somehow into a coherent story, or they become warnings to prevent making mistakes in the future or are simply forgotten.

Therefore, meaning can be continuously modified and grown through pursuing our objectives in life, producing a satisfactory life narrative, and nurturing our social and broader life experiences. However, the development of meaning does not necessarily proceed in a steady and continuous manner. Some specific events, such as traumatic ones, can shake up our existence, thus changing life meaning, perhaps even radically. Some strong life experiences may have the capacity to challenge our assumptions about life, our social relationships, objectives, and overall life story. The passing away of a very important person in our life, a life-threatening disease such as cancer can alter our view of life and life meaning (e.g. Thompson and Janigian 1988). The more satisfactory the new meaning is, the faster we can overcome the uncertainties caused by the traumatic experience and therefore the greater our well-being will be.

Once our life acquires a specific meaning, such meaning imbues life itself with purpose. That is, meaning becomes the motor that propels us towards the future, towards embarking in specific experiences. If science or art gives meaning to my life, my life is bound to move in the direction of scientific understanding and discovery, or creation of new works of art and explorations of new dimensions of artistic creativity, and so forth. Wisdom will tell us which goals are attainable at any point in time; and even if ultimate goals are not likely to be achieved, wisdom will tell us whether the effort was nonetheless worthwhile so that others may be helped in the task of pursuing such goals in the future, or whether it is better to abandon the project and move in a different direction altogether. Without wisdom the inability to attain some specific goals in life may lead to depression. A category of individuals, that Thompson and Janician (1988) call "hardy" may pig-headedly persist in the pursuit of those goals that give meaning to their life. Those who succeed against unspeakable odds become the object of much admiration, sometimes even legend, which biases our evaluation of "hardiness" in a positive direction. But the many people who fail despite their uncommon effort should be also taken into consideration to understand how to proceed in a wise manner in any given circumstance. Being regarded as "hardy" but a "failure" is not a combination that can uplift the life of many people. A fulfilling life may be aided by a degree of hardiness, but it is not helped by declaring a life to be a "failure".

The unsatisfaction with our current personal system of meaning triggers the search for a new meaning, lest our life becomes meaningless or at least unsatisfactory. Understanding the reasons for the collapse of our previous system of meaning may help in the construction of a new one. Some of our beliefs may have proven ineffective once tested in real life (e.g. religious beliefs) and hence the new system of meaning must be constructed using different tools and building blocks (e.g. new principles). In Thompson and Janician's (1988, p. 270) view: "Thus the criterion for finding meaning is not just that the life scheme be able to explain one's experiences, but also that the world and self views be compatible with setting and obtaining personal goals". Meaning making may provide new goals that make our life meaningful under the changed circumstances (see the Sect. 8.2.5 in this chapter).

Creating meaning, maintaining meaning once it has been acquired, and recovering meaning if it is lost can have important effects on health (e.g. Wong 1989), as a meaningless life may be more susceptible to become a stressful life. Therefore, life meaning may not only protect ego integrity, promote well-being and self-esteem (Becker 1971), and therefore protect us against despair, but it can also protect us against stress-related health problems.

Wong (1989) proposed several strategies that can help enhance life meaning in old people, these include *reminiscence*, *commitment*, *optimism*, and *religiosity*. *Reminiscence* as expressed through memories, life review, writing diaries may help maintain a sense of continuity in life, give some coherence to life, strengthen the awareness of past achievements, and perhaps provide an opportunity to tackle unresolved conflicts (e.g. with close family members). Reminiscence can also acquire a practical purpose through sharing life experiences with the younger generations (generativity). Reminiscence, to be constructive and to serve the purpose of meaning making should avoid becoming obsessive and plunge into continuous rumination, especially about what the past could have been had different choices been made. Watt and Wong (1991) have listed the different types of reminiscence that include both those that are positive and negative (Table 8.1).

Table 8.1 Types of reminiscence

Type	Function
Obsessive	Ruminating over upsetting events from the past
Escapist	Finding refuge in "the good old times" to escape the present
Narrative	Story telling
Instructive or transmissive	Sharing instructive stories, life lessons
Instrumental	Identifying coping and problem-solving strategies used in the past; applying those to current life situations, favouring self-control/efficacy
Integrative	Re-evaluating past events; solving past conflicts ("unfinished business"); reconciling ideals with reality; identifying a pattern of continuity between past and present; finding meaning and worth in life as it is lived

From Watt and Wong (1991)

Commitment is the resolve and determination used in pursuing meaning in life in the specific way that one has established for him/herself. It involves being consistent with the patterns of behaviour, activities, values, traditions, ideals, and significant relationships that give meaning to our life. Through *optimism* life meaning can have positive effects on health, by decreasing stress, and by more speedily overcoming negative or even traumatic experiences that could jeopardise our sense of living a good life. Finally, spiritual well-being can be cultivated through *religiosity* and other means that can help maintain self-coherence, purpose, and hence meaning in life.

Meaning in old age is also affected by the realisation that the body has changed, the mind and accumulated experience too, and that a degree of adaptation and adjustment in the life objectives is required if fulfilment in the late phase of life is going to be experienced (Dittmann-Kohli 1990). This requires adopting an attitude of constraint towards self-criticism: change what you can change, calmly accept what you cannot, and use your wisdom to distinguish between those two options.

8.2.2 The Study of Meaning

Reker and Guppy (1988) devised the *Sources of Meaning Profile* (SMP) with the objective to study meaning. The SMP is a questionnaire instrument that includes the following 16 sources, covering both individual and social dimensions of life:

- participation in leisure activities.
- meeting basic and everyday needs.
- taking part in creative activities.
- engaging in personal relationships with family and/or friends.
- being acknowledged for personal achievements.
- experiencing personal growth.
- taking part in religious activities.
- interest in social causes.

- being of service to others.
- preserving human values and ideals.
- preserving culture and tradition.
- leaving a legacy for the next generation.
- feeling financially secure.
- interest in human rights (humanistic concerns).
- participating in hedonistic activities.
- acquiring material possessions to enjoy the good life.

In other words, just about any aspect of our life can be a potential source of meaning. The specific importance of each factor may be variable among individuals but in the end our life acquires meaning through the contribution of all dimensions of our existence. The relevance of each source of meaning may also vary as we become older. According to Bar-Tur and Prager (1996) throughout old age the types of sources of meaning tend not to qualitatively change in each individual but there is a gradual level of disengagement with life in the old-old. As shown in Table 8.2, personal growth, pleasure, and acquiring material goods significantly decrease in importance in the old-old.

Wong's (2012a) positive psychology and existential approaches to life meaning, that we mentioned previously in this chapter, have been synthesised by him into his *Dual System Model* (DSM) which addresses: (a) what people really want and the ways they can use to achieve their life goals; (b) what people fear and how they can overcome their anxieties; and (c) how people make sense of their predicaments and the paradoxes of their life.

The basic postulates of the *DSM* include:

a. Positive and negative conditions for meaning construction coexist. This duality leads to the *Duality hypothesis* that emphasises discovering the positive potential in negative life events (a concept derived from Frankl's logotherapy). In the words of Paul Wong: "The duality hypothesis posits that optimal outcomes depend on the interaction between the approach and avoidance systems… a combination of positive and negative interventions is more likely to yield better outcomes than either positive and negative methods alone… Similarly, enhancing one's strengths and correcting one's weaknesses will lead to greater improvement than focusing on only strengths or only weaknesses." (Wong 2012a, p. 16). Thus, the duality hypothesis suggests that meaning can emerge from the combined contribution of both positive and negative experiences.

Table 8.2 Sources of personal meaning that differ significantly between the young-old and old-old

Sources of meaning	Young-old (n = 104) M	SD	Old-old (n = 106) M	SD	P
Experiencing wisdom, maturity, and personal growth	5.28	1.70	4.49	2.38	≤0.001
Taking part in pleasurable activities	4.03	1.72	3.33	2.21	≤0.01
Acquiring possessions and material goods	4.43	1.40	3.57	2.25	≤0.001

From Bar-Tur and Prager (1996)

b. Approach and avoidance systems coexist in meaning making. Positive affects are approach, whereas defensive mechanisms against threats and negative emotions are avoidance.
c. The model takes into account emotions, cognition, awareness regulation, meaning, individual differences, resources, and motivations whether they are personal or cultural.
d. Meaning is crucial in this model, playing a role in life protection and expansion, enhancing well-being, and buffering against negative experiences.
e. Self-regulation systems are affected by individual differences and resources, alongside cultural and contextual variables.

Therefore, Wong's *DSM* is aimed at achieving personal well-being by enhancing the positive aspects of our life and by also finding meaning in the negative aspects of it.

Wong subsequently expanded the above basic *DSM* into a more general model that encapsulates the *PURE principle*. *PURE* includes four major components: Purpose (motivational), Understanding (cognitive), Responsibility (social and moral), Enjoyment/Evaluation (affective).

Purpose
Motivational dimension of life meaning: Motivationally, life is given meaning to the extent that we have goals and that we are striving in order to achieve such goals, whatever the goals may be and no matter how much they may be revised and adjusted to changing physical and mental circumstances due to ageing, and also to changing external environmental/social conditions. Meaning helps in providing resilience to the individual in the face of negative experiences in life.

Understanding
Cognitive dimension of life meaning: Cognitively we give meaning to life to the extent that we find unity in it, with all its components making unitary sense. In this context narratives can help transform the negative in life into positive by emphasising positive aspects and ignoring/diminishing the negative ones, thus helping in the construction of meaning (Wong 2012). *Post-traumatic growth* is a meaning-making exercise where new narratives help overcome the negative effects of trauma.

Enjoyment/Evaluation
Affective dimension of life meaning: Life meaning increases the more an individual is in a better mood although mood can be variable and the more variable it is the less stable life meaning will also become. Maintaining positive affect through experiences of suffering is a challenge but if it is achieved life meaning can be increased (Frankl 2004; Wong 2012).

Responsibility
Social/Moral dimension of life meaning: Life can acquire meaning to the extent that we fulfil moral obligations towards others and towards society at large. When we are both part and co-contributors to the fulfillment of a bigger project, our existence extends its meaning beyond the self and, to an extent, beyond the present. The moral dimension of life meaning also blends with the religious/spiritual dimension.

Religion and spirituality can help provide life meaning to old people through a worldview that can make sense of "perceived ambiguities, puzzles, and paradoxes in human experience" (Fortuin et al. 2018, p. 1).

The components of the *PURE* principle define what Wong (2012a, p. 18) describes as the *Meaning Mindset*:

> The meaning mindset refers to one's overarching motivations and worldviews. The meaning mindset focuses on the person… as meaning-seeking and meaning-making creatures. It also capitalizes on the human capacity for reflection and awakening … Joy comes from living meaningfully and authentically….

An *ABCDE* intervention strategy was derived by Wong from the *DSM* model and the *PURE* principle:

a. *A*ccept and confront reality (reality principle).
b. *B*elieve that life is worth living (faith principle).
c. *C*ommit to goals and action (action principle).
d. *D*iscover the meaning and significance of self and situations (Aha! Principle).
e. *E*valuate the foregoing (self-regulation principle).

PURE and *ABCDE* are meaning-based self-regulation systems. Self-regulation should prevent positive-feedback runaway processes from taking place; for instance, runaway processes that may lead to despair and disillusionment with life on the one hand, or blind pursue of success as a merely hedonistic endeavour on the other. In addition, Wong also included the *Mindful Awareness* hypothesis in his view of meaning. Mindful awareness is comprised of Openness, Compassion, Empathy, Acceptance, and Non-judgment (OCEAN).

Therefore, following on the path laid down by Viktor Frankl, Wong's *Dual System Model* "affirms that one can still squeeze out meaning and satisfaction from negative situations, which people instinctively avoid, but also guards against the excesses and pitfalls in the sole pursuit of success and happiness" (2012, p. 13–14) (see Fig. 8.3).

The *DSM*, as illustrated in Fig. 8.3, represents a very comprehensive framework for both understanding and acting on personal meaning, it is a fundamentally optimistic approach and one that recruits all our mental and social resources towards living life in a fulfilling manner.

Some negative experiences in life include our frustrations, and the *DSM* suggests finding ways to overcome frustrations by looking for manners in which our negative experiences may contribute to personal growth. For instance, Wong (2012a, p. 17) mentions what he calls the *Deep-and-Wide hypothesis* which suggests that:

1. Setbacks or failures can arouse an individual or team members to give their best to achieve a specific goal in life.
2. Desperation can lead to ingenuity and resourcefulness.
3. Prolonged frustration can result in a broader search for alternatives and motivate a major change in life direction.
4. The experience of overcoming progressively more difficult tasks or enduring increasing levels of frustration can result in higher frustration tolerance or a higher breaking point.

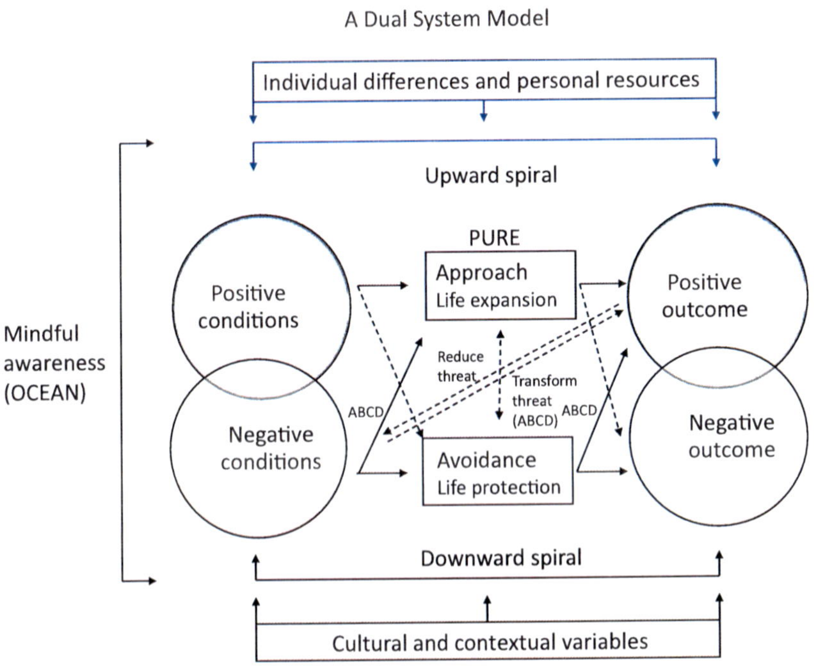

Fig. 8.3 Interactions and meaning-based regulations within the *Dual System Model*. Redrawn from Wong (2012a)

However, as suggested by Wong, there is also a personal final *breaking point* beyond which the negative aspects of life are no longer beneficial.

8.2.3 Meaning and Life Satisfaction

Greater meaning in life is found in the old who are in better physical and mental conditions. In general, the greater the elder's well-being, the more likely it is that she/he may find life more meaningful (e.g. Takkinen and Ruoppila 2001); the same is the case with regard to human relations: better human relations are associated with greater meaning in life (Takkinen and Ruoppila 2001). Personality traits also affect the ability to find meaning in life. High levels of Extraversion, Conscientiousness, Openness to experience, and low levels of Neuroticism, tend to be associated with higher meaning and purpose in life (McAdams 2012; Wong 2012). Wong (2012) has also proposed a "meaning mindset" personality trait, which can help develop a positive psychological attitude in those who are going through a rough patch in life. Conversely, developing greater life meaning can improve our well-being and life satisfaction through increasing: positive affect such as feelings of high morale, love, joy, pleasure, and vitality; and *global happiness*, which is a more positive perception of the world and life in general. Positive affect can help sustain our life meaning from an emotional perspective, contributing to make our life enjoyable no matter what turn it may take (e.g. Park et al. 2010; King and Hicks 2012).

Greater life meaning can also improve psychological adjustment, quality of life, self-esteem, self-acceptance, self-actualisation, positive self-regard, self-worth, self-confidence, self-empowerment, ego strength, internal health locus of control, personal growth, sense of mastery (Park et al. 2010; Steger 2013 and references therein); and it is associated with various additional self-empowering traits such as ego resilience, autonomy, ambition, personal responsibility, optimism, better capacity to manage life's challenges (Steger 2012).

However, meaning can be also found in suffering, as we have mentioned with regard to the works of Viktor Frankl, Paul Wong, and others (see also Steger 2013), and by finding meaning in suffering the individual may be able to cope better.

Life satisfaction improves once meaning is found, but the process of searching for meaning is often associated with depression, either because the search for meaning is triggered by traumatic experiences in life or the search process becomes frustrating when no end point is found (Park et al. 2010).

Figure 8.4 shows that satisfaction with life increases as presence of meaning also increases. For levels of presence of meaning lower than 75%, however, life satisfaction tends to decrease as we search for meaning more intensely and perhaps experience frustration; whereas satisfaction is positively correlated with search for meaning at the highest levels of presence of meaning (76–100%). In other words, life satisfaction decreases when search for meaning is intense but relatively unproductive.

Life satisfaction can be measured through the construct of *subjective well-being* (SWB), which can be defined as: "cognitive and affective evaluations that people make about their lives. The former evaluations include long-term judgments of one's general satisfaction with life, whereas the latter include reports on one's immediate experiences of positive and negative affect" (Shmotkin and Shrira 2012, p. 144). SWB includes hedonistic concerns but especially eudaimonic concerns and, in general, the greater the SWB experienced by a person the more meaningful the life of that person tends to be (Steger 2012).

Even though an individual may experience high life meaning and SWB, it is still adaptive to remain conscious and alert about potential *hostile-world scenarios* that may lurk in the future (Shmotkin 2005). Life meaning and SWB could be maintained even in the face of such hostile-world scenarios for as long as a broadly optimistic stance in our capacity to overcome adversity is adopted. In the words of Shmotkin and Shrira (2012, p. 148): "SWB lets people benefit from the favorable psychological environment while remaining vigilant for hostile conditions". One adaptive benefit of "unifying frameworks of meaning" such as philosophy, spirituality, and religion, is that of empowering the individual to better confront hostile-world scenarios (e.g. Moremen 2005).

Setting oneself some goals in life—that may change as we become older—working towards the achievement of such goals and perhaps fulfilling our aspirations in the context of those goals can increase our level of happiness, well-being, and meaning in life. The achievement of goals and other positive subjective experiences

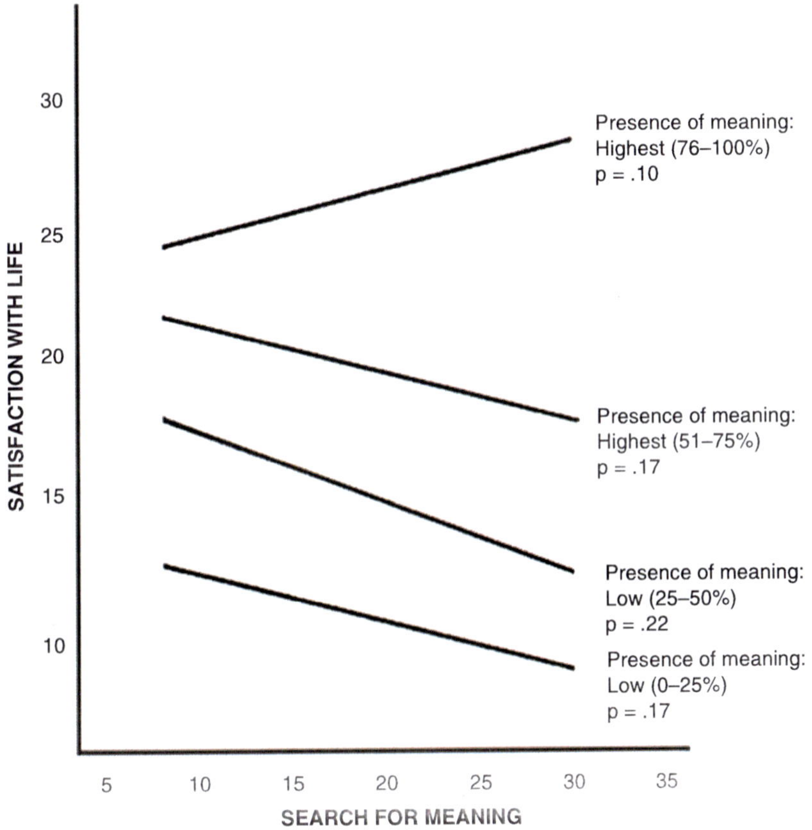

Fig. 8.4 Relationship between life satisfaction and search for meaning among respondents differing in the presence of meaning. Partial correlations (ρs) estimate the association between search for meaning and life satisfaction within each sub-sample, after controlling for age, gender, education, and presence of meaning (from Park et al. 2010)

can be of great help in giving shape to life meaning. Such positive experiences in life are causes of happiness, life satisfaction, flow, contentment, optimism, and hope (Emmons 2003), and they form the basic platform for the program of *positive psychology* (e.g. Seligman and Csikszentmihalyi 2000). Emmons (2003) points to three main goals in life that can contribute to increasing SWB in old age: intimacy, generativity, and spirituality, which are in accordance with Tornstam's *three levels of gerotranscendence* mentioned in the introductory section of this chapter.

Positive psychology can help older people keep an optimistic view of their life throughout the final stages of such life. It helps build the strengths to live a fulfilling life in the face of life challenges, by developing courage, interpersonal skills, rationality, insight, creativity, coping skills, hope, perseverance, realism, capacity to experience pleasure, putting troubles into perspective, self-esteem, self-efficacy and personal control, future-mindedness and finding purpose and optimism (Hewitt

2002; Maddux 2002; Seligman 2002; Simonton 2002; Snyder et al. 2002; Stanton et al. 2002; Thompson 2002).

A level of life meaning that can bring personal satisfaction to the individual, no matter what the stage of his/her life is, can be better developed the more optimistic the individual is, especially when the individual goes through many experiences in life that can be challenging. We have already seen how Wong (1989) included optimism as one of four major strategies that can improve life meaning in old people. Reker (1997, p. 710) states that optimism "reflects the human capacity to transcend temporal boundaries, to anticipate a positive future, to hope. Optimism is defined as looking forward to many desirable events and being very confident that these will take place". That is, optimism is a psychological tool that helps us overcome those life challenges that, without a capacity to transcend them could overwhelm us, thus negatively affecting our well-being.

Dispositional optimism, on the other hand, is an enduring characteristic of personality, rather than an occasional response to life events (Scheier and Carver 1987; Carver and Scheier 2002), which tends to make individuals more likely to experience happiness, satisfaction in life, and overall SWB (Diener et al. 2002; Langer 2002). Dispositional optimism tends to be associated with better mental health in older people (Reker 1997 and references therein). That is, dispositional optimism increases the level of resilience of the individual to the challenges of life (Masten and Reed 2002).

Dispositional optimism and also *positive affectivity*—a psychological trait defined by cheerfulness, enthusiasm, energy, self-confidence, and alertness, Watson (2002)—are expected to favour the experiencing of *positive emotions* (Fredrickson 2002) and states of *flow* by the individual, in which case the individual is more easily absorbed in whatever task, in the conviction that such task is not only meaningful but it is generally worthwhile pursuing and will hopefully produce positive results whatever they might be (Nakamura and Csikszentmihalyi 2002).

Greater life meaning can also help our health and general well-being, as already noted, to the point of reducing the risk of death (Sone et al. 2008; Boyle et al. 2009). Meaning and meaning making may affect health through the mediation of stress control. Less meaning may lead to more stress (What's the point of living? Where am I heading to?) and more stress may negatively affect health; but higher stress levels may also lead to less meaning (What's the point of all this suffering? No time to think about anything else but work hard, with all the stress involved). In the face of stress, meaning may be restored through social help, for instance, whereas once acquired, meaning may provide a buffer against further stress (Krause 2012).

Life meaning in old age also varies across cultures but there are some cross-cultural commonalities on how to achieve satisfaction and meaning in life, especially with regard to the centrality of physical health and functioning, material security, family, and sociality (Fry et al. 2007). In Indian culture, for instance, following the Hindu tradition, life is divided into four stages: the *brahmacārya* or studenthood, *gṛhasta* or the stage of the householder, *vānaprastha* or stage of the forest-dweller, and *sannyāsa* or the stage of the renouncer, the old age (Ram-Prasad 1995). During the life course the individual fulfils a specific set of purposes:

appropriate conduct/responsibility/duty (*dharma*), wealth/material well-being (*artha*), desire/pleasure (*kdma*), and liberation/realisation (*mokṣa*). Meaning is given to life through material well-being (wealth and health) and moral regulation, a conception that will also sound familiar to many Westerners.

8.2.4 Life Meaning and Life Traumas

Traumatic events in life can shatter our view of the meaning of life, making us question what is life for, and whether it is worth living at all. But at the same time traumatic events represent a challenge, an opportunity to come up with new meanings and with a transformed view of our life and its direction. Trauma may be an opportunity for personal growth. A trauma can be defined as any event or series of events that are "spectacular, horrifying, and just deeply disturbing experiences" (Wheaton 1994, p. 90), and yet from trauma hope can emerge.

Viktor Frankl's contribution to the idea of finding meaning in traumatic life experiences has been already mentioned in Chap. 1 and previously in this chapter as well, including in references to Frankl's influence on the work of Paul Wong and others, but here we want to dig deeper into Frankl's ideas, with a special focus on the end-of-life process.

Finding meaning in life springs spontaneously from our own human nature, being expressed in what Frankl calls *will to meaning*: "It refers to the fact which reveals itself to a phenomenological analysis, namely, that man is basically striving to find and fulfil meaning and purpose in life" (Frankl 1967, p. 140). Meaning making is the expression of what Frankl regarded as the *defiant power of the human spirit* (Frankl 1967), the capacity that each one of us has to transcend the contingent, even when it involves suffering. When will to meaning fails, the individual may experience existential frustration or an *existential vacuum*, that is expressed through a "sense of futility and meaninglessness or emptiness and void" (Frankl 1967, p. 140). Frankl in fact goes as far as describing the existential vacuum as "the mass neurosis of our age". Existential vacuum is a state of boredom and life meaninglessness which, if it persists, may lead to depression, a situation that, as we have seen in Chap. 4, can well arise after retirement.

Although Frankl developed these concepts in the context of psychological therapy to treat neuroses, they apply more generally to all individuals, especially those confronting the imminent end to their life. Finding meaning in our own existence is a way to avoid existential loneliness and prevent the "horror vacui" of a life without purpose. In our view a fulfilling life is a meaningful life that, however, must be detached from any concept of "success", which is usually externally imposed and controlled, whereas life fulfilment is determined and judged by each individual (e.g. Kimble and Ellor 2001).

In Frankl's perspective, the prospect of death charges life with meaning and with the responsibility of living each moment to the full, in accordance to our personal choices, because "life holds meaning, in whatever circumstance, until the last breath" (Kimble and Ellor 2001, p. 16). Meaning allows us to constructively adapt to the basic challenges of life, such as what Frankl (1963) referred to as the *tragic triad*: guilt, suffering, and death. None of the components of the triad can be fully eliminated from our life, especially death, but it is entirely up to us to live our life in a fulfilling manner in spite of the inevitability of some suffering, some guilt about acts done in the past and decisions made, and the inescapable end of life.

Frankl also contributed a specific psychological intervention known as *logotherapy*, which emerged from his experience not only as a psychiatrist and psychotherapist, but especially as an internee in the Theresienstadt, Auschwitz, Kaufering, and Türkheim Nazi concentration camps from 1942 until his liberation in 1945 (Frankl 2004). Through the analysis of his own experience and that of his fellow inmates in the camps, Frankl ultimately adopted an optimistic view of life—rather than succumbing to despair and existential nihilism—a view that involves a continuous search for our personal path to a fulfilling life. As expressed by Martin Gilbert in the Foreword to the 2004 edition of Frankl's *Man's search for meaning*:

> 'Man is ultimately self-determining,' he tells us. The ability to decide is at the centre of our being: 'Man does not simply exist but always decides what his existence will be, what he will become in the next moment.' The power to choose is a power every individual can use, and which gives each individual his own personal place and space: 'Everyone has his own specific vocation or mission in life to carry out a concrete assignment which demands fulfilment. Therein he cannot be replaced, nor can his life be repeated. Thus, everyone's task is as unique as his specific opportunity to implement it' (p. xii)

The camps experience drove Frankl to conclude that personal fulfilment can only be developed through love and a rich internal life: "This intensification of inner life helped the prisoner find a refuge from the emptiness, desolation and spiritual poverty of his existence, by letting him escape into the past… As the inner life of the prisoner tended to become more intense, he also experienced the beauty of art and nature as never before" (Frankl 2004, p. 32).

Therefore, meaning is central to Frankl's view of the human mind and the human condition. His logotherapy approach: "healing through meaning", stands on three basic pillars: (a) freedom of will, or the capacity of the individual to make decisions about his/her life; (b) the will to meaning or the human pursue of meaning in life; and (c) finding the meaning of our own life. His view of the human condition is therefore fundamentally optimistic, always in search for a good reason to find contentment in life even from experiences that could far more easily lead to depression and despair. He realises that often our situation may not be easily changed (e.g. in the case of a terminal disease) and the past can never be changed, but what it can be changed is our attitude towards both.

The search for meaning in life is obviously a personal journey, but such journey benefits from our interactions with others. Recounting a phone-call he received early one morning from a person who intended to commit suicide, Frankl explains how he tried to persuade her from such action. Later-on he met her in the hospital where she went for a visit with him and upon asking her the reason why she changed her mind it turned out that it was not what he suggested to her that made a difference:

> The only reason she had decided not to commit suicide was the fact that, rather than growing angry because of having been disturbed in my sleep in the middle of the night, I had patiently listened to her and talked with her for half an hour, and a world – she found – in which this can happen, must be a world worth living in. (Frankl 1988, p. xvii)

Life meaning and social bonds establish the basic foundations from which the individual can build self-transcendence: we humans can transcend ourselves "either toward another human being or toward meaning" (Frankl 1988, p. 6); and finding meaning in life leads to fulfilment and a better mental health. On the other hand, lack of meaning in life produces an existential vacuum that could lead to unhappiness and, potentially, mental illness. Finding meaning in life is an individual responsibility: "meaning cannot be given arbitrarily but must be found responsibly" (Frankl 1988, p. 43); but the social milieu can certainly help in the task.

We have seen in Fig. 1.7 how Frankl points to the possibilities of finding success and personal fulfilment if life has meaning or, alternatively, failure and despair if meaning is lost or never found. But we can also find life fulfilment despite failure, or we may experience despair despite success whenever our life is beset by existential vacuum. In fact, this strengthens our view of the ultimate uselessness of the concepts of "success" and "failure" as they are applied to the life of any person. Meaning and fulfilment in life can be experienced by anyone: famous or unknown, non-disabled or disabled, young or old, and so on.

The logotherapeutic approach is less retrospective and less introspective than other forms of psychoanalysis as it focuses on the future, "that is to say, on the meanings to be fulfilled by the patient in his future… At the same time, logotherapy defocuses all the vicious-circle formations and feedback mechanisms which play such a great role in the development of neuroses. Thus, the typical self-centredness of the neurotic is broken up instead of being continually fostered and reinforced" (Frankl 2004, p. 80). That is, the objective of logotherapy is to focus the individual on the construction of life meaning and to drive him/her away from hedonism and power control: the obsession with "success". According to Frankl, life meaning can be developed in at least three ways (Frankl 2004, p. 90):

a. "By creating a work or doing a deed".
b. "By experiencing something or encountering someone".
c. "By the attitude we take towards unavoidable suffering".

Life meaning brings fulfilment in life through a process of personal development in interaction with others. A life without meaning may lead the individual to experience, in Frankl's words, *existential frustration*, *existential distress*, an *existential vacuum*.

Given the virtual inevitability of experiencing suffering in life, Frankl (2004, p. 111) introduces his concept of *tragic optimism* as a way of coping with losses of all kinds, losses being more common as we become older:

> … hence the reason I speak of tragic optimism, that is, an optimism in the face of tragedy and in view of the human potential which at its best always allows for: (1) turning suffering into a human achievement and accomplishment; (2) deriving from guilt the opportunity to change oneself for the better; and (3) deriving from life's transitoriness an incentive to take responsible action.

Through tragic optimism meaning can be found even in the most negative of experiences. There is meaning even in suffering. Quoting Frankl, Kimble and Ellor (2001, p. 15) stated that we "give meaning to life through *what we give* (in terms of creative works) ... by *what we take* from the world (in terms of our experiencing values) ... through the *stand we take* toward a fate we no longer can change (an incurable disease, an inoperable cancer, or the like)". On the other hand, without optimism existential vacuum can be experienced even when the individual has other aspects of life perfectly under control.

Frankl invites older people to engage in a constructive life review at any time, including as they approach the end of their life, to project their existence into the future. The future can be personal for as long as we stay alive, beyond death it can branch out, in one way or another, if we have devoted some effort to generativity. The life review approach was also supported by Charlotte Bühler, a colleague of Frankl (Bühler 1964). Such process of synthesis through a life review, however, needs to be done before we decay to such a point where we can only focus on the basics of life and may lack the strength, motivation, or perhaps the mental capacity to engage in more complex cognitive tasks. To be constructive, the life review must avoid become bogged down in regrets and sadness, thinking about the missed opportunities in life. Quite on the contrary, the idea is to look forward to the fulfilment of one's own current projects whilst standing on the positive experiences of the past; or simply the fulfilment of a plan to live the last years of our life in calm reflection and enjoyment of the simple things in life (Bühler 1964). In both cases our past positive accomplishments, whichever they may be, are the building blocks that we use to construct our life review. In the words of Charlotte Bühler (1964, p. 6):

> *Accomplishments* may be defined in this context as either the creating of a product or the bringing about of changes in the world which were conceived of as *constructive*. The term 'constructive' I defined previously as 'having a beneficial intent.' A truly constructive accomplishment would come about through individuals' bringing their 'best potentials' to realization. This in turn would be experienced by them as 'fulfilling.'

A fulfilling life that is experienced, in Bühler's words, "through the realization of the individual's best potentials" is at the core of what is known as *Humanistic Psychology*.

Neal Krause, just like Frankl, Bühler, and others, also focused on the effects that experiencing traumatic events has on life meaning. Life meaning can be affected by traumatic events in very different ways, depending on how stressful or not so stressful those events are. For instance, some individuals may cope better than others, being more capable of seeing the positive in the negative due to individual characteristics such as personality, and thus they would be better able to find meaning in traumatic experiences. Personal resilience to trauma is not necessarily easy to build, but it can be achieved, and it is indeed a frequent response in people affected by trauma (e.g. Bonanno 2005). Social coping mechanisms can also buffer the individual from the stresses of life, and they may promote a switch of life meaning from a mere individualistic focus to one where integration with others becomes more relevant (e.g. Krause 2005). More generally: "significant others may help stressed elders rediscover, renew and re-establish goals and plans, by helping them develop a clear sense of what needs to be done" (Krause 2005, p. 504).

In his empirical study carried out in a population of English-speaking, 65 years old and older retirees in the USA, Krause (2005) found that greater life exposure to trauma is associated with a lower sense of meaning in life in older age. This is especially true for the specific case of greater exposure to trauma at ages between 18 and 30 years old which negatively affects the sense of meaning in the oldest-old. On the other hand, the young-old and the old-old found a sense of meaning in life even when they had experienced traumatic events. However, greater meaning could be developed despite trauma the more social support the individual received from family and friends.

Although traumatic stresses can negatively affect life meaning, could a well-developed sense of life meaning buffer the individual from the effects of those traumas, as suggested by Viktor Frankl (1963)? This idea was empirically tested and supported by Krause (2007) who found that although traumatic events experienced throughout life are associated with more symptoms of depression in old age, this is not true for all old people. Some old people are better able to cope with life traumas, and the reason for that is because they have a deeper sense of life meaning. Krause (2009) went even further, demonstrating that a stronger sense of life meaning in old age may decrease the probability of dying during the study period: from 1992 to 2007. This seems to be due to the positive effects of sense of meaning—having a strong sense of purpose in life in particular—on health, after the effects of attendance to religious services and emotional support were controlled statistically.

Frankl also emphasised the spiritual aspects of meaning. He referred to the spiritual dimension of meaning as the *noological dimension*—from the Greek *noös*, which means *mind*—to avoid restricting the concept to religion. Spirituality needs only imply a broad sense of transcendence and of connection with something that is larger than oneself (Steger 2013).

Ultimately, we can also find meaning in death. The inevitability of death does not just render "life more meaningful" (Tomer 2012), but death itself may acquire its own meaning that will affect our attitudes towards it. Paul Wong and his colleagues, for instance, have developed the *Death Attitude Profile-revisited* (DAP-R; Wong et al. 1994) to study our approach to death. The widespread view of death as an opening to a new life or at least as a fact of life that can have a transcendent dimension (e.g. through legacy) seems to decrease the fear of death and help achieve a sense of peace, whereas a nihilistic view of death that offers no hope and no continuity is more likely to lead to fear. According to Frankl (1963), dying is not just a fact of life but also a task towards which we should work if we want to die a peaceful death. Facing death with a sense of peace becomes easier for as long as we do so after relieving our self from *existential guilt*, that is after shedding any regret that "you owe to yourself" (van Deurzen 2002, p. 212).

8.2.5 Meaning Making

In his analysis of meaning, Wong (2012) distinguished two major aspects of it: (a) what gives us meaning, the *content of meaning*: "experiences, activities, goals, and emotional states that imbue life with meaning" (p. xxxi); here he suggests eight specific sources of meaning: happiness, achievement, intimacy, relationship, self-transcendence, self-acceptance, and fairness. We have already mentioned several components of the *content of meaning*. (b) The second aspect of meaning is meaning making, or the *processes of meaning*: "psychological mechanisms or adaptive efforts in the quest for meaning" (p. xxxi) which Wong sees unfolding through six stages that we reproduce here verbatim:

a. the *inertia stage*, in which individuals have not yet embarked on the quest for meaning;
b. the *exploratory stage*, in which individuals are struggling to find meaning but have not yet discovered one;
c. the *discovery stage*, in which individuals have already experienced some success in finding meaning in the major domains of life;
d. the *completion stage*, in which individuals cease their quests for meaning because they have found satisfactory answers to all their existential concerns;
e. the *emergency stage*, in which something horrible happens and shatters the assumptive world of individuals, triggering a quest for meaning; and
f. the *stagnant stage*, in which individuals get stuck in their search because they ask the wrong questions or come to conclusions that do not provide any closure or satisfaction (p. xxxi, italics ours)

Thus, the journey towards life meaning may come to a fulfilling conclusion or not according to the specific individual meaning-making experience.

We are constantly engaged in meaning making as we try to fit the events of life into our current system of meaning and as we restructure such system following specific experiences and traumatic events (e.g. Heine et al. 2006). That is, meaning making emerges from our need to create coherence in our world and to develop some purpose and direction in life, in order to exert some degree of control on life (Sommer et al. 2012), and in order to restore emotional stability that may be unsettled by unsatisfaction with some aspects of our life.

Meaning making "involves an individual's search for a clear understanding of his or her experiences" (Weinstein et al. 2012, p. 87; see also Reker and Wong 1988) leading to regular questioning and perhaps reframing and re-evaluating of our worldview. Thus, if we are restlessly meaning making it may either signify that we currently lack some sense of meaning, or that we are unsatisfied with the one we have (e.g. Steger et al. 2009).

Although meaning making is a continuous process that spans the entire life course, it becomes especially pressing in old age, starting from retirement (e.g. Malette and Oliver 2006) and gaining in importance especially closer to the end of life.

Ultimately, any activity that can promote fulfilment in life and life satisfaction can give meaning to such life (see the *outcome approach* of Dennis et al. 1993), but which specific activity or goal in life will give meaning to the life of each one of us is strictly personal. As stated by Weinstein et al. (2012, p. 94): "not all goals are created equal".

At some point in time our system of meaning may stabilise, giving raise to what has been called our *assumptive world* (Parkes 1971), or the worldview emerging from our preconceptions that leads us to select those relationships we understand and to manipulate them so that they conform to our expectations. Seeking meaning, making meaning, and finally finding meaning in life lead to greater life satisfaction, less anxiety, and less probability of developing depression (Weinstein et al. 2012).

Losing the assumptive world during a crisis threatens our life meaning in such a way that "the whole organization of reality from which the sense of life derived is undermined" (Marris 2002, p. 18), which challenges us to construct a new world view that restores life meaning. In old age, for instance, traumas such as widowhood are followed by grief leading to a search and to the reconstruction of meaning in one's life in bereavement (Neimeyer et al. 2006; Koren and Lowenstein 2008).

Tomer (2012, p. 210) makes the distinction between *global meaning* and *situational meaning* in the process of meaning making:

> The first includes people's basic goals, assumptions and expectations from the world, others, and oneself. The situational meaning refers to particular events and interaction between the person and the environment. The person brings the basic beliefs or global meanings to these interactions and uses them to appropriate the meaning of an event in processes of primary and secondary appraisal…, as well as to search for meaning in an effort to reduce the incongruence between situational meaning and basic beliefs or global meaning … This process of meaning making might involve, therefore, a reappraisal of the event but also… revisions of one's basic beliefs and life goals.

Thus, if global meaning is meaning of life in general (long time-span perspective) and situational meaning is meaning of specific activities or life events (short time-span perspective) then meaning making becomes an effort to reduce the discrepancy between situational meaning and global meaning (Reker and Wong 2012; Slattery and Park 2012; Tomer 2012). When the discrepancy becomes too great after experiencing a series of specific life events, then a new global meaning may emerge that accommodates the changes (Slattery and Park 2012). Recovering a sense of global meaning after a crisis is important to maintain well-being (Carver and Scheier 1981; Carver and Scheier 1998), physical health (Wrosch et al. 2012) and to experience fulfilment in life. Wrosch et al. (2012, p. 539) suggest that to recover global meaning:

> a person needs to engage in adaptive self-regulation processes that allow disengagement from the unattainable goal and reengagement in other meaningful goals and activities. Those people who are better able to disengage from unattainable goals and engage in other new goals should be more likely to maintain a sense of purpose in life, and thereby prevent distress and physical health problems, than people who have more difficulty with goal disengagement and goal reengagement.

To achieve disengagement and re-engagement with life goals the individual must recur to the plastic capacities of his/her mind, perhaps aided by external inputs such as a good advice.

According to Reker and Wong's (1988) *Three-component model* of the structure of personal global meaning—which comprises cognitive, motivational, and affective components— a person who is high in global meaning should display a clear life purpose, a specific sense of direction; he or she should strive for goals consistent with such life purpose, and is expected to feel both satisfied with past achievements and determined to project towards a meaningful future. The *Three-component model* is illustrated in Fig. 8.5.

Global meaning and situational meaning have been integrated into a *Contextual model of personal meaning* by Reker and Wong (1988) (see Fig. 8.6). The model comprises four levels: (a) self-preoccupation with hedonistic pleasure and personal comfort; (b) devotion of time and energy to the realisation of a personal potential; (c) service to others and commitment to larger societal and political causes; and (d) transcendent values that encompass cosmic meaning and ultimate purpose. Global meaning serves as a framework to both creating meaning from and discovering meaning in specific situations in life, and in doing so global meaning functions as a buffer between the stresses of life and personal well-being.

Meaning making is thus a process that recruits the three components of meaning at the individual level (cognitive, motivational, and affective) to maintain coherence between situational and global meaning. Although meaning making in old age springs fundamentally from the individual and his/her personal agency, it can take inspiration and strength from social sources, and it could be also limited and

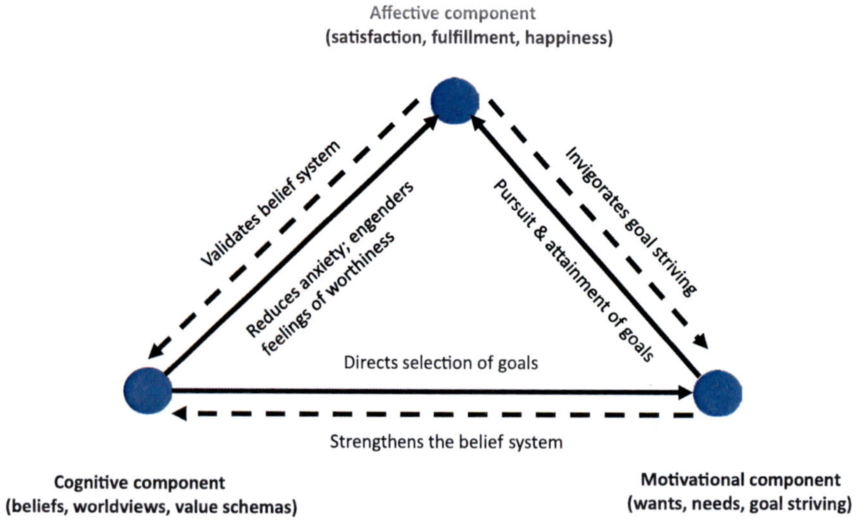

Fig. 8.5 The *Three-component model* of the structure of personal global meaning. Solid arrows represent the direction of influence, whereas dashed arrows represent feedback. Redrawn from Reker and Wong (2012)

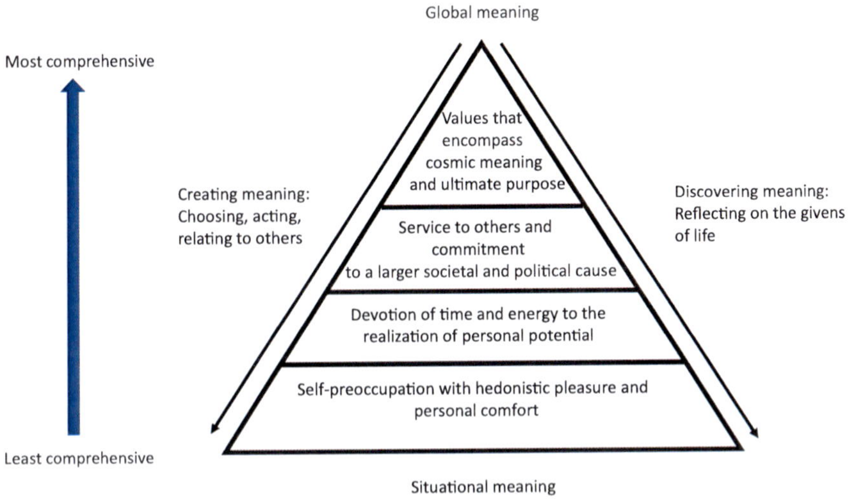

Fig. 8.6 Contextual model of personal meaning in life. Redrawn from Reker and Wong (2012)

constrained by social influences. Social influences that can shape meaning can be found in culture, the characteristics of the elder's cohort, demographic change, work experience, education, leisure activities (including volunteering), personal relationships, interactions with members of the local community of residence; but we can also go beyond, expanding the social influences to participation at ever broader scales of social complexity, even to the point where an elder may be an active member of an international organisation pursuing global objectives (Settersen 2002, see also Sherman et al. 2011).

The place we live in also has an important role in meaning making and the maintenance of meaning in old age (Rowles and Ravdal 2002). Meaning is linked to personal history and personal history is rooted, among other things, into the places we have lived in throughout our life. *Meaning of place* can become especially critical after relocation into a nursing home or aged care facility. Individuals develop a sense of physical intimacy and comfort with settings they routinely inhabit or traverse over long periods of time. Emotional intimacy and comfort combine with cognition to transform our conception of living space, for instance, from *house* to *home*, which is achieved by imbuing the house with meaning. In addition, places also imply a specific dimension of social immersion: a home can be both a personal and a social space, providing an overall sense of *experiential place integration* to the elder (Cutchin 1997). Changing place, therefore, is not an easy decision for elders as it can lead to a sense of *placelessness* (Relph 1976) that may affect their well-being and life meaning. For an old person at the end of life it is also important to be able to choose the appropriate place to die.

Maddi (2012) suggests that meaning making is also affected by *hardiness* and *courage*. We introduced the concept of hardiness in Sect. 8.2.1 in this chapter when we mentioned the work of Thompson and Janician (1988). Hardiness is found in people who "welcome challenges in life, exert control over difficult events, and aim to make lasting commitments amid uncertainty and change. A hardy disposition should promote healthy behavior and the ability to cope well with stress" (McAdams 2012, p. 109).

Displaying hardiness involves showing: *commitment*, or a tendency to become involved in whatever happens around; *control*, or a belief in that direct action can influence the course of events; and *challenge*, or the conviction that what is ultimately most fulfilling is to continue to grow in wisdom through learning from experiences, whether such experiences are positive or negative. Hardiness decreases anxiety in the face of stressful experiences and indeed, through hardiness a stressful experience can be transformed into an opportunity for personal growth. Maddi regards such individual capacity to overcome stressful experiences through hardiness as a positive aspect of "individualism", but he unfortunately sees the social aspects of meaning making only in a somewhat negative way, through *conformism* (see also our critique of Jung in the introductory section of this chapter). His point is to emphasise the proactive stance of individualists, whereas conformists would show a passive stance in life. However, enjoying the benefits of, and actively participating in the life of a community does not necessarily lead to passive conformism. That is why, contrary to Maddi's view, stressors may not necessarily have greater negative effects on conformists' life meaning than on individualists' life meaning. If "conformists" are socially well integrated and enjoy greater social support during periods of personal crisis, then they are likely to better overcome the effects of stressors.

Therefore, there is an important social dimension to meaning making. Through our interactions with others we may learn and grow, thus improving our life meaning. Social interactions are a very important source of meaning through love and affection. This is true whether we receive affection or whether we give affection to others through providing care and support (Aron and Aron 2012).

Another important source of meaning making is our memories of the past, especially the fond and *nostalgic memories*. Nostalgia can be associated with positive mood, greater self-esteem and social connectedness alongside increased life meaning for as long as memories that are recalled by the individual are positive, filling the individual with a sense of satisfaction in life. Memories of "momentous life events" (Wildschut et al. 2006) can obviously elicit nostalgic thoughts that can confer meaning to life and strengthen a sense of fulfilment: the birth of children and grandchildren, marriage, a prize received, the completion of a project, and so forth.

Routledge et al. (2011) studied the role of nostalgia in shaping meaning in life and described how nostalgia increases life meaning, especially through social connectedness, and helps improve psychological well-being when meaning is threatened. The authors conclude that the "provision of existential meaning is a pivotal function of nostalgia" (p. 638).

More broadly, Roy Baumeister (1991) has proposed a structure of meaning making that is grounded on:

1. Sense of purpose in life (finding meaning in objectives for the future and outcomes).
2. Feelings of efficacy (perception that you are the master of your destiny).
3. Positive value (performing morally justified actions).
4. Positive self-worth (striving to feel that you are a good and worthy person, a target of admiration).

If life meaning derives from finding a purpose in life, we should expect it to also lead to experiencing states of life fulfilment, as we have argued. However, Baumeister (1991, see also Baumeister and Vohs 2002) is sceptical about the role of fulfilment in the process of meaning making, instead referring to the *Myth of fulfilment*. In a book chapter co-authored by Baumeister, the *Myth of fulfilment* was described so:

> It appears that actual achievement is less important for a sense of purpose than is the process of working toward one's goals. People do not become permanently satisfied upon reaching some goal or fulfillment state… the term *myth of fulfillment* … refer(s) to the contrast between popular notions of fulfillment, which are generally permanent, and actual experiences of fulfillment, which tend to be transient. In Western culture, the popular notion of passionate, romantic love is a striking example of the myth of fulfillment because the image celebrated by novels, movies, poetry, and other sources promises undying love, whereas the empirical evidence suggests that such feelings typically *drop off substantially* after a few months or years… The mythical nature of fulfillment is suggested by the fact that the process of approaching it is often equally important as, or more important than, the fleeting moments of bliss (Sommer et al. 2012, pp. 301–302; italics ours).

Such criticism is grounded in a reductive view of fulfilment. Sommer et al. seem to regard fulfilment as an ephemeral emotional experience coming and going like the waves on a seashore. But fulfilment also has a long-term existential dimension which may span over a lifetime, alongside a short-term dimension that can build up as we go through diverse experiences in life. Fulfilment can indeed be experienced at the end of life, as we stop and look backwards to the perhaps meandering path we followed throughout our life, the little (or not so little) things we have built along the way and how such accumulated experience has made us who we are at that specific point in time. Fulfilment is not equivalent to ultimate "success"/"achievement", as we argue in this book, instead fulfilment is equivalent to calm satisfaction (more eudaimonia than hedonism), which can be experienced even when the dreams of youth may have been shattered by the harsh reality of life, or when we have gone through a life course that was perhaps unexpected but that it can be quite satisfying nonetheless. Fulfilment is both a matter of process, in agreement with Sommer et al., but also a matter of satisfaction with whatever that process has led our life into. In old age such satisfaction can be enduring rather than being just a "fleeting moment of bliss".

Through meaning making life can be fulfilling in whichever way we want, for as long as our attitude to life remains constructive. For instance, in his book *The Ascent of Man* (Bronowski 1976), Jacob Bronowski recounts the story of the *Watts Towers* built by Sabato Rodia—an Italian immigrant who made a living as a tile mason and construction worker in Los Angeles—over a period of 33 years (1921–1954). Rodia wanted to build something big that could endure over time and be remembered, and he did so by building a large, gothic-looking structure made up of scrap metal, pieces of tiles and glass all held together by concrete. What is the meaning of the *Watts Towers*? It is just the life fulfilment of a single man, who gave material expression to his aesthetic perceptions. The structure survived attempts to demolish it and today there are guided tours to visit the towers that are organised by the adjacent *Watts Towers Arts Center*. In this case, one man's life fulfilment has endured over time and it may have helped others develop their own meaning in life. We can only wonder how long it will take for the *Watts Towers* to "drop off substantially".

Meaning can also emerge from negative experiences, as we already mentioned with reference to Frankl's and others' works. Such experiences may be of such intensity that they may challenge our worldview, leading to a profound transformation. Traumatic events, for instance, can lead to a process of personal development known as *post-traumatic growth* (Tedeschi and Calhoun 1996; see also Wong's *PURE principle* in this chapter). Some of the variables affecting post-traumatic growth include: (a) Cognitive processing of traumatic events; (b) expression of concerns about such traumatic events; (c) supportive reactions of others to the acts of self-disclosure of people who suffered traumas; (d) effects of sociocultural contexts affecting both traumas and the attempts to process, disclose, and resolve them; (e) the personal dispositions and resilience of the survivors of traumas; (f) the extent to which circumstances favour or suppress the above process (Tedeschi and Calhoun 2012). Thus, the chances of post-traumatic growth increase the better the trauma is processed cognitively and resolved emotionally by the individual, and the more effective social support is.

When the capacity for meaning making is lost, we may fall into a form of "existential sickness" such as *vegetativeness*: "At the cognitive level, individuals suffering from vegetativeness cannot find anything they are doing or can imagine doing that seems interesting or worthwhile." (Maddi 2012, p. 69). Vegetativeness leads to apathy, aimlessness, and the depletion of the will to live.

Another form of "existential sickness", according to Maddi, is *nihilism* which, paradoxically, finds "meaning" in lack of meaning. Maddi also includes *adventurousness* as a form of "existential sickness", whenever the individual seeks meaning through extreme and risky activities. However, not all forms of adventurousness are likely to be maladaptive, or an expression of sickness, existential or otherwise. In 2019 we watched the documentary *Free Solo*, starring the American climber Alex Honnold, who was able to reach the summit of the *El Capitan*, a vertical granite rock formation situated within the *Yosemite National Park*, USA. Honnold achieved the feat without any safety aid whatever (no rope, just shoes, and chalk) and he did so with apparent ease and complete self-control. If anything, it was the viewers who felt rather sick, so scary the adventure looked to non-climbers: one little mistake and he would have fallen to his death. At some point in the documentary it was revealed that Honnold conforms to some characteristics of Asperger's syndrome. Should Honnold's adventurousness be described as sickness? Or should his psychological and physical characteristics just put him in a position where the experiencing of specific forms of meaning is possible, whereas the same forms of meaning are out of reach to most other people? Perhaps Asperger's syndrome may simply allow some individuals to achieve exceptional levels of flow in the performance of specific activities that can help in the fulfilment of specific existential objectives.

The process of meaning making also involves a constant effort of *meaning repairing*, where we tend to restore the old meaning after a challenge, if we can (Steger 2012; Proulx and Inzlicht 2012). The capacity of meaning repairing varies according to how resilient our meaning-making system is to the challenges of life. The resilience of our meaning-making system to external challenges has been encapsulated by Proulx and Inzlicht (2012) in their *Meaning maintenance model* (MMM). The more our life meaning is resilient to the disruptions from life events, the easier the recovery of meaning will be, until contradictions pile up to such an extent that a breaking point is reached and a more serious effort of meaning making, involving a radical reconstruction of new meaning in life is required.

Meaning making can be also understood from the perspective of additional theoretical models. *Self-determination theory* (SDT) (Ryan and Deci 2000), for instance, propounds that "humans share at least three *basic psychological needs*—the needs for competence, autonomy, and relatedness—that must be satisfied for the healthy functioning of these inherent growth and integrative tendencies." (Weinstein et al. 2012, p. 83). Thus, SDT points to the evolved capacity (or *intrinsic motivation*) of humans to effectively operate in the world as both individuals and members of an integrated group. The emergence of new meanings can be favoured (or prevented) by both individual (e.g. a drive to self-organisation) and social factors. *Competence* is seen in SDT as a necessity to be and feel effective in the world; *autonomy* is the necessity to experience own behaviour as self-organised and driven by personal motivations. *Relatedness*, on the other hand, refers to our evolved social nature that leads us to seek connectedness with others, and engaging in cooperative acts.

Personal frustration in the meaning-making process may lead to *amotivation*, *passivity,* and *helplessness* and hence existential vacuum (Weinstein et al. 2012), whereas finding meaning in life leads to experiencing life satisfaction, positive affect, subjective well-being, and better mental and general health.

As we have already seen, Frankl, Wong, and others regard the life review to be a very important aspect of the meaning-making process. Hétu (2000) has proposed a series of stages and categories for the life review. The stages describe how a life review unfolds as a process (Table 8.3), whereas the categories describe the

Table 8.3 Stages of life review

Stage	Function
Priming	Evoking the past, using a question or a recent event
Immersion	Recalling past events and their associated affect and reviewing them in detail
Reaction	Recalling not only past events and their associated affect, but also reacting emotionally to them
Reframing	Using a new frame of interpretation to review past events, and possibly finding new meaning
Integration	Reconciling with oneself, one's life, one's past

From Hétu (2000)

Table 8.4 Categories of life review

Category	Function
Priming	Recalling an event without exploring the context or associated affect
Defensive	Putting aside difficult past events or minimising their impact (Life Review is not blocked but is slowed down)
Blocked	Avoidance of the past, and/or struggling with "unfinished business" or obsessiveness
Ongoing	Going to and fro between the stages of immersion and reaction
Consolidated	Experiencing a certain peace regarding the past concomitant with a need to review certain events which are no longer troubling
Ceilinged	Reaching a certain level of peace due to relating to the past, due to the conclusion that all that could be done was done
Terminated	Integrating one's past and one's losses peacefully (full integration)

From Hétu (2000)

relationship of a person with his/her past (Table 8.4). Thus, we can recall past events cognitively, emotionally, or more likely a combination of both and from such recalling we may draw some conclusions and reframe our worldview, creating new meanings. By going through a process of life review we may also reconcile ourselves with some events of our past. Life review categories include straightforward recalling of a past event (*priming*), dismissing other events (*defensive*), struggling with unfinished businesses (*blocked*), continuously shifting between stages of immersion and reaction (*ongoing*), reconciliation with our past (*consolidated*), achievement of peace (*ceilinged*), and full integration of our past within our self (*terminated*).

Meaning making could be used in psychological therapy to foster not only *personal consolidation*, or the pulling together of the various loose threads in our life, but also *personal transcendence* and *personal transformation* that can offer a creative way forward to the direction of our life, through a diversity of autobiographical methods, such as life review, reminiscence, and guided autobiography (Reker et al. 2012).

Broadly inspired on Frankl's logotherapy, Wong proposed the *Meaning-centered Counseling and Therapy* (MCCT), which is a therapeutic approach based on meaning making that is delivered through the use of various social-psychological methods such as life review, cognitive reframing, existential and spiritual coping, attribution, and stress appraisal (Wong 2012b, p. 627):

> Consistent with most faith traditions and the tenets of logotherapy, MCCT believes that the terminal value of self-centered pursuits of personal happiness and success often lead to disillusion and misery, whereas the ultimate concern of actualizing one's mission leads to authentic happiness and fulfillment.

Such happiness and fulfilment can be experienced through meaning making, according to Wong (2007), by promoting goal striving (agency), intimacy and family relationships (love), social relationships more generally (community), self-transcendence (larger cause), religion (spirituality), self-acceptance (maturity), and fair treatment (justice and morality). In fact, following Frankl (2019), happiness should not be regarded as a goal but as an outcome of the fulfilment of what makes our life meaningful to us.

William Breitbart designed an *Individual Meaning Centred Psychotherapy* (IMCP) (Breitbart 2002), that is also based on the works of Viktor Frankl on logotherapy, and that can be used with old people approaching the end of their life. Breitbart roots his own meaning-making therapy on the concept of self-transcendence that, as we have seen, was also included by Wong in his MCCT. Self-transcendence can be achieved through assertive communication, constructive thinking, clarification of own values, management of feelings and ability to solve problems, for instance. The process of achieving self-transcendence is aided by a logotherapist who guides the individual through a journey of self-discovery. IMCP involves the following steps:

1. Enhancing the rapport with the therapist.
2. Eliciting sources that have provided meaning in the elder's life.
3. Focusing on the impact of illness.
4. Dealing with the fear of dying using the technique of *dereflection*: taking the focus of thoughts off oneself—the issue of one's death in this case—decrease excessive self-reflection that may interfere with action, loosen up and thus be better able to perform specific tasks. That is, dereflection aims at limiting the tendency to ruminate.
5. Enhancing a sense of closure with significant others in life as death approaches.

Steps 4 and 5 of IMCP specifically refer to strategies aimed at intervening on individuals who are close to their death. Therefore, it is appropriate to ask: Can meaning making be used therapeutically also in palliative care? Palliative care is the preamble to the end of life for old people diagnosed with terminal illnesses. At this stage meaning can mainly come from the satisfaction of past achievements, including having a good family that shares that last period in life with the dying elder, and a conception of the future found in transcendence, spirituality, religion, and legacy, the latter can be constructed in its details after a period of life review (e.g. Romanoff and Thompson 2006). Through this process, an old patient in palliative care can find fulfilment in preparation to the end of life. Breitbart's IMCP, for instance, has been already shown to be useful in terminal patients to decrease anxiety, mood disturbance, and distress derived from physical problems.

To recap, seeking meaning in life is a basic characteristic of our humanity. A meaningful life is one that is worth living, whereas meaninglessness may plunge us into existential despair. Old age is usually associated with greater life meaning, but it does not preclude the search for meaning if meaning is shaky or lost. Understanding life meaning in old age requires an appreciation of the cognitive, emotional, motivational, and social factors that affect meaning production, maintenance, and change. The more meaningful our life is, the greater satisfaction we can experience in life, the more capable we are of overcoming life traumas and the better our health may be. Following Frankl's logotherapy approach, meaning can be derived even from traumatic experiences when optimism prevails. Through optimism we can productively engage in a process of meaning making that can help us find new meaning when the old one is lost or is no longer satisfactory. Help from others can also assist in the meaning-making process. Finding fulfilment in a meaningful life as we approach death should better guide us towards the last stage of our existence. Beyond that, our past life may retain some meaning through our legacy.

In the next section we focus on the *Legacy* sub-theme of our analysis of *Life meaning and legacy*.

8.2.6 Legacy

> I once was sitting in a restaurant with Rabbi Zalman Schachter talking about the search for meaning in old age. Suddenly Reb Zalman turned to me and said, 'You know, this search for meaning comes down to a simple question: Are you saved?' Hearing this, I was puzzled, and I thought to myself, well, he is a clergyman 'Is this perhaps a theological reference?' I asked. 'No,' said Reb Zalman, 'I don't mean it in a theological sense but in a computer sense. Are you saved? Have you downloaded your life experience for coming generations? Have you started doing your legacy work?'
>
> Moody (2009/2010, p. 70)

We briefly mentioned the issue of legacy in previous chapters with reference to our preparation for retirement, life in retirement, and its role in generativity (Chap. 1). In Chap. 5 we also mentioned how legacy can improve life satisfaction in old age. In this section we will mainly focus on legacy in the context of the final stages of our life.

The drive for legacy in old age is ancient in the human species and linked to our desire to extend ourselves through time and beyond death. This drive can be experienced by everybody, not just leaders or people in the position to be more broadly influential (Zacher et al. 2011). Myths of immortality abound across cultures but in fact the easiest way to achieve an "immortality" of sorts is through our transmission of knowledge to the next generation. Payne (2015, p. 19–20) quotes Plato who in his *Laws* points to the importance of producing children who can carry on with the legacy of their forebears:

> A man shall marry between the ages of thirty and thirty-five, considering that in a manner the human race naturally partakes of immortality, which every man is by nature inclined to desire to the utmost; for the desire of every man that he may become famous, and not lie in the grave without a name, is only the love of continuance. Now mankind are coeval with all time, and are ever following, and will ever follow, the course of time; and so they are immortal, because they leave children's children behind them, and partake of immortality in the unity of generation. And for a man voluntarily to deprive himself of this gift, as he deliberately does who will not have a wife or children, is impiety.

Of course, the legacy work does not necessarily rely on an elder having his/her own children and grandchildren, but obviously someone must have children, otherwise generativity stops.

The search for identity beyond the grave is what Butler (1970) called *historicity* and that we will refer to as *legacy*. Legacy is not supposed to be a mere narcissistic endeavour, but rather a "genuine concern for posterity", a "healthy expression of the progression of generations" (Butler 1970, p. 123). All people can become more aware of their own personal legacy after undertaking a process of life review (Butler

1963). Such life review is based on reminiscence and it is especially common the closer one gets to the end of life, as we have seen in the previous sections.

Legacy has been defined by Hunter and Rowles (2005, p. 328) as *a mechanism for transmitting a resilient and enduring image of what we stood for*; or, more generally *the process of passing one's self through generations, creating continuity from the past through the present to the future* (Hunter 2008, p. 314). This drive to give more meaning to life by extending our influence on the world beyond death, can take expression through legacy in the transmission of (a) influence (e.g. behaviours, intellectual achievements, political initiatives, and so forth) and/or (b) material entities (e.g. our belongings, our creations, our children, even our own body).

Just about anything we have done in life can be part of our legacy, but what will endure is mainly not up to us to control, but to the next generations to appreciate, treasure, and make use of it as they wish. In fact, most of our deeds are likely to be forgotten and have no lasting consequence whatever, but some of them may endure.

We have seen in Chap. 1 Erikson's (1997) concept of generativity: a process used to achieve meaningfulness and transcendence through the ability to guide the next generation. For as long as our legacy is generative it is more likely to endure in time, and we agree with Hunter and Rowles (2005) that legacy is a "mutualistic" concept that benefits both the elder (in terms of being remembered) and the younger generations (in terms of benefitting from the experience of the elders).

Hunter and Rowles (2005) have proposed a model of legacy expressed in terms of the inter-generational transmission of biological traits (e.g. through reproduction), material products (e.g. wealth, artefacts), and values (e.g. personal ones) (see Fig. 8.7, and Hunter 2008). But we can also transmit specific pieces of knowledge (e.g. cooking recipes) and ways of expressing our emotions (e.g. music). In a study

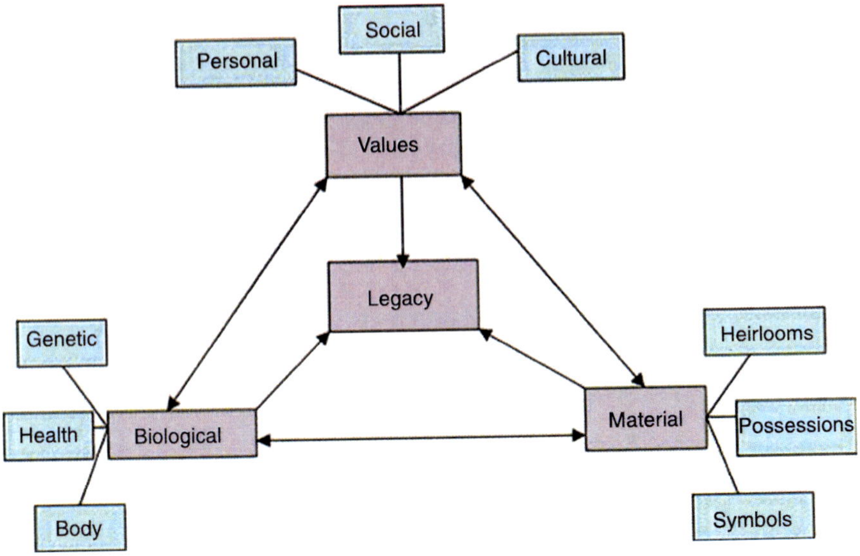

Fig. 8.7 A model of legacy as it was derived from interviews with elderly people. In addition, we can also transmit specific pieces of knowledge (cognitive transmission) and ways of expressing our emotions (emotional transmission). From Hunter and Rowles (2005)

carried out in the USA, Savishinsky (2007) listed legacy initiatives taken by elderly participants that included composing autobiographies, drawing genealogies, writing oral histories, organising and distributing possessions, family photographs and also practical initiatives such as finalising their wills and funeral arrangements that would lighten the burden on relatives upon the individual's death.

Ultimately, it is the life of the individual that can become a legacy for the next generation: a *legacy by example* (Savishinsky 2007). In the words of Savishinsky (2007, p. 85): "In that sense, a legacy offers a proof of life, a hope that personal identity will outlive the person" or, we may also say that legacy offers a form of "symbolic immortality" (Lifton 1973). Although referring to legacy as "symbolic immortality" may seem rather optimistic, as most legacies have a relatively limited time span in personal or collective memories. But in a sense, it is true that, as Poiani's old father used to say: You really die only when nobody remembers you anymore.

Hunter (2008) studied perceptions about legacy in a sample of 38 women (31–94 years old) in the USA, addressing a specific aspect of legacy: "transmission of self through personal actions and personal artifacts" (see Fig. 8.8). Hunter described five major actions categories that were included in the interviewees' understanding of behavioural legacy:

- *Personal behaviour* (being a role model): "The legacy I'd like to leave behind is that I know that I cared for people, that I treated people well and that I respected people" (p. 50).
- *Spiritual and/or religious practices* (share personal foundations for living a moral life): "belief in kindness, honesty and love" (p. 50; see also the specific perception of legacy in old nuns (Wolf 1990)).
- *Career roles* (making a difference to people's lives): "The people I do counseling with may not know when I die, but hopefully many of them will remember me or remember the change that I helped facilitate in them" (p. 50–51).
- *Medical decision-making* (transmitting lessons learned from experiences of illness): "fighter or not being a burden" (p. 51).
- *Health behaviours* (influence disease prevention through advice coming from personal experience with health issues): "(hope) that their own fight with cancer

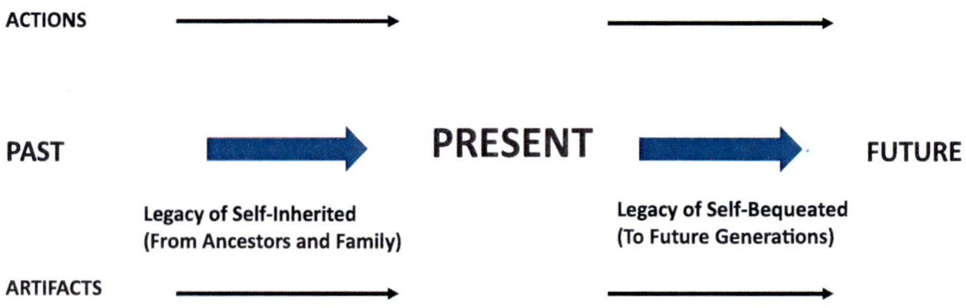

Fig. 8.8 Historical perspective of legacy through inter-generational transmission of actions and artefacts. Redrawn from Hunter (2008)

would influence their family member's health behaviors educating them about the disease and encouraging them to seek preventive testing in order to avoid the problem or detect it early" (p. 51).

Examples of legacy of artefacts or, more generally, material legacy that was recorded by Hunter included:

- *Autobiographies* (written or recorded; in this regard see the use of digital media (Manchester and Facer 2015)); *funerals*, which are correctly described by Hunter as "concrete rituals that allow one to present oneself, one final time" (p. 51); *wills*, which allow the individual to pass on cherished possessions (*belongings*) to people or institutions they hold dear; and *body donation* (e.g. to medicine).

Legacy can be also used in specific psychological interventions to help old people who are approaching the end of their life. For instance, encouraging older people to tell the story of their family to the younger members, including their own personal experiences within the family, can help build a sense of identity in the young and at the same time give meaning to the elder's life through the experience of "passing on of the baton" (Randall and Martin 2003) and generativity. The elders can also pass on specific historical knowledge from outside the confine of the family, as they may have gone through experiences that marked specific historical periods (e.g. the Great Depression, a war). The stories can attract the interest and curiosity of the young and represent an important instance of learning for them.

Conscious legacy building is also used to help old terminal patients come to terms with their end of life, by alleviating the effects of negative emotions (Otis-Green 2003). Writing documents based on a life review can be useful, not just in terms of transmission of specific knowledge about the family or historical events, as mentioned above, but also in terms of passing on wisdom and personal reflections.

In extreme cases, especially when the elder lives in a nursing home and has no family members coming to visit, legacy could still be used to help improve the elder's life meaning through visits by *volunteer listeners* (McNees 2009). Although in this case the objective is to help the elder, it is quite possible that even a volunteer, a stranger, may be personally moved by some stories and life experiences, thus learning from them and perhaps being transformed by them. In such a case, the elder would have in fact been generative and left a legacy.

Reminiscence and legacy have also been used as a form of meaning reconstruction to help palliative care patients alleviate their symptoms of depression (Allen et al. 2014).

Harvey Chochinov introduced *Dignity Therapy* that includes aspects of legacy to alleviate suffering, especially existential suffering, in terminal patients (Chochinov and Cann 2005, Chochinov 2006; see also other approaches used by Kearney and Mount 2000; Rousseau 2000). In this intervention the therapist follows a specific protocol to guide participants to reflect on their feelings, thoughts, hopes, past, present, and future. The conversations may be made available to the family as part of the participant's legacy. More specifically, the protocol follows these nine steps:

1. Identify mentally competent, terminally ill patients who wish to take part in dignity therapy.
2. Assisted by a therapist, patients are guided through a dignity therapy interview that uses the questions protocol [see below] to direct its content.
3. Depending on the patient's ability and wishes, offer one or two sessions in as close proximity as possible (1 to 3-day interval).
4. Transcribe conversations within 2–3 days of the final interview.
5. Over the next 24–48 h, edit the transcripts to convert the dialogue format into a polished narrative (it involves removing colloquialisms, correcting time sequences, eliminating material not intended for generativity purposes).
6. When editing is complete, revisit the patient to read the entire manuscript, ensuring that all changes are correct and meet the patient's specifications.
7. Any necessary corrections to the document are made within 24–48 h.
8. When corrections are completed, the patient is provided a hard copy of the generativity document.
9. When they wish to do so, patients may share or bequeath the document to friends, family, or loved ones of their choosing. Slightly adapted from McClement et al. (2007, p. 1077).

The conversation with the patient is guided and motivated by a series of recommended questions, as part of a questions protocol (McClement et al. 2007, p. 1077):

- Tell me a little about your life history; particularly the parts that you either remember most or think are the most important.
- When did you feel most alive?
- Are there specific things that you would want your family to know about you, and are there particular things you would want them to remember?
- What are the most important roles you have played in life (family roles, vocational roles, community service roles, and so forth)?
- Why were they so important to you and what do you think you accomplished in those roles?
- What are your most important accomplishments, and what do you feel most proud of?
- Are there particular things that you feel still need to be said to your loved ones, or things that you would want to take the time to say once again?
- What are your hopes and dreams for your loved ones?
- What have you learned about life that you would want to pass along to others?
- What advice or words of guidance would you wish to pass along to your [son, daughter, husband, wife, parents, other(s)]?
- Are there words or perhaps even instructions you would like to offer your family, to help prepare them for the future?
- In creating this permanent record, are there other things that you would like included?

The effectiveness of *Dignity Therapy* was evaluated by Chochinov et al. (2005) in Canada and Australia using the following dignity themes:

– Generativity.
– Continuity of self.
– Role preservation.
– Maintenance of pride.
– Hopefulness.
– Aftermath concerns (concerns about the burden that patient's death may impose on others).
– Care tenor (ways in which others interact with the patient that may foster or constrain dignity).

Dignity Therapy was especially useful to patients who were initially in greater psychosocial despair and it helped them increase their will to live and their life meaningfulness.

In a subsequent work Chochinov et al. (2011) compared the effectiveness of *Dignity Therapy* with that of *Client Centred Care* (CCC) which is based on conversations between the patient and a nurse-therapist focusing on issues that are currently relevant to the patient. Thus, CCC does not have a generative focus. An additional set of participants was also available for comparison who just received standard palliative care, where generative issues are also not addressed. A total of 326 participants completed this study: 50.6% women, who were mainly receiving home based palliative care (60.1%). Upon completion of the intervention, patients in the *Dignity Therapy* group were more likely to express the opinion that the study was useful and that it had improved both their quality of life and sense of dignity. *Dignity Therapy* patients also reported that the experience had positively affected their relationship with family, they enjoyed greater spiritual well-being, decreased sadness and depression, and they improved the feeling of personal satisfaction.

Dignity Therapy has also been successfully used to help frail elderly people not in palliative care (Chochinov et al. 2012). As it is also the case for patients in palliative care, dignity therapy in this circumstance helped diminish elderly people's suffering whilst increasing life meaning and dignity. Such benefits were also extended to the family of the participants.

Allen et al. (2014) evaluated an intervention program for palliative care patients in the USA that involves the participation of retired senior volunteers (RSVs) delivering a three-session reminiscence and creative activity. The authors measured the effects of the intervention on emotional and spiritual functioning in a group of 55 years old or older palliative care patients who were living in the community or in assisted living, they had an advanced illness or a combination of chronic illnesses, and were at the most only mildly cognitively impaired. After controlling for the effect of patient health, patients in the intervention group displayed greater reduction in the number of emotional symptoms and emotional bother than patients in the control group, the latter increased both emotional symptoms and bother. In addition, intervention patients expressed more forgiveness than controls and they also displayed less negative religious coping, such as feelings of abandonment by God, through the period from baseline to post-intervention. Therefore, this intervention delivered by RSVs had a positive effect on emotional and spiritual well-being in this group of palliative care patients.

So far, we have mainly stressed the legacy projections of old people towards their self-identity and their family members. However, there is a more global dimension of legacy that transcends the self and the immediate family, to extend towards the future of society, the environment, and indeed life on Earth itself. Increasing environmental consciousness calls for a consideration of our own legacy in terms of bequeathing an environment to future generations that is as good or better than the one we have lived in. Moody (2009/2010) used the term *eco-elders* that he characterised as people with legacy motivations that focus on our collective future. For instance, he lists some "types" of eco-elders such as the *Gatekeepers of the Future*, or those who "cultivate a 'long view' about the future"; the *Memory Keepers* or the "custodians of memory of the past"; and those engaging in *Grand Generativity*, or generativity initiatives directed to grandchildren and greatgrandchildren. The conservation of our natural environment is a generative activity that involves all those types of eco-elders.

Elders can also act as *keepers-of-the-culture* for the benefit of future generations (McNees 2009) and can teach the young about the past of their community or ethnic group. Reminiscing and transmitting cultural knowledge are also useful to the elder and their sense of life meaning.

In sum, older people at the end of their life can find transcendence through a work of generativity by building and then transmitting their legacy to younger generations. Our values, our deeds, our material possessions and creations, our children can all be integrated into a legacy that can bring a sense of fulfilment in life. Elders who may lack the tools to think about and build a coherent legacy can be helped with appropriate interventions if they so wish.

Although approaching the end of life in a positive manner and with a strong sense of meaning and legacy is expected to decrease our fear of death and death anxiety, a degree of fear and anxiety may linger on. Individuals will have to find the resources to cope with their fear of death and death anxiety but, as usual, others may also help. This is the focus of the next section.

8.3 Fear of Death and Death Anxiety

> Dying is nothing to fear...it all depends on how you have lived
>
> Elisabeth Kübler-Ross (1997, p. 286).

In the last song composed by the Canadian singer, songwriter and poet Leonard Cohen before he died at the age of 82 (*You want it darker*; the album was released nineteen days before his death) he provided a less than optimistic view of life, and yet his refrain *Hineni, hineni*, a Hebrew word meaning "here I am" denotes a preparedness, a readiness to abandon himself and accept death. Is that attitude towards death common? Or are fear of death and anxiety about death more prevalent among elderly people? What kind of life should we live in order not to fear death when the moment comes?

In this section we start by introducing the factors that affect fear of death and death anxiety, to then analyse some approaches used to understand them: psychological, evolutionary. We will see how the belief in literal immortality or the construction of some form of symbolic immortality can give transcendence to old people at the end of their life that can help them cope with fear and anxiety. Coping with fear of death and death anxiety requires using all the tools we have at our disposal: personal (religious, spiritual) and social (emotional support, legacy, generativity).

The human mind had to come to grips with death since we evolved consciousness. We do not have many records of the understanding of death in prehistoric times, but we know that dead bodies were intentionally buried in the Palaeolithic, that is, since at least 120,000 years before present (Pettit 2010). Since then humans have produced a large variety of funerary rites and views of death and the afterlife (see Kastenbaum and Costa 1977; Davis 1999; Insoll 2011), with a common feature among them being a widespread desire for death not be a final stage of total annihilation of being, a belief that has the immediate effect of decreasing fear and anxiety about death and loss.

Nash (1980) described four major aspects of death: pain, loneliness, intrusiveness (or lack of control), and loss. All those aspects can be sources of anxiety and fear. Beliefs in death as a mere passage into a different state of existence can fulfil the immediate function of decreasing anxiety and fear in the dying individual, his/her family, and acquaintances.

Once people become aware that death is near, they may adopt a variety of responses:

> (a) closed (don't talk about it), (b) suspicion (no communication but there is suspicion that something is not going well), (c) mutual pretense (pretend that everything is fine), and (d) open (openly talk about death and dying) (Nussbaum et al. 2000, p. 296).

The different responses are idiosyncratic strategies to manage death anxiety and fear of death.

The concept of *death anxiety* refers to a "dread of complete annihilation" (Lehto and Stein 2009, p. 31), a dread of death that "resides in the unconscious, a dread that is formed early in life at a time prior to the development of precise conceptual formation, a dread that is terrible and inchoate and exists outside of language and image" (Yalom 1980, p. 189). *Fear of death*, on the other hand, is a "more concrete belief that death is frightening" (Lehto and Stein 2009, p. 31; see also Momeyer 1988; Cicirelli 2006).

Further distinctions have also been made regarding different types of death anxiety. For instance, *unknown death anxiety* or *existential death anxiety* is caused by concerns about what happens after one die. On the other hand, *known death anxiety* or *tangible death anxiety* results from worries about what happens to the body during the process of dying or after death (Benton et al. 2007).

The precise expression of death anxiety and fear of death can be affected by multiple factors, including our emotions, cognition, motivation, and sociocultural influences that vary over time (Florian and Mikulincer 1997; Lehto and Stein 2009).

Attig (1979) and Wong (2007) described various aspects of fear of death, including seeing death as a final stage that cannot be avoided or reversed; the fear of uncertainty about what follows after death; the fear that death is just a state of annihilation of being; the fear of losing everything we value as a result of death, including the relationship with loved ones. The possibility of death causes fear of disruption of the broader "flow of life" and the fear of abruptly ending those life projects that remain unfinished. Confronted with the end of life we may fear indignity, loss of control, pain, a violent death, and the loneliness of dying; fear of being a burden to family during the dying process, fear of judgement and retributions.

Fear of death tends to draw people away from those who are dying, and such response seems to be more widespread in Western societies at present. Kastenbaum and Aisenberg (1972) argue that there may be at least three reasons for this: (a) attributing social inferiority to dying people; (b) incapacity to deal with dying people; and (c) avoiding psychological perturbation caused by dying people.

Death anxiety derives from an increased awareness of *death salience*, or reminders of own mortality. Such reminders may spring from experiencing major stresses in life such as those coming from living in stressful environments (e.g. war, natural disasters), the diagnosis of a life-threatening illness (e.g. cancer), and personal experiences with death and dying (e.g. the passing away of family members, loved ones, friends) (Lehto and Stein 2009).

More specifically, six attributes of death anxiety have been identified that refer to its emotional, cognitive, experiential, developmental, sociocultural, and motivational aspects (Lehto and Stein 2009):

Emotional: Death anxiety is closely related to fear of annihilation of one's existence. We have seen in Chap. 2 how Erikson described old age as a stage in life that tends to be characterised not only by greater wisdom, but also a tension between integrity and despair (Erikson 1982). Such despair can emerge from death anxiety.

Cognitive: Cognitive factors determine individual differences in the need to put order and give structure to our world, whilst establishing a degree of tolerance for ambiguity and capacity to manage uncertainty, modulate the impact of perceptions of threat, and therefore modulate the degree of death anxiety that we experience. Some relevant cognitive aspects of death anxiety include: "attitudes, conceptual ability to predict and anticipate the future, and awareness of the salience of death… beliefs about death, images and beliefs regarding the death experience, and the self as dead or no longer existing" (Lehto and Stein 2009, p. 26).

Experiential: Death anxiety tends to be repressed as an adaptive strategy to reduce fear and terror that may jeopardise our survival. Self-regulation processes such as self-esteem are used in such repression or inhibition of death anxiety.

Developmental: Death anxiety tends to change with age showing a curvilinear relationship, with a maximum observed at around middle age and young-old, and subsequently declining at older ages.

Sociocultural: Culture affects different aspects of death anxiety and it can also exert a protective effect on the individual that can decrease death anxiety and fear of death (see also Florian and Mikulincer 1997).

Motivational: Death anxiety can be a source of motivation in artistic creativity and in the accumulation of resources; some of these initiatives may be aimed at overcoming the fear of death.

Death anxiety has been analysed from a variety of theoretical perspectives. On the one hand, it has been interpreted negatively as evidence of lack of self-realisation and fulfilment in life; on the other hand, it has been seen positively as evidence of openness and ego-strength as we confront a radical aspect of human existence (Wink 2006). A so-called healthy view of death anxiety suggests that improvement of satisfaction in life leads to lower death anxiety in old age, whereas the "morbid view" propounds that death anxiety "is a given of human existence because we are the only species to realize our mortality, an insight that contributes to the dread of non-existence" (Wink 2006, p. 95). Put in simpler terms, death anxiety emerges from a conflict between being aware that death is inevitable and wishing to extend our life for as long as possible (Yalom 1980).

From an existential perspective, death anxiety is seen as a result of realising that our life is finite and therefore it is necessarily incomplete, ultimately leading to non-existence and meaninglessness (Sartre 1956); alternatively, it is such realisation of life incompleteness and finitude that should ultimately free us from anxiety (Heidegger 1963): "While for Heidegger death is the foundation of one's freedom, for Sartre … death prevents an individual from realizing his or her possibilities" (Tomer 1992, p. 478).

Greater death anxiety can emerge in individuals who regard being alive as a fundamental condition for the realisation of meaning but possess poor self-worth. Thus, death anxiety may decrease in old age the more we are satisfied that our life has been meaningful and productive (Frankl 1963, see also Erikson 1963), or the old individual still has a purpose in life and is looking towards the future, enjoying good health (Rappaport et al. 1993).

In addition, death anxiety may be affected by specific cultural constructs of "immortality", as seen in Becker's (1973) and Greenberg et al.'s (1986) *Terror Management Theory* that we will analyse in more detail below. The *illusion of immortality* can have the positive effect of decreasing death anxiety and therefore increasing our well-being (Tomer 1992).

Apart from the issue of increased life meaningfulness in old age and beliefs about immortality, a decrease of death anxiety in old age has also been explained by individuals being in denial about ageing, or simply by coming to terms with mortality, thus decreasing fear of dying (Chopik 2017, and references therein).

From an evolutionary perspective, fear of death and death anxiety are adaptive traits that may prevent individuals from taking unnecessary risks that may jeopardise their survival. The traits may *proximally* derive from fear of separation or "object loss" and the fear of the process of dying itself, which may involve pain, suffocation and, psychologically, shame and dependency (Stolorow 1973).

Ultimately we are living organisms and unless we have some form of "fear of annihilation of life" that is expressed starting from basic reflexes to then also involve complex cognitive and emotional capacities that allow us to escape threatening situations, our career as a living organism may not last long. That is, we are confronted with the challenge of both decreasing fear of death (e.g. by denying annihilation through religious views of the afterlife), lest we become paralysed in our capacity for action, and at the same time we must fear the reality of death if we want to prevent it from happening too soon (Dumont and Foss 1972).

When fear and anxiety become obsessive, they may have a negative effect on us. At that point, we should focus on controlling the proximate physiological and psychological causes of fear and anxiety—through a medical intervention if necessary— and to coping with their consequences (Berger 1967). Forms of coping with fear of death and death anxiety at the end of life include enjoying the close contact of loved ones, engaging in generativity in order to contain the fear of separation and loss, and receiving appropriate palliative care in order to decrease any pain derived from the dying process. We may respond adaptively to our death anxiety by strengthening our social relationships, better securing our material well-being, modulating our emotions, reinforcing transcendental beliefs, and restructuring our philosophy of life. Lehto and Stein (2009, and references therein), however, also mention some maladaptive responses to death anxiety such as post-traumatic stress reactions, eating and self-mutilation disorders, anxiety disorders, the expression of negative emotions such as distress, insecurity, worry, tension, and uneasiness.

Seen from an evolutionary adaptive perspective, we would expect death anxiety to slowly increase at young ages and until puberty, to then increase faster after we reach our sexual maturity, as death anxiety may protect us from harmful experiences that may jeopardise our survival and current ability to reproduce. Death anxiety may continue to increase into adulthood as we improve our physical and behavioural chances of attracting a partner and therefore transmit our genes through our offspring. At one point, however, death anxiety may start to decrease as our reproductive capacity decreases (faster in females than males; with older males perhaps continuing to slowly increase death anxiety until they also reach a point where their chance of reproduction becomes negligible), but the rate of decrease should be initially slow, as young-older individuals, even if they are not reproductive (or are less reproductive), can still support the reproductive effort of their younger relatives, through grandparenting for instance. Death would deprive close relatives of important help and therefore it should be feared. Death anxiety is expected to fall even faster, however, once we reach such ages when our body and mind can do little to sustain ourselves in our daily activities, let alone helping others (see Fig. 8.9 for a representation of the model, Fig. 3.9 for some data, and Fortner and Neimeyer (1999) and Missler et al. (2012) for reviews). In fact, there may be even a point in time when death anxiety may take "negative values", that is the old person may not only not fear death but in fact he or she may desire to hasten death in order to avoid unbearable pain and a life that has become meaningless (see Sect. 8.5 in this chapter). The curvilinear relationship between death anxiety and age expected from Fig. 8.9, that is based on evolutionary considerations, has been observed in many empirical studies (Gesser et al. 1987; Neimeyer et al. 1988; Neimeyer and Van Brunt 1995; Wong et al. 1994; Thorson and Powell 1994; Fortner and Neimeyer 1999).

Nussbaum et al. (2000, p. 296) specifically stated: "…fear of death is strongest between the ages of 40 and 55, and that fear of death is lower for elderly people."

Evolutionarily, the curvilinear relationship between death anxiety and age is expected under conditions of ancestral living, where older people remained within multi-generational social groups for as long as they could keep pace with the movements of the group. Currently, however, many old people are institutionalised, living in retirement villages or nursing homes, being mainly surrounded by other old individuals. In such a condition the recurrent reminder of death may tend to increase death anxiety in old age (Missler et al. 2012, see also Mullins and Lopez 1982). Although even in institutionalised settings, some elders may feel less anxious to think about death than others (Portal Moreno et al. 2008). In Israel, however, Ron (2010) found no differences in death and dying anxiety among elderly living in nursing homes and those living in the community, a similar non-significant difference was also found by Sridevi and Swathi (2014).

Some additional data broadly consistent with the evolutionary model illustrated in Fig. 8.9 are available in Chopik (2017). Figure 8.10 shows how death-related thoughts reach peaks at around 38 and then 58 years old to then decline afterwards. The linear regression with negative slope seems to us to be too simplistic, missing a greater complexity in the data although Chopik did test the data for quadratic and cubic effects of age and found that neither was significant. Why do death-related thoughts seem to decline in the 40s and then increase again in the 50s is unclear, one possible explanation may be found in different proportions of males and females in the different age groups. But in this study gender did not have a significant effect on death-related thoughts. Alternatively, the trough may reflect life commitments at that age that are so intense that death thoughts are not frequently contemplated.

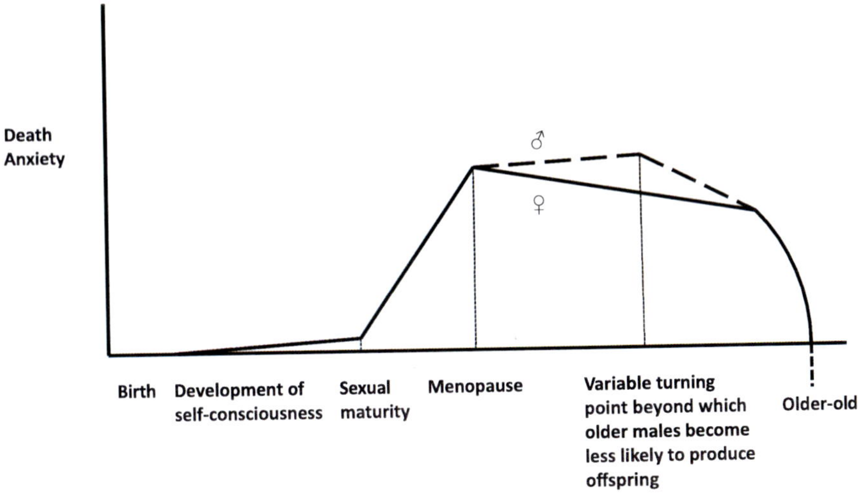

Fig. 8.9 Evolutionary model of death anxiety across the lifespan

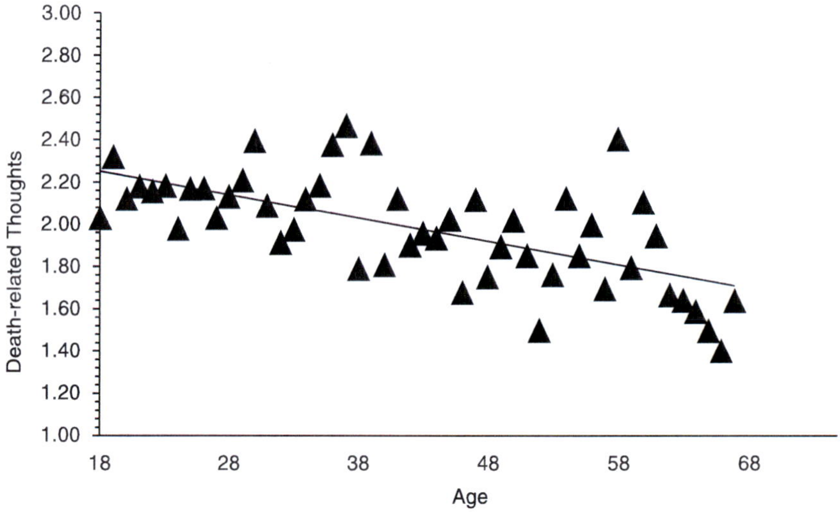

Fig. 8.10 Age differences in death-related thoughts for individuals aged 18–67 years old (from Chopik 2017)

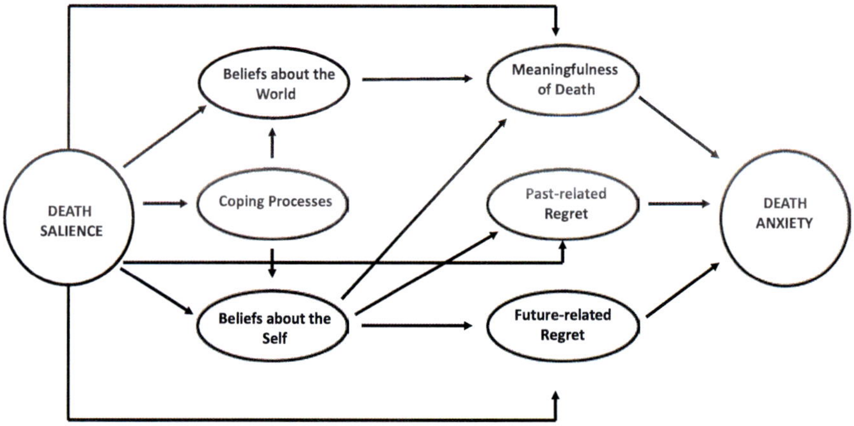

Fig. 8.11 *Comprehensive model of death anxiety*. Death anxiety is proximately affected by past-regrets (regrets about things done or not done in the past), future-regrets (regrets about thing that will not be done in the future) and how meaningful death is. Those three factors are in turn affected by death salience both directly and indirectly via beliefs and coping. Redrawn from Tomer and Eliason (1996)

Tomer and Eliason (1996) produced a *Comprehensive model of death anxiety* (Fig. 8.11) which is grounded on three direct causes of death anxiety: past-related regrets (having done something that one regrets to have done, or regretting not having done something), future-related regrets (concerns about not achieving some important goals in life), and personal conceptualisation of death (i.e. the personal meaning of death). An individual is expected to experience lower levels of anxiety the more meaningful death is and the fewer regrets he or she has. The three factors: meaningfulness of death, past regrets, and future regrets, can be in turn modulated directly or indirectly by *death salience* or "the extent to which the individual contemplates or ponders his or her mortality" (Tomer and Eliason 1996, p. 346).

We can protect ourselves from the anxiety that comes with thinking about death through a series of coping strategies (see also Fig. 8.11). For instance, a reassuring worldview and also self-esteem derived from fulfilment in life, especially when such fulfilment includes a transcendental dimension that projects our life achievements through time (e.g. family, community, deeds, creations of whatever kind whether it is in the artistic, scientific, technology, business, or any other domain) can be protective. Such protective strategies have been articulated by *Terror Management theory* TMT (Greenberg et al. 1986), which we already mentioned in Chaps. 3 and 6.

TMT points to our human need to put some order in our perception of the universe and in our lives. Order allows us to feel more in control and control decreases our anxiety. TMT is often studied using *death-thought accessibility* (DTA) techniques (Hayes et al. 2010).

On the other hand, *Post-traumatic growth theory* (PTG, Tedeschi and Calhoun 1996; Tedeschi and Calhoun 2004) suggests that confronting a life crisis, such as the possibility of our own death or that of a loved one "can result in positive changes, such as a greater appreciation for life, a shift in priorities toward intrinsic goals, and improved interpersonal relationships" (Furer and Walker 2008, p. 168). Such post-traumatic growth also helps individuals decrease death anxiety. Lykins et al. (2007) have put forward a suggestion to reconcile TMT and PTG. Terror management involves an immediate response to cope with the threatening possibility of death; but this initial response may be followed, over time, by post-traumatic growth and a reconstruction of our worldview that incorporates, in a meaningful manner, the initial experience that caused death anxiety. Our previous book *Behavioural oncology* was not only an intellectual journey and a critical synthesis of current scientific research on the relationship between cancer and behaviour; but also an occasion to constructively manage the terror of death and provide an opportunity for post-traumatic growth for both M. Cordella, who prematurely lost her mother to cancer, and A. Poiani who was himself diagnosed with cancer.

Greenberg and Arndt (2012, p. 398) offer the following description of the basic tenets of TMT:

> The theory posits that to manage the potential terror engendered by the awareness of mortality, humans sustain faith in worldviews which provide a sense that they are significant beings in an enduring, meaningful world rather than mere material animals fated only to obliteration upon death.

In the words of Victor Cicirelli (Cicirelli 2002, p. 358):

> all humans are instinctively driven toward survival and continued existence, while at the same time have knowledge of their inevitable mortality. As a result, there is a potential for them to consciously experience a terror of death. The core of death fear is fear of annihilation, which refers to the extinction of mind, spirit, and soul, as well as the destruction of the body—that is, total nonexistence. According to these theorists, the experience of such death

terror would be paralyzing without some means of suppressing it from awareness. The basic task of the theory is to identify the factors helping to maintain this suppression.

Some of such factors that decrease fear of death and death anxiety include views that emphasise the rationality, predictability, and permanence of our world and that contribute meaning to life; and in so doing such views promote self-esteem and life fulfilment, with projections into the future through generativity and legacy. Developing a perception that we can project ourselves towards the future, beyond death, helps in creating a sense of "immortality" that is, by definition, a defeat of death. We can see immortality as *literal immortality* or "the belief that a noncorporeal aspect of the individual will live on indefinitely in some way" which usually relates to religious beliefs, or as *symbolic immortality* or "the belief that individuals are represented by something or someone other than themselves that will continue to exist after they are dead" (Cicirelli 2002, p. 359) which is broader. Generativity is a way of achieving some form of symbolic immortality, producing works that can endure is another. If we can "defeat death", death is not "terrifying" anymore.

Since antiquity, the fear of inevitable death, and the existential "horror vacui" coming with it, has spurred a search for some form of immortality. The fear of emptiness after a period full of life provides a strong motivation to somehow fill that gap, to eliminate the feeling that death is equivalent to falling off a cliff and into the depth of nothingness. This is especially true in the case when life has been experienced in a positive manner. In other contexts, however, death could be regarded as relief from extreme suffering; but even in such dramatic circumstances there remains some motivation to make our life and death meaningful. Even in the worst moments of the experience in a Nazi concentration camp, when all seems lost and death inevitable, a final effort to achieve lasting meaning can emerge, as in the poem left by a prisoner moments before she entered the gas chamber in Auschwitz. The poem is entitled *We, the Dead, Accuse!* (Dov Kulka 2013, p. 52) and it ends with these words:

And then we will emerge, in awful ranks,
A skull on our skulls and bony shanks;
And we'll roar in the faces of all the people
We, the dead, accuse!

In this case, the poem is a strong call for posthumous justice, a claim to meaning in the face of the terror of inevitable annihilation of life, a life that in the end was characterised by intense suffering.

TMT is based on two considerations: first, we, as animals, have evolved behaviours that are aimed at keeping us alive; and second, that humans have also evolved the self-conscious ability to realise that life inevitably comes to an end. Thus, the realisation that we are mortal clashes with our inbuilt drive to stay alive, such clash produces anxiety and even terror that we must learn to manage to make life bearable. This has led to various strategies to manage such terror: from religion to spirituality, cult of the ancestors, legacy, and so forth. In modern psychology the aim is to strengthen those psychological aspects of self that limit the negative effects of thoughts about mortality, this is what Greenberg and Arndt (2012) refer to as the *existential shield*.

According to TMT death concerns may be decreased by using defence strategies as an existential shield, in the form of distraction, distancing, and denial that simply avoid thinking about death. Those strategies may well remove death concerns from *consciousness*, but they do not prevent death concerns from lingering on in our *sub-conscious/unconscious* and therefore resurfacing regularly on occasions. Therefore, we are still confronted with the need to manage sub-conscious and unconscious death concerns.

Pyszczynski et al. (1999, p. 837) explain the role of *Dual-process theory* in the management of both conscious and unconscious death concerns:

> Dual-process theory posits that conscious thoughts of death are defended against with proximal defenses: rational threat-focused attempts to remove conscious death-related thoughts from focal attention either by suppressing such thoughts with distractions or by pushing the problem of death into the distant future by biasing rational inferential processes to deny one's vulnerability. Unconscious thoughts of death, on the other hand, are defended against with distal defenses: symbolic conceptions of self and reality specified by one's culture as a way of dealing with the unconscious knowledge (as opposed to current conscious awareness) of the inevitability of death.

Given that death is unavoidable, only the construction of some sense of "immortality" would be able to decrease our fear of death at both conscious and unconscious levels. Fortunately, for those who do not believe in literal immortality, symbolic immortality can be accessed in very many ways through children, passing our knowledge on, specific products of our creativity, helping others, and so forth. Symbolic immortality cannot be guaranteed, of course, even to individuals capable of producing high levels of legacy, but it can be better achieved to the extent that people are contributors to something that is bigger than themselves: membership of movements, political entities, institutions. Pyszczynski et al. (1999, p. 838) may have been thinking along the same lines when they wrote that conscious and unconscious death concerns could be removed: "through the use of distal terror management defenses that enable the individual to conceive of himself or herself as a person of value in an eternal world of meaning".

Of course, fear of death and death anxiety may never be an issue to the nihilist, once the process of dying is made painless, but then one can only wonder whether we would want to live in a world inhabited by nihilists, who may be at the most concerned with the present but rather uninterested about the future.

TMT has also been the target of some criticisms. For instance, Kirkpatrick and Navarrete (2006) provide a critique of TMT from an evolutionary perspective. They start by identifying four major tenets of TMT:

1. Humans (and other species) possess a survival instinct.
2. The evolution of advanced cognitive abilities in humans led to paralysing and incapacitating fear of death that created a new adaptive problem for our ancestors.
3. The adaptation produced through natural selection in response to this adaptive problem was an anxiety-reducing terror-management system.
4. A central process incorporated in this terror-management system involves the defence of our worldview.

Kirkpatrick and Navarrete criticised the "survival instinct" on the ground that what is selected in evolutionary terms is genes—not individuals—and the information they contain, information that is eventually expressed into a phenotype. They correctly point out that inclusive fitness is what matters evolutionarily in terms of transmission of genes, and inclusive fitness involves both the direct reproductive output of an individual through offspring and an indirect output through helping close relatives produce their own offspring. Thus, lower probability of survival may still be adaptive if it increases inclusive fitness. However, even though the argument is correct, it misses the point that inclusive fitness should be measured over the lifetime of the individual and, everything else being the same, a longer life may be still associated with higher inclusive fitness than a shorter life. Moreover, after menopause, old women can continue to contribute to their inclusive fitness if they stay alive and well and help in the reproductive effort of children and grandchildren. From this we can conclude that longevity should remain a selected trait in humans and fear of death would help increase longevity. Longevity as a selected trait can be under ongoing natural selection only if it possesses some degree of heritability (e.g. Mousseau and Roff 1987). In fact, we have seen in Chap. 4 that heritability of longevity ranges between 0.20 and 0.30, and it increases after the age of 65 to 0.36, to reach a value of 0.40 after 85 years old.

The evolution of high cognitive abilities presumably leading to "a paralyzing and incapacitating fear of death" is criticised by Kirkpatrick and Navarrete on the ground that fear of death may be just an epiphenomenon of our brain activity; but their statement that "It is unclear why fear of death is incapacitating" is puzzling, as fear of death can indeed make the individual so cautious to lead to an excessive decrease of risk-taking. Such conservatism is unlikely to be adaptive in a changing or complex environment. Adaptively, one could argue that individuals should be able to dynamically balance risk and cautiousness, any inappropriate swing to one or the other may lead to a maladaptive response in any specific environment.

Kirkpatrick and Navarrete further dispute the evolution by natural selection of an anxiety-reducing terror-management system on the ground that reducing the ability to experience anxiety and fear may not be adaptive. We obviously agree with this point, but we must also note that fear and anxiety carry a cost—both physiological and psychological—through the effects of stress. Controlling our fear and anxiety becomes adaptive in reducing the long-term negative effects of chronic stress.

Finally, the defence of our worldview as a terror-management mechanism espoused by TMT is criticised by Kirkpatrick and Navarrete on the ground that defence of a worldview may be an exaptation. But again, defence of a specific worldview may be proximally adaptive as a stress-reducing mechanism. In the long term, however, persisting with a worldview that is not helpful in terms of survival and reproduction, no matter how much it relaxes us is arguably not adaptive.

Other views such as *Denial of death* (Becker 1973) imply reducing death anxiety through the development of some sense of immortality, as we have already mentioned. This is usually achieved through spirituality and religion, but legacy could also serve the same purpose.

Death anxiety can be also decreased by greater capacity for self-determination, as contemplated by *Self-determination theory* (SDT), which may lead to less death anxiety in older people who retain a reasonably good health, compared to people in their middle ages (Tomer 2012).

Finally, Wong (2008) have proposed the *Meaning Management theory* (MMT) to control death anxiety. The theory is rooted in existential-humanism and cognitive behavioural processes inspired by Frankl's work. In this model terror management is achieved through meaning seeking, meaning making, and meaning reconstruction. Those processes are especially useful after a loss (e.g. of a beloved one, a long-term job or career, and so forth) but they can be of help also in the case of an imminent loss, such as the approaching end of own life. The more life is meaningful and perceived as complete and fulfilled, the more death is likely to be accepted in old age and less anxiety experienced.

MMT is grounded on some basic principles such as that humans are complex beings embedded in a multidimensional reality: biological, psychosocial, and spiritual. Some basic tenets of MMT include (Wong 2007):

a. Humans are meaning-seeking and meaning-making beings, living in a social world of culturally constructed meanings.
b. They constantly engage in meaning making to make sense of life, and they react to such perceived meanings in their response to life events.
c. Unity and coherence of our personal worldview can be achieved despite the often contradictory and fragmented nature of life experiences; producing such unity is helped by our capacities to produce symbolic meaning and storytelling.
d. MMT predicts that a sense of meaning and purpose not only offers the best protection against the terrors of life and death, but also contributes the most to healing and well-being as compared to other psychological variables such as internal control and self-efficacy.
e. Meaning making is necessary in our quest for survival as it can awaken in us defence mechanisms against the terror of death, and attitudes that spur purpose in life and the desire to stay alive. MMT also predicts that when survival is fraught with suffering, it will motivate a quest for reasons to live despite the pain.
f. Meaning can be found in all situations, including the most hopeless and horrific experiences.
g. MMT predicts that meaning is essential for maintaining hope and happiness in the face of suffering, pain, and death.
h. The positive and proactive tendency to create a happy and meaningful life serves a growth-oriented function, facilitating life fulfilment. MMT also predicts that the best way to reduce death anxiety is to facilitate death acceptance and positive tendencies.

Meaning management involves the management of our inner life; that is, our combined feelings, desires, perceptions, thoughts, hopes and dreams, and all the complexities of our consciousness. Through meaning management we can control our fears, anxieties and hopes, hates and loves, memories and dreams, doubts and beliefs, regrets and celebrations, and the various meanings we attach to events and people; so that we can experience life fulfilment, happiness, hope, and equanimity in spite of setbacks, sufferings, and the closeness to death. Whenever we live embedded in an environment where ambiguity, uncertainty, and social change are prevalent, meaning management can be even more important for our existential well-being (Wong 2007).

8.3.1 Factors Affecting Fear of Death and Death Anxiety

Mellor and Shilling (1993) argued that, given our temporal finitude, an obsessive focus on the self and the body may predispose us to experience greater fear of death and death anxiety. On the other hand, making death more meaningful and hence less stressful from an existential perspective may decrease fear of death and death anxiety. We have seen that legacy and generativity also help in projecting the life of an individual through time, via the influence that such life has on others, which also may decrease fear of death and death anxiety; whereas religiosity and spirituality may provide a response to fear and anxiety by telling us that our being can be extended into eternity.

Fear of death and death anxiety can be modulated by various factors, including: age, education, gender, physical appearance, financial situation, ethnicity, marital status, personality, level of ego integrity and purpose in life, self-esteem, health status including psychological health (e.g. depression); living alone or in an institution or with family, religion/spirituality, political ideology, a recent loss and the prospect of losses that death entails, social support, place of residence (e.g. rural vs urban), sense of legacy, and previous experience with dying people (Wass and Sisler 1979; Myers et al. 1980; Thorson and Powell 1988; Thorson and Powell 1993; Nussbaum et al. 2000; Benton et al. 2007; Castano et al. 2011; Kawano 2011; Missler et al. 2012).

The effects that each one of the above factors has on fear and anxiety may vary according to circumstances. For instance, greater religiosity but of a superficial nature may lead to greater fear of death, whereas a more profound form of religiosity may lead to less fear of death (e.g. Wagner and Lorion 1984).

Older people tend to display lower levels of death anxiety compared to younger ones (e.g. Thorson and Powell 1988), this result seems to be resilient to the methodology used to assess death anxiety (De Raedt and Van Der Speeten 2008). Lower death anxiety may in turn make older people less likely to go through all of Elizabeth Kübler-Ross's *five stages of grief* if they are diagnosed with a terminal illness:

1. Shock and denial.
2. Anger.
3. Bargaining ("it must be a mistake", "let's see another doctor").
4. Depression.
5. Acceptance.

The older the person is—and therefore the less anxious she or he is about death—the less shocking a diagnosis of a terminal illness is likely to be, and therefore the less likely it is that such diagnosis will be confronted with an attitude of denial. Anger may also be less, especially if the elder has lived a fulfilling life, not so if he or she has too many regrets. Regrets may lead to a state of depression upon being diagnosed with a terminal illness, but with some help the older person may reach a peaceful state of acceptance, especially if a sense of legacy is developed, perhaps with the help of family and friends. The same conclusions apply to Pattison's (1978) three phases of the *living-dying process*: (a) the acute phase, where a crisis unfolds; (b) the chronic living-dying phase which is the longest one and it may be experienced by the individual in a manner that integrates his/her existence (ego integrity) or that it leads to disintegration (ego dissolution); and (c) the terminal phase, which is characterised by withdrawal and little hope (see also Samarel 1995). Older people should be better able to feel more integrated and accept the terminal phase without experiencing hopelessness if they have lived a fulfilling life.

Therefore, older people displaying less death anxiety and who have reached a point where they perceive their life to be fulfilled may simply respond with "acceptance" upon being diagnosed with a terminal illness. That is, both family and health professionals should not expect, let alone instigate the passage of the old person through a predetermined series of stages (Belsky 1984), each elder will walk to the end of life following his/her own path.

Fear of death is higher in older people who are in worse health condition (Wass and Sisler 1979; Cicirelli 2006), but the better educated ones tend to score lower on death anxiety compared to the less educated ones (Thorson and Powell 1988). Better education can modulate fear of death and death anxiety through cognition. Neimeyer et al. (1986) used the *Templer Death Anxiety Scale* (DAS) to study cognitive aspects of death anxiety in a group of 35 participants (62.8% women) who were between 19 and 68 years old. Female participants tended to score higher on death anxiety in this test. They also applied the *Death Attitude Repertory Test* (DART) which measures the process of formation of the individual's views about death. Older people had a more coherent and integrated view of death and those with better education were also more flexible, actualised and subtle in their conception of death. Cognitively, death anxiety can also affect the way in which old individuals visualise death with specific imagery: low levels of death anxiety tend to be associated with the production of more positives images of death (McDonald and Hilgendorf 1986).

In addition, the greater the level of ego integrity the lower the level of death anxiety (Fortner and Neimeyer 1999), but death anxiety in old people tends to increase following the recent loss of a dear one, when social support is low, and when purpose in life is also low (Missler et al. 2012).

Death concerns are higher among rural residents than small town or urban residents. Living alone also increases death concerns as compared with living with family; but being in a better financial situation decreases death concerns. Not surprisingly, simply not thinking about death is a good approach to decreasing death anxiety. This does not require true denial of death, which would be delusional, but simply avoid ruminating about death matters. According to Wass and Sisler (1979) those with a higher sense of legacy also tend to show higher levels of "death concerns". In this case, however, the concern is positive, and it is shown through taking legacy seriously, reflecting "a need for some kind of sense of immortality or continuity" (p. 79).

Perhaps paradoxically, exposure to death may decrease death anxiety: "Experiencing the deaths of others may… have some positive effects… the experience of loss provides a rehearsal for one's death. One gets used to the idea of death and becomes somewhat more accepting of mortality by having friends and family members die and by seeing that life goes on" (Nussbaum et al. 2000, p. 294). Incorporating the dead in the life of the living (e.g. through ceremonies, celebrations, feasts) may likewise normalise death and therefore decrease death anxiety.

Chopik (2017) studied the effects of social support received from spouses, children, family, and friends on death anxiety in old people. Greater social support was associated with lower levels of death anxiety over time. The decline in death anxiety due to social support was mainly affected by quality spousal and friend relationships. Support received from children was only marginally associated with lower death anxiety, whereas relationships with other family members were unrelated to changes in death anxiety. The social sphere, however, is not always a source of support to old people. Kalish's (1976), for instance, described the phenomenon of *social death*, which is treating an old person (whether she or he is dying or not) as if he/she is already dead, before death eventuates. Death anxiety experienced by family and friends may even enhance the negative effects of ageism, as older people may remind us of our own mortality (Bodner et al. 2015).

In many studies men tend to display a relatively lower level of fear of death and death anxiety than women (e.g. Wass and Sisler 1979; Neimeyer et al. 1986; Depaola et al. 2003; Russac et al. 2007; Singh Madnawat and Singh Kachhawa 2007; Chan and Yap 2009; Azaiza et al. 2010); this result is not expected from our evolutionary model (see Fig. 8.9), but not all studies support such pattern. Missler et al. (2012) carried out a study of death anxiety in elderly women and men living in assisted accommodation in The Netherlands (66.6% women, 60-96 years old). The results indicate that men display greater death anxiety regarding "fear of the unknown" than women, whereas women showed greater death anxiety than men regarding "fear for significant others", and "fear of the dying process". It is possible that old men are less cognitively prepared to death than older women, as expected from our evolutionary model, whereas women are more socially oriented (a common characteristic of females in social mammals; Poiani 2010) and are also less prepared to endure pain in the dying process (in many societies males are expected to show resilience in the face of pain—e.g. Mancini and Bonanno 2006, being educated into it). Troll (1982) described women being more accepting of death than men.

Dattel and Neimeyer (1990) found that women display greater death concerns than men on the *Death Anxiety Scale* (DAS) but both genders scored similarly on the *Threat Index*, which is a more cognitive measure of death concern. On the other hand, Thorson and Powell (1993) found no statistically significant difference between men and women on death anxiety using the *Revised Death Anxiety Scale*. In Thorson and Powell's (1993) work, men lower on affiliation scored higher on death anxiety; in the case of women death anxiety was negatively correlated with deference, order, change, and endurance, whereas death anxiety was positively correlated with exhibition, heterosexuality, and aggression. Suhail and Akram (2002) measured death anxiety using the DAS and fear of death using the *Collett-Lester Fear of Death Scale* (CLS). Women scored higher on both measures, but the difference between genders was more pronounced when the CLS was used. Russac et al. (2007) also used the CLS to find that women scored significantly higher on fear of death than men. In addition, they found that although death anxiety for men was greatest during their 20s and then declined gradually thereafter, women displayed a bimodal pattern, with a major peak in death anxiety during their 20s and a second, smaller peak during the early 50s. This lends support to the idea that the peaks in death anxiety that we pointed out in Fig. 8.10: one at 38 and the other at 58 years old may in fact be real, as 70.2% of participants in Chopik's (2017) work, from which the figure was taken, were women.

Depaola et al. (2003) also found that women display higher levels of death anxiety than men although only in the *Fear of Dead* subscale of the *Multidimensional Fear of Death Scale* (MFODS).

Old women not only seem to display higher death anxiety directed to self, but they also show higher anxiety about the potential death of their partner (Momtaz et al. 2015). Although this result could be proximately explained by some of those women fearing loss of material support if the partner dies, a more basic cause could be found in the higher sociability of women than men, which may make women more sensitive to the loss of a partner. In fact, in a study carried out in New Zealand (MacLeod et al. 2016), women displayed more anxiousness when thinking not only about own death but also about the death of somebody they care about (see also Missler et al. 2012).

Differences in personality traits can also affect death anxiety. For instance, Neuroticism is positively associated with death anxiety (Thorson and Powell 1993, and references therein).

Despite some technical difficulties confronting cross-cultural research on death anxiety (Beshai 2008), ethnicity and culture have been also shown to affect the level of death anxiety experienced by elderly people. In general, there seems to be a trend for Asian cultures to be associated with lower levels of death anxiety than Western cultures (Wu et al. 2002 and references therein). Religious beliefs that include reincarnation (e.g. Buddhism) may in part explain this trend, but in Japan there is also the added effect of beliefs that death is only a transformation from one state of being to another, the state of ancestor. Ancestors are the subject of worship in Japan. In the words of Kawano (2011, p. 372):

> Participating in the system of ancestor worship allowed people to see their personal death as a transitional point leading to another state of social existence and attaining a sense of the assurance that they, too, would become ancestors. The two cultural devices central to the gradual transformation of deceased family members from their state as dangerous, unstable newly dead into stable ancestors were a domestic altar and a family grave, both of which contributed to keeping the dead alive as social beings. A domestic altar accommodated ancestral tablets of the deceased family dead, and received offerings regularly … The living

took care of ancestors ritually and ensured their peaceful rest, while ancestors were also supposed to protect their descendants and watch over them. The family grave, meanwhile, often provided the interment site that housed generations of the family dead. Survivors were expected to visit the grave regularly, make offerings, and have veneration rites performed, usually by a Buddhist priest.

A similar cult of the ancestors was also common in the West in antiquity, in Etruscan and Roman cultures, for instance (see Chap. 2).

Both spirituality and religiosity can be regarded as sources of *transcendental meaning* that can help to *cognitively reframe* life and death and decrease death anxiety (Tomer 2012). In a study carried out by Wagner and Lorion (1984) in the USA, they investigated correlates of death anxiety in 122 elders ranging in age between 60 and 89 years old who were residents in institutions or lived in the community. The authors applied the DAS to the group of participants, and their analysis suggests that death anxiety was more common among less religious people. Achievements can also give transcendental meaning to the life of a person and thus they can buffer against death anxiety, especially if such achievements are expected to endure in the memory of the living.

Given the various factors that may affect, alone or in combination, fear of death and death anxiety, in the next section we provide a more thorough analysis of some specific approaches that could be used to decrease fear of death and death anxiety in elderly people.

8.3.2 Decreasing Fear of Death and Death Anxiety

Death anxiety ends as soon as we accept death not only as a reality but also as a stage of our existence that we are ready to embrace. This may be because death has acquired more meaning than life, or simply because we want to end our constant suffering and pain.

In the *Death Attitude Profile* (DAP) (Gesser et al. 1987, see also the more recent DAP-R of Wong et al. 1994, and Wong 2007 for an introduction), death acceptance is classified into three types:

1. *Neutral death acceptance*—facing death rationally as an inevitable end of life;
2. *Approach acceptance*—accepting death as a gateway to a better afterlife; and
3. *Escape acceptance*—choosing death as a preferable alternative to a painful existence. In this case the terror of death is superseded by the "terror of living" (Cicirelli 2001).

According to Wong's MMT, the best way to promote death acceptance is to simply focus on how to live a meaningful life (Wong 2007), and we should add: how to live a life that is fulfilling. Life satisfaction should improve our readiness to accept the end of our life, when life carries a sense of mission accomplished and there is little else that we can contribute given our age and condition, whereas unsatisfaction with life makes us anxious about reaching the end prematurely. When anxiety prevails, specific psychological interventions have been proposed to decrease death anxiety such as various *cognitive behavioural therapies* (CBT) (e.g. Furer and Walker 2008).

Although an old person may have accepted that he or she is dying, family and friends should continue to interact with the person in as much a normal way as possible. It is not helpful to engage in *anticipatory grieving*, or "experiencing sorrow about a person's death before the death has occurred" (Nussbaum et al. 2000, p. 306). Although anticipatory grieving does make it easier for survivors to recover faster after the death of a loved elder (Troll 1982), one of its negative consequences is that the elder may experience a *social death* in life. Surprising as it may sound, some family members may even become irritated that the death of their elder relative is taking too long (Kalish 1976); needless to say, such attitude is not helpful.

Fulton and Fulton (1980) suggested that anticipatory grieving can become a "double-edged sword" helping to protect the well-being of the survivors on the one hand, but on the other it breaks the remaining social bond between the dying person and the survivors.

From the perspective of the elders, coping with death anxiety and fear of death can be achieved with the use of internal individual resources or with the help of external support. One important internal individual resource is engaging in a life review, which has been seen to decrease death anxiety in the elderly (Fishman 1992).

We also try to abate—or perhaps hide—our fear and anxiety behind metaphoric language, when confronted with death. Nussbaum et al. (2000, p. 292) mention, with a tone of sarcasm, the ambiguity of such terms for death such as "'gone' (where to?); 'eternal rest'; 'lost' (the battle or cannot be found?); and the medicalized 'coded'". However, frankly talking about death may help bring about death acceptance and therefore decrease death anxiety, but care must be taken about the cultural background of the elder, as in some cultures direct talk about death is seen as hastening death (see Sect. 8.6 in this chapter).

Other individual strategies to decrease death anxiety include enhanced self-esteem and even denial of death, especially when one feels that is not ready to die yet (Firestone 1993). Humour can be also used to decrease death anxiety, even in extreme cases when death is predictably about to happen. Richman (2006, p. 44) reports the remarkable example of a criminal sentenced to death and about to be electrocuted:

> A criminal by the name of Wood was about to be electrocuted. He was strapped into the electric chair, and as the switch was about to be thrown, said to the onlookers, 'You are about to see the effect of electricity upon Wood.'

In this case humour may have decreased death anxiety with a mix of defiance. But of course, humour can be useful to decrease death anxiety only if the elder is fully engaged in it and all persons involved share such sense of humour, otherwise it may be counterproductive. Richman drives the following conclusions from his experience with using humour in psychotherapy to decrease anxiety in his patients:

a. Humour usually emerges spontaneously during conversations.
b. The timing of the humour is essential, and it is best used when it is most relevant and meaningful.

c. The humour therapy experience fosters social cohesion, especially when everyone in the meeting laughs together (see also Cordella and Poiani 2014).
d. Death is less feared when you do not face it alone.
e. Humour possesses the power to reduce stress and affirm life which, in the case of a person who is close to death, can be associated with legacy.

Another aspect of our psychology that can help us cope with death anxiety is *emotional complexity*. Emotional complexity has been variably described in the psychological literature, with Lindquist and Feldman Barrett (2008) mentioning that such complexity can be seen and measured in the precision and discursive complexity of self-reports of emotional experiences, the explicit knowledge about emotions in various situations, and each individual self-characterisation of the complexity of her/his emotional experiences. Emotional complexity can have emotional regulation functions, and it has been also suggested as a source of coping against fear of death and death anxiety. Bodner et al. (2015a) have shown that emotional complexity—that can be measured with the *Scale of Positive and Negative Experience* (SPANE) and the *Positive and Negative Affect Schedule* instruments, which consider both high and low arousal emotions—can moderate the relationship between anxiety of ageing, anxiety of death, and emotional distress. Low emotional complexity is associated with a positive relationship between those two forms of anxiety and psychological distress, whereas greater emotional complexity simply affords more psychological tools to the individual to buffer the potential negative effects of anxiety. The more tools we have, the more capable we are of coping with anxiety if our initial response fails.

Lau and Cheng (2011) point out that *gratitude* in a life review may also help decrease death anxiety and fear of death: "By reexamining life events with a thankful attitude, people may become less fearful of death due to a sense that life has been well-lived." (p. 169). Gratitude can prevent the life review from becoming negatively biased towards regrets, the latter are more likely to increase death anxiety through feelings of a "wasted life" and "missed opportunities". Increased life satisfaction, sense of fulfilment in life, and ego integrity are also positive effects of a life review marked by gratitude.

Social resources and the support they provide can also help old people cope with death anxiety. Such social resources usually start with the family (e.g. Chokami and Razavi 2015) and they are particularly useful when they include contact with younger people, babies in particular. Zhou et al. (2009) suggest that young members of the family, as the expression of new life, may reduce death anxiety in the elderly by providing a "representation of symbolic immortality" and by fostering self-esteem: "having children provides a sense of security, fulfillment and value, which may result in a heightened sense of self-esteem" (Zhou et al. 2009, p. 41; see also Fritsche et al. 2007). Such symbolic immortality is not only based on the transmission of genes, but also culture and personal influence; therefore, a childless elder may reduce death anxiety in this way by transmitting his/her legacy to others, whoever they are.

Given the important social support received by dying elders in professional settings, death anxiety should not only be decreased in the elderly, but also in nursing home and hospital staff. Staff members may display more negative attitudes towards the elderly when they experience high levels of death anxiety (Eakes 1985).

We conclude our analysis of the approaches that may be used to decrease death anxiety and fear of death by mentioning the specific role of spirituality and religiosity, which can have effects on both the personal and the social dimensions of fear and anxiety.

We already defined spirituality and religiosity, and analysed them from the psychological perspective of the elders' mental health and well-being in Sect. 4.3 of Chap. 4. Here we specifically focus on their role in coping with anxiety and fear of death.

Although still of uncertain origin, the word "religion" may have derived from the Latin word *religare* (to bind) (Hoyt 1912), perhaps alluding to the social bonds that are created among individuals on the ground of sharing a common set of beliefs. Affiliative social bonds can help decrease anxiety (e.g. MacDonald 1970). In addition, religion has the capacity of decreasing anxiety and fear of death through a more individual route: the belief in an extension of being after death and into eternity. As stated by Bronislaw Malinowski (1948, p. 32–33) in words worth quoting in full, the human being is

> intensely afraid of death, probably as the result of some deep-seated instincts common to man and animals. He does not want to realize it as an end, he cannot face the idea of complete cessation, of annihilation. The idea of spirit and of spiritual existence is near at hand… Grasping at it, man reaches the comforting belief in spiritual continuity and in the life after death. Yet this belief does not remain unchallenged in the complex, double-edged play of hope and fear which sets in always in the face of death. To the comforting voice of hope, to the intense desire of immortality, to the difficulty, in one's own case, almost the impossibility, of facing annihilation there are opposed powerful and terrible forebodings. The testimony of the senses, the gruesome decomposition of the corpse, the visible disappearance of the personality—certain apparently instinctive suggestions of fear and horror seem to threaten man at all stages of culture with some idea of annihilation, with some hidden fears and forebodings. And here into this play of emotional forces, into this supreme dilemma of life and final death, religion steps in, selecting the positive creed, the comforting view, the culturally valuable belief in immortality, in the spirit independent of the body, and in the continuance of life after death. In the various ceremonies at death, in commemoration and communion with the departed, and worship of ancestral ghosts, religion gives body and form to the saving beliefs. Thus the belief in immortality is the result of a deep emotional revelation, standardized by religion, rather than a primitive philosophic doctrine. Man's conviction of continued life is one of the supreme gifts of religion, which judges and selects the better of the two alternatives suggested by self-preservation—the hope of continued life and the fear of annihilation. The belief in spirits is the result of the belief in immortality. The substance of which the spirits are made is the full-blooded passion and desire for life, rather than the shadowy stuff which haunts his dreams and illusions. Religion saves man from a surrender to death and destruction, and in doing this it merely makes use of the observations of dreams, shadows, and visions. The real nucleus of animism lies in the deepest emotional fact of human nature, the desire for life.

That religion has deep roots into—among other things—our fear of death and our love for life is well understood. The thanatological—from *Thanatos*, the god of death in ancient Greek mythology—aspect of religion has a long history in western culture. A well-known quote attributed to Petronius (c. 27–66 c.e.): *primus in orbe deos fecit timor* (fear first made gods in the world) (Highet 1941, p. 185) points exactly to the association between religion and fear; and if fear springs from confronting the insecurity of the unknown, there is nothing more unknown than what lies beyond death (see Jong et al. 2018, for a brief historical review).

Religiosity can relieve us from death anxiety and fear of death, but the relationship is not simple and straightforward, and it is not linear (Leming 1980). Previous research has found that religiosity can be either positively, negatively, or not associated with fear of death (Ochsman 1984, and references therein). In the words of Michael Leming (1980, p. 347): "religiosity may serve the dual function of afflicting the comforted and of comforting the afflicted".

Leming (1980) suggested a hump-up curvilinear relationship between religiosity and death anxiety/fear of death, with values of anxiety/fear being lower for low and high levels of religiosity, whereas medium levels of religiosity are associated with the highest levels of anxiety/fear of death. Leming (1980, p. 356) provides the following interpretation for this pattern:

> Religiosity, for the highly committed individual, not only promises a reward in the afterlife but it diminishes the fear ascribed to death from secular society… The non-religiously committed individual need not fear an afterlife or divine judgment but only the disruptive effects of death upon his social life. The moderate religionist fears both the afterlife as well as the secular crisis which death presents.

The curvilinear relationship between fear of death and religiosity has been further supported by later studies as shown in Fig. 8.12 (see also Ochsman 1984; Wink and Scott 2005; Jong et al. 2018; and references therein for additional examples). Henrie and Patrick (2014) studied the effects of religious doubt on death anxiety and

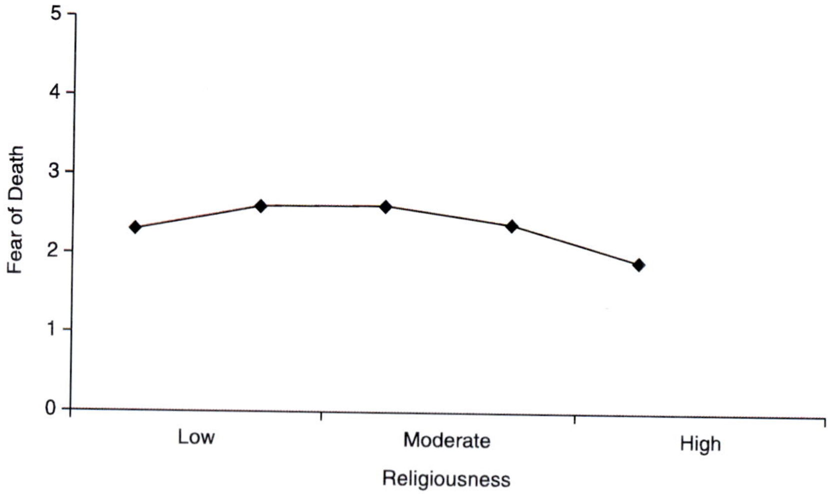

Fig. 8.12 Relationship between fear of death and religiosity in late adulthood (from Wink and Scott 2005)

found religious doubt to be positively associated with death anxiety. Religious doubters would fall in the "moderate" region for religiosity in Fig. 8.12.

Wink (2006) interviewed old people in their late 60s to mid-70s ($n = 181$; 53% women) in the USA to understand the relationship between religiosity, spirituality, and attitudes towards death. Whilst religiosity was positively associated with an accepting identity, identification with social norms, and life satisfaction; spirituality was positively associated with a searching identity, creative temperament, and the positive use of reminiscence. Although neither religiosity nor spirituality was correlated with fear of death, a more sophisticated analysis unraveled various trends. Religiosity increases fear of death if the individual has low belief in the afterlife; this may explain the hump in Fig. 8.12 that characterises people who are experiencing a religious conflict. However, religiosity decreases fear of death with high belief in the afterlife, as also shown in Fig. 8.12. With regard to spirituality, higher spirituality is associated with lower fear of death but only in those who have high belief in the afterlife, suggesting that the projection of being beyond death plays an important role in decreasing fear of death, whether such projection is framed in a religious, spiritual or, presumably, any other way (e.g. generativity).

From an individualistic perspective, adherence to specific religious beliefs can help decrease death anxiety through seeing death just as a transition between one state of being and another, with the overall process extending into eternity, either in a linear or in a cyclical fashion. Anxiety can be further decreased following the promise of re-conjunction with loved ones in the afterlife (e.g. Henrie and Patrick 2014, and references therein). Religion can also give meaning to the life of an individual thus decreasing anxiety in general, but the relationship between religiosity, meaning, and death anxiety is complex (Henrie and Patrick 2014), just as it is for the case of spirituality. Hui and Coleman (2012) have shown that intrinsic religiosity (see Chapter 1) can decrease death anxiety through the combined effect of a positive view of the afterlife and greater ego integrity. Moreover, various forms of death anxiety (e.g. towards own or significant others' death or dying) can be decreased by intrinsic religiosity either directly or, as in the case of anxiety towards own dying, through the mediation of "purpose in life" (Hui and Fung 2008, see Fig. 8.13). Extrinsic religiosity, on the other hand, tends to be positively associated with death anxiety (Cohen et al. 2005).

Socially, religiosity can help older people decrease death anxiety by providing them with social contact and a sense of belonging, through their participation in religious services and other activities of their church community (e.g. Henrie and Patrick 2014, and references therein). On the other hand, social relationships can themselves be a source of death fear and anxiety, as it could be the case in some retirement communities, where the high concentration of old people may create "a sense of doom, a thanatophobic atmosphere in the entire community" (Duff and Hong 1995, p. 20). In this case, religiosity could help counterbalance such potential negative effects.

From a methodological perspective, mere cross-sectional and correlational studies of religiosity and death anxiety may give confusing results, as people experiencing greater death anxiety may tend to be more assiduous participants to religious services, as they seek to cope with their anxiety; or, alternatively, people who participate to religious rites may have achieved greater peace of mind regarding life and

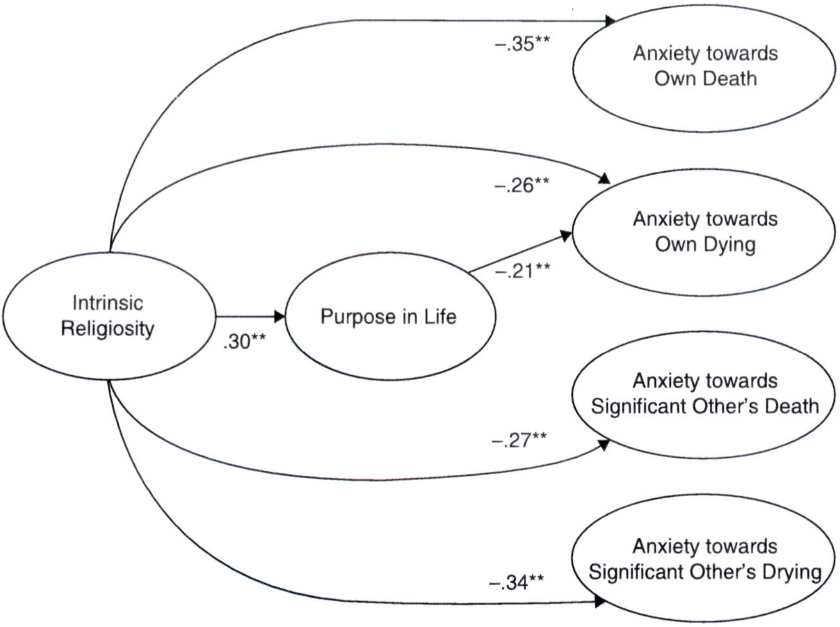

Fig. 8.13 Direct and indirect effects of intrinsic religiosity on decreasing various types of death and dying anxiety. An indirect effect was observed between intrinsic religiosity and a decrease in anxiety towards own dying that is mediated by a positive correlation between intrinsic religiosity and purpose in life (from Hui and Fung 2008)

death and thus they may experience less death anxiety. If both types of believers are studied in the same sample, then no correlation between religiosity and death anxiety may be found.

Religious doubt can also lead to increased death anxiety for as long as it represents an unresolved conflict. However, it could lead to decreased death anxiety as soon as the conflict is resolved and the individual is happy and fulfilled with a new, non-religious existential state of being, whilst keeping the "doubt" only as an admission of personal lack of knowledge about God and its existence. This attitude is commonly found among agnostics (Streib and Klein 2013). Pooling together samples of participants who are at different stages of this process may lead to confusing results.

Therefore, fear of death and death anxiety can be decreased, to an extent at least, through living a fulfilling life that produces personal satisfaction and legacy, and through beliefs about the afterlife or the continuation of being beyond death. This has sparked the creation of numerous spiritual belief-systems and religions, some of the most ancient of which are likely to be the various cults of the ancestors. The dead, whether they died old or young, can persist over time in the mind of the living through the memory of their past existence within a causal chain linking all generations. Moreover, ancestors do not only persist in memory as faces and bodies, they also persist through their teaching, examples, and perhaps practical achievements; that is, through their cultural legacy. The genetic legacy also provides a degree of continuity across generations. This has currently sparked much interest in genealogies and family trees in some countries. In ancient times the preservation of the body of ancestors through often complex processes of mummification was intended to help the individual continue his/her journey in the afterlife and thus decrease death anxiety and fear of death, but the actual preservation of the body of a deceased relative also reinforced the reality of the historical continuity between generations. This was the case in the Chinchorro culture of South America, where mummification concerned all people (Sanz et al. 2014) and the classical example of ancient Egypt where human mummification could be mainly afforded by wealthy families (Millet et al. 1998). The tradition of mummification started in the Neolithic, at around the same time in both Egypt and South America.

In the next section we further expand the theme of religion and spirituality at the end of life beyond the issues of fear of death and death anxiety.

8.4 Religion and Spirituality

We already defined religiosity and spirituality in Sect. 4.3 of Chap. 4, and from those definitions we should keep in mind that spirituality is a far broader concept than religiosity although both ultimately aim at providing a cosmic view or a transcendent view of life to the individual (Braam et al. 2006). More specifically, we defined spirituality in Chap. 4 as *the search a human being undertakes to find a sense of peace, satisfaction and meaning in life, it is an existential journey that engages the whole of our self*. The journey may be transcendent in many ways: through the effects it has on others, through leaving material testimonies, and through believing in its continuation into an afterlife.

In this comment on religion, spirituality, and their role in old age we start from Popper's (1959) line of demarcation between physics and metaphysics. That line is the principle of *falsifiability*: a hypothesis is scientific, and it therefore falls within the realm of physics, only if it can be disproven through empirical evidence. Given that our approach to the study of old age is scientific, we will have to leave the realm of metaphysics aside. Therefore, we see religious and spiritual beliefs as mind processes that involve our bodily functions in interaction with the rest of our environment, especially our social environment and the cultural knowledge that is transmitted through it. Any reference to the afterlife and its metaphysical implications is simply understood here as a belief expressed by a real person in real time in this world. In addition, we see the dualism body-soul as a psychological construct (Bloom 2007) of a brain that can reflect on itself. No matter what our emphasis in the study of religion might be: biological, psychological, linguistic, anthropological, social, historical, philosophical, and so on (Dow 2006; Stausberg 2010), we will remain within the physical side of the line of demarcation. Any mention of God and the afterlife is intended to be a mention of the "human belief in God and the afterlife".

To a great extent, religiosity and spirituality are well suited for old people, as the "essential rhythms of religious life" blend smoothly with the natural rhythms of old age: "(1) conversion or transition from one style of life to another…; (2) death-resurrection, or loss and gain …; (3) transcendence, or a realization of a life beyond the here and now" (Kollar 1985, p. 55). But above all, religiosity and spirituality provide a basic scaffolding from which some people can build life meaning as they age (Langer 2000), and such scaffolding allows old people to better focus on and to give more *salience*, or centrality, to some aspects of life than others (Ward 1984). Hence the tendency towards increasing spirituality and religiosity in old age (e.g. Levin and Taylor 1997; Argue et al. 1999; Wink and Dillon 2002; Manning 2012; Zimmer et al. 2016; see also Fig. 8.14 for spirituality and Fig. 8.15 for religiosity, the latter being measured as level of prayer). In later adulthood, women increase their spirituality more rapidly than men (Wink and Dillon 2002), the same with religiosity which is higher in older women than older men (Levin and Taylor 1997; Argue et al. 1999).

Manning (2012) recorded three major themes referring to spirituality in the discourse of old women: being profoundly grateful, engaging in complete acceptance, and having a strong sense of assuredness. That is, complementing Kollar's dynamic notion of the essential rhythms of religious life that include transcendence, participants in Manning's work were focusing more on the state of being achieved in old age rather than change. This diversity of focus is a recurrent theme in religious/spiritual studies, reflecting a constant interplay between *being* and *becoming*. Old people obviously *were*, and they still *are*, but what about their *becoming*? Religiosity and spirituality aim at providing full continuity to all those states of existence, to the extent that they also offer an opportunity to *become* even when we are right on the threshold of death.

The broad themes of *being* and *becoming* also emerge from the metaphors and symbols inspired by religion and spirituality that may be used by old people towards

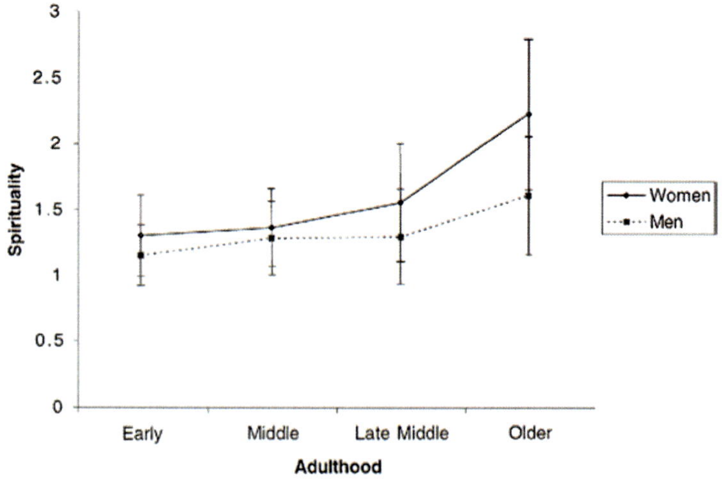

Fig. 8.14 Level of spirituality for four adult age periods (from Wink and Dillon 2002)

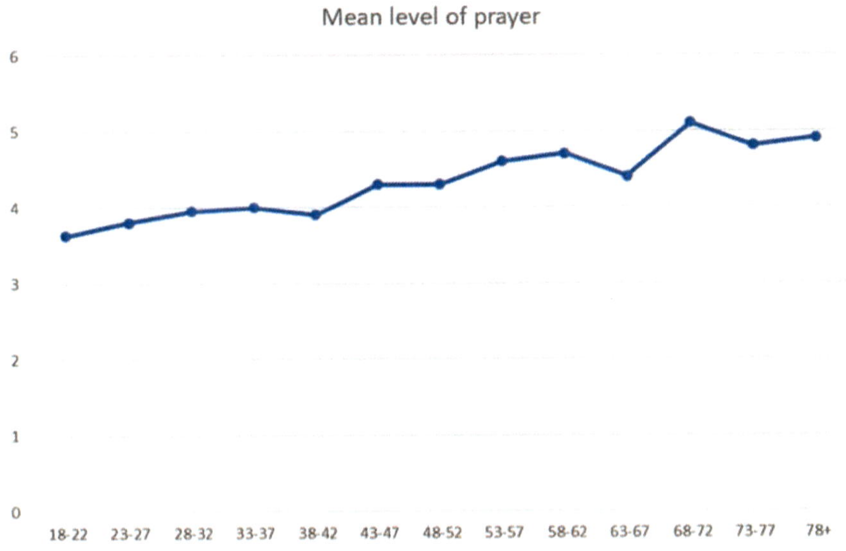

Fig. 8.15 Increase in mean level of prayer with age cohort (years). Redrawn from Levin et al. (1997)

the end of their life and that may help them find meaning. For instance, Marcoen (1994, p. 530–531) quotes various such metaphors that are commonly recorded in the elderly, and many more could be added:

> elderly people in the final stage of life may identify themselves with the suffering Christ, the seed which has to die, the artist who is going to deliver his work of art, the traveller or pilgrim coming home, the actor leaving the scene, a piece of nature merging into the cosmic reality, a leaf falling from the tree of life, the child going back to the Father, etc. These and other metaphors may provide the old person with a frame of reference which inspires the person to a certain way of life and helps him or her to cope with difficult experiences.

Both spiritual and religious experiences share some commonalities in terms of producing wonder and clarity of ideas, compassion towards others, perceptive stillness, and appreciation of silence, but also the development of a sense of expansiveness of the self towards the exterior, as a form of broader integration with life, the world, and beyond. Such experiences involve the use of all our faculties, be they sensorial, emotional, or cognitive (e.g. Atchley 2008).

Spirituality and religiosity are complex constructs that are better described in a multidimensional manner. For instance, as summarised by Marcoen (1994), Elkins et al. (1988) described nine major components of spirituality that include a transcendent dimension that gives meaning, purpose, and a mission in life, and which confers sacredness to life, including the material aspects of it; and social components that focus on the relationships with others via idealism and altruism. In addition, our values can help us adopt a clear stance that can produce a coherent spiritual view of our entire life:

(1) *Transcendent dimension* (the spiritual person has an experientially based belief that there is a transcendent dimension to life). (2) *Meaning and purpose in life* (the spiritual person has known the quest for meaning and purpose and has emerged from this quest with the confidence that life is deeply meaningful and that one's own existence has purpose). (3) *Mission in life* (the spiritual person has a sense of 'vocation'). (4) *Sacredness of life* (the spiritual person believes life is infused with sacredness and he or she often experiences a sense of awe, reverence, and wonder even in 'non-religious' settings). (5) *Material values* (the spiritual person can appreciate material goods such as money and possessions but does not seek ultimate satisfaction from them nor attempts to use them as a substitute for frustrated spiritual needs). (6) *Altruism* (the spiritual person believes we are our 'brother's keeper' and is touched by the pain and suffering of others). (7) *Idealism* (the spiritual person is a visionary committed to the betterment of the world). (8) *Awareness of the tragic* (the spiritual person is solemnly conscious of the tragic realities of human existence). (9) *Fruits of spirituality* (the spiritual person is one whose spirituality has born fruit in his or her life)." (Marcoen 1994, p. 528)

Religiosity builds on those aspects of spirituality and adds an emphasis on a transcendental concept of God or the divine more broadly. In this regard, Marcoen (1994, p. 528–529) quotes Genia's (1990) description of the concept of *transcendental faith* which includes experiencing a transcendent relationship to something greater than oneself (God, a divine power); adopting a style of living, including moral behaviour, that is consistent with religious values; the maintenance of a degree of doubt which is part of the person's quest for higher levels of spiritual growth; openness to religiously diverse viewpoints; a strong social interest and humanitarian concern; infusion of a sense of meaning and purpose.

Religiosity and spirituality can therefore be particularly important to our conception of death, beyond the immediate aspects of fear and anxiety that we have seen in the previous section. A religious/spiritual worldview provides meaning and coherence to the life of old people that help make sense of their experiences, including the process of dying and of death itself (Dow 2006). Adherence to religious or spiritual beliefs may also help find fulfilment in life past, as that life is coming to an end, through ego integrity and transcendence.

If religiosity and spirituality are mental, social, and cultural phenomena that emerge from our interactions with the world, it is appropriate to ask: What are the evolutionary roots of religiosity and spirituality? And can such evolutionary knowledge be useful to better understand the role of religiosity and spirituality in old age?

From an evolutionary perspective, religion—along with the ritualised behaviours expressed through it—and spirituality have been seen as *adaptations* (characteristics that evolved as a result of natural selection), *exaptations* (characteristics that did not originally evolve as adaptations, but that were subsequently co-opted in adaptive ways), or as non-adaptive (or even maladaptive) *by-products* of adaptations.

Religion and spirituality as adaptations: Traditionally, religiosity and spirituality have been associated either with positive health outcomes (as in the views of William James), which would support the idea that religion is an adaptation; or with negative health outcomes (a view promoted by Sigmund Freud) which would make religion maladaptive (Oman and Thorensen 2005). In fact, religiosity can positively affect health through decreasing stress, by giving the individual a source of reassurance in situations of uncertainty, and it could also benefit health through promoting specific healthy behaviours.

Religiosity and spirituality can have various adaptive consequences that encompass both individual and social dimensions (Sosis 2009). From the perspective of the individual, religiosity and spirituality can improve body health through the prescription of specific eating habits and hygienic behaviours; but they can also promote reproduction, positive attitudes towards life and death, a strong work ethics, prohibition of cannibalism and incest, to name a few (Lachman 2010). In addition, religiosity and spirituality can also improve mental health through emotional regulation and cognitive integrity, encouraging positive emotions and cooperation that can alleviate the burden of life in demanding (e.g. competitive) environments, and they can increase social cohesion and pro-social behaviours such as truthfulness and honesty.

We have previously analysed the links between religiosity/spirituality and health—mental health in particular—and well-being in Chap. 4, whereas a more specific relationship between religiosity/spirituality and cancer, and some potential mechanisms linking the two that involve the biology of stress were addressed in our book *Behavioural oncology: Psychological, communicative, and social dimensions* (Cordella and Poiani 2014). There we quoted George Vaillant (Vaillant 2008) who pointed out the adaptive role of spirituality in terms of being a conduit for the expression of positive emotions (e.g. hope, joy, compassion, gratitude) that promote cooperation among members of a group, thus decreasing stress. Such social-cohesiveness role of spirituality and religiosity can be extended as far as embracing the whole of humanity (Lahti 2009).

Both young and old people can experience stress caused by losses of loved ones or losses of personal capacities to operate effectively in life (Marcoen 1994; Fry 2000; Puchalski 2002; Moberg 2005; Wilkinson and Coleman 2010; Hafeez and Rafique 2014; Noronha 2015). The positive effects of spirituality on reducing stress and recovering psychological well-being are especially evident in frail elders (Kirby et al. 2004, see Fig. 8.16), that is in those people who are in more need of help. In fact, such positive effects of spirituality and religiosity remain valid even as we approach the point of death (e.g. Puchalski 2002).

However, this does not mean that religiosity and spirituality are the only strategies that old people can use to cope with stresses, or even that they are the best strategies. What is best will depend on any specific individual, his/her preferences, and life choices. For instance, Wilkinson and Coleman (2010) tested the comparative ability of atheists and religious old people (60 years old and older) to cope with the challenges of ageing and old age, and they found that both groups were equally capable of coping effectively.

The ability to cope with adversity is dependent on how *resilient* we are. More specifically, resilience can be seen in a diversity of ways: as a "(a) process of *recovery* (how well individuals are able to bounce back from adversity), (b) *sustainability* (the capacity to continue to move forward in the face of adversity), and (c) *growth* (the ability to further develop as a response to adversity)" (Manning 2013, p. 569;

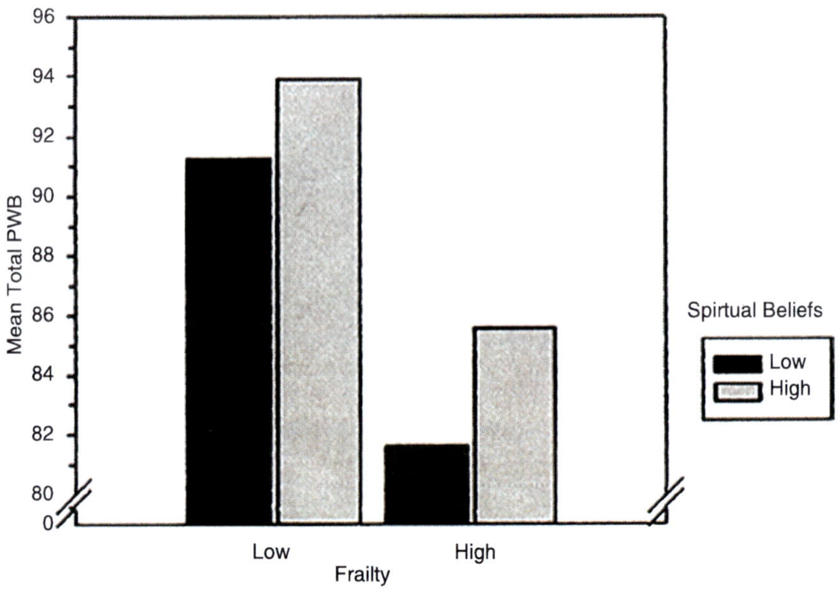

Fig. 8.16 Greater psychological well-being (PWB) is observed in elders who also have stronger spiritual beliefs, but the positive effect of spiritual beliefs is accentuated in the case of more frail elders (from Kirby et al. 2004)

based on Reich et al. 2010, see also Ramsey 2012). Religiosity and spirituality help develop resilience in the old against the mounting difficulties of life (Manning 2013) by acting as a stimulating source of hope and optimism, that can help in recovering from stress or in withstanding the effects of adversity first and then personally grow from the experience.

Pargament et al. (2005) identified a diversity of forms of religious coping (see Table 8.5). Coping strategies include finding meaning in a stressful experience. Such meaning can be positive, as in seeing the learning benefits of experiencing stressful situations, as if they were the acts of a benevolent God. But the meaning could be also negative, such as seeing the stress as punishment from God for past sins. In this case, the individual is expected to redeem him/herself through expiating the sin. Here coping relies on the hope for improvement in life that is expected from repentance and the belief in a forgiving God.

Religious coping can also operate through gaining mastery and control. In this case the control can be left in the hands of God, as it is expected from intercessory prayers, for instance, or it can be taken in the own hands. Religious coping can also rely on getting comfort from the closeness to God, which to an extent is a substitute for social support, and on finding a new path in life through religion.

Both bereavement and grief can be better sustained if their negative effects on old widows and widowers are attenuated by religious and spiritual beliefs, and the support of equally minded people. Adjustment to widowhood can be achieved through spirituality and religiosity by reconstructing life meaning, receiving social

Table 8.5 The variety of forms of religious coping

Religious methods of coping to find meaning:
Benevolent religious reappraisal—redefining the stressor through religion as potentially beneficial
Punishing God reappraisal—redefining the stressor as a punishment from God for the individual's sins
Demonic reappraisal—redefining the stressor as an act of the devil
Reappraisal of God's powers—redefining God's power to influence the stressful situation
Religious methods of coping to gain mastery and control:
Collaborative religious coping—seeking control through a partnership with God in problem solving
Passive religious deferral—passive waiting for God to control the situation
Active religious surrender—active giving up of control to God
Pleading for direct intercession—seeking control indirectly by pleading to God for a miracle or divine intervention
Self-directing religious coping—seeking control through individual initiative rather than help from God
Religious methods of coping to gain comfort and closeness to God:
Seeking spiritual support—searching for comfort and reassurance through God's love and care
Religious focus—engaging in religious activities to shift focus from the stressor
Religious purification—searching for spiritual cleansing through religious actions
Spiritual connection—seeking a sense of connectedness with forces that transcend the self
Spiritual discontent—expressing confusion and dissatisfaction with God' relationship to the individual in the stressful situation
Marking religious boundaries—clearly demarcating acceptable from unacceptable religious behaviour and remaining within religious boundaries
Religious methods of coping to gain intimacy with others and closeness to God:
Seeking support from clergy or members—searching for intimacy and reassurance through the life and care of congregation members and clergy
Religious helping—attempting to provide spiritual support and comfort to others. *Interpersonal religious discontent*—expressing confusion and dissatisfaction with the relationship of clergy or members to the individual in the stressful situation
Religious methods of coping to achieve a life transformation:
Seeking religious direction—looking to religion for assistance in finding a new direction for living
Religious conversion—looking to religion for a radical change in life
Religious forgiving—looking to religion for help in shifting from anger, hurt, and fear associated with an offense to peace

From Pargament et al. (2005)

and material support, giving a reason for the death of the beloved one ("It was God's plan"), and maintaining some contact with the deceased whether it is through spiritual beliefs or, more basically, through remembering or "keeping the loved one alive" (Michael et al. 2003). Given that death is irreversible, inability to cope with the death of a loved one can be associated with health deterioration, including deterioration of mental health. Such deterioration is usually short-term (Wilcox et al. 2003), as after an initial period of pain and grieving there is a point when the emotional bond with the deceased one is severed and the bereaved one moves on (Stroebe 1992). But in some cases, deterioration may extend over the longer-term (Stroebe et al. 2007).

Better coping is also conducive to a better quality of life. Religiosity and spirituality can improve both elders' quality of life in general and quality of the end of life, in particular (Molzahn 2007). One may wonder: what is the point of referring to "quality of life" when life is ending? In fact, the concern about the quality of the end of life is quite relevant, as the end of life should not be marked by pain and anguish if they can be avoided. After all, "the period at the end of life is still life, and not death" (Idler et al. 2009, p. 528). Adaptively, a more peaceful death may help in the grieving process of the bereaved family members.

Religiosity can improve quality of life as we approach the end of it through mechanisms that are both individual and social and that can vary across different cohorts of old people (Idler et al. 2009; Jacó Chaves and Aranha Gil 2015). We have mentioned in previous chapters that we cannot study old people without a thorough understanding of cohort effects that can explain some of their behaviours. Old people who have gone through wars, famine, natural disasters, revolutions, or other momentous national or world events that marked a generation will necessarily develop their spiritual life and their religiosity in a specific manner that is likely to differ from that of other generations who had a different life experience (e.g. McFadden 2005). Cohort experiences can profoundly affect the extent in which either religiosity or, alternatively, spirituality is used (Sutherland et al. 2003), or whether any of them is even necessary in the process of coping with the stresses of life.

Starting from the work of David Moberg carried out in the 1950s there has been an increasing interest in the links between spirituality/religiosity and health and well-being in old people. Several studies have found positive associations between physical health and spirituality/religiosity in the areas of global self-ratings of health, functional health and limitations of disability, physical symptomatology, incidence and prevalence of coronary heart disease, hypertension, cerebrovascular disease, health-related behaviours such as smoking, drug abuse, alcohol consumption leading to a decrease in mortality (see Levin and Chatters 2008 for a review).

Religiosity and spirituality can improve longevity and foster reproductive success through the combined adaptive effect of both individual and social factors (e.g. Cohen and Koenig 2003; Daaleman et al. 2004; Coleman 2005; Sanderson 2008; Lavretsky 2010; Sijuwade 2013; Zimmer et al. 2016). Indeed, religiosity and spirituality have been associated with decreased mortality in the elderly population in various studies (e.g. Zhang 2008; Chida et al. 2009 and references therein). This effect is mediated by both psychological factors (e.g. stress reduction) and healthy behaviours (e.g. exercise, diet). In reviewing the effects of spirituality and religiosity on health, Musick et al. (2000) pointed out that empirical evidence indicates that religious old people tend to be in better health condition and live longer than non-religious ones.

The concept of *spiritual well-being* has been introduced to encapsulate the various effects of spirituality and religiosity that can have fitness benefits to the individual. An elder's spiritual well-being can increase as a result of a satisfactory quest for meaning and the establishment of fulfilling relationships with other people, the rest of the natural environment, and for religious people in particular, God (Canda 1988). From the perspective of the individual, spiritual well-being is expressed in terms of heightened sense of identity and greater "integrity, satisfaction, beauty, love, respect, positive attitudes, inner peace and harmony, a goal, and a direction in life" (Velasco-Gonzalez and Rioux 2013, p. 1125). Greater spiritual well-being can lead to broader psychological well-being.

Psychological well-being achieved through religiosity and spirituality can be mediated by emotional regulation and cognitive integrity that help the individual operate more effectively in life without fear, doubt, and uncertainty. The individual can still learn from experience but without necessarily going through such phases of continuous self-doubt and instability that may lead to long-term inaction and therefore incapacity to operate effectively in life, especially when environmental and social circumstances change. In this regard, religiosity and spirituality may be regarded as brain adaptations to cope with the evolution of brain plasticity, the latter being adaptive in a variable environment (see also Rappaport 1974). When brain plasticity overshoots its immediate adaptive functions, it may lead to emotional dysregulation and cognitive incoherence, whereas spirituality, religiosity, philosophy, a coherent world view (a *Weltanschauung*, e.g. Koltko-Rivera 2004) provide a broad and more stable framework.

Religiosity and spirituality can therefore be important components of the identity of an individual throughout his or her life (Marcia 1966, Kiesling et al. 2006). This means that supporting the religious and spiritual needs of old people can be a way of maintaining their ego integrity, despite some radical changes occurring in their lives. This is especially important as they approach death, in which case integrity is maintained in the psychological reassurance that some aspect of the self will endure beyond death in one way or another, according to specific beliefs.

Throughout history religion has also provided a cognitive frame that has been used not only to give meaning to life (Park 2005) but also to encapsulate a series of accumulated traditional experiences that have been applied to solving specific practical problems: the cycles of sowing and harvesting, animal husbandry, care of pregnant women, care of sick people, delivery of babies and how to care for newborns, understanding the weather, activities associated with lunar cycles and the seasons, and so forth, all wrapped up in rites, ceremonies, dances, and chants.

Eckart Voland, refers to a *cognitive imperative* of religion:

> The cognitive imperative forces one to constantly reflect on the regularities and rules of one's experiences. The cognitive imperative compels a plausible and coherent design of the portrayal of world happenings, without any gaps in explanations, without any islands of irrationality. Human beings obviously cannot stand contingencies, irrationality or causal uncertainty, because what is not understood generates fear. To avoid this, reasons and causes are seen, even where there aren't any (Voland 2009, p. 12–13; see also the various chapters in Voland and Schiefenhövel 2009).

Thus, religion can be a way of making sense of our world, of achieving a degree of coherence and understanding of what happens around us, and such understanding is not only emotionally reassuring but it will be eventually used to guide action.

Cognitively, religiosity can be negatively affected by ageing through a decrease in memory, for instance, which may impinge on remembering sacred texts, songs, specific aspects of ritual that can be important for proper religious living. In addition, increased physical frailty can difficult attendance to services, but help from the community of believers may assist older people in adapting their relationship with their religion and God according to their changing circumstances (McFadden 2005).

Emotionally, religion can relieve death anxiety and fear of death as already mentioned.

Specifically, a strong belief in God and the afterlife tend to be positively associated with death acceptance (e.g. Harding et al. 2005). In addition, religious rituals can afford cathartic experiences that may release stress whilst strengthening social bonds (McFadden 2005, Bloom 2012). Although there is a long psychoanalytic tradition associating religiosity with neuroses, religiosity has also been associated with better mental health and improved capacity to cope with stress (Koenig 2009, Koenig et al. 2001). Older believers may find comfort in their religion, a sense of belonging, ethical guidance, communion in the participation in rituals, and emotional reassurance in the belief of the immortality of the soul. Such positive emotional aspects of religion are consistent with an emotionally positive valence of God in some religious traditions (Thagard 2005), but we should remind that God can be also an object of fear, even within the same religion (e.g. Christianity). The association of God with both positive and negative emotions suggests a potential role of religion in emotional and more broadly behavioural regulation. In fact, religion has been suggested to be a method of achieving self-control, or the "capability for bringing one's behavior into line with meaningful rules and standards" (Geyer and Baumeister 2005, p. 412). Self-control is obviously necessary to live a life according to rules or, in other words, to live a *moral life*. But of course, control through religion can be also social (e.g. Raven 1999) and in this respect control can be either oppressive or liberating (e.g. McSweeney 2005).

If religious and spiritual beliefs are adaptive in regulating our emotions, providing a source of cognitive integrity, and gathering social support, how can we explain religious rituals that can be painful if not overtly dangerous? For instance, initiation rites, fasting, mutilations. From an evolutionary adaptive perspective such rituals could be seen as *handicaps* (Zahavi 1975; Zahavi and Zahavi 1997), that are used to prove the worth and commitment of individuals to be collaborative members of a group, or contributors in solving any challenge the group may confront (e.g. Boyer and Bergstrom 2008). Showing to be able to withstand physical and psychological challenges is also proof of personal fitness.

We have repeatedly mentioned sociality as a major adaptive aspect of religiosity and spirituality. Seen from the perspective of sociality, religiosity and spirituality are readily regarded as adaptive in fostering cooperation, promoting empathy and enhancing the ability of the group to not only satisfy the daily living needs of the members (find food, build shelter) but also to better compete against other groups.

Morality, as guided by religion, may have adaptive consequences within both individual and social contexts (e.g. Bloom 2012), with the moral codes enshrined in religious books, documents, and traditions regulating social conduct and behaviours (e.g. do not lie, do not steal), alongside some personal habits that may benefit the individual (e.g. prohibition to eat specific kinds of food). Moreover, even the association of personality traits such as Agreeableness and Conscientiousness with religiosity can be regarded as links between religiosity and adaptation, in this case adaptation would be in the context of social living and cooperation (Saroglou 2010).

Socially, religiosity may also confer some benefits in terms of better policing the behaviour of the members of the group, thus decreasing the probability of cheating (Bulbulia 2004; Bulbulia and Frean 2010; this is also a function of religious rituals as handicaps, that we mentioned above). *Commitment theory* suggests that "people signal each other by making irrational commitments to unverifiable truths. This tends to create trust within the group. Trust is always a problem in human groups, because the syntactic symbolic communication that creates human culture also creates a potential for deception and autopredation" (Dow 2006, p. 70, referring to Richard Sosis' and others' work).

Greater levels of trust may also enhance the ability of the group to coordinate actions against common threats—including those coming from other groups—thus finding more efficient solutions to the problems that individuals confront. This approach harks back to the theoretical work of David Sloan Wilson on *interdemic selection* (Wilson 1975; not to be confused with the discredited "group selection" theory of Wynne-Edwards (1962)), suggesting that cooperation within groups benefits the members of the group in competition with other groups, thus resulting in the spread of any allele that contributes to the expression of cooperative behaviour. In this regard, Bulbulia (2004) suggests that religious coalitions may be more effective in inter-group competition than secular coalitions.

Nevertheless, the benefits of empathy and cooperation that derive from belonging to a religious group can be also enjoyed by the members of the group without any direct reference to inter-group competition (Hill 2006). For instance, church members may enjoy specific financial benefits if they are in troubles, or if they are looking for a job, another member of the church may be keen to come to their aid (e.g. Iannaccone 1995).

To recap, religiosity and spirituality can be regarded as adaptations given their positive effects on health, survival, stress reduction, reproduction, psychological well-being, increased life meaning, cooperation. The positive psychological and social effects of religiosity and spirituality could be also helpful to some old people throughout their end-of-life process. Seeing their dying relative content and at peace may in turn provide psychological relief to family and friends.

Religion and spirituality as exaptations: In an influential article published in 1979, Stephen J. Gould and Richard Lewontin criticised the "adaptationist program" or the tendency to always look for an adaptive function in each trait, be it anatomical, physiological, or behavioural. In their critique they used the metaphor of an architectural feature, the spandrel that inevitably appears when two rounded arches meet at right angle. As such the spandrel is not the product of architectural design but an inevitable outcome of designing two arches and positioning them perpendicular to each other. Once the spandrel is available, however, it offers a new surface upon which some figures may be drawn, or frescoes may be painted. Gould and Vrba (1982) gave the name of *exaptation* to this phenomenon. Thus, exaptations are initially non-adaptive traits that may have resulted from the evolution of adaptations, and that are subsequently co-opted for adaptive purposes. In the words of Gould and Vrba (1982, p. 4): exaptations are "features that now enhance fitness but were not built by natural selection for their current role".

It has been proposed that religion may be an exaptation that emerged from pre-existing, naturally selected cognitive capacities, and that subsequently may have evolved adaptive functions in social interactions (e.g. promoting cooperation) (Richerson and Boyd 2005; Pyysiäinen 2006). Thagard (2005) has also considered the possibility that religion may be an exaptation of emotion-cognition, dismissing it. In our view, however, Thagard's arguments are weak as the evolutionary link between emotion-cognition and religiosity is not necessarily direct. Broader aspects of religiosity and spirituality that derive from consciousness and then meaning-seeking/meaning making are likely to be part of the causal link extending from basic emotion-cognition to the arguably more complex religiosity/spirituality.

Religion and spirituality as non-adaptive (neutral) or even maladaptive by-products: Lee Kirkpatrick (2005, p. 108; see also Pyysiäinen and Hauser 2010) regards religion as an adaptively neutral by-product of:

> evolved systems dedicated to understanding the physical, biological, and interpersonal worlds… specific forms of religious belief emerge as by-products of psychological mechanisms dedicated to processing information about functionally distinct kinds of interpersonal relationships—attachments, kinships, dominance and status competitions, social exchange relationships, friendships, coalitions, and so forth—that whir into action to shape specific beliefs and expectations about these beings and guide behavior toward them. Thus, for example, gods might be perceived as attachment figures, dominant or high-status individuals, or social exchange partners, with each possibility leading to a different set of expectations and inferences about those gods' behavior and decisions about how to best interact with them—processes emerging from functionally distinct psychological systems designed to solve such adaptive problems in human relations.

In this view, religion has no adaptive function per se, being a neutral by-product of adaptive behaviours, social behaviours in particular:

> The ability to construct abstract philosophical explanations about the world was almost certainly not directly related to reproductive success in Stone Age environments (Kirkpatrick 2005, p. 112).

An explanation for the emergence of religion as a by-product of our evolved capacity to infer causality and agency was suggested early on by Stewart Guthrie (Guthrie 1993). In the words of Bulbulia (2004, p. 658):

> What makes religion plausible, for Guthrie, is our experience of the world as filled with animate beings. Contrary to the methodological assumptions of late twentieth century anthropology, Guthrie didn't think that anthropomorphic tendencies could be explained solely as products of local culture and context. You don't 'learn' to read gods into the fabric

of reality. Rather god-mongering is a panhuman tendency; even secularists do it, for example when we perceive faces in clouds…

Causality and agency are adaptive traits that can also lead to the adaptively neutral development of the concept of God (e.g. Atran 2002; Bloom 2007). It is certainly adaptive to look for a cause given a specific effect: vomiting may be the result of some food item we have just eaten, a roar in the savannah may be evidence for the presence of a dangerous predator, and therefore responding by increasing our alertness is adaptive. But there is nothing specifically adaptive in suggesting that God is the cause of everything. At the most God may fill the gap of our ignorance and provide temporary psychological relief from the anxiety of not knowing, but reality is bound to come back and hit us hard if we get our causality wrong.

Gods are therefore the agents causing things to happen in the world, whether causation is proximate or ultimate; they artificially fill the gaps of our ignorance. The agency of God can explain the immediate and the ultimate cause of things, thus ending a potentially never-ending causative regression. God as the ultimate cause of everything stops our brain running forever into a *regression ad infinitum*, but it does not provide any useful knowledge of the world we live in. That knowledge comes from our direct experience, with the superimposed religious structure being rather superfluous.

Neutrally adaptive religious experiences can emerge from the action of a *hyperactive agent detection device* (HADD) (Barrett 2000). Detecting causative agents is obviously adaptive, but we may lose more by dismissing an agent that is real only because we want to cleanse our perception from agents that are not. Therefore, given that seeing animals or faces in clouds carries no real cost to us, even if we infer that those faces are real and we call them Gods, dismissing the shape of a potential enemy or predator seen through a dense vegetation could be fatal.

Other aspects of religiosity that have no obvious adaptive value include mystical experiences, awe, visions, trance (Boyer and Bergstrom 2008, Bloom 2012). In fact, such experiences tend to be confined to specialised individuals (e.g. a shaman, a witch doctor) rather than being widespread in the population. Meditation, on the other hand, could be adaptive in emotion regulation and it can be of more general utility.

Religion can be maladaptive when all members of the church decrease their fitness following their beliefs. Below in Sect. 8.5.3 we will mention one such case: *apocalyptic suicide*.

Therefore, religiosity and spirituality display some aspects of adaptations, which may have ultimately derived from previously evolved traits associated with social living, emotion, and cognition. The adaptive value of religiosity and spirituality is retained into old age. But as far as specific religious practices are concerned, they could be adaptive, neutral, or even maladaptive.

In addition to the above evolutionary considerations, we have seen in Chap. 5 how greater integration of the self and life fulfilment can be achieved through wisdom. It has been suggested that greater wisdom could be facilitated in old age by spirituality and religiosity (Alexander et al. 1990, see also Le 2008), but the mechanisms that have been put forward are in fact not peculiar to spirituality and religiosity. For instance, Mihaly Csikszentmihályi (1993, p. 241) suggests that "the essence of spirituality seems to consist in an effort to free consciousness from the thrall of genetic instructions" but this is just another way of saying that consciousness and spirituality are ultimately products of plastic mechanisms, and we have seen that our nervous system displays a remarkable degree of plasticity. That neural plasticity is likely to be a core proximate mechanism underlying the complexities of religiosity, spirituality and practical wisdom is emerging from neurobiological studies that suggest the involvement of highly plastic brain centres in spiritual experiences (Mohandas 2008, see also Sect. 4.2 of Chap. 4). It is also likely due to the plasticity of our brain that specific religious practices can be adaptive, neutral, or maladaptive, as we mentioned in the previous paragraph, even though religiosity and spirituality as such may be ultimately grounded on adaptive mechanisms.

Transcendent wisdom (see Chap. 5), however, may benefit from the control of our brain plasticity, to achieve a better "emptying or quieting the mind, and … stopping the mental chatter" (Le 2008, p. 388). "Quieting the mind" is associated with contemplative religiosity and spirituality.

Supporting dying elders in their spiritual and religious needs is something that their caregivers in both the family and in professional institutions must be aware of and should be capable to facilitate (e.g. Phillips 2003; Murdock 2005). As stated by Christina Puchalski (2002, p. 291): "We need systems of care where people are able to find their deep sense of meaning and purpose in the midst of suffering, where they can find peace". Spiritual and religious support may be important to help some terminal elders in their *healing* process (Nelson-Becker et al. 2015). Through healing, older people who are at the end of their life could find "solace, comfort, connection, meaning, and purpose in the midst of suffering, disarray, and pain." (Puchalski et al. 2009, p. 890).

In a study carried out in the UK, Narayanasamy et al. (2004) described a series of spiritual/religious needs expressed by older patients to their nurses in health care centres: allowing the expression of religious beliefs and practice, absolution, connectedness, comfort and reassurance, healing, and meaning and purpose. In turn, nurses responded to such old patients' needs by being sensitive to patients' religious beliefs and practices, utilising personal religious beliefs to assist patients, expressing respect for privacy and dignity, helping patients to connect, helping patients to complete unfinished businesses, listening to patients concerns, and being comforting and reassuring.

Sometimes older people receiving care in nursing homes express the need for religious support to come directly from nursing staff (e.g. Rykkje et al. 2013). Religious and spiritual support provided by healthcare professionals and others become especially relevant as older people are approaching death. In the words of Phillips (2003, p. 254): "The physician who listens thoughtfully to the patient's reminiscences, welcomes discussion of spiritual struggles, and respects the wisdom the dying patient has to offer will grant the patient the opportunity to approach death with dignity."

If medical practitioners, nursing, and other healthcare staff lack the adequate training to be of assistance in the case of elders requiring religious and spiritual support at the end of their life, they should contact appropriately trained people to come and become directly involved (Puchalski 2002).

Medical and nursing staff working in *palliative care* are expected to be especially aware of the spiritual/religious needs of elders who are at the end of their life. Following the definition of the *World Health Organisation*, palliative care is

> an approach that improves the quality of life of patients and their families facing the problem associated with life-threatening illness, through the prevention and relief of suffering by means of early identification and impeccable assessment and treatment of pain and other problems, physical, psychosocial and spiritual. (https://www.who.int/cancer/palliative/definition/en/; downloaded on 12 June 2019)

The WHO definition of palliative care reflects an approach to assisting old people towards the end of their life that is known as the *biopsychosocial-spiritual model*. This model stresses interventions on dying people that recognise the interplay among various factors: biological, social, psychological, and spiritual. Those factors represent sources of specific needs for the elder that should be addressed and satisfied as life is ending (Richardson 2014).

Finally, there is a phenomenon that is sometimes framed in the context of spirituality and religion and that is known as *premonition of death*. It is not uncommon to hear stories about old people correctly predicting their own imminent death in circumstances where there is no clinical indication suggesting that their life is coming to an end as yet; or examples of individuals living only just long enough to accomplish a specific objective, or dying after rather than before an important event that they wanted to attend; or the apparent ability of predicting own death on the ground of self-assessment of health (Idler et al. 2009, and references therein). Can old people at the end of their life determine that the end is imminent, against the opinion of medical personnel? The phenomenon of *premonition of death* has been reported in old individuals and there is plenty of anecdotal evidence to support it (Ngeh 2003, and references therein), but to the best of our knowledge it has not received attention in systematic studies. Premonition of death is often wrapped up in religious or spiritual imagery experienced in dreams. On the other hand, there is scientific work done on the relationship between will to live and death. For instance, Karppinen et al. (2012) have shown that, after controlling for age, gender, and comorbidities, survival was higher among old people displaying a stronger will to live. Will to live was also positively related to survival in 70+ years old Israeli elders over a period of 7.5 years, the pattern was observed in women but not in men (Carmel et al. 2007). Conversely, after analysing data from the 16-year longitudinal *Berlin Aging Study* (BASE), Kotter-Grühn et al. (2009) found that feeling old and being dissatisfied with one's age and ageing at baseline, and also experiencing negative changes in self-perceptions of ageing over time, were associated with an increased mortality risk, after controlling for age, gender, socio-economic status, number of diagnosed illnesses, and dementia.

To recap, religiosity and spirituality can be important aspects of the end of life process in some elders, increasing in relevance with age, especially in women. Religiosity and spirituality allow the dying elder to project him/herself into the future, thus providing existential reassurance and a better ability to cope with the dying process. Religiosity and spirituality have been mainly seen as adaptive responses in both personal and social contexts, although there are also views that interpret them as exaptations, and some specific expressions of religiosity and spirituality can be even non-adaptive or maladaptive by-products. Providing religious/spiritual support to dying elders who request it is an important way of helping them dying with dignity.

In the next three sections we explore the reasons why some old people may decide that life has reached a point where it is not worth living anymore, thus opting to hasten their death through euthanasia or suicide.

8.5 Hastening Own Death: Euthanasia and Suicide

We tend not to complain about going on living into a very advanced old age for as long as we feel well and satisfied with our life. The ancient Greek philosopher Gorgias (483–375 b.c.e.) kept on working up to the end, dying at the advanced age of 107: "When someone asked him why he chose to stay alive so long, he replied: 'I have no complaint to make about old age'" (Nussbaum and Levmore 2017, p. 75). But at some point in time complaints about old age may pile up to such an extent that we may desire to hasten the end of our life, a life that has lost its meaning, perhaps being beset by pain and suffering.

At the age of 104 the Australian academic David Goodall was on his way to Switzerland to undergo assisted suicide, which is legal in that country, but it is not so in Australia for individuals in his situation. He was still in a reasonably good condition for his age, he had no terminal illness, and his mind was sharp, but he had decided that life was not worth living anymore. Currently euthanasia is legal in the Australian states of Victoria and Western Australia, but only for patients with an advanced terminal condition that is likely to lead to death within 6–12 months and that is causing them unacceptable suffering. In a public statement, Goodall expressed his conviction about the step he was taking and frustration towards a system that did not allow him to do the same in his own country:

> I don't want to go to Switzerland, though it's a nice country. But I have to do that in order to get the opportunity of suicide which the Australian system does not permit. I feel very resentful ... [My family realises] how unsatisfactory my life here is, unsatisfactory in almost every respect. The sooner it comes to an end the better ... My feeling is that an old person like myself should have full citizenship rights including the right to assisted suicide (https://www.abc.net.au/news/2018-05-01/david-goodall-to-travel-to-switzerland-for-voluntary-euthanasia/9714292)

Is "life-at-any-cost" (Burt 2002) a principle worth subscribing to? Or is there a point in time when we may decide that "enough is enough" (Cohen 2003)? Doesn't the Hippocratic plea to doctors *to do no harm*:

> The physician must be able to tell the antecedents, know the present, and foretell the future—must mediate these things, and have two special objects in view with regard to disease, namely, to do good or *to do no harm* (Hippocrates, *Of the Epidemics*; http://classics.mit.edu/Hippocrates/epidemics.mb.txt, italics ours)

stand in contradiction with the fact that some procedures used to prolong life at any cost can cause harm to the patient in terms of pain and suffering? Is there a point where providing care to extend life becomes a futile exercise (Misak et al. 2014)?

Before we get into the complexities of death hastening decisions, we should clarify some of the key concepts.

Death hastening behaviour can be defined as: "killing, abandoning and forsaking of the elderly; it is defined as nonsupportive treatment that leads directly to the death of aged individuals" (Glascock 2009, p. 84). Similar concepts include *gerontocide* and *gericide*. Forsaking can occur in the form of denial of food or denial of all support. It is seen in Western societies consequent to elder abuse (see also Chap. 6), but also in traditional societies such as the *Bororo* of central Brazil or the *Yakut* of north-central Siberia (Glascock 2009). Killing of the old has been reported among the *Chuckchee* of northeastern Siberia, whereas abandoning occurred among the *Lau* of the central Pacific Ocean (Glascock 2009) (see Sect. 3.6 in Chap. 3). Table 8.6 shows how death hastening has been observed in one form or another in 51% of traditional societies studied. In the above definition, death hastening behaviour is a result of others' decisions rather than an explicit request from the elder. In modern societies, however, hastening own death is often a personal decision of the elder, who may withdraw from treatment or voluntarily refuse basic life support such as food and drink (Quill et al. 2018).

In a survey of doctors carried out in the UK, Seale (2009) reported a tendency of doctors in intensive care units to be more open to the possibility to hasten the end of life of a patient "lacking the capacity to discuss these decisions", whereas palliative care specialists were less keen to decide in favour of hastening death. Such decision, when it was taken, was usually made in consultation with colleagues, family and when possible the patient him/herself. In Australia, Douglas et al. (2001, p. 51) reported that 36.2% of the general surgeons that they surveyed gave "drugs in doses that they perceived to be *greater* than those required to relieve symptoms with the *intention* of hastening death" (italics original), and more than half of those surgeons

Table 8.6 Hastening death in the elderly in traditional societies

Treatment of the elderly	Number of societies	%
Only support	12	29
Support and non-threatening	4	10
Only non-threatening	4	10
Only death hastening	6	14
Death hastening and non-threatening	4	10
Death hastening and support	11	27
Total	41	100

From Glascock (2009)

stated that "they had never received an unambiguous request for a lethal dose of medication". In previous surveys carried out in the USA, Denmark, England, and Australia, a variable percentage of doctors (2.2–12.3%) have reported assisting patients dying as a result of an explicit request from the patient, whereas Willems et al. (2000) have reported that 2% of North American and 15% of Dutch doctors stated that they proceeded to end the life of a patient without an explicit request from him/her. In Australia, the practice has been suggested to account for 3.5% of all deaths (Kuhse et al. 1997).

Euthanasia can be defined as "an easy death or inducing one, or the act or practice of painlessly putting to death persons suffering from incurable conditions or diseases" (Scheper and Duursma 1994, p. 4). Although euthanasia has been traditionally subdivided into *passive*, *active*, *voluntary*, and *involuntary* (e.g. Meier and Cassel 1983), Lundberg (1988) provides a more detailed subdivision of different types of euthanasia: *passive* (choosing not to treat a terminal disease or medical condition); *semi-passive* (medical withholding of nutrition and fluids from a patient who is in coma caused by a terminal progressive disease); *semi-active* (disconnecting a ventilator from a patient who has suffered a massive cerebral lesion and who is regarded as incapable of recovering consciousness); *accidental* (administration of a narcotic intended to relieve severe pain in a terminal patient with the narcotic suppressing respiration leading to death); *suicidal* (the doctor provides drugs for a terminal patient with the patient taking an overdose of such drugs leading to death, also referred to as *physician-assisted suicide*); and *active* euthanasia (a doctor administering a fatal dose of a drug to a patient in the terminal stage of a disease).

Aid-in-dying (AID) is a legal term used in the USA since the first medically assisted euthanasia statute (the *Death with Dignity Act*) was introduced in that country by the state of Oregon in 1997 (Orentlicher et al. 2016). In spite of this legal framework being in place in the USA, findings from *The Vermont Study on Aid-in-Dying* suggest that even in jurisdictions where the act applies, terminally ill people confront a series of barriers to access euthanasia which include safeguards, cost of medication, access to physicians, and access to information (Buchbinder 2018).

Active shortening of the dying process can be defined as performing an act with the specific intent of speeding up dying such as administering an intentional overdose of narcotics, anaesthetics, or potassium chloride (Sprung et al. 2003). This approach is more open than the *accidental euthanasia* mentioned above.

Intensified alleviation of symptoms with the intention of shortening life has been described as a "grey area" in the care of terminal patients. In this case medical doctors and nurses decide to use drugs intended to alleviate specific symptoms but in doses that could hasten death (Löfmark et al. 2008), this is known as the *rule of double effect* (Billings 2011). It is often the case that the decision is taken on the ground of "personal ethical norms rather than legal rules, professional guidelines or shared moral values" (van Bruchem-van de Scheur et al. 2008, p. 682). In a review of end-of-life decisions in the Netherlands spanning the period between 1990 and 2015, van der Heide et al. (2017) pointed out that the use of morphine in a way that it provides an intensified alleviation of symptoms increased from 19% to 36% during that period.

Withdrawing/withholding life-prolonging treatment involves refusing further medical treatment that is not indicated and is regarded not to benefit the patient (withholding); or the removal of medical therapy that is regarded to be of no benefit (withdrawing) (Vincent et al. 2004). When both withdrawing and withholding are decisions originated from medical personnel, they generate a series of ethical issues, especially how to respond to the pleas of some patients not to "give up" on them.

Continuous deep sedation applied without the administration of artificial nutrition/hydration has been considered as a form of "terminal sedation", given its intended or foreseen life-shortening effects (Miccinesi et al. 2006). The use of continuous deep sedation has increased in the Netherlands from 8.2% in 2005 to 18.3% in 2015 (van der Heide et al. 2017).

Suicide is the "act of intentionally ending one's own life" (Nock et al. 2008, p. 134). The practice has been condemned, but sometimes also justified across cultures throughout history (Leenaars et al. 2001). Following Turecki and Brent (2016, p. 1228), *active suicide ideation* involves having "thoughts about taking action to end one's life, identifying a method, having a plan, or having intent to act". *Passive suicide ideation* is expressed in "thoughts about death or wanting to be dead without any plan or intent". Following a period of suicide ideation, *preparatory acts* may be undertaken towards an imminent suicidal action; whereas *non-suicidal self-injury* is "self-injurious behaviour with no intent to die" and it mostly "consists of repetitive cutting, rubbing, burning, or picking; the main motivations are either to relieve distress, to 'feel something', to induce self-punishment, to get attention, or to escape a difficult situation". It differs from suicide attempt in motivation, non-familial transmission, younger age of onset, psychopathology, and lower level of functional impairment. A *suicide attempt* is "a potentially self-injurious behaviour associated with at least some intent to die"; on the other hand, *deliberate self-harm* is defined as "any type of self-injurious behaviour, including suicide attempts and non-suicidal self-injury". Self-harm can be *direct* (e.g. taking an overdose of medication) or *indirect* (e.g. refusing to eat or drink) (Draper et al. 2002). Finally, *suicidal events* are the worsening of suicide ideation leading to a suicide attempt.

Why would somebody, whether young or old, want to end his/her life through various forms of euthanasia or suicide? Deciding to end one's life is neither an easy decision nor is something that one would normally expect from a living being, whose body and functions have evolved to stay alive and reproduce. However, life can reach a point when it may become a burden. This point can be crossed when the individual is suffering unbearable physical and/or psychological pain. When life does not offer anything else that can justify staying alive, a life not worth living may as well end sooner than later. Life may become not worth living when the individual goes through life-changing experiences that destroy the concept of self without replacement, without a self we may as well not be (e.g. Girard 1993).

Whenever medical personnel are involved in hastening the end of life of an old person, both legal and ethical principles necessarily apply. Each country has a legal framework to deal with death hastening initiatives. A broad ethical framework has been also put forward—originally by Beauchamp and Childress (1979)—that includes adhering to the basic principles of *Beneficence* (i.e. acting in a way that is in the best interest of the patient); *Nonmaleficence* (i.e. an obligation to not intentionally inflict harm); *Autonomy* (i.e. acknowledging "the right and ability of a well-informed and competent individual to choose their own management plan, including end of life care, based on their values and beliefs. Importantly, a patient may opt for an approach that differs from the advised course of care, and this should be respected" Vincent et al. 2004, p. 116); *Distributive justice* (i.e. fairness in the administration of care at the end of life).

We should point out that, from an ethical point of view, although hastening own death could be seen as a right (*right to die*), it should not be framed as a "duty", as some people have been tempted to do:

> ex-Governor Richard Lamm of Colorado and ex-Senator Jacob Javits of New York raised the sensitive issue of the costs associated with the medical care of the terminally ill elderly and even suggested that terminally ill people may have the 'duty to die' (Glascock 2009, p. 79).

A "duty to die" to help the government's budget bottomline? Whilst the same government may perhaps be obsessed with decreasing tax rates? One may only wonder whether anyone would like to live and grow old in such a society. The idea, however, is not new, as it was already espoused by Euripides in the fifth century b.c.e. (quoted in Battin 1987, p. 317):

> I hate the men who would prolong their lives
> By foods and drinks and charms of magic art
> Perverting nature's course to keep off death
> They ought, when they no longer serve the land
> To quit this life, and clear the way for youth

Duty to die aside, the right-to-die concept initially emerged as a result of the significant medical developments achieved since the mid-twentieth century, that have enabled the medical profession to extend the life of many individuals but at the cost of often keeping alive a body that is seriously impaired and deteriorating. Many such patients have responded to their worsening conditions, pain, helplessness, and hopelessness with a desire to actively end their life and suffering in a manner that could be legal, painless, and professionally delivered.

An emblematic case in the development of the right-to-die movement in the USA was that of Karen Anne Quinlan. She had been:

> severely injured in a motor vehicle crash and could not breathe on her own. When the physicians refused to remove the ventilator, the family went to court. Even though the court decided that the parents had the right to refuse treatment, the hospital refused to remove the ventilator. Instead, they slowly weaned Quinlan from it over a period of months. She lived for 10 additional years without the ventilator in an unconscious state, finally dying from pneumonia, for which the parents had refused treatment. The Quinlan case illustrates a key concern for advocates of the right to die. They fear the technological imperative: the belief that doctors are obligated to use whatever medical treatment and technology is available to prolong life ... A growing number of patients now feel that continuing their life may cause more harm to them than allowing their life to end… (McCormick 2011, p. 120)

Modern medical technology is increasingly capable of extending life, but it is not always capable of matching its life-extending capacities with an equal extension of the quality of life. It is such constraint that sets a limit to the desire for life extension, thus providing a good argument in favour of conferring to the individual the legal ability to decide when enough is enough, and end his/her life painlessly with the help of other advances in the medical science.

When the biological life of an old person is prolonged but without the maintenance of a satisfactory level of self-identity (Giddens 1991), then such life extension may be called into question by the affected person. As stated by Seale (1998), since the twentieth century there has been a push towards a right to a medically hastened death, reflecting a will to maintain a narrative of self-identity up until the end (see also Richards 2012).

The idea of introducing the right-to-die has a complex legal, ethical, and academic history rooted in early enquiries dating back to the first half of the twentieth century, that were further developed following the movements for civil and other rights of the 1960s and 1970s to finally become a target of systematic investigation within the new field of *bioethics*. The term *bioethics* was introduced by Van Rensselaer Potter in the 1970s (Potter 1971) to "describe his proposal that we need an ethic that can incorporate our obligations, not just to other humans, but to the biosphere as a whole" (Kuhse and Singer 2006, p. 1). The term, however, is currently used mainly in a more restricted sense to refer to ethical issues associated with activities in the biological and medical sciences (Kuhse and Singer 2006).

According to Beauchamp (2006, p. 644), since the 1970s the specific bioethical aspects of the right-to-die have risen to prominence as a subject of inquiry and public controversy, becoming "an impressive example of the triumph of *autonomy* in bioethics" (italics ours). Such triumph of autonomy means that the will of the individual has risen to greater prominence in the decisions concerning his/her own end of life.

The principle of autonomy simply stresses the right of the individual to *self-determination* in such important matters as the continuation of own life (McCormick 2011). In a healthcare setting, such right is grounded on individual choice, liberty rights, privacy, prevailing culture, and freedom of the will (Beauchamp and Childress 2001), but it is constrained by what the law allows the medical and nursing personnel to do.

The principle of autonomy is clearly respected in Speijer and Diesktra's (1980) rules for assisted suicide. Leenaars et al. (2001, p. 35) describes the rules in this manner:

1. Request is made voluntarily and directly by the individual.
2. Individual is mentally competent at time of request.
3. Wish to end life is longstanding.
4. Presence of unbearable suffering.
5. No reasonable perspective of improvement.
6. Remaining treatment alternatives uncertain/only palliative (offered but rejected).
7. Helper is an acknowledged professional.
8. Helper has consulted with other professionals.

9. Avoidance of preventable harm/damage to others.
10. Decision-making process and steps taken are documented for professional and legal evaluation.

People who are confronting the end of their life should not be regarded as passive participants in a natural and ultimately inevitable process, but they should be considered active agents retaining rights up to the end. On the one hand, wishing to hasten own death is a profoundly personal decision (Monforte-Royol et al. 2011), on the other such personal decision should be contextualised within a framework of rights. Finn (1979; see also Gwyther et al. 2009), for instance, suggested a list of rights of dying people that include (a) the right of choice, (b) the right of communication, (c) the right of companionship, and (d) the right of relief from pain and anxiety.

Baron (2004) calls for greater respect—enshrined in legislation—for Beauchamp and Childress (1979) principle of autonomy, to allow terminally ill people—elders in particular—to end their suffering. The right-to-die movement that has developed in support of euthanasia is grounded in the principle of autonomy and self-determination that should allow any individual the chance of ending unnecessary and unbearable suffering and pain that make life not worth living anymore (Richards 2012). Not legislating ways of humanely ending the life of a terminal patient exposes society to what Charles Baron has called the "seven deadly sins of the status quo": inhumanity, paternalism, utilitarianism, hypocrisy, lawlessness, injustice, and the deadly risk of error and abuse (Baron 2004). In short, denying the possibility of medical help in ending one's life may inhumanely enforce suffering on a terminal patient, it is also an act of paternalism as it assumes that society knows better about what the patient needs than the patient him/herself, and it puts some vague principles above the specific interests and needs of the individual, when the decision of the individual exclusively implies the end of his/her own life. In addition, it can be argued that it is hypocritical of some sectors of the healthcare system to reject a hastening the end of life legislation, when it is common practice to assist in ending the life of a patient in a hospital when circumstances call for it; such action being justified on humanitarian grounds. Appropriate legislation would protect medical personnel and family members from potential legal action, and ultimately, anyone can use it or not at their discretion.

Contemplating the possibility of hastening own death is a direct consequence of a decay in the *will to live*. The will to live can fluctuate temporally as we approach the end of our life, being affected by anxiety, depression, and physical problems such as dyspnoea, or breathing difficulties (Chochinov et al. 1999). Will to live also decreases as older people feel that they are becoming a burden to others, to their family, in particular (McPherson et al. 2007). However, will to live may also simply decay because the elder has concluded that there is nothing else left to do in life.

Religiosity and values that aim at preserving life inhibit decisions to hasten death. But this reluctance may be lost in old people suffering a terminal illness who experience an unacceptably low quality of life. Such old people tend to be more open to hasten their own death (Cicirelli et al. 2000).

In sum, for as long as life in old age is worth living old people will remain motivated to live, but when the conditions of old age make life painful, unbearable, becoming a burden, then the desire to shed such burden and end life may gather momentum. Society should come to the assistance of such elders in a way that respects their autonomy, dignity, and desire for a peaceful and painless death.

In the next section we analyse in greater detail the issues concerning euthanasia from the perspective of old people.

8.5.1 Euthanasia

The word *euthanasia* comes from the Greek *eu*, meaning "good", and *thanatos*, "death", so euthanasia is the "good death". Euthanasia was already in use during the Hellenistic period (fourth to first century b.c.e.) in ancient Greece, where it was understood as a "serene death that seals a virtuous life, the heroic death on the battlefield or for the defense of supreme ideals, the death that comes at the right time, and the non-tortuous death" (Kraniodiotis 2018, p. 533; see also van Hooff 2004; Mystakidou et al. 2005).

According to Mystakidou et al. (2005) euthanasia was associated in ancient Greek culture with the concept of *eugiria* (from *eu*, "good" and *giras*, "old age"). Therefore, dying well in old age was considered an important virtue, an appropriate culmination of the process of living well and ageing well. Aeschylus (523–456 b.c.e.) referred to death as "the only one who can heal all incurable diseases" (Mystakidou et al. 2005, p. 95), this conception of death is consistent with our modern view of euthanasia, as we will see below, as euthanasia can be seen as a way out of a life marked by unbearable pain and suffering. The same view of euthanasia as an end to suffering was supported by Sophocles (fifth century b.c.e.) who, in his play *The Women of Trachis*, wrote about Heracles being in the grip of unbearable pain and asking his son, Hyllus, to assist in ending his life:

> 'Lay my body thereupon and kindle it with flaming pine-torch. And let no tear of mourning be seen there.' Hyllus complains that in so doing, he will 'become a murderer' and will be showing disrespect to the gods. 'What a deed dost thou require of me my father that I should become the murderer guilty of thy blood.' 'Father, father, how can you? You are asking me to be your murderer, polluted with your blood.' And Heracles replies: 'No, I am not. I ask you to be my healer,' or '(be) healer of my sufferings, sole physician of my pain.' (Papadimitriou et al. 2007, p. 26)

Epicurus (c. 341–270 b.c.e.) also supported hastening the end of life if we are suffering constant and unbearable pain (Papadimitriou et al. 2007). He did not believe in the afterlife and so death was devoid of any fear of what may be awaiting us after death; therefore, it did not make any sense to the Epicureans to prolong life if life was marked by constant and unbearable pain (van Hooff 2004). Life without quality is not life.

Hippocrates, on the other hand, was against euthanasia: "I will not prescribe a deadly drug to please someone, nor give advice that may cause his death"

(Mystakidou et al. 2005, p. 97); as it was Aristotle who refers to hastened death in both the *Eudemian Ethics* and the *Nichomachean Ethics*: "In the former, he states that people seeking death are weak and depraved: 'The base among mankind, by toil o'ercome conceive a love of death,' and in the latter, he writes, 'But to seek death in order to escape from poverty, or the pangs of love or from pain or sorrow is not the act of courageous man, but rather of a coward.' (Papadimitriou et al. 2007, p. 27).

According to Kraniodiotis (2018) the word *euthanasia* was introduced into the English language by Francis Bacon in the seventeenth century, to refer to the "painless death that is accomplished by the application of medical means of pain relief" (p. 533). The current meaning of the word, as we defined it in the previous section following Scheper and Duursma (1994, p. 4): "an easy death or inducing one, or the act or practice of painlessly putting to death persons suffering from incurable conditions or diseases" came into use in the nineteenth century.

The practice of euthanasia as we understand it today has found much resistance throughout history, but since the late nineteenth century more consideration has been given to the possibility of ending a life that is not worth living anymore. Although we will focus on euthanasia intended as an intervention to end suffering, it is important to mention that the term was also used in the context of *eugenics* (from the Greek word *eugenes*: *eu*, meaning "well/good" and *genos* "born"), a practice aimed at "improving" the race through selective breeding and survival of individuals possessing specific and desirable traits. The eugenics movement led to practices such as the killing of people who lived an "unworthy life": terminal patients (to be killed against their will), the mentally handicapped, the disabled, psychotic, and old people (Garver and Garver 1991). The eugenics movement directly contributed to the establishment of the extermination camps in Nazi Germany and their occupied territories, and to the Nazi program of the so-called mercy killings directed to "any child with congenital deformities- such as idiocy or mongolism, microencephaly, hydrocephaly, missing limbs, malformations of the head, spina bifida, spastics, etc." (Garver and Garver 1991, p. 1113). But the eugenics movement also led to discriminatory laws in the USA in the first half of the twentieth century affecting immigration and involuntary sterilisation. Eugenics also has precedents in antiquity. For instance, it was practiced by the Spartans in ancient Greece to dispose of babies showing visible deformities (Mystakidou et al. 2005).

Quite differently from its use in eugenics, euthanasia as a mean to achieve a "good death" is a perfectly appropriate end to a fulfilling life. We have seen in Chap. 1 how the ancient Greek concept of *eudaimonia* encapsulates the ideal of a life lived in a fulfilling manner, a life in which our *daimon* or true nature is allowed full expression. The decision to end our pain, taken in the exercise of our right not to suffer unnecessarily, can also be an expression of our *daimon*. In the words of Shaw (2009, p. 530): "While death is to be accepted as part of life, it should not be left to nature to dictate the way we die, and it is fundamentally unjust to grant people liberal latitude in how they live their lives, while granting them little control over the conclusion of their life narratives."

By accepting euthanasia as an appropriate way of ending a fulfilling life, we depart from Viktor Frankl with whom we are otherwise much in agreement. For Frankl (2019, p. 46–47):

> … fate is part of our lives and so is suffering. Therefore, if life has meaning, suffering also has meaning. Consequently, suffering, as long as it is necessary and unavoidable, also holds the possibility of being meaningful… in the final analysis, it is not a question of either achievement or endurance—rather, in some cases, endurance itself is the greatest achievement.

But unavoidable pain and suffering at the end of life need not be endured, nor should elders in such a situation be asked to withstand unbearable distress and instead look for some meaning hidden in it. There is a breaking point, a very personal breaking point, when enough is enough.

Some form of euthanasia or other is currently legal in Canada, Belgium, Luxembourg, The Netherlands, Colombia, Switzerland, in some USA states such as Oregon, Washington, Vermont, Montana, Colorado, California, Hawai'i, and Washington DC, and it is also legal in the Australian states of Victoria (Young et al. 2019) and Western Australia (https://www.theguardian.com/society/2019/dec/11/western-australia-legalises-voluntary-assisted-dying-after-momentous-process).

Although we may have some intuitive idea of what we may wish a "good death" to be, it may help to have a more specific guidance when it comes to take actions that affect real people. In January 2000, the editor of the *British Medical Journal* (BMJ) Richard Smith published a set of *Principles of a good death*:

- To know when death is coming, and to understand what can be expected.
- To be able to retain control of what happens.
- To be afforded dignity and privacy.
- To have control over pain relief and other symptom control.
- To have choice and control over where death occurs (at home or elsewhere).
- To have access to information and expertise of whatever kind is necessary.
- To have access to any spiritual or emotional support required.
- To have access to hospice care in any location, not only in hospital.
- To have control over who is present and who shares the end.
- To be able to issue advance directives which ensure wishes are respected.
- To have time to say goodbye, and control over other aspects of timing.
- To be able to leave when it is time to go, and not to have life prolonged pointlessly (Smith 2000, p. 129).

Thus, in Smith's view a process of good death is grounded on the retention of agency and control, and it is defined by support and hopefully painlessness. In fact, across cultures what is regarded as "good death" tends to be characterised by a "mild, painless, peaceful, dignified, brief" dying process, that is experienced in a social context defined by "family involvement, good relationships, communication, conflicts resolved", where the dying person, for as long as it is possible, retains "autonomy and control over the dying process" (Pool 2004).

There are several moral principles that have guided the debate on euthanasia in recent decades and that give substance to its ethical foundations: *autonomy*, *respect for life*, and *utility* (Meier and Cassel 1983).

Autonomy refers to the respect for the individual and his/her free choices. Thus, the wishes of a fully competent person to end his/her suffering by ending life should be given appropriate consideration, and the fulfilment of the wish should be facilitated by the medical profession. In this regard, the law is expected to provide both guidance and protection to medical practitioners who facilitate the ending-of-life wishes of an individual. Whenever the wishes of the patient conflict with the moral/religious principles of the doctor, the doctor should be compelled to leave the case in the hands of a colleague who shares the same principles as the patient under the law (e.g. Kouwenhoven et al. 2014).

Although *respect for life* is a principle that emerges from granting human life an intrinsic value, we must also consider the situation in which life is not worth living anymore. In this specific situation respect for life becomes respect for the life-decision of the individual, which may include hastening the end of his/her life under specific circumstances. In other words, respect for life is subsumed under the principle of respect for autonomy:

> … the principle of autonomy can and should inform our concept of respect for life. Especially in the elderly, as the prospects of cure diminish, the responsibility to care and comfort grows in importance. This often demands even greater skill, stamina, and attention to detail from health professionals… Perhaps the principle of respect for life should be redefined to direct terminal care away from futile attempts to prolong life and toward efforts to ease pain, maximize function, and lessen the sense of abandonment and loneliness so often cited by the dying. This shift in emphasis in no way weakens the moral foundation of the physician's role. It requires compassion and courage… (Meier and Cassel 1983, p. 297)

The principle of *utility* points to the requirement to maximise the welfare of all concerned, so that the "greatest good is achieved" (Meier and Cassel 1983, p. 297). This principle extends the decision-making process on euthanasia beyond the individual, involving also family and medical personnel at least, but only for as long as the elder is unconscious and there is no will explicitly ruling in favour or against euthanasia. In such a circumstance, others may intervene within the limits of their legal authority and decide on the ground of what is best for all involved. In our view, utility should not overrule autonomy if the elder is in full control of his/her mental faculties.

The practice of *verstervern*, a Dutch word indicating "abstaining from giving food and liquid to patients who refuse or who do not experience sensations of hunger and thirst due to old age or illness" (Pool 2004, p. 956) points to the challenges of amalgamating autonomy, respect for life and utility in the achievement of a good death. The term *verstervern* was introduced in the euthanasia debate by the Dutch psychiatrist Boudewijn Chabot in 1996 to indicate an ancient practice of "good death". In fact, a slow decay of the body due to loss of weight and dehydration is normally observed in people experiencing a death trajectory (Pool 2004). Even if verstervern is regarded as a natural "mild death" there are still issues about how to regulate it in institutional settings and how to involve relatives. Moreover, it is not always possible to allow the patient to exert self-determination when the mental condition has deteriorated excessively.

Various arguments have been also put forward against euthanasia. A common argument against the legalisation of euthanasia is that it would expose vulnerable people in society to abuse and to premature death against their will, this is the so-called *slippery slope argument*. The *slippery slope* is a *consequentialist* argument that stresses unwanted and undesirable but allegedly inevitable negative consequences of a decision, in this case that of legalising euthanasia (e.g. Muller et al. 1998). A *deontological* argument has been also advanced against euthanasia. Deontology does not focus on consequences of actions but on their morality, based on a set of a priori principles of right/wrong or good/evil. For instance, the conception that *life is sacred* would obviously run against any action intended to hasten the end of life. *Clinical pragmatic* arguments, on the other hand, would oppose euthanasia on the ground of some clinical observations such as if good medical care is provided, then a request to end life may not eventuate; whereas a euthanasia request is more likely to be made when medical care is inadequate or deficient. The euthanasia debate continues because none of those positions is final and alternatives can be found. For instance, abuse of euthanasia can be prevented by checks, balances, protocols, and legal penalties; conflicts between competing principles can be resolved by giving priority to the wishes of the elder; the same for the clinical pragmatic argument, as a terminal patient may still request euthanasia in spite of being offered the best medical care available.

In general, old age is expected to increase the chances of accepting euthanasia as a personal decision, but being a religious person makes people less likely to accept euthanasia, even among the elders, whereas higher levels of education and greater belief in individual rights and autonomy increase the level of acceptance of voluntary euthanasia (Ward 1980). In a study of attitudes towards euthanasia carried out in the USA, Ward (1980) found that euthanasia was better accepted by younger people, males, whites, and by less religious and more educated people. Interestingly, the results suggest that the attitude towards euthanasia becomes more positive the more distant the person is from being in the real situation of taking a decision (the young), the less constrained the person is by external pressures (the less religious and the more educated) and the more control of their destiny they may feel they have (males and whites in the 1970s USA). Control may give greater reassurance to the person that nothing will be done without their consent.

Regarding religion, churches retain a degree of influence on some individuals despite an increasing level of secularisation observed in society. Referring to the situation in the Netherlands, where euthanasia is legal, Fortuin et al. (2019) reported results of 26 interviews with 79 years old and older Dutch people. The authors subdivided the participants on the ground of their responses into three groups: a *refraining group*, who were not keen to allow euthanasia to determine the moment of their death; a *depending group*, who were in favour of assigning the authority to decide on euthanasia mainly to medical professionals; and a *self-determining group*, who preferred to give older adults the right to personally determine the moment of their death through euthanasia. Did those three groups differ in terms of their religiosity?

The least religious group was the *depending group*, this means that although they did not have religious reasons to oppose euthanasia, they were keen to shift the responsibility of the decision to the medical profession. Interestingly, the other two groups: *refraining* and *self-determining* were equally religious, differing only in the relative percentage of Catholics and Protestants. The Catholics were relatively more numerous in the *self-determining group*, which may be explained by some of them "having united their religious beliefs with science or scholarly learning" (Fortuin et al. 2019, p. 15); whereas Protestants were relatively more numerous in the *refraining group*. These results suggest that at least in the Netherlands, religion as such is not always an impediment to accept euthanasia even though, in principle, Christian religions are fundamentally against euthanasia, regarding life to be sacred and leaving any action on end of life matters to the will of God. This is also the view of other Abrahamic religions such as Hebraism and Islam. Hinduism and Buddhism also stress the sanctity of life, but they are open to the possibility of euthanasia whenever it can be seen "as a compassionate act or a 'mercy killing' for a dying person striving to the highest purpose of human destiny, liberation" (Campbell 2000, p. 39).

Other studies have reported increasingly positive attitudes towards euthanasia as people become older (Leinbach 1993). Fok and Chong (2003), for instance, have shown how increasingly frail elders can become more favourable towards euthanasia not only because of fear of personal pain and suffering, but also motivated by social considerations such as fear of being a burden to the family and feeling useless in old age.

The actual association of euthanasia with age can be complex, as it relies on the health condition of individuals at different ages. Those who are diagnosed with cancer at the age of 60 may tend to show a more positive stance towards euthanasia than older but healthier individuals. The latter, however, may change their mind towards euthanasia as they become even older and more frail, infirm, and less capable of living a fulfilling life (e.g. Onwuteaka-Philipsen et al. 1997; Muller et al. 1998).

Euthanasia as the expression of a good death should simply allow the elder to exit life in a dignified and fulfilling manner. Euthanasia affords an opportunity to old people to die with dignity at a moment and in conditions of their choosing. It is thus appropriate to end this section with the image of Alexis Zorba's good death, famously created by the mind of Nikos Kazantzakis in his novel *Zorba the Greek*. Kazantzakis represents the last moment in the life of his character as an instance of supreme self-affirmation. A life lived in full has no regrets and the individual has the right to be in control up to the very end:

> He then sat up in his bed, threw back the sheets and tried to get up. We ran to prevent him—Lyuba, his wife, and I, along with several sturdy neighbours. But he brushed us all roughly aside, jumped out of bed and went to the window. There, he gripped the frame, looked out far into the mountains, opened wide his eyes and began to laugh, then to whinny like a horse. It was thus, standing, with his nails dug into the window-frame, that death came to him.

Kazantzakis died in 1957 and he is buried in Heraklion, on the Island of Crete. Perhaps he was thinking about Zorba when he decided on the epitaph for his own tomb:

> I hope for nothing. I fear nothing. I am free.

8.5.2 Euthanasia, Life Fatigue, and Existential Suffering

In October 1991, the Dutch Supreme Court judge Huib Drion published the essay *The self-chosen end of old people* ("Het zelfgewilde einde van oudere mensen"), re-published in 1992 as a book that includes others' reactions to Drion's idea (Drion 1992). Based on the principles of autonomy and self-determination of older adults, Drion suggested to make the medical means for euthanasia freely available to all people 75 years old and older, hence the proposal being renamed "Drion's pill". In October 2016, a proposal for supplementary legislation was introduced in the Dutch Parliament by the health and justice ministers to "legalise euthanasia for older adults with a persistent death wish not stemming from a medical condition" (Fortuin et al. 2019, p. 2). Fortuin et al. (2019, p. 2) explain that, in this Dutch debate:

> a persistent death wish in older people who are tired of living is referred to as a 'completed life' … This term evokes the connection with a perceived 'social death', or the 'cessation of the individual person as an active agent in other's lives'… Supplementary legislation legalising euthanasia without medical grounds for older adults upon their own, persistent and voluntary request would enable a reduction of the time gap between perceived social and physical death by bringing forward physical death …

Although in the above quote, Fortuin et al. emphasise social aspects of the existential conflicts of old people at the end of life, existential suffering can be rooted in a diversity of personal experiences, not just the social ones.

Unbearable physical suffering that cannot be relieved tends to be broadly seen as a justifiable reason for euthanasia, but there is much resistance to apply the same logic to cases of *existential suffering* (i.e., distress, helplessness and grief emerging from enduring threats to life) or *life fatigue* (i.e., being tired of living). However, existential suffering resulting from loneliness, emptiness, and fear of further physical and mental decline can be as unbearable as current physical suffering, potentially leading to a desire to hasten the end of life (e.g. Huxtable and Mölle 2007, van Baarsen 2009). Commenting on the case of two old Australians who had attempted suicide and requested euthanasia, Wand et al. (2016, p. 4) pointed out that:

> Some of these individuals may be chronically bored, restricted in activity, and have little or no avenues to experience any pleasure in life, such that they are 'waiting to die.'

If we accept that euthanasia can relieve existential suffering and life fatigue, then it becomes a *beneficent* response (*sensu* Beauchamp and Childress 1979, see above) to such suffering and fatigue, strengthening the case for the use of euthanasia in situations of unbearable suffering (Huxtable and Mölle 2007). In addition, respect for the autonomy of the individual has been put forward as an argument in favour of administering euthanasia in cases of unbearable existential suffering (Buiting et al. 2012).

The link between existential suffering and wish for euthanasia, however, may sometimes be potentiated by the burden of financial difficulties, in which case it is the latter problem that should be solved first, before considering any end-of-life solution (Verhofstadt et al. 2017). Getting rid of "uncompetitive" and "unproductive" poor may as well be a neoliberal ideal, but it is certainly not a humanistic and democratic one.

In spite of the arguments in favour of including existential suffering and life fatigue as acceptable criteria to administer euthanasia, there is much resistance to the idea from the medical profession, being seen as "stretching the borders of euthanasia too far" (Sheldon 2001, p. 1384). Varelius (2014), however, has provided a counter-argument that if euthanasia "based on suffering arising from illness or injury is taken to fall within the domain of medical expertise, it is consistent that [euthanasia] cannot be denied in cases of purely existential suffering by maintaining that physicians' professional expertise does not extend to existential questions" (p. 107). Whenever suffering becomes unbearable it should take a central place in any debate on euthanasia, irrespective of its specific origin.

The case of Edward Brongersma has become emblematic of the use of euthanasia for cases of existential suffering. Cohen-Almagor (2017, p. 188) provides a summary of this case:

> In 2000, a court in Haarlem in The Netherlands considered the case of 86-year-old Edward Brongersma who had expressed his wish to die to his general practitioner, Dr. Philip Sutorius, claiming that death had 'forgotten' him, his friends and relatives were dead, and he experienced 'a pointless and empty existence'. After repeated requests, Dr. Sutorius euthanized his insisting patient and was then put on trial. The public prosecution recognized that Dr. Sutorius fulfilled all the legal criteria but one: 'hopeless and unbearable suffering.' Therefore, the patient's request should have been refused. The court did not discipline Dr. Sutorius, saying that the patient was obsessed with his 'physical decline' and 'hopeless existence' and therefore was suffering 'hopelessly and unbearably'. A spokesman for the Royal Dutch Medical Association reacted to the court judgment by saying that the definition of 'unbearable suffering' had been stretched too far and that 'what is new is that it goes beyond physical or psychiatric illness to include social decline'. The then Justice Minister Benk Korthals said that being 'tired of life' is not sufficient reason for euthanasia. Since then, the debate as to whether physicians should comply with euthanasia requests of people who are 'tired of life' has been widened and many people in Belgium and in The Netherlands are calling for the law to be expanded in order to include similar patients.

In a quote above, Fortuin et al. (2019) mention the concept of *completed life*. In fact, the idea of a *completed life* adds a new dimension to the end-of-life process, a dimension that goes beyond the somewhat negative perspective offered by *life fatigue* and *existential suffering*. Reaching the conclusion that your life is completed and that there is nothing else that you can contribute either to yourself or to others given your current limitations, it is not necessarily a view reflecting negativity, but a simple matter of fact (e.g. Satalkar and van der Geest 2019). At that point in time, is it appropriate to decide to end our completed life?

One may argue that, not matter the circumstances, there is always something that one can contribute (to family, friends, or society at large), but when all our energy and time are devoted to the simple process of just staying alive, then it may be appropriate to ask the question whether living is still worth the pain. The story of David Goodall recounted at the beginning of Sect. 8.5 is a good example of a request for euthanasia following an acknowledgment of *completed life*.

The issue of a *completed life* has been investigated by Satalkar and van der Geest (2019) in the Dutch context. The concept has recently risen to prominence in the euthanasia debate after a Dutch group calling themselves *Uit Vrije Wil* (Of Free Will) launched a petition with the objective of extending the current euthanasia legislation to older people who regard their life to be completed. Before we proceed any further, we must clarify that a "completed life", as intended here, is one that has nothing else to offer to the individual, and therefore it may as well come to an end. We use the concept of *life fulfilment* in this book not to mean that life is completed and has nothing else to offer, but simply that we have found the meaning that makes our life worth living. We can go on living our fulfilling life for as long as we can or want, but an elder who feels fulfilled in life may independently reach a point in time when she or he concludes that life is completed and is ready to embrace the end of it.

Satalkar and van der Geest (2019) found that among the supporters of ending own life on the ground that such life is declared completed, there was a strong sense of autonomy and self-determination as leading values in their life. The idea that they could reach a point where their life would deteriorate to such an extent that they would lose their dignity and independence, made life unworthy of living from that point on. In addition, support for euthanasia often resulted from the experience of having witnessed the extended suffering of relatives and friends before they died:

> They said that they had been closely involved in the care of family members and friends who had grown progressively dependent on their caregivers. They had witnessed their gradual decline and ultimate reduction to a living 'thing' confined to a bed. They also emphasized that they had been confronted by what they experienced as the minimal assistance one gets in such situations to achieve a dignified death. The respondents considered such living conditions and state of dependency unacceptable, and against their values of autonomy and self-determination. Thus they 'decided' that they did not want to get into such a situation themselves. (Satalkar and van der Geest 2019, p. 11)

Finally, others have argued that in fact euthanasia is not needed for the purpose of ending unbearable existential suffering. Non-lethal alternatives have been proposed such as sedation, which is used in the case of terminal patients close to death, but in the case of people only expressing existential suffering the practice of sedation requires close scrutiny and monitoring (Taylor and McCann 2005), alongside more debate to understand its applicability and limits (Bruce and Boston 2011) before consensus on its use can be reached (Lam et al. 2017). In our view, sedation is unlikely to eliminate suffering at the existential level, if anything it may make it worse, as we find it difficult to see a medically "sedated life" becoming a fulfilling life.

To recap, euthanasia may be contemplated by people at the end of their life who wish to put an end to their physical suffering and unbearable pain. It is especially relevant when the individual is old and approaching the end of life. In old age, euthanasia should be seen in the context of achieving the ideal of a good death at the end of a fulfilling life. It should be applied in accordance to the principles of autonomy (free choice), respect for life (prevent suffering and respect the individual's decision to end own life) and utility (in case a dying individual cannot decide on his/her own). Although euthanasia remains the focus of much public debate, the gradual trend has been to introduce it into various countries. In addition, euthanasia may be requested by an old person in order to put an end to life fatigue and existential suffering, and it can be also requested following the conclusion that life is now complete and there is no point in further extending it.

In the next section we shift our analysis to suicide, a proactive way of ending our life that involves the individual not only in terms of decision-making but also action.

8.5.3 Suicide

The phenomenology of suicide is complex, involving a diversity of behaviours. We already defined *suicide* in Sect. 8.5 as the "act of intentionally ending one's own life" and listed various stages of the process leading to suicide, starting from active or passive suicide ideation, preparatory acts, suicide attempts, and suicidal events, eventually culminating in a successful act of suicide.

Suicide, in the elderly or otherwise, is not a new phenomenon in the history of humankind. Ancient historical antecedents of suicide are available, especially from the classical Greco-Roman world, suggesting that the practice was mainly followed in response to unbearable physical suffering/pain, shame/dishonour, grief, mental disease, desperate love, advanced age, devotion, faith, and *taedium vitae* (the meaning of the Latin word *taedium* can range from weariness, boredom, and disgust, to aversion, repugnance, loathing, and sadness; in other words, *taedium vitae* implies a rejection of life); whilst the most common methods used to commit suicide were starvation, cutting, and self-poisoning (van Hooff 1993; Lykouras et al. 2013).

Ancient Greeks made the distinction between *honourable* and *cowardly* suicide. The so-called cowardly suicide mainly conformed to the characteristics mentioned above, whereas "honourable" suicide (also known as *heroic suicide*) resulted from a desire for honour, fear of shame, and self-sacrifice for the benefit of the group (Seidel 1995). Suicide for the benefit of the group can be considered a case of *altruistic suicide* (van Hoof 2004a). Anthony Giddens (Giddens 1970) pointed out that the earliest reference to suicide in the classical Greek literature seems to be that of Epikaste (Jocasta, mother of Oedipus) in Homer's *Odyssey*. Suicides are common events in the Homeric works, being mostly of the heroic type. Heroic suicide was well regarded in classical antiquity, as it is also today (e.g. self-sacrifice in times of war).

Most ancient Greek philosophers did not support the so-called cowardly suicide, but they (e.g. Plato, Aristotle) and Romans such as Seneca (a Stoic) justified suicide in various cases, including infirmity in old age, madness, senile mental decline, incurable disease, pain, poverty, insufferable distress caused by shame and the tyranny of a despot, grief, and insolvency (Giddens 1970; Seidel 1995; Lykouras et al. 2013). Although Seneca recommended against suicide if one is still useful to relatives and/or friends (Seidel 1995). Epicurus articulated his view on suicide by counselling against it to extend the pleasurable experiences of life but accepting it when life becomes unbearably painful. The so-called *Epicurean epitaph* illustrates this open mindedness and almost nonchalance about life and death: *Non fui, fui, non sum, non curo* (I was not, I was, I am not, I do not care).

Sometimes suicide was legally *imposed* on a person accused and found guilty of a crime. Socrates was condemned to death by committing suicide, the reason being his authoritarian stance that threatened Athenian democracy: Socrates did not support the democratic rule of the majority, but the rule of the "knowledgeable" (Stone 1989).

Group suicides also occurred. For instance, what Durheim calls "suicide of the besieged" (1897/1951) is exemplified by the siege of Masada (Netzer 1991). Masada is a hill in Southern Israel, close to the Dead Sea, on top of which Herod the Great built a palace fortress in the first century b.c.e. and where a group of Jewish people made a last stand against the Roman occupation in the first century c.e. After the fall of Jerusalem a Roman army under the command of Flavius Silva besieged the fortress and in spite of a tenacious defence Masada fell in 73/74 c.e.. The occupants, rather than being taken prisoners by the Romans, decided to commit mass suicide.

Another form of group suicide of the *heroic* kind was reported for the period between the fourth and third centuries b.c.e. on the island of Kea in the Aegean Sea, where some people:

> used to end their lives after gaining the consent of their co-citizens … Rather than growing older and ill, they preferred a mass exodus from life by drinking hemlock, in a festive atmosphere like that of the ancient symposia, when they drank wine in the name of their gods. The eagerness to terminate their lives in this way might stem from the fact that the act was considered heroic… (Mystakidou et al. 2005, p. 97–98).

The so-called *apocalyptic suicide* is yet another form of group suicide which may result from the confluence of various factors that include a strong dualistic philosophy, a leader with total control over the followers and who spreads apocalyptic teachings, and a relative isolation of the group from other social and cultural influences (Dein and Littlewood 2000). A copy-cat effect quickly spreads the suicide behaviour throughout the members of the group (Durkheim 1897). A well-known modern example of apocalyptic suicide is represented by the case of *Jonestown*, a settlement in northwestern Guyana where in November 1978 a total of 918 members of the *People's Temple*, a cult led by Jim Jones, committed suicide following the directions of their leader (Black 1990).

Although justified under specific circumstances in the ancient world, the practice of suicide had to be kept under control, lest it destabilised the entire society. As an example of this limit, Lykouras et al. (2013, p. 549) recount the story of Hegesias:

> Hegesias, known as 'Peisithanatios' ('Death persuader'), taught that death is better than life… After a great number of suicide acts, considered as dangerous for the society, the governor, Ptolemy I Lagos banished him from Alexandria, whilst banning the propagation of his ideas.

The Christian era introduced a total rejection of suicide in the West, being considered a sin of diabolical inspiration and equivalent to murder. In fact, since the *Council of Prague* of 563, survivors of a suicide attempt were expected to be penalised (Avrami 2002). Probably because of this there is agreement that, in Europe, suicide was low in the Middle Ages. However, suicide increased during the Renaissance and then it continued to increase throughout the 18th century and afterwards. This happened in concomitance with a relative decline in the importance of religiosity in Europe. In the 19th century, suicide was more frequently linked to mental illness (Giddens 1970).

The scientific study of suicide started in the second half of the nineteenth century especially with the works of Enrico Morselli, Thomas Masaryk, and Emile Durkheim (Kushner 1993). Masaryk (1881) ascribed a central role to pain as a cause for committing suicide, and his observations pointed to a gender bias in the frequency of suicides, with men being more inclined to commit suicide than women.

Regarding the relationship between age and suicide, Masaryk suggested that the tendency to commit suicide increases in direct proportion to age, with a maximum variably observed between 40 and 80 years old, depending on the country. He cited psychoses, boredom with life, physical suffering, and weakness as the most common causes of suicide, with depression being also commonly diagnosed. In addition, people living in urban environments were more likely to commit suicide than those living in rural areas. He concluded that "*the social mass phenomenon of suicide is the fruit of progress, of education, of civilization*" (p. 144, italics original) and, ultimately, "irreligiosity of the masses" (p.141). Consequently, as a remedy to suicide in society, he suggested to increase the level of religiosity: "We need a religion; we need to be religious" (p. 224). This may sound a bit surprising to many of us in the twenty-first century, given the increasing level of secularisation observed around the world (e.g. Bullard 2016), but in fact, in spite of the historical decline of religion, in the nineteenth century religiosity was still a major venue to find meaning and to connect or re-connect with self, life, and other members of the community, given the social disruptions caused by classical modernity (Kushner 1993).

Emile Durkheim's book *Suicide: A study in sociology* (Durkheim 1897) is arguably the most influential of the nineteenth century scientific studies on suicide. Durkheim's definition of suicide states that "*suicide is applied to all causes of death resulting directly or indirectly from a positive or negative act of the victim himself, which he knows will produce this result.* An attempt is an act thus defined but falling short of actual death" (p. 44, italics original).

Durkheim, just like Masaryk and Morselli, pointed to a bias towards men in the frequency of suicides and he also stressed the geographic variability observed in Europe at the time. In France suicide rate increased for men across the entire age spectrum, whereas for women it increased steadily from young to 70–80 years old to then decrease at older ages. In Prussia (an area that included parts of modern Germany and Poland), frequency of suicides steadily increased with age in men, but in women there was a drop in the 30–40 age interval, to then increase afterwards. Saxony (a region in Germany) displayed a steady increase of suicides with age for both men and women. In Italy there was a broad increase with age in men until the

60–70 years old interval, but with a drop in the 30–40 years old interval in between, and a steady decline at ages older than 70; for women there was a steady increase from young ages until 60–70 years old, a drop at 70–80 and then an increase again at ages older than 80. Contemporary data from Denmark that combined men and women showed an increase in the rate of suicide from young ages until 70–80 years old and a decline from then on. In other words, in nineteenth century Europe suicide broadly tended to increase with age, starting to decline between the ages of 60 and 80 years old, especially in women but sometimes also in men.

Durkheim identified three major types of suicide: *egoistic*, *altruistic*, and *anomic*. We have already seen cases of egoistic suicide (e.g. the so-called "cowardly" suicide) which is caused by the desires and priorities of the individual; and of altruistic (e.g. "honourable") suicide, which is outwardly focused on the needs of the group. Anomic suicide occurs during periods of economic, social, or political change, which lead to "deregulation of desires and suffering", as stated by Hodwitz and Frey (2016, p. 236). Such suffering in turn causes an increase in the frequency of suicides, suicide being the ultimate relief from suffering. Hodwitz and Frey (2016) also provided recent evidence for anomic suicide, showing that suicide rates increased following the global financial markets crash of 2008 in the European Union and decreased once the economic conditions improved in 2010. The pattern was found in males only.

Durkheim was of the opinion that suicidal tendencies decrease decrease to the extent that egoistic, altruistic, and anomic tendencies are in equilibrium within each individual or moral agent—a moral agent is a person capable of fulfilling any personal aspiration in communion with the group, whilst being resilient to the stressful effects of changes in society.

Subsequent studies carried out on suicide patterns and suicidal intentions since the twentieth century have provided a more precise picture of population trends and individual diversity.

Cantor et al. (1996) pointed out that since the beginning of the twentieth century suicide rates in westernised countries have been increasing with age, although relatively less so in females than males and with variability observed among countries. The authors carried out a comparative study of suicide rates in English speaking countries divided between "New World" countries: United States of America (USA), Canada (CAN), Australia (AUS), and New Zealand (NZ); and "Old World" countries: England and Wales (E&W), Scotland (SCOT), and Northern Ireland (NIRE). The logic of this grouping is that "New World" countries, being mainly countries of immigration, may be characterised by greater levels of social isolation, alienation, and by reduced cohesion; such countries also contain minority indigenous populations that may suffer high levels of stress due to social injustice, discrimination, and racism. Following the combined effects of all those factors it is expected "New World" countries to be characterised by higher levels of suicide. The analyses were based on data supplied by the *World Health Organisation* (WHO). Table 8.7 shows their main results.

Consistent with previous work on suicide carried out by Masaryk, Durkheim, and others in the nineteenth century, Table 8.7 indicates that the male bias in the rate of suicide has continued throughout the twentieth century (e.g. Harwood et al. 2000). Older people are also more inclined to commit suicide than younger people,

Table 8.7 Mean suicide rate of sex and age groups from seven localities: four "New World" and three "Old World"

Sex/age	CAN	USA	AUS	NZ	E&W	NIRE	SCOT	Total
Males (age groups)								
15–19	14.4	11.1	10.0	10.0	3.8	3.9	5.5	8.4
20–24	25.2	21.2	21.5	19.4	9.9	9.6	11.6	16.9
25–29	23.6	21.7	22.6	17.2	11.7	8.9	12.5	16.9
30–54	24.9	23.6	25.8	20.2	14.8	11.1	17.0	19.7
55 +	28.1	35.0	30.4	28.2	21.3	13.1	19.0	25.0
Females (age groups)								
15–19	3.1	2.9	3.1	3.1	1.6	1.9	2.1	2.5
20–24	5.4	5.1	6.3	6.1	3.6	2.7	4.0	4.7
25–29	7.0	6.7	8.2	6.4	4.5	3.4	5.3	5.9
30–54	9.7	9.6	11.9	8.6	8.4	6.2	9.1	9.1
55 +	8.6	8.5	12.1	12.6	12.8	6.8	10.5	10.3

Suicide rate is per 100,000 people. *CAN* Canada, *USA* United States of America, *AUS* Australia, *NZ* New Zealand, *E&W* England and Wales, *NIRE* Northern Ireland, *SCOT* Scotland. Adapted from Cantor et al. (1996)

but the trend is more consistent across countries in men than women. In addition, as expected, "New World" countries had higher suicide rates than "Old World" countries, except for females aged 55+ years, where suicide rates are very similar.

Gulbinat (1996) also analysed WHO data, but in this case a comparison with Asia was included. Overall, suicide rates are also higher in men than women in Asia, as it is the case in other regions of the world, but variability between countries is high. Although attempted suicide tends to be higher in women than men and the gender difference in suicide is lower in Asia than elsewhere, mainly due to high suicide rates found among older women in that region of the world. From his analyses, Gulbinat (1996) draws the following conclusions:

- There is great suicide variation among countries, with some countries reporting sharp increases of male suicide over the last three decades, and others showing a clear decline in such rates.
- Countries with an increase in old age suicide outnumber those with a decrease by ratios between 2:1 (Europe) and 5:1 (Asia).
- Increases in old age suicide are much more pronounced than decreases (up to 150% increases, but not more than 60% decreases).
- The above phenomena can be observed across continents (i.e. Americas, Asia, and Europe).

More recent epidemiological data on suicide indicate that the rate is higher in low-income and middle-income countries, which account for 84% of all suicides around the world according to WHO (Phillips and Cheng 2012); but the variability between countries is high and, as one might expect, the prevalence of non-fatal suicidal behaviours (ideation and attempts) is higher than that of suicides (Fig. 8.17). Sinyor et al. (2017) have produced an updated report on the global trends of suicide noting that increases in suicide frequency have been associated with the effects of

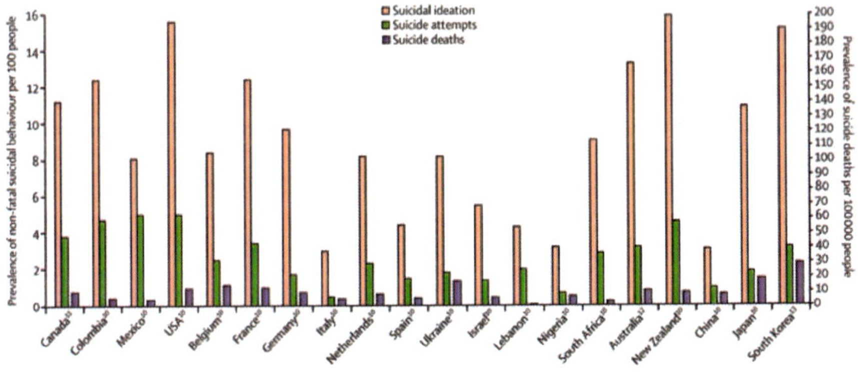

Fig. 8.17 Prevalence values of suicidal behaviours (ideation, attempts, and deaths) in various countries. From Turecki and Brent (2016)

major social disruptions such as the global financial crisis of 2008—as already noted above in our discussion of anomic suicide—psychological stresses, natural disasters, and environmental pollution.

Risk factors for suicide have been categorised as: *population-level factors* such as social fragmentation, economic crises leading to unemployment, living in rural areas—this contrasts with the opposite trend found in the nineteenth century by Masaryk, who observed that urban communities suffered greater rates of suicides than rural communities—and the effect of the media; *distal individual-level factors* such as family history of suicide, adversity in early life, and personality characteristics; and *proximal individual-level factors* such as major depression, psychosis, substance abuse, and stressful life events (Sinyor et al. 2017).

Turecki and Brent (2016) have proposed a comprehensive model of risk factors for suicide that includes different levels of analysis (see Fig. 8.18). Population factors reflected in disruption of social cohesion, economic turmoil, and social isolation can modulate causes of suicide operating at other levels. For instance, a predisposition to suicide that derives from family history can be triggered by changes in economic circumstances. The same occurs when the predisposition to suicide is developmental. Other proximate factors for suicide, such as substance abuse, could also act on their own. Ultimately, different factors can operate alone or in synergy, simultaneously or in sequence, thus increasing the likelihood that a person may commit suicide.

Regarding the specific relationship between old age and suicide, we have mentioned that suicide has been reported to broadly increase with age since the early works of Thomas Masaryk and Emile Durkheim. The pattern has been confirmed to these days (O'Connell et al. 2004), its consistency being stronger across countries in the case of men (Cantor et al. 1996) who tend to commit suicide at higher frequency than women; although, as we have already noted, in Asia the gender difference of suicides in old age is smaller than elsewhere as a result of high suicide rates observed among old women (Gulbinat 1996). Despite the general trend of increased suicide rates with ageing, there is some variability across countries: some countries

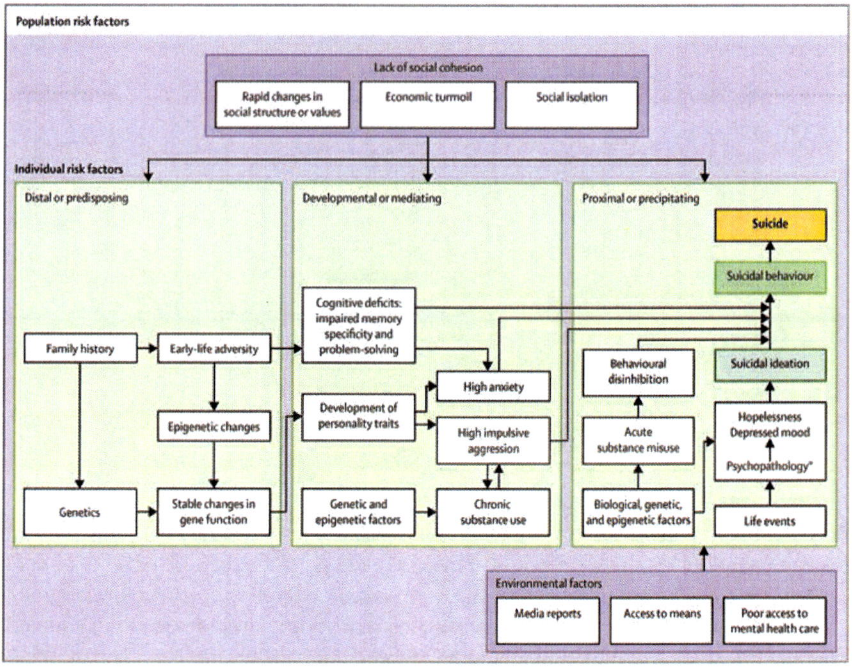

Fig. 8.18 Turecki and Brent's (2016) model for suicide risk. Risk of suicide is affected by a variety of factors operating at both the population and individual levels. Individual risk factors can be grouped into distal (or predisposing), developmental (or mediating), and proximal (or precipitating) factors, and many of these factors interact to contribute to the risk of developing suicidal behaviours. *Any single or a combination of mental illnesses associated with suicide risk, such as major depressive disorder, bipolar disorder, schizophrenia, and personality disorders

peak in frequency of suicides at younger ages, whereas others display low levels of suicidality across all ages (Shah 2007).

From a temporal perspective, it should be noted that suicide among older people declined in many countries in the late 1980s and early 1990s (Shah and Ganesvaran 1994). In a study of data for the period between 1998 and 2003, Shah et al. (2008) focused on suicide trends among 65–74 and 75+ years old across the world. Their main conclusions were that, over the study period: (1) in most countries, elderly suicide rates either declined or remained unchanged; (2) suicide rates declined in both sexes in several European countries; and (3) suicide rates in both sexes and in both age groups remained unchanged in several Central and South American countries, Eastern European countries, and countries that were part of the former Soviet Union. Figure 8.19 shows results for the USA also displaying the same decline in suicide rate in the population of 65–85+ old people that started in the 1980s and extended up to 2007, whereas for 35 to 64 year olds there was a slight increase starting in the late 1990s.

Therefore, the period comprising the 1980s and until the early 2000s seems to have been a benign period as far as the elderly population is concerned, at least in terms of decreased or unchanged levels of suicide. But then, in 2008 the *Global*

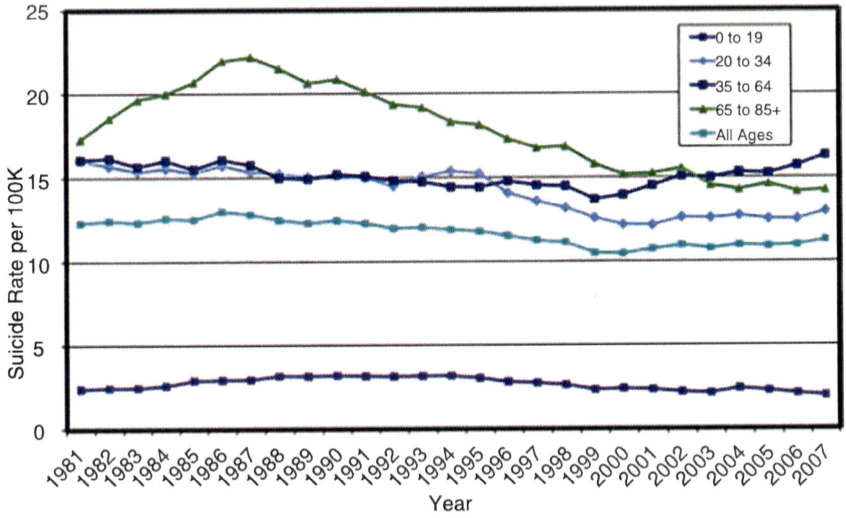

Fig. 8.19 Rates of suicides per 100,000 population in the USA (all races, both sexes) from 1981 to 2007. Data are taken from the *Centers for Disease Control and Prevention*, 2010 (Conwell et al. 2011)

Financial Crisis (GFC) struck, sending the world into a serious economic turmoil. In which ways did the social instability that followed the GFC affect the levels of suicides in the elderly population across the world? Alicandro et al. (2019) studied the effects of the GFC on suicide and concluded that populations in different countries responded differently. In some countries the frequency of suicides increased (e.g. Greece, the Netherlands, and the UK), whereas in others the downward trend observed before the GFC stopped (e.g. Italy, Spain, Germany). The surge in the levels of unemployment during that period is likely to have contributed to the increase in suicides, but other factors may have also played a role. In general, suicide mortality rates were higher in the older population compared to the young during this period; but in some countries suicides were more frequent among middle-age people. Alicandro et al. (2019) suggest that the economic instability caused by the GFC may have especially affected working-age people through unemployment, whilst the elderly population was more affected by the stress of social exclusion and chronic diseases (see also Breuer 2015; Yur'yev et al. 2010). Ultimately, however, an economic crisis undermines the capacity of all (middle-aged and old) for survival, thus increasing the effect of suicide risk factors on everybody, no matter what proximal mechanism explains suicide in each individual case.

Several suicide risk factors have been identified in the population. Suicide tends to be more frequent not only in old people, especially men, as already noted, but also those with physical disorders, depression, and anxiety although suicide attempts are more likely in women (Turecki and Brent 2016). Dombrovski et al. (2008), however, studied 125 men and women who were 65 years old on average (range = 50–91 years) and who had been hospitalised after a suicide attempt, whilst also having been diagnosed with major depression. In this group, *suicidal intent*—measured as behaviour and thoughts before a recent suicide attempt using the *Suicide Intent Scale* (SIS)—was strongest in older men, as it also was *attempt lethality*, measured using the *Beck's Lethality Scale* (LS).

Turecki and Brent (2016) also noted that the male bias in suicides is higher in high-income countries than in low-to-middle-income countries and it is also higher in Europe and the American continent than in Asian Pacific countries. An important socio-economic factor that may predispose to suicide is poverty (Iemmi et al. 2016), but it seems that being poor in a rich country may have stronger effects on increasing suicide rate than being poor in a poor country.

Over the decades and to these days, there has been an insistence to focus on psychopathological causes of suicide in old age, with depression being recurrently at the centre of attention (e.g. Cattell 2000; Conwell et al. 2002; Waern et al. 2003; Almeida et al. 2012; Dombrovski et al. 2013). However, suicidal feelings and suicide itself also occur in mentally healthy old people, not just those affected by depression, and most old people diagnosed with a mental disorder do not commit suicide, a pattern that takes some shine off the centrality of the psychopathology hypothesis (Duberstein et al. 2004). Social factors have also been targeted as causes for the emergence of suicidal tendencies among old people (O'Connell et al. 2004) and life stressors can contribute to making depression even worse (Brilman and Ormel 2001).

The symptomatology of depression could be present in people who have simply lost their *will to live* due to plummeting quality of life—caused by chronic or terminal medical conditions, for instance—or lack of life meaningfulness, as we mentioned in Sect. 8.5.2. In fact, O'Connell et al. (2004) report that the ratio between parasuicides and completed suicides is much lower in elderly people than in adolescents and the general population: 4:1 in elderly people vs. 200:1 in adolescents, and 8:1–33:1 for the general population concluding that "Suicidal behaviour among elderly people is therefore more likely to carry a higher degree of intent. This is further supported by the reported increased use of lethal means by older people, such as firearms and hanging" (p. 896).

If rational intentionality based on poor quality of life and no further prospects for the future is at the root of suicide in most old people, we would expect that a diagnosis of dementia would lead to a spike of suicides before cognitive capacity is lost than afterwards. This has been shown to be the case in a study of dementia and suicide in 50+ years-old people living in Denmark. Erlangsen et al. (2008) found that most suicides in men (63%) occurred within the first 3 years post-diagnosis of dementia, for women it was 59%. Moreover, an additional 18% of total suicides occurred in men in the period between 3 and 10 years from diagnosis, whereas 29% of suicides occurred in women during the same period. Osvath et al. (2005) studied hospital in-patients in Hungary who were 65 years old or older and who were admitted into a psychiatric ward after they attempted to commit suicide. Most elders (about 70%) were diagnosed with hypertension and cardiovascular disorders; depression and adjustment disorders were also commonly diagnosed. Dementia, however, was more common among elders who did not attempt to commit suicide, and who belonged to a control group comprising general elderly inpatients who had been hospitalised in the same period in the same hospital. This suggests that the decision to commit suicide requires functioning cognitive capabilities. In fact, increasing levels of cognitive impairment carried a lower risk of attempting suicide in this study.

A further exploration of the role of *existential willingness to live* as a deeper cause of suicide in the elderly can be provided by a study of suicides carried out among old people in Greece (Paraschakis et al. 2012). It is well known that Greece is one of the countries with the highest proportion of old people, including centenarians and ultracentenarians (see Sect. 4.7 in Chap. 4) and yet Greece is also a country with one of the lowest rates of suicides in the EU (Paraschakis et al. 2012, and references therein). Why would that be? Is it because the level of clinical depression is lower in Greece than elsewhere in the EU? That does not seem to be the case, quite on the contrary, in the EU there is a North-South gradient of depression among the elderly, with older people in Southern Europe being more susceptible to develop depression, especially among the more poorly educated, and presumably less well-off ones; this trend is also valid for Greece (Ladin 2008). It is possible that elders in Greece are better socially integrated—within their family in particular—and that may decrease suicidal ideation, attempts and events (Yur'yev et al. 2010). Social support, however, may not be enough to stop some elders from committing suicide once the willingness to live is lost.

Paraschakis et al. (2012) studied 118 cases of suicide among elders who were mainly living in the Athens area (69.5%; mean age at death: 73.6 years old, range 60–97; 30.5% women). Importantly, 70.8% of the elderly who committed suicide were not living alone. Recent psychiatric hospitalisation (within the past 12 months) was only recorded in 12.4% of the group although 65.2% had a past psychiatric history, mainly due to depression, with the diagnoses being provided by relatives. However, "(a)s many as 81.6% had a known physical illness and were receiving treatment. Additionally, 41.4% had recent hospitalization for a physical ailment" (p. 137). Therefore, the majority of those elders who committed suicide fit more closely to a description of individuals who have run out of personal reasons to stay alive, given their decaying physical condition and in spite of not living alone.

Osvath et al. (2005) reported that old suicide attempters were more likely to take anxiolytic medication or anti-depressants than elders who did not attempt to commit suicide, suggesting that other factors may play a greater role in their suicidal attempts other than depression, if the medication was effective. In fact, factors such as somatic deterioration (55%), interpersonal conflicts (32.3%), and death of close relatives (9.7%) were frequently reported, that could have caused a decrease in the will to live.

Osvath et al. (2005) point to a confluence of multiple factors that, in combination, may push an old person over the limit and induce him/her to commit suicide, in what they call a "catastrophic reaction":

> The high proportion of psychosocial factors raises the possibility of a catastrophic reaction ..., manifesting under stressful circumstances and marked by anxiety and agitation secondary to the intellectual deficits, playing a role in the development of suicidal behavior (p. 224)

In a study carried out in Sweden, Waern et al. (2003) analysed forensic reports and semi-structured interviews with informants about 85 cases of suicide (71 certain and 14 strongly suspected). Obtaining information from informants about cases of suicide is a methodological approach known as *psychological autopsy*, in which the events and circumstances of a suicide are reconstructed through interviewing people who knew the deceased person well.

The deceased individuals were categorised as "old elderly" (age range 75–97 years old; there were also 71 controls for this age group) and "young elderly" (65–74 years old; this age group had 82 controls). Requiring assistance with daily living was a strong predictor of suicide in the old elderly age group, whereas most young elderly did not require home assistance. Members of the older group who had a lower suicide risk also benefitted from family support. Conversely, family discord was a strong predictor of suicide in both old and young elderly. Economic problems also increased the probability of committing suicide, but this was confined to the young elderly.

Other factors predisposing to suicide in old people include personality types such as Neuroticism, whereas Openness to experience seems to be protective (Duberstein et al. 1994). Timidity and shy seclusiveness, hypochondriasis, hostility, and a rigid and independent style have also been indicated as personality risk factors for suicide in old people (Conwell et al. 2002).

Kjølseth et al. (2009) carried out a study of elderly suicide in Norway using the methodology of *psychological autopsy*. Through the testimonials of close relatives, family doctors, and home-based nurses the authors analysed 23 cases of suicide (17.4% women) in elders who were between 65 and 90 years old at the time of death. Overall, the elders were described as people with a strong-willed personality, obstinate, controllers, introverts, and non-expressive of their feelings, with a difficult childhood and adolescence (poverty, loss of protective carers). In spite of that, the individuals had shown to be resilient enough to become "achievers" in life, being characterised by an action-oriented attitude although they also showed little flexibility to adapt to changing circumstances in life, including ageing. Kjølseth et al. (2009) suggest that those personality characteristics made those elders highly dependent on work, and the achievements at work were the fundamental source of pleasurable experiences and life fulfilment. Such obstinacy and stubbornness are defined as *obsessive-compulsive* or *anankastic personality traits* that are expressed through controlling behaviours towards own life and the life of others: "Those around them viewed them as strong-willed individuals capable of taking action and with a considerable ability to exert control, thus gaining confirmation of themselves and their own self image" (Kjølseth et al. 2009, p. 908). Such strong ability to control own life, however, can lead to despair when control is lost and attempts to recover it fail, leading to loss of self-esteem. More flexibility in life can save the individual from committing suicide, but when flexibility is lacking changing direction becomes more difficult. Their introversion also likely contributed to them being disconnected from others, thus becoming isolated from potential sources of help.

Deteriorating physical health (e.g. due to illness) and disability may also affect suicidality on their own (Cattell 2000; Conwell et al. 2002; O'Connell et al. 2004;

Duberstein et al. 2004; Erlangsen et al. 2005; Hawton and Harriss 2006; Kaplan et al. 2007, see Fiske et al. 2008 for a review; although not all studies have found such association, e.g. Fiske et al. 2008 and references therein), as they may put the elder in the situation to ask him/herself whether life is worth living under those circumstances. In a study of 1329 cases of suicide, spanning over a period of nine years and involving 66-years-old and older Canadians resident in the province of Ontario, Juurlink et al. (2004) found that although depression and other mental conditions were considered important as potential causative factors for suicide, what best predicted suicide was the accumulation of medical conditions. In fact, the risk of suicide increased proportionally with the number of illnesses diagnosed in the individual. The five most common diagnoses made in the week before suicide included anxiety, unspecified gastrointestinal symptoms, depression, unspecified cardiac symptoms, and hypertension. In Waern et al.'s (2003) study, the diagnosis of a somatic illness was associated with an increase in suicide risk that was more than three-fold higher among 75+ years old people compared to a control group of 65–74 years old. This suicide-sensitive age group was mainly affected by illnesses of the eyes, nose, throat, and ears. Serious visual impairment in this group was associated with an eight-fold increase in the risk of suicide, whereas a three-fold increase in suicide risk was linked to malignant illnesses.

In terms of emotions, hopelessness, anxiety, pessimism, self-reproach, anger, guilt, and worthlessness have been identified as risk factors for suicidal ideation and suicide in old people (Kiosses et al. 2014). Maladaptive emotional regulation strategies are also a risk factor for suicide in older adults. Such strategies include ruminating, blaming oneself, expressing negative emotions, avoidance coping, thought suppression. Conversely, reducing negative emotions and increasing emotional coping improve the reasons for living and therefore are associated with lower suicidal ideation (Kiosses et al. 2014).

It has been also suggested that suicidality in old people may result from some disruption of cognitive control mechanisms, deficits in social processing, and impulsivity that alter the adaptiveness of decision-making (Kiosses et al. 2014). The tendency to commit suicide may be enhanced by an abnormal response to social interactions and/or an incapacity to solve social problems, thus leading to social isolation.

Social factors such as bereavement and loneliness may contribute to suffering and to develop a sense of life meaninglessness that together can predispose an old person to commit suicide (Cattell 2000; Conwell et al. 2002; O'Connell et al. 2004). Loss of direction in life after retirement is also a factor that may contribute to suicidality in old people (Juurlink et al. 2004). Social and other factors may more broadly affect the levels of psychological stress in older people thus triggering a suicidal response in extreme cases (Conwell et al. 2002), these include financial problems and family discord (Duberstein et al. 2004).

Erlangsen et al. (2004) carried out a study in Denmark from 1994 to 1998 to test the effects of loss of partner on the risk of suicide in 50 years old and older Danish residents. During that period 2323 persons (1494 men and 829 women) committed suicide. Suicide risk increased among all those who experienced a loss although oldest-old men showed the highest absolute increase in suicide risk. Men are also more affected by a loss of partner than women across ages. The suicide risk of people aged 80+ years increased almost six-fold during the first year after the death of a partner, whereas for women the increase was about three-fold. Recovering from the loss also takes longer for the oldest-old men than for women. Oldest-old men do not significantly reduce their risk of suicide after the first year from their loss, whereas for women the risk is eventually reduced to a quarter of the first-year risk.

Accumulated unfavourable life circumstances can also contribute to increase the probability of committing suicide, especially in the oldest-old. Knowing to be transferred to a nursing home may finally tip the balance in some elders to the point where they may regard life not worth living anymore. Radical transitions, such as retirement, that find old people unprepared may have similar negative effects on them (De Leo and Spathonis 2003), as already noted.

A summary of major risk factors for suicide in the elderly is provided in Table 8.8.

Van Orden et al. (2010) linked social factors to suicide in old people into an *Interpersonal theory of suicide*, suggesting that social disconnection and the stress coming with it can increase the tendency to commit suicide, perhaps through the mediation of depression and hopelessness. Personality factors such as Neuroticism just make the individual more sensitive to life stresses (see Fig. 8.20). Two social proximal causes are identified in the theory that can increase the willingness to commit suicide: *thwarted belongingness* and *perceived burdensomeness* (Conwell et al. 2011). Such factors may act independently or in conjunction to push the individual beyond the limit and commit suicide if the individual is already under unbearable pain and stress. We are social organisms and we have an inbuilt "need to belong" (Baumeister and Leary 1995) but if we become unwillingly socially isolated or develop a limited degree of social connectedness, the probability of committing suicide tends to increase (Mellqvist Fässberg et al. 2012). On the other hand, elders may not be socially isolated; e.g. they may be well integrated into their family, but they may perceive that they are becoming an excessive burden (e.g. economic) to their loved ones as their physical and mental conditions deteriorate, thus increasing their willingness to commit suicide (Conwell et al. 2011). Conversely, strong social connections in an environment that is cooperative and both capable and willing to provide material support to the elderly in case of need, will increase the feeling of belongingness in the elderly thus protecting them from suicide, at least to an extent. However, the more advanced the age of the elder is, the more likely it is that he or she may reach a point in time when the will to live collapses, independently of any other consideration. At this point, the existential factors listed in Table 8.8: a sense of completed life or a feeling that staying alive has lost meaning, may acquire greater prominence.

The results so far strongly indicate the role of accumulated, significant, and persistent stressors in tipping the balance towards suicide in some elderly individuals. Such accumulation of stressors may push the individual beyond the point where life is not worth living anymore. Once that point is crossed, the end of life is not seen as a loss but as a gain, a gain in ultimate tranquillity and peace.

Table 8.8 Risk factors for suicide in the elderly

Age	The older the individual, the higher the risk
Socio-demographic factors	Male
	Caucasian
	Lower socio-economic status
	Social disconnection
Psychopathologic factors	Psychological stress
	Anxiety
	Depressive disorder
	Depressive symptomatology
	Schizophrenia or paraphrenia
	Dementia (only before they completely lose their competence and agency)
	Personality disorder
	Alcohol abuse disorder
	Comorbid depressive/alcohol abuse disorders
	Comorbid depressive/anxiety disorders
	Psychiatric hospital admission within previous year
Physical illness	Physical illness (acute, chronic, painful)
	Fear of illness
	Comorbid physical illness/depressive disorder
	Comorbid physical illness/alcohol abuse disorder
Personality traits	Strong, frustrated need to be active and independent
	Neuroticism (more openness decreases risk of suicide)
	Introverts
	Obsessive-compulsive
	Hostility
	Hopelessness and helplessness
	Lower openness to experience
	Inability to tolerate change
	Inability to verbally express psychological pain
	Loss of control
	Dependency on others
Life conditions and events	Broad life adversities leading to low quality of life
	Widowed, single, divorced
	Family conflict, loneliness, despair, loss of social ties
	Cumulative recent losses
	Death of spouse
	Loss of relatives, friends, pets
	Unresolved grief and bereavement
	Abrupt, complete, and involuntary retirement
	Inadequate income, poverty (especially if living in a wealthy country)
	Loss of physical and mental capabilities
	Living in rural areas
Existential factors	Completed life, low will to live, low life meaningfulness

From: De Leo and Spathonis (2003), with modifications

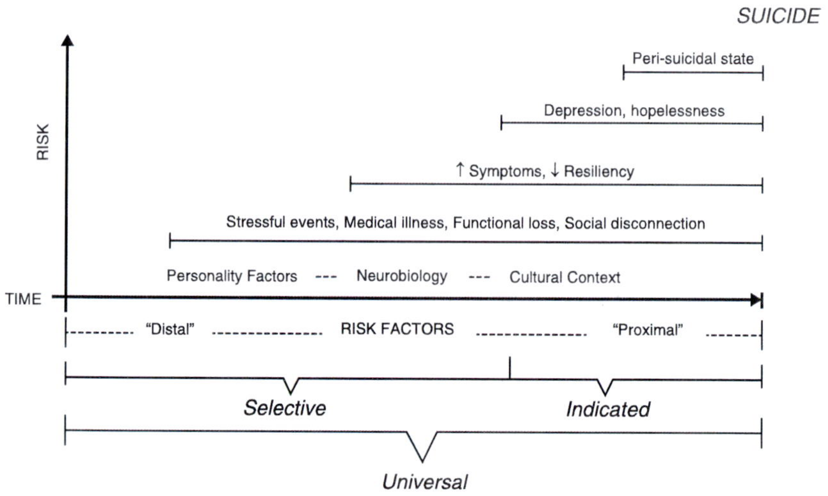

Fig. 8.20 Van Orden et al.'s (2010) *Interpersonal theory of suicide.* From Conwell et al. (2011)

To recap, the tendency to commit suicide may be increased in old people by mood disorders such as depression and anxiety, a recent psychiatric hospital admission, a limited or inefficient social network (e.g. family conflicts, loneliness), an accumulation of physical illnesses (especially those causing functional limitations), and other unfavourable life circumstances; being male, personality traits such as neuroticism, a prevalence of negative emotions, maladaptive emotional regulation, and coping difficulties; disruption of cognitive control mechanisms, but not to the extent observed in advanced dementia which is associated with a lower probability of committing suicide; bereavement, and other factors contributing to develop a sense of meaninglessness. Those are the most relevant risk factors for suicide in the elderly which may become especially important when they act in synergy. Will to live, however, may plunge and suicidality increases at some point in time no matter what, consequent to very advanced age.

8.5.4 Will to Live and Suicide

Both the emotional and cognitive approaches to understanding suicidality in older people have tended to emphasise issues of mental health and physical limitations. Of course, it cannot be denied that mental health problems and physical constraints play an important role, but the decision to commit suicide made by a person who is already approaching the end of his/her life does not necessarily require the diagnosis of a mental disorder. Life fatigue, low existential willingness to live a life marked by pain, the feeling that there is nothing else one can contribute, may be enough reasons for a collapse in the will to live and to want to end one's life and suffering.

We believe that it is important to further consider how the will to live affects suicidality in old people.

As stated by Kjølseth et al. (2009, based on the works of Kjølseth and Ekeberg 1997, Crocker et al. 2006, among others) referring to suicide in old age in a group of elderly Norwegians:

> …it was not the risk factors that were of greatest importance but the extent to which these people saw them as a 'turning point'—a warning of functional decline, dependence and isolation. The elderly talked about loss of control and a feeling of being 'invisible', of being disconnected from others. Among elderly Norwegians who had attempted suicide, self-esteem was strongly associated with the ability to achieve ... When aging involved the loss of this ability, they perceived themselves to be of no value. (p. 904)

Attempted suicide in those elders was consequent to a decay in life meaning and the will to live, rather than a way to escape from the clutches of depression (see also Kjølseth et al. 2010 and the Sect. 8.5.2 in this chapter).

If a sense of *completed life* or even *life fatigue* and *existential suffering* play a role in the desire to commit suicide by plummeting the will to live, they are likely to be even more relevant in the case of the oldest-old. Although the number of oldest-old is increasing in the world, their overall condition is not improving, as it is often the case that many frail and infirm oldest-old people are kept alive by medical interventions, whilst their quality of life is fast deteriorating. Given this trend, it could be also expected that the frequency of suicides in the oldest-old may be higher than at any other old age, as suggested by Table 8.7 and Fig. 8.19.

Erlangsen et al. (2003) studied cases of suicide that occurred in Denmark between 1972 and 1998, including all people aged 50 years old or older and compared the cases of the middle-aged (50–64 years old), the old (65–79 years old), and the oldest-old (80 years old and older). Over the 26 years included in the study, suicide rates for oldest-old men remained relatively higher than those for men of the middle-aged and old age groups and those for women of any age. They also remained relatively constant throughout the entire study period. For the younger men, however, there was a drop in the suicide rates that started in the 1980s and continued throughout the 1990s. For oldest-old women suicide rates were relatively constant throughout the period, just like for oldest-old men, but at a much lower rate (fluctuating roughly between 15 and 30 suicides/100,000 people vs. 40 and 100 for oldest-old men), and they were mostly lower than the suicide rates of women in the middle-aged and old age classes. Suicide rates in women of the younger age classes (i.e. middle-age and old) remained relatively constant until the 1980s and then started to decline throughout the rest of the period, just like it was the case for men in the same-age classes. We have already seen how suicide rates in old people declined during the earlier neoliberal phase of economic reforms in the western world, but with a change of direction for the worse that followed the global financial crisis of 2008. Such period was obviously not included in Erlangsen et al.'s (2003) work. Therefore, from this work we may conclude that the oldest-old are more likely to commit suicide than younger age groups, and they are more resilient to cohort effects and to changes in their overall environment, at least in terms of their decision to commit suicide. This is consistent with a view that their decision was based more on personal circumstances that were likely affected by the ageing process itself.

Sinyor et al. (2015) carried out a study in Canada where they collected information on old people who had committed suicide in Toronto from 1998 to 2011. They compared two age groups: 65–79 years old and 80+ years old (i.e. the oldest-old). A larger percentage of people in the oldest-old group were widowed, they were more likely to be living in a nursing home and were more likely to have a physical problem compared to the rest. However recent stresses (e.g. interpersonal and financial/employment) were higher in the youngest group. Importantly, the two groups were indistinguishable in terms of depression and past suicide attempts. This suggest that the oldest-old may be especially affected in their willingness to commit suicide by a life that has lost purpose, whereas recent stresses may be more likely to be the trigger for suicide in younger old people. Sinyor et al. (2015) also carried out a cluster analysis using data for the 80+ years old group, finding three clear groups: (a) the largest cluster included oldest-old people who were widowed/married, had more health issues, and were also depressed; (b) the second cluster included almost all the single individuals, the majority of whom were living alone and displayed the lowest rate of mental illness; (c) the third cluster was comprised of mainly single individuals who were living in a nursing home or retirement residence, had the highest rates of mental illness, including depression and dementia and they had more past attempts at committing suicide. Therefore, within the oldest-old group there is a diversity of reasons why elders may commit suicide, some of which involve deteriorating mental health, but others simply suggest a deteriorating will to live (e.g. see cluster (b)).

Will to live and life fatigue may also vary in old age among cohorts, following specific life experiences that are broadly shared by the members of the cohort. In commenting on Erlangsen et al.'s (2003) work, we mentioned that, in terms of suicide, the oldest-old seem to be more resilient to cohort effects. But what about other old age classes and the findings of other researchers on cohort effects and suicide? Do the studies show some cohort effects, whereby individuals who are born in a particular year (or a small range of consecutive years) have suicide rates when they are old that are specific to that cohort and that differ from the suicide rates of other cohorts?

Snowdon and Hunt (2002) analysed data from the *Australian Bureau of Statistics* (ABS) about age and sex of people dying by suicide between 1919 and 1999. Data were analysed for successive 5-year periods. As expected also from previous studies, the rate of suicides was higher for men than women, and it increased as individuals grew older, until a peak was reached somewhere between 40 and 64 years old, to then decrease at older ages. Interestingly, as shown in Fig. 8.21, each cohort peaked in suicide frequency at a different age, with peak-suicide age shifting towards younger ages in cohorts born in successive 5-year intervals starting in 1899 and ending in 1923. In this specific population, it is possible that the effect of having spent the early years of life during the Great Depression may have negatively affected the ability of those men to overcome life stresses in old age, and the effect seems to be stronger the younger the boy was during the Great Depression.

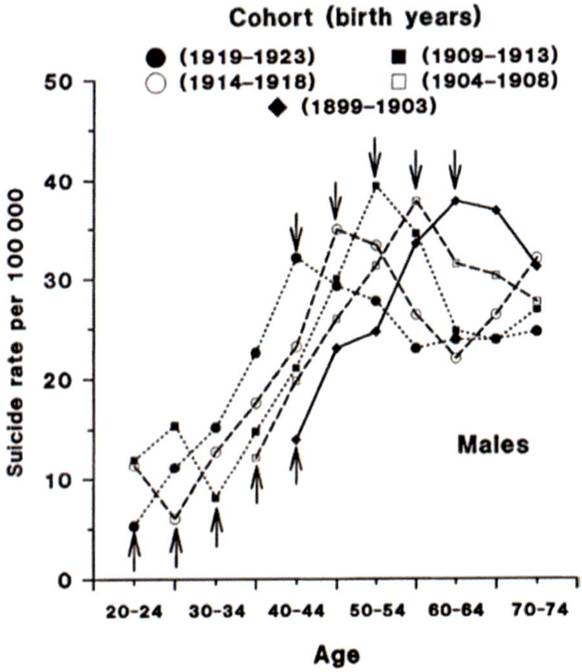

Fig. 8.21 Period effect on suicide in different cohorts: suicide rates between the early 1940s and late 1990s. Upward arrows indicate the age of each cohort during WWII, when rates were low, whereas downward arrows indicate the peak rate for each cohort. From Snowdon and Hunt (2002)

In Australia, unemployment—which was already high at 10% in mid-1929 (the year of the October 1929 Wall Street stock market crash)—was 21% by mid-1930, and reached almost 32% in mid-1932, when the boys of the latest cohort were between 12 and 16 years old. In addition, the peak suicide rate observed at younger ages for the cohort born between 1919 and 1923 may have been also affected by WWII.

Therefore, loss of will to live is an important factor, among others, that can explain suicidality in the old. Will to live is variable not only individually, but it could also differ between cohorts. Cohort differences in suicidality can be affected by specific historical events.

8.5.5 *Attempted Suicide, Protective Factors, and Interventions*

As we noted before, both attempted suicide and suicidal ideation are more common than actual suicide (see Fig. 8.17), which is not surprising, as many suicide attempts may not be carried out with the real intention to take own life; rather, they may be ways of attracting attention or cries for help in cases where solutions to solve a personal problem are running out and despair takes over. Early studies on attempted suicides in old people interpreted such behaviour as an effect of a current mental disorder, with other factors such as a family and personal history of psychiatric problems, the effect of broken homes, vulnerable personalities that may lead to social isolation, depression, significant physical disease also contributing. Batchelor and Napier (1953, p. 1190), in their concluding remarks, point to the likely multicausality of attempted suicides in the elderly:

> Attention is drawn to the multiplicity of precipitating factors and motives, and stress is laid on the importance of feelings of loneliness and physical incapacity, fear, and identification with a dead person. Adverse social factors of a material kind were found to be less important than community attitudes, which tend to make the old person feel useless and unwanted.

Feeling useless and unwanted are factors that will undoubtedly affect the will to live negatively. Although attempted suicides may not necessarily indicate an intention to die, it is worth noting that the ratio of attempted suicide to successful suicide has been estimated at 4:1 for the elderly, whereas for the general population it is between 8:1 and 20:1 (McIntosh 1992, see also Draper 1996). This suggests that attempted suicides in the elderly carry a stronger intention to end their life than attempted suicides in the broader population.

Actions that cause self-harm can be a prelude to an attempted suicide (Wand et al. 2018). In fact, the intent to commit suicide has been reported to be higher among old people who have self-harmed (Dennis et al. 2007). Self-harm often follows loss of life meaning. Wand et al. (2018, p. 294) report how the narratives of some old people who self-harmed:

> reflected the perception of being no longer able to give to others or to achieve anything more in life, and a desire to feel useful and needed ... Some older people feared becoming a burden upon family due to increasing dependency with nothing to offer in return ... In the absence of meaning, life was an obligation …, or boring and rudderless …, and some people searched for a reason to live after their attempt ... One participant particularly highlighted this point 'If there was anything good in my life, I wouldn't want to die'

An accumulated life of suffering has been also reported as a predisposing factor to commit self-harm and eventually suicide (Wand et al. 2018, and references therein).

Once risk factors for attempted suicide have been identified, protective measures could be implemented. For instance, if social isolation is a risk factor, then improving at least the quality but perhaps also the size of the supportive social network may become a protective factor (Wand et al. 2018). Other protective factors may include strengthening the will to live and improving the ability to solve problems (Sinyor et al. 2017), improving health in general, and being more open to experience (Heisel 2006). Religion and spirituality can be also a protective factor for as long as they help the individual at times of crisis, providing meaning, hope, and resilience (McClain et al. 2003; Wu et al. 2015).

Specific suicide-prevention strategies have been also implemented, such as pharmacological interventions especially targeting depression, and psychological therapies such as *Cognitive Behavioural Therapy* and *Dialectical Behaviour Therapy* (Heisel 2006; Zalsman et al. 2016). Lapierre et al. (2011) have reviewed the main interventions currently used to prevent suicide in the elderly.

Some primary care intervention programs that target depression include the IMPACT program (*Improving Mood—Promoting Access to Collaborative Treatment*) and the PROSPECT program (*Prevention of Suicide in Primary Care Elderly: Collaborative Trial*), such programs seem to have some benefits especially for elders diagnosed with major depression. Depression is also treated with pharmacotherapy and *interpersonal psychotherapy* sessions with apparently some success in decreasing suicidal ideation. Nevertheless, even before any psychological or pharmacological intervention is adopted to decrease depression in the elderly, some root social causes of depression that can lead to suicide ideation and then suicide should be addressed. First among such social causes are providing adequate financial support to the elderly to relieve them from their current material hardships and insecurity (e.g. Chi and Chou 2000).

Community-based outreach programs have also been used with some success to improve depression symptoms in elderly at suicide risk (see *Beyond Blue* in Australia, for instance; https://www.beyondblue.org.au/). Telephone counselling support services, such as *Tele-Help* and *Tele-Check*, have also produced some positive results in reducing suicide among elderly people at risk, being more effective in women. Another telephone counselling intervention, *Link-Plus*, has been proven effective in reducing symptoms of depression. On the other hand, the telephone services *The Friendship Line* and the *Geriatric Outreach* programs were able to reduce the levels of hopelessness but not depression (Lapierre et al. 2011).

Those programs notwithstanding, we have already argued that targeting depression alone may not be the answer to helping people who are at risk of committing suicide. Improving resilience in older adults has also been the target of interventions. Some of such programs are based on cognitive behavioural approaches and are especially targeted to elders experiencing some problems of adaptation to retirement. They are "aimed at increasing meaning in life by helping participants set, plan, pursue, and realize meaningful, concrete personal goals" (Lapierre et al. 2011, p. 6). Other programs use *interpersonal psychotherapy* to "improve their social functioning and skills in order to enhance social support and satisfaction of interpersonal needs" (p. 7). Various strategies have also been used to enhance positive ageing and quality of life in order to prevent suicide in the elders: empowerment, coping and adaptive behaviour, flexibility, social skills, self-esteem, sense of belonging, reasons for living, hope, meaning in life, religion/spirituality, and humour (Lapierre et al. 2011).

All those programs that are aimed at preventing suicide in the elderly are obviously important and necessary, but we must keep in mind that old people may reach a point in their life when death is not a threat anymore but, in the words of Sophocles: a healer.

In sum, several factors increase the tendency of old people to commit suicide, including older age itself and being male, physical deterioration, psychological decay, social isolation, and loss of life meaning. But suicide may simply be a decision of an elder who has come to the conclusion that his/her life journey has been completed, he/she feels fulfilled and ready to end the negative aspects of old age that are making living a less than enjoyable experience. In that case, we should not prevent that person from fulfilling his/her desire to end life, and we should not put him/her in the situation where committing suicide is the only option either. Instead, society should provide the means to access humane euthanasia, to be delivered under controlled and professional supervision in the environment of choice of the elder. Again, just as we did for the case of euthanasia, here we depart from the views of Frankl (2019, p. 48) who was of the opinion that "(s)uicide is in no way the answer to any question; suicide is never able to solve a problem". But in fact, suicide can be the solution to the problem of pain and suffering at the end of life, if euthanasia is not available. Contrary to Frankl, we do not believe that suicide is the hallmark of the lazy one, or perhaps the coward in life. We are unconvinced that "suicide also flouts the rules of the game of life; these rules do not require us to win at all costs, but they do demand from us that we never give up the fight" (Frankl 2019, p. 49). We are all for fighting to fulfill our objectives in life, fighting to stay alive also comes naturally from our evolved biology, but fighting to just endure the pain of life, for endurance sake, when life is coming to an end seems to us to be a rather meaningless objective.

In the next section we analyse the specific issue of the end of life in multicultural societies, where cultural diversity can be both an asset and a challenge, in terms of the capacity to provide appropriate end-of-life care in a fair and humane manner.

8.6 End of Life in Multicultural Societies

Modern societies contain large numbers of people that can range into the millions of citizens and non-citizens, natives, and immigrants, and with people it comes an inevitable degree of diversity, including cultural diversity. Such cultural diversity represents an asset but also a challenge to society at large, especially to the providers of healthcare who may have to deal with some elders' conception of the end of life that may be quite alien to them (e.g. Crawley et al. 2002).

Cultural diversity has motivated a large amount of research on the need for more *cultural sensitivity* (i.e. being open to issues derived from cultural diversity), *cultural awareness* (i.e. capacity to recognise and acknowledge culturally specific needs), and *cultural competence* (i.e. ability to act in a culturally appropriate manner), among providers of care for dying elders (Meñaca et al. 2012). Moreover, cultural diversity in society inevitably confronts us with the issue of *multiculturalism*.

Multiculturalism is arguably one of the most misunderstood concepts that have been the target of media, social and political interest for some time now. Specific views about multiculturalism range from the trivial and positive (e.g. enjoying a diversity of cuisines) to the negative and paranoid (e.g. a dangerous threat to national identity). Multiculturalism, however, is something different and it can be seen as "a social-intellectual movement that promotes the value of diversity as a core principle and insists that all cultural groups be treated with respect and as equals" (Fowers and Richardson 1996, p. 609).

Of course, in a democratic society there is an implication intrinsic to multiculturalism that all cultural groups must also obey the law, whereas citizens from all backgrounds are entitled to vote and therefore alter the configuration of parliament with the objective of supporting a specific legislative agenda. Moreover, multiculturalism involves not only the respect for the cultural identity of others but also open mindedness to learn from them and perhaps incorporate such learning into our own identity. Thus, multiculturalism starts with respect for cultural diversity under a common law, but it can lead to a degree of cultural syncretism in some individuals.

Will Kymlicka provides an interesting perspective on multiculturalism that goes to the core of the concept and its utility in the past, the present, and arguably the future. Kymlicka (2010, p. 99) sees *multiculturalism* as a "political project that attempts to redefine the relationship between ethno-cultural minorities and the state through the adoption of new laws, policies or institutions" and we add that those laws, policies, and institutions should ensure the harmonious and democratic co-existence of diverse cultural traditions within the same country. This definition of multiculturalism defends not only the basic freedoms of ethnic minorities, but also the basic freedoms of all citizens, such as the freedom of religion, association, choice of partner, expression, and political allegiance, which are core aspects not just of multiculturalism but of democracy itself, and that therefore are of relevance to all the citizens, regardless of their ethnic background. In our view it is impossible to live in a democratic country unless that country is also multicultural. Moreover, following the above definition, we agree with Kymlicka that multiculturalism is "as old as humanity—different cultures have always found ways of co-existing and respect for diversity was a familiar feature of many empires throughout history" (p. 100).

The phenomenon of multiculturalism that we see today around the world, however, has been specifically affected by recent historical events such as the post-World War II fall of colonialism and the increased movement of people around the world due to the combined needs to work (by immigrants) and for manpower (by host countries), that follow from economic development. Such movement of people is facilitated by the speedy and safe means of transportation over long distances that are currently available. Out of this migratory process has emerged a phenomenon that Kymlicka (2010) calls *citizenisation*, which involves:

> turning the earlier catalogue of hierarchical relations into relationships of liberal-democratic citizenship, both in terms of the vertical relationship between the members of minorities and the state and the horizontal relationships amongst the members of different groups. In the past it was often assumed that the only way to engage in this process of citizenisation was to impose a single undifferentiated model of citizenship on all individuals. But the ideas and policies of multiculturalism that emerged from the 1960s start from the assumption that this complex history inevitably and appropriately generates group differentiated ethno-political claims. The key to citizenisation is not to suppress these differential claims but to filter and frame them through the language of human rights, civil liberties and democratic accountability. This is what multiculturalist movements have aimed to do. (p. 100)

Kymlicka also points to three major manifestations of multiculturalism that have emerged as a result of *citizenisation* in Western democracies: (a) the empowerment of indigenous people (e.g. in New Zealand and Canada); (b) new forms of autonomy and power-sharing for internal ethnic groups (e.g. the Basques in Spain); (c) new policies of *multicultural citizenship* aimed at immigrant groups (e.g. constitutional recognition of multiculturalism and introduction of laws that are consistent with it).

Increasingly since the global financial crisis of 2008 and following the serious effects that it had on unemployment and on declining living standards in the West in particular, multiculturalism has been the target of media, popular, and political attack. But we agree with Kymlicka that such attacks have been largely misguided and we should add that they also look like a subterfuge to save the neoliberal system threatened by the global financial crisis by creating scapegoats: the immigrants. In reality, if we see multiculturalism as being inspired by human rights norms, and as a system that "seeks to deepen relations of democratic citizenship", we can only conclude that "reports of multiculturalism's death are very much exaggerated" (Kymlicka 2010, p. 104).

Although, in one way or another, all countries are multicultural, in this section we will mainly focus on countries that have received many immigrants from around the world, the end-of-life experiences of those immigrants, and the challenges of approaching the end of life in old age in a multicultural society.

In previous chapters we have seen how older immigrants may experience a variable degree of acculturative stress and culture shock upon arriving in their new country, and how the local communities of the same ethnicity can help them feel less isolated (e.g. see Sect. 6.8 in Chap. 6). As the end of life approaches, the old immigrant will need a special kind of support. The experience for the immigrant elders—and their family—at such crucial point in life can be rather traumatic, unless the local community, health care system, and social services provide much needed assistance. Such support must be culturally appropriate if it is going to be effective. For instance, across cultures, there are different rituals and beliefs that aim at decreasing fear of death and death anxiety, such rituals and beliefs are taken by immigrants to their host country and local services providers should be aware of them (see Sect. 3.10 of Chap. 3). In order to better understand the needs of dying old immigrants, we have seen in Sect. 7.7 of Chap. 7 that translation and interpreting services are crucial for old immigrants who do not speak the official language/s of their country of residence.

For as long as we accept the principle of autonomy in decision-making at the end of life, we need to be open to the specific requirements of old people from different ethnic and cultural backgrounds (Macklin 1998), and such openness is more likely to be found in societies that take multiculturalism seriously. Autonomy, of course, does not mean freedom to be above the law, but it certainly means to be listened to with respect and specific requests accommodated when possible. As stated by Ruth Macklin, after all we can respect cultural diversity without being compelled to accept every aspect and ritual of various cultural traditions. The issue therefore is to make sure that the elderly from different cultural backgrounds can make informed decisions regarding their end of life process, rather than the decision being imposed on them by the medical system (e.g. Levin and Sprung 2003).

The effect of racial and ethnic diversity on end-of-life decision-making was reviewed by Kwak and Haley (2005), mainly based on studies carried out in the USA. They pointed out that Caucasians tend to be more knowledgeable about advance care directives than other groups, in part because Caucasians seem to have a more positive attitude towards advance care planning. Caucasians also tend not to ask for care support to prolong life, whereas African Americans are keener to try all forms of care that is available with the hope of extending life, even through aggressive treatment; Hispanics are also keen to use life support. Likewise, African Americans tend to be less likely to adopt *Do Not Resuscitate* (DNR) orders than Caucasians, the same with Hispanics, whereas Native Americans are more willing to opt for DNR orders than African Americans, Hispanics, and Asians. In other studies Asians (the Japanese in particular) have been recorded to be more likely to give directions not to use life support than African Americans, Hispanics, and Caucasians. The same for the case of Chinese elders diagnosed with a terminal illness, they were of the opinion that "use of life support or prolonging life indefinitely would bring burden and suffering on the individual, family, and society" (Kwak and Haley 2005, p. 638), a view that springs from the broadly collectivistic cultures of Asia.

We have mentioned in previous chapters the contrast between *individualistic* and *collectivistic* cultures/societies. Although describing individualism and collectivism as a dichotomy is simplistic, we can broadly differentiate those two forms of society, following Albert Bandura, according to the sources of agency. Individualism assumes that agency is fundamentally rooted in the capacity of individuals to decide about their life, whereas collectivism spreads agency across the group. In the end, however, we agree with Bandura that "(s)uccessful functioning requires an agentic blend of these different modes of agency" (Bandura 2002, p. 270).

Collectivistic cultures tend to protect and shield the dying elder from further stress, but they may also tend to protect and shield the family of the dying elder. Protection of the elder in such cultures may be expressed in terms of not disclosing a medical diagnosis of serious illness and a terminal prognosis. For instance, elders from more collectivistic cultures, such as Hispanics in the USA, tend to be keener to defer end-of-life decisions to a family member whenever they fear becoming incapacitated as they approach death (Kwak and Haley 2005). In Europe, Meñaca et al. (2012) compared end-of-life care in still relatively collectivistic Italy, Portugal, and Spain. In Spain and Italy there is an ongoing tradition of only partial disclosure of information for terminal elderly patients and when information is given it is more frequently about diagnosis than prognosis. The situation, however, is evolving and it seems that an expectation of open disclosure is on the rise, especially in Portugal, and the attitude may also change among migrants from those countries through acculturation into a more individualistic society.

Chattopadhyay and Simon (2008) provided a specific comparison of end-of-life decisions between the relatively more individualistic West (Germany) and the more collectivistic East (India). Understanding cross-cultural differences in the bioethics of the end of life process allows us to also understand that there is not a single way of acting ethically. If the objective is to make the dying individual as comfortable as possible with the process, then different attitudes and strategies will have to be adopted according to the characteristics of the individual, including cultural characteristics. For instance, with respect to how much information should be given to the patient, the Western tradition is to fully and frankly inform the patient, whereas the Eastern tradition is to inform the family and protect the patient as much as possible from the psychological effects of a negative prognosis (see also Chap. 3).

We agree with Chattopadhyay and Simon's (2008, p. 172) conclusion that:

> Cross-cultural perspective in end-of-life issues also serves to create new possibilities for opening a dialogue and a more enriched understanding of moral values and meaning of life, illness, suffering, and death across cultures, rather than to circumvent it. Further, this wider perspective on the dynamics of end-of-life decision making offers a broader range of approaches and possibilities available for formulation, analysis, and implementation of healthcare policy in end-of-life care.

Thus, understanding the views of immigrants from other cultures about the end of life does not only serve the purpose of helping those immigrants but it also becomes an opportunity to expand our own personal horizon. Perhaps paradoxically, in the more individualistic West end-of-life care will have to consider the specific collectivistic cultural needs of older immigrants, as such cultural needs are undeniably part of their own individuality. In so doing, we may learn (or re-learn) some of the benefits of collectivism.

The bioethics of end of life in a multicultural world calls for greater flexibility and understanding of specific cultural requirements (Turner 2003). For instance, the meaning of key ethical principles such as Beauchamp and Childress' (1979) principle of individual autonomy may require some revisiting. Individual autonomy is well suited for a situation of end-of-life in a Western setting, but "autonomy" in more collectivistic cultures involves the autonomy of the family group, not just of the individual concerned.

An individual-centred idea of autonomy is embedded within the broader concept of self-determination, which gives the individual control over decision-making, an expectation of truth telling, and a broader control over the dying process (Ersek et al. 1998). However, the concepts of autonomy and self-determination at the end of life have their own limitations. As sarcastically put by Mrs. Kim, a 79 years old Korean American immigrant, quoted by Frank et al. (1998, p. 404): "I am the one who is going to die; so I don't control the situation". Autonomy is limited by the constraints and inevitabilities of the end-of-life experience. If full autonomy is already an individualistic ideal that can be difficult to achieve when we are young and in good health, it is even more constrained when we are old, frail, and close to our end of life. Still, some regions of autonomous action, especially in decision-making, are available to elders at the end of their life, for as long as their mental faculties remain good and they cherish such autonomy. For those who fear that they may not be able to decide when time comes, providing advance care directives is an option.

Frank et al. (1998) reported on the concept of autonomy in decision-making at the end of life comparing four ethnic groups living in the USA: African Americans,

European Americans, Korean Americans, and Mexican Americans. Decision-making was highly influenced by ethnicity, with the Korean Americans being the ethnic group least supportive of telling the truth to the dying elder about his/her situation, whereas they were the most supportive of shifting the responsibility of decision-making to the family and away from the elder with regard to the use of life-sustaining technology. This was seen as a family responsibility for the care of the elderly to the very end (*hyodo* or "filial piety"), rather than fear of death per se. Across all ethnicities studied by Frank et al., Korean Americans and Mexican Americans showed the clearest evidence of collectivism (*familism*) in information and decision-making, whereas the responses of European Americans and African Americans were more individualistic.

In addition, full autonomy may put an additional burden on elders and their families in areas of decision-making they may have little knowledge of or experience about. As always, one would expect wisdom to prevail in such circumstances, where all concerned can navigate the complexities of an end-of-life situation through communication. At one point, as suggested by Bandura, perhaps a mixture of "autonomy" and "paternalism" may be the best way forward for any specific individual.

Among collectivistic cultures autonomy, as expressed in truth telling about a life-threatening diagnosis, may find some resistance. Candib (2002) reported such conflicts in examples taken from family medicine practice in the USA. The case of the 55-year-old Mr. Nguyen, an immigrant from Vietnam, is one in which the family recommended not to tell the patient about his cancer diagnosis as "he was already sick enough, and it would make him give up hope" (p. 215). In fact, old people from collectivistic cultures often expect to be protected from bad news and rely on the family to absorb the full emotional brunt of such news and take decisions for them. An emblematic example is given by Candib regarding Mr. Sokolov, a Russian man in his seventies diagnosed with pancreatic cancer. He was usually attended by a Russian doctor who knew how to handle the communication and the need not to disclose the full diagnosis, this shocked the doctor's supervisor who visited the patient and disclosed the diagnosis to him. It is worthwhile quoting the full reaction of the patient as reported by Candib (2002, p. 215–216):

> Initially he listened carefully with a rather stoic demeanor. I was certain that my approach would again be the right one. However, as the news sank in, he became rather agitated and appeared quite angry. This prompted me to have the interpreter explore his feelings:
>
> MD: You look very upset and angry. Can you tell me what you are thinking?
>
> Pt: I am very angry that you have given me such terrible news. Where I come from, doctors give bad news to the family.
>
> MD: So you are angry because I told you about the likelihood of having cancer.
>
> Pt (appearing more angry): Yes, and I don't want to talk with you about it any more. You are telling me that I have a terrible disease without chance of cure. How am I to get better without having hope? Please just decide what you need to do and talk with my family and Dr. Danilova.
>
> They will tell you what to do.
>
> MD: (rather remorseful and shocked): I am sorry if I have upset you. I believe that it is important for people to be in charge of their healthcare and only wanted to provide you with that opportunity.

The patient acknowledged the explanation with a dismissive 'yes' and made no further eye contact.

Meñaca et al. (2012, p. 391) explain that in the still somewhat collectivistic culture of Italy, non-disclosure of a life-threatening illness:

> ...was also understood as a way to maintain hope and continuity of social life... An anthropological study described life in Tuscany as an unpredictable flow of events shared by all members of society. Amidst this flow of events, the most vulnerable were protected and cared for. Avoiding the certainty inherent in doctors' diagnosis/prognosis was considered to avert the separation of the individual from social life and, therefore, an anticipated social death.

Interestingly, in more collectivistic cultures it is not just the patient who expects to be protected from bad news through the active participation of the family, but the family itself is also protected by the elderly patient, in other words, protection is mutual. To illustrate this point Candib (2002, p. 216) provides the case of Mr. Do, a Vietnamese patient diagnosed with metastatic squamous cell carcinoma who, upon knowing the diagnosis and prognosis, told the doctor: "Do not tell my family. They have enough on their minds already. They do not need to worry about this."

A diagnosis of life-threatening illness may also clash with other cultural characteristics of old people such as optimism. For instance, Candib provides another example, this time from the Native American Navajo culture that is contrary to advance directives. The reason for this is because:

> In the Navajo belief system, thought and language are believed to have the power to shape reality and control events. It is important 'think and speak in a positive way' within *hozho*, a way of life that combines beauty, goodness, harmony, and all that is ideal in how one thinks, speaks, and relates to others and the world.... Bringing up in conversation the possibility of death is tantamount to wishing death upon the person. This belief is not a random idiosyncrasy of Navajo belief but rather an integral part of an entire way of life. Such a view makes the discussion of advance directives a direct assault upon a Navajo person's well-being. While physicians need not agree that talking about death can bring it about, once physicians are aware of this belief, it becomes ethically necessary to arrive at an alternative approach (Candib 2002, p. 219)

Some authors have criticised the extension of the concept of autonomy to the acceptance of elders' decisions that are dictated by specific cultural beliefs or traditions (e.g. Fagan 2004). The criticism is grounded in the assumption that the elder, by acting in accordance to such traditions, is not exercising his/her autonomy but is being coerced into behaving in particular ways by external pressure. This criticism would be valid if the decision of the elder runs in favour of some cultural expectation, and against the personal wishes of the individual; but it is not valid if the elder is fully aware of the cultural framework of his/her decision, accepts it and feels far better by operating within such a framework than outside it (see the case of Mr. Sokolov above). In other words, acting autonomously simply means acting in a way that is personally satisfactory or personally fulfilling, not in a way that reinforces a view of individualism as an alternative to collectivism.

Each culture is embedded in a *web of significance* (Geertz 1973) that provides meaning to the life of individuals and a framework to interpret and understand their world. The challenge of bioethics in multicultural societies is that of distinguishing between *tolerable* and *intolerable* cultural practices (Turner 2003). However, what is tolerable is not necessarily set in stone, even when we refer to current laws and behavioural expectations, as what is invalid (illegal/unethical) now may become valid (legal/ethical) tomorrow. The history of homosexuality in western culture, for instance, may teach us one or two lessons in this regard (e.g. Poiani 2010).

Common views of gender roles in the care of older adults may also vary inter-culturally, alongside marriage relationships. The patterns of communication may also differ across cultures, from a direct style to an indirect style; the relative emphasis on nonverbal communication may differ as may the specific meanings of nonverbal gestures. Importantly, elders may differ in their religion or specific spiritual-belief system, which may affect their views on suffering and the afterlife (see Table 8.9 for a comparison of some principles and issues in geriatric bioethics among the major monotheistic religions). Advance care planning may be also interpreted differently across cultures (Volker 2005) as it may the need for specific approaches to palliative care (Chater and Tsai 2009).

Given the reality of cultural diversity, doctors working in multicultural societies should be specifically trained to be able to better handle end-of-life situations involving old people from diverse ethnic backgrounds (e.g. McNamara et al. 1997; Searight and Gafford 2005). Such training should incorporate the development of *cultural humility*: "a lifelong process of self-reflection and self-critique whereby the individual not only learns about another's culture, but one starts with an examination of her/his own beliefs and cultural identities" (Yeager and Bauer-Wu 2013, p. 2; see also Tervalon and Murray-Garcia 1998). Importantly, training in *cultural humility* may allow health professionals to assist dying elders from a diversity of ethnicities in a way that helps the elder experience a fulfilling end of life, rather than making the health professional feel better or more in tune with his/her own views and priorities.

Table 8.9 Comparison of three monotheistic religions with respect to several principles and relevant issues in geriatric bioethics (from Clarfield et al. 2003)

Principle/issue	Judaism	Catholicism	Islam
Autonomy	+	++	+
Beneficence	+++	+++	+++
Non-maleficence	++++	+++	+++
Justice	+	++*	+
Role of family/community	+++	+++*	+++
Second religious opinion	*	*	±
Withholding truth	**	*	*
Moral equivalence of withholding or withdrawing treatment	*	***	**
Withholding nutrition	*	**	*
Use of advance directive	*	***	±
Autopsy	*	***	*

+ to ++++ = least to most important; * to *** = least to most acceptable; ± = not well formulated

To recap, modern societies are increasingly multicultural, in part due to the high flow of migrants seen around the world. Therefore, assisting old people at the end of their life requires knowledge of their cultural background. In particular, elders coming from an individualistic cultural tradition will need a different approach to palliative care than elders coming from a collectivistic tradition. Providing culturally appropriate care at the end of life requires developing a degree of cultural humility in all caregivers.

In the previous section we analysed some issues of end-of-life care in the specific context of multicultural societies. In the final section of this chapter we expand on end-of-life care in a broader perspective.

8.7 End-of-Life Care

> Physical contact provides both a mental and physical representation of the fact that you are there for a person, and there is no better example of this than the warmth and comfort that a person can receive from a cuddle. And this is why oversized 'cuddle beds' are having such a massive impact for both the families and patients in Palliative Care (https://hellocaremail.com.au/oversized-cuddle-beds-bringing-families-closer-palliative-care-patients/)

In early 2019 news spread in Australia that two hospital beds had been provided in the state of Queensland for the purpose of allowing family and friends to share time with their terminally ill relative whilst keeping close physical contact on an oversized "cuddle bed". This is an image of end-of-life care that exudes humanity and warmth, but is that the experience of most dying people?

Meier and Morrison (1999, p. 6) provide a vignette of a more common end-of-life experience:

> In our society, the overwhelming majority of people who die are elderly. They typically die slowly of chronic diseases, over long periods of time, with multiple coexisting problems, progressive dependency on others, and heavy care needs met mostly by family members. They spend the majority of their final months and years at home but, in most parts of the country, actually die in the hospital or nursing home surrounded by strangers. Many of these deaths become protracted and negotiated processes, with healthcare providers and family members making difficult, often wrenching, decisions about the use or discontinuation of such life-prolonging technologies as feeding tubes, ventilators, and intravenous fluids. There is abundant evidence that the quality of life during the dying process is often poor, characterized by inadequately treated physical distress, fragmented care systems, poor to absent communication between doctors and patients and families, and enormous strains on family caregiver and support systems.

An even greater departure from the "cuddle beds" practice than the one mentioned by Meier and Morrison was implemented just one year after the publication of the HelloCare article quoted above in this section, not just in Australia but also around the world, as social distancing measures were introduced in response to a coronavirus (COVID-19) pandemic (https://www.who.int/emergencies/diseases/novel-coronavirus-2019/advice-for-public). In pandemic-affected areas, seriously ill elderly people showing the symptoms of the disease were taken to hospital and remained isolated, many dying alone, away from their family and loved ones. In this case the intervention was medically justified, to protect others from also being infected, but this should not discourage us from pursuing better end-of-life care practices when possible.

Therefore, most death experiences seem to depart, quite considerably, from the ideals of a "death with dignity" or a "good death". The final stage of life is usually associated with pain, physical distress that come after a period of intense dependency from others such as family and/or healthcare professionals (Meier and Morrison 1999; Carr and Khodyakov 2007), and multiple co-morbidities (Rashidi et al. 2011). Serious physical and mental decay accumulates and develops into a *frailty syndrome* which in the elderly at the end of their life indicates the need for palliative care. The frailty syndrome "is associated with a state of extreme vulnerability and poor clinical outcomes. Frailty is typified with loss of weight and strength, altered balance, poor endurance, inactivity and fatigue" (Lo and Woo 2006, p. 36).

There are different approaches to care at the end of life. On the one hand, the "heroic" view pushes medical intervention as far as it goes and for as long as there are new therapies to try, including the use of *Intensive Care Units* (e.g. Galzerano et al. 2009; Flaatten and Garrouste-Orgeas 2015). Adopting the heroic approach may be an initiative of the patient, the medical team, the family, or a combination of those. On the other hand, the "humanistic" perspective focuses on the dying person's specific needs and it is usually adopted in hospice and palliative care centres (e.g. Kaufman 1998).

Dependency at the end of life increases the older the individuals are, due to their accumulated disabilities (Zhao et al. 2010). Given that pain, distress and dependency are unlikely to be optimal experiences for any old person, discussing end-of-life care options with elders is therefore essential as, on average, they are still expected to live about 30 months after receiving a terminal diagnosis (Old 2008). If the elder is cognitively incapacitated the family will have to play a greater role in decision-making, in conjunction with health care professionals.

Controlling pain in terminal people is one of the most important objectives of the humanistic approach to palliative care. The *International Association for the Study of Pain* (IASP) defines pain as "an unpleasant sensory and emotional experience associated with actual or potential tissue damage, or described in terms of such damage" (https://www.iasp-pain.org/Education/Content.aspx?ItemNumber=1698#Pain; downloaded on 3 July 2019). It should be noted from the definition that pain has not only a sensory component but also a psychological one. Lo and Woo (2006) note that elderly persons at the end of their life may associate pain with a worsening of their condition, which may cause distress. Thus, alleviating pain is a matter of priority in these elders. If pain is controlled, then finding meaning in the last period of life becomes easier to achieve as the mind can be occupied on transmitting legacy, developing spiritual growth, and finding transcendence.

Minimising pain and discomfort is a key aspect of a "good death", alongside avoiding prolongation of dying, achieving a sense of control, relieving the burden placed on family, and strengthening of relationships (Detering et al. 2010). This is also consistent with the list of characteristics of a "good death" provided by Gott et al. (2008) and Lloyd et al. (2011) in the context of palliative care that we mentioned in the introduction to this chapter. However, different individuals may have different approaches to their end of life. In a study of 91 elders—80 years old and older—carried out in the UK, Lloyd and Cameron (2005) reported how the elders themselves expressed a diversity of views about death and dying that the authors classified into four major groups:

1. Only think about the day-to-day: This group is described as "stoic", living what time is left in their life as it comes, without much planning for the future. They obviously realise that their life is coming to an end, but are not too focused on detailed planning, usually going no further than writing their will.
2. Pragmatic or practical: Unlike the previous group, these elders have made detailed plans for their death, including making funeral arrangements and finalising legal matters.
3. Religious or spiritual: These elders trust their life and future in the hands of God and their belief in the afterlife.
4. Desire to determine own death: The members of this group are fearful of losing their independence and becoming a burden to their family. For them death is preferable.

Dying with dignity is an important aspect of the dying process for many elders. There are several aspects of a dying body that involve loss of dignity, humiliation, disempowerment, threat to autonomy, respect, and self-esteem and that add further distress to a situation that instead should require full focus on personal fulfilment and transcendence. Such aspects include toileting and bathing, aesthetic issues of the body, odorous wounds, incontinence, helplessness (Lo and Woo 2006; see also Lo et al. 2010), but also the prospect of dying alone and in an unfamiliar environment (e.g. Frances 2015). Lo and Woo (2006) also point out that invasive and futile treatment may further undermine dignity, which may be worsened by the insistence of family members that new, experimental treatments be tried, believing that there is "always hope" and that "you never give up".

Dying with dignity should indeed be considered a human right, being a means of protecting the individual against unnecessary and futile suffering.

As part of best terminal care practice that focuses on dignity, it is always recommended that the elders themselves write *advance care directives* to clearly explain their wishes, otherwise they may find themselves at the will of healthcare personnel and/or family (e.g. Hesse 1995; Wu et al. 2008). This may include a specific list of actions (*instructive advance care directives*) or the nomination of an individual (a *proxy*) who will take decisions for the elder in case of incapacitation. Instructive advance care directives may include a living will, a do not resuscitate order ("no code" order, or no cardiopulmonary resuscitation in the eventuality of cardiac arrest), a personal letter describing the patient's end-of-life wishes, a verbal statement, and organ donor card, whereas proxy advance care directives may include a durable power of attorney for health care and financial decisions, a common law directive or "common sense order" (Old 2008). Advance care planning should benefit the dying person, providing peace of mind in preparation for the end of life, and

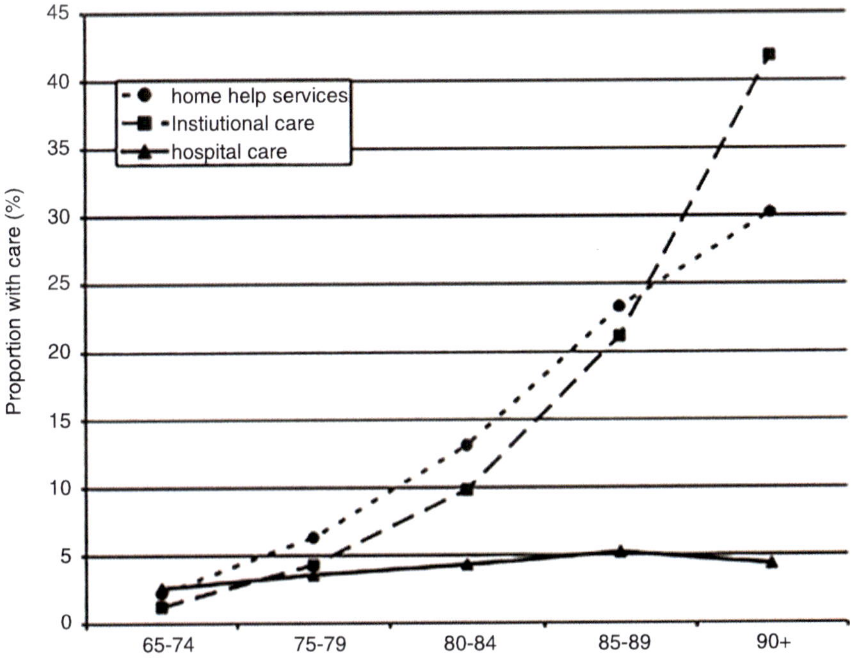

Fig. 8.22 Proportion of older people in Sweden, by age, using home help services, living in institutional care facilities, and in hospital. Data are for October 2003. From Larsson et al. (2008)

it has been also shown to decrease anxiety, stress, and depression in the members of the family (Detering et al. 2010).

End-of-life care can be provided in a variety of settings, but for the older age groups the most common one is a nursing home (e.g. Bravell et al. 2010; Lindskog et al. 2015; see Fig. 8.22 for an example from Sweden) even though they may desire to be cared for in a familiar environment such as home (Abarshi et al. 2010; Hunt et al. 2014). When care is provided at home, it is usually a responsibility of a female member of the family (Morgan et al. 2016).

Unfortunately, care is not always of the appropriate standard for elderly people at the end of their life, even in economically advanced countries such as Sweden. For instance, Lindskog et al. (2015) reported that the likelihood of dying in the presence of a family member or a caregiver decreases with age (Table 8.10), whereas the same-age group also showed the least use of palliative consultation services in end-of-life care (Table 8.11).

Many programs have been introduced to help healthcare staff and family acquire the technical and emotional tools to cope with the burden of caring for a dying elder (Bolmsjo 2008).

As far as the wishes of the elders are concerned, there is a convergence of opinions that when the end comes it should be hopefully peaceful and painless, preferably occurring during sleep (Fleming et al. 2016). The ideal death as seen from the perspective of a dying person often clashes with the capacity of family and the health care system to provide the means and the environment that can facilitate such

Table 8.10 Likelihood of dying in the presence of a family member or care giver

Group	Age (y)	%	Odds ratios	95% confidence interval	*P*
1	18–39	96.3	6.0	3.2–11.0	<0.001
2	40–59	92.1	2.8	2.2–3.2	<0.001
3	60–74	89.4	1.9	1.7–2.1	<0.001
4	75–84	85.1	1.2	1.1–1.4	<0.001
5	≥85	83.1	1		

Data are from the *Swedish Register of Palliative Care*, odds ratios and *p*-values were calculated using the oldest group as reference. From Lindskog et al. (2015)

Table 8.11 Use of palliative consultation services in end-of-life care

Group	Age (y)	%	Odds ratios	95% confidence interval	*P*
1	18–39	34.9	4.3	3.3–5.7	<0.001
2	40–59	28.0	2.9	2.5–3.3	<0.001
3	60–74	24.5	2.1	1.9–2.3	<0.001
4	75–84	21.5	1.4	1.2–1.5	<0.001
5	≥85	19.7	1		

From Lindskog et al. (2015)

preferred death process. For instance, although most elders may like to die at home, surrounded by the people they love, in many cases they actually die in hospital (Andersson et al. 2007; Perrels et al. 2014) and the reason for doing so is not difficult to understand, as any crisis will inevitably trigger a response in the family caregivers to hospitalise the elder. One of those crises will be final and hence the elder will pass away in hospital. On the other hand, if death occurs in circumstances that are not preceded by a need for hospitalisation—or hospitalisation is required but no beds are available—one would expect that the elder may most likely die at home or in a retirement institution (e.g. Temkin-Greener and Mukamel 2002; Andersson et al. 2007).

There are different approaches to meeting the needs of a dying elder. On the one hand, professionals and the community at large may press the appropriate authorities to make the services needed to assist a dying person available to all, even at home if possible. On the other, both professionals and community may accept that the ideal will not be achieved all the time and may settle for a second-best option: a "good enough death" (Lloyd 2000). Whichever approach is followed, a concern should remain all along: it is easy to, step-by-step, slip down a path of decaying and de-humanising healthcare services pushed by the mantra that there are not enough resources, a mantra that is rather incomprehensible in economically advanced countries. Thus, if we build a healthcare system that is only as ambitious as to provide a "good enough death" to the elders, at least it should be the elders and their families who define what is "good" and what is "enough". After all, as Liz Lloyd (2000, p. 185) put it: "The ways in which older people are supported and cared for through the process of dying provides a clear indication of broader social values".

In this chapter we have seen how life meaning and the existential dimension of life take a special place as we embark on the last stretch of our life journey. Dobríková et al. (2016) investigated life meaning in a group of elders who had received a medical prognosis for a life expectancy of less than one month. The study was carried out in two hospices in the Slovak Republic. They used the *Logo-Test* (Lukas 1986) that is based on Viktor Frankl's logotherapy. Life meaning increased and sense of loneliness decreased in this group of elders the more social support they received from members of their family, but also friends and even hospice staff (see also Sand and Strang 2006). Moreover, the greater the fulfilment of their biological, psychological, social, and spiritual needs was, the higher their purpose in life also was.

Life fulfilment and having lived a meaningful life better prepare an elder at the end of such life. In an ideal world, once we have reached fulfilment in life, we should be able to carry on with it to the very end, and when the end comes it should be hopefully painless, and we may hope to be surrounded by those we love the most. Understanding and then preparing for such an experience requires some clarity of ideas about how to die. Alas, Lloyd (2000, p. 171), quoting Howarth's (1998) views, reminded us that: "since most people spend their lives avoiding thoughts of death, contrary to popular opinion, old people 'do not know 'how to die'''. But knowing how to die is rooted in our knowing how to live, and we are of the opinion that having lived a fulfilling life prepares us to better accept the end of it, when the end comes, in spite of unpredictable circumstances. Indeed, a "good death" (or even a "good enough death") should just be a natural extension, the final stage, of a fulfilling and meaningful life.

Eventually, death will come, unleashing a series of consequences: the grieving of family and friends, the immediate management of the corpse and its final mortuary treatment.

The term *bereavement* refers to the factual loss of a dear one to death, whereas *grief* describes the "emotional, cognitive, functional and behavioral responses to the death" (Zisook and Shear 2009, p. 67). *Mourning* refers to the "behavioral manifestations of grief, which are influenced by social and cultural rituals, such as funerals, visitations, or other customs" (Zisook and Shear 2009, p. 67).

The death of a loved one defines a state of bereavement in the survivors, leading to grief that may last for a variable amount of time depending on how close the relationship with the deceased elder was. When interacting and comforting bereaved people, empathy is obviously recommended.

Bereavement tends to rob "survivors of love and companionship, as well as future hopes and plans that they expected to share with the person who died" (Boerner et al. 2015, p. 1). This produces a break in the continuity of attachment of the survivor with the deceased one, filling the survivor with a sense of loss. Freud (1917/1957) described the bereaved person as one suffering a traumatic experience, being initially trapped between the absence of a loved one and the yearning for reunion. This conflict is eventually resolved over a period of time as emotional ties with the deceased gradually weaken, or at least the absence of a loved person becomes more bearable, thus allowing the survivor to overcome the initial trauma and resume a more normal life. Bowlby (1969, 1973, 1980) also focused on the conflict that a bereaved person experiences between the loss of a beloved one and the initial desire for reunion. Such desire is a natural response of attachment functions. The conflict is resolved, according to Bowlby, by going through various stages of grieving.

In general, bereavement is associated with an increased fear of death and *dying anxiety* (Azaiza et al. 2011, and references therein), which may lead to the use of coping strategies such as religiosity and spirituality (e.g. Ghufran and Ansari 2008). Later in the process of bereavement, family and friends may continue to nurture a relationship with the deceased one through helping maintain his/her legacy and keeping the memory of the deceased alive. In about 8–16% of bereaved people grief can become complicated, affecting their mental health (e.g. chronic grief, chronic depression) and therefore requiring professional intervention (Bonanno et al. 2002). For more details on the diversity of experiences of bereavement found across cultures, the work of Parker et al. (2015) is a very good start.

Grief can sometimes remain unresolved, leading to trauma and psychological complications affecting social relationships, work, and even health. Grief can manifest differently in different individuals, varying in psychological effects, duration, and intensity according to differences in personality, age, health condition, attachment to the deceased and other factors (Zisook and Shear 2009). Normal experiences of grief tend to be a mixture of feelings of pain and loss but also relief and peace of mind, knowing that a loved person is not suffering anymore. The dynamics of grief can be described through an initial *acute* phase that occurs soon after the death of a loved one, and that is characterised by intense pain, sadness, crying, recurrent thoughts about the dead one, difficulties in maintaining concentration, and loss of interest in daily life. Later, grief may take the shape of *integrated* or *abiding grief* which is characterised by sadness and longing. Bowlby (1980) described the process of grieving unfolding in four stages: (a) an initial period of numbness, disbelief, or shock; (b) followed by yearning or searching for the deceased, a process that can also come with anger and protest; (c) despair and disorganisation, as the bereaved gives up the search, during this period depression and hopelessness may manifest themselves; and finally, (d) reorganisation or recovery as the loss of the deceased is accepted and there is a gradual return to previous activities and normal life. In Sect. 8.3.1 of this chapter, we also mentioned Elizabeth Kübler-Ross's *Five stages of grief*: Shock and denial, anger, bargaining, depression, and acceptance which approximately parallel Bowlby's stages. Those stages, however, are experienced differently by different individuals in different cultures.

Acceptance of the reality of the loss of a dear person may eventually lead to the integration of the new status of the deceased into a fulfilling life for the living. In other words, the dead become assimilated into the life of those who care about them. In this regard, Zisook and Shear (2009, p. 68) point out that:

> During the transition from acute to integrated grief, usually beginning within the first few months of the death, the wounds begin to heal, and the bereaved person finds his or her way back to a fulfilling life. The reality and meaning of the death are assimilated and the bereaved are able to engage once again in pleasurable and satisfying relationships and activ-

ities. Even though the grief has been integrated, they do not forget the people they lost, relinquish their sadness nor do they stop missing their loved ones. The loss becomes integrated into autobiographical memory and the thoughts and memories of the deceased are no longer preoccupying or disabling.

According to Neimeyer (2001, 2006), the construction of a new meaning throughout the process of grieving is achieved thanks to the production of a coherent life story, a narrative, that can return life to a path of fulfilment. Such process of meaning making is not necessarily easy and it may benefit from social support.

In Sect. 3.10 of Chap. 3 we mentioned how bereavement, grief, and mourning can be helped by religious funerary rituals, or other rites of aggregation, that may provide socio-emotional support to the participants. We quoted the work of Worden (1982) who described four tasks of mourning: (1) accepting that the loss is a reality; (2) experiencing the emotional pain produced by grief; (3) adjusting to the new reality where the loved one is no longer present; (4) achieving emotional detachment from the lost loved one. Mourning rituals help in the re-construction of life meaning, that was lost with the death of a loved one, by serving integrative and regulatory functions, and they achieve this by "providing a structure for the emotional chaos of grief, conferring a symbolic order on events" (Neimeyer et al. 2002, p. 237). According to Neimeyer et al. (2002), ritual practices are used during mourning in order to: (a) help the mourners reflect on their own sense of self and their attachment to the deceased; (b) favour the transition of both deceased and bereaved to a new social status, with the former becoming a "spirit" or a memory as ancestor and the latter transitioning from wife/husband to widow/widower, for instance; (c) retain some connection with the deceased through sharing with others thoughts and memories in ceremonies or informal conversations. Neimeyer et al. (2002) also note that this process is made more difficult in cases where the bereaved is socially disenfranchised (e.g. an extra-marital lover).

The relatives of the deceased have the responsibility to decide and supervise the proper treatment of the body according to their customs, but within the legal restrictions of the country of residence. In this area there may be some clashes between cultural/religious practices and medical/legal requirements. For instance, some deaths may call for the performance of an autopsy, but the relatives may object it on religious or other grounds (e.g. Renteln 2001, and references therein). Funeral directors may be asked to treat the body of a deceased person according to specific traditions, which they may be unfamiliar with (e.g. embalming, transportation). A failure to proceed in accordance to custom may be highly distressing to the family and it could lead to lawsuits (Renteln 2001).

The body of the deceased goes through a final journey in accordance to the will of the deceased, the mortuary practices of the family, or the decision of those who are responsible to act on this matter. Mortuary practices are affected by philosophical, spiritual, religious, and broader cultural/social factors (Carr 1995) and they may involve various alternative treatments of the body. Burial or cremation are two well-known alternatives to treat the body of the deceased, but in fact across cultures there is a very large diversity of treatments for the dead body. White et al. (2017, p. 147) mention some of those practices in a somewhat colourful way:

> corpses are washed, embalmed, anointed, pickled, dismantled, painted, adorned with jewellery, clothed, wrapped, placed in a container, moved, viewed extensively, touched, embraced, wept over, shouted at, danced over, and force-fed food, among other practices.

Specific mortuary treatments also vary according to social status—the higher the status the more energy and resources are usually spent in mortuary practices, age (a distinction is usually made between child and adult), gender, and personal identity.

In their cross-cultural study of mortuary rituals, White et al. (2017) point out that most such rituals involve a degree of exposure of the corpse to kin, and some physical contact such as washing and dressing. Some benefits of mortuary rituals have been postulated such as control of fear of mortality, strengthening social order/bonds, reinforcing group values and "redefining social relationships at a time of social uncertainty and change" (White et al. 2017, p. 157). In addition, White et al. (2017) suggest a function of mortuary rituals in the reduction of the suffering caused by grief, which if it is prolonged it could impair the functional life of the survivors and the activities of the family group. From an evolutionary perspective, it is expected that death of a loved one would trigger, in the short-term, mechanisms aimed at seeking the re-conjunction with the lost person, only for such response to gradually fade away and be replaced by disengagement and re-orientation, after coming to terms with the fact that the loss is permanent. Disengagement with the dead person may be also helped by distancing from the corpse, which is also an adaptive response that helps protect against potential contagion. In fact, practices that involve the direct manipulation of the corpse, such as mummification or cannibalism are quite rare (White et al. 2017).

In sum, end-of-life care should be ideally marked by an effort to facilitate a dignified dying process. There is no need for dying to be heroic and unnecessarily painful. The passing away of an old person leads the family to start a period of grief and mourning that will hopefully give way to a return to a more normal life after the acute phase is over and disengagement and re-orientation are completed.

8.8 Conclusions

The end of life can be very different to different people, both in terms of wishes and above all reality. But in general, we may say that it is a period of reminiscence, synthesis, and of projection of the self to the future in the form of transcendence. A meaningful life can also give meaning to death and it can help overcome the existential loneliness of dying. Transcendence can be achieved not just through spirituality and religious beliefs, but also and perhaps above all through legacy and generativity work. A "good death" may come at the completion of a fulfilling life, which may decrease death anxiety and fear of death. However, when suffering and meaninglessness beset life in old age the will to live may plunge and an earlier exit may be desired. In that case society should make the legal and medical facilities available to those old people who want to opt for euthanasia or assisted suicide. When the end is approaching, we may wish to die with dignity, without pain and hopefully in the company of our loved ones.

We started this chapter with a quote from the Italian poet Salvatore Quasimodo that emphasised how life comes to an end all too soon, but we want to conclude it with more uplifting words from the Bengali poet Rabindranath Tagore, who in his poem *Gitanjali* reminds us of continuity and legacy, that give meaning to life and death:

> On the day when death will knock at thy door what wilt thou offer to him?
>
> Oh, I will set before my guest the full vessel of my life - I will never let him go with empty hands.
>
> All the sweet vintage of all my autumn days and summer nights, all the earnings and gleanings of my busy life will I place before him at the close of my days when death will knock at my door.

Through memory and legacy, the earnings and gleanings of a busy life, Tagore reminds us that the life fulfilment of one may also contribute to the existential "dawn" and the fulfilment of many:

> Death is not extinguishing the light. It is putting out the lamp because the dawn has come.

References

Abarshi, E., Echteld, M. A., van den Block, L., Donker, G., Deliens, L., & Onwuteaka-Philipsen, B. (2010). The oldest old and GP end-of-life care in the Dutch community: A nationwide study. *Age and Ageing, 39*, 716–722.

Adler, A. (1927). *Understanding human nature: A key to self-knowledge*. Greenwich, CT: Fawcett Publications.

Alexander, C. N., Davies, J. L., Dixon, C. A., Dillbeck, M. C., Druker, S. M., & Oetzel, R. A. (1990). Growth of higher stages of consciousness: Maharishi's vedic psychology of human development. In C. N. Alexander & E. J. Langer (Eds.), *Higher stages of human development: Perspectives on adult growth* (pp. 286–341). New York: Oxford University Press.

Alicandro, G., Malvezzi, M., Gallus, S., La Vecchia, C., Negri, E., & Bertuccio, P. (2019). Worldwide trends in suicide mortality from 1990 to 2015 with a focus on the global recession time frame. *International Journal of Public Health*. https://doi.org/10.1007/s00038-019-01219-y

Allen, R. S., Harris, G. M., Burgio, L. D., Azuero, C. B., Miller, L. A., Shin, J., et al. (2014). Can senior volunteers deliver reminiscence and creative activity interventions? Results of the legacy intervention family enactment randomized controlled trial. *Journal of Pain and Symptom Management, 48*, 590–601.

Almeida, O. P., Draper, B., Snowdon, J., Lautenschlager, N. T., Pirkis, J., Byrne, G., et al. (2012). Factors associated with suicidal thoughts in a large community study of older adults. *The British Journal of Psychiatry, 201*, 466–472.

Andersson, M., Hallberg, I. R., & Edberg, A.-K. (2007). Health care consumption and place of death among old people with public home care or in special accommodation in their last year of life. *Aging Clinical and Experimental Research, 19*, 228–239.

Antonovsky, A. (1979). *Health, stress and coping*. San Francisco: Jossey-Bass.

Antonovsky, A. (1987). *Unraveling the mystery of health: How people manage stress and stay well*. San Francisco: Jossey-Bass.

Argue, A., Johnson, D. R., & White, L. K. (1999). Age and religiosity: Evidence from a three-wave panel analysis. *Journal for the Scientific Study of Religion, 38*, 423–435.

Arnett, J. J. (2000). Emerging adulthood: A theory of development from the late teens through the twenties. *American Psychologist, 55*, 469–480.
Aron, A., & Aron, E. N. (2012). The meaning of love. In P. T. P. Wong (Ed.), *The human quest for meaning: Theories, research, and applications* (pp. 185–208). London: Routledge.
Atchley, R. C. (2008). Spirituality, meaning, and the experience of aging. *Generations, 32*, 12–16.
Atran, S. (2002). *In gods we trust: The evolutionary landscape of religion*. New York: Oxford University Press.
Attig, T. (1979). Death, respect and vulnerability. In A. de Vries & A. Carmi (Eds.), *The dying human* (pp. 3–15). Ramat Gan: Turtledove Publishing.
Avrami, S. (2002). "I wish he had died in the war": Suicide survivors - The Israeli case. *Omega, 46*, 273–286.
Azaiza, F., Ron, P., Shoham, M., & Gigini, I. (2010). Death and dying anxiety among elderly Arab Muslims in Israel. *Death Studies, 34*, 351–364.
Azaiza, F., Ron, P., Shoham, M., & Tinsky-Roimi, T. (2011). Death and dying anxiety among bereaved and nonbereaved elderly parents. *Death Studies, 35*, 610–624.
Baars, J. (2013). A Deepening involvement in life with others. Towards a philosophy of aging. *Research on Ageing and Social Policy, 1*, 6–26.
Baars, J. (2017). Aging: Learning to live a finite life. *The Gerontologist, 57*, 969–976.
Baird, R. M. (1985). Meaning in life: Discovered or created? *Journal of Religion and Health, 24*, 117–124.
Bakan, D. (1966). *The duality of human existence*. Chicago: Rand McNally.
Bandura, A. (2002). Social cognitive theory in cultural context. *Applied Psychology: An International Review, 51*, 269–290.
Baron, C. H. (2004). Hastening death: The seven deadly sins of the status quo. In T. E. Quill & M. P. Battin (Eds.), *Psysician-assisted dying: The case for palliative care and patient choice* (pp. 309–321). Baltimore: Johns Hopkins University Press.
Barrett, J. L. (2000). Exploring the natural foundations of religion. *Trends in Cognitive Sciences, 4*, 29–34.
Bar-Tur, L., & Prager, E. (1996). Sources of personal meaning in a sample of young-old and old-old Israelis. *Activities, Adaptation and Aging, 21*, 59–75.
Batchelor, I. R. C., & Napier, M. B. (1953). Attempted suicide in old age. *British Medical Journal, 28*, 1186–1190.
Battin, M. P. (1987). Age rationing and the just distribution of health care: Is there a duty to die? *Ethics, 97*, 317–340.
Battista, J., & Almond, R. (1973). The development of meaning in life. *Psychiatry, 36*, 409–427.
Baumeister, R. F. (1991). *Meanings of life*. New York: Guilford Press.
Baumeister, R. F., & Leary, M. R. (1995). The need to belong: Desire for interpersonal attachments as a fundamental human motivation. *Psychological Bulletin, 117*, 497–529.
Baumeister, R. F., & Vohs, K. D. (2002). The pursuit of meaningfulness in life. In C. R. Snyder & S. J. Lopez (Eds.), *Handbook of positive psychology* (pp. 608–618). Oxford: Oxford University Press.
Beauchamp, T. L. (2006). The right to die as the triumph of autonomy. *Journal of Medicine and Philosophy, 31*, 643–654.
Beauchamp, T. L., & Childress, J. F. (1979). *Principles of biomedical ethics*. New York: Oxford University Press.
Beauchamp, T. L., & Childress, J. F. (2001). *Principles of biomedical ethics*. New York: Oxford University Press.
Becker, E. (1971). *The birth and death of meaning: An interdisciplinary perspective on the problem of man*. New York: Free Press.
Becker, E. (1973). *The denial of death*. New York: Macmillan.
Beike, D. R., & Crone, T. S. (2012). Autobiographical memory and personal meaning: Stable versus flexible meanings of remembered life experiences. In P. T. P. Wong (Ed.), *The human quest for meaning: Theories, research, and applications* (pp. 315–334). London: Routledge.
Belsky, J. (1984). *The psychology of aging*. Monterrey, CA: Brooks/Cole.

Benton, J. P., Christopher, A. N., & Walter, M. I. (2007). Death anxiety as a function of aging anxiety. *Death Studies, 31*, 337–350.
Berger, P. (1967). *The sacred canopy: Elements of a sociological theory of religion*. New York: Doubleday.
Beshai, J. A. (2008). Are cross-cultural comparisons of norms on death anxiety valid? *Omega, 57*, 299–313.
Billings, J. A. (2011). Double effect: A useful rule that alone cannot justify hastening death. *Journal of Medical Ethics, 37*, 437–440.
Black Jr., A. (1990). Jonestown - Two faces of suicide: A Durkheimian analysis. *Suicide and Life-threatening Behavior, 20*, 285–306.
Bloom, P. (2007). Religion is natural. *Developmental Science, 10*, 147–151.
Bloom, P. (2012). Religion, morality, evolution. *Annual Review of Psychology, 63*, 179–199.
Boerner, K., Stroebe, M., Schut, H., & Wortman, C. B. (2015). Theories of grief and bereavement. In N. Pachana (Ed.), *Encyclopedia of geropsychology* (pp. 1–10). Singapore: Springer.
Bodner, E., Shrira, A., Bergman, Y. S., & Cohen-Fridel, S. (2015a). Anxieties about aging and death and psychological distress: The protective role of emotional complexity. *Personality and Individual Differences, 83*, 91–96.
Bodner, E., Shrira, A., Bergman, Y. S., Cohen-Fridel, S., & Grossman, E. S. (2015). The interaction between aging and death anxieties predicts ageism. *Personality and Individual Differences, 86*, 15–19.
Bolmsjo, I. A. (2008). End-of-life care for old people: A review of the literature. *American Journal of Hospice and Palliative Medicine, 25*, 328–338.
Bonanno, G. A. (2005). Resilience in the face of potential trauma. *Current Directions in Psychological Science, 14*, 135–138.
Bonanno, G. A., Wortman, C. B., Lehman, D. R., Tweed, R. G., Haring, M., Sonnega, J., et al. (2002). Resilience to loss and chronic grief: A prospective study from pre-loss to 18 months post-loss. *Journal of Personality and Social Psychology, 83*, 1150–1164.
Bowlby, J. (1969). *Attachment. Attachment and loss* (Vol. 1). New York: Basic Books.
Bowlby, J. (1973). *Separation: Anxiety and anger. Attachment and loss* (Vol. 2). New York: Basic Books.
Bowlby, J. (1980). *Attachment and loss: Vol. 3: Loss: Sadness and depression*. London: Hogarth Press.
Boyer, P., & Bergstrom, B. (2008). Evolutionary perspectives on religion. *Annual Review of Anthropology, 37*, 111–130.
Boyle, P. A., Barnes, L. L., Buchman, A. S., & Bennett, D. A. (2009). Purpose in life is associated with mortality among community-dwelling older persons. *Psychosomatic Medicine, 71*, 574–579.
Braam, A. W., Bramsen, I., van Tilburg, T. G., van der Ploeg, H. M., & Deeg, D. J. H. (2006). Cosmic transcendence and framework of meaning in life: Patterns among older adults in The Netherlands. *Journal of Gerontology, 61B*, S121–S128.
Bravell, M. E., Malmberg, B., & Berg, S. (2010). End-of-life care in the oldest old. *Palliative and Supportive Care, 8*, 335–344.
Breitbart, W. (2002). Spirituality and meaning in supportive care: Spirituality- and meaning- centered broup psychotherapy interventions in advanced cancer. *Supportive Care in Cancer, 10*, 272–286.
Breuer, C. (2015). Unemployment and suicide mortality: Evidence from regional panel data in Europe. *Health Economics, 24*, 936–950.
Brilman, E. I., & Ormel, J. (2001). Life events, difficulties and onset of depressive episodes in later life. *Psychological Medicine, 31*, 859–869.
Bronowski, J. (1976). *The ascent of man*. Cambridge, MA: MIT Press.
Bruce, A., & Boston, P. (2011). Relieving existential suffering through palliative sedation: Discussion of an uneasy practice. *Journal of Advanced Nursing, 67*, 2732–2740.
Buchbinder, M. (2018). Access to aid-in-dying in the United States: Shifting the debate from rights to justice. *American Journal of Public Health, 108*, 754–759.

Bühler, C. (1964). The human course of life in its goal aspects. *Journal of Humanistic Psychology, 4*, 1–17.

Buiting, H. M., Deeg, D. J. H., Knol, D. L., Ziegelmann, J. P., Pasman, H. R. W., Widdershoven, G. A. M., et al. (2012). Older peoples' attitudes towards euthanasia and an end-of-life pill in The Netherlands: 2001-2009. *Journal of Medical Ethics, 38*, 267–273.

Bulbulia, J. (2004). The cognitive and evolutionary psychology of religion. *Biology and Philosophy, 19*, 655–686.

Bulbulia, J., & Frean, M. (2010). The evolution of charismatic cultures. *Method and Theory in the Study of Religion, 22*, 254–271.

Bullard, G. (2016, April 22). The World's newest major religion: No religion. *National Geographic*. https://news.nationalgeographic.com/2016/04/160422-atheism-agnostic-secular-nones-rising-religion/

Burt, R. (2002). *Death is that man taking names: Intersections of American medicine, law, and culture*. Berkeley, CA: The University of California Press.

Butler, R. N. (1963). The life review: An interpretation of reminiscence in the aged. *Psychiatry, 26*, 65–76.

Butler, R. N. (1970). Looking forward to what? The life review, legacy, and excessive identity versus change. *The American Behavioral Scientist, 14*, 121–128.

Campbell, C. (2000). Euthanasia and religion. *UNESCO Courier, 53*, 37–39.

Canda, E. (1988). Spirituality, diversity and social work practice. *Social Casework, 69*, 238–247.

Candib, L. M. (2002). Truth telling and advance planning at the end of life: Problems with autonomy in a multicultural world. *Families, Systems & Health, 20*, 213–228.

Cantor, C. H., Leenaars, A. A., Lester, D., Slater, P. J., Wolanowski, A. M., & O'Toole, B. (1996). Suicide trends in eight predominantly English-speaking countries 1960-1989. *Social Psychiatry and Psychiatric Epidemiology, 31*, 364–373.

Cappeliez, P., & O'Rourke, N. (2002). Personality traits and existential concerns as predictors of the functions of reminiscence in older adults. *Journal of Gerontology, 57B*, P116–P123.

Carmel, S., Baron-Epel, O., & Shemy, G. (2007). The will-to-live and survival at old age: Gender differences. *Social Science and Medicine, 65*, 518–523.

Carpentier, N., & Van Brussel, L. (2012). On the contingency of death: A discourse- theoretical perspective on the construction of death. *Critical Discourse Studies, 9*, 99–115.

Carr, C. (1995). Mortuary practices: Their social, philosophical-religious, circumstantial, and physical determinants. *Journal of Archaeological Method and Theory, 2*, 105–200.

Carr, D., & Khodyakov, D. (2007). End-of-life health care planning among young-old adults: An assessment of psychosocial influences. *Journal of Gerontology, 62B*, S135–S141.

Carver, C. S., & Scheier, M. F. (1981). *Attention and self-regulation: A control-theory approach to human behavior*. New York: Springer.

Carver, C. S., & Scheier, M. F. (1998). *On the self regulation of behavior*. New York: Cambridge University Press.

Carver, C. S., & Scheier, M. F. (2002). Optimism. In C. R. Snyder & S. J. Lopez (Eds.), *Handbook of positive psychology* (pp. 231–243). Oxford: Oxford University Press.

Castano, E., Leidner, B., Bonacossa, A., Nikkah, J., Perrulli, R., Spencer, B., et al. (2011). Ideology, fear of death, and death anxiety. *Political Psychology, 32*, 601–621.

Cattell, H. (2000). Suicide in the elderly. *Advances in Psychiatric Treatment, 6*, 102–108.

Chan, L. C., & Yap, C. C. (2009). Age, gender, and religiosity as related to death anxiety. *Sunway Academic Journal, 6*, 1–16.

Chater, K., & Tsai, C.-T. (2009). Palliative care in a multicultural society: A challenge for western ethics. *Australian Journal of Advanced Nursing, 26*, 95–100.

Chattopadhyay, S., & Simon, A. (2008). East meets West: Cross-cultural perspective in end-of-life decision making from Indian and German viewpoints. *Medicine, Health Care and Philosophy, 11*, 165–174.

Chi, I., & Chou, K.-L. (2000). Financial strain and depressive symptoms among Hong Kong Chinese elderly: A longitudinal study. *Journal of Gerontological Social Work, 32*, 41–60.

Chida, Y., Steptoe, A., & Powell, L. H. (2009). Religiosity/spirituality and mortality: A systematic quantitative review. *Psychotherapy and Psychosomatics, 78*, 81–90.
Chochinov, H. M. (2006). Dying, dignity, and new horizons in palliative end-of-life care. *CA: A Cancer Journal for Clinicians, 56*, 84–103.
Chochinov, H. M., Cann, B., Cullihall, K., Kristjanson, L., Harlos, M., McClement, S. E., et al. (2012). Dignity therapy: A feasibility study of elders in long-term care. *Palliative and Supportive Care, 10*, 3–15.
Chochinov, H. M., & Cann, B. J. (2005). Interventions to enhance the spiritual aspects of dying. *Journal of Palliative Medicine, 8*, S103–S115.
Chochinov, H. M., Hack, T., Hassard, T., Kristjanson, L. J., McClement, S., & Harlos, M. (2005). Dignity therapy: A novel psychotherapeutic intervention for patients near the end of life. *Journal of Clinical Oncology, 23*, 5520–5525.
Chochinov, H. M., Kristjanson, L. J., Breitbart, W., McClement, S., Hack, T. F., Hassard, T., et al. (2011). The effect of dignity therapy on distress and end-of-life experience in terminally ill patients: A randomised controlled trial. *Lancet Oncology, 12*, 753–762.
Chochinov, H. M., Tataryn, D., Clinch, J. J., & Dudgeon, D. (1999). Will to live in the terminally ill. *The Lancet, 354*, 816–819.
Chokami, A. A., & Razavi, V. (2015). Relationship between social support and hope and death anxiety among the old people of Tehran Omid Cultural Center. *International Journal of Life Sciences, 9*, 65–70.
Chopik, W. J. (2017). Death across the lifespan: Age differences in death-related thoughts and anxiety. *Death Studies, 41*, 69–77.
Cicirelli, V. G. (2001). Personal meanings of death in older adults and young adults in relation to their fears of death. *Death Studies, 25*, 663–683.
Cicirelli, V. G. (2002). Fear of death in older adults: Predictions from terror management theory. *Journal of Gerontology, 57B*, P358–P366.
Cicirelli, V. G. (2006). Fear of death in mid-old age. *Journal of Gerontology, 61*, 75–81.
Cicirelli, V. G., MacLean, A. P., & Cox, L. S. (2000). Hastening death: A comparison of two end-of-life decisons. *Death Studies, 24*, 401–419.
Clarfield, A. M., Gordon, M., Markwell, H., & Alibhai, S. M. H. (2003). Ethical issues in end-of-life geriatric care: The approach of three monotheistic religions - Judaism, Catholicism, and Islam. *Journal of the American Geriatrics Society, 51*, 1149–1154.
Cohen, A. B., & Koenig, H. G. (2003). Religion, religiosity and spirituality in the biopsychosocial model of health and ageing. *Ageing International, 28*, 215–241.
Cohen, L. M. (2003). A medical revolution in death-hastening decisions. *Palliative and Supportive Care, 1*, 377–379.
Cohen, A. B., Pierce Jr., J. D., Chambers, J., Meade, R., Gorvine, B. J., & Koenig, H. G. (2005). Intrinsic and extrinsic religiosity, belief in the afterlife, death anxiety, and life satisfaction in young Catholics and Protestants. *Journal of Research in Personality, 39*, 307–324.
Cohen-Almagor, R. (2017). Euthanizing people who are 'tired of life' in Belgium. In D. A. Jones, C. Gastmans, & C. MacKellar (Eds.), *Euthanasia and assisted suicide: Lessons from Belgium* (pp. 188–201). Cambridge: Cambridge University Press.
Coleman, P. G. (2005). Spirituality and ageing: The health implications of religious belief and practice. *Age and Ageing, 34*, 318–319.
Collins, A. L., Sarkisian, N., & Winner, E. (2009). Flow and happiness in later life: An investigation into the role of daily and weekly flow experiences. *Jornal of Happiness Studies, 10*, 703–719.
Conwell, Y., Duberstein, P. R., & Caine, E. D. (2002). Risk factors for suicide in later life. *Biological Psychiatry, 52*, 193–204.
Conwell, Y., Van Orden, K., & Caine, E. D. (2011). Suicide in older adults. *Psychiatric Clinics of North America, 34*, 451–468.
Cordella, M., & Poiani, A. (2014). *Behavioural oncology: Psychological , communicative, and social dimensions*. New York: Springer.

Crawley, L. M., Marshall, P. A., Lo, B., Koenig, B. A., & for the End-of-Life Care Consensus Panel. (2002). Strategies for culturally effective end-of-life care. *Annals of Internal Medicine, 136*, 673–679.

Crocker, L., Clare, L., & Evans, K. (2006). Giving up or finding a solution? The experience of attempted suicide in later life. *Aging and Mental Health, 10*, 638–647.

Csikszentmihályi, M. (1990). *Flow: The psychology of optimal experience*. New York: Harper & Row.

Csikszentmihályi, M. (1993). *The evolving self*. New York: Harper Perennial.

Cutchin, M. P. (1997). Community and self: Concepts for rural physician integration and retention. *Social Science and Medicine, 44*, 1661–1674.

Daaleman, T. P., Perera, S., & Studenski, S. A. (2004). Religion, spirituality, and health status in geriatric outpatients. *Annals of Family Medicine, 2*, 49–53.

Damon, W., Menon, J., & Bronk, K. C. (2003). The development of purpose during adolescence. *Applied Developmental Science, 7*, 119–128.

Dattel, A. R., & Neimeyer, R. A. (1990). Sex differences in death anxiety: Testing the emotional expressiveness hypothesis. *Death Studies, 14*, 1–11.

Davis, J. (1999). *Death, burial and rebirth in the relions of antiquity*. London: Routledge.

De Leo, D., & Spathonis, K. (2003). Suicide and euthanasia in late life. *Aging Clinical and Experimental Research, 15*, 99–110.

De Raedt, R., & Van Der Speeten, N. (2008). Discrepancies between direct and indirect measures of death anxiety disappear in old age. *Depression and Anxiety, 25*, E11–E17.

Dein, S., & Littlewood, R. (2000). Apocalyptic suicide. *Mental Health, Religion and Culture, 3*, 109–114.

Dennis, R. E., Williams, W., Giangreco, M. F., & Cloninger, C. J. (1993). Quality of life as context for planning and evaluation of services for people with disabilities. *Exceptional Children, 59*, 499–512.

Dennis, M. S., Wakefield, P., Molloy, C., Andrews, H., & Friedman, T. (2007). A study of self-harm in older people: Mental disorder, social factors and motives. *Aging and Mental Health, 11*, 520–525.

Depaola, S. J., Griffin, M., Young, J. R., & Neimeyer, R. A. (2003). Death anxiety and attitudes toward the elderly among older adults: The role of gender and ethnicity. *Death Studies, 27*, 335–354.

Detering, K. M., Hancock, A. D., Reade, M. C., & Silvester, W. (2010). The impact of advance care planning on end of life care in elderly patients: Randomised controlled trial. *BMJ, 340*, c1345. https://doi.org/10.1136/bmj.c1345

Diener, E., Lucas, R. E., & Oishi, S. (2002). Subjective well-being: The science of happiness and life satisfaction. In C. R. Snyder & S. J. Lopez (Eds.), *Handbook of positive psychology* (pp. 63–73). Oxford: Oxford University Press.

Dittmann-Kohli, F. (1990). The construction of meaning in old age: Possibilities and constraints. *Ageing and Society, 10*, 279–294.

Dobríková, P., Macková, J., Pavelek, L., Miller, A., & West, D. J. (2016). The effect of social and existential aspects during end of life care. *Nursing and Palliative Care, 1*, 47–51.

Dombrovski, A. Y., Szanto, K., Clark, L., ReynoldsIII, C. F., & Siegle, G. J. (2013). Reward signals, attempted suicide, and impulsivity in late-life depression. *JAMA Psychiatry, 70*, 1020–1030.

Dombrovski, A. Y., Szanto, K., Duberstein, P., Conner, K. R., Houck, P. R., & Conwell, Y. (2008). Sex differences in correlates of suicide attempt lethality in late life. *American Journal of Geriatric Psychiatry, 16*, 905–913.

Douglas, C. D., Kerridge, I. H., Rainbird, K. J., McPhee, J. R., Hancock, L., & Spigelman, A. D. (2001). The intention to hasten death: A survey of attitudes and practices of surgeons in Australia. *Medical Journal of Australia, 175*, 511–515.

Dov Kulka, O. (2013). *Landscapes of the metropolis of death*. London: Allen Lane.

Dow, J. W. (2006). The evolution of religion: Three anthropological approaches. *Method and Theory in the Study of Religion, 18*, 67–91.
Draper, B. (1996). Attempted suicide in old age. *International Journal of Geriatric Psychiatry, 11*, 577–587.
Draper, B., Brodaty, H., & Low, L. F. (2002). Types of nursing home residents with self- destructive behaviours: Analysis of the harmful behaviours scale. *International Journal of Geriatric Psychiatry, 17*, 670–675.
Drion, H. (1992). *Het zelfgewilde einde van oude mensen*. Amsterdam: Balans.
Duberstein, P. R., Conwell, Y., & Caine, E. D. (1994). Age differences in the personality characteristics of suicide completers: Preliminary findings from a psychological autopsy study. *Psychiatry, 57*, 213–224.
Duff, R. W., & Hong, L. K. (1995). Age density, religiosity and death anxiety in retirement communities. *Review of Religious Research, 37*, 19–32.
Dumont, R., & Foss, D. (1972). *The American view of death: Acceptance or denial?* Cambridge, MA: Schenkman.
Durkheim, E. (1951). *Suicide: A study in sociology*. New York: The Free Press. (Original work published 1897).
Eakes, G. G. (1985). The relationship between death anxiety and attitudes toward the elderly among nursing staff. *Death Studies, 9*, 163–172.
Ebersole, P., & Quiring, G. (1991). Meaning in life depth: The MILD. *Journal of Humanistic Psychology, 31*, 113–124.
Eliason, G. T., Samide, J. L., Williams, G., & Lepore, M. F. (2010). Existential theory and our search for spirituality. *Journal of Spirituality in Mental Health, 12*, 86–111.
Elkins, D. N., Hedstrom, L. J., Hughes, L. L., Leaf, J. A., & Saunders, C. (1988). Toward a humanistic-phenomenological spirituality: Definition, description, and measurement. *Journal of Humanistic Psychology, 28*, 5–18.
Emanuel, E. J., & Emanuel, L. L. (1998). The promise of a good death. *Lancet, 351*, SII21–SII29.
Emmons, R. A. (1986). Personal strivings: An approach to personality and subjective well- being. *Journal of Personality and Social Psychology, 51*, 1058–1068.
Emmons, R. A. (2003). Personal goals, life meaning, and virtue: Well springs of a positive life. In C. L. M. Keyes & J. Haidt (Eds.), *Flourishing: Positive psychology and the life well-lived* (pp. 105–128). Washington, DC: American Psychological Association.
Erikson, E. H. (1963). *Childhood and society* (2nd ed.). New York: Norton.
Erikson, E. H. (1968). *Identity: Youth and crisis*. New York: W.W. Norton and Company.
Erikson, E. H. (1982). *The life cycle completed*. New York: Norton.
Erikson, E. H. (1997). *The life cycle completed*. New York: Norton.
Erlangsen, A., Bille-Brahe, U., & Jeune, B. (2003). Differences in suicide between the old and the oldest old. *Journal of Gerontology, 58B*, S314–S322.
Erlangsen, A., Jeune, B., Bille-Brahe, U., & Vaupel, J. W. (2004). Loss of partner and suicide risks among oldest old: A population-based register study. *Age and Ageing, 33*, 378–383.
Erlangsen, A., Vach, W., & Jeune, B. (2005). The effect of hospitalization with medical illnesses on the suicide risk in the oldest old: A population-based register study. *Journal of the American Geriatric Society, 53*, 771–776.
Erlangsen, A., Zarit, S. H., & Conwell, Y. (2008). Hospital-diagnosed dementia and suicide: A longitudinal study using prospective, nationwide register data. *American Journal of Geriatric Psychiatry, 16*, 220–228.
Ersek, M., Kagawa-Singer, M., Barnes, D., Blackhall, L., & Koenig, B. A. (1998). Multicultural considerations in the use of advance directives. *Oncology Nursing Forum, 25*, 1683–1690.
Fagan, A. (2004). Challenging the bioethical application of the autonomy principle within multicultural societies. *Journal of Applied Philosophy, 21*, 15–31.
Finn, W. F. (1979). The rights of the dying human. In A. de Vries & A. Carmi (Eds.), *The dying human* (pp. 187–189). Ramat Gan: Turtledove Publishing.
Firestone, R. W. (1993). Individual defenses against death anxiety. *Death Studies, 17*, 497–515.

Fishman, S. (1992). Relationships among an older adult's life review, ego integrity, and death anxiety. *International Psychogeriatrics, 4*, 267–277.
Fiske, A., O'Riley, A. A., & Widoe, R. K. (2008). Physical health and suicide in late life: An evaluative review. *Clinical Gerontologist, 31*, 31–50.
Flaatten, H., & Garrouste-Orgeas, M. (2015). The very old ICU patient: A never-ending story. *Intensive Care Medicine, 41*, 1996–1998.
Fleming, J., Farquhar, M., Cambridge City over-75s Cohort (CC75C) Study Collaboration, Brayne, C., & Barclay, S. (2016). Death and the oldest old: Attitudes and preferences for end-of-life care—Qualitative research within a population-based cohort study. *PLoS One, 11*(4), e0150686. https://doi.org/10.1371/journal.pone.0150686
Florian, V., & Mikulincer, M. (1997). Fear of death and the judgment of social transgressions: A multidimensional test of terror management theory. *Journal of Personality and Social Psychology, 73*, 369–380.
Fok, S.-Y., & Chong, A. M.-L. (2003). Euthanasia and old age: The case of Hong Kong. *Hallym International Journal of Aging, 5*, 41–53.
Fortner, B. V., & Neimeyer, R. A. (1999). Death anxiety in older adults: A quantitative review. *Death Studies, 23*, 387–411.
Fortuin, N. P. M., Schilderman, J. B. A. M., & Venbrux, E. (2018). The search for meaning in later life: On the connection between religious narratives and narratives of aging. *Journal of Religion, Spirituality and Aging*. https://doi.org/10.1080/15528030.2018.1553225
Fortuin, N. P. M., Schilderman, J. B. A. M., & Venbrux, E. (2019). Determining the end of life: A qualitative study of religion and euthanasia among older Dutch adults. *Mortality*. https://doi.org/10.1080/13576275.2019.1584165
Fowers, B. J., & Richardson, F. C. (1996). Why is multiculturalism good? *American Psychologist, 51*, 609–621.
Frances, A. (2015, April 17). Dying well means dying at home. *Psychiatric Times*. http://www.psychiatrictimes.com/blogs/couch-crisis/dying-well- means-dying-home?GUID=097EB3EEB7AC-46D2-AF0F- 8CC6E4A7E556&rememberme=1&ts=30042015
Frank, G., Blackhall, L. J., Michel, V., Murphy, S. T., Azen, S. P., & Park, K. (1998). A discourse of relationships in bioethics: Patient autonomy and end-of-life decision making among elderly Korean Americans. *Medical Anthropology Quarterly, 12*, 403–423.
Frankl, V. E. (1963). *Man's search for meaning: An introduction to logotherapy*. New York: Washington Square Press.
Frankl, V. E. (1967). Logotherapy and existentialism. *Psychotherapy: Theory, Research and Practice, 4*, 138–142.
Frankl, V. E. (1988). *The will to meaning: Foundations and applications of logotherapy*. New York: A Plume Book.
Frankl, V. E. (2004). *Man's search for meaning*. London: Rider.
Frankl, V. E. (2019). *Yes to life: In spite of everything*. London: Rider.
Fredrickson, B. L. (2002). Positive emotions. In C. R. Snyder & S. J. Lopez (Eds.), *Handbook of positive psychology* (pp. 120–134). Oxford: Oxford University Press.
Freud, S. (1917/1957). Mourning and melancholia. In: J. Strachey (Ed. and trans.), *Standard edition of the complete psychological works of Sigmund Freud* (Vol. 14, pp. 237–258). London: Hogarth Press.
Fritsche, I., Jonas, E., Fischer, P., Koranyi, N., Berger, N., & Fleischmann, B. (2007). Mortality salience and the desire for offspring. *Journal of Experimental Social Psychology, 43*, 753–762.
Fry, C. L., Dickerson-Putman, J., Draper, P., Ikels, C., Keith, J., Glascock, P., et al. (2007). Culture and the meaning of a good old age. In J. Sokolovsky (Ed.), *The cultural context of aging: Worldwide perspectives* (pp. 99–123). Westport, CT: Bergin and Garvey.
Fry, P. S. (2000). Religious involvement, spirituality and personal meaning for life: Existential predictors of psychological wellbeing in community-residing and institutional care elders. *Aging and Mental Health, 4*, 375–387.

Fry, P. S. (2001). The unique contribution of key existential factors to the prediction of psychological well-being of older adults following spousal loss. *The Gerontologist, 41*, 69–81.
Fulton, R., & Fulton, J. (1980). A psychological aspect of terminal care: Anticipatory grief. *Perspectives on Death and Dying, 2*, 87–96.
Furer, P., & Walker, J. R. (2008). Death anxiety: A cognitive-behavioral approach. *Journal of Cognitive Psychotherapy, 22*, 167–182.
Galzerano, A., Sabatini, E., Durì, D., Rossi, C., Silvestri, L., Mazzon, D., et al. (2009). Old patients in intensive care unit (ICU): What decisions to make? *Archives of Gerontology and Geriatrics, 49*, 294–297.
Garrow, S., & Walker, J. A. (2001). Existential group therapy and death anxiety. *Adultspan Journal, 3*, 77–87.
Garver, K. L., & Garver, B. (1991). Eugenics: Past, present, and the future. *American Journal of Human Genetics, 49*, 1109–1118.
Geertz, C. (1973). *The interpretation of cultures*. New York: Basic Books.
Genia, V. (1990). Religious development: A synthesis and reformulation. *Journal of Religion and Health, 29*, 85–99.
Gerstorf, D., Ram, N., Estabrook, R., Schupp, J., Wagner, G. G., & Lindenberger, U. (2008). Life satisfaction shows terminal decline in old age: Longitudinal evidence from the German Socioeconomic Panel Study. *Developmental Psychology, 44*, 1148–1159.
Gesser, G., Wong, P. T. P., & Reker, G. T. (1987). Death attitudes across the life-span: The development and validation of the death attitude profile. *Omega, 18*, 113–128.
Geyer, A. L., & Baumeister, R. F. (2005). Religion, morality, and self-control: Values, virtues, and vices. In R. F. Paloutzian & C. L. Park (Eds.), *Handbook of the psychology of religion and spirituality* (pp. 412–432). New York: The Guilford Press.
Ghufran, M., & Ansari, S. (2008). Impact of widowhood on religiosity and death anxiety among senior citizens. *Journal of the Indian Academy of Applied Psychology, 34*, 175–180.
Giddens, A. (1970). Introduction. In T. G. Masaryk (Ed.), *1881/1970 Suicide and the meaning of civilization* (pp. xix–xli). Chicago: The University of Chicago Press.
Giddens, A. (1991). *Modernity and self-identity: Self and society in the late modern age*. Stanford, CA: Stanford University Press.
Girard, C. (1993). Age, gender, and suicide: A cross-national analysis. *American Sociological Review, 58*, 553–574.
Glascock, A. (2009). Is killing necessarily murder? Moral questions surrounding assisted suicide and death. In J. Sokolovsky (Ed.), *The cultural context of aging: Worldwide perspectives* (pp. 77–92). Westport, CT: Praeger.
Gott, M., Small, N., Barnes, S., Payne, S., & Seamark, D. (2008). Older people's views of a good death in heart failure: Implications for palliative care provision. *Social Science and Medicine, 67*, 1113–1121.
Gould, S. J., & Vrba, E. S. (1982). Exaptation - A missing term in the science of form. *Paleobiology, 8*, 4–15.
Greenberg, J., & Arndt, J. (2012). Terror management theory. In P. A. M. Van Lange, A. W. Kruglanski, & E. T. Higgins (Eds.), *Handbook of theories of social psychology* (Vol. 1, pp. 398–415). Thousand Oaks, CA: Sage.
Greenberg, J., Pyszczynski, T., & Solomon, S. (1986). The causes and consequences of a need for self-esteem: A terror management theory. In R. F. Baumeister (Ed.), *Public self and private self* (pp. 189–212). New York: Springer.
Gulbinat, W. H. (1996). The epidemiology of suicide in old age. *Archives of Suicide Research, 2*, 31–42.
Guthrie, S. (1993). *Faces in the clouds: A new theory of religion*. New York: Oxford University Press.
Guttmann, D. (2001). A logotherapeutic approach to the quest for meaningful old age. *Social Thought, 20*, 117–128.
Gwyther, L., Brennan, F., & Harding, R. (2009). Advancing palliative care as a human right. *Journal of Pain and Symptom Management, 38*, 767–774.

Hafeez, A., & Rafique, R. (2014). Spirituality and religiosity as predictors of psychological well-being in residents of old homes. *The Dialogue, 8*, 285–301.
Hallberg, I. R. (2004). Death and dying from old people's point of view. A literature review. *Aging Clinical and Experimental Research, 16*, 87–103.
Harwood, D. M. J., Hawton, K., Hope, T., & Jacoby, R. (2000). Suicide in older people: Mode of death, demographic factors, and medical contact before death. *International Journal of Geriatric Psychiatry, 15*, 736–743.
Harding, S. R., Flannelly, K. J., Weaver, A. J., & Costa, K. G. (2005). The influence of religion on death anxiety and death acceptance. *Mental Health, Religion and Culture, 8*, 253–261.
Hawton, K., & Harriss, L. (2006). Deliberate self-harm in people aged 60 years and over: Characteristics and outcome of a 20-year cohort. *International Journal of Geriatric Psychiatry, 21*, 572–581.
Hayes, J., Schimel, J., Arndt, J., & Faucher, E. H. (2010). A theoretical and empirical review of the death-thought accessibility concept in terror management research. *Psychological Bulletin, 136*, 699–739.
Heidegger, M. (1963). *Sein und Zeit*. Tubingen: Niemeyer.
Heine, S. J., Proulx, T., & Vohs, K. D. (2006). The meaning maintenance model: On the coherence of social motivations. *Personality and Social Psychology Review, 10*, 88–110.
Heisel, M. J. (2006). Suicide and its prevention among older adults. *Canadian Journal of Psychiatry, 51*, 143–154.
Henrie, J., & Patrick, J. H. (2014). Religiousness, religious doubt, and death anxiety. *International Journal of Aging and Human Development, 78*, 203–227.
Heo, J., Lee, Y., Pedersen, P. M., & McCormick, B. P. (2010). Flow experience in the daily lives of older adults: An analysis of the interaction between flow, individual difference serious leisure, location, and social context. *Canadian Journal on Aging, 29*, 411–423.
Hermans, H. J. M. (1989). The meaning of life as an organized process. *Psychotherapy, 26*, 11–22.
Hesse, K. A. (1995). Terminal care of the very old: Changes in the way we die. *Archives of Internal Medicine, 155*, 1513–1518.
Hétu, J.-L. (2000). *Bilan de vie : Quand le passé nous rattrape*. Montréal, QC: Fides.
Hewitt, J. P. (2002). The social construction of self-esteem. In C. R. Snyder & S. J. Lopez (Eds.), *Handbook of positive psychology* (pp. 135–147). Oxford: Oxford University Press.
Highet, G. (1941). Petronium the moralist. *Transactions and Proceedings of the American Philological Association, 72*, 176–194.
Hill, T. D. (2006). Religion, spirituality, and healthy cognitive aging. *Southern Medical Journal, 99*, 1176–1177.
Hirao, K., Kobayashi, R., Okishima, K., & Tomokuni, Y. (2012). Flow experience and health-related quality of life in community dwelling elderly Japanese. *Nursing and Health Sciences, 14*, 52–57.
Hodwitz, O., & Frey, K. (2016). Anomic suicide: A Durkheimian analysis of European normlessness. *Sociological Spectrum, 36*, 236–254.
Howarth, G. (1998). 'Just live for today': Living, caring, ageing and dying. *Ageing and Society, 18*, 673–689.
Hoyt, S. F. (1912). The etymology of religion. *Journal of the American Oriental Society, 32*, 126–129.
Hui, V. K.-Y., & Coleman, P. G. (2012). Afterlife beliefs and ego integrity as two mediators of the relationship between intrinsic religiosity and personal death anxiety among older adult British Christians. *Research on Aging, 35*, 144–162.
Hui, V. K.-Y., & Fung, H. H. (2008). Mortality anxiety as a function of intrinsic religiosity and perceived purpose in life. *Death Studies, 33*, 30–50.
Hunt, K. J., Shlomo, N., & Addington-Hall, J. (2014). End-of-life care and preferences for place of death among the oldest old: Results of a population-based survey using VOICES–short form. *Journal of Palliative Medicine, 17*, 176–182.
Hunter, E. G. (2008). Legacy: The occupational transmission of self through actions and artifacts. *Journal of Occupational Science, 15*, 48–54.

Hunter, E. G., & Rowles, G. D. (2005). Leaving a legacy: Toward a typology. *Journal of Aging Studies, 19*, 327–347.
Huxtable, R., & Mölle, M. (2007). Setting a principled boundary'? Euthanasia as a response to 'life fatigue. *Bioethics, 21*, 117–126.
Iannaccone, L. R. (1995). Voodoo economics? Defending the rational approach to religion. *Journal for the Scientific Study of Religion, 34*, 76–88.
Idler, E. L., McLaughlin, J., & Kasl, S. (2009). Religion and the quality of life in the last year of life. *Journal of Gerontology, 64B*, 528–537.
Iemmi, V., Bantjes, J., Coast, E., Channer, K., Leone, T., McDaid, D., et al. (2016). Suicide and poverty in low-income and middle- income countries: A systematic review. *Lancet Psychiatry, 3*, 774–783.
Insoll, T. (Ed.). (2011). *The Oxford handbook of the archaeology of ritual and religion*. Oxford: Oxford University Press.
Jacó Chaves, L., & Aranha Gil, C. (2015). Older people's concepts of spirituality, related to aging and quality of life. *Ciência & Saúde Coletiva, 20*, 3641–3652.
Jong, J., Ross, R., Philip, T., Chang, S.-H., Simons, N., & Halberstadt, J. (2018). The religious correlates of death anxiety: A systematic review and metaanalysis. *Religion, Brain and Behavior, 8*, 4–20.
Jung, C. G. (1989). *Memories, dreams, reflections*. New York: Vintage Books. (Original work published 1963).
Jung, C. G. (2006). *The undiscovered self*. New York: Signet. (Original work published 1958).
Juurlink, D. N., Herrmann, N., Szalai, J. P., Kopp, A., & Redelmeier, D. A. (2004). Medical illness and the risk of suicide in the elderly. *Archives of Internal Medicine, 164*, 1179–1184.
Kalish, R. A. (1976). Death and dying in social context. In R. H. Binstock & E. Shanas (Eds.), *Handbook of aging and the social sciences* (pp. 483–507). New York: Van Nostrand.
Kaplan, M. S., McFarland, B. H., Huguet, N., & Newsom, J. T. (2007). Physical illness, functional limitations, and suicide risk: A population-based study. *American Journal of Orthopsychiatry, 77*, 56–60.
Karppinen, H., Laakkonen, M.-L., Strandberg, T. E., Tilvis, R. J., & Pitkälä, K. H. (2012). Will-to-live and survival in a 10-year follow-up among older people. *Age and Ageing, 41*, 789–794.
Kastenbaum, R., & Aisenberg, R. (1972). *The psychology of death*. New York: Springer.
Kastenbaum, R., & Costa Jr., P. T. (1977). Psychological perspectives on death. *Annual Review of Psychology, 28*, 225–249.
Kaufman, S. R. (1998). Intensive care, old age, and the problem of death in America. *The Gerontologist, 38*, 715–725.
Kawano, S. (2011). A sociocultural analysis of death anxiety among older Japanese urbanites in a citizens' movement. *Omega, 62*, 369–386.
Kearney, M., & Mount, B. (2000). Spiritual care of the dying patient. In H. M. Cochinov & W. Breitbart (Eds.), *Handbook of psychiatry in palliative medicine* (pp. 357–373). New York: Oxford University Press.
Kellehear, A. (2009). Dying old - and preferably alone? Agency, resistance and dissent at the end of life. *International Journal of Ageing and Later Life, 4*, 5–21.
Kiesling, C., Sorell, G. T., Montgomery, M. J., & Colwell, R. K. (2006). Identity and spirituality: A psychosocial exploration of the sense of spiritual self. *Developmental Psychology, 42*, 1269–1277.
Kimble, M. A. (1991). Aging and the search for meaning. *Journal of Religious Gerontology, 7*, 111–129.
Kimble, M. A., & Ellor, J. W. (2001). Logotherapy: An overview. *Journal of Religious Gerontology, 11*, 9–24.
King, L. A., & Hicks, J. A. (2012). Positive affect and meaning in life: The intersection of hedonism and eudaimonia. In P. T. P. Wong (Ed.), *The human quest for meaning: Theories, research, and applications* (pp. 125–141). London: Routledge.

Kiosses, D. N., Szanto, K., & Alexopoulos, G. S. (2014). Suicide in older adults: The role of emotions and cognition. *Current Psychiatry Reports, 16*(11), 495. https://doi.org/10.1007/s11920-014-0495-3

Kirby, S. E., Coleman, P. G., & Daley, D. (2004). Spirituality and well-being in frail and nonfrail older adults. *Journal of Gerontology, 59B*, P123–P129.

Kirkpatrick, L. A. (2005). Evolutionary psychology: An emerging new foundation for the psychology of religion. In R. F. Paloutzian & C. L. Park (Eds.), *Handbook of the psychology of religion and spirituality* (pp. 101–119). New York: The Guilford Press.

Kirkpatrick, L. A., & Navarrete, C. D. (2006). Reports of my death anxiety have been greatly exaggerated: A critique of terror management theory from an evolutionary perspective. *Psychological Inquiry, 17*, 288–298.

Kjølseth, I., & Ekeberg, Ø. (1997). Suicide attempts among the elderly. *Tidsskrift for den Norske Lægeforening, 117*, 3988–3991.

Kjølseth, I., Ekeberg, Ø., & Steihaug, S. (2009). "Why do they become vulnerable when faced with the challenges of old age?" Elderly people who committed suicide, described by those who knew them. *International Psychogeriatrics, 21*, 903–912.

Kjølseth, I., Ekeberg, Ø., & Steihaug, S. (2010). Why suicide? Elderly people who committed suicide and their experience of life in the period before their death. *International Psychogeriatrics, 22*, 209–218.

Klinger, E. (2012). The search for meaning in evolutionary goal-theory perspective and its clinical applications. In P. T. P. Wong (Ed.), *The human quest for meaning: Theories, research, and applications* (pp. 23–56). London: Routledge.

Koenig, H. G. (2009). Research on religion, spirituality, and mental health: A review. *Canadian Journal of Psychiatry, 54*, 283–291.

Koenig, H. G., Larson, D. B., & Larson, S. S. (2001). Religion and coping with serious medical illness. *Annals of Pharmacotherapy, 35*, 352–359.

Kollar, N. R. (1985). Towards a spirituality of aging and old age. *Journal of Religion and Aging, 1*, 49–59.

Koltko-Rivera, M. E. (2004). The psychology of worldviews. *Review of General Psychology, 8*, 3–58.

Koren, C., & Lowenstein, A. (2008). Late-life widowhood and meaning in life. *Ageing International, 32*, 140–155.

Kotter-Grühn, D., Kleinspehn-Ammerlahn, A., Gerstorf, D., & Smith, J. (2009). Self- perceptions of aging predict mortality and change with approaching death: 16- year longitudinal results from the Berlin Aging Study. *Psychology and Aging, 24*, 654–667.

Kouwenhoven, P. S. C., van Thiel, G. J. M. W., Raijmakers, N. J. H., Rietjens, J. A. C., van der Heide, A., & van Delden, J. J. M. (2014). Euthanasia or physician-assisted suicide? A survey from the Netherlands. *The European Journal of General Practice, 20*, 25–31.

Kraniodiotis, G. (2018). Euthanasia: A historical review of meanings, ideas and practices. *Archives of Hellenic Medicine, 35*, 527–534.

Krause, N. (2005). Traumatic events and meaning in life: Exploring variations in three age cohorts. *Ageing and Society, 25*, 501–524.

Krause, N. (2007). Evaluating the stress-buffering function of meaning in life among older people. *Journal of Aging and Health, 19*, 792–812.

Krause, N. (2009). Meaning in life and mortality. *Journal of Gerontology, 64B*, 517–527.

Krause, N. (2012). Meaning in life and healthy aging. In P. T. P. Wong (Ed.), *The human quest for meaning: Theories, research, and applications* (pp. 409–432). London: Routledge.

Kübler-Ross, E. (1997). *The wheel of life: A memoir of living and dying*. New York: Simon and Schuster.

Kuhse, H., & Singer, P. (2006). Introduction. In H. Kuhse & P. Singer (Eds.), *Bioethics: An anthology* (2nd ed., pp. 1–7). Malden, MA: Blackwell Publishing.

Kuhse, H., Singer, P., Baume, P., & Clark, M. (1997). End-of-life decisions in Australian medical practice. *The Medical Journal of Australia, 166*, 191–196.

Kushner, H. I. (1993). Suicide, gender, and the fear of modernity in nineteenth-century Medical and social thought. *Journal of Social History, 26*, 461–490.
Kwak, J., & Haley, W. E. (2005). Current research findings on end-of-life decision making among racially or ethnically diverse groups. *The Gerontologist, 45*, 634–641.
Kwee, M. G. T. (2012). Relational Buddhism: A psychological quest for meaning and sustainable happiness. In P. T. P. Wong (Ed.), *The human quest for meaning: Theories, research, and applications* (pp. 249–273). London: Routledge.
Kymlicka, W. (2010). The rise and fall of multiculturalism? New debates on inclusion and accommodation in diverse societies. *International Social Science Journal, 61*, 97–112.
Laceulle, H., & Baars, J. (2014). Self-realization and cultural narratives about later life. *Journal of Aging Studies, 31*, 34–44.
Lacey, H. P., Smith, D. M., & Ubel, P. A. (2006). Hope I die before I get old: Mispredicting happiness across the adult lifespan. *Journal of Happiness Studies, 7*, 167–182.
Ladin, K. (2008). Risk of late-life depression across 10 European Union countries: Deconstructing the education effect. *Journal of Aging and Health, 20*, 653–670.
Lahti, D. C. (2009). The correlated history of social organization, morality, and religion. In E. Voland & W. Schiefenhövel (Eds.), *The biological evolution of religious mind and behavior* (pp. 67–88). Berlin: Springer.
Lam, M., Lam, H. R., Agarwal, A., Chow, R., Chow, S., Chow, E., et al. (2017). Clinicians' views on palliative sedation for existential suffering: A systematic review and thematic synthesis of qualitative studies. *Journal of Pain Management, 10*, 31–40.
Langer, E. (2002). Well-being: Mindfulness versus positive evaluation. In C. R. Snyder & S. J. Lopez (Eds.), *Handbook of positive psychology* (pp. 202–213). Oxford: Oxford University Press.
Langer, N. (2000). The importance of spirituality in later life. *Gerontology and Geriatrics Education, 20*, 41–50.
Längle, A. (2001). Old age from an existential-analytical perspective. *Psychological Reports, 89*, 211–215.
Lapierre, S., Erlangsen, A., Waern, M., De Leo, D., Oyama, H., Scocco, P., et al. (2011). A systematic review of elderly suicide prevention programs. *Crisis, 32*, 88–98.
Larsson, K., Kåreholt, I., & Thorslund, M. (2008). Care utilisation in the last years of life in relation to age and time to death: Results from a Swedish urban population of the oldest old. *European Journal of Ageing, 5*, 349–357.
Lau, R. W. L., & Cheng, S.-T. (2011). Gratitude lessens death anxiety. *European Journal of Ageing, 8*, 169–175.
Lavretsky, H. (2010). Spirituality and aging. *Aging Health, 6*, 749–769.
Le, T. N. (2008). Age differences in spirituality, mystical experiences and wisdom. *Ageing and Society, 28*, 383–411.
Leenaars, A., Connolly, J., & on behalf of the International Ethical and Legal Working Group in Suicidology. (2001). Suicide, assisted suicide and euthanasia: International perspectives. *Irish Journal of Psychological Medicine, 18*, 33–37.
Lehto, R. H., & Stein, K. F. (2009). Death anxiety: An analysis of an evolving concept. *Research and Theory for Nursing Practice, 23*, 23–41.
Leinbach, R. M. (1993). Euthanasia attitudes of older persons: A cohort analysis. *Research on Aging, 15*, 433–448.
Leming, M. R. (1980). Religion and death: A test of Homan's thesis. *Omega, 10*, 347–364.
Levi, P. (1958). *Se questo è un uomo*. Torino: Giulio Einaudi editore s.p.a.
Levin, J. S., & Taylor, R. J. (1997). Age differences in patterns and correlates of the frequency of prayer. *The Gerontologist, 37*, 75–88.
Levin, J., & Chatters, L. M. (2008). Religion, aging, and health: Historical perspectives, current trends, and future directions. *Journal of Religion, Spirituality and Aging, 20*, 153–172.
Levin, P. D., & Sprung, C. L. (2003). Cultural differences at the end of life. *Critical Care Medicine, 31*(Suppl), S354–S357.

Lifton, R. J. (1973). *Home from the war: Vietnam veterans: Neither victims or executioners*. New York: Simon and Schuster.
Linden, M., & Barnow, S. (1997). The wish to die in very old persons near the end of life: A psychiatric problem? Results from the Berlin Aging Study. *International Psychogeriatrics, 9*, 291–307.
Lindquist, K. A., & Feldman Barrett, L. (2008). Emotional complexity. In M. Lewis, J. M. Haviland-Jones, & L. Feldman Barrett (Eds.), *Handbook of emotions* (3rd ed., pp. 513–530). New York: The Guilford Press.
Lindskog, M., Tavelin, B., & Lundström, S. (2015). Old age as risk indicator for poor end-of-life care quality - A population-based study of cancer deaths from the Swedish Register of Palliative Care. *European Journal of Cancer, 51*, 1331–1339.
Lloyd, L. (2000). Dying in old age: Promoting well-being at the end of life. *Mortality, 5*, 171–188.
Lloyd, L., & Cameron, A. (2005). Significant life events: Developing knowledge for care at the end of life in old age. *Journal of Integrated Care, 13*, 34–39.
Lloyd, L., White, K., & Sutton, E. (2011). Researching the end-of-life in old age: Cultural, ethical and methodological issues. *Ageing and Society, 31*, 386–407.
Lo, R. S. K., Kwan, B. H. F., Lau, K. P. K., Kwan, C. W. M., Lam, L. M., & Woo, J. (2010). The needs, current knowledge, and attitudes of care staff toward the implementation of palliative care in old age homes. *American Journal of Hospice and Palliative Medicine, 27*, 266–271.
Lo, R. S. K., & Woo, J. (2006). Palliative care in old age. *Reviews in Clinical Gerontology, 16*, 35–44.
Löfmark, R., Nilstun, T., Cartwright, C., Fischer, S., van der Heide, A., Mortier, F., et al. (2008). Physicians' experiences with end-of-life decision-making: Survey in 6 European countries and Australia. *BMC Medicine, 6*, 4. https://doi.org/10.1186/1741-7015-6-4
Lukas, E. (1986). LOGO-Test. In *Test zur Messung von "innerer Sinnerfüllung" und "existentieller Frustration"*. Wien: Deuticke.
Lundberg, G. D. (1988). 'It's over, Debbie' and the euthanasia debate. *JAMA, 259*, 2142–2143.
Lykins, E. L. B., Segerstrom, S. C., Averill, A. J., Evans, D. R., & Kemeny, M. E. (2007). Goal shifts following reminders of mortality: Reconciling posttraumatic growth and terror management theory. *Personality and Social Psychology Bulletin, 33*, 1088–1099.
Lykouras, L., Poulakou-Rebelakou, C., Tsiamis, C., & Ploumpidis, D. (2013). Suicidal behavior in the ancient Greek and Roman world. *Asian Journal of Psychiatry, 6*, 548–551.
Macklin, R. (1998). Ethical relativism in a multicultural society. *Kennedy Institute of Ethics Journal, 8*, 1–22.
MacDonald Jr., A. P. (1970). Anxiety, affiliation, and social isolation. *Developmental Psychology, 3*, 242–254.
MacLeod, R., Crandall, J., Wilson, D., & Austin, P. (2016). Death anxiety among New Zealanders: The predictive role of gender and marital status. *Mental Health, Religion and Culture, 19*, 339–349.
Macquarrie, J. (1973). *Existentialism*. New York: Penguin Books.
Maddi, S. R. (2012). Creating meaning through making decisions. In P. T. P. Wong (Ed.), *The human quest for meaning: Theories, research, and applications* (pp. 57–80). London: Routledge.
Maddux, J. E. (2002). Self-efficacy: The power of believing you can. In C. R. Snyder & S. J. Lopez (Eds.), *Handbook of positive psychology* (pp. 277–287). Oxford: Oxford University Press.
Malette, J., & Oliver, L. (2006). Retirement and existential meaning in the older adult: A qualitative study using life review. *Counselling, Psychotherapy, and Health, 2*, 30–49.
Malinowski, B. (1948). *Magic, science and religion and other essays*. Glencoe, IL: The Free Press.
Manchester, H., & Facer, K. (2015). Digital curation: Learning and legacy in later life. *E- Learning and Digital Media, 12*, 242–258.
Mancini, A. D., & Bonanno, G. A. (2006). Resilience in the face of potential trauma: Clinical practices and illustrations. *Journal of Clinical Psychology, 62*, 971–985.
Manning, L. K. (2012). Spirituality as a lived experience: Exploring the essence of spirituality for women in late life. *International Journal of Aging and Human Development, 75*, 95–113.

Manning, L. K. (2013). Navigating hardships in old age: Exploring the relationship between spirituality and resilience in later life. *Qualitative Health Research, 23*, 568–575.
Marcia, J. E. (1966). Development and validation of ego-identity status. *Journal of Personality and Social Psychology, 3*, 551–558.
Marcoen, A. (1994). Spirituality and personal well-being in old age. *Ageing and Society, 14*, 521–536.
Marris, P. (2002). Holding onto meaning through the life cycle. In R. S. Weiss & S. A. Bass (Eds.), *Challenges of the third age: Meaning and purpose in later life* (pp. 13–28). Oxford: Oxford University Press.
Maslow, A. H. (1943). A theory of human motivation. *Psychological Review, 50*, 370–396.
Masaryk, T. G. (1881/1970). *Suicide and the meaning of civilization*. Chicago: The University of Chicago Press.
Mascarenhas Fonseca, L., & Testoni, I. (2011–2012). The emergence of thanatology and current practice in death education. *Omega, 64*, 157–169.
Masten, A. S., & Reed, M.-G. J. (2002). Resilience in development. In C. R. Snyder & S. J. Lopez (Eds.), *Handbook of positive psychology* (pp. 74–88). Oxford: Oxford University Press.
McAdams, D. P. (2012). Meaning and personality. In P. T. P. Wong (Ed.), *The human quest for meaning: Theories, research, and applications* (pp. 107–123). London: Routledge.
McClain, C. S., Rosenfeld, B., & Breitbart, W. (2003). Effect of spiritual wellbeing on end of life despair in terminally ill cancer patients. *Lancet, 361*, 1603–1607.
McClement, S., Chochinov, H. M., Hack, T., Hassard, T., Kristjanson, L. J., & Harlos, M. (2007). Dignity therapy: Family member perspectives. *Journal of Palliative Medicine, 10*, 1076–1082.
McCormick, A. J. (2011). Self-determination, the right to die, and culture: A literature review. *Social Work, 56*, 119–128.
McDonald, R. T., & Hilgendorf, W. A. (1986). Death imagery and death anxiety. *Journal of Clinical Psychology, 42*, 87–91.
McFadden, S. H. (2005). Points of connection: Gerontology and the psychology of religion. In R. F. Paloutzian & C. L. Park (Eds.), *Handbook of the psychology of religion and spirituality* (pp. 162–176). New York: The Guilford Press.
McIntosh, J. L. (1992). Suicide of the elderly. In B. Bongar (Ed.), *Suicide: Guidelines for assessment, management, and treatment*. New York: Oxford University Press.
McNamara, B., Martin, K., Waddell, C., & Yuen, K. (1997). Palliative care in a multicultural society: Perceptions of health care professionals. *Palliative Medicine, 11*, 359–367.
McNees, P. (2009). The beneficial effects of life story and legacy activities. *Journal of Geriatric Care Management, Spring, 2009*, 15–19.
McPherson, C. J., Wilson, K. G., & Murray, M. A. (2007). Feeling like a burden to others: A systematic review focusing on the end of life. *Palliative Medicine, 21*, 115–128.
McSweeney, J. (2005). Foucault and theology. *Foucault Studies, 2*, 117–144.
Meier, D. E., & Cassel, C. K. (1983). Euthanasia in old age: A case study and ethical analysis. *Journal of the American Geriatrics Society, 31*, 294–298.
Meier, D. E., & Morrison, R. S. (1999). Old age and care near the end of life. *Generations, 23*, 6–11.
Mellor, P. A., & Shilling, C. (1993). Modernity, self-identity and the sequestration of death. *Sociology, 27*, 411–431.
Mellqvist Fässberg, M., van Orden, K. A., Duberstein, P., Erlangsen, A., Lapierre, S., Bodner, E., et al. (2012). A systematic review of social factors and suicidal behavior in older adulthood. *International Journal of Environmental Research and Public Health, 9*, 722–745.
Meñaca, A., Evans, N., Andrew, E. V. W., Toscani, F., Finetti, S., Gómez-Batiste, X., et al. (2012). End-of-life care across Southern Europe: A critical review of cultural similarities and differences between Italy, Spain and Portugal. *Critical Reviews in Oncology/Hematology, 82*, 387–401.
Miccinesi, G., Rietjens, J. A. C., Deliens, L., Paci, E., Bosshard, G., Nilstun, T., et al. (2006). Continuous deep sedation: Physicians' experiences in six European countries. *Journal of Pain and Symptom Management, 31*, 122–129.

Michael, S. T., Crowther, M. R., Schmid, B., & Allen, R. S. (2003). Widowhood and Spirituality: Coping Responses to Bereavement. *Journal of Women and Aging, 15*, 145–165.
Millet, N. B., Hart, G. D., Reyman, T. A., Zimmerman, M. R., & Lewin, P. K. (1998). ROMI: Mummification for the common people. In A. Cockburn, E. Cockburn, & T. A. Reyman (Eds.), *Mummies, disease and ancient cultures* (pp. 91–105). Cambridge: Cambridge University Press.
Misak, C. J., White, D. B., & Truog, R. D. (2014). Medical futility a new look at an old problem. *Chest, 146*, 1667–1672.
Missler, M., Stroebe, M., Geurtsen, L., Mastenbroek, M., Chmoun, S., & van der Houwen, K. (2012). Exploring death anxiety among elderly people: A literature review and empirical investigation. *Omega, 64*, 357–379.
Moberg, D. O. (2005). Research in spirituality, religion, and aging. *Journal of Gerontological Social Work, 45*, 11–40.
Mohandas, E. (2008). Neurobiology of spirituality. *Mens Sana Monographs, 6*, 63–80.
Molzahn, A. E. (2007). Spirituality in later life: Effect on quality of life. *Journal of Gerontological Nursing, 33*, 32–39.
Momeyer, R. W. (1988). *Confronting death*. Indianapolis, IL: Indiana University Press.
Momtaz, Y. A., Haron, S. A., Ibrahim, R., & Hamid, T. A. (2015). Spousal death anxiety in old age: Gender perspective. *Omega, 72*, 69–80.
Monforte-Royol, C., Villavicencio-Chávez, C., Tomás-Sábado, J., & Balaguer, A. (2011). The wish to hasten death: A review of clinical studies. *Psycho-Oncology, 20*, 795–804.
Moody, H. R. (2002). The changing meaning of aging. In R. S. Weiss & S. A. Bass (Eds.), *Challenges of the third age: Meaning and purpose in later life* (pp. 41–54). Oxford: Oxford University Press.
Moody, H. R. (2009/2010). Eco-elders: Legacy and environmental advocacy. *Generations, 33*, 70–74.
Moremen, R. D. (2005). What is the meaning of life? Women's spirituality at the end of the life span. *Omega, 50*, 309–330.
Morgan, J. H. (2012). Geriatric logotherapy: Exploring the psychotherapeutics of memory in treating the elderly. *Psychological Thought, 5*, 99–105.
Morgan, T., Williams, L. A., Trussardi, G., & Gott, M. (2016). Gender and family caregiving at the end-of-life in the context of old age: A systematic review. *Palliative Medicine, 30*, 616–624.
Mousseau, T. A., & Roff, D. A. (1987). Natural selection and the heritability of fitness components. *Heredity, 59*, 181–197.
Moustakas, C. (1972). *Loneliness and love*. Englewood Cliffs, NJ: Prentice-Hall.
Muller, M. T., Kimsma, G. K., & van der Wal, G. (1998). Euthanasia and assisted suicide: Facts, figures and fancies with special regard to old age. *Drugs and Aging, 13*, 185–191.
Mullins, L. C., & Lopez, M. A. (1982). Death anxiety among nursing home residents: A comparison of the young-old and the old-old. *Death Education, 6*, 75–86.
Murdock, V. (2005). Guided by ethics. *Journal of Gerontological Social Work, 45*, 131–154.
Musick, M. A., Traphagan, J. W., Koenig, H. G., & Larson, D. B. (2000). Spirituality in physical health and aging. *Journal of Adult Development, 7*, 73–86.
Myers, J. E., Wass, H., & Murphey, M. (1980). Ethnic differences in death anxiety among the elderly. *Death Studies, 4*, 237–244.
Mystakidou, K., Parpa, E., Tsilika, E., Katsouda, E., & Vlahos, L. (2005). The evolution of euthanasia and its perceptions in Greek culture and civilization. *Perspectives in Biology and Medicine, 48*, 95–104.
Nakamura, J., & Csikszentmihalyi, M. (2002). The concept of flow. In C. R. Snyder & S. J. Lopez (Eds.), *Handbook of positive psychology* (pp. 89–105). Oxford: Oxford University Press.
Narayanasamy, A., Clissett, P., Parumal, L., Thompson, D., Annasamy, S., & Edge, R. (2004). Responses to the spiritual needs of older people. *Journal of Advanced Nursing, 48*, 6–16.
Nash, M. L. (1980). Dignity of person in the final phase of life: An exploratory study. *Perspectives on Death and Dying, 2*, 62–70.

Neimeyer, R. A. (2001). *Meaning, reconstruction and the experience of loss*. Washington, DC: American Psychological Association.
Neimeyer, R. A. (2002). Traumatic loss and the reconstruction of meaning. *Journal of Palliative Medicine, 5*, 935–942.
Neimeyer, R. A. (2006). Widowhood, grief and the quest for meaning: A narrative perspective on resilience. In D. Carr, R. M. Nesse, & C. B. Wortman (Eds.), *Spousal bereavement in late life* (pp. 227–252). New York: Springer.
Neimeyer, R. A., Bagley, K. J., & Moore, M. K. (1986). Cognitive structure and death anxiety. *Death Studies, 10*, 273–288.
Neimeyer, R. A., Baldwin, S. A., & Gillies, J. (2006). Continuing bonds and reconstructing meaning: Mitigating complications in bereavement. *Death Studies, 30*, 715–738.
Neimeyer, R. A., Moore, M. K., & Bagley, K. (1988). A preliminary factor structure for the Threat Index. *Death Studies, 12*, 217–225.
Neimeyer, R. A., Prigerson, H. G., & Davies, B. (2002). Mourning and meaning. *American Behavioral Scientist, 46*, 235–251.
Neimeyer, R. A., & Van Brunt, D. (1995). Death anxiety. In H. Wass & R. A. Neimeyer (Eds.), *Dying: Facing the facts* (3rd ed., pp. 49–88). Washington, D.C.: Taylor and Francis.
Nelson-Becker, H., Ai, A. L., Hopp, F. P., McCormick, T. R., Schlueter, J. O., & Camp, J. K. (2015). Spirituality and religion in end-of-life care ethics: The challenge of interfaith and cross-generational matters. *British Journal of Social Work, 45*, 104–119.
Netzer, E. (1991). The last days and hours at Masada. *Biblical Archaeology Review, 17*, 20–32.
Ngeh, J. K. T. (2003). The phenomenon of premonition of death in older patient. *Journal of the American Geriatrics Society, 51*, 1672–1673.
Nietzsche, F. (1968). *The will to power*. New York: Vintage.
Nock, M. K., Borges, G., Bromet, E. J., Cha, C. B., Kessler, R. C., & Lee, S. (2008). Suicide and suicidal behavior. *Epidemiologic Reviews, 30*, 133–154.
Noronha, K. J. (2015). Impact of religion and spirituality on older adulthood. *Journal of Religion, Spirituality and Aging, 27*, 16–33.
Nussbaum, B. (2003). African culture and *Ubuntu*: Reflections of a South African in America. *Perspectives, 17*, 1–12.
Nussbaum, J. F., Pecchioni, L. L., Robinson, J. D., & Thompson, T. L. (2000). *Communication and aging* (2nd ed.). Mahwah, NJ: Lawrence Erlbaum Associates Publishers.
Nussbaum, M. C., & Levmore, S. (2017). *Aging thoughfully: Conversations about retirement, romance, wrinkles, & regret*. New York: Oxford University Press.
O'Connell, H., Chin, A.-V., Cunningham, C., & Lawlor, B. A. (2004). Recent developments: Suicide in older people. *BMJ, 329*, 895–899.
Ochsman, R. (1984). Belief in afterlife as a moderator of fear of death? *European Journal of geSocial Psychology, 14*, 53–67.
Old, J. L. (2008). Discussing end-of-life care with your patients. *Family Practice Management, 2008*, 18–22.
Olde Scheper, T. M. J. J., & Duursma, S. A. (1994). Euthanasia: The Dutch experience. *Age and Ageing, 23*, 3–8.
Olson, L. K. (1994). *The graying of the world: Who will care for the frail elderly*. New York: The Haworth Press.
Oman, D., & Thorensen, C. E. (2005). Do religion and spirituality influence health? In R. F. Paloutzian & C. L. Park (Eds.), *Handbook of the psychology of religion and spirituality* (pp. 435–459). New York: The Guilford Press.
Onwuteaka-Philipsen, B. D., Muller, M. T., & van der Wal, G. (1997). Euthanasia and old age. *Age and Ageing, 26*, 487–492.
Orentlicher, D., Pope, T. M., & Rich, B. (2016). Clinical criteria for physician aid in dying. *Journal of Palliative Medicine, 19*, 259–262.
Osborne, J. W. (2012). Existential and psychological aspects of the transition to retirement. *European Journal of Psychotherapy and Counselling, 2012*, 1–15.

Osvath, P., Kovacs, A., Voros, V., & Fekete, S. (2005). Risk factors of attempted suicide in the elderly: The role of cognitive impairment. *International Journal of Psychiatry in Clinical Practice, 9*, 221–225.
Otis-Green, S. (2003). Legacy building. *Smith College Studies in Social Work, 73*, 395–404.
Papadimitriou, J. D., Skiadas, P., Mavrantonis, C. S., Polimeropoulos, V., Papadimitriou, D. J., & Papacostas, K. J. (2007). Euthanasia and suicide in antiquity: Viewpoint of the dramatists and philosophers. *Journal of the Royal Society of Medicine, 100*, 25–28.
Paraschakis, A., Douzenis, A., Michopoulos, I., Christodoulou, C., Vassilopoulou, K., Koutsaftis, F., et al. (2012). Late onset suicide: Distinction between "young-old" vs. "old-old" suicide victims. How different populations are they? *Archives of Gerontology and Geriatrics, 54*, 136–139.
Pargament, K. I., Ano, G. G., & Wachholtz, A. B. (2005). The religious dimension of coping: Advances in theory, research, and practice. In R. F. Paloutzian & C. L. Park (Eds.), *Handbook of the psychology of religion and spirituality* (pp. 479–495). New York: The Guilford Press.
Park, C. L. (2005). Religion and meaning. In R. F. Paloutzian & C. L. Park (Eds.), *Handbook of the psychology of religion and spirituality* (pp. 295–314). New York: The Guilford Press.
Park, N., Park, M., & Peterson, C. (2010). When is the search for meaning related to life satisfaction? *Applied Psychology: Health and Well-Being, 2*, 1–13.
Parker, C., Laungani, P., & Young, B. (Eds.). (2015). *Death and bereavement across cultures* (2nd ed.). London: Routledge.
Parkes, C. M. (1971). Psycho-social transitions: A field for study. *Social Science and Medicine, 5*, 101–115.
Pattison, E. M. (1978). On death and dying. In F. J. Berghorn & D. E. Schafer (Eds.), *The dynamics of aging* (pp. 83–99). Boulder, CO: Westview Press.
Payne, B. R., Jackson, J. J., Noh, S. R., & Stine-Morrow, E. A. L. (2011). In the zone: Flow state and cognition in older adults. *Psychology and Aging, 26*, 738–743.
Payne, T. (2015). *The ancient art of growing old*. London: Vintage Books.
Perlman, D., & Peplau, L. A. (1981). Toward a social psychology of loneliness. In S. W. Duck & R. Gilmour (Eds.), *Personal relationships in disorder* (pp. 31–56). London: Academic Press.
Perls, F., Hefferline, R. F., & Goodman, P. (1958). *Gestalt therapy*. New York: Julian Press.
Perrels, A. J., Fleming, J., Zhao, J., Barclay, S., Farquhar, M., Buiting, H. M., et al. (2014). Place of death and end-of-life transitions experienced by very old people with differing cognitive status: Retrospective analysis of a prospective population-based cohort aged 85 and over. *Palliative Medicine, 28*, 220–233.
Peterson, C., & Park, N. (2012). Character strengths and life of meaning. In P. T. P. Wong (Ed.), *The human quest for meaning: Theories, research, and applications* (pp. 277–295). London: Routledge.
Peterson, C., Park, N., & Seligman, M. E. P. (2005). Orientations to happiness and life satisfaction: The full life versus the empty life. *Journal of Happiness Studies, 6*, 25–41.
Peterson, J. B. (1999). *Maps of meaning: The architecture of belief*. New York: Routledge.
Pettitt, P. (2010). *The Palaeolithic origins of human burial*. London: Routledge.
Phillips, I. (2003). Infusing spirituality into geriatric health care: Practical applications from the literature. *Topics in Geriatrics Rehabilitation, 19*, 249–256.
Phillips, M. R., & Cheng, H. G. (2012). The changing global face of suicide. *The Lancet, 379*, 2318–2319.
Poiani, A. (2010). *Animal homosexuality: A biosocial perspective*. Cambridge: Cambridge University Press.
Pool, R. (2004). "You're not going to dehydrate mom, are you?": Euthanasia, *versterving*, and good death in the Netherlands. *Social Science and Medicine, 58*, 955–966.
Popper, K. R. (1959). *The logic of scientific discovery*. New York: Routledge.
Portal Moreno, R., De La Fuente Solana, E. I., Rico, M. A., & Lozano Fernández, L. M. (2008). Death anxiety in institutionalized and non-institutionalized elderly people in Spain. *Omega, 58*, 61–76.

Potter, V. R. (1971). *Bioethics: Bridge to the future*. Englewood Cliffs, NJ: Prentoce-Hall.
Proulx, T., & Inzlicht, M. (2012). The five "A"s of meaning maintenance: Finding meaning in the theories of sense-making. *Psychological Inquiry, 23*, 317–335.
Puchalski, C., Ferrell, B., Virani, R., Otis-Green, S., Baird, P., Bull, J., et al. (2009). Improving the quality of spiritual care as a dimension of palliative care: The report of the consensus conference. *Journal of Palliative Medicine, 12*, 885–904.
Puchalski, C. M. (2002). Spirituality and end-of-life care: A time for listening and caring. *Journal of Palliative Medicine., 5*, 289–294.
Pyszczynski, T., Greenberg, J., & Solomon, S. (1999). A dual-process model of defense against conscious and unconscious death-related thoughts: An extension of terror management theory. *Psychological Review, 106*, 35–845.
Pyysiäinen, I. (2006). Amazing grace: Religion and the evolution of the human mind. In P. McNamara (Ed.), *Where God and Science meet: How brain and evolutionary studies alter our understanding of religion* (Vol. I, pp. 209–225). Santa Barbara, CA: Praeger.
Pyysiäinen, I., & Hauser, M. (2010). The origins of religion: Evolved adaptation or by- product? *Trends in Cognitive Sciences, 14*, 104–109.
Quill, T., Ganzini, L., Truog, R., & Pope, T. (2018). Voluntarily stopping eating and drinking among patients with serious advanced illness-clinical, ethical, and legal aspects. *JAMA Internal Medicine, 178*, 123–127.
Ram-Prasad, C. (1995). A classical Indian philosophical perspective on ageing and the meaning of life. *Ageing and Society, 15*, 1–36.
Ramsey, J. (2012). Spirituality and aging: Cognitive, affective, and relational pathways to resiliency. *Annual Review of Gerontology and Geriatrics, 32*, 131–150.
Randall, G. K., & Martin, P. (2003). Developing and using stories or narratives to transmit values and legacy. *Organization Development Journal, 21*, 44–50.
Rappaport, H., Fossler, R. J., Bross, L. S., & Gilden, D. (1993). Future time, death anxiety, and life purpose among older adults. *Death Studies, 17*, 369–379.
Rappaport, R. A. (1974). *Pigs for the ancestors: Ritual in the ecology of a New Guinea people* (2nd ed.). New Haven: Yale.
Rashidi, N. M., Zordan, R. D., Flynn, E., & Philip, J. A. M. (2011). The care of the very old in the last three days of life. *Journal of Palliative Medicine, 14*, 1339–1344.
Raven, B. H. (1999). Kurt Lewin address: Influence, power, religion, and the mechanisms of social control. *Journal of Social Issues, 55*, 161–186.
Reich, J. W., Zautra, A., & Hall, J. S. (2010). *Handbook of adult resilience*. New York: Guilford Press.
Reker, G. T. (1997). Personal meaning, optimism, and choice: Existential predictors of depression in community and institutional elderly. *The Gerontologist, 37*, 709–716.
Reker, G. T., Birren, J. E., & Svensson, C. M. (2012). Restoring, maintaining, and enhancing personal meaning in life through autobiographical methods. In P. T. P. Wong (Ed.), *The human quest for meaning: Theories, research, and applications* (pp. 383–407). London: Routledge.
Reker, G. T., & Guppy, B. (1988). *Sources of personal meaning among young, middle- aged, and older adults*. Paper presented at the Canadian Association on Gerontology Meeting, Halifax, NS.
Reker, G. T., & Wong, P. T. P. (1988). Aging as an individual process: Toward a theory of personal meaning. In J. E. Birren & V. L. Bengston (Eds.), *Emergent theories of aging* (pp. 214–246). New York: Springer.
Reker, G. T., & Wong, P. T. P. (2012). Personal meaning in life and psychological adaptation in the later years. In P. T. P. Wong (Ed.), *The human quest for meaning: Theories, research, and applications* (pp. 433–456). London: Routledge.
Relph, E. (1976). *Place and placelessness*. London: Pion Limited.
Renteln, A. D. (2001). The rights of the dead: Autopsies and corpse mismanagement in multicultural societies. *The South Atlantic Quarterly, 100*, 1005–1027.

Richards, N. (2012). The fight-to-die: Older people and death activism. *International Journal of Ageing and Later Life, 7*, 7–32.

Richardson, P. (2014). Spirituality, religion and palliative care. *Annals of Palliative Medicine, 3*, 150–159.

Richerson, P., & Boyd, R. (2005). *Not by genes alone: How culture transformed human evolution*. Chicago: University of Chicago Press.

Richman, J. (2006). The role of psychotherapy and humor for death anxiety, death wishes, and aging. *Omega, 54*, 41–51.

Romanoff, B. D., & Thompson, B. E. (2006). Meaning construction in palliative care: The use of narrative, ritual, and the expressive arts. *American Journal of Hospice and Palliative Medicine, 23*, 309–316.

Ron, P. (2010). Elderly people's death and dying anxiety: A comparison between elderly living within the community and in nursing homes in Israel. *Illness, Crisis and Loss, 18*, 3–17.

Rousseau, P. (2000). Spirituality and the dying patient. *Journal of Clinical Oncology, 18*, 2000–2002.

Routledge, C., Arndt, J., Wildschut, T., Sedikides, C., Hart, C., Juhl, J., et al. (2011). The past makes the present meaningful: Nostalgia as an existential resource. *Journal of Personality and Social Psychology, 101*, 638–652.

Rowles, G. D., & Ravdal, H. (2002). Aging, place, and meaning in the face of changing circumstances. In R. S. Weiss & S. A. Bass (Eds.), *Challenges of the third age: Meaning and purpose in later life* (pp. 81–114). Oxford: Oxford University Press.

Russac, R. J., Gatliff, C., Reece, M., & Spottswood, D. (2007). Death anxiety across the adult years: An examination of age and gender effects. *Death Studies, 31*, 549–561.

Russell, B. (1967). A free man's worship. In P. Edwards (Ed.), *Why I am not a Christian and other essays on religion and related subjects*. New York: A Touchstone Book.

Ryan, R. M., & Deci, E. L. (2000). Self-determination theory and the facilitation of intrinsic motivation, social development, and well-being. *American Psychologist, 55*, 68–78.

Ryff, C. D. (2012). Existential well-being and health. In P. T. P. Wong (Ed.), *The human quest for meaning: Theories, research, and applications* (pp. 233–247). London: Routledge.

Rykkje, L. L. R., Eriksson, K., & Raholm, M.-B. (2013). Spirituality and caring in old age and the significance of religion – A hermeneutical study from Norway. *Scandinavian Journal of Caring Sciences, 27*, 275–284.

Samarel, N. (1995). The dying process. In H. Wass & R. A. Neimeyer (Eds.), *Dying: Facing the facts* (pp. 89–116). Washington, DC: Taylor and Francis.

Sand, L., & Strang, P. (2006). Existential loneliness in a palliative home care setting. *Journal of Palliative Medicine, 9*, 1376–1387.

Sanderson, S. K. (2008). Adaptation, evolution, and religion. *Religion, 38*, 141–156.

Sanz, N., Arriaza, B. T., & Standen, V. G. (Eds.). (2014). *The Chinchorro culture: A comparative perspective. The archaeology of the earliest human mummification*. Paris: UNESCO.

Saroglou, V. (2010). Religiousness as a cultural adaptation of basic traits: A five-factor model perspective. *Personality and Social Psychology Review, 14*, 108–125.

Sartre, J.-P. (1956). *Being and nothingness*. New York: The Philosophical Library.

Satalkar, P., & van der Geest, S. (2019). Escaping the 'unprepared generation' trap: Discussions about euthanasia after a 'completed life' in the Netherlands. *Mortality*, 1–15. https://doi.org/10.1080/13576275.2019.1585780

Savishinsky, J. S. (2007). The quest for legacy in later life. *Journal of Intergenerational Relationships, 4*, 75–90.

Scheier, M. E., & Carver, C. S. (1987). Dispositional optimism and physical well-being: The influence of generalized outcome expectancies on health. *Journal of Personality, 55*, 169–210.

Scheper, T.M.J.J. olde & Duursma, S.A. (1994). Euthanasia: The Dutch experience. Age and Ageing 23: 3–8

Seale, C. (1998). *Constructing death: The sociology of dying and bereavement*. Cambridge: Cambridge University Press.

Seale, C. (2009). Hastening death in end-of-life care: A survey of doctors. *Social Science and Medicine, 69*, 1659–1666.
Searight, H. R., & Gafford, J. (2005). Cultural diversity at the end of life: Issues and guidelines for family physicians. *American Family Physician, 71*, 515–522.
Seidel, G. (1995). Suicide in the elderly in antiquity. *International Journal of Geriatric Psychiatry, 10*, 1077–1084.
Seligman, M., & Csikszentmihalyi, M. (2000). Positive psychology: An introduction. *American Psychologist, 55*, 5–14.
Seligman, M. E. P. (2002). *Authentic happiness*. New York: Free Press.
Settersten Jr., R. A. (2002). Social sources of meaning in later life. In R. S. Weiss & S. A. Bass (Eds.), *Challenges of the third age: Meaning and purpose in later life* (pp. 55–79). Oxford: Oxford University Press.
Shah, A. (2007). The relationship between suicide rates and age: An analysis of multinational data from the World Health Organization. *International Psychogeriatrics, 19*, 1141–1152.
Shah, A., Bhat, R., MacKenzie, S., & Koen, C. (2008). Elderly suicide rates: Cross-national comparisons of trends over a 10-year period. *International Psychogeriatrics, 20*, 673–686.
Shah, A. K., & Ganesvaran, T. (1994). Suicide in the elderly. In E. Chiu & D. Ames (Eds.), *Functional psychiatric disorders of the elderly* (pp. 221–244). Cambridge: Cambridge University Press.
Shaw, D. M. (2009). Euthanasia and eudaimonia. *Journal of Medical Ethics, 35*, 530–533.
Sheldon, T. (2001). "Existential" suffering not a justification for euthanasia. *BMJ: British Medical Journal, 323*, 1384.
Sherman, E., Michel, R., Rybak, C., Randall, G. K., & Davidson, J. (2011). Meaning in life and volunteerism in older adults. *Adultspan Journal, 10*, 78–90.
Shmotkin, D. (2005). Happiness in face of adversity: Reformulating the dynamic and modular bases of subjective well-being. *Review of General Psychology, 9*, 291–325.
Shmotkin, D., & Shrira, A. (2012). Happiness and suffering in the life story: An inquiry into conflicting expectations concerning the association of perceived past with present subjective well-being in old age. *Journal of Happiness Studies, 13*, 389–409.
Sijuwade, P. O. (2013). Spiritual intelligence, living status and general health of the elderly. *Pakistan Journal of Social Sciences, 10*, 135–138.
Simonton, D. K. (2002). Creativity. In C. R. Snyder & S. J. Lopez (Eds.), *Handbook of positive psychology* (pp. 189–201). Oxford: Oxford University Press.
Singh Madnawat, A. V., & Singh Kachhawa, P. (2007). Age, gender, and living circumstances: Discriminating older adults on death anxiety. *Death Studies, 31*, 763–769.
Sinyor, M., Tan, L. P., Schaffer, A., Gallagher, D., & Shulman, K. (2015). Suicide in the oldest old: An observational study and cluster analysis. In A. Sheils, J. Ashmore, K. Kõlves, & D. De Leo (Eds.), *Suicide Research: Selected readings* (Vol. 13, pp. 76–78). Samford, QLD: Australian Academy Press.
Sinyor, M., Tse, R., & Pirkis, J. (2017). Global trends in suicide epidemiology. *Current Opinion in Psychiatry, 30*, 1–6.
Sjöberg, M., Beck, I., Rasmussen, B. H., & Edberg, A.-K. (2018). Being disconnected from life: Meanings of existential loneliness as narrated by frail older people. *Aging and Mental Health, 22*, 1357–1364.
Slattery, J. M., & Park, C. L. (2012). Clinical approaches to discrepancies in meaning: Conceptualization, assessment, and treatment. In P. T. P. Wong (Ed.), *The human quest for meaning: Theories, research, and applications* (pp. 497–520). London: Routledge.
Smith, R. (2000). A good death: An important aim for health services and for us all. *BMJ, 320*, 129–130.
Snowdon, J., & Hunt, G. E. (2002). Age, period and cohort effects on suicide rates in Australia, 1919-1999. *Acta Psychiatrica Scandinavica, 105*, 265–270.

Snyder, C. R., Rand, K. L., & Sigmo, D. R. (2002). Hope theory: A member of the positive psychology family. In C. R. Snyder & S. J. Lopez (Eds.), *Handbook of positive psychology* (pp. 257–276). Oxford: Oxford University Press.

Sommer, K. L., Baumeister, R. F., & Stillman, T. F. (2012). The construction of meaning from life events: Empirical studies of personal narratives. In P. T. P. Wong (Ed.), *The human quest for meaning: Theories, research, and applications* (pp. 297–313). London: Routledge.

Sone, T., Nakaya, N., Ohmori, K., Shimazu, T., Higashiguchi, M., Kakizaki, M., et al. (2008). Sense of life worth living (*ikigai*) and mortality in Japan: Ohsaki Study. *Psychosomatic Medicine, 70*, 709–715.

Sosis, R. (2009). The adaptationist-byproduct debate on the evolution of religion: Five misunderstandings of the adaptationist program. *Journal of Cognition and Culture, 9*, 315–332.

Speijer, N., & Diesktra, R. (1980). *Assisted suicide: A study of the problems related to self-chosen death*. Peventer: van Loghun Slaterur.

Sprung, C. L., Cohen, S. L., Sjokvist, P., Baras, M., Bulow, H.-H., Hovilehto, S., et al. (2003). End-of-life practices in European intensive care units: The Ethicus study. *JAMA, 290*, 790–797.

Sridevi, G., & Swathi, P. (2014). Death anxiety, death depression, geriatric depression and suicidal ideation among institutionalized and non-institutionalized elders. *International Journal of Scientific and Research Publications, 4*, 1–8.

Stanton, A. L., Parsa, A., & Austenfeld, J. L. (2002). The adaptive potential of coping through emotional approach. In C. R. Snyder & S. J. Lopez (Eds.), *Handbook of positive psychology* (pp. 148–158). Oxford: Oxford University Press.

Stausberg, M. (2010). Prospects in theories of religion. *Method and Theory in the Study of Religion, 22*, 223–238.

Steger, M. F. (2012). Experiencing meaning in life: Optimal functioning at the nexus of well-being, psychopathology, and spirituality. In P. T. P. Wong (Ed.), *The human quest for meaning: Theories, research, and applications* (pp. 165–184). London: Routledge.

Steger, M. F. (2013). Experiencing meaning in life: Optimal functioning at the nexus of well-being, psychopathology, and spirituality. In P. T. P. Wong (Ed.), *The human quest for meaning: Theory, research, and application* (2nd ed., pp. 165–184). New York: Routledge.

Steger, M. F., Oishi, S., & Kashdan, T. B. (2009). Meaning in life across the life span: Levels and correlates of meaning in life from emerging adulthood to older adulthood. *The Journal of Positive Psychology, 4*, 43–52.

Stolorow, R. D. (1973). Perspectives on death anxiety: A review. *Psychiatric Quarterly, 47*, 473–486.

Stone, I. F. (1989). *The trial of Socrates*. New York: Anchor Books.

Streib, H., & Klein, C. (2013). Atheists, agnostics, and apostates. In K. I. Pargament, J. J. Exline, & J. W. Jones (Eds.), *APA handbook of psychology, religion, and spirituality (Vol. 1): Context, theory, and research* (pp. 713–728). Washington, DC: American Psychological Association.

Stroebe, M. (1992). Coping with bereavement: A review of the grief work hypothesis. *Omega, 26*, 19–42.

Stroebe, M., Schut, H., & Stroebe, W. (2007). Health outcomes of bereavement. *Lancet, 370*, 1960–1973.

Suhail, K., & Akram, S. (2002). Correlates of death anxiety in Pakistan. *Death Studies, 26*, 39–50.

Suri, R. (2010). Working with the elderly: An existential-humanistic approach. *Journal of Humanistic Psychology, 50*, 175–186.

Sutherland, J.-A., Poloma, M. M., & Pendleton, B. F. (2003). Religion, spirituality, and alternative health practices: The Baby Boomer and Cold War cohorts. *Journal of Religion and Health, 42*, 315–338.

Takkinen, S., & Ruoppila, I. (2001). Meaning in life as an important component of functioning in old age. *International Journal of Aging and Human Development, 53*, 211–231.

Taylor, B. R., & McCann, R. M. (2005). Controlled sedation for physical and existential suffering? *Journal of Palliative Medicine, 8*, 144–147.

Tedeschi, R. G., & Calhoun, L. G. (1996). The posttraumatic growth inventory: Measuring the positive legacy of trauma. *Journal of Traumatic Stress, 9*, 455–471.
Tedeschi, R. G., & Calhoun, L. G. (2004). Posttraumatic growth: Conceptual foundations and empirical evidence. *Psychological Inquiry, 15*, 1–18.
Tedeschi, R. G., & Calhoun, L. G. (2012). Pathways to personal transformation: Theoretical and empirical developments. In P. T. P. Wong (Ed.), *The human quest for meaning: Theories, research, and applications* (pp. 559–572). London: Routledge.
Temkin-Greener, H., & Mukamel, D. B. (2002). Predicting place of death in the Program of All-inclusive Care for the Elderly (PACE): Participant versus program characteristics. *Journal of the American Geriatrics Society, 50*, 125–135.
Tervalon, M., & Murray-Garcia, J. (1998). Cultural humility versus cultural competence: A critical distinction in defining physician training outcomes in multicultural education. *Journal of Health Care for the Poor and Underserved, 9*, 117–125.
Thagard, P. (2005). The emotional coherence of religion. *Journal of Cognition and Culture, 5*, 58–74.
Thompson, S. C. (2002). The role of personal control in adaptive functioning. In C. R. Snyder & S. J. Lopez (Eds.), *Handbook of positive psychology* (pp. 202–213). Oxford: Oxford University Press.
Thompson, S. C., & Janigian, A. S. (1988). Life schemes: A framework for understanding the search for meaning. *Journal of Social and Clinical Psychology, 7*, 260–280.
Thorson, J. A., & Powell, F. C. (1988). Elements of death anxiety and meanings of death. *Journal of Clinical Psychology, 44*, 691–701.
Thorson, J. A., & Powell, F. C. (1993). Personality, death anxiety, and gender. *Bulletin of the Psychonomic Society, 3*, 589–590.
Thorson, J. A., & Powell, F. C. (1994). A revised death anxiety scale. In R. A. Neimeyer (Ed.), *Death anxiety handbook* (pp. 303–381). Washington, DC: Taylor and Francis.
Tomer, A. (1992). Death anxiety in adult life - Theoretical perspectives. *Death Studies, 16*, 475–506.
Tomer, A. (2012). Meaning and death attitudes. In P. T. P. Wong (Ed.), *The human quest for meaning: Theories, research, and applications* (pp. 209–231). London: Routledge.
Tomer, A., & Eliason, G. (1996). Toward a comprehensive model of death anxiety. *Death Studies, 20*, 343–365.
Tornstam, L. (1996). Gerotranscendence – A theory about maturing into old age. *Journal of Aging and Identity, 1*, 37–50.
Tornstam, L. (2011). Maturing into gerotranscendence. *The Journal of Transpersonal Psychology, 43*, 166–180.
Troll, L. E. (1982). *Continuous adult development and aging*. Brooks and Cole, CA: Monterey.
Turecki, G., & Brent, D. A. (2016). Suicide and suicidal behavior. *Lancet, 387*, 1227–1239.
Turner, L. (2003). Bioethics in a multicultural world: Medicine and morality in pluralistic settings. *Health Care Analysis, 11*, 99–117.
Vaillant, G. E. (2008). *Spiritual evolution: How we are wired for faith, hope, and love*. New York: Broadway Books.
van Hooff, A. J. L. (2004). Ancient euthanasia: 'good death' and the doctor in the Graeco-Roman world. *Social Science and Medicine, 58*, 975–985.
van Hooff, A. (2004a). Paetus, it does not hurt: Altruistic suicide in the Greco-Roman world. *Archives of Suicide Research, 8*, 43–56.
van Baarsen, B. (2009). Suffering, loneliness, and the euthanasia choice: An explorative study. *Journal of Social Work in End-Of-Life and Palliative Care, 4*, 189–213.
Van Bruchem-van de Scheur, G., van der Arend, A. J., Abu-Saad, H. H., van Wijmen, F. C. B., Spreeuwenberg, C., & ter Meulen, R. H. J. (2008). Alleviation of pain and symptoms with a life shortening intention. *Nursing Ethics, 15*, 682–696.
van der Heide, A., van Delden, J. J. M., & Onwuteaka-Philipsen, B. D. (2017). End-of-life decisions in the Netherlands over 25 Years. *The New England Journal of Medicine, 377*, 492–494.

van Deurzen, E. (2002). *Existential counseling and psychotherapy in practice*. London: Sage.
van Hooff, A. J. L. (1993). Suicide and parasuicide in ancient personal testimonies. *Crisis, 14*, 76–82.
Van Orden, K. A., Witte, T. K., Cukrowicz, K. C., Braithwaite, S. R., Selby, E. A., & Joiner, T. E., Jr. (2010). The interpersonal theory of suicide. *Psychological Review, 117*, 575–600.
Varelius, J. (2014). Medical expertise, existential suffering and ending life. *Journal of Medical Ethics, 40*, 104–107.
Velasco-Gonzalez, L., & Rioux, L. (2013). The spiritual well-being of elderly people: A study of a French sample. *Journal of Religion and Health, 53*, 1123–1137.
Verhofstadt, M., Thienpont, L., & Ygram Peters, G.-J. (2017). When unbearable suffering incites psychiatric patients to request euthanasia: Qualitative study. *The British Journal of Psychiatry, 211*, 238–245.
Vincent, J.-L., Berre, J., & Creteur, J. (2004). Withholding and withdrawing life prolonging treatment in the intensive care unit: A current European perspective. *Chronic Respiratory Disease, 1*, 115–120.
Voland, E. (2009). Evaluating the evolutionary status of religiosity and religiousness. In E. Voland & W. Schiefenhövel (Eds.), *The biological evolution of religious mind and behavior* (pp. 9–24). Berlin: Springer.
Voland, E., & Schiefenhövel, W. (Eds.). (2009). *The biological evolution of religious mind and behavior*. Berlin: Springer.
Volker, D. L. (2005). Control and end-of-life care: Does ethnicity matter? *American Journal of Hospice and Palliative Care, 22*, 442–446.
Wadensten, B. (2005). Introducing older people to the theory of gerotranscendence. *Journal of Advanced Nursing, 52*, 381–388.
Waern, M., Rubenowitz, E., & Wilhelmson, K. (2003). Predictors of suicide in the old elderly. *Gerontology, 49*, 328–334.
Wagner, K. D., & Lorion, R. P. (1984). Correlates of death anxiety in elderly persons. *Journal of Clinical Psychology, 40*, 1235–1241.
Wand, A. P. F., Peisah, C., Draper, B., & Brodaty, H. (2018). Understanding self-harm in older people: A systematic review of qualitative studies. *Aging and Mental Health, 22*, 289–298.
Wand, A. P. F., Peisah, C., Draper, B., Jones, C., & Brodaty, H. (2016). Rational suicide, euthanasia, and the very old: Two case reports. *Case Reports in Psychiatry, 2016*, 4242064, 5 pages. https://doi.org/10.1155/2016/4242064
Ward, R. A. (1980). Age and acceptance of euthanasia. *Journal of Gerontology, 35*, 421–431.
Ward, R. A. (1984). The marginality and salience of being old. *The Gerontologist, 24*, 227–232.
Wass, H., & Sisler, H. H. (1979). Death concern and views on various aspects of dying among elderly persons. In A. de Vries & A. Carmi (Eds.), *The dying human* (pp. 71–86). Ramat Gan: Turtledove Publishing.
Watson, D. (2002). Positive affectivity: The disposition to experience pleasurable emotional states. In C. R. Snyder & S. J. Lopez (Eds.), *Handbook of positive psychology* (pp. 106–119). Oxford: Oxford University Press.
Watt, L., & Wong, P. T. P. (1991). A taxonomy of reminiscence and therapeutic implications. *Journal of Gerontological Social Work, 16*, 37–57.
Weinstein, N., Ryan, R. M., & Deci, E. L. (2012). Motivation, meaning, and wellness: A self-determination perspective on the creation and internalization of personal meanings and life goals. In P. T. P. Wong (Ed.), *The human quest for meaning: Theories, research, and applications* (pp. 81–106). London: Routledge.
Weiss, R. S., & Bass, S. A. (2002). Introduction. In R. S. Weiss & S. A. Bass (Eds.), *Challenges of the third age: Meaning and purpose in later life* (pp. 3–12). Oxford: Oxford University Press.
Wheaton, B. (1994). Sampling the stress universe. In W. R. Avison & I. H. Gotlieb (Eds.), *Stress and mental health: Contemporary issues and prospects for the future* (pp. 77–114). New York: Plenum.

Whitaker, A. (2010). The body as existential midpoint—The aging and dying body of nursing home residents. *Journal of Aging Studies, 24*, 96–104.

White, C., Marin, M., & Fessler, D. M. T. (2017). Not just dead meat: An evolutionary account of corpse treatment in mortuary rituals. *Journal of Cognition and Culture, 17*, 146–168.

Wiesel, E. (1958). *Night*. London: Penguin Books.

Wilcox, S., Evenson, K. R., Aragaki, A., Wassertheil-Smoller, S., Mouton, C. P., & Loevinger, B. L. (2003). The effects of widowhood on physical and mental health, health behaviors, and health outcomes: The Women's Health Initiative. *Health Psychology, 22*, 513–522.

Wildschut, T., Sedikides, C., Arndt, J., & Routledge, C. D. (2006). Nostalgia: Content, triggers, functions. *Journal of Personality and Social Psychology, 91*, 975–993.

Wilkinson, P. J., & Coleman, P. G. (2010). Strong beliefs and coping in old age: A case- based comparison of atheism and religious faith. *Ageing and Society, 30*, 337–361.

Willems, D. L., Daniels, E. R., van der Wal, G., van der Maas, P. J., & Emanuel, E. J. (2000). Attitudes and practices concerning the end of life: A comparison between physicians from the United States and from The Netherlands. *Archives of Internal Medicine, 160*, 63–68.

Wilson, D. S. (1975). A theory of group selection. *Proceedings of the National Academy of Sciences of the United States of America, 72*, 143–146.

Wink, P. (2006). Who is afraid of death? Religiousness, spirituality, and death anxiety in late adulthood. *Journal of Religion, Spirituality and Aging, 18*, 93–110.

Wink, P., & Dillon, M. (2002). Spiritual development across the adult life course: Findings from a longitudinal study. *Journal of Adult Development, 9*, 79–94.

Wink, P., & Scott, J. (2005). Does religiousness buffer against the fear of death and dying in late adulthood? Findings from a longitudinal study. *Journal of Gerontology, 60B*, P207–P214.

Wolf, M. A. (1990). The crisis of legacy: Life review interviews with elderly women religious. *Journal of Women and Aging, 2*, 67–79.

Wong, P. T., Reker, G. T., & Gesser, G. (1994). Death attitude profile revised. In R. A. Neimeyer (Ed.), *Death anxiety handbook* (pp. 121–148). New York: Taylor and Francis.

Wong, P. T. P. (1989). Personal meaning and successful aging. *Canadian Psychology/Psychologie Canadienne, 30*, 516–525.

Wong, P. T. P. (2007). *Meaning management theory and death acceptance*. Downloaded on March 26, 2019, from http://www.meaning.ca/archives/archive/art_MMT-death-acceptance_P_Wong.htm

Wong, P. T. P. (2008). Meaning management theory and death acceptance. In A. Toner, G. T. Eliason, & P. T. P. Wong (Eds.), *Existential and spiritual issues in death attitudes* (pp. 65–87). New York: Erlbaum.

Wong, P. T. P. (2012). Introduction: A roadmap for meaning research and applications. In P. T. P. Wong (Ed.), *The human quest for meaning: Theories, research, and applications* (pp. xxix–xlvi). London: Routledge.

Wong, P. T. P. (2012a). Toward a dual-system model of what makes life worth living. In P. T. P. Wong (Ed.), *The human quest for meaning: Theories, research, and applications* (pp. 3–22). London: Routledge.

Wong, P. T. P. (2012b). From logotherapy to meaning-centered counseling and therapy. In P. T. P. Wong (Ed.), *The human quest for meaning: Theories, research, and applications* (pp. 619–647). London: Routledge.

Worden, J. W. (1982). *Grief counselling and grief therapy: A handbook for the mental health practitioner*. New York: Springer.

Wrosch, C., Scheier, M. F., Miller, G. E., & Carver, C. S. (2012). When meaning is threatened: The importance of goal adjustment for psychological and physical health. In P. T. P. Wong (Ed.), *The human quest for meaning: Theories, research, and applications* (pp. 539–557). London: Routledge.

Wu, A., Wang, J.-Y., & Jia, C.-X. (2015). Religion and completed suicide: A meta- analysis. *PLoS One, 10*, e0131715. https://doi.org/10.1371/journal.pone.0131715

Wu, A. M. S., Tang, C. S. K., & Kwok, T. C. Y. (2002). Death anxiety among Chinese elderly people in Hong Kong. *Journal of Aging and Health, 14*, 42–56.

Wu, P., Lorenz, K. A., & Chodosh, J. (2008). Advance care planning among the oldest old. *Journal of Palliative Medicine, 11*, 152–157.

Wynne-Edwards, V. C. (1962). *Animal dispersion in relation to social behaviour*. Edinburgh: Oliver and Boyd.

Yalom, I. D. (1980). *Existential psychotherapy*. New York: Basic Books.

Yeager, K. A., & Bauer-Wu, S. (2013). Cultural humility: Essential foundation for clinical researchers. *Applied Nursing Research, 26*(4). https://doi.org/10.1016/j.apnr.2013.06.008

Young, J., Egan, R., Walker, S., Graham-DeMello, A., & Jackson, C. (2019). The euthanasia debate: Synthesising the evidence on New Zealander's attitudes. *Kōtuitui: New Zealand Journal of Social Sciences Online, 14*(1), 1–21. https://doi.org/10.1080/1177083X.2018.1532915

Yur'yev, A., Leppik, L., Tooding, L. M., Sisask, M., Värnik, P., Wu, J., et al. (2010). Social inclusion affects elderly suicide mortality. *International Psychogeriatrics, 22*, 1337–1343.

Zacher, H., Rosing, K., & Frese, M. (2011). Age and leadership: The moderating role of legacy beliefs. *The Leadership Quarterly, 22*, 43–50.

Zahavi, A. (1975). Mate selection: Selection for a handicap. *Journal of Theoretical Biology, 53*, 205–214.

Zahavi, A., & Zahavi, A. (1997). *The handicap principle: A missing piece of Darwin's puzzle*. Oxford: Oxford University Press.

Zalsman, G., Hawton, K., Wasserman, D., van Heeringen, K., Arensman, E., Sarchiapone, M., et al. (2016). Suicide prevention strategies revisited: 10-year systematic review. *Lancet Psychiatry, 3*, 646–659.

Zhang, W. (2008). Religious participation and mortality risk among the oldest old in China. *Journal of Gerontology, 63B*, S293–S297.

Zhao, J., Barclay, S., Farquhar, M., Kinmonth, A. L., Brayne, C., Fleming, J., et al. (2010). The oldest old in the last year of life: Population-based findings from Cambridge city over-75s cohort study participants aged 85 and older at death. *Journal of the American Geriatrics Society, 58*, 1–11.

Zhou, X., Lei, Q., Marley, S. C., & Chen, J. (2009). Existential function of babies: Babies as a buffer of death-related anxiety. *Asian Journal of Social Psychology, 12*, 40–46.

Zimmer, Z., Jagger, C., Chiu, C.-T., Ofstedal, M. B., Rojo, F., & Saito, Y. (2016). Spirituality, religiosity, aging and health in global perspective: A review. *SSM - Population Health, 2*, 373–381.

Zisook, S., & Shear, K. (2009). Grief and bereavement: What psychiatrists need to know. *World Psychiatry, 8*, 67–74.

Chapter 9
Ageing in Evolutionary Perspective

Abstract *Ageing in evolutionary perspective* delves into an area of inquiry that is still in development but that it promises to be of great utility in the study of ageing. In this chapter we analyse several evolutionary aspects of longevity and ageing and aim to lay down the foundations for an *evolutionary gerontology*.

… old age has already conquered my once delicate skin
and my hair has turned white instead of black;

And my heart has grown heavy, and my knees will not support [me],
that once were nimble for the dance as young fawns.

I often lament these things; but what do I do?
Being human, it is impossible to be ageless.

For people used to say that the rose-armed Dawn,
overtaken by love, once carried off Tithonus to the world's end,

since he was handsome and young then, but in time grey age
overtook him, although he had an immortal wife.

Tithonus Song, Sappho (c. 630–c. 570 b.c.e.; Claudio 2018, pp. 107–108)

This is how the ancient Greek poet Sappho recounts the story of Tithonus, a story that first appeared in Homer's *Hymn to Aphrodite*. As we mentioned in Chap. 6, in a quote from Baltes and Smith (2003), Tithonus was granted immortality by Zeus but not eternal youth. Thus, he lived forever but suffering the ever-increasing decay of old age. Tithonus' story points to the difference between individual *longevity* (or *lifespan*): the duration of the life of an individual from birth to death, and *ageing* (or *senescence*) which is a decline later in life in the fertility and probability of survival of an individual (Partridge and Barton 1993). The distinction between longevity and ageing reminds us that the mere passage of time as expressed in the ticking of a clock, that marks chronological age, may not be related to the physical and psychological decay that are seen in biological senescence in a simple and straightforward manner.

M. Cordella, A. Poiani, *Fulfilling Ageing*, International Perspectives on Aging 30,
https://doi.org/10.1007/978-3-030-60071-6_9

We should also state at the outset that in the evolutionary literature on ageing the words *ageing* and *senescence* are usually regarded as synonyms: (e.g. Charnov 1993; Williams et al. 2006; Freeman and Herron 2007; Monaghan et al. 2008; Ricklefs 2008; Maklakov and Immler 2016). Here we will follow such tradition and therefore use both words interchangeably. The word *longevity* will be also used as a synonym for the *lifespan*.

The evolutionary biologist George C. Williams took the story of Tithonus as a metaphorical starting point for his evolutionary analysis of ageing, referring to it as *The Tithonus Error* (Williams 1999). Williams was of the opinion that senescence does not follow from the wearing-out or depletion of our biological structures (the *Wear-and-tear* hypothesis), but from their deficient maintenance or replacement. He argued that evolution must be biased in favour of youth over old age simply because death at a young age means a lower probability of having left any descendant, leading to the extinction of the lineage, whereas death at old age, after reproduction has taken place, would still allow the transmission of genes to the next generation. Thus, in this view, deleterious or semi-deleterious alleles may be expressed in old age with no consequence for their transmission. This assumes that (1) older people do not reproduce, which is the case for women post-menopause but, at least biologically, it is not necessarily so for all older men; and (2) that there are no indirect ways of transmitting own genes in old age, which, again, it is not true as we will see in the section *Kin selection and ageing* in this chapter, and as also noted by Williams. Williams also remarked that in most species the probability of living to old age is low in the wild due to ecological factors such as diseases, parasitism, predation, and starvation, thus there is usually no population of old age individuals upon which natural selection may exert an effect. This, however, is variable among species, as we will see in this chapter.

From an evolutionary point of view, ageing can be defined as "a persistent decline in the age-specific fitness components of an organism due to internal physiological deterioration" (Rose et al. 2012, p. 1; see also Rose 1991), where fitness, or more precisely *Darwinian fitness* is, from an organismal perspective, the lifetime production of reproductive offspring. Thus, ageing is the process that underlies the increasing physiological inability of individuals to stay alive and reproduce. *Darwinian fitness* can be encapsulated within the broader concept of *inclusive fitness* (Hamilton 1964) which involves two major components: *direct fitness* and *indirect fitness*:

$$\text{Inclusive fitness} = \text{Direct fitness} + \text{Indirect fitness}$$

Direct fitness is accrued through the production of own offspring, whereas indirect fitness is gained through helping genetically related individuals produce their own offspring. Therefore, increasing inclusive fitness involves increasing own direct reproductive success and helping increase the reproductive success of genetic relatives, in any combination.

Some evolutionary issues of ageing were addressed opportunistically in several of the previous chapters. In Table 1.2, we listed some evolutionary theories of ageing and in Chap. 5 we briefly mentioned adaptive aspects of emotions, whereas in Chap. 6 we mentioned evolutionary aspects of social behaviour. In Chap. 8 both death anxiety and religiosity/spirituality were also discussed taking into consideration their evolutionary context. Here we analyse the evolutionary aspects of ageing in a more detailed and systematic manner.

The research field of evolutionary gerontology is currently developing fast, so much so that Partridge and Gems (2006) have coined the term *evo-gero* to encapsulate the discipline, but the interest of evolutionary biologists in longevity and ageing is not new.

Rose et al. (2008, see also Rauser et al. 2009) briefly reviewed the development of evolutionary thinking on ageing since Charles Darwin. The authors indicated August Weismann as the first evolutionist to analyse senescence and longevity from the perspective of selection; he suggested that longevity was determined by "the needs of the species" (Weismann 1891, p. 9), a group-selectionist approach that has been subsequently discredited. However, Weismann did also provide a more traditional natural selectionist approach to old age by suggesting that hereditary aspects of old age could be transmitted to the next generation for as long as living longer is selectively advantageous.

The modern gene-based selectionist approach to old age was introduced by J.B.S. Haldane in the first half of the twentieth century. In his book *New paths in genetics* (Haldane 1941), referring to the age of onset of hereditary diseases, he wrote that:

> Above all, in many cases the age of onset is very variable, and this makes a great practical difference. It does not much matter if a man develops the first symptoms of 'creeping paralysis' at the age of sixty. A life is wrecked if a child does so at the age of six (p. 187).

Referring to Huntington's disease more specifically, a genetically heritable neurodegenerative disorder, Haldane stated that:

> The present age of onset of that disease may merely mean that primitive men and women seldom lived much beyond forty, so postponement of onset beyond this age had no selective advantage. If the unfavourable modifiers are not disadvantageous, they will spread until the disease becomes one of old age (p. 193).

Thus, Haldane argued that genes whose deleterious action is delayed until the post-reproductive ages can be maintained in the population and will not be wiped out by natural selection. This line of thinking was further developed by Peter Medawar who in his article *Old age and natural death* (Medawar 1946) argued that ageing evolved through selection of alleles that confer an advantage early in life, when the organism reproduces, but that can have pleiotropically harmful effects later in life, once the organism is not reproductive anymore. Alleles that express themselves at post-reproductive ages cannot be eliminated by natural selection. This has become known as the *Mutation accumulation* theory of ageing (see also Medawar 1952, and for more recent developments, see Maklakov et al. 2015). The argument is valid for as long as there are no consequences of the expression of such alleles on the *Darwinian fitness* of close relatives of the individual. If old individuals do help improve the reproductive success of close relatives, then alleles that negatively affect the performance and survival of old individuals could be eliminated by selection, resulting in directional selection towards an increased lifespan.

We already mentioned George C. Williams with reference to *The Tithonus Error*. In the 1950s, he stressed the trade-offs between current reproductive success and later survival in the evolution of ageing: natural selection would favour genes that may decrease survival over time, provided that they increase reproductive success earlier on. This process was later named *antagonistic pleiotropy* by Michael Rose

(1982; see also Abrams 1991; Leroi et al. 2005; Maklakov et al. 2017, for empirical evidence, and Gaillard and Lemaître 2017, for a review of Williams' influence on evolutionary thinking about senescence).

William Hamilton theoretically formalised the evolutionary aspects of ageing in his famous article *The moulding of senescence by natural selection* (Hamilton 1966). Hamilton provided a more precise theoretical framework to the arguments on ageing put forward by Haldane, Medawar, and Williams. Hamilton also contributed the idea that the mismatch between the long human lifespan and the end of effective reproduction (especially in women) could be maintained by natural selection if post-reproductive individuals help in the care of younger generations (e.g. grandchildren):

> It is, moreover, a matter of common knowledge that the post-menopausal woman normally remains a useful and healthy member of the community for some time. A woman does sometimes live to twice the age of her menopause… The vigour of the post-reproductive adult can be attributed to beneficial effects of continued survival on the survival and reproduction of descendants. In fact, the 15 or so years of comparatively healthy life of the post-reproductive woman is so long in itself and so conspicuously better than the performance of the male that it inevitably suggests a special value of the old woman as mother or grandmother during a long ancestral period… (Hamilton 1966, p. 37).

The process alluded to by Hamilton was subsequently named by John Maynard-Smith *kin selection* (Maynard-Smith 1964), and its specific extension to grandparenting by old women is now known as the *Grandmother hypothesis* (Hawkes et al. 1998). Traits can evolve through kin selection if they increase inclusive fitness via the specific effect of increasing indirect fitness.

Subsequent work carried out by Brian Charlesworth indicated that both antagonistic pleiotropy and accumulation of mutations could result in the evolution of ageing, due to the decreasing effect of natural selection at post-reproductive ages (Charlesworth 1980). Therefore, ageing is a by-product of those processes, rather than an adaptation. In the words of Stephen Stearns (1992, p. 200):

> Thus ageing should follow the onset of reproduction with widespread, diffuse erosion of physiological and biochemical functions caused by many genes of relatively small effect that produce ageing as a by-product, not as an adaptation.

The general decay coming with ageing would eventually decrease the direct fitness component of inclusive fitness, as reproduction decreases over time until it stops, but for as long as the individual remains alive and in reasonably good condition, the indirect component of his/her inclusive fitness could be increased. Old age then can become a "spandrel", an *exaptation*, upon which old people can "paint the images" of additional grandchildren (see Gould and Lewontin 1979 and Chap. 8).

Studies carried out in the 1990s on a diversity of organisms have also shown that mortality rates decelerate at very old ages. This may occur when late life does not involve the expression of deleterious alleles or the constant accumulation of genetic errors, thus in some organisms the expected continuous deterioration of physiological functions with age may not occur (Rose et al. 2012). In other words, groups of individuals who survive to older ages may do so because they are more robust and therefore their death rate would be lower (Vaupel et al. 1979, see also Charlesworth and Partridge 1997; Partridge and Mangel 1999). This is an important issue when studying the processes of ageing across species. As far as humans are concerned, however, even though the improvements of health, nutrition, and the effectiveness of medical interventions achieved in the broader human population over the past decades have been associated with an increase in the mean length of the lifespan (Vaupel 2010), Jeanne Calment's age record of 122 years and 164 days has not been officially broken since 1997 (see Chap. 4) and our ageing process is marred by obvious general decay.

Despite our hope to be smarter than Tithonus and therefore be able to extend our lifespan, but in good health conditions, Gawande (2015, p. 35) offers some sobering thoughts:

> I spoke to Felix Silverstone, who for twenty-four years was the senior geriatrician at the Parker Jewish Institute, in New York, and who has published more than a hundred studies on aging. There is, he told me, 'no single, common cellular mechanism to the aging process'. Our bodies accumulate lipofuscin and oxygen free-radical damage and random DNA mutations and numerous other microcellular problems. The process is gradual and unrelenting. I asked Silverstone whether gerontologists have discerned any particular, reproducible pathway to aging. 'No,' he said. 'We just fall apart'.

We have argued in section *Reaching very old ages* of Chap. 4 that available evidence suggests that there is a current limit to human longevity. There we mentioned recent work carried out by Leonid Gavrilov and Natalia Gavrilova. Gavrilov and Gavrilova have applied *Reliability theory* to the study of ageing and longevity. The theory has evolutionary implications, offering a complement to other evolutionary models of ageing that are consistent with Silverstone's "falling apart" mechanism (Gavrilov and Gavrilova 2001, 2003, 2004, 2006, 2011).

Reliability theory (e.g. Birolini 2007) was originally developed within the field of engineering with the objective of modelling systems *failure*—defined as "the event when a required function is terminated" (Gavrilov and Gavrilova 2006, p. 45), given a specified structure of a system and reliability of its components. Failures can be either due to *degradation* of a system, in which case some of the system's components do not function properly anymore, or they can be *catastrophic* or *fatal failures*, in which case the component's or the entire system's life abruptly comes to an end (Gavrilov and Gavrilova 2006). Catastrophic failures do occur with ageing in some vertebrate species (Ricklefs 2008).

Gavrilov and Gavrilova (2001, p. 527) provided a succinct summary of the theory:

> … even those systems that are entirely composed of non-aging elements (with a constant failure rate) will nevertheless deteriorate (fail more often) with age, if these systems are *redundant* in irreplaceable elements. Aging, therefore, is a direct consequence of systems redundancy. Reliability theory also predicts the late-life mortality deceleration with subsequent leveling-off, as well as the late-life mortality plateaus, as an inevitable consequence of *redundancy exhaustion* at extreme old ages. The theory explains why mortality rates increase exponentially with age (the Gompertz law) in many species, by taking into account the *initial flaws* (*defects*) in newly formed systems…The theory explains why relative differences in mortality rates of compared populations (within a given species) vanish with age, and mortality convergence is observed due to the exhaustion of initial differences in redundancy levels. (italics original).

Therefore, according to *Reliability theory*, ageing is not just the passing of chronological time but the degradation of the body, resulting into functions failure ("we just fall apart"), which is expressed in terms of adverse health outcomes, frailty, and finally death. Despite the accumulation of body elements failures with ageing, the individual may still remain alive for a period of time due to system redundancy, but with accumulating probability of mortality over time, until a point is reached when redundancy is exhausted, and mortality becomes maximal, reaching a plateau (e.g. Pletcher and Curtsinger 1998). This pattern had been already envisaged by Benjamin Gompertz in the early nineteenth century (Gavrilov and Gavrilova 2001), although recent studies of human mortality may not support the idea of a mortality plateau (Gavrilova and Gavrilov 2019). In this theory, differences between individuals in a population and differences between species can be explained by structural differences in the redundancy of body components. Redundancy prevents the individual from immediately "falling off a cliff" as soon as it reaches an age when reproduction ceases; and given that redundancy is lost with ageing, mortality rates of different populations or even species become less different at older ages.

As an obvious corollary to the association of ageing with loss of redundancy over time in biological systems, it is expected that individuals who lose system redundancy at early ages, e.g. due to disease, are comparatively more likely to die younger (Gavrilov and Gavrilova 2004).

Reliability theory points to the tear and wear of material systems, and from a genetic perspective it has been also referred to as the *Damage accumulation hypothesis* (Charlesworth 2000).

Thus, *Reliability theory* adds a structural element to the evolutionary approach to ageing by identifying factors that can influence longevity: initial redundancy levels in the functions of the organism, rate of redundancy loss over time, and repair potential that can slow down the ageing process (Gavrilov and Gavrilova 2004). The theory is also consistent with the antagonistic pleiotropy theory of ageing and it has received some support from empirical studies (Boonekamp et al. 2015).

In this chapter we explore various evolutionary aspects of old age in more detail. We start with a review of the evolution of lifespan to then focus more specifically on the ageing process. We also devote a specific section to sex and ageing, to be followed by a review of ageing and kin selection. We conclude this chapter with an analysis of the evolutionary process of neoteny and its potential relevance to human ageing. We hope that a critical review of the published literature on evolution and ageing may help clarify the foundations for an *evolutionary gerontology* or *evo-gero*.

9.1 Evolution and Lifespan

> It is indeed remarkable that after a seemingly miraculous feat of morphogenesis a complex metazoan should be unable to perform the much simpler task of merely maintaining what is already formed
>
> George C. Williams (1957, p. 398).

The lifespan or longevity is simply the time interval between birth and death. Lifespan is a life-history variable that describes a crucial aspect of the life of an organism. Other life-history variables include senescence, age at maturity, age at last reproduction, juvenile survival, adult survival, fertility, size at birth, growth pattern, size at maturity; number, size, and sex of offspring, and more (e.g. Stearns 1992).

Williams' conundrum points to the question of why has not immortality evolved in metazoan organisms? To respond to this question, evolutionary biologists have offered a basic argument stating that organisms will be affected by life events, accidents, predation, parasitism, diseases, environmental fluctuations in some crucial factor (e.g. temperature), food availability that may eventually kill the organism no matter what. Hence, to quote Kirkwood (1987, p. 213):

> There is therefore little advantage to be gained from investment in potential somatic immortality when in practice the return from this investment may not be realized.

However, organisms can still transmit their genes to future generations through reproduction, and the process can be sustainable even at the expense of shortening the lifespan or undergoing senescence as a result.

Our species went through a period of significant increase in longevity throughout most of its evolution, but especially since the Early Upper Palaeolithic (the Upper Palaeolithic period spans from 40,000 to 10,000 years before present) (Straus 1995; Caspari and Lee 2004). Longevity can evolve if there is genetic heritable variation for it (e.g. Fox et al. 2011).

We have mentioned in previous chapters (Chaps. 1 and 4) that there is some heritable variation in human populations to survive to old ages, such heritability tends to be lower than 50%, usually ranging between 20 and 30%, but with increasing values at older ages: 36% after the age of 65 and 40% after the age of 85 years old. Heritability of the lifespan is about 25% in identical twins (Herskind et al. 1996; Finch and Tanzi 1997). Estimates of gender-specific heritability of longevity are very variable, however, as indicated in Chap. 4.

Therefore, given that there is heritable variation for longevity, the evolution of human lifespan could proceed through natural selection when the *response to selection* (R) takes positive values ($R > 0$), given that *heritability* (h^2) is larger than zero, and there is variability in the population in terms of lifespan (*selection differential*, S), following the equation (e.g. Maynard-Smith 1989):

$$R = h^2 S$$

Some selective conditions that favour a longer lifespan include low environmentally caused mortality rate and a higher tissue-repair efficiency (Cichoń 1997).

A common approach used to study the evolution of life-history variables such as the lifespan, is the so-called comparative methods. For an introduction to such methods the reader may like to consult Garamszegi (2014). In short, comparisons are carried out among taxa (e.g. a group of species) with regard to a number of variables of interest, in order to test evolutionary hypotheses whilst taking into account the evolutionary relationships among those taxa, as expressed in a phylogenetic tree, for instance. Similarities and differences among taxa in various traits may be due to phylogenetic relationships (close or distant), convergent/divergent adaptive evolution, drift. Comparative methods aim at unravelling those different evolutionary possibilities.

The evolution of the lifespan in primates, including humans, has been studied by several authors using comparative methods. Judge and Carey (2000) used a simple comparative approach that partially controls for phylogenetic effects, by regarding genera as evolutionarily independent datapoints. The value of continuous traits, such as the lifespan, is calculated for each genus as the arithmetic mean of the values for congeneric species. This is known as the *higher node method* in comparative analyses (Harvey and Pagel 1991), and even though it is much better than just using the values of traits for species—Harvey and Pagel also provide a statistical criterion to determine what higher node to use—, it does not fully solve the problem of lack of independence among the data used to test evolutionary hypotheses. Nonetheless, Judge and Carey compared values of lifespan, body size, and brain size (all log-transformed) for 36 genera of anthropoid primates and 16 genera of prosimians (data are from females) and found that apes in general display lifespans that tend to exceed the expected lifespan from their body mass and brain mass. In addition, residual longevity after the effects of both body mass and brain mass are controlled is greatest in humans and in the genus *Cebus*. This analysis predicts a human lifespan of 72–91 years (Gurven and Kaplan (2007) suggest a modal adult lifespan of 68–78 years) and it indicates that primates in general are characterised by long lives, late age at first reproduction, and low fertility (Bronikowski et al. 2011, Holland Jones 2011).

Humans, however, do show some peculiarities compared to other primates. Bronikowski et al. (2011) indicated that humans are among the primates with the lowest combined values of risk of death at onset of adulthood and rate of increase in age-specific mortality with advancing age. Alberts et al. (2013) carried out a comparative analysis using long-term, individual-based data on reproductive lifespan in natural populations of primates, including a population of humans experiencing natural fertility and mortality (the*!Kung*, also known as *Ju/'hoansi*), and showed that female humans live much longer beyond the age at last reproduction than females of other primates (see Fig. 9.1), as also suggested by Judge and Carey's work, which also used data for females. As remarked by Alberts et al. (2013, p. 13443): "among the nonhuman primate species, 1–6% of adult-female-years are lived by postreproductive individuals, whereas in the *!Kung* this metric reaches 42.5%". A similar life-history pattern of longevity, that allows females to survive beyond the age when their ability to reproduce is nil due to menopause, has been described in natural populations of killer whales (*Orcinus orca*) by Ward et al. (2009, see also Brent et al. 2015; Croft et al. 2017).

Lower mortality and an extended lifespan beyond reproduction have likely co-evolved with sociality, brain complexity, and learning capabilities in the human species, as suggested by a comparative analysis of non-human primates carried out by Street et al. (2017) using a more sophisticated comparative analysis approach (i.e. Bayesian phylogenetic generalised linear mixed models; Hadfield 2010).

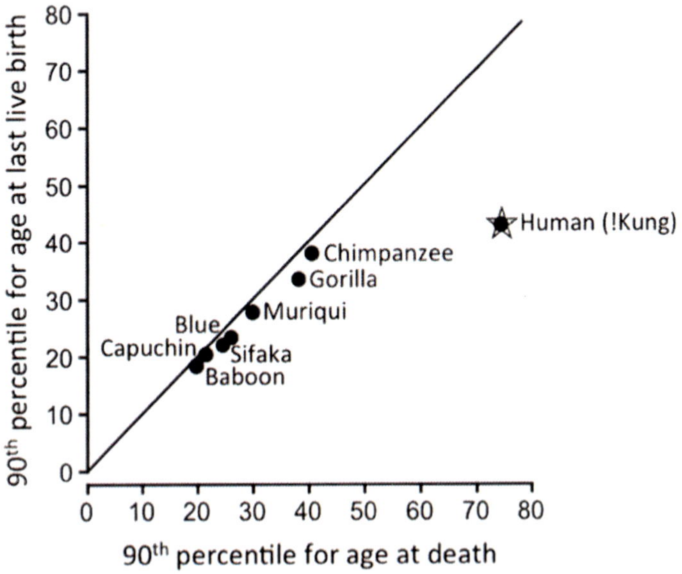

Fig. 9.1 The 90th percentiles of the distributions for age at death (in years) are plotted against the 90th percentiles of the distributions for age at last live birth (in years). The line of identity between the two distributions is represented by the black line. Non-human primates fall very close to the line of identity, not so for the *!Kung* population of humans (reproduced, with permission, from Alberts et al. 2013)

Sociality in particular does not just offer greater protection to the individual, thus increasing survival and longevity, but also access to valuable food items, such as large preys, that are shared by the members of the group and that can be better found and defended if in scavenging mode, or captured and killed if in hunting mode (Kaplan et al. 2000), and it can also provide learning opportunities to the young (Gurven et al. 2006).

Sociality, brain complexity, and learning capabilities also shift the physiology of organisms more towards investment in maintenance and an extended lifespan, what in ecology was traditionally known as *K*-selection (in contrast to the short lifespan, high reproductive rate of *r*-selected populations). *r*/*K* selection has now been enriched with other views of life-histories such as a focus on age-specific mortality (Reznick et al. 2002) and *CSR* selection, that focuses on the specific selective pressures experienced by populations: competition (C-selection), stress tolerance (S-selection), and regeneration (R-selection) (Cassill 2019).

Organisms in populations subject to low extrinsic mortality (e.g. low competition, low predation, lower environmental stress, low levels of infections) are expected to invest more on the maintenance of body functions that will increase longevity, thus pacing the reproductive effort over a longer lifespan and slowing down the rate of ageing. This is, in a nutshell, what Thomas Kirkwood introduced as the *Disposable soma* theory (Kirkwood 1977, 1981), which states that:

> longevity is determined through the setting of longevity assurance mechanisms so as to provide an optimal compromise between investments in somatic maintenance (including stress resistance) and in reproduction. A corollary is that species with low extrinsic mortality are predicted to invest relatively more effort in maintenance, resulting in slower intrinsic ageing, than species with high extrinsic mortality (Kirkwood et al. 2000, p. 587).

Therefore, the costs of maintaining the germline functional for as long as possible can be selectively advantageous even at the expense of somatic senescence later in life (Maklakov and Immler 2016) in organisms with low mortality rates and therefore an extended lifespan.

The more the organism is capable of repairing internal damage to body parts and functions, that is caused by external factors, including those causing stress, the longer its lifespan is expected to be. Kirkwood et al. (2000) tested this hypothesis from an evolutionary perspective by carrying out a comparative analysis of stress resistance in primary skin fibroblasts from rat, hamster, marmoset, rabbit, sheep, pig, cow, and human and confirmed that cells from long-living species are more resistant to stressors, as expected from the *Disposable soma* theory.

The stress response can be improved through calorie restriction, which may slow down the process of ageing, and extend the lifespan (Harrison and Archer 1989; Kirkwood et al. 2000; Haigis and Yankner 2010). General food restriction upregulates mechanisms that protect against somatic damage, thus increasing longevity (Kirkwood and Shanley 2005); this can be regarded as a straightforward adaptation to survive periods of food scarcity, an eventuality that a long-living individual is more likely to experience. Food restriction may also lead to decreased reproduction (e.g. Martin et al. 2008). Thus, under environmental conditions of food scarcity leading to restriction in food intake, life-history patterns shift from (1) reproduction now, to (2) survival now and reproduction later.

More recently, Kyriazis (2019) has further refined the *Disposable soma* theory by offering the *Indispensable soma* hypothesis which states that lower and delayed reproduction in humans is an evolutionary result of reallocation of resources to cognitive mechanisms, somatic structures, and repair of such structures that are needed for survival over a long period of time. Kyriazis especially emphasises the possibility of an important trade-off between neurons and germline components.

Caleb Finch more specifically focused on those evolutionary aspects of the lifespan that are affected by infections, nutrition, and the activity of the immune system as expressed in the inflammatory response (Finch 2010). As seen in Fig. 9.2 mortality is not only lower, and therefore the percentage of survivors is higher in human populations living traditional lives in pre-industrial conditions than in wild chimpanzees, but mortality also starts accelerating in humans in old age about 30 years later than in chimpanzees, which is consistent with Fig. 9.1. A longer lifespan in humans is helped by an efficient immune system that protects us against infections and diseases, but that it can also become immunosenescent over time—as seen in the depletion of the limited number of naïve T-cells, for instance—, being negatively affected by chronic diseases that have inflammatory consequences. The evolution of an active immune system that contributes to extending the lifespan may, therefore, have the consequence of making life in old age problematic if the immune

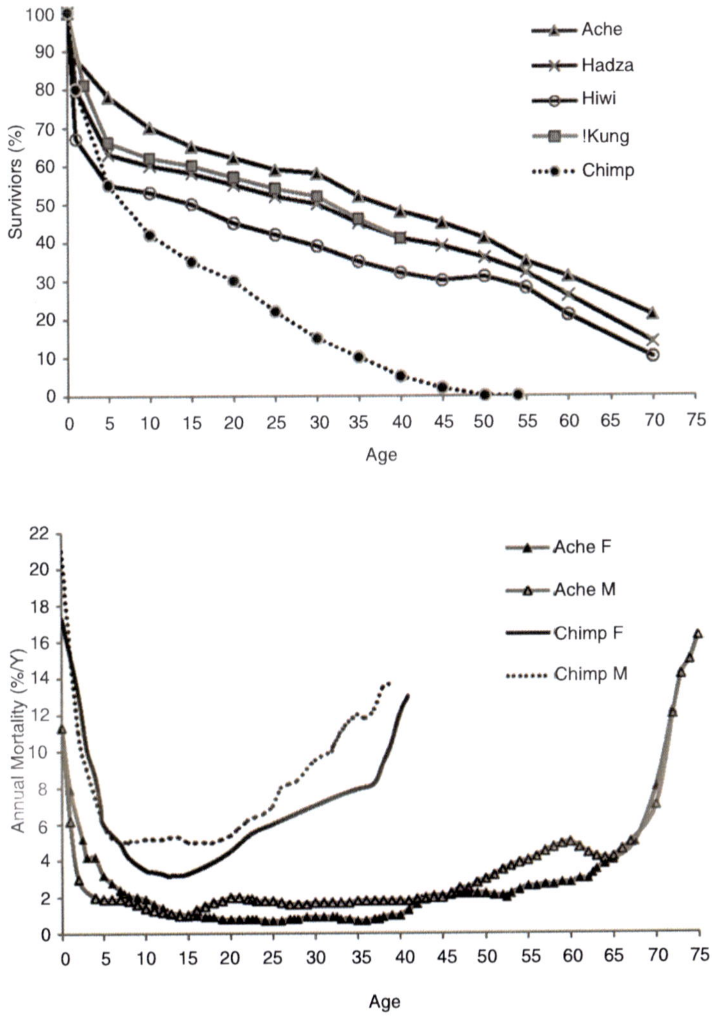

Fig. 9.2 Demographic comparisons of wild chimpanzees with human populations living traditional lives in pre-industrial conditions and with little access to medicine. (Top) Survival curves. (Bottom) Age-specific mortality. At all ages, chimpanzees have higher mortality than the Aché and lower survival than the Aché, Hazda, Hiwi, and Kung, and they show acceleration of mortality about 30 years earlier (reproduced, with permission, from Kaplan et al. 2000)

system becomes dysregulated in old age, thus causing chronic diseases of inflammatory origin (Finch 2010, 2012). This may explain the evolution of genes such as the apolipoprotein E (*ApoE*) that we mentioned in Chap. 4, that can modulate chronic inflammation and various aspects of ageing in the brain and arteries and that could increase human lifespan in some of its forms (e.g. *ApoE2*), but with other alleles contributing to the onset of disease, such as dementia, at older ages (e.g. *ApoE4*). Raichlen and Alexander (2014) have suggested that humans were able to survive to older ages despite the negative effects of alleles such as *ApoE4* later in life, thanks to their high levels of physical activity (aerobic exercise). Physical activity may enhance neurogenesis and decrease amyloid deposition in the brain. That for *Homo sapiens* physical activity has been an important aspect of our evolutionary history is easily seen from even a cursory analysis of our anatomy. Our bipedalism is a good adaptation that not only frees our hands to manipulate various aspects of our environment, and it gives us a better 360° vigilance ability, but it also allows us to efficiently walk over long distances to find food (e.g. pursuing wounded large preys), water, and to disperse.

We mentioned in Chaps. 1 and 4 that shortening of telomeres in our DNA is associated with ageing (see also Milewski 2010). Shorter telomeres are also associated with higher mortality and therefore a shorter lifespan, but the relationship between telomere length and mortality diminishes in older ages (Boonekamp et al. 2013). The association between telomere length and lifespan is widespread among various species (Haussmann et al. 2003) and Tricola et al. (2018) have recently found that longer-living bird species lose fewer telomeric repeats over time compared with species with shorter lifespans, suggesting that there are evolutionary factors that may modulate telomeric length. At the cellular level, a long lifespan is associated not only with longer telomeres, but also with less overall DNA damage, lower mitochondrial deterioration, and less oxidative stress.

Finally, we would like to mention studies that have associated phenotypic plasticity—which in humans is clear in the plastic capacities of our nervous system, for instance—with increased longevity. It is usually assumed that the evolution of longevity should drag the evolution of plasticity, as longer-living organisms are more likely to experience variable environmental conditions (e.g. Gonzalez-Lagos et al. 2010). On the other hand, in plants it seems that it is greater plasticity that leads to greater longevity, as plastic individuals have more capacity to stay alive longer despite changing environmental conditions (e.g. Borges 2009). In the end, it seems more plausible to think that both lifespan and plasticity can coevolve (Ratikainen and Kokko 2019).

In conclusion, humans' long lifespan is a trait broadly shared with other anthropoid primates, but unlike our closest living relatives, human longevity far exceeds that expected from our body and brain mass, presumably as an evolutionary consequence of sociality, brain complexity, capacity of repairing body damage and low extrinsic mortality.

In the next section we move beyond longevity to focus on the process of senescence.

9.2 Evolution and Senescence

In an influential article published in 1954, Alex Comfort (1954) put forward the idea that senescence (or ageing) results from the combined action of the cumulative effects of injuries incurred throughout life and the growing incapacity of the organism to repair such damage, plus the effect of accumulating metabolites that may interfere with physiological processes over time, and the "exhaustion of stored irreplenishable reserves" (p. 290) of important metabolites used in vital processes. He summarised these ideas in his concept of *morphogenetic senescence*:

> This senescence appears to arise directly from the operation of the processes which determine shape and size characteristic of the species; it expresses itself as a decline in the capacity to regenerate or maintain structures or conditions which during growth and a post-growing period of variable duration, are normally regenerated and maintained. Morphogenetic senescence is a major, cumulative failure of homeostasis, to which coincident or dependent mechanical failure or accumulative processes may contribute, but which appears to be continuous with the processes which control cell-differentiation and regulation. It is particularly well seen in man. The chief evidence that this morphogenetic senescence is more than the 'sum of environmental insult' which was formerly invoked to explain it, is the existence in many organisms of specific age, analogous to specific size and apparently related to it, and displaying little environmental, but market inter-race and interspecific, variation (p. 291).

Therefore, senescence is a process that is ingrained in our biology and it is also one that eventually leads to death. How can senescence be able to withstand the effects of natural selection which should theoretically oppose it? Old group-selectionist views of senescence, that regarded it as an adaptation that benefits the population: removing the old to benefit the young (e.g. Weismann 1882; Emerson 1960), were speedily dismissed by Williams (1966). However, Williams could not conceive of senescence as a "selectively irrelevant" process and he proposed an alternative mechanism:

> I shall assume initially, therefore, that senescence is an unfavorable character, and that its development is opposed by selection. To account for its prevalence, therefore, it is necessary to postulate another force that favors its development in such a way that the observed variations in senescence reflect variations in the balance between these two forces. I believe that this other force is an indirect effect of selection, and results from the selection of genes that have different effects on fitness at different ages (Williams 1957, p. 399).

> Obviously a gene that conveyed a slight increase in fitness with a high *p*-value might be favorably selected in spite of a serious decrease in fitness with a low *p*-value. This, I believe, is the key to understanding the evolution of senescence, because any genic effect that arises late in life will have a low *p*-value. Natural selection may be said to be biased in favor of youth over old age whenever a conflict of interests arises (p. 401).

Therefore, in Williams' view ageing results from negative pleiotropic effects of genes in old age, with those genes being selected due to their positive effects on fitness in youth; this is an idea that was first proposed by Peter Medawar (1952). This antagonistic pleiotropic mechanism of ageing and the age-specific effects of genes on fitness are major evolutionary views on senescence (Rose and Graves 1989; see also Partridge 2001).

Senescence affects and is affected by multiple aspects of our biology at the molecular level such as cellular detoxification, autophagy, antioxidant defences, role of heath shock proteins, role of reactive oxygen species and mitochondria, and more (Bowles 1998; Partridge and Gems 2006).

Thomas Kirkwood further developed the views on old age put forward by Comfort, Williams, and Medawar stressing the molecular mechanisms that could underlie the ageing process:

> …there is good reason to believe that ageing may be due primarily to a breakdown in the accuracy of macromolecular synthesis within cells and we shall consider the implications of this theory for the evolution of longevity (Kirkwood and Holliday 1979, p. 536).

Starting from a suggestion first put forward by Leslie Orgel in 1963 and then amended in 1970 (Orgel 1963, 1970) whereby accumulated molecular errors over time could lead to senescence either in a continuous or a catastrophic manner (see also Laird and Sherratt 2009), Kirkwood proposed that although accumulation of molecular errors is a possibility, there are molecular mechanisms (e.g. proofreading and scavenging enzymes) that can correct such molecular errors and that can therefore slow down the processes of senescence. The longer living the organism is, the more effective such molecular repair mechanisms are expected to be and therefore the later in life the evidence of senescence should be in such organisms (Kirkwood 2005).

Kirkwood's *Disposable soma theory* of senescence states that species will evolve an optimum level of accuracy in their molecular processes and repair capacity:

> If errors were too frequent, it would be impossible to develop, sustain and reproduce a complex organism. On the other hand, a completely error-free organism, with no possibility of mutation, would no longer be capable of evolution. In the initial evolution of increasing accuracy the law of diminishing returns must apply and there will always be an optimum point where there is no further advantage in investing energy and cellular resources to reduce the error levels. Since organisms have very different environments and life styles we would certainly expect them to evolve to different optimum error levels, and our basic hypothesis is that these will be directly related to the time the species must survive in the natural environment in order to reproduce itself successfully (Kirkwood and Holliday 1979, p. 544).

Thus, from an evolutionary perspective, although errors can put a limit to efficiency, they can open up the possibility for adaptation and therefore survival under changing environmental conditions. We encountered the same principle in Chap. 4 when we addressed the issue of brain plasticity: although a plastic nervous system can lead to errors, it is through nervous system plasticity that novel behavioural solutions to the challenges of survival and reproduction can fast emerge.

Beyond such age when reproduction is no longer feasible, accumulation of molecular errors can cause senescence without risking the possibility of extinction. This point was further developed by Kirkwood and Rose in their article *Evolution of senescence: Late survival sacrificed for reproduction* (Kirkwood and Rose 1991):

> the optimum level of investment in somatic maintenance for an iteroparous species is less than the level required for indefinite survival. In other words, natural selection should favour a strategy that results in the progressive accumulation of unrepaired somatic defects, which in turn must result in senescence (p. 17).

Such optimum level of investment in somatic maintenance that is lower than what is required for indefinite survival is simply the result of a trade-off between survival and iteroparous reproduction (see Fig. 9.3). Excessive investment in somatic maintenance and therefore decreased senescence (non-senescing side of Fig. 9.3) will

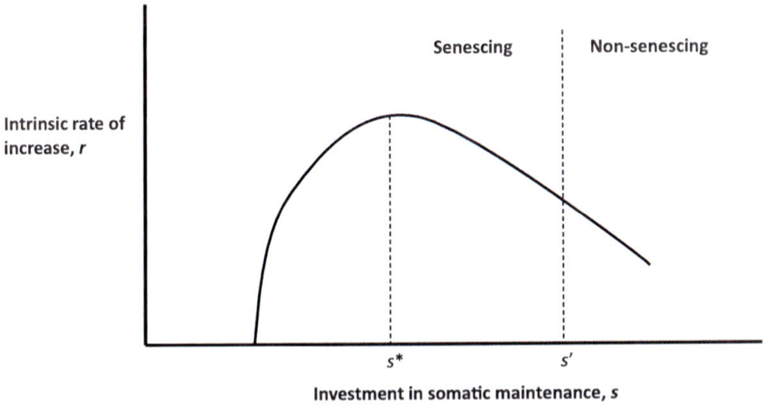

Fig. 9.3 Relation between fitness, measured as intrinsic rate of population increase r, and level of investment in somatic maintenance s. For $s > s'$, senescence does not occur due to the high investment in somatic maintenance. The optimum $s = s^*$ is predicted to be always less than s' given that reproduction compromises somatic maintenance and therefore survival, whilst reproduction decreases the chances of extinction (redrawn from Kirkwood and Rose 1991)

detract resources otherwise devoted to reproduction. If reproduction (r in Fig. 9.3) does not occur often enough, the death of the individual due to environmental factors may mark the extinction of the lineage. The optimal investment in somatic maintenance is expected to maximise r, at the cost of stabilising senescence ($s^* < s'$).

Somatic maintenance is required due to the natural wear and tear of organic tissues. Investment in somatic maintenance also links back to an original idea put forward by Leo Szilard who suggested that ageing is a process of accumulation of somatic mutations (Szilard 1959) resulting from the decreasing capacity of the organism to repair such mutations over time. This process may also play an important role in ageing (Kirkwood 1989, 1997; see also Partridge and Gems 2006; Kenyon 2010). As stated by Kirkwood (2002, p. 744): "we are not programmed to die, merely insufficiently programmed to survive".

Therefore, the pattern of senescence observed within a species, although somewhat variable among individuals, can be broadly described through a general process. The damage accumulated over time implicit in the *Disposable soma theory* (see also Gavrilov and Gavrilova's *Reliability theory* and Charlesworth's *Damage accumulation hypothesis* introduced above in this chapter) is likely the result of a combined effect of genetic and environmental influences, that can explain both the broader similarities and the individual differences in the patterns of ageing observed across individuals within and between species (Gladyshev 2013).

The somatic mutations mentioned by Szilard that lead to senescence can be caused by the accumulation of free radicals, especially the reactive oxygen species (ROS) that we mentioned in Chap. 4. ROS accumulate with age causing negative effects on physiology such as oxidising membrane lipids, changing the structure of proteins, damaging nucleic acids, causing deficits in synaptic plasticity, and neurodegenerative disorders (Beckhauser et al. 2016).

The specific role of free radicals as causative agents of ageing was initially suggested by Denham Harman in his *Free radical theory of ageing* (Harman 1956). This theory proposes that the physiological aspects of ageing are caused by the cumulative effects of free radicals over time (see also Bonsall 2006 and Chap. 1). In turn, Bonsall (2006) linked Harman's theory to the *Rate of living theory* (Loeb and Northrop 1916; Pearl 1928) that suggests that shorter lifespans are associated with higher rates of energy expenditure and therefore higher oxygen consumption, and oxygen provides the element for the production of ROS; hence the idea of extending the lifespan through calorie restriction that would decrease expenditure of energy and reduce the accumulation of ROS (see also Calder 1985).

Given that mitochondria display an array of important cellular functions in energy and ROS production, homeostasis, iron processing, steroid synthesis, apoptosis (e.g. Murphy 2009; Gonzalez-Freire et al. 2015), they have been a special focus of research on ageing processes (e.g. Wallace 2005; Gonzalez-Freire et al. 2015). The mitochondrial-DNA (mtDNA) has been associated with various aspects of ageing (Zapico and Ubelaker 2013; Swerdlow et al. 2017; Haas 2019; Reynolds et al. 2020), and some specific mtDNA haplotypes have been also linked to higher longevity in various studies carried out on Italian, Irish, Finnish, and Japanese centenarians (Wallace 2005). Such high longevity results from better protection against diseases, including neurodegenerative diseases.

Consistent with the *Disposable soma theory*, mitochondria fulfil a biological role in providing the energy to repair damaged biological functions that if left unrepaired could lead to senescence. Senescence is characterised by progressive mitochondrial dysfunction which affects especially tissues that require high energy: brain, heart, skeletal muscle, kidney, liver (Gonzalez-Freire et al. 2015). The hypothesis that mitochondria have an important role in the process of senescence has withstood the test of time (Gonzalez-Freire et al. 2015).

Insulin-like growth factor-1 (IGF-1) has also been the focus of evolutionary and physiological studies on senescence. IGF-1 mediates the action of growth hormone (GH) in mammals, thus controlling the organism's growth via effects on cell reproduction and negative effects on apoptosis. Reduced GH in elderly people contributes to ageing symptoms such as loss of muscle mass, increased adiposity, decreased bone mineral density, and lower energy levels, among others (Bartke et al. 2003). The role of IGF-1 on ageing and lifespan extension seems to be conserved across invertebrates and vertebrates (van Heemst 2010).

Genes that increase resistance to stress such as those encoding for heat shock proteins and detoxification systems can also decrease the level of senescence (e.g. Ricklefs 2008, and references therein).

Kirkwood (2005) suggested a network approach to understand the interconnected molecular mechanisms of senescence. For instance, mtDNA mutations accumulated over time (e.g. years) might lead to a steady increase in ROS production and a gradual decline in energy production, both of which may contribute to the death of the cell and therefore tissue malfunction, thus leading to organismic senescence.

If senescence is characterised by an increase in the probability of death over time, we may as well describe the periods in the human lifespan when the death rate decreases, as periods of "negative senescence" (Vaupel et al. 2004), or at least "negligible senescence" (Finch 1998, 2007, 2009) (see Fig. 9.2, bottom). Vaupel et al. (2004, p. 339) define *negative senescence* as a "decline in mortality with age after reproductive maturity, generally accompanied by an increase in fecundity". Negligible senescence, however, is expected to carry a large long-term cost resulting from the side effects of activating molecular mechanisms that simultaneously increase survival and the reproductive effort (e.g. Kirkwood and Holliday 1979); therefore, negligible senescence is more likely to characterise animal species with determinate growth that are very short-lived, or relatively short periods in the lifetime of long-living species such as humans.

In sum, our current understanding of senescence suggests that it results not just from the accumulation of tissue insult and redundancy exhaustion over time, but also from the incapacity of the organism to continuously repair accumulated molecular and cellular damage. The process is affected by our genetic makeup and environmental factors, displaying some general patterns across individuals, but also inter-individual differences. Senescence can persist, thus escaping natural selection, for as long as it is expressed during post-reproductive periods, and mechanisms that cause senescence later in life have positive effects earlier on in terms of increased reproduction. This may explain why there is no selective advantage in investing energy into continuously repairing tissue damage over time and therefore eliminate senescence, for as long as such investment detracts from the capacity of the organism to reproduce. However, the longer and more effectively such damage is repaired, the lower the level of senescence of the organism will be over time, thus an optimum age at senescence is expected. The specific molecular mechanisms involved in senescence that can be affected by evolutionary processes include accumulation of somatic mutations, the effects of ROS, stress, IGF-1, mitochondrial function among others. Calorie restriction, on the other hand, may help in slowing down the rate of ageing.

In the next section we focus on evolutionary aspects of inter-sexual differences in ageing.

9.2.1 Evolution, Sex, and Ageing

It is well known that women currently live longer than men (see Chap. 2). On this ground alone, one would expect that, from an evolutionary and adaptive perspective, ageing would differ between human males and females; but other factors also contribute to sex-specific ageing. For instance, the biological costs of reproduction are more taxing on the female body than on the male body, from this we would expect females to behave more like *K*-strategists, which involves a higher emphasis on investment in body maintenance, an extended lifespan, and a reproductive effort represented by a small number of offspring per pregnancy. Therefore, females are expected to compete for resources such as food, shelter, and so forth. Males, on the other hand, produce relatively cheap sperm which may predispose them to adopt an *r*-strategy, involving a comparatively shorter lifespan, higher reproductive rate, and competition to access mates. Therefore, we may expect males to age faster than females, following the trade-off between investment in development (and repair of somatic structures/functions) and reproductive effort suggested by the *Disposable soma theory*. That is, males are expected to conform to the so-called "live fast, die young" life-history strategy (O'Dor and Webber 1986; Promislow and Harvey 1990).

Although the suggested "*r*-strategy male" vs "*K*-strategy female" life-history patterns have been observed in nature across various taxa (Bonduriansky et al. 2008 and references therein), the scenario is somewhat more complex in both humans and other species.

Bonduriansky et al. (2008) reviewed the role of sexual selection and sexual conflict in the evolution of ageing, concluding that in addition to the *r*-*K* strategies mentioned above, ageing may be differentially expressed in the sexes in response to other factors such as the mating system: greater similarity in ageing patterns between males and females may be expected in socially monogamous species with biparental care. In sex-role reversed species, a situation that has been observed in some bird and other vertebrate species, it is females who conform to an *r*-strategy pattern, whereas males fit the expectations of a *K*-strategy life history. In addition, males may be under strong selection to increase lifespan and delay senescence if they can only reproduce after reaching a high social status that requires the development of a strong body and advanced skills (e.g. Graves 2007).

Bonduriansky et al. (2008) have proposed a model linking together lifespan duration, ageing, and *sexual selection*. Sexual selection is the selection of traits that increase the capacity to attract mates, copulate, and produce offspring. Bonduriansky et al.'s model is summarised in Fig. 9.4. Overall, there is a general ageing trend towards decreasing both reproductive rate and the size of secondary sexual traits in older individuals. However, (Fig. 9.4a) after considering environmental variation in extrinsic mortality risk: if life expectancy is brief (illustrated by vertical line I), then individuals that invest more heavily in reproduction in early life (solid line) will achieve higher

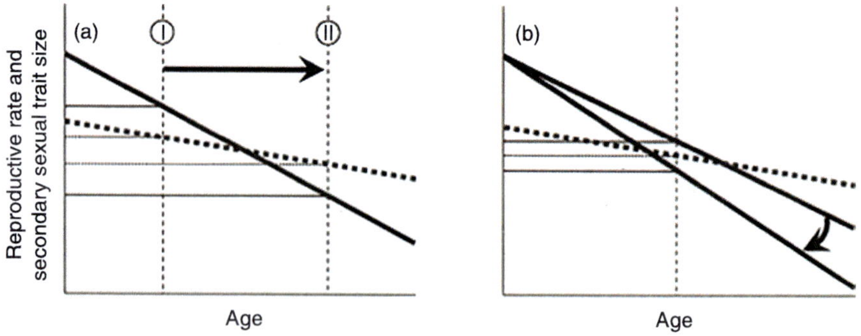

Fig. 9.4 Ageing can mediate the effects of environmental factors on sexual selection when phenotypes vary in lifespan or ageing rate. Reproduced, with permission, from Bonduriansky et al. (2008)

lifetime fitness than individuals that invest less in reproduction (dotted line), and the traits that mediate variation in reproductive rate will be under positive sexual selection. If extrinsic mortality rate is reduced (arrow), extending life expectancy from vertical line I to II, then individuals with a lower investment in reproduction at younger ages will achieve higher lifetime fitness. Figure 9.4b models the environmental effects on the expression of ageing: An environmental change may result in the acceleration of ageing (arrow), resulting in faster decrease of the reproductive rate over time (lower solid line) in individuals that used to show high reproductive rate thanks to a large investment in secondary sexual traits (upper solid line); but the same environmental change does not significantly affect individuals exhibiting a lower reproductive rate, as they invested less in a secondary sexual trait (dotted line). Prior to the environmental change (dotted vertical line), high-investing individuals achieved higher lifetime fitness and there was a positive sexual selection on the secondary sexual trait; whereas after the change, the sign of sexual selection is reversed. This may explain the evolutionary response of our species confronted with hard environmental conditions, that may have produced the selection of bi-parental care, inter-sexual collaboration and division of labour, and lower reproductive rate spread over a long lifespan.

In principle, one would expect that given that the benefits of reproduction are accrued by both the female producing the ovum and the male producing the sperm, both sexes would collaborate for as long as both are reproductively active. But in fact, male–female sexual conflicts abound in nature and such conflicts may affect life-histories, including the patterns of ageing, leading to the evolution of *sexually antagonistic traits* (Chapman et al. 2003). For instance, inter- and intra-sexual competition leading to shortening of the lifespan of one sex may lead to increased rates of ageing (Promislow 2003), but the affected sex may also evolve adaptations against such effects. On the other hand, monogamy with bi-parental care would be expected to decrease the rate of ageing in both sexes (Promislow 2003).

Across species, there is an interesting pattern of sex differences in longevity that follows not the male/female divide but the homogametic/heterogametic sex divide. In humans, females are the homogametic sex (XX), whereas males are the heterogametic sex (XY) and females tend to live longer than males. In most species of birds, it is males who live longer than females and in this case the males are the homogametic sex (ZZ), whereas females are the heterogametic sex (ZW). Although a similar pattern has been found in other organisms (Tower and Arbeitman 2009; Xirocostas et al. 2020), the sex difference in lifespan among mammals and birds can be also explained by their broadly different mating systems, with mammals being largely socially polygynous, displaying highly costly male–male competition for access to copulations, whereas birds are mainly socially monogamous. In socially polygynous bird species, it is males that tend to display higher mortality than females (Liker and Szekely 2005). This suggests that there are chromosomal, but also other factors that can contribute to sex differences in lifespan and senescence. Maklakov and Lummaa (2013) noted that although old women are biologically prevented from reproducing due to menopause and they reach older ages than men, old men can continue to have children into old age and some do so, thus old age involves, at least in principle, different direct fitness opportunities for men and women.

In theory, the homogametic sex may have an advantage due to a lower potential for the expression of deleterious recessive alleles (e.g. Liker and Szekely 2005). The X-chromosome advantage in terms of senescence in humans also seems to be linked to having longer telomeres, which we have seen are associated with lower levels of ageing (Barrett and Richardson 2011, and references therein). Moreover, female mitochondria produce lower amounts of ROS than males, thus favouring females in terms of lower senescence and a more extended longevity (Viña et al. 2011).

Mitochondria are inherited maternally and have been also suggested to increase male ageing and mortality due to the accumulation of male-specific deleterious mutations, a mechanism that has been metaphorically labelled the "mother's curse" (Gemmell et al. 2004; Camus et al. 2012).

Differential longevity between men and women could also result from differential levels of *immunosenescence*. In this case, we would expect that immune competence to be higher in older women than men. Caruso et al. (2013) investigated this possibility by studying the levels of various types of T-cells and B-cells in a group of healthy, 20–90 years old Japanese. They found that the rate of decline with age of such components of the immune system was generally higher in men than women. For those cell types that showed an increase with ageing, the rate of increase was higher in women than men. T-cell proliferation capacity and immune score also decline with age and the rate of decline is higher in men. Thus, this work suggests that older women remain relatively more immunocompetent as they age than older men. Table 9.1 summarises a list of similarities and differences in various aspects of the immune systems of old men and women. For instance, old women (≥60 years old) tend to have higher NK cell numbers, relatively better proliferative and cytokine-secretion capacity of T-cells and also better ability of monocytes to secrete IFN-γ-inducible protein 10 (IP-10), which is an attractant to activated T-cells. Women also retain their levels of IL-10 compared to a decrease in old age for men, the same for IL-17. Women also decrease their number of B- and T-cells in old age less dramatically than men, but they increase their IFN-γ production relative to old men, or at least the decrease is less dramatic than in old men.

The above trends suggest that old women are better able to maintain the integrity of their immune system with ageing. After puberty, women are more susceptible to develop autoimmune conditions but are less prone to exhibit infections and chronic inflammatory diseases, suggesting that some components of their immune system may be highly active, presumably as a defence against pathogens during the reproductive years. Ageing involves a decrease in the competence and self-regulatory capacity of the immune system in both sexes, but such changes are more significant in old men than women:

> Men and women experience the same types of aging-related changes to the immune system, but males experience them earlier or more dramatically than women (Gubbels Bupp 2015, p. 102).

To recap, harsh environmental conditions confronted by our ancestors may have established the selective ground for the evolution of bi-parental care—therefore inter-sexual collaboration—, and lower reproductive rate spread over a long

Table 9.1 Ageing-related immune phenotypes observed in males and females

Immune phenotype	Sex-specific change with age	Age and condition
Innate		
NK cell numbers	↑ in both, but more dramatic ↑ in F	Humans ≥60 years old
Monocytes		
IL-6 production (LPS)	No change in either M or F with age	Healthy humans ≥65 years old
TNF-α production (LPS)	No change in either M or F with age	Healthy humans ≥65 years old
IP-10 production (IFN-γ)	↓ in both, but more dramatic ↓ in M	Healthy humans ≥65 years old
Serum concentrations of		
C-reactive protein	↑ in M, ↑ in F (M and F not compared in same study)	Humans ≥60 years old
IL-6	↑ In M, ↑ in F (M and F not compared in same study)	Males ≥60 years old; females post-menopause or ≥ 60 years old
TNF-α	↑ in M, ↑ in F (M and F not compared in same study)	Males ≥60 years old; females post-menopause or ≥60 years old
IL-10	↓ in M, no change in F	Humans ≥60 years old
Adaptive		
B-cell numbers	↓ in both, but more dramatic ↓ in M	Humans ≥66 years old
T-cell numbers	↓ in both, but more dramatic ↓ in M	Humans ≥66 years old
Proliferative capacity	↓ in both, but more dramatic ↓ in M	Humans ≥66 years old
IL-2 production (anti-CD3 + anti-CD28)	↓ in both	Healthy humans ≥65 years old
IFN-γ production (anti-CD3 + anti-CD28)	↓ in M but ↑ in F	Healthy humans ≥65 years old
IFN-γ production (IL-18 + IL-12)	↓ in both, but more dramatic ↓ in M	Healthy humans ≥65 years old
IL-17 production (anti-CD3 + anti-CD28)	↓ in M but no change in F	Healthy humans ≥65 years old
CD8+ CD28− (T_{EMRA})	↑ in both, but more dramatic ↑ in M	Healthy humans 20–90 years old
Inverted CD4/8 ratio	Both, but more prevalent in M	Humans ≥66 years old
CD4+ overall	↓ in both, but more dramatic ↓ in M	Humans ≥66 years old
Mucosal associated invariant T-cells	↓ In both, but more dramatic ↓ in M	Humans ≥60 years old

NK natural Killer cells, *IL* interleukin, *TFN* tumour necrosis factor, *LPS* lipopolysaccharide (used to induce production of inflammatory cytokines), *IP* inducible protein, *IFN* interferon, *CD* cell receptors, T_{EMRA} subset of differentiated T-cells. Reproduced from Gubbels Bupp (2015) with only minor modifications lifespan, with the lifespan being especially extended in women. Intra- and inter-sexual competition, on the other hand, may lead to the shortening of the lifespan and increased reproductive effort. Inter-sexual differences in lifespan and senescence are also affected by genetic factors, including which sex is homogametic and which is heterogametic. Females can have an additional advantage in terms of longevity compared to males thanks to immune and mitochondrial effects.

In the next section we explore the potential role of kin selection in explaining the ageing patterns of humans, with a special focus on women.

9.2.2 *Kin Selection and Ageing*

One of the most striking aspects of human life-histories is that women's reproduction sharply ends biologically at about 50 years old, whilst they can survive post-menopause to over 70 years, even in conditions that approach ancestral hunter-gatherer living (Fig. 9.1). What evolutionary mechanisms could maintain survival post-reproduction with continued senescence? As we already mentioned in the introduction to this chapter, one of the mechanisms that have been proposed is *kin selection*.

It was William Hamilton who first theoretically formalised the evolutionary aspects of ageing proposed by Medawar and Williams, and he also introduced what is now known as the *Grandmother hypothesis*, which follows from the more general *Kin selection theory* that Hamilton also contributed so much to develop (Hamilton 1963, 1964). *Kin selection theory* is grounded on what is now known as *Hamilton's rule*, formally derived by Hamilton in the 1960s:

$$rB > C$$

Where in an interaction between a donor and a recipient of help, B stands for the added benefits accrued by the recipient (e.g. in terms of a marginal increase in the number of reproductive offspring), C are the costs paid by the donor of help (e.g. in terms of the marginal decrease in the number of reproductive offspring caused by forfeiting reproduction), and r is the coefficient of genetic relatedness between donor and recipient. That is, evolution is expected to result in the selection of genes—and the phenotype they contribute to express—that lead to incurring costs when individuals help others, to the extent that the recipient benefits more than what the donor loses, relative to their genetic relatedness. In other words, the closest the individuals are related, the greater the reproductive benefits to the recipient and the lower the reproductive cost to the donor of help, the more likely it is that the donor will act altruistically and help. In humans, grandparents are the classical example of such helping individuals, as they are not reproductive but could help raise their grand-offspring, thus increasing their inclusive fitness.

Before we get into the specific matter of kin selection and senescence, we will review some important evolutionary aspects of menopause, which is a major turning point in women's ageing and a key aspect of the mechanism linking survival post-reproduction, continued senescence, and kin selection.

Among primates, menopause, or the end of ovulation, seems to be restricted to humans (Rovirosa-Hernández et al. 2017), although it is a process shared with the short-finned pilot whale (*Globicephala macrorhynchus*) in which females experience menopause between the ages of 30 and 40 years, but are capable of surviving 14 or more years beyond menopause (Marsh and Kasuya 1984); and resident killer whales (*Orcinus orca*) a species in which females tend to reproduce between the ages of 12 and 40 years but they can survive beyond 90 years old (Photopoulou et al. 2017; Croft et al. 2015; see Fig. 9.5). More recently, Photopoulou et al. (2017) have added the false killer whale (*Pseudorca crassidens*) to the list, as females in this species cease reproduction at about 55 years old but they can survive up to 62.5-63.5 years old.

Menopause is an evolutionary conundrum that we have highlighted before, and that is loudly asking for a solution (see also Austad 1994). Men, on the other hand, are not biologically impeded from reproducing into old age, at least not in the same manner as women are; although, in practice most men stop siring offspring well before they reach very advanced ages. In traditional societies male fertility remains positive at ages between 55–75 years old (Tuljapurkar et al. 2007); that is, men can extend their reproductive life to much older ages than women.

The most basic hypothesis that has been put forward to explain human menopause is that it is an epiphenomenon of a recent evolutionary trend to increased lifespan (e.g. Peccei 2001, and references therein). A longer lifespan that evolved without changing the ancestral reproductive physiology would produce menopause as a side effect. In fact, anthropoid primates that keep reproducing until the end of their life, live to about the age when humans start menopause (*Gorilla gorilla*, 60.1 years; *Pan paniscus*, 55 years; *Pan troglodytes*, 59.4 years; *Pongo pygmaeus*, 59 years; http://genomics.senescence.info/species/search_list.php; see also Emery Thompson et al. 2007). The pattern of an extended lifespan that exceeds the ability to reproduce is also expected from the evolutionary process of neoteny that we will analyse in the Sect. 9.3.

Williams (1957, see also Peccei 1995a, b, 2001), however, also suggested that the end of pregnancies as the body becomes relatively more frail with ageing has a *direct fitness* advantage, as it would allow mothers to avoid the potentially negative effects of such pregnancies on her own survival and therefore devote her energy to raise her still dependent children. As stated by Williams (1957, p. 408):

> Menopause, although apparently a cessation of reproduction, may have arisen as a reproductive adaptation to a life-cycle already characterized by senescence, unusual hazards in pregnancy and childbirth, and a long period of juvenile dependence.

This is known as the *Mother hypothesis* for the evolution of menopause (e.g. Peccei 1995a).

In fact, maternal mortality during pregnancy and delivery significantly increases in women who are 35 years old and older (see Fig. 9.6). Menopause eliminates such risk, thus increasing the chances of the mother to complete parental care for the children who are still growing up.

In addition, older mothers are also more likely to give birth to children displaying birth defects that may decrease the probability of those children surviving and/or

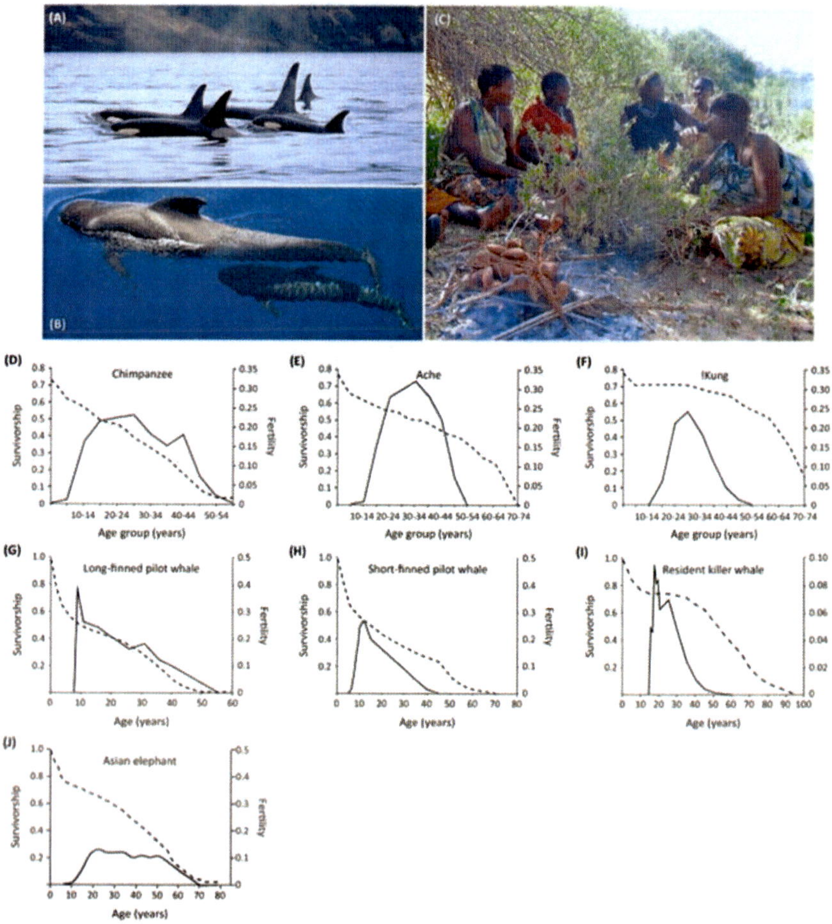

Fig. 9.5 (**a–c**) Vertebrate species that display prolonged post-reproductive lifespans. (**A**) Resident killer whale (*Orcinus orca*); (**B**) short-finned pilot whale (*Globicephala macrorhynchus*); and (**C**) *Homo sapiens*, the image shows Hadza women. (**D–J**) Survivorship and fertility plots for vertebrate species that do (humans (**E**, **F**), short-finned pilot whale (**H**), resident killer whale (**I**)) and do not (chimpanzee (**D**), long-finned pilot whale (**G**), Asian elephant (**J**)) have prolonged post-reproductive lifespans. Fertility (continuous line), survivorship (dashed line). Reproduced from Croft et al. (2015), with permission

producing their own offspring. This is a cost of ageing that is also found in other species (e.g. Priest et al. 2002). For instance, the increased risk of chromosomal malformations with maternal age is well known, especially for mothers older than 40 (see Fig. 9.7).

Offspring also experience increased risks of non-chromosomal malformations with increased maternal age. Hollier et al. (2000) reported an additional age-related risk of non-chromosomal malformations of 1% in children of women 35 years old or older, after studying 102,728 pregnancies. An increase in non-chromosomal

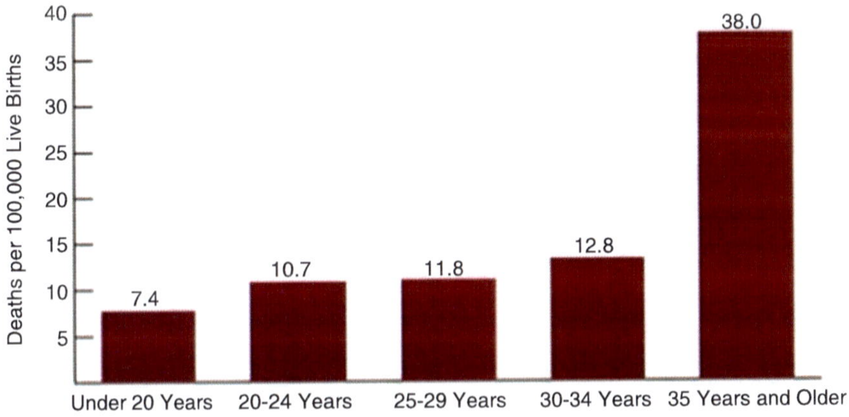

Fig. 9.6 Maternal mortality rates in the USA in 2005 as reported to the *Centers for Disease Control and Prevention*. Reproduced from Sauer (2015) with permission

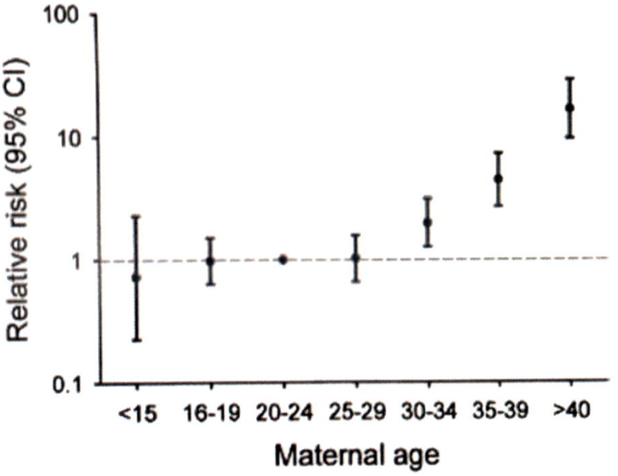

Fig. 9.7 Relative risk (expressed semi-logarithmically) for abnormal karyotypes compared with baseline risks at age 20–24 (reproduced with permission from Hollier et al. 2000)

malformations was also detected by Reefhuis and Honein (2004) in the offspring of both very young and advanced-age mothers (see also Gill et al. 2012).

Perinatal mortality, intrauterine foetal death, and neonatal death increase with maternal age (e.g., from 20–29 years old to 40–44 years, and >45 years in Jacobsson et al. 2004). Moreover, increased risk of foetal death with maternal age has remained despite an overall trend for deaths at birth to decrease over time (e.g. in a comparison between 1961 and 1974 vs. 1978 and 1993 in the study carried out by Fretts et al. 1995).

In Canada, however, Baird et al. (1991) found that the risk of children being born with birth defects of unknown aetiology tends to decrease with advanced maternal age, but only after excluding cases of chromosomal anomalies and birth defects of known aetiology. Women giving birth after the age of 34 are also at higher risk of pre-term birth than younger women (Mousiolis et al. 2013).

What about paternal agc and birth defects? Although men can biologically produce offspring at older ages than women, the offspring of older men are also more likely to display birth defects than offspring of younger men (McIntosh et al. 1995). Although men younger than 20 years old are also at an elevated risk of fathering children with birth defects, as it is seen in very young women as well. After controlling for maternal age, Fisk Green et al. (2010) found that the odds of some birth defects increase in offspring of older fathers. These results can be explained by a decline in sperm quality in old age and an increase in the number of new mutations in spermatozoa (Paul and Robaire 2013; Sharma et al. 2015).

Potentially, the combined effects of increased birth defects in the offspring of both older women and men may contribute to the selection of menopause in women—the sex that is burdened with the highest cost of reproduction—well before the age of death. Perhaps oocyte depletion is adjusted to maximise the viability of the offspring at birth, rather than the survival of the mother and therefore the survival of the offspring still needing parental care, as suggested by Williams? In fact, those two selective mechanisms can act synergistically, as they are not mutually exclusive. Pregnancy in old women carries risks for both mother and her offspring, both those already born and in need of parental care and the one in gestation (Schoen and Rosen 2009).

Therefore, the combined effects of risks of pregnancies in older ages plus diverting time and resources that could be used to care for the last offspring, who may be still in need of parental care, could explain the adaptive value of menopause. However, women (and men) survive to post-menopausal ages that far exceed the age of independence of their last offspring (e.g. Pavelka and Fedigan 1991). What could explain such extended post-reproductive lifespan? Could old humans be increasing their inclusive fitness benefits, as suggested by Hamilton (1966) in the quote reproduced in the introductory section of this chapter?

In 1997 Kristen Hawkes and her collaborators published an article addressing the above questions in a study of Hazda women from Northern Tanzania. There, the authors pointed out that women's foraging not only positively affects their own children's well-being, and that a newborn child can negatively affect the contribution of the mother to the nutrition of a weaned sibling, but also that reduction of maternal care for the weaned child was somewhat compensated by an increase in care from grandmothers (Hawkes et al. 1997). Grandmothers can provide not only material support to their grandchildren but also valuable knowledge. Transmission of knowledge by grandmothers has been reported not just in humans but also in another species displaying end of reproduction several years before death: killer whales (*Orcinus orca*) (Brent et al. 2015).

Gurven and Hill (1997), commenting on Hawkes et al.'s article, argued that provisioning by grandfathers may also be important, not just grandmothers, in order to explain the long life of males and their decreased reproductive success as they age. To this, Hawkes et al. replied that male provisioning of meat from large game animals, which is what men tend to provide to the family group in the Hazda, does not seem to make a great difference to the group in terms of food intake, as compared to capturing small animals or the gathering activity of women. The following year Hawkes and collaborators published an article in the *Proceedings of the National Academy of Science of the USA* where they officially introduced the *Grandmother hypothesis* as an evolutionary explanation for the long human, and especially women, lifespan that extends well beyond menopause (Hawkes et al. 1998).

The *Grandmother hypothesis* derives directly from *Hamilton's rule* and *kin selection* that we already introduced in this chapter. It states that the long post-menopausal life of women and slowed down senescence could be adaptive if those old women help increase the fitness of their daughters/sons through providing additional care for their young grandchildren. Such adaptive strategy could be selected if the costs of producing and caring for own offspring become too high as women age, compared with the costs of simply helping their daughters/sons with some aspects of childcare.

Grandmothering may have reinforced the adaptive value of a natural process of senescence that is already slowed down in response to an evolutionary shift in life-history strategies from *r*-selected to *K*-selected (e.g. Hawkes 2003).

Since its introduction in the late 1990s, the *Grandmother hypothesis* has been subject to various empirical tests. Lahdenperä et al. (2004) studied complete multi-generational demographic records from pre-modern populations in Finland and Canada describing that women with a lengthy post-reproductive lifespan also had more grandchildren, mainly because they increased the lifetime reproductive success of their adult offspring by providing alloparental care to their grandchildren. Shanley et al. (2007) also supported the hypothesis after analysing a long-term dataset coming from The Gambia. Maternal grandmothers have a significant additional positive effect on the survival of their grandchildren during their initial 2 years of life. More recently, Engelhardt et al. (2019) used a very detailed dataset of pre-industrial French settlers in the St. Lawrence Valley of Canada to test the *Grandmother hypothesis*. The geographic distance between family members was used as a proxy for help. The vicinity of grandmothers allowed daughters to increase their number of offspring by 2.1, and to increase the number of offspring surviving to 15 years old by 1.1, compared to situations where the grandmother was dead. Moreover, the farther away the grandmother lived, the lower the number of children born was and the lower the overall lifetime reproductive success of the daughters also was. A statistically weaker result, due to a small sample size, but that also goes in support of the *Grandmother hypothesis* was obtained by Sorenson Jamison et al. (2002). They studied population registers from a village in central Japan for the period 1671–1871 and found that after the presence of other household members was accounted for, the grandparent who had a consistently positive effect on child's survival was the maternal grandmother.

Hawkes and Coxworth (2013, see also Hawkes 2020) and Blell (2018) have reviewed the empirical standing of the *Grandmother hypothesis* suggesting that although some studies have supported the hypothesis others have not been so clear.

Controlling for a diversity of confounding variables is necessary before an appropriate test of the hypothesis can be regarded as valid. For instance, in some modern societies, grandparents may treat their grandchildren with tasty but unhealthy food, thus negatively affecting their grandchildren's health. The grandparents' intent supports the hypothesis, but the end-result of their action falsifies it. In a recent comparative analysis, Péron et al. (2019) have described a broader trend across non-human species exhibiting grandparental care that indicates that females of those species have a relatively longer lifespan than males, and they also have a longer lifespan than females in polygynous species without grandparental care, a pattern that is consistent with the hypothesis.

Therefore, although more empirical studies that control for potential confounding variables are needed, grandmothering seems to be promising as a potential explanation for, or at least a contributing factor to explain the extension of human female longevity beyond menopause.

We mentioned that the *Grandmother hypothesis* is a special case of *kin selection*. *Kin selection* was originally proposed as an evolutionary mechanism to explain *altruism*, or the provision of help to others at a cost to the donor of help. Figure 9.8 summarises four major categories of social interactions classified according to the benefits or costs (disadvantage) accrued by the interacting individuals.

From the perspective of older people, they may engage in *cooperation* with others whereby both parties benefit from the interaction (e.g. see inter-generational programmes that involve positive interactions between young and old; e.g. Cordella and Huang 2016; Teater 2016). Sometimes, however, social interactions may not be positive at all to both parties, as a consequence of spiteful acts or simply bad luck; whereas on other occasions one of the interactants benefits and the other pays a cost. The latter case may involve a straightforward *exploitative* relationship. On the

		Recipient of help	
		Benefits	**Disadvantaged**
Donor of help	**Benefits**	**Cooperation, Mutualism**	**Exploitation/ Altruism**
	Disadvantaged	**Exploitation/ Altruism**	**Spite, Bad luck**

Fig. 9.8 Various kinds of social interactions described in terms of the benefits and costs (disadvantage) accrued by the interacting individuals

other hand, participants who incur a cost may do so consciously and voluntarily, as in the case of grandparents helping their adult children through taking care of grandchildren or providing financial assistance; in this case the older person may be acting *altruistically*. If at some later stage the grandparent does receive some form of compensation for the help provided, altruism may in fact be reciprocal (Trivers 1971). *Reciprocal altruism* may involve the same interactants or it may be more diffuse as in the case when an elder may help a grandchild and the grandchild, in return, feels the responsibility to help his/her own grandchildren when time comes. This is also known as the *demonstration effect* (Cox and Stark 2005), which is equivalent to the *Arrow of Time reciprocation* that we mentioned in Chap. 6, and the *transitional reciprocity* mentioned in Chap. 7. Culture also modulates these different patterns of altruism.

Kin selection is more likely to operate when the apparently altruistic elder does not receive a major direct benefit from his/her helping acts performed towards close kin. If the helping elder enjoys emotional satisfaction by supporting closely related individuals, and such satisfaction (happiness, fulfilment) results in health benefits for the elder (Post 2005)—which may even translate into an extended lifespan (e.g. Brown et al. 2005)—, more than "altruism", the helper may be engaged in some form of mutualism. The situation becomes a bit more complicated when the elder only benefits from cooperation through a more extended lifespan, with no consequence on the elder ability to reproduce. Living longer may be regarded as beneficial by an individual, but from an evolutionary perspective it is the transmission of genes through the next generation that matters, hence if a non-reproductive elder only benefits from cooperating with kin via an increased lifespan, such increased lifespan is adaptively significant mainly in terms of additional opportunities to keep helping kin (i.e. increasing the indirect component of inclusive fitness). The adaptive link breaks at ages when the elders are so old (e.g. centenarians, supercentenarians) that they may only be a burden to their relatives. Here, cultural duties towards such elders can be interpreted as a form of delayed reciprocal altruism, as mentioned above, but the system is obviously unstable at very advanced ages, as demonstrated by the many cases observed across cultures of elder neglect, abandonment, and even abuse to the point of hastening the elder's death (geronticide). This can explain why some elders may contemplate euthanasia or suicide at some stage, as a way of easing the burden of care on the family.

The operation of kin selection relies on a close genetic relationship between interacting members of a group. In humans, where females tend to be the dispersing sex, women may initially have low genetic relatedness with the members of the extended family (the "in-laws"), but relatedness will increase over time, as the older she is the more children, grandchildren and great-grandchildren are added to the family (e.g. Koster et al. 2019).

Bourke (2007) extended the classification of social interactions to specifically focus on the effects of changes in the lifespan, ageing, and altruism (Table 9.2), noting that as far as kin selection is concerned, what matters with regard to an extension of the lifespan is what old people actually do:

Table 9.2 Classification of hypothesised social effects on lifespan and rate of ageing

	(*i*) Recipient's fitness is increased	(*ii*) Recipient's fitness is decreased
(*a*) Focal individual increases lifespan	Kin-selected delayed ageing in caregiving parent or grandparent	Longer-lived individual denies inherited resources to potential successor; corollary of case $b \times i$
(*b*) Focal individual decreases lifespan or dies	Kin-selected accelerated ageing to release inherited resources to potential successor	Death of caregiver curtails care provided to dependent individuals; corollary of case $a \times i$

From Bourke (2007)

> For example, extending one's lifespan could be altruistic (if there is post-reproductive care of others) or selfish (if resources are denied to others) (p. 109).

Most of us can probably think about real life examples of both situations.

Kin selection may explain an extended lifespan and delayed somatic ageing in spite of declining reproductive capacity, and eventually menopause in women, if help provided by the elder increases the fitness (lifetime reproductive success) of relatives who are recipient of such help ($a \times i$ quadrant in Table 9.2). On the other hand, an extended lifespan could also be selected if the fitness of close relatives is decreased ($a \times ii$ quadrant), for as long as the elder remains reproductively active. This is more likely to be the case in old men than in old post-menopausal women, in which case those old men would be acting in a selfish manner. This possibility, however, may invite retaliatory actions from the part of the offspring, that in some cases may even jeopardise the probability of survival of the elder (see Chap. 6). The latter situation would take the relationship to quadrant $b \times i$ which may involve an actual acceleration of biological ageing if the elder is mentally prepared to give up life to favour a successor, or loss of life is beyond the control of the elder, whether death is accidental or caused by others. Kin selection may potentially modulate the lifespan: dying too late or too early may be detrimental to inclusive fitness according to circumstances. The situation where an elder dies too early, thus decreasing his/her inclusive fitness, is exemplified in quadrant $b \times ii$ which includes accidental cases of elder's death, a case of "bad luck" (see also Fig. 9.8).

In sum, menopause, or the termination of reproduction in women well before their death, is an evolutionary conundrum. Menopause has been explained as an epiphenomenon of increased lifespan in humans, and as an adaptation to ageing that would maximise the survival of the last offspring. However, humans survive to post-menopausal ages that exceed the age of independence of their last offspring. Kin selection, in the form of the *Grandmother hypothesis*, may offer a comprehensive explanation for human's extended lifespan, delayed somatic ageing, declining reproductive capacity, and eventually menopause in women if the help provided by the elder increases the lifetime reproductive success of relatives, whilst avoiding the high costs of pregnancy incurred by an ageing body.

The *Grandmother hypothesis*, however, may be missing an important aspect of our evolution that could also contribute to explain human patterns of senescence. In the next and final section of this chapter we explore the relationship between neoteny and ageing.

9.3 Neoteny

What a tenuous position for the crown of creation! An ape arrested in its development; holding the spark of divinity only through a chemical brake placed upon its glandular development; retaining a corporeal reminder, so to speak, of original sin – the potential of sinking again into the Tertiary abyss, should that break ever be released

Stephen J. Gould *Ontogeny and phylogeny* (p. 361).

It is not with the prolongation of life that we should be so much concerned as with the prolongation of our pedomorphic drives, the prolongation of our youthfulness, the prolongation of the quality of life, the realization of our destiny.

Ashley Montagu *Growing young* (p. 174).

In an article published in 2015, Darren Croft and collaborators stated that "… in natural populations of humans as well as some cetaceans and insects, reproductive senescence occurs much faster than somatic aging and females exhibit prolonged post-reproductive lifespans" (Croft et al. 2015, p. 407). Although in relative terms reproductive senescence does occur before somatic ageing in humans (women in particular), we have seen that reproductive senescence in humans occurs, in fact, at a similar age than in other anthropoid primates, whereas somatic senescence occurs at a later age. In other words, in humans there is a development of reproductive and somatic tissues that is out of synchrony, with somatic development being slowed down, or delayed, compared to reproductive development. This is the definition of the evolutionary process known as *neoteny*.

In this section we will try to answer the question whether the patterns of ageing seen in humans are simply the result of our species having evolved through a neotenic process. Skulachev et al. (2017) have provided a recent review of neoteny as a possible evolutionary mechanism to explain human ageing.

Neoteny could explain both the extended lifespan in humans—compared to other anthropoid primates—and some traits of old humans. But before we get into addressing those issues, we need to explain in more detail what neoteny is.

In his book *Ontogeny and Phylogeny*, Stephen J. Gould (1977, p. 482) defined *heterochrony*, following De Beer (1930), as a process whereby new species may evolve through mutations that alter the rate of development of some traits relative to others, in comparison to the ancestral state. The array of potential processes that fit the definition of heterochrony is diverse as we illustrate in Fig. 9.9, and they include *paedomorphosis* (*neoteny*, *progenesis*) and *peramorphosis* (*acceleration*, *hypermorphosis*).

Paedomorphosis describes an evolutionary process in which the development of individuals in a derived species is slowed down compared to the ancestor, whereas *peramorphosis* is an acceleration of development (Gariépy et al. 2001). In the context of somatic vs. reproductive structures and functions, peramorphic evolutionary processes that involve speeding up the rate of development of somatic structures, with the timing of gonadal maturation remaining unchanged, are called *acceleration*, whereas those involving an unchanged somatic development throughout speciation but a delayed or slowed down gonadal maturation are called *hypermorphosis*.

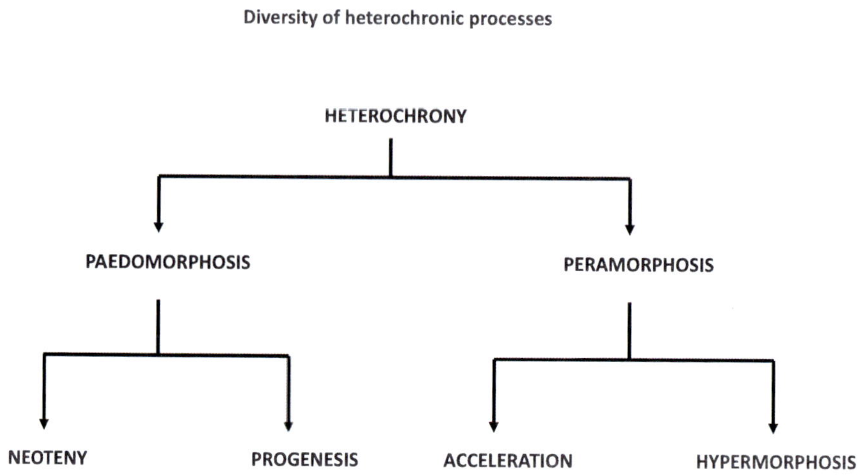

Fig. 9.9 Neoteny is one of the two evolutionary heterochronic processes of paedomorphosis

It can be seen that peramorphosis will be involved in evolutionary changes in a derived species that delay the onset of reproductive ability relative to the development of other physiological and morphological characteristics compared with ancestors, in other words species that evolved through a process of peramorphosis will have adult-looking developmental stages (compared with the ancestor) that nevertheless are slower at developing full reproductive capability. Paedomorphosis, on the other hand, may involve either an accelerated gonadal maturation with rate of somatic development remaining unchanged, a process known as *progenesis*; or delayed somatic development with gonadal development unchanged, a process known as *neoteny* (Raff and Wray 1989; Gariépy et al. 2001). In both cases, paedomorphosis results in a derived species with individuals attaining reproductive capacity at a developmental stage in which they look like a juvenile of the ancestral species.

Paedomorphosis is consistent with the evolutionary patterns of several morphological structures and behaviours found in various groups of mammals, including primates (Hafner and Hafner 1984; Doran 1992; German et al. 1994; Berge 1998; Chaline et al. 1998; Gariépy et al. 2001; Cubo et al. 2002; Goldberg 2003; Mitteroecker et al. 2004).

The idea that humans evolved through neoteny is not new. Gould (1977) traces the first indication of thoughts about adult humans resembling young of great apes for evolutionary reasons rather than chance, back to the first half of the nineteenth century.

The term neoteny was introduced by the Swiss zoologist Julius Kollman in the late 1800s (Kollman 1885), who suggested that humans evolved from ancestral anthropoid apes through neoteny. However, it was the Dutch anatomist Louis Bolk (1926) who explicitly made the link between what we now recognise as neoteny and human evolution with his theory of *fetalization* of human anatomy.

De Beer further developed Bolk's theory, with the neotenic roots of human evolution recurrently surfacing in evolutionary debates in subsequent decades. Konrad Lorenz (1971) extended the idea to the evolution of human behaviour. He observed that old animals are less flexible in their ability to learn, whereas humans retain the "juvenile characteristic of investigative curiosity" (p. 179) into older ages. Lorenz could see "no cogent reason for regarding the general juvenescence of man as anything other than a special case of true *neoteny*" (p. 180, italics original).

Stephen J. Gould expressed the opinion that:

> human beings are 'essentially' neotenous, not because I can enumerate a list of important paedomorphic features, but because a *general, temporal retardation of development has clearly characterized human evolution. This retardation established a matrix within which all trends in the evolution of human morphology must be assessed.* This matrix does not in itself guarantee a central role for paedomorphosis, but it certainly provides a mechanism for such a result, if this result be of selective value (Gould 1977, p. 365; italics original).

The idea had been also supported by Ashley Montagu in the 1950s (Montagu 1955) and further developed in the 1980s (Montagu 1989). Montagu stressed how, thanks to neoteny, we may retain youthful traits of curiosity and exploration into old age, that we can use to build wisdom. The neotenic characteristics of humans have been more recently popularised in a book by Clive Bromhall *The eternal child* (Bromhall 2004).

Neoteny could extend juvenile-like biological processes in humans through the slowing down of the development of somatic tissues, whilst maintaining the ancestral rate of development of reproductive tissues. This could explain a long lifespan beyond the age of reproduction (Buffenstein et al. 2020) and also the expression of characteristics in the adult that are typical of juvenile stages in both humans and other closely related species; alongside a pattern of development of reproductive tissues that is equivalent to that found in closely related species and that could therefore be closer to the ancestral state.

Shostak (2006) has suggested that a longer lifespan in humans may be maintained by the evolution, through neoteny, of a state where more stem cells are acquired: "The lynchpin connecting neoteny with an expanded life span would seem to be the embryonic addition of SR [self-renewing] cells to tissues, thus tipping the scale in favor of extended cellular replacement in adult organisms" (p. 141).

Vladimir Skulachev and collaborators have provided the most comprehensive work so far arguing for the human patterns of extended longevity and slowed down senescence post-sexual maturity, to be a result of evolutionary neotenic processes (Skulachev et al. 2017). It is important to note that in order to show evidence of neotenic effects on human senescence, what must be proven is not just the slowing down of somatic senescence, but also the retention of the ancestral rate of development of reproductive tissues. We have suggested that this is the case, on the ground of available comparative evidence illustrated, for instance, in Fig. 9.5, where chimpanzees, the closest living relatives of humans, display female reproduction that stops at around 50 years old, which coincides with the end of life in this species; whereas in the case of humans living in ancestral hunter-gatherer conditions (e.g. Aché and!Kung) female reproductive senescence is also achieved at about 50 years old with menopause, but female lifespan extends well beyond the end of reproduction. Thus, the expected neotenic pattern of delayed somatic development with retention of ancestral reproductive development, seems to be fulfilled in humans. This pattern, however, could be attributed to mere coincidence if it wasn't for the fact that adult humans also show other evidence of neoteny, such as juvenile-like somatic characteristics.

Skulachev et al. (2017), following Bolk (1926, 1927) and more recent authors, compiled a preliminary list of neotenic traits in humans that we reproduce here in a reduced form:

1. "flat faced", orthognathy;
2. considerable reduction of body hair;
3. loss of pigmentation in the skin, eyes, and hair;
4. epicanthic eye fold;
5. central position of the foramen magnum (it migrates backward during the ontogeny of most other primates);
6. high relative brain weight and other neotenic aspects of brain development (e.g. synaptic plasticity; younger metabolic brain age compared to chronological age, especially in women Goyal et al. 2019);
7. labia majora of women;
8. structure of the hands and feet;
9. form of the pelvis;
10. ventrally directed position of the sexual canal in women;
11. absence of brow ridges;
12. absence of cranial crests;
13. thinness and shape of the skull;
14. position of the orbits under the cranial cavity;
15. brachycephaly (flattened rear side of the skull);
16. small teeth;
17. small nose;
18. late eruption of teeth;
19. absence of a baculum (penis bone);
20. no rotation of the big toe;
21. longer legs than arms;
22. upright stance;
23. prolonged period of infantile dependency;
24. prolonged period of growth;
25. long lifespan.

Humans are not the only mammalian species thought to have evolved through neoteny. Another example that has been extensively studied is the naked mole-rat *Heterocephalus glaber*, an African subterranean rodent characterised by high levels of sociality and a hairless body that resembles that of newborns of other rodent species (Fig. 9.10). Naked mole-rats were first indicated as a possible neotenic species by Richard Alexander in his work *Some unanswered questions about naked mole-rats* (Alexander 1991).

Fig. 9.10 Digging chain of naked mole-rats. Reproduced, with permission, from Griffin (2008; Photo: Justin O'Riain

Skulachev et al. (2017) also provided evidence for the neotenic aspects of the biology of the naked mole-rat, some of which are species-specific, whereas others are shared with humans, the "naked ape". The following is a partial list of that evidence:

1. Much longer gestation than is expected based on body weight.
2. Absence of auricles, as in newborns and in contrast to other adult mammals.
3. Absence of scrotum, as in newborns.
4. Very small vomeronasal organ, as in newborns.
5. Limited ability to maintain a constant body temperature, as in newborns.
6. Age-linked diseases such as cancer, diabetes, and cardiovascular, brain, liver, and some kidney pathologies, as well as many infections are very rare or absent as causes of death of naked mole-rats (as is the case in mammalian juveniles).
7. Absence of age-linked declines in cognitive functions.
8. Very high resistance of brain neurons to anoxia and subsequent oxidative stress during reoxygenation (as in newborns).
9. A very extended time frame for brain development compared with that of other rodents.
10. Maintenance of the capacity for the regeneration and elongation of neurons during adulthood.
11. Delayed maturation of the skeleton system.
12. No age-linked decrease in metabolic rate.
13. Very long maximal lifespan; compared with other rodents, a very low mortality rate that increases only slightly at least during the first 24 year.
14. Absence of hair (unique among adult rodents).

15. Much longer time to maturity than for other rodents. For this trait to be neotenic it is required that the ancestor of the Bathyergidae, the rodent family to which the naked mole-rat belongs, to have also displayed a long time to maturity.

Regarding behaviour, we mentioned the work of Konrad Lorenz pointing to psychological traits of adult humans that resemble the behaviour of juvenile primates, such as "investigative curiosity". Brüne (2000) suggested the following list of neotenic human behaviours based on Lorenz (1973), Gould (1977) and Montagu (1989): Persisting learning abilities, curiosity, play, flexibility, explorative behaviour, behavioural adaptability, prolonged dependency period, and persistent emotional attachment. Several of those adult behaviours rely on our evolved brain plasticity, which is also a characteristic of juvenile vertebrates (Happel and Frischknecht 2016).

Humans have the capacity to retain such juvenile-like behaviours not only into adulthood but, to some extent, well into old age with variability observed among individuals; and additional physiological characteristics that have been described as neotenic can also persist into old age (e.g. brain metabolic rate, Goyal et al. 2019; capacity for rejuvenation of the haematopoietic system, Melamed and Scott 2012).

Following *Reliability theory*, we may speculate that neoteny offers greater redundancy to an organism in the form of increased plasticity, that is a characteristic of embryonic tissues. More plasticity confers more adaptability to our species, but we may pay the cost of higher probability of developing cancer in old age (see Chap. 1 of Cordella and Poiani 2014, for a brief introduction).

The hypothesis that a longer lifespan and a delayed development of some somatic traits and behaviours have evolved in humans (and in some other mammalian species) through neoteny stands on good supportive empirical evidence, but there are also some critics.

Brian Shea, for instance, although accepting that "(h)eterochrony in general has undoubtedly played a central role in the evolutionary transformations characterizing our lineage, and neotenic paedomorphism in particular can probably be implicated in several of these shifts" (Shea 1989, p. 97), also stresses that other features of the human biology do not fit the neoteny model and are more likely to have evolved through different processes (see also Bogin 1997). We agree that neoteny may not explain all aspects of human biology, but it looks like that it can explain enough of them to be considered an important mechanism to understand human evolution and the patterns of ageing in our species.

To recap, human lifespan may extend well beyond the age when our last offspring has become independent and started reproducing. This life-history pattern may result from our species undergoing an evolutionary process of neoteny. The age at menopause is consistent with the age at last reproduction for females in our closest living primate relatives, thus suggesting that it may be a conserved ancestral state. Neoteny could explain the retention of such pattern of reproductive development with a slowing down in the development of somatic structures, that may also affect behaviour, and that make adults of our species resemble, to an extent, a juvenile state of our closest living relatives. Not all aspects of our biology may have been affected by neoteny, but several human patterns of ageing are consistent with it.

9.4 Conclusions

Human ageing is expressed in biological processes that have been affected by our evolution. Ageing involves a limit in the capacity of our body to repair the ongoing deterioration of its components at the molecular, cellular, and systemic levels. In general, natural selection on traits that are expressed beyond the age of reproduction is rather weak, but through natural selection the age at last reproduction can be extended, often through an overall restructuring of life-history traits that conform to a *K*-selected pattern. In addition, human females can survive well beyond their age at last reproduction, the latter being biologically determined by menopause. Menopause has been explained as an epiphenomenon of human evolution and extended lifespan, or as a specific adaptation to the costs of reproduction (e.g. the *Mother hypothesis*). Women's lifespan, however, extends so far beyond the age of sexual maturation of their last offspring that a more comprehensive model has been proposed to account for the evolution of menopause, the *Grandmother hypothesis*. The *Grandmother hypothesis* is grounded on kin selection. In addition, the age at menopause, long lifespan, and patterns of senescence in humans may also conform to a scenario where neoteny may have been central to many aspects of the evolution of our species, as evidenced by the retention of juvenile-like traits into adulthood and, to an extent, also into old age.

References

Abrams, P. A. (1991). The fitness costs of senescence: The evolutionary importance of events in early adult life. *Evolutionary Ecology, 5*, 343–360.

Alberts, S. C., Altmann, J., Brockman, D. K., Cords, M., Fedigan, L. M., Pusey, A., et al. (2013). Reproductive aging patterns in primates reveal that humans are distinct. *Proceedings of the National Academy of Sciences of the United States of America, 110*, 13440–13445.

Alexander, R. D. (1991). Some unanswered questions about naked mole-rats. In P. W. Sherman, J. U. M. Jarvis, & R. D. Alexander (Eds.), *The biology of the naked mole-rat* (pp. 446–465). Princeton, NJ: Princeton University Press.

Austad, S. N. (1994). Menopause: An evolutionary perspective. *Experimental Gerontology, 29*, 255–263.

Baird, P. A., Sadovnick, A. D., & Yee, I. M. L. (1991). Maternal age and birth defects: A population study. *The Lancet, 337*, 527–530.

Baltes, P. B., & Smith, J. (2003). New frontiers in the future of aging: From successful aging of the young old to the dilemmas of the fourth age. *Gerontology, 49*, 123–135.

Barrett, E. L. B., & Richardson, D. S. (2011). Sex differences in telomeres and lifespan. *Aging Cell, 10*, 913–921.

Bartke, A., Chandrashekar, V., Dominici, F., Turyn, D., Kinney, B., Steger, R., et al. (2003). Insulin-like growth factor 1 (IGF-1) and aging: Controversies and new insights. *Biogerontology, 4*, 1–8.

Beckhauser, T. F., Francis-Oliveira, J., & De Pasquale, R. (2016). Reactive oxygen species: Physiological and physiopathological effects on synaptic plasticity. *Journal of Experimental Neuroscience, 10*(Suppl. 1), 23–48.

Berge, C. (1998). Heterochronic processes in human evolution: An ontogenetic analysis of the hominid pelvis. *American Journal of Physical Anthropology, 105*, 441–459.

Birolini, A. (2007). *Reliability engineering: Theory and practice* (5th ed.). Berlin: Springer.
Blell, M. (2018). Grandmother hypothesis, grandmother effect, and residence patterns. In H. Callan (Ed.), *The international encyclopedia of anthropology* (pp. 1–5). New York: Wiley.
Bogin, B. (1997). Evolutionary hypotheses for human childhood. *Yearbook of Physical Anthropology, 40*, 63–89.
Bolk, L. (1926). On the problem of anthropogenesis. *Proceedings of the Section of Sciences Koninklijke Akademie van Wetenschappen te Amsterdam, 29*, 465–475.
Bolk, L. (1927). On the origin of human races. *Proceedings of the Koninklijke Nederlandse Akademie van Wetenschappen—Amsterdam, 30*, 320–328.
Bonduriansky, R., Maklakov, A., Zajitschek, F., & Brooks, R. (2008). The evolutionary ecology of senescence: Sexual selection, sexual conflict and the evolution of ageing and life span. *Functional Ecology, 22*, 443–453.
Bonsall, M. B. (2006). Longevity and ageing: Appraising the evolutionary consequences of growing old. *Philosophical Transactions of the Royal Society B., 361*, 119–135.
Boonekamp, J. J., Briga, M., & Verhulst, S. (2015). The heuristic value of redundancy models of aging. *Experimental Gerontology, 71*, 95–102.
Boonekamp, J. J., Simons, M. J. P., Hemerik, L., & Verhulst, S. (2013). Telomere length behaves as biomarker of somatic redundancy rather than biological age. *Aging Cell, 12*, 330–332.
Borges, R. M. (2009). Phenotypic plasticity and longevity in plants and animals: Cause and effect? *Journal of Biosciences, 34*, 605–611.
Bourke, A. F. G. (2007). Kin selection and the evolutionary theory of aging. *Annual Review of Ecology, Evolution and Systematics, 38*, 103–128.
Bowles, J. T. (1998). The evolution of aging: A new approach to an old problem of biology. *Medical Hypotheses, 51*, 179–221.
Brent, L. J. N., Franks, D. W., Foster, E. A., Balcomb, K. C., Cant, M. A., & Croft, D. P. (2015). Ecological knowledge, leadership, and the evolution of menopause in killer whales. *Current Biology, 25*, 746–750.
Bromhall, C. (2004). *The eternal child: How evolution has made children of us all*. London: Ebury Press.
Bronikowski, A. M., Altmann, J., Brockman, D. K., Cords, M., Fedigan, L. M., Pusey, A., et al. (2011). Aging in the natural world: Comparative data reveal similar mortality patterns across primates. *Science, 331*, 1325–1328.
Brown, W. M., Consedine, N. S., & Magai, C. (2005). Altruism relates to health in an ethnically diverse sample of older adults. *Journal of Gerontology, 60B*, P143–P152.
Brüne, M. (2000). Neoteny, psychiatric disorders and the social brain: Hypotheses on heterochrony and the modularity of the mind. *Anthropology & Medicine, 7*, 301–318.
Buffenstein, R., Lewis, K. N., Gibney, P. A., Narayan, V., Grimes, K. M., Smith, M., et al. (2020). Probing pedomorphy and prolonged lifespan in naked mole-rats and dwarf mice. *Physiology, 35*, 96–111.
Calder, W. A. I. (1985). The comparative biology of longevity and lifetime energetics. *Experimental Gerontology, 20*, 161–170.
Camus, M. F., Clancy, D. J., & Dowling, D. K. (2012). Mitochondria, maternal inheritance, and male aging. *Current Biology, 22*, 1717–1721.
Caruso, C., Accardi, G., Virruso, C., & Candore, G. (2013). Sex, gender and immunosenescence: A key to understand the different lifespan between men and women? *Immunity and Ageing, 10*, 20. http://www.immunityageing.com/content/10/1/20
Caspari, R., & Lee, S.-H. (2004). Older age becomes common late in human evolution. *Proceedings of the National Academy of Sciences of the United States of America, 101*, 10895–10900.
Cassill, D. L. (2019). Extending r/K selection with a maternal risk-management model that classifies animal species into divergent natural selection categories. *Nature, 9*, 6111. https://doi.org/10.1038/s41598-019-42562-7
Chaline, J., David, B., Magniez-Jannin, F., Dambricourt Malassé, A., Marchand, D., Courant, F., et al. (1998). Quantification de l'évolution morphologique du crâne des hominidés et hété-

rochronies. *Comptes rendus de L'Academie des Sciences Series II Sciences de la Terre et des Planetes, 326*, 291–298.
Chapman, T., Arnqvist, G., Bangham, J., & Rowe, L. (2003). Sexual conflict. *Trends in Ecology and Evolution, 18*, 41–47.
Charlesworth, B. (1980). *Evolution in age-structured populations*. London: Cambridge University Press.
Charlesworth, B. (2000). Fisher, Medawar, Hamilton and the evolution of aging. *Genetics, 156*, 927–931.
Charlesworth, B., & Partridge, L. (1997). Ageing: Levelling of the grim reaper. *Current Biology, 7*, R440–R442.
Charnov, E. L. (1993). *Life history invariants: Some explorations of symmetry in evolutionary ecology*. Oxford: Oxford University Press.
Cichoń, M. (1997). Evolution of longevity through optimal resource allocation. *Proceedings of the Royal Society of London B, 264*, 1383–1388.
Claudio, B. L. (2018). *The study of historical and philological papyrology: Case studies "new Sappho" and "newest Sappho"*. Master of Arts Thesis, Faculty of Graduate Studies, The University of Manitoba. Department of Classics, University of Manitoba, Winnipeg, Manitoba, ON.
Comfort, A. (1954). Biological aspects of senescence. *Biological Reviews, 29*, 284–329.
Cordella, M., & Huang, H. (Eds.). (2016). *Rethinking second language learning: Using intergenerational community resources*. Bristol: Multilingual Matters.
Cordella, M., & Poiani, A. (2014). *Behavioural oncology: Psychological , communicative, and social dimensions*. New York: Springer.
Cox, D., & Stark, O. (2005). On the demand for grandchildren: Tied transfers and the demonstration effect. *Journal of Public Economics, 89*, 1665–1697.
Croft, D. P., Brent, L. J. N., Franks, D. W., & Cant, M. A. (2015). The evolution of prolonged life after reproduction. *Trends in Ecology and Evolution, 30*, 407–416.
Croft, D. P., Johnstone, R. A., Ellis, S., Nattrass, S., Franks, D. W., Brent, L. J. N., et al. (2017). Reproductive conflict and the evolution of menopause in killer whales. *Current Biology, 27*, 298–304.
Cubo, J., Berge, C., Quilhac, A., de Margerie, E., & Castanet, J. (2002). Heterochronic patterns in primate evolution : Evidence from endochondral ossification. *European Journal of Morphology, 40*, 81–88.
De Beer, G. R. (1930). *Embriology and evolution*. Oxford: Clarendon Press.
de Rovirosa-Hernández, M. J., Hernández González, M., Guevara-Pérez, M. A., García-Orduña, F., de los Aguilar-Tirado, A., Puga-Olguín, A., et al. (2017). Menopause in nonhuman primates: A comparative study with humans. In J. F. Rodríguez-Landa (Ed.), *A multidisciplinary look at menopause* (pp. 28–48). Intech Open. https://doi.org/10.5772/intechopen.69657
Doran, D. M. (1992). The ontogeny of chimpanzee and pigmy chimpanzee locomotor behaviour: A case study of paedomorphism and its behavioural correlates. *Journal of Human Evolution, 23*, 139–157.
Emerson, A. E. (1960). The evolution of adaptation in population systems. In S. Tax (Ed.), *Evolution after Darwin* (pp. 307–348). Chicago: University of Chicago Press.
Emery Thompson, M., Jones, J. H., Pusey, A. E., Brewer-Marsden, S., Goodall, J., Marsden, D., et al. (2007). Aging and fertility patterns in wild chimpanzees provide insights into the evolution of menopause. *Current Biology, 17*, 2150–2156.
Engelhardt, S. C., Bergeron, P., Gagnon, A., Dillon, L., & Pelletier, F. (2019). Using geographic distance as a potential proxy for help in the assessment of the grandmother hypothesis. *Current Biology, 29*, 651–656.
Finch, C. (1998). Variations in senescence and longevity include the possibility of negligible senescence. *Journal of Gerontology, 53A*, B235–B239.
Finch, C. E. (2007). *The biology of human longevity: Inflammation, nutrition, and aging in the evolution of life spans*. Amsterdam: Elsevier.

Finch, C. E. (2009). Update on slow aging and negligible senescence - A mini-review. *Gerontology, 55*, 307–313.

Finch, C. E. (2010). Evolution of the human lifespan and diseases of aging: Roles of infection, inflammation, and nutrition. *Proceedings of the National Academy of Sciences of the United States of America, 107*(Suppl 1), 1718–1724.

Finch, C. E. (2012). Evolution of the human lifespan, past, present, and future: Phases in the evolution of human life expectancy in relation to the inflammatory load. *Proceedings of the American Philosophical Society, 156*, 9–44.

Finch, C. E., & Tanzi, R. E. (1997). Genetics of aging. *Science, 278*, 407–411.

Fisk Green, R., Devine, O., Crider, K. S., Olney, R. S., Archer, N., Olshan, A. F., et al. (2010). Association of paternal age and risk for major congenital anomalies from the National Birth Defects Prevention Study, 1997 to 2004. *Annals of Epidemiology, 20*, 241–249.

Fox, C. W., Wagner, J. D., Cline, S., Thomas, F. A., & Messina, F. J. (2011). Rapid evolution of lifespan in a novel environment: Sex-specific responses and underlying genetic architecture. *Evolutionary Biology, 38*, 182–196.

Freeman, S., & Herron, J. C. (2007). *Evolutionary analysis* (4th ed.). San Francisco, CA: Pearson Benjamin Cummings.

Fretts, R. C., Schmittdiel, J., McLean, F. H., Usher, R. H., & Goldman, M. B. (1995). Increased maternal age and the risk of fetal death. *The New England Journal of Medicine, 333*, 953–957.

Gaillard, J.-M., & Lemaître, J.-F. (2017). The Williams' legacy: A critical reappraisal of his nine predictions about the evolution of senescence. *Evolution, 71*, 2768–2785.

Garamszegi, L. Z. (Ed.). (2014). *Modern phylogenetic comparative methods and their application in evolutionary biology: Concepts and practice*. Berlin: Springer.

Gariépy, J.-L., Bauer, D. J., & Cairns, R. B. (2001). Selective breeding for differential aggression in mice provides evidence for heterochrony in social behaviours. *Animal Behaviour, 61*, 933–947.

Gavrilov, L. A., & Gavrilova, N. S. (2001). The reliability theory of aging and longevity. *Journal of Theoretical Biology, 213*, 527–545.

Gavrilov, L. A., & Gavrilova, N. S. (2003). The quest for a general theory of aging and longevity. *Science of Aging Knowledge Environment, 2003*, RE5.

Gavrilov, L. A., & Gavrilova, N. S. (2004). The reliability-engineering approach to the problem of biological aging. *Annals of the New York Academy of Science, 1019*, 509–512.

Gavrilov, L. A., & Gavrilova, N. S. (2006). Models of systems failure in aging. In P. M. Conn (Ed.), *Handbook of models for human aging* (pp. 45–68). Burlington, MA: Elsevier.

Gavrilov, L. A., & Gavrilova, N. S. (2011). Reliability theory of aging and longevity. In *2nd LEPAS Conference on the Biology and Economics of Aging*, June 16–17, 2011 University of Alicante, Spain.

Gavrilova, N. S., & Gavrilov, L. A. (2019). Are we approaching a biological limit to human longevity? *Journals of Gerontology Series A, Biological Sciences and Medical Sciences, 75*(6), 1061–1067. https://doi.org/10.1093/gerona/glz164

Gawande, A. (2015). *Being mortal: Illness, medicine, and what matters in the end*. London: Wellcome Collection.

Gemmell, N. J., Metcalf, V. J., & Allendorf, F. W. (2004). Mother's curse: The effect of mtDNA on individual fitness and population viability. *Trends in Ecology and Evolution, 19*, 238–244.

German, R. Z., Hertweck, D. W., Sirianni, J. E., & Swindler, D. R. (1994). Heterochrony and sexual dimorphism in the pigtail macaque (*Macaca nemestrina*). *American Journal of Physical Anthropology, 93*, 373–380.

Gill, S. K., Broussard, C., Devine, O., Fisk Green, R., Rasmussen, S. A., Reefhuis, J., et al. (2012). Association between maternal age and birth defects of unknown etiology—United States, 1997–2007. *Clinical and Molecular Teratology, 94*, 1010–1018.

Gladyshev, V. N. (2013). The origin of aging: Imperfectness-driven non-random damage defines the aging process and control of lifespan. *Trends in Genetics, 29*, 506–512.

Goldberg, J. (2003). Domestication et comportement. *Bulletin de la Société Zoologique de France, 128*, 175–181.

Gonzalez-Freire, M., de Cabo, R., Bernier, M., Sollott, S. J., Fabbri, E., Navas, P., et al. (2015). Reconsidering the role of mitochondria in aging. *Journal of Gerontology, 70*, 1334–1342.

Gonzalez-Lagos, C., Sol, D., & Reader, S. M. (2010). Large-brained mammals live longer. *Journal of Evolutionary Biology, 23*, 1064–1074.

Gould, S. J. (1977). *Ontogeny and phylogeny*. Cambridge, MA: Belknap Press.

Gould, S. J., & Lewontin, R. C. (1979). The spandrels of San Marco and the Panglossian paradigm: A critique of the adaptationist programme. *Proceedings of the Royal Society of London, Series B, 205*, 581–598.

Goyal, M. S., Blazey, T. M., Su, Y., Couture, L. E., Durbin, T. J., Bateman, R. J., et al. (2019). Persistent metabolic youth in the aging female brain. *Proceedings of the National Academy of Sciences of the United States of America, 116*, 3251–3255.

Graves, B. M. (2007). Sexual selection effects on the evolution of senescence. *Evolution and Ecology, 21*, 663–668.

Griffin, A. S. (2008). Naked mole-rat. *Current Biology, 18*, R844–R845.

Gubbels Bupp, M. R. (2015). Sex, the aging immune system, and chronic disease. *Cellular Immunology, 294*, 102–110.

Gurven, M., & Hill, K. (1997). Comment on: Hadza women's time allocation, offspring provisioning, and the evolution of long postmenopausal life spans, by K. Hawkes, J.F. O'Connell, N.G. Blurton Jones. *Current Anthropology, 38*, 566–567.

Gurven, M., Kaplan, H., & Gutierrez, M. (2006). How long does it take to become a proficient hunter? Implications for the evolution of delayed growth. *Journal of Human Evolution, 51*, 454–470.

Haas, R. H. (2019). Mitochondrial dysfunction in aging and diseases of aging. *Biology, 8*, 48. https://doi.org/10.3390/biology8020048

Hadfield, J. D. (2010). MCMC methods for multi-response generalized linear mixed models: The MCMCglmm R package. *Journal of Statistical Software, 33*, 1–22.

Hafner, M. S., & Hafner, J. C. (1984). Brain size, adaptation and heterochrony in geomyoid rodents. *Evolution, 38*, 1088–1098.

Haigis, M. C., & Yankner, B. A. (2010). The aging stress response. *Molecular Cell, 40*, 333–344.

Haldane, J. B. S. (1941). *New paths in genetics*. London: George Allen and Unwin Ltd.

Hamilton, W. D. (1963). The evolution of altruistic behaviour. *American Naturalist, 97*, 354–356.

Hamilton, W. D. (1964). The genetical evolution of social behaviour: I, II. *Journal of Theoretical Biology, 7*, 1–52.

Hamilton, W. D. (1966). The moulding of senescence by natural selection. *Journal of Theoretical Biology, 12*, 12–45.

Happel, M. F., & Frischknecht, R. (2016). Neuronal plasticity in the juvenile and adult brain regulated by the extracellular matrix. In F. Travascio (Ed.), *Composition and function of the extracellular matrix in the human body* (pp. 143–158). Rijeka: Intech.

Harman, D. (1956). Aging: A theory based on free radical and radiation chemistry. *Journal of Gerontology, 2*, 298–300.

Harrison, D. E., & Archer, J. R. (1989). Natural selection for extended longevity from food restriction. *Growth Development and Aging, 53*, 3–6.

Harvey, P. H., & Pagel, M. D. (1991). *The comparative method in evolutionary biology*. Oxford: Oxford University Press.

Haussmann, M., Winkler, D. W., O'Reilly, K. M., Huntington, C. E., Nisbet, I. C. T., & Vleck, C. M. (2003). Telomeres shorten more slowly in long-lived birds and mammals than in short-lived ones. *Proceedings of the Royal Society of London B, 270*, 1387–1392.

Hawkes, K. (2003). Grandmothers and the evolution of human longevity. *American Journal of Human Biology, 15*, 380–400.

Hawkes, K. (2020). Reproduction: The female perspective from an integrative and comparative framework. *Integrative and Comparative Biology, 9*, icaa029. https://doi.org/10.1093/icb/icaa029

Hawkes, K., & Coxworth, J. E. (2013). Grandmothers and the evolution of human longevity: A review of findings and future directions. *Evolutionary Anthropology, 22*, 294–302.

Hawkes, K., O'Connell, J. F., & Blurton Jones, N. G. (1997). Hadza women's time allocation, offspring provisioning, and the evolution of long postmenopausal life spans. *Current Anthropology, 38*, 551–577.
Hawkes, K., O'Connell, J. F., Blurton Jones, N. G., Alvarez, H., & Charnov, E. L. (1998). Grandmothering, menopause, and the evolution of human life histories. *Proceedings of the National Academy of Sciences of the United States of America, 95*, 1336–1339.
Herskind, A. M., McGue, M., Holm, N. V., Sorensen, T. I., Harvald, B., & Vaupel, J. W. (1996). The heritability of human longevity: A population-based study of 2872 Danish twin pairs born 1870–1900. *Journal of Human Genetics, 97*, 319–323.
Holland Jones, J. (2011). Primates and the evolution of long, slow life histories. *Current Biology, 21*, R708–R717.
Hollier, L. M., Leveno, K. J., Kelly, M. A., McIntire, D. D., & Cunningham, F. G. (2000). Maternal age and malformations in singleton births. *Obstetrics and Gynecology, 96*, 701–706.
Jacobsson, B., Ladfors, L., & Milsom, I. (2004). Advanced maternal age and adverse perinatal outcome. *Obstetrics and Gynecology, 104*, 727–733.
Judge, D. S., & Carey, J. R. (2000). Postreproductive life predicted by primate patterns. *Journal of Gerontology, 55A*, B201–B209.
Kaplan, H., Hill, J., Lancaster, J., & Hurtado, A. M. (2000). A theory of human life history evolution: Diet, intelligence, and longevity. *Evolutionary Anthropology, 9*, 156–185.
Kaplan, M. S., McFarland, B. H., Huguet, N., & Newsom, J. T. (2007). Physical illness, functional limitations, and suicide risk: A population-based study. *American Journal of Orthopsychiatry, 77*, 56–60.
Kenyon, C. J. (2010). The genetics of ageing. *Nature, 464*, 504–512.
Kirkwood, T. B. L. (1977). Evolution of aging. *Nature, 270*, 301–304.
Kirkwood, T. B. L. (1981). Repair and its evolution: Survival versus reproduction. In C. R. Townsend & P. Calow (Eds.), *Physiological ecology: An evolutionary approach to resource use* (pp. 165–189). Oxford: Blackwell Scientific.
Kirkwood, T. B. L. (1987). Immortality of germ-line versus disposability of the soma. In A. D. Woodhead & K. H. Thompson (Eds.), *Evolution of longevity in animals: A comparative approach* (pp. 209–218). New York: Plenum Press.
Kirkwood, T. B. L. (1989). DNA, mutations and aging. *Mutation Research, 219*, 1–7.
Kirkwood, T. B. L. (1997). The origins of human ageing. *Philosophical Transactions of the Royal Society of London B, 352*, 1765–1772.
Kirkwood, T. B. L. (2002). Evolution of ageing. *Mechanisms of Ageing and Development, 123*, 737–745.
Kirkwood, T. B. L. (2005). Understanding the odd science of aging. *Cell, 120*, 437–447.
Kirkwood, T. B. L., & Holliday, R. (1979). The evolution of ageing and longevity. *Proceedings of the Royal Society of London Series B, Biological Sciences, 205*, 531–554.
Kirkwood, T. B. L., Kapahi, P., & Shanley, D. P. (2000). Evolution, stress, and longevity. *Journal of Anatomy, 197*, 587–590.
Kirkwood, T. B. L., & Rose, M. R. (1991). Evolution of senescence: Late survival sacrificed for reproduction. *Philosophical Transactions: Biological Sciences, 332*, 15–24.
Kirkwood, T. B. L., & Shanley, D. P. (2005). Food restriction, evolution and ageing. *Mechanisms of Ageing and Development, 126*, 1011–1016.
Kollman, J. (1885). Das Uberwintern von europaischen Frosch- und Tritonlarven und die Umwandlung des mexikanischen Axolotl. *Verhandlungen der Naturforschenden Gesellschaft in Basel, 7*, 387–398.
Koster, J., Lukas, D., Nolin, D., Power, E., Alvergne, A., Mace, R., et al. (2019). Kinship ties across the lifespan in human communities. *Philosophical Transactions of the Royal Society of London, B, 374*(1780), 20180069.
Kyriazis, M. (2019). Ageing throughout history: The evolution of human lifespan. *Journal of Molecular Evolution*. https://doi.org/10.1007/s00239-019-09896-2
Lahdenperä, M., Lummaa, V., Helle, S., Tremblay, M., & Russell, A. F. (2004). Fitness benefits of prolonged post-reproductive lifespan in women. *Nature, 428*, 178–181.

Laird, R. A., & Sherratt, T. N. (2009). The evolution of senescence through decelerating selection for system reliability. *Journal of Evolutionary Biology, 22*, 974–982.
Leroi, A. M., Bartke, A., De Benedictis, G., Franceschi, C., Gartner, A., Gonos, E., et al. (2005). What evidence is there for the existence of individual genes with antagonistic pleiotropic effects? *Mechanisms of Ageing and Development, 126*, 421–429.
Liker, A., & Szekely, T. (2005). Mortality costs of sexual selection and parental care in natural populations of birds. *Evolution, 59*, 890–897.
Loeb, J., & Northrop, J. H. (1916). Is there a temperature coefficient for the duration of life? *Proceedings of the National Academy of Sciences of the United States of America, 2*, 456–457.
Lorenz, K. (1971). *Studies in animal and human behaviour* (Vol. 2). London: Methuen and Co., Ltd..
Lorenz, K. (1973). *Die Rückseite des Spiegels*. München: Piper.
Maklakov, A. A., Carlsson, H., Denbaum, P., Lind, M. I., Mautz, B., Hinas, A., et al. (2017). Antagonistically pleiotropic allele increases lifespan and late-life reproduction at the cost of early-life reproduction and individual fitness. *Proceedings of the Royal Society B, 284*(1856), 20170376.
Maklakov, A. A., & Immler, S. (2016). The expensive germline and the evolution of ageing. *Current Biology, 26*, R577–R586.
Maklakov, A. A., & Lummaa, V. (2013). Evolution of sex differences in lifespan and aging: Causes and constraints. *BioEssays, 35*, 717–724.
Maklakov, A. A., Rowe, L., & Friberg, U. (2015). Why organisms age: Evolution of senescence under positive pleiotropy? *BioEssays, 37*, 802–807.
Marsh, H., & Kasuya, T. (1984). Changes in the ovaries of the short-finned pilot whale, *Globicephala macrorhychus*, with age and reproductive activity. In W. F. Perrin, R. L. Brownell, & D. P. DeMaster (Eds.), *Reproduction in whales, dolphins and porpoises* (pp. 311–335). Cambridge, MA: Report of the International Whaling Commission.
Martin, B., Golden, E., Carlson, O. D., Egan, J. M., Mattson, M. P., & Maudsley, S. (2008). Caloric restriction: Impact upon pituitary function and reproduction. *Ageing Research Reviews, 7*, 209–224.
Maynard-Smith, J. (1964). Group selection and kin selection. *Nature, 201*, 1145–1147.
Maynard-Smith, J. (1989). *Evolutionary genetics*. Oxford: Oxford University Press.
McIntosh, G. C., Olshan, A. F., & Baird, P. A. (1995). Paternal age and the risk of birth defects in offspring. *Epidemiology, 6*, 282–288.
Medawar, P. B. (1946). Old age and natural death. *Modern Quarterly, 1*, 30–56.
Medawar, P. B. (1952). *An unsolved problem of biology*. London: H.K. Lewis.
Melamed, D., & Scott, D. W. (2012). Aging and neoteny in the B lineage. *Blood, 120*, 4143–4149.
Milewski, L. A. K. (2010). The evolution of ageing. *Bioscience Horizons, 3*, 77–84.
Mitteroecker, P., Gunz, P., Bernhard, M., Schaefer, K., & Bookstein, F. L. (2004). Comparison of cranial ontogenetic trajectories among great apes and humans. *Journal of Human Evolution, 46*, 679–698.
Monaghan, P., Charmantier, A., Nussey, D. H., & Ricklefs, R. E. (2008). The evolutionary ecology of senescence. *Functional Ecology, 22*, 371–378.
Montagu, A. (1989). *Growing young*. Massachusetts: Bergin and Garvey Publishers, Inc..
Montagu, M. F. A. (1955). Time, morphology, and neoteny in the evolution of man. *American Anthropologist, 57*, 13–27.
Mousiolis, A., Baroutis, G., Sindos, M., Costalos, C., & Antsaklis, A. (2013). Maternal age as a predictive factor of pre-term birth. An epidemiological study from 1999 to 2008 in Greece. *Journal of Obstetrics and Gynaecology, 33*, 28–31.
Murphy, M. P. (2009). How mitochondria produce reactive oxygen species. *Biochemical Journal, 417*, 1–13.
O'Dor, R. K., & Webber, D. M. (1986). The constraints on cephalopods: Why squid aren't fish. *Canadian Journal of Zoology, 64*, 1591–1605.

Orgel, L. E. (1963). The maintenance of the accuracy of protein synthesis and its relevance to ageing. *Proceedings of the National Academy of Sciences of the United States of America, 49*, 517–521.
Orgel, L. E. (1970). The maintenance of the accuracy of protein synthesis and its relevance to ageing: A correction. *Proceedings of the National Academy of Sciences of the United States of America, 67*, 1476.
Partridge, L. (2001). Evolutionary theories of ageing applied to long-lived organisms. *Experimental Gerontology, 36*, 641–650.
Partridge, L., & Barton, N. H. (1993). Optimality, mutation, and the evolution of ageing. *Nature, 362*, 305–311.
Partridge, L., & Gems, D. (2006). Beyond the evolutionary theory of ageing, from functional genomics to evo-gero. *Trends in Ecology and Evolution, 21*, 334–340.
Partridge, L., & Mangel, M. (1999). Messages from mortality: The evolution of death rates in the old. *Trends in Ecology and Evolution, 14*, 438–442.
Paul, C., & Robaire, B. (2013). Ageing of the male germ line. *Nature Reviews Urology, 10*, 227–234.
Pavelka, M. S. M., & Fedigan, L. M. (1991). Menopause: A comparative life history perspective. *Yearbook of Physical Anthropology, 34*, 13–38.
Pearl, R. (1928). *The rate of living: Being an account of some experimental studies on the biology of life duration*. New York: Alfred A. Knopf.
Peccei, J. S. (1995a). A hypothesis for the origin and evolution of menopause. *Maturitas, 21*, 83–89.
Peccei, J. S. (1995b). The origin and evolution of menopause: The altriciality-lifespan hypothesis. *Ethology and Sociobiology, 16*, 425–449.
Peccei, J. S. (2001). Menopause: Adaptation or epiphenomenon? *Evolutionary Anthropology, 10*, 43–57.
Péron, G., Bonenfant, C., Lemaitre, J.-F., Ronget, V., Tidiere, M., & Gaillard, J.-M. (2019). Does grandparental care select for a longer lifespan in non-human mammals? *Biological Journal of the Linnean Society, 128*(2), 360–372. https://doi.org/10.1093/biolinnean/blz078
Photopoulou, T., Ferreira, I. M., Kasuya, T., Best, P. B., & Marsh, H. (2017). Evidence for a postreproductive phase in female false killer whales *Pseudorca crassidens*. *Frontiers in Zoology, 14*, 1–24.
Pletcher, S. D., & Curtsinger, J. W. (1998). Mortality plateaus and the evolution of senescence: Why are old-age mortality rates so low? *Evolution, 52*, 454–464.
Post, S. G. (2005). Altruism, happiness, and health: It's good to be good. *International Journal of Behavioral Medicine, 12*, 66–77.
Priest, N. K., Mackowiak, B., & Promislow, D. E. L. (2002). The role of parental age effects on the evolution of aging. *Evolution, 56*, 927–935.
Promislow, D. (2003). Mate choice, sexual conflict, and evolution of senescence. *Behavior Genetics, 33*, 191–201.
Promislow, D. E. L., & Harvey, P. H. (1990). Living fast and dying young: A comparative analysis of life-history variation among mammals. *Journal of Zoology, 220*, 417–437.
Raff, R. A., & Wray, G. A. (1989). Heterochrony: Developmental mechanisms and evolutionary results. *Journal of Evolutionary Biology, 2*, 409–434.
Raichlen, D. A., & Alexander, G. E. (2014). Exercise, APOE genotype, and the evolution of the human lifespan. *Trends Neuroscience, 37*, 247–255.
Ratikainen, I. I., & Kokko, H. (2019). The coevolution of lifespan and reversible plasticity. *Nature Communications, 10*, 538. https://doi.org/10.1038/s41467-019-08502-9
Rauser, C. L., Mueller, L. D., Travisano, M., & Rose, M. R. (2009). Evolution of aging and late life. In T. Garland Jr. & M. R. Rose (Eds.), *Experimental evolution: Concepts, methods, and applications of selection experiments* (pp. 551–584). Berkeley, CA: California University Press.
Reefhuis, J., & Honein, M. A. (2004). Maternal age and non-chromosomal birth defects, Atlanta—1968–2000: Teenager or thirty-something, who is at risk? *Clinical and Molecular Teratology, 70*, 572–579.
Reynolds, J. C., Bwiza, C. P., & Lee, C. (2020). Mitonuclear genomics and aging. *Human Genetics, 139*, 381–399.

Reznick, D., Bryant, M. J., & Bashey, F. (2002). r- and K-selection revisited: The role of population regulation in life-history evolution. *Ecology, 83*, 1509–1520.
Ricklefs, R. E. (2008). The evolutionary ecology of senescence: The evolution of senescence from a comparative perspective. *Functional Ecology, 22*, 379–392.
Rose, M. R. (1982). Antagonistic pleiotropy, dominance, and genetic variation. *Heredity, 48*, 63–78.
Rose, M. R. (1991). *Evolutionary biology of ageing*. Oxford: Oxford University Press.
Rose, M. R., Burke, M. K., Shahrestani, P., & Mueller, L. D. (2008). Evolution of ageing since Darwin. *Journal of Genetics, 87*, 363–371.
Rose, M. R., Flatt, T., Graves, J. L., Greer, L. F., Martinez, D. E., Matos, M., et al. (2012). What is aging? *Frontiers in Genetics, 3*, 134.
Rose, M. R., & Graves, J. L., Jr. (1989). What evolutionary biology can do for gerontology. *Journal of Gerontology, 44*, B27–B29.
Sauer, M. V. (2015). Reproduction at an advanced maternal age and maternal health. *Fertility and Sterility, 103*, 1136–1143.
Schoen, C., & Rosen, T. (2009). Maternal and perinatal risks for women over 44 - A review. *Maturitas, 64*, 109–113.
Shanley, D. P., Sear, R., Mace, R., & Kirkwood, T. B. L. (2007). Testing evolutionary theories of menopause. *Proceedings of the Royal Society B, Biological Sciences, 274*, 2943–2949.
Sharma, R., Agarwal, A., Rohra, V. K., Assidi, M., Abu-Elmagd, M., & Turki, R. F. (2015). Effects of increased paternal age on sperm quality, reproductive outcome and associated epigenetic risks to offspring. *Reproductive Biology and Endocrinology, 13*, 35. https://doi.org/10.1186/s12958-015-0028-x
Shea, B. T. (1989). Heterochrony in human evolution: The case for neoteny reconsidered. *Yearbook of Physical Anthropology, 32*, 69–101.
Shostak, S. (2006). *The evolution of death: Why we are living longer*. Albany, NY: University of New York.
Skulachev, V. P., Holtze, S., Vyssokikh, M. Y., Bakeeva, L. E., Skulachev, M. V., Markov, A. V., et al. (2017). Neoteny, prolongation of youth: From naked mole rats to "naked apes" (humans). *Physiological Reviews, 97*, 699–720.
Sorenson Jamison, C., Cornell, L. L., Jamison, P. L., & Nakazato, H. (2002). Are all grandmothers equal? A review and a preliminary test of the "grandmother hypothesis" in Tokugawa Japan. *American Journal of Physical Anthropology, 119*, 67–76.
Stearns, S. C. (1992). *The evolution of life histories*. Oxford: Oxford University Press.
Straus, L. G. (1995). The upper paleolithic of Europe: An overview. *Evolutionary Anthropology, 4*, 4–16.
Street, S. E., Navarrete, A. F., Reader, S. M., & Laland, K. N. (2017). Coevolution of cultural intelligence, extended life history, sociality, and brain size in primates. *Proceedings of the National Academy of Sciences of the United States of America, 114*, 7908–7914.
Swerdlow, R. H., Koppel, S., Weidling, I., Hayley, C., Ji, Y., & Wilkins, H. M. (2017). Mitochondria, cybrids, aging, and Alzheimer's disease. *Progress in Molecular Biology and Translational Science, 146*, 259–302.
Szilard, L. (1959). On the nature of the aging process. *Proceedings of the National Academy of Sciences of the United States of America, 45*, 35–45.
Teater, B. (2016). Intergenerational programs to promote active aging: The experiences and perspectives of older adults. *Activities, Adaptation and Aging, 40*, 1–19.
Tower, J., & Arbeitman, M. (2009). The genetics of gender and life span. *Journal of Biology, 8*, 38. https://doi.org/10.1186/jbiol141
Tricola, G. M., Simons, M. J. P., Atema, E., Boughton, R. K., Brown, J. L., Dearborn, D. C., et al. (2018). The rate of telomere loss is related to maximum lifespan in birds. *Philosophical Transactions of the Royal Society of London B Biological Sciences, 373*(1741), 20160445.
Trivers, R. L. (1971). The evolution of reciprocal altruism. *Quarterly Review of Biology, 46*, 35–57.

Tuljapurkar, S. D., Puleston, C. O., & Gurven, M. D. (2007). Why men matter: Mating patterns drive evolution of human lifespan. *PLoS One, 2*(8), e785. https://doi.org/10.1371/journal.pone.0000785

van Heemst, D. (2010). Insulin, IGF-1 and longevity. *Aging and Disease, 1*, 147–157.

Vaupel, J. W. (2010). Biodemography of human ageing. *Nature, 464*, 536–542.

Vaupel, J. W., Baudisch, A., Dölling, M., Roach, D. A., & Gampe, J. (2004). The case for negative senescence. *Theoretical Population Biology, 65*, 339–351.

Vaupel, J. W., Manton, K. G., & Stallard, E. (1979). The impact of heterogeneity in individual frailty on the dynamics of mortality. *Demography, 16*, 439–454.

Viña, J., Gambini, J., Lopez-Grueso, R., Abdelaziz, M. K., Jove, M., & Borrás, C. (2011). Females live longer than males: Role of oxidative stress. *Current Pharmaceutical Design, 17*, 3959–3965.

Wallace, D. C. (2005). A mitochondrial paradigm of metabolic and degenerative diseases, aging, and cancer: A dawn for evolutionary medicine. *Annual Review of Genetics, 39*, 359–407.

Ward, E. J., Parsons, K., Holmes, E. E., Balcomb, K. C. I. I. I., & Ford, J. K. B. (2009). The role of menopause and reproductive senescence in a long-lived social mammal. *Frontiers in Zoology, 6*, 4.

Weismann, A. (1882). *Essays upon heredity and kindred biological problems*. Oxford: Oxford University Press.

Weismann, A. (1891). *Essays upon heredity and kindred biological problems, volume I* (2nd ed.). Oxford: Clarendon Press.

Williams, B. A., Lindquist, K., Sudore, R. L., Strupp, H. M., Willmott, D. J., & Walter, L. C. (2006). Being old and doing time: Functional impairment and adverse experiences of geriatric female prisoners. *Journal of the American Geriatrics Society, 54*, 702–707.

Williams, G. C. (1957). Pleiotropy, natural selection, and the evolution of senescence. *Evolution, 11*, 398–411.

Williams, G. C. (1966). *Adaptation and natural selection: A critique of some current evolutionary thought*. Princeton: Princeton University Press.

Williams, G. C. (1999). The Tithonus error in modern gerontology. *The Quarterly Review of Biology, 74*, 405–415.

Xirocostas, Z. A., Everingham, S. E., & Moles, A. T. (2020). The sex with the reduced sex chromosome dies earlier: A comparison across the tree of life. *Biology Letters, 16*, 20190867. https://doi.org/10.1098/rsbl.2019.0867

Zapico, S. C., & Ubelaker, D. H. (2013). mtDNA mutations and their role in aging, diseases and forensic sciences. *Aging and Disease, 4*, 364–380.

Chapter 10
Fulfilling Ageing

Abstract *Fulfilling ageing* provides our synthesis of old age and here we also review some approaches to helping older people live a more fulfilling life. We emphasise the intermingled individual and social dimensions of fulfilling ageing, through the idea of finding personal meaning in the experience of living with others.

Although old age is the outcome of biological processes that affect every human being, the way we age is not fully determined at birth and it is not the same for every one of us. The manners in which we relate to older people are not fixed either. On the other hand, old people were once young and most retain their memory of past experiences; therefore, they do not expect to be defined only by superficial views of how they look now.

Elders are agentic individuals, capable of thinking, creating, and doing, even though they are aware that their agency is continuously changing and subject to various constraints. Objective constraints can be either constructively and creatively incorporated into a positive new view of life, or they can be regarded as an obstacle, an impediment to achieve an ideal where ageing is perhaps seen as part of a lifelong race to success. The constraints of old age stand on a line of demarcation between ageing that can lead to happiness, and ageing that can lead to frustration. The idealistic notion of "successful ageing" carries with it the seeds of frustration and disappointment when our constraints make us less "competitive" and therefore less able to "succeed". Elders do not control the criteria for success, the criteria are set socially, and elders who do not fit within those criteria are more limited in their capacity for action, control fewer resources and are, therefore, more likely to be seen as "losers" in the race to success, their life being tainted by defeat. The problem is, the life course is walked only once, there is no second chance, if you are "defeated" when you reach the end of the course, that is it, you cannot turn back and start all over again in the hope to do better next time around. The current neoliberal obsession with individualism, competition, and success that puts the concept of

M. Cordella, A. Poiani, *Fulfilling Ageing*, International Perspectives on Aging 30,
https://doi.org/10.1007/978-3-030-60071-6_10

successful ageing squarely into the realm of ideologies, ignores the fate of the "losers". But in fact, there is no need to create "winners" and "losers" in the first place, if we prioritise life fulfilment over success. Fulfilment is not a commodity that is only available in short supply, we do not need to compete for it, we can all be satisfied with life in old age and feel fulfilled in our own way, nobody needs to miss out.

Therefore, as we cross the line of demarcation and leave the concept of "successful ageing" behind, we encounter *fulfilling ageing*. Fulfilling ageing—or the ability to effectively express capability and agency, leading to a personally satisfactory level of well-being, contentment, and adaptation within the self and with our surrounding environment as we become older—is not a new concept, the ancient Greeks already distinguished between the pleasure-obsessed hedonism and the life-fulfilling eudaimonia, but it is a concept that has been obscured in recent decades, under the influence of the "winners"/"losers" dichotomy.

Fulfilling ageing is open to dynamic change and adaptation as we age and accumulate more experiences in life. It can fit just about any worldview within a common set of general rules—as established in a national constitution or common law, for instance—, provided that those rules are not too restrictive and we do not negatively interfere with the right of others to also live their life in a fulfilling manner; and if we can help instead, so much the better.

10.1 Experiencing Fulfilling Ageing

Life fulfilment develops with the aid of our physical, cognitive, emotional, social, and spiritual capacities, and it gives meaning to our life past and future. A fulfilling life is not just a source of personal satisfaction but also a good source of legacy for future generations, a source of generativity. Our legacy can be as variegated as we wish it to be and it may affect many people or not so many, for a short or a long time, and through legacy we can find transcendence. In the old say, people are expected to leave a legacy by "planting a tree, having a child, and writing a book" before they die, in this way life may transcend the here and now to leave a trace that can endure into the future. In reality, we are free to be as imaginative as we like in the generative search for fulfilment, but there are some underlying objectives hidden in the old say that can have more general validity: preserve the environment for future generations, maintain our species through love and consideration for our family and beyond, produce knowledge and transmit the accumulated knowledge to others.

It is natural to hope for a fulfilling life in old age marked by good health, but life fulfilment can be experienced no matter what our physical conditions are, provided that we are not abandoned to the ravages of current contingencies and we stay largely in control of what course our life takes to the very end.

Although the path to life fulfilment is diverse and personal there are some general tools that can be developed through life that can help in building fulfilment. A better education may certainly help as it gives the elder greater mental flexibility and knowledge; financial resources are undoubtedly not to be dismissed as they help the elder to be more independent; we obviously hope for as good a body and mental health condition as we can master and this can be achieved through maintaining an active and engaged life; extending the plastic capabilities of our brain in particular requires continuous mental engagement with life, curiosity, and learning. Finding spiritual contentment and wisdom that help us make better sense of our life and the world we live in, leading to gerotranscendence, and nurturing social connectedness at all levels may be something that many elders would certainly appreciate, especially if they feel unsatisfied being alone. Loneliness is one of the most common impediments to fulfilling ageing that we hear from old people, especially when they are institutionalised. In fact, the integration of older people into society, or greater social embeddedness in a well-integrated social network, can allow them to create their own niche, an eldership, from which to relate with other members of society.

In the pursue of our fulfilment in life we may combine both ever-developing personal needs and interests, and our connectedness with others. In the end, it is not a matter of self vs the other; rather, fulfilment is experienced through a very personal blend of our individual and social experiences.

Therefore, living a fulfilling old age is not just in the hands of the elder, society at large also has a specific responsibility through providing appropriate care, preventing discrimination and prejudice based on senseless stereotypes that lead to stigma; fostering social integration of the elders, and making sure that government also accomplishes its role as the supporter of all people when people require support, including support for both elders and their caregivers, the latter may regularly require relief from their burden of care.

Individual differences in the way we live fulfilling ageing can be also affected by our cultural and ethnic background, which is something that contemporary societies must urgently consider given the current levels of globalisation and migratory movements around the world; but also the ever-increasing diversity of lifestyles, philosophies of life, world and political views that abound even within the same ethnic group living in the same country. Fulfilling ageing experienced within a diverse social matrix requires an improved sense of acceptance, compassion, and upholding not only the law but also the basic principles of democracy. Fulfilling ageing is a walk along a personal path in our common human journey.

Although ageing is a continuous process, there are some events that occur in life that can trace specific boundaries between before and after. One such event is undoubtedly retirement. There is no reason why retirement should mark the end or even a significant disruption of a life that has been fulfilling at work. In principle, retirement can simply be a change of ways to experience fulfilment or, in cases where life at work was stressful and frustrating, it could be the stage in life when fulfilment can finally start being seriously addressed. Self-identity can go through dynamic processes of change through retirement, in part retaining the old, in part acquiring the new, and always synthesising both into a coherent and dynamic view of the self.

Retirement can be highly fulfilling to the extent that older people engage in the activities of their choice, experiencing a state of flow and personal satisfaction. Planning for retirement whilst still at work can also help in smoothing the transition. Some people may like a bridge employment during the transition period, others may be ready for a full change in direction. The main issue is that retirement may last many years, and so the personal and social resources of the individual should be such that life fulfilment can be sustained and nurtured throughout the entire duration of the experience.

Living a fulfilling life in old age should also prepare us to experience a fulfilling death, for as long as we live in the awareness that life is finite and that each moment counts, even as we become increasingly vulnerable with ageing. Remaining optimistic through our old age helps in lifting our morale, strengthening our resolve to better enjoy each moment. Our optimism is in turn increased the more meaningful and fulfilling our life is to us. This may prepare us for a good death, free form fears, anxieties, and regrets.

Elders help themselves by living a fulfilling old age through initiatives that are grounded on personal activities and choices, and that are naturally incorporated in a meaningful way into their daily life. Such personal initiatives are as diverse as the interests, background, sensibilities, knowledge, and imagination of the elders can be. In the next section, we provide a brief summary of spontaneous personal and community-based activities for the elders.

10.2 Spontaneous Personal and Community Initiatives

Our biological functions ultimately deteriorate with ageing and therefore many personal and community initiatives have been spontaneously adopted to decrease the rate of decay of such functions in the elders, including the decay in our mental capabilities. Cognitive aspects such as memory are usually a prime concern. Mental and physical exercise can improve cognition through spontaneous activities, but more specific cognitive training programmes based on the use of brain plasticity, cognitive reserve, and scaffolding such as meditation and mindfulness are also available.

Spontaneous mentally and physically stimulating activities that involve social interactions, inter-generational experiences (with children of various ages or young adults), language "boosting" (e.g. spelling games), language-learning initiatives (learning a second language), life review, creative writing, memory training programmes, artistic activities such as theatre, dance; guided visits to museums, cinema; writing poetry, song writing and music performances, are all initiatives that can be easily implemented, usually at low cost. Such programmes must be specifically tailored to the needs of the group of elders involved, taking into consideration not only their age variability, gender diversity, and educational background, but also ethnicity and other relevant inter-individual differences. Broad physical health can be tackled not just through physical exercise and social engagement but also suggestions for a healthier diet, including calorie restriction (see Hartman-Stein and La Rue 2011, for details on those various initiatives).

A major issue for older people is loneliness, which may interfere with life fulfilment and ultimately lead to loss of life meaning, increased depression, and in the worst-case scenario loss of willingness to live, and potentially suicide. Loneliness can be tackled through many personal and community initiatives that do not require great organisation, logistics, and external resources. For start, older people who can access the Internet may be able to create and then nurture networks of relationships that can be meaningful (e.g. with friends, family) using that medium. The use of Skype, for instance, can add a greater level of realism and immediacy to communication at all spatial scales, including the intercontinental one. Zoom, a cloud platform for video and audio conferencing, allows more than one elder to engage in group conversations. Free Internet access can be available in public libraries in some countries, along with some introductory courses on how to use it. Use of technology aside, visits from relatives, and engaging with trusted neighbours and local community organisations can also make an important difference. Loneliness is better tackled when the relationship between the elder and others is ongoing and available whenever it is needed. Extended intervals of solitude only punctuated by occasional encounters may not be an effective solution for loneliness in all cases.

Loneliness is often the result of entrenched ageist stereotypes, prejudice, and discrimination. The elder may feel lonely simply because he/she is ostracised, nobody is interested in engaging in a conversation, nobody is available to offer support. But spontaneous community initiatives may suffice to decrease loneliness in the elders: family and friends networks inviting the elder to celebrations and gatherings along with keeping regular contact may be already of great help. Visits by grandchildren, whether they are young or not so young, are usually highly appreciated by elders interested in generativity and legacy.

When there is no immediate family that could provide support, these days most communities, whether they are urban or rural, provide access to opportunities to volunteer in activities that help others or the environment. Volunteering groups can offer services to the elders through assisting such elders in their activities of daily living, but also opportunities to the elders themselves as volunteers to increase social engagement, physical, and cognitive capabilities which can improve functioning and life meaning (Anderson et al. 2014).

Group activities can also focus more specifically on physical fitness and mental well-being, such as the popular laughing therapy and Tai Chi groups that can spontaneously gather in public parks, being open to members of the local community.

In Chap. 3 we mentioned *naturally occurring retirement communities* (NORC) that can allow groups of elders to share activities that can be organised through regular meetings and then delivered, and even promote engagement in initiatives with others within the community at large. Local self-organised clubs of elderly people can be an important natural community for participation in social activities, such as regular lunches, singing in a chorus, playing instruments, excursions, annual celebrations of the religious or secular kind that are typical of specific groups or ethnicities. In this case some financial assistance from local governments or in-kind support such as providing a safe meeting venue can further help strengthen any spontaneous initiative of the elders.

Through voluntarism elders can become a social capital for the rest of the community, and this is one of the reasons why some communities can and do organise themselves to be more elders-friendly (Stafford 2009). Elder-friendly communities self-organise in order to help address elders' basic needs, optimise the physical and mental health and well-being of the elders, maximise independence of the frail and those with disability whilst providing necessary help and support, and promote voluntary social and civic engagement of the elders (e.g. Shenk and Mahon 2009; Binstock et al. 2009–2010).

Ultimately, local communities of elders can constitute an *eldertopia*, or a:

> community that improves the quality of life for people of all ages by strengthening and improving the means by which the community protects, sustains, and nurtures its elders, and the elders contribute to the well-being and foresight of the community. An Eldertopia that is blessed with a large number of older people is acknowledged to be 'elder rich' and uses this richness to the advantage of all (Thomas 2009, p. 535).

Eldertopias are not necessarily static, they can be changeable in composition and dynamic in their activities, as they displace themselves across the spatial and population landscape, thus defining specific *elderscapes* (Katz 2009a, b) or *landscapes of ageing* (Cohen 1988). A classic example of a dynamic eldertopia is that of the *Grey Nomads*, *RVers*, or *Snowbirds* that we mentioned in Chap. 2. The landscape, however, can be also cultural and the cultural landscapes of the elders can change over time. Elders may embrace such changes, adapt to them, and incorporate them into their own sense of life fulfilment.

Broader communities can also integrate local elders by making their skills available to the residents for their benefit and for the improvement of their quality of life; see, for instance, the *Transition Towns* grassroots movement which is organised around the concept of environmental sustainability (Barnett and Mordey 2010).

10.3 Institutional and Professionally Guided Initiatives

Beyond the spontaneous initiatives of each individual and the self-organisation of natural communities, elders can be also helped in their journey to life fulfilment through professionally guided programmes that are offered by institutions, groups, or government, often in collaboration with families (e.g. Stuart and Boll Hansen 2009). We mentioned in previous chapters several such programmes, some of which are specifically aimed at supporting the elders in their cognitive and emotional experiences and in developing their capabilities (e.g. positive narrative groups, Ruini et al. 2014; emotional processing and regulation programmes, Scheibe and Carstensen 2010), other initiatives provide sensory stimulation that can have effects on cognition and emotions (e.g. music therapy, Hays et al. 2002); opportunities to develop social bonds and connectedness can be also included in specifically structured activities (e.g. animal-assisted therapy, Banks and Banks 2002; discussion groups, Cattan et al. 2005; participating in conservation programmes, Caissie and Halpenny 2003).

A first step for any intervention is to identify groups of elders who are particularly vulnerable and at risk of experiencing any kind of decline in their functions (Saliba et al. 2001). *Vulnerability* can be defined as "the exposure to contingencies and stress, and difficulty coping with them" (Chambers 1989, p. 1). Social exclusion is certainly a factor that can strongly affect the vulnerability of older persons, and greater vulnerability can compromise the ability to live life in a fulfilling manner (e.g. Schröder-Butterfill and Marianti 2006). Therefore, interventions that help the elderly to remain socially engaged in a meaningful manner are an important area of research and action (Morrow-Howell et al. 2003).

Inter-generational programmes have been vastly used around the world that allow older people to establish a relationship with younger generations, especially when elders are socially isolated or do not have a family or, worse still, their family members show little interest to stay in touch. Inter-generational programmes have demonstrated two-way benefits for the elder and for the young participants: the elder receives warmth and emotional support, an opportunity to develop greater meaningfulness, perform generative work and perhaps re-energise the journey to life fulfilment by staying in touch with the new world of younger generations; the young benefit from the experience of the elders, some specific skills they may have, their memories and their witnessing of a world long gone which often represents the roots of current international phenomena and concerns (wars, revolutions, environmental and political issues) (e.g. Lundgren 2010). Marisa Cordella has been involved in the implementation of one of such programmes in Australia based on interactions between older non-institutionalised immigrants and school-level language students. Both interacting generations not only learnt from each other in terms of their different life experiences, but the elders also became teachers of their native language that the younger participants were learning at school (Cordella and Huang 2016).

Many such inter-generational programmes are available around the world; see, for instance, *NewBridge*, a retirement community in Boston, USA, where elders share experiences with school children (Gawande 2015). A recent television programme: *Old People's Home For 4 Year Olds*, was transmitted by the Australian Broadcasting Corporation in 2019 describing an initiative: "that brings together elderly people in a retirement community with a group of 4-year-olds" (https://iview.abc.net.au/show/old-people-s-home-for-4-year-olds).

These initiatives allow retirees to live in as much an autonomous way as possible even if they are based in an institution. They foster the creation of an elders' own community, whilst enjoying the benefits of professionally organised and guided activities.

Whatever their residential arrangement is, the elders can also access programmes such as the *University of the Third Age* (U3A) an initiative that began in France in the 1970s which aims to:

1. Encourage and enable older people no longer in full-time paid employment to help each other to share their knowledge, skills, interests, and experience.
2. Demonstrate the benefits and enjoyment to be gained and the new horizons to be discovered in learning throughout life.
3. Celebrate the capabilities and potential of older people and their value to society.
4. U3A members regard themselves as both learners and teachers.
   (Adapted from Hubble and Tew 2013, p. 58).

The U3A extends the creative, active, and productive life experience into older ages in an environment that is both intellectually "challenging" and socially supportive, and that fosters social interactions among people who are equally motivated to keep growing as a person. Personal fulfilment is a central priority of this initiative.

Institutional care for older people is also fraught with problems that require specific professional attention, sometimes even legal intervention (Clarke et al. 2003). An environment of neglect and perhaps abuse is obviously useless to help in developing life fulfilment in old age. Best state-of-the-art assisted living should be not only affordable but also provide a safe and stimulating environment to the resident elders who can live as independently as they can. Support staff should be highly qualified and be always available to provide assistance when required.

Our evolutionary history has endowed us with the capacity for self-reflection and therefore the ability to think about ourselves, our life, and the meaning of it. We have also evolved the potential to live a long life, and now our challenge is what do we make of it. It is up to us, but also the social and institutional environment we live in, to either waste such life in old age or transform it into a fulfilling experience. We can only hope that reading this book may be a source of ideas and inspiration for others, just as writing it was for us.

References

Anderson, N. D., Damianakis, T., Kröger, E., Wagner, L. M., Dawson, D. R., Binns, M. A., et al. (2014). The benefits associated with volunteering among seniors: A critical review and recommendations for future research. *Psychological Bulletin, 140*, 1505–1533.

Banks, M. R., & Banks, W. A. (2002). The effects of animal-assisted therapy on loneliness in an elderly population in long-term care facilities. *Journal of Gerontology, 57A*, M428–M432.

Barnett, R., & Mordey, M. (2010). Honouring the elders: A transition model of inclusion for older people. *Working with Older People, 14*, 12–16.

Binstock, R. H., Sykes, K., & Reilly, S. (2009–2010). Imagining the American community environmental services: A vision for environmentalists and elders. *Generations, 33*, 75–81.

Caissie, L. T., & Halpenny, E. A. (2003). Volunteering for nature: Motivations for participating in a biodiversity conservation volunteer program. *World Leisure Journal, 45*, 38–50.

Cattan, M., White, M., Bond, J., & Learmouth, A. (2005). Preventing social isolation and loneliness among older people: A systematic review of health promotion interventions. *Ageing and Society, 25*, 41–67.

Chambers, R. (1989). Vulnerability, coping and policy. *Institute of Development Studies Bulletin, University of Sussex, 20*, 1–7.

Clarke, A., Hanson, E. J., & Ross, H. (2003). Seeing the person behind the patient: Enhancing the care of older people using a biographical approach. *Journal of Clinical Nursing, 12*, 697–706.

Cohen, G. D. (1988). Do health science concepts influence care?: The case for a new landscape of aging. *The American Journal of Geriatric Psychiatry, 6*, 273–276.

Cordella, M., & Huang, H. (Eds.). (2016). *Rethinking second language learning: Using intergenerational community resources*. Bristol: Multilingual Matters.

Gawande, A. (2015). *Being mortal: Illness, medicine, and what matters in the end*. London: Wellcome Collection.

Hartman-Stein, P. E., & La Rue, A. (Eds.). (2011). *Enhancing cognitive fitness in adults: A guide to the use and development of community-based programs*. New York: Springer.

Hays, T., Bright, R., & Minichiello, V. (2002). The contribution of music to positive aging: A review. *Journal of Aging and Identity, 7*, 165–175.

Hubble, N., & Tew, P. (2013). *Ageing, narrative and identity: New qualitative social research*. Hampshire: Palgrave Macmillan.

Katz, R. (2009a). Intergenerational family relations and subjective well-being in old age: A cross-national study. *European Journal of Ageing, 6*, 79–90.

Katz, S. (2009b). Spaces of age, snowbirds and the gerontology of mobility: The elderscapes of Charlotte County, Florida. In J. Sokolovsky (Ed.), *The cultural context of aging: Worldwide perspectives* (pp. 463–486). Westport, CT: Praeger.

Lundgren, A. S. (2010). "In the good old days": Insidious nostalgia and the constitution of old age identity. *Journal of Aging Studies, 24*, 248–256.

Morrow-Howell, N., Hinterlong, J., Rozario, P. A., & Tang, F. (2003). Effects of volunteering on the well-being of older adults. *Journal of Gerontology, 58B*, S137–S145.

Ruini, C., Masoni, L., Ottolini, F., & Ferrari, S. (2014). Positive Narrative Group Psychotherapy: The use of traditional fairy tales to enhance psychological well- being and growth. *Psychology of Well-Being: Theory, Research and Practice, 4*, 13. http://www.psywb.com/content/4/1/13

Saliba, D., Elliott, M., Rubenstein, L. Z., Solomon, D. H., Young, R. T., Kamberg, C. J., et al. (2001). The vulnerable elders survey: A tool for identifying vulnerable older people in the community. *Journal of the American Geriatrics Society, 49*, 1691–1699.

Scheibe, S., & Carstensen, L. L. (2010). Emotional aging: Recent findings and future trends. *Journal of Gerontology, 65B*, 135–144.

Schröder-Butterfill, E., & Marianti, R. (2006). A framework for understanding old-age vulnerabilities. *Ageing Society, 26*, 9–35.

Shenk, D., & Mahon, J. (2009). Standing up for others: The AMIGOS volunteer model for working with elders in Peru. In J. Sokolovsky (Ed.), *The cultural context of aging: Worldwide perspectives* (pp. 259–265). Westport, CT: Praeger.

Stafford, P. B. (2009). Aging in the hood: Creating and sustaining elder-friendly environments. In J. Sokolovsky (Ed.), *The cultural context of aging: Worldwide perspectives* (pp. 441–452). Westport, CT: Praeger.

Stuart, M., & Boll Hansen, E. (2009). Danish home care policy and the family: Implications for the United States. In J. Sokolovsky (Ed.), *The cultural context of aging: Worldwide perspectives* (pp. 266–276). Westport, CT: Praeger.

Thomas, W. (2009). Eldertopia: A vision for old age in a new world. In J. Sokolovsky (Ed.), *The cultural context of aging: Worldwide perspectives* (pp. 533–535). Westport, CT: Praeger.

Index

A

Abuse, 45, 122, 141, 155–159, 375, 379, 415, 487–493, 525, 550, 564, 570, 573, 574, 576, 664, 670, 813, 818, 823, 841, 919, 944
ageism, 488, 491
alcoholism, 490
Applied Ecological Bi-Focal Intergenerational Model, 491
caregivers, 45, 156, 158, 375, 488–493
caregiver Stress theory, 488
"cognitive mask", 492
conflicts, 488, 823
cultures, 121, 122, 141, 156, 488, 490, 919
dementia, 156, 157, 375, 489, 490
deprivation, 488
discrimination, 122, 155, 159, 488, 492, 525
fear of retaliation, 492
financial, 156–158, 416, 487, 489
granny battering, 487, 488
habits, 488
helplines, 492
infantilising the elder, 488
injuries, 492
institutionalised elders, 489
International Network for the Prevention of Elder Abuse (INPEA), 525
material, 157, 487, 488, 493
media, 489, 492
multiples, 156, 491
neglect, 158, 487, 488, 919
nursing homes, 157, 488–490
oppression, 491
perpetrators, 157, 158, 375
physical, 45, 122, 156, 158, 375, 379, 416, 487–490, 492, 493, 570, 574, 841
Positioning theory, 488
power, 488
Power and Control theory, 488
power imbalances, 491
powerlessness, 159, 488
prevent, 158, 159, 492
psychological, 45, 156–158, 487–489, 492, 493, 573, 574
sadistic acts, 157
sexual, 156, 416, 487, 489, 574, 576
shame, 159, 488, 492
signs, 492, 493
social ecology theory, 488
social exchange theory, 488
social learning theory, 488
stresses, 488, 489, 570, 574, 823
suffering, 492, 550
violence, 158, 375, 488, 489, 491, 493, 570, 574
Acculturation, 145, 147, 166, 168, 169, 548, 550, 552, 553, 851
Achenbaum, W.A., 44, 58, 91, 95, 96, 421, 532
Activities of daily living (ADL), 2, 16, 17, 25, 160, 267, 292, 294, 295, 310, 311, 313, 316, 390, 483, 487, 497, 511, 519, 542, 543, 580, 581, 584, 585, 589, 598, 599, 655, 941
Activity theories, 31, 33, 463
Adrenal glands, 14, 279, 415
Advanced Cognitive Training for Independent and Vital Elderly (ACTIVE) program, 263

M. Cordella, A. Poiani, *Fulfilling Ageing*, International Perspectives on Aging 30,
https://doi.org/10.1007/978-3-030-60071-6

Aegidius Romanus, 83
Aeschylus, 819
Afterlife, 122, 148, 186, 187, 679, 682, 722, 775, 778, 790, 794–797, 819, 858
Age
ages of life, 59–63, 65, 90, 122
chronological, 39, 60, 64, 120, 316, 360, 388, 469, 534, 538, 658, 896, 924
fourth, 39, 104, 241
ideal, 85, 652, 937
integrations, 20, 124, 386, 809
pensions, 1–3, 7, 60, 67–69, 98, 104, 105, 182, 420, 465
social, 7, 25, 33, 35, 38, 46, 57, 60, 64–66, 68, 70, 73, 77, 80, 81, 85, 86, 91, 98, 99, 101, 110–113, 121, 124, 128, 129, 131, 150, 157, 159, 178, 181, 216, 218, 226, 231, 232, 238, 269, 270, 274, 277–279, 283, 284, 286, 294–296, 301, 305, 310, 311, 316–319, 360, 372, 377, 378, 380, 381, 388, 405, 407, 413, 414, 419–421, 426, 462–464, 466, 468, 473, 476, 480, 487, 501, 503, 506, 508, 509, 515, 520, 523, 530, 531, 534, 543, 559, 561, 564, 565, 577, 596, 599, 635, 638, 656, 663, 664, 676, 678, 680, 684, 685, 690, 724, 729, 737, 742, 760, 761, 767, 779, 797, 830, 840, 864, 940, 944
subjective, 37, 215, 285, 288, 360, 378, 388, 392, 400, 414, 679
Aged care, 7, 130, 480, 490, 564, 571, 583, 596, 656, 723, 761
Ageing (senescence)
active ageing, 33, 162
"age as adventure", 108
ageing in place, 33, 506, 513, 546, 570, 581
ageing well, 27, 127–129, 819
apoptosis, 9, 11, 211, 212, 252, 906
bladder, 294
blood clots (thromboses), 203
blood pressure, 25, 36, 203, 235
bones, 202, 203, 415
bystander damage, 202
cancer, 11, 17, 18, 202, 203, 225, 297, 300, 309, 311, 312, 415, 528, 926
cardiovascular system, 204
codon restriction, 9
cognitive decline, 36, 203, 217, 220, 236, 239, 244, 246, 247, 253, 257, 260, 276, 278, 292, 317, 319, 415, 417
differential ageing, 541
digestive system, 164, 203, 298
discourse on, 39, 43, 681
DNA damage, 417, 902
dysdifferentiation, 9
epigenomics of ageing, 314
error accumulation, 9
error catastrophe, 9
free radicals, 9, 10, 202, 205, 261, 895, 905, 906
free radical theory of ageing, 906
fulfilling ageing, 8, 18, 26–35, 38, 40, 43, 46, 73, 110, 121, 127–137, 175, 188, 213, 215, 216, 222, 292, 299, 300, 310, 314, 317–319, 322, 355, 361, 362, 367, 391, 406, 408, 412, 465, 475, 513, 520, 538, 543, 547, 556, 559, 561, 586, 676, 679, 684, 691, 692, 694, 708, 728, 937–944
gene regulation, 9, 316
genomic instability, 9
hair, 203, 274, 300, 527, 891, 924, 925
hearing, 203, 235, 265, 295, 297, 298, 308, 392, 529, 637, 641
heritability, 10, 250, 251, 303, 314, 315, 897
increased glial cells activation, 205
instant, 664
irreversible protein glycation, 205
kidneys, 203, 260, 906, 925
landscapes of ageing, 107, 942
life course paradigm, 23
muscles, 12, 14, 16, 25, 203, 204, 206, 234, 235, 245, 246, 295
neuroerosion, 356, 358
oxidative damage, 10, 205, 249, 316
pain, 41, 43, 203, 286, 291, 292, 294, 295, 313, 316, 577
primary ageing, 314
programmed ageing, 8, 314, 407, 905
raka, 123
rate-of-living, 906
respiratory system, 203
secondary ageing, 273, 314
skin, 14, 203, 204, 274, 309, 417, 891, 900, 924
smell, 203, 376
somatic mutation, 9, 905, 906
stroke, 203, 214, 251, 262, 309, 312
subjective ageing, 388, 392
successful ageing, 26, 27, 30, 35, 465, 562, 679, 681, 740, 937
taste, 203, 265, 295, 520
telomere theory, 9
temperature perception, 203

touch (sensitivity to), 133, 203, 204, 265, 283, 300, 515, 637, 943
vascular complications, 205
vision, 203, 235, 308
wear-and-tear, 9, 148, 205, 314, 892, 905
wrinkles, 203, 273, 417, 527
Ageism, 27, 152, 153, 530, 532–539, 599, 705, 708
ageist stereotypes, 533, 535
benign, 533
changes, 535, 538, 539, 708
cognitive dissonance, 539
commonsense reality, 535, 539
compassionate, 533, 534, 539
discrimination, 68, 100, 532–536, 538, 539, 565, 708
gerontophobia, 533
greedy geezers, 534, 535
implicit, 533, 535
inhibiting, 539
institutionalised, 535, 539
internalisation, 535
intra-generational, 535
malignant, 533, 534, 539
positive image, 492, 537
positive self-perception, 529, 537
self-disclosure, 538
social identity theory (SIT), 536
"spoiled identity", 99
terror management theory, 536
theories of ageism, 537
tit-for-tat, 537
Agency, 23, 24, 175, 363, 364, 464, 578, 634, 653, 677, 683, 690, 728, 729, 740, 808, 809, 821, 851, 937
agentic process, 34
existential, 728
identities, 363, 634, 653, 690
life course, 23, 363, 364, 937
pragmatic, 363, 633
Ages of life
ages of men, 60, 63
first age, 60, 61, 65
four ages of man, 88
fourth age, 60
Generation X, 66, 644
G.I. Generation, 66
maturity, 60, 63
millennials, 66, 676
old ages, 60–62, 64, 65, 122
older boomers, 66
old-old age, 60
old testament, 61
pythagorean seasonal metaphor, 61
sannyasi, 129
second age, 314
senility, 61
silent generation, 66
third age, 469, 474, 943
vanaprastha, 129, 752
Younger Boomers, 66
young-old age, 60
youth, 60, 61, 63
Aging in Manitoba (AIM) study, 145
Alexander, R., 901, 924
Almshouse, 94, 106
Aloneness, 36, 478, 486
Alzheimer, A., 243
Alzheimer's disease (AD)
aggression, 245
β-amyloid (Aβ), 249–251, 398
β-amyloid accumulation, 247
Amyloid Cascade Hypothesis, 249
β-amyloid plaques (AβP), 248, 249, 251
amyloid precursor protein (APP), 251
ApoE2 allele (ε2), 315
ApoE3 allele (ε3), 398
ApoE4 allele (ε4), 243, 315, 398, 399
ApoE gene, 398
ApoE peptide, 250, 251, 398
apolipoprotein E (ApoE), 398
apoptosis, 315
Attentional Network Task (ANT), 259
bilingualism, 260
brain blood flow, 251
cell microtubules, 251
cerebral amyloid angiopathy (CAA), 248
cholesterol, 398
cholesteryl ester transferase protein (CETP), 250
chromosome
1, 315
9, 250
10, 250
14, 246
21, 246, 251
chronic depression, 269
chronic infections, 251
chronic psychological stress, 248
cognitive declines, 217
cognitive/emotion-oriented interventions, 263
cognitive reserve, 269
confusion, 245
cortical thinning, 247
Dementia Australia, 267
dietary antioxidants, 261
disorientation, 244, 245

Alzheimer's disease (AD) (*cont.*)
double-strand breaks in DNA, 246
down syndrome, 251
early-onset AD, 249, 250
emotional dysregulation, 252
endoplasmic reticulum, 250
entorhinal cortex, 252, 253, 255
environmental manipulation interventions, 266
forgetfulness, 150, 244, 245
granulin gene (*GRN*), 251
grey matter, 252, 257
hallucinations, 245
hippocampus, 212, 248, 251–253, 255, 261
HIV-AIDS, 251
hyper-phosphorylated tau, 251
incontinence, 245
inflammation, 14, 202, 203, 231, 248, 315, 417
late-onset AD, 250, 251
memories, 269, 398, 399, 472
microtubule-associated protein tau gene (*MAPT*), 251
mitochondria-associated ER membranes (MAMs), 250
multilingualism, 260
muscle atrophy, 245
narrative dissolution, 258
necrosis, 252
neocortex, 251, 252
neural compensation, 257
neural reserve, 484
neuritic plaques (β-amyloid plaques), 244, 248
neurodegeneration, 245, 253
neurofibrillary tangles, 247–249, 251, 484
obstructive sleep apnoea, 251
polyunsaturated fatty acids, 261
presenilins, 251
prevalence, 239
PSEN gene, 251
retrogenesis ("last in, first out") hypothesis, 253
risk factors, 269
segmental progeroid syndromes, 245
sensory stimulation interventions, 263, 266
sleeplessness, 245
SORL1, 246
synapses, 248, 249, 251, 253
synaptic loss, 253
trilingualism, 260
use it or lose it approach, 255
vascular abnormalities, 251
American Psychological Association, 404
American Society on Aging, 532
Amici principis, 77
Ancestors, 71, 76, 120, 122, 134, 183, 310, 475, 536, 782, 783, 789, 790, 796, 797, 863, 910, 921, 922, 926
Anorexia, 12, 272
Antagonistic pleiotropic effects, 11
Anthropometric measurements
body mass index (BMI), 17
body weight, 17
fat-free mass, 17
fat mass, 17
height, 17
hip circumference, 17
waist-to-hip ratio, 17
Anthropophagy, 78
Antioxidants, 205, 248, 261, 310
catalase, 205
glutathione peroxidase, 205
superoxide dismutase, 205
Antonucci, T., 133, 137, 141, 283, 318, 472, 495, 508
Anxiety, 34, 121, 125, 144, 147, 148, 185, 186, 217, 228, 230, 234–238, 266, 267, 270, 284, 297, 299, 371, 374, 377, 399, 418, 425, 476–478, 487, 560, 562, 575, 591, 638, 656, 726, 762, 774, 775, 778, 781, 782, 786, 790, 791, 793, 796, 800, 809, 835, 837, 862
chronic, 236, 784, 862
death, 184, 185, 187, 189, 226, 577, 578, 697, 726, 727, 774–779, 782–796, 806, 864, 892
existential, 189, 223, 477, 727, 775, 777, 782, 786, 796
Apolipoprotein E (APOE) gene, 10
Apoptosis, 9, 11, 211, 212, 252, 315, 906
Appetite
ghrelin, 12
reduction of appetite, 12
Archetypes, 724
Area Agencies on Aging (AAA), 571
Aristotle, 28, 60, 72–74, 83, 88, 96, 378, 820, 828
Arteries
thickening of arterial walls, 18
Asperger's syndrome, 764, 765
Atherosclerosis, 231, 244
Attachment, 35, 36, 74, 106, 137, 218, 220, 236, 283, 365, 467–469, 475, 502, 542, 544, 565, 688, 693, 727, 808, 861–863
figures, 35, 469, 470, 478, 808

Attention restoration theory, 511
Attention span, 19
Auschwitz, 754, 782
Australian
  aboriginal, 2, 123
  Australian Pensioners League, 104
Authenticity, 41, 42, 379, 730
  accommodation, 42
  assimilation, 42
Autoimmunity, 9, 12
Autonomy
  autonomous, 541, 852
  heteronomous, 28
Autophagy, 11, 903
Awareness, 28, 183, 186, 239, 356, 378, 394, 522, 528, 531, 533, 537, 539, 572, 637, 646, 661, 689, 706, 722, 726, 731, 744, 747, 748, 776, 781–783, 800, 848, 940

B

Bacon, F., 820
Baltes, P., 20, 22, 32, 39, 42, 175, 203, 213, 308, 317, 356, 357, 386, 502, 540, 891
Baltimore Longitudinal Study of Aging, 276
Bandura, A., 34, 271, 363, 364, 488, 575, 851, 853
Basel Prospective Study, 400
Baumeister, R., 217, 736, 738, 739, 762, 763, 806, 840
Beauvais, Vincent of, 81
Belongingness, 479, 556, 565
  connectedness, 479, 485
  embeddedness, 479
Bereavement, 183, 231, 232, 236, 285, 417, 480, 481, 571, 577, 657, 658, 722, 737, 759, 802, 839, 841, 842, 861–863
Berlin aging study, 237, 303, 321, 383, 542, 811
Betula project, 277
Big Five, 274, 275, 406–410
Bioethics, 516, 817, 851, 852, 855
  intolerable cultural practices, 855
  tolerable cultural practices, 855
Biological clocks, 12
Blood
  blood vessels, 18
  capillaries, 18
  pressure, 25, 36, 203, 235, 262, 399, 485, 595
Blue Zones, 309, 313
Body
  images, 3, 273, 274, 289, 294, 520
  mass, 12, 17, 235, 262, 402, 898, 902
  mass index, 17, 96, 235, 402
Bolk, L., 922–924
Bone
  fractures, 14
  mineral density, 14, 233, 906
Brain
  amygdala, 210, 212, 231, 252, 374
  anterior pituitary gland, 279
  astrocytes, 18
  attractors, 358
  auditory cortex, 210
  axo-dendritic synapses, 211
  basal ganglia, 210, 253, 256
  blood oxygenation level dependent (BOLD) response, 211
  brain reserve capacity, 20, 381, 385
  calcarine cortex, 19
  caudatum, 19
  cerebellar cortex, 210
  cerebellum, 19, 205, 256
  corpus callosum, 19
  de-differentiation, 20
  default mode network, 213, 255
  dentate gyrus, 211, 212, 252
  diaschisis, 212
  entorhinal cortices, 19, 205, 210, 252, 253, 255
  frontal lobe, 20, 358
  fronto-executive network, 213
  fronto-parietal network, 213
  fusiform cortex, 19
  glial cells, 19, 211, 252
  hemodynamic response, 211
  hippocampus, 19, 211, 212, 214, 231, 248, 251, 253
  hypothalamus, 279
  neuroerosion, 356, 358
  neuronal loss, 19, 204, 249
  nucleus accumbens, 214
  olfactory bulb, 211
  plasticity, 19, 20, 22, 44, 66, 201, 206, 207, 210–215, 263, 315, 382, 385, 393, 399, 414, 729, 742, 805, 810, 904, 924, 926, 940
  prefrontal cortex, 19, 22, 212, 214, 252, 256, 257
  primary somatosensory cortex, 210
  reorganization, 20
  selection, 20, 257, 278
  size, 20, 96, 204, 214, 898
  tertiary association cortices, 19

Brain (*cont.*)
upregulation, 20
visual cortex, 19, 210, 214
white matter, 19, 204, 231, 252, 257, 385
Brain plasticity
adaptiveness, 215
adaptive plasticity, 207
apoptosis, 211, 212
axonal branches, 210
bilateral activation, 208
cognitive flexibility, 212
cognitive reserve, 212, 213, 215
compensatory masquerade (compensation), 212
"compensatory-related utilization of neuronal circuits" (CRUNCH), 208
dedifferentiation, 212
dendritic spines, 212
diet, 212
elasticity, 203, 207
exercise, 212
experience-dependent neuroplasticity, 212
extra-region recruitment, 212
functional connectivity, 212
functionally redundant, 207
functional plasticity, 207
Hebb, D., 207
Hebb's law, 207
"hemispheric asymmetry reduction in old age" (HAROLD), 208
inter-neuronal connectivity, 210
Lugaro, E., 207
maladaptive plasticity, 208
map expansion, 207, 208
mirror region takeover, 208
neurogenesis, 212, 213
neuroplasticity, 215
"plastic paradox", 208
reactive neuroplasticity, 212
recovery, 212
redundancy, 208
reorganisation, 212
restitution, 212
robustness (stability), 207
sensory reassignment, 207
stresses, 212, 215, 382
structural plasticity, 207
synaptogenesis, 212
Tanzi, E., 207
Vicariance, 208
Brain reserve capacity
brain plasticity, 20, 382, 385
brain reserve capacity hypothesis, 20, 23
scaffolding, 20, 23, 381
Scaffolding Theory of Ageing and Cognition (*STAC*), 20, 23, 382
Breitbart, W., 222, 223, 767
Bridge employment, 173, 177, 430
Bridge, W., 89, 90
British Household Panel Study (BHPS), 407
Bromhall, C., 923
Bronowski, J., 763
Bühler, C., 756
Bulbulia, J., 807, 808
Burden of dependency, 100, 101, 103

C

Calcium
metabolism, 19
Caloric
intake, 402
restrictions, 10, 310
Camus, A., 735
Cancer
tumour, 17
Capabilities
crystallised abilities, 8, 393
fluid abilities, 8, 393
Capillaries, 18
Carbohydrate intolerance, 12
Caregiving, 65, 123, 160, 161, 164, 269, 270, 278, 510, 581, 583, 584, 587, 589, 590, 592–599, 655, 695, 696, 920
arrow of time reciprocation, 589, 663, 919
burden, 150, 151, 160, 161, 164, 270, 587–595, 597, 599, 695
burnout, 587, 592, 599
care provided by the family, 551, 581
caregivers, 149, 151, 160, 164, 266, 269, 510, 584, 587–599, 655, 695
dementia patients, 149, 151, 266, 269, 591, 592
depression, 593, 598, 599
exhaustion, 589
family caregiver alliance, 581, 582
formal caregiver, 581, 584, 585
informal caregiver, 581, 584, 597
institutionalised elders, 170, 581, 586, 599
interdependence, 139, 463, 503, 508, 588, 666
person-centred care, 599
seeking employment, 590
skills training, 592
social exchange theory, 588, 589
stress/health model, 591
types of help, 581

Carstensen, L.L., 22, 33, 35, 175, 370–372, 374, 379, 386, 395, 396, 495, 502, 942
Catlin, G., 96
Cato Maior de Senectute, 74
Cell death, 9, 10, 211, 212
Celsus, 73, 240
Centenarians, 2, 10, 44, 201, 203, 250, 286, 301–304, 306–313, 316–321, 385, 393, 540, 837, 919
Charlesworth, B., 894, 896, 905
Chaucer, 80
China Health and Retirement Longitudinal Study (CHARLS), 277
Chinese culture
- *bao-da*, 500
- *cao gio*, 703
- *danxin*, 163
- *fengshui*, 150
- *giac hoi*, 703
- *guanxi*, 150
- *juk*, 169
- *laodao*, 653, 667
- *lao-nien*, 170
- *qi*, 144, 146, 150, 169, 182
- *qi gong*, 144
- *taijiquan*, 144
- *yang*, 144, 146, 169
- *yin*, 144, 169

Chinese Longitudinal Healthy Longevity Survey (CLHLS), 277, 319, 543, 545
Chochinov, H.M., 184, 223, 723, 771–773, 818
Cholecystokinin, 14
Cholesterol, 10, 166, 208, 247, 250, 262, 295, 315
Chromosomes, 10, 246, 250, 251, 315, 316
Chronic anxiety, 213
Chronic fatigue, 213
Cicero, 61, 72–74, 76–78, 205, 402, 683
Cicirelli, V.G., 35, 185, 377, 469, 470, 495, 503, 775, 781, 782, 787, 790, 818
Circadian rhythms, 36, 264, 269
Client Centred Care (CCC), 773
Cognition
- active process, 385, 386
- adaptation, 29, 368, 370, 635, 783, 805, 809, 847
- Advanced Cognitive Training for Independent and Vital Elderly (ACTIVE), 45, 266, 278, 318
- affective-cognitive structure, 369
- autonoetic awareness, 394
- bilingualism, 385
- change connections, 366
- cognitive appraisals, 368
- cognitive behavioural therapies (CBT), 234, 237, 238, 791, 847
- cognitive capacities, 22, 390, 391
- cognitive consonance, 524
- cognitive dissonance, 524, 539
- cognitive declines, 278, 319
- cognitive discrepancy, 477
- cognitive flexibility, 212, 257, 387, 415
- cognitive functions, 266, 278, 417
- cognitive integration, 386
- cognitive plasticity, 391
- cognitive processing speed, 387, 392, 635
- cognitive reframing, 766
- cognitive reserve, viii, 34, 212, 213, 215, 255–257, 260, 262, 263, 269, 319–321, 381, 385, 386, 388, 390, 392, 439, 940
- cognitive skills, 3, 386, 389, 392, 701
- compensation, 23
- confirmatory biases, 524
- creative thinking, 387
- creativity, 255
- crystallised intelligence, 383, 384, 392
- delayed recall, 265
- *differential-preservation* hypothesis, 205, 206, 320
- environmental complexity, 388, 429
- episodic memory, 275, 401, 403, 417
- event-event connections, 366, 367
- executive functions, 265
- *expertise preservation* hypothesis, 385
- explanatory connections, 367
- fluid intelligence, 257, 383, 384, 386, 392
- functional plasticity in cognitive ageing hypothesis, 23
- generational effects, 382
- illusionary correlations, 524
- inductive reasoning, 382
- information overload, 387
- inhibition, 387
- interference control, 387
- mastery, 152
- memories, 45, 215, 385, 388, 484, 940
- multilingualism, 385, 390
- numeric ability, 382
- passive process, 385
- perceptual speed, 388
- planning, 387
- *preserved-differentiation* hypothesis (*Preservation differentiation* hypothesis), 206

Cognition (*cont.*)
problem solving, 358
processing of connections, 366
processing speed, 205, 382, 387, 390, 392, 417, 635
Scaffolding Theory of Ageing and Cognition (STAC), 20, 23, 208, 381
schemata, 381, 652
selective optimisation with compensation (SOC) model, 22
self-event connections, 367
self-fulfilling prophecies, 524
self-regulation, 388, 739, 747
short-term recall, 19
spatial memory, 19
spatial navigation ability, 19
spatial orientation, 388
speech comprehension, 205, 382
thematic coherence, 367
threshold model, 385
use it/lose it hypothesis, 206, 255, 320, 382, 383, 429
verbal ability, 253
verbal memory, 278, 388
visuospatial skills, 205, 382
vocabulary, 205
wear-and-tear hypothesis, 892
wholesome knowledge, 367
working memory, 257, 387
Cognitive behavioural therapies (CBT), 791
Cognitive impairment
dementia, 205, 231, 246, 489, 836
mild cognitive impairment (MCI), 245, 246, 260, 320
Cognitive reserve, 34, 255–257, 260, 262, 263, 319–321, 381, 385, 386, 388, 390, 392, 437, 940
Cognitive social learning theory, 271
Cognitive vulnerability-stress model, 281
Co-habitation, 129
Cohen, L., 119
Cohorts, 7, 16, 39, 112, 113, 141, 181, 182, 221, 227, 242, 306, 311, 314, 319, 366, 367, 378, 382, 388, 389, 397, 405, 413, 463, 471, 473, 559, 569, 640, 661, 688, 707, 761, 799, 804, 843–845
Cold war, 7, 97, 112
Comfort, A., 902, 904
Commedia dell'Arte, 87
Communication
accommodation, 142, 640, 644, 649–651, 668
accommodative moves, 642
across ages, 670
age-irrelevant positivity, 648
approaches, 635, 644, 647, 649, 655, 674, 702, 703, 707, 855
asymmetrical, 642
attitudes, 537, 538, 639, 647, 649, 662, 675, 704
attuning, 645, 708
avoidance, 649, 653
baby talk, 638
Communication Accommodation theory (CAT), 641, 647, 652, 655
Communication Predicament of Ageing model (CPA), 638–640, 647, 655
Communicative Enhancement of Ageing model (CEA), 639, 640
confirming statements, 636
connectedness, 667, 675, 676
contextualisation, 657
convergence, 640, 642, 646, 648, 649, 657, 662, 665, 668
counter-attuning, 645
deference, 649, 653
deficit paradigm, 637, 647
developmental stake hypothesis, 640, 647
directive talk, 638
dissatisfaction, 653
divergence, 435, 642
doctor-patient communication, 704–707
downward, 636, 639, 652
elderspeak, 638
enhanced predicament model (EPM), 640, 647, 655
ethical-moral adviser, 673
face compensation, 658
face-threatening act, 657, 658, 660, 668
facilitators, 661
generational consciousness, 661
generational intelligence model, 661–663
gerontophilic, 652
gerontophobic, 652
grandfather effect., 663
grandmother hypothesis, 894, 912, 917
helping, 497, 513, 652, 653, 672
ideal age, 662
ideological matching, 657
imperviousness, 636
incommensurability, 661
incompatibility, 661
instant ageing, 664
inter-generational, 45, 138, 141, 142, 537, 633–636, 638, 639, 643–646, 648–654, 658, 659, 662, 666–669, 671, 673–676, 686–688, 708

inter-generational communication
schemata (ICS), 652
Inter-generational programs (IPs), 675–676
Intergenerational Work/Study Program
(IWSP), 670
interpersonal conflict management
styles, 668
intimacy, 435, 477, 640, 666
intra-generational, 45, 142, 636, 646,
654–656, 676
inversion, 658
labelled age-irrelevant positivity, 648
Lakoff's rules of rapport, 673
landed learning, 670
language attitudes, 660
language instructor, 673
laodao, 653, 667
leniency hypotheses, 640, 647
maintenance, 124, 133, 213, 278, 538, 636,
642, 690
masquerades, 662, 686
media, 634, 676, 698, 941
memory line project, 671
minimisation, 658
mixed feelings, 556, 653
multimodal, 642
mutual, 152, 538, 635, 640, 643, 645, 647,
655, 660, 666, 673, 674, 708, 775
mutually satisfying, 653
mutually unpleasant, 653
mutual reciprocity, 663
negative, 142, 152, 375, 538, 546, 633,
638, 639, 643–645, 647, 651,
653–655, 657, 664, 666–668, 675,
704, 708
non-accommodation, 643, 648, 649
non-verbal, 633, 637, 659, 676
not listening, 634, 646
null, 635
overaccommodation, 658
overly personal talk, 638
painful self-disclosure (PSD), 657, 658,
660, 676, 708
patronising, 638, 658, 659, 664
persona, 662
perspectives, 266, 639, 647–654, 659, 666,
673, 675, 677
politeness, 649, 653
positive, 142, 152, 265, 266, 278, 538,
638–640, 643–648, 651, 653, 655,
657, 662, 664, 666, 674, 675,
699, 704
pragmatic, 633, 651, 668
presbycusis, 637
prestige dialects, 660
proxemics, 660
rationalisation, 658
relational competence, 660, 676
reminiscence, 634
respect, 125, 648, 651
respect norms, 665
schemata, 653
selective assertiveness, 639, 674
selective assertiveness model, 675
self-deprecation, 657
self-handicapping, 657, 659
self-stereotyping, 659
situational theory of publics, 644–646
small talk, 653
social age, 662
social identity theory (SIT), 635, 646,
652, 655
social masks, 662
sociocultural guide, 673
staircase pattern, 635, 644–646, 649, 654
stances, 673
stereotypes, 142, 152, 634, 638–640, 651,
655, 658, 659, 664, 665, 707
stereotype-sensitive model, 655
subjective age, 662
superficial talk, 638
symmetrical, 642
symmetrical convergence, 642
tangential remarks, 636
telescoping, 663
touch, 141, 656
uncertainty reduction theory (URT), 646
underaccommodation, 643, 652, 658, 708
unimodal, 642
upward, 636
transitional reciprocity, 663
verbal, 266, 633, 634, 664, 676
verbosity, 375, 651
Communication predicament of ageing model
(CPA), 638
Communism, 723
Community
absenting elders, 511
ageing in place, 506, 513, 546
antagonizing elders, 511
attention restoration theory, 511
being elders, 591
belonging elders, 509, 707
bestowing elders, 135
bounded relationships, 510
Cantor, M.H., 37, 466, 504, 560, 831–833
casual relationships, 510
committed relationships, 37, 561

Community (*cont.*)
communing elders, 511
community organisations, 504, 562, 941
company of elders, 512
diabetes club, 513
Disengagement theory, 31, 110, 427, 463, 464, 508
downtowners, 107
elderscape, 107
engagement, 107, 128, 306, 473, 508, 561, 562, 673, 941
exchange theory, 508
fictive kin, 510, 542
Grey Nomads, 107, 108
incorporative relationships, 510
intentional communities, 506, 508
kindness, 512
landscapes of ageing, 107, 942
neighbourhoods, 471
neighbours, 87, 164, 504, 508, 512, 531, 585, 941
networks, 129, 131, 472, 473, 508, 548, 561, 693, 941
restorative activities, 511
retirement village, 108, 129, 433, 600, 655, 657
RVers, 107, 170, 179, 942
snowbirds, 108
social capital, 467, 506, 511–513, 669
social care model (SCM), 504
social connectedness, 689
Social Convoy model, 318, 472, 508
socialisation, 24, 470, 503, 531, 561, 595
spontaneous community, 508, 941
Compadrazgo, 139
Comparison of Longitudinal European Studies on Aging (CLESA), 144
Compassion, 36, 216, 358, 501, 680, 724, 748, 799, 801, 822, 939
Concentration, 11, 14, 25, 36, 202, 240, 264, 555, 638, 754, 782, 795, 862, 911
Conditioning
trace eyeblink, 19
Confucian philosophy
Analects, 121
Chong, 121
Li, 121
Ren, 121
Xiao, 120, 121
Yi, 121
Confucius, 121
Connectedness, 258, 471, 677, 732, 803, 810
Consciousness, 65, 111, 358, 359, 394, 568, 661, 662, 687, 722, 727, 739, 774, 775, 783, 785, 808, 810, 814
Consumerism, 7, 129, 499, 681
Contentment, 28, 31, 319, 751, 754, 938, 939
Contextual embeddedness, 35
Continuity theory, 33, 427
Coping
acceptance, 281
altruism, 281
belief in a just world, 26
cognitive, 224, 234, 285, 403, 416, 418, 842, 847
cognitive reassessment, 281
corrective, 26
denial, 281, 418, 635
distortion of reality, 281
emotion-focused, 403, 418
externalising, 281
humour, 281, 418
instrumental, 403
internalising, 281, 285
mechanisms, 26, 413, 415, 420, 576, 583, 596, 599, 635, 756, 804, 842
preventive, 26
proactive, 26, 234
problem-focused, 418
resilience, 159, 296, 420, 756, 847
Cornaro, Alvise (Luigi), 86
Cortisol, 215, 247, 278, 279, 403
Country
economically developed, 5, 6, 241, 364, 721
Couplehood
couplehood, 36, 501, 502
living apart together (LAT), 501, 502
re-marriage, 501
social regression, 503
Creativity, 36, 110, 255, 274, 385, 387, 422, 462, 511, 512, 519, 668, 687, 751, 777, 783
Critical, Cultural and Humanistic Gerontology (CCHG), 464
Croft, D.P., 898, 913, 914, 921
Cross-cultural
similarities, 44, 126, 143
Cross-Sectional and Longitudinal Aging Study (CALAS), 390
Csikszentmihalyi, M., 36, 110, 422, 742, 751, 810
Culturally and linguistically diverse (CALD), 159, 513
Culture, 28, 60, 71, 88, 90, 108, 119, 122, 128, 135, 136, 143, 148, 152, 153, 164, 165, 168, 179, 364, 405, 417, 461, 468, 474, 505, 547, 548, 555, 565, 650, 672, 673, 686–690, 701, 724, 726, 746, 777, 783, 789, 792, 793, 797, 807, 808, 819, 850, 854, 919

acculturation, 548
afterlife, 679, 778
Asians, 136, 650, 789
"bad blood", 148
beliefs, 119, 165, 548, 650, 789, 793, 796, 850
collectivistic, 135, 854
cultural adaptation, 125
cultural awareness, 848
cultural bifocality, 180
cultural changes, 153
cultural continuity, 690
cultural diversity, 38, 120, 183, 468, 848–850, 855
cultural frame, 119, 218, 854
cultural humility, 855, 856
culturally and linguistically diverse (CALD), 159, 513
culturally syncretic identities, 180
cultural networks, 129
cultural scripts, 474
cultural sensitivity, 848
cultural tailoring, 151
cultural values, 505
doreh, 131
eastern, 121, 134, 536, 648
elders, 71, 122, 135, 143, 153, 168, 179, 364, 418, 505, 650, 671, 672, 688–690, 792, 850, 919
ethnicities, 462, 789, 850
familism, 151, 270, 365, 480, 505, 583
festivals, 475
filial piety, 120, 121, 125, 142, 150, 153, 158, 644, 649, 652
gerontocratic, 71
individualistic, 137, 365, 536, 648, 853
intolerable cultural practices, 855
jivanmukta, 187
karmas, 187
machismo, 186
multiculturalism, 548
myth, 37, 60, 100–101, 230–231, 494, 519, 678, 735, 739, 763, 768
nu, 123
num, 123
tolerable cultural practices, 855
transculturality, 107
web of significance, 854
western, 28, 60, 122, 136, 364, 690, 789, 855
"worration", 148
Cumming, E., 31, 110, 111, 175, 463, 464, 508
Cumulative disadvantage, 101
Custodial grandparents, 140, 416
Cytokines, 14, 18, 202, 315, 911

D

Daimon, 28, 820
Darwin, C.
The Expression of the Emotions in Man and Animals, 370
Death
abandoning, 155, 156, 813
accabadora, 156
acceptance, 65, 184, 186, 225, 787, 790, 791, 806, 862
advance directives, 854
afterlife, 73, 122, 186, 775, 778, 794–797, 806, 819
appropriate death, 722
autonomy, 577, 722, 816, 821, 822, 825, 828, 852
bad death, 722
bereavement, 577, 722, 861, 862
concerns, 26, 45, 78, 83, 278, 577, 729, 773, 775, 780, 783, 787, 788, 804
death anxiety, 406, 537, 723, 774–797, 850
Death Attitude Profile (DAP), 184, 185, 790
Death Attitude Profile-revisited (DAP-R), 757, 790
death hastening behaviours, 155, 156, 376
death salience, 776, 780, 781
denial of death, 787
Día de los Muertos, 536
die alone, 726
dignity, 722, 726, 727, 810, 857, 864
do not resuscitate (DNR) order, 851, 858
duty to die, 85
ego dissolution, 787
fear of death, 73, 184–187, 225, 406, 725, 732, 757, 774–797, 850, 853, 862
forsaking, 155, 156, 813
funerary rites, 775
gerotranscendence, 43, 726, 727
good death, 722, 723, 732, 821, 857
grief, 183, 186, 802, 861–864
hastening behaviour, 813
ideational convergence, 188
intrusiveness, 775
killing, 155, 156, 813
life review, 74, 184, 756, 791, 792
life span perspective, 184
loneliness, 775

Death (*cont.*)
lonely death, 727
losses, 12, 73, 187, 203, 285, 376, 577, 722, 725, 775–778, 785–789, 801, 822, 839, 857, 862–864, 920
pain, 43, 75, 78, 577, 722, 726, 775–778, 788, 790, 803, 804, 814, 818–820, 827, 857, 862–864
pokkuri, 161
premonition of death, 811
Principles of a good death, 821
reincarnation, 187, 789
remembrance, 187
respect for life, 822, 828
right to die, 816
rites of aggregation, 183, 863
rites of liminality, 183
rites of passage, 59
rites of separation, 183
social death, 788, 791, 825
su jualeddu, 156
su mazzolu, 156
symbolic immortality, 775, 782, 783
Terminal Decline hypothesis, 187
Terror Management Theory, 186, 536, 777, 781
thanatology, 183
"unfinished business", 184
utility, 822, 828
Death anxiety, 184, 185, 187, 189, 226, 406, 537, 577, 578, 697, 723, 726, 727, 774–797, 806, 850, 864, 892
comprehensive model of death anxiety, 780
death acceptance, 785, 790
death-thought accessibility (DTA), 781
denial of death, 784, 791
Dual-process theory, 783
ego dissolution, 787
fear of death, 185, 187, 189, 377, 697, 727, 774, 775, 777–779, 782, 783, 786–794, 796, 806, 864
"healthy view", 777
illusion of immortality, 777
literal immortality, 775, 782
Meaning Management theory (MMT), 785, 790
"morbid view", 777
Post-traumatic growth theory, 781
religiosity, 189, 225, 727, 790, 793–795, 892
Self-determination theory (SDT), 785
symbolic immortality, 775, 782, 783, 792
Terror Management Theory, 186, 777, 781
Death hastening behaviour
autonomy, 765, 816
beneficence, 815
bioethics, 817, 851, 852
clinical pragmatic arguments, 823
deontological argument, 823
distributive justice, 816
euthanasia, 78, 812–848
existential suffering, 825–828, 843
gericide, 813
gerontocide, 813
life fatigue, 825–828, 842–844
nonmaleficence, 816
right to die, 86, 816–818
self-determination, 817, 818, 823, 825, 827
slippery slope argument, 823
suicide, 85, 812–848
verstervern, 822
will-to-live, 317, 818, 836, 840–845
Deguileville, Guillaume de, 83
Dementia
Alzheimer's disease, 147, 148, 163, 241, 243, 260, 584
caregiving, 149, 151, 591, 592
cultural beliefs, 148
elder population, 34
forgetfulness, 240, 244, 245
"long goodbye", 149
neurofibrillary plaques, 19
prevalence, 34, 227, 241, 242, 244, 250, 260, 269, 297, 575, 842
tangles, 19, 249, 484
vascular dementia, 34, 148, 241, 262, 263
Democracy
Athenian, 71
Old man Demos, 71
Demography
cohort, 761
demographic transition, 4, 131, 138
Demoralisation, 35, 36, 43, 220, 489, 529, 728
Denial, 108, 155, 239, 242, 267, 526, 535, 634, 652, 693, 777, 783, 813
"denial of age", 110
Deoxyribonucleic acid (DNA)
methylation, 10, 16, 316
oxidative damage, 10, 205, 316
Depression, 25, 31, 35, 36, 43, 68, 106, 124–126, 140, 146–148, 150, 164, 176, 184, 212, 217, 220, 224, 228–238, 241, 247, 269, 270, 272, 274, 280–282, 284–286, 291–294, 298, 299, 311, 313, 317, 371, 374, 375, 399, 406, 411, 412, 414, 415,

422, 423, 425, 429, 433, 435, 472, 484, 489–492, 502, 527, 529, 530, 552, 553, 560–565, 570, 573, 574, 576, 587, 589–591, 593–596, 598, 599, 635, 685, 728, 736, 744, 750, 753, 754, 757, 759, 771, 773, 786, 787, 818, 830, 833, 835–837, 839, 840, 842–844, 846, 847, 859, 862
De Senectute, 206
Developmental imperative, 109, 111
Diabetes, 14, 202, 247, 261, 262, 297, 298, 312, 513, 925
Día de los Muertos, 536
Dignity, 77, 101, 134, 419, 487, 514–517, 539, 553, 577, 592, 723, 821, 824, 827, 858
  therapy, 771–773
Diocletian, 67
Diogenes syndrome, 364, 412
Disability
  acquired, 292, 293
  assistive technologies, 295
  care, 292, 293, 296, 297
  cerebral palsy, 296, 300
  combat-related disabilities, 293
  depression, 231, 232, 280, 299
  developmental, 291–293, 299, 498
  disabled, 291, 293–296, 300
  diseases, 30, 44, 292, 297, 300
  down syndrome, 296, 299, 300
  epilepsy, 296, 300
  fulfilling ageing, 30, 292, 299, 300, 322
  gastrointestinal function, 294
  genitourinary complications, 294
  haemorrhoids, 294
  Healthy Ageing and Intellectual Disabilities (HA-ID) project, 299
  hearing impairment, 295, 297, 637
  intellectual, 291, 296–300
  life expectancy, 296, 300
  life satisfaction, 293
  living conditions, 300
  multi-morbidity, 299, 308
  multiple, 300, 530
  muscle atrophy, 245, 295
  neuropathies, 382, 392, 484
  oppression, 291
  optimism, 296, 308, 414
  osteoporosis, 297, 298
  paralytic poliomyelitis, 294
  physical, 30, 44, 236, 280, 282, 284, 291–297, 299, 300, 308, 319, 322, 838, 941
  prejudice, 291, 299
  psychiatric issues, 296
  psychological support, 293
  respiratory complications, 298
  rights, 291
  spinal cord injury (SCI), 294
  stigma, 291, 529, 530
  tetraplegic, 294
  triple jeopardy, 295
Discourse
  on ageing, 39, 43, 680, 681
  agentive, 677
  analysis, 39, 672, 679
  face-threatening act, 657, 658
  narrative gerontology, 679
  narratives, 39, 637, 680, 686, 708
  natural, 39
  normal, 39
  normative, 39
Discrimination
  ageism, 37, 68, 99, 100, 492, 517, 532–536, 538, 539, 565
  machismo, 186
  negative, 27, 37, 68, 99, 122, 153, 154, 188, 376, 517, 525, 532–535, 537–539, 549, 562, 565
  prejudice, 188, 535
Disease
  arthritis, 295, 483
  autoimmune, 14
  avoidance, 30
  cancer, 143, 203, 309, 743, 770
  degenerative, 202
  dementia, 34, 147, 148, 163, 203, 243, 245, 250, 259, 262, 267, 269, 398, 399, 409, 473, 584, 593, 696
  diabetes, 295, 483
  folk medicine, 144
  infectious, 5, 251, 298, 415
  non-infectious, 5
  osteoporosis, 295
  vascular, 202
Disengagement, 32, 33, 111, 170, 465, 519, 658, 683, 746, 759, 760, 864
  mutual, 36, 478
  social, 36, 221
  theory, 110, 175, 427, 463, 464, 508
Distress
  distressing experiences, 26
Divorce, 36, 97, 367, 478, 498, 501, 502
Doctor-patient communication
  advocate, 705
  antagonist, 706

Doctor-patient communication (*cont.*)
 Communication Enhancement model, 707
 companion, 705, 706
 metaphors, 707
 partnership builder, 706
 passive participant, 705
 shield from bad news, 706
 social communicator, 706
 telecare, 707
 triadic medical consultations, 705, 706
Downtowners, 107
Drion, H., 825
Durkheim, E., 469, 829–831, 833

E

Education, 4, 5, 8, 16, 17, 23, 24, 31, 65, 72, 101, 106, 125, 136, 139, 144, 145, 151, 160, 163, 165, 171, 172, 175, 177, 179, 224, 231, 232, 236, 242, 247, 255, 256, 258, 259, 262, 265, 271, 280, 294, 311, 318–321, 379, 385, 388, 389, 392, 400, 413, 423, 480, 483, 489, 492, 498, 508, 509, 539, 545, 554, 556, 570, 579, 590, 591, 594, 596, 597, 652, 657, 683, 684, 751, 761, 786, 787, 823, 830, 857, 938
Ego integrity, 65, 185, 360, 697, 744, 786, 787, 792, 795, 805
E-health, 517
Elder
 definition, 487, 811, 813
 family historian, 140
 healthcare services, 860
 housing, 295, 531
 na, 160
 public services, 76, 420, 513
Elderscapes, 107, 137, 942
Eldership, 128, 939
Emotional regulation, 22, 235, 357, 374, 495, 792, 801, 805, 839
Emotions, 19, 22, 29, 45, 144, 164, 252, 357, 368–375, 380, 381, 397, 412, 437, 501, 521, 556, 646, 736, 741, 747, 769, 771, 778, 792, 801, 806, 839, 942
 affect, 45, 282, 368, 369, 371, 372, 792
 affective-cognitive structure, 369
 affective response, 520
 affective solidarity, 544
 anger, 164, 274, 369, 371, 372, 375, 381, 839
 apathetic acceptance, 478
 attachments, 371–373
 authentic pride, 381
 aversive emotional state, 35
 avoidance, 144, 839
 avoidant relationships, 372
 contempt, 369, 372, 375, 381
 curiosity, 274, 380, 381
 demoralisation, 35
 devaluated pride, 378
 differential-emotions theory, 369
 disgust, 369, 370, 372, 381
 disinhibition, 375
 dismissingness, 372
 dismissive relationships, 372
 dynamic integration theory, 374
 emotional complexity, 372, 792
 emotional cues, 375
 emotional isolation, 478
 emotionality, 19, 272, 274, 374, 741
 emotional support, 36, 161, 183, 224, 365, 465, 466, 471, 487, 497, 498, 501, 542, 543, 551, 574, 671, 757, 775, 821, 943
 emotional stability, 19, 22, 371, 372, 374, 381
 empathy, 274
 excessive pride, 378
 fearful avoidance, 372
 fears, 238, 272, 369, 381, 806
 guilt, 164, 368, 369, 372, 381, 839
 happiness, 282, 381, 646
 hopelessness, 839
 hubristic pride, 381
 insecure attachment, 372
 insecure relationships, 372
 interests, 29, 272, 368, 369, 372, 380
 irritability, 146, 147, 221, 235, 375, 404, 552, 593
 joy, 29, 369, 372, 736, 741, 801
 Life-span theory of control, 22, 374
 loneliness, 164
 maturational dualism, 374
 moods, 369
 negative, 29, 237, 274, 282, 357, 371, 372, 374, 376, 380, 381, 412, 521, 736, 741, 747, 771, 792, 806, 839
 neurological change hypothesis, 374
 positive, 29, 144, 274, 282, 371, 372, 374, 380, 381, 437, 521, 736, 747, 792, 801, 806, 942
 positive effects, 372
 positive reminiscence, 379
 pride, 29
 recognition, 374

regulation, 252, 368, 371, 372, 374, 806, 942
sadness, 29, 369, 372, 381, 646, 736
secure attachment, 372
security, 372
Selective Optimisation with Compensation (SOC) model, 22
self-regulation, 36, 363, 387, 429, 525, 739, 747–748, 776
sentiment, 375
shame, 369, 372, 381
shyness, 369
situation modification, 374
situation selection, 374
Socioemotional selectivity theory, 22, 35, 372, 502
suffering, 252, 375
surprise, 369, 372, 381, 556
verbosity, 375
violence, 282
wonder, 110, 370, 729
Employment, 100, 104, 172, 421, 422, 429, 433, 437, 563, 569, 590, 943
sub-employment, 8, 97, 422
unemployment, 100, 104
Endocrine systems, 11, 14, 369
The end of history illusion, 202
End-of-life care
advance care directives, 851, 852, 858
"cuddle beds", 856
dying with dignity, 858
frailty syndrome, 857
good death, 857, 858, 861
heroic view, 857
humanistic perspective, 857
intensive care units, 857
International Association for the Study of Pain (IASP), 857
pain, 857
palliative care, 856–860
stoic, 858
Enlightenment, 91, 680
Entrapment
veiled, 129
Epicurus, 28, 819, 829
Epigenetic mechanisms
DNA methylation, 10
Erasmus of Rotterdam, 85
Erikson, E.H.
"ego-dystonic" factors, 65
ego integrity, 65
"ego-syntonic" factors, 65
ESRC/MRC HSRC Quality of Life Survey, 280
Ethics
accessibility, 516, 517
assistive technology, 516
autonomy, 514
bioethics, 516, 817, 851, 852, 855
biogerontology, 516
conduct, 137
confidentiality, 516, 703
consent, 516, 517
dignity, 514
equality, 515
equity, 515
Eudemian Ethics, 820
fairness, 515
familism, 499, 515
filialism, 515
honour, 90
informed consent, 517
intersectionality, 516
intrusive, 517
justice, 514
moral distress, 515
moral obligations, 514, 515
Nichomachean Ethics, 820
privacy, 517
protection of privacy, 517
resourcefulness, 516
risk of confinement, 517
Ethnicity, 119, 120, 124, 130, 145, 151, 165, 488, 493, 558, 597, 940
aboriginal, 123
Aché, 155, 901
African Americans, 597
Aleut, 128
Asian American, 561
Athabascan, 128
Aztec, 124
Bororo, 155
Chamorros, 144
Chukchee, 155
Compadrazgo, 139
compensation, 120, 130
cubanidad, 131
deprivation, 120
Druse, 124
entrapment, 129, 130
Eskimo, 128
ethnic diversity, 120
ethnogeriatrics, 119
ethnogerontology, 119
European, 145
European American, 157, 561
Fulani, 170
Hazda, 901, 916, 917

Ethnicity (*cont.*)
- Hispanic, 130, 140, 147, 163, 165, 494
- Hiwi, 901
- Hua, 123
- Innu, 3
- *Ju/'hoansi* (*!Kung*), 155, 898
- Kaska, 690
- *Kpelle*, 4
- *!Kung* (*Ju/'hoansi*), 155, 898
- Lau, 155
- *Machismo*, 186
- Maori, 158
- Mapuche, 159
- Mayan, 147
- *Métis*, 689, 690
- *Minangkabau*, 133
- Native Americans, 145
- Navajo, 187, 854
- Noongar, 690
- *Nuu-chah-nulth*, 689
- *Samburu*, 122
- *Samia*, 170
- Tiwi, 123, 155
- *Yakut*, 155
- *Yup'ik*, 128

Eudaimonia
- *model of eudaimonic living*, 28
- *Nicomachean ethics*, 28

Eugenics, 95, 820

Eugiria, 819

Eukrates' motion stele, 72

Euthanasia, 43, 45, 221, 727, 812–848, 864, 919
- accidental, 814
- active, 814, 825
- *Aid-in-dying* (AID), 814
- beneficent response, 825
- Brongersma, E., 826
- completed life, 825, 827
- continuous deep sedation, 815
- *Death with Dignity Act*, 814
- "Drion's pill", 825
- existential suffering, 825–828
- good death, 45, 819, 820, 822, 824, 828, 864
- intensified alleviation of symptoms, 814
- involuntary, 814, 820
- life fatigue, 825–828
- passive, 814
- *rule of double effect*, 814
- semi-active, 814
- semi-passive, 814
- suicidal, 814
- *The Vermont Study on Aid-in-Dying*, 814
- voluntary, 814, 823, 825
- will to live, 864
- withdrawing life-prolonging treatment, 815
- withholding life-prolonging treatment, 815

Evolution
- acceleration, 901, 909, 920, 921
- adaptation, 369, 805, 808, 894, 902, 909, 927
- age-specific mortality, 898, 899, 901
- altruism, 918, 919
- antagonistic pleiotropy theory, 896
- *Arrow of Time reciprocation*, 919
- by-product, 894
- catastrophic failure, 895
- comparative methods, 897, 898
- cooperation, 369
- *CSR* selection, 899
- *Damage accumulation hypothesis*, 896, 905
- *Darwinian fitness*, 892, 893
- *degradation* of a system, 895
- demonstration effect, 919
- direct fitness, 892, 894, 909, 913
- *Disposable soma* theory, 899, 900, 904–906, 908
- error accumulation, 9
- evo-gero, 896
- evolutionary gerontology, 896
- exaptations, 807, 808
- exploitative relationship, 918
- fatal failure, 895
- fetalization, 922
- *Free radical theory of ageing*, 906
- *Gorilla gorilla*, 913
- *Grandmother hypothesis*, 920, 927
- *Hamilton's rule*, 912, 917
- Handicaps, 806
- heritability (h^2), 897
- heritable variation, 897
- heterochrony, 921
- heterogametic sex, 909
- homogametic sex, 909, 910, 912
- hyperactive agent detection device (HADD), 809
- hypermorphosis, 921
- immunosenescence, 910
- inclusive fitness, 912
- indirect fitness, 892, 894
- *Indispensable Soma* hypothesis, 900
- interdemic selection, 807
- killer whales (*Orcinus orca*), 898, 913, 916
- kin selection, 588, 892, 894, 896, 912–920
- *Kin selection theory*, 912
- *K*-selection, 899

life histories, 897, 908, 909, 927
maladaptive by-product, 800
menopause, 912, 913, 917, 920, 926, 927
morphogenetic senescence, 903
Mother hypothesis, 913, 927
mutation, 894, 904
Mutation accumulation theory, 893
naked mole-rat (*Heterocephalus glaber*), 924–926
natural selection, 94, 369, 784, 808, 892–894, 897, 903, 927
negative senescence, 907
neoteny, 896, 920, 922, 923, 926, 927
paedomorphosis, 923
Pan paniscus, 913
Pan troglodytes, 913
Peramorphosis, 921, 922
Pongo pygmaeus, 913
progenesis, 921, 922
Rate of living theory, 906
reciprocal altruism, 919
redundancy, 895, 896, 926
redundancy exhaustion, 895, 907
reliability theory, 895, 896, 905, 926
response to selection (*R*), 897
r/K selection, 899
r-selection, 899
selection differential (*S*), 897
sexually antagonistic traits, 909
sexual selection, 908
short-finned pilot whale (*Globicephala macrorhynchus*), 913, 914
systems failure, 895
transitional reciprocity, 919
X-chromosome, 910
Exceptional longevity, 201, 303, 315, 316, 321, 322
Exercise, 16, 17, 72–74, 81, 111, 144, 146, 162, 176, 178, 179, 182, 184, 203, 205, 212–217, 234, 248, 253, 255, 261, 265, 266, 269, 294, 302, 307, 310, 313, 322, 368, 379, 386, 389, 391, 392, 399, 402, 429, 432, 485, 525, 554, 576, 592, 678–680, 697, 704, 747, 804, 813, 820, 902, 940
Existential, 22, 29, 30, 33, 34, 41–43, 45, 65, 74, 109, 128, 215, 221–223, 321, 363, 418, 577, 633, 658, 722, 723, 725, 727–736, 746, 754, 758, 763, 764, 766, 767, 771, 782, 783, 786, 797, 812, 825–828, 837, 840, 842, 861, 865
adventurousness, 764
continuity, 33, 188, 727, 757
dimensions, 42, 215, 222, 321, 722, 727–734, 757, 763, 826, 861
distress, 222, 478, 722
frustration, 753, 755
fulfillment, 29, 30, 33, 42, 188
guilt, 757
identities, 41, 128, 633, 658
integrity, 722
life fatigue, 825–828, 842, 843
life meaning, 31, 43, 45, 109, 215, 283, 727, 730, 735, 741, 746, 762, 767, 861
life review, 75
loneliness, 477, 486, 487, 727, 741, 825, 861
meanings, 732, 754, 758, 764, 785
narrative foreclosure, 734
nihilism, 734, 735, 754, 764
perspectives, 741, 777, 786, 826
shield, 782, 783
sickness, 764
sufferings, 786
tragic optimism, 755, 756
vacuum, 31, 728, 753, 755, 756, 765
vegetativeness, 764
wisdom, 42, 65, 75, 728

F

Familism, 151, 241, 270, 365, 496, 583, 695
Family
ambivalence model, 495
ambivalent grandparents, 496
apathetic siblings, 496
autonomous and highly engaged grandparents, 496
babysitting, 139, 494
binuclear, 501
blended, 493, 494
children, 2, 7, 76, 82, 94, 98, 99, 129–132, 139, 149, 160, 161, 163, 283, 319, 409, 435, 473, 483, 484, 490–495, 498–501, 545, 546, 550, 552, 557, 560, 561, 563, 578, 581, 583, 593, 594, 597, 651, 654, 688, 690, 699, 701, 788, 792, 917, 919
children-in-law, 583
chosen kin, 542
commuter marriages, 501
congenial siblings, 496
conjugal, 82
constructed kin, 542
contradictions, 495, 654
co-parenthood, 139

Family (*cont.*)
co-residential, 129, 494
custodial grandparents, 498
daughter, 93, 98, 151, 314, 319, 469, 497, 515, 578, 581, 583, 592, 593, 917
dilemmas, 495, 557
divorce, 36, 97, 498, 501, 502
duteous grandparents, 496
emotional regulation, 495, 842
extended, 2, 91, 98, 129, 139, 159, 163, 493, 494, 496, 499, 503, 561, 563, 583, 663, 694, 773, 919
familism, 130, 139, 151, 480, 496, 499, 505, 515, 583, 651, 695, 853
family-independent grandparents, 496
fictive kin, 550
filial piety, 120, 121, 130, 143, 158, 161, 163, 181, 188, 475, 500, 550, 644, 645, 647, 651, 652, 853
Filial Piety Prize, 500
grandchildren, 98, 105, 139, 140, 283, 475, 490, 495, 496, 498, 501, 502, 551, 557, 579, 688, 699, 919, 941
hostile siblings, 496
individualism, 97, 129, 136, 143, 499, 510, 695
in-law, 319
inter-generational ambivalence, 500, 501
inter-generational conflicts, 82, 495, 663, 671
inter-generational solidarity, 82, 143, 495
integrated grandparents, 496
intimate siblings, 496
kinship network, 542
lending money, 494
living apart together (LAT), 501–503
loyal siblings, 496
modernisation, 129, 136, 163, 480, 494, 497, 499–501, 503, 505, 550
multi-generational co-residential, 129
nuclear, 97, 129, 130, 469, 493, 494, 503
orphans, 494
parents, 2, 7, 25, 36, 76, 82, 91, 129, 130, 139, 150, 160, 161, 259, 319, 435, 483, 493, 494, 497–500, 546, 550, 552, 560, 578, 586, 651, 652, 654, 663, 689, 816
"patchwork family", 499
patriarchal, 82
principle of substitution, 503
relationships, 4, 36, 45, 121, 134, 135, 148, 150, 151, 161, 186, 218, 239, 267, 435, 465, 472, 474, 486, 493–496, 499–501, 503, 515, 545, 550, 560, 563, 589, 596, 599, 660, 662, 663, 694, 701, 745, 776, 862, 919, 943
reverence, 500
rosy family, 494
siblings, 314, 491, 560, 561, 563, 594
social capital, 466, 467, 494, 510
social regression, 503
solidarity-conflict paradigm, 495
stability, 35, 139, 409
Father-of-moran, 122
Fatigue, 292, 731
chronic, 213
Fear, 38, 75, 89, 99, 125, 126, 150, 185, 217, 230, 234, 266, 313, 370, 372, 376, 377, 521, 526, 527, 531, 533, 534, 536, 553, 556, 564, 569, 570, 573, 583, 656, 668, 682, 690, 735, 746, 767, 775–778, 781, 782, 784, 786, 789, 791, 793–795, 800, 803, 816, 819, 824, 825, 828, 851, 852, 864
ego dissolution, 787
fear of death, 185, 635, 734, 793
religiosity, 225, 786, 793–795, 862
Feminism, 7
Fertility, 4, 5, 891, 897, 898, 913, 914
Fête de la vieillesse, 90
Ficino, M., 84
Filialism, 515
Filial piety, 120, 121, 142, 150, 153, 172, 500, 649, 708
authoritarian, 121
eastern, 120
hyodo, 853
pietas, 121
reciprocal, 121
Western, 120, 121, 142, 163, 182, 647, 649
Finch, C., 10, 202, 250–252, 261, 897, 900, 901, 907
Flow, 36, 107, 110, 120, 133, 139, 146, 171, 176, 178, 211, 251, 261, 422, 483, 511, 546, 637, 700, 742, 751, 752, 765, 776, 854, 856, 939
Folk medicine, 144
Food
alcohol, 402, 689
carbohydrates, 86, 166
choices, 164–167, 169
cholesterol, 10, 166, 208, 247, 250, 262, 295, 315, 398
complex carbohydrates, 166
diet, 86, 165, 168–170, 260, 261, 269
energies, 167
fibre, 166
intake, 12, 164, 166, 167, 260, 900, 917

macronutrients, 166
memory schemata, 165
polyunsaturated fatty acids, 166, 261
prestige, 165
satiety, 165
smell, 164, 376
sugar, 166, 167
taste, 164, 165, 168–170, 376
tolerance, 165
Forgetfulness, 72, 81, 244, 245, 704
The Fountain of Age, 110
Fountain of Youth, 83, 84, 88, 110, 517
Frailty, 12, 35, 39, 63, 105, 120, 186, 376, 508, 540, 682, 705, 806
frailty syndrome, 12, 18, 148, 857
Framingham Heart Study, 242
Frankl, V., 27, 29–31, 217, 306, 681, 728–730, 735, 739, 742, 746–748, 750, 753–757, 764–767, 777, 785, 821, 848, 861
Freud, S., 207, 800, 861
Friedan, B., 99, 108, 110, 111
Friendly Societies Act of 1875, 2, 67
Frustration, 27, 125, 140, 148, 184, 220, 224, 236, 239, 245, 269, 375, 412, 488, 523, 553, 653, 695, 696, 699, 748, 750, 765, 812, 937
Fulfilling ageing, 8, 27–35, 38, 40, 46, 73, 110, 121, 127–129, 131–134, 137, 175, 188, 213, 215, 216, 222, 300, 317, 355, 362, 367, 391, 465, 475, 520, 538, 543, 547, 676, 684, 691, 694, 728, 938, 939
accommodative flexibility, 33
activity theory, 31
ageing well, 127, 129
agency, 8, 938
authentically (life lived), 41, 730
authenticity, 41
autonoetic awareness, 394
cognitive flexibility, 212
cognitive integration, 386
cognitive reserve, viii, 213, 215, 255, 269, 940
connectedness, 31, 110, 513, 669, 939, 942
continuity theory, 33, 425, 427
dignity, 33, 134, 135, 419, 722, 727, 824
disabilities, 30, 299, 300
disengagement theory, 31
emotional regulation, 805
eudaimonia, 28, 29, 938
euthanasia, 819–825
existential well-being, 41–43, 215, 221, 223, 419, 728–729, 732, 786
flow, 133, 422, 939
generativity, 412
gerotranscendence, 939
hardiness, 434, 744, 762
inner strength, 693, 708
legacy, 30, 189, 412, 691
life review, 730, 940
life satisfaction index, 29
life span perspective, 406
meaningfulness, 691
meanings, 46, 362
optimism, 413, 751, 940
paradox of well-being, 39
plasticity, 213, 322
possibilities of the self, 684
possible selves, 684
psychological well-being, 27, 127, 132, 135, 173, 217, 372, 432, 466, 594, 635, 762, 801, 805
purpose in life, 29, 43, 216, 217
religiosity, 221
resilience, 216
resourcefulness, 748
"San Quentin", 30
satisfaction, 30–33, 131
selective optimisation with compensation (SOC) model, 32
self-transcendence, 217
sense of coherence, 736
sense of purpose, 224, 468, 512, 554, 739–740, 757, 763
social cohesion, 467
social-environmental theory, 34
socioemotional selectivity theory, 33, 35
spiritual intelligence, 43
spirituality, 109, 221
subjective well-being, 26, 42, 127, 132, 177, 215, 225, 270, 282, 285, 317, 319, 378, 411, 415, 510, 552, 750, 765
thematic coherence, 367
transcendence, 216, 427, 727, 864, 938
wholesome knowledge, 367
wisdom, 368, 939
Functional magnetic resonance imaging (fMRI), 21, 208, 212, 213
Funeral, 134, 183, 770, 771, 858, 861, 863

G

Galen
cooling, 73
drying, 73, 83
gerocomic art, 73

Galen (*cont.*)
harpastum, 73
humours, 73
Gavrilova, N., 305–308, 314, 895, 896, 905
Gavrilov, L., 305–308, 314, 895, 896, 905
Geist, L., 96
Gender
androgyny, 273, 566
bisexuals, 558, 559, 569
body image, 270, 273
cognitive capacities, 388
cognitive social learning theory, 271
cultural stereotypes, 274
"degendering" of identities, 273
"double standard of ageing", 274
evolutionary theories, 271, 278
expectancy-value theory, 271
femininity, 273, 566
gay, 559, 565, 569, 693
gender roles, 5, 97, 123, 273, 435, 566, 569, 595, 855
gender undifferentiated, 273
heterosexual, 97, 558, 565, 668
homosexual, 559, 668
intersex, 558, 569
lesbians, 559, 569, 668, 693
LGBTI, 558, 564, 569, 668
masculinity, 273, 566
sociocultural theory, 271
trans, 570
transsexual/transgender (trans), 45, 558–572
Gene
APOE2, 10, 315
APOE4, 10, 251, 315
mutations, 10, 251
Generation
baby boomers, 96, 97, 112, 113, 567, 644
downward skipped generation, 636
generational conflicts, 636
generational consciousness, 661
generational experience, 661, 940
generational imprinting, 688
generational intelligence model, 661
generational solidarity, 636
Generation Next, 645
Generation X, 66, 644
generativity, 41, 737
G.I. Generation, 66
leading edge boomers, 96
Lucky Generation, 96, 97
Millennials, 66, 676
needy generation, 636
older boomers, 66
options generation, 97
pivotal generation, 663
sandwich generation, 636
seniors, 644
shadow boomers, 96
Silent Generation, 66
skipped generation, 636
stoic, 96
upward skipped generation, 636
younger boomers, 66
Generativity, 65, 110, 409, 430, 496, 680, 688, 768, 772, 774, 778, 782, 786, 864, 941
eco-elders, 774
gatekeepers of the future, 774
generative needs, 34
grand, 774
keepers-of-the-culture, 774
legacy, 768, 774, 782, 786, 864, 941
memory keepers, 774
role-modelling, 65
Genetics
genomic instability, 9
heritable variations, 897
Genius, 76, 736
Geriatrics, 73, 95, 96, 227, 529, 664, 705, 855
Geroanthropology, 119
Gerontocracy, 70, 72
council of the elders, 70, 72
gerontes, 71
Gerousia, 72
Kenbet, 70
The Republic, 72
Roman Catholic popes, 84
Sparta, 71
Venetian *doges*, 84
Venetian Republic, 84
Zekenim, 70
Gerontological Society, 98, 564
Gerontology, 633, 893
Critical, Cultural and Humanistic Gerontology (CCHG), 464
critical gerontology, 35
gerontology of mobility, 107
narrative gerontology, 39, 184, 679
new gerontology, 35
Gerontophobia, 99, 521, 526
Gerotranscendence
cosmic, 725
gerotranscendence, 111, 216, 691, 725, 726, 730, 751
self, 112, 691, 725, 726, 730
social, 112, 691, 725, 726, 730
three levels, 725, 751
Ghrelin, 12

Glial cells
astrocytes, 208, 211, 252
oligodendrocytes, 208, 252
Global AgeWatch Index, 6, 7
Global Financial Crisis (GFC), 306, 546, 833, 835, 843, 850
Globalisation, 120, 124, 142, 364, 546, 939
centrifugal reactive force, 124
centripetal homogenising force, 124
global cities, 133
glocalisation, 124
world cities, 133
Glucocorticoids, 12, 14, 216, 231, 232, 415
Glucose
metabolism, 14
God
Jewish God, 72
Zeus, 72
Goffman, E., 38, 99, 131, 149, 152, 292, 363, 522, 526–529, 687
Goldberg, E.
Wisdom Paradox, 65, 356
Gonadotropins, 14
Goodall, D., 812, 827
Gorgias, 812
Gould, S.J., 446, 807, 808, 894, 921–923, 926
Government, 2, 3, 31, 37, 44, 66–68, 91, 101, 102, 131–133, 139, 140, 159, 163, 164, 171, 172, 182, 218, 221, 227, 259, 260, 300, 301, 362, 420, 427, 428, 431, 433, 437, 462, 464, 466, 467, 500, 504, 505, 510, 512, 513, 515, 525, 534, 552, 557, 561, 563, 570, 578, 585, 596, 600, 661, 692, 699, 724, 816, 939, 941, 942
Grandchildren, 3, 41, 77, 89, 98, 105, 139, 140, 157, 283, 361, 396, 416, 475, 490, 495, 496, 498, 501, 502, 551, 557, 578, 579, 589, 636, 647, 651, 666, 667, 688, 689, 699, 732, 762, 768, 774, 784, 894, 916–919, 941
Grandmothers of Plaza de Mayo, 578
Grandparents
ambivalent grandparents, 496
autonomous, 496
custodial, 140, 416, 498
duteous grandparents, 496
family-independent grandparents, 496
integrated grandparents, 496
role captivity, 140
Great depression, 103, 106, 112, 366, 688, 771, 844
Grey Nomads, 107, 108, 942
Grief
abiding grief, 862
acute grieving period, 285
anticipatory grieving, 791
five stages of grief, 786, 862
grief, 36, 164, 553, 558, 560, 571, 635, 759, 841, 861, 864
integrated grief, 862
Growth factor system, 10, 315
Growth hormone (GH), 12, 906
Guthrie, S., 808

H

Hā Kūpuna: National Resource Center for Native Hawaiian Elders, 468
Haldane, J.B.S., 893, 894
Hamilton, W., 588, 892, 894, 912, 916, 917
Happiness, 28, 42, 81, 111, 215, 291, 319, 379, 464, 473, 646, 652, 766, 785
frustration, 27
unhappiness, 27, 552, 755
Harman, D., 10, 906
Harmonious coexistence, 467, 468
Hawkes, K., 894, 916, 917
Healing, 144, 147, 160, 204, 224, 301, 682, 683, 693, 754, 785, 810
Health
arthritis, 3, 416
auditory acuity, 17
autoimmune conditions, 16
back and neck problems, 16
circadian rhythms, 36, 264, 269
diabetes, 16, 286, 309, 399
expiratory volume, 17
folk medicine, 144
general frailty syndrome, 148
grip, 17
healthcare services, 5, 7, 860
health-related quality of life (HRQL), 472
healthy ageing, 144, 201, 274
heart pathologies, 16
hypertension, 16, 297, 309, 311, 804, 839
illnesses, 303, 811, 839
incontinence, 16, 311, 530
limbs strength, 17
loss of muscle mass (sarcopenia), 16
mental, 3, 18, 32–34, 38, 44, 77, 98, 122, 143–145, 201, 215–218, 220–223, 226, 227, 262, 267, 277, 278, 293, 296, 300, 304, 308, 310, 311, 317, 318, 321, 355, 375, 409, 414, 416, 432–435, 472, 473, 481, 483, 485, 511, 542, 545, 549, 552–554, 560, 561, 569, 570, 574–576, 578, 580, 697, 728, 752, 793, 801, 803, 806, 839, 842, 844, 862, 938, 941

Health (*cont.*)
motor functions, 17
neuromuscular problems, 16
normal function, 143
obesity, 241, 261, 270, 286, 298, 316
osteoporosis, 3, 16, 297, 309, 311
physical functional limitations, 16
self-rated health (SRH), 144–146, 233
senility, 148, 149
sensorimotor capacities, 17
tactile abilities, 17
thermal regulation, 16
visual acuity, 637
Health and lifestyle survey, 411
Health and retirement study, 400
Healthcare, 5, 7, 34, 83, 172, 292, 360, 530, 554, 562, 587, 633, 702, 707, 810, 811, 817, 818, 848, 852, 853, 856–860
Health-Related Quality of Life (HRQL), 234, 282, 283, 472
Hearing impairment, 16, 295, 298, 637
Hebb's law, 207
Hedonism, 28, 97, 755, 938
aristippus, 28
Hellenistic period, 72, 819
Helplessness, 148, 492, 638, 647, 664, 682, 727, 765, 816, 841, 858
Heracles, 819
Heritability
to survive (to oldest ages), 897
Herodotus, 83
Hierokles, 76
Hippocrates, 72, 73, 88, 813, 819
Hispanic paradox, 166, 167
History (old age in)
Black Death, 57
Great Depression, 103, 106, 366
Greek world, 59
Hellenistic period, 72
Middle Ages, 59, 61, 62, 67, 79–87, 94, 96, 100, 104, 111
Renaissance, 59, 62, 72, 79–86
Roman Egypt, 67
Roman Empire, 57, 72, 75, 77
Roman Senate, 67, 76
Romanticism, 63, 94
Roman world, 59, 76, 78
social democratic phase, 104
World War II, 58, 96, 100, 104, 112
Holocaust
concentration camp, 25, 735
post-traumatic stress disorder (PTSD), 24
survivors, 24, 25, 735
trauma, 25, 682
Holstein, M., 27, 35, 40, 101–103, 359, 679
Homeostasis, 9, 10, 903, 906
Homer, 71, 489, 491, 828, 891
Honnold, A., 764
Honores, 67
Hope, 8, 29, 43, 65, 74, 81, 83, 148, 218, 224, 296, 360, 361, 372, 375, 425, 467, 539, 668, 688, 696, 739, 751–753, 757, 770–772, 785–787, 793, 801, 802, 825, 846, 847, 851, 853, 854, 858, 861, 895, 896, 937, 938, 944
Horace, 60
Hormones
adrenocorticotropic hormone (ACTH), 279
corticosterone, 211, 213
corticotropin releasing hormone (CRH), 279, 415
cortisol, 14
glucocorticoids, 14, 18
oestrogens, 208
prolactin, 211
prostaglandins, 18
Hospices, 91, 92, 821, 857, 861
Household
multi-generational, 130, 917
Hozho, 854
H70 Longitudinal Gerontological and Geriatric Population Study, 132
Human development index (HDI), 7
Human gut microbiota (GM)
Actinobacteria, 203
Bacteroidetes, 203
Firmicutes, 203
mutualistic symbionts, 203
pathobionts, 203
probiotic bacteria, 203
Proteobacteria, 203
Verrucomicrobia, 203
Humanistic Psychology, 263, 756
Humour, 357, 371, 380, 403, 413, 419, 420, 791, 792, 847
happiness, 380
Huntington's disease, 240, 893
Hymn to Aphrodite, 891
Hyodo, 853
Hypertension, 12, 19, 247, 251, 261, 295, 483, 836
Hyponatraemia, 12
Hypothalamo-pituitary-adrenal axis, 12
Hypothalamo-pituitary-gonadal axis, 12

I
Identity
auto-biographical affinity, 365
assimilation, 361
"complex identity", 362
cultural bifocality, 180
culturally syncretic identities, 180
dissociative identity disorder, 687
environmental, 365
environmental self, 365
"fixity", 40, 684, 686, 687
"flux", 40, 367, 684
insecurity, 364
insideness, 365
intersectionality, 511
labelled, 526, 531
liminality, 362
"multiple identity", 362
narrative of identity, 677, 684, 686, 692, 742
self-identity, 361, 362, 364, 635, 684, 685, 688, 724
social identity threat, 521
stereotypes, 40, 364, 522, 655, 668, 685
threats, 521, 668
Identity theory, 216, 427
Ikigai, 177
Illnesses, 70, 72, 73, 78, 87, 124, 143, 144, 149, 150, 163, 222, 225, 231, 232, 236, 309, 311, 431, 490, 502, 530, 552, 553, 556, 569, 575, 577, 659, 683, 696, 701, 704, 755, 767, 770, 773, 776, 786, 787, 811, 812, 818, 822, 826, 830, 834, 837–839, 841, 842, 844, 851, 852, 854
Immigration, 38, 58, 120, 124, 130, 151, 181, 364, 547, 553, 694, 699, 820
adaptation with continuity, 144
cultural tailoring, 151
Immune system
adaptive, 202
antibodies, 15–17
antigen presenting cells (APC), 14
antioxidants, 202
autoantibodies, 16
B-lymphocytes, 14
bone marrow, 14, 415
CD4 "helper" T cells, 14
CD8 cytotoxic T cells, 14
cytokines, 14, 18, 202, 315
dendritic cells, 14
dysregulated, 5, 14, 901
granulocytes, 14
IL-6, 14, 18
innate, 202
interleukin-1 (IL-1), 14, 18
leukocytes (white blood cells), 14
lymph glands, 14
lymphocytes, 14, 415
macrophages, 202
mast cells, 14
monocytes, 14, 910
mucus-associated lymphoid tissue, 14
naïve T-cells, 202, 900
natural killer (NK) cells, 14
neurotrophins, 18
oxidases, 202
polymorphonuclear granulocytes, 14
primary lymphoid organs, 14
probiotic bacteria, 203
reactive oxygen species (ROS), 202
secondary lymphoid organs, 14
spleen, 14
thymus, 14, 415
T-lymphocytes, 14
tumour necrosis factor (TNF), 14
Immunity
acquired immune response, 14
adaptive, 15
antibody affinity, 16
antigens, 15
autoimmunity, 9, 12
autoimmune disorders, 16
cellular, 14, 15
chronic inflammation, 231
cytotoxicity, 16
dysregulation, 16, 202, 231
human gut microbiota (GM), 203
humoral, 14–16
immunosenescence, 14
inflamm-ageing, 14, 203
inflammation, 202
innate immune response, 14, 15
molecular inflammation hypothesis, 202
redox balance, 202, 203
Indian culture
artha, 753
brahmacārya, 752
dharma, 752
gṛhasta, 752
kdma, 752
mokṣa, 752
sannyāsa, 752
vānaprastha, 129, 752
Individual
diversity of individualities, 40
individualism, 92, 129, 218, 535, 762
individuality, 131, 520, 678
possible selves, 40

Industrial Revolution, 91, 92, 546
Inequality
  wealth, 7, 104
Instrumental activities of daily living (IADL), 16, 243, 584, 585, 598, 655
Instruments
  Anxiety about Ageing scale, 125
  Attitudes Towards Old People Questionnaire, 153
  Auditory verbal learning test, 214
  Bakas Caregiving Outcomes Scale (BCO), 588
  Beck's Lethality Scale (*LS*), 836
  Benton Visual Retention Test (*BVRT*), 276
  Boston Naming Test, 276
  Brunstein's five-item scale, 26
  California Verbal Learning Test (*CVLT*), 276
  Card Rotations Test, 276
  Caregiver Burden Inventory (CBI), 588
  Caregiver Burden Scale (CBS), 588
  Caregivers Reaction Assessment (CRA), 588
  Caregiver Strain Index (CSI), 588
  Carer Quality of Life (CarerQol), 588
  Collett-Lester Fear of Death Scale (CLS), 789
  Cost of Care Index (CCI), 588
  Death Anxiety Scale (DAS), 787, 788
  Death Attitude Profile (DAP), 184, 185, 757, 790
  Death Attitude Profile-revisited (DAP-R), 757, 790
  Death Attitude Repertory Test (DART), 787
  Digit Symbol task, 276
  Geriatric Depression Scale (GDS), 529
  Global Perceptions of Intergenerational Communication (GPIC), 650, 666
  Life Satisfaction Index, 29
  Mattis Dementia Rating Scale, 214
  Mini Mental State Examination (MMSE), 276, 388
  Multidimensional Fear of Death Scale (MFODS), 789
  Perceptions of Intergenerational Communication (PIC), 648
  Personal Belief in a Just World Scale, 26
  Positive and Negative Affect Scales, 26
  Positive and Negative Affect Schedule, 792
  Project Talent Personality Inventory (PTPI), 410
  Purpose in Life Test (PIL), 217
  Relatives Stress Scale (RSS), 588
  Resilience Scale (RS), 217
  Revised Death Anxiety Scale, 788
  Revised NEO Personality Inventory (NEO-PI-R), 126
  Scale of Positive and Negative Experience (SPANE), 792
  Self-Rated Burden Scale (SRB), 588
  Self-Transcendence Scale (STS), 217
  Sense of Coherence Scale (SOC), 217
  Sense of Competence Questionnaire (SCQ), 588
  SF-36 Health Survey (SF-36), 217
  Sources of Meaning Profile (SMP), 745
  Suicide Intent Scale (SIS), 836
  Templer Death Anxiety Scale (DAS), 787
  Threat Index, 788, 880
  Trail Making test (A and B), 214, 276
  Trier Social Stress Test (TSST), 279
  Zarit Burden Interview (ZBI), 588
Insulin
  IGF-II, 10
  insulin-like growth factor-I (IGF-I), 10
Inter-generational
  cooperation, 112
  exchange, 82, 131, 161
  exchanges, 635, 663
  reciprocity, 500, 589, 669, 672
  solidarity, 137, 138, 142, 539, 645, 646, 655, 662, 666
  tensions, 4, 71, 82, 100, 112, 534, 583, 648
Interpreting
  Australian Institute of Interpreters and Translators, 704
  informal interpreters, 701
  language brokers, 701
  source language, 699
  target languages, 699
  transcultural care, 702
  transcultural perspective-taking, 701
Intersectionality, 516
Inter-systemic communication, 18
Iowa Youth and Families Project (IYFP), 360
Isolation, 18, 122, 125, 220, 224, 282, 309, 468, 478, 480, 481, 485, 490, 516, 519, 527, 530, 549, 558, 564, 570, 573, 577, 593, 656, 724, 730, 737, 829, 843
Izard, C., 369, 370

J

James, W., 359, 404, 800
Jerusalem Longitudinal Cohort Study, 24
Johnson, L.B., 104
Juncus, 73

Jung, C.G., 110, 723
Just-world, 26, 152, 524
Juvenal, 73, 74, 81

K

Kazantzakis, N., 824, 825
Kin, 92, 106, 123, 126, 128, 129, 133, 139, 157, 161, 369, 470, 475, 493, 494, 503, 505, 509, 510, 542, 549, 550, 577, 864, 919
Kinscripts, 128
Kin selection, 588, 892, 894, 896, 912–920, 927
Kirkwood, T., 899, 904
Kollman, J., 922
Krause, N, 132, 216, 218–220, 224, 225, 279, 411, 415, 420, 740, 752, 756, 757
Kübler-Ross, E., 774, 786, 862
Kung Fu, 162
Kungsholmen Project, 262, 545
Kymlicka, W., 849, 850

L

Laissez-faire capitalism, 92
Landscape of ageing, 107, 942
Language
  bilinguals, 256–260, 656, 701
  first language reversion, 656
  L1, 656, 672, 674
  L2, 258, 259, 656, 672, 674, 699
  multilingualism, 390
  prestige dialects, 660
  second, 136, 256, 258–260, 672, 673, 940
  second language attrition, 656
  trilinguals, 320
Lares, 76
Law, 38, 67, 103, 104, 151, 157, 362, 393, 468, 548, 589, 817, 822, 826, 849, 850, 858, 895, 904, 938, 939
Learning, 18, 22, 36, 37, 99, 107, 110, 128, 136, 145, 172, 180, 204, 206, 207, 211, 212, 214, 215, 220, 247, 248, 259, 260, 271, 276, 277, 320, 362, 369, 374, 380, 383, 386, 388, 392, 394, 430, 468, 519, 535, 546, 549, 552, 558, 653, 661, 669–674, 693, 699, 724, 741, 762, 771, 802, 849, 898, 899, 926, 943
  concentration, 36, 276
Legacy, 35, 80, 322, 531, 569, 579, 679, 726–728, 746, 768–771, 774, 787, 788, 792, 796, 862, 938
  concern for posterity, 768
  conscious legacy building, 771
  by example, 770
  generativity, 322, 726, 728, 769, 938
  historicity, 768
  symbolic immortality, 531, 770, 792
Leisure world, 108
Leptin, 14
Levy, B., 522
LGBTI people
  accelerated ageing theory, 568, 580
  ageing in place, 570
  American Society on Aging's LGBT Aging Issues Network (LAIN), 565
  androgyny, 273, 566
  assisted living facilities, 571
  birds of a feather, 571
  bisexual, 558, 559, 565, 568, 569
  care support networks, 561
  chosen family, 566
  community of practice, 561
  Continuing Care Retirement Communities (CCRC), 570
  crisis competence theory, 568
  cultural imperialism, 565
  discrimination, 558, 559, 562–564, 571, 572
  disenfranchised grief, 571
  disorders of sex development (DSD), 570
  family of choice, 561, 571
  family of origin, 560, 561
  femininity, 566
  gay, 558, 559, 564
  Gay and Lesbian Association of Retiring Persons (GLARP), 571
  Gays and Lesbians Older and Wiser (GLOW), 565
  healthcare, 562
  heteronormativity, 558
  heterosexuals, 560, 561, 563, 571
  hierarchical compensatory theory, 560
  HIV/AIDS epidemic, 559
  homophobia, 564
  homosexuals, 564
  intersex, 558, 565, 569, 570
  lesbians, 558, 564
  minstrelisation, 565
  Naturally Occurring Retirement Communities (NORC), 570
  non-hegemonic patterns of masculinity, 566
  nursing homes, 570
  palms of manasota, 571
  psychological ambivalence, 564

LGBTI people (*cont.*)
rainbow vision, 571
Senior Action in a Gay Environment (SAGE), 564
sexual fluidity, 566
social isolation, 563
social support networks, 561
stereotypes, 564, 599
stigma, 559, 562, 572
stonewall *communities*, 571
tolerant societies, 566
trans, 558–572
transgender, 558–572
transphobia, 569
transsexual, 558–572
Life
accommodative flexibility, 33
adversities, 26, 217, 682, 690, 693, 750, 801, 833, 841
belief in a just world, 26
coherence, 216, 217, 223, 280, 317, 321, 366–368, 437, 687, 693, 743, 744, 758, 785, 800
competent living, 32
costs of living, 7
"fluid life cycle", 65
life cycle, 65, 68, 113, 474
life expectancy, 3–6, 16, 61, 82, 90, 591, 721, 861, 908
life journey, 27, 40, 848, 861
life script, 109
meaning, 27, 30, 34, 41, 42, 78, 89, 105, 108, 109, 121, 157, 164, 173, 175–178, 180, 181, 183, 184, 186, 189, 215, 216, 222, 226, 281, 321, 357, 361, 365, 409, 419, 422, 425, 430, 434, 438, 468, 474, 543, 553, 599, 669, 677, 678, 684, 686–688, 692, 697, 708, 722, 723, 726–732, 734–741, 743–759, 762, 763, 766–769, 771, 774, 777, 782, 785, 795, 797, 799, 800, 802, 805, 807, 810, 812, 820, 821, 827, 840, 846, 847, 852, 854, 857, 863–865, 938, 940, 944
quality of life, 6, 8, 25, 26, 33, 37, 43, 63, 78, 100, 237, 238, 265, 291, 295, 313, 314, 378, 415, 483, 510, 515, 516, 722, 728, 750, 773, 804, 811, 817, 818, 836, 841, 843, 847, 856, 921, 942
reviews, 42, 121, 183, 218, 265, 285, 300, 405, 431, 462, 566, 580, 591, 707, 727, 740, 765, 766, 778, 794, 814, 839, 940
satisfaction, 26, 27, 39, 42, 128, 174, 176, 184, 187, 188, 224, 282–284, 294, 295, 317, 318, 357, 379, 380, 418, 423, 427, 428, 430, 434, 437, 471–473, 500, 504, 542, 543, 554, 560, 577, 591, 596, 635, 645, 723, 726, 728, 729, 740, 749–753, 762, 763, 767, 768, 773, 777, 790, 792, 795–797, 800, 847, 938, 939
success, 28, 30, 31, 79, 569, 681, 733, 748, 753, 755, 763, 916, 920, 937
Life course
age-based model, 24
generation-based model, 24
historical contingencies, 23
life course paradigm, 23
life journey, 27, 40, 848, 861
past traumas, 24, 25
Life review, 109, 634, 678, 692, 730, 740, 765, 768, 771
blocked, 766
ceilinged, 766
consolidated, 766
defensive, 766
ongoing, 766
priming, 766
terminated, 766
Lifespan, 10, 12, 60, 64, 68, 75, 102, 182, 183, 221, 250, 270, 300, 302, 304, 305, 359, 378, 432, 663, 894–900, 902, 906–909, 912, 913, 917–921, 923, 926
healthspan, 314
Lifestyle, 2, 31, 32, 34, 63, 68, 101, 103, 107, 108, 144, 159, 160, 170, 179–182, 214, 223, 224, 242, 256, 302, 306, 309, 314, 320, 355, 389–392, 402, 414, 421–423, 437, 508, 535, 545, 573, 574, 645, 939
Liminality, 362, 430, 437
Lipids, 9, 10, 202, 203, 205, 235, 250, 315, 316, 398, 905
Living arrangements
co-habitation, 129
couplehood, 36, 501, 502
living apart together (LAT), 479, 501, 503
Locus of control
internal, 224
Logotherapy, 754, 755, 766
Loneliness, 35, 36, 38, 106, 111, 129, 181, 318, 420, 432, 462, 464, 471, 476–487, 543–545, 547–549, 552, 553, 564, 565, 569, 572, 639, 682, 776, 822, 839, 846, 939–941
aloneness, 36, 478, 486

apathetic acceptance, 478
attributional factors, 478
cathectic deprivation, 485
cognitive discrepancy, 477
dementia, 483–485
depression, 36, 111, 220, 234, 297, 432, 483–485, 487, 552, 553, 565, 940
emotional isolation, 478
emotional loneliness, 35, 478, 484, 485
empty nest syndrome, 483
existential loneliness, 477, 478, 486, 487, 728, 731–734, 753, 864
hopelessness, 483
isolation, 477, 480, 564, 727
loneliness threshold, 479
mortality rates, 471, 483, 487, 545
negative loneliness, 478
parasuicide, 484
perceived isolation, 485
positive loneliness, 478, 486
privacy, 478, 487
reminiscence therapy, 485
social disconnectedness, 220, 485
social isolation, 106, 129, 220, 477, 478, 480, 485, 487, 548, 569
social loneliness, 35, 478, 484, 543
stereotypes, 564, 941
stigma, 129, 462, 564, 565
suicide ideation, 484
Longevity, 10, 68, 70, 84, 128, 250, 270, 301, 302, 314–316, 379, 412, 413, 639, 784, 804, 891–893, 897, 898, 900, 902, 906, 909, 910, 912, 918, 923
adaptive behaviour, 310, 808, 847
"ageing cliff", 305
blue zones, 309, 313
Calment, J., 301, 302, 305, 895
centenarians, 10, 301, 302, 306, 310, 314, 315, 321, 322, 906
Chinese Healthy Longevity Survey, 313
cognitive decline, 149, 204, 217, 220, 244, 247, 253, 276, 292, 319, 382, 389–390, 416, 473
cognitive reserve, 34, 212–213, 255–257, 260, 262, 269, 320, 385–388
"compression of mortality", 302
delayers, 312
differential-preservation hypothesis, 320
disuse hypothesis, 320
emotional health, 319
epigenetic mechanisms, 316
escapers, 312
exceptional longevity, 301, 305, 311, 315, 322
fulfilling ageing, 412
Gavrilov, L.A., 305–308, 895, 896, 905
Gavrilova, N., 305–308, 895, 896
genetic factors, 302, 314, 912
"good luck", 37, 310
health capital, 310
healthspan, 314
healthy habits, 302
Heidelberg *Centenarian Study*, 317, 320
heritability, 10, 314, 784, 897
Hokusai, K., 310
'individual Gompertz slope', 302
International Database on Longevity (IDL), 301
life expectancy, 302
lifespans, 891, 892
maximum lifespan, 302, 305, 306
maximum reported age at death (MRAD), 306
medical treatments, 302, 322
New England *Centenarian Study* (NECS), 309
paradox of ageing, 292, 317
100-plus Study, 321
semi-supercentenarians, 301, 302, 314, 315, 322
social connectivity, 302
supercentenarians, 302, 305, 306
survivors, 312
technologies, 302
ultracentenarians, 301
USA Georgia Centenarian Study, 317
well-being, 8, 22, 26, 28, 43, 109, 128, 132, 175, 178, 215–217, 221, 224–225, 282–284, 288, 291, 318, 369, 378, 411, 434–435, 512, 543, 554, 594, 637, 728, 747, 805, 941
Longitudinal Aging Study Amsterdam (LASA), 236, 471, 509
Lord of the castle of Goet's, 84
Lorenz, K., 923, 926
Louisiana healthy aging study, 542
Love, 29, 36, 42, 74, 75, 93, 121, 129, 140, 142, 143, 157, 218, 269, 289, 310, 357, 358, 364, 492, 494, 495, 501, 540, 564, 579, 583, 588, 592, 688, 723, 728, 737, 749, 754, 762, 763, 766, 768, 770, 785, 793, 803, 805, 820, 828, 860, 861, 891, 938
Lysosomes, 11

M

Maastricht aging study, 277, 388
Machiavelli, N., 86

Magnetic resonance imaging (MRI), 19
Marriage, 113, 136, 145, 170, 496, 501, 559, 855
  commuter, 501
  divorce, 16, 36, 97, 106, 236, 286, 367, 478, 498, 501, 502, 841
  social regression, 503
Masaryk, T., 830, 831, 833
Masquerade, 40, 41, 207, 361, 686, 687
Massachusetts Hispanic Elders Study (MAHES), 166
Mastery
  passive, 124
  stereotypes, 152
Mater familias, 75, 76
Maximian (Maximianus), 67, 75
Maynard-Smith, J., 894, 897
Meaning
  ABCDE intervention strategy, 748
  adventurousness, 764
  affective components, 740, 760
  autotelic (activity), 175, 177, 182, 226, 741, 759
  avoidance systems, 746, 747
  become yourself, 741
  cognitive component, 737, 740
  commitments, 172, 361, 599, 745, 760, 762
  created, 108, 741, 759
  creativity, 30, 736, 743, 744
  deep-and-*wide hypothesis*, 748
  defensive mechanisms, 747
  dispositional optimism, 321, 752
  duality hypothesis, 746
  Dual System Model (DSM), 747, 748
  duality hypothesis, 746
  ego resilience, 750
  ego strength, 409, 750
  emotional well-being, 128, 372, 543, 736, 741
  existential distress, 755
  existential frustration, 753, 755
  existential guilt, 757
  existential meaning, 741, 762
  existential perspectives, 741
  existential vacuum, 31, 728, 753, 755, 756, 765
  flow, 742
  global happiness, 749
  global meanings, 759, 760
  gratitude, 739
  hostile-world, 750
  implicit, 741
  know thyself, 741
  life meaning, 26, 31, 217, 596, 708, 728, 734–774, 798, 843, 941
  life satisfaction, 293, 741, 750, 751, 759, 765
  logotherapy, 746, 754, 755, 766, 767, 861
  mastery, 750
  meaningfulness, 41, 224, 321, 781
  meaning making, 216, 677, 690, 727, 734, 744, 747, 748, 752, 753, 758–767, 785, 863
  meaning-mindset, 748, 749
  meaning of place, 761
  mindful awareness, 748, 749
  motivational component, 739, 740
  myth of fulfillment, 763
  narrative of identity, 692, 742
  noological dimension, 757
  nostalgia, 762
  OCEAN, 748, 749
  optimism, 224, 697, 744, 745, 750–752, 767
  positive affects, 749, 765
  positive psychology, 741, 751
  post-traumatic growth, 747, 764, 778, 781
  presence of, 468, 732, 737, 750, 751
  psychological closure, 419, 743
  PURE principle, 748
  purpose in life, 109, 216, 739, 749, 753, 757, 763, 777, 795, 800, 861
  religiosity, 223, 224, 226, 744, 745, 797, 798, 800, 804, 830
  reminiscences, 730, 737, 744, 771, 864
  search for, 723, 736, 737, 741, 744, 750, 751, 754, 758, 759, 767, 768, 782, 938
  seeking, 42, 163, 434, 736, 743, 759, 767, 785
  sense of coherence, 217, 321, 363, 736, 739
  sense of purpose, 224, 468, 739, 740, 757, 763
  situational meaning, 741, 759, 760
  Sources of Meaning Profile (SMP), 745
  spirituality, 42, 45, 222–224, 226, 596, 727, 736, 739, 748, 750, 751, 757, 766, 767, 790, 795, 797, 798, 800, 804, 847, 864
  striving for self-enhancement, 742
  tragic triad, 754
  transcendence, 41, 216, 217, 321, 727, 728, 766, 767, 769, 864
  transcendental meaning, 790
  unifying frameworks of meaning, 750
  will to meaning, 753, 754
  wisdom, 357, 680, 723, 736, 744, 762

Meaning making
amotivation, 765
assumptive world, 758, 759
challenges, 762, 765
commitments, 744, 762
completion stage, 758
conformism, 762
content of meaning, 758
contextual model of personal meaning, 760, 761
controls, 752, 758, 762
courage, 761
dereflection, 767
discovery stage, 758
emergency stage, 758
exploratory stage, 758
helplessness, 765
global meaning, 759–761
hardiness, 761, 762
Individual Meaning Centred Psychotherapy (IMCP), 767
inertia stage, 758
intrinsic motivation, 765
life review, 744, 766, 767
meaning-centered counseling and therapy (MCCT), 766
meaning maintenance model (MMM), 765
meaning management theory (MMT), 785
meaning of place, 761
meaning repairing, 765
momentous life events, 762
nostalgic memories, 762
outcome approach, 759
passivity, 765
personal consolidation, 766
personal transcendence, 766
personal transformation, 357, 766
post-traumatic growth, 747, 764, 781
processes of meaning, 758
self-determinatyion theory (SDT), 765, 785
situational meaning, 759
stagnant stage, 758
three-component model, 760
vegetativeness, 764
Medawar, P.B., 9, 893, 894, 903, 904, 912
Media, 106, 396, 489, 492, 534, 537, 539, 634, 638, 676, 685, 698, 771
mainstream, 698
Medicaid, 104
Medicare, 104
Memory
alzheimer's disease, 150, 212, 259, 269, 398, 399, 473
amnesia, 393, 395, 396
anterograde amnesia, 395
antioxidants, 269, 402
autobiographical memories, 213, 394, 395, 397, 400, 743, 863
autonoetic awareness, 394
brain-derived neurotrophic factor (BDNF), 399
caloric restriction, 402
catechol-o-methyltransferase (COMT), 399
consolidation, 393
crystallised abilities, 393
declarative memory, 266, 393, 394, 397, 417
delayed recall, 19, 277, 401
dementia, 240, 242, 252, 263, 266, 385, 392, 396, 398, 683
ε4 allele, 250, 398, 399
education, 243, 259, 276–278, 385, 397, 400, 401, 404
encodes, 260, 393, 395, 398
episodic memory, 20, 21, 205, 253, 255, 275–278, 393–395, 397–401, 403, 417
explicit memory, 394, 397
false memories, 21, 397, 404
fluid abilities, 393
foils, 392
forgetfulness, 150, 240
gender, 276–278, 388, 389, 397, 400, 404, 940
haptoglobin system (genetic marker), 398
hippocampus, 19, 21, 205, 210, 252, 255, 392, 398
human brain-derived neurotrophic factor (BDNF) protein, 398
implicit memory, 394, 397
infantile amnesia, 395
intrusions, 392
lifestyles, 107, 278, 397, 402, 404, 473
long-term memory, 19, 259, 393, 395, 397
memory consolidation, 213, 245, 393
memory training, 403, 940
multi-tasking, 21
negative stereotypes, 152
nostalgic memories, 762
osteocalcin, 205
perceptual representational system (PRS), 393
personalities, 45, 238, 397, 400, 404, 406, 742
positive stereotyping, 402
positive reminiscence, 379
primary memory, 21, 394, 395

Memory (*cont.*)
priming, 393, 394, 402
procedural memory, 240, 266, 393, 397
prospective memory, 21, 397
reminiscence bumps, 395, 397
remote memory, 395
retrieve, 260, 393–395, 402
retrograde amnesia, 395
Ribot's law, 393
schemata, 165
self-defining memories, 743
semantic memory, 276, 393, 395–399, 402, 404, 416
semanticised, 394
short-term memory, 21, 259, 393, 395, 397
sociality, 397, 400, 404
spatial memory, 19, 205
stores, 260, 393, 394, 404
temporally graded retrograde amnesia, 393
tip-of-the-tongue experience (TOT), 394, 395
vocabulary, 205, 656
working memory, 21, 205, 255, 276, 277, 387, 393, 394, 397–400, 403, 416
Menopause, 2, 14, 60, 93, 143, 146, 147, 286, 287, 405, 909, 913, 916, 917, 920, 926, 927
climacteric, 146
depression, 35, 126, 147, 212, 220, 228, 230–234, 237–238, 247, 269, 280–281, 294, 299, 415, 483–484, 490, 530, 552, 560, 576, 591, 594, 635, 786, 836–837, 844, 847
gengnianqi, 146
hot flashes, 146, 147
kônenki, 146
menopausal syndrome, 146
symptoms, 146, 147
Mental
concentrations, 276, 279, 485
disorders, 34, 215, 226–230, 238, 240, 270, 296, 406, 559, 575, 697, 834, 836, 842, 846
handicaps, 32
health, 3, 18, 32–34, 37, 44, 98, 122, 143–145, 159, 182, 201–322, 355, 375, 406, 409, 414, 416, 434, 472, 473, 483, 485, 511, 519, 545, 549, 552–554, 560, 562, 569, 570, 574–578, 580, 677, 697, 728, 752, 755, 765, 793, 801, 803, 806, 838, 842, 852, 862, 939, 941
Mental disorders
alzheimer's disease, 34
anxiety, 228, 238
cognitive behavioural therapy (CBT), 238
cognitive restructuring, 237
delirium, 239, 317
dementias, 227
depression, 228–236
dysthymic disorder, 233
frontotemporal NCD, 240, 272
generalised anxiety disorder (GAD), 238
major depressive disorder, 228, 229, 233, 834
major neurocognitive disorders, 227, 238
mild cognitive impairment (MCI), 205, 242, 245, 260, 266, 320, 403
mild neurocognitive disorder, 266
minor depression, 233
mood disorders, 227, 228
NCD due to Parkinson's disease, 240
NCD with Lewy bodies, 240, 241, 272
neurocognitive disorders (NCD), 228
other NCDs, 240
relaxation training, 237
vascular NCD, 239, 251, 272
Mental health
anger, 221, 375, 552
anxiousness, 221
dementias, 317
depression, 35, 126, 147, 212, 220, 228, 230–234, 237–238, 247, 269, 280–281, 294, 299, 415, 483–484, 490, 530, 552, 560, 576, 591, 594, 635, 786, 836–837, 844, 847
difficulty concentrating, 221
drinking, 221
drug dependence, 221
existential anguish, 221
irritability, 221, 375, 552
mental well-being, 217, 222, 941
sleep disturbance, 220
social support, 216, 218, 224, 355, 433, 485, 552, 554, 561, 569
vitality, 217
Metabolism, 9, 10, 12, 211, 315, 316
Migrant, 138, 181, 548, 551, 552, 557
acculturation, 549, 550, 553, 554
acculturative stress, 417, 547, 850
assimilation, 552
bao-da, 550
belonging, 418, 479, 509, 549, 556, 795, 806
caregiver burden, 36, 492, 563, 580, 587–588, 590, 593, 695
culture shock, 547, 548, 850
discrimination, 549
emigration, 158
emotional vacuum, 555

ethnic neighbourhood, 555
Fourfold theory of acculturation, 548
ghorbat, 553
han, 552
health selection effect, 551
homeland, 555
hwabyung, 552
immigrants, 138, 181, 548, 549, 551, 552, 558
immigration, 38, 44, 58, 124, 125, 130, 145, 151, 154, 181, 364, 547, 550, 553, 694, 699, 820
institutional support, 38, 180
integration, 548
international retirement migrants (IRM), 546
lifeworld, 556
longing, 556
marginalisation, 548
migration, 138, 181, 551
migratory grief, 553
oong-uh-ri, 552
refugees, 553
rootedness, 549
schengen area, 557
senior social centres, 554
sense of purpose, 468, 512, 554, 739–740, 757, 763
separation, 548, 557
services, 548
shinggyongshaeyak, 552
snowbirds, 107, 108, 170, 179, 546, 942
social network of choice, 549
social network of necessity, 549
transnational caregiving, 551
transnational grandparents, 546
traumatised, 553
uprootedness, 553
zero generation, 551
Mild cognitive impairment (MCI), 205, 242, 246, 266, 403
Mind
active, 33, 39, 74, 201, 247, 307, 400
awareness, 689
delusional disorders, 34, 298
forgetfulness, 73, 150, 240, 244, 245, 704
mental disorders, 201
mindful, 28, 111, 748, 749
mindfulness, 28, 111, 213, 216, 404, 420, 940
mood disorders, 227–230, 233, 575, 589, 842
stress-related disorders, 34
Mindfulness, 28, 111, 216, 403, 404, 940
Minois, G., 60, 62, 71–77, 80–86
Mitochondria, 10, 316, 903, 906, 910
Modernisation, 44, 91, 120, 124, 129, 130, 136, 137, 142–144, 153, 163, 188, 480, 494, 497, 499–501, 503, 505, 536, 550, 574, 647, 649, 654, 708
Montagu, A., 356, 921, 923, 926
Mood, 29, 228, 230, 231, 238, 240, 244, 368, 420, 486, 747, 767
disorders, 228, 230, 233, 575, 589, 842
The Monongahela Valley Independent Elders Survey (MoVIES project), 390
Moran, 122
Morbidities, 8, 16, 156, 220, 284, 285, 309, 311–313, 465, 591, 592
Morselli, E., 830
Mortality, 3, 4, 8, 25, 36, 38, 43, 58, 59, 61, 82, 91, 93, 144, 145, 178, 186, 203, 220, 225, 233, 234, 237, 270, 281, 285, 296, 301, 302, 308, 311, 372, 375, 390, 412–415, 465, 471, 473, 485, 516, 545, 551, 590–592, 776, 777, 781, 782, 788, 804, 811, 835, 864, 894–902, 907–910, 913, 915, 925
Mourning, 186, 564, 819, 861, 863, 864
four tasks of mourning, 863
Multiculturalism, 848–850
citizenisation, 849, 850
multicultural citizenship, 850
Munera
munera of decurion, 67
munera personalia, 67
Muscle
muscle mass, 14, 203, 204, 906
muscle function, 14
skeletal muscle strength, 12
Mutation
somatic, 9, 905, 907
Myth
eternal youth, 80, 86, 541, 891
Fountain of Youth, 83
greedy geezers, 100
Herakles and Geras, 78
Island of Jupiter, 83
leaners, 100
lifters, 100
myths of rejuvenation, 70

N
Narrative, 27, 40, 256, 258, 366, 367, 658, 677–681, 683, 684, 686, 690, 692, 693, 695–698, 734, 742, 743, 772, 817, 942
accrual, 678
anchor scripts, 682, 683

Narrative (*cont.*)
biographical reasoning, 684
counterfactuality, 679
deconstruction, 696
discourse, 256, 258, 366, 658, 677, 679–681, 696
emplotment, 684
environment, 680, 681, 691, 696
fiction and cultural mediation of ageing project (FCMAP), 690
fiction narratives, 690
fixity, 686
fluidity, 686
flux, 684
folk narratives, 678
foreclosure, 734
generational imprinting, 366
gerontology, 680
habitus, 677
heroes, 685, 686
heroins, 685, 686
intersubjectivity, 678
life reviews, 697
lifestory, 680
life world, 180, 678
masquerade, 40, 686, 687
narrative of decline, 679
narrative of identity, 677, 684, 686, 692, 742
narrative of suffering, 683
natural discourse, 39, 680
normal discourse, 39, 680
normative discourse, 681
poetical dimension., 680
possibilities of the self, 686
possible selves, 684
reconstruction, 366, 684, 696
self-integration, 691, 692
situation models, 677
spiritual dimension, 680
storytelling, 678
storytime, 679
therapies, 696, 697, 942
transactional inter-individuation, 687
U-meek, the first whaler, 689
Villains, 685, 686
wisdom environment, 680
Nascher, I.L., 95
National Social Life, Health, and Aging Project (NSHAP), 220
National Study on Aging and Care in Kungsholmen (SNACK), 399
National Survey of Psychiatric Morbidity, 378
Naturally occurring retirement community (NORC), 129
Nazism, 723
Neglect, 136, 141, 155, 157–159, 369, 416, 489, 536, 550, 944
psychological, 158, 489, 536
Neoliberalism, 27, 43, 68, 95, 97, 102, 105, 129, 154, 172, 360, 500, 515, 516, 534
Neruda, P., vii, viii
Nervous system
axonal atrophy, 18
brain, 11, 18, 207, 810, 904
central nervous systems, 18
glial cells, 18
motoneurons, 18
myelination, 18
myelin sheath, 18
nerve fibres, 18
nerve impulse, 18
nerve regeneration, 18
neural plasticity, 18, 810
neurogenesis, 18
neurotransmitters, 18
peripheral nervous system, 369
spinal cord, 18
Neurodegeneration, 16, 248, 253
Neuroendocrine system, 9, 218
Neuron
apoptosis, 211, 252
axons, 210, 251
dendrites, 210
dendritic arborisation, 210
dopaminergic receptors, 19
impulse, 18, 19
myelin, 18, 19
neurofibrillary plaques, 19
neurogenesis, 207, 211, 212
neuronal size, 204
neurotrophins, 18, 211
spine density, 204
synapsis, 210, 211, 251
synaptic plasticity, 19
synaptogenesis, 210
tangles, 251
Neurotransmitter
dopamine, 208
Nietzsche, F., 734
Nihilism, 23, 97, 322, 734
Nimrod, G., 175, 176, 432
Nottenbohm, F., 211

O

Obesity, 14, 17, 247, 261, 270, 286, 297, 316, 415
Oestradiol, 12, 14

Offline seniors, 513
Old age
  as "disease", 96
  winter of life, 61, 88
Oldest-old, 2, 3, 5, 22, 38, 39, 45, 63, 64, 177, 184, 185, 277–280, 293, 301, 304, 313, 318–321, 462, 481, 509, 539–545, 586, 599, 693, 757, 840, 843, 844
Old-old, 2, 3, 64, 177, 214, 509, 746, 757
Old Poor Law, 90
On the Decline of Life, 63
Optimism, 22, 97, 181, 224, 226, 308, 363, 382, 411, 413, 587, 685, 686, 751, 752, 940
  dispositional, 321, 752
  dynamic integration theory, 22
  flow, 751, 752
  happiness, 317, 752
  illusory optimism, 412
  positive affectivity, 752
  positive emotions, 752
  positive psychology, 751
  satisfaction, 554
  tragic optimism, 755, 756
Orgel, L., 904
Osteocalcin, 205
Otium honestum, 77, 111, 178, 407, 431

P
Palliative care, 572, 577, 722, 767, 771, 773, 774, 778, 811, 813, 855–858, 860
Paradox of ageing, 292, 317
Paradox of well-being, 39
Parathyroid hormone, 14
Parkinson's disease, 212, 217, 240, 241, 309, 593
Partners, 25, 33, 36, 45, 65, 97, 124, 149, 175, 184, 272, 280, 283, 285, 288, 289, 362, 372, 375, 379, 419, 432, 435, 473, 481, 485, 493, 494, 499, 501–503, 510, 525, 531, 542, 560, 561, 563, 566, 569, 571, 581, 588, 639, 642, 644–646, 655, 660, 668, 671, 674, 675, 691, 778, 789, 808, 839, 840, 849
Passive mastery, 124
Pater familias, 76
Pathogens, 15, 473, 910
Patria potestas, 121
Pavle, K., 101–103
Penates, 76
Pensions
  accumulation, 68
  age pensions, 2, 68, 69, 420
  Beveridgean model, 68
  Bismarckian model, 68
  Danish Act on Old Age Relief, 67
  defined benefits, 68, 102
  defined contributions, 68, 102
  "deserving" poor, 67
  friendly societies, 2, 67
  The Friendly Societies Act, 2, 67
  government pensions, 67, 101, 421
  Old Age Pensions Act, 67
  pensionable age, 2
  pension systems, 2, 104
  "undeserving" poor, 104
Pericles, 75, 77
Peroxisomes, 10
Personality, 33, 36, 42, 126, 148, 151, 185, 228, 231, 235, 236, 270, 272, 274, 275, 285, 299, 321, 355, 369, 404–414, 434, 437, 479, 483, 491, 541, 544, 575, 587, 687, 730, 749, 789, 807, 833, 834, 838, 840, 842
  agreeableness, 126, 274, 275, 285, 404, 406–410, 413, 434, 437, 807
  Alzheimer's disease (AD), 409
  Big Five, 274, 275, 406–410
  catastrophising, 412
  changes, 33, 151, 231, 404–411, 413, 414, 434, 686
  conscientiousness, 126, 274, 321, 404, 406–410, 413, 437, 749, 807
  coping, 235, 400, 411, 587, 842
  extraversion, 42, 126, 274, 321, 400, 404, 406–410, 413, 434, 730, 749
  gender patterns, 410
  hardiness, 185
  illusory optimism, 412
  illusory pessimism, 412
  intra-individual consistency, 409
  ipsative stability, 409
  La dolce vita effect, 405
  life satisfaction, 409, 749
  major changes, 406
  masquerade, 40, 207, 361, 662, 686
  maturity principle, 404, 410
  mean-level consistency, 409
  neuroticism, 126, 185, 231, 235, 236, 274, 275, 285, 400, 404–411, 413, 434, 437, 483, 730, 749, 789, 838, 840, 842
  normative, 405

Personality (*cont.*)
openness, 126, 274, 406–410, 413, 434, 730, 749, 838
plaster hypothesis, 404
rank-order consistency, 409, 410
resolution, 29
Self-identified psychological turning points (SPTPs), 406
social support, 318, 411, 414, 541, 544, 587
temperament, 272, 404
turning points, 406
zest, 29–30, 739
Pessimism, 95, 102, 388, 412, 686, 839
Petrarch, 81, 82
Phenomenological hermeneutical method, 693
Phenotypes, 9, 314, 784, 908, 911, 912
Philosophy
nihilism, 735
therapeutic nihilism, 23
Pietas, 76
Plasticity
Functional Plasticity in Cognitive Ageing hypothesis, 23
neuronal, 12, 19, 207, 208, 210, 212, 263, 391
synaptic, 19, 207, 399
Plato, 72, 74, 241, 768, 828
Pliny, 73, 77
Plutarch, 72, 74–76
Politics of ageing
conservative, 132, 153, 645, 653
neoliberalism, 43
pensions, 44
political participation, 474, 543
social democracy, 102, 104, 112, 132, 171, 420, 505
superannuation, 3, 421, 427
welfare system, 132
Poor
"deserving poor", 67, 92, 104
"undeserving poor", 92, 104
workhouses, 92
Pope Innocent III, 81
Population
global, 4, 243, 464
growth, 5, 101, 103, 211
Positron emission tomography (PET), 208, 253, 254
Post-traumatic growth, 747, 764, 781
Post-traumatic stress disorder (PTSD), 25, 230, 236, 293, 416
Prejudice, 79, 99, 100, 154, 227, 260, 282, 291, 364, 478, 518, 521, 523–526, 530, 532, 534, 535, 538, 547, 586, 638, 639, 663, 664, 666, 675, 939
cognitive consonance, 524
cognitive dissonance, 524
confirmatory biases, 524
overaccommodation, 99
self-fulfilling prophecies, 524
self-regulation, 525
stereotypes, 79, 99, 154, 521, 523, 525, 526, 533–535, 638, 639, 663, 666, 675, 939
Prison inmates
depersonalisation, 573
dying in prison, 577
functional impairment, 575
institutional dependency, 573
lifetime traumas, 574
major illnesses, 575
mental conditions, 575
recidivists, 572
Reintegration Effort for Long-term Infirm and Elderly Federal Offenders (RELIEF), 578
Resettlement and Care for Older Ex-Offenders (Recoop), 578
RESTORE 50+ Program, 578
self-efficacy, 575, 576
Senior Ex-Offender Program (SEOP), 578
victimisation, 574
Probiotic bacteria, 203
Project Talent, 410
Proteasomes, 11
Protein
apolipoprotein E, 250
Psychoanalysis, 65, 419, 724, 755
Psychological closure, 419, 743
Pythagoras, 60, 74

Q

Quality of life
activities of daily living (ADL), 2, 24, 160, 267, 292, 294, 295, 310, 311, 313, 316, 390, 483, 487, 497, 511, 519, 542, 543, 580, 581, 589, 599, 941
instrumental activities of daily living (IADL), 16, 243, 585, 598, 655
subjective experience, 25
Quasimodo, S., 721, 727, 865
Queensland Association for Healthy Communities (QAHC), 563
Quetelet, A., 96
Quinlan, K.A., 816

R

Race, 27, 532, 562, 768, 820, 937
racism, 95, 150, 682, 831
Reactive oxygen species (ROS), 10, 11, 202, 246, 903, 905–907, 910
Reciprocation, 283, 406, 475, 588, 589, 644
arrow of time reciprocation, 589, 663, 919
Reciprocity
mutual, 508, 663
transitional, 663, 919
Racism, 95, 150, 682, 831
Rawls, J., 101
Relationships
positive, 4, 42, 131, 149, 220, 402, 514, 560, 564, 745, 805
Religion
afterlife, 81, 795, 797, 806, 855
Buddhism, 187, 824
Christianity, 187, 806
cognitive imperative, 805
extrinsic, 187, 225
Hindu, 129, 130, 475, 752
intrinsic, 187, 224, 225, 795, 796
Islam, 187, 824
Judaism, 187, 855
old testament, 61, 89
religiosity, 187, 222, 410, 795, 800, 806, 808
rituals, 222, 806
Taoism, 182, 184
traditional, 222, 552, 805
Religiosity
cognitive imperative, 805
commitment theory, 807
extrinsic, 42, 187, 222, 224–226
healing, 224
hyperactive agent detection device (HADD), 809
intrinsic, 42, 187, 222, 224–226, 795, 796
moral life, 806
"positive illusory beliefs", 223
religion, aging, and health survey, 225–226
transcendental faith, 800
Resilience, 23, 185, 206, 217, 257, 320, 321, 356, 364, 385, 405, 682, 683, 690, 693, 747, 752, 765, 801, 802
Resourcefulness, 516, 748
Respect, 67, 73, 74, 79, 82, 90, 93, 120, 121, 125, 128, 134–137, 143, 146, 150, 152, 153, 180, 188, 257, 375, 468, 475, 490, 497, 498, 500, 536, 552, 569, 574, 580, 644, 647, 649, 650, 653, 654, 665, 703, 727, 806, 810, 822, 825, 849, 850, 852, 855
disrespect, 136, 644
neglect, 158
Restorative activities, 511
Retirement
active, 25, 67–69, 77, 170, 172, 177–179, 182, 188, 355, 389, 420, 428, 429, 431, 436
adjustment process, 422, 423
Australian *masters games*, 181
bridge employment, 69, 172–174, 422, 430, 431, 939
"busy ethics", 432
changes, 25, 41, 69, 93, 95, 102, 107, 113, 120, 129, 154, 171, 173–175, 177, 188, 221, 286, 378, 401, 421, 423, 425, 426, 430, 432, 435, 437, 939
concentrators, 176
continuity, 42, 171, 173–177, 426, 429, 430, 437
continuity theory, 174, 425
deciding the time to let go, 427
decision-making process, 422, 426, 428
depression, 68, 176, 423, 425, 429, 432, 433, 435
diffusers, 176
disenchantment, 174, 423
expanders, 176
financial security, 170, 172, 182, 421, 426
gender, 185, 277, 282, 303, 426, 434–437
gerontomigration, 180
gerotranscendence, 175
"goal instability", 434
Grey Nomads, 107, 170, 179
happiness, 378, 432, 437
hardiness, 434
health, 25, 34, 77, 104, 128, 154, 170–173, 175, 181, 182, 188, 221, 282, 421, 423, 426, 428, 429, 432–435, 437, 438, 464, 570, 657, 667, 844, 859
healthy lifestyle, 379, 432
historical aspects, 69, 421
hobbyist, 178
healthcare, 34, 172
identities, 42, 173, 174, 420, 431, 560, 657
imagining the possibility of retirement, 426
innovation theory, 176
involuntary, 173, 434, 463
leisure, 170, 173, 176–179, 181, 182, 189, 277, 429–432, 435
leisure-time physical activity (LTPA), 178
life course perspective, 174, 425
life satisfaction, 173, 175–177, 421, 423, 425, 431–434
liminality, 431, 437

Retirement (*cont.*)
loneliness, 657
newbridge, 943
passive, 177, 178, 389
pensions, 1, 2, 34, 68, 93, 95, 104, 120, 170–172, 546
planning for, 34, 171, 172, 429, 430, 939
post-retirement behaviour, 176
planning for, 171, 172, 430, 431, 939
putting plans into action to transition into retirement, 429
reducers, 176
relationships, 25, 34, 175, 286, 401, 426, 430, 432, 433, 435, 463, 667, 779, 795
residential tourism, 180
resource *perspective*, 174
retrenchment, 173
role theory, 173, 174, 425
RVers, 107, 170, 179
self-esteem, 427, 432
self-fulfillment, 421, 430
serious leisure, 178
Silver Human Resource Centres (SHRCs), 177
snowbirds, 107, 170, 179, 546
social roles, 173, 174, 432
stage in career development, 422, 423
stage theories, 174, 427
stresses, 175, 176, 421, 427, 431, 432, 434, 438, 844
subjective wellbeing, 177, 282, 433
superannuation, 3, 421, 427
transitions, 69, 95, 172–175, 221, 420, 421, 423, 425, 426, 428, 430, 431, 434, 435, 437, 730, 939
University of the Third Age (U3A), 943
un-retire, 422
USA senior *games*, 181
voluntary, 3, 67, 173, 421, 423, 432, 463
Ribot, T.-A., 393
Rodia, S., 763
Roosevelt, F.D., 104
Rose, M.R., 892–894, 903–905
Ruminating, 730, 745, 787, 839
Rural Missouri Elders longitudinal study, 132
Russell, B., 736
RVers, 942

S
Sappho, 891
Scaffolding Theory of Ageing and Cognition (STAC), 20, 23, 208, 381
Seattle Longitudinal Study, 383
"Second childhood", 111
Second Duke Longitudinal Study, 410
Selective Optimisation with Compensation (SOC) model, 22, 175, 386
Self, 26, 28–30, 40, 41, 59, 65, 109, 178, 359–362, 365, 366, 372, 374, 378, 381, 393, 394, 422, 431, 437, 464, 486, 517, 527, 536, 639, 684, 686, 687, 689, 691–693, 722–724, 727, 737, 740, 742, 744, 747, 748, 757, 766, 769, 770, 776, 783, 786, 799, 805, 809, 815, 863, 864, 938, 939
autonomy, 596, 722, 750
coherent, 367, 692
crystallised self-identity, 40
disengagement theory, 427
environmental mastery, 29
personal growth, 178, 743
positive relationships, 29
possibilities of the self, 691, 708
possible selves, 40
purpose in life, 431, 743
self-acceptance, 27
self-actualisation, 504, 736, 750
self-affirmation, 359
self-attribution, 359
self-awareness, 359, 527
self-centred, 74, 142, 405, 407, 642
self-concept, 29, 359, 743
self-confidence, 100, 289, 294, 376, 512, 564, 675, 750, 752
self-criticism, 109, 357, 745
self-defining, 366
self-deprecation, 657
self-disclosure, 471, 537, 538, 646, 657, 658, 660, 661, 666, 676, 708, 764
self-efficacy, 359, 378, 528
self-empowerment, 750
self-esteem, 359, 638, 838
self-evaluation, 359
self-event connections, 366
self-exception, 652
self-handicapping, 657, 659
self-identity, 359–361, 364–367, 437, 684, 686, 691, 723, 724, 774
self-ideology, 359
self-image, 359, 838
self-inclusion, 652
selfishness, 41
self-perception, 60, 271, 388, 411, 529, 537, 570, 639, 662, 811
self-protection, 359
self-realisation, 109, 777
self-reflection, 36, 357, 363, 366, 687, 739, 767, 855, 944

self-regard, 750
self-regulation, 34, 36, 363, 387, 429, 525, 739, 747, 748, 759, 776
self-reliance, 44, 137, 143, 218, 366, 475, 480, 499, 598, 688
self-stereotyping, 362, 522, 523, 659
self-sufficient, 41
self-worth, 359
suffering, 681
Sen, A., 7, 8
Seneca, 828, 829
Senescence, *see* Ageing
Senicide, 78
Senility, 111, 148
boke, 149
Sex
intersex, 570
ratio, 5, 59, 86, 270, 665
sexual orientation, 461, 528, 560, 562, 564, 568, 569, 571, 693
Sexual life
asexuality, 286
attitudes about sex, 288
body image, 270, 273, 274, 289
depression, 270, 272, 274, 280–282, 284–286, 291
desire, 75, 93, 286
ejaculatory functions, 286
erectile dysfunctions, 285–287
erection, 286, 287, 291
"gender-equal sexual regime", 289, 290
interest in sex, 286–288, 291
lubrication, 291
"male-centred sexual regime", 289, 290
"marital duty", 286
masturbation, 272, 288
menopause, 93, 270, 286, 287
orgasm, 286
pain during intercourse (*dyspareunia*), 286, 291
pleasure, 289, 290
pronatalist views, 287
relational satisfaction, 291
satisfaction, 289, 290
self-confidence, 289
sexual activity, 286
sexual compatibility, 286
sexual function, 286, 287, 289, 290
sexual functioning, 288, 291
"sexual retirement", 285
sexual stimulation, 288
subjective sexual well-being, 288–291
vaginal dryness, 286
vaginal soreness, 286
Viagra© (sildenafil citrate), 287
Shakespeare, W., 85, 520, 735
Silver Human Resource Centres (SHRCs), 177
Silver surfers, 513
Simmons, L.W., 122
Sisyphus, The myth of, 735
Skin, 14, 203, 204, 274, 309, 417, 891, 900, 924
Skulachev, V.P., 921, 923–925
Sleeping, 170, 247, 374, 473, 592
disrupted, 473
limb movements, 473
snoring, 473
Smoking, 16, 247, 297, 302, 402, 463, 560, 592, 804
Snowbirds, 107, 108, 179, 942
Social
accommodation, 91, 93, 635, 643, 655, 676, 708, 788
acculturation, 38, 140, 163, 232, 418, 548, 549, 554
activity theory, 463
actors, 661
affective solidarity, 544
age integration, 469, 663
ageism, 110, 547, 663, 724, 788
age segregation, 530, 663
age stratification theory, 463
anticipated support, 218, 220
antisocial, 372
asocial, 221, 372, 511
attachments, 724, 725
biopsychosocial model, 463
biopsychosocial-spiritual model, 463
bonding capital, 467
bridging capital, 467
capital, 35, 462, 464, 467, 468, 475, 510, 552, 941
civic engagement, 468, 942
class, 35, 81, 82, 86, 92, 170, 178, 378, 431, 497, 534
coexistence, 131, 462, 467, 468
cohesion, 91, 133, 224, 371, 464, 467, 468, 475, 480, 502, 554, 642, 792, 801, 831, 833
collective efficacy, 133
community, 7, 33, 35, 38, 45, 87, 91, 97, 107, 108, 124, 128, 131, 133, 137, 140, 150, 151, 157–160, 180, 218, 220, 224, 226, 291, 301, 306, 362, 420, 429, 431, 433, 462, 464, 466–469, 472, 475, 484, 488, 493, 494, 497, 503, 504, 506, 508, 510–513, 516, 526, 530, 531, 534, 537, 543, 546, 548, 549, 552–556, 558, 560, 561, 563–565, 570, 571, 578, 583, 597, 599, 633, 657, 663, 667–673, 689, 690, 693, 698, 699, 708, 724, 761, 762, 766, 779, 795, 830, 833, 846, 850, 860, 941, 942

Social (*cont.*)
compadrazgo, 139
connectedness, 31, 110, 216, 262, 278, 467, 472, 479, 485, 513, 669, 675, 676, 693, 698, 762, 765, 840, 939, 942
constraints, 37, 38, 41, 113, 120, 129, 137, 140, 181, 278, 288, 319, 464, 502, 514, 515, 544, 545, 556, 557, 563, 564, 646, 685
contextual embeddedness, 35
conventions, 371
convoy model, 472, 508
cooperation, 29, 138, 139, 159, 220, 371, 414, 495, 505, 663, 668, 676, 724, 726, 801, 806–808, 918
co-parenthood, 139
co-residence, 133, 139, 480, 543
death, 854
diogenes syndrome, 221, 364, 463
disconnectedness, 221
discrimination, 37, 93, 98, 100, 151, 152, 232, 291, 376, 478, 517, 528, 533, 549, 563, 565, 569, 571, 583, 668, 682, 690, 693, 831, 939
disengagement, 34, 464, 465, 475, 579, 864
disengagement theory, 427, 463, 464, 508
dismissing attachment style, 35
display rules, 371
disrespect, 136, 644, 660
division of labour, 271, 465
doreh, 131
downtowners, 107
engagement, 30, 137, 159, 175, 255, 262, 269, 318, 322, 379, 380, 385, 388, 405, 471, 474, 484, 485, 529, 542, 576, 939–941
exchange, 469, 808
festivals, 90, 475, 555, 558
formal public system of support, 466
funerai, 161
gated communities, 468
government, 31, 68, 91, 101, 218, 221, 301, 534, 724
identity threat, 521
ihsan, 475
illusionary correlations, 524
inequality, 35, 100, 112, 152, 161, 172, 282, 305, 420, 464, 467, 480, 526, 536
informal support system, 466, 467
integration, 35, 38, 45, 221, 233, 291, 299, 301, 319, 358, 361, 368, 372, 401, 462–464, 467–469, 471–475, 478, 496, 531, 542, 546–550, 554, 555, 558, 564, 565, 569, 578, 662, 666, 669, 670, 676, 698, 756, 761, 939
interconnectivity, 66
inter-generational network, 36, 89
interpersonal conflict management styles, 668
isolated, 93, 220, 510, 513, 693, 840, 850, 857
isolation, 164, 220, 231, 300, 415, 478, 489, 494, 527, 546, 554, 571–572, 590, 669, 699, 831, 833, 839, 846, 847
justice, 35, 514, 515, 579, 766, 826
kinship network, 542
Life Course theory, 463
linking capital, 467
living apart together (LAT), 479, 502, 503
machismo, 186
masks, 100, 662, 686
masquerade, 662, 686
matrilineal system, 133
multilingual communities, 468
neglect, 549
neighbourhood, 107, 109, 466, 467, 508, 555, 571, 656, 724
network, 6, 33–38, 45, 82, 93, 98, 100, 110, 124, 129, 131–133, 137, 143, 148–150, 160, 175, 180–182, 185, 220, 221, 234, 236, 262, 267, 277, 280, 283, 292, 294, 317–319, 372, 405, 414, 419, 428, 432, 462, 464–469, 471–473, 476, 477, 479, 480, 483, 485–487, 494, 496, 498, 503–506, 508, 509, 513, 530, 533, 535, 541–543, 545, 549–551, 555, 558, 560–562, 565, 566, 569–571, 582, 597, 598, 635, 667, 676, 723, 726, 737, 842, 846
networks of choice, 549
networks of necessity, 549
normative climate, 480
normative solidarity, 544
objective social integration, 464
overaccommodation, 643, 659, 708
patronising, 536, 647, 664, 667, 676, 708
persona, 662
positive social relationships, 37, 66, 122, 465, 466
power, 72, 96, 270, 434, 516, 653
prejudice, 60, 68, 70, 94, 98, 151, 299, 376, 518, 524, 528, 569, 698
proximity seeking, 469

received support, 218, 220
reciprocity, 76, 129, 133, 466, 475, 589, 663, 672, 919
relationships, 26, 27, 29, 30, 35–38, 40, 45, 57, 82, 100, 119, 121, 131, 133, 139, 141, 147–149, 178, 221, 222, 224, 231, 262, 280, 283, 288, 303, 314, 317–319, 363, 368, 402, 403, 411, 418, 462–464, 467–469, 471, 474, 475, 477, 479, 480, 484, 487, 488, 493, 495, 498, 501–504, 508, 510, 517, 527, 530, 531, 538, 541, 542, 544, 545, 554, 556, 560–564, 566, 587, 588, 598, 635, 636, 640, 661–663, 690, 694, 695, 708, 724, 725, 729, 733, 742, 743, 761, 766, 778, 788, 795, 799–801, 808, 821, 830, 862, 864, 918
resilience, 29, 216, 218, 317, 466, 588, 682, 698, 708, 764, 788, 846
respect, 37, 94, 121, 139, 598, 690
restricted social networks, 463, 464, 481
safe haven, 469
saudara angkat, 475
secure attachment, 372
secure base, 469
segregation, 469, 527, 530, 531
skinship, 161
Social ecology theory, 488
Social-environmental theory, 34
social exchange theory, 141, 588, 589
"Social Support Bank", 133
socioemotional selectivity theory (SST), 35, 372, 464, 502
solidarity, 132, 467, 494, 503, 541, 544, 583, 662, 695
solipsism, 41
stereotypes, 37, 40, 72, 74, 89, 98, 122, 151, 152, 286, 402, 478, 517, 518, 522, 523, 528, 534, 564–566, 579, 599, 653, 655, 657, 659, 675, 676, 708
stigma, 37, 38, 40, 45, 60, 92, 98, 149–152, 291, 295, 364, 376, 517, 526–528, 530, 531, 565, 569, 570, 599, 663, 664, 668, 724, 939
structural component of social capital, 542
subjective social integration, 464
support, 6, 26, 27, 31, 34–36, 38, 41, 69, 76, 82, 91, 93, 94, 101, 102, 110, 113, 120, 121, 124, 128, 131–133, 137, 139, 143, 149, 150, 162, 164, 171, 178, 180–183, 220, 221, 234, 236, 239, 263, 266, 270, 278, 279, 283–285, 292, 294–296, 299–301, 316–318, 362–364, 374, 380, 428, 431, 432, 462, 464–466, 471, 472, 475, 476, 480, 483–487, 492–494, 496–498, 500, 502–506, 508, 510–513, 515–517, 529, 531, 534, 536, 538, 541–545, 547–553, 555–558, 560, 561, 563, 565, 566, 569, 571, 577, 578, 583, 596, 597, 599, 640, 645, 658, 663, 674, 685, 692, 693, 695, 708, 724, 726, 757, 762, 764, 775, 786–788, 792, 793, 802, 806, 812, 821, 837, 840, 847, 850, 861, 863, 939, 941
tolerance, 38, 291, 294, 468
underaccommodation, 643
"universal brotherhood", 475
voluntary organisations, 469, 505
volunteering, 401, 430, 472, 761
wealth inheritance, 133
welfare, 58, 100, 103, 158, 430, 464, 468, 480, 505, 556, 557
welfare agencies, 466
well-integrated social network, 483, 939
Social Darwinism, 94, 95
Social Ecology theory
exo-system, 488
macro-system, 488
meso-system, 488
micro-system, 488
Social-environmental theory, 34
Social exchange theory, 588
Social network
choice, 549
necessity, 549
primary children and relatives, 132
primary friendships, 132
secondary formal, 132
Social relationships
adopted, 139
adoptive kin, 475
asymmetrical, 133
compadrazgo, 139
connectedness, 479
couplehood, 501
custodial grandparent, 140
daughter, 515
direct family, 45
divorce, 36, 286
friends, 45, 318, 411, 433, 462, 541, 554, 561, 563, 667
grandfather, 667
grandmother, 667
grandparent, 363, 496, 499

Social relationships (*cont.*)
in-laws, 496
inter-generational exchange, 161
intimacy, 766
kinship network, 542
neighbour, 401, 420, 467, 472, 509, 510, 542, 563, 597
partner, 33, 36, 175, 289, 291, 501, 531, 566
positive, 37, 502, 544, 554
saudara angkat, 475
social networks, 37, 319, 462, 465, 467, 541, 542, 544, 563, 667
son, 121, 515
spouses, 493, 542
widow, 285
widower, 285
Social Security Act of 1935, 103
Social Support Bank model, 137
Society
gerontocratic, 153, 155
multicultural, 136, 548, 672, 727
multi-ethnic, 45, 136, 151, 159, 475
traditional, 4, 72, 85, 110, 124, 125, 129, 137, 153, 170, 188, 500, 550
Sociocultural theory, 271
Socio-Economic Panel Study (SOEP), 187
Socioemotional selectivity theory, 22, 33, 175
Sokolovsky, J., 5, 87, 101, 105, 113, 119, 120, 123, 124, 129–132, 139, 155, 536
Solidarity, 80, 138, 468
associational, 138
cooperation, 137, 138, 142, 495, 544
functional, 138
normative, 138
Solipsism, 41
Somatization, 25
Sophocles, 819, 847
Sparta
gerontes, 71
Spinal cord
calcareous deposits, 18
degeneration, 18
Spirituality
ancestor, 122
attitudes towards death, 795
becoming, 93, 561, 689, 798
being, 42, 81, 222, 223, 511, 554, 561, 689, 729, 786, 795, 797, 798, 811, 846
biopsychosocial-spiritual model, 811
healing, 224, 810
moral life, 770
oosumch, 689
salience, 798
spiritual intelligence, 43
spiritual suffering, 223
spiritual well-being, 42, 137, 728, 745, 773, 774, 804, 805
transcendent dimension, 799, 800
Spouse, 36, 91, 98, 105, 106, 108, 149, 150, 161, 184, 278, 281, 283, 319, 465, 469, 470, 483, 490, 491, 494, 581, 583, 591, 593–595, 597, 655, 788, 841
Standing, Guy
grinners, 97
groaners, 97
The Precariat, 97
St Bernard, 81
Stearns, S.C., 894, 897
Stereotype, 37, 38, 78, 80, 87, 92, 100, 153, 154, 417, 518, 520, 522, 525, 529, 664
admiration, 521
affective response, 520
automatic perpetuation, 518
change, 142, 154, 417, 518, 520, 522
cognitive consonance, 524
cognitive dissonance, 524
commonsense reality, 519, 535, 539
confirmatory biases, 524
consensus, 152, 518, 520
contemptuous, 521
correction, 525
curmudgeon, 154
diffusion, 520
disconfirmation, 524
disidentification, 525
distinction, 520
embodiment, 417, 522
envious, 521
evolutionary root, 518
generalizations, 37
gerontophobia, 99, 526, 533
golden agers, 664
hegemonic, 152, 154
identity threat, 522
illusionary correlations, 524
"incompetent", 153
individuation, 525
inner self, 517
inter-generational conflict, 153
internalise, 522
irresponsible, 142
just-world, 152
mask of ageing, 517
myths, 37, 78

naïve, 142
negative, 72, 78, 80, 154, 286, 417, 518, 521–523, 533
oppositional stereotype enhancement effect, 152
paternalistic stereotype, 521
persona, 518
pigeon-holing, 518
positive, 92, 154, 155, 417, 521, 523, 533
prejudice, 99, 151
refractoriness to change, 520
self-fulfilling prophecies, 522, 524
self-regulation, 525
self-reinforcing process, 153
self-stereotyping, 522, 659
simplifications, 518
social identity threat, 521
social prejudice, 518
stability, 520
Stereotype Content model (SCM), 521
Stereotype-sensitive model, 655
Suppression, 525
System-justification theory, 152
"victimism" ("playing the victim"), 153
visibility, 518
"warmth", 152, 153
Stigma, 38, 88, 99, 106, 148, 153, 227, 526–529, 531, 532, 565, 577
aesthetics, 529
concealability, 529
course, 149, 565, 664
courtesy, 149, 527
discrimination, 38, 99, 526, 529, 532
disruptiveness, 529
double, 530, 547, 559, 566
double living, 527
enacted, 527
expected, 129, 150, 299, 526, 529
Goffman, E., 38, 99, 149, 152, 526–529
Goffman's "recursive process", 527
gossip, 526
immortality through stigma, 531
labelled identities, 531
"marks", 529
minstrelisation, 528, 565
"normal", 526, 529, 565
origin, 148, 529
peril, 529
resistance, 38, 532
runaway process, 99
self-awareness, 528
self-esteem, 299, 527–529
self-stigma, 528, 529
social death, 527
stigmatisation, 100, 150, 531, 569
stigmatised people, 526, 527, 531, 565
"why try" effect, 528
"wise", 527, 529, 565
Stoicism, 75, 78, 551
Stress, 9, 20, 33, 34, 40, 41, 68, 74, 94, 97, 98, 107, 111, 123, 140, 147, 148, 150, 151, 178, 181, 205, 216–219, 226, 227, 233, 239, 246, 261, 262, 270, 274, 278, 279, 284, 285, 292, 293, 310, 315, 316, 318, 319, 322, 372, 403, 413–421, 423, 425, 434, 463, 465, 470, 490, 491, 498, 528, 537, 554, 556, 577, 585, 587, 591, 593, 597–599, 683, 697, 723, 752, 766, 778, 792, 801, 802, 804, 806, 807, 824, 831, 835, 840, 846, 899, 900, 906, 925
acculturative, 417, 547, 850
acute, 403, 412, 416
adrenal glands, 279, 415
adenocorticotropic hormone, 415
attention, 25, 107, 218, 293, 416, 585, 846
avoidant style, 419
cardiovascular problems, 415
chronic, 215, 403, 415–418, 589
cognitive appraisals, 416
cognitive decline, 239, 382, 415, 416
cognitive vulnerability-stress model, 281
confrontational style, 419
coping, 159, 218, 220, 270, 281, 282, 293, 403, 413, 415, 416, 418–420, 438, 599, 766, 778, 802
corticotropin-releasing hormone (CRH), 415
custodial grandparents, 416
dance movement therapy, 420
dementia, 34, 148, 151, 239, 247, 262, 382, 412, 416, 591
denial style, 419
depression, 150, 587, 593, 859
diseases, 215, 224, 261, 310, 315, 413, 415, 438, 589, 593, 835
distress, 247, 282, 403, 416, 418, 421, 423, 434, 597, 778
emotion-focused coping, 418
emotion-oriented, 417–418
glucocorticoids, 215, 415
gratitude interventions, 420
humour, 683, 792
humour and laughing therapy, 420
identities, 40, 41, 123, 159, 178, 216, 414
immune dysregulation, 231, 415
immunosenescence, 415
long-term effects, 415
management, 412, 597

Stress (*cont.*)
memories, 215, 218, 239, 247, 355, 382, 403, 414–416, 437, 683, 697
mindfulness-based stress reduction (MBSR), 213, 420
music therapy, 420
mortality, 591
natural disasters, 416
negative effects, 159, 215, 220, 232, 274, 281, 414–417, 419, 420, 438, 589, 591, 778
obesity, 415
personality, 412, 414, 470, 599
pituitary, 415
problem-focused coping, 403
psychological, 45, 159, 247, 355, 414–418, 420, 437, 839
quality of life, 33, 465
reminiscence interviews, 420
rumination, 403, 417
social support, 140, 159, 218, 220, 224, 226, 232, 233, 262, 281, 285, 293, 318, 322, 415, 418, 420, 570, 587, 599, 723
stereotype embodiment, 417
stress-related disorders, 34
subjective well-being, 270, 416
task-oriented, 417
unconstructive repetitive thinking (URT), 417
vasopressin, 415
vulnerability, 281, 942
weakness, 415
yoga, 420, 512
Stress management
arts, 216
connectedness to others, 216
exercise, 216
meditation, 216, 404
mindfulness, 216, 404
relaxation, 216
spirituality, 216
Subjective well-being, 26, 42, 127, 132, 177, 215, 225, 270, 282, 285, 317, 319, 378, 411, 415, 433, 510, 552, 596, 598, 750, 765
Successful ageing, 27, 465, 681, 937
achievement, 27
ageing well, 27
competitiveness, 27
failures, 27
healthy ageing, 681
hedonism, 28
optimal ageing, 27
productive ageing, 27
successful ageing 2.0, 27
Suffering, 1, 41, 43, 73, 78, 95, 149, 155, 180, 203, 213, 223, 231, 244, 252, 287, 294, 374, 382, 429, 485, 492, 493, 550, 559, 565, 570, 574, 591–593, 596, 599, 659, 677, 681–684, 691, 695–698, 708, 722, 726, 727, 731, 734, 735, 741, 747, 750, 752–756, 764, 771, 773, 782, 785, 790, 799, 800, 810–822, 824–828, 831, 851, 852, 855, 858, 861, 862, 864, 891
Suicide
active suicide ideation, 815
altruistic, 828, 831, 919
anankastic personality traits, 838
anomic, 831, 833
apocalyptic suicide, 829
attempt, 230, 484, 815, 828, 830, 835, 844, 845
attempted suicides, 825, 832, 843, 845, 846
attempt lethality, 836
beyond blue, 847
cognitive behavioural therapy, 847
community-based outreach programs, 847
completed life, 840, 843
Council of Prague, 830
cowardly, 828, 831
deliberate self-harm, 815
depression, 234
Dialectical Behaviour Therapy, 847
dishonour, 828
egoistic, 831
epicurean epitaph, 829
existential suffering, 825, 843
The Friendship Line, 847
geriatric outreach, 847
hegesias, 829
heroic, 828, 829
honourable, 828, 831
hopelessness, 840, 847
IMPACT program, 847
indirect self-harm, 815
interpersonal psychotherapy, 847
Interpersonal theory of suicide, 840
legally imposed, 829
life fatigue, 844
Link-Plus, 847
loneliness, 842
Masada, 829
non-suicidal self-injury, 815
obsessive-compulsive personality traits, 838, 841
passive suicide ideation, 815, 828
perceived burdensomeness, 840
preparatory acts, 815, 828

PROSPECT program, 847
psychological autopsy, 838
risk factors, 833–835, 838–842, 846
suffering, 43, 577, 727, 825, 828, 830, 831, 839, 842, 846, 848
suicidal events, 815, 828
suicidal ideation, 230, 837, 839, 845, 847
suicidal intent, 836
suicide of the besieged, 829
taedium vitae, 828
Tele-Check, 847
Tele-Help, 847
thwarted belongingness, 840
unfavourable life circumstances, 842
will to live, 837, 840, 842–844, 846
worthlessness, 839
Superannuation
defined benefits system, 3
Superoxide dismutase (SOD), 10, 11
Survey in Europe on Nutrition and the Elderly, a Concerted Action (SENECA), 61, 72, 75, 78, 96, 167
Survival, 5, 16, 18, 34, 58, 80, 94, 95, 100, 123–125, 132, 143, 152, 166, 178, 211, 216, 244, 262, 284, 303, 311, 312, 368, 370, 390, 404, 430, 435, 471, 510, 545, 683, 689, 690, 729, 776–778, 781, 783–785, 807, 811, 820, 835, 891, 893, 894, 897, 899–901, 904, 905, 907, 912, 913, 916, 917, 920
Sushruta samhita, 70, 96
Swedish OCTO-Twin-study, 542
Sweetser's Rule, 98
Synaptogenesis, 208, 210, 212
Szilard, L., 9, 905

T
Tagore, R., 865
Tai Chi, 145, 162, 214, 420, 512, 941
Taoist philosophy, 182
Telemedicine, 517
Telomerase, 10
Telomere, 9, 10, 16, 316, 902, 910
Telomere shortening, 10, 16, 902
Terence, 96
Terman Life-Cycle Study, 413
Terminal Decline hypothesis, 187
Terror Management Theory, 186, 536, 726, 777, 781
Testosterone, 12, 14
Thane, P., 58, 60, 63, 67, 74, 82, 98, 105
Thyroid, 12, 247, 297, 298, 309
Tithonos (also known as Tithonus), 540, 541, 891–893, 895
Tithonus Song, 891
Tolerance, 165, 748, 776, 899
Tornstam, L., 42, 111, 175, 216, 321, 725, 751
Trace eyeblink conditioning, 19
Transcendence, 41, 216, 321, 361, 380, 427, 680, 725–728, 734, 757, 766, 767, 769, 774, 775, 798, 800, 857, 858, 864, 938
Transculturality, 107
Transition Towns, 942
Translating
Australian Institute of Interpreters and Translators, 704
decentering, 700
language brokering, 699, 701
natural translation, 701
Seven-Step Translation Model, 700
source language, 699, 700
target languages, 701
Trauma, 693, 753, 756, 757, 764, 861, 862
life course, 25, 433, 466, 568, 692
life meaning, 753–757
Tulving, E., 393–394

U
Ubuntu, 724
Ultra-centenarians, 2, 301, 837
Umeå 85+ study, 217, 310, 542
Uncertainty Reduction theory (URT), 636
Unemployment, 2, 8, 97, 100, 103, 105, 112, 421, 426, 435, 483, 833, 835, 845, 850
United Nations, 2, 58
Universal brotherhood, 475
University of the Third Age (U3A), 469, 474, 943
U.S. Department of Veteran Affairs, 293
1996 U.S. General Social Survey, 371

V
Vaillant, G.
"developmental tasks", 65
Vascular dementia, 34, 148, 241, 262, 263
Victoria Longitudinal Study, 276
Villanova, Arnoldus de, 84
Virgil, 76
Vision impairment, 16, 235
Vitamins, 11, 165, 204, 261, 274, 402

Volunteering, 172, 176, 177, 180, 220, 363, 470, 512, 586, 941
volunteer listeners, 771
von Bismarck, Chancellor Otto, 91
Vulnerability, 12, 124, 126, 204, 281, 515, 547, 558, 583, 704, 734, 783, 857, 942

W
Wagga Wagga Aboriginal Elders Group Incorporated (WWAEGI), 467
Wallace, A.R., 94
Walsperger map, 83
Watts Towers, 763, 764
Wear-and-tear, 148, 205, 314, 382, 892, 905
Web of significance, 854
Weight
gain, 164, 230
loss, 17, 164, 203, 228, 230, 245, 247, 590
Welfare
system, 8, 100, 132, 480, 500
Well-being, 6, 22, 26, 28, 42, 109, 127, 132, 178, 215, 220, 224–225, 282, 288–289, 364, 378, 411, 435, 461, 512, 543, 594, 635, 728, 805, 916, 938
existential, 41–43, 45, 215, 221, 223, 418, 728, 729, 732, 786
meaning in life, 176, 728, 750
paradox of well-being, 39
psychological, 27, 127, 132, 135, 170, 173, 217, 296, 372, 434, 466, 475, 594, 635, 762, 801, 802, 805, 807
resilience, 29, 217, 218, 321
self-transcendence, 217, 321
sense of coherence, 217, 321, 693
spiritual, 42, 137, 728, 745, 773, 774, 804, 805
subjective, 26, 42, 127, 132, 177, 215, 225, 270, 282, 285, 317, 319, 378, 411, 416, 433, 510, 552, 596, 598, 750, 765
wellness, 31, 508
Widowhood
loneliness, 478, 481
principle of substitution, 503
re-marriage, 498, 501
"societies of widows", 285
widowers, 285, 802
widows, 285, 802
Williams, G.C., 892, 893, 896, 897, 903, 904, 912, 913, 916
Will to live, 223, 284, 285, 313, 314, 317, 321, 764, 773, 811, 836, 841–845
Wilson, D.S., 807
Wisdom, 57, 62, 70–74, 81, 111, 125, 270, 356–359, 371, 437, 672, 697, 745, 771, 809, 810, 853
attractors, 358
emotional regulation, 357
environment, 680
fundamental pragmatics of life, 356, 357
practical, 358, 359, 810
self-esteem, 357
transcendent, 358, 359, 810
Wisdom Paradox, 65, 356
Witch, 75, 80, 664, 678, 809
Women's Health and Aging Study (WHAS), 146
Wong, P.T.P., 712, 736, 739–741, 744–750, 753, 757–761, 765–767, 778, 785, 786, 790
Workhouse, 88, 91, 92
World Health Organisation (WHO), 2, 160, 487, 531, 811, 831, 832
World Psychiatric Association, 531
World War II, 7, 96, 104, 112, 171, 420
Worden, J.W., 183, 863

Y
Yoga, 145, 420, 512
Young-old, 2, 3, 64, 294, 538, 746, 757, 776

Z
Zerbi, G., 83, 84